ELECTRONIC CIRCUIT ANALYSIS AND DESIGN

ELECTRONICS AND VLSI CIRCUITS

Allison: *Electronic Engineering: Semiconductors and Devices*

Chang and Sze: *ULSI Technology*

Chen: *Elements of Optoelectronics and Fiber Optics*

Colclaser and Diehl-Nagel: *Materials and Devices for Electrical Engineers and Physicists*

DeMicheli: *Synthesis and Optimization of Digital Circuits*

Fabricius: *Introduction to VLSI Design*

Ferendici: *Physical Foundations of Solid-State and Electron Devices*

Fonstad: *Microelectronic Devices and Circuits*

Franco: *Design with Operational Amplifiers and Analog Integrated Circuits*

Geiger, Allen, and Strader: *VLSI Design Techniques for Analog and Digital Circuits*

Gopalan: *Introduction to Digital Microelectronic Circuits*

Grinich and Jackson: *Introduction to Integrated Circuits*

Hodges and Jackson: *Analysis and Design of Digital Integrated Circuits*

Huelsman: *Active and Passive Analog Filter Design: An Introduction*

Ismail and Fiez: *Analog VLSI: Signal and Information Processing*

Jaeger: *Microelectronic Circuit Design*

Kang and Leblebici: *CMOS Digital Integrated Circuits Analysis and Design*

Kasap: *Principles of Electrical Engineering Materials and Devices*

Kovacs: *Micromachined Transducers Sourcebook*

Laker and Sansen: *Design of Analog Integrated Circuits and Systems*

Massobrio and Antogentti: *Semiconductor Device Modeling with SPICE*

Millman and Grabel: *Microelectronics*

Millman and Taub: *Pulse, Digital, and Switching Waveforms*

Neamen: *Electronic Circuit Analysis and Design 2nd ed.*

Neamen: *Semiconductor Physics and Devices*

Ng: *Complete Guide to Semiconductor Devices*

Pollock: *Fundamentals of Optoelectronics*

Roulston: *Bipolar Semiconductor Devices*

Ruska: *Microelectronic Processing: An Introduction to the Manufacture of Integrated Circuits*

Sandige: *Modern Digital Design*

Sarrafzadeh: *An Introduction to VLSI Physical Design*

Schilling and Belove: *Electronic Circuits: Discrete and Integrated*

Seraphim: *Principles of Electronic Packaging*

Singh: *Optoelectronics: An Introduction to Materials and Devices*

Singh: *Physics of Semiconductors and Their Heterostructures*

Singh: *Semiconductor Devices: An Introduction*

Singh: *Semiconductor Optoelectronics: Physics and Technology*

Smith: *Modern Communication Circuits*

Smith: *Modern Optical Engineering: The Design of Optical Systems*

Sze: *VLSI Technology*

Taub: *Digital Circuits and Microprocessors*

Taub and Schilling: *Digital Integrated Electronics*

Tsividis: *Mixed Analog-Digital VLSI Devices and Technology*

Tsividis: *Operation and Modeling of the MOS Transistor*

Vithayathil: *Power Electronics: Principles and Applications*

Wait, Huelsman, and Korn: *Introduction to Operational and Amplifier Theory Application*

Walker: *Optical Engineering Fundamentals*

Watson: *Mastering Electronics*

Yang: *Microelectronic Devices*

Zambuto: *Semiconductor Devices*

ELECTRONIC CIRCUIT ANALYSIS AND DESIGN

SECOND EDITION

DONALD A. NEAMEN
University of New Mexico

Boston   Burr Ridge, IL   Dubuque, IA   Madison, WI   New York   San Francisco   St. Louis
Bangkok   Bogotá   Caracas   Lisbon   London   Madrid
Mexico City   Milan   New Delhi   Seoul   Singapore   Sydney   Taipei   Toronto

To my wife, Mimi,
for her extreme patience, understanding,
and support during this project.

McGraw-Hill Higher Education
*A Division of The **McGraw-Hill** Companies*

ELECTRONIC CIRCUIT ANALYSIS AND DESIGN
Published by McGraw-Hill, an imprint of The McGraw-Hill Companies, Inc., 1221 Avenue of the Americas, New York, NY, 10020. Copyright © 2001, 1996, by The McGraw-Hill Companies, Inc. All rights reserved. No part of this publication may be reproduced or distributed in any form or by any means, or stored in a database or retrieval system, without the prior written consent of The McGraw-Hill Companies, Inc., including, but not limited to, in any network or other electronic storage or transmission, or broadcast for distance learning.
Some ancillaries, including electronic and print components, may not be available to customers outside the United States.

This book is printed on acid-free paper.

1 2 3 4 5 6 7 8 9 0 DOW/DOW 0 9 8 7 6 5 4 3 2 1 0

ISBN 0256261156

Publisher:	*Thomas Casson*
Executive editor:	*Betsy Jones*
Sponsoring editor:	*Catherine Fields*
Developmental editor:	*Kelley Butcher*
Marketing manager:	*John Wannemacher*
Project editor:	*Karen J. Nelson*
Production supervisor:	*Rose Hepburn*
Coordinator freelance design:	*Kiera Cunningham*
Supplement coordinator:	*Carol Bielski*
New media:	*Christopher Styles*
Cover image:	*© 2000 Jed & Kaoru Share/Tony Stone Images*
Interior design:	*Diane Beasley*
Compositor:	*Keyword Publishing Services*
Typeface:	*10/12 Times Roman*
Printer:	*R. R. Donnelley & Sons Company*

Library of Congress Cataloging-in-Publication Data
Neamen, Donald A.
   Electronic circuit analysis and design / Donald A. Neamen.—2nd ed.
     p. cm.
   ISBN 0-256-26115-6 (Electronic and VLSI circuits)
   Includes index.
   1. Electronic circuit design. 2. Semiconductors. I. Title. II. Series.
TK7867.N412 2001
621.3815dc—21   00-044787

www.mhhe.com

Donald Neamen is a professor and associate chairman for the Department of Electrical and Computer Engineering at The University of New Mexico. He received his Ph.D. from The University of New Mexico and then became an electronics engineer at the Solid State Sciences Laboratory at Hanscom Air Force Base. In 1976, he became an Assistant Professor in the EECE department at The University of New Mexico, where he teaches the semiconductor physics and devices course and electronics courses.

In 1980, Professor Neamen received the Outstanding Teacher Award for The University of New Mexico. In 1983 and 1985, he was recognized as Outstanding Teacher in the College of Engineering by Tau Beta Pi. In 1990 and 1994 to 1999, he received the Faculty Recognition Award, presented by graduating EECE students. He was also honored with the Teaching Excellence Award in the College of Engineering for 1994.

In addition to his teaching, Professor Neamen has worked in industry with Martin Marietta, Sandia National Laboratories, and Raytheon Company. He has published many papers and is the author of *Semiconductor Physics and Devices: Basic Principles*, 2nd ed.

PHILOSOPHY AND GOALS

Electronic Circuit Analysis and Design is intended as a core text in electronics for undergraduate electrical and computer engineering students. The purpose of the second edition of the book is to provide a foundation for analyzing and designing both analog and digital electronic circuits.

A majority of electronic circuits today are designed as integrated circuits (ICs), in which the entire circuit is fabricated on a single piece of semiconductor material. The integrated circuit may contain over a million semiconductor devices and other elements, and may perform many complex functions. The microprocessor is an example of such a circuit. The ultimate objective is to understand the operation, characteristics, and limitations of these integrated circuits.

Initially, discrete transistor circuits are analyzed and designed. The complexity of the circuits studied is then increased. Eventually the reader should be able to analyze and design the basic elements of integrated circuits, such as digital logic gates.

This text is an introduction to the complex subject of electronic circuits. Therefore, more advanced material is not included. Specific technologies, such as gallium arsenide, which are used in special applications, are also not included, although reference may be made to a few specialized applications. Finally, the layout and fabrication of ICs are not covered, since these topics alone can warrant entire texts.

COMPUTER-AIDED ANALYSIS AND DESIGN (PSPICE)

Computer analysis and computer-aided design (CAD) are significant factors in electronics. One of the most prevalent electronic circuit simulation programs is Simulation Program with Integrated Circuit Emphasis (SPICE), developed at the University of California. A version of SPICE tailored for the personal computer is PSpice. A comprehensive appendix on the PSpice circuit modeling program is included in this text. Example programs are also given in Appendix B. Instructors may introduce PSpice at any point in the course.

Electronics Workbench (EWB) simulates over 100 circuits taken from

the figures, examples, and problems marked by the icon . These

circuit files are on the CD-ROM that accompanies this text. Current users of EWB have access to all 100 files. Readers unfamiliar with EWB can use the free demonstration to use 20 unlocked files. If you wish to purchase the Electronics Workbench program, it is provided on the CD and can be unlocked for $79 by contacting Interactive Image Technologies, the maker of Electronics Workbench.

It is our belief that the use of computer analysis must be handled with care in an introductory electronics course. The objective in this introductory course is to learn the fundamental aspects of device models, circuit configurations, and basic circuit properties. We believe this learning is enhanced by going through the hand analysis and calculations. In this course, then, the computer analysis can be used as a check of the hand analysis results.

In several places in the text, PSpice results are included and are correlated with the hand analysis results. The PSpice capture schematics are included, as well as the computer simulation results. Specific computer simulation problems are included at the end of most chapters. However, at the instructor's discretion, PSpice can be used for any exercise or problem, to verify the hand analysis.

In some chapters, particularly the chapters on frequency response and feedback, computer analysis is used more heavily. Even in these situations, however, computer analysis is considered only after the fundamental properties of the circuit have been covered. The computer is a tool that can aid in the analysis and design of electronic circuits, but it is not a substitute for a thorough understanding of the basic concepts of circuit analysis.

DESIGN EMPHASIS

Design is the heart of engineering. Good design evolves out of considerable experience with analysis. In this text, we point out various characteristics and properties of circuits as we go through the analysis. The objective is to develop an intuition that can be applied to the design process.

Many design examples, design exercise problems, and end-of-chapter design problems are included. These problems are designated with an RD or a D. The RD is for redesign and generally refers to a circuit that has already been designed. The redesign problems are generally less complicated and may lead to a unique solution. Meeting an explicit set of design criteria is a first step in the design process. The D is for design problems that generally do not have unique solutions. A separate section in the end-of-chapter problems contains open-ended design problems.

PREREQUISITES

This book is intended for junior undergraduates in electrical and computer engineering. The prerequisites for understanding the material include dc analysis and steady-state sinusoidal analysis of electric circuits and the transient analysis of *RC* circuits. Various network concepts, such as Thevenin's and Norton's theorems, are used extensively. Some background in Laplace transform techniques may also be useful. Prior knowledge of semiconductor device physics is not required.

ORGANIZATION

The book is divided into three parts. Part I, consisting of the first eight chapters, covers semiconductor devices and basic circuit applications. Chapter 1 introduces semiconductor material and the diode, which leads to the application of diodes in circuits, given in Chapter 2. Chapter 3 covers the bipolar junction transistor, with basic linear amplifier applications given in Chapter 4. Chapter 5 discusses the field-effect transistor (FET), with strong emphasis on the metal-oxide-semiconductor FET (MOSFET), and Chapter 6 covers basic FET linear amplifier circuits. Although the bipolar transistor is discussed before the FET, these chapters are written to stand alone, so the FET and its circuits can be covered before the bipolar transistor. The frequency response of transistors and transistor circuits is covered in Chapter 7. Finally, Chapter 8 addresses special applications, including power amplifiers.

Although many of the circuit configurations for bipolar transistors and FETs are very similar, we feel that the emphasis in Part I should be on analysis and design techniques. We therefore feel that mixing the two types of transistors in the same chapter would introduce unnecessary confusion. In Part II, however, both bipolar and FET circuits are evaluated in the same chapters.

Part I in essence presents the basic components of electronics. Part II covers more advanced topics in analog electronics, and Part III considers digital electronic circuits. Again, these two parts are written to stand alone, so the order of study of advanced analog and digital electronics can be interchanged, although it may be somewhat disconcerting to both students and instructors to jump from Chapter 6 to Chapter 16.

Chapters 9 through 15 are included in Part II, which addresses more advanced analog electronics. In this part, the emphasis is placed on the operational amplifier. The ideal operational amplifier and ideal op-amp circuits are covered in Chapter 9. Chapter 10 analyzes constant-current source biasing circuits and introduces the active load, both of which are used extensively in integrated circuits (ICs). The differential amplifier, the heart of the op-amp, is discussed in Chapter 11, and feedback is considered in Chapter 12. Chapter 13 presents the analysis and design of various circuits that form operational amplifiers. Nonideal effects in analog ICs are addressed in Chapter 14, and applications, such as active filters and oscillators, are covered in Chapter 15.

Part III, consisting of Chapters 16 and 17, covers the basics of digital electronics, including both FET and bipolar logic gates. Chapter 16 analyzes MOS digital electronic circuits. Since MOSFETs are used extensively in digital electronics, they are covered before bipolar circuits. Chapter 17 presents bipolar digital circuits, including emitter-coupled logic and classic transistor–transistor logic circuits.

Six appendixes are included at the end of the text. Appendix A contains physical constants and conversion factors. Appendix B is a discussion of PSpice, including analysis types. Several examples are presented in which the PSpice circuit schematic is given as well as the output response. This will allow the reader to get started with PSpice. Several available texts provide a complete discussion of SPICE and PSpice.

Manufacturers' data sheets for several devices and circuits are included in Appendix C. References are listed in Appendix E, and answers to selected end-of-chapter problems are given in Appendix F.

FEATURES OF THE SECOND EDITION

- A preview section introduces each chapter. This preview links the chapter to previous chapters and states the chapter's goals, i.e., what the reader should gain from the chapter.

- An extensive number of worked examples are used throughout the text to reinforce the theoretical concepts being developed. These examples contain all the details of the analysis or design, so the reader does not have to fill in missing steps.

- Exercise or drill problems are included throughout the chapter. These problems are generally placed immediately after an example problem, rather than at the end of a long section, so that readers can immediately test their understanding of the material just covered. Answers are given for each drill problem so readers do not have to search for an answer at the end of the book. These problems will reinforce readers' grasp of the material before they move on to the next section.

- Problem Solving Techniques are given throughout each chapter to assist the reader in analyzing circuits. Although there may be more than one method of solving a problem, these Problem Solving Techniques are intended to help the reader get started in the analysis of a circuit.

- A Summary section follows the text of each chapter. This section summarizes the overall results derived in the chapter and reviews the basic concepts developed.

- A Checkpoint section follows the Summary section. This section states the goals that should have been met and states the abilities the reader should have gained. The Checkpoints will help assess progress before moving to the next chapter.

- A list of review questions is included at the end of each chapter. These questions serve as a self-test to help the reader determine how well the concepts developed in the chapter have been mastered.

- A large number of problems are given at the end of each chapter, organized according to the subject of each section. Approximately 300 new problems have been included in the second edition. Design oriented problems are included as well as problems with varying degrees of difficulty. Again, RD indicates redesign-type problems, D indicates design-type problems, and an asterisk (*) indicates more difficult problems. Separate computer simulation problems and open-ended design problems are also included.

- An Electronics Workbench CD-ROM is included with the second edition of the text. The CD-ROM contains a copy of Electronics Workbench software and approximately 100 circuit simulations from the text.

- Answers to selected problems are given in an appendix. Knowing the answer to a problem can aid and reinforce the problem solving.

- Manufacturers' data sheets for selected devices and circuits are given in another appendix. These data sheets should allow the reader to relate the basic concepts and circuit characteristics studied to real circuit characteristics and limitations.

SUPPLEMENTS

The book is supported by the following supplements:

- A CD-ROM containing a copy of Electronics Workbench software and circuit simulations is provided with each book.
- Solutions Manual, available to instructors in paper and electronic form.
- Transparency masters of important figures.
- A Web site containing PowerPoint slides of 200 important figures, a solutions manual, links to Electronics Workbench and other important sites, and text updates. The URL for the Web site is http://www.mhhe.com/neamen.

ACKNOWLEDGMENTS

I am indebted to the many students I have taught over the years who have helped in the evolution of this text. Their enthusiasm and constructive criticism have been invaluable, and their delight when they think they have found an error their professor may have made is priceless. I also want to acknowledge Professors Cox, Hawkins, and McNeil at the University of New Mexico who have taught from the first edition and who have made excellent suggestions for improvement.

The University of New Mexico has my appreciation for providing an atmosphere conducive to writing this book.

I want to thank the many people at McGraw-Hill for their tremendous support. To Tom Casson, publisher, and Catherine Fields, sponsoring editor, I am grateful for their encouragement and support. A special thanks to Kelley Butcher, senior developmental editor at McGraw-Hill. Her attention to the plethora of details, her unwavering support, and her tremendous enthusiasm throughout the entire project are especially recognized and appreciated. I also appreciate the efforts of Karen Nelson, project manager, who guided the work through its final phase toward publication. This effort included gently, but firmly, pushing me through proofreading.

Four special groups of people deserve my thanks. These are the reviewers who read the original manuscript in its various phases, a focus group who spent an entire precious weekend discussing and evaluating the original project, and the accuracy checkers who worked the examples, exercises, and problems to minimize any errors I may have introduced. The fourth group consists of those individuals who have reviewed the first edition and the manuscript for the second edition. The efforts and suggestions made by all of these individuals have made this a better book. These people are recognized for their valuable contributions.

Reviewers for the First Edition

Timothy F. Darling
*University of California—
Santa Barbara*
Daniel J. Moore
*Rose Hulman Institute of
Technology*
R. G. Deshmukh
Florida Institute of Technology
Khalil Najafi
University of Michigan—Ann Arbor
Godi Fischer
University of Rhode Island
Bruce Johnson
University of Nevada—Reno
Dennis Polla
University of Minnesota
Donnie K. Reinhard
Michigan State University
Raymond S. Winton
Mississippi State University
Wesley G. Lawson
University of Maryland
Eugene D. Fabricius
*California Polytechnic State
University—San Luis Obispo*
Glen C. Gerhard
University of Arizona
David J. Dumin
Clemson University
James C. Gottling
Ohio State University
Daniel W. Hart
Valparaiso University
William Wilson
Rice University
Anura P. Jayasumana
Colorado State University
Charles E. Smith
University of Mississippi
Robert J. Krueger
*University of Wisconsin—
Milwaukee*
Philip C. Munro
Youngstown State University
Thomas Wong
Illinois Institute of Technology

Focus Group Participants

Donnie K. Reinhard
Michigan State University
Richard Hester
Texas Instruments
Peter E. Engler
New Jersey Institute of Technology
Mahmood Navi
*California Polytechnic State
University—San Luis Obispo*
Stuart M. Wentworth
Auburn University
Farid Tranjan
*University of North Carolina—
Charlotte*
Arthur F. Witulski
University of Arizona
Michael Hassul
*California State University—
Long Beach*
Glen C. Gerhard
University of Arizona
Daniel J. Moore
*Rose Hulman Institute of
Technology*
Ronald S. Gyurcsik
North Carolina State University

Accuracy Checkers

Daniel J. Moore
*Rose Hulman Institute of
Technology*
William Davis
*Virginia Polytechnic Institute and
State University*
Paul Weston
University of Illinois—Urbana
Hongyan Diao
University of Houston
Calvin L. Finn
University of Idaho
Tony King
University of Houston

George Aliftiras
Virginia Polytechnic Institute and State University
Howard Hao Wu
University of Houston
Montanez Wade
Tennessee State University
William Schneider
University of Houston
Carl Erickson
Messiah College
William Kuhn
Virginia Polytechnic Institute and State University
Maritza Kozicki
Sam Stone

Reviewers for the Second Edition

Kay D. Baker
Utah State University
Paul Benkeser
Georgia Institute of Technology

John Brews
University of Arizona
Steven M. Durbin
Florida State University
Martin Feldman
Louisiana State University
Jack Lee
University of Texas at Austin
Daniel J. Moore
Rose Hulman Institute of Technology
Farid Najm
University of Illinois— Urbana-Champaign
Mehmet Ozturk
North Carolina State University
Donald Parker
Texas A&M University
Andrew Rys
Kansas State University
Mahmoud Wagdy
California State University— Long Beach

Please forward any comments, suggestions, or corrections concerning the book to me in care of McGraw-Hill, 1333 Burr Ridge Parkway, Burr Ridge, IL 60521.

Donald A. Neamen

Industry Insight

RICHARD J. VALENTINE
Principal Staff Engineer
Motorola, Inc.

''Discrete semiconductor devices form the basis of most electronic circuits. Even complex integrated devices like microcomputer chips are designed from individual diodes and transistors. Understanding how these individual devices function is essential before an engineer can start a circuit design, whether it is a modest home computer power supply or a 5-million-transistor microprocessor integrated circuit. Future electronic designs will incorporate more semiconductor devices to achieve higher efficiency levels. Discrete power semiconductor and microcontroller devices allow motors to perform smarter with less power consumption, which is significant when you realize that half of the world's electric energy supply is consumed just by electric motors.

Each type of semiconductor device has unique properties that are optimized for different circuit requirements. Learning these properties enables the circuit designer to select the correct units quickly. The time spent learning how each discrete semiconductor device type operates and interacts with other components is always rewarded. For example, our engineering team was given a task of designing a motor control unit for electric vehicles. We quickly determined the best power transistor and rectifier types, based on our education in semiconductor technology. After building and testing the electric vehicle motor controller, we focused on areas of improvement for the semiconductor device design, again based on our knowledge of semiconductor theory.

Any career in electronics will involve the use, and probably the design, of semiconductor devices. Mastering the material presented in the following chapters is important for both the technician, who will be testing diodes, transistors, and microchips, and the engineer, who will be designing these semiconductor devices into electronic equipment.''

Semiconductor Devices and Basic Applications

In the first part of the text, we introduce the physical characteristics and operation of the major semiconductor devices and the basic circuits in which they are used, to illustrate how the device characteristics are utilized in switching, digital, and amplification applications

Chapter 1 briefly discusses semiconductor material characteristics and then introduces the semiconductor diode. Chapter 2 looks at various diode circuits that demonstrate how the nonlinear characteristics of the diode itself are used in switching and waveshaping applications. Chapter 3 introduces the bipolar transistor, presents the dc analysis of bipolar transistor circuits, and discusses basic applications of the transistor. In Chapter 4, we design and analyze fundamental bipolar transistor circuits, including amplifiers.

Chapter 5 introduces the field-effect transistor (FET), and FET circuits are analyzed and designed in Chapter 6. Chapter 7 considers the frequency response of both bipolar and field-effect transistor circuits. Finally, Chapter 8 discusses the designs and applications of these basic electronic circuits, including power amplifiers and various output stages.

1

Semiconductor Materials and Diodes

1.0 PREVIEW

This text deals with the analysis and design of circuits containing electronic devices, such as diodes and transistors. These electronic devices are fabricated using semiconductor materials, so we begin Chapter 1 with a brief discussion of the properties and characteristics of semiconductors. The intent of this brief discussion is to become familiar with some of the semiconductor material terminology.

A basic electronic device is the pn junction diode. One of the more interesting characteristics of the diode is its nonlinear current–voltage properties. The resistor, for example, has a linear relation between the current through it and the voltage across the element. The diode is also a two-terminal device, but the *i*–*v* relationship is nonlinear. The current is an exponential function of voltage in one direction and is essentially zero in the other direction. As we will see, this nonlinear characteristic makes possible the generation of a dc voltage from an ac voltage source and the design of digital logic circuits, for example.

Since the diode is a nonlinear element, the analysis of circuits containing diodes is not as straightfoward as is the analysis of simple resistor circuits. A mathematical model of the diode, describing the nonlinear *i*–*v* properties, is developed. However, the circuit cannot be analyzed, in general, by direct mathematical calculations. In many engineering problems, approximate "back-of-the-envelope" solutions replace difficult complex solutions. We develop one such approximation technique using the piecewise linear model of the diode. In this case, we replace the nonlinear diode properties by linear characteristics that are approximately valid over a limited region of operation. This concept is used throughout the study of electronics.

Besides the pn junction diode, we consider five other types of diodes that are used in specialized electronic applications. These include the solar cell, photodiode, light-emitting diode, Schottky barrier diode, and the Zener diode.

The general properties of the diode are considered in this chapter. Simple diode circuits are analyzed with the intent of developing a basic understanding of analysis techniques and diode circuit characteristics. Chapter 2 then considers applications of diodes in circuits that perform various electronic functions.

1.1   SEMICONDUCTOR MATERIALS AND PROPERTIES

Most electronic devices are fabricated by using semiconductor materials along with conductors and insulators. To gain a better understanding of the behavior of the electronic devices in circuits, we must first understand a few of the characteristics of the semiconductor material. Silicon is by far the most common semiconductor material used for semiconductor devices and integrated circuits. Other semiconductor materials are used for specialized applications. For example, gallium arsenide and related compounds are used for very-high-speed devices and optical devices.

1.1.1   Intrinsic Semiconductors

An atom is composed of a nucleus, which contains positively charged protons and neutral neutrons, and negatively charged electrons that, in the classical sense, orbit the nucleus. The electrons are distributed in various "shells" at different distances from the nucleus, and electron energy increases as shell radius increases. Electrons in the outermost shell are called **valence electrons**, and the chemical activity of a material is determined primarily by the number of such electrons.

Elements in the period table can be grouped according to the number of valence electrons. Table 1.1 shows a portion of the periodic table in which the more common semiconductors are found. Silicon (Si) and germanium (Ge) are in group IV and are **elemental semiconductors**. In contrast, gallium arsenide is a group III–V **compound semiconductor**. We will show that the elements in group III and group V are also important in semiconductors.

Table 1.1   A portion of the periodic table

III	IV	V
B	C	
Al	Si	P
Ga	Ge	As

Figure 1.1(a) shows five noninteracting silicon atoms, with the four valence electrons of each atom shown as dashed lines emanating from the atom. As silicon atoms come into close proximity to each other, the valence electrons interact to form a crystal. The final crystal structure is a tetrahedral configuration in which each silicon atom has four nearest neighbors, as shown in Figure 1.1(b). The valence electrons are shared between atoms, forming what are called **covalent bonds**. Germanium, gallium arsenide, and many other semiconductor materials have the same tetrahedral configuration.

Figure 1.1(c) is a two-dimensional representation of the lattice formed by the five silicon atoms in Figure 1.1(a). An important property of such a lattice is that valence electrons are always available on the outer edge of the silicon crystal so that additional atoms can be added to form very large single-crystal structures.

A two-dimensional representation of a silicon single crystal is shown in Figure 1.2, for $T = 0\,°K$, where $T =$ temperature. Each line between atoms

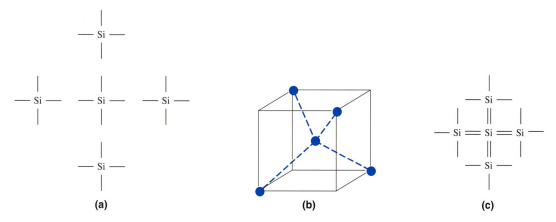

Figure 1.1 Silicon atoms in a crystal matrix: (a) five noninteracting silicon atoms, each with four valence electrons, (b) the tetrahedral configuration, (c) a two-dimensional representation showing the covalent bonding

represents a valence electron. At $T = 0\,°K$, each electron is in its lowest possible energy state, so each covalent bonding position is filled. If a small electric field is applied to this material, the electrons will not move, because they will still be bound to their individual atoms. Therefore, at $T = 0\,°K$, silicon is an **insulator**; that is, no charge flows through it.

If the temperature increases, the valence electrons will gain thermal energy. Any such electron may gain enough thermal energy to break the covalent bond and move away from its original position (Figure 1.3). The electron will then be free to move within the crystal.

Since the net charge of the material is neutral, if a negatively charged electron breaks its covalent bond and moves away from its original position, a positively charged "empty state" is created at that position (Figure 1.3). As the temperature increases, more covalent bonds are broken and more free electrons and positive empty states are created.

In order to break the covalent bond, a valence electron must gain a minimum energy, E_g, called the **bandgap energy**. Materials that have large bandgap energies, in the range of 3 to 6 electron–volts[1] (eV), are insulators because, at room temperature, essentially no free electrons exist in these materials. In contrast, materials that contain very large numbers of free electrons at room temperature are **conductors**.

In a *semiconductor,* the bandgap energy is on the order of 1 eV. The net flow of free electrons in a semiconductor causes a current. In addition, a valence electron that has a certain thermal energy and is adjacent to an empty state may move into that position, as shown in Figure 1.4 making it appear as if a positive charge is moving through the semiconductor. This positively charged "particle" is called a **hole**. In semiconductors, then, two types of charged particles contribute to the current: the negatively charged free electron, and the positively charged hole. (This description of a hole is

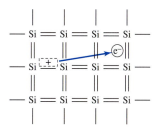

Figure 1.2 Two-dimensional representation of the silicon crystal at $T = 0\,°K$

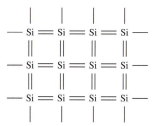

Figure 1.3 The breaking of a covalent bond for $T > 0\,°K$

Figure 1.4 A two-dimensional representation of the silicon crystal showing the movement of the positively charged hole

[1]An electron–volt is the energy of an electron that has been accelerated through a potential difference of 1 volt, and $1\,eV = 1.6 \times 10^{-19}$ joules.

greatly oversimplified, and is meant only to convey the concept of the moving positive charge.)

The concentrations ($\#/cm^3$) of electrons and holes are important parameters in the characteristics of a semiconductor material, because they directly influence the magnitude of the current. An **intrinsic semiconductor** is a single-crystal semiconductor material with no other types of atoms within the crystal. In an intrinsic semiconductor, the densities of electrons and holes are equal, since the thermally generated electrons and holes are the only source of such particles. Therefore, we use the notation n_i as the **intrinsic carrier concentration** for the concentration of the free electrons, as well as that of the holes. The equation for n_i is as follows:

$$n_i = BT^{3/2}e^{\left(\frac{-E_g}{2kT}\right)} \tag{1.1}$$

where B is a constant related to the specific semiconductor material, E_g is the bandgap energy (eV), T is the temperature (°K), and k is Boltzmann's constant (86×10^{-6} eV/°K). The values for B and E_g for several semiconductor materials are given in Table 1.2. The bandgap energy is not a strong function of temperature.

Table 1.2 Semiconductor constants

Material	E_g (eV)	B (cm^{-3}°K$^{-3/2}$)
Silicon (Si)	1.1	5.23×10^{15}
Gallium arsenide (GaAs)	1.4	2.10×10^{14}
Germanium (Ge)	0.66	1.66×10^{15}

Example 1.1 **Objective:** Calculate the intrinsic carrier concentration in silicon at $T = 300$ °K.

Solution: For silicon at $T = 300$ °K, we can write

$$n_i = BT^{3/2}e^{\left(\frac{-E_g}{2kT}\right)}$$

$$= (5.23 \times 10^{15})(300)^{3/2}e^{\left(\frac{-1.1}{2(86 \times 10^{-6})(300)}\right)}$$

or

$$n_i = 1.5 \times 10^{10} \text{ cm}^{-3}$$

Comment: An intrinsic electron concentration of 1.5×10^{10} cm^{-3} may appear to be large, but it is relatively small compared to the concentration of silicon atoms, which is 5×10^{22} cm^{-3}.

The intrinsic concentration n_i is an important parameter that appears often in the current–voltage equations for semiconductor devices.

Test Your Understanding

1.1   Calculate the intrinsic carrier concentration in gallium arsenide and germanium at $T = 300\,°\mathrm{K}$. (Ans. GaAs, $n_i = 1.80 \times 10^6\,\mathrm{cm}^{-3}$; Ge, $n_i = 2.40 \times 10^{13}\,\mathrm{cm}^{-3}$)

1.2   Determine the intrinsic carrier concentration in silicon, gallium arsenide, and germanium at $T = 400\,°\mathrm{K}$. (Ans. Si, $n_i = 4.76 \times 10^{12}\,\mathrm{cm}^{-3}$; GaAs, $n_i = 2.44 \times 10^9\,\mathrm{cm}^{-3}$; Ge, $n_i = 9.06 \times 10^{14}\,\mathrm{cm}^{-3}$)

1.1.2   Extrinsic Semiconductors

Because the electron and hole concentrations in an intrinsic semiconductor are relatively small, only very small currents are possible. However, these concentrations can be greatly increased by adding controlled amounts of certain impurities. A desirable impurity is one that enters the crystal lattice and replaces (i.e., substitutes for) one of the semiconductor atoms, even though the impurity atom does not have the same valence electron structure. For silicon, the desirable substitutional impurities are from the group III and V elements (see Table 1.1).

The most common group V elements used for this purpose are phosphorus and arsenic. For example, when a phosphorus atom substitutes for a silicon atom, as shown in Figure 1.5, four of its valence electrons are used to satisfy the covalent bond requirements. The fifth valence electron is more loosely bound to the phosphorus atom. At room temperature, this electron has enough thermal energy to break the bond, thus being free to move through the crystal and contribute to the electron current in the semiconductor.

The phosphorus atom is called a **donor impurity**, since it donates an electron that is free to move. Although the remaining phosphorus atom has a net positive charge, the atom is immobile in the crystal and cannot contribute to the current. Therefore, when a donor impurity is added to a semiconductor, free electrons are created without generating holes. This process is called **doping**, and it allows us to control the concentration of free electrons in a semiconductor.

A semiconductor that contains donor impurity atoms is called an **n-type semiconductor** (for the negatively charged electrons).

The most common group III element used for silicon doping is boron. When a boron atom replaces a silicon atom, its three valence electrons are used to satisfy the covalent bond requirements for three of the four nearest silicon atoms (Figure 1.6). This leaves one bond position open. At room temperature, adjacent silicon valence electrons have sufficient thermal energy to move into this position, thereby creating a hole. The boron atom then has a net negative charge, but cannot move, and a hole is created that can contribute to a hole current.

Because the boron atom has accepted a valence electron, the boron is therefore called an **acceptor impurity**. Acceptor atoms lead to the creation of holes without electrons being generated. This process, also called doping, can be used to control the concentration of holes in a semiconductor.

Figure 1.5   Two-dimensional representation of a silicon lattice doped with a phosphorus atom

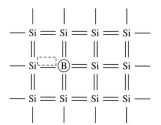

Figure 1.6   Two-dimensional representation of a silicon lattice doped with a boron atom

A semiconductor that contains acceptor impurity atoms is called a **p-type semiconductor** (for the positively charged holes created).

The materials containing impurity atoms are called **extrinsic semiconductors**, or **doped semiconductors**. The doping process, which allows us to control the concentrations of free electrons and holes, determines the conductivity and currents in the material.

A fundamental relationship between the electron and hole concentrations in a semiconductor *in thermal equilibrium* is given by

$$n_o p_o = n_i^2 \qquad\qquad\qquad (1.2)$$

where n_o is the thermal equilibrium concentration of free electrons, p_o is the thermal equilibrium concentration of holes, and n_i is the intrinsic carrier concentration.

At room temperature ($T = 300\,°K$), each donor atom donates a free electron to the semiconductor. If the donor concentration N_d is much larger than the intrinsic concentration, we can approximate

$$n_o \cong N_d \qquad\qquad\qquad (1.3)$$

Then, from Equation (1.2), the hole concentration is

$$p_o = \frac{n_i^2}{N_d} \qquad\qquad\qquad (1.4)$$

Similarly, at room temperature, each acceptor atom accepts a valence electron, creating a hole. If the acceptor concentration N_a is much larger than the intrinsic concentration, we can approximate

$$p_o \cong N_a \qquad\qquad\qquad (1.5)$$

Then, from Equation (1.2), the electron concentration is

$$n_o = \frac{n_i^2}{N_a} \qquad\qquad\qquad (1.6)$$

Example 1.2   Objective: Calculate the thermal equilibrium electron and hole concentrations.

Consider silicon at $T = 300\,°K$ doped with phosphorus at a concentration of $N_d = 10^{16}\,\text{cm}^{-3}$. Recall from Example 1.1 that $n_i = 1.5 \times 10^{10}\,\text{cm}^{-3}$.

Solution: Since $N_d \gg n_i$, the electron concentration is

$$n_o \cong N_d = 10^{16}\,\text{cm}^{-3}$$

and the hole concentration is

$$p_o = \frac{n_i^2}{N_d} = \frac{(1.5 \times 10^{10})^2}{10^{16}} = 2.25 \times 10^4\,\text{cm}^{-3}$$

Comment: In an extrinsic semiconductor, the electron and hole concentrations normally differ by many orders of magnitude.

In an n-type semiconductor, electrons are called the **majority carrier** because they far outnumber the holes, which are termed the **minority carrier**. The results obtained in Example 1.2 clarify this definition. In contrast, in a p-type semiconductor, the holes are the majority carrier and the electrons are the minority carrier.

Test Your Understanding

1.3 Calculate the majority and minority carrier concentrations in silicon at $T = 300\,^\circ K$ if (a) $N_a = 10^{17}\,\text{cm}^{-3}$, and (b) $N_d = 5 \times 10^{15}\,\text{cm}^{-3}$. (Ans. (a) $p_o = 10^{17}\,\text{cm}^{-3}$, $n_o = 2.25 \times 10^3\,\text{cm}^{-3}$, (b) $n_o = 5 \times 10^{15}\,\text{cm}^{-3}$, $p_o = 4.5 \times 10^4\,\text{cm}^{-3}$)

1.1.3 Drift and Diffusion Currents

The two basic processes which cause electrons and holes to move in a semiconductor are: (a) **drift**, which is the movement caused by electric fields; and (b) **diffusion**, which is the flow caused by variations in the concentration, that is, concentration gradients. Such gradients can be caused by a nonhomogeneous doping distribution, or by the injection of a quantity of electrons or holes into a region, using methods to be discussed later in this chapter.

To understand drift, assume an electric field is applied to a semiconductor. The field produces a force that acts on free electrons and holes, which then experience a net drift velocity and net movement. Consider an n-type semiconductor with a large number of free electrons (Figure 1.7(a)). An electric field E applied in one direction produces a force on the electrons in the *opposite* direction, because of the electrons' negative charge. The electrons acquire a drift velocity v_{dn} (in cm/s) which can be written as

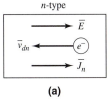

(a)

$$v_{dn} = -\mu_n E \tag{1.7}$$

where μ_n is a constant called the electron mobility and has units of $\text{cm}^2/\text{V--s}$. For low-doped silicon, the value of μ_n is typically 1350 $\text{cm}^2/\text{V--s}$. The mobility can be thought of as a parameter indicating how well an electron can move in a semiconductor. The negative sign in Equation (1.7) indicates that the electron drift velocity is opposite to that of the applied electric field as shown in Figure 1.7(a). The electron drift produces a drift current density J_n (A/cm^2) given by

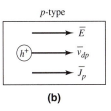

(b)

Figure 1.7 Applied electric field, carrier drift velocity, and drift current density in (a) an n-type semiconductor and (b) a p-type semiconductor

$$J_n = -env_{dn} = -en(-\mu_n E) = +en\mu_n E \tag{1.8}$$

where n is the electron concentration ($\#/\text{cm}^3$) and e is the magnitude of the electronic charge. The conventional drift current is in the opposite direction from the flow of negative charge, which means that the drift current in an n-type semiconductor is in the same direction as the applied electric field.

Next consider a p-type semiconductor with a large number of holes (Figure 1.7(b)). An electric field E applied in one direction produces a force on the holes in the *same* direction, because of the positive charge on the holes. The holes acquire a drift velocity v_{dp} (in cm/s) which can be written as

$$v_{dp} = +\mu_p E \tag{1.9}$$

where μ_p is a constant called the hole mobility, and again has units of $cm^2/V\text{--}s$. For low-doped silicon, the value of μ_p is typically 480 $cm^2/V\text{--}s$, which is slightly less than half the value of the electron mobility. The positive sign in Equation (1.9) indicates that the hole drift velocity is in the same direction as the applied electric field as shown in Figure 1.7(b). The hole drift produces a drift current density J_p (A/cm^2) given by

$$J_p = +epv_{dp} = +ep(+\mu_p E) = +ep\mu_p E \tag{1.10}$$

where p is the hole concentration ($\#/cm^3$) and e is again the magnitude of the electronic charge. The conventional drift current is in the same direction as the flow of positive charge, which means that the drift current in a p-type material is also in the same direction as the applied electric field.

Since a semiconductor contains both electrons and holes, the total drift current density is the sum of the electron and hole components. The total drift current density is then written as

$$J = en\mu_n E + ep\mu_p E = \sigma E \tag{1.11(a)}$$

where

$$\sigma = en\mu_n + ep\mu_p \tag{1.11(b)}$$

and where σ is the **conductivity** of the semiconductor in $(\Omega\text{--cm})^{-1}$. The conductivity is related to the concentration of electrons and holes. If the electric field is the result of applying a voltage to the semiconductor, then Equation (1.11(a)) becomes a linear relationship between current and voltage and is one form of Ohm's law.

From Equation (1.11(b)), we see that the conductivity can be changed from strongly n-type, $n \gg p$, by donor impurity doping to strongly p-type, $p \gg n$, by acceptor impurity doping. **Being able to control the conductivity of a semiconductor by selective doping is what allows us to fabricate the variety of electronic devices that are available.**

With diffusion, particles flow from a region of high concentration to a region of lower concentration. This is a statistical phenomenon related to kinetic theory. To explain, the electrons and holes in a semiconductor are in continuous motion, with an average speed determined by the temperature, and with the directions randomized by interactions with the lattice atoms. Statistically, we can assume that, at any particular instant, approximately half of the particles in the high-concentration region are moving *away* from that region toward the lower-concentration region. We can also assume that, at the same time, approximately half of the particles in the lower-concentration region are moving *toward* the high-concentration region. However, by definition, there are fewer particles in the lower-concentration region than there are in the high-concentration region. Therefore, the net result is a flow of particles away from the high-concentration region and toward the lower-concentration region. This is the basic diffusion process.

For example, consider an electron concentration that varies as a function of distance x, as shown in Figure 1.8(a). The diffusion of electrons from a high-concentration region to a low-concentration region produces a flow of electrons in the negative x direction. Since electrons are negatively charged, the conventional current direction is in the positive x direction.

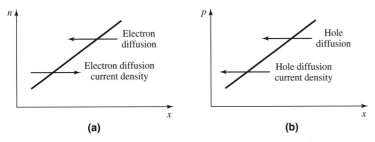

Figure 1.8 Current density caused by concentration gradients: (a) electron diffusion and corresponding current density and (b) hole diffusion and corresponding current density

In Figure 1.8(b), the hole concentration is a function of distance. The diffusion of holes from a high-concentration region to a low-concentration region produces a flow of holes in the negative x direction.

The *total* current density is the sum of the drift and diffusion components. Fortunately, in most cases only one component dominates the current at any one time in a given region of a semiconductor.

1.1.4 Excess Carriers

Up to this point, we have assumed that the semiconductor is in thermal equilibrium. In the discussion of drift and diffusion currents, we implicitly assumed that equilibrium was not significantly disturbed. Yet, when a voltage is applied to, or a current exists in, a semiconductor device, the semiconductor is really not in equilibrium. In this section, we will discuss the behavior of nonequilibrium electron and hole concentrations.

Valence electrons may acquire sufficient energy to break the covalent bond and become free electrons if they interact with high-energy photons incident on the semiconductor. When this occurs, both an electron and a hole are produced, thus generating an electron–hole pair. These additional electrons and holes are called **excess electrons** and **excess holes**.

When these excess electrons and holes are created, the concentrations of free electrons and holes increase above their thermal equilibrium values. This may be represented by

$$n = n_o + \delta n \tag{1.12(a)}$$

and

$$p = p_o + \delta p \tag{1.12(b)}$$

where n_o and p_o are the thermal equilibrium concentrations of electrons and holes, and δn and δp are the excess electron and hole concentrations.

If the semiconductor is in a steady-state condition, the creation of excess electrons and holes will not cause the carrier concentration to increase indefinitely, because a free electron may recombine with a hole, in a process called **electron–hole recombination**. Both the free electron and the hole disappear causing the excess concentration to reach a steady-state value. The mean time over which an excess electron and hole exist before recombination is called the **excess carrier lifetime**.

Test Your Understanding

1.4 Consider silicon at $T = 300\,°\text{K}$. Assume that $\mu_n = 1350\,\text{cm}^2/\text{V–s}$ and $\mu_p = 480\,\text{cm}^2/\text{V–s}$. Determine the conductivity if (a) $N_d = 5 \times 10^{16}\,\text{cm}^{-3}$ and (b) $N_a = 5 \times 10^{16}\,\text{cm}^{-3}$. (Ans. (a) $10.8\,(\Omega\text{–cm})^{-1}$, (b) $3.84\,(\Omega\text{–cm})^{-1}$.

1.5 A sample of silicon at $T = 300\,°\text{K}$ is doped to $N_d = 8 \times 10^{15}\,\text{cm}^{-3}$. (a) Calculate n_o and p_o. (b) If excess holes and electrons are generated such that their respective concentrations are $\delta p = \delta n = 10^{14}\,\text{cm}^{-3}$, determine the total concentrations of holes and electrons. (Ans. (a) $n_o = 8 \times 10^{15}\,\text{cm}^{-3}$, $p_o = 2.81 \times 10^4\,\text{cm}^{-3}$; (b) $n_o = 8.1 \times 10^{15}\,\text{cm}^{-3}$, $p_o \cong 10^{14}\,\text{cm}^{-3}$)

1.6 The conductivity of silicon is $\sigma = 10\,(\Omega\text{–cm})^{-1}$. Determine the drift current density if an electric field of $E = 15\,\text{V/cm}$ is applied. (Ans. $J = 150\,\text{A/cm}^2$)

1.2   THE pn JUNCTION

In the preceding sections, we looked at characteristics of semiconductor materials. The real power of semiconductor electronics occurs when p- and n-regions are directly adjacent to each other, forming a **pn junction**. One important concept to remember is that in most integrated circuit applications, the entire semiconductor material is a single crystal, with one region doped to be p-type and the adjacent region doped to be n-type.

1.2.1   The Equilibrium pn Junction

Figure 1.9(a) is a simplified block diagram of a pn junction. Figure 1.9(b) shows the respective p-type and n-type doping concentrations, assuming uniform doping in each region, as well as the minority carrier concentrations in each region, assuming thermal equilibrium.

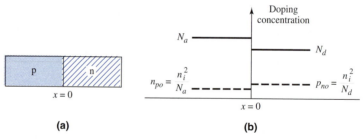

| (a) | (b) |

Figure 1.9   The pn junction: (a) simplified geometry of a pn junction and (b) doping profile of an ideal uniformly doped pn junction

The interface at $x = 0$ is called the **metallurgical junction**. A large density gradient in both the hole and electron concentrations occurs across this junction. Initially, then, there is a diffusion of holes from the p-region into the n-region, and a diffusion of electrons from the n-region into the p-region (Figure 1.10). The flow of holes from the p-region uncovers negatively charged acceptor ions, and the flow of electrons from the n-region uncovers positively

(a)

(b)

Figure 1.10 Initial diffusion of electrons and holes at the metallurgical junction, establishing thermal equilibrium

Figure 1.11 The pn junction in thermal equilibrium: (a) the space-charge region and electric field and (b) the potential through the junction

charged donor ions. This action creates a charge separation (Figure 1.11(a)), which sets up an electric field oriented in the direction from the positive charge to the negative charge.

If no voltage is applied to the pn junction, the diffusion of holes and electrons must eventually cease. The direction of the induced electric field will cause the resulting force to repel the diffusion of holes from the p-region and the diffusion of electrons from the n-region. Thermal equilibrium occurs when the force produced by the electric field and the "force" produced by the density gradient exactly balance.

The positively charged region and the negatively charged region comprise the **space-charge** region, or **depletion region**, of the pn junction, in which there are essentially no mobile electrons or holes. Because of the electric field in the space-charge region, there is a potential difference across that region (Figure 1.11(b)). This potential difference is called the **built-in potential barrier**, or built-in voltage, and is given by

$$V_{bi} = \frac{kT}{e} \ln\left(\frac{N_a N_d}{n_i^2}\right) = V_T \ln\left(\frac{N_a N_d}{n_i^2}\right) \tag{1.13}$$

where $V_T \equiv kT/e$, k = Boltzmann's constant, T = absolute temperature, e = the magnitude of the electronic charge, and N_a and N_d are the net acceptor and donor concentrations in the p- and n-regions, respectively. The parameter V_T is called the **thermal voltage** and is approximately $V_T = 0.026$ V at room temperature, $T = 300\,°$K.

Example 1.3 **Objective:** Calculate the built-in potential barrier of a pn junction.

Consider a silicon pn junction at $T = 300\,°$K, doped at $N_a = 10^{16}\,cm^{-3}$ in the p-region and $N_d = 10^{17}\,cm^{-3}$ in the n-region.

Solution: From the results of Example 1.1, we have $n_i = 1.5 \times 10^{10}\,cm^{-3}$ for silicon at room temperature. We then find

$$V_{bi} = V_T \ln\left(\frac{N_a N_d}{n_i^2}\right) = (0.026) \ln\left[\frac{(10^{16})(10^{17})}{(1.5 \times 10^{10})^2}\right] = 0.757\,V$$

Comment: Because of the log function, the magnitude of V_{bi} is not a strong function of the doping concentrations. Therefore, the value of V_{bi} for silicon pn junctions is usually within 0.1 to 0.2 V of this calculated value.

The potential difference, or built-in potential barrier, across the space-charge region cannot be measured by a voltmeter because new potential barriers form between the probes of the voltmeter and the semiconductor, canceling the effects of V_{bi}. In essence, V_{bi} maintains equilibrium, so no current is produced by this voltage. However, the magnitude of V_{bi} becomes important when we apply a forward-bias voltage, as discussed later in this chapter.

Test Your Understanding

1.7 Determine V_{bi} for a silicon pn junction at $T = 300\,°K$ for (a) $N_a = 10^{15}\,\text{cm}^{-3}$, $N_d = 10^{17}\,\text{cm}^{-3}$, and for (b) $N_a = N_d = 10^{17}\,\text{cm}^{-3}$. (Ans. (a) $V_{bi} = 0.697\,\text{V}$, (b) $V_{bi} = 0.817\,\text{V}$)

1.8 Calculate V_{bi} for a GaAs pn junction at $T = 300\,°K$ for $N_a = 10^{16}\,\text{cm}^{-3}$ and $N_d = 10^{17}\,\text{cm}^{-3}$. (Ans. $V_{bi} = 1.23\,\text{V}$)

1.2.2   Reverse-Biased pn Junction

Assume a positive voltage is applied to the n-region of a pn junction, as shown in Figure 1.12. The applied voltage V_R induces an applied electric field, E_A, in the semiconductor. The direction of this applied field is the same as that of the E-field in the space-charge region. Since the electric fields in the areas outside the space-charge region are essentially zero, the magnitude of the electric field *in* the space-charge region increases above the thermal equilibrium value. This increased electric field holds back the holes in the p-region and the electrons in the n-region, so there is essentially no current across the pn junction. By definition, this applied voltage polarity is called **reverse bias**.

When the electric field in the space-charge region increases, the number of positive and negative charges also increases. If the doping concentrations are not changed, the increases in the charges can only occur if the width W of the

Figure 1.12   A pn junction with an applied reverse-bias voltage, showing the direction of the electric field induced by V_R and of the space-charge electric field

space-charge region increases. Therefore, with an increasing reverse-bias voltage V_R, space-charge width W also increases.

Because of the additional positive and negative charges in the space-charge region, a capacitance is associated with the pn junction when a reverse-bias voltage is applied. This **junction capacitance**, or depletion layer capacitance, can be written in the form

$$C_j = C_{jo}\left(1 + \frac{V_R}{V_{bi}}\right)^{-1/2} \qquad\qquad (1.14)$$

where C_{jo} is the junction capacitance at zero applied voltage.

The capacitance–voltage characteristics make the pn junction useful for electrically tunable resonant circuits. Junctions fabricated specifically for this purpose are called **varactor diodes**. Varactor diodes can be used in electrically tunable oscillators, such as a Hartley oscillator, discussed in Chapter 15, or in tuned amplifiers, considered in Chapter 8.

Example 1.4   Objective: Calculate the junction capacitance of a pn junction.

Consider a silicon pn junction at $T = 300\,^\circ\text{K}$, with doping concentrations of $N_a = 10^{16}\,\text{cm}^{-3}$ and $N_d = 10^{15}\,\text{cm}^{-3}$. Assume that $n_i = 1.5 \times 10^{10}\,\text{cm}^{-3}$ and let $C_{jo} = 0.5\,\text{pF}$. Calculate the junction capacitance at $V_R = 1\,\text{V}$ and $V_R = 5\,\text{V}$.

Solution: The built-in potential is determined by

$$V_{bi} = V_T \ln\left(\frac{N_a N_d}{n_i^2}\right) = (0.026)\ln\left[\frac{(10^{16})(10^{15})}{(1.5 \times 10^{10})^2}\right] = 0.637\,\text{V}$$

The junction capacitance for $V_R = 1\,\text{V}$ is then found to be

$$C_j = C_{jo}\left(1 + \frac{V_R}{V_{bi}}\right)^{-1/2} = (0.5)\left(1 + \frac{1}{0.637}\right)^{-1/2} = 0.312\,\text{pF}$$

For $V_R = 5\,\text{V}$

$$C_j = (0.5)\left(1 + \frac{5}{0.637}\right)^{-1/2} = 0.168\,\text{pF}$$

Comment: The magnitude of the junction capacitance is usually at or below the picofarad range, and it decreases as the reverse-bias voltage increases.

As implied in the previous section, the magnitude of the electric field in the space-charge region increases as the reverse-bias voltage increases, and the maximum electric field occurs at the metallurgical junction. However, neither the electric field in the space-charge region nor the applied reverse-bias voltage can increase indefinitely because at some point, breakdown will occur and a large reverse bias current will be generated. This concept will be described in detail later in this chapter.

Test Your Understanding

1.9 A silicon pn junction at $T = 300\,°K$ is doped at $N_d = 10^{16}\,\text{cm}^{-3}$ and $N_a = 10^{17}\,\text{cm}^{-3}$. The junction capacitance is to be $C_j = 0.8\,\text{pF}$ when a reverse-bias voltage of $V_R = 5\,\text{V}$ is applied. Find the zero-biased junction capacitance C_{jo}. (Ans. $C_{jo} = 2.21\,\text{pF}$)

1.2.3   Forward-Biased pn Junction

To review briefly, the n-region contains many more free electrons than the p-region; similarly, the p-region contains many more holes than the n-region. With zero applied voltage, the built-in potential barrier prevents these majority carriers from diffusing across the space-charge region; thus, the barrier maintains equilibrium between the carrier distributions on either side of the pn junction.

If a positive voltage v_D is applied to the p-region, the potential barrier decreases (Figure 1.13). The electric fields in the space-charge region are very large compared to those in the remainder of the p- and n-regions, so essentially all of the applied voltage exists across the pn junction region. The applied electric field, E_A, induced by the applied voltage is in the opposite direction from that of the thermal equilibrium space-charge E-field. The net result is that the electric field in the space-charge region is lower than the equilibrium value. This upsets the delicate balance between diffusion and the E-field force. Majority carrier electrons from the n-region diffuse into the p-region, and majority carrier holes from the p-region diffuse into the n-region. The process continues as long as the voltage v_D is applied, thus creating a current in the pn junction. This process would be analogous to lowering a dam wall slightly. A slight drop in the wall height can send a large amount of water (current) over the barrier.

Figure 1.13   A pn junction with an applied forward-bias voltage, showing the direction of the electric field E_A induced by v_D and of the net space-charge electric field E

This applied voltage polarity (i.e., bias) is known as **forward bias**. The forward-bias voltage v_D must always be less than the built-in potential barrier V_{bi}.

As the majority carriers cross into the opposite regions, they become minority carriers in those regions, causing the minority carrier concentrations to increase. Figure 1.14 shows the resulting excess minority carrier concentrations

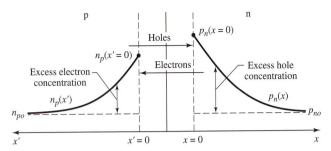

Figure 1.14   Steady-state minority carrier concentration in a pn junction under forward bias

at the space-charge region edges. These excess minority carriers diffuse into the neutral n- and p-regions, where they recombine with majority carriers, thus establishing a steady-state condition, as shown in Figure 1.14.

1.2.4   Ideal Current–Voltage Relationship

As shown in Figure 1.14, an applied voltage results in a gradient in the minority carrier concentrations, which in turn causes diffusion currents. The theoretical relationship between the voltage and the current in the pn junction is given by

$$i_D = I_S\left[e^{\left(\frac{v_D}{nV_T}\right)} - 1\right] \tag{1.15}$$

The parameter I_S is the **reverse-bias saturation current**. For silicon pn junctions, typical values of I_S are in the range of 10^{-15} to 10^{-13} A. The actual value depends on the doping concentrations and the cross-sectional area of the junction. The parameter V_T is the thermal voltage, as defined in Equation (1.13), and is approximately $V_T = 0.026$ V at room temperature. The parameter n is usually called the emission coefficient or ideality factor, and its value is in the range $1 \leq n \leq 2$.

The emission coefficient n takes into account any recombination of electrons and holes in the space-charge region. At very low current levels, recombination may be a significant factor and the value of n may be close to 2. At higher current levels, recombination is less a factor, and the value of n will be 1. Unless otherwise stated, we will assume the emission coefficient is $n = 1$.

Example   1.5   **Objective:**   Determine the current in a pn junction.

Consider a pn junction at $T = 300\,°K$ in which $I_S = 10^{-14}$ A and $n = 1$. Find the diode current for $v_D = +0.70$ V and $v_D = -0.70$ V.

Solution: For $v_D = +0.70$ V, the pn junction is forward-biased and we find

$$i_D = I_S\left[e^{\left(\frac{v_D}{V_T}\right)} - 1\right] = (10^{-14})\left[e^{\left(\frac{+0.70}{0.026}\right)} - 1\right] \Rightarrow 4.93\,mA$$

For $v_D = -0.70$ V, the pn junction is reverse-biased and we find

$$i_D = I_S\left[e^{\left(\frac{v_D}{V_T}\right)} - 1\right] = (10^{-14})\left[e^{\left(\frac{-0.70}{0.026}\right)} - 1\right] \cong -10^{-14}\,\text{A}$$

Comment: Although I_S is quite small, even a relatively small value of forward-bias voltage can induce a moderate junction current. With a reverse-bias voltage applied, the junction current is virtually zero.

Test Your Understanding

1.10 A silicon pn junction diode at $T = 300\,^\circ\text{K}$ has a reverse-saturation current of $I_S = 10^{-14}\,\text{A}$. (a) Determine the forward-bias diode current for (i) $v_D = 0.5\,\text{V}$, (ii) $v_D = 0.6\,\text{V}$, and (iii) $v_D = 0.7\,\text{V}$. (b) Find the reverse-bias diode current for (i) $v_D = -0.5\,\text{V}$, and (ii) $v_D = -2\,\text{V}$. (Ans. (a) (i) 2.25 µA, (ii) 105 µA, (iii) 4.93 mA; (b) (i) $10^{-14}\,\text{A}$, (ii) $10^{-14}\,\text{A}$)

1.11 A silicon pn junction diode at $T = 300\,^\circ\text{K}$ has a reverse-saturation current of $I_S = 10^{-13}\,\text{A}$. The diode is forward-biased with a resulting current of 1 mA. Determine v_D. (Ans. $v_D = 0.599\,\text{V}$)

1.2.5   pn Junction Diode

Figure 1.15 is a plot of the derived current–voltage characteristics of a pn junction. For a forward-bias voltage, the current is an exponential function

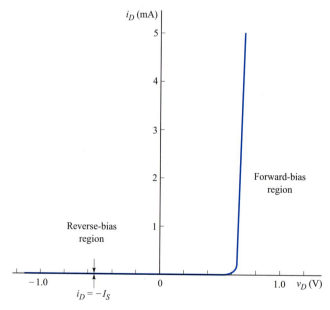

Figure 1.15   Ideal *I–V* characteristics of a pn junction diode for $I_S = 10^{-14}\,\text{A}$

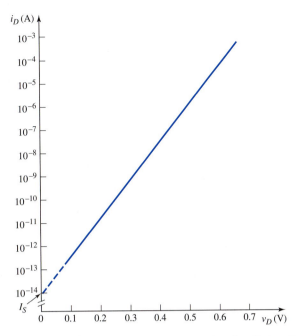

Figure 1.16   Ideal forward-biased *I–V* characteristics of a pn junction diode, with the current plotted on a log scale for $I_S = 10^{-14}$ A and $n = 1$

of voltage. Figure 1.16 depicts the forward-bias current plotted on a log scale. With only a small change in the forward-bias voltage, the corresponding forward-bias current increases by orders of magnitude. For a forward-bias voltage $v_D > +0.1$ V, the (-1) term in Equation (1.15) can be neglected. In the reverse-bias direction, the current is almost zero.

The semiconductor device that displays these *I–V* characteristics is called a **pn junction diode**. Figure 1.17 shows the diode circuit symbol and the conventional current direction and voltage polarity. The diode can be thought of and used as a voltage controlled switch that is "off" for a reverse-bias voltage and "on" for a forward-bias voltage. In the forward-bias or "on" state, a relatively large current is produced by a fairly small applied voltage; in the reverse-bias, or "off" state, only a very small current is created.

When a diode is reverse-biased by at least 0.1 V, the diode current is $i_D = -I_S$. The current is in the reverse direction and is a constant, hence the name reverse-bias saturation current. Real diodes, however, exhibit reverse-bias currents that are considerably larger than I_S. This additional current is called a generation current and is due to electrons and holes being generated within the space-charge region. Whereas a typical value of I_S may be 10^{-14} A, a typical value of reverse-bias current may be 10^{-9} A or 1 nA. Even though this current is much larger than I_S, it is still small and negligible in most cases.

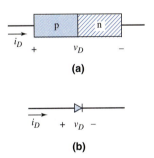

Figure 1.17   The basic pn junction diode: (a) simplified geometry and (b) circuit symbol, and conventional current direction and voltage polarity

Temperature Effects

Since both I_S and V_T are functions of temperature, the diode characteristics also vary with temperature. The temperature-related variations in forward-bias characteristics are illustrated in Figure 1.18. For a given current, the required

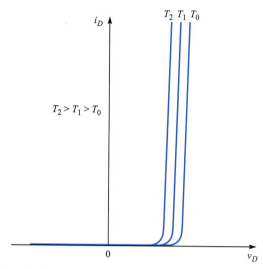

Figure 1.18   Forward-bias characteristics versus temperature

forward-bias voltage decreases as temperature increases. For silicon diodes, the change is approximately $2\,\text{mV}/°\text{C}$.

The parameter I_S is a function of the intrinsic carrier concentration n_i, which in turn is strongly dependent on temperature. Consequently, the value of I_S approximately doubles for every $5\,°\text{C}$ increase in temperature. The actual reverse-bias diode current, as a general rule, doubles for every $10\,°\text{C}$ rise in temperature. As an example of the importance of this effect, in germanium, the relative value of n_i is large, resulting in a large reverse-saturation current in germanium-based diodes. Increases in this reverse current with increases in the temperature make the germanium diode highly impractical for most circuit applications.

Breakdown Voltage

When a reverse-bias voltage is applied to a pn junction, the electric field in the space-charge region increases. The electric field may become large enough that covalent bonds are broken and electron-hole pairs are created. Electrons are swept to the n-region and holes to the p-region by the electric field generating a reverse-bias current. This breakdown mechanism is called the **Zener effect**. Another breakdown mechanism is called **avalanche breakdown**, which occurs when minority carriers crossing the space-charge region gain sufficient kinetic energy to be able to break covalent bonds during a collision process. The generated electron-hole pairs can themselves be involved in a collision process generating additional electron-hole pairs, thus, the avalanche process. The reverse-bias current for each breakdown mechanism will be limited by the external circuit.

The voltage at which breakdown occurs depends on fabrication parameters of the pn junction, but is usually in the range of 50 to 200 V for discrete devices, although breakdown voltages outside this range are possible—in excess of 1000 V, for example. A pn junction is usually rated in terms of its

peak inverse voltage or **PIV**. The PIV of a diode must never be exceeded in circuit operation if reverse breakdown is to be avoided.

Zener diodes are fabricated with a specifically designed breakdown voltage and are designed to operate in the breakdown region. These diodes are discussed later in this chapter.

Switching Transient

Since the pn junction diode can be used as an electrical switch, an important parameter is its transient response, that is, its speed and characteristics, as it is switched from one state to the other. Assume, for example, that the diode is switched from the forward-bias "on" state to the reverse-bias "off" state. Figure 1.19 shows a simple circuit that will switch the applied voltage at time $t = 0$. For $t < 0$, the forward-bias current i_D is

$$i_D = I_F = \frac{V_F - v_D}{R_F} \tag{1.16}$$

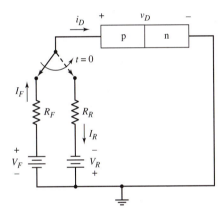

Figure 1.19   Simple circuit for switching a diode from forward to reverse bias

The minority carrier concentrations for an applied forward-bias voltage and an applied reverse-bias voltage are shown in Figure 1.20. Here, we neglect the change in the space charge region width. When a forward-bias voltage is applied, excess minority carrier charge is stored in both the p- and n-regions. The excess charge is the difference between the minority carrier concentrations for a forward-bias voltage and those for a reverse-bias voltage as indicated in the figure. This charge must be removed when the diode is switched from the forward to the reverse bias.

As the forward-bias voltage is removed, relatively large diffusion currents are created in the reverse-bias direction. This happens because the excess minority carrier electrons flow back across the junction into the n-region, and the excess minority carrier holes flow back across the junction into the p-region.

The large reverse-bias current is initially limited by resistor R_R to approximately

$$i_D = -I_R \cong \frac{-V_R}{R_R} \tag{1.17}$$

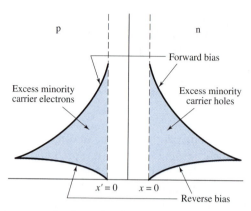

Figure 1.20   Stored excess minority carrier charge under forward bias compared to reverse bias

The junction capacitances do not allow the junction voltage to change instantaneously. The reverse current I_R is approximately constant for $0^+ < t < t_s$, where t_s is the **storage time**, which is the length of time required for the minority carrier concentrations at the space-charge region edges to reach the thermal equilibrium values. After this time, the voltage across the junction begins to change. The fall time t_f is typically defined as the time required for the current to fall to 10 percent of its initial value. The total **turn-off time** is the sum of the storage time and the fall time. Figure 1.21 shows the current characteristics as this entire process takes place.

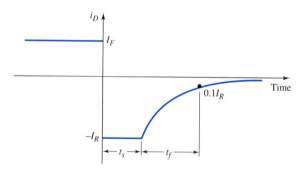

Figure 1.21   Current characteristics versus time during diode switching

In order to switch a diode quickly, the diode must have a small excess minority carrier lifetime, and we must be able to produce a large reverse current pulse. Therefore, in the design of diode circuits, we must provide a path for the transient reverse-bias current pulse. These same transient effects impact the switching of transistors. For example, the switching speed of transistors in digital circuits will affect the speed of computers.

The turn-on transient occurs when the diode is switched from the "off" state to the forward-bias "on" state, which can be initiated by applying a forward-bias current pulse. The transient **turn-on time** is the time required to establish the forward-bias minority carrier distributions. During this time, the voltage across the junction gradually increases toward its steady-state value. Although the turn-on time for the pn junction diode is not zero, it is usually less than the transient turn-off time.

Test Your Understanding

1.12   Recall that the forward-bias diode voltage decreases approximately by $2 \, \text{mV}/°\text{C}$ for silicon diodes with a given current. If $V_D = 0.650 \, \text{V}$ at $I_D = 1 \, \text{mA}$ for a temperature of $25 °\text{C}$, determine the diode voltage at $I_D = 1 \, \text{mA}$ for $T = 125 °\text{C}$. (Ans. $V_D = 0.450 \, \text{V}$)

1.3   DIODE CIRCUITS: DC ANALYSIS AND MODELS

In this section, we begin to study the diode in various circuit configurations. As we have seen, the diode is a two-terminal device with nonlinear i–v characteristics, as opposed to a two-terminal resistor, which has a linear relationship between current and voltage. The analysis of nonlinear electronic circuits is not as straightforward as the analysis of linear electric circuits. However, there are electronic functions that can be implemented only by nonlinear circuits. Examples include the generation of dc voltages from sinusoidal voltages and the implementation of logic functions.

Mathematical relationships, or **models**, that describe the current–voltage characteristics of electrical elements allow us to analyze and design circuits without having to fabricate and test them in the laboratory. An example is Ohm's law, which describes the properties of a resistor. In this section, we will develop the dc analysis and modeling techniques of diode circuits.

To begin to understand diode circuits, consider a simple diode application. The current–voltage characteristics of the pn junction diode were given in Figure 1.15. An **ideal diode** (as opposed to a diode with ideal I–V characteristics) has the characteristics shown in Figure 1.22(a). When a reverse-bias voltage is applied, the current through the diode is zero (Figure 1.22(b)); when current through the diode is greater than zero, the voltage across the diode is zero (Figure 1.22(c)). An external circuit connected to the diode must be designed to control the forward current through the diode.

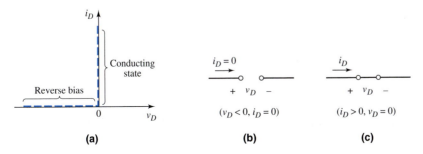

Figure 1.22   The ideal diode: (a) I–V characteristics, (b) equivalent circuit under reverse bias, and (c) equivalent circuit in the conducting state

One diode circuit is the **rectifier** circuit shown in Figure 1.23(a). Assume that the input voltage v_I is a sinusoidal signal, as shown in Figure 1.23(b), and the diode is an ideal diode (see Figure 1.22(a). During the positive half-cycle of the sinusoidal input, a forward-bias current exists in the diode and the voltage across the diode is zero. The equivalent circuit for this condition is shown in Figure 1.23(c). The output voltage v_O is then equal to the input voltage. During the negative half-cycle of the sinusoidal input, the diode is reverse biased. The equivalent circuit for this condition is shown in Figure 1.23(d). In this part of the cycle, the diode acts as an open circuit, the current is zero, and the output voltage is zero. The output voltage of the circuit is shown in Figure 1.23(e).

Over the entire cycle, the input signal is sinusoidal and has a zero average value; however, the output signal contains only positive values and therefore has a positive average value. Conseqently, this circuit is said to **rectify** the input signal, which is the first step in generating a dc voltage from a sinusoidal (ac) voltage. A dc voltage is required in virtually all electronic circuits.

As mentioned, the analysis of nonlinear circuits is not as straightforward as that of linear circuits. In this section, we will look at four approaches to the dc analysis of diode circuits: (a) iteration; (b) graphical techniques; (c) a piece-wise linear modeling method; and (d) a computer analysis. Methods (a) and (b) are closely related and are therefore presented together.

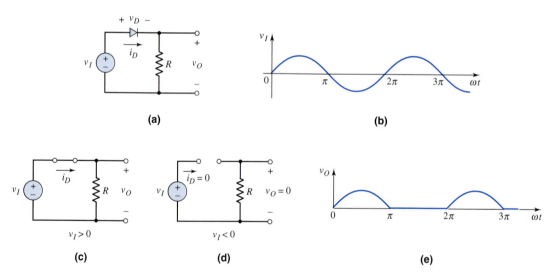

Figure 1.23 The diode rectifier: (a) circuit, (b) sinusoidal input signal, (c) equivalent circuit for $v_I > 0$, (d) equivalent circuit for $v_I < 0$, and (e) rectified output signal

1.3.1   Iteration and Graphical Analysis Techniques

Iteration means using trial and error to find a solution to a problem. The graphical analysis technique involves plotting two simultaneous equations and locating their point of intersection, which is the solution to the two equations. We will use both techniques to solve the circuit equations, which include the diode equation. These equations are difficult to solve by hand because they contain both linear and exponential terms.

Consider, for example, the circuit shown in Figure 1.24, with a dc voltage V_{PS} applied across a resistor and a diode. **Kirchhoff's voltage law** applies both to nonlinear and linear circuits, so we can write

$$V_{PS} = I_D R + V_D \qquad (1.18(a))$$

which can be rewritten as

$$I_D = \frac{V_{PS}}{R} - \frac{V_D}{R} \qquad (1.18(b))$$

Figure 1.24 A simple diode circuit

[Note: In the remainder of this section in which dc analysis is emphasized, the dc variables are denoted by uppercase letters and uppercase subscripts.]

The diode voltage V_D and current I_D are related by the ideal diode equation as

$$I_D = I_S \left[e^{\left(\frac{V_D}{V_T} \right)} - 1 \right] \qquad (1.19)$$

where I_S is assumed to be known for a particular diode.

Combining Equations (1.18(a)) and (1.19), we obtain

$$V_{PS} = I_S R \left[e^{\left(\frac{V_D}{V_T} \right)} - 1 \right] + V_D \qquad (1.20)$$

which contains only one unknown, V_D. However, Equation (1.20) is a transcendental equation and cannot be solved directly. The use of iteration to find a solution to this equation is demonstrated in the following example.

Example 1.6 **Objective:** Determine the diode voltage and current for the circuit shown in Figure 1.24.

Consider a diode with a given reverse-saturation current of $I_S = 10^{-13}$ A.

Solution: We can write Equation (1.20) as

$$5 = (10^{-13})(2 \times 10^3) \left[e^{\left(\frac{V_D}{0.026} \right)} - 1 \right] + V_D \qquad (1.21)$$

If we first try $V_D = 0.60$ V, the right side of Equation (1.21) is 2.7 V, so the equation is not balanced and we must try again. If we next try $V_D = 0.65$ V, the right side of Equation (1.21) is 15.1 V. Again, the equation is not balanced, but we can see that the solution for V_D is between 0.6 and 0.65 V. If we continue refining our guesses, we will be able to show that, when $V_D = 0.619$ V, the right side of Equation (1.21) is 4.99 V, which is essentially equal to the value of the left side of the equation.

The current in the circuit can then be determined by dividing the voltage difference across the resistor by the resistance, or

$$I_D = \frac{V_{PS} - V_D}{R} = \frac{5 - 0.619}{2} = 2.19 \, \text{mA}$$

Comment: Once the diode voltage is known, the current can also be determined from the ideal diode equation. However, dividing the voltage difference across a resistor by the resistance is usually easier, and this approach is used extensively in the analysis of diode and transistor circuits.

To use a graphical approach to analyze the circuit, we go back to Kirchhoff's voltage law, as expressed by Equation (1.18(b)), which produces a straight-line relationship between current I_D and voltage V_D for a given V_{PS} and R. This equation is referred to as the circuit **load line**, which can be plotted on a graph with I_D and V_D as the axes. From Equation (1.18(b)), we see that if $I_D = 0$, then $V_D = V_{PS}$. Also from this equation, if $V_D = 0$, then $I_D = V_{PS}/R$. The load line can be drawn between these two points. Using the values given in Example (1.6), we can plot the straight line shown in Figure 1.25. The second plot in the figure is that of Equation (1.19), which is the ideal diode equation relating the diode current and voltage. The intersection of the load line and the device characteristics curve provides the dc current $I_D \approx 2.2\,\text{mA}$ through the diode and the dc voltage $V_D \approx 0.62\,\text{V}$ across the diode. This point is referred to as the **quiescent point**, or the **Q-point**.

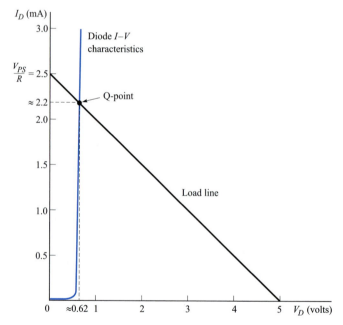

Figure 1.25   The diode and load line characteristics for the circuit shown in Figure 1.24

The graphical analysis method can yield accurate results, but it is somewhat cumbersome. However, the concept of the load line and the graphical approach are useful for "visualizing" the response of a circuit, and the load line is used extensively in the evaluation of electronic circuits.

Test Your Understanding

*1.13   Consider the circuit in Figure 1.24. Let $V_{PS} = 4\,\text{V}$, $R = 40\,\text{k}\Omega$, and $I_S = 10^{-12}\,\text{A}$. Determine V_D and I_D, using the ideal diode equation and the iteration method. (Ans. $V_D = 0.535\,\text{V}$, $I_D = 0.864\,\text{mA}$)

1.14   Consider the diode and circuit in Exercise 1.13. Determine V_D and I_D, using the graphical technique. (Ans. $V_D \approx 0.54\,\text{V}$, $I_D \approx 0.87\,\text{mA}$)

1.3.2 Piecewise Linear Model

Another, simpler way to analyze diode circuits is to *approximate* the diode's current–voltage characteristics, using linear relationships or straight lines. Figure 1.26, for example, shows the ideal current–voltage characteristics and two linear approximations.

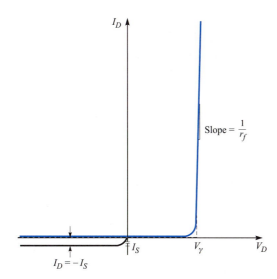

Figure 1.26   The ideal diode I–V characteristics and two linear approximations

For $V_D \geq V_\gamma$, we assume a straight-line approximation whose slope is $1/r_f$, where V_γ is the **turn-on**, or **cut-in**, **voltage** of the diode, and r_f is the **forward diode resistance**. The equivalent circuit for this linear approximation is a constant-voltage source in series with a resistor (Figure 1.27(a)).[2] For $V_D < V_\gamma$, we assume a straight-line approximation parallel to the V_D axis at the zero current level. In this case, the equivalent circuit is an open circuit (Figure 1.27(b)).

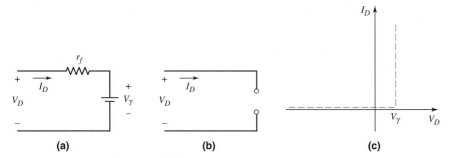

Figure 1.27   The diode equivalent circuit (a) in the "on" condition when $V_D \geq V_\gamma$, (b) in the "off" condition when $V_D < V_\gamma$, and (c) piecewise linear approximation when $r_f = 0$

---

[2]It is important to keep in mind that the voltage source in Figure 1.27(a) only represents a voltage drop for $V_D \geq V_\gamma$. When $V_D < V_\gamma$, the V_γ source does *not* produce a negative diode current. For $V_D < V_\gamma$, the equivalent circuit in Figure 1.27(b) must be used.

This method models the diode with segments of straight lines; thus the name **piecewise linear model**. If we assume $r_f = 0$, the piecewise linear diode characteristics are shown in Figure 1.27(c).

Example   1.7   **Objective:**   Determine the diode voltage and current in the circuit shown in Figure 1.24, using a piecewise linear model.

Assume piecewise linear diode parameters of $V_\gamma = 0.6\,\text{V}$ and $r_f = 10\,\Omega$.

Solution: With the given input voltage polarity, the diode is forward biased or "turned on," so $I_D > 0$. The equivalent circuit is shown in Figure 1.27(a). The diode current is determined by

$$I_D = \frac{V_{PS} - V_\gamma}{R + r_f} = \frac{5 - 0.6}{2 \times 10^3 + 10} \Rightarrow 2.19\,\text{mA}$$

and the diode voltage is

$$V_D = V_\gamma + I_D r_f = 0.6 + (2.19 \times 10^{-3})(10) = 0.622\,\text{V}$$

Comment: This solution, obtained using the piecewise linear model, is nearly equal to the solution obtained in Example 1.6, in which the ideal diode equation was used. However, the analysis using the piecewise-linear model in this example is by far easier than using the actual diode I–V characteristics as was done in Example 1.6. In general, we are willing to accept some slight analysis inaccuracy for ease of analysis.

Because the forward diode resistance r_f in Example 1.7 is much smaller than the circuit resistance R, the diode current I_D is essentially independent of the value of r_f. In addition, if the cut-in voltage is 0.7 V instead of 0.6 V, the calculated diode current will be 2.15 mA, which is not significantly different from the previous results. Therefore, the calculated diode current is not a strong function of the cut-in voltage. Consequently, we will often assume a cut-in voltage of 0.7 V for silicon pn junction diodes.

The concept of the load line and the piecewise linear model can be combined in diode circuit analyses. Using Kirchhoff's voltage law, expressed as Equation 1.14(b), and the circuit in Figure 1.24, assume $V_\gamma = 0.7\,\text{V}$, $r_f = 0$, $V_{PS} = +5\,\text{V}$, and $R = 2\,\text{k}\Omega$. Figure 1.28(a) shows the resulting load line and the piecewise linear characteristic curves of the diode. The two curves intersect

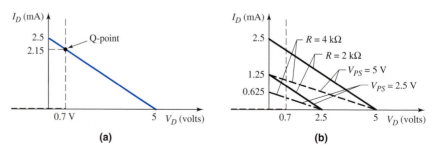

Figure 1.28   Piecewise linear approximation (a) load line for $V_{PS} = 5\,\text{V}$ and $R = 2\,\text{k}\Omega$ and (b) several load lines

at the Q-point, or diode current, $I_{DQ} \cong 2.15\,\text{mA}$, which is essentially a function of only V_{PS} and R. Figure 1.28(b) shows the same piecewise linear characteristics of the diode but with four different load lines, corresponding to: $V_{PS} = 5\,\text{V}$, $R = 4\,\text{k}\Omega$; $V_{PS} = 5\,\text{V}$, $R = 2\,\text{k}\Omega$; $V_{PS} = 2.5\,\text{V}$, $R = 4\,\text{k}\Omega$; and $V_{PS} = 2.5\,\text{V}$, $R = 2\,\text{k}\Omega$. The Q-point changes for each load line.

The load line concept is also useful when the diode is reverse biased. Figure 1.29(a) shows the same diode circuit as before, but with the direction of the diode reversed. The diode current I_D and voltage V_D shown are the usual forward-biased parameters. Applying Kirchhoff's voltage law, we can write

$$V_{PS} = I_{PS}R - V_D = -I_D R - V_D \qquad (1.22(a))$$

or

$$I_D = -\frac{V_{PS}}{R} - \frac{V_D}{R} \qquad (1.22(b))$$

where $I_D = -I_{PS}$. Equation (1.22(b)) is the load line equation. The two end points are found by setting $I_D = 0$, which yields $V_D = -V_{PS} = -5\,\text{V}$, and by setting $V_D = 0$, which yields $I_D = -V_{PS}/R = -5/2 = -2.5\,\text{mA}$. The diode characteristics and the load line are plotted in Figure 1.29(b). We see that the load is now in the third quadrant, where it intersects the diode characteristics curve at $V_D = -5\,\text{V}$ and $I_D = 0$, demonstrating that the diode is reverse biased.

Although the piecewise linear model may yield solutions that are less accurate than those obtained with the ideal diode equation, the analysis is much easier.

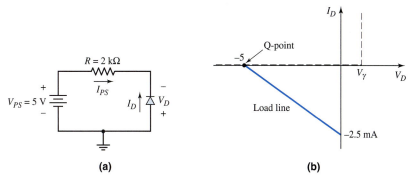

(a) **(b)**

Figure 1.29 Reverse-biased diode (a) circuit and (b) piecewise linear approximation and load line

Test Your Understanding

1.15 (a) Consider the circuit in Figure 1.24. Let $V_{PS} = 5\,\text{V}$, $R = 4\,\text{k}\Omega$, and $V_\gamma = 0.7\,\text{V}$. Assume $r_f = 0$. Determine I_D. (b) If V_{PS} is increased to 8 V, what must be the new value of R such that I_D is the same value as in part (a)? (c) Draw the diode characteristics and load lines for parts (a) and (b). (Ans. (a) $I_D = 1.08\,\text{mA}$, (b) $R = 6.79\,\text{k}\Omega$)

1.16   The power supply (input) voltage in the circuit of Figure 1.24 is $V_{PS} = 10\,\text{V}$ and the diode cut-in voltage is $V_\gamma = 0.7\,\text{V}$ (assume $r_f = 0$). The power dissipated in the diode is to be no more than $1.05\,\text{mW}$. Determine the maximum diode current and the minimum value of R to meet the power specification. (Ans. $I_D = 1.5\,\text{mA}$, $R = 6.2\,\text{k}\Omega$)

1.3.3   Computer Simulation and Analysis

Today's computers are capable of using detailed simulation models of various components and performing complex circuit analyses quickly and relatively easily. Such models can factor in many diverse conditions, such as the temperature dependence of various parameters. One of the earliest, and now the most widely used, circuit analysis programs is the *s*imulation *p*rogram with *i*ntegrated *c*ircuit *e*mphasis (SPICE). This program, developed at the University of California at Berkeley, was first released about 1973, and has been continuously refined since that time. One outgrowth of SPICE is PSpice, which is designed for use on personal computers.

Example   1.8   **Objective:**   Determine the diode current and voltage characteristics of the circuit shown in Figure 1.24 using a PSpice analysis.

Solution: Figure 1.30(a) is the PSpice circuit schematic. A standard 1N4002 diode from the PSpice library was used in the analysis. The input voltage $V1$ was varied (dc sweep) from 0 to 5 V. Figure 1.30(b) and (c) shows the diode voltage and diode current characteristics versus the input voltage.

(a)

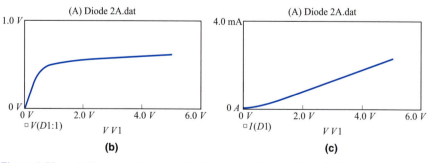

(b)   (c)

Figure 1.30   (a) PSpice circuit schematic, (b) diode voltage, and (c) diode current for Example 1.8

Discussion: Several observations may be made from the results. The diode voltage increases at almost a linear rate up to approximately 400 mV without any discernible (mA) current being measured. For an input voltage greater than approximately 500 mV, the diode voltage increases gradually to a value of about 610 mV at the maximum input voltage. The current also increases to a maximum value of approximately 2.2 mA at the maximum input voltage. The piecewise linear model predicts quite accurate results at the maximum input voltage. However, these results show that there is definitely a non-linear relation between the diode current and diode voltage. We must keep in mind that the piecewise linear model is an approximation technique that works very well in many applications.

1.3.4   Summary of Diode Models

The two dc diode models used in the hand analysis of diode circuits are: the ideal diode equation and the piecewise linear approximation. For the ideal diode equation, the reverse-saturation current I_S must be specified. For the piecewise linear model, the cut-in voltage V_γ and forward diode resistance r_f must be specified. In most cases, however, r_f is assumed to be zero unless otherwise given.

1.4   DIODE CIRCUITS: AC EQUIVALENT CIRCUIT

Up to this point, we have only looked at the dc characteristics of the pn junction diode. When semiconductor devices with pn junctions are used in linear amplifier circuits, the time-varying, or ac, characteristics of the pn junction become important, because sinusoidal signals may be superimposed on the dc currents and voltages. The following sections examine these ac characteristics.

1.4.1   Sinusoidal Analysis

In the circuit shown in Figure 1.31(a), the voltage source v_i is assumed to be a sinusoidal, or time-varying, signal. The total input voltage v_I is composed of a dc component V_{PS} and an ac component v_i superimposed on the dc value. To investigate this circuit, we will look at two types of analyses: a dc analysis involving only the dc voltages and currents, and an ac analysis involving only the ac voltages and currents. (We should point out that the circuit in the figure is not a practical circuit, since it is not desirable to have a dc current flowing through an ac signal source. However, the circuit is useful for a discussion of dc and ac analyses.)

Current–Voltage Relationships

Since the input voltage contains a dc component with an ac signal superimposed, the diode current will also contain a dc component with an ac signal superimposed, as shown in Figure 1.31(b). Here, I_{DQ} is the dc quiescent diode current. In addition, the diode voltage will contain a dc value with an ac signal superimposed, as shown in Figure 1.31(c). For this analysis, assume that the ac

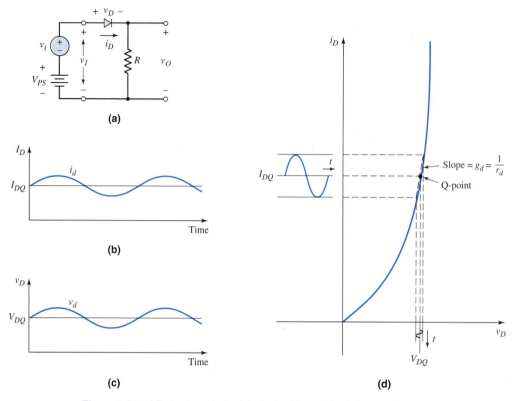

Figure 1.31 AC circuit analysis: (a) circuit with combined dc and sinusoidal input voltages, (b) sinusoidal diode current superimposed on the quiescent current, (c) sinusoidal diode voltage superimposed on the quiescent value, and (d) forward-biased diode *I–V* characteristics with a sinusoidal current and voltage superimposed on the quiescent values

signal is small compared to the dc component, so that a linear ac model can be developed from the nonlinear diode.

The relationship between the diode current and voltage can be written as

$$i_D \cong I_S e^{\left(\frac{v_D}{V_T}\right)} = I_S e^{\left(\frac{V_{DQ}+v_d}{V_T}\right)} \tag{1.23}$$

where V_{DQ} is the dc quiescent voltage and v_d is the ac component. We are neglecting the -1 term in the diode equation. Equation (1.23) can be rewritten as

$$i_D = I_S \left[e^{\left(\frac{V_{DQ}}{V_T}\right)} \right] \cdot \left[e^{\left(\frac{v_d}{V_T}\right)} \right] \tag{1.24}$$

If the ac signal is "small," then $v_d \ll V_T$, and we can expand the exponential function into a linear series, as follows:

$$e^{\left(\frac{v_d}{V_T}\right)} \cong 1 + \frac{v_d}{V_T} \tag{1.25}$$

We may also write the quiescent diode current as

$$I_{DQ} = I_S e^{\left(\frac{V_{DQ}}{V_T}\right)} \tag{1.26}$$

The diode current–voltage relationship from Equation (1.24) can then be written as

$$i_D = I_{DQ}\left(1 + \frac{v_d}{V_T}\right) = I_{DQ} + \frac{I_{DQ}}{V_T} \cdot v_d = I_{DQ} + i_d \tag{1.27}$$

where i_d is the ac component of the diode current. The relationship between the ac components of the diode voltage and current is then

$$i_d = \left(\frac{I_{DQ}}{V_T}\right) \cdot v_d = g_d \cdot v_d \tag{1.28(a)}$$

or

$$v_d = \left(\frac{V_T}{I_{DQ}}\right) \cdot i_d = r_d \cdot i_d \tag{1.28(b)}$$

The parameters g_d and r_d, respectively, are the diode **small-signal incremental conductance** and **resistance**, also called the **diffusion conductance** and **diffusion resistance**. We see from these two equations that

$$r_d = \frac{1}{g_d} = \frac{V_T}{I_{DQ}} \tag{1.29}$$

This equation tells us that the incremental resistance is a function of the dc bias current I_{DQ} and is inversely proportional to the slope of the $I-V$ characteristics curve, as shown in Figure 1.31(d).

Circuit Analysis

To analyze the circuit shown in Figure 1.31(a), we can use the piecewise linear model for the dc calculations and Equation (1.29) for the ac calculation.

Example 1.9   Objective: Analyze the circuit shown in Figure 1.31(a).

Assume circuit and diode parameters of $V_{PS} = 5\,\text{V}$, $R = 5\,\text{k}\Omega$, $V_\gamma = 0.6\,\text{V}$, and $v_i = 0.1 \sin \omega t\,(\text{V})$.

Solution: Divide the analysis into two parts: the dc analysis and the ac analysis.

For *the dc analysis,* we set $v_i = 0$ and then determine the dc quiescent current as

$$I_{DQ} = \frac{V_{PS} - V_\gamma}{R} = \frac{5 - 0.6}{5} = 0.88\,\text{mA}$$

The dc value of the output voltage is

$$V_O = I_{DQ}R = (0.88)(5) = 4.4\,\text{V}$$

For *the ac analysis,* we consider only the ac signals and parameters in the circuit. In other words, we effectively set $V_{PS} = 0$. The ac Kirchhoff voltage law (KVL) equation becomes

$$v_i = i_d r_d + i_d R = i_d (r_d + R)$$

where r_d is again the small-signal diode diffusion resistance. From Equation (1.29), we have

$$r_d = \frac{V_T}{I_{DQ}} = \frac{0.026}{0.88} = 0.0295 \, \text{k}\Omega$$

The ac diode current is

$$i_d = \frac{v_i}{r_d + R} = \frac{0.1 \sin \omega t}{0.0295 + 5} \Rightarrow 19.9 \sin \omega t (\mu\text{A})$$

The ac component of the output voltage is

$$v_o = i_d R = 0.0995 \sin \omega t (\text{V})$$

Comment: Throughout the text, we will divide the circuit analysis into a dc analysis and an ac analysis. To do so, we will use separate equivalent circuit models for each analysis.

Frequency Response

In the previous analysis, we implicitly assumed that the frequency of the ac signal was small enough that capacitance effects in the circuit would be negligible. If the frequency of the ac input signal increases, the **diffusion capacitance** associated with a forward-biased pn junction becomes important. The source of the diffusion capacitance is shown in Figure 1.32, which displays the dc values of the minority carrier concentrations and the changes caused by an ac component being superimposed. The ΔQ charge is alternately being charged and discharged through the junction as the voltage across the junction changes. The diffusion capacitance is the change in the stored minority carrier charge that is caused by a change in the voltage, or

$$C_d = \frac{dQ}{dV_D} \tag{1.30}$$

The diffusion capacitance C_d is normally much larger than the junction capacitance C_j, because of the magnitude of the charges involved.

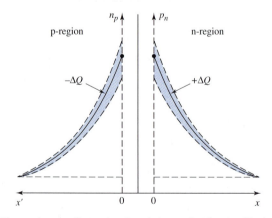

Figure 1.32 Change in minority carrier stored charge, leading to diffusion capacitance

1.4.2 Small-Signal Equivalent Circuit

The small-signal equivalent circuit of the forward-biased pn junction is shown in Figure 1.33 and is developed partially from the equation for the **admittance**, which is given by

$$Y = g_d + j\omega C_d \tag{1.31}$$

where g_d and C_d are the diffusion conductance and capacitance, respectively. We must also add the junction capacitance, which is in parallel with the diffusion resistance and capacitance, and a series resistance, which is required because of the finite resistances in the neutral n- and p-regions.

The small-signal equivalent circuit of the pn junction is used to obtain the ac response of a diode circuit subjected to ac signals superimposed on the Q-point values. Small-signal equivalent circuits of pn junctions are also used to develop small-signal models of transistors, and these models are used in the analysis and design of transistor amplifiers.

(a)

(b)

Figure 1.33 Small-signal equivalent circuit of the diode: (a) simplified version and (b) complete circuit

Test Your Understanding

1.17   Determine the diffusion conductance of a pn junction diode at $T = 300\,°\text{K}$ and biased at a current of 0.8 mA. (Ans. $g_d = 30.8\,\text{mS}$)

1.18   The diffusion resistance of a pn junction diode at $T = 300\,°\text{K}$ is determined to be $r_d = 50\,\Omega$. What is the quiescent diode current? (Ans. $I_{DQ} = 0.52\,\text{mA}$)

1.5   OTHER DIODE TYPES

Other types of diodes with specialized characteristics include the solar cell, photodiode, light-emitting diode, Schottky diode, and Zener diode. The solar cell, photodiode, light-emitting diode, and Zener diode are types of pn junction diodes with specific characteristics that make them useful in particular circuit applications.

1.5.1   Solar Cell

A **solar cell** is a pn junction device with no voltage directly applied across the junction. The pn junction, which converts solar energy into electrical energy, is connected to a load as indicated in Figure 1.34. When light hits the space-charge region, electrons and holes are generated. They are quickly separated and swept out of the space-charge region by the electric field, thus creating a **photocurrent**. The generated photocurrent will produce a voltage across the load, which means that the solar cell has supplied power. Solar cells are usually fabricated from silicon, but may be made from GaAs or other III–V compound semiconductors.

Solar cells have long been used to power the electronics in satellites and space vehicles, and also as the power supply to some calculators. Solar cells are also used to power race cars in a Sunrayce event. Collegiate teams in the United States design, build and drive the race cars. Typically, a Sunrayce car

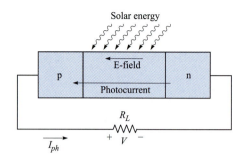

Figure 1.34   A pn junction solar cell connected to load

has $8\,\mathrm{m}^2$ of solar cell arrays that can produce 800 W of power on a sunny day at noon. The power from the solar array can be used either to power an electric motor or to charge a battery pack.

1.5.2   Photodiode

Photodetectors are devices that convert optical signals into electrical signals. An example is the **photodiode**, which is similar to a solar cell except that the pn junction is operated with a reverse-bias voltage. Incident photons or light waves create excess electrons and holes in the space-charge region. These excess carriers are quickly separated and swept out of the space-charge region by the electric field, thus creating a "photocurrent." This generated photocurrent is directly proportional to the incident photon flux.

1.5.3   Light-Emitting Diode

The **light-emitting diode (LED)** converts current to light. As previously explained, when a forward-bias voltage is applied across a pn junction, electrons and holes flow across the space-charge region and become excess minority carriers. These excess minority carriers diffuse into the neutral semiconductor regions, where they recombine with majority carriers. If the semiconductor is a **direct bandgap material**, such as GaAs, the electron and hole can recombine with no change in momentum, and a photon or light wave can be emitted. Conversely, in an **indirect bandgap material**, such as silicon, when an electron and hole recombine, both energy and momentum must be conserved, so the emission of a photon is very unlikely. Therefore, LEDs are fabricated from GaAs or other compound semiconductor materials. In an LED, the diode current is directly proportional to the recombination rate, which means that the output light intensity is also proportional to the diode current.

Monolithic arrays of LEDs are fabricated for numeric and alphanumeric displays, such as the readout of a digital voltmeter.

An LED may be integrated into an optical cavity to produce a coherent photon output with a very narrow bandwidth. Such a device is a laser diode, which is used in optical communications applications.

The LED can be used in conjunction with a photodiode to create an optical system such as that shown in Figure 1.35. The light signal created

Figure 1.35 Basic elements in an optical transmission system

may travel over relatively long distances through the optical fiber, because of the low optical absorption in high-quality optical fibers.

1.5.4 Schottky Barrier Diode

A **Schottky barrier diode**, or simply a Schottky diode, is formed when a metal, such as aluminium, is brought into contact with a *moderately* doped n-type semiconductor. Figure 1.36(a) shows the metal-semiconductor contact, and Figure 1.36(b) shows the circuit symbol with the current direction and voltage polarity.

The current–voltage characteristics of a Schottky diode are very similar to those of a pn junction diode. The same ideal diode equation can be used for both devices. However, there are two important differences between the two diodes that directly affect the response of the Schottky diode.

First, the current mechanism in the two devices is different. The current in a pn junction diode is controlled by the diffusion of minority carriers. The current in a Schottky diode results from the flow of majority carriers over the potential barrier at the metallurgical junction. This means that there is no minority carrier storage in the Schottky diode, so the switching time from a forward bias to a reverse bias is very short compared to that of a pn junction diode. The storage time, t_s, for a Schottky diode is essentially zero.

Second, the reverse-saturation current I_S for a Schottky diode is larger than that of a pn junction diode for comparable device areas. This property means that the current in a Schottky diode is larger than that in a pn junction diode for the same forward-bias voltage.

Figure 1.37 compares the characteristics of the two diodes. Applying the piecewise linear model, we can determine that the Schottky diode has a smaller turn-on voltage than the pn junction diode. In later chapters, we will see how this lower turn-on voltage and the shorter switching time make the Schottky diode useful in integrated-circuit applications.

(a)

(b)

Figure 1.36 Schottky barrier diode: (a) simplified geometry and (b) circuit symbol

Figure 1.37 Comparison of the forward-bias *I–V* characteristics of a pn junction diode and a Schottky barrier diode

Figure 1.38 Simple circuit with both a pn junction diode and a Schottky barrier diode

Example 1.10   Objective: Calculate the currents in a circuit containing both a pn junction diode and a Schottky diode.

Consider the circuit shown in Figure 1.38. Assume the cut-in voltages for the pn junction diode and the Schottky diode are $V_\gamma = 0.7\,$V and $V_\gamma = 0.3\,$V, respectively. Let $r_f = 0$ for both diodes.

Solution: The current I_1 is the voltage difference across R_1 divided by the resistance R_1, or

$$I_1 = \frac{4 - 0.7}{4} = 0.825\,\text{mA}$$

Similarly, the current I_2 is the voltage difference across R_2 divided by the resistance R_2, or

$$I_2 = \frac{4 - 0.3}{4} = 0.925\,\text{mA}$$

Comment: The dc calculations for a circuit containing a Schottky diode are the same as those for a circuit containing a pn junction diode.

Another type of metal–semiconductor junction is also possible. A metal applied to a heavily doped semiconductor forms, in most cases, an *ohmic contact*: that is, a contact that conducts current equally in both directions, with very little voltage drop across the junction. Ohmic contacts are used to connect one semiconductor device to another on an IC, or to connect an IC to its external terminals.

Test Your Understanding

1.19 The reverse-saturation currents of a pn junction diode and a Schottky diode are $I_S = 10^{-12}$ A and 10^{-8} A, respectively. Determine the forward-bias voltages required to produce 1 mA in each diode. (Ans. pn diode, $V_D = 0.539$ V; Schottky diode, $V_D = 0.299$ V)

1.20 A pn junction diode and a Schottky diode both have forward-bias currents of 1.2 mA. The reverse-saturation current of the pn junction diode is $I_S = 4 \times 10^{-15}$ A. The difference in forward-bias voltages is 0.265 V. Determine the reverse-saturation current of the Schottky diode. (Ans. $I_S = 1.07 \times 10^{-10}$ A)

1.5.5   Zener Diode

As mentioned earlier in this chapter, the applied reverse-bias voltage cannot increase without limit. At some point, breakdown occurs and the current in the reverse-bias direction increases rapidly. The voltage at this point is called the breakdown voltage. The diode I–V characteristics, including breakdown, are shown in Figure 1.39.

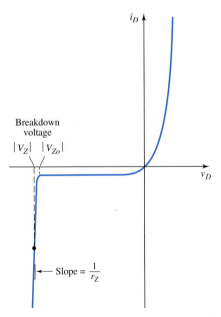

Figure 1.39   Diode I–V characteristics showing breakdown effects

Diodes, called **Zener diodes**, can be designed and fabricated to provide a specified breakdown voltage V_{Zo}. (Although the breakdown voltage is on the negative voltage axis (reverse-bias), its value is given as a positive quantity.) The large current that may exist at breakdown can cause heating effects and catastrophic failure of the diode due to the large power dissipation in the device. However, diodes can be operated in the breakdown region by limiting the current to a value within the capabilities of the device. Such a diode can be used as a constant-voltage reference in a circuit. The diode breakdown voltage is essentially constant over a wide range of currents and temperatures. The slope of the I–V characteristics curve in breakdown is quite large, so the incremental resistance r_z is small. Typically, r_z is in the range of a few ohms or tens of ohms.

The circuit symbol of the Zener diode is shown in Figure 1.40. (Note the difference between this symbol and the Schottky diode symbol.) The voltage V_Z is the Zener breakdown voltage, and the current I_Z is the reverse-bias current when the diode is operating in the breakdown region.

Figure 1.40 Circuit symbol of the Zener diode

Design Example 1.11 **Objective:** Consider a simple constant-voltage reference circuit and design the value of resistance required to limit the current in this circuit.

Consider the circuit shown in Figure 1.41. Assume that the Zener diode breakdown voltage is $V_Z = 5.6$ V and the Zener resistance is $r_z = 0$. The current in the diode is to be limited to 3 mA.

Solution: As before, we can determine the current from the voltage difference across R divided by the resistance. That is,

$$I = \frac{V_{PS} - V_Z}{R}$$

The resistance is then

$$R = \frac{V_{PS} - V_Z}{I} = \frac{10 - 5.6}{3} = 1.47 \text{ k}\Omega$$

Figure 1.41 Simple circuit containing a Zener diode

Comment: The resistance external to the Zener diode limits the current when the diode is operating in the breakdown region. In the circuit shown in the figure, the output voltage will remain constant at 5.6 V, even though the power supply voltage and the resistance may change over a limited range. Hence, this circuit provides a constant output voltage. We will see further applications of the Zener diode in the next chapter.

Test Your Understanding

1.21 Consider the circuit shown in Figure 1.41. Determine the value of resistance R required to limit the power dissipated in the Zener diode to 10 mW. (Ans. $R = 2.46$ kΩ)

1.22 A Zener diode has an equivalent series resistance of 20 Ω. If the voltage across the Zener diode is 5.20 V at $I_Z = 1$ mA, determine the voltage across the diode at $I_Z = 10$ mA. (Ans. $V_Z = 5.38$ V)

1.6 SUMMARY

- We initially considered some of the characteristics and properties of semiconductor materials. We discussed the concept of electrons (negative charge) and holes (positive charge) as two distinct charge carriers in a semiconductor. The doping of pure semiconductor crystals with specific types of impurity atoms produces either n-type materials, which have a preponderance of electrons, or p-type materials, which have a preponderance of holes. The concepts of n-type and p-type materials are used throughout the text.

- A pn junction diode is formed when an n-doped region and a p-doped region are directly adjacent to each other. The current–voltage characteristics of the diode are nonlinear: The current is an exponential function of voltage in the forward-bias condition, and is essentially zero in the reverse-bias condition.

- Since the i–v relationship of the diode is nonlinear, the analysis of circuits containing diodes is not as straightfoward as that of linear circuits that contain only linear resistors. A piecewise-linear model of the diode was developed so that approximate hand calculation results can be easily obtained. The i–v characteristics of the diode are broken into linear segments, which are valid over particular regions of operation. The concept of a diode turn-on voltage was introduced as part of the piecewise linear model.

- Time-varying, or ac signals, may be superimposed on a dc diode current and voltage. A small-signal linear equivalent circuit was developed and is used to determine the relationship between the ac current and ac voltage. This same equivalent circuit will be applied extensively when the frequency response of transistors is discussed.

CHECKPOINT

After studying this chapter, the reader should have the ability to:

✔ Understand the concept of intrinsic carrier concentration, the difference between n-type and p-type materials, and the concept of drift and diffusion currents. (Section 1.1)

✔ Analyze a simple diode circuit using the ideal diode current–voltage characteristics and using the iteration analysis technique. (Section 1.3)

✔ Analyze a diode circuit using the piecewise linear approximation model for the diode. (Section 1.3)

✔ Determine the small-signal characteristics of a diode using the small-signal equivalent circuit. (Section 1.4)

✔ Understand the general characteristics of a solar cell, light-emitting diode, Schottky barrier diode, and Zener diode. (Section 1.5)

REVIEW QUESTIONS

1. Describe an intrinsic semiconductor material. What is meant by the intrinsic carrier concentration?

2. Describe the concept of an electron and a hole as charge carriers in the semiconductor material.

3. Describe an extrinsic semiconductor material. What is the value of the electron concentration in an n-type material, and what is the value of the hole concentration in a p-type material?

4. Describe the concepts of drift current and diffusion current in a semiconductor material.

5. How is a pn junction formed? What is meant by a built-in potential barrier, and how is it formed?

6. How is a junction capacitance created in a reverse-biased pn junction diode?

7. Write the ideal diode current–voltage relationship. Describe the meaning of I_S and V_T.

8. Describe the iteration method of analysis and when it must be used to analyze a diode circuit.

9. Describe the piecewise linear model of a diode and why it is useful. What is the diode turn-on voltage?

10. Define a load line in a simple diode circuit.

11. Under what conditions is the small-signal model of a diode used in the analysis of a diode circuit?

12. Describe the operation of a simple solar cell circuit.

13. How do the i–v characteristics of a Schottky barrier diode differ from those of a pn junction diode?

14. What characteristic of a Zener diode is used in the design of a Zener diode circuit?

PROBLEMS

[Note: Unless otherwise specified, assume that $T = 300\,°K$ in the following problems. Also, assume the emission coefficient is $n = 1$ unless otherwise stated.]

Section 1.1   Semiconductor Materials and Properties

1.1 (a) Calculate the intrinsic carrier concentration in silicon at (i) $T = 275\,°K$ and (ii) $T = 325\,°K$. (b) Repeat part (a) for gallium arsenide.

1.2 (a) The intrinsic carrier concentration in silicon is to be no larger than $n_i = 10^{12}\,\text{cm}^{-3}$. Determine the maximum allowable temperature. (b) Repeat part (a) for $n_i = 10^9\,\text{cm}^{-3}$.

1.3 (a) Find the concentrations of electrons and holes in a sample of silicon that has a concentration of donor atoms equal to $5 \times 10^{15}\,\text{cm}^{-3}$. Is the semiconductor n-type or p-type? (b) Repeat part (a) for gallium arsenide.

1.4 (a) Calculate the concentration of electrons and holes in a silicon semiconductor sample that has a concentration of acceptor atoms equal to $10^{16}\,\text{cm}^{-3}$. Is the semiconductor n- or p-type? (b) Repeat part (a) for germanium.

1.5 The electron concentration in silicon at $T = 300\,°K$ is $n_o = 5 \times 10^{15}\,\text{cm}^{-3}$. (a) Determine the hole concentration. (b) Is the material n-type or p-type? (c) What is the impurity doping concentration?

D1.6 (a) A silicon semiconductor material is to be designed such that the majority carrier electron concentration is $n_o = 7 \times 10^{15}\,\text{cm}^{-3}$. Should donor or acceptor impurity atoms be added to intrinsic silicon to achieve this electron concentration? What concentration of dopant impurity atoms is required? (b) In this silicon material, the minority carrier hole concentration is to be no larger than $p_o = 10^6\,\text{cm}^{-3}$. Determine the maximum allowable temperature.

1.7 The applied electric field in p-type silicon is $E = 15\,\text{V/cm}$. The semiconductor conductivity is $\sigma = 2.2\,(\Omega\text{–cm})^{-1}$ and the cross-sectional area is $A = 10^{-4}\,\text{cm}^2$. Determine the drift current in the semiconductor.

1.8 A drift current density of 85 A/cm^2 is established in n-type silicon with an applied electric field of $E = 12$ V/cm. Determine the conductivity of the semiconductor.

1.9 In GaAs, the mobilities are $\mu_n = 8500$ cm^2/V–s and $\mu_p = 400$ cm^2/V–s. (a) Determine the range in conductivity for a range in donor concentration of $10^{15} \leq N_d \leq 10^{19}$ cm^{-3}. (b) Using the results of part (a), determine the range in drift current density if the applied electric field is $E = 0.10$ V/cm.

1.10 GaAs is doped to $N_a = 10^{17}$ cm^{-3}. (a) Calculate n_o and p_o. (b) Excess electrons and holes are generated such that $\delta n = \delta p = 10^{15}$ cm^{-3}. Determine the total concentration of electrons and holes.

Section 1.2   The pn Junction

1.11 Calculate V_{bi} in a silicon pn junction for: (a) $N_d = N_a = 10^{15}$ cm^{-3}; (b) $N_d = 10^{15}$ cm^{-3}, $N_a = 10^{18}$ cm^{-3}; and (c) $N_d = N_a = 10^{18}$ cm^{-3}.

1.12 Repeat Problem 1.11 for gallium arsenide.

1.13 The donor concentration in the n-region of a silicon pn junction is $N_d = 10^{16}$ cm^{-3}. Plot V_{bi} versus N_a over the range $10^{15} \leq N_a \leq 10^{18}$ cm^{-3} where N_a is the acceptor concentration in the p-region.

1.14 Consider a uniformly doped GaAs pn junction with doping concentrations of $N_a = 5 \times 10^{18}$ cm^{-3} and $N_d = 5 \times 10^{16}$ cm^{-3}. Plot the built-in potential barrier V_{bi} versus temperature for $200\,°K \leq T \leq 500\,°K$.

1.15 A silicon pn junction has zero-bias junction capacitance of $C_{jo} = 1$ pF and doping concentrations of $N_a = 2 \times 10^{16}$ cm^{-3} and $N_d = 2 \times 10^{15}$ cm^{-3}. Calculate the junction capacitance at: (a) $V_R = 1$ V and (b) $V_R = 5$ V.

*1.16 The zero-bias capacitance of a silicon pn junction diode is $C_{jo} = 0.02$ pF and the built-in potential is $V_{bi} = 0.80$ V. The diode is reverse biased through a 47-kΩ resistor and a voltage source. (a) For $t < 0$, the applied voltage is 5 V and, at $t = 0$, the applied voltage drops to zero volts. Estimate the time it takes for the diode voltage to change from 5 V to 1.5 V. (As an approximation, use the average diode capacitance between the two voltage levels.) (b) Repeat part (a) for an input voltage change from 0 V to 5 V and a diode voltage change from 0 V to 3.5 V. (Use the average diode capacitance between these two voltage levels.)

1.17 A silicon pn junction is doped at $N_a = 10^{18}$ cm^{-3} and $N_d = 10^{15}$ cm^{-3}. The zero-bias junction capacitance is $C_{jo} = 0.25$ pF. An inductance of 2.2 mH is placed in parallel with the pn junction. Calculate the resonant frequency f_o of the circuit for reverse-bias voltages of: (a) $V_R = 1$ V, and (b) $V_R = 10$ V.

1.18 (a) At what reverse bias voltage does the reverse-bias current in a silicon pn junction diode reach 90 percent of its saturation value? (b) What is the ratio of the current for a forward-bias voltage of 0.2 V to the current for a reverse-bias voltage of 0.2 V?

1.19 (a) Determine the current in a silicon pn junction diode for forward-bias voltages of 0.5, 0.6, and 0.7 V if the reverse-saturation current is $I_S = 10^{-11}$ A. (b) Repeat part (a) for $I_S = 10^{-13}$ A.

1.20 For a pn junction diode, what must be the forward-bias voltage to produce a current of 150 μA if (a) $I_S = 10^{-11}$ A and (b) $I_S = 10^{-13}$ A.

1.21 A silicon pn junction diode has an emission coefficient of $n = 2$. The diode current is 1 mA when $V_D = 0.7$ V. (a) Find the reverse-saturation current. (b) Determine the diode current when the voltage is increased to 0.8 V. (c) Repeat parts (a) and (b) when the emission coefficient is $n = 1$.

1.22   The reverse-saturation current of a silicon pn junction diode at $T = 300\,°K$ is $I_S = 10^{-12}$ A. Determine the temperature range over which I_S varies from 0.5×10^{-12} A to 50×10^{-12} A.

1.23   A silicon pn junction diode has an applied forward-bias voltage of 0.6 V. Determine the ratio of current at $100\,°C$ to that at $-55\,°C$.

1.24   (a) Consider a silicon pn junction diode operating in the forward-bias region. Determine the increase in forward-bias voltage that will cause a factor of 10 increase in current. (b) Repeat part (a) for a factor of 100 increase in current.

Section 1.3   DC Diode Analysis

1.25   A pn junction diode is in series with a 10 MΩ resistor and a 1.5 V power supply. The reverse-saturation current of the diode is $I_S = 30$ nA. (a) Determine the diode current and voltage if the diode is forward biased. (b) Repeat part (a) if the diode is reverse biased.

*1.26   (a) The diode in the circuit shown in Figure P1.26 has a reverse-saturation current of $I_S = 5 \times 10^{-13}$ A. Determine the diode voltage and current. (b) Repeat part (a) with a computer simulation analysis.

Figure P1.26

1.27   The reverse-saturation current of each diode in the circuit shown in Figure P1.27 is $I_S = 2 \times 10^{-13}$ A. Determine the input voltage V_I required to produce an output voltage of $V_O = 0.60$ V.

Figure P1.27   **Figure P1.28**

1.28   (a) In the circuit shown in Figure P1.28, find the diode voltage V_D and the supply voltage V such that the current is $I = 0.50$ mA. Assume the reverse-saturation current is $I_S = 5 \times 10^{-12}$ A. (b) From the results of part (a), determine the power dissipated in the diode.

1.29   (a) Consider the circuit shown in Figure P1.26. The value of R_1 is reduced to $R_1 = 10$ kΩ and the cut-in voltage of the diode is $V_\gamma = 0.7$ V. Determine I_D and V_D. (b) Repeat part (a) if $R_1 = 50$ kΩ. (c) Repeat parts (a) and (b) with a computer simulation analysis.

1.30 The cut-in voltage for each diode in the circuits shown in Figure P1.30 is $V_\gamma = 0.6\,\text{V}$. For each circuit, determine the diode current I_D and the voltage V_O (measured with respect to ground potential).

Figure P1.30

*D1.31 The cut-in voltage of the diode shown in the circuit in Figure P1.31 is $V_\gamma = 0.7\,\text{V}$. The diode is to remain biased "on" for a power supply voltage in the range $5 \le V_{PS} \le 10\,\text{V}$. The minimum diode current is to be $I_D(\text{min}) = 2\,\text{mA}$. The maximum power dissipated in the diode is to be no more than $10\,\text{mW}$. Determine appropriate values of R_1 and R_2.

Figure P1.31

1.32 Assume each diode in the circuit shown in Figure P1.32 has a cut-in voltage of $V_\gamma = 0.65\,\text{V}$. The input voltage is $V_I = 5\,\text{V}$. Determine the value of R_1 required such that I_{D1} is one-half the value of I_{D2}. What are the values of I_{D1} and I_{D2}?

Figure P1.32

1.33   The voltage V in Figure P1.28 is $V = 1.7\,\text{V}$. If the cut-in voltage of the diode is $V_\gamma = 0.65\,\text{V}$, determine the new value of R required to limit the power dissipation in the diode to no more than $0.20\,\text{mW}$.

1.34   Repeat Problem 1.25 if the diode cut-in voltage is $V_\gamma = 0.7\,\text{V}$. Compare these results to those obtained in Problem 1.25. Discuss any discrepancies.

Section 1.4   Small-Signal Diode Analysis

1.35   (a) Consider a pn junction diode biased at $I_{DQ} = 1\,\text{mA}$. A sinusoidal voltage is superimposed on V_{DQ} such that the peak-to-peak sinusoidal current is $0.05I_{DQ}$. Find the value of the applied peak-to-peak sinusoidal voltage. (b) Repeat part (a) if $I_{DQ} = 0.1\,\text{mA}$.

*1.36   The diode in the circuit shown in Figure P1.36 is biased with a constant current source I. A sinusoidal signal v_s is coupled through R_S and C. Assume that C is large so that it acts as a short circuit to the signal. (a) Show that the sinusoidal component of the diode voltage is given by

$$v_o = v_s\left(\frac{V_T}{V_T + IR_S}\right)$$

(b) If $R_S = 260\,\Omega$, find v_o/v_s for $I = 1\,\text{mA}$, $I = 0.1\,\text{mA}$, and $I = 0.01\,\text{mA}$.

Figure P1.36

Section 1.5   Other Types of Diodes

1.37   The reverse-saturation currents of a pn junction diode and a Schottky diode are $I_S = 10^{-14}\,\text{A}$ and $10^{-9}\,\text{A}$, respectively. Determine the forward-bias voltages required to produce a current of $100\,\mu\text{A}$ in each diode.

1.38   A pn junction diode and a Schottky diode have equal cross-sectional areas and have forward-bias currents of $0.5\,\text{mA}$. The reverse-saturation current of the Schottky diode is $I_S = 5 \times 10^{-7}\,\text{A}$. The difference in forward-bias voltages between the two diodes is $0.30\,\text{V}$. Determine the reverse-saturation current of the pn junction diode.

1.39   Consider the circuit shown in Figure P1.39. The reverse-saturation currents of the Schottky diode and pn junction diode are $10^{-8}\,\text{A}$ and $10^{-12}\,\text{A}$, respectively. Determine the value of R such that the currents in the diodes are equal. What is the voltage across each diode?

Figure P1.39

1.40   The reverse-saturation currents of a Schottky diode and a pn junction diode are $I_S = 5 \times 10^{-8}\,\text{A}$ and $10^{-12}\,\text{A}$, respectively. (a) The diodes are connected in parallel and the parallel combination is driven by a constant current of $0.5\,\text{mA}$. (i) Determine the current in each diode. (ii) Determine the voltage across each diode. (b) Repeat part (a) for the diodes connected in series, with a voltage of 0.90 V connected across the series combination.

*1.41 Consider the Zener diode circuit shown in Figure P1.41. The Zener breakdown voltage is $V_Z = 5.6\,\text{V}$ at $I_Z = 0.1\,\text{mA}$, and the incremental Zener resistance is $r_z = 10\,\Omega$. (a) Determine V_O with no load ($R_L = \infty$). (b) Find the change in the output voltage if V_{PS} changes by $\pm 1\,\text{V}$. (c) Find V_O if $V_{PS} = 10\,\text{V}$ and $R_L = 2\,\text{k}\Omega$.

Figure P1.41

1.42 A voltage regulator consists of a 6.8 V Zener diode in series with a $200\,\Omega$ resistor and a 9 V power supply. (a) Neglecting r_z, calculate the diode current and power dissipation. (b) If the power supply is increased to 12 V, calculate the percentage increase in diode current and power dissipation.

*1.43 Consider the Zener diode circuit shown in Figure P1.41. The Zener diode voltage is $V_Z = 6.8\,\text{V}$ at $I_Z = 0.1\,\text{mA}$ and the incremental Zener resistance is $r_z = 20\,\Omega$. (a) Calculate V_O with no load ($R_L = \infty$). (b) Find the change in the output voltage when a load resistance of $R_L = 1\,\text{k}\Omega$ is connected.

COMPUTER SIMULATION PROBLEMS

1.44 Use a computer simulation to generate the ideal current–voltage characteristics of a diode from a reverse-bias voltage of 5 V to a forward-bias current of 10 mA, for an I_S parameter value of: (a) 10^{-14} A and (b) 10^{-10} A. Use the default values for all other parameters.

1.45 Use a computer simulation to generate the I–V characteristics of a diode with $I_S = 10^{-12}$ A at temperatures of: (a) $T = 0\,°\text{C}$, (b) $T = 25\,°\text{C}$, (c) $T = 75\,°\text{C}$, and (d) $T = 125\,°\text{C}$. Plot the characteristics from a reverse-bias voltage of 5 V to a forward-bias current of 10 mA.

1.46 Consider the circuit shown in Figure 1.31(a) with $V_{PS} = 5\,\text{V}$. Let $I_S = 10^{-14}$ A and assume that v_i is a sinusoidal source with a peak value of 0.25 V. Choose values of R to generate quiescent diode currents of approximately 0.1, 1.0, and 10 mA. From a computer simulation analysis, determine the peak values of the sinusoidal diode current and sinusoidal diode voltage for each dc diode current. Compare the relationship between the ac diode current and voltage to Equation (1.28(b)), where r_d is given by Equation (1.29). Do the computer simulation results compare favorably with the theoretical predictions?

1.47 Repeat Problem 1.16 using the actual C versus V_R characteristics.

DESIGN PROBLEMS

[Note: Each design should be verified by a computer simulation.]

*D1.48   Design a circuit to produce the characteristics shown in Figure P1.48, where i_D is the diode current and v_I is the input voltage. Assume the diode has piecewise linear parameters of $V_\gamma = 0.7\,\text{V}$ and $r_f = 0$.

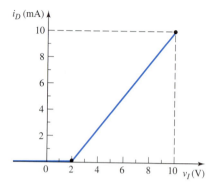

Figure P1.48

*D1.49   Design a circuit to produce the characteristics shown in Figure P1.49 where v_I is the input voltage and i_I is the current supplied by v_I. Assume any diodes in the circuit have piecewise linear parameters of $V_\gamma = 0.7\,\text{V}$ and $r_f = 0$.

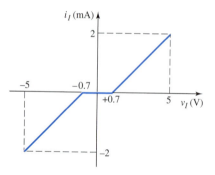

Figure P1.49

*D1.50   Design a circuit to produce the characteristics shown in Figure P1.50, where v_O is an output voltage and v_I is the input voltage.

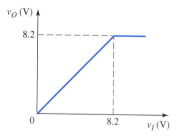

Figure P1.50

2

Diode Circuits

2.0 PREVIEW

In the last chapter, we discussed some of the properties of semiconductor materials, and introduced the diode. We presented the ideal current–voltage relationship, and considered the piecewise linear model, which simplifies the dc analysis of diode circuits. In this chapter, the techniques and concepts developed in Chapter 1 are used to analyze and design electronic circuits containing diodes. A general goal of this chapter is to develop the ability to use the piecewise linear model and approximation techniques in the hand analysis and design of various diode circuits.

Each circuit to be considered accepts an input signal at a set of input terminals and produces an output signal at a set of output terminals. This process is called **signal processing**. The circuit "processes" the input signal and produces an output signal that is a different shape or a different function compared to the input signal. We will see in this chapter how diodes are used to perform these various signal processing functions.

Circuits to be considered perform functions such as rectification, clipping, and clamping. These functions are possible only because of the nonlinear properties of the pn junction diode. The conversion of an ac voltage to a dc voltage, such as for a dc power supply, is called rectification. Clipper diode circuits clip portions of a signal that are above or below some reference level. Clamper circuits shift the entire signal by some dc value.

Zener diodes, which operate in the reverse-bias breakdown region, have the advantage that the voltage across the diode in this region is nearly constant over a wide range of currents. Such diodes are used in voltage reference or voltage regulator circuits. Finally, we look at the circuits of two special diodes: the light-emitting diode (LED) and the photodiode. An LED circuit is used in visual displays, such as the seven-segment numerical display. The photodiode circuit is used to detect the presence or absence of light and convert this information into an electrical signal.

Although diodes are useful electronic devices, we will begin to see the limitations of these devices and the desirability of having some type of "amplifying" device.

2.1   RECTIFIER CIRCUITS

One important application of diodes is in the design of rectifier circuits. A diode rectifier forms the first stage of a dc power supply as shown in Figure 2.1. As we will see throughout the text, a dc power supply is required to bias all electronic circuits. The dc output voltage v_O will usually be in the range of 3 to 24 V depending on the particular electronics application. Throughout the first part of this chapter, we will analyze and design the various stages in the power supply circuit.

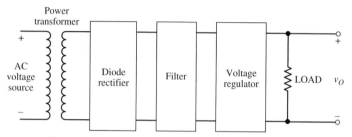

Figure 2.1   Block diagram of an electronic power supply

Rectification is the process of converting an alternating (ac) voltage into one that is limited to one polarity. The diode is useful for this function because of its nonlinear characteristics, that is, current exists for one voltage polarity, but is essentially zero for the opposite polarity. Rectification is classified as **half-wave** or **full-wave**, with half-wave being the simplest.

2.1.1   Half-Wave Rectification

Figure 2.2(a) shows a power transformer with a diode and resistor connected to the secondary of the transformer. We will use the piecewise linear approach in analyzing this circuit, assuming the diode forward resistance is $r_f = 0$ when the diode is "on."

The input signal, v_I, is, in general, a 120 V(rms), 60 Hz ac signal. Recall that the secondary voltage, v_S, and primary voltage, v_I, of an ideal transformer are related by

$$\frac{v_I}{v_S} = \frac{N_1}{N_2} \tag{2.1}$$

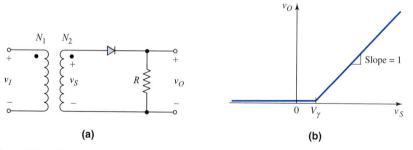

Figure 2.2   Diode in series with ac power source: (a) circuit and (b) voltage transfer characteristics

where N_1 and N_2 are the number of primary and secondary turns, respectively. The ratio N_1/N_2 is called the **transformer turns ratio**. The transformer turns ratio will be designed to provide a particular secondary voltage, v_S, which in turn will produce a particular output voltage v_O.

Problem-Solving Technique: Diode Circuits

In using the piecewise linear model of the diode, the first objective is to determine the linear region (conducting or not conducting) in which the diode is operating. To do this, we can:

1. Determine the input voltage condition such that a diode is conducting (on). Then find the output signal for this condition.
2. Determine the input voltage condition such that a diode is not conducting (off). Then find the output signal for this condition.

[Note: Item 2 can be performed before item 1 if desired.]

Figure 2.2(b) shows the voltage transfer characteristics, v_O versus v_S, for the circuit. For $v_S < 0$, the diode is reverse biased, which means that the current is zero and the output voltage is zero. As long as $v_S < V_\gamma$, the diode will be nonconducting, so the output voltage will remain zero. When $v_S > V_\gamma$, the diode becomes forward biased and a current is induced in the circuit. In this case, we can write

$$i_D = \frac{v_S - V_\gamma}{R} \qquad\qquad\qquad (2.2(a))$$

and

$$v_O = i_D R = v_S - V_\gamma \qquad\qquad\qquad (2.2(b))$$

For $v_S > V_\gamma$, the slope of the transfer curve is 1.

If v_S is a sinusoidal signal, as shown in Figure 2.3(a), the output voltage can be found using the voltage transfer curve in Figure 2.2(b). For $v_S \leq V_\gamma$ the output voltage is zero; for $v_S > V_\gamma$, the output is given by Equation (2.2(b)), or

$$v_O = v_S - V_\gamma$$

and is shown in Figure 2.3(b). We can see that while the input signal v_S alternates polarity and has a time-average value of zero, the output voltage v_O is unidirectional and has an average value that is not zero. The input signal is therefore rectified. Also, since the output voltage appears only during the positive cycle of the input signal, the circuit is called a **half-wave rectifier**.

When the diode is cut off and nonconducting, no voltage drop occurs across the resistor R; therefore, the entire input signal voltage appears across the diode (Figure 2.3(c)). Consequently, the diode must be capable of handling the peak current in the forward direction and sustaining the largest peak inverse voltage (PIV) without breakdown. For the circuit shown in Figure 2.2(a), the value of PIV is equal to the peak value of v_S.

The load line concept can help in visualizing the operation of the half-wave rectifier circuit. Figure 2.4(a) shows the sine wave input. Figure 2.4(b) shows the piecewise linear characteristics of the diode, along with the load lines at

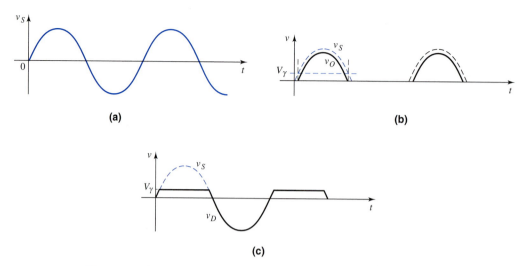

Figure 2.3 Half-wave rectifier circuit: (a) sinusoidal input voltage, (b) output voltage, and (c) diode voltage

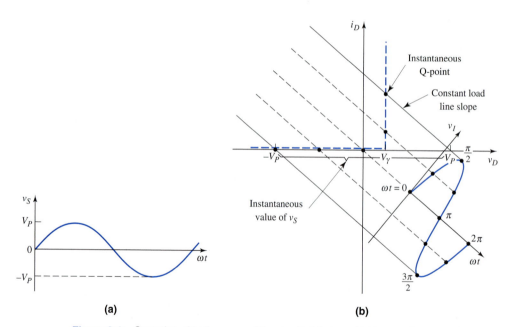

Figure 2.4 Operation of half-wave rectifier circuit: (a) sinusoidal input voltage and (b) diode piecewise linear characteristics and circuit load line at various times

various times. Because the resistance R is a constant, the slope of the load lines remains constant. However, since the magnitude of the power supply voltage varies with time, the magnitude of load line also changes with time. As the load line sweeps across the diode I–V characteristics, the output voltage, diode voltage, and diode current can be determined as a function of time.

We can use a half-wave rectifier circuit to charge a battery as shown in Figure 2.5(a). Charging current exists whenever the instantaneous ac source voltage is greater than the battery voltage plus the diode cut-in voltage as

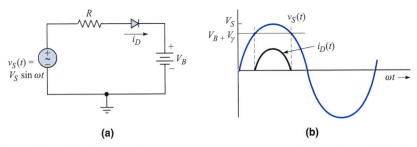

Figure 2.5 (a) Half-wave rectifier used as a battery charger; (b) input voltage and diode current waveforms

shown in Figure 2.5(b). The resistance R in the circuit is to limit the current. When the ac source voltage is less than V_B, the current is zero. Thus current flows only in the direction to charge the battery. One disadvantage of the half-wave rectifier is that we "waste" the negative half-cycles. The current is zero during the negative half-cycles, so there is no energy dissipated, but at the same time, we are not making use of any possible available energy.

Test Your Understanding

2.1 Figure 2.5(a) shows a simple circuit for charging a battery. Assume $V_B = 12\,\text{V}$ and $R = 100\,\Omega$. Also assume that v_S is a sinusoidal signal with a peak amplitude of 24 V and that the diode has piecewise linear parameters of $V_\gamma = 0.6\,\text{V}$ and $r_f = 0$. Determine: (a) the peak diode current; (b) the maximum reverse-bias diode voltage; and (c) the fraction (percent) of the cycle over which the diode conducts. (Ans. (a) 114 mA, (b) 36 V, (c) 32.4%)

2.1.2 Full-Wave Rectification

The full-wave rectifier inverts the negative portions of the sine wave so that a unipolar output signal is generated during both halves of the input sinusoid. One example of a full-wave rectifier circuit appears in Figure 2.6(a). The input to the rectifier consists of a power transformer, in which the input is normally a 120 V (rms), 60 Hz ac signal, and the two outputs are from a center-tapped secondary winding that provides equal voltages v_S, with the polarities shown. When the input line voltage is positive, both output signal voltages v_S are also positive.

The primary winding connected to the 120 V ac source has N_1 windings, and each half of the secondary winding has N_2 windings. The value of the v_S output voltage is $120\,(N_2/N_1)$ volts (rms). The **turns ratio** of the transformer, usually designated (N_1/N_2) can be designed to "step down" the input line voltage to a value that will produce a particular dc output voltage from the rectifier.

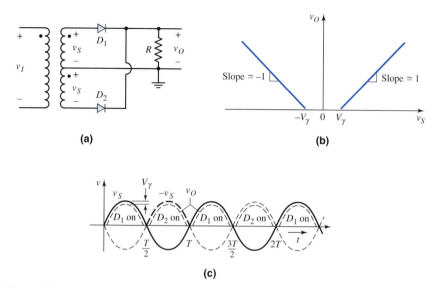

Figure 2.6   Full-wave rectifier: (a) circuit with center-tapped transformer, (b) voltage transfer characteristics, and (c) input and output waveforms

The input power transformer also provides electrical isolation between the powerline circuit and the electronic circuits to be biased by the rectifier circuit. This isolation reduces the risk of electrical shock.

During the positive half of the input voltage cycle, both output voltages v_S are positive; therefore, diode D_1 is forward biased and conducting and D_2 is reverse biased and cut off. The current through D_1 and the output resistance produce a positive output voltage. During the negative half cycle, D_1 is cut off and D_2 is forward biased, or "on," and the current through the output resistance again produces a positive output voltage. If we assume that the forward diode resistance r_f of each diode is small and negligible, we obtain the voltage transfer characteristics, v_O versus v_S, shown in Figure 2.6(b).

For a sinusoidal input voltage, we can determine the output voltage versus time by using the voltage transfer curve shown in Figure 2.6(b). When $v_S > V_\gamma$, D_1 is on and the output voltage is $v_O = v_S - V_\gamma$. When v_S is negative, then for $v_S < -V_\gamma$ or $-v_S > V_\gamma$, D_2 is on and the output voltage is $v_O = -v_S - V_\gamma$. The corresponding input and output voltage signals are shown in Figure 2.6(c). Since a rectified output voltage occurs during both the positive and negative cycles of the input signal, this circuit is called a **full-wave rectifier**.

Another example of a full-wave rectifier circuit appears in Figure 2.7(a). This circuit is a **bridge rectifier**, which still provides electrical isolation between the input ac powerline and the rectifier output, but does not require a center-tapped secondary winding. However, it does use four diodes, compared to only two in the previous circuit.

During the positive half of the input voltage cycle, v_S is positive, D_1 and D_2 are forward biased, D_3 and D_4 are reverse biased, and the direction of the current is as shown in Figure 2.7(a). During the negative half-cycle of the input voltage, v_S is negative, and D_3 and D_4 are forward biased. The direction of the current, shown in Figure 2.7(b), produces the same output voltage polarity as before.

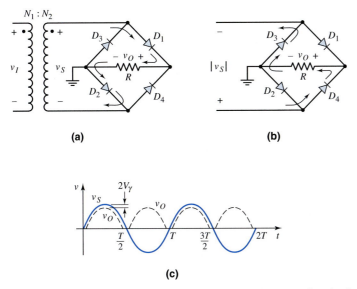

Figure 2.7 A full-wave bridge rectifier: (a) circuit showing the current direction for a positive input cycle, (b) current direction for a negative input cycle, and (c) input and output voltage waveforms

Figure 2.7(c) shows the sinusoidal voltage v_S and the rectified output voltage v_O. Because two diodes are in series in the conduction path, the magnitude of v_O is two diode drops less than the magnitude of v_S.

One difference to be noted in the bridge rectifier circuit in Figure 2.7(a) and the rectifier in Figure 2.6(a) is the ground connection. Whereas the center tap of the secondary winding of the circuit in Figure 2.6(a) is at ground potential, the secondary winding of the bridge circuit (Figure 2.7(a)) is not directly grounded. One side of the load R is grounded, but the secondary of the transformer is not.

Example 2.1 **Objective:** Compare voltages and the transformer turns ratio in two full-wave rectifier circuits.

Consider the rectifier circuits shown in Figures 2.6(a) and 2.7(a). Assume the input voltage is from a 120 V (rms), 60 Hz ac source. The desired peak output voltage v_O is 9 V, and the diode cut-in voltage is assumed to be $V_\gamma = 0.7$ V.

Solution: For the center-tapped transformer circuit shown in Figure 2.6(a), a peak voltage of $v_O(\max) = 9$ V means that the peak value of v_S is

$$v_S(\max) = v_O(\max) + V_\gamma = 9 + 0.7 = 9.7\,\text{V}$$

For a sinusoidal signal, this produces an rms value of

$$v_{S,\text{rms}} = \frac{9.7}{\sqrt{2}} = 6.86\,\text{V}$$

The turns ratio of the primary to each secondary winding must then be

$$\frac{N_1}{N_2} = \frac{120}{6.86} \cong 17.5$$

For the bridge circuit shown in Figure 2.7(a), a peak voltage of $v_O(\text{max}) = 9\,\text{V}$ means that the peak value of v_S is

$$v_S(\text{max}) = v_O(\text{max}) + 2V_\gamma = 9 + 2(0.7) = 10.4\,\text{V}$$

For a sinusoidal signal, this produces an rms value of

$$v_{S,\text{rms}} = \frac{10.4}{\sqrt{2}} = 7.35\,\text{V}$$

The turns ratio should then be

$$\frac{N_1}{N_2} = \frac{120}{7.35} \cong 16.3$$

For the center-tapped rectifier, the peak inverse voltage (PIV) of a diode is

$$\text{PIV} = v_R(\text{max}) = 2v_S(\text{max}) - V_\gamma = 2(9.7) - 0.7 = 18.7\,\text{V}$$

For the bridge rectifier, the peak inverse voltage of a diode is

$$\text{PIV} = v_R(\text{max}) = v_S(\text{max}) - V_\gamma = 10.4 - 0.7 = 9.7\,\text{V}$$

Comment: These calculations demonstrate the advantages of the bridge rectifier over the center-tapped transformer circuit. First, only half as many turns are required for the secondary winding in the bridge rectifier. This is true because only half of the secondary winding of the center-tapped transformer is utilized at any one time. Second, for the bridge circuit, the peak inverse voltage that any diode must sustain without breakdown is only half that of the center-tapped transformer circuit.

Because of the advantages demonstrated in Example 2.1, the bridge rectifier circuit is used more often than the center-tapped transformer circuit.

2.1.3   Filters, Ripple Voltage, and Diode Current

If a capacitor is added in parallel with the load resistor of a half-wave rectifier to form a simple filter circuit (Figure 2.8(a)), we can begin to transform the half-wave sinusoidal output into a dc voltage. Figure 2.8(b) shows the positive half of the output sine wave, and the beginning portion of the voltage across the capacitor, assuming the capacitor is initially uncharged. If we assume that the diode forward resistance is $r_f = 0$, which means that the $r_f C$ time constant is zero, the voltage across the capacitor follows this initial portion of the signal voltage. When the signal voltage reaches its peak and begins to decrease, the voltage across the capacitor also starts to decrease, which means the capacitor starts to discharge. The only discharge current path is through the resistor. If the RC time constant is large, the voltage across the capacitor discharges exponentially with time (Figure 2.8(c)). During this time period, the diode is cut off.

A more detailed analysis of the circuit response when the input voltage is near its peak value indicates a subtle difference between actual circuit operation and the qualitative description. If we assume that the diode turns off immediately when the input voltage starts to decrease from its peak value, then the output voltage will decrease exponentially with time, as previously indicated. An exaggerated sketch of these two voltages is shown in Figure 2.8(d). The output voltage decreases at a faster rate than the input voltage,

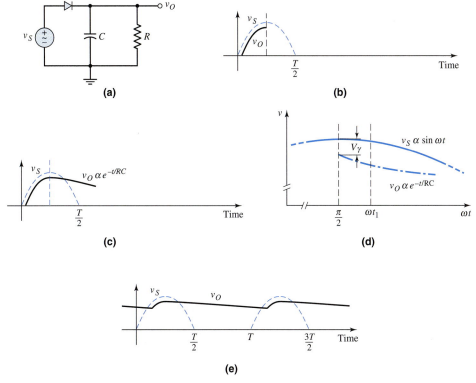

Figure 2.8 Simple filter circuit: (a) half-wave rectifier with an RC filter, (b) positive input voltage and initial portion of output voltage, (c) output voltage resulting from capacitor discharge, (d) expanded view of input and output voltages assuming capacitor discharge begins at $\omega t = \pi/2$, and (e) steady-state input and output voltages

which means that at time t_1, the difference between v_I and v_O, that is, the voltage across the diode, is greater than V_γ. However, this condition cannot exist. Therefore, the diode does not turn off immediately. If the RC time constant is large, there is only a small difference between the time of the peak input voltage and the time the diode turns off. In this situation, a computer analysis may provide more accurate results than an approximate hand analysis.

During the next positive cycle of the input voltage, there is a point at which the input voltage is greater than the capacitor voltage, and the diode turns back on. The diode remains on until the input reaches its peak value and the capacitor voltage is completely recharged.

Since the capacitor filters out a large portion of the sinusoidal signal, it is called a **filter capacitor**. The steady-state output voltage of the RC filter is shown in Figure 2.8(e).

The ripple effect in the output from a full-wave filtered rectifier circuit can be seen in the output waveform in Figure 2.9. The capacitor charges to its peak voltage value when the input signal is at its peak value. As the input decreases, the diode becomes reverse biased and the capacitor discharges through the output resistance R. Determining the ripple voltage is necessary for the design of a circuit with an acceptable amount of ripple.

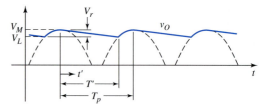

Figure 2.9   Output voltage of a full-wave rectifier with an *RC* filter

To a good approximation, the output voltage, that is, the voltage across the capacitor or the *RC* circuit, can be written as

$$v_O(t) = V_M e^{-t'/\tau} = V_M e^{-t'/RC} \tag{2.3}$$

where t' is the time after the output has reached its peak value, and RC is the time constant of the circuit.

The smallest output voltage is

$$V_L = V_M e^{-T'/RC} \tag{2.4}$$

where T' is the discharge time, as indicated in the figure.

The **ripple voltage** V_r is defined as the difference between V_M and V_L, and is determined by

$$V_r = V_M - V_L = V_M \left(1 - e^{-T'/RC} \right) \tag{2.5}$$

Normally, we will want the discharge time T' to be small compared to the time constant, or $T' \ll RC$. Expanding the exponential in a series and keeping only the linear terms of that expansion, we have the approximation

$$e^{-T'/RC} \cong 1 - \frac{T'}{RC} \tag{2.6}$$

The ripple voltage can now be written as

$$V_r \cong V_M \left(\frac{T'}{RC} \right) \tag{2.7}$$

Since the discharge time T' depends on the RC time constant, Equation (2.7) is difficult to solve. However, if the ripple effect is small, then as an approximation, we can let $T' = T_p$, so that

$$V_r \cong V_M \left(\frac{T_p}{RC} \right) \tag{2.8}$$

where T_p is the time between peak values of the output voltage. For a full-wave rectifier, T_p is one-half the signal period. Therefore, we can relate T_p to the signal frequency,

$$f = \frac{1}{2T_p}$$

The ripple voltage is then

$$V_r = \frac{V_M}{2fRC} \tag{2.9}$$

For a half-wave rectifier, the time T_p corresponds to a full period (not a half period) of the signal, so the factor 2 does not appear in Equation (2.9).

Equation (2.9) can be used to determine the capacitor value required for a particular ripple voltage.

Example 2.2   Objective: Determine the capacitance required to yield a particular ripple voltage.

Consider a full-wave rectifier circuit with a 60 Hz input signal and a peak output voltage of $V_M = 10\,\text{V}$. Assume the output load resistance is $R = 10\,\text{k}\Omega$ and the ripple voltage is to be limited to $V_r = 0.2\,\text{V}$.

Solution: From Equation (2.9), we can write

$$C = \frac{V_M}{2fRV_r} = \frac{10}{2(60)(10 \times 10^3)(0.2)} \Rightarrow 41.7\,\mu\text{F}$$

Comment: If the ripple voltage is to be limited to a smaller value, a larger filter capacitor must be used.

The diode in a filtered rectifier circuit conducts for a brief interval Δt near the peak of the sinusoidal input signal (Figure 2.10(a)). The diode current supplies the charge lost by the capacitor during the discharge time. Figure 2.11 shows the equivalent circuit of the full-wave rectifier during the charging time. We see that

$$i_D = i_C + i_R = C\frac{dv_O}{dt} + \frac{v_O}{R} \tag{2.10}$$

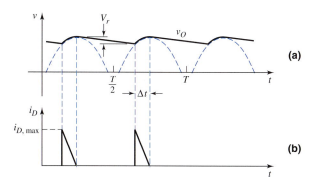

Figure 2.10   Output of a full-wave rectifier with an *RC* filter: (a) diode conduction time and (b) diode current

Figure 2.11   Equivalent circuit of a full-wave rectifier during capacitor charging cycle

If the ripple voltage is small, then the resistor or load current is $i_R \cong V_M/R$. If we neglect the diode cut-in voltage, then

$$V_M \cos(\omega\Delta t) = V_M - V_r \tag{2.11}$$

If the ripple voltage is small, then $\omega\Delta t$ is small. Therefore, $\cos(\omega\Delta t)$ $\cong 1 - \frac{1}{2}(\omega\Delta t)^2$. Using Equation (2.11), we find that

$$\omega\Delta t = \sqrt{\frac{2V_r}{V_M}} \tag{2.12}$$

The charge supplied to the capacitor through the diode is

$$Q_{\text{sup}} = i_{C,\text{avg}}\Delta t \tag{2.13(a)}$$

The charge lost by the capacitor during the discharge time is

$$Q_{\text{lost}} = CV_r \tag{2.13(b)}$$

To find the average diode current $i_{D,\text{avg}}$ during conduction, we equate these two equations, which yields

$$i_{C,\text{avg}} = \frac{CV}{\Delta t} \tag{2.14}$$

where $i_{C,\text{avg}}$ is the average current through the capacitor during the charging cycle.

The average currents in the diode and capacitor during the charging cycle are related by

$$i_{D,\text{avg}} = i_{C,\text{avg}} + \frac{V_M}{R} \tag{2.15}$$

The average diode current during the diode conduction time in a full-wave rectifier circuit is then

$$i_{D,\text{avg}} = \frac{V_M}{R}\left(1 + \pi\sqrt{\frac{V_M}{2V_r}}\right) \tag{2.16}$$

where we have used the frequency from Equation (2.9). The average capacitor current $i_{C,\text{avg}}$ is zero if the ripple voltage is zero. From Equation (2.15), the average diode current in this ideal case is then V_M/R, and does not become infinite as Equation (2.16) might suggest.

The peak diode current is found to be

$$i_{D,\text{max}} = \frac{V_M}{R}\left(1 + 2\pi\sqrt{\frac{V_M}{2V_r}}\right) \tag{2.17}$$

During the diode conduction time, for small ripple voltages, the current through the capacitor is much larger than the load current. Comparing Equations (2.16) and (2.17), we see that

$$i_{D,\text{max}} \cong 2i_{D,\text{avg}}$$

The resulting diode current approximates a triangular wave, as shown in Figure 2.10(b). The average diode current over the entire input signal period is then

$$i_D(\text{avg}) = \frac{1}{2\pi}\sqrt{\frac{2V_r}{V_M}}\left[\left(\frac{V_M}{R}\right)\left(1 + \pi\sqrt{\frac{V_M}{2V_r}}\right)\right] \tag{2.18}$$

Design Example 2.3 **Objective:** Design a full-wave rectifier to meet particular specifications.

A full-wave rectifier is to be designed to produce a peak output voltage of 12 V, deliver 120 mA to the load, and produce an output with a ripple of not more than 5 percent. An input line voltage of 120 V (rms), 60 Hz is available.

Solution: A full-wave bridge rectifier will be used, because of the advantages previously discussed. The effective load resistance is

$$R = \frac{V_O}{I_L} = \frac{12}{0.12} = 100 \, \Omega$$

Assuming a diode cut-in voltage of 0.7 V, the peak value of v_S is

$$v_S(\text{max}) = v_O(\text{max}) + 2V_\gamma = 12 + 2(0.7) = 13.4 \, \text{V}$$

For a sinusoidal signal, this produces an rms voltage value of

$$v_{S,\text{rms}} = \frac{13.4}{\sqrt{2}} = 9.48 \, \text{V}$$

The transformer turns ratio is then

$$\frac{N_1}{N_2} = \frac{120}{9.48} = 12.7$$

For a 5 percent ripple, the ripple voltage is

$$V_r = (0.05)V_M = (0.05)(12) = 0.6 \, \text{V}$$

The required filter capacitor is found to be

$$C = \frac{V_M}{2fRV_r} = \frac{12}{2(60)(100)(0.6)} \Rightarrow 1667 \, \mu\text{F}$$

The peak diode current is

$$i_{D,\text{max}} = \frac{12}{100}\left[1 + 2\pi\sqrt{\frac{12}{2(0.6)}}\right] = 2.50 \, \text{A}$$

and the average diode current over the entire signal period is

$$i_D(\text{avg}) = \frac{1}{2\pi}\sqrt{\frac{2(0.6)}{12}}\left[\left(\frac{12}{100}\right)\left(1 + \pi\sqrt{\frac{12}{2(0.6)}}\right)\right] \Rightarrow 66 \, \text{mA}$$

Finally, the peak inverse voltage that each diode must sustain is

$$\text{PIV} = v_R(\text{max}) = v_S(\text{max}) - V_\gamma = 13.4 - 0.7 = 12.7 \, \text{V}$$

Comment: The minimum specifications for the diodes in this full-wave rectifier circuit are: a peak current of 2.50 A, an average current of 66 mA, and a peak inverse voltage of 12.7 V. In order to meet the desired ripple specification, the required filter capacitance must be large, since the effective load resistance is small.

Design Pointer: (1) A particular turns ratio was determined for the transformer. However, this particular transformer design is probably not commercially available. This means an expensive custom transformer design would be required, or if a standard transformer is used, then additional circuit design is required to meet the output voltage specification. (2) A constant 120 V (rms) input voltage is assumed to be available. However, this voltage can fluctuate, so the output voltage will also fluctuate.

We will see later how more sophisticated designs will solve these two problems.

Computer Verification: Since we simply used an assumed cut-in voltage for the diode and used approximations in the development of the ripple voltage equations, we can use PSpice to give us a more accurate evaluation of the circuit. The PSpice circuit schematic and the steady-state output voltage are shown in Figure 2.12. We see that the peak output voltage is 11.6 V, which is close to the desired 12 V. One reason for the slight discrepancy is that the diode voltage drop for the maximum input voltage is slightly greater than 0.8 V rather than the assumed 0.7 V. The ripple voltage is approximately 0.5 V, which is within the 0.6 V specification.

(a)

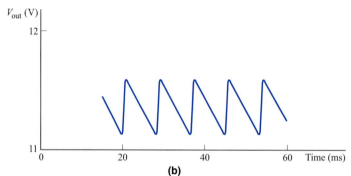

(b)

Figure 2.12 (a) PSpice circuit schematic of diode bridge circuit; (b) Steady-state output voltage of PSpice analysis of diode bridge circuit for a 60 Hz input sine wave with a peak value of 13.4 V

Discussion: In the PSpice simulation, a standard diode, 1N4002, was used. In order for the computer simulation to be valid, the diode used in the simulation and in the actual circuit must match. In this example, to reduce the diode voltage and increase the peak output voltage, a diode with a larger cross-sectional area should be used.

Test Your Understanding

2.2 The input voltage to the half-wave rectifier in Figure 2.8(a) is $v_S = 75 \sin [2\pi(60)t]$ V. Assume a diode cut-in voltage of $V_\gamma = 0$. The ripple voltage is to be no more than $V_r = 4$ V. If the filter capacitor is 50 μF, determine the minimum load resistance that can be connected to the output. (Ans. $R = 6.25$ kΩ)

2.3 The circuit in Figure 2.6(a) is used to rectify a sinusoidal input signal with a peak voltage of 120 V and a frequency of 60 Hz. A filter capacitor is connected in parallel with

R. If the output voltage cannot drop below 100 V, determine the required value of the capacitance *C*. The transformer has a turns ratio of $N_1 : N_2 = 1 : 1$, where N_2 is the number of turns on each of the secondary windings. Assume the diode cut-in voltage is 0.7 V and the output resistance is 2.5 kΩ. (Ans. $C = 20.6 \, \mu\text{F}$)

2.4 The secondary transformer voltage of the rectifier circuit shown in Figure 2.7(a) is $v_S = 50 \sin[2\pi(60)t] \, \text{V}$. Each diode has a cut-in voltage of $V_\gamma = 0.7 \, \text{V}$, and the load resistance is $R = 10 \, \text{k}\Omega$. Determine the value of the filter capacitor that must be connected in parallel with *R* such that the ripple voltage is no greater than $V_r = 2 \, \text{V}$. (Ans. $C = 20.3 \, \mu\text{F}$)

2.5 Determine the fraction (percent) of the cycle that each diode is conducting in (a) Exercise 2.2, (b) Exercise 2.3, and (c) Exercise 2.4. (Ans. (a) 5.2%, (b) 18.1%, (c) 9.14%)

2.1.4 Voltage Doubler Circuit

A **voltage doubler circuit** is very similar to the full-wave rectifier, except that two diodes are replaced by capacitors, and it can produce a voltage equal to approximately twice the peak output of a transformer (Figure 2.13).

Figure 2.14(a) shows the equivalent circuit when the voltage polarity at the "top" of the transformer is negative; Figure 2.14(b) shows the equivalent circuit for the opposite polarity. In the circuit in Figure 2.14(a), the forward diode resistance of D_2 is small; therefore, the capacitor C_1 will charge to almost the peak value of v_S. Terminal 2 on C_1 is positive with respect to terminal 1. As the magnitude of v_S decreases from its peak value, C_1 discharges through R_L

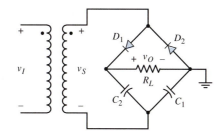

Figure 2.13 A voltage doubler circuit

| (a) | (b) |

Figure 2.14 Equivalent circuit of the voltage doubler circuit: (a) negative input cycle and (b) positive input cycle

and C_2. We assume that the time constant $R_L C_2$ is very long compared to the period of the input signal.

As the polarity of v_S changes to that shown in Figure 2.14(b), the voltage across C_1 is essentially constant at V_M, with terminal 2 remaining positive. As v_S reaches its maximum value, the voltage across C_2 essentially becomes V_M. By Kirchhoff's voltage law, the peak voltage across R_L is now essentially equal to $2V_M$, or twice the peak output of the transformer. The same ripple effect occurs as in the output voltage of the rectifier circuits, but if C_1 and C_2 are relatively large, then the ripple voltage V_r, is quite small.

There are also voltage tripler and voltage quadrupler circuits. These circuits provide a means by which multiple dc voltages can be generated from a single ac source and power transformer.

2.2   ZENER DIODE CIRCUITS

In Chapter 1, we saw that the breakdown voltage of a Zener diode was nearly constant over a wide range of reverse-bias currents. This makes the Zener diode useful in a **voltage regulator**, or a constant-voltage reference circuit. In this chapter, we will look at an ideal voltage reference circuit, and the effects of including a nonideal **Zener resistance**.

The results of this section will then complete the design of the electronic power supply in Figure 2.1. We should note that in actual power supply designs, the voltage regulator will be a more sophisticated integrated circuit rather than the simpler Zener diode design that will be developed here. One reason is that a standard Zener diode with a particular desired breakdown voltage may not be available. However, this section will provide the basic concept of a voltage regulator.

2.2.1   Ideal Voltage Reference Circuit

Figure 2.15 shows a Zener voltage regulator circuit. For this circuit, the output voltage should remain constant, even when the output load resistance varies over a fairly wide range, and when the input voltage varies over a specific range.

Figure 2.15   A Zener diode voltage regulator circuit

We determine, initially, the proper input resistance R_i. The resistance R_i limits the current through the Zener diode and drops the "excess" voltage between V_{PS} and V_Z. We can write

$$R_i = \frac{V_{PS} - V_Z}{I_I} = \frac{V_{PS} - V_Z}{I_Z + I_L} \qquad (2.19)$$

which assumes that the Zener resistance is zero for the ideal diode. Solving this equation for the diode current, I_Z, we get

$$I_Z = \frac{V_{PS} - V_Z}{R_i} - I_L \qquad (2.20)$$

where $I_L = V_Z/R_L$, and the variables are the input voltage source V_{PS} and the load current I_L.

For proper operation of this circuit, the diode must remain in the breakdown region and the power dissipation in the diode must not exceed its rated value. In other words:

1. The current in the diode is a minimum, $I_Z(\min)$, when the load current is a maximum, $I_L(\max)$, and the source voltage is a minimum, $V_{PS}(\min)$.
2. The current in the diode is a maximum, $I_Z(\max)$, when the load current is a minimum, $I_L(\min)$, and the source voltage is a maximum, $V_{PS}(\max)$.

Inserting these two specifications into Equation (2.19), we obtain

$$R_i = \frac{V_{PS}(\min) - V_Z}{I_Z(\min) + I_L(\max)} \qquad (2.21(a))$$

and

$$R_i = \frac{V_{PS}(\max) - V_Z}{I_Z(\max) + I_L(\min)} \qquad (2.21(b))$$

Equating these two expressions, we then obtain

$$\begin{aligned} &[V_{PS}(\min) - V_Z] \cdot [I_Z(\max) + I_L(\min)] \\ &= [V_{PS}(\max) - V_Z] \cdot [I_Z(\min) + I_L(\max)] \end{aligned} \qquad (2.22)$$

Reasonably, we can assume that we know the range of input voltage, the range of output load current, and the Zener voltage. Equation (2.22) then contains two unknowns, $I_Z(\min)$ and $I_Z(\max)$. Further, as a minimum requirement, we can set the minimum Zener current to be one-tenth the maximum Zener current, or $I_Z(\min) = 0.1 I_Z(\max)$. (More stringent design requirements may require the minimum Zener current to be 20 to 30 percent of the maximum value.) We can then solve for $I_Z(\max)$, using Equation (2.22), as follows:

$$I_Z(\max) = \frac{I_L(\max) \cdot [V_{PS}(\max) - V_Z] - I_L(\min) \cdot [V_{PS}(\min) - V_Z]}{V_{PS}(\min) - 0.9 V_Z - 0.1 V_{PS}(\max)} \qquad (2.23)$$

Using the maximum current thus obtained from Equation (2.23), we can determine the maximum required power rating of the Zener diode. Then, combining Equation (2.23) with either Equation (2.21(a)) or (2.21(b)), we can determine the required value of the input resistance R_i.

Design Example 2.4   **Objective:**   Design a voltage regulator using the circuit in Figure 2.15.

The voltage regulator is to power a car radio at $V_L = 9\,\text{V}$ from an automobile battery whose voltage may vary between 11 and 13.6 V. The current in the radio will vary between 0 (off) to 100 mA (full volume).

The equivalent circuit is shown in Figure 2.16.

Figure 2.16   Circuit for Design Example 2.4

Solution: The maximum Zener diode current can be determined from Equation (2.23) as

$$I_Z(\text{max}) = \frac{(100)[13.6 - 9] - 0}{11 - (0.9)(9) - (0.1)(13.6)} \cong 300\,\text{mA}$$

The maximum power dissipated in the Zener diode is then

$$P_Z(\text{max}) = I_Z(\text{max}) \cdot V_Z = (300)(9) \Rightarrow 2.7\,\text{W}$$

The value of the current-limiting resistor R_i, from Equation (2.21(b)), is

$$R_i = \frac{13.6 - 9}{0.3} = 15.3\,\Omega$$

The maximum power dissipated in this resistor is

$$P_{Ri}(\text{max}) = \frac{(V_{PS}(\text{max}) - V_Z)^2}{R_i} = \frac{(13.6 - 9)^2}{15.3} \cong 1.4\,\text{W}$$

 Comment: From this design, we see that the minimum power ratings of the Zener diode and input resistor are 2.7 W and 1.4 W, respectively. The minimum Zener diode current occurs for $V_{PS}(\text{min})$ and $I_L(\text{max})$. We find $I_Z(\text{min}) = 30.7\,\text{mA}$, which is approximately 10 percent of $I_Z(\text{max})$ as specified by the design equations.

Design Pointer: (1) The variable input in this example was due to a variable battery voltage. However, referring back to Example 2.3, the variable input could also be a function of using a standard transformer with a given turns ratio as opposed to a custom design with a particular turns ratio and/or having a 120 V (rms) input voltage that is not exactly constant.

(2) The 9 V output is a result of using a 9 V Zener diode. However, a Zener diode with exactly a 9 V breakdown voltage may also not be available. We will again see later how more sophisticated designs can solve this problem.

Test Your Understanding

2.6 The Zener diode regulator circuit shown in Figure 2.15 has an input voltage that varies between 10 and 14 V, and a load resistance that varies between $R_L = 20$ and $100\,\Omega$. Assume a 5.6 Zener diode is used, and assume $I_Z(\min) = 0.1 I_Z(\max)$. Find the value of R_i required and the minimum power rating of the diode. (Ans. $P_Z = 3.31$ W, $R_i \cong 13\,\Omega$)

2.7 Suppose the current-limiting resistor in Example 2.4 is replaced by one whose value is $R_i = 20\,\Omega$. Determine the minimum and maximum Zener diode current. Does the circuit operate "properly"?

2.8 Suppose the power supply voltage in the circuit shown in Figure 2.16 drops to $V_{PS} = 10$ V. Let $R_i = 15.3\,\Omega$. What is the maximum load current in the radio if the minimum Zener diode current is to be maintained at $I_Z(\min) = 30$ mA?

2.2.2 Zener Resistance and Percent Regulation

In the ideal Zener diode, the Zener resistance is zero. In actual Zener diodes, however, this is not the case. The result is that the output voltage is a function of the Zener diode current or the load current.

Figure 2.17 shows the equivalent circuit of the voltage regulator in Figure 2.15. Because of the Zener resistance, the output voltage will not remain constant. We can determine the minimum and maximum values of output voltage. A figure of merit for a voltage regulator is called the **percent regulation**, and is defined as

$$\% \text{ Regulation} = \frac{V_L(\max) - V_L(\min)}{V_L(\text{nom})} \times 100 \qquad (2.24)$$

where $V_L(\text{nom})$ is the nominal value of the output voltage. As the percent regulation approaches zero percent, the circuit approaches that of an ideal voltage regulator.

Figure 2.17 A Zener diode voltage regulator circuit with a nonzero Zener resistance

Example 2.5 **Objective:** Determine the percent regulation of a voltage regulator.

Consider the circuit described in Example 2.4 and assume a Zener resistance of $r_z = 4\,\Omega$. The nominal output voltage is $V_L(\text{nom}) = 9$ V.

Solution: As a first approximation, we can assume that the output voltage does not change significantly. Therefore, the minimum and maximum Zener diode currents will be the same, as in Example 2.4. Then,

$$V_L(\text{max}) = V_L(\text{nom}) + I_Z(\text{max})r_z = 9 + (0.30)(4) = 10.20 \text{ V}$$

and

$$V_L(\text{min}) = V_L(\text{nom}) + I_Z(\text{min})r_z = 9 + (0.030)(4) = 9.12 \text{ V}$$

The percent regulation is then

$$\% \text{ Regulation} = \frac{V_L(\text{max}) - V_L(\text{min})}{V_L(\text{nom})} \times 100 = \frac{10.2 - 9.12}{9} \times 100 = 12\%$$

Comment: Because of the relatively high current levels in this example, the percent regulation is fairly high. The percent regulation can be improved significantly by using amplifiers in conjunction with the voltage reference circuit. We will see examples of these circuits in later chapters.

Test Your Understanding

2.9   If the diode in Exercise 2.6 has a Zener resistance of $r_z = 1.5\,\Omega$, determine the percent regulation. (Ans. 14.3%)

2.3   CLIPPER AND CLAMPER CIRCUITS

In this section, we continue our discussion of nonlinear circuit applications of diodes. Diodes can be used in waveshaping circuits that either limit or "clip" portions of a signal, or shift the dc voltage level. The circuits are called **clippers** and **clampers**, respectively.

2.3.1   Clippers

Clipper circuits, also called **limiter circuits**, are used to eliminate portions of a signal that are above or below a specified level. For example, the half-wave rectifier is a clipper circuit, since all voltages below zero are eliminated. A simple application of a clipper is to limit the voltage at the input to an electronic circuit so as to prevent breakdown of the transistors in the circuit. The circuit may be used to measure the frequency of the signal, so the amplitude is not an important part of the signal.

Figure 2.18 shows the general voltage transfer characteristics of a limiter circuit. The limiter is a linear circuit if the input signal is in the range $V_O^-/A_v \leq v_I \leq V_O^+/A_v$, where A_v is the slope of the transfer curve. If $A_v \leq 1$, as in diode circuits, the circuit is a **passive limiter**. If $v_I > V_O^+/A_v$, the output is limited to a maximum value of V_O^+. Similarly, if $v_I < V_O^-/A_v$, the output is limited to a minimum value of V_O^-. Figure 2.18 shows the general transfer curve of a double limiter, in which both the positive and negative peak values of the input signal are clipped.

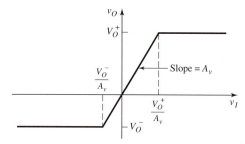

Figure 2.18   General voltage transfer characteristics of a limiter circuit

Various combinations of V_O^+ and V_O^- are possible. Both parameters may be positive, both negative, or one may be positive while the other negative, as indicated in the figure. If either V_O^- approaches minus infinity or V_O^+ approaches plus infinity, then the circuit reverts to a single limiter.

Figure 2.19(a) is a single-diode clipper circuit. The diode D_1 is off as long as $v_I < V_B + V_\gamma$. With D_1 off, the current is zero, the voltage drop across R is zero, and the output voltage follows the input voltage. When $v_I > V_B + V_\gamma$, the diode turns on, the output voltage is clipped, and v_O equals $V_B + V_\gamma$. The output signal is shown in Figure 2.19(b). In this circuit, the output is clipped above $V_B + V_\gamma$.

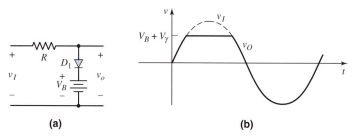

(a)   **(b)**

Figure 2.19   Single-diode clipper: (a) circuit and (b) output response

Other clipping circuits can be constructed by reversing the diode, the polarity of the voltage source, or both. Figures 2.20(a), (b), and (c) show these circuits, along with the corresponding input and output signals.

Positive and negative clipping can be performed simultaneously by using a double limiter or a **parallel-based clipper**, such as the circuit shown in Figure 2.21. The input and output signals are also shown in the figure. The parallel-based clipper is designed with two diodes and two voltage sources oriented in opposite directions.

Example 2.6   **Objective:**   Find the output of the parallel-based clipper in Figure 2.22(a).

For simplicity, assume that $V_\gamma = 0$ and $r_f = 0$ for both diodes.

Solution: For $t = 0$, we see that $v_I = 0$ and both D_1 and D_2 are reverse biased. For $0 < v_I \le 2\,\text{V}$, D_1 and D_2 remain off; therefore, $v_O = v_I$. For $v_I > 2\,\text{V}$, D_1 turns on and

$$i_1 = \frac{v_I - 2}{10 + 10}$$

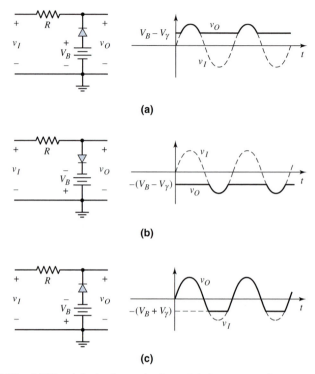

(a)

(b)

(c)

Figure 2.20 Additional diode clipper circuits and their corresponding output responses

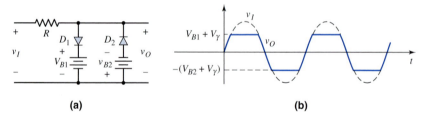

(a)   **(b)**

Figure 2.21 A parallel-based diode clipper circuit and its output response

Also,

$$v_O = i_1 R_2 + 2 = \tfrac{1}{2}(v_I - 2) + 2 = \tfrac{1}{2}v_I + 1$$

If $v_I = 6\,\text{V}$, then $v_O = 4\,\text{V}$.

For $-4 < v_I < 0\,\text{V}$, both D_1 and D_2 are off and $v_O = v_I$. For $v_I \le -4\,\text{V}$, D_2 turns on and the output is constant at $v_O = -4\,\text{V}$. The input and output waveforms are plotted in Figure 2.22(b).

Comment: If we assume that $V_\gamma \ne 0$, the output will be very similar to the results calculated here. The only difference will be the points at which the diodes turn on.

Diode clipper circuits can also be designed such that the dc power supply is in series with the input signals. Figure 2.23 shows various circuits based on this design. The battery in series with the input signal causes the input signal to be

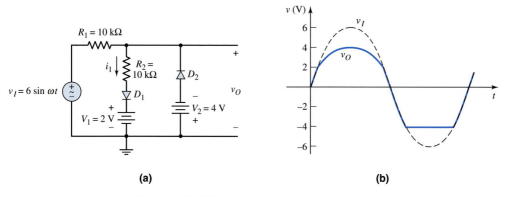

(a)   **(b)**

Figure 2.22   Figure for Example 2.6

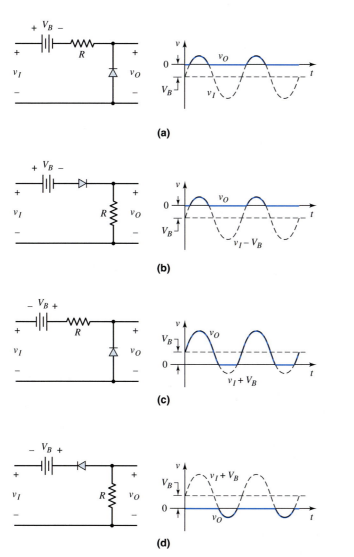

Figure 2.23   Series-based diode clipper circuits and their corresponding output responses

superimposed on the V_B dc voltage. The resulting conditioned input signals and corresponding output signals are also shown in Figure 2.23.

Test Your Understanding

2.10   Plot the voltage transfer characteristics (v_O versus v_I) for the circuit in Figure 2.22(a). Assume each diode cut-in voltage is $V_\gamma = 0.7\,\text{V}$. (Ans. For $-4.7 \le v_I \le 2.7\,\text{V}$, $v_O = v_I$; for $v_I > 2.7\,\text{V}$, $v_O = (\frac{1}{2})v_I + 1.35$; for $v_I < -4.7\,\text{V}$, $v_O = -4.7\,\text{V}$)

D2.11   Design a parallel-based clipper that will yield the voltage transfer function shown in Figure 2.24. Assume diode cut-in voltages of $V_\gamma = 0.7\,\text{V}$. (Ans. For Figure 2.22(a), $V_2 = 4.3$, $V_1 = 1.8\,\text{V}$, and $R_1 = 2R_2$)

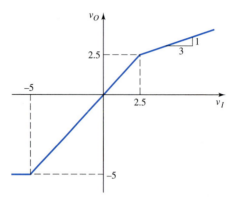

Figure 2.24   Figure for Exercise 2.11

2.3.2   Clampers

Clamping shifts the entire signal voltage by a dc level. In steady state, the output waveform is an exact replica of the input waveform, but the output signal is shifted by a dc value that depends on the circuit. The distinguishing feature of a clamper is that it adjusts the dc level without needing to know the exact waveform.

An example of clamping is shown in Figure 2.25(a). The sinusoidal input voltage signal is shown in Figure 2.25(b). Assume that the capacitor is initially uncharged. During the first 90 degrees of the input waveform, the voltage across the capacitor follows the input, and $v_C = v_I$ (assuming that $r_f = 0$ and $V_\gamma = 0$). After v_I and v_C reach their peak values, v_I begins to decrease and the diode becomes reverse biased. Ideally, the capacitor cannot discharge, so the voltage across the capacitor remains constant at $v_C = V_M$. By Kirchhoff's voltage law

$$v_O = -v_C + v_I = -V_M + V_M \sin \omega t \qquad (2.25(\mathbf{a}))$$

or

$$v_O = V_M(\sin \omega t - 1) \qquad (2.25(\mathbf{b}))$$

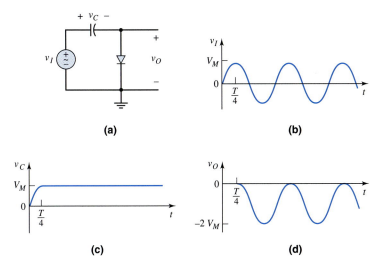

(a) **(b)**

(c) **(d)**

Figure 2.25 Action of a diode clamper circuit: (a) a typical diode clamper circuit, (b) the sinusoidal input signal, (c) the capacitor voltage, and (d) the output voltage

The capacitor and output voltages are shown in Figures 2.25(c) and (d). The output voltage is "clamped" at zero volts, that is, $v_O \leq 0$. In steady state, the waveshapes of the input and output signals are the same, and the output signal is shifted by a certain dc level compared to the input signal.

A clamping circuit that includes an independent voltage source V_B is shown in Figure 2.26(a). In this circuit, the $R_L C$ time constant is assumed to be large, where R_L is the load resistance connected to the output. If we assume, for simplicity, that $r_f = 0$ and $V_\gamma = 0$, then the output is clamped at V_B. Figure 2.26(b) shows an example of a sinusoidal input signal and the resulting output voltage signal. When the polarity of V_B is as shown, the output is shifted in a negative voltage direction. Similarly, Figure 2.26(c) shows a square-wave input signal and the resulting output voltage signal. For the square-wave signal, we neglect the diode capacitance effects and assume the voltage can change instantaneously.

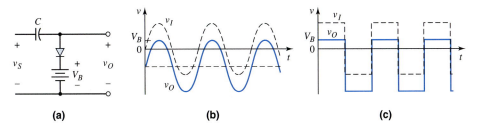

(a) **(b)** **(c)**

Figure 2.26 Action of a diode clamper circuit with a voltage source: (a) the circuit, (b) steady-state sinusoidal input and output signals, and (c) steady-state square-wave input and output signals

Electronic signals tend to lose their dc levels during signal transmission. For example, the dc level of a TV signal may be lost during transmission, so that the dc level must be restored at the TV receiver. The following example illustrates this effect.

Example 2.7   **Objective:**   Find the steady-state output of the diode-clamper circuit shown in Figure 2.27(a).

The input v_I is assumed to be a sinusoidal signal whose dc level has been shifted with respect to a receiver ground by a value V_B during transmission. Assume $V_\gamma = 0$ and $r_f = 0$ for the diode.

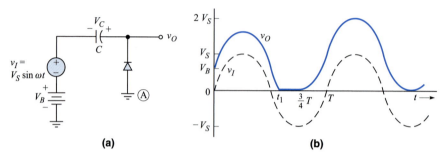

Figure 2.27   (a) Circuit for Example 2.7; (b) input and output waveforms

Solution: Figure 2.27(b) shows the sinusoidal input signal. If the capacitor is initially uncharged, then the output voltage is $v_O = V_B$ at $t = 0$ (diode reverse-biased). For $0 \le t \le t_1$, the effective RC time constant is infinite, the voltage across the capacitor does not change, and $v_O = v_I + V_B$.

At $t = t_1$, the diode becomes forward biased; the output cannot go negative, so the voltage across the capacitor changes (the $r_f C$ time constant is zero).

At $t = (\frac{3}{4})T$, the input signal begins increasing and the diode becomes reverse biased, so the voltage across the capacitor now remains constant at $V_S - V_B$ with the polarity shown. The output voltage is now given by

$$v_O = (V_S - V_B) + v_I + V_B = (V_S - V_B) + V_S \sin \omega t + V_B$$

or

$$v_O = V_S(1 + \sin \omega(t - (\tfrac{3}{4})T))$$

Comment: For $t > (\frac{3}{4})T$, steady state is reached. The output signal waveform is an exact replica of the input signal waveform and is now measured with respect to the reference ground at terminal A.

Test Your Understanding

2.12   Sketch the steady-state output voltage for the input signal given for the circuit in Figure 2.28. (Ans. Square wave between -2 and -10 V.)

2.13   Determine the steady-state output voltage v_O for the circuit in Figure 2.29(a), if the input is as shown in Figure 2.29(b). Assume the diode cut-in voltage is $V_\gamma = 0$. (Ans. Output is a square wave between $+5$ V and $+35$ V)

Figure 2.28 Figure for Exercise 2.12

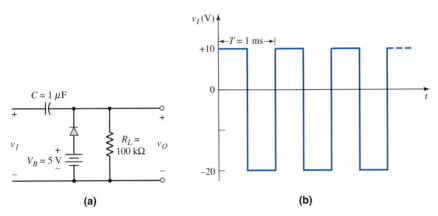

Figure 2.29 Figure for Exercise 2.13

2.4   MULTIPLE-DIODE CIRCUITS

Since a diode is a nonlinear device, part of the analysis of a diode circuit involves determining whether the diode is "on" or "off." If a circuit contains more than one diode, the analysis is complicated by the various possible combinations of "on" and "off."

In this section, we will look at several multiple-diode circuits. We will see, for example, how diode circuits can be used to perform logic functions. This section serves as an introduction to digital logic circuits that will be considered in detail in Chapters 16 and 17.

2.4.1   Example Diode Circuits

To review briefly, consider two single-diode circuits. Figure 2.30(a) shows a diode in series with a resistor. A plot of voltage transfer characteristics, v_O versus v_I, shows the piecewise linear nature of this circuit (Figure 2.30(b)). The diode does not begin to conduct until $v_I = V_\gamma$. Consequently, for $v_I \leq V_\gamma$, the output voltage is zero; for $v_I > V$, the output voltage is $v_O = v_I - V_\gamma$.

Figure 2.31(a) shows a similar diode circuit, but with the input voltage source explicitly included to show that there is a path for the diode current. The voltage transfer characteristic is shown in Figure 2.31(b). In this circuit, the diode remains conducting for $v_I < V_S - V_\gamma$, and the output voltage is $v_O =$

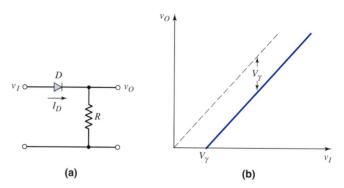

Figure 2.30   Diode and resistor in series: (a) circuit and (b) voltage transfer characteristics

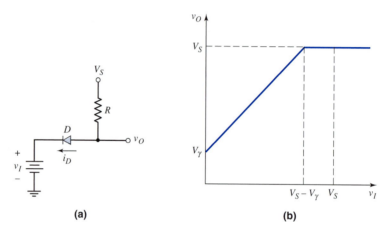

Figure 2.31   Diode with input voltage source: (a) circuit and (b) voltage transfer characteristics

$v_I + V_\gamma$. When $v_I > V_S - V_\gamma$, the diode turns off and the current through the resistor is zero; therefore, the output remains constant at V_S.

These two examples demonstrate the piecewise linear nature of the diode and the diode circuit. They also demonstrate that there are regions where the diode is "on," or conducting, and regions where the diode is "off," or nonconducting.

In multidiode circuits, each diode may be either "on" or "off." Consider the two-diode circuit in Figure 2.32. Since each diode may be either on or off, the circuit has four possible states. However, some of these states may not be feasible because of diode directions and voltage polarities.

If we assume that $V^+ > V^-$ and that $V^+ - V^- > V_\gamma$, there is at least a possibility that D_2 can be turned on. First, v' cannot be less than V^-. Then, for $v_I = V^-$, diode D_1 must be off. In this case, D_2 is on, $i_{R1} = i_{D2} = i_{R2}$, and

$$v_O = V^+ - i_{R1} R_1 \tag{2.26}$$

where

$$i_{R1} = \frac{V^+ - V_\gamma - V^-}{R_1 + R_2} \tag{2.27}$$

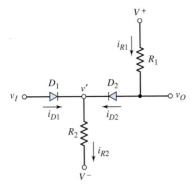

Figure 2.32　A two-diode circuit

Voltage v' is one diode drop below v_O, and D_1 remains off as long as v_I is less than the output voltage. As v_I increases and becomes equal to v_O, both D_1 and D_2 turn on. This condition or state is valid as long as $v_I < V^+$. When $v_I = V^+$, $i_{R1} = i_{D2} = 0$, at which point D_2 turns off and v_O cannot increase any further.

Figure 2.33 shows the resulting plot of v_O versus v_I. Three distinct regions, $v_O{}^{(1)}$, $v_O{}^{(2)}$, and $v_O{}^{(3)}$, correspond to the various conducting states of D_1 and D_2. The fourth possible state, corresponding to both D_1 and D_2 being off, is not feasible in this circuit.

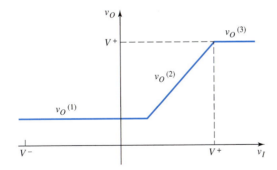

Figure 2.33　Voltage transfer characteristics for the two-diode circuit in Figure 2.32

Example 2.8 **Objective:** Determine the output voltage and diode currents for the circuit shown in Figure 2.32, for two values of input voltage.

Assume the circuit parameters are $R_1 = 5\,\text{k}\Omega$, $R_2 = 10\,\text{k}\Omega$, $V_\gamma = 0.7\,\text{V}$, $V^+ = +5\,\text{V}$, and $V^- = -5\,\text{V}$. Determine v_O, i_{D1}, and i_{D2} for $v_I = 0$ and $v_I = 4\,\text{V}$.

Solution: For $v_I = 0$, assume initially that D_1 is off. The currents are then

$$i_{R1} = i_{D2} = i_{R2} = \frac{V^+ - V_\gamma - V^-}{R_1 + R_2} = \frac{5 - 0.7 - (-5)}{5 + 10} = 0.62\,\text{mA}$$

The output voltage is

$$v_O = V^+ - i_{R1}R_1 = 5 - (0.62)(5) = 1.9\,\text{V}$$

and v' is

$$v' = v_O - V_\gamma = 1.9 - 0.7 = 1.2\,\text{V}$$

From these results, we see that diode D_1 is indeed cut off, $i_{D1} = 0$, and our analysis is valid.

For $v_I = 4\,\text{V}$, we see from Figure 2.33 that $v_O = v_I$; therefore, $v_O = v_I = 4\,\text{V}$. In this region, both D_1 and D_2 are on, and

$$i_{R1} = i_{D2} = \frac{V^+ - v_O}{R_1} = \frac{5 - 4}{5} = 0.2\,\text{mA}$$

Note that $v' = v_O - V_\gamma = 4 - 0.7 = 3.3\,\text{V}$. Thus,

$$i_{R2} = \frac{v' - V^-}{R_2} = \frac{3.3 - (-5)}{10} = 0.83\,\text{mA}$$

The current through D_1 is found from $i_{D1} + i_{D2} = i_{R2}$ or

$$i_{D1} = i_{R2} - i_{D2} = 0.83 - 0.2 = 0.63\,\text{mA}$$

Comment: For $v_I = 0$, we see that $v_O = 1.9\,\text{V}$ and $v' = 1.2\,\text{V}$. This means that D_1 is reverse biased, or off, as we initially assumed. For $v_I = 4\,\text{V}$, we have $i_{D1} > 0$ and $i_{D2} > 0$, indicating that both D_1 and D_2 are forward biased, as we assumed.

Computer Analysis: For multidiode circuits, a PSpice analysis may be useful in determining the conditions under which the various diodes are conducting or not conducting. This avoids guessing the conducting state of each diode in a hand analysis. Figure 2.34 is the PSpice circuit schematic of the diode circuit in Figure 2.32. Figure 2.34 also shows the output voltage and the two diode currents as the input is varied between $-1\,\text{V}$ and $+7\,\text{V}$. From these curves, we can determine when the diodes turn on and off.

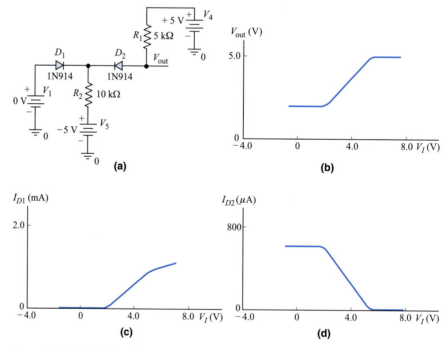

Figure 2.34   (a) PSpice circuit schematic; (b) output voltage; (c) current in diode 1, and (d) current in diode 2 for the diode circuit in Example 2.8

Comment: The hand analysis results, based on the piecewise linear model for the diode, agree very well with the computer simulation results. This gives us confidence in the piecewise linear model when quick hand calculations are made.

Problem-Solving Technique: Multiple Diode Circuits

Analyzing multidiode circuits requires determining if the individual devices are "on" or "off." In many cases, the choice is not obvious, so we must initially guess the state of each device, then analyze the circuit to determine if we have a solution consistent with our initial guess. To do this, we can:

1. Assume the state of a diode. If a diode is assumed "on," the voltage across the diode is assumed to be V_γ. If a diode is assumed to be "off," the current through the diode is assumed to be zero.

2. Analyze the "linear" circuit with the assumed diode states.

3. Evaluate the resulting state of each diode. If the initial assumption were that a diode is "off" and the analysis shows that $I_D = 0$ and $V_D \leq V_\gamma$, then the assumption is correct. If, however, the analysis actually shows that $I_D > 0$ and/or $V_D > V_\gamma$, then the initial assumption is incorrect. Similarly, if the initial assumption were that a diode is "on" and the analysis shows that $I_D \geq 0$ and $V_D = V_\gamma$, then the initial assumption is correct. If, however, the analysis shows that $I_D < 0$ and/or $V_D < V_\gamma$, then the initial assumption is incorrect.

4. If any initial assumption is proven incorrect, then a new assumption must be made and the new "linear" circuit must be analyzed. Step 3 must then be repeated.

Example 2.9 **Objective:** Demonstrate how inconsistencies develop in a solution with incorrect assumptions.

For the circuit shown in Figure 2.32, assume that parameters are the same as those given in Example 2.8. Determine v_O, i_{D1}, i_{D2}, and i_{R2} for $v_I = 0$.

Solution: Assume initially that both D_1 and D_2 are conducting (i.e., on). Then, $v' = -0.7\,\text{V}$ and $v_O = 0$. The two currents are

$$i_{R1} = i_{D2} = \frac{V^+ - v_O}{R_1} = \frac{5 - 0}{5} = 1.0\,\text{mA}$$

and

$$i_{R2} = \frac{v' - V^-}{R^2} = \frac{-0.7 - (-5)}{10} = 0.43\,\text{mA}$$

Summing the currents at the v' node, we find that

$$i_{D1} = i_{R2} - i_{D2} = 0.43 - 1.0 = -0.57\,\text{mA}$$

Since this analysis shows the D_1 current to be negative, which is an impossible or inconsistent solution, our initial assumption must be incorrect. If we go back to Example 2.8, we will see that the correct solution is D_1 off and D_2 on when $v_I = 0$.

Comment: We can perform linear analyses on diode circuits, using the piecewise linear model. However, we must first determine if each diode in the circuit is operating in the "on" linear region or the "off" linear region.

Test Your Understanding

2.14   Consider the circuit shown in Figure 2.35, in which the diode cut-in voltages are $V_\gamma = 0.6\,\text{V}$. Plot v_O versus v_I for $0 \le v_I \le 10\,\text{V}$. (Ans. For $0 \le v_I \le 3.5\,\text{V}$, $v_O = 4.4\,\text{V}$; for $v_I > 3.5\,\text{V}$, D_2 turns off; and for $v_I \ge 9.4\,\text{V}$, $v_O = 10\,\text{V}$)

2.15   Determine V_O, I_{D1}, I_{D2}, and I in the circuit shown in Figure 2.36. Assume $V_\gamma = 0.6\,\text{V}$ for each diode. (Ans. $V_O = -0.6\,\text{V}$, $I_{D1} = 0$, $I_{D2} = I = 4.27\,\text{mA}$)

Figure 2.35   Figure for Exercise 2.14

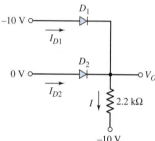

Figure 2.36   Figure for Exercise 2.15

2.4.2   Diode Logic Circuits

Diodes in conjunction with other circuit elements can perform certain **logic functions**, such as AND and OR. The circuit in Figure 2.37 is an example of a diode logic circuit. The four conditions of operation of this circuit depend on various combinations of input voltages, as follows:

$V_1 = V_2 = 0$: There is no excitation to the circuit; therefore, $V_O = 0$ and both diodes are off.

$V_1 = 5\,\text{V}$, $V_2 = 0$: Diode D_1 becomes forward biased and D_2 is reverse biased. Assuming a diode cut-in voltage of $V_\gamma = 0.7\,\text{V}$, the output voltage is $V_O = 4.3\,\text{V}$. The currents are $I_{D2} = 0$ and $I_{D1} = I = V_O/R$.

$V_1 = 0$, $V_2 = 5\,\text{V}$: Diode D_2 turns on, diode D_1 is cut off, and the output voltage is $V_O = 4.3\,\text{V}$. The currents are $I_{D1} = 0$ and $I_{D2} = I = V_O/R$.

$V_1 = V_2 = 5\,\text{V}$: Both diodes are forward biased, so the output is again $V_O = 4.3\,\text{V}$. The current in the resistor is $I = V_O/R$. Since both diodes are on, we assume that the current I splits evenly between the two diodes; therefore, $I_{D1} = I_{D2} = I/2$.

These results are shown in Table 2.1. By definition, in a positive logic system, a voltage near zero corresponds to a logic 0 and a voltage close to

Table 2.1 Two-diode OR logic circuit response

V_1(V)	V_2(V)	V_O(V)
0	0	0
5	0	4.3
0	5	4.3
5	5	4.3

the supply voltage of 5 V corresponds to a logic 1. The results shown in Table 2.1 indicate that this circuit performs the OR logic function. The circuit of Figure 2.37, then, is a two-input diode OR logic circuit.

Figure 2.37 A two-input diode OR logic circuit

Figure 2.38 A two-input diode AND logic circuit

Next, consider the circuit in Figure 2.38. Assume a diode cut-in voltage of $V_\gamma = 0.7$ V. Again, there are four possible states, depending on the combination of input voltages, as follows:

$V_1 = V_2 = 0$: Both diodes are forward biased, and the output voltage is $V_O = 0.7$ V. The current in the resistor is $I = (5 - 0.7)/R$, which we assume splits evenly between the two diodes, so that $I_{D1} = I_{D2} = I/2$.

$V_1 = 5$ V, $V_2 = 0$: In this case, diode D_1 is off, D_2 is on, and the output voltage is $V_O = 0.7$ V. The currents are: $I_{D1} = 0$, and $I_{D2} = I = (5 - 0.7)/R$.

$V_1 = 0$, $V_2 = 5$ V: In this situation, D_1 is on, D_2 is off, and the output voltage is $V_O = 0.7$ V. The currents are: $I_{D1} = I = (5 - 0.7)/R$, and $I_{D2} = 0$.

$V_1 = V_2 = 5$ V: Since there is no potential difference between the supply voltage and the input voltages, all currents are zero and both diodes are off. Also, since there is no potential drop across the resistor R, the output voltage is $V_O = 5$ V.

These results are shown in Table 2.2. This circuit performs the AND logic function. The circuit of Figure 2.38 is a two-input diode AND logic circuit.

If we examine Tables 2.1 and 2.2, we see that the input "low" and "high" voltages may not be the same as the output "low" and "high" voltages. As an example, for the AND circuit (Table 2.2), the input "low" is 0 V, but the output "low" is 0.7 V. This can create a problem because the output of one logic gate is often the input to another logic gate. Another problem occurs when diode logic circuits are connected in cascade; that is, the output of one

Table 2.2 Two-diode AND logic circuit response

V_1(V)	V_2(V)	V_O(V)
0	0	0.7
5	0	0.7
0	5	0.7
5	5	5

OR gate is connected to the input of a second OR gate. The logic 1 levels of the two OR gates are not the same (see Problems 2.42 and 2.43). The logic 1 level degrades or decreases as additional logic gates are connected. However, these problems may be overcome with the use of amplifying devices (transistors) in digital logic systems.

Test Your Understanding

2.16 Consider the OR logic circuit shown in Figure 2.37. Assume a diode cut-in voltage of $V_\gamma = 0.6$ V. (a) Plot V_O versus V_1 for $0 \leq V_1 \leq 5$ V, if $V_2 = 0$. (b) Repeat part (a) if $V_2 = 3$ V. (Ans. (a) $V_O = 0$ for $V_1 \leq 0.6$ V, $V_O = V_1 - 0.6$ for $0.6 \leq V_1 \leq 5$ V; (b) $V_O = 2.4$ V for $0 \leq V_1 \leq 3$ V, $V_O = V_1 - 0.6$ for $3 \leq V_1 \leq 5$ V)

2.17 Consider the AND logic circuit shown in Figure 2.38. Assume a diode cut-in voltage of $V_\gamma = 0.6$ V. (a) Plot V_O versus V_1 for $0 \leq V_1 \leq 5$ V, if $V_2 = 0$. (b) Repeat part (a) if $V_2 = 3$ V. (Ans. (a) $V_O = 0.6$ V for all V_1, (b) $V_O = V_1 + 0.6$ for $0 \leq V_1 \leq 3$ V, $V_O = 3.6$ V for $V_1 \geq 3$ V)

2.5   PHOTODIODE AND LED CIRCUITS

A photodiode converts an optical signal into an electrical current, and a light-emitting diode (LED) transforms an electrical current into an optical signal.

2.5.1   Photodiode Circuit

Figure 2.39 A photodiode circuit

Figure 2.39 shows a typical photodiode circuit in which a reverse-bias voltage is applied to the photodiode. If the photon intensity is zero, the only current through the diode is the reverse-saturation current, which is normally very small. Photons striking the diode create excess electrons and holes in the space-charge region. The electric field quickly separates these excess carriers and sweeps them out of the space-charge region, thus creating a **photocurrent** in the reverse-bias direction. The photocurrent is

$$I_{ph} = \eta e \Phi A \qquad (2.28)$$

where η is the quantum efficiency, e is the electronic charge, Φ is the photon flux density (#/cm²–s), and A is the junction area. This linear relationship between photocurrent and photon flux is based on the assumption that the reverse-bias voltage across the diode is constant. This in turn means that the voltage drop across R induced by the photocurrent must be small, or that the resistance R is small.

Example 2.10 **Objective:** Calculate the photocurrent generated in a photodiode.

For the photodiode shown in Figure 2.39 assume the quantum efficiency is 1, the junction area is $10^{-2}\,\text{cm}^2$, and the incident photon flux is $5 \times 10^{17}\,\text{cm}^{-2}\text{--}\,\text{s}^{-1}$.

Solution: From Equation (2.28), the photocurrent is

$$I_{ph} = \eta e \Phi A = (1)(1.6 \times 10^{-19})(5 \times 10^{17})(10^{-2}) \Rightarrow 0.8\,\text{mA}$$

Comment: The incident photon flux is normally given in terms of light intensity, in lumens, foot-candles, or W/cm^2. The light intensity includes the energy of the photons, as well as the photon flux.

Test Your Understanding

2.18 (a) Photons with an energy of $h\nu = 2\,\text{eV}$ are incident on the photodiode shown in Figure 2.39. The junction area is $A = 0.5\,\text{cm}^2$, the quantum efficiency is $\eta = 0.8$, and the light intensity is $6.4 \times 10^{-2}\,\text{W/cm}^2$. Determine the photocurrent I_{ph}. (b) If $R = 1\,\text{k}\Omega$, determine the minimum power supply voltage V_{PS} needed to ensure that the diode is reverse biased. (Ans. (a) $I_{ph} = 12.8\,\text{mA}$, (b) $V_{PS}(\text{min}) = 12.8\,\text{V}$)

2.5.2 LED Circuit

A light-emitting diode (LED) is the inverse of a photodiode; that is, a current is converted into an optical signal. If the diode is forward biased, electrons and holes are injected across the space-charge region, where they become excess minority carriers. These excess minority carriers diffuse into the neutral n- and p-regions, where they recombine with majority carriers, and the recombination can result in the emission of a photon.

LEDs are fabricated from compound semiconductor materials, such as gallium arsenide or gallium arsenide phosphide. Because these materials have higher bandgap energies than silicon, the forward-bias junction voltage is larger than that in silicon-based diodes.

It is common practice to use a seven-segment LED for the numeric readout of digital instruments, such as a digital voltmeter. The **seven-segment display** is sketched in Figure 2.40. Each segment is an LED normally controlled by IC logic gates.

Figure 2.41 shows one possible circuit connection, known as a common-anode display. In this circuit, the anodes of all LEDs are connected to a 5 V source and the inputs are controlled by logic gates. If V_{I1} is "high," for example, D_1 is off and there is no light output. When V_{I1} goes "low," D_1 becomes forward biased and produces a light output.

Figure 2.40 Seven-segment LED display

Figure 2.41   Control circuit for the seven-segment LED display

---

Example   2.11   **Objective:** Determine the value of R required to limit the current in the circuit in Figure 2.41 when the input is in the low state.

Assume that a diode current of 10 mA produces the desired light output, and that the corresponding forward-bias voltage drop is 1.7 V.

Solution: If $V_I = 0.2$ V in the "low" state, then the diode current is

$$I = \frac{5 - V_\gamma - V_I}{R}$$

The resistance R is then determined as

$$R = \frac{5 - V_\gamma - V_I}{I} = \frac{5 - 1.7 - 0.2}{10} \Rightarrow 310\,\Omega$$

Comment: Typical LED current-limiting resistor values are in the range of 300 to 350 Ω.

One application of LEDs and photodiodes is in **optoisolators**, in which the input signal is electrically decoupled from the output (Figure 2.42). An input signal applied to the LED generates light, which is subsequently detected by the photodiode. The photodiode then converts the light back to an electrical signal. There is no electrical feedback or interaction between the output and input portions of the circuit.

Figure 2.42   Optoisolator using an LED and a photodiode

Test Your Understanding

2.19 Determine the value of resistance R required to limit the current in the circuit shown in Figure 2.41 to $I = 15\,\text{mA}$. Assume $V_\gamma = 1.7\,\text{V}$, $r_f = 15\,\Omega$, and $V_I = 0.2\,\text{V}$ in the "low" state. (Ans. $R = 192\,\Omega$)

2.6 SUMMARY

- In this chapter, we analyzed several classes of diode circuits that can be used to produce various desired output signals. The resulting characteristics of each of the circuits considered rely on the nonlinear i–v relationship of the diode. We continued to use the piecewise linear model and approximation techniques in our hand analyses. Computer simulation can be used to obtain more accurate results when actual diode properties are known.

- Half-wave and full-wave rectifier circuits convert a sinusoidal (i.e., ac) signal to an approximate dc signal. A dc power supply, which is used to bias electronic circuits and systems, utilize these types of circuits. An RC filter can be connected to the output of the rectifier circuit to reduce the ripple effect. The ripple voltage in the output signal was determined as a function of the RC filter and other circuit parameters.

- Zener diodes operate in the reverse breakdown region. Since the breakdown voltage is nearly constant over a wide range of currents, these devices are useful in voltage reference or regulator circuits. The percent regulation, a figure of merit for the circuit, is a function of the range of input voltage and load resistance values, and of the individual device parameters.

- Techniques used to analyze multidiode circuits, which are used in various signal-processing applications, were discussed. The technique requires making assumptions as to whether a diode is conducting (on) or not conducting (off). After analyzing the circuit using these assumptions, we must go back and verify that the assumptions made were valid. This analysis technique is obviously not as straightfoward as the one for linear circuits.

- Diode circuits can be designed to perform basic digital logic functions. We considered the circuit that performs the OR logic function and the circuit that performs the AND logic function. However, we noted some inconsistencies between input and output logic values, which will limit the use of diode logic gates.

- The light-emitting diode (LED) converts an electrical current to light and is used extensively in such applications as the seven-segment alphanumeric display. Conversely, the photodiode detects an incident light signal and transforms it into an electrical current. Examples of these types of circuits were analyzed.

CHECKPOINT

After studying this chapter, the reader should have the ability to:

✔ In general, apply the diode piecewise linear model in the analysis of diode circuits.

✔ Analyze diode rectifier circuits, including the calculation of ripple voltage. (Section 2.1)

✔ Analyze Zener diode circuits, including the effect of a Zener resistance. (Section 2.2)

✔ Determine the output signal for a given input signal of diode clipper and clamper circuits. (Section 2.3)

✔ Analyze circuits with multiple diodes by making initial assumptions and then verifying these initial assumptions (Section 2.4)

REVIEW QUESTIONS

1. What characteristic of a diode is used in the design of diode signal processing circuits?

2. Describe a simple half-wave diode rectifier circuit and sketch the output voltage versus time.

3. Describe a simple full-wave diode rectifier circuit and sketch the output voltage versus time.

4. What is the advantage of connecting an RC filter to the output of a diode rectifier circuit?

5. Define ripple voltage. How can the magnitude of the ripple voltage be reduced?

6. Describe a simple Zener diode voltage reference circuit.

7. What effect does the Zener diode resistance have on the voltage reference circuit operation?

8. What are the general characteristics of diode clipper circuits?

9. Describe a simple diode clipper circuit to limit the negative portion of a sinusoidal input voltage to a specified value.

10. What are the general characteristics of diode clamper circuits?

11. What one circuit element, besides a diode, is present in all diode clamper circuits?

12. Describe the procedure used in the analysis of a circuit containing two diodes. How many initial assumptions concerning the state of the circuit are possible?

13. Describe a diode OR logic circuit. Compare a logic 1 value at the output compared to a logic 1 value at the input. Are they the same value?

14. Describe a diode AND logic circuit. Compare a logic 0 value at the output compared to a logic 0 value at the input. Are they the same value?

15. Describe a simple circuit that can be used to turn an LED on or off with a high or low input voltage.

PROBLEMS

[Note: In the following problems, assume $r_f = 0$ unless otherwise specified.]

Section 2.1   Rectifier Circuits

2.1   Assume the input to the circuit in Figure P2.1 is a triangular wave of 20 V peak-to-peak amplitude with a zero time-average value. Let $R = 1\,k\Omega$ and assume piecewise linear diode parameters of $V_\gamma = 0.6\,V$ and $r_f = 20\,\Omega$. Sketch the output voltage versus time over one cycle and label all appropriate voltages.

2.2   For the circuit shown in Figure P2.1, show that for $v_I \geq 0$, the output voltage is approximately given by

$$v_O = v_I - V_T \ln\left(\frac{v_O}{I_S R}\right)$$

Figure P2.1

2.3   Consider the half-wave rectifier circuit in Figure 2.2(a). The input voltage is $v_I = 160 \sin[2\pi(60)t]$ V and the transformer turns ratio is $N_1/N_2 = 4$. If $V_\gamma = 0$ and $r_f = 0$, determine (a) the peak diode current, (b) the value of PIV, (c) the average value of the output voltage, and (d) the fraction (percent) of a cycle that $v_O > 0$.

RD2.4 The input signal voltage to the full-wave rectifier circuit in Figure 2.7(a) in the text is $v_I = 160 \sin[2\pi(60)t]$ V. Assume $V_\gamma = 0.7$ V for each diode. Determine the required turns ratio of the transformer to produce a peak output voltage of (a) 25 V, and (b) 100 V. What must be the diode PIV rating for each case? Verify the results with a computer simulation analysis.

D2.5 The output resistance of the full-wave rectifier in Figure 2.7(a) in the text is $R = 150\,\Omega$. A filter capacitor is connected in parallel with R. Assume $V_\gamma = 0.7$ V. The peak output voltage is to be 24 V and the ripple voltage is to be no more than 0.5 V. The input frequency is 60 Hz. (a) Determine the required rms value of v_S. (b) Determine the required filter capacitance value. (c) Determine the peak current through each diode.

RD2.6 Repeat Problem 2.5 for the half-wave rectifier in Figure 2.2(a).

2.7 The circuit in Figure P2.7 is a complementary output rectifier. If $v_s = 26 \sin [2\pi(60)t]$ V, sketch the output waveforms v_o^+ and v_o^- versus time, assuming $V_\gamma = 0.6$ V for each diode.

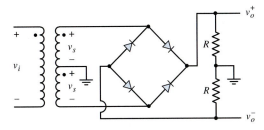

Figure P2.7

D2.8 Consider the battery charging circuit in Figure 2.5(a). Let $V_B = 12$ V, $V_\gamma = 0.7$ V, $V_S = 24$ V, and $\omega = 2\pi(60)$. The average battery charging current is to be $i_D = 2$ A. Determine the required value of R and find the fraction of time the diode is conducting. What must be the power rating of the resistor R?

D2.9 The full-wave rectifier in Figure 2.6(a) is to deliver 0.1 A and 15 V (average) to a load. The ripple voltage is to be no larger than 0.4 V peak-to-peak. The input signal is 120 V (rms) at 60 Hz. Assume diode cut-in voltages of 0.7 V. Determine the required turns ratio, the filter capacitance value, and the diode PIV rating. Verify the design with a computer simulation analysis.

*2.10 Sketch v_o versus time for the circuit in Figure P2.10 with the input shown. Assume $V_\gamma = 0$.

Figure P2.10

$R_1 = 2.2\,\text{k}\Omega$

D_1

$-\ v_o\ +$

v_i

$R_L = 6.8\,\text{k}\Omega$

D_2

$R_2 = 2.2\,\text{k}\Omega$

Figure P2.11

*2.11 (a) Sketch v_o versus time for the circuit in Figure P2.11. The input is a sine wave given by $v_i = 10 \sin \omega t$ V. Assume $V_\gamma = 0$. (b) Determine the rms value of the output voltage.

Section 2.2   Zener Diode Circuits

2.12 In the voltage regulator circuit in Figure P2.12, let $V_I = 6.3$ V, $R_i = 12\,\Omega$, and $V_z = 4.8$ V. The Zener diode current is to be limited to the range $5 \le I_Z \le 100$ mA. (a) Determine the range of possible load currents and load resistances. (b) Determine the power rating required for the Zener diode and the load resistor.

R_i

$+$ I_I I_Z $+$ I_L $+$

V_I V_Z R_L V_L

$-$ $-$ $-$

Figure P2.12

*2.13 In the voltage regulator circuit in Figure P2.12, $V_I = 20$ V, $V_Z = 10$ V, $R_i = 222\,\Omega$, and $P_Z(\text{max}) = 400$ mW. (a) Determine I_L, I_Z, and I_I, if $R_L = 380\,\Omega$. (b) Determine the value of R_L that will establish $P_Z(\text{max})$ in the diode. (c) Repeat part (b) if $R_i = 175\,\Omega$.

D2.14 A Zener diode is connected in a voltage regulator circuit as shown in Figure P2.12. The Zener voltage is $V_Z = 10$ V and the Zener resistance is assumed to be $r_z = 0$. (a) Determine the value of R_i such that the Zener diode remains in breakdown if the load current varies from $I_L = 50$ to 500 mA and if the input voltage varies from $V_I = 15$ to 20 V. Assume $I_Z(\text{min}) = 0.1 I_Z(\text{max})$. (b) Determine the power rating required for the Zener diode and the load resistor.

2.15 Reconsider Problem D2.14. (a) Determine the maximum variation in the output voltage if the Zener resistance is $r_z = 2\,\Omega$. (b) Calculate the percent regulation.

2.16 The percent regulation of the Zener diode regulator shown in Figure 2.17 is 5 percent. The Zener voltage is $V_{ZO} = 6$ V and the Zener resistance is $r_z = 3\,\Omega$. Also, the load resistance varies between 500 and 1000 Ω, the input resistance is $R_i = 280\,\Omega$, and the minimum power supply voltage is $V_{PS}(\text{min}) = 15$ V. Determine the maximum power supply voltage allowed. (Ans. $V_{PS}(\text{max}) = 41.3$ V)

*D2.17 A voltage regulator is to have a nominal output voltage of 10 V. The specified Zener diode has a rating of 1 W, has a 10 V drop at $I_Z = 25$ mA, and has a Zener resistance of $r_z = 5\,\Omega$. The input power supply has a nominal value of $V_{PS} = 20$ V and can vary by ± 25 percent. The output load current is to vary between $I_L = 0$ and 20 mA. (a) If the minimum Zener current is to be $I_Z = 5$ mA, determine the required R_i. (b) Determine the maximum variation in output voltage. (c) Determine the percent regulation.

*D2.18 Consider the circuit in Figure P2.18. Let $V_\gamma = 0$. The secondary voltage is given by $v_s = V_s \sin \omega t$, where $V_S = 24$ V. The Zener diode has parameters $V_Z = 16$ V at $I_Z = 40$ mA and $r_z = 2\,\Omega$. Determine R_i such that the load current can vary over the range $40 \le I_L \le 400$ mA with $I_Z(\text{min}) = 40$ mA and find C such that the ripple voltage is no larger than 1 V.

Figure P2.18

*2.19 The secondary voltage in the circuit in Figure P2.18 is $v_s = 12 \sin \omega t$ V. The Zener diode has parameters $V_Z = 8$ V at $I_Z = 100$ mA and $r_z = 0.5\,\Omega$. Let $V_\gamma = 0$ and $R_i = 3\,\Omega$. Determine the percent regulation for load currents between $I_L = 0.2$ and 1 A. Find C such that the ripple voltage is no larger than 0.8 V.

Section 2.3   Clipper and Clamper Circuits

2.20 Consider the circuit in Figure P2.20. Let $V_\gamma = 0$. (a) Plot v_O versus v_I over the range $-10 \le v_I \le +10$ V. (b) Plot i_1 over the same input voltage range as part (a).

Figure P2.20   **Figure P2.21**

2.21 For the circuit in Figure P2.21, (a) plot v_O versus v_I for $0 \le v_I \le 15$ V. Assume $V_\gamma = 0.7$ V. Indicate all breakpoints. (b) Plot i_D over the same range of input voltage. (c) Compare the results of parts (a) and (b) with a computer simulation.

2.22 For the circuit in Figure P2.22, let $V_\gamma = 0.7$ V and assume the input voltage varies over the range $-10 \le v_I \le +10$ V. Plot (a) v_O versus v_I and (b) i_D versus v_I over the input voltage range indicated.

*2.23 The diode in the circuit of Figure P2.23(a) has piecewise linear parameters $V_\gamma = 0.7$ V and $r_f = 10\,\Omega$. (a) Plot v_O versus v_I for $-30 \le v_I \le 30$ V. (b) If the triangular wave, shown in Figure P2.23(b), is applied, plot the output versus time.

Figure P2.22

(a)   (b)

Figure P2.23

Figure P2.24

2.24 Consider the circuit in Figure P2.24. Sketch v_O versus time if $v_I = 15 \sin \omega t$ V. Assume $V_\gamma = 0.6$ V.

2.25 Plot v_O for each circuit in Figure P2.25 for the input shown. Assume (a) $V_\gamma = 0$ and (b) $V_\gamma = 0.6$ V.

(a)   (b)

Figure P2.25

***D2.26** A car's radio may be subjected to voltage spikes induced by coupling from the ignition system. Pulses on the order of ± 250 V and lasting for $120\,\mu s$ may exist. Design a clipper circuit using resistors, diodes, and Zener diodes to limit the input voltage between $+14$ V and -0.7 V. Specify power ratings of the components.

2.27 Sketch v_O versus time for each circuit with the input shown in Figure P2.27. Assume $V_\gamma = 0$ and assume the RC time constant is large.

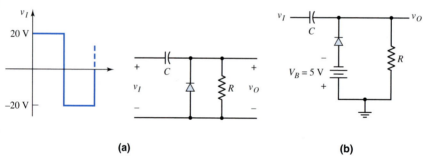

(a)   (b)

Figure P2.27

D2.28 Design a diode clamper to generate the output v_O from the input v_I shown in Figure P2.28 if (a) $V_\gamma = 0$, and (b) $V_\gamma = 0.7$ V.

(a)   (b)

Figure P2.28

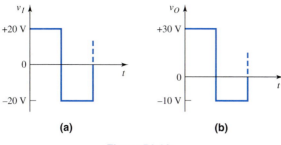

Figure P2.29

D2.29 Design a diode clamper to generate the output v_O from the input v_I in Figure P2.29 if $V_\gamma = 0$.

2.30 For the circuit in Figure 2.27(a), let $V_\gamma = 0$ and $v_I = 10 \sin \omega t$. Plot v_O versus time over 3 cycles of input voltage for (a) $V_B = +3$ V and (b) $V_B = -3$ V.

2.31 Repeat Problem 2.30 for the case when the direction of the diode in the circuit shown in Figure 2.27(a) is reversed.

Section 2.4   Multiple Diode Circuits

2.32 The diodes in the circuit in Figure P2.32 have piecewise linear parameters of $V_\gamma = 0.6$ V and $r_f = 0$. Determine the output voltage V_O and the diode currents I_{D1} and I_{D2} for the following input conditions: (a) $V_1 = 10$ V, $V_2 = 0$; (b) $V_1 = 5$ V, $V_2 = 0$; (c) $V_1 = 10$ V, $V_2 = 5$ V; and (d) $V_1 = V_2 = 10$ V. (e) Compare the results of parts (a) through (d) with a computer simulation analysis.

Figure P2.32   **Figure P2.33**

2.33 In the circuit in Figure P2.33 the diodes have the same piecewise linear parameters as described in Problem 2.32. Calculate the output voltage V_O and the currents I_{D1}, I_{D2}, and I for the following input conditions: (a) $V_1 = V_2 = 10$ V; (b) $V_1 = 10$ V, $V_2 = 0$; (c) $V_1 = 10$ V, $V_2 = 5$ V; and (d) $V_1 = V_2 = 0$.

2.34 The diodes in the circuit in Figure P2.34 have the same piecewise linear parameters as described in Problem 2.32. Determine the output voltage V_O and the currents I_{D1}, I_{D2}, I_{D3}, and I for the following input conditions: (a) $V_1 = V_2 = 0$; (b) $V_1 = V_2 = 5$ V; (c) $V_1 = 5$ V, $V_2 = 0$; and (d) $V_1 = 5$ V, $V_2 = 2$ V.

Figure P2.34   Figure P2.35

2.35 The cut-in voltage for each diode in Figure P2.35 is $V_\gamma = 0.6$ V. (a) Find V_1, V_2, and each diode current for $R_1 = 2\,\text{k}\Omega$, $R_2 = 6\,\text{k}\Omega$, and $R_3 = 2\,\text{k}\Omega$. (b) Repeat part (a) for $R_1 = 6\,\text{k}\Omega$, $R_2 = R_3 = 5\,\text{k}\Omega$. (c) Determine R_1, R_2, and R_3 such that each diode current is 0.5 mA.

2.36 (a) For the circuit in Figure P2.36, each diode has $V_\gamma = 0.6$ V. Plot v_O versus v_I over the range $0 \le v_I \le 10$ V. (b) Compare the results of part (a) with a computer simulation analysis.

Figure P2.36   Figure P2.37

*2.37 Assume $V_\gamma = 0.7$ V for each diode in the circuit in Figure P2.37. Plot v_O versus v_I for $-10 \le v_I + 10$ V.

2.38 Let $V_\gamma = 0.7$ V for each diode in the circuit in Figure P2.38. (a) Find I_{D1} and V_O for $R_1 = 5\,\text{k}\Omega$ and $R_2 = 10\,\text{k}\Omega$. (b) Repeat part (a) for $R_1 = 10\,\text{k}\Omega$ and $R_2 = 5\,\text{k}\Omega$.

2.39 If $V_\gamma = 0.7$ V for the diode in the circuit in Figure P2.39, determine I_D and V_O.

Figure P2.38   Figure P2.39   Figure P2.40

2.40 Let $V_\gamma = 0.6\,\text{V}$ for the diode in the circuit in Figure P2.40. Determine I_D and V_D if: (a) $V_1 = 15\,\text{V}$, $V_2 = 10\,\text{V}$; and (b) $V_1 = 10\,\text{V}$, $V_2 = 15\,\text{V}$.

2.41 (a) Each diode in the circuit in Figure P2.41 has piecewise linear parameters of $V_\gamma = 0$ and $r_f = 0$. Plot v_O versus v_I for $0 \le v_I \le 30\,\text{V}$. Indicate the breakpoints and give the state of each diode in the various regions of the plot. (b) Compare the results of part (a) with a computer simulation analysis.

Figure P2.41   Figure P2.42

2.42 Consider the circuit in Figure P2.42. The output of a diode OR logic gate is connected to the input of a second diode OR logic rate. Assume $V_\gamma = 0.6\,\text{V}$ for each diode. Determine the outputs V_{O1} and V_{O2} for: (a) $V_1 = V_2 = 0$; (b) $V_1 = 5\,\text{V}$, $V_2 = 0$; and (c) $V_1 = V_2 = 5\,\text{V}$. What can be said about the relative values of V_{O1} and V_{O2} in their "high" state?

2.43 Consider the circuit in Figure P2.43. The output of a diode AND logic gate is connected to the input of a second diode AND logic gate. Assume $V_\gamma = 0.6\,\text{V}$ for each

Figure P2.43

diode. Determine the outputs V_{O1} and V_{O2} for: (a) $V_1 = V_2 = 5\,\text{V}$; (b) $V_1 = 0$, $V_2 = 5\,\text{V}$; and (c) $V_1 = V_2 = 0$. What can be said about the relative values of V_{O1} and V_{O2} in their "low" state?

2.44   Determine the Boolean expression for V_O in terms of the four input voltages for the circuit in Figure P2.44.

Figure P2.44

Section 2.5   LED and Photodiode Circuits

2.45   Consider the circuit shown in Figure P2.45. The forward-bias cut-in voltage of the diode is 1.5 V and the forward-bias resistance is $r_f = 10\,\Omega$. Determine the value of R required to limit the current to $I = 12\,\text{mA}$ when $V_I = 0.2\,\text{V}$.

Figure P2.45

2.46   The light-emitting diode in the circuit shown in Figure P2.45 has parameters $V_\gamma = 1.7\,\text{V}$ and $r_f = 0$. Light will first be detected when the current is $I = 8\,\text{mA}$. If $R = 750\,\Omega$, determine the value of V_I at which light will first be detected.

2.47   If the resistor in Example 2.10 is $R = 2\,\text{k}\Omega$ and the diode is to be reverse biased by at least 1 V, determine the minimum power supply voltage required.

2.48   Consider the photodiode circuit shown in Figure 2.39. Assume the quantum efficiency is 1. A photocurrent of 0.6 mA is required for an incident photon flux of $\Phi = 10^{17}\,\text{cm}^{-2}\text{–s}^{-1}$. Determine the required cross-sectional area of the diode.

COMPUTER SIMULATION PROBLEMS

2.49   Correlate the results of Example 2.1 with a computer simulation.

2.50   Consider the voltage doubler circuit shown in Figure 2.13. Assume a 60 Hz sinusoidal input signal and a 1:1 transformer turns ratio. Let $R = 5\,\text{k}\Omega$ and $C_1 = C_2 = 100\,\mu\text{F}$. Plot the steady-state output voltage over two cycles of input voltage.

2.51   Correlate the design results of Example 2.4 with a computer simulation.

2.52   Perform a computer simulation analysis of Exercise 2.13. How much does the output voltage change during each half-cycle?

DESIGN PROBLEMS

[Note: Each design should be correlated with a computer simulation.]

*D2.53   Design a full-wave rectifier to produce a peak output voltage of 9 V, deliver 150 mA to the load, and produce an output with a ripple of not more than 5 percent. A line voltage of 120 V (rms), 60 Hz is available. The only transformers available have turns ratios of $N_1/N_2 = 10$, 15, and 20.

*D2.54   Design a full-wave rectifier to provide a dc output of 28 V at a current of 4 A, with a ripple of not more than 3 percent. A line voltage of 120 V (rms), 60 Hz is available.

*D2.55   Design a full-wave regulated power supply using a 5 : 1 center-tapped transformer and a 7.5 V, 1 W Zener diode. The power supply must provide a constant 7.5 V to a load varying from 120 to 450 Ω. The input voltage is 120 V (rms), 60 Hz.

3

The Bipolar Junction Transistor

3.0 PREVIEW

In the last chapter, we saw that the rectifying current–voltage characteristics of the diode are useful in electronic switching and waveshaping circuits. However, diodes are not capable of amplifying currents or voltages. The electronic device that is capable of current and voltage amplification, or gain, in conjunction with other circuit elements, is the transistor, which is a three-terminal device. The development of the silicon transistor by Bardeen, Brattain, and Schockley at Bell Telephone Laboratories in the late 1940s started the first electronics revolution of the 1950s and 1960s. This invention led to the development of the first integrated circuit in 1958 and to the operational transistor amplifier (op-amp), which is one of the most widely used electronic circuits.

The bipolar transistor, which is introduced in this chapter, is the first of two major types of transistors. The second type of transistor, the field-effect transistor (FET), is introduced in Chapter 5 and has led to the second electronics revolution in the 1970s and 1980s. These two device types are the basis of modern day microelectronics. Each device type is equally important and each has particular advantages for specific applications.

We begin this chapter with a look at the physical structure and operation of the bipolar transistor. The chapter deals mainly with the transistor characteristics and with the dc analysis and design of bipolar circuits. We continue to use the piecewise linear approximation techniques, developed for the diode, in the bipolar transistor calculations. We discuss how the transistor can be used in switch, digital, and linear amplifier applications.

Much of the material in this chapter may appear to be skewed toward discrete transistor biasing. However, the principal goal of the chapter is to ensure that readers become familiar with transistor characteristics and are able to quickly analyze and design the dc response of bipolar transistor circuits. Integrated circuit biasing is discussed toward the end of the chapter and is emphasized to a greater extent in the later chapters.

3.1 BASIC BIPOLAR JUNCTION TRANSISTOR

The **bipolar junction transistor (BJT)** has three separately doped regions and contains two pn junctions. A single pn junction has two modes of operation—

forward bias and reverse bias. The bipolar transistor, with two pn junctions, therefore has four possible modes of operation, depending on the bias condition of each pn junction, which is one reason for the versatility of the device. With three separately doped regions, the bipolar transistor is a three-terminal device. The basic transistor principle is that *the voltage between two terminals controls the current through the third terminal.*

Our discussion of the bipolar transistor starts with a description of the basic transistor structure and a qualitative description of its operation. To describe its operation, we use the pn junction concepts presented in Chapter 1. However, the two pn junctions are sufficiently close together to be called interacting pn junctions. The operation of the transistor is therefore totally different from that of two back-to-back diodes.

Current in the transistor is due to the flow of both electrons and holes, hence the name **bipolar**. Our discussion covers the relationship between the three terminal currents. In addition, we present the circuit symbols and conventions used in bipolar circuits, the bipolar transistor current-voltage characteristics, and finally, some nonideal current–voltage characteristics.

3.1.1   Transistor Structures

Figure 3.1 shows simplified block diagrams of the basic structure of the two types of bipolar transistor: npn and pnp. The **npn bipolar transistor** contains a thin p-region between two n-regions. In contrast, the **pnp bipolar transistor** contains a thin n-region sandwiched between two p-regions. The three regions and their terminal connections are called the **emitter**, **base**, and **collector**. The operation of the device depends on the two pn junctions being in close proximity, so the width of the base must be very narrow, normally in the range of tenths of a micrometer (10^{-6} m).

The actual structure of the bipolar transistor is considerably more complicated than the block diagrams of Figure 3.1. For example, Figure 3.2 is the cross section of a classic npn bipolar transistor fabricated in an integrated circuit. One important point is that the device is not symmetrical electrically. This asymmetry occurs because the geometries of the emitter and collector regions are not the same, and the impurity doping concentrations in the

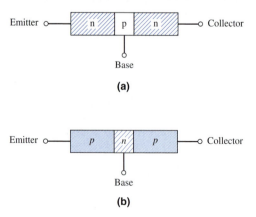

Figure 3.1   Simple geometry of bipolar transistors: (a) npn and (b) pnp

Figure 3.2   Cross section of a conventional integrated circuit npn bipolar transistor

three regions are substantially different. For example, the impurity doping concentrations in the emitter, base, and collector may be on the order of 10^{19}, 10^{17}, and 10^{15} cm^{-3}, respectively. Therefore, even though both ends are either p-type or n-type on a given transistor, switching the two ends makes the device act in drastically different ways.

Although the block diagrams in Figure 3.1 are highly simplified, they are still useful for presenting the basic transistor characteristics.

3.1.2   npn Transistor: Forward-Active Mode Operation

Since the transistor has two pn junctions, four possible bias combinations may be applied to the device, depending on whether a forward or reverse bias is applied to each junction. For example, if the transistor is used as an amplifying device, the **base–emitter (B–E) junction** is forward biased and the **base–collector (B–C) junction** is reverse biased, in a configuration called the **forward-active operating mode**, or simply the **active region**. The reason for this bias combination will be illustrated as we look at the operation of such transistors and the characteristics of circuits that use them.

Transistor Currents

Figure 3.3 shows an idealized npn bipolar transistor biased in the forward-active mode. Since the B–E junction is forward biased, electrons from the emitter are injected across the B–E junction into the base, creating an excess minority carrier concentration in the base. Since the B–C junction is reverse

Figure 3.3   An npn bipolar transistor biased in the forward-active mode

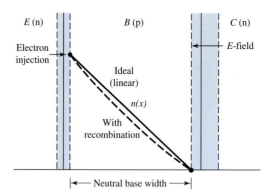

Figure 3.4   Minority carrier electron concentration across the base region of an npn bipolar transistor biased in the forward-active mode

biased, the electron concentration at the edge of that junction is approximately zero.

The electron concentration in the base region is shown in Figure 3.4. Because of the large gradient in this concentration, electrons injected from the emitter diffuse across the base into the B–C space-charge region, where the electric field sweeps them into the collector region. Ideally, as many electrons as possible will reach the collector without recombining with majority carrier holes in the base. Figure 3.4 shows the ideal case in which no recombination occurs, so that the electron concentration is a linear function of distance across the base. However, if some carrier recombination does occur in the base, the electron concentration will deviate from the ideal linear curve, as shown in the figure. To minimize recombination effects, the width of the neutral base region must be small compared to the minority carrier diffusion length.

Emitter Current   Since the B–E junction is forward biased, we expect the current through this junction to be an exponential function of B–E voltage, just as we saw that the current through a pn junction diode was an exponential function of the forward-biased diode voltage. We can then write the current at the emitter terminal as

$$i_E = I_S \left[e^{v_{BE}/V_T} - 1 \right] \cong I_S e^{v_{BE}/V_T} \tag{3.1}$$

where the approximation of neglecting the (−1) term is usually valid since $v_{BE} \gg V_T$ in most cases. The parameter V_T is the usual thermal voltage. The emission coefficient n that multiplies V_T is assumed to be 1, as we discussed in Chapter 1 in considering the ideal diode equation. The flow of the negatively charged electrons is through the emitter into the base and is opposite to the conventional current direction. The conventional emitter current is therefore out of the emitter terminal.

The multiplying constant, I_S, contains electrical parameters of the junction, but in addition is directly proportional to the active B–E cross-sectional area. Therefore, if two transistors are identical except that one has twice the area of the other, then the emitter currents will differ by a factor of two for the same applied B–E voltage. Typical values of I_S are in the range of 10^{-12} to 10^{-15} A, but may, for special transistors, vary outside of this range.

Collector Current   Since the doping concentration in the emitter is much larger than that in the base region, the vast majority of emitter current is due to the injection of electrons into the base. The number of these injected electrons reaching the collector is the major component of collector current.

The number of electrons reaching the collector per unit time is proportional to the number of electrons injected into the base, which in turn is a function of the B–E voltage. To a first approximation, the collector current is proportional to e^{v_{BE}/V_T} and is independent of the reverse-biased B–C voltage. The device therefore looks like a **constant-current source**. The collector current is controlled by the B–E voltage; in other words, the current at one terminal (the collector) is controlled by the voltage across the other two terminals. *This control is the basic transistor action.*

The collector current is proportional to the emitter current, so we can write the collector current as[1]

$$i_C = \alpha_F i_E = \alpha_F I_S e^{v_{BE}/V_T} \tag{3.2}$$

where α_F is a constant less than 1 but very close to 1. This parameter is called the **common-base current gain**. The reason for this name will become clearer as we proceed through the chapter.

Base Current   Since the B–E junction is forward biased, holes from the base flow across the B–E junction into the emitter. However, because these holes do not contribute to the collector current, they are not part of the transistor action. Instead, the flow of holes forms one component of the base current. This component is also an exponential function of the B–E voltage, because of the forward-biased B–E junction. We can write

$$i_{B1} \propto e^{v_{BE}/V_T} \tag{3.3(a)}$$

A few electrons recombine with majority carrier holes in the base. The holes that are lost must be replaced through the base terminal. The flow of such holes is a second component of the base current. This "recombination current" is directly proportional to the number of electrons being injected from the emitter, which in turn is an exponential function of the B–E voltage. We can write

$$i_{B2} \propto e^{v_{BE}/V_T} \tag{3.3(b)}$$

The total base current is the sum of the two components from Equations (3.3(a)) and (3.3(b)):

$$i_B \propto e^{v_{BE}/V_T} \tag{3.4}$$

Figure 3.5 shows the flow of electrons and holes in an npn bipolar transistor, as well as the terminal currents.[2] (Reminder: the conventional current

---

[1]In many cases, the multiplying constant in the collector current, i_C, equation is written as I_S, which means that the multiplying constant for the emitter current would be I_S/α_F. The important point of this discussion is that the currents are an exponential function of the B–E voltage. Which multiplying parameter is used in each equation is simply a matter of preference.

[2]A more thorough study of the physics of the bipolar transistor shows that there are other current components, in addition to the ones mentioned. However, these additional currents do not change the basic properties of the transistor and can be neglected for our purposes.

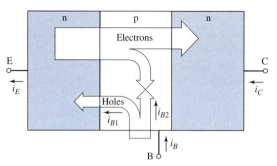

Figure 3.5   Electron and hole currents in an npn transistor biased in the forward-active mode

direction is the same as the flow of positively charged holes and opposite to the flow of negatively charged electrons.)

If the concentration of electrons in the n-type emitter is much larger than the concentration of holes in the p-type base, then the number of electrons injected into the base will be much larger than the number of holes injected into the emitter. This means that the i_{B1} component of the base current will be much smaller than the collector current. In addition, if the base width is small, then the number of electrons that recombine in the base will be small, and the i_{B2} component of the base current will also be much smaller than the collector current.

Common-Emitter Current Gain

In the transistor, the rate of flow of electrons and the resulting collector current are an exponential function of the B–E voltage, as is the resulting base current. This means that the collector current and the base current are linearly related. Therefore, we can write

$$\frac{i_C}{i_B} = \beta_F \tag{3.5}$$

or

$$i_B = \frac{\alpha_F I_S}{\beta_F} e^{v_{BE}/V_T} \tag{3.6}$$

The parameter β is the **common-emitter current gain** and is a key parameter of the bipolar transistor. In this idealized situation, β is considered to be a constant for any given transistor. The value of β is usually in the range of $50 < \beta < 300$, but it can be smaller or larger for special devices.

Figure 3.6 shows an npn bipolar transistor in a circuit. Because the emitter is the common connection, this circuit is referred to as a **common-emitter configuration**. When the transistor is biased in the forward-active mode, the B–E junction is forward biased and the B–C junction is reverse biased. Using the piecewise linear model of a pn junction, we assume that the B–E voltage is equal to $V_{BE}(\text{on})$, the junction turn-on voltage. Since $V_{CC} = v_{CE} + i_C R_C$, the power supply voltage must be sufficiently large to keep the B–C junction reverse biased. The base current is established by V_{BB} and R_B, and the resulting collector current is $i_C = \beta i_B$.

Figure 3.6   An npn transistor circuit in the common-emitter configuration

If we set $V_{BB} = 0$, the B–E junction will have zero applied volts; therefore, $i_B = 0$, which implies that $i_C = 0$. This condition is called **cutoff**.

Current Relationships

If we treat the bipolar transistor as a single node, then, by Kirchhoff's current law, we have

$$i_E = i_C + i_B \tag{3.7}$$

If the transistor is biased in the forward-active mode, then

$$i_C = \beta i_B \tag{3.8}$$

Substituting Equation (3.8) into (3.7), we obtain the following relationship between the emitter and base currents:

$$i_E = (1 + \beta)i_B \tag{3.9}$$

Solving for i_B in Equation (3.8) and substituting into Equation (3.9), we obtain a relationship between the collector and emitter currents, as follows:

$$i_C = \left(\frac{\beta}{1 + \beta}\right)i_E \tag{3.10}$$

From Equation (3.2), we had $i_C = \alpha_F i_E$ so

$$\alpha_F = \frac{\beta_F}{1 + \beta_F} \tag{3.11}$$

The parameter α_F is called the common-base current gain and is always slightly less than 1. We may note that if $\beta_F = 100$, then $\alpha_F = 0.99$, so α_F is indeed close to 1. From Equation (3.11), we can state the common-emitter current gain in terms of the common-base current gain:

$$\beta_F = \frac{\alpha_F}{1 - \alpha_F} \tag{3.12}$$

Summary of Transistor Operation

We have presented a first-order model of the operation of the npn bipolar transistor biased in the forward-active region. The forward-biased B–E voltage, v_{BE}, causes an exponentially related flow of electrons from the emitter into the base where they diffuse across the base region and are collected in the collector region. The collector current, i_C, is independent of the B–C voltage as long as the B–C junction is reverse biased. The collector, then, behaves as an ideal current source. The collector current is a fraction α_F of the emitter current, and the base current is a fraction $1/\beta_F$ of the collector current. If $\beta_F \gg 1$, then $\alpha_F \cong 1$ and $i_C \cong i_E$.

Example 3.1   Objective: Calculate the collector and emitter currents, given the base current and current gain.

Assume a common-emitter current gain of $\beta = 150$ and a base current of $i_B = 15\,\mu A$. Also assume that the transistor is biased in the forward-active mode.

Solution: The relation between collector and base currents gives

$$i_C = \beta i_B = (150)(15\,\mu A) \Rightarrow 2.25\,mA$$

and the relation between emitter and base currents yields

$$i_E = (1 + \beta)i_B = (151)(15\,\mu A) \Rightarrow 2.27\,mA$$

From Equation (3.11), the common-base current gain is

$$\alpha = \frac{\beta}{1 + \beta} = \frac{150}{151} = 0.9934$$

Comment: For reasonable values of β, the collector and emitter currents are nearly equal, and the common-base current gain is nearly 1.

Test Your Understanding

3.1   Transistors of a particular type have common-base current gains in the range of $0.980 \leq \alpha \leq 0.995$. Find the corresponding range of β. (Ans. $49 \leq \beta \leq 199$)

3.2   The common-emitter current gains of two transistors are $\beta = 75$ and $\beta = 125$. Determine the common-base current gains. (Ans. $\alpha = 0.9868$, $\alpha = 0.9921$)

3.1.3   pnp Transistor: Forward-Active Mode Operation

We have discussed the basic operation of the npn bipolar transistor. The complementary device is the pnp transistor. Figure 3.7 shows the flow of holes and electrons in a pnp device biased in the forward-active mode. Since the B–E junction is forward biased, the p-type emitter is positive with respect to the n-type base, holes flow from the emitter into the base, the holes diffuse across the base, and they are swept into the collector. The collector current is a result of this flow of holes.

Again, since the B–E junction is forward biased, the emitter current is an exponential function of the B–E voltage. Noting the direction of emitter current and the polarity of the foward-biased B–E voltage, we can write

$$i_E = I_S e^{v_{EB}/V_T} \tag{3.13}$$

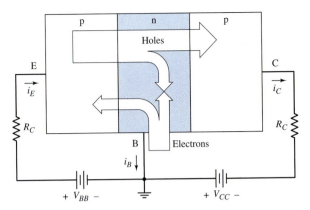

Figure 3.7 Electron and hole currents in a pnp transistor biased in the forward-active mode

where we are again assuming the (-1) term in the ideal diode equation is negligible.

The collector current is an exponential function of the E–B voltage, and the direction is out of the collector terminal, which is opposite to that in the npn device. We can now write

$$i_C = \alpha_F i_E = \alpha_F I_S e^{v_{EB}/V_T} \tag{3.14}$$

where α_F is again the common-base current gain.

The base current in a pnp device is the sum of two components. The first component comes from electrons flowing from the base into the emitter as a result of the forward-biased E–B junction. The second component comes from the flow of electrons supplied through the base terminal to replace those lost by recombination with holes in the base. The direction of the base current is out of the base terminal. The base current in the pnp device is also an exponential function of the E–B voltage, as follows:

$$i_B = \frac{i_C}{\beta_F} = \frac{\alpha_F I_S}{\beta_F} e^{v_{EB}/V_T} \tag{3.15}$$

The parameter β is also the common-emitter current gain of the pnp bipolar transistor.

The relationships between the terminal currents of the pnp transistor are exactly the same as those of the npn transistor and are summarized in Table 3.1 in the next section. Also the relationships between β_F and α_F are the same as given in Equations (3.11) and (3.12).

3.1.4 Circuit Symbols and Conventions

The block diagram and conventional circuit symbol of an npn bipolar transistor are shown in Figures 3.8(a) and 3.8(b). The arrowhead in the circuit symbol is always placed on the emitter terminal, and it indicates the direction of the emitter current. For the npn device, this direction is out of the emitter. The simplified block diagram and conventional circuit symbol of a pnp bipolar transistor are shown in Figures 3.9(a) and 3.9(b). Here, the arrowhead on the

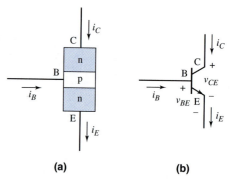

Figure 3.8 npn bipolar transistor: (a) simple block diagram and (b) circuit symbol

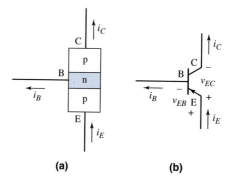

Figure 3.9 pnp bipolar transistor: (a) simple block diagram and (b) circuit symbol

emitter terminal indicates that the direction of the emitter current is into the emitter.

Referring to the circuit symbols given for the npn (Figure 3.8(b)) and pnp (Figure 3.9(b)) transistors showing current directions and voltage polarities, we can summarize the current–voltage relationships as given in Table 3.1. Figure 3.10(a) shows a common-emitter circuit with an npn transistor. The figure includes the transistor currents, and the base-emitter (B–E) and collector–emitter (C–E) voltages. Figure 3.10(b) shows a common-emitter circuit with a

Table 3.1 Summary of the bipolar current–voltage relationships in the active region

npn	pnp
$i_E = I_S e^{v_{BE}/V_T}$	$i_E = I_S e^{v_{EB}/V_T}$
$i_C = \alpha_F i_E = \alpha_F I_S e^{v_{BE}/V_T}$	$i_C = \alpha_F i_E = \alpha_F I_S e^{v_{EB}/V_T}$
$i_B = \dfrac{i_C}{\beta_F} = \dfrac{\alpha_F I_S}{\beta_F} e^{v_{BE}/V_T}$	$i_B = \dfrac{i_C}{\beta_F} = \dfrac{\alpha_F I_S}{\beta_F} e^{v_{EB}/V_T}$

For both transistors

$i_C = \beta_F i_B$	
$i_E = (1 + \beta_F) i_B$	$\alpha_F = \dfrac{\beta_F}{1 + \beta_F}$
$i_C = \left(\dfrac{\beta_F}{1 + \beta_F}\right) i_E = \alpha_F i_E$	$\beta_F = \dfrac{\alpha_F}{1 - \alpha_F}$

Figure 3.10 Common-emitter circuits: (a) with an npn transistor, (b) with a pnp transistor, and (c) with a pnp transistor biased with a positive voltage source

pnp bipolar transistor. Note the different current directions and voltage polarities in the two circuits. A more usual circuit configuration using the pnp transistor is shown in Figure 3.10(c). This circuit allows positive voltage supplies to be used.

Test Your Understanding

3.3 An npn transistor is biased in the forward-active mode. The base current is $I_B = 9.60\,\mu A$ and the emitter current is $I_E = 0.780\,\text{mA}$. Determine β, α, and I_C. (Ans. $\beta = 80.3$, $\alpha = 0.9877$, $I_C = 0.771\,\text{mA}$)

3.4 The emitter current in a pnp transistor biased in the forward-active mode is $I_E = 2.15\,\text{mA}$. The common-base current gain of the transistor is $\alpha = 0.990$. Determine β, I_B, and I_C. (Ans. $\beta = 99$, $I_B = 21.5\,\mu A$, $I_C = 2.13\,\text{mA}$)

3.1.5   Current–Voltage Characteristics

Figures 3.11(a) and 3.11(b) are **common-base circuit configurations** for an npn and a pnp bipolar transistor, respectively. The current sources provide the emitter current. Previously, we stated that the collector current i_C was nearly independent of the C–B voltage as long as the B–C junction was reverse biased. When the B–C junction becomes forward biased, the transistor is no longer in the forward-active mode, and the collector and emitter currents are no longer related by $i_E = \alpha_F i_C$.

Figure 3.12 shows the typical common-base current–voltage characteristics. When the collector–base junction is reverse biased, then for constant values of emitter current, the collector current is nearly equal to i_E. These characteristics show that the common-base device is nearly an ideal constant-current source.

The C–B voltage can be varied by changing the V^+ voltage (Figure 3.11(a)) or the V^- voltage (Figure 3.11(b)). When the collector–base junction becomes forward biased in the range of 0.2 and 0.3 V, the collector current i_C is still essentially equal to the emitter current i_E. In this case, the transistor is still

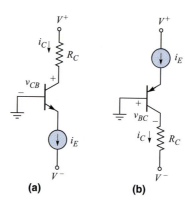

Figure 3.11 Common-base circuit configurations: (a) an npn transistor and (b) a pnp transistor

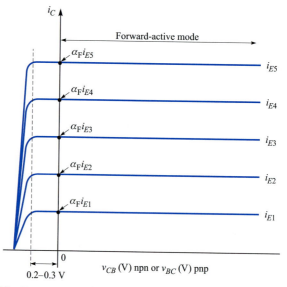

Figure 3.12 Transistor current–voltage characteristics of the common-base circuit

basically biased in the forward-active mode. However, as the forward-bias C–B voltage increases, the linear relationship between the collector and emitter currents is no longer valid, and the collector current very quickly drops to zero.

The common-emitter circuit configuration provides a slightly different set of current–voltage characteristics, as shown in Figure 3.13. For these curves, the collector current is plotted against the collector–emitter voltage, for various constant values of the base current. These curves are generated from the common-emitter circuits shown in Figure 3.10. In this circuit, the V_{BB} source forward biases the B–E junction and controls the base current i_B. The C–E voltage can be varied by changing V_{CC}.

In the npn device, in order for the transistor to be biased in the forward-active mode, the B–C junction must be zero or reverse biased, which means

Figure 3.13   Transistor current–voltage characteristics of the common-emitter circuit

that V_{CE} must be greater than approximately $V_{BE}(\text{on})$.[3] For $V_{CE} > V_{BE}(\text{on})$, there is a finite slope to the curves. If, however, $V_{CE} < V_{BE}(\text{on})$, the B–C junction becomes forward biased, the transistor is no longer in the forward-active mode, and the collector current very quickly drops to zero.

Figure 3.14 shows an exaggerated view of the current–voltage characteristics plotted for constant values of the B–E voltage. The curves are theoretically linear with respect to the C–E voltage in the forward-active mode. When the curves are extrapolated to zero current, they meet at a point on the negative voltage axis, at $v_{CE} = -V_A$. The voltage V_A is a positive quantity called the **Early voltage**, after J. M. Early, who first predicted these characteristics. Typical values of V_A are in the range $50 < V_A < 300$ V.

For a given value of v_{BE}, if v_{CE} increases, the reverse-bias voltage on the collector–base junction increases, which means that the width of the B–C

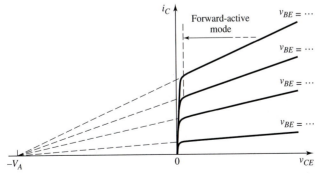

Figure 3.14   Current–voltage characteristics for the common-emitter circuit, showing the Early voltage

---

[3] Even though the collector current is essentially equal to the emitter current when the B–C junction becomes slightly forward biased, as was shown in Figure 3.12, the transistor is said to be biased in the forward-active mode when the B–C junction is zero or reverse biased.

space-charge region also increases. This in turn reduces the neutral base width W (see Figure 3.4). A decrease in the base width causes the gradient in the minority carrier concentration to increase, which increases the diffusion current through the base. The collector current then increases as the C–E voltage increases.

The linear dependence of i_C versus v_{CE} in the forward-active mode can be described by

$$i_C = \alpha_F I_S \left(e^{v_{BE}/V_T}\right) \cdot \left(1 + \frac{v_{CE}}{V_A}\right) \tag{3.16}$$

where I_S and α_F are assumed to be constant.

In Figure 3.14, the nonzero slope of the curves indicates that the **output resistance** r_o looking into the collector is finite. This output resistance is determined from

$$\frac{1}{r_o} = \left.\frac{\partial i_C}{\partial v_{CE}}\right|_{v_{BE}=\text{const.}} \tag{3.17}$$

Using Equation (3.16), we can show that

$$r_o \cong \frac{V_A}{I_C} \tag{3.18}$$

where I_C is the quiescent collector current when v_{BE} is a constant and v_{CE} is small compared to V_A.

In most cases, the dependence of i_C on v_{CE} is not critical in the dc analysis or design of transistor circuits. However, the finite output resistance r_o may significantly affect the amplifier characteristics of such circuits. This effect is examined more closely in Chapter 4 of this text.

Test Your Understanding

3.5 Find the output resistance r_o of a bipolar transistor for which $V_A = 150\,\text{V}$ at collector currents of $I_C = 0.1$, 1.0, and 10 mA. (Ans. $r_o = 1.5\,\text{M}\Omega$, 150 k$\Omega$, 15 k$\Omega$)

3.6 Assume that $I_C = 1\,\text{mA}$ at $V_{CE} = 1\,\text{V}$, and that V_{BE} is held constant. Determine I_C at $V_{CE} = 10\,\text{V}$ if: (a) $V_A = 75\,\text{V}$; and (b) $V_A = 150\,\text{V}$. (Ans. $I_C = 1.12\,\text{mA}$, 1.06 mA)

3.1.6 Nonideal Transistor Leakage Currents and Breakdown Voltage

In discussing the current–voltage characteristics of the bipolar transistor in the previous sections, two topics were ignored: leakage currents in the reverse-biased pn junctions and breakdown voltage effects.

Leakage Currents

In the common-base circuits in Figure 3.11, if we set the current source $i_E = 0$, transistors will be cut off, but the B–C junctions will still be reverse biased. A reverse-bias leakage current exists in these junctions, and this current corresponds to the reverse-bias saturation current in a diode, as described in Chapter 1. The direction of these reverse-bias leakage currents is the same as that of the collector currents. The term I_{CBO} is the collector leakage current in the common-base configuration, and is the collector-base leakage current when the emitter is an open circuit.

Another leakage current can exist between the emitter and collector with the base terminal an open circuit. Figure 3.15 is a block diagram of an npn transistor in which the base is an open circuit ($i_B = 0$). The current component I_{CBO} is the normal leakage current in the reverse-biased B–C pn junction. This current component causes the base potential to increase, which forward biases the B–E junction and induces the B–E current I_{CEO}. The current component αI_{CEO} is the normal collector current resulting from the emitter current I_{CEO}. We can write

$$I_{CEO} = \alpha I_{CEO} + I_{CBO} \tag{3.19(a)}$$

or

$$I_{CEO} = \frac{I_{CBO}}{1 - \alpha} \cong \beta I_{CBO} \tag{3.19(b)}$$

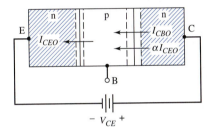

Figure 3.15   Block diagram of an npn transistor in an open-base configuration

This relationship indicates that the open-base configuration produces different characteristics than the open-emitter configuration.

When the transistors are biased in the forward-active mode, the leakage currents still exist. The common-emitter and common-base current gain parameters, β_A and α_F, are dc parameters and can be written as $\beta_F = I_C/I_B$ and $\alpha_F = I_C/I_E$, where I_C includes the leakage current component. In the next chapter, we will discuss ac current gain factors. In most instances in this text, leakage currents will be completely negligible.

Breakdown Voltage: Common-Base Characteristics

The common-base current–voltage characteristics shown in Figure 3.12 are ideal in that breakdown is not shown. Figure 3.16 shows the same i_C versus v_{CB} characteristics with the breakdown voltage.

Consider the curve for $i_E = 0$ (the emitter terminal is effectively an open circuit). The collector–base junction breakdown voltage is indicated as BV_{CBO}. This is a simplified figure in that it shows breakdown occurring abruptly at BV_{CBO}. For the curves in which $i_E > 0$, breakdown actually begins earlier. The carriers flowing across the junction initiate the breakdown avalanche process at somewhat lower voltages.

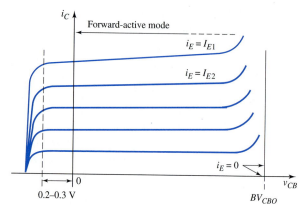

Figure 3.16   The i_C versus v_{CB} common-base characteristics, showing the collector–base junction breakdown

Breakdown Voltage: Common-Emitter Characteristics

Figure 3.17 shows the i_C versus v_{CE} characteristics of an npn transistor, for various constant base currents, and an ideal breakdown voltage of BV_{CEO}. The value of BV_{CEO} is less than the value of BV_{CBO} because BV_{CEO} includes the effects of the transistor action, while BV_{CBO} does not. This same effect was observed in the I_{CEO} leakage current.

The breakdown voltage characteristics for the two configurations are also different. The breakdown voltage for the open-base case is given by

$$BV_{CEO} = \frac{BV_{CBO}}{\sqrt[n]{\beta}}$$

(3.20)

where n is an empirical constant usually in the range of 3 to 6.

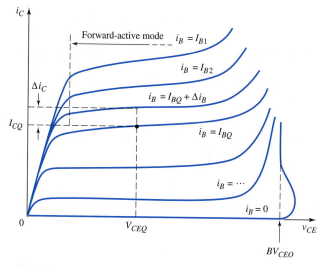

Figure 3.17   Common-emitter characteristics showing breakdown effects

Example 3.2 **Objective:** Calculate the breakdown voltage of a transistor connected in the open-base configuration.

Assume that the transistor current gain is $\beta = 100$ and that the breakdown voltage of the B–C junction is $BV_{CBO} = 120\,\text{V}$.

Solution: If we assume an empirical constant of $n = 3$, we have

$$BV_{CEO} = \frac{BV_{CBO}}{\sqrt[n]{\beta}} = \frac{120}{\sqrt[3]{100}} = 25.9\,\text{V}$$

Comment: The breakdown voltage of the open-base configuration is substantially less than that of the C–B junction. This represents a worst-case condition, which must be considered in any circuit design.

Design Pointer: The designer must be aware of the breakdown voltage of the specific transistors used in a circuit, since this will be a limiting factor in the size of the dc bias voltages that can be used.

Breakdown may also occur in the B–E junction if a reverse-bias voltage is applied to that junction. The junction breakdown voltage decreases as the doping concentrations increase. Since the emitter doping concentration is usually substantially larger than the doping concentration in the collector, the B–E junction breakdown voltage is normally much smaller than that of the B–C junction. Typical B–E junction breakdown voltage values are in the range of 6 to 8 V.

Test Your Understanding

3.7 The open-emitter breakdown voltage is $BV_{CBO} = 200\,\text{V}$, the current gain is $\beta = 120$, and the empirical constant is $n = 3$. Determine BC_{CEO}. (Ans. 40.5 V)

3.8 A particular transistor circuit requires a minimum open-base breakdown voltage of $BV_{CEO} = 30\,\text{V}$. If $\beta = 100$ and $n = 3$, determine the minimum required value of BV_{CBO}. (Ans. 139 V)

3.2   DC ANALYSIS OF TRANSISTOR CIRCUITS

We've considered the basic transistor characteristics and properties. We can now start analyzing and designing the dc biasing of bipolar transistor circuits. A primary purpose of the rest of the chapter is to become familiar and comfortable with the bipolar transistor and transistor circuits. The dc biasing of transistors, the focus of this chapter, is an important part of designing bipolar amplifiers, the focus of the next chapter.

The piecewise linear model of a pn junction can be used for the dc analysis of bipolar transistor circuits. We will first analyze the common-emitter circuit and introduce the load line for that circuit. We will then look at the dc analysis of other bipolar transistor circuit configurations. Since a transistor in a linear amplifier must be biased in the forward-active mode, we emphasize, in this section, the analysis and design of circuits in which the transistor is biased in this mode.

3.2.1 Common-Emitter Circuit

One of the basic transistor circuit configurations is called the **common-emitter circuit**. Figure 3.18(a) shows one example of a common-emitter circuit. The emitter terminal is obviously at ground potential. This circuit configuration will appear in many amplifiers that will be considered in Chapter 4.

Figure 3.18(a) shows a common-emitter circuit with an npn transistor, and Figure 3.18(b) shows the dc equivalent circuit. We will assume that the B–E junction is forward biased, so the voltage drop across that junction is the cut-in or turn-on voltage $V_{BE}(\text{on})$. When the transistor is biased in the forward-active mode, the collector current is represented as a dependent current source that is a function of the base current. We are neglecting the reverse-biased junction leakage current and the Early effect in this case. In the following circuits, we will be considering dc currents and voltages, so the dc notation for these parameters will be used.

(a)   **(b)**

Figure 3.18   (a) Common-emitter circuit with an npn transistor and (b) dc equivalent circuit, with piecewise linear parameters

The base current is

$$I_B = \frac{V_{BB} - V_{BE}(\text{on})}{R_B} \tag{3.21}$$

Implicit in Equation (3.21) is that $V_{BB} > V_{BE}(\text{on})$, which means that $I_B > 0$. When $V_{BB} < V_{BE}(\text{on})$, the transistor is cut off and $I_B = 0$.

In the collector–emitter portion of the circuit, we can write

$$I_C = \beta I_B \tag{3.22}$$

and

$$V_{CC} = I_C R_C + V_{CE} \tag{3.23(a)}$$

or

$$V_{CE} = V_{CC} - I_C R_C \tag{3.23(b)}$$

In Equation (3.23(b)), we are also implicitly assuming that $V_{CE} > V_{BE}(\text{on})$, which means that the B–C junction is reverse biased and the transistor is biased in the forward-active mode.

Example 3.3 **Objective:** Calculate the base, collector, and emitter currents and the C–E voltage for a common-emitter circuit.

For the circuit shown in Figure 3.18(a), the parameters are: $V_{BB} = 4\,\text{V}$, $R_B = 220\,\text{k}\Omega$, $R_C = 2\,\text{k}\Omega$, $V_{CC} = 10\,\text{V}$, $V_{BE}(\text{on}) = 0.7\,\text{V}$, and $\beta = 200$. Figure 3.19(a) shows the circuit without explicitly showing the voltage sources.

(a) **(b)**

Figure 3.19 Circuit for Example 3.3

Solution: Referring to Figure 3.19(b), the base current is found as

$$I_B = \frac{V_{BB} - V_{BE}(\text{on})}{R_B} = \frac{4 - 0.7}{220} \Rightarrow 15\,\mu\text{A}$$

The collector current is

$$I_C = \beta I_B = (200)(15\,\mu\text{A}) \Rightarrow 3\,\text{mA}$$

and the emitter current is

$$I_E = (1 + \beta) \cdot I_B = (201)(15\,\mu\text{A}) \Rightarrow 3.02\,\text{mA}$$

From Equation (3.23(b)), the collector-emitter voltage is

$$V_{CE} = V_{CC} - I_C R_C = 10 - (3)(2) = 4\,\text{V}$$

Comment: Since $V_{BB} > V_{BE}(\text{on})$ and $V_{CE} > V_{BE}(\text{on})$, the transistor is indeed biased in the forward-active mode. As a note, in an actual circuit, the voltage across a B–E junction may not be exactly 0.7 V, as we have assumed using the piecewise linear approximation. This may lead to slight inaccuracies between the calculated currents and voltages and the measured values. Also note that, if we take the difference between I_E and I_C, which is the base current, we obtain $I_B = 20\,\mu\text{A}$ rather than $15\,\mu\text{A}$. The difference is the result of roundoff error in the emitter current.

Figure 3.20(a) shows a common-emitter circuit with a pnp bipolar transistor, and Figure 3.20(b) shows the dc equivalent circuit. In this circuit, the emitter is at ground potential, which means that the polarities of the V_{BB}

Figure 3.20 (a) Common-emitter circuit with a pnp transistor and (b) dc equivalent circuit using piecewise linear parameters

and V_{CC} power supplies must be reversed compared to those in the npn circuit. The analysis proceeds exactly as before, and we can write

$$I_B = \frac{V_{BB} - V_{EB}(\text{on})}{R_B} \tag{3.24}$$

$$I_C = \beta I_B \tag{3.25}$$

and

$$V_{EC} = V_{CC} - I_C R_C \tag{3.26}$$

We can see that Equations (3.24), (3.25), and (3.26) for the pnp bipolar transistor in the common-emitter configuration are exactly the same as Equations (3.21), (3.22), and (3.23(b)) for the npn bipolar transistor in a similar circuit, if we properly define the current directions and voltage polarities.

In many cases, the pnp bipolar transistor will be reconfigured in a circuit so that positive voltage sources, rather than negative ones, can be used. We see this in the following example.

Example 3.4 **Objective:** Analyze the common-emitter circuit with a pnp transistor.

For the circuit shown in Figure 3.21(a), the parameters are: $V_{BB} = 1.5\,\text{V}$, $R_B = 580\,\text{k}\Omega$, $V_{CC} = 5\,\text{V}$, $V_{EB}(\text{on}) = 0.6\,\text{V}$, and $\beta = 100$. Find I_B, I_C, I_E, and R_C such that $V_{EC} = \left(\frac{1}{2}\right)V_{CC}$.

Solution: Writing a Kirchhoff voltage law equation around the E–B loop, we find the base current to be

$$I_B = \frac{V_{CC} - V_{EB}(\text{on}) - V_{BB}}{R_B} = \frac{5 - 0.6 - 1.5}{580} \Rightarrow 5\,\mu\text{A}$$

The collector current is

$$I_C = \beta I_B = (100)(5\,\mu\text{A}) \Rightarrow 0.5\,\text{mA}$$

and the emitter current is

$$I_E = (1 + \beta)I_B = (101)(5\,\mu\text{A}) \Rightarrow 0.505\,\text{mA}$$

Figure 3.21   Circuit for Example 3.4

For a C–E voltage of $V_{EC} = \frac{1}{2}V_{CC} = 2.5\,\text{V}$, R_C is

$$R_C = \frac{V_{CC} - V_{EC}}{I_C} = \frac{5 - 2.5}{0.5} = 5\,\text{k}\Omega$$

Comment: In this case, the difference between V_{CC} and V_{BB} is greater than the transistor turn-on voltage, or $(V_{CC} - V_{BB}) > V_{EB}(\text{on})$. Also, because $V_{EC} > V_{EB}(\text{on})$, the pnp bipolar transistor is biased in the forward-active mode.

Discussion: In this example, we used an emitter-base turn-on voltage of $V_{EB}(\text{on}) = 0.6\,\text{V}$, whereas previously we used a value of $0.7\,\text{V}$. We must keep in mind that the turn-on voltage is an approximation and the actual base–emitter voltage will depend on the type of transistor used and the current level. In most situations, choosing a value of $0.6\,\text{V}$ or $0.7\,\text{V}$ will make only minor differences. However, most people tend to use the value of $0.7\,\text{V}$.

The dc equivalent circuits, such as those given in Figures 3.18(b) and 3.20(b), are useful initially in analyzing transistor circuits. From this point on, however, we will not explicitly draw the equivalent circuit. We will simply analyze the circuit using the transistor circuit symbols, as in Figures 3.19 and 3.21.

3.2.2   Load Line and Modes of Operation

The load line can help us visualize the characteristics of a transistor circuit. For the common-emitter circuit in Figure 3.19(a), we can use a graphical technique for both the B–E and C–E portions of the circuit. Figure 3.22(a) shows the piecewise linear characteristics for the B–E junction and the input load line. The input load line is obtained from Kirchhoff's voltage law equation around the B–E loop, written as follows:

$$I_B = \frac{V_{BB}}{R_B} - \frac{V_{BE}}{R_B} \tag{3.27}$$

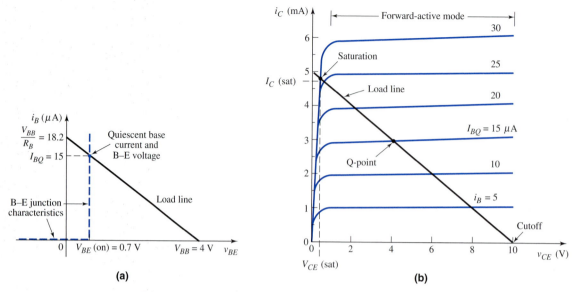

Figure 3.22   (a) Base–emitter junction characteristics and the input load line and (b) common-emitter transistor characteristics and the collector–emitter load line

Both the load line and the quiescent base current change as either or both V_{BB} and R_B change. The load line in Figure 3.22(a) is essentially the same as the load line characteristics for diode circuits, as shown in Chapter 1.

For the C–E portion of the circuit in Figure 3.19(a), the load line is found by writing Kirchhoff's voltage law equation around the C–E loop. We obtain

$$V_{CE} = V_{CC} - I_C R_C \qquad\qquad (3.28(a))$$

which can be written in the form

$$I_C = \frac{V_{CC}}{R_C} - \frac{V_{CE}}{R_C} = 5 - \frac{V_{CE}}{2}\ (\text{mA}) \qquad\qquad (3.28(b))$$

Equation (3.28(b)) is the load line equation, showing a linear relationship between the collector current and collector–emitter voltage. Since we are considering the dc analysis of the transistor circuit, this relationship represents the dc load line. The ac load line is presented in the next chapter.

Figure 3.22(b) shows the transistor characteristics for the transistor in Example 3.3, with the load line superimposed on the transistor characteristics. The two end points of the load line are found by setting $I_C = 0$, yielding $V_{CE} = V_{CC} = 10\,\text{V}$, and by setting $V_{CE} = 0$, yielding $I_C = V_{CC}/R_C = 5\,\text{mA}$.

The quiescent point, or Q-point, of the transistor is given by the dc collector current and the collector–emitter voltage. The Q-point is the intersection of the load line and the I_C versus V_{CE} curve corresponding to the appropriate base current. The Q-point also represents the simultaneous solution to two expressions. The load line is useful in visualizing the bias point of the transistor. In the figure, the Q-point shown is for the transistor in Example 3.3.

As previously stated, if the power supply voltage in the base circuit is smaller than the turn-on voltage, then $V_{BB} < V_{BE}(\text{on})$ and $I_B = I_C = 0$, and the transistor is in the cutoff mode. In this mode, all transistor currents are zero, neglecting leakage currents, and for the circuit shown in Figure 3.19(a), $V_{CE} = V_{CC} = 10\,\text{V}$.

As V_{BB} increases ($V_{BB} > V_{BE}(\text{on})$), the base current I_B increases and the Q-point moves up the load line. As I_B continues to increase, a point is reached where the collector current I_C can no longer increase. At this point, the transistor is biased in the **saturation mode**; that is, the transistor is said to be in saturation. The B–C junction becomes forward biased, and the relationship between the collector and base currents is no longer linear. The transistor C–E voltage in saturation, $V_{CE}(\text{sat})$, is less than the B–E cut-in voltage. The forward-biased B–C voltage is always less than the forward-biased B–E voltage, so the C–E voltage in saturation is a small positive value. Typically, $V_{CE}(\text{sat})$ is in the range of 0.1 to 0.3 V.

Example 3.5   Objective: Calculate the currents and voltages in a circuit when the transistor is driven into saturation.

For the circuit shown in Figure 3.23, the transistor parameters are: $\beta = 100$, and $V_{BE}(\text{on}) = 0.7\,\text{V}$. If the transistor is biased in saturation, assume $V_{CE}(\text{sat}) = 0.2\,\text{V}$.

Solution: Since $+8\,\text{V}$ is applied to the input side of R_B, the base–emitter junction is certainly forward biased, so the transistor is turned on. The base current is

$$I_B = \frac{V_{BB} - V_{BE}(\text{on})}{R_B} = \frac{9 - 0.7}{220} \Rightarrow 33.2\,\mu\text{A}$$

If we first assume that the transistor is biased in the active region, then the collector current is

$$I_C = \beta I_B = (100)(33.2\,\mu\text{A}) \Rightarrow 3.32\,\text{mA}$$

The collector–emitter voltage is then

$$V_{CE} = V_{CC} - I_C R_C = 10 - (3.32)(4) = -3.28\,\text{V}$$

However, the collector–emitter voltage of the npn transistor in the common-emitter configuration shown in Figure 3.23(a) cannot be negative. Therefore, our initial assumption of the transistor being biased in the forward-active mode is incorrect. Instead, the transistor must be biased in saturation.

As given in the "objective" statement, set $V_{CE}(\text{sat}) = 0.2\,\text{V}$. The collector current is

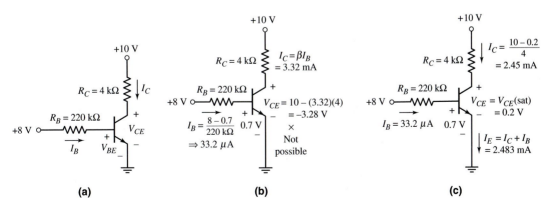

Figure 3.23   Circuit for Example 3.5

$$I_C = I_C(\text{sat}) = \frac{V_{CC} - V_{CE}(\text{sat})}{R_C} = \frac{10 - 0.2}{4} = 2.45\,\text{mA}$$

Assuming that the B–E voltage is still equal to $V_{BE}(\text{on}) = 0.7\,\text{V}$, the base current is $I_B = 33.2\,\mu\text{A}$, as previously determined. If we take the ratio of collector current to base current, then

$$\frac{I_C}{I_B} = \frac{2.45}{0.0332} = 74 < \beta$$

The emitter current is

$$I_E = I_C + I_B = 2.45 + 0.033 = 2.48\,\text{mA}$$

Comment: When a transistor is driven into saturation, we use $V_{CE}(\text{sat})$ as another piecewise linear parameter. In addition, when a transistor is biased in the saturation mode, we have $I_C < \beta I_B$. This condition is very often used to prove that a transistor is indeed biased in the saturation mode.

Problem-Solving Technique: Bipolar DC Analysis

Analyzing the dc response of a bipolar transistor circuit requires knowing the mode of operation of the transistor. In some cases, the mode of operation may not be obvious, which means that we have to guess the state of the transistor, then analyze the circuit to determine if we have a solution consistent with our initial guess. To do this, we can:

1. Assume that the transistor is biased in the forward-active mode in which case $V_{BE} = V_{BE}(\text{on})$, $I_B > 0$, and $I_C = \beta I_B$.
2. Analyze the "linear" circuit with this assumption.
3. Evaluate the resulting state of the transistor. If the initial assumed parameter values and $V_{CE} > V_{CE}(\text{sat})$ are true, then the initial assumption is correct. However, if $I_B < 0$, then the transistor is probably cut off, and if $V_{CE} < 0$, the transistor is likely biased in saturation.
4. If the initial assumption is proven incorrect, then a new assumption must be made and the new "linear" circuit must be analyzed. Step 3 must then be repeated.

Because it is not always clear whether a transistor is biased in the forward-active or saturation mode, we may initially have to make an educated guess as to the state of the transistor and then verify our initial assumption. This is similar to the process we used for the analysis of multidiode circuits. For instance, in Example 3.5, we assumed a forward-active mode, performed the analysis, and showed that $V_{CE} < 0$. However, a negative V_{CE} for an npn transistor in the common-emitter configuration is not possible. Therefore, our initial assumption was disproved, and the transistor was biased in the saturation mode. Using the results of Example 3.5, we also see that when a transistor is in saturation, the ratio of I_C to I_B is always less than β, or

$$I_C/I_B < \beta$$

This condition is true for both the npn and the pnp transistor biased in the saturation mode.

Another mode of operation for a bipolar transistor is the **inverse-active mode**. In this mode, the B–E junction is reverse biased and the B–C junction is forward biased. In effect, the transistor is operating "upside down"; that is, the emitter is acting as the collector and the collector is operating as the emitter. We will postpone discussions on this operating mode until we discuss digital electronic circuits later in this text.

To summarize, the four modes of operation for an npn transistor are shown in Figure 3.24. The four possible combinations of B–E and B–C voltages determine the modes of operation. If $v_{BE} > 0$ (forward-biased junction) and $v_{BC} < 0$ (reverse-biased junction), the transistor is biased in the forward-active mode. If both junctions are zero or reverse biased, the transistor is in cutoff. If both junctions are forward biased, the transistor is in saturation. If the B–E junction is reverse biased and the B–C junction is forward biased, the transistor is in the inverse-active mode.

Figure 3.24   Bias conditions for the four modes of operation of an npn transistor

Test Your Understanding

3.9   For the circuit shown in Figure 3.25, assume $\beta = 50$. Determine V_O, I_B, and I_C for: (a) $V_I = 0.2\,\text{V}$, and (b) $V_I = 3.6\,\text{V}$. Then, calculate the power dissipated in the transistor for the two conditions. (Ans. (a) $I_B = I_C = 0$, $V_O = 5\,\text{V}$, $P = 0$; (b) $I_B = 4.53\,\text{mA}$, $I_C = 10.9\,\text{mA}$, $P = 5.35\,\text{mW}$)

3.10   For the circuit shown in Figure 3.25, let $\beta = 50$, and determine V_I such that $V_{BC} = 0$. Calculate the power dissipated in the transistor. (Ans. $V_I = 0.825\,\text{V}$, $P = 6.98\,\text{mW}$)

Figure 3.25   Figure for Exercises 3.9 and 3.10

3.2.3   Common Bipolar Circuits: DC Analysis

There are a number of other bipolar transistor circuit configurations in addition to the common-emitter circuits shown in Figures 3.19 and 3.21. Several examples of such circuits are presented in this section. BJT circuits tend to be very similar in terms of dc analysis procedures, so that the same basic analysis approach will work regardless of the appearance of the circuit. We continue our dc analysis and design of bipolar circuits to increase our proficiency and to become more comfortable with these types of circuits.

Example   3.6   Objective: Calculate the characteristics of a circuit containing an emitter resistor.

For the circuit shown in Figure 3.26(a), let $V_{BE}(\text{on}) = 0.7\,\text{V}$ and $\beta = 75$.

(a)   (b)

Figure 3.26   Circuit for Example 3.6

Solution:
Q-Point Values:
Writing the Kirchhoff's voltage law equation around the B–E loop, we have

$$V_{BB} = I_B R_B + V_{BE}(\text{on}) + I_E R_E \tag{3.29}$$

Assuming the transistor is biased in the forward-active mode, we can write $I_E = (1 + \beta)I_B$. We can then solve Equation (3.29) for the base current:

$$I_B = \frac{V_{BB} - V_{BE}(\text{on})}{R_B + (1 + \beta)R_E} = \frac{6 - 0.7}{25 + (76)(0.6)} \Rightarrow 75.1\,\mu\text{A}$$

The collector and emitter currents are

$$I_C = \beta I_B = (75)(75.1\,\mu\text{A}) \Rightarrow 5.63\,\text{mA}$$

and

$$I_E = (1 + \beta)I_B = (76)(75.1\,\mu\text{A}) \Rightarrow 5.71\,\text{mA}$$

Referring to Figure 3.26(b), the collector–emitter voltage is

$$V_{CE} = V_{CC} - I_C R_C - I_E R_E = 12 - (5.63)(0.4) - (5.71)(0.6)$$

or

$$V_{CE} = 6.32\,\text{V}$$

Solution:
Load Line:
We again use Kirchhoff's voltage law around the C–E loop. From the relationship between the collector and emitter currents, we find

$$V_{CE} = V_{CC} - I_C\left[R_C + \left(\frac{1 + \beta}{\beta}\right)R_E\right] = 12 - I_C\left[0.4 + \left(\frac{76}{75}\right)(0.6)\right]$$

or

$$V_{CE} = 12 - I_C(1.01)$$

The load line and the calculated Q-point are shown in Figure 3.27. A few transistor characteristics of I_C versus V_{CE} are superimposed on the figure.

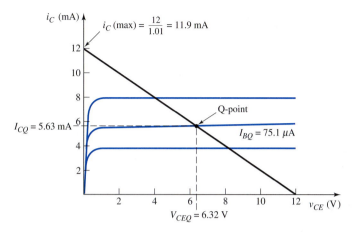

Figure 3.27 Load line for the circuit in Figure 3.26

Comment: Since the C–E voltage is 6.32 V, $V_{CE} > V_{BE}$(on) and the transistor is biased in the forward-active mode, as initially assumed. We will see, later in the chapter, the value of including an emitter resistor in a circuit.

Example   3.7 **Objective:** Calculate the characteristics of a circuit containing both a positive and a negative power supply voltage.

For the circuit shown in Figure 3.28, let V_{BE}(on) $= 0.65$ V and $\beta = 100$. Even though the base is at ground potential, the B–E junction is forward biased through R_E and V^-.

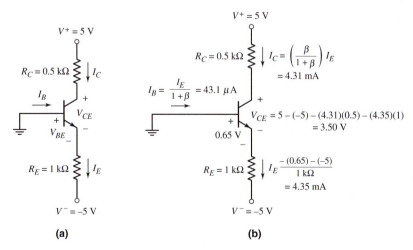

Figure 3.28 Circuit for Example 3.7

Solution:
Q-Point Values:
Writing the Kirchhoff's voltage law equation around the B–E loop, we have

$$0 = V_{BE}(\text{on}) + I_E R_E + V^-$$

which yields

$$I_E = \frac{-V^- - V_{BE}(\text{on})}{R_E} = \frac{-(-5) - 0.65}{1} = 4.35\,\text{mA}$$

The base current is

$$I_B = \frac{I_E}{1 + \beta} = \frac{4.35}{101} \Rightarrow 43.1\,\mu\text{A}$$

and the collector current is

$$I_C = \left(\frac{\beta}{1+\beta}\right)I_E = \left(\frac{100}{101}\right) \cdot (4.35) = 4.31\,\text{mA}$$

Referring to Figure 3.28(b), the C–E voltage is

$$V_{CE} = V^+ - I_C R_C - I_E R_E - V^-$$

or

$$V_{CE} = 5 - (4.31)(0.5) - (4.35)(1) - (-5) = 3.50\,\text{V}$$

Solution:
Load Line:
The load line equation is

$$V_{CE} = (V^+ - V^-) - I_C\left[R_C + \left(\frac{1+\beta}{\beta}\right)R_E\right] = (5 - (-5)) - I_C\left[0.5 + \left(\frac{101}{100}\right)(1)\right]$$

or

$$V_{CE} = 10 - I_C(1.51)$$

The load line and the calculated Q-point are shown in Figure 3.29.

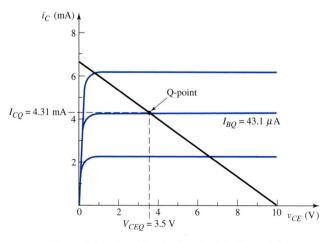

Figure 3.29   Load line for the circuit in Figure 3.28

Comment: The B–E junction is forward biased, even though V_{BB} is at ground potential. The forward-bias voltage is a result of the negative potential V^- applied at the "bottom" of the emitter resistor R_E. The transistor is biased in the forward-active mode.

Test Your Understanding

(Note: In the following exercises, assume the B–E cut-in voltage is 0.7 V for both the npn and pnp transistors. Also assume that the C–E saturation voltage is 0.2 V for each type of transistor.)

RD3.11   Redesign the circuit shown in Figure 3.30 such that $I_{CQ} = 1.5$ mA and $V_C = +4$ V. Assume $\beta = 100$. (Ans. $R_C = 4$ kΩ, $R_E = 6.14$ kΩ)

3.12   For the circuit shown in Figure 3.31, the measured value of V_C is $V_C = +6.34$ V. Determine I_B, I_E, I_C, V_{CE}, β, and α (Ans. $I_C = 0.915$ mA, $I_E = 0.930$ mA, $\alpha = 0.9839$, $I_B = 15.0\,\mu$A, $\beta = 61$, $V_{CE} = 7.04$ V)

3.13   Determine I_B, I_C, I_E, and V_{EC}, assuming $\beta = 50$ for the circuit shown in Figure 3.32. (Ans. $I_E = 1.16$ mA, $I_B = 22.7\,\mu$A, $I_C = 1.14$ mA, $V_{EC} = 6.14$ V)

Figure 3.30   Figure for Exercise 3.11

Figure 3.31   Figure for Exercise 3.12

Figure 3.32   Figure for Exercise 3.13

Design Example   3.8   **Objective:**   Design a pnp bipolar transistor circuit.

For the circuit shown in Figure 3.33(a), let $V_{EB}(\text{on}) = 0.6$ V and $\beta = 60$. Design the circuit such that $V_{ECQ} = 2.5$ V.

Solution:

Q-Point Values:

Writing the Kirchhoff's voltage law equation around the E–C loop, we obtain

$$V^+ = I_E R_E + V_{EC}$$

or

$$5 = I_E(2) + 2.5$$

which yields $I_E = 1.25$ mA. The collector current is

$$I_C = [\beta/(1+\beta)]I_E = [(60)/61](1.25) = 1.23 \text{ mA}$$

The base current is

Figure 3.33   Circuit for Example 3.8

$$I_B = I_E/(1 + \beta) = 1.25/61 = 0.0205 \, \text{mA}$$

Writing the Kirchhoff's voltage law equation around the E–B loop, we find

$$V^+ = I_E R_E + V_{EB}(\text{on}) + I_B R_B + V_{BB}$$

or

$$5 = (1.25)(2) + 0.6 + (0.0205)R_B + (-2)$$

which yields $R_B = 190 \, \text{k}\Omega$.

Solution:
Load Line:
The load line equation is

$$V_{EC} = V^+ - I_E R_E = V^+ - I_C \left(\frac{1 + \beta}{\beta}\right) R_E$$

or

$$V_{EC} = 5 - I_C \left(\frac{61}{60}\right)(2) = 5 - I_C(2.03)$$

The load line and calculated Q-point are shown in Figure 3.34.

Figure 3.34   Load line for the circuit in Figure 3.33

Comment: Even though there is no collector resistor, there is still a collector current. Also, the transistor is biased in the forward-active mode.

Computer Simulation: It is often desirable to verify a transistor circuit design with a computer simulation. This verification becomes more important as the complexity of the circuit increases and as the complexity of the transistor model increases.

As an introduction to computer simulation, a PSpice analysis of the circuit design shown in Figure 3.33(b) was performed. Figure 3.35 shows the PSpice circuit schematic. A standard 2N3906 transistor from the circuit library was used. Shown below are the schematic's netlist that was created, a partial listing of the transistor parameters, and the resulting Q-point values.

Figure 3.35 PSpice circuit schematic for Example 3.8

```
              *Schematics Netlist*
    Q_Q1   0 $N_0001 $N_0002 Q2N3906
    V_V2   $N_0003 0 -2V
    V_V1   $N_0004 0 5V
    R_R3   $N_0003 $N_0001 190K
    R_R1   $N_0004 $N_0002 2K
```

```
    Q2N3906            **** BIPOLAR JUNCTION TRANSISTORS
    PNP
IS   1.410000E-15
BF  180.7          NAME           Q_Q1
NF   1             MODEL          Q2N3906
VAF  18.7          IB             -1.15E-05
IKF   .08          IC             -2.04E-03
BR   4.977         VBE            -7.25E-01
NR   1             VBC            1.77E-01
RB  10             VCE            -9.01E-01
RBM 10             BETADC         1.78E+02
RC   2.5
```

We see that the current gain β_F of the 2N3906 is approximately 180 compared to the assumed value of 60 used in the design calculations. This important difference produces an emitter–collector voltage of $V_{EC} = 0.901 \text{ V}$ compared to the desired value of 2.5 V. Using the value of $\beta_F = 180$, a new value of R_B would need to be determined to produce the desired V_{EC} value.

Discussion: This example illustrates one extremely important point. In order for the computer simulation to accurately predict the circuit response, the transistor parameters must be known. In the above design, if the circuit is to actually function properly with

$\beta_F = 60$, then the transistor in the PSpice analysis would need to be changed and a different transistor with the proper parameters would need to be used. However, for a given device model, PSpice can determine the value of β_F depending on operating point and temperature.

Example 3.9 Objective: Calculate the characteristics of an npn bipolar circuit with a load resistance. The load resistance can represent a second transistor stage connected to the output of a transistor circuit.

For the circuit shown in Figure 3.36(a), the transistor parameters are: $V_{BE}(\text{on}) = 0.7\,\text{V}$, and $\beta = 100$.

(a)

(b)   (c)

Figure 3.36   Circuit for Example 3.9

Solution:

Q-Point Values:
Kirchhoff's voltage law equation around the B–E loop yields

$$I_B R_B + V_{BE}(\text{on}) + I_E R_E + V^- = 0$$

Again assuming $I_E = (1 + \beta)I_B$, we find

$$I_B = \frac{-(V^- + V_{BE}(\text{on}))}{R_B + (1 + \beta)R_E} = \frac{-(-5 + 0.7)}{10 + (101)(5)} \Rightarrow 8.35 \,\mu\text{A}$$

The collector and emitter currents are

$$I_C = \beta I_B = (100)(8.35 \,\mu\text{A}) \Rightarrow 0.835 \,\text{mA}$$

and

$$I_E = (1 + \beta)I_B = (101)(8.35 \,\mu\text{A}) \Rightarrow 0.843 \,\text{mA}$$

At the collector node, we can write

$$I_C = I_1 - I_L = \frac{V^+ - V_O}{R_C} - \frac{V_O}{R_L}$$

or

$$0.835 = \frac{12 - V_O}{5} - \frac{V_O}{5}$$

Solving for V_O, we get $V_O = 3.91 \,\text{V}$. The currents are then $I_1 = 1.62 \,\text{mA}$ and $I_L = 0.782 \,\text{mA}$. Referring to Figure 3.36(b), the collector–emitter voltage is

$$V_{CE} = V_O - I_E R_E - (-5) = 3.91 - (0.843)(5) - (-5) = 4.70 \,\text{V}$$

Solution:

Load Line:
The load line equation for this circuit is not as straightforward as for previous circuits. The easiest approach to finding the load line is to make a "Thevenin equivalent circuit" of R_L, R_C, and V^+, as indicated in Figure 3.36(a). (Thevenin equivalent circuits are also covered later in this chapter, in Section 3.4.) The Thevenin equivalent resistance is

$$R_{TH} = R_L \| R_C = 5 \| 5 = 2.5 \,\text{k}\Omega$$

and the Thevenin equivalent voltage is

$$V_{TH} = \left(\frac{R_L}{R_L + R_C}\right) \cdot V^+ = \left(\frac{5}{5 + 5}\right) \cdot (12) = 6 \,\text{V}$$

The equivalent circuit is shown in Figure 3.36(c). The Kirchhoff voltage law equation around the C–E loop is

$$V_{CE} = (6 - (-5)) - I_C R_{TH} - I_E R_E = 11 - I_C(2.5) - I_C\left(\frac{101}{100}\right) \cdot (5)$$

or

$$V_{CE} = 11 - I_C(7.55)$$

The load line and the calculated Q-point values are shown in Figure 3.37.

Comment: Remember that the collector current, determined from $I_C = \beta I_B$, is the current into the collector terminal of the transistor; it is not necessarily the current in the collector resistor R_C.

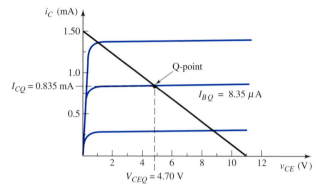

Figure 3.37   Load line for the circuit in Figure 3.36(a)

Test Your Understanding

RD3.14   For the transistor shown in the circuit of Figure 3.38, the common-base current gain is $\alpha = 0.9920$. Determine R_E such that the emitter current is limited to $I_E = 1.0$ mA. Also determine I_B, I_C, and V_{BC}. (Ans. $R_E = 3.3$ kΩ, $I_C = 0.992$ mA, $I_B = 8.0$ µA, $V_{BC} = 4.01$ V)

Figure 3.38   Figure for Exercise 3.14

3.15   For the circuit shown in Figure 3.39, determine I_E, I_B, I_C, and V_{CE}, if $\beta = 75$. (Ans. $I_B = 15.1$ µA, $I_C = 1.13$ mA, $I_E = 1.15$ mA, $V_{CE} = 6.03$ V)

Figure 3.39   Figure for Exercise 3.15

RD3.16   Let $\beta = 100$ for the circuit shown in Figure 3.40. Determine R_E such that $V_{CE} = 2.5$ V. (Ans. $R_E = 138$ Ω)

3.17   For the circuit shown in Figure 3.41, assume $\beta = 50$ and determine V_{BB} such that $I_E = 2.2$ mA. Then, find I_C and V_{EC}. (Ans. $I_C = 2.16$ mA, $V_{BB} = 5.06$ V, $V_{EC} = 2.8$ V)

Figure 3.40 Figure for
Exercise 3.16

Figure 3.41 Figure for
Exercise 3.17

3.3 BASIC TRANSISTOR APPLICATIONS

Transistors can be used to: switch currents, voltages, and power; perform digital logic functions; and amplify time-varying signals. In this section, we consider the switching properties of the bipolar transistor, analyze a simple transistor digital logic circuit, and then show how the bipolar transistor is used to amplify time-varying signals.

3.3.1 Switch

Figure 3.42 shows a bipolar circuit called an inverter, in which the transistor in the circuit is switched between cutoff and saturation. The load, for example, could be a motor, a light-emitting diode (see Exercise 3.21), or some other electrical device. If $v_I < V_{BE}(\text{on})$, then $i_B = i_C = 0$ and the transistor is cut off. Since $i_C = 0$, the voltage drop across R_C is zero, so the output voltage is $v_O = V_{CC}$. Also, since the currents in the transistor are zero, the power dissipation in the transistor is zero. If the load were a motor, the motor would be off with zero current. Likewise, if the load were a light-emitting diode, the light output would be zero with zero current.

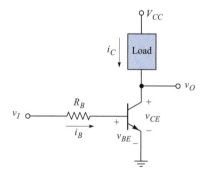

Figure 3.42 An npn bipolar inverter circuit used as a switch

If we let $v_I = V_{CC}$ and if the ratio of R_B to R_C, where R_C is the effective resistance of the load, is less than β, then the transistor is usually driven into saturation, which means that

$$i_B \cong \frac{v_I - V_{BE}(\text{on})}{R_B} \tag{3.30}$$

$$i_C = I_C(\text{sat}) = \frac{V_{CC} - V_{CE}(\text{sat})}{R_C} \tag{3.31}$$

and

$$v_O = V_{CE}(\text{sat}) \tag{3.32}$$

In this case, a collector current is induced that would turn on the motor or the LED, depending on the type of load.

Equation (3.30) assumes that the B–E voltage can be approximated by the turn-on voltage. This approximation will be modified slightly when we discuss bipolar digital logic circuits.

Example 3.10   Objective: Calculate the currents, output voltage, and power dissipation in the transistor for the bipolar inverter shown in Figure 3.42.

Assume the circuit and transistor parameters are: $R_B = 240\,\Omega$, $V_{CC} = 12\,\text{V}$, $V_{BE}(\text{on}) = 0.7\,\text{V}$, $V_{CE}(\text{sat}) = 0.1\,\text{V}$, and $\beta = 75$. Assume the load is a motor with an effective resistance of $R_C = 5\,\Omega$.

Solution: For $v_I = 0$, the transistor is cut off, $i_B = i_C = 0$, $v_O = V_{CC} = 12\,\text{V}$, and the power dissipated in the transistor is zero. For $v_I = 12\,\text{V}$,

$$i_B = \frac{v_I - V_{BE}(\text{on})}{R_B} = \frac{12 - 0.7}{240} \Rightarrow 47.1\,\text{mA}$$

Assuming the transistor is in saturation, we find that

$$i_C = \frac{V_{CC} - V_{CE}(\text{sat})}{R_C} = \frac{12 - 0.1}{5} = 2.38\,\text{A}$$

If we take the ratio of collector current to base current, we have

$$\frac{i_C}{i_B} = \frac{2.38}{0.0471} = 50.6 < \beta$$

Since $i_C/i_B < \beta$, the transistor is indeed in saturation, as we initially assumed. Also, since the transistor is in saturation, the output voltage is

$$v_O = V_{CC}(\text{sat}) = 0.1\,\text{V}$$

The power dissipated in the transistor is

$$P = i_C v_{CE} + i_B v_{BE} = (2.38)(0.1) + (0.0471)(0.7)$$

or

$$P = 0.271\,\text{W}$$

Comment: With a collector current of 2.38 A, the transistor would have to be a power transistor. However, from the results we see that we can "switch" a relatively large collector current ($i_C = 2.38\,\text{A}$) using only a relatively small base current ($i_B = 47\,\text{mA}$).

Design Pointer: Motors tend to be inductive, so that during start-up and shutdown a relatively large di/dt voltage could be induced in the circuit. This voltage, especially during shutdown, could cause the transistor to go into breakdown and be damaged.

When a transistor is biased in saturation, the relationship between the collector and base currents is no longer linear. Consequently, this mode of operation cannot be used for linear amplifiers. On the other hand, switching a transistor between cutoff and saturation produces the greatest change in output voltage, which is especially useful in digital logic circuits, as we will see in the next section.

3.3.2 Digital Logic

In the simple transistor inverter circuit shown in Figure 3.43(a), if the input is approximately zero volts, the transistor is in cutoff and the *output* is high and equal to V_{CC}. If, on the other hand, the *input* is high and equal to V_{CC}, the transistor is driven into saturation, and the output is low and equal to $V_{CE}(\text{sat})$.

Figure 3.43   A bipolar (a) inverter circuit and (b) NOR logic gate

Now consider the case when a second transistor is connected in parallel, as shown in Figure 3.43(b). When the two inputs are zero, both transistors Q_1 and Q_2 are in cutoff, and $V_O = 5\,\text{V}$. When $V_1 = 5\,\text{V}$ and $V_2 = 0$, transistor Q_1 can be driven into saturation, and Q_2 remains in cutoff. With Q_1 in saturation, the output voltage is $V_O = V_{CE}(\text{sat}) \cong 0.2\,\text{V}$. If we reverse the input voltages so that $V_1 = 0$ and $V_2 = 5\,\text{V}$, then Q_1 is in cutoff, Q_2 can be driven into saturation, and $V_O = V_{CE}(\text{sat}) \cong 0.2\,\text{V}$. If both inputs are high, meaning $V_1 = V_2 = 5\,\text{V}$, then both transistors can be driven into saturation, and $V_O = V_{CE}(\text{sat}) \cong 0.2\,\text{V}$.

Table 3.2 shows these various conditions for the circuit in Figure 3.43(b). In a **positive logic system**, meaning that the larger voltage is a logic 1 and the lower voltage is a logic 0, this circuit performs the **NOR logic function**. The circuit of Figure 3.43(b) is then a two-input bipolar NOR logic circuit.

Table 3.2   The bipolar NOR logic circuit response

V_1 (V)	V_2 (V)	V_O (V)
0	0	5
5	0	0.2
0	5	0.2
5	5	0.2

Example 3.11   **Objective:** Determine the currents and voltages in the circuit shown in Figure 3.43(b).

Assume the transistor parameters are: $\beta = 50$, $V_{BE}(\text{on}) = 0.7\,\text{V}$, and $V_{CE}(\text{sat}) = 0.2\,V$. Let $R_C = 1\,\text{k}\Omega$ and $R_B = 20\,\text{k}\Omega$. Determine the currents and output voltage for various input conditions.

Solution: The following table indicates the equations and results for this example.

Condition	V_O	I_R	Q_1	Q_2
$V_1 = 0$, $V_2 = 0$	5 V	0	$I_{B1} = I_{C1} = 0$	$I_{B2} = I_{C2} = 0$
$V_1 = 5\,V$, $V_2 = 0$	0.2 V	$\dfrac{5 - 0.2}{1} = 4.48\,\text{mA}$	$I_{B1} = \dfrac{5 - 0.7}{20} = 0.215\,\text{mA}$ $I_{C1} = I_R = 4.8\,\text{mA}$	$I_{B2} = I_{C2} = 0$
$V_1 = 0$, $V_2 = 5\,V$	0.2 V	4.8 mA	$I_{B1} = I_{C1} = 0$	$I_{B2} = 0.215\,\text{mA}$ $I_{C2} = I_R = 4.8\,\text{mA}$
$V_1 = 5\,V$, $V_2 = 5\,V$	0.2 V	4.8 mA	$I_{B1} = 0.215\,\text{mA}$ $I_{C1} = \dfrac{I_R}{2} = 2.4\,\text{mA}$	$I_{B2} = 0.215\,\text{mA}$ $I_{C2} = \dfrac{I_R}{2} = 2.4\,\text{mA}$

Comment: In this example, we see that whenever a transistor is conducting, the ratio of collector current to base current is always less than β. This shows that the transistor is in saturation, which occurs when either V_1 or V_2 is 5 V.

This example and the accompanying discussion illustrate that bipolar transistor circuits can be configured to perform logic functions. In Chapter 17, we will see that this circuit can experience loading effects when load circuits or other digital logic circuits are connected to the output. Therefore, logic circuits must be designed to minimize or eliminate such loading effects.

3.3.3   Amplifier

The bipolar inverter circuit shown in Figure 3.43(a) can also be used as an amplifier. We will initially develop the voltage transfer characteristics of a specific inverter circuit and then superimpose a time-varying signal on a dc input voltage.

Example 3.12   **Objective:** Determine the dc voltage transfer characteristics and then the amplification factor of the circuit shown in Figure 3.44(a).

Assume the transistor parameters are: $\beta_F = 100$, $V_A = \infty$, $V_{BE}(\text{on}) = 0.7\,\text{V}$, and $V_{CE}(\text{sat}) = 0.2\,\text{V}$.

DC Solution: For $v_I \leq 0.7\,\text{V}$, Q is cut off and $v_O = 5\,\text{V}$. For $v_I > 0.7\,\text{V}$, Q turns on and is biased in the active region, so that

$$i_B = \frac{v_I - V_{BE}(\text{on})}{R_B} = \frac{v_I - 0.7}{100\,\text{k}\Omega}$$

The output voltage is

$$v_O = V^+ - i_C R_C = V^+ - \beta_F i_B R_C$$

or

$$v_O = 5 - (100)\left[\frac{v_I - 0.7}{100\,\text{k}\Omega}\right](4\,\text{k}\Omega) = 7.8 - 4v_I$$

This equation is valid for $v_I \geq 0.7\,\text{V}$ and $v_O \geq V_{CE}(\text{sat}) = 0.2\,\text{V}$. The input voltage for $v_O = 0.2\,\text{V}$ is found to be $v_I = 1.9\,\text{V}$. Now, for $v_I > 1.9\,\text{V}$, the transistor is biased in

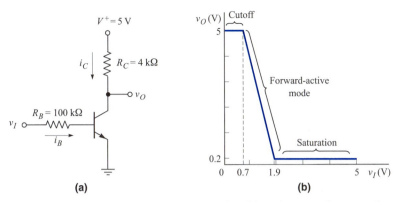

Figure 3.44   (a) A bipolar inverter used as an amplifier; (b) the inverter voltage transfer characteristics

saturation and the output voltage is constant at 0.2 V. The voltage transfer characteristics are shown in Figure 3.44(b).

AC Solution: Now bias the transistor in the center of the active region with an input voltage of $v_I = V_{BB} = 1.3$ V. Also include a second input voltage source, denoted as Δv_I in Figure 3.45(a). The dc output voltage is 2.6 V, which is the Q-point of the transistor.

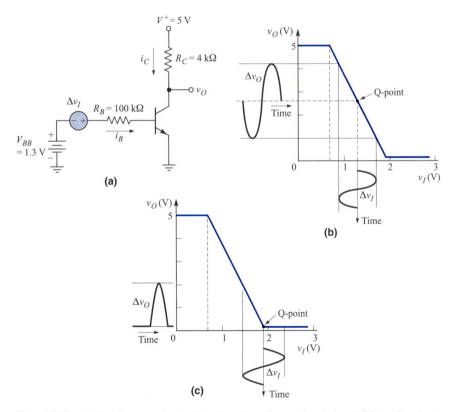

Figure 3.45   (a) The inverter circuit with both a dc and an ac input signal; (b) the dc voltage transfer characteristics, Q-point, and sinusoidal input and output signals; (c) the transfer characteristics showing improper dc biasing

From Figure 3.45(a), we see that the base current is

$$i_B = \frac{V_{BB} + \Delta v_I - V_{BB}(\text{on})}{R_B} = \frac{(1.3 - 0.7) + \Delta v_I}{100 \, \text{k}\Omega}$$

where the B–E voltage is assumed to be constant at 0.7 V. The output voltage is

$$v_O = V^+ - i_C R_C = V^+ - \beta_F i_B R_C$$

which can be written as

$$v_O = 5 - (100) \left[\frac{0.6 + \Delta v_I}{100 \, \text{k}\Omega} \right] (4 \, \text{k}\Omega) = 2.6 - 4(\Delta v_I)$$

The change in the output voltage due to the change in the input voltage can be written as

$$\Delta v_O = -4(\Delta v_I)$$

The change in output voltage per change in input voltage is the amplification, or

$$A_v = \frac{\Delta v_O}{\Delta v_I} = -4$$

Comment: As the input voltage changes, we move along the voltage transfer characteristics as shown in Figure 3.35(b). The negative sign occurs because of the inverting property of the circuit.

Discussion: In this example, we have biased the transistor in the center of the active region. If the input signal Δv_I is a sinusoidal function as shown in Figure 3.45(b), then the output signal Δv_O is also a sinusoidal signal, which is the desired response for an analog circuit. (This assumes the magnitude of the sinusoidal input signal is not too large.) If the Q-point, or dc biasing, of the transistor were at $v_I = 1.9$ V and $v_O = 0.2$ V, as in Figure 3.45(c), the output response changes. Shown in the figure is a symmetrical sinusoidal input signal. When the input sinusoidal signal is on its positive cycle, the transistor remains biased in saturation and the output voltage does not change. During the negative half of the input signal, the transistor becomes biased in the active region, so a half sinusoidal output response is produced. The output signal is obviously not a replication of the input signal.

This discussion emphasizes the importance of properly biasing the transistor for analog or amplifier applications. The primary objective of this chapter, as stated previously, is to help readers become familiar with transistor circuits, but it is also to enable them to design the dc biasing of transistor circuits that are to be used in analog applications.

Test Your Understanding

[Note: In the following exercises, assume the B–E cut-in voltage is 0.7 V for both the npn and pnp transistors. Also assume the C–E saturation voltage is 0.2 V for both types of transistor.]

*3.18   Consider the circuit in Figure 3.46. Determine I_B, I_C, I_E, and V_{CE} for $\beta = 80$. (Ans. $I_B = 0.402$ mA, $I_C = 0.880$ mA, $I_E = 1.28$ mA, $V_{CE} = 0.2$ V)

3.19   For the circuit in Figure 3.47, assume $\beta = 50$ and determine V_I such that $I_C/I_B = 2$. Determine the values of I_B, I_C, and V_{EC}. (Ans. $V_{EC} = 0.2$ V, $I_C = 0.48$ mA, $I_B = 0.24$ mA, $V_I = -5.5$ V)

Figure 3.46 Figure for Exercise 3.18

Figure 3.47 Figure for Exercise 3.19

Figure 3.48 Figure for Exercise 3.20

*3.20 For the circuit in Figure 3.48, let $\beta = 75$, and determine I_C, I_E, and V_{CE} for:
(a) $V_I = -4.5\,\text{V}$; (b) $V_I = -3.5\,\text{V}$; and (c) $V_I = +3.5\,\text{V}$. (Ans. (a) $I_B = I_C = I_E = 0$,
$V_{CE} = 10\,\text{V}$; (b) $I_B = 2.55\,\mu\text{A}$, $I_C = 0.191\,\text{mA}$, $I_E = 0.194\,\text{mA}$, $V_{CE} = 8.84\,\text{V}$;
(c) $I_B = 0.112\,\text{mA}$, $I_C = 1.56\,\text{mA}$, $I_E = 1.67\,\text{mA}$, $V_{CE} = 0.2\,\text{V}$)

D3.21 The transistor in the circuit in Figure 3.49 is used to turn the LED on and off.
Assume $\beta_F = 50$ for the transistor and $V_\gamma = 1.5\,\text{V}$ for the LED. Determine the value of
R to limit the diode current to 15 mA when the transistor is driven into saturation.
Determine R_B such that $I_C/I_B = 20$ when the transistor is driven into saturation for
$v_I = 5\,\text{V}$.

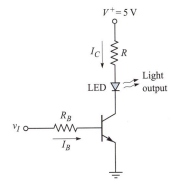

Figure 3.49 Figure for Exercise 3.21

3.22 The transistor parameters in the circuit in Figure 3.43(b) are: $\beta = 40$,
$V_{BE}(\text{on}) = 0.7\,\text{V}$, and $V_{CE}(\text{sat}) = 0.2\,\text{V}$. Let $R_C = 600\,\Omega$ and $R_B = 950\,\Omega$. Determine
the currents and output voltage for: (a) $V_1 = V_2 = 0$; (b) $V_1 = 5\,\text{V}$, $V_2 = 0$; and
(c) $V_1 = V_2 = 5\,\text{V}$. (Ans. (a) The currents are zero, $V_O = 5\,\text{V}$; (b) $I_{B2} = I_{C2} = 0$,
$I_{B1} = 4.53\,\text{mA}$, $I_{C1} = I_R = 8\,\text{mA}$, $V_O = 0.2\,\text{V}$; (c) $I_{B1} = I_{B2} = 4.53\,\text{mA}$, $I_{C1} = I_{C2} = 4\,\text{mA} = I_R/2$, $V_O = 0.2\,\text{V}$)

D3.23 Consider the inverter amplifier shown in Figure 3.45(a). Redesign the circuit
such that the voltage amplification is $\Delta v_O / \Delta v_I = -5$. Determine the Q-point values so
that the transistor is biased in the center of the active region.

3.4   BIPOLAR TRANSISTOR BIASING

As mentioned in the previous section, in order to create a linear amplifier, we must keep the transistor in the forward-active mode, establish a Q-point near the center of the load line, and couple the time-varying input signal to the base. The circuit in Figure 3.45(a) is impractical for two reasons: (1) the signal source is not connected to ground, and (2) the dc base current flows through the signal source. In this section, we will examine several alternative biasing schemes. These basic biasing circuits illustrate some desirable and some undesirable biasing characteristics. More sophisticated biasing circuits that use additional transistors and that are used in integrated circuits are discussed in Chapter 10.

3.4.1   Single Base Resistor Biasing

The circuit shown in Figure 3.50(a) is one of the simplest transistor circuits. There is a single dc power supply, and the quiescent base current is established through the resistor R_B. The **coupling capacitor** C_C acts as an open circuit to dc, isolating the signal source from the dc base current. If the frequency of the input signal is large enough and C_C is large enough, the signal can be coupled through C_C to the base with little attenuation. Figure 3.50(b) is the dc equivalent circuit; the Q-point values are indicated by the additional subscript Q.

Figure 3.50   (a) Common-emitter circuit with a single bias resistor in the base and (b) dc equivalent circuit

Design Example   3.13   **Objective:**   Design the circuit shown in Figure 3.50(b) to yield a given I_{CQ} and V_{CEQ}.

    Assume that $V_{CC} = 12\,$V, $\beta = 100$, and $V_{BE}(\text{on}) = 0.7\,$V. The Q-point values are to be $I_{CQ} = 1\,$mA and $V_{CEQ} = 6\,$V.

Solution: The collector resistance can be found from

$$R_C = \frac{V_{CC} - V_{CEQ}}{I_{CQ}} = \frac{12 - 6}{1} = 6\,\text{k}\Omega$$

The base current must then be

$$I_{BQ} = \frac{I_{CQ}}{\beta} = \frac{1\,\text{mA}}{100} \Rightarrow 10\,\mu\text{A}$$

and the base resistance is determined to be

$$R_B = \frac{V_{CC} - V_{BE}(\text{on})}{I_{BQ}} = \frac{12 - 0.7}{10\,\mu\text{A}} = 1.13\,\text{M}\Omega$$

Comment: Although a value of $1.13\,\text{M}\Omega$ for R_B will establish the required base current, this resistance is too large to be used in an integrated circuit.

Figure 3.51(a) shows the transistor characteristics and load line for the circuit in Example 3.13. Although we assumed a current gain of $\beta = 100$, a given transistor type may exhibit a range of values for β as a result of slight variations in the fabrication process. For example, a second circuit with the same configuration as Figure 3.50(a) could be fabricated using a transistor with

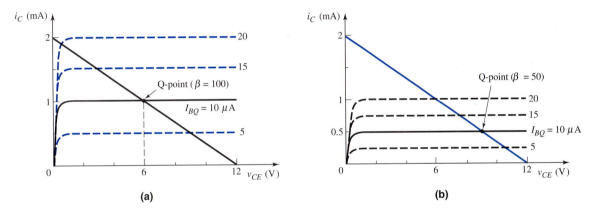

Figure 3.51 Transistor characteristics and load line for the circuit in Example 3.13 when (a) $\beta = 100$ and (b) $\beta = 50$

a current gain of $\beta = 50$. Using the R_C and R_B values previously determined, we would find new quiescent values, as follows:

$$I_{BQ} = \frac{V_{CC} - V_{BE}(\text{on})}{R_B} = 10\,\mu\text{A (unchanged)} \tag{3.33}$$

$$I_{CQ} = \beta I_{BQ} = (50)(10\,\mu\text{A}) \Rightarrow 0.5\,\text{mA} \tag{3.34}$$

and

$$V_{CEQ} = V_{CC} - I_{CQ}R_C = 12 - (0.5)(6) = 9\,\text{V} \tag{3.35}$$

Figure 3.51(b) shows the new transistor characteristics, load line, and Q-point. We can see that the Q-point has shifted substantially. In this circuit, then, the Q-point is not stabilized against variations in β; as β changes, the Q-point varies significantly. We noted in Example 3.12 that the position of the Q-point is significant in a circuit design.

The Q-point is also influenced by variations in resistance values. Tolerances in discrete resistance and integrated circuit resistance values result from process variations and material property variations. For example, if a discrete resistor has a ±5 percent tolerance, a 6 kΩ resistor may actually have a value between 5.7 and 6.3 kΩ; similarly, a 1.13 MΩ resistor may actually be between 1.07 and 1.19 MΩ.

Assuming $\beta = 100$, as in the last example, a ±5 percent variation in R_B means that the base current may be in the range of 9.5 to 10.6 µA. The collector current would then be in the range of 0.95 to 1.06 mA. The variation in collector current, in conjunction with a ±5 percent tolerance in R_C, means that the C–E voltage may range from 5.32 to 6.59 V. Our design, then, may specify a value of $V_{CE} = 6$ V, but the actual value of V_{CE} may vary considerably.

Figure 3.52 Figure for Exercises 3.24 and 3.25

Test Your Understanding

[Note: In the following exercises, assume the B–E cut-in voltage is 0.7 V for both the npn and pnp transistors. Also assume the C–E saturation voltage is 0.2 V for both types of transistor.]

3.24 Consider the circuit shown in Figure 3.52. (a) If $\beta = 100$, determine R_B such that $V_{CEQ} = 2.5$ V. (b) Determine the minimum and maximum allowed values of β if the quiescent collector–emitter voltage is to be in the range $1 \leq V_{CEQ} \leq 4$ V. (Ans. (a) $R_B = 344$ kΩ; (b) $40 \leq \beta \leq 160$)

*D3.25 For the circuit shown in Figure 3.52, let $R_B = 800$ kΩ. If the range of β is between 75 and 150, determine a new value of R_C such that the Q-point will always be in the range $1 \leq V_{CEQ} \leq 4$ V. What will be the actual range of V_{CEQ} for the new value of R_C? (Ans. For $V_{CEQ} = 2.5$ V, $R_C = 4.14$ kΩ; (b) $1.66 \leq V_{CEQ} \leq 3.33$ V)

3.4.2   Voltage Divider Biasing and Bias Stability

The circuit in Figure 3.53(a) is a classic example of discrete transistor biasing. The single bias resistor R_B in the previous circuit is replaced by a pair of

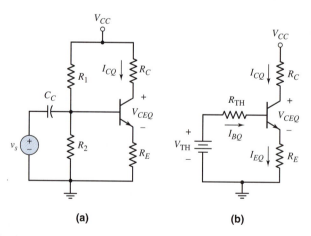

(a)   (b)

Figure 3.53   (a) A common-emitter circuit with an emitter resistor and voltage divider bias circuit in the base; (b) the dc circuit with a Thevenin equivalent base circuit resistors R_1 and R_2, and an emitter resistor R_E is added. The ac signal is still coupled to the base of the transistor through the coupling capacitor C_C.

The circuit is most easily analyzed by forming a **Thevenin equivalent circuit** for the base circuit. The coupling capacitor acts as an open circuit to dc. The equivalent Thevenin voltage is

$$V_{TH} = [R_2/(R_1 + R_2)]V_{CC}$$

and the equivalent Thevenin resistance is

$$R_{TH} = R_1 \| R_2$$

where the symbol $\|$ indicates the parallel combination of resistors. Figure 3.53(b) shows the equivalent dc circuit. As we can see, this circuit is similar to those we have previously considered.

Applying Kirchhoff's law around the B–E loop, we obtain

$$V_{TH} = I_{BQ}R_{TH} + V_{BE}(\text{on}) + I_{EQ}R_E \tag{3.36}$$

If the transistor is biased in the forward-active mode, then

$$I_{EQ} = (1 + \beta)I_{BQ}$$

and the base current, from Equation (3.36), is

$$I_{BQ} = \frac{V_{TH} - V_{BE}(\text{on})}{R_{TH} + (1 + \beta)R_E} \tag{3.37}$$

The collector current is then

$$I_{CQ} = \beta I_{BQ} = \frac{\beta(V_{TH} - V_{BE}(\text{on})}{R_{TH} + (1 + \beta)R_E} \tag{3.38}$$

Example 3.14 **Objective:** Analyze a circuit using a voltage divider bias circuit, and determine the change in the Q-point with a variation in β when the circuit contains an emitter resistor.

For the circuit given in Figure 3.53(a), let $R_1 = 56\,\text{k}\Omega$, $R_2 = 12.2\,\text{k}\Omega$, $R_C = 2\,\text{k}\Omega$, $R_E = 0.4\,\text{k}\Omega$, $V_{CC} = 10\,\text{V}$, $V_{BE}(\text{on}) = 0.7\,\text{V}$, and $\beta = 100$.

Solution: Using the Thevenin equivalent circuit in Figure 3.53(b), we have

$$R_{TH} = R_1 \| R_2 = 56 \| 12.2 = 10.0\,\text{k}\Omega$$

and

$$V_{TH} = \left(\frac{R_1}{R_1 + R_2}\right) \cdot V_{CC} = \left(\frac{12.2}{56 + 12.2}\right)(10) = 1.79\,\text{V}$$

Writing the Kirchhoff voltage law equation around the B–E loop, we obtain

$$I_{BQ} = \frac{V_{TH} - V_{BE}(\text{on})}{R_{TH} + (1 + \beta)R_E} = \frac{1.79 - 0.7}{10 + (101)(0.4)} \Rightarrow 21.6\,\mu\text{A}$$

The collector current is

$$I_{CQ} = \beta I_{BQ} = (100)(21.6\,\mu\text{A}) \Rightarrow 2.16\,\text{mA}$$

and the emitter current is

$$I_{EQ} = (1 + \beta)I_{BQ} = (101)(21.6\,\mu\text{A}) \Rightarrow 2.18\,\text{mA}$$

The quiescent C–E voltage is then

$$V_{CEQ} = V_{CC} - I_{CQ}R_C - I_{EQ}R_E = 10 - (2.16)(2) - (2.18)(0.4) = 4.81 \text{ V}$$

These results show that the transistor is biased in the active region.

If the current gain of the transistor were to decrease to $\beta_F = 50$ or increase to $\beta_F = 150$, we obtain the following results:

β_F	$I_{BQ}(\mu A)$	$I_{CQ}(mA)$	$I_{EQ}(mA)$	$V_{CEQ}(V)$
50	35.9	1.80	1.83	5.67
100	21.6	2.16	2.18	4.81
150	15.5	2.32	2.34	4.40

For a 3 : 1 ratio in β_F, the collector current and collector-emitter voltage change by only a 1.29 : 1 ratio.

Comment: The voltage divider circuit of R_1 and R_2 can bias the transistor in its active region using resistor values in the low kilohm range. In contrast, single resistor biasing requires a resistor in the megohm range. In addition, the change in I_{CQ} and V_{CEQ} with a change in β_F has been substantially reduced compared to the change shown in Figure 3.51. Including an emitter resistor R_E has tended to **stabilize** the Q-point. This means that including the emitter resistor helps to stabilize the Q-point with respect to variations in β.

The design requirement for bias stability is $R_{TH} \ll (1 + \beta)R_E$. Consequently, the collector current, from Equation (3.38), becomes approximately

$$I_{CQ} \cong \frac{\beta(V_{TH} - V_{BE}(\text{on}))}{(1 + \beta)R_E} \tag{3.39}$$

Normally, $\beta \gg 1$; therefore, $\beta/(1 + \beta) \cong 1$, and

$$I_{CQ} \cong \frac{(V_{TH} - V_{BE}(\text{on}))}{R_E} \tag{3.40}$$

Now the quiescent collector current is essentially a function of only the dc voltages and the emitter resistance, and the Q-point is stabilized against β variations. However, if R_{TH} is too small, then R_1 and R_2 are small, and excessive power is dissipated in these resistors. The general rule is that a circuit is considered **bias stable** when

$$R_{TH} \cong 0.1(1 + \beta)R_E \tag{3.41}$$

Design Example 3.15   Objective: Design a bias-stable circuit.

Consider the circuit shown in Figure 3.53(a). Let $V_{CC} = 5\text{ V}$, $R_C = 1\text{ k}\Omega$, $V_{BE}(\text{on}) = 0.7\text{ V}$, and $\beta = 120$. Choose R_E and determine R_1 and R_2 such that the circuit is bias stable and that $V_{CEQ} = 3\text{ V}$.

Design Pointer: Typically, the voltage across R_E should be on the same order of magnitude as $V_{BE}(\text{on})$. Larger voltage drops may mean the supply voltage V_{CC} has to be increased in order to get the required voltage across the collector-emitter and across R_C.

Solution: With $\beta = 120$, $I_{CQ} \approx I_{EQ}$. Then, choosing a standard value of $0.51\,k\Omega$ for R_E, we find

$$I_{CQ} \cong \frac{V_{CC} - V_{CEQ}}{R_C + R_E} = \frac{5 - 3}{1 + 0.51} = 1.32\,mA$$

The voltage drop across R_E is now $(1.32)(0.51) = 0.673\,V$, which is approximately the desired value. The base current is found to be

$$I_{BQ} = \frac{I_{CQ}}{\beta} = \frac{1.32}{120} \Rightarrow 11.0\,\mu A$$

Using the Thevenin equivalent circuit in Figure 3.53(b), we find

$$I_{BQ} = \frac{V_{TH} - V_{BE}(\text{on})}{R_{TH} + (1 + \beta)R_E}$$

For a bias-stable circuit, $R_{TH} = 0.1(1 + \beta)R_E$, or

$$R_{TH} = (0.1)(121)(0.51) = 6.17\,k\Omega$$

Then,

$$I_{BQ} = 11.0\,\mu A \Rightarrow \frac{V_{TH} - 0.7}{6.17 + (121)(0.51)}$$

which yields

$$V_{TH} = 0.747 + 0.70 = 1.45\,V$$

Now

$$V_{TH} = \left(\frac{R_2}{R_1 + R_2}\right)V_{CC} = \left(\frac{R_2}{R_1 + R_2}\right)5 = 1.45\,V$$

or

$$\left(\frac{R_2}{R_1 + R_2}\right) = \frac{1.45}{5} = 0.288$$

Also,

$$R_{TH} = \frac{R_1 R_2}{R_1 + R_2} = 6.05\,k\Omega = R_1\left(\frac{R_2}{R_1 + R_2}\right) = R_1(0.288)$$

which yields

$$R_1 = 21\,k\Omega$$

and

$$R_2 = 8.5\,k\Omega$$

From Appendix D, we can choose standard resistor values of $R_1 = 20\,k\Omega$ and $R_2 = 8.2\,k\Omega$.

Comment: The Q-point in this example is now considered stabilized against variations in β, and the voltage divider resistors R_1 and R_2 have reasonable values in the kilohm range.

Computer Simulation: Figure 3.54 shows the PSpice circuit schematic with the standard resistor values and with a standard 2N2222 transistor from the PSpice library for the circuit designed in this example. A dc analysis was performed and the resulting transister Q-point values are shown. The collector–emitter voltage is $V_{CE} = 2.80\,V$, which is close to the design value of $3\,V$. One reason for the difference is that the standard-valued resistors are not exactly equal to the design values. Another reason

Figure 3.54   PSpice circuit schematic for Example 3.15

for the slight difference is that the effective β_F of the 2N2222 is 157 compared to the assumed value of 120.

```
**** BIPOLAR JUNCTION TRANSISTORS

NAME        Q_Q1
MODEL       Q2N2222
IB          9.25E-06
IC          1.45E-03
VBE         6.55E-01
VBC         -2.15E+00
VCE         2.80E+00
BETADC      1.57E+02
```

Another advantage of including an emitter resistor is that it stabilizes the Q-point with respect to temperature. To explain, we noted in Figure 1.18 that the current in a pn junction increases with increasing temperature, for a constant junction voltage. We then expect the transistor current to increase as the temperature increases. If the current in a junction increases, the junction temperature increases (because of I^2R heating), which in turn causes the current to increase, thereby further increasing the junction temperature. This phenomenon can lead to thermal runaway and to device destruction. However, from Figure 3.53(b), we see that as the current increases, the voltage drop across R_E increases. The Thevenin equivalent voltage and resistance are assumed to be essentially independent of temperature, and the temperature-induced change in the voltage drop across R_{TH} will be small. The net result is that the increased voltage drop across R_E reduces the B–E junction voltage, which then tends to stabilize the transistor current against increases in temperature.

Test Your Understanding

3.26   For the circuit shown in Figure 3.53(a), let $V_{CC} = 5$, $R_1 = 9\,\text{k}\Omega$, $R_2 = 2.25\,\text{k}\Omega$, $R_E = 200\,\Omega$, $R_C = 1\,\text{k}\Omega$, and $\beta = 150$. (a) Determine R_{TH} and V_{TH}. (b) Find I_{BQ}, I_{CQ}, and V_{CEQ}. (c) Repeat part (b) if β changes to $\beta = 75$. (Ans. (a) $R_{TH} = 1.8\,\text{k}\Omega$,

$V_{TH} = 1.0\,\text{V}$; (b) $I_{BQ} = 9.38\,\mu\text{A}$, $I_{CQ} = 1.41\,\text{mA}$, $V_{CEQ} = 3.31\,\text{V}$; (c) $I_{BQ} = 17.6\,\mu\text{A}$, $I_{CQ} = 1.32\,\text{mA}$, $V_{CEQ} = 3.41\,\text{V}$)

*3.27 In the circuit shown in Figure 3.53(a), let $V_{CC} = 5\,\text{V}$, $R_E = 0.2\,\text{k}\Omega$, $R_C = 1\,\text{k}\Omega$, and $\beta = 150$. If $R_1 + R_2 = 11.25\,\text{k}\Omega$, determine R_1 and R_2 such that the Q-point is in the center of the load line. (Ans. $R_1 = 8.67\,\text{k}\Omega$, $R_2 = 2.58\,\text{k}\Omega$)

D3.28 Consider the circuit shown in Figure 3.55. Let $\beta = 150$, $R_E = 0.2\,\text{k}\Omega$, and $R_C = 1\,\text{k}\Omega$. Design a dc bias-stable circuit such that the quiescent output voltage is zero. (Ans. $R_1 = 16.7\,\text{k}\Omega$, $R_2 = 3.68\,\text{k}\Omega$)

D3.29 In the circuit shown in Figure 3.55, assume $\beta = 120$, $R_C = 1.2\,\text{k}\Omega$, and $R_E = 0.3\,\text{k}\Omega$. Design a bias-stable circuit such that $V_{CEQ} = 5\,\text{V}$. (Ans. $R_1 = 20.1\,\text{k}\Omega$, $R_2 = 4.44\,\text{k}\Omega$)

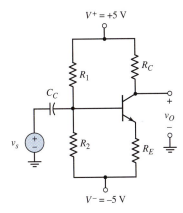

Figure 3.55 Figure for Exercises 3.28 and 3.29

3.4.3 Integrated Circuit Biasing

The resistor biasing of transistor circuits considered up to this point is primarily applied to discrete circuits. For integrated circuits, we would like to eliminate as many resistors as possible since, in general, they require a larger surface area than transistors.

A bipolar transistor can be biased by using a constant-current source I_Q, as shown in Figure 3.56. The advantages of this circuit are that the emitter current is independent of β and R_B, and the collector current and C–E voltage are essentially independent of transistor current gain, for reasonable values of β. The value of R_B can be increased, thus increasing the input resistance at the base, without jeopardizing the bias stability.

The constant-current source can be implemented by using transistors, as shown in Figure 3.57. Transistors Q_1 and Q_2 and resistor R_1 form the constant-current source. Even though transistor Q_1 is connected as a diode, it operates as a transistor in the active region.

Current I_1 is called the **reference current** and is found by writing Kirchhoff's voltage law equation around the R_1–Q_1 loop, as follows:

$$0 = I_1 R_1 + V_{BE}(\text{on}) + V^- \qquad (3.42(\text{a}))$$

Figure 3.56 Bipolar transistor biased with a constant-current source

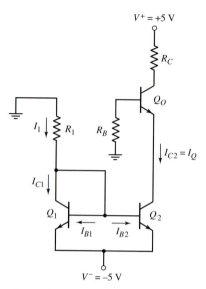

Figure 3.57 Constant-current source biasing

which yields

$$I_1 = \frac{-(V^- + V_{BE}(\text{on}))}{R_1} \qquad \qquad (3.42(\text{b}))$$

Summing the currents at the collector of Q_1 gives

$$I_1 = I_{C1} + I_{B1} + I_{B2} \qquad \qquad (3.43)$$

Since the B–E voltages of Q_1 and Q_2 are equal, if Q_1 and Q_2 are identical transistors and are held at the same temperature, then $I_{B1} = I_{B2}$ and $I_{C1} = I_{C2}$. Equation (3.43) can then be written as

$$I_1 = I_{C1} + 2I_{B2} = I_{C2} + \frac{2I_{C2}}{\beta} = I_{C2}\left(1 + \frac{2}{\beta}\right) \qquad \qquad (3.44)$$

Solving for I_{C2}, we find

$$I_{C2} = I_Q = \frac{I_1}{\left(1 + \dfrac{2}{\beta}\right)} \qquad \qquad (3.45)$$

This current biases the transistor Q_O in the active region.

Example 3.16   **Objective:** Determine the currents in a two-transistor current source.

For the circuit in Figure 3.57, the circuit and transistor parameters are: $R_1 = 10\,\text{k}\Omega$, $\beta = 50$, and $V_{BE}(\text{on}) = 0.7\,\text{V}$.

Solution: The reference current is

$$I_1 = \frac{-(V^- + V_{BE}(\text{on}))}{R_1} = \frac{-((-5) + 0.7)}{10} = 0.43\,\text{mA}$$

From Equation (3.45), the bias current I_Q is

$$I_{C2} = I_Q = \frac{I_1}{\left(1 + \dfrac{2}{\beta}\right)} = \frac{0.43}{\left(1 + \dfrac{2}{50}\right)} = 0.413 \, \text{mA}$$

The base currents are then

$$I_{B1} = I_{B2} = \frac{I_{C2}}{\beta} = \frac{0.413}{50} \Rightarrow 8.26 \, \mu\text{A}$$

Comment: For relatively large values of current gain β, the bias current I_Q is essentially the same as the reference current I_1.

As mentioned, constant-current biasing is used almost exclusively in integrated circuits. As we will see in Part II of the text, circuits in integrated circuits use a minimum number of resistors, and transistors are often used to replace these resistors. Transistors take up much less area than resistors on an IC chip, so it's advantageous to minimize the number of resistors.

Test Your Understanding

$V^+ = 5 \, \text{V}$

*3.30 (a) For the circuit in Figure 3.56, the parameters are: $I_Q = 1 \, \text{mA}$, $V^+ = 10 \, \text{V}$, $V^- = -10 \, \text{V}$, $R_B = 50 \, \text{k}\Omega$, and $R_C = 5 \, \text{k}\Omega$. For the transistor, $\beta = 100$, and $I_S = 3 \times 10^{-14} \, \text{A}$. Determine the dc voltage at the base and V_{CEQ}. (b) Repeat part (a) if $\beta = 50$. (Ans. (a) $V_B = -0.495 \, \text{V}$, $V_{CEQ} = 6.18 \, \text{V}$; (b) $V_B = -0.98 \, \text{V}$, $V_{CEQ} = 6.71 \, \text{V}$)

3.31 The circuit shown in Figure 3.58 is biased with a constant-current source I_Q. For the transistor, $\beta = 120$, and the E–B turn-on voltage is $V_{EB}(\text{on}) = 0.7 \, \text{V}$. Determine I_Q such that $V_{ECQ} = 3 \, \text{V}$. (Ans. $I_Q = 0.710 \, \text{mA}$)

Figure 3.58 Figure for Exercise 3.31

3.5 MULTISTAGE CIRCUITS

Most transistor circuits contain more than one transistor. We can analyze and design these **multistage circuits** in much the same way as we studied single-transistor circuits. As an example, Figure 3.59 shows an npn transistor, Q_1, and a pnp bipolar transistor, Q_2, in the same circuit.

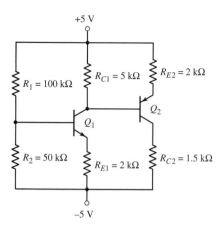

Figure 3.59   A multistage transistor circuit

Example   3.17   Objective: Calculate the dc voltages at each node and the dc currents through the elements in a multistage circuit.

For the circuit in Figure 3.59, assume the B–E turn-on voltage is 0.7 V and $\beta = 100$ for each transistor.

Solution: The Thevenin equivalent circuit of the base circuit of Q_1 is shown in Figure 3.60. The various currents and nodal voltages are defined as shown. The Thevenin resistance and voltage are

$$R_{TH} = R_1 \| R_2 = 100 \| 50 = 33.3 \, \text{k}\Omega$$

and

$$V_{TH} = \left(\frac{R_2}{R_1 + R_2}\right)(10) - 5 = \left(\frac{50}{150}\right)(10) - 5 = -1.67 \, \text{V}$$

Kirchhoff's voltage law equation around the B–E loop of Q_1 is

$$V_{TH} = I_{B1} R_{TH} + V_{BE}(\text{on}) + I_{E1} R_{E1} - 5$$

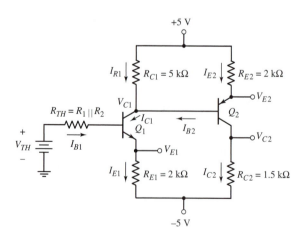

Figure 3.60   Multistage transistor circuit with a Thevenin equivalent circuit in the base of Q_1

Noting that $I_{E1} = (1 + \beta)I_{B1}$, we have

$$I_{B1} = \frac{-1.67 + 5 - 0.7}{33.3 + (101)(2)} \Rightarrow 11.2\,\mu\text{A}$$

Therefore,

$$I_{C1} = 1.12\,\text{mA}$$

and

$$I_{E1} = 1.13\,\text{mA}$$

Summing the currents at the collector of Q_1, we obtain

$$I_{R1} + I_{B2} = I_{C1}$$

which can be written as

$$\frac{5 - V_{C1}}{R_{C1}} + I_{B2} = I_{C1} \tag{3.46}$$

Then, the base current I_{B2} can be written in terms of the emitter current I_{E2}, as follows:

$$I_{B2} = \frac{I_{E2}}{1 + \beta} = \frac{5 - V_{E2}}{(1 + \beta)R_{E2}} = \frac{5 - (V_{C1} + 0.7)}{(1 + \beta)R_{E2}} \tag{3.47}$$

Substituting Equation (3.47) into (3.46), we obtain

$$\frac{5 - V_{C1}}{R_{C1}} + \frac{5 - (V_{C1} + 0.7)}{(1 + \beta)R_{E2}} = I_{C1} = 1.12\,\text{mA}$$

which can be solved for V_{C1} to yield

$$V_{C1} = -0.482\,\text{V}$$

Then,

$$I_{R1} = \frac{5 - (-0.482)}{5} = 1.10\,\text{mA}$$

To find V_{E2}, we have

$$V_{E2} = V_{C1} + V_{EB}(\text{on}) = -0.482 + 0.7 = 0.218\,\text{V}$$

The emitter current I_{E2} is

$$I_{E2} = \frac{5 - 0.218}{2} = 2.39\,\text{mA}$$

Then,

$$I_{C2} = \left(\frac{\beta}{1 + \beta}\right)I_{E2} = \left(\frac{100}{101}\right)(2.39) = 2.37\,\text{mA}$$

and

$$I_{B2} = \frac{I_{E2}}{1 + \beta} = \frac{2.39}{101} \Rightarrow 23.7\,\mu\text{A}$$

The remaining nodal voltages are

$$V_{E1} = I_{E1}R_{E1} - 5 = (1.13)(2) - 5 \Rightarrow V_{E1} = -2.74\,\text{V}$$

and

$$V_{C2} = I_{C2}R_{C2} - 5 = (2.37)(1.5) - 5 \Rightarrow V_{C2} = -1.45\,\text{V}$$

We then find that

$$V_{CE1} = -0.482 - (-2.74) = 2.26\,\text{V}$$

and that

$$V_{EC2} = 0.218 - (-1.45) = 1.67\,\text{V}$$

Comment: These results show that both Q_1 and Q_2 are biased in the forward-active mode, as originally assumed. However, when we consider the ac operation of this circuit as an amplifier in the next chapter, we will see that a better design would increase the value of V_{EC2}.

Test Your Understanding

[Note: In the following exercises, assume the B–E cut-in voltage is 0.7 V and that $\beta = 100$ for both the npn and pnp transistors.]

*RD3.32   In the circuit shown in Figure 3.59, determine new values of R_{C1} and R_{C2} such that $V_{CEQ1} = 3.25\,\text{V}$ and $V_{ECQ2} = 2.5\,\text{V}$. (Ans. $R_{C1} = 4.08\,\text{k}\Omega$, $R_{C2} = 1.97\,\text{k}\Omega$)

*3.33   For the circuit shown in Figure 3.59, change the +5 V bias to +12 V and the −5 V bias to ground potential. Determine all new currents and node voltages. (Ans. $I_{B1} = 14.0\,\mu\text{A}$, $I_{C1} = 1.40\,\text{mA}$, $I_{E1} = 1.42\,\text{mA}$, $V_{B1} = 3.53\,\text{V}$, $V_{E1} = 2.83\,\text{V}$, $V_{C1} = V_{B2} = 5.15\,\text{V}$, $V_{E2} = 5.85\,\text{V}$, $I_{E2} = 3.08\,\text{mA}$, $I_{C2} = 3.04\,\text{mA}$, $I_{B2} = 30.4\,\mu\text{A}$, $V_{C2} = 4.56\,\text{V}$)

3.6   SUMMARY

- In this chapter, we considered the basic characteristics and properties of the bipolar transistor, which is a three-terminal device that has three separately doped semiconductor regions and two pn junctions. The three terminals are called the base (B), emitter (E), and collector (C). Both npn and pnp complementary bipolar transistors can be formed. The defining transistor action is that the voltage across two terminals (base and emitter) controls the current in the third terminal (collector).

- The modes of operation of a bipolar transistor are determined by the biases applied to the two junctions. The four modes are: forward active, cutoff, saturation, and inverse active. In the forward-active mode, the B–E junction is forward biased and the B–C junction is reverse biased, and the collector and base currents are related by the common-emitter current gain β. The relationship is the same for both npn and pnp transistors, as long as the conventional current directions are maintained. When a transistor is cut off, all currents are zero. In the saturation mode, the collector current is no longer a function of base current.

- The dc analysis and the design of dc biasing of bipolar transistor circuits were emphasized in this chapter. We continued to use the piecewise linear model of the pn junction in these analyses and designs. Techniques to design a transistor circuit with a stable Q-point were developed.

- Basic applications of the transistor were discussed. These include switching currents and voltages, performing digital logic functions, and amplifying time-varying signals. The amplifying characteristics will be considered in detail in the next chapter.

CHECKPOINT

After studying this chapter, the reader should have the ability to:

✔ Understand and describe the general current–voltage characteristics for both the npn and pnp bipolar transistors. (Section 3.1)

✔ Define the four modes of operation of a bipolar transistor. (Section 3.1)

✔ Apply the piecewise linear model to the dc analysis and design of various bipolar transistor circuits, including the understanding of the load line. (Section 3.2)

✔ Qualitatively understand how a transistor circuit can be used to switch currents and voltages, to perform digital logic functions, and to amplify time-varying signals. (Section 3.3)

✔ Design the dc biasing of a transistor circuit to achieve specified dc currents and voltages, and to stabilize the Q-point against transistor parameter variations. (Section 3.4)

✔ Apply the dc analysis and design techniques to multistage transistor circuits. (Section 3.5)

REVIEW QUESTIONS

1. What are the bias voltages that need to be applied to an npn bipolar transistor such that the transistor is biased in the forward-active mode?

2. Define the conditions for cutoff, forward-active mode, and saturation mode for a pnp bipolar transistor.

3. Define common-base current gain and common-emitter current gain.

4. Describe the current components that contribute to the collector current and to the base current.

5. Define Early voltage and collector output resistance.

6. Describe a simple common-emitter circuit with an npn bipolar transistor and discuss the relation between collector–emitter voltage and input base current.

7. Define Q-point.

8. Describe the parameters that define a load line.

9. What are the steps used to analyze the dc response of a bipolar transistor circuit?

10. Describe how an npn transistor can be used to switch an **LED** diode on and off.

11. Describe a bipolar transistor **NOR** logic circuit.

12. Describe how a transistor can be used to amplify a time-varying voltage.

13. Discuss the advantages of using resistor voltage divider biasing compared to a single base resistor.

14. How can the Q-point be stabilized against variations in transistor parameters?

15. What is the principal difference between biasing techniques used in discrete transistor circuits and integrated circuits?

PROBLEMS

[Note: In the following problems, let $V_{BE}(\text{on}) = 0.7\,\text{V}$ and $V_{CE}(\text{sat}) = 0.2\,\text{V}$ for npn transistors, and let $V_{EB}(\text{on}) = 0.7\,\text{V}$ and $V_{EC}(\text{sat}) = 0.2\,\text{V}$ for pnp transistors.]

Section 3.1   Basic Bipolar Junction Transistor

3.1   (a) In a bipolar transistor biased in the forward-active mode, the base current is $i_B = 6.0\,\mu A$ and the collector current is $i_C = 510\,\mu A$. Determine β_F, α_F, and i_E. (b) Repeat part (a) if $i_B = 50\,\mu A$ and $i_C = 2.65\,mA$.

3.2   (a) The range of β for a particular type of transistor is $110 \leq \beta \leq 180$. Determine the corresponding range of α. (b) If the base current is $50\,\mu A$, determine the range of collector current.

3.3   An npn transistor with $\alpha = 0.982$ is connected in a common-base configuration. The collector is connected to ground through a 5 V source and a $2\,k\Omega$ resistor. The emitter is driven with a constant-current source. Determine the maximum emitter current such that $V_{CE} \geq 1.0\,V$.

3.4   An npn bipolar transistor with $\beta_F = 120$ is connected in the same circuit configuration described in Problem 3.3. The collector-emitter voltage is $v_{CE} = 2\,V$. Determine i_C, i_B, and i_E.

3.5   A pnp transistor with $\beta = 60$ is connected in a common-base configuration and is biased in the forward-active mode. The collector current is $I_C = 0.85\,mA$. Determine α, I_E, and I_B.

3.6   An npn transistor has a reverse-saturation current of $I_S = 10^{-13}\,A$ and a current gain of $\beta_F = 90$. The transistor is biased at $v_{BE} = 0.685\,V$. Determine I_E, I_C, and I_B.

3.7   Two pnp transistors, fabricated with the same technology, have different junction areas. Both transistors are biased with an emitter-base voltage of $v_{EB} = 0.650\,V$ and have emitter currents of 0.50 and 12.2 mA. Find I_S for each device. What are the relative junction areas?

3.8   A BJT has an Early voltage of 250 V. What is the output resistance for (a) $I_C = 1\,mA$ and (b) $I_C = 0.10\,mA$.

3.9   The open-emitter breakdown voltage of a B–C junction is $BV_{CBO} = 60\,V$. If $\beta = 100$ and the empirical constant is $n = 3$, determine the C–E breakdown voltage in the open-base configuration.

3.10   In a particular circuit application, the minimum required breakdown voltages are $BV_{CBO} = 220\,V$ and $BV_{CEO} = 56\,V$. If $n = 3$, determine the maximum allowed value of β.

3.11   A particular transistor circuit design requires a minimum open-base breakdown voltage of $BV_{CEO} = 50\,V$. If $\beta = 50$ and $n = 3$, determine the minimum required value of BV_{CBO}.

Section 3.2   DC Analysis of Transistor Circuits

3.12   For all the transistors in Figure P3.12, $\beta = 75$. The results of some measurements are indicated on the figures. Find the values of the other labeled currents and voltages.

3.13   The collector resistor values in the circuits shown in Figures P3.12(c) and (d) may vary by ± 5 percent from the given value. Determine the range of calculated parameters.

RD3.14   For the circuit shown in Figure 3.20(a), $V_{BB} = 2.5\,V$, $V_{CC} = 5\,V$, and $\beta_F = 70$. Redesign the circuit such that $I_{BQ} = 15\,\mu A$ and $V_{ECQ} = 2.5\,V$.

3.15   In the circuits shown in Figure P3.15, the values of measured parameters are shown. Determine β, α, and the other labeled currents and voltages. Sketch the dc load line and plot the Q-point.

Figure P3.12

Figure P3.15

3.16 For the transistor in the circuit shown in Figure P3.16, $\beta = 200$. Determine I_E and V_C for: (a) $V_B = 0$, (b) $V_B = 1\,\text{V}$, and (c) $V_B = 2\,\text{V}$.

Figure P3.16   **Figure P3.17**

3.17 (a) The current gain of the transistor in Figure P3.17 is $\beta = 75$. Determine V_O for: (i) $V_{BB} = 0$, (ii) $V_{BB} = 1\,\text{V}$, and (iii) $V_{BB} = 2\,\text{V}$. (b) Verify the results of part (a) with a computer simulation.

3.18 (a) The transistor shown in Figure P3.18 has $\beta_F = 100$. Determine V_O for (i) $I_Q = 0.1\,\text{mA}$, (ii) $I_Q = 0.5\,\text{mA}$, and (iii) $I_Q = 2\,\text{mA}$. (b) Determine the percent change in V_O for the conditions in part (a) if the current gain increases to $\beta_F = 150$.

Figure P3.18   **Figure P3.21**   **Figure P3.22**

3.19 For the circuit in Figure P3.16, determine V_B and I_E such that $V_B = V_C$. Assume $\beta = 50$.

3.20 For the circuit shown in Figure P3.18, determine the value of I_Q such that $V_{CB} = 0.5\,\text{V}$. Assume $\beta_F = 100$.

3.21 Consider the circuit shown in Figure P3.21. The measured value of the emitter voltage is $V_E = 2\,\text{V}$. Determine I_E, I_C, β, α, and V_{EC}. Sketch the dc load line and plot the Q-point.

3.22 The transistor in the circuit shown in Figure P3.22 is biased with a constant current in the emitter. If $I_Q = 1\,\text{mA}$, determine V_C and V_E. Assume $\beta = 50$.

3.23 In the circuit in Figure P3.22, the constant current is $I = 0.5\,\text{mA}$. If $\beta = 50$, determine the power dissipated in the transistor and the power supplied by the constant-current source.

3.24 (a) For the circuit shown in Figure P3.22, calculate and plot the power dissipated in the transistor for $I_Q = 0$, 0.5, 1.0, 1.5, 2.0, 2.5, and 3.0 mA. Assume $\beta_F = 50$. (b) Verify the results of part (a) with a computer simulation.

Figure P3.25   **Figure P3.26**

3.25 For the circuit shown in Figure P3.25, if $\beta = 200$ for each transistor, determine: (a) I_{E1}, (b) I_{E2}, (c) V_{C1}, and (d) V_{C2}.

D3.26 (a) For the circuit shown in Figure P3.26, the Q-point is $I_{CQ} = 2\,\text{mA}$ and $V_{CEQ} = 12\,\text{V}$ when $\beta = 60$. Determine the values of R_C and R_B. (b) If the transistor is replaced by a new one with $\beta = 100$, find the new values of I_{CQ} and V_{CEQ}. (c) Sketch the load line and Q-point for both parts (a) and (b).

Figure P3.27   **Figure P3.28**   **Figure P3.29**

3.27 Consider the common-base circuit shown in Figure P3.27. Assume the transistor alpha is $\alpha = 0.9920$. Determine I_E, I_C, and V_{BC}.

3.28 For the transistor in Figure P3.28, $\beta = 30$. Determine V_1 such that $V_{CEQ} = 6\,\text{V}$.

3.29 Let $\beta = 25$ for the transistor in the circuit shown in Figure P3.29. Determine the range of V_1 such that $1.0 \leq V_{CE} \leq 4.5\,\text{V}$. Sketch the load line and show the range of the Q-point values.

Section 3.4   Bipolar Transistor Biasing

3.30   For the transistor in the circuit shown in Figure P3.30, $\beta = 50$. Determine I_{CQ} and V_{CEQ}.

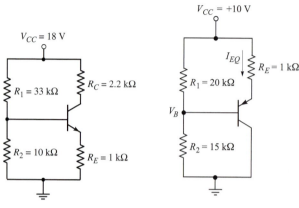

Figure P3.30   **Figure P3.32**

*RD3.31   For the circuit shown in Figure P3.30, let $V_{CC} = 18\,$V, $R_E = 1\,$kΩ, and $\beta_F = 80$. Redesign the circuit such that $I_{CQ} = 1.2\,$mA and $V_{CEQ} = 9\,$V. Let $R_{TH} = 50\,$kΩ. Correlate the design with a computer simulation.

3.32   The current gain of the transistor shown in the circuit of Figure P3.32 is $\beta = 100$. Determine V_B and I_{EQ}.

3.33   For the circuit shown in Figure P3.33, let $\beta = 125$. (a) Find I_{CQ} and V_{CEQ}. (b) If each resistor can vary by ± 5 percent, determine the range in I_{CQ} and V_{CEQ}.

3.34   Consider the circuit shown in Figure P3.34. Determine I_{BQ}, I_{CQ}, and V_{CEQ} for: (a) $\beta = 75$, and (b) $\beta = 150$.

Figure P3.33   **Figure P3.34**

RD3.35   (a) Redesign the circuit shown in Figure P3.30 using $V_{CC} = 9\,$V such that the voltage drop across R_C is $(\frac{1}{3})V_{CC}$ and the voltage drop across R_E is $(\frac{1}{3})V_{CC}$. Assume $\beta_F = 100$. The quiescent collector current is to be $I_{CQ} = 0.4\,$mA, and the current through R_1 and R_2 should be approximately $0.2 I_{CQ}$. (b) Replace each resistor in part (a) with the closest standard value (Appendix D). What is the value of I_{CQ} and what are the voltage drops across R_C and R_E? (c) Verify the design with a computer simulation.

3.36 For the circuit shown in Figure P3.36, let $\beta = 100$. (a) Find R_{TH} and V_{TH} for the base circuit. (b) Determine I_{CQ} and V_{CEQ}. (c) Draw the load line for the circuit. (d) If each resistor can vary by ± 5 percent, determine the range in I_{CQ} and V_{CEQ}.

Figure P3.36   **Figure P3.37**

D3.37 In the circuit shown in Figure P3.37, find R_C such that the Q-point is in the center of the load line. Let $\beta = 75$. What are the values of I_{CQ} and V_{ECQ}?

D3.38 (a) For the circuit shown in Figure P3.38, design a bias-stable circuit such that $I_{CQ} = 0.8\,\text{mA}$ and $V_{CEQ} = 5\,\text{V}$. Let $\beta_F = 100$. (b) Using the results of part (a), determine the percentage change in I_{CQ} if β_F is in the range $75 \le \beta_F \le 150$. (c) Repeat parts (a) and (b) if $R_E = 1\,\text{k}\Omega$.

Figure P3.38   **Figure P3.40**

*D3.39 Design a bias-stable circuit in the form of Figure P3.38 with $\beta = 120$ such that $I_{CQ} = 0.8\,\text{mA}$, $V_{CEQ} = 5\,\text{V}$, and the voltage across R_E is approximately 0.7 V. The current through the bias resistors R_1 and R_2 should be no larger than 0.1 mA.

D3.40 Using the circuit in Figure P3.40, design a bias-stable amplifier such that the Q-point is in the center of the load line. Let $\beta = 125$. Determine I_{CQ}, V_{CEQ}, R_1, and R_2.

D3.41 For the circuit shown in Figure P3.40, the quiescent collector current is to be $I_{CQ} = 1\,\text{mA}$. (a) Design a bias-stable circuit for $\beta = 80$. Determine V_{CEQ}, R_1, and R_2. (b) If each resistor can vary by ± 5 percent, determine the range in I_{CQ} and V_{CEQ}.

*D3.42 A bias-stable circuit with the configuration shown in Figure P3.40 is to be designed such that $I_{CQ} = (3 \pm 0.1)\,\text{mA}$ and $V_{CEQ} \cong 5\,\text{V}$ using a transistor with $75 \leq \beta \leq 150$. Verify the design with a computer simulation.

3.43 (a) For the circuit shown in Figure P3.43, assume $\beta = 75$. Determine I_{BQ}, I_{CQ}, and V_{ECQ}. (b) Determine the values of I_{BQ}, I_{CQ}, and V_{ECQ} if $\beta = 100$.

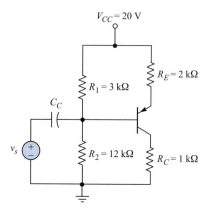

Figure P3.43

D3.44 The dc load line and Q-point of the circuit in Figure P3.44(a) are shown in Figure P3.44(b). For the transistor, $\beta = 120$. Find R_E, R_1, and R_2 such that the circuit is bias stable.

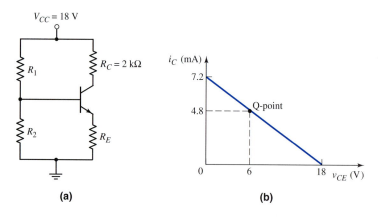

(a)   (b)

Figure P3.44

*D3.45 The range of β for the transistor in the circuit in Figure P3.45 is $50 \leq \beta \leq 90$. Design a bias-stable circuit such that the nominal Q-point is $I_{CQ} = 2\,\text{mA}$ and $V_{CEQ} = 10\,\text{V}$. The value of I_C must fall in the range $1.75 \leq I_C \leq 2.25\,\text{mA}$.

*D3.46 The nominal Q-point of the circuit in Figure P3.46 is $I_{CQ} = 1\,\text{mA}$ and $V_{CEQ} = 5\,\text{V}$, for $\beta = 60$. The current gain of the transistor is in the range $45 \leq \beta \leq 75$. Design a bias-stable circuit such that I_{CQ} does not vary by more than 10 percent from its nominal value.

Figure P3.45   **Figure P3.46**   **Figure P3.48**

RD3.47 (a) For the circuit in Figure P3.46, the value of V_{CC} is changed to 3 V. Let $R_C = 5R_E$ and $\beta_F = 120$. Redesign a bias-stable circuit such that $I_{CQ} = 100\,\mu A$ and $V_{CEQ} = 1.4$ V. (b) Using the results of part (a), determine the dc power dissipation in the circuit. (c) Verify the design with a computer simulation.

D3.48 For the circuit in Figure P3.48, let $\beta = 100$ and $R_E = 3\,k\Omega$. Design a bias-stable circuit such that $V_E = 0$.

3.49 For the circuit in Figure P3.49, let $R_C = 2.2\,k\Omega$, $R_E = 2\,k\Omega$, $R_1 = 10\,k\Omega$, $R_2 = 20\,k\Omega$, and $\beta = 60$. (a) Find R_{TH} and V_{TH} for the base circuit. (b) Determine I_{BQ}, I_{CQ}, V_E, and V_C.

*D3.50 Design the circuit in Figure P3.49 to be bias stable and to provide nominal Q-point values of $I_{CQ} = 0.5\,mA$ and $V_{ECQ} = 8$ V. The maximum current in R_1 and R_2 is to be limited to $40\,\mu A$.

3.51 In the circuit in Figure P3.51, $\beta = 75$. (a) Find V_{TH} and R_{TH} for the base circuit. (b) Determine I_{CQ} and V_{CEQ}. (c) If each resistor can vary by ± 5 percent, determine the range in I_{CQ} and V_{CEQ}.

3.52 For the circuit in Figure P3.52, let $\beta = 100$. (a) Find V_{TH} and R_{TH} for the base circuit. (b) Determine I_{CQ} and V_{CEQ}.

3.53 Find I_{CQ} and V_{CEQ} for the circuit in Figure P3.53, if $\beta = 100$.

Figure P3.49

Figure P3.51   **Figure P3.52**

Section 3.5   Multistage Circuits

*3.54 For each transistor in the circuit in Figure P3.54, $\beta = 120$ and the B–E turn-on voltage is 0.7 V. Determine the quiescent base, collector, and emitter currents in Q_1 and Q_2. Also determine V_{CEQ1} and V_{CEQ2}.

*3.55 The parameters for each transistor in the circuit in Figure P3.55 are $\beta = 80$ and $V_{BE}(\text{on}) = 0.7\,\text{V}$. Determine the quiescent values of base, collector, and emitter currents in Q_1 and Q_2.

3.56 Consider the circuit used in Example 3.16. Determine the power supplied by the $+5\,\text{V}$ source and by the $-5\,\text{V}$ source.

*3.57 (a) For the transistors in the circuit shown in Figure P3.57, the parameters are: $\beta = 100$ and $V_{BE}(\text{on}) = 0.7\,\text{V}$. Determine R_{C1}, R_{E1}, R_{C2}, and R_{E2} such that $I_{C1} = I_{C2} = 0.8\,\text{mA}$, $V_{ECQ1} = 3.5\,\text{V}$, and $V_{CEQ2} = 4.0\,\text{V}$. (b) Correlate the results of part (a) with a computer simulation.

Figure P3.53   **Figure P3.54**   **Figure P3.55**

COMPUTER ANALYSIS PROBLEMS

3.58 Generate the i_C versus v_{CE} characteristics for an npn silicon bipolar transistor at $T = 300\,^\circ\text{K}$, using a saturation current of $I_S = 10^{-14}\,\text{A}$. Limit the characteristics to $v_{CE}(\text{max}) = 10\,\text{V}$ and $i_C(\text{max}) = 10\,\text{mA}$. Plot curves for: (a) $\beta_F = 100$, $V_A = \infty$ (default value); (b) $\beta_F = 50$, $V_A = \infty$; and (c) $\beta_F = 100$, $V_A = 50\,\text{V}$.

3.59 Correlate the results of Example 3.4 with a computer simulation.

3.60 The circuit shown in Figure P3.18 is driven by a constant-current source. Using a computer simulation, investigate the B–E and C–E voltages as the transistor is driven into saturation.

3.61 For Exercise 3.28, use a computer simulation to obtain a plot of the Q-point values versus temperature over the range $0\,^\circ\text{C} \le T \le 125\,^\circ\text{C}$.

3.62 For Exercise 3.28, use a computer simulation to obtain a plot of the Q-point values versus β over the range $50 \le \beta \le 200$.

DESIGN PROBLEMS

[Note: Each design should be correlated with a computer simulation.]

*D3.63 Consider a common-emitter circuit with the configuration shown in Figure 3.53(a). The circuit parameters are: $V_{CC} = 10\,\text{V}$, $R_E = 0.5\,\text{k}\Omega$, and $R_C = 4\,\text{k}\Omega$. The transistor B–E turn-on voltage is 0.7 V and the current gain is in the range $80 \leq \beta \leq 120$. Design the circuit such that the nominal Q-point is in the center of the load line and the Q-point parameters do not deviate from the nominal value by more than ± 10 percent. In addition, the dc currents in R_1 and R_2 should be at least a factor of ten larger than the quiescent base current.

*D3.64 (a) For the transistors in the circuit shown in Figure 3.57, the parameters are: $V_{BE}(\text{on}) = 0.7\,\text{V}$, and $\beta = 80$. If $R_B = 10\,\text{k}\Omega$ and $R_C = 2\,\text{k}\Omega$, design the circuit such that the quiescent collector–emitter voltage of Q_O is $V_{CEQ}(Q_O) = 3\,\text{V}$. (b) If the three transistors have the same value of β, but the value is in the range $60 \leq \beta \leq 100$, determine the maximum tolerance in R_1 such that the C–E voltage of Q_O remains in the range $2.7 \leq V_{CEQ} \leq 3.3\,\text{V}$.

*D3.65 Design a discrete circuit using the configuration shown in Figure P3.65, given that $V^+ = 15\,\text{V}$, $V^- = -15\,\text{V}$, $V_{CEQ} \cong 8\,\text{V}$, and $I_{CQ} \cong 5\,\text{mA}$. The transistor parameters are: $V_{BE}(\text{on}) = 0.7\,\text{V}$, and $100 \leq \beta \leq 400$. Select standard 5 percent tolerance resistance values.

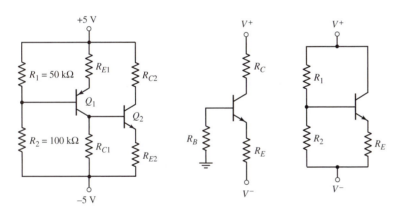

Figure P3.57   Figure P3.65   Figure P3.66

*D3.66 Design a discrete emitter–follower with the configuration shown in Figure P3.66, given that $V^+ = 10\,\text{V}$, $V^- = -10\,\text{V}$, $V_{CEQ} \cong \left(\frac{1}{2}\right)(V^+ - V^-)$, and $I_{CQ} \cong 100\,\text{mA}$. The transistor parameters are: $V_{BE}(\text{on}) = 0.7\,\text{V}$, and $80 \leq \beta \leq 160$. Select standard 5 percent tolerance resistance values.

*D3.67 Redesign the multistage circuit shown in Figure 3.59 such that $V_{CE1} > 3\,\text{V}$ and $V_{EC2} \cong 5\,\text{V}$. Assume transistor turn-on voltages of 0.7V and nominal transistor gains of $\beta = 100$.

4

Basic BJT Amplifiers

4.0 PREVIEW

In the last chapter, we described the operation of the bipolar junction transistor, and analyzed and designed the dc response of circuits containing these devices. In this chapter, we emphasize the use of the bipolar transistor in linear amplifier applications.

Linear amplifiers imply that we are dealing with analog signals. The magnitude of an analog signal may have any value and may vary continuously with respect to time. A linear amplifier then means that the output signal is equal to the input signal multiplied by a constant, where the magnitude of the constant of proportionality is, in general, greater than unity. A linear amplifier is capable of producing signal power gain; that is, the power in the output signal is greater than the power in the input signal. We will investigate the source of this "extra" power.

We examine the properties of three basic single-stage, or single-transistor, amplifier circuits. These circuits are the common-emitter, emitter-follower, and common-base configurations. These configurations form the building blocks for more complex amplifiers, so gaining a good understanding of these three amplifier circuits is an important goal of this chapter.

We introduce a few of the many possible multistage configurations in which multiple amplifiers are connected in series, or cascade, to increase the overall small-signal voltage gain or to provide a particular combination of voltage gain and output resistance. Our discussion includes the method of analysis required for these types of circuits and a synopsis of their properties.

4.1 ANALOG SIGNALS AND LINEAR AMPLIFIERS

In this chapter, we will be considering **signals**, **analog** circuits, and **amplifiers**. A signal contains some type of information. For example, sound waves produced by a speaking human contain the information the person is conveying to another person. Our physical senses, such as hearing, vision, and touch, are naturally analog. Analog signals can represent parameters such as temperature, pressure, and wind velocity. Here, we are interested in electrical signals, such as the output signal from a compact disc, a signal from a microphone, or a signal from a heart rate monitor. The electrical signals are in the form of time-varying currents and voltages.

The magnitude of an **analog signal** can take on any value and may vary continuously with time. Electronic circuits that process analog signals are called **analog circuits**. One example of an analog circuit is a linear amplifier. A **linear amplifier** magnifies an input signal and produces an output signal that is larger and directly proportional to the input signal.

Time-varying signals from a particular source very often need to be amplified before the signal is capable of being "useful." For example, Figure 4.1 shows a signal source that is the output of a compact disc system. That signal consists of a small time-varying voltage and current, which means the signal power is relatively small. The power required to drive the speakers is larger than the output signal from the compact disc, so the compact disc signal must be amplified before it is capable of driving the speakers in order that sound can be heard. Other examples of signals that must be amplified before they are capable of driving loads include the output of a microphone, voice signals received on earth from an orbiting manned shuttle, and video signals from an orbiting weather satellite.

Also shown in Figure 4.1 is a dc voltage source connected to the amplifier. The amplifier contains transistors that must be biased in the forward-active region so that the transistors can act as amplifying devices. We want the output signal to be linearly proportional to the input signal so that the output of the speakers is an exact (as much as possible) reproduction of the signal generated from the compact disc. Therefore, we want the amplifier to be a **linear** amplifier.

Figure 4.1   Block diagram of a compact disc player system

Figure 4.1 suggests that there are two types of analyses of the amplifier that we must consider. The first is a dc analysis because of the applied dc voltage source, and the second is a time-varying or ac analysis because of the time-varying signal source. A linear amplifier means that the superposition principle applies. The principle of superposition states: *The response of a linear circuit excited by multiple independent input signals is the sum of the responses of the circuit to each of the input signals alone.* For the linear amplifier, then, the dc analysis can be performed with the ac source set to zero, the ac analysis can be performed with the dc source set to zero, and the total response is the sum of the two individual responses.

4.2   THE BIPOLAR LINEAR AMPLIFIER

The transistor is the heart of an amplifier. In this chapter, we will consider bipolar transistor amplifiers. Bipolar transistors have traditionally been used in linear amplifier circuits because of their relatively high gain. In Chapter 6, we will consider the field-effect transistor amplifier, and will compare those results with the bipolar amplifier characteristics developed in this chapter.

We begin our discussion by considering the same bipolar circuit that was discussed in the last chapter. Figure 4.2(a) shows the circuit and Figure 4.2(b) shows the voltage transfer characteristics that were developed in Chapter 3. To use the circuit as an amplifier, the transistor needs to be biased with a dc voltage at a quiescent point (Q-point), as shown in the figure, such that the transistor is biased in the forward-active region. This dc analysis or design of the circuit was the focus of our attention in Chapter 3. If a time-varying (e.g., sinusoidal) signal is superimposed on the dc input voltage, V_{BB}, the output voltage will change along the transfer curve producing a time-varying output voltage. If the time-varying output voltage is directly proportional to and larger than the time-varying input voltage, then the circuit is a linear amplifier. From this figure, we see that if the transistor is not biased in the active region (biased either in cutoff or saturation), the output voltage does not change with a change in the input voltage. Thus, we no longer have an amplifier.

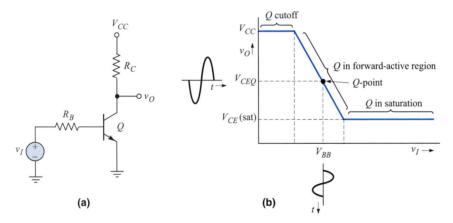

Figure 4.2   (a) Bipolar transistor inverter circuit; (b) inverter transfer characteristics

In this chapter, we are interested in the ac analysis and design of bipolar transistor amplifiers, which means that we must determine the relationships between the time-varying output and input signals. We will initially consider a graphical technique that can provide an intuitive insight into the basic operation of the circuit. We will then develop a small-signal equivalent circuit that will be used in the mathematical analysis of the ac signals. In general, we will be considering a steady-state, sinusoidal analysis of circuits. We will assume that any time-varying signal can be written as a sum of sinusoidal signals of different frequencies and amplitudes (Fourier series), so that a sinusoidal analysis is appropriate.

We will be dealing with time-varying as well as dc currents and voltages in this chapter. Table 4.1 gives a summary of notation that will be used. This

Table 4.1 Summary of notation

Variable	Meaning
i_B, v_{BE}	Total instantaneous values
I_B, V_{BE}	DC values
i_b, v_{be}	Instantaneous ac values
I_b, V_{be}	Phasor values

notation was discussed in the Prologue, but is repeated here for convenience. A lowercase letter with an uppercase subscript, such as i_B or v_{BE}, indicates *total instantaneous values*. An uppercase letter with an uppercase subscript, such as I_B or V_{BE}, indicates *dc quantities*. A lowercase letter with a lowercase subscript, such as i_b or v_{be}, indicates instantaneous values of *ac signals*. Finally, an uppercase letter with a lowercase subscript, such as I_b or V_{be}, indicates *phasor quantities*. The phasor notation, which was reviewed in the Prologue becomes especially important in Chapter 7 during the discussion of frequency response. However, the phasor notation will be generally used in this chapter in order to be consistent with the overall ac analysis.

4.2.1 Graphical Analysis and AC Equivalent Circuit

Figure 4.3 shows the same basic bipolar inverter circuit that has been discussed, but now includes a sinusoidal signal source in series with the dc source. (This circuit is not practical and should not be built in the lab since a dc current flows through the sinusoidal signal source. However, the circuit can be used to conveniently illustrate the basic circuit operation and the mechanism of amplification.)

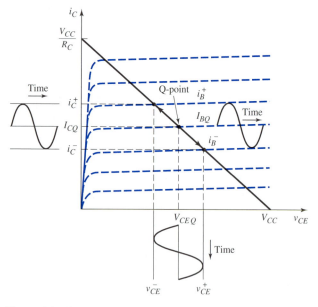

Figure 4.3 A common-emitter circuit with a time-varying signal source in series with the base dc source

Figure 4.4 Common-emitter transistor characteristics, dc load line, and sinusoidal variation in base current, collector current, and collector–emitter voltage

Figure 4.4 shows the transistor characteristics, the dc load line, and the Q-point. The sinusoidal signal source, v_s, will produce a time-varying or ac base current superimposed on the quiescent base current as shown in the figure. The time-varying base current will induce an ac collector current superimposed on the quiescent collector current. The ac collector current then produces a time-varying voltage across R_C, which induces an ac collector–emitter voltage as shown in the figure. The ac collector–emitter voltage, or output voltage, in general, will be larger than the sinusoidal input signal, so that the circuit has produced signal amplification—that is, the circuit is an amplifier.

We need to develop a mathematical method or model for determining the relationships between the sinusoidal variations in currents and voltages in the circuit. As already mentioned, a linear amplifier implies that superposition applies so that the dc and ac analyses can be performed separately. To obtain a linear amplifier, the time-varying or ac currents and voltages must be small enough to ensure a linear relation between the ac signals. To meet this objective, the time-varying signals are assumed to be *small signals,* which means that the amplitudes of the ac signals are small enough to yield linear relations. The concept of "small enough," or small signal, will be discussed further as we develop the small-signal equivalent circuits.

A time-varying signal source, v_s, in the base of the circuit in Figure 4.3 generates a time-varying component of base current, which implies there is also a time-varying component of base–emitter voltage. Figure 4.5 shows the exponential relationship between base-current and base–emitter voltage. If the magnitudes of the time-varying signals that are superimposed on the dc quiescent point are small, then we can develop a linear relationship between the ac base–emitter voltage and ac base current. This relationship corresponds to the slope of the curve at the Q-point.

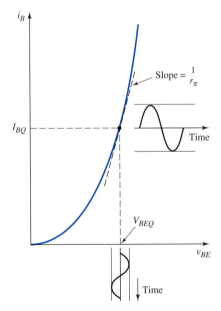

Figure 4.5 Base current versus base–emitter voltage characteristic with superimposed sinusoidal signals

Using Figure 4.5, we can now determine one quantitative definition of small signal. From the discussion in Chapter 3, the relation between base–emitter voltage and base current can be written as

$$i_B = \frac{I_S}{1 + \beta_F} \cdot \exp\left(\frac{v_{BE}}{V_T}\right) \tag{4.1}$$

If v_{BE} is composed of a dc term with a sinusoidal component superimposed, i.e., $v_{BE} = V_{BEQ} + v_{be}$, then

$$i_B = \frac{I_S}{1 + \beta_F} \cdot \exp\left(\frac{V_{BEQ} + v_{be}}{V_T}\right) = \frac{I_S}{1 + \beta_F} \cdot \exp\left(\frac{V_{BEQ}}{V_T}\right) \cdot \exp\left(\frac{v_{be}}{V_T}\right) \tag{4.2}$$

where V_{BEQ} is normally referred to as the base–emitter turn-on voltage, $V_{BE}(\text{on})$. The term $[I_S/(1 + \beta_F)] \cdot \exp(V_{BEQ}/V_T)$ is the quiescent base current, so we can write

$$i_B = I_{BQ} \cdot \exp\left(\frac{v_{be}}{V_T}\right) \tag{4.3}$$

The base current, given in this form, cannot be written as an ac current superimposed on a dc quiescent value. However, if $v_{be} \ll V_T$, then we can expand the exponential term in a Taylor series, keeping only the **linear term**. This approximation is what is meant by **small signal**. We then have

$$i_B \cong I_{BQ}\left(1 + \frac{v_{be}}{V_T}\right) = I_{BQ} + \frac{I_{BQ}}{V_T} \cdot v_{be} = I_{BQ} + i_b \tag{4.4(a)}$$

where i_b is the time-varying (sinusoidal) base current given by

$$i_b = \left(\frac{I_{BQ}}{V_T}\right)v_{be} \tag{4.4(b)}$$

The sinusoidal base current, i_b, is linearly related to the sinusoidal base–emitter voltage, v_{be}. In this case, the term small-signal refers to the condition in which v_{be} is sufficiently small for the linear relationships between i_b and v_{be} given by Equation (4.4(b)) to be valid. As a general rule, if v_{be} is less than $10\,\text{mV}$, then the exponential relation given by Equation (4.3) and its linear expansion in Equation (4.4(a)) agree within approximately 5 percent.

From the concept of small signal, all the time-varying signals shown in Figure 4.4 will be linearly related and are superimposed on dc values. We can write (refer to notation given in Table 4.1)

$$i_B = I_{BQ} + i_b \tag{4.5(a)}$$

$$i_C = I_{CQ} + i_c \tag{4.5(b)}$$

$$v_{CE} = V_{CEQ} + v_{ce} \tag{4.5(c)}$$

and

$$v_{BE} = V_{BEQ} + v_{be} \tag{4.5(d)}$$

If the signal source, v_s, is zero, then the base–emitter and collector–emitter loop equations are,

$$V_{BB} = I_{BQ}R_B + V_{BEQ} \tag{4.6(a)}$$

and

$$V_{CC} = I_{CQ}R_C + V_{CEQ} \qquad \textbf{(4.6(b))}$$

Taking into account the time-varying signals, we find the base–emitter loop equation is

$$V_{BB} + v_s = i_B R_B + v_{BE} \qquad \textbf{(4.7(a))}$$

or

$$V_{BB} + v_s = (I_{BQ} + i_b)R_B + (V_{BEQ} + v_{be}) \qquad \textbf{(4.7(b))}$$

Rearranging terms, we find

$$V_{BB} - I_{BQ}R_B - V_{BEQ} = i_b R_B + v_{be} - v_s \qquad \textbf{(4.7(c))}$$

From Equation (4.6(a)), the left side of Equation (4.7(c)) is zero. Equation (4.7(c)) can then be written as

$$v_s = i_b R_B + v_{be} \qquad \textbf{(4.8)}$$

which is the base–emitter loop equation with all dc terms set equal to zero.

Taking into account the time-varying signals, the collector–emitter loop equation is

$$V_{CC} = i_C R_C + v_{CE} = (I_{CQ} + i_c)R_C + (V_{CEQ} + v_{ce}) \qquad \textbf{(4.9(a))}$$

Rearranging terms, we find

$$V_{CC} - I_{CQ}R_C - V_{CEQ} = i_c R_C + v_{ce} \qquad \textbf{(4.9(b))}$$

From Equation (4.6(b)), the left side of Equation (4.9(b)) is zero. Equation (4.9(b)) can be written as

$$i_c R_C + v_{ce} = 0 \qquad \textbf{(4.10)}$$

which is the collector–emitter loop equation with all dc terms set equal to zero.

Equations (4.8) and (4.10) relate the ac parameters in the circuit. These equations can be obtained directly by setting all dc currents and voltages equal to zero, so the dc voltage sources become short circuits and any dc current sources would become open circuits. *These results are a direct consequence of applying superposition to a linear circuit.* The resulting BJT circuit, shown in Figure 4.6, is called the *ac equivalent circuit,* and all currents and voltages shown are time-varying signals. We should stress that this circuit is an

Figure 4.6   The ac equivalent circuit of the common-emitter circuit with an npn transistor

equivalent circuit. We are implicitly assuming that the transistor is still biased in the forward-active region with the appropriate dc voltages and currents.

Another way of looking at the ac equivalent circuit is as follows. In the circuit in Figure 4.3, the base and collector currents are composed of ac signals superimposed on dc values. These currents flow through the V_{BB} and V_{CC} voltage sources, respectively. Since the voltages across these sources are assumed to remain constant, the sinusoidal currents do not produce any sinusoidal voltages across these elements. Then, since the sinusoidal voltages are zero, the equivalent ac impedances are zero, or short circuits. In other words, the dc voltage sources are ac short circuits in an equivalent ac circuit. We say that the node connecting R_C and V_{CC} is at signal ground.

4.2.2 Small-Signal Hybrid-π Equivalent Circuit of the Bipolar Transistor

We developed the ac equivalent circuit shown in Figure 4.6. We now need to develop a **small-signal equivalent circuit** for the transistor. One such circuit is the **hybrid-π** model, which is closely related to the physics of the transistor. This effect will become more apparent in Chapter 7 when a more detailed hybrid-π model is developed to take into account the frequency response of the transistor.

We can treat the bipolar transistor as a two-port network as shown in Figure 4.7. One element of the hybrid-π model has already been described. Figure 4.5 showed the base current versus base–emitter voltage characteristic, with small time-varying signals superimposed at the Q-point. Since the sinusoidal signals are small, we can treat the slope at the Q-point as a constant, which has units of conductance. The inverse of this conductance is the small-signal resistance defined as r_π. We can then relate the small-signal input base current to the small-signal input voltage by

Figure 4.7 The BJT as a small-signal, two-port network

$$v_{be} = i_b r_\pi \tag{4.11}$$

where $1/r_\pi$ is equal to the slope of the i_B–v_{BE} curve, as shown in Figure 4.5. We then find r_π from

$$\frac{1}{r_\pi} = \left.\frac{\partial i_B}{\partial v_{BE}}\right|_{Q\text{-pt}} = \left.\frac{\partial}{\partial v_{BE}}\left[\frac{I_S}{1+\beta_F}\cdot\exp\left(\frac{v_{BE}}{V_T}\right)\right]\right|_{Q\text{-pt}} \tag{4.12(a)}$$

or

$$\frac{1}{r_\pi} = \frac{1}{V_T}\cdot\left.\left[\frac{I_S}{1+\beta_F}\cdot\exp\left(\frac{v_{BE}}{V_T}\right)\right]\right|_{Q\text{-pt}} = \frac{I_{BQ}}{V_T} \tag{4.12(b)}$$

Then

$$\frac{v_{be}}{i_b} = r_\pi = \frac{V_T}{I_{BQ}} = \frac{\beta_F V_T}{I_{CQ}} \tag{4.13}$$

The resistance r_π is called the **diffusion resistance** or base–emitter input resistance. We note that r_π is a function of the Q-point parameters.

We can consider the output terminal characteristics of the bipolar transistor. If we initially consider the case in which the output collector current is independent of the collector–emitter voltage, then the collector current is a function only of the base–emitter voltage, as discussed in Chapter 3. We can then write

$$\Delta i_C = \left. \frac{\partial i_C}{\partial v_{BE}} \right|_{Q\text{-}pt} \cdot \Delta v_{BE} \qquad\qquad (4.14(a))$$

or

$$i_c = \left. \frac{\partial i_C}{\partial v_{BE}} \right|_{Q\text{-}pt} \cdot v_{be} \qquad\qquad (4.14(b))$$

From Chapter 3, we had written

$$i_C = \alpha I_S \exp\left(\frac{v_{BE}}{V_T}\right) \qquad\qquad (4.15)$$

Then

$$\left. \frac{\partial i_C}{\partial v_{BE}} \right|_{Q\text{-}pt} = \frac{1}{V_T} \cdot \alpha I_S \exp\left(\frac{v_{BE}}{V_T}\right)\Bigg|_{Q\text{-}pt} = \frac{I_{CQ}}{V_T} \qquad\qquad (4.16)$$

The term $\alpha I_S \exp(v_{BE}/V_T)$ evaluated at the Q-point is just the quiescent collector current. The term I_{CQ}/V_T is a conductance. Since this conductance relates a current in the collector to a voltage in the B–E circuit, the parameter is called a **transconductance** and is written

$$g_m = \frac{I_{CQ}}{V_T} \qquad\qquad (4.17)$$

The small-signal transconductance is also a function of the Q-point parameters and is directly proportional to the dc bias current.

Using these new parameters, we can develop a simplified small-signal hybrid-π equivalent circuit for the npn bipolar transistor, as shown in Figure 4.8. The phasor components are given in parentheses. This circuit can be inserted into the ac equivalent circuit previously shown in Figure 4.6.

We can develop a slightly different form for the output of the equivalent circuit. We can relate the small-signal collector current to the small-signal base current. We can write

$$\Delta i_C = \left. \frac{\partial i_C}{\partial i_B} \right|_{Q\text{-}pt} \cdot \Delta i_B \qquad\qquad (4.18(a))$$

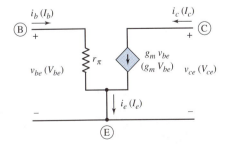

Figure 4.8 A simplified small-signal hybrid-π equivalent circuit for the npn transistor

or

$$i_c = \left.\frac{\partial i_C}{\partial i_B}\right|_{Q\text{-}pt} \cdot i_b \tag{4.18(b)}$$

where

$$\left.\frac{\partial i_C}{\partial i_B}\right|_{Q\text{-}pt} \equiv \beta \tag{4.18(c)}$$

and is called an incremental or ac common-emitter current gain. We can then write

$$i_c = \beta i_b \tag{4.19}$$

The small-signal equivalent circuit of the bipolar transistor in Figure 4.9 uses this parameter. The parameters in this figure are also given as phasors. This circuit can also be inserted in the ac equivalent circuit given in Figure 4.6. Either equivalent circuit, Figure 4.8 or 4.9, may be used. We will use both circuits in the examples that follow in this chapter.

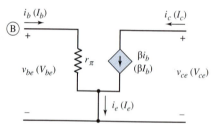

Figure 4.9   BJT small-signal equivalent circuit using common-emitter current gain

Relation between β_F and β

At this point, we need to pause and discuss the relationship between β_F and β. The difference between the two terms is illustrated in Figure 4.10. The term β_F is the ratio of dc collector current to dc base current. These currents include any leakage currents that might exist. As just mentioned, the term β is the ratio of the incremental change in collector current to the incremental change in base current. In an ideal BJT, these terms are identical.

In the derivation of r_π and g_m, the ideal exponential relation between current and base–emitter voltage was assumed. This implies that leakage currents are negligible. If we multiply r_π and g_m, we find

$$r_\pi g_m = \left(\frac{\beta_F V_T}{I_{CQ}}\right) \cdot \left(\frac{I_{CQ}}{V_T}\right) = \beta_F \tag{4.20}$$

Since leakage currents were neglected, and assuming β_F is independent of collector current, the β_F term is actually equivalent to the ac β. In general, we will assume that β and β_F are equivalent throughout the remainder of the text. However, we must keep in mind that β may vary from one device to another and that β does vary with collector current. This variation with I_C will be specified on data sheets for specific transistors.

Figure 4.10 Transistor characteristics showing definitions of β_F and β

Continuing our discussion of equivalent circuits, we may now insert the bipolar equivalent circuit in Figure 4.8, for example, into the ac equivalent circuit in Figure 4.6. The result is shown in Figure 4.11. Note that we are using the phasor notation.

Figure 4.11 The small-signal equivalent circuit of the common-emitter circuit using the npn transistor hybrid-π model

The **small-signal voltage gain**, $A_v = V_o/V_s$, of the circuit is defined as the ratio of output signal voltage to input signal voltage. The dependent current $g_m V_{be}$ flows through R_C, producing a negative collector–emitter voltage, or

$$V_o = V_{ce} = -(g_m V_{be})R_C \tag{4.21}$$

and, from the input portion of the circuit, we find

$$V_{be} = \left(\frac{r_\pi}{r_\pi + R_B}\right) \cdot V_s \tag{4.22}$$

The small-signal voltage gain is then

$$A_v = \frac{V_o}{V_s} = -(g_m R_C) \cdot \left(\frac{r_\pi}{r_\pi + R_B}\right) \tag{4.23}$$

Example 4.1 **Objective:** Calculate the small-signal voltage gain of the bipolar transistor circuit shown in Figure 4.3.

Assume the transistor and circuit parameters are: $\beta = 100$, $V_{CC} = 12\,\text{V}$, $V_{BE} = 0.7\,\text{V}$, $R_C = 6\,\text{k}\Omega$, $R_B = 50\,\text{k}\Omega$, and $V_{BB} = 1.2\,\text{V}$.

DC Solution: We first do the dc analysis to find the Q-point values. We obtain

$$I_{BQ} = \frac{V_{BB} - V_{BE}(\text{on})}{R_B} = \frac{1.2 - 0.7}{50} \Rightarrow 10\,\mu\text{A}$$

so that

$$I_{CQ} = \beta I_{BQ} = (100)(10\,\mu\text{A}) \Rightarrow 1\,\text{mA}$$

Then,

$$V_{CEQ} = V_{CC} - I_{CQ}R_C = 12 - (1)(6) = 6\,\text{V}$$

Therefore, the transistor is biased in the forward-active mode.

AC Solution: The small-signal hybrid-π parameters are

$$r_\pi = \frac{\beta V_T}{I_{CQ}} = \frac{(100)(0.026)}{1} = 2.6\,\text{k}\Omega$$

and

$$g_m = \frac{I_{CQ}}{V_T} = \frac{1}{0.026} = 38.5\,\text{mA/V}$$

The small-signal voltage gain is determined using the small-signal equivalent circuit shown in Figure 4.11. From Equation (4.23), we find

$$A_v = \frac{V_o}{V_s} = -(g_m R_C) \cdot \left(\frac{r_\pi}{r_\pi + R_B} \right)$$

or

$$= -(38.5)(6)\left(\frac{2.6}{2.6 + 50} \right) = -11.4$$

Comment: We see that the magnitude of the sinusoidal output voltage is 11.4 times the magnitude of the sinusoidal input voltage. We will see that other circuit configurations result in even larger small-signal voltage gains.

Discussion: We may consider a specific sinusoidal input voltage. Let

$$v_s = 0.25 \sin \omega t\,\text{V}$$

The sinusoidal base current is given by

$$i_b = \frac{v_s}{R_B + r_\pi} = \frac{0.25 \sin \omega t}{50 + 2.6} \rightarrow 4.75 \sin \omega t\,\mu\text{A}$$

The sinusoidal collector current is

$$i_c = \beta i_b = (100)(4.75 \sin \omega t) \rightarrow 0.475 \sin \omega t\,\text{mA}$$

and the sinusoidal collector-emitter voltage is

$$v_{ce} = -i_c R_C = -(0.475)(6) \sin \omega t = -2.85 \sin \omega t\,\text{V}$$

Figure 4.12 shows the various currents and voltages in the circuit. These include the sinusoidal signals superimposed on the dc values. Figure 4.12(a) shows the sinusoidal input voltage, and Figure 4.12(b) shows the sinusoidal base current superimposed on the quiescent value. The sinusoidal collector current superimposed on the dc quiescent value is shown in Figure 4.12(c). Note that, as the base current increases, the collector current increases.

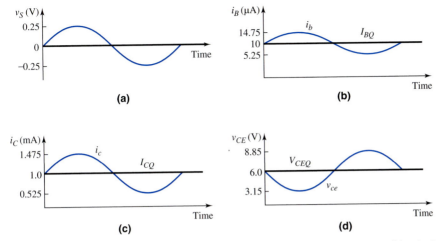

Figure 4.12 The common-emitter circuit: (a) input voltage, (b) input base current, (c) output collector current, and (d) output collector–emitter voltage

Figure 4.12(d) shows the sinuosidal component of the C–E voltage superimposed on the quiescent value. As the collector current increases, the voltage drop across R_C increases so that the C–E voltage decreases. Consequently, the sinusoidal component of the output voltage is 180 degrees out of phase with respect to the input signal voltage. The minus sign in the voltage gain expression represents this 180-degree **phase shift**.

Analysis Method: To summarize, the analysis of a BJT amplifier proceeds as shown in the box "Problem Solving Method: Bipolar AC Analysis."

Problem-Solving Technique: Bipolar AC Analysis

Since we are dealing with linear amplifier circuits, superposition applies, which means that we can perform the dc and ac analyses separately. The analysis of the BJT amplifier proceeds as follows:

1. Analyze the circuit with only the dc sources present. This solution is the dc or quiescent solution, which uses the dc signal models for the elements, as listed in Table 4.2. The transistor must be biased in the forward-active region in order to produce a linear amplifier.

2. Replace each element in the circuit with its small-signal model, as shown in Table 4.2. The small-signal hybrid-π model applies to the transistor although it is not specifically listed in the table.

3. Analyze the small-signal equivalent circuit, setting the dc source components equal to zero, to produce the response of the circuit to the time-varying input signals only.

In Table 4.2, the dc model of the resistor is a resistor, the capacitor model is an open circuit, and the inductor model is a short circuit. The forward-biased diode model includes the cut-in voltage V_γ and the forward resistance r_f.

Table 4.2 Transformation of elements in dc and small-signal analysis

Element	I–V relationship	DC model	AC model
Resistor	$I_R = \dfrac{V}{R}$	R	R
Capacitor	$I_C = sCV$	Open —o o—	C
Inductor	$I_L = \dfrac{V}{sL}$	Short —o—o—	L
Diode	$I_D = I_S(e^{v_D/V_T} - 1)$	$+V_\gamma - r_f$	$r_d = V_T/I_D$ —ww—
Independent voltage source	$V_S = $ constant	$+ V_S -$	Short —o o—
Independent current source	$I_S = $ constant	I_S	Open —o o—

Table suggested by Richard Hester of Iowa State University.

The small-signal models of R, L, and C remain the same. However, if the signal frequency is sufficiently high, the impedance of a capacitor can be approximated by a short circuit. The small-signal, low-frequency model of the diode becomes the diode diffusion resistance r_d. Also, the independent dc voltage source becomes a short circuit, and the independent dc current source becomes an open circuit.

4.2.3 Hybrid-π Equivalent Circuit, Including the Early Effect

So far in the small-signal equivalent circuit, we have assumed that the collector current is independent of the collector–emitter voltage. We discussed the Early effect in the last chapter in which the collector current does vary with collector–emitter voltage. Equation (3.16) in the previous chapter gives the relation

$$i_C = \alpha I_S\left[\exp\left(\frac{v_{BE}}{V_T}\right)\right]\cdot\left(1 + \frac{v_{CE}}{V_A}\right) \tag{3.16}$$

where V_A is the Early voltage. The equivalent circuits in Figures 4.8 and 4.9 can be expanded to take into account the Early voltage.

The output resistance r_o is defined as

$$r_o = \left.\frac{\partial v_{CE}}{\partial i_C}\right|_{Q\text{-}pt} \tag{4.24}$$

Using Equations (3.16) and (4.24), we can write

$$\frac{1}{r_o} = \left.\frac{\partial i_C}{\partial v_{CE}}\right|_{Q\text{-}pt} = \left.\frac{\partial}{\partial v_{CE}}\left\{\alpha I_S\left[\exp\left(\frac{v_{BE}}{V_T}\right)\left(1 + \frac{v_{CE}}{V_A}\right)\right]\right\}\right|_{Q\text{-}pt} \tag{4.25(a)}$$

or

$$\frac{1}{r_o} = \left.\alpha I_S\left[\exp\left(\frac{v_{BE}}{V_T}\right)\right]\cdot\frac{1}{V_A}\right|_{Q\text{-}pt} \cong \frac{I_{CQ}}{V_A} \tag{4.25(b)}$$

Then

$$r_o = \frac{V_A}{I_{CQ}}$$ (4.26)

and is called the **small-signal transistor output resistance**.

This resistance can be thought of as an equivalent Norton resistance, which means that r_o is in parallel with the dependent current sources. Figure 4.13(a) and (b) show the modified bipolar equivalent circuits including the output resistance r_o.

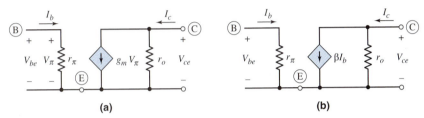

Figure 4.13 Expanded small-signal model of the BJT, including the Early effect, for the case when the circuit contains the (a) transconductance and (b) the current gain parameters

---

Example 4.2 **Objective:** Determine the small-signal voltage gain, including the effect of the transistor output resistance r_o.

Reconsider the circuit shown in Figure 4.1, with the parameters given in Example 4.1. In addition, assume the Early voltage is $V_A = 50\,\text{V}$.

Solution: The small-signal output resistance r_o is determined to be

$$r_o = \frac{V_A}{I_{CQ}} = \frac{50}{1\,\text{mA}} = 50\,\text{k}\Omega$$

Using the small-signal eqivalent circuit in Figure 4.11, we see that the output resistance r_o is in parallel with R_C. The small-signal voltage gain is therefore

$$A_v = \frac{V_o}{V_s} = -g_m(R_C \| r_o)\left(\frac{r_\pi}{r_\pi + R_B}\right)$$

$$= -(38.5)(6\|50)\left(\frac{2.6}{2.6 + 50}\right) = -10.2$$

Comment: Comparing this result to that of Example 4.1, we see that r_o reduces the magnitude of the small-signal voltage gain. In many cases, the magnitude of r_o is much larger than that of R_C, which means that the effect of r_o is negligible.

Test Your Understanding

4.1 A BJT with $\beta = 120$ and $V_A = 150\,\text{V}$ is biased such that $I_{CQ} = 0.25\,\text{mA}$. Determine g_m, r_π, and r_o. (Ans. $g_m = 9.62\,\text{mA/V}$, $r_\pi = 12.5\,\text{k}\Omega$, $r_o = 600\,\text{k}\Omega$)

4.2   The Early voltage of a BJT is $V_A = 75\,V$. Determine the minimum required collector current such that the output resistance is at least $r_o = 200\,k\Omega$. (Ans. $I_{CQ} = 0.375\,mA$)

The hybrid-π model derives its name, in part, from the hybrid nature of the parameter units. The four parameters of the equivalent circuits shown in Figures 4.13(a) and 4.13(b) are: input resistance r_π (ohms), current gain β (dimensionless), output resistance r_o (ohms), and transconductance g_m (mhos).

Up to this point, we have considered only circuits with npn bipolar transistors. However, the same basic analysis and equivalent circuit also applies to the pnp transistor. Figure 4.14(a) shows a circuit containing a pnp transistor. Here again, we see the change of current directions and voltage polarities compared to the circuit containing the npn transistor. Figure 4.14(b) is the ac equivalent circuit, with the dc voltage sources replaced by an ac short circuit, and all current and voltages shown are only the sinusoidal components.

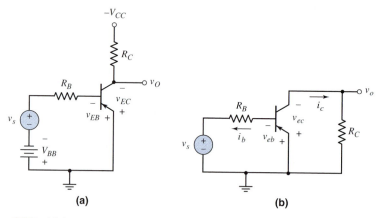

(a)   **(b)**

Figure 4.14   (a) A common-emitter circuit with a pnp transistor and (b) the corresponding ac equivalent circuit

The transistor in Figure 4.14(b) can now be replaced by either of the hybrid-π equivalent circuits shown in Figure 4.15. The hybrid-π equivalent circuit of the pnp transistor is the same as that of the npn device, except that again all current directions and voltage polarities are reversed. The hybrid-π parameters are determined by using exactly the same equations as

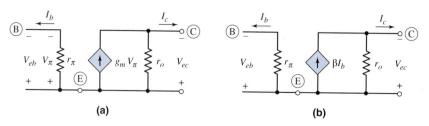

(a)   **(b)**

Figure 4.15   The hybrid-π model of the pnp transistor with (a) the transconductance parameter and (b) the current gain parameter for the npn device; that is, Equation (4.13) for r_π, Equation (4.17) for g_m, and Equation (4.26) for r_o.

We can note that, in the small-signal equivalent circuits in Figure 4.15, if we define currents of opposite direction and voltages of opposite polarity, the equivalent circuit model is exactly the same as that of the npn bipolar transistor. However, the author prefers to use the models shown in Figure 4.15 because the current directions and voltage polarities are consistent with the pnp device.

Combining the hybrid-π model of the pnp transistor (Figure 4.15(a)) with the ac equivalent circuit (Figure 4.14(b)), we obtain the small-signal equivalent circuit shown in Figure 4.16. The output voltage is given by

$$V_o = (g_m V_\pi)(r_o \| R_C) \tag{4.27}$$

The control voltage V_π can be expressed in terms of the input signal voltage V_s using a voltage divider equation. Taking into account the polarity, we find

$$V_\pi = -\frac{V_s r_\pi}{R_B + r_\pi} \tag{4.28}$$

Combining Equations (4.27) and (4.28), we obtain the small-signal voltage gain:

$$A_v = \frac{V_o}{V_s} = \frac{-g_m r_\pi}{R_B + r_\pi}(r_o \| R_C) = \frac{-\beta}{R_B + r_\pi}(r_o \| R_C) \tag{4.29}$$

The expression for the small-signal voltage gain of the circuit containing a pnp transistor is exactly the same as that for the npn transistor circuit. Taking into account the reversed current directions and voltage polarities, the voltage gain still contains a negative sign indicating a 180-degree phase shift between the input and output signals.

Figure 4.16 The small-signal equivalent circuit of the common-emitter circuit with a pnp transistor

Test Your Understanding

4.3 For the circuit in Figure 4.3 let $\beta = 150$, $V_A = 200$ V, $V_{CC} = 7.5$ V, $V_{BE}(\text{on}) = 0.7$ V, $R_C = 15$ kΩ, $R_B = 100$ kΩ, and $V_{BB} = 0.92$ V. (a) Determine the small-signal hybrid-π parameters r_π, g_m, and r_o. (b) Find the small-signal voltage gain $A_v = V_o/V_s$. (Ans. (a) $g_m = 12.7$ mA/V, $r_\pi = 11.8$ kΩ, $r_o = 606$ kΩ (b) $A_v = -19.6$)

4.4 For the circuit in Figure 4.14(a), let $\beta = 90$, $V_A = 120$ V, $V_{CC} = 5$ V, $V_{EB}(\text{on}) = 0.7$ V, $R_C = 2.5$ kΩ, $R_B = 50$ kΩ, and $V_{BB} = 1.145$ V. (a) Determine the small-signal hybrid-π parameters r_π, g_m, and r_o. (b) Find the small-signal voltage gain $A_v = V_o/V_s$. (Ans. (a) $g_m = 30.8$ mA/V, $r_\pi = 2.92$ kΩ, $r_o = 150$ kΩ (b) $A_v = -4.17$)

4.2.4   Expanded Hybrid-π Equivalent Circuit

Figure 4.17 shows an expanded hybrid-π equivalent circuit, which includes two additional resistances, r_b and r_μ.

The parameter r_b is the **series resistance** of the semiconductor material between the external base terminal B and an idealized internal base region B$'$. Typically, r_b is a few tens of ohms and is usually much smaller than r_π; therefore, r_b is normally negligible at low frequencies. However, at high frequencies, r_b may not be negligible, since the input impedance becomes capacitive, as we will see in Chapter 7.

Figure 4.17   Expanded hybrid-π equivalent circuit

The parameter r_μ is the **reverse-biased diffusion resistance** of the base–collector junction. This resistance is typically on the order of megohms and can normally be neglected (an open circuit). However, the resistance does provide some feedback between the output and input, meaning that the base current is a slight function of the collector–emitter voltage.

In this text, when we use the hybrid-π equivalent circuit model, we will neglect both r_b and r_μ, unless they are specifically included.

4.2.5   Other Small-Signal Parameters and Equivalent Circuits

Other small-signal parameters can be developed to model the bipolar transistor or other transistors described in the following chapters.

One common equivalent circuit model for bipolar transistor uses the **h-parameters**, which relate the small-signal terminal currents and voltages of a two-port network. These parameters are normally given in bipolar transistor data sheets, and are convenient to determine experimentally at low frequency.

Figure 4.18(a) shows the small-signal terminal current and voltage phasors for a common-emitter transistor. If we assume the transistor is biased at a Q-

(a)   (b)

Figure 4.18   (a) Common-emitter npn transistor and (b) the *h*-parameter model of the common-emitter bipolar transistor point in the forward-active region, the linear relationships between the small-signal terminal currents and voltages can be written as

$$V_{be} = h_{ie}I_b + h_{re}V_{ce} \tag{4.30(a)}$$

$$I_c = h_{fe}I_b + h_{oe}V_{ce} \tag{4.30(b)}$$

These are the defining equations of the common–emitter h-parameters, where the subscripts are: i for input, r for reverse, f for forward, o for output, and e for common emitter.

These equations can be used to generate the small-signal h-parameter equivalent circuit, as shown in Figure 4.18(b). Equation (4.30(a)) represents a Kirchhoff voltage law equation at the input, and the resistance h_{ie} is in series with a dependent voltage source equal to $h_{re}V_{ce}$. Equation (4.30(b)) represents a Kirchhoff current law equation at the output, and the conductance h_{oe} is in parallel with a dependent current source equal to $h_{fe}I_b$.

Since both the hybrid-π and h-parameters can be used to model the characteristics of the same transistor, these parameters are not independent. We can relate the hybrid-π and h-parameters using the equivalent circuit shown in Figure 4.17. The **small-signal input resistance** h_{ie}, from Equation (4.30(a)), can be written as

$$h_{ie} = \left.\frac{V_{be}}{I_b}\right|_{V_{ce}=0} \tag{4.31}$$

where the small-signal C–E voltage is held zero. With the C–E voltage equal to zero, the circuit in Figure 4.17 is transformed to the one shown in Figure 4.19. From this figure, we see that

$$h_{ie} = r_b + r_\pi \| r_\mu \tag{4.32}$$

In the limit of a very small r_b and a very large r_μ, $h_{ie} \cong r_\pi$.

The parameter h_{fe} is the **small-signal current gain**. From Equation (4.30(b)), this parameter can be written as

$$h_{fe} = \left.\frac{I_c}{I_b}\right|_{V_{ce}=0} \tag{4.33}$$

Since the collector–emitter voltage is again zero, we can use Figure 4.19, for which the short-circuit collector current is

$$I_c = g_m V_\pi \tag{4.34}$$

If we again consider the limit of a very small r_b and a very large r_μ, then

$$V_\pi = I_b r_\pi$$

Figure 4.19 Expanded hybrid-π equivalent circuit with the output short-circuited

and

$$h_{fe} = \left.\frac{I_c}{I_b}\right|_{V_{ce}=0} = g_m r_\pi = \beta \tag{4.35}$$

Consequently, at low frequency, the small-signal current gain h_{fe} is essentially equal to β in most situations.

The parameter h_{re} is called the **voltage feedback ratio**, which, from Equation (4.30(a)), can be written as

$$h_{re} = \left.\frac{V_{be}}{V_{ce}}\right|_{I_b=0} \tag{4.36}$$

Since the input signal base current is zero, the circuit in Figure 4.17 is transformed to that shown in Figure 4.20, from which we can see that

$$V_{be} = V_\pi = \left(\frac{r_\pi}{r_\pi + r_\mu}\right) \cdot V_{ce} \tag{4.37(a)}$$

and

$$h_{re} = \left.\frac{V_{be}}{V_{ce}}\right|_{I_b=0} = \frac{r_\pi}{r_\pi + r_\mu} \tag{4.37(b)}$$

Since $r_\pi \ll r_\mu$, this can be approximated as

$$h_{re} \cong \frac{r_\pi}{r_\mu} \tag{4.38}$$

Since r_π is normally in the kilohm range and r_μ is in the megohm range, the value of h_{re} is very small and can usually be neglected.

The fourth h-parameter is the **small-signal output admittance** h_{oe}. From Equation (4.30(b)), we can write

$$h_{oe} = \left.\frac{I_c}{V_{ce}}\right|_{I_b=0} \tag{4.39}$$

Since the input signal base current is again set equal to zero, the circuit in Figure 4.20 is applicable, and a Kirchhoff current law equation at the output node produces

$$I_c = g_m V_\pi + \frac{V_{ce}}{r_o} + \frac{V_{ce}}{r_\pi + r_\mu} \tag{4.40}$$

where V_π is given by Equation (4.37(a)). For $r_\pi \ll r_\mu$, Equation (4.40) becomes

$$h_{oe} = \left.\frac{I_c}{V_{ce}}\right|_{I_b=0} = \frac{1+\beta}{r_\mu} + \frac{1}{r_o} \tag{4.41}$$

In the ideal case, r_μ is infinite, which means that $h_{oe} = 1/r_o$.

Figure 4.20   Expanded hybrid-π equivalent circuit with the input open-circuited

The *h*-parameters for a pnp transistor are defined in the same way as those for an npn device. Also, the small-signal equivalent circuit for a pnp transistor using *h*-parameters is identical to that of an npn device, except that the current directions and voltage polarities are reversed.

Example 4.3   **Objective:** Determine the *h*-parameters of a specific transistor.

The 2N2222A transistor is a commonly used npn transistor. Data for this transistor are shown in Figure 4.21. Assume the transistor is biased at $I_C = 1\,\text{mA}$ and let $T = 300\,^\circ\text{K}$.

Solution: In Figure 4.21, we see that the small-signal current gain h_{fe} is generally in the range $100 < h_{fe} < 170$ for $I_C = 1\,\text{mA}$, and the corresponding value of h_{ie} is generally

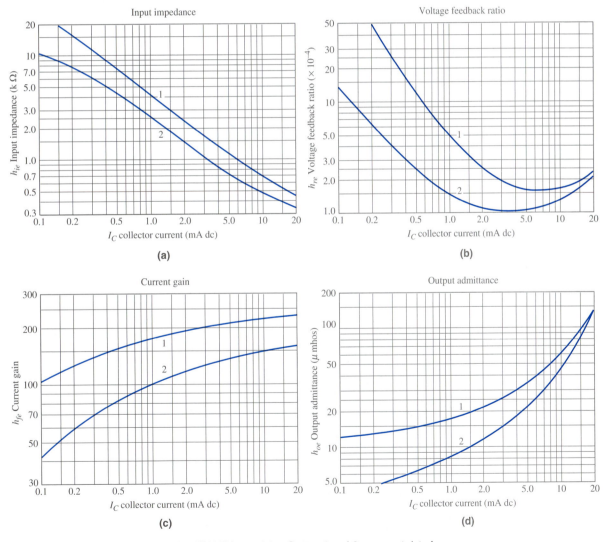

Figure 4.21   *h*-parameter data for the 2N2222A transistor. Curves 1 and 2 represent data from high-gain and low-gain transistors, respectively

between 2.5 and 5 kΩ. The voltage feedback ratio h_{re} varies between 1.5×10^{-4} and 5×10^{-4}, and the output admittance h_{oe} is in the range $8 < h_{oe} < 18 \, \mu$mhos.

Comment: The purpose of this example is to show that the parameters of a given transistor type can vary widely. In particular, the current gain parameter can easily vary by a factor of two. These variations are due to tolerances in the initial semiconductor properties and in the production process variables.

Design Pointer: This example clearly shows that there can be a wide variation in transistor parameters. Normally, a circuit is designed using nominal parameter values, but the allowable variations must be taken into account. In Chapter 3, we noted how a variation in β affects the Q-point. In this Chapter, we will see how the variations in small-signal parameters affect the small-signal voltage gain and other characteristics of a linear amplifier.

In the previous discussion, we indicated that the h-parameters h_{ie} and $1/h_{oe}$ are essentially equivalent to the hybrid-π parameters r_π and r_o, respectively, and that h_{fe} is essentially equal to β. The transistor circuit response is independent of the transistor model used. This reinforces the concept of a relationship between hybrid-π parameters and h-parameters. In fact, this is true for any set of small-signal parameters; that is, any given set of small-signal parameters is related to any other set of parameters.

Data Sheet

In the previous example, we showed some data for the 2N2222 discrete transistor. Figure 4.22 shows additional data from the data sheet for this transistor. Data sheets contain a lot of information, but we can begin to discuss some of the data at this time.

The first set of parameters pertains to the transistor in cutoff. The first two parameters listed are $V_{(BR)CEO}$ and $V_{(BR)CBO}$, which are the collector–emitter breakdown voltage with the base terminal open and the collector–base breakdown voltage with the emitter open. These parameters were discussed in Section 3.1.6 in the last chapter. In that section, we argued that $V_{(BR)CBO}$ was larger than $V_{(BR)CEO}$, which is supported by the data shown. These two voltages are measured at a specific current in the breakdown region. The third parameter, $V_{(BR)EBO}$, is the emitter–base breakdown voltage, which is substantially less than the collector–base or collector–emitter breakdown voltages.

The current I_{CBO} is the reverse-biased collector–base junction current with the emitter open ($I_E = 0$). This parameter was also discussed in Section 3.1.6. In the data sheet, this current is measured at two values of collector–base voltage and at two temperatures. The reverse-biased current increases with increasing temperature, as we would expect. The current I_{EBO} is the reverse-biased emitter–base junction current with the collector open ($I_C = 0$). This current is also measured at a specific reverse-bias voltage. The other two current parameters, I_{CEX} and I_{BL}, are the collector current and base current measured at given specific cutoff voltages.

The next set of parameters applies to the transistor when it is turned on. As was shown in Example 4.3, the data sheets give the h-parameters of the transistor. The first parameter, h_{FE}, is the dc common-emitter current gain and is measured over a wide range of collector current. We discussed, in Section 3.4.2, stabilizing the Q-point against variations in current gain. The data presented in the data sheet show that the current gain for a given transistor can vary significantly, so that stabilizing the Q-point is indeed an important issue.

We have used $V_{CE}(\text{sat})$ as one of the piecewise linear parameters when a transistor is driven into saturation and have always assumed a particular value in our analysis or design. This parameter, listed in the data sheet, is not a constant but varies with collector current. If the collector becomes relatively large, then the collector–emitter saturation voltage also becomes relatively large. The larger $V_{CE}(\text{sat})$ value would need to be taken into account in large-current situations. The base–emitter voltage for a transistor driven into saturation, $V_{BE}(\text{sat})$, is also given. Up to this point in the text, we have not been concerned with this parameter; however, the data sheet shows that the base–emitter voltage can increase significantly when a transistor is driven into saturation at high current levels.

The other parameters listed in the data sheet become more applicable later in the text when the frequency response of transistors is discussed. The intent of this short discussion is to show that we can begin to read through data sheets even though there are a lot of data presented.

4.3   BASIC TRANSISTOR AMPLIFIER CONFIGURATIONS

As we have seen, the bipolar transistor is a three-terminal device. Three basic single-transistor amplifier configurations can be formed, depending on which of the three transistor terminals is used as signal ground. These three basic configurations are appropriately called **common emitter**, **common collector (emitter follower)**, and **common base**. Which configuration or amplifier is used in a particular application depends to some extent on whether the input signal is a voltage or current and whether the desired output signal is a voltage or current. The characteristics of the three types of amplifiers will be determined to show the conditions under which each amplifier is most useful.

The input signal source can be modeled as either a Thevenin or Norton equivalent circuit. Figure 4.23(a) shows the Thevenin equivalent source that would represent a voltage signal, such as the output of a microphone. The voltage source v_s represents the voltage generated by the microphone. The resistance R_S is called the output resistance of the source and takes into account the change in output voltage as the source supplies current. Figure 4.23(b) shows the Norton equivalent source that would represent a current signal, such as the output of a photodiode. The current source i_s represents the current generated by the photodiode and the resistance R_S is the output resistance of this signal source.

Each of the three basic transistor amplifiers can be modeled as a two-port network in one of four configurations as shown in Table 4.3. We will determine the gain parameters, such as A_{vo}, A_{io}, G_{mo}, and R_{mo}, for each of the three transistor amplifiers. These parameters are important since they determine the amplification of the amplifier. However, we will see that the input and output resistances, R_i and R_o, are also important in the design of these amplifiers. Although one configuration shown in Table 4.3 may be preferable for a given application, any one of the four can be used to model a given amplifier. Since each configuration must produce the same terminal characteristics for a

2N2222
2N2222A

TO–18

PN2222
PN2222A

TO–92

MMBT2222
MMBT2222A

TO–236
(SOT–23)

MPQ2222

TO–116

NPN General Purpose Amplifier

Electrical Characteristics T_A = 25 °C unless otherwise noted

Symbol	Parameter		Min	Max	Units
OFF CHARACTERISTICS					
$V_{(BR)CEO}$	Collector-Emitter Breakdown Voltage (Note 1) (I_C = 10 mA, I_B = 0) 2222 2222A	2222 2222A	30 40		V
$V_{(BR)CBO}$	Collector-Base Breakdown Voltage (I_C = 10 μA, I_E = 0)	2222 2222A	60 75		V
$V_{(BR)EBO}$	Emitter Base Breakdown Voltage (I_E = 10 μA, I_C = 0)	2222 2222A	5.0 6.0		V
I_{CEX}	Collector Cutoff Current (V_{CE} = 60 V, V_{EB}(off) = 3.0 V)	2222A		10	nA
I_{CBO}	Collector Cutoff Current (V_{CB} = 50 V, I_E = 0) (V_{CB} = 60 V, I_E = 0) (V_{CB} = 50 V, I_E = 0, T_A = 150 °C) (V_{CB} = 60 V, I_E = 0, T_A = 150 °C)	2222 2222A 2222 2222A		0.01 0.01 10 10	μA
I_{EBO}	Emitter Cutoff Current (V_{EB} = 3.0 V, I_C = 0)	2222A		10	nA
I_{BL}	Base Cutoff Current (V_{CE} = 60 V, V_{EB}(off) = 3.0)	2222A		20	nA
ON CHARACTERISTICS					
h_{FE}	DC Current Gain (I_C = 0.1 mA, V_{CE} = 10 V) (I_C = 1.0 mA, V_{CE} = 10 V) (I_C = 10 mA, V_{CE} = 10 V) (I_C = 10 mA, V_{CE} = 10 V, T_A = –55 °C) (I_C = 150 mA, V_{CE} = 10 V) (Note 1) (I_C = 150 mA, V_{CE} = 1.0 V) (Note 1) (I_C = 500 mA, V_{CE} = 10 V) (Note 1)	2222 2222A	35 50 75 35 100 50 30 40	300	

Note 1: Pulse Test: Pulse Width ≤ 300 μs, Duty Cycle ≤ 2.0%.

2N2222/PN2222/MMBT2222/MPQ2222/2N2222A/PN2222A/MMBT2222A NPN General Purpose Amplifier

Figure 4.22   Basic data sheet for the 2N2222 bipolar transistor

2222/PN2222/MMBT2222/MPQ2222/2N2222/2N2222A/PN2222A/MMBT2222A NPN General Purpose Amplifier

NPN General Purpose Amplifier (Continued)

Electrical Characteristics $T_A = 25\,°C$ unless otherwise noted (Continued)

Symbol	Parameter		Min	Max	Units
ON CHARACTERISTICS (Continued)					
V_{CE} (sat)	Collector-Emitter Saturation Voltage (Note 1) ($I_C = 150$ mA, $I_B = 15$ mA) ($I_C = 500$ mA, $I_B = 50$ mA)	2222 2222A 2222 2222A		0.4 0.3 1.6 1.0	V
V_{BE} (sat)	Base-Emitter Saturation Voltage (Note 1) ($I_C = 150$ mA, $I_B = 15$ mA) ($I_C = 500$ mA, $I_B = 50$ mA)	2222 2222A 2222 2222A	0.6 0.6	1.3 1.2 2.6 2.0	V
SMALL-SIGNAL CHARACTERISTICS					
f_T	Current Gain—Bandwidth Product (Note 3) ($I_C = 20$ mA, $V_{CE} = 20$ V, $f = 100$ MHz)	2222 2222A	250 300		MHz
C_{obo}	Output Capacitance (Note 3) ($V_{CB} = 10$ V, $I_E = 0$, $f = 100$kHz)			8.0	pF
C_{ibo}	Input Capacitance (Note 3) ($V_{EB} = 0.5$ V, $I_C = 0$, $f = 100$ kHz)	2222 2222A		30 25	pF
$rb'C_C$	Collector Base Time Constant ($I_E = 20$ mA, $V_{CB} = 20$ V, $f = 31.8$ MHz)	2222A		150	ps
NF	Noise Figure ($I_C = 100\,\mu$A, $V_{CE} = 10$V, $R_S = 1.0$ kΩ, $f = 1.0$ kHz)	2222A		4.0	dB
$Re(h_{ie})$	Real Part of Common-Emitter High Frequency Input Impedance ($I_C = 20$ mA, $V_{CE} = 20$ V, $f = 300$ MHz)			60	Ω
SWITCHING CHARACTERISTICS					
t_D	Delay Time	($V_{CC} = 30$ V, V_{BE}(off) = 0.5 V, $I_C = 150$ mA, $I_{B1} = 15$ mA) except MPQ2222		10	ns
t_R	Rise Time			25	ns
t_S	Storage Time	($V_{CC} = 30$ V, $I_C = 150$ mA, $I_{B1} = I_{B2} = 15$ mA) except MPQ2222		225	ns
t_F	Fall Time			60	ns

Note 1: Pulse Test: Pulse Width < 300 μs, Duty Cycle ≤ 2.0%.
Note 2: For characteristics curves, see Process 19.
Note 3: f_T is defined as the frequency at which h_{fe} extrapolates to unity.

Figure 4.22 Continued

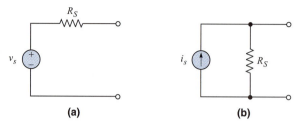

Figure 4.23 Input signal source modeled as (a) Thevenin equivalent circuit and (b) Norton equivalent circuit

Table 4.3 Four equivalent two-port networks

Type	Equivalent circuit	Gain property
Voltage amplifier	R_o, i_o, v_{in}, R_i, $A_{vo}v_{in}$, v_o	Output voltage proportional to input voltage
Current amplifier	i_{in}, i_o, R_i, $A_{is}i_{in}$, R_o, v_o	Output current proportional to input current
Transconductance amplifier	i_o, v_{in}, R_i, $G_{ms}v_{in}$, R_o, v_o	Output current proportional to input voltage
Transresistance amplifier	i_{in}, R_o, i_o, R_i, $R_{mo}i_{in}$, v_o	Output voltage proportional to input current

given amplifier, the various gain parameters are not independent, but are related to each other.

If we wish to design a voltage amplifier (preamp) so that the output voltage of a microphone, for example, is amplified, the total equivalent circuit may be that shown in Figure 4.24. The input voltage to the amplifier is given by

Figure 4.24   Equivalent preamplifier circuit

$$v_{\text{in}} = \frac{R_i}{R_i + R_S} \cdot v_s \qquad\qquad (4.42)$$

In general, we would like the input voltage v_{in} to the amplifier to be as nearly equal to the source voltage v_s as possible. This means, from Equation (4.42), that we need to design the amplifier such that the input resistance R_i is much larger than the signal source output resistance R_S. (The output resistance of an ideal voltage source is zero, but is not zero for most practical voltage sources.) To provide a particular voltage gain, the amplifier must have a gain parameter A_{vo} of a certain value. The output voltage supplied to the load (where the load may be a second power amplifier stage) is given by

$$v_o = \frac{R_L}{R_L + R_o} \cdot A_{vo} v_{\text{in}} \qquad\qquad (4.43)$$

Normally, we would like the output voltage to the load to be equal to the Thevenin equivalent voltage generated by the amplifier. This means that we need $R_o \ll R_L$ for the voltage amplifier. So again, for a voltage amplifier, the output resistance should be very small. The input and output resistances are significant in the design of an amplifier.

For a current amplifier, we would like to have $R_i \ll R_S$ and $R_o \gg R_L$. We will see as we proceed through the chapter that each of the three basic transistor amplifier configurations exhibits characteristics that are desirable for particular applications.

We should note that, in this chapter, we will be primarily using the two-port equivalent circuits shown in Table 4.3 to model single-transistor amplifiers. However, these equivalent circuits are also used to model multitransistor circuits. This will become apparent as we get into Part II of the text.

4.4   COMMON-EMITTER AMPLIFIERS

In this section, we consider the first of the three basic amplifiers—the **common-emitter** circuit. We will apply the equivalent circuit of the bipolar transistor that was previously developed. In general, we will use the hybrid-π model throughout the text.

4.4.1   Basic Common-Emitter Amplifier Circuit

Figure 4.25 shows the basic common-emitter circuit with voltage-divider biasing. We see that the emitter is at ground potential—hence the name common emitter. The signal from the signal source is coupled into the base of the transistor through the coupling capacitor C_C, which provides dc isolation between the amplifier and the signal source. The dc transistor biasing is established by R_1 and R_2, and is not disturbed when the signal source is capacitively coupled to the amplifier.

Figure 4.25   A common-emitter circuit with a voltage-divider biasing circuit and a coupling capacitor

If the signal source is a sinusoidal voltage at frequency f, then the magnitude of the capacitor impedance is $|Z_c| = [1/(2\pi f C_C)]$. For example, assume that $C_C = 10\,\mu\text{F}$ and $f = 2\,\text{kHz}$. The magnitude of the capacitor impedance is then

$$|Z_c| = \frac{1}{2\pi f C_C} = \frac{1}{2\pi(2 \times 10^3)(10 \times 10^{-6})} \cong 8\,\Omega \tag{4.44}$$

The magnitude of this impedance is much less than the Thevenin resistance at the capacitor terminals, which in this case is $R_1 \| R_2 \| r_\pi$. We can therefore assume that the capacitor is essentially a short circuit to signals with frequencies greater than 2 kHz. We are also neglecting any capacitance effects within the transistor. Using these results, our analyses in this chapter assume that the signal frequency is sufficiently high that any coupling capacitance acts as a perfect short circuit, and is also sufficiently low that the transistor capacitances can be neglected. Such frequencies are in the midfrequency range, or simply at the midband of the amplifier.

The small-signal equivalent circuit in which the coupling capacitor is assumed to be a short circuit is shown in Figure 4.26. The small-signal variables, such as the input signal voltage and input base current, are given in phasor form. The control voltage V_π is also given as a phasor.

Example   4.4   Objective: Determine the small-signal voltage gain of the circuit shown in Figure 4.25.

Assume the transistor parameters are: $\beta = 100$, $V_{BE}(\text{on}) = 0.7\,\text{V}$, and $V_A = 100\,\text{V}$.

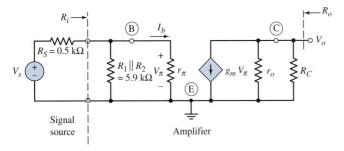

Figure 4.26 The small-signal equivalent circuit, assuming the coupling capacitor is a short circuit

DC Solution: We first do a dc analysis to find the Q-point values. We find that $I_{CQ} = 0.95\,\text{mA}$ and $V_{CEQ} = 6.31\,\text{V}$, which shows that the transistor is biased in the forward-active mode.

AC Solution: The small-signal hybrid-π parameters for the equivalent circuit are

$$r_\pi = \frac{V_T \beta}{I_{CQ}} = \frac{(0.026)(100)}{(0.95)} = 2.74\,\text{k}\Omega$$

$$g_m = \frac{I_{CQ}}{V_T} = \frac{0.95}{0.026} = 36.5\,\text{mA/V}$$

and

$$r_o = \frac{V_A}{I_{CQ}} = \frac{100}{0.95} = 105\,\text{k}\Omega$$

Assuming that C_C acts as a short circuit, Figure 4.26 shows the small-signal equivalent circuit. The small-signal output voltage is

$$V_o = -(g_m V_\pi)(r_o \| R_C)$$

The dependent current $g_m V_\pi$ flows through the parallel combination of r_o and R_C, but in a direction that produces a negative output voltage. We can relate the control voltage V_π to the input voltage V_s by a voltage divider. We have

$$V_\pi = \left(\frac{R_1 \| R_2 \| r_\pi}{R_1 \| R_2 \| r_\pi + R_S} \right) \cdot V_s$$

We can then write the small-signal voltage gain as

$$A_v = \frac{V_o}{V_s} = -g_m \left(\frac{R_1 \| R_2 \| r_\pi}{R_1 \| R_2 \| r_\pi + R_S} \right) (r_o \| R_C)$$

or

$$A_v = -(36.5)\left(\frac{5.9 \| 2.74}{5.9 \| 2.74 + 0.5} \right)(105 \| 6) = -163$$

We can also calculate R_i, which is the input resistance to the amplifier. From Figure 4.26, we see that

$$R_i = R_1 \| R_2 \| r_\pi = 5.9 \| 2.74 = 1.87\,\text{k}\Omega$$

The output resistance R_o is found by setting the independent source V_s equal to zero. In this case, there is no excitation to the input portion of the circuit so $V_\pi = 0$, which implies that $g_m V_\pi = 0$ (an open circuit). The output resistance looking back into the output terminals is then

$$R_o = r_o \| R_C = 105 \| 6 = 5.68\,\text{k}\Omega$$

Comment: In this circuit, the effective series resistance between the voltage source V_s and the base of the transistor is much less than that given in Example 4.1. For this reason, the magnitude of the voltage gain for the circuit given in Figure 4.25 is much larger than that found in Example 4.1.

Discussion: The two-port equivalent circuit along with the input signal source for the common-emitter amplifier analyzed in this example is shown in Figure 4.27. We can determine the effect of the source resistance R_S in conjunction with the amplifier input resistance R_i. Using a voltage-divider equation, we find the input voltage to the amplifier is

$$V_{in} = \left(\frac{R_i}{R_i + R_S}\right)V_s = \left(\frac{1.87}{1.87 + 0.5}\right)V_s = 0.789V_s$$

Because the input resistance to the amplifier is not very much greater than the signal source resistance, the actual input voltage to the amplifier is reduced to approximately 80 percent of the signal voltage. This is called a **loading effect**. The voltage V_{in} is a function of the amplifier connected to the source. In other amplifier designs, we will try to minimize the loading effect, or make $R_i \gg R_S$, which means that $V_{in} \cong V_s$.

Figure 4.27   Two-port equivalent circuit for the amplifier in Example 4.4

4.4.2   Circuit with Emitter Resistor

For the circuit in Figure 4.25, the bias resistors R_1 and R_2 in conjunction with V_{CC} produce a base current of $9.5 \,\mu\text{A}$ and a collector current of $0.95 \,\text{mA}$, when the B–E turn-on voltage is assumed to be $0.7 \,\text{V}$. If the transistor in the circuit is replaced by a new one with slightly different parameters so that the B–E turn-on voltage is $0.6 \,\text{V}$ instead of $0.7 \,\text{V}$, then the resulting base current is $26 \,\mu\text{A}$, which is sufficient to drive the transistor into saturation. Therefore, the circuit shown in Figure 4.25 is not practical. An improved dc biasing design includes an emitter resistor.

In the last chapter, we found that the Q-point was stabilized against variations in β if an emitter resistor were included in the circuit, as shown in Figure 4.28. We will find a similar property for the ac signals, in that the voltage gain of a circuit with R_E will be less dependent on the transistor current gain β. Even though the emitter of this circuit is not at ground potential, this circuit is still referred to as a common-emitter circuit.

Assuming that C_C acts as a short circuit, Figure 4.29 shows the small-signal hybrid-π equivalent circuit. In this case, we are using the equivalent circuit with the current gain parameter β, and we are assuming that the Early voltage is infinite so the transistor output resistance r_o can be neglected (an open circuit). The ac output voltage is

$$V_o = -(\beta I_b)R_C \tag{4.45}$$

Figure 4.28 A bipolar circuit with an emitter resistor, a voltage-divider biasing circuit, and a coupling capacitor

Figure 4.29 The small-signal equivalent circuit with an emitter resistor

To find the small-signal voltage gain, it is worthwhile finding the input resistance first. The resistance R_{ib} is the input resistance looking into the base of the transistor. We can write the following loop equation

$$V_{\text{in}} = I_b r_\pi + (I_b + \beta I_b)R_E \tag{4.46}$$

The input resistance R_{ib} is then defined as, and found to be,

$$R_{ib} = \frac{V_{\text{in}}}{I_b} = r_\pi + (1 + \beta)R_E \tag{4.47}$$

In the common-emitter configuration that includes an emitter resistance, the small-signal input resistance looking into the base of the transistor is r_π plus the emitter resistance multiplied by the factor $(1 + \beta)$. This effect is called the **resistance reflection rule**. We will use this result throughout the text without further derivation.

The input resistance to the amplifier is now

$$R_i = R_1 \| R_2 \| R_{ib} \tag{4.48}$$

We can again relate V_{in} to V_s through a voltage-divider equation as

$$V_{\text{in}} = \left(\frac{R_i}{R_i + R_S}\right) \cdot V_s \tag{4.49}$$

Combining Equations (4.45), (4.47), and (4.49), we find the small-signal voltage gain is

$$A_v = \frac{V_o}{V_s} = \frac{-(\beta I_b)R_C}{V_s} = -\beta R_C \left(\frac{V_{\text{in}}}{R_{ib}}\right) \cdot \left(\frac{1}{V_s}\right) \tag{4.50(a)}$$

or

$$A_v = \frac{-\beta R_C}{r_\pi + (1 + \beta)R_E}\left(\frac{R_i}{R_i + R_S}\right) \tag{4.50(b)}$$

From this equation, we see that if $R_i \gg R_S$ and if $(1 + \beta)R_E \gg r_\pi$, then the small-signal voltage gain is approximately

$$A_v \cong \frac{-\beta R_C}{(1 + \beta)R_E} \cong \frac{-R_C}{R_E} \tag{4.51}$$

Equations (4.50(b)) and (4.51) show that the voltage gain is less dependent on the current gain β than in the previous example, which means that there is a smaller change in voltage gain when the transistor current gain changes. The circuit designer now has more control in the design of the voltage gain, but this advantage is at the expense of a smaller gain.

In Chapter 3, we discussed the variation in the Q-point with variations or tolerances in resistor values. Since the voltage gain is a function of resistor values, it is also a function of the tolerances in those values. This must be considered in a circuit design.

Example 4.5   Objective: Determine the small-signal voltage gain of a common-emitter circuit with an emitter resistor.

For the circuit in Figure 4.28, the transistor parameters are: $\beta = 100$, $V_{BE}(\text{on}) = 0.7\,\text{V}$, and $V_A = \infty$.

DC Solution: From a dc analysis of the circuit, we can determine that $I_{CQ} = 2.16\,\text{mA}$ and $V_{CEQ} = 4.81\,\text{V}$, which shows that the transistor is biased in the forward-active mode.

AC Solution: The small-signal hybrid-π parameters are determined to be

$$r_\pi = \frac{V_T \beta}{I_{CQ}} = \frac{(0.026)(100)}{(2.16)} = 1.20\,\text{k}\Omega$$

$$g_m = \frac{I_{CQ}}{V_T} = \frac{2.16}{0.026} = 83.1\,\text{mA/V}$$

and

$$r_o = \frac{V_A}{I_{CQ}} = \infty$$

The input resistance to the base can be determined as

$$R_{ib} = r_\pi + (1 + \beta)R_E = 1.20 + (101)(0.4) = 41.6\,\text{k}\Omega$$

and the input resistance to the amplifier is now found to be

$$R_i = R_1 \| R_2 \| R_{ib} = 10 \| 41.6 = 8.06\,\text{k}\Omega$$

Using the exact expression for the voltage gain, we find

$$A_v = \frac{-(100)(2)}{1.20 + (101)(0.4)} \left(\frac{8.06}{8.06 + 0.5} \right) = -4.53$$

If we use the approximation given by Equation (4.51), we obtain

$$A_v = \frac{-R_C}{R_E} = \frac{-2}{0.4} = -5.0$$

Comment: The magnitude of the small-signal voltage gain is substantially reduced when an emitter resistor is included. Also, Equation (4.51) gives a good first approximation for the gain, which means that it can be used in the initial design of a common-emitter circuit with an emitter resistor.

Discussion: The amplifier gain is nearly independent of changes in the current gain parameter β. This fact is shown in the following calculations:

β	A_v
50	-4.41
100	-4.53
150	-4.57

In addition to gaining an advantage in stability by including an emitter resistance, we also gain an advantage in the loading effect. We see that, for $\beta = 100$, the input voltage to the amplifier is

$$V_{in} = \left(\frac{R_i}{R_i + R_S}\right) \cdot V_s = (0.942)V_s$$

We see that V_{in} is much closer in value to V_s than in the previous example. There is less loading effect because the input resistance to the base of the transistor is higher when an emitter resistor is included.

The same equivalent circuit as shown in Figure 4.27 applies to this example also. The difference in the two examples is the values of input resistance and gain parameter.

Test Your Understanding

4.5   For the circuit in Figure 4.30, let $R_E = 0.6\,\text{k}\Omega$, $R_C = 5.6\,\text{k}\Omega$, $\beta = 120$, $V_{BE}(\text{on}) = 0.7\,\text{V}$, $R_1 = 250\,\text{k}\Omega$, and $R_2 = 75\,\text{k}\Omega$. (a) For $V_A = \infty$, determine the small-signal voltage gain A_v. (b) Determine the input resistance looking into the base of the transistor. (Ans. (a) $A_v = -8.27$, (b) $R_{ib} = 80.1\,\text{k}\Omega$)

*D4.6   For the circuit shown in Figure 4.30, let $\beta = 100$, $V_{BE}(\text{on}) = 0.7\,\text{V}$, and $V_A = \infty$. Design a bias-stable circuit such that $I_{CQ} = 0.5\,\text{mA}$, $V_{CEQ} = 2.5\,\text{V}$, and $A_v = -8$. (Ans. To a good approximation: $R_C = 4.54\,\text{k}\Omega$, $R_E = 0.454\,\text{k}\Omega$, $R_1 = 24.1\,\text{k}\Omega$, and $R_2 = 5.67\,\text{k}\Omega$)

4.7   Assume a 2N2907A transistor is used in the circuit in Figure 4.31 and that the nominal dc transistor parameters are $\beta = 100$ and $V_{EB}(\text{on}) = 0.7\,\text{V}$. Determine the small-voltage gain, using the h-parameter model of the transistor. Find the minimum and maximum values of gain corresponding to the minimum and maximum h-parameter values. See Appendix C. For simplicity, assume $h_{re} = h_{oe} = 0$. (Ans. $A_v = -2.54$ for both cases)

Figure 4.30   Figure for Exercises 4.5 and 4.6

Figure 4.31   Figure for Exercise 4.7

4.4.3   Circuit with Emitter Bypass Capacitor

There may be times when the emitter resistor must be large for the purposes of dc design, but degrades the small-signal voltage gain too severely. We can use an emitter bypass capacitor to effectively short out a portion or all of the emitter resistance as seen by the ac signals. Consider the circuit shown in Figure 4.32 biased with both positive and negative voltages. Both emitter resistors R_{E1} and R_{E2} are factors in the dc design of the circuit, but only R_{E1} is part of the ac equivalent circuit, since C_E provides a short circuit to ground for the ac signals.

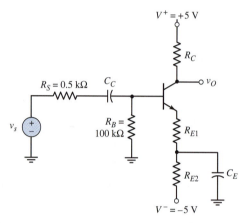

Figure 4.32   A bipolar circuit with an emitter resistor and an emitter bypass capacitor

Design Example 4.6   **Objective:**   An amplifier with the configuration in Figure 4.32 is to be designed such that a 12 mV sinusoidal signal from a microphone is amplified to a 0.4 V sinusoidal output signal. Standard resistor values are to be used in the final design.

Initial Design Approach: The magnitude of the voltage gain of the amplifier needs to be

$$|A_v| = \frac{0.4\,\text{V}}{12\,\text{mV}} = 33.3$$

From Equation (4.51), the approximate voltage gain of the amplifier is

$$|A_v| \cong \frac{R_C}{R_{E1}}$$

Noting from the last example that this value of gain produces an optimistically high value, we can set $R_C/R_{E1} = 40$ or $R_C = 40R_{E1}$.

The dc base-emitter loop equation is

$$5 = I_B R_B + V_{BE}(\text{on}) + I_E(R_{E1} + R_{E2})$$

Assuming $\beta = 100$ and $V_{BE}(\text{on}) = 0.7\,\text{V}$, we can design the circuit to produce a quiescent emitter current of, for example, 0.20 mA. We then have

$$5 = \frac{(0.20)}{(101)}(100) + 0.70 + (0.20)(R_{E1} + R_{E2})$$

which yields

$$R_{E1} + R_{E2} = 20.5 \, \text{k}\Omega$$

Assuming $I_E \cong I_C$ and designing the circuit such that $V_{CEQ} = 4 \, \text{V}$, the collector–emitter loop equation produces

$$5 + 5 = I_C R_C + V_{CEQ} + I_E(R_{E1} + R_{E2}) = (0.2)R_C + 4 + (0.2)(20.5)$$

or

$$R_C = 9.5 \, \text{k}\Omega$$

Then

$$R_{E1} = \frac{R_C}{40} = \frac{9.5}{40} = 0.238 \, \text{k}\Omega$$

and $R_{E2} = 20.3 \, \text{k}\Omega$.

From Appendix D, we can pick standard resistor values of $R_{E1} = 240 \, \Omega$, $R_{E2} = 20 \, \text{k}\Omega$, and $R_C = 10 \, \text{k}\Omega$.

Computer Simulation: Since we used approximation techniques in our design, we can use PSpice to give us a more accurate evaluation of the circuit for the standard resistor values that were chosen. Figure 4.33 shows the PSpice circuit schematic.

Using the standard resistor values and the 2N3904 transistor, the output signal voltage produced by a 12 mV input signal is 323 mV. A frequency of 2 kHz and capacitor values of 100 μF were used in the simulation. The magnitude of the output signal is slightly less than the desired value of 400 mV. The principal reason for the difference is that the r_π parameter of the transistor was neglected in the design. For a collector current of approximately $I_C = 0.2 \, \text{mA}$, r_π can be significant.

In order to increase the small-signal voltage gain, a smaller value of R_{E1} is necessary. For $R_{E1} = 160 \, \Omega$, the output signal voltage is 410 mV, which is very close to the desired value.

Design Pointer: Approximation techniques are extremely useful in an initial electronic circuit design. A computer simulation, such as PSpice, can then be used to verify the design. Slight changes in the design can then be made to meet the required specifications.

Figure 4.33   PSpice circuit schematic

Test Your Understanding

*D4.8   Design the circuit in Figure 4.34 such that it is bias stable and the small-signal voltage gain is $A_v = -8$. Let $I_{CQ} = 0.6\,\text{mA}$, $V_{ECQ} = 3.75\,\text{V}$, $\beta = 100$, $V_{EB}(\text{on}) = 0.7\,\text{V}$, and $V_A = \infty$. (Ans. To a good approximation: $R_C = 5.62\,\text{k}\Omega$, $R_E = 0.624\,\text{k}\Omega$, $R_1 = 7.40\,\text{k}\Omega$, and $R_2 = 42.4\,\text{k}\Omega$)

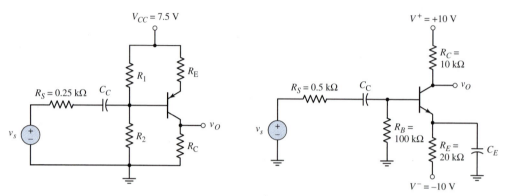

Figure 4.34   Figure for Exercise 4.8   **Figure 4.35**   Figure for Exercise 4.9

4.9   For the circuit in Figure 4.35, let $\beta = 100$, $V_{BE}(\text{on}) = 0.7\,\text{V}$, and $V_A = 100\,\text{V}$. (a) Determine the small-signal voltage gain. (b) Determine the input resistance seen by the signal source and the output resistance looking back into the output terminal. (Ans. (a) $A_v = -148$ (b) $R_{\text{in}} = 6.09\,\text{k}\Omega$, $R_o = 9.58\,\text{k}\Omega$)

4.10   For the circuit in Figure 4.28, the small-signal voltage gain is given approximately by $-R_C/R_E$. For the case of $R_C = 2\,\text{k}\Omega$, $R_E = 0.4\,\text{k}\Omega$, and $R_S = 0$, what must be the value of β such that the approximate value is within 5 percent of the actual value? (Ans. $\beta = 76$)

4.11   For the circuit in Figure 4.36, let $\beta = 125$, $V_{BE}(\text{on}) = 0.7\,\text{V}$, and $V_A = 200\,\text{V}$. (a) Determine the small-signal voltage gain A_v. (b) Determine the output resistance R_o. (Ans. (a) $A_v = -50.5$ (b) $R_o = 2.28\,\text{k}\Omega$)

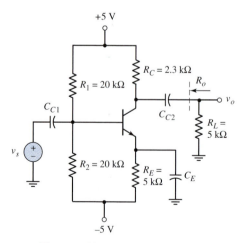

Figure 4.36   Figure for Exercise 4.11

4.4.4 Advanced Common-Emitter Amplifier Concepts

Our previous analysis of common-emitter circuits assumed constant load or collector resistances. The common-emitter circuit shown in Figure 4.37(a) is biased with a constant-current source and contains a nonlinear, rather than a constant, collector resistor. Assume the current–voltage characteristics of the nonlinear resistor are described by the curve in Figure 4.37(b). Neglecting base currents, assume that the collector resistor is biased at I_Q and V_{RQ}. At the Q-point of the resistor, assume the incremental resistance $\Delta v_R / \Delta i_C$ is r_c.

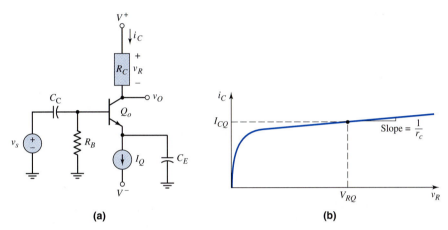

(a)   **(b)**

Figure 4.37 (a) A common-emitter circuit with current source biasing and a nonlinear load resistor and (b) current–voltage characteristics of the nonlinear load resistor

The small-signal equivalent circuit of the common-emitter amplifier circuit in Figure 4.37(a) is shown in Figure 4.38. The collector resistor R_C is replaced by the small-signal equivalent resistance r_c that exists at the Q-point. The small-signal voltage gain is then, assuming an ideal voltage signal source,

$$A_v = \frac{V_o}{V_s} = -g_m(r_o \| r_c) \tag{4.52}$$

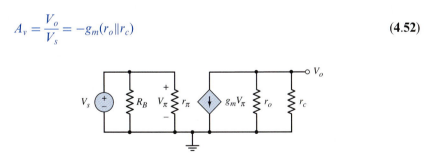

Figure 4.38 Small-signal equivalent circuit of the circuit in Figure 4.37(a)

Example 4.7   **Objective:** Determine the small-signal voltage gain of a common-emitter circuit with a nonlinear load resistance.

Assume the circuit shown in Figure 4.37(a) is biased at $I_Q = 0.5\,\text{mA}$, and the transistor parameters are $\beta = 120$ and $V_A = 80\,\text{V}$. Also assume that nonlinear small-signal collector resistance is $r_c = 120\,\text{k}\Omega$.

Solution: For a transistor current gain of $\beta = 120$, $I_{CQ} \cong I_{EQ} = I_Q$, and the small-signal hybrid-π parameters are

$$g_m = \frac{I_{CQ}}{V_T} = \frac{0.5}{0.026} = 19.2 \, \text{mA/V}$$

and

$$r_o = \frac{V_A}{I_{CQ}} = \frac{80}{0.5} = 160 \, \text{k}\Omega$$

The small-signal voltage gain is therefore

$$A_v = -g_m(r_o \| r_c) = -(19.2)(160 \| 120) = -1317$$

Comment: As we will see in Part II of this text, the nonlinear resistor R_C is produced by the I–V characteristics of another bipolar transistor. Because the resulting effective load resistance is large, a very large small-signal voltage gain is produced. A large effective load resistance r_c means that the output resistance r_o of the amplifying transistor cannot be neglected; therefore, the loading effects must be taken into account.

4.5   AC LOAD LINE ANALYSIS

A dc load line gives us a way of visualizing the relationship between the Q-point and the transistor characteristics. When capacitors are included in a transistor circuit, a new effective load line, called an **ac load line**, may exist. The ac load line helps in visualizing the relationship between the small-signal response and the transistor characteristics. The ac operating region is on the ac load line.

4.5.1   AC Load Line

The circuit in Figure 4.33 has emitter resistors and an emitter bypass capacitor. The dc load line is found by writing a Kirchhoff voltage law (KVL) equation around the collector–emitter loop, as follows:

$$V^+ = I_C R_C + V_{CE} + I_E(R_{E1} + R_{E2}) + V^- \tag{4.53}$$

Noting that $I_E = [(1 + \beta)/\beta]I_C$, Equation (4.53) can be written as

$$V_{CE} = (V^+ - V^-) - I_C\left[R_C + \left(\frac{1+\beta}{\beta}\right)(R_{E1} + R_{E2})\right] \tag{4.54}$$

which is the equation of the dc load line. For the parameters and standard resistor values found in Example 4.6, the dc load line and the Q-point are plotted in Figure 4.39. If $\beta \gg 1$, then we can approximate $(1 + \beta)/\beta \cong 1$.

From the small-signal analysis in Example 4.6, the KVL equation around the collector–emitter loop is

$$i_c R_C + v_{ce} + i_e R_{E1} = 0 \tag{4.55(a)}$$

or, assuming $i_c \cong i_e$, then

$$v_{ce} = -i_c(R_C + R_{E1}) \tag{4.55(b)}$$

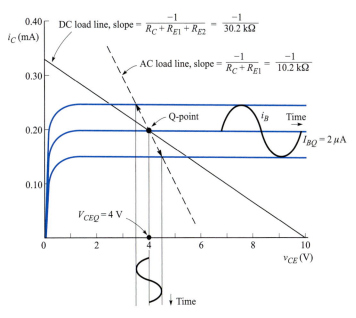

Figure 4.39 The dc and ac load lines for the circuit in Figure 4.33, and the signal responses to input signal

This equation is the ac load line. The slope is given by

$$\text{Slope} = \frac{-1}{R_C + R_{E1}}$$

The ac load line is shown in Figure 4.39. When $v_{ce} = i_c = 0$, we are at the Q-point. When ac signals are present, we deviate about the Q-point on the ac load line.

The slope of the ac load line differs from that of the dc load line because the emitter resistor is not included in the small-signal equivalent circuit. The small-signal C–E voltage and collector current response are functions of the resistor R_C and R_{E1} only.

Example 4.8 **Objective:** Determine the dc and ac load lines for the circuit shown in Figure 4.40.

Assume the transistor parameters are: $V_{EB}(\text{on}) = 0.7\,\text{V}$, $\beta = 150$, and $V_A = \infty$.

DC Solution: The dc load line is found by writing a KVL equation around the C–E loop, as follows:

$$V^+ = I_E R_E + V_{EC} + I_C R_C + V^-$$

The dc load line equation is then

$$V_{EC} = (V^+ - V^-) - I_C \left[R_C + \left(\frac{1 + \beta}{\beta} \right) R_E \right]$$

Assuming that $(1 + \beta)/\beta \cong 1$, the dc load line is plotted in Figure 4.41.

Figure 4.40 Circuit for Example 4.8

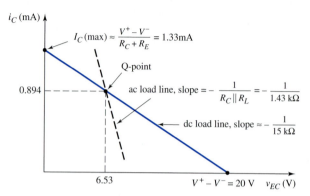

Figure 4.41 Plots of dc and ac load lines for Example 4.8

To determine the Q-point parameters, write a KVL equation around the B–E loop, as follows:

$$V^- = (1 + \beta)I_{BQ}R_E + V_{EB}(\text{on}) + I_{BQ}R_B$$

or

$$I_{BQ} = \frac{V^+ - V_{EB}(\text{on})}{R_B + (1 + \beta)R_E} = \frac{10 - 0.7}{50 + (151)(10)} \Rightarrow 5.96\,\mu A$$

Then,

$$I_{CQ} = \beta I_{BQ} = (150)(5.96\,\mu A) \Rightarrow 0.894\,\text{mA}$$
$$I_{EQ} = (1 + \beta)I_{BQ} = (151)(5.96\,\mu A) \Rightarrow 0.90\,\text{mA}$$

and

$$V_{ECQ} = (V^+ - V^-) - I_{CQ}R_C - I_{EQ}R_E$$
$$= [10 - (-10)] - (0.894)(5) - (0.90)(10) = 6.53\,\text{V}$$

The Q-point is also plotted in Figure 4.41.

AC Solution: Assuming that all capacitors act as short circuits, the small-signal equivalent circuit is shown in Figure 4.42. Note that the current directions and voltage polarities in the hybrid-π equivalent circuit of the pnp transistor are reversed compared to those of the npn device. The small-signal hybrid-π parameters are

$$r_\pi = \frac{V_T \beta}{I_{CQ}} = \frac{(0.026)(150)}{0.894} = 4.36\,\text{k}\Omega$$

$$g_m = \frac{I_{CQ}}{V_T} = \frac{0.894}{0.026} = 34.4\,\text{mA/V}$$

Figure 4.42 The small-signal equivalent circuit for Example 4.8

and

$$r_o = \frac{V_A}{I_{CQ}} = \frac{\infty}{I_{CQ}} = \infty$$

The small-signal output voltage, or C–E voltage, is

$$v_o = v_{ce} = +(g_m v_\pi)(R_C \| R_L)$$

where

$$g_m v_\pi = i_c$$

The ac load line, written in terms of the E–C voltage, is defined by

$$v_{ec} = -i_c(R_C \| R_L)$$

The ac load line is also plotted in Figure 4.41.

Comment: In the small-signal equivalent circuit, the large $10\,k\Omega$ emitter resistor is effectively shorted by the bypass capacitor C_E, the load resistor R_L is in parallel with R_C as a result of the coupling capacitor C_{C2}, so that the slope of the ac load line is substantially different than that of the dc load line.

4.5.2 Maximum Symmetrical Swing

When symmetrical sinusoidal signals are applied to the input of an amplifier, symmetrical sinusoidal signals are generated at the output, as long as the amplifier operation remains linear. We can use the ac load line to determine the **maximum output symmetrical swing**. If the output exceeds this limit, a portion of the output signal will be clipped and signal distortion will occur.

Example 4.9 **Objective:** Determine the maximum symmetrical swing in the output voltage of the circuit given in Figure 4.40.

Solution: The ac load line is given in Figure 4.41. The maximum negative swing in the collector current is from 0.894 mA to zero; therefore, the maximum possible symmetrical peak-to-peak ac collector current is

$$\Delta i_c = 2(0.894) = 1.79\,\text{mA}$$

The maximum symmetrical peak-to-peak output voltage is given by

$$|\Delta v_{ec}| = |\Delta i_c|(R_C \| R_L) = (1.79)(5\|2) = 2.56\,\text{V}$$

Therefore, the maximum instantaneous collector current is

$$i_C = I_{CQ} + \tfrac{1}{2}|\Delta i_c| = 0.894 + 0.894 = 1.79\,\text{mA}$$

Comment: Considering the Q-point and the maximum swing in the C–E voltage, the transistor remains biased in the forward-active region. Note that the maximum instantaneous collector current, 1.79 mA, is larger than the maximum dc collector current, 1.33 mA, as determined from the dc load line. This apparent anomaly is due to the different resistance in the C–E circuit for the ac signal and the dc signal.

Problem-Solving Technique: Maximum Symmetrical Swing

Again, since we are dealing with linear amplifier circuits, superposition applies so that we can add the dc and ac analysis results. To design a BJT amplifier for maximum symmetrical swing, we perform the following steps.

1. Write the dc load line equation that relates the quiescent values I_{CQ} and V_{CEQ}.
2. Write the ac load line equation that relates the ac values i_c and v_{ce} : $v_{ce} = -i_c R_{eq}$ where R_{eq} is the effective ac resistance in the collector-emitter circuit.
3. In general, we can write $i_c = I_{CQ} - I_C(\text{min})$, where $I_C(\text{min})$ is zero or some other specified minimum collector current.
4. In general, we can write $v_{ce} = V_{CEQ} - V_{CE}(\text{min})$, where $V_{CE}(\text{min})$ is some specified minimum collector-emitter voltage.
5. The above four equations can be combined to yield the optimum I_{CQ} and V_{CEQ} values to obtain the maximum symmetrical swing in the output signal.

Test Your Understanding

4.12   For the circuit in Figure 4.30, use the parameters given in Exercise 4.5. If the total instantaneous current must always be greater than 0.1 mA and the total instantaneous C–E voltage must be in the range $0.5 \leq v_{CE} \leq 5$ V, determine the maximum symmetrical swing in the output voltage. (Ans. 3.82 V peak-to-peak)

4.13   Reconsider the circuit in Figure 4.31. Let $r_o = \infty$, $\beta = 120$, and $V_{EB}(\text{on}) = 0.7$ V. (a) Plot the dc and ac load lines on the same graph. (b) Determine the maximum symmetrical swing in the output voltage, for $i_c > 0$ and $0.5 \leq v_{EC} \leq 12$ V. (Ans. (b) 6.58V peak-to-peak)

*D4.14   For the circuit in Figure 4.35, assume the transistor parameters are: $\beta = 100$, $V_{BE}(\text{on}) = 0.7$ V, and $V_A = \infty$. Determine a new value of R_E that will achieve a maximum symmetrical swing in the output voltage, for $i_C > 0$ and $0.7 \leq v_{CE} \leq 19.5$ V. What is the maximum symmetrical swing that can be achieved? (Ans. $R_E = 16.4$ kΩ, 10.6 V peak-to-peak)

4.15   For the circuit in Figure 4.36, let $\beta = 125$, $V_{BE}(\text{on}) = 0.7$ V, and $V_A = 200$ V. (a) Plot the dc and ac load lines on the same graph. (b) Determine the maximum symmetrical swing in the output voltage for $i_C > 0$ and $0.5 \leq v_{CE} \leq 9.5$ V. (Ans. (b) 2.66 V)

*D4.16   For the circuit shown in Figure 4.43, let $\beta = 120$, $V_{EB}(\text{on}) = 0.7$ V, and $r_o = \infty$. (a) Design a bias-stable circuit such that $I_{CQ} = 1.6$ mA. Determine V_{ECQ}. (b) Determine the value of R_L that will produce the maximum symmetrical swing in the output voltage and collector current for $i_C \geq 0.1$ mA and $0.5 \leq v_{EC} \leq 11.5$ V. (Ans. (a) $R_1 = 15.24$ kΩ, $R_2 = 58.7$ kΩ, $V_{ECQ} = 3.99$ V (b) $R_L = 5.56$ kΩ)

Figure 4.43   Figure for Exercise 4.16

4.6   COMMON-COLLECTOR (EMITTER-FOLLOWER) AMPLIFIER

The second type of transistor amplifier to be considered is the **common-collector circuit**. An example of this circuit configuration is shown in Figure 4.44. As seen in the figure, the output signal is taken off of the emitter with respect to ground and the collector is connected directly to V_{CC}. Since V_{CC} is at signal ground in the ac equivalent circuit, we have the name common-collector. The more common name for this circuit is **emitter follower**. The reason for this name will become apparent as we proceed through the analysis.

4.6.1   Small-Signal Voltage Gain

The dc analysis of the circuit is exactly the same as we have already seen, so we will concentrate on the small-signal analysis. The hybrid-π model of the bipolar transistor can also be used in the small-signal analysis of this circuit. Assuming the coupling capacitor C_C acts as a short circuit, Figure 4.45 shows the small-signal equivalent circuit of the circuit shown in Figure 4.44. The collector terminal is at signal ground and the transistor output resistance r_o is in parallel with the dependent current source.

Figure 4.46 shows the equivalent circuit rearranged so that all signal grounds are at the same point.

Figure 4.44   Emitter-follower circuit

Figure 4.45   Small-signal equivalent circuit of the emitter-follower

Figure 4.46   Another small-signal equivalent circuit for the emitter-follower

We see that

$$I_o = (1 + \beta)I_b \tag{4.56}$$

so the output voltage can be written as

$$V_o = I_b(1 + \beta)(r_o \| R_E) \tag{4.57}$$

Writing a KVL equation around the base-emitter loop, we obtain

$$V_{in} = I_b[r_\pi + (1 + \beta)(r_o \| R_E)] \tag{4.58(a)}$$

or

$$R_{ib} = \frac{V_{in}}{I_b} = r_\pi + (1 + \beta)(r_o \| R_E) \tag{4.58(b)}$$

We can also write

$$V_{in} = \left(\frac{R_i}{R_i + R_S}\right) \cdot V_s \tag{4.59}$$

where $R_i = R_1 \| R_2 \| R_{ib}$.

Combining Equations (4.57), (4.58(b)), and (4.59), the small-signal voltage gain is

$$A_v = \frac{V_o}{V_s} = \frac{(1 + \beta)(r_o \| R_E)}{r_\pi + (1 + \beta)(r_o \| R_E)} \cdot \left(\frac{R_i}{R_i + R_S}\right) \tag{4.60}$$

Example 4.10   Objective: Calculate the small-signal voltage gain of an emitter-follower circuit.

For the circuit shown in Figure 4.44, assume the transistor parameters are: $\beta = 100$, $V_{BE}(on) = 0.7\,V$, and $V_A = 80\,V$.

Solution: The dc analysis shows that $I_{CQ} = 0.793\,mA$ and $V_{CEQ} = 3.4\,V$. The small-signal hybrid-π parameters are determined to be

$$r_\pi = \frac{V_T\beta}{I_{CQ}} = \frac{(0.026)(100)}{0.793} = 3.28\,k\Omega$$

$$g_m = \frac{I_{CQ}}{V_T} = \frac{0.793}{0.026} = 30.5\,mA/V$$

and

$$r_o = \frac{V_A}{I_{CQ}} = \frac{80}{0.793} \cong 100\,\text{k}\Omega$$

We may note that

$$R_{ib} = 3.28 + (101)(100\|2) = 201\,\text{k}\Omega$$

and

$$R_i = 50\|50\|201 = 22.2\,\text{k}\Omega$$

The small-signal voltage gain is then

$$A_v = \frac{(101)(100\|2)}{3.28 + (101)(100\|2)} \cdot \left(\frac{22.2}{22.2 + 0.5}\right)$$

or

$$A_v = +0.962$$

Comment: The magnitude of the voltage gain is slightly less than 1. An examination of Equation (4.60) shows that this is always true. Also, the voltage gain is positive, which means that the output signal voltage at the emitter is in phase with the input signal voltage. The reason for the terminology emitter-follower is now clear. The output voltage at the emitter is essentially equal to the input voltage.

At first glance, a transistor amplifier with a voltage gain essentially of 1 may not seem to be of much value. However, the input and output resistance characteristics make this circuit extremely useful in many applications, as we will show in the next section.

4.6.2 Input and Output Impedance

The input impedance, or small-signal input resistance, of the emitter-follower is determined in the same manner as for the common-emitter circuit. For the circuit in Figure 4.44, the input resistance looking into the base is denoted R_{ib} and is indicated in the small-signal equivalent circuit shown in Figure 4.46.

The input resistance R_{ib} was given by Equation (4.58(b)) as

$$R_{ib} = r_\pi + (1 + \beta)(r_o \| R_E) \tag{4.58(b)}$$

Since the emitter current is $(1 + \beta)$ times the base current, the effective impedance in the emitter is multiplied by $(1 + \beta)$. We saw this same effect when an emitter resistor was included in a common-emitter circuit. This multiplication by $(1 + \beta)$ is again called the resistance reflection rule. The input resistance at the base is r_π plus the effective resistance in the emitter multiplied by the $(1 + \beta)$ factor. This resistance reflection rule will be used extensively throughout the remainder of the text.

The circuit in Figure 4.47 can be used to determine the output resistance of the emitter-follower circuit looking back into the output terminals. The independent voltage source is set equal to zero ($v_s = 0$), which means v_s acts as a short circuit. A test voltage V_x is applied to the output terminals, and the resulting test current is I_x. The output resistance, R_o, is given by

Figure 4.47 Small-signal equivalent circuit for emitter-follower output resistance calculations

$$R_o = \frac{V_x}{I_x} \tag{4.61}$$

In this case, the control voltage V_π is not zero, but is a function of the test voltage V_x. The dependent current source must be included in this analysis. Summing currents at the output node, we have

$$I_x + g_m V_\pi = \frac{V_x}{R_E} + \frac{V_x}{r_o} + \frac{V_x}{r_\pi + R_1 \| R_2 \| R_S} \tag{4.62}$$

The control voltage can be written in terms of the test voltage by a voltage divider equation as

$$V_\pi = -\left(\frac{r_\pi}{r_\pi + R_1 \| R_2 \| R_S}\right) \cdot V_x \tag{4.63}$$

Equation (4.62) can then be written as

$$I_x = \left(\frac{g_m r_\pi}{r_\pi + R_1 \| R_2 \| R_S}\right) \cdot V_x + \frac{V_x}{R_E} + \frac{V_x}{r_o} + \frac{V_x}{r_\pi + R_1 \| R_2 \| R_S} \tag{4.64}$$

Noting that $g_m r_\pi = \beta$, we find

$$\frac{I_x}{V_x} = \frac{1}{R_o} = \frac{1 + \beta}{r_\pi + R_1 \| R_2 \| R_S} + \frac{1}{R_E} + \frac{1}{r_o} \tag{4.65}$$

or

$$R_o = \left(\frac{r_\pi + R_1 \| R_2 \| R_S}{1 + \beta}\right) \| R_E \| r_o \tag{4.66}$$

Equation (4.66) says that the output resistance looking back into the output terminals is the effective resistance in the emitter, $R_E \| r_o$, in parallel with the resistance looking back into the emitter. In turn, the resistance looking into the emitter is the total resistance in the base circuit divided by $(1 + \beta)$. This is an important result and is the inverse of the resistance reflection rule looking to the base.

Example 4.11 Objective: Calculate the input and output resistance of the emitter-follower circuit shown in Figure 4.44.

The small-signal parameters, as determined in Example 4.10, are $r_\pi = 3.28\,\text{k}\Omega$, $\beta = 100$, and $r_o = 100\,\text{k}\Omega$.

Solution: Input Resistance. The input resistance looking into the base was determined in Example 4.10 as

$$R_{ib} = r_\pi + (1 + \beta)(r_o \| R_E) = 3.28 + (101)(100\|2) = 201 \text{ k}\Omega$$

and the input resistance seen by the signal source R_i is

$$R_i = R_1 \| R_2 \| R_{ib} = 50\|50\|201 = 22.2 \text{ k}\Omega$$

Comment: The input resistance of the emitter-follower looking into the base is substantially larger than that of the simple common-emitter circuit because of the $(1 + \beta)$ factor. This is one advantage of the emitter-follower circuit. However, in this case, the input resistance seen by the signal source is dominated by the bias resistors R_1 and R_2. To take advantage of the large input resistance of the emitter-follower circuit, the bias resistors must be designed to be much larger.

Solution: Output Resistance. The output resistance is found from Equation (4.66) as

$$R_o = \left(\frac{r_\pi + R_1\|R_2\|R_S}{1 + \beta}\right) \| R_E \| r_o = \left(\frac{3.28 + 50\|50\|0.5}{101}\right)\|2\|100$$

or

$$R_o = 0.0373\|2\|100 = 0.0366 \text{ k}\Omega \Rightarrow 36.6 \,\Omega$$

The output resistance is dominated by the first term that has $(1 + \beta)$ in the denominator.

Comment: The emitter-follower circuit is sometimes referred to as an **impedance transformer**, since the input impedance is large and the output impedance is small. The very low output resistance makes the *emitter-follower act almost like an ideal voltage source,* so the output is not loaded down when used to drive another load. Because of this, the emitter-follower is often used as the output stage of a multistage amplifier.

Test Your Understanding

4.17 For the circuit shown in Figure 4.44, let $V_{CC} = 5$ V, $\beta = 120$, $V_A = 100$ V, $R_E = 1$ kΩ, V_{BE}(on) $= 0.7$ V, $R_1 = 25$ kΩ, and $R_2 = 50$ kΩ. (a) Determine the small-signal voltage gain $A_v = V_o/V_s$. (b) Find the input resistance looking into the base of the transistor. (c) Find the output resistance looking into the output terminals. (Ans. (a) $A_v = 0.956$ (b) $R_{ib} = 120$ kΩ (c) $R_o = 15\,\Omega$)

D4.18 Assume the transistor parameters for the circuit in Figure 4.44 are: $\beta = 100$, V_{BE}(on) $= 0.7$ V, and $V_A = \infty$. Assume the circuit parameters are: $R_E = 2$ kΩ, and $V_{CC} = 10$ V. Design the circuit such that $V_{CEQ} = 5$ V and the resistance seen by the signal source is at least 65 kΩ. What is the resulting small-signal voltage gain and output resistance? (Ans. $I_E = 2.5$ mA, $R_1 = 118$ kΩ, $R_2 = 504$ kΩ, $A_v = 0.995$, $R_o = 10.3\,\Omega$, neglecting R_S)

4.6.3 Small-Signal Current Gain

We can determine the small-signal current gain of an emitter-follower by using the input resistance and the concept of current dividers. For the small-signal emitter-follower equivalent circuit shown in Figure 4.46, the small signal current gain is defined as

$$A_i = \frac{I_e}{I_i} \tag{4.67}$$

where I_e and I_i are the output and input current phasors.

Using a current divider equation, we can write the base current in terms of the input current, as follows:

$$I_b = \left(\frac{R_1 \| R_2}{R_1 \| R_2 + R_{ib}}\right) I_i \tag{4.68}$$

Since $g_m V_\pi = \beta I_b$, then,

$$I_o = (1 + \beta) I_b = (1 + \beta)\left(\frac{R_1 \| R_2}{R_1 \| R_2 + R_{ib}}\right) I_i \tag{4.69}$$

Writing the load current in terms of I_o produces

$$I_e = \left(\frac{r_o}{r_o + R_E}\right) I_o \tag{4.70}$$

Combining Equations (4.69) and (4.70), we obtain the small-signal current gain, as follows:

$$A_i = \frac{I_e}{I_i} = (1 + \beta)\left(\frac{R_1 \| R_2}{R_1 \| R_2 + R_{ib}}\right)\left(\frac{r_o}{r_o + R_E}\right) \tag{4.71}$$

If we assume that $R_1 \| R_2 \gg R_{ib}$ and $r_o \gg R_E$, then

$$A_i \cong (1 + \beta) \tag{4.72}$$

which is the current gain of the transistor.

Although the small-signal voltage gain of the emitter follower is slightly less than 1, the small-signal current is normally greater than 1. Therefore, the emitter-follower circuit produces a small-signal power gain.

Although we did not explicitly calculate a current gain in the common-emitter circuit previously, the analysis is the same as that for the emitter-follower and in general the current gain is also greater than unity.

Design Example   4.12   Objective: Consider the output signal of the amplifier designed in Example 4.6. We now want to design an emitter-follower circuit such that the output signal from this circuit does not vary by more than 5 percent when a load in the range $R_L = 4\,\text{k}\Omega$ to $R_L = 20\,\text{k}\Omega$ is connected to the output.

Discussion: The output resistance of the common-emitter circuit designed in Example 4.6 is $R_o = R_C = 10\,\text{k}\Omega$. Connecting a load resistance between $4\,\text{k}\Omega$ and $20\,\text{k}\Omega$ will load down this circuit, so that the output voltage will change substantially. For this reason, an emitter-follower circuit with a low output resistance must be designed to minimize the loading effect. The Thevenin equivalent circuit is shown in Figure 4.48. The output voltage can be written as

Figure 4.48 Thevenin equivalent of the output of an amplifier.

$$v_o = \left(\frac{R_L}{R_L + R_o}\right) \cdot v_{TH}$$

where v_{TH} is the ideal voltage generated by the amplifier. In order to have v_o change by less than 5 percent as a load resistance R_L is added, we must have R_o less than or equal to approximately 5 percent of the minimum value of R_L. In this case, then, we need R_o to be approximately 200 Ω.

Initial Design Approach: Consider the emitter-follower circuit shown in Figure 4.49. Note that the source resistance is $R_S = 10\,\text{k}\Omega$, corresponding to the output resistance of the circuit designed in Example 4.6. Assume that $\beta = 100$, $V_{BE}(\text{on}) = 0.7$ V, and $V_A = 80$ V.

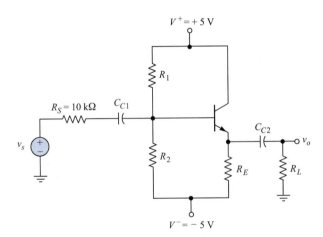

Figure 4.49   Figure for Example 4.12 and Exercises 4.19 and 4.20

The output resistance, given by Equation (4.66), is

$$R_o = \left(\frac{r_\pi + R_1 \| R_2 \| R_S}{1 + \beta}\right) \| R_E \| r_o$$

The first term, with $(1 + \beta)$ in the denominator, dominates, and if $R_1 \| R_2 \| R_S \cong R_S$, then we have

$$R_o \cong \frac{r_\pi + R_S}{1 + \beta}$$

For $R_o = 200\,\Omega$, we find

$$0.2 = \frac{r_\pi + 10}{101}$$

or $r_\pi = 10.2\,\text{k}\Omega$. Since $r_\pi = (\beta V_T)/I_{CQ}$, the quiescent collector current must be

$$I_{CQ} = \frac{\beta V_T}{r_\pi} = \frac{(100)(0.026)}{10.2} = 0.255\,\text{mA}$$

Assuming $I_{CQ} \cong I_{EQ}$ and letting $V_{CEQ} = 5\,\text{V}$, we find

$$R_E = \frac{V^+ - V_{CEQ} - V^-}{I_{EQ}} = \frac{5 - 5 - (-5)}{0.255} = 19.6\,\text{k}\Omega$$

The term $(1 + \beta)R_E$ is

$$(1 + \beta)R_E = (101)(19.6) \Rightarrow 1.98\,\text{M}\Omega$$

With this large resistance, we can design a bias-stable circuit as defined in Chapter 3 and still have large values for bias resistances. Let

$$R_{TH} = (0.1)(1 + \beta)R_E = (0.1)(101)(19.6) = 198\,\text{k}\Omega$$

The base current is

$$I_B = \frac{V_{TH} - V_{BE}(\text{on}) - V^-}{R_{TH} + (1 + \beta)R_E}$$

where

$$V_{TH} = \left(\frac{R_2}{R_1 + R_2}\right)(10) - 5 = \frac{1}{R_1}(R_{TH})(10) - 5$$

We can then write

$$\frac{0.255}{100} = \frac{\dfrac{1}{R_1}(198)(10) - 5 - 0.7 - (-5)}{198 + (101)(19.6)}$$

We find $R_1 = 344\,\text{k}\Omega$ and $R_2 = 467\,\text{k}\Omega$.

☞ **Comment:** The quiescent collector current $I_{CQ} = 0.255\,\text{mA}$ establishes the required r_π value which in turn establishes the required output resistance R_o.

Computer Simulation: We again used approximation techniques in our design. For this reason, it is useful to verify our design with a PSpice analysis, since the computer simulation will take into account more details than our hand design.

Figure 4.50 shows the PSpice circuit schematic. A 1 mV sinusoidal signal source is capacitively coupled to the output of the emitter follower. The input signal source has been set equal to zero. The current from the output signal source was found to be

Figure 4.50   PSpice circuit schematic for Example 4.12

5.667 μA. The output resistance of the emitter follower is then $R_o = 176\,\Omega$, which means that we have met our desired specification that the output resistance should be less than $200\,\Omega$.

```
    BJT MODEL PARAMETERS         **** BIPOLAR JUNCTION TRANSISTORS

            Q2N3904              NAME      Q_Q1
            NPN                  MODEL     Q2N3904
    IS    6.734000E-15           IB        2.08E-06
    BF  416.4                    IC        2.39E-04
    NF    1                      VBE       6.27E-01
    VAF  74.03                   VBC      -4.65E+00
    IKF    .06678                VCE       5.28E+00
    ISE   6.734000E-15           BETADC    1.15E+02
    NE    1.259                  GM        9.19E-03
    BR     .7371                 RPI       1.47E+04
    NR    1                      RX        1.00E+01
    RB   10                      RO        3.30E+05
    RBM  10                      CBE       9.08E-12
    RC    1                      CBC       1.98E-12
    CJE   4.493000E-12           CJS       0.00E+00
    MJE    .2593                 BETAAC    1.35E+02
    CJC   3.638000E-12           CBX       0.00E+00
    MJC    .3085                 FT        1.32E+08
    TF  301.200000E-12
    XTF   2
    VTF   4
    ITF    .4
    TR  239.500000E-09
    XTB    1.5
```

Discussion: The transistor Q-point values from the PSpice analysis are also listed. From the computer simulation, the quiescent collector current is $I_{CQ} = 0.239\,\text{mA}$ compared to the designed value of $0.255\,\text{mA}$. The principal reason for the difference in value is the difference in base-emitter voltage and current gain between the hand analysis and computer simulation.

Test Your Understanding

*D4.19 Design a bias-stable emitter-follower circuit, using the circuit shown in Figure 4.49, such that $I_{CQ} = 0.75\,\text{mA}$ and the small-signal gain is $A_i = i_o/i_i = 15$. Assume $R_S = 0$. The transistor parameters are: $\beta = 100$, $V_{BE}(\text{on}) = 0.7\,\text{V}$, and $V_A = 125\,\text{V}$. (Ans. $R_E = 2.0\,\text{k}\Omega$, $R_1 = 85.2\,\text{k}\Omega$, and $R_2 = 26.5\,\text{k}\Omega$)

*D4.20 For the circuit in Figure 4.49, the transistor parameters are: $\beta = 100$, $V_{BE}(\text{on}) = 0.7\,\text{V}$ and $V_A = 125\,\text{V}$. Assume $R_S = 0$. (a) Design a bias-stable circuit such that $I_{CQ} = 1.25\,\text{mA}$ and $V_{CEQ} = 4\,\text{V}$. (b) What is the small-signal current gain $A_i = i_o/i_i$? (c) What is the output resistance looking back into the output terminals? (d) What is the maximum symmetrical swing in the output voltage if the total collector–emitter voltage is to remain in the range $0.5 \leq v_{CE} \leq 9.5\,\text{V}$? (Ans. (a) $R_E = 4.76\,\text{k}\Omega$, $R_1 = 65.8\,\text{k}\Omega$, $R_2 = 178.8\,\text{k}\Omega$ (b) $A_i = 30.0$ (c) $R_o = 20.4\,\Omega$ (d) 2.07 V peak-to-peak)

4.21 Assume the circuit in Figure 4.51 uses a 2N2222A transistor. Assume a nominal dc current gain $\beta = 130$. Using the average h-parameter values (assume $h_{re} = 0$) given in the data sheets, determine $A_v = v_o/v_s$, $A_i = i_o/i_i$, R_{ib}, and R_o for:

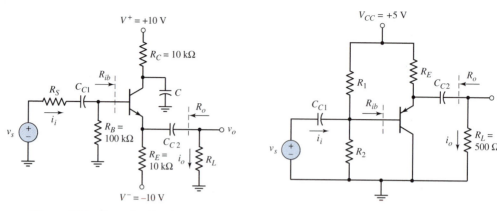

Figure 4.51   Figure for Exercise 4.21   **Figure 4.52**   Figure for Exercises 4.22 and 4.23

(a) $R_S = R_L = 10\,\text{k}\Omega$, and (b) $R_S = 1\,\text{k}\Omega$, $R_L = 10\,\text{k}\Omega$. (Ans. (a) $A_v = 0.891$, $A_i = 8.59$, $R_{ib} = 641\,\text{k}\Omega$, $R_o = 96.0\,\Omega$ (b) $A_v = 0.982$, $R_o = 36.8\,\Omega$)

4.22   For the circuit in Figure 4.52, $R_E = 2\,\text{k}\Omega$, $R_1 = 50\,\text{k}\Omega$, $R_2 = 50\,\text{k}\Omega$ and the transistor parameters are: $\beta = 100$, $V_{EB}(\text{on}) = 0.7\,\text{V}$, and $V_A = 125\,\text{V}$. (a) Determine the small-signal voltage gain $A_v = v_o/v_s$. (b) Find the resistances R_{ib} and R_o. (c) Determine the range of R_o if each resistor varies by ± 5 percent. (Ans. (a) $A_v = 0.925$ (b) $R_{ib} = 4.37\,\text{k}\Omega$, $R_o = 32.0\,\Omega$ (c) $28.6 \le R_o \le 35.7\,\Omega$)

D4.23   For the circuit in Figure 4.52, the transistor parameters are: $\beta = 75$, $V_{EB}(\text{on}) = 0.7\,\text{V}$, and $V_A = 75\,\text{V}$. The small-signal current gain is to be $A_i = i_o/i_i = 10$. Assume $V_{ECQ} = 2.5\,\text{V}$. Determine the values of the elements required if: (a) $R_E = R_L$, and (b) $R_E = 4R_L$. (Ans. (a) $R_1 = 26.1\,\text{k}\Omega$, $R_2 = 9.55\,\text{k}\Omega$ (b) $R_1 = 18.6\,\text{k}\Omega$, $R_2 = 9.6\,\text{k}\Omega$)

4.7   COMMON-BASE AMPLIFIER

A third amplifier circuit configuration is the **common-base circuit**. To determine the small-signal voltage and current gains, and the input and output impedances, we will use the same hybrid-π equivalent circuit for the transistor that was used previously. The dc analysis of the common-base circuit is essentially the same as for the common-emitter circuit.

4.7.1   Small-Signal Voltage and Current Gains

Figure 4.53 shows the basic common-base circuit, in which the base is at signal ground and the input signal is applied to the emitter. Assume a load is connected to the output through a coupling capacitor C_{C2}.

Figure 4.54(a) again shows the hybrid-π model of the npn transistor, which the output resistance r_o assumed to be infinite. Figure 4.54(b) shows the small-signal equivalent circuit of the common-base circuit, including the hybrid-π model of the transistor. As a result of the common-base configuration, the hybrid-π model in the small-signal equivalent circuit may look a little strange.

The small signal output voltage is given by

Figure 4.53   Basic common-base circuit

(a)   **(b)**

Figure 4.54   (a) Simplified hybrid-π model of the npn transistor and (b) small-signal equivalent circuit of the common-base circuit

$$V_o = -(g_m V_\pi)(R_C \| R_L) \tag{4.73}$$

Writing a KCL equation at the emitter node, we obtain

$$g_m V_\pi + \frac{V_\pi}{r_\pi} + \frac{V_\pi}{R_E} + \frac{V_s - (-V_\pi)}{R_S} = 0 \tag{4.74}$$

Since $\beta = g_m r_\pi$, Equation (4.74) can be written

$$V_\pi \left(\frac{1+\beta}{r_\pi} + \frac{1}{R_E} + \frac{1}{R_S} \right) = -\frac{V_s}{R_S} \tag{4.75}$$

Then,

$$V_\pi = -\frac{V_s}{R_S} \left[\left(\frac{r_\pi}{1+\beta} \right) \middle\| R_E \| R_S \right] \tag{4.76}$$

Substituting Equation (4.76) into (4.73), we find the small-signal voltage gain, as follows:

$$A_v = \frac{V_o}{V_s} = +g_m \left(\frac{R_C \| R_L}{R_S} \right) \left[\left(\frac{r_\pi}{1+\beta} \right) \middle\| R_E \| R_S \right] \tag{4.77}$$

We can show that as R_S approaches zero, the small-signal voltage gain becomes

$$A_v = g_m(R_C \| R_L) \tag{4.78}$$

Figure 4.54(b) can also be used to determine the small-signal current gain. The current gain is defined as $A_i = I_o / I_i$. Writing a KCL equation at the emitter node, we have

$$I_i + \frac{V_\pi}{r_\pi} + g_m V_\pi + \frac{V_\pi}{R_E} = 0 \tag{4.79}$$

Solving for V_π, we obtain

$$V_\pi = -I_i \left[\left(\frac{r_\pi}{1 + \beta} \right) \| R_E \right] \qquad (4.80)$$

The load current is given by

$$I_o = -(g_m V_\pi) \left(\frac{R_C}{R_C + R_L} \right) \qquad (4.81)$$

Combining Equations (4.80) and (4.81), we obtain an expression for the small-signal current gain, as follows:

$$A_i = \frac{I_o}{I_i} = g_m \left(\frac{R_C}{R_C + R_L} \right) \left[\left(\frac{r_\pi}{1 + \beta} \right) \| R_E \right] \qquad (4.82)$$

If we take the limit as R_E approaches infinity and R_L approaches zero, then the current gain becomes the short-circuit current gain given by

$$A_{io} = \frac{g_m r_\pi}{1 + \beta} = \frac{\beta}{1 + \beta} = \alpha \qquad (4.83)$$

where α is the common-base current gain of the transistor.

Equations (4.77) and (4.83) indicate that, for the common-base circuit, the small-signal voltage gain is usually greater than 1 and the small-signal current gain is slightly less than 1. However, we still have a small-signal power gain. The applications of a common-base circuit take advantage of the input and output resistance characteristics.

4.7.2   Input and Output Impedance

Figure 4.55 shows the small-signal equivalent circuit of the common-base configuration looking into the emitter. In this circuit, for convenience only, we have reversed the polarity of the control voltage, which reverses the direction of the dependent current source.

Figure 4.55   Common-base equivalent circuit for input resistance calculations

The input resistance looking into the emitter is defined as

$$R_{ie} = \frac{V_\pi}{I_i} \qquad (4.84)$$

If we write a KCL equation at the input, we obtain

$$I_i = I_b + g_m V_\pi = \frac{V_\pi}{r_\pi} + g_m V_\pi = V_\pi \left(\frac{1 + \beta}{r_\pi} \right) \qquad (4.85)$$

Therefore,

$$R_{ie} = \frac{V_\pi}{I_i} = \frac{r_\pi}{1+\beta} \equiv r_e \tag{4.86}$$

The resistance looking into the emitter, with the base grounded, is usually defined as r_e and is quite small, as already shown in the analysis of the emitter-follower circuit. When the input signal is a current source, a small input resistance is desirable

Figure 4.56 shows the circuit used to calculate the output resistance. The independent source v_s has been set equal to zero. Writing a KCL equation at the emitter, we find

$$g_m V_\pi + \frac{V_\pi}{r_\pi} + \frac{V_\pi}{R_E} + \frac{V_\pi}{R_S} = 0 \tag{4.87}$$

This implies that $V_\pi = 0$, which means that the independent source $g_m V_\pi$ is also zero. Consequently, the output resistance looking back into the output terminals is then

$$R_o = R_C \tag{4.88}$$

Because we have assumed r_o is infinite, the output resistance looking back into the collector terminal is essentially infinite, which means that the common-base circuit looks almost like an ideal current source.

Figure 4.56 Common-base equivalent circuit for output resistance calculations

Discussion

The common-base circuit is very useful when the input signal is a current. We will see this type of application when we discuss the cascode circuit in Section 4.9.

Test Your Understanding

4.24 For the circuit shown in Figure 4.57, the transistor parameters are: $\beta = 100$, $V_{EB}(\text{on}) = 0.7\,\text{V}$, and $r_o = \infty$. (a) Calculate the quiescent values of I_{CQ} and V_{ECQ}. (b) Determine the small-signal current gain $A_i = i_o/i_i$. (c) Determine the small-signal voltage gain $A_v = v_o/v_s$. (d) Determine the maximum symmetrical swing in the output voltage if the emitter–collector voltage is in the range $0.5 \le v_{EC} \le 19.5\,\text{V}$. (Ans. (a) $I_{CQ} = 0.921\,\text{mA}$, $V_{ECQ} = 6.1\,\text{V}$ (b) $A_i = 0.987$ (c) $A_v = 177$ (d) 9.21 V peak-to-peak)

Figure 4.57   Figure for Exercise 4.24

Figure 4.58   Figure for Exercises 4.25 and 4.26

4.25   For the circuit shown in Figure 4.58, the parameters are: $R_B = 100\,\text{k}\Omega$, $R_E = 10\,\text{k}\Omega$, $R_C = 10\,\text{k}\Omega$, $V_{CC} = V_{EE} = 10\,\text{V}$, $R_L = 1\,\text{k}\Omega$, $R_S = 1\,\text{k}\Omega$, $V_{BE}(\text{on}) = 0.7\,\text{V}$, $\beta = 100$, and $V_A = \infty$. (a) Determine the small-signal transistor parameters g_m, r_π, and r_o. (b) Find the small-signal current gain $A_i = i_o/i_i$ and the small-signal voltage gain $A_v = v_o/v_s$. (c) Determine the input resistance R_i and the output resistance R_o. (Ans. (a) $r_\pi = 3.1\,\text{k}\Omega$, $g_m = 32.23\,\text{mA/V}$, $r_o = \infty$ (b) $A_v = 0.870$, $A_i = 0.90$ (c) $R_i = 0.0307\,\text{k}\Omega$, $R_o = 10\,\text{k}\Omega$)

D4.26   For the circuit shown in Figure 4.58, let $R_S = 0$, $C_B = 0$, $R_C = R_L = 2\,\text{k}\Omega$, $V_{CC} = V_{EE} = 5\,\text{V}$, $\beta = 100$, $V_{BE}(\text{on}) = 0.7\,\text{V}$, and $V_A = \infty$. Design R_E and R_B for a dc quiescent collector current of 1 mA and a small-signal voltage gain of 20. (Ans. $R_B = 2.4\,\text{k}\Omega$, $R_E = 4.23\,\text{k}\Omega$)

4.8   THE THREE BASIC AMPLIFIERS: SUMMARY AND COMPARISON

The basic small-signal characteristics of the three single-stage amplifier configurations are summarized in Table 4.4.

Table 4.4   Characteristics of the three BJT amplifier configurations

Configuration	Voltage gain	Current gain	Input resistance	Output resistance
Common emitter	$A_v > 1$	$A_i > 1$	Moderate	Moderate to high
Emitter follower	$A_v \cong 1$	$A_i > 1$	High	Low
Common base	$A_v > 1$	$A_i \cong 1$	Low	Moderate to high

For the common-emitter circuit, the voltage and current gains are generally greater than 1. For the emitter-follower, the voltage gain is slightly less than 1, while the current gain is greater than 1. For the common-base circuit, the voltage gain is greater than 1, while the current gain is less than 1.

The input resistance looking into the base terminal of a common-emitter circuit may be in the low kilohm range; in an emitter follower, it is generally in the 50 to 100 kΩ range. The input resistance looking into the emitter of a common-base circuit is generally on the order of tens of ohms.

The overall input resistance of both the common-emitter and emitter-follower circuits can be greatly affected by the bias circuitry.

The output resistance of the emitter follower is generally in the range of a few ohms to tens of ohms. In contrast, the output resistance looking into the collector terminal of the common-emitter and common-base circuits is very high. In addition, the output resistance looking back into the output terminal of the common-emitter and common-base circuits is a strong function of the collector resistance. For these circuits, the output resistance can easily drop to a few kilohms.

The characteristics of these single-stage amplifiers will be used in the design of multistage amplifiers.

4.9   MULTISTAGE AMPLIFIERS

In most applications, a single transistor amplifier will not be able to meet the combined specifications of a given amplification factor, input resistance, and output resistance. For example, the required voltage gain may exceed that which can be obtained in a single transistor circuit. We also saw an illustration of this effect in Example 4.12, in which a low output resistance was required in a particular design.

Transistor amplifier circuits can be connected in series, or **cascaded**, as shown in Figure 4.59. This may be done either to increase the overall small-signal voltage gain or to provide an overall voltage gain greater than 1, with a very low output resistance. The overall voltage or current gain, in general, is not simply the product of the individual amplification factors. For example, the gain of stage 1 is a function of the input resistance of stage 2. In other words, loading effects may have to be taken into account.

There are may possible multistage configurations; we will examine a few here, in order to understand the type of analysis required.

Figure 4.59   A generalized three-stage amplifier

4.9.1   Multistage Analysis: Cascade Configuration

In Figure 4.60, the circuit is a cascade configuration of two common-emitter circuits. The dc analysis of this circuit, done in Example 3.17 of Chapter 3, showed that both transistors are biased in the forward-active mode. Figure 4.61 shows the small-signal equivalent circuit, assuming all capacitors act as short circuits and each transistor output resistance r_o is infinite.

We may start the analysis at the output and work back to the input, or start at the input and work toward the output.

The small-signal voltage gain is

$$A_v = \frac{V_o}{V_s} = g_{m1}g_{m2}(R_{C1}\|r_{\pi2})(R_{C2}\|R_L)\left(\frac{R_i}{R_i + R_S}\right) \qquad \textbf{(4.89)}$$

Figure 4.60 A two-stage amplifier in a cascade configuration

Figure 4.61 Small-signal equivalent circuit of the cascade configuration

The input resistance of the amplifier is

$$R_{is} = R_1 \| R_2 \| r_{\pi 1}$$

which is identical to that of a single-stage common-emitter amplifier. Similarly, the output resistance looking back into the output terminals is $R_o = R_{C2}$. To determine the output resistance, the independent source V_s is set equal to zero, which means that $V_{\pi 1} = 0$. Then $g_{m1} V_{\pi 1} = 0$, which gives $V_{\pi 2} = 0$ and $g_{m2} V_{\pi 2} = 0$. The output resistance is therefore R_{C2}. Again, this is the same as the output resistance of a single-stage common-emitter amplifier.

Computer Example 4.13   **Objective:** Determine the small-signal voltage gain of the multitransistor circuit shown in Figure 4.60 using a PSpice analysis.

The dc and ac analyses of a multitransistor circuit become more complex compared to those for a single-transistor circuit. In this situation, a computer simulation of the circuit, without a hand analysis, is extremely useful.

The PSpice circuit schematic is shown in Figure 4.62. Also given are the Q-point values of the transistors. The ac voltage at the collector of the npn transistor is $51\,\mu V$ and that at the collector of the pnp transistor is $4.79\,mV$. Since the input voltage was assumed to be $1\,\mu V$, this result shows that a significant voltage gain can be achieved in a two-stage amplifier.

Figure 4.62 PSpice circuit schematic for Example 4.13

```
****  BIPOLAR JUNCTION TRANSISTORS

NAME        Q_Q1          Q_Q2
MODEL       Q2N3906       Q2N3904
IB          -1.42E-05     8.59E-06
IC          -2.54E-03     1.18E-03
VBE         -7.30E-01     6.70E-01
VBC         3.68E-01      -1.12E+00
VCE         -1.10E+00     1.79E+00
BETADC      1.79E+02      1.37E+02
GM          9.50E-02      4.48E-02
RPI         1.82E+03      3.49E+03
RX          1.00E+01      1.00E+01
RO          7.52E+03      6.37E+04
CBE         3.11E-11      2.00E-11
CBC         7.75E-12      2.74E-12
CJS         0.00E+00      0.00E+00
BETAAC      1.73E+02      1.57E+02
CBX         0.00E+00      0.00E+00
FT          3.89E+08      3.14E+08
```

Comment: We can see from the Q-point values that the collector–emitter voltage of each transistor is quite small. This implies that the maximum symmetrical swing in the output voltage is limited to a fairly small value. These Q-point values can be increased by a slight redesign of the circuit.

Discussion: The transistors used in this PSpice analysis of the circuit were standard bipolar transistors from the PSpice library. We must keep in mind that, for the computer simulation to be valid, the models of the devices used in the simulation must match those of the actual devices used in the circuit. If the actual transistor characteristics were substantially different from those used in the computer simulation, then the results of the computer analysis would not be accurate.

In some applications, it would be desirable to have a bipolar transistor with a much larger current gain than can normally be obtained. Figure 4.63(a) shows a multitransistor configuration, called a Darlington pair or a Darlington configuration, that provides increased current gain.

Figure 4.63   (a) A Darlington pair configuration; (b) small-signal equivalent circuit

The small-signal equivalent in which the input signal is assumed to be a current source, is shown in Figure 4.63(b). We will use the input current source to determine the current gain of the circuit. To determine the small-signal current gain $A_i = I_o/I_i$, we see that

$$V_{\pi 1} = I_i r_{\pi 1} \tag{4.90}$$

Therefore,

$$g_{m1} V_{\pi 1} = g_{m1} r_{\pi 1} I_i = \beta_1 I_i \tag{4.91}$$

Then,

$$V_{\pi 2} = (I_i + \beta_1 I_i) r_{\pi 2} \tag{4.92}$$

The output current is

$$I_o = g_{m1} V_{\pi 1} + g_{m2} V_{\pi 2} = \beta_1 I_i + \beta_2 (1 + \beta_1) I_i \tag{4.93}$$

where $g_{m2} r_{\pi 2} = \beta_2$. The overall current gain is then

$$A_i = \frac{I_o}{I_i} = \beta_1 + \beta_2 (1 + \beta_1) \cong \beta_1 \beta_2 \tag{4.94}$$

From Equation (4.94), we see that the overall small-signal current gain of the Darlington pair is essentially the product of the individual current gains.

The input resistance is $R_i = V_i/I_i$. We can write that

$$V_i = V_{\pi 1} + V_{\pi 2} = I_i r_{\pi 1} + I_i (1 + \beta_1) r_{\pi 2} \tag{4.95}$$

so that

$$R_i = r_{\pi 1} + (1 + \beta_1) r_{\pi 2} \tag{4.96}$$

The base of transistor Q_2 is connected to the emitter of Q_1, which means that the input resistance to Q_2 is multiplied by the factor $(1 + \beta_1)$, as we saw in circuits with emitter resistors. We can write

$$r_{\pi 1} = \frac{\beta_1 V_T}{I_{CQ1}} \qquad (4.97)$$

and

$$I_{CQ1} \cong \frac{I_{CQ2}}{\beta_2} \qquad (4.98)$$

Therefore,

$$r_{\pi 1} = \beta_1 \left(\frac{\beta_2 V_T}{I_{CQ2}} \right) = \beta_1 r_{\pi 2} \qquad (4.99)$$

From Equation (4.96), the input resistance is then approximately

$$R_i \cong 2\beta_1 r_{\pi 2} \qquad (4.100)$$

We see from these equations that the overall gain of the Darlington pair is large. At the same time, the input resistance tends to be large, because of the β multiplication.

4.9.2 Cascode Configuration

A slightly different multistage configuration, called a **cascode configuration**, is shown in Figure 4.64(a). The input is into a common-emitter amplifier (Q_1), which drives a common-base amplifier (Q_2). The ac equivalent circuit is shown in Figure 4.64(b). We see that the output signal current of Q_1 is the input signal of Q_2. We mentioned previously that, normally, the input signal of a common-base configuration is to be a current. One advantage of this circuit is that the output resistance looking into the collector of Q_2 is much larger than the output

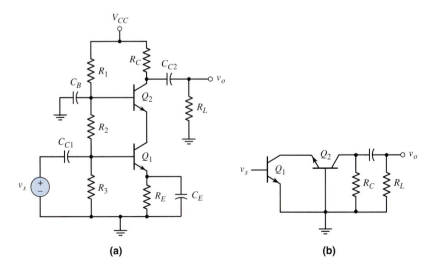

(a)   (b)

Figure 4.64   (a) Cascode amplifier and (b) the ac equivalent circuit of a common-emitter circuit. Another important advantage of this circuit is in the frequency response, as we will see in Chapter 7.

The small-signal equivalent circuit is shown in Figure 4.65 for the case when the capacitors act as short circuits. We see that $V_{\pi1} = V_s$ since we are assuming an ideal signal voltage source. Writing a KCL equation at E_2, we have

$$g_{m1} V_{\pi1} = \frac{V_{\pi2}}{r_{\pi2}} + g_{m2} V_{\pi2} \tag{4.101}$$

Figure 4.65   Small-signal equivalent circuit of the cascode configuration

Solving for the control voltage $V_{\pi2}$, we find

$$V_{\pi2} = \left(\frac{r_{\pi2}}{1 + \beta_2}\right)(g_{m1} V_s) \tag{4.102}$$

where $\beta_2 = g_{m2} r_{\pi2}$. The output voltage is

$$V_o = -(g_{m2} V_{\pi2})(R_C \| R_L) \tag{4.103(a)}$$

or

$$V_o = -g_{m1} g_{m2} \left(\frac{r_{\pi2}}{1 + \beta_2}\right)(R_C \| R_L) V_s \tag{4.103(b)}$$

Therefore, the small-signal voltage gain is

$$A_v = \frac{V_o}{V_s} = -g_{m1} g_{m2} \left(\frac{r_{\pi2}}{1 + \beta_2}\right)(R_C \| R_L) \tag{4.104}$$

An examination of Equation (4.104) shows

$$g_{m2}\left(\frac{r_{\pi2}}{1 + \beta_2}\right) = \frac{\beta_2}{1 + \beta_2} \cong 1 \tag{4.105}$$

The gain of the cascode amplifier is then approximately

$$A_v \cong -g_{m1}(R_C \| R_L) \tag{4.106}$$

which is the same as for a single-stage common-emitter amplifier. This result is to be expected since the current gain of the common-base circuit is essentially unity.

Test Your Understanding

*4.27 For each transistor in the circuit in Figure 4.66, the parameters are: $\beta = 125$, $V_{BE}(\text{on}) = 0.7\,\text{V}$, and $r_o = \infty$. (a) Determine the Q-points of each transistor. (b) Find the overall small-signal voltage gain $A_v = V_o/V_s$. (c) Determine the input resistance R_i and the output resistance R_o. (Ans. (a) $I_{CQ1} = 0.366\,\text{mA}$, $V_{CEQ1} = 7.90\,\text{V}$, $I_{CQ2} = 4.81\,\text{mA}$, $V_{CEQ2} = 2.72\,\text{V}$ (b) $A_v = -17.7$ (c) $R_i = 4.76\,\text{k}\Omega$, $R_o = 43.7\,\Omega$)

*4.28 Consider the circuit in Figure 4.63(a). Let $\beta = 100$, $V_{BE}(\text{on}) = 0.7\,\text{V}$, and $V_A = \infty$ for each transistor. Assume $R_B = 10\,\text{k}\Omega$, $R_C = 4\,\text{k}\Omega$, $I_{Eo} = 1\,\text{mA}$, $V^+ = 5\,\text{V}$, and $V^- = -5\,\text{V}$. (a) Determine the Q-point values for each transistor. (b) Calculate the small-signal hybrid-π parameters for each transistor. (c) Find the overall small-signal voltage gain $A_v = V_o/V_s$. (d) Find the input resistance R_i. (Ans. (a) $I_{CQ1} = 0.0098\,\text{mA}$, $V_{CEQ1} = 1.7\,\text{V}$, $I_{CQ2} = 0.990\,\text{mA}$, $V_{CEQ2} = 2.4\,\text{V}$ (b) $r_{\pi1} = 265\,\text{k}\Omega$, $g_{m1} = 0.377\,\text{mA/V}$, $r_{\pi2} = 2.63\,\text{k}\Omega$, $g_{m2} = 38.1\,\text{mA/V}$ (c) $A_v = -76.9$ (d) $R_i = 531\,\text{k}\Omega$)

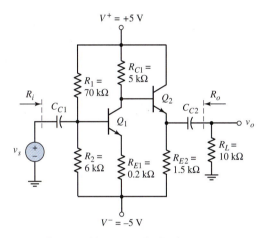

Figure 4.66 Figure for Exercise 4.27

4.29 For the circuit shown in Figure 4.67, let $\beta = 100$, $V_{BE}(\text{on}) = 0.7\,\text{V}$, and $V_A = \infty$ for each transistor. Repeat parts (a)–(d) of Exercise 4.28. (Ans. (a) $I_{CQ1} = 0.464\,\text{mA}$, $V_{CEQ1} = 8.18\,\text{V}$, $I_{CQ2} = 0.842\,\text{mA}$, $V_{CEQ2} = 8.88\,\text{V}$ (b) $r_{\pi1} = 5.60\,\text{k}\Omega$, $g_{m1} = 17.8\,\text{mA/V}$, $r_{\pi2} = 3.09\,\text{k}\Omega$, $g_{m2} = 32.4\,\text{mA/V}$, $r_{o1} = r_{o2} = \infty$ (c) $A_v = -0.298$ (d) $R_i = 1.99\,\text{M}\Omega$)

Figure 4.67 Figure for Exercise 4.29

D4.30   Consider the cascode circuit in Figure 4.64(a). Let $\beta = 100$, $V_{BE}(\text{on}) = 0.7\,\text{V}$, and $V_A = \infty$ for each transistor. Assume $V_{CC} = 12\,\text{V}$, $R_L = 2\,\text{k}\Omega$, and $R_E = 0.5\,\text{k}\Omega$. (a) Find R_C, R_1, R_2, and R_3 such that $I_{CQ2} = 0.5\,\text{mA}$ and $V_{CE1} = V_{CE2} = 4\,\text{V}$. Let $R_1 + R_2 + R_3 = 100\,\text{k}\Omega$. (Hint: Neglect the dc base currents and assume $I_C = I_E$ in both Q_1 and Q_2.) (b) Determine the small-signal hybrid-π parameters for each transistor. (c) Determine the small-signal voltage gain $A_v = V_o/V_s$. (Ans. (a) $R_C = 7.5\,\text{k}\Omega$, $R_3 = 7.92\,\text{k}\Omega$, $R_2 = 33.3\,\text{k}\Omega$, $R_1 = 58.8\,\text{k}\Omega$ (b) $r_{\pi 1} = r_{\pi 2} = 5.2\,\text{k}\Omega$, $g_{m1} = g_{m2} = 19.23$ mA/V, $r_{o1} = r_{o2} = \infty$ (c) $A_v = -30.1$)

4.10   POWER CONSIDERATIONS

As mentioned previously, an amplifier produces a **small-signal power gain**. Since energy must be conserved, the question naturally arises as to the source of this "extra" signal power. We will see that the "extra" signal power delivered to a load is a result of a redistribution of power between the load and the transistor.

Consider the simple common-emitter circuit shown in Figure 4.68 in which an ideal signal voltage source is connected at the input. The dc power supplied by the V_{CC} voltage source P_{CC}, the dc power dissipated or supplied to the collector resistor P_{RC}, and the dc power dissipated in the transistor P_Q are given, respectively, as

$$P_{CC} = I_{CQ}V_{CC} + P_{\text{Bias}} \qquad\qquad (4.107(\text{a}))$$

$$P_{RC} = I_{CQ}^2 R_c \qquad\qquad (4.107(\text{b}))$$

and

$$P_Q = I_{CQ}V_{CEQ} + I_{BQ}V_{BEQ} \cong I_{CQ}V_{CEQ} \qquad\qquad (4.107(\text{c}))$$

The term P_{Bias} is the power dissipated in the bias resistors R_1 and R_2. Normally in a transistor $I_{CQ} \gg I_{BQ}$, so the power dissipated is primarily a function of the collector current and collector–emitter voltage.

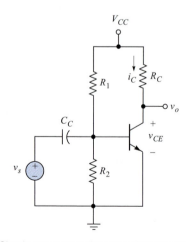

Figure 4.68   Simple common-emitter amplifier for power calculations

If the signal voltage is given by

$$v_s = V_p \cos \omega t \tag{4.108}$$

then the total base current is given by

$$i_B = I_{BQ} + \frac{V_p}{r_\pi} \cos \omega t = I_{BQ} + I_b \cos \omega t \tag{4.109}$$

and the total collector current is

$$i_C = I_{CQ} + \beta I_b \cos \omega t = I_{CQ} + I_c \cos \omega t \tag{4.110}$$

The total instantaneous collector-emitter voltage is

$$v_{CE} = V_{CC} - i_C R_C = V_{CC} - (I_{CQ} + I_c \cos \omega t) R_C = V_{CEQ} - I_c R_C \cos \omega t \tag{4.111}$$

The average power, including ac signals, supplied by the voltage source V_{CC} is given by

$$\begin{aligned}
\bar{p}_{cc} &= \frac{1}{T} \int_0^T V_{CC} \cdot i_C \, dt + P_{\text{Bias}} \\
&= \frac{1}{T} \int_0^T V_{CC} \cdot [I_{CQ} + I_c \cos \omega t] \, dt + P_{\text{Bias}} \\
&= V_{CC} I_{CQ} + \frac{V_{CC} I_c}{T} \int_0^T \cos \omega t \, dt + P_{\text{Bias}}
\end{aligned} \tag{4.112}$$

Since the integral of the cosine function over one period is zero, the average power supplied by the voltage source is the same as the dc power supplied. The dc voltage source does not supply additional power.

The average power delivered to the load R_C is found from

$$\begin{aligned}
\bar{p}_{RC} &= \frac{1}{T} \int_0^T i_C^2 R_C \, dt = \frac{R_C}{T} \int_0^T [I_{CQ} + I_c \cos \omega t]^2 \, dt \\
&= \frac{I_{CQ}^2 R_C}{T} \int_0^T dt + \frac{2 I_{CQ} I_c}{T} \int_0^T \cos \omega t \, dt + \frac{I_c^2 R_C}{T} \int_0^T \cos^2 \omega t \, dt
\end{aligned} \tag{4.113}$$

The middle term of this last expression is again zero, so

$$\bar{p}_{RC} = I_{CQ}^2 R_C + \tfrac{1}{2} I_c^2 R_C \tag{4.114}$$

The average power delivered to the load has increased because of the signal source. This is expected in an amplifier.

Now, the average power dissipated in the transistor is

$$\bar{p}_Q = \frac{1}{T}\int_0^T i_C \cdot v_{CE}\, dt$$

$$= \frac{1}{T}\int_0^T [I_{CQ} + I_c \cos \omega t] \cdot [V_{CEQ} - I_c R_C \cos \omega t]\, dt \qquad \textbf{(4.115)}$$

which produces

$$\bar{p}_Q = I_{CQ} V_{CEQ} - \frac{I_c^2 R_C}{T}\int_0^T \cos^2 \omega t\, dt \qquad \textbf{(4.116(a))}$$

or

$$\bar{p}_Q = I_{CQ} V_{CEQ} - \tfrac{1}{2} I_c^2 R_C \qquad \textbf{(4.116(b))}$$

From Equation (4.116(b)), we can deduce that the average power dissipated in the transistor decreases when an ac signal is applied. The V_{CC} source still supplies all of the power, but the input signal changes the relative distribution of power between the transistor and the load.

Test Your Understanding

4.31 In the circuit in Figure 4.69 the transistor parameters are: $\beta = 80$, $V_{BE}(\text{on}) = 0.7\,\text{V}$, and $V_A = \infty$. Determine the average power dissipated in R_C, R_L, and Q for: (a) $v_s = 0$, and (b) $v_s = 18 \cos \omega t\,\text{mV}$. (Ans. (a) $\bar{p}_{RC} = 8\,\text{mW}$, $\bar{p}_{RL} = 0$, $\bar{p}_Q = 14\,\text{mW}$ (b) $\bar{p}_Q = 13.0\,\text{mW}$, $\bar{p}_{RL} = 0.479\,\text{mW}$, $\bar{p}_{RC} = 8.48\,\text{mW}$)

4.32 For the circuit in Figure 4.70, the transistor parameters are: $\beta = 100$, $V_{BE}(\text{on}) = 0.7\,\text{V}$, and $V_A = \infty$. (a) Determine R_C such that the Q-point is in the center of the load line. (b) Determine the average power dissipated in R_C and Q for $v_s = 0$. (c) Considering the maximum symmetrical swing in the output voltage, determine the ratio of maximum signal power delivered to R_C to the total power dissipated in R_C and the transistor. (Ans. (a) $R_C = 2.52\,\text{k}\Omega$ (b) $\bar{p}_{RC} = \bar{p}_Q = 2.48\,\text{mW}$ (c) 0.25)

Figure 4.69 Figure for Exercise 4.31   **Figure 4.70** Figure for Exercise 4.32

4.11 SUMMARY

- This chapter emphasized the application of bipolar transistors in linear amplifier circuits. A small-signal equivalent circuit for the transistor was developed that is used in the analysis and design of linear amplifiers.

- Three basic circuit configurations were considered: the common–emitter, emitter–follower, and common–base. These three configurations form the basic building blocks for more complex integrated circuits. The small-signal voltage gains, current gains, and input and output resistances for these circuits were analyzed. Table 4.4 compares the circuit characteristics of the three circuits.

- The cascode configuration of two common-emitter circuits was analyzed as one example of a multistage transistor circuit. The overall small-signal voltage gain is increased compared to that of a single stage amplifier. A Darlington pair configuration was also analyzed. In this multistage circuit, the overall current gain increases and the input resistance increases compared to a single transistor circuit.

CHECKPOINT

After studying this chapter, the reader should have the ability to:

✔ Explain graphically the amplification process in a simple bipolar amplifier circuit. (Section 4.2)

✔ Describe the small-signal hybrid-π equivalent circuit of the bipolar transistor and to determine the values of the small-signal hybrid-π parameters. (Section 4.2)

✔ Apply the small-signal hybrid-π equivalent circuit to various bipolar amplifier circuits to obtain the time-varying circuit characteristics.

✔ Characterize the small-signal voltage and current gains and the input and output resistances of a common-emitter amplifier. (Section 4.4)

✔ Characterize the small-signal voltage and current gains and the input and output resistances of an emitter-follower amplifier. (Section 4.6)

✔ Characterize the small-signal voltage and current gains and the input and output resistances of a common-base amplifier. (Section 4.7)

✔ Apply the bipolar small-signal equivalent circuit in the analysis of multistage amplifier circuits

REVIEW QUESTIONS

1. Discuss, using the concept of a load line superimposed on the transistor characteristics, how a simple common-emitter circuit can amplify a time-varying signal.

2. Why can the analysis of a transistor circuit be split into a dc analysis, with all ac sources set equal to zero, and then an ac analysis, with all dc sources set equal to zero?

3. What are the physical meanings of the hybrid-π parameters r_π and r_o?

4. What does the term small-signal imply?

5. Discuss the difference between the ac and dc common-emitter current gains.

6. Discuss similarities and differences between the hybrid-π and h-parameter models of the bipolar transistor.

7. Discuss the four equivalent two-port networks and discuss the conditions under which each equivalent circuit would be used.

8. Sketch a simple common-emitter amplifier circuit and discuss the general ac circuit characteristics (voltage gain, current gain, input and output resistances).

9. Discuss the general conditions under which a common-emitter amplifier would be used.

10. What are the changes in the ac characteristics of a common-emitter amplifier when an emitter resistor and an emitter bypass capacitor are incorporated in the design?

11. Discuss the concepts of a dc load line and an ac load line.

12. Sketch a simple emitter-follower amplifier circuit and discuss the general ac circuit characteristics (voltage gain, current gain, input and output resistances).

13. Discuss the general conditions under which an emitter-follower amplifier would be used.

14. Sketch a simple common-base amplifier circuit and discuss the general ac circuit characteristics (voltage gain, current gain, input and output resistances).

15. Discuss the general conditions under which a common-base amplifier would be used.

16. Compare the ac circuit characteristics of the common-emitter, emitter-follower, and common-base circuits.

17. State at least two reasons why a multistage amplifier circuit would be required in a design rather than a single-stage circuit.

18. If a transistor circuit provides signal power gain, discuss the source of this additional signal power.

PROBLEMS

[Note: In the following problems, assume that the B–E turn-on voltage is 0.7 V for both npn and pnp transistors and that $V_A = \infty$ unless otherwise stated. Also assume that all capacitors act as short circuits to the signal.]

Section 4.2   The Bipolar Linear Amplifier

4.1   (a) If the transistor parameters are $\beta = 180$ and $V_A = 150\,\text{V}$, and it is biased at $I_{CQ} = 2\,\text{mA}$, determine the values of g_m, r_π, and r_o. (b) Repeat part (a) if $I_{CQ} = 0.5\,\text{mA}$.

4.2   The transistor parameters are $\beta = 125$ and $V_A = 200\,\text{V}$. A value of $g_m = 200\,\text{mA/V}$ is desired. Determine the collector current required, and then find r_π and r_o.

4.3   For the circuit in Figure 4.3, the transistor parameters are $\beta = 120$ and $V_A = \infty$, and the circuit parameters are $V_{CC} = 5\,\text{V}$, $R_C = 4\,\text{k}\Omega$, $R_B = 250\,\text{k}\Omega$, and $V_{BB} = 2.0\,\text{V}$. (a) Determine the hybrid-π parameter values of g_m, r_π, and r_o. (b) Find the small-signal voltage gain $A_v = v_o/v_s$. (c) If the time-varying output signal is given by $v_o = 0.8\sin(100t)\,\text{V}$, what is v_s?

4.4   The nominal quiescent collector current of a transistor is 1.2 mA. If the range of β for this transistor is $80 \leq \beta \leq 120$ and if the quiescent collector current changes by ± 10 percent, determine the range in values for g_m and r_π.

D4.5   For the circuit in Figure 4.3, $\beta = 120$, $V_{CC} = 5\,\text{V}$, $V_A = 100\,\text{V}$, and $R_B = 25\,\text{k}\Omega$. (a) Determine V_{BB} and R_C such that $r_\pi = 5.4\,\text{k}\Omega$ and the Q-point is in the center of the load line. (b) Find the resulting small-signal voltage gain $A_v = v_o/v_s$.

D4.6   For the circuit in Figure 4.14, $\beta = 100$, $V_A = \infty$, $V_{CC} = 10\,\text{V}$, and $R_B = 50\,\text{k}\Omega$. (a) Determine V_{BB} and R_C such that $I_{CQ} = 0.5\,\text{mA}$ and the Q-point is in the center of the load line. (b) Find the small-signal parameters g_m, r_π, and r_o. (c) Determine the small-signal voltage gain $A_v = v_o/v_s$.

Section 4.4   Common-Emitter Amplifier

4.7   The parameters of the transistor in the circuit in Figure P4.7 are $\beta = 100$ and $V_A = 100\,\text{V}$. (a) Find the dc voltages at the base and emitter terminals. (b) Find R_C such that $V_{CEQ} = 3.5\,\text{V}$. (c) Assuming C_C and C_E act as short circuits, determine the small-signal voltage gain $A_v = v_o/v_s$. (d) Repeat part (c) if a $500\,\Omega$ source resistor is in series with the v_s signal source.

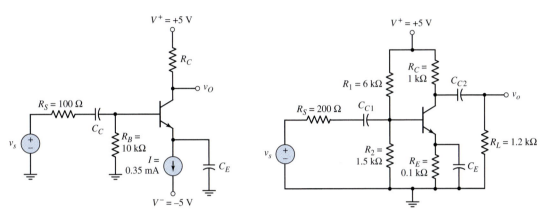

Figure P4.7   **Figure P4.8**

4.8   For the circuit in Figure P4.8, the transistor parameters are $\beta = 180$ and $r_o = \infty$. (a) Determine the Q-point values. (b) Find the small-signal hybrid-π parameters. (c) Find the small-signal voltage gain $A_v = v_o/v_s$.

*D4.9   The parameters of the transistor in the circuit in Figure P4.9 are $\beta = 150$ and $V_A = \infty$. (a) Determine R_1 and R_2 to obtain a bias-stable circuit with the Q-point in the center of the load line. (b) Determine the small-signal voltage gain $A_v = v_o/v_s$.

Figure P4.9   **Figure P4.10**

4.10   Assume that $\beta = 100$, $V_A = \infty$, $R_1 = 10\,\text{k}\Omega$, and $R_2 = 50\,\text{k}\Omega$ for the circuit in Figure P4.10. (a) Plot the Q-point on the dc load line. (b) Determine the small-signal voltage gain. (c) Determine the range of voltage gain if each resistor value varies by ± 5 percent.

D4.11 The transistor parameters for the circuit in Figure P4.10 are $\beta = 100$ and $V_A = \infty$. (a) Design the circuit such that it is bias stable and that the Q-point is in the center of the load line. (b) Determine the small-signal voltage gain of the designed circuit.

D4.12 For the circuit in Figure P4.12, the transistor parameters are $\beta = 100$ and $V_A = \infty$. Design the circuit such that $I_{CQ} = 0.25\,\text{mA}$ and $V_{CEQ} = 3\,\text{V}$. Find the small-signal voltage gain $A_v = v_o/v_s$. Find the input resistance seen by the signal source v_s.

Figure P4.12   Figure P4.13

D4.13 For the circuit in Figure P4.13, the transistor parameters are $\beta = 80$ and $V_A = 80\,\text{V}$. (a) Determine R_E such that $I_{EQ} = 0.75\,\text{mA}$. (b) Determine R_C such that $V_{ECQ} = 7\,\text{V}$. (c) For $R_L = 10\,\text{k}\Omega$, determine the small-signal voltage gain $A_v = v_o/v_s$. (d) Determine the impedance seen by the signal souce v_s.

D4.14 Assume the transistor in the circuit in Figure P4.14 has parameters $\beta = 120$ and $V_A = 100\,\text{V}$. (a) Design the circuit such that $V_{CEQ} = 3.75\,\text{V}$. (b) Determine the small-signal transresistance $R_m = v_o/i_s$.

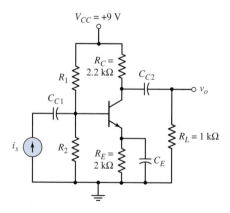

Figure P4.14

D4.15 For transistor parameters $\beta = 65$ and $V_A = 75\,\text{V}$, (a) design the circuit in Figure P4.15 such that the dc voltages at the base and collector terminals are 0.30 V and −3 V, respectively. (b) Determine the small-signal transconductance $G_f = i_o/v_s$.

Figure P4.15   Figure P4.16

4.16 The transistor in the circuit in Figure P4.16 is a 2N2907A with a nominal dc current gain of $\beta = 100$. Assume the range of h_{fe} is $80 \le h_{fe} \le 120$ and the range of h_{oe} is $10 \le h_{oe} \le 20\,\mu\text{S}$. For $h_{re} = 0$ determine: (a) the range of small-signal voltage gain $A_v = v_o/v_s$, and (b) the range in the input and output resistances R_i and R_o.

*D4.17 Design a one-transistor common-emitter preamplifier that can amplify a 10 mV (rms) microphone signal and produce a 0.5 V (rms) output signal. The source resistance of the microphone is 1 kΩ.

*4.18 For the transistor in the circuit in Figure P4.18, the parameters are $\beta = 100$ and $V_A = \infty$. (a) Determine the Q-point. (b) Find the small-signal parameters g_m, r_π, and r_o. (c) Find the small-signal voltage gain $A_v = v_o/v_s$ and the small-signal current gain $A_i = i_o/i_s$. (d) Find the input resistances R_{ib} and R_{is}. (e) Repeat part (c) if a 1 kΩ source resistor is in series with the v_s signal source.

Figure P4.18

4.19 If the collector of a transistor is connected to the base terminal, the transistor continues to operate in the forward-active mode, since the B–C junction is not reverse biased. Determine the small-signal resistance, $r_e = v_{ce}/i_e$, of this two-terminal device in terms of g_m, r_π, and r_o.

D4.20 Design an amplifier with the configuration similar to that shown in Figure 4.28. The source resistance is $R_S = 100\,\Omega$ and the voltage gain should be approximately -10. The total power dissipated in the circuit should be no more than approximately $0.12\,\text{mW}$.

D4.21 An ideal signal voltage source is given by $v_s = 5\sin(5000t)$ (mV). The peak current that can be supplied by this source is $0.2\,\mu\text{A}$. The desired output voltage across a $10\,\text{k}\Omega$ load resistor is $v_o = 100\sin(5000t)$ (mV). Design a one-transistor common-emitter amplifier to meet this specification.

***D4.22** Design a bias-stable common-emitter circuit that has a minimum open-circuit small-signal voltage gain of $|A_v| = 10$. The circuit is to be biased from a single power supply $V_{CC} = 10\,\text{V}$ that can supply a maximum current of 1 mA. The available transistors are pnp's with $\beta = 80$ and $V_A = \infty$. Minimize the number of capacitors required in the circuit.

***D4.23** Design a common-emitter circuit whose output is capacitively coupled to a load resistor $R_L = 10\,\text{k}\Omega$. The minimum small-signal voltage gain is to be $|A_v| = 20$. The circuit is to be biased at ± 5 V and each voltage source can supply a maximum of $0.5\,\text{mA}$. The parameters of the available transistors are $\beta = 120$ and $V_A = \infty$.

Section 4.5   AC Load Line Analysis

4.24 For the circuit in Figure P4.10 with circuit and transistor parameters as described in Problem 4.10, determine the maximum undistorted swing in the output voltage if the total instantaneous E–C voltage is to remain in the range $1 \le v_{EC} \le 11$ V.

4.25 For the circuit in Figure P4.12, let $\beta = 100$, $V_A = \infty$, $R_E = 12.9\,\text{k}\Omega$, and $R_C = 6\,\text{k}\Omega$. Determine the maximum undistorted swing in the output voltage if the total instantaneous C–E voltage is to remain in the range $1 \le v_{CE} \le 9$ V and if the total instantaneous collector current is to remain greater or equal to $50\,\mu\text{A}$.

4.26 Consider the circuit in Figure P4.7 with transistor parameters given in Problem 4.7. Assume $R_C = 6\,\text{k}\Omega$. (a) Determine the maximum undistorted swing in the output voltage if the total instantaneous C–E voltage is to remain in the range $0.5 \le v_{CE} \le 4.5$ V. (b) Using the results of part (a), determine the range of collector current.

4.27 Consider the circuit in Figure P4.15. Let $\beta = 100$, $V_A = \infty$, $R_B = 10\,\text{k}\Omega$, and $R_C = 4\,\text{k}\Omega$. Determine the maximum undistorted swing in the output current i_o if the total instantaneous collector current is $i_C \ge 0.08\,\text{mA}$ and the total instantaneous E–C voltage is in the range $1 \le v_{EC} \le 9$ V.

4.28 Consider the circuit in Figure P4.18 with transistor parameters described in Problem 4.18. Determine the maximum undistorted swing in the output current i_C if the total instantaneous collector current is $i_C \ge 0.1\,\text{mA}$ and the total instantaneous C–E voltage is in the range $1 \le v_{CE} \le 21$ V.

***D4.29** For the circuit in Figure P4.14, the transistor parameters are $\beta = 150$ and $V_A = \infty$. Design a bias-stable circuit to achieve the maximum undistorted swing in the output voltage if the total instantaneous C–E voltage is to remain in the range $1 \le v_{CE} \le 8$ V.

***RD4.30** In the circuit in Figure P4.8 with transistor parameters $\beta = 180$ and $V_A = \infty$, redesign the bias resistors R_1 and R_2 to achieve maximum symmetrical swing in the output voltage and to maintain a bias-stable circuit. The total instantaneous C–E voltage is to remain in the range $0.5 \leq v_{CE} \leq 4.5\,\text{V}$ and the total instantaneous collector current is to be $i_C \geq 0.25\,\text{mA}$.

4.31 For the circuit in Figure P4.16, the transistor parameters are $\beta = 100$ and $V_A = \infty$. (a) Determine the maximum undistorted swing in the output voltage if the total instantaneous E–C voltage is to remain in the range $1 \leq v_{EC} \leq 9\,\text{V}$. (b) Using the results of part (a), determine the range of collector current.

Section 4.6 Common-Collector (Emitter-Follower) Amplifier

4.32 The transistor parameters for the circuit in Figure P4.32 are $\beta = 180$ and $V_A = \infty$. (a) Find I_{CQ} and V_{CEQ}. (b) Plot the dc and ac load lines. (c) Calculate the small-signal voltage gain. (d) Determine the input and output resistances R_{ib} and R_o.

Figure P4.32   **Figure P4.33**

4.33 Consider the circuit in Figure P4.33. The transistor parameters are $\beta = 120$ and $V_A = \infty$. Repeat parts (a)–(d) of Problem 4.32.

*D4.34 For the circuit in Figure P4.34, the transistor current gain is $\beta = 80$ and $R_L = 500\,\Omega$. Design the circuit to obtain a small-signal current gain of $A_i = i_o/i_s = 8$. Let $V_{CC} = 10\,\text{V}$. Find R_1, R_2, and the output resistance R_o if $R_E = 500\,\Omega$. What is the current gain if $R_L = 2000\,\Omega$?

Figure P4.34

D4.35 Design an emitter-follower circuit with the configuration shown in Figure 4.44 such that the input resistance R_i, as defined in Figure 4.46, is $120 \, k\Omega$. Assume transistor parameters of $\beta = 120$ and $V_A = \infty$. Let $V_{CC} = 5 \, V$ and $R_E = 2 \, k\Omega$. Find new values of R_1 and R_2. The Q-point should be approximately in the center of the load line.

D4.36 (a) For the emitter-follower circuit in Figure P4.34, assume $V_{CC} = 24 \, V$, $\beta = 75$, and $A_i = i_o/i_s = 8$. Design the circuit to drive an $8 \, \Omega$ load. (b) Determine the maximum undistorted swing in the output voltage. (c) Determine the output resistance R_o.

4.37 For the circuit shown in Figure P4.37, let $V_{CC} = 5 \, V$, $R_L = 4 \, k\Omega$, $R_E = 3 \, k\Omega$, $R_1 = 60 \, k\Omega$, and $R_2 = 40 \, k\Omega$. The transistor parameters are $\beta = 50$ and $V_A = 80 \, V$. (a) Determine I_{CQ} and V_{ECQ}. (b) Plot the dc and ac load lines. (c) Determine $A_v = v_o/v_s$ and $A_i = i_o/i_s$. (d) Determine R_{ib} and R_o. (e) Determine the range in current gain if each resistor value varies by ± 5 percent.

V_{CC}

Figure P4.37

*D4.38 The output of an amplifier can be represented by $v_s = 4 \sin \omega t (V)$ and $R_S = 4 \, k\Omega$. An emitter-follower circuit, with the configuration shown in Figure 4.49, is to be designed such that the output signal does not vary by more than 5 percent when a load in the range $R_L = 4$ to $10 \, k\Omega$ is connected to the output. The transistor current gain is in the range $90 \le \beta \le 130$ and the Early voltage is $V_A = \infty$. For your design, find the minimum and maximum possible value of the output voltage.

4.39 (a) In the circuit shown in Figure P4.39, determine the range in small-signal voltage gain $A_v = v_o/v_s$ and current gain $A_i = i_o/i_s$ if β is in the range $75 \le \beta \le 150$.

$V_{CC} = +10 \, V$

Figure P4.39

Figure P4.40

(b) Repeat part (a) if a $5\,\text{k}\Omega$ source resistor is in series with the v_s signal source.

4.40 For the transistor in Figure P4.40, $\beta = 80$ and $V_A = 150\,\text{V}$. (a) Determine the dc voltages at the base and emitter terminals. (b) Calculate the small-signal parameters g_m, r_π, and r_o. (c) Determine the small-signal voltage gain and current gain. (d) Repeat part (c) if a $2\,\text{k}\Omega$ source resistor is in series with the v_s signal source.

4.41 Consider the circuit shown in Figure P4.40. The transistor current gain is in the range $100 \leq \beta \leq 180$ and the Early voltage is $V_A = 150\,\text{V}$. Determine the range in small-signal voltage gain if the load resistance varies from $R_L = 0.5\,\text{k}\Omega$ to $R_L = 500\,\text{k}\Omega$.

*D4.42 For the transistor in Figure P4.42, the parameters are $\beta = 100$ and $V_A = \infty$. (a) Design the circuit such that $I_{EQ} = 1\,\text{mA}$ and the Q-point is in the center of the dc load line. (b) If the peak-to-peak sinusoidal output voltage is $4\,\text{V}$, determine the peak-to-peak sinusoidal signals at the base of the transistor and the peak-to-peak value of v_s. (c) If a load resistor $R_L = 1\,\text{k}\Omega$ is connected to the output through a coupling capacitor, determine the peak-to-peak value in the output voltage, assuming v_s is equal to the value determined in part (b).

Figure P4.42

*D4.43 An emitter-follower amplifier, with the configuration shown in Figure 4.49, is to be designed such that an audio signal given by $v_s = 5\sin(3000t)\,\text{V}$ but with a source resistance of $R_S = 10\,\text{k}\Omega$ can drive a small speaker. Assume the supply voltages are $V^+ = +12\,\text{V}$ and $V^- = -12\,\text{V}$. The load, representing the speaker, is $R_L = 12\,\Omega$. The amplifier should be capable of delivering approximately $1\,\text{W}$ of average power to the load. What is the signal power gain of your amplifier?

Section 4.7   Common-Base Amplifier

4.44 For the circuit shown in Figure P4.44, $\beta = 125$, $V_A = \infty$, $V_{CC} = 18$ V, $R_L = 4\,k\Omega$, $R_E = 3\,k\Omega$, $R_C = 4\,k\Omega$, $R_1 = 25.6\,k\Omega$, and $R_2 = 10.4\,k\Omega$. The input signal is a current. (a) Determine the Q-point values. (b) Determine the transresistance $R_m = v_o/i_s$. (c) Find the small-signal voltage gain $A_v = v_o/v_s$.

Figure P4.44

*RD4.45 For the common-base circuit shown in Figure P4.44, let $\beta = 100$, $V_A = \infty$, $V_{CC} = 12$ V, $R_L = 12\,k\Omega$, and $R_E = 500\,\Omega$. (a) Redesign the circuit such that the small-signal voltage gain is $A_v = v_o/v_s = 10$. (b) What are the Q-point values? (c) What is the small-signal voltage gain if R_2 is bypassed by a large capacitor?

*D4.46 A photodiode in an optical transmission system, such as shown in Figure 1.35, can be modeled as a Norton equivalent circuit with i_s in parallel with R_S as shown in Figure P4.44. Assume that the current source is given by $i_s = 2.5\sin\omega t\,\mu A$ and $R_S = 50\,k\Omega$. Design the common-base circuit of Figure P4.44 such that the output voltage is $v_o = 5\sin\omega t\,mV$. Assume transistor parameters of $\beta = 120$ and $V_A = \infty$. Let $V_{CC} = 5$ V.

4.47 For the circuit shown in Figure P4.47, the transistor parameters are $\beta = 100$ and $V_A = \infty$. (a) Determine the dc voltages at the collector, base, and emitter terminals. (b) Determine the small-signal voltage gain $A_v = v_o/v_s$. (c) Find the input resistance R_i.

4.48 In the common-base circuit shown in Figure P4.48, the transistor is a 2N2907A, with a nominal dc current gain of $\beta = 80$. (a) Determine I_{CQ} and V_{ECQ}. (b) Using the h-parameters (assuming $h_{re} = 0$), determine the range in small-signal voltage gain $A_v = v_o/v_s$. (c) Determine the range in input and output resistances R_i and R_o.

Figure P4.47

Figure P4.48

*D4.49   In the circuit of Figure P4.48, let $V_{EE} = V_{CC} = 5\,\text{V}$, $\beta = 100$, $V_A = \infty$, $R_L = 1\,\text{k}\Omega$, and $R_S = 0$. (a) Design the circuit such that the small-signal voltage gain is $A_v = v_o/v_s = 25$ and $V_{ECQ} = 3\,\text{V}$. (b) What are the values of the small-signal parameters g_m, r_π, and r_o?

Section 4.9   Multistage Amplifiers

*4.50   The parameters for each transistor in the circuit shown in Figure P4.50 are $\beta = 100$ and $V_A = \infty$. (a) Determine the small-signal parameters g_m, r_π, and r_o for both transistors. (b) Determine the small-signal voltage gain $A_{v1} = v_{o1}/v_s$, assuming v_{o1} is connected to an open circuit, and determine the gain $A_{v2} = v_o/v_{o1}$. (c) Determine the overall small-signal voltage gain $A_v = v_o/v_s$. Compare the overall gain with the product $A_{v1} \cdot A_{v2}$, using the values calculated in part (b).

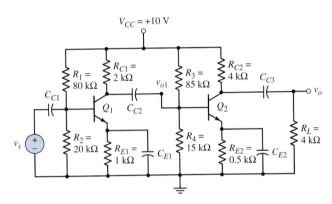

Figure P4.50

*4.51   Consider the circuit shown in Figure P4.51 with transistor parameters $\beta = 120$ and $V_A = \infty$. (a) Determine the small-signal parameters g_m, r_π, and r_o for both transistors. (b) Plot the dc and ac load lines for both transistors. (c) Determine the overall small-signal voltage gain $A_v = v_o/v_s$. (d) Determine the input resistance R_{is} and the output resistance R_o. (e) Determine the maximum undistorted swing in the output voltage.

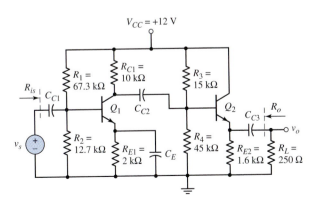

Figure P4.51

4.52 For the circuit shown in Figure P4.52, assume transistor parameters of $\beta = 100$ and $V_A = \infty$. (a) Determine the dc collector current in each transistor. (b) Find the small-signal voltage gain $A_v = v_o/v_s$. (c) Determine the input and output resistances R_{ib} and R_o.

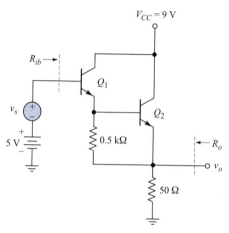

Figure P4.52

*4.53 For each transistor in Figure P4.53, the parameters are $\beta = 100$ and $V_A = \infty$. (a) Determine the Q-point values for both Q_1 and Q_2. (b) Determine the overall small-signal voltage gain $A_v = v_o/v_s$. (c) Determine the input and output resistances R_{is} and R_o.

Figure P4.53   **Figure P4.54**

4.54 An equivalent ac circuit of a Darlington pair configuration is shown in Figure P4.54. (a) Derive the expression for the output resistance R_o as a function of I_{Bias} and I_{C2}. Take into account the transistor output resistances r_{o1} and r_{o2}. (b) Assuming transistor parameters of $\beta = 100$ and $V_A = 100\,\text{V}$, determine R_o for (i) $I_{C2} = I_{Bias} = 1\,\text{mA}$ and (ii) $I_{C2} = 1\,\text{mA}$, $I_{Bias} = 0$.

Section 4.10   Power Considerations

4.55 The transistor in the circuit shown in Figure 4.25 has parameters $\beta = 100$ and $V_A = 100\,\text{V}$. (a) Determine the average power dissipated in the transistor and R_C, for $v_s = 0$. (b) Determine the maximum undistorted signal power that can be delivered to R_C.

4.56 Consider the circuit shown in Figure 4.35. The transistor parameters are $\beta = 120$ and $r_o = \infty$. (a) Calculate the average power dissipated in the transistor, R_E, and R_C, for $v_s = 0$. (b) Determine the maximum undistorted signal power that can be delivered to R_L.

4.57 For the circuit shown in Figure 4.40, use the circuit and transistor parameters described in Example 4.8. (a) Calculate the average power dissipated in the transistor, R_E, and R_C, for $v_s = 0$. (b) Determine the maximum signal power that can be delivered to R_L. What are the signal powers dissipated in R_E and R_C, and what is the average power dissipated in the transistor in this case?

4.58 For the circuit shown in Figure 4.51, the transistor parameters are $\beta = 100$ and $V_A = 100\,\text{V}$, and the source resistor is $R_S = 0$. Determine the maximum undistorted signal power that can be delivered to R_L if: (a) $R_L = 1\,\text{k}\Omega$, and (b) $R_L = 10\,\text{k}\Omega$.

4.59 Consider the circuit shown in Figure 4.58 with parameters given in Exercise 4.25. (a) Calculate the average power dissipated in the transistor and R_C, for $v_s = 0$. (b) Determine the maximum undistorted signal power that can be delivered to R_L, and the resulting average power dissipated in the transistor and R_C.

COMPUTER SIMULATION PROBLEMS

4.60 Consider Example 4.2. Using a computer simulation analysis, investigate the effect of the Early voltage on the small-signal characteristics of the circuit.

4.61 The circuit in Figure P4.61 can be used to simulate the circuit shown in Figure 4.37(a). Assume Early voltages of $V_A = 60\,\text{V}$. (a) Plot the voltage transfer characteristics, v_O versus V_{BB}, over the range $0 \leq V_{BB} \leq 1\,\text{V}$. (b) Set V_{BB} such that the dc value of the output voltage is $v_O \cong 2.5\,\text{V}$. Determine the small-signal voltage gain at this Q-point. Compare the results to those found in Example 4.7.

4.62 Verify the results of Example 4.8 with a computer simulation analysis.

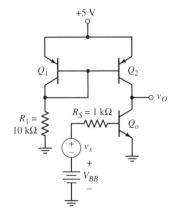

Figure P4.61

4.63   Verify the input and output resistances of the emitter-follower circuit described in Example 4.11.

4.64   Perform a computer simulation analysis of the common-base circuit described in Exercise 4.25. In addition, assume $V_A = 80$ V and determine the output resistance looking into the collector of the transistor. How does this value compare to $r_o = V_A/I_{CQ}$?

DESIGN PROBLEMS

[Note: Each design should be correlated with a computer simulation.]

*D4.65   Design a common-emitter circuit with a small-signal voltage gain of $|A_v| = 50$ while driving a load $R_L = 5$ kΩ. The source signal is $v_s = 0.02 \cos \omega t$ V and the source resistance is $R_S = 1$ kΩ. Bias the circuit at ± 5 V, and use transistors with a maximum collector current rating of 10 mA and current gains in the range $80 \le \beta \le 150$.

*D4.66   For the circuit in Figure P4.37, let $V_{CC} = 10$ V and $R_L = 1$ kΩ. The transistor parameters are $\beta = 120$ and $V_A = \infty$. (a) Design the circuit such that the current gain is $A_i = 18$. (b) Determine R_{ib} and R_o. (c) Find the maximum undistorted swing in the output voltage.

*D4.67   Design a common-base amplifier with the general configuration shown in Figure 4.58. The available power supplies are ± 10 V. The output resistance of the signal source is $50\ \Omega$, and the input resistance of the amplifier should match this value. The output resistance is $R_L = 2$ kΩ, and the output voltage is to have the largest possible symmetrical swing. In order to maintain linearity, the peak value of the B–E signal voltage should be limited to 15 mV. Assume that transistors with $\beta = 150$ are available. Specify the current and power ratings of the transistors.

*D4.68   A microphone puts out a peak voltage of 1 mV and has an output resistance of 10 kΩ. Design an amplifier system to drive an 8 Ω speaker, producing 2 W of signal power. Use a 24 V power supply to bias the circuit. Assume a current gain of $\beta = 50$ for the available transistors. Specify the current and power ratings of the transistors.

*D4.69   Redesign the two-stage amplifier in Figure 4.60 such that a symmetrical sine wave with a peak value of ± 3 V can be obtained at the output. The load resistor is still $R_L = 5$ kΩ. To avoid distortion, the minimum C–E voltage should be at least 1 V and the maximum C–E voltage should be no more than 9 V. Assume the transistor current gains are $\beta = 100$. If the Early voltage for each transistor is $V_A = \infty$, calculate the resulting overall small-signal voltage gain. State the value of each resistor and the quiescent values of each transistor.

5

The Field-Effect Transistor

5.0 PREVIEW

In the last two chapters, we looked at the bipolar junction transistor and BJT circuits. In this chapter, we introduce the second major type of transistor, the field-effect transistor (FET). There are two general classes of FETs: the metal-oxide-semiconductor FET (MOSFET) and the junction FET (JFET). The MOSFET has led to the second electronics revolution in the 1970s and 1980s, in which the microprocessor has made possible powerful desktop computers and sophisticated hand-held calculators. The MOSFET can be made very small, so high-density VLSI circuits and high-density memories are possible.

We begin the chapter with a look at the physical structure and operation of the MOSFET. The current–voltage characteristics of the device are developed, and then the dc analysis and design of MOSFET circuits are considered. We will see how the MOSFET can be used in place of resistors in a circuit, so that circuits containing only MOSFETs can be designed.

In the junction FET, the junction may be a pn junction, which forms a pn JFET, or a Schottky barrier junction, which forms a metal-semiconductor FET, or MESFET. MESFETs are used in very high speed or high-frequency applications, such as microwave amplifiers. Since JFETs are specialized devices, the discussion of JFET circuits is brief.

Although the emphasis of this chapter is on dc circuits, we discuss how the FET can be used in switch, digital, and linear amplifier applications. A major goal of this chapter is to enable the reader to become very familiar and comfortable with the MOSFET properties and to be able to quickly analyze and design the dc response of FET circuits.

5.1 MOS FIELD-EFFECT TRANSISTOR

The **metal-oxide-semiconductor field-effect transistor (MOSFET)** became a practical reality in the 1970s. The MOSFET, compared to BJTs, can be made very small (that is, it occupies a very small area on an IC chip). Since digital circuits can be designed using only MOSFETs, with essentially no resistors or diodes required, high-density VLSI circuits, including microprocessors and memories, can be fabricated. The MOSFET has made possible the hand-held calculator and the powerful personal computer. MOSFETs can also be used in analog circuits, as we will see in the next chapter.

In the MOSFET, the current is controlled by an electric field applied perpendicular to both the semiconductor surface and to the direction of current. The phenomenon used to modulate the conductance of a semiconductor, or control the current in a semiconductor, by applying an electric field perpendicular to the surface is called the **field effect**. Again, the basic transistor principle is that the voltage between two terminals controls the current through the third terminal.

In the following two sections, we will discuss the various types of MOSFETs, develop the i–v characteristics, and then consider the dc biasing of various MOSFET circuit configurations. After studying these sections, you should be familiar and comfortable with the MOSFET and MOSFET circuits.

5.1.1   Two-Terminal MOS Structure

The heart of the MOSFET is the metal-oxide-semiconductor capacitor shown in Figure 5.1. The metal may be aluminum or some other type of metal. In many cases, the metal is replaced by a high-conductivity polycrystalline silicon layer deposited on the oxide. However, the term metal is usually still used in referring to MOSFETs. In the figure, the parameter t_{ox} is the thickness of the oxide and ϵ_{ox} is the oxide permittivity.

Figure 5.1   The basic MOS capacitor structure

The physics of the MOS structure can be explained with the aid of a simple parallel-plate capacitor.[1] Figure 5.2(a) shows a parallel-plate capacitor with the top plate at a negative voltage with respect to the bottom plate. An insulator material separates the two plates. With this bias, a negative charge exists on the top plate, a positive charge exists on the bottom plate, and an electric field is induced between the two plates, as shown.

A MOS capacitor with a p-type semiconductor substrate is shown in Figure 5.2(b). The top metal terminal, also called the **gate**, is at a negative voltage with respect to the semiconductor substrate. From the example of the parallel-plate capacitor, we can see that a negative charge will exist on the top metal plate and an electric field will be induced in the direction shown in the figure. If the electric field penetrates the semiconductor, the holes in the p-type semiconductor will experience a force toward the oxide-semiconductor inter-

---

[1]The capacitance of a parallel plate capacitor, neglecting fringing fields, is $C = \epsilon A/d$, where A is the area of one plate, d is the distance between plates, and ϵ is the permittivity of the medium between the plates.

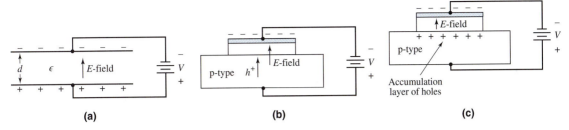

Figure 5.2 (a) A parallel-plate capacitor, showing the electric field and conductor charges, (b) a corresponding MOS capacitor with a negative gate bias, showing the electric field and charge flow, and (c) the MOS capacitor with an accumulation layer of holes

face. The equilibrium distribution of charge in the MOS capacitor with this particular applied voltage is shown in Figure 5.2(c). An accumulation layer of positively-charged holes in the oxide-semiconductor junction corresponds to the positive charge on the bottom "plate" of the MOS capacitor.

Figure 5.3(a) shows the same MOS capacitor, but with the polarity of the applied voltage reversed. A positive charge now exists on the top metal plate and the induced electric field is in the opposite direction, as shown. In this case, if the electric field penetrates the semiconductor, holes in the p-type material will experience a force away from the oxide-semiconductor interface. As the holes are pushed away from the interface, a negative space-charge region is created, due to the fixed acceptor impurity atoms. The negative charge in the induced depletion region corresponds to the negative charge on the bottom "plate" of the MOS capacitor. Figure 5.3(b) shows the equilibrium distribution of charge in the MOS capacitor with this applied voltage.

Figure 5.3 The MOS capacitor with p-type substrate: (a) effect of positive gate bias, showing the electric field and charge flow, (b) the MOS capacitor with an induced space-charge region due to a moderate gate bias, and (c) the MOS capacitor with an induced space-charge region and electron inversion layer due to a larger gate bias

When a larger positive voltage is applied to the gate, the magnitude of the induced electric field increases. Minority carrier electrons are attracted to the oxide-semiconductor interface, as shown in Figure 5.3(c). This region of minority carrier electrons is called an **electron inversion layer**. The magnitude of the charge in the inversion layer is a function of the applied gate voltage.

The same basic charge distributions can be obtained in a MOS capacitor with an n-type semiconductor substrate. Figure 5.4(a) shows this MOS capacitor structure, with a positive voltage applied to the top gate terminal. A positive charge is created on the top gate and an electric field is induced in the

Figure 5.4   The MOS capacitor with n-type substrate for: (a) a positive gate bias, (b) a moderate negative bias, and (c) a larger negative bias

direction shown. In this situation, an accumulation layer of electrons is induced in the n-type semiconductor.

Figure 5.4(b) shows the case when a negative voltage is applied to the gate terminal. A positive space-charge region is induced in the n-type substrate by the induced electric field. When a larger negative voltage is applied, a region of positive charge is created at the oxide-semiconductor interface, as shown in Figure 5.4(c). This region of minority carrier holes is called a **hole inversion layer**. The magnitude of the positive charge in the inversion layer is a function of the applied gate voltage.

The term **enhancement mode** means that a voltage must be applied to the gate to create an inversion layer. For the MOS capacitor with a p-type substrate, a positive gate voltage must be applied to create the electron inversion layer; for the MOS capacitor with an n-type substrate, a negative gate voltage must be applied to create the hole inversion layer.

5.1.2   n-Channel Enhancement-Mode MOSFET

We will now apply the concepts of an inversion layer charge in a MOS capacitor to create a transistor.

Transistor Structure

Figure 5.5(a) shows a simplified cross section of a MOS field-effect transistor. The gate, oxide, and p-type substrate regions are the same as those of a MOS capacitor. In addition, we now have two n-regions, called the **source terminal** and **drain terminal**. The current in a MOSFET is the result of the flow of charge in the inversion layer, also called the **channel region**, adjacent to the oxide–semiconductor interface.

The channel length L and channel width W are defined on the figure. The channel length of a typical integrated circuit MOSFET is less than 1 μm (10^{-6} m), which means that MOSFETs are small devices. The oxide thickness t_{ox} is typically on the order of 400 angstroms, or less.

The diagram in Figure 5.5(a) is a simplified sketch of the basic structure of the transistor. Figure 5.5(b) shows a more detailed cross section of a MOSFET fabricated into an integrated circuit configuration. A thick oxide, called the **field oxide**, is deposited outside the area in which the metal interconnect lines are formed. The gate material is usually heavily doped polysilicon. Even though the actual structure of a MOSFET may be fairly complex, the simplified diagram may be used to develop the basic transistor characteristics.

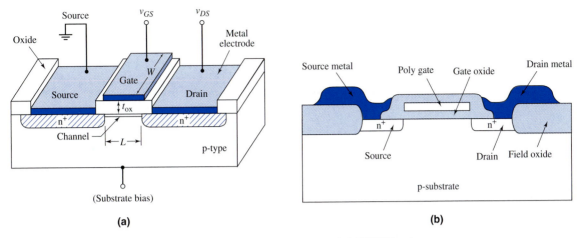

Figure 5.5   (a) Schematic diagram of an n-channel enhancement mode MOSFET and (b) an n-channel MOSFET, showing the field oxide and polysilicon gate

Basic Transistor Operation

With zero bias applied to the gate, the source and drain terminals are separated by the p-region, as shown in Figure 5.6(a). This is equivalent to two back-to-back diodes, as shown in Figure 5.6(b). The current in this case is essentially zero. If a large enough positive gate voltage is applied, an electron inversion layer is created at the oxide–semiconductor interface and this layer "connects" the n-source to the n-drain, as shown in Figure 5.6(c). A current can then be generated between the source and drain terminals. Since a voltage must be applied to the gate to create the inversion charge, this transistor is called an **enhancement-mode MOSFET**. Also, since the carriers in the inversion layer are electrons, this device is also called an **n-channel MOSFET**.

The source terminal supplies carriers that flow through the channel, and the drain terminal allows the carriers to drain from the channel. For the n-channel MOSFET, or **NMOS transistor**, electrons flow from the source to the drain with an applied drain-to-source voltage, which means the conventional

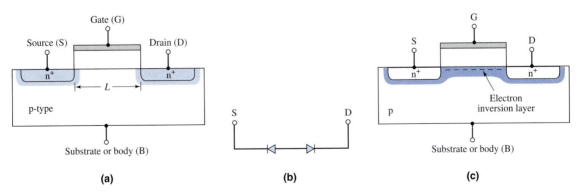

Figure 5.6   (a) Cross section of the n-channel MOSFET prior to the formation of an electron inversion layer, (b) equivalent back-to-back diodes between source and drain when the transistor is in cutoff, and (c) cross section after the formation of an electron inversion layer

current enters the drain and leaves the source. The magnitude of the current is a function of the amount of charge in the inversion layer, which in turn is a function of the applied gate voltage. Since the gate terminal is separated from the channel by an oxide or insulator, there is no gate current. Similarly, since the channel and substrate are separated by a space-charge region, there is essentially no current through the substrate.

5.1.3 Ideal MOSFET Current–Voltage Characteristics

The **threshold voltage** of the n-channel MOSFET is denoted as V_{TN} and is defined[2] as the applied gate voltage needed to create an inversion charge in which the density is equal to the concentration of majority carriers in the semiconductor substrate. In simple terms, we can think of the threshold voltage as the gate voltage required to "turn on" the transistor.

For the n-channel enhancement-mode MOSFET, the threshold voltage is positive because a positive gate voltage is required to create the inversion charge. If the gate voltage is less than the threshold voltage, the current in the device is essentially zero. If the gate voltage is greater than the threshold voltage, a drain-to-source current is generated as the drain-to-source voltage is applied. The gate and drain voltages are measured with respect to the source.

Figure 5.7(a) shows an n-channel enhancement-mode MOSFET with the source and substrate terminals connected to ground. The gate-to-source voltage is less than the threshhold voltage, and there is a small drain-to-source voltage. With this bias configuration, there is no electron inversion layer, the drain-to-substrate pn junction is reverse biased, and the drain current is zero (neglecting pn junction leakage currents).

Figure 5.7(b) shows the same MOSFET with an applied gate voltage greater than the threshold voltage. In this situation, an electron inversion layer is created and, when a small drain voltage is applied, electrons in the inversion layer flow from the source to the positive drain terminal. The conven-

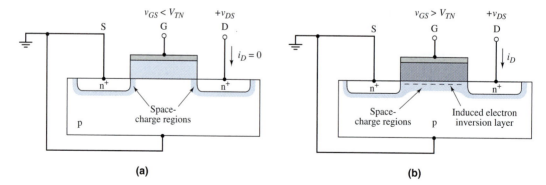

Figure 5.7   The n-channel enhancement-mode MOSFET (a) with an applied gate voltage $v_{GS} < V_{TN}$, and (b) with an applied gate voltage $v_{GS} > V_{TN}$

[2]The usual notation for threshold voltage is V_T. However, since we have defined the thermal voltage as $V_T = kT/q$, we will use V_{TN} for the threshold voltage of the n-channel device.

tional current enters the drain terminal and leaves the source terminal. Note that a positive drain voltage creates a reverse-biased drain-to-substrate pn junction, so current flows through the channel region and not through a pn junction.

The i_D versus v_{DS} characteristics for small values of v_{DS} are shown in Figure 5.8. When $v_{DS} < V_{TN}$, the drain current is zero. When v_{GS} is greater than V_{TN}, the channel inversion charge is formed and the drain current increases with v_{DS}. Then, with a larger gate voltage, a larger inversion charge density is created, and the drain current is greater for a given value of v_{DS}.

Figure 5.9(a) shows the basic MOS structure for $v_{GS} > V_{TN}$ and a small applied v_{DS}. In the figure, the thickness of the inversion channel layer qualitatively indicates the relative charge density, which for this case is essentially constant along the entire channel length. The corresponding i_D versus v_{DS} curve is also shown in the figure.

Figure 5.8 Plot of i_D versus v_{DS} characteristic for small values of v_{DS} at three v_{GS} voltages

Figure 5.9(b) shows the situation when v_{DS} increases. As the drain voltage increases, the voltage drop across the oxide near the drain terminal decreases, which means that the induced inversion charge density near the drain also decreases. The incremental conductance of the channel at the drain then decreases, which causes the slope of the i_D versus v_{DS} curve to decrease. This effect is shown in the i_D versus v_{DS} curve in the figure.

As v_{DS} increases to the point where the potential difference across the oxide at the drain terminal is equal to V_{TN}, the induced inversion charge density at the drain terminal is zero. This effect is shown schematically in Figure 5.9(c). For this condition, the incremental channel conductance at the drain is zero, which means that the slope of the i_D versus v_{DS} curve is zero. We can write

$$v_{GS} - v_{DS}(\text{sat}) = V_{TN} \qquad\qquad \textbf{(5.1(a))}$$

or

$$v_{DS}(\text{sat}) = v_{GS} - V_{TN} \qquad\qquad \textbf{(5.1(b))}$$

where $v_{DS}(\text{sat})$ is the drain-to-source voltage that produces zero inversion charge density at the drain terminal.

When v_{DS} becomes larger than $v_{DS}(\text{sat})$, the point in the channel at which the inversion charge is just zero moves toward the source terminal. In this case, electrons enter the channel at the source, travel through the channel toward the drain, and then, at the point where the charge goes to zero, are injected into the space-charge region, where they are swept by the E-field to the drain contact. This process is similar to electrons being swept across a reverse-biased B–C junction in a bipolar transistor. In the ideal MOSFET, the drain current is constant for $v_{DS} > v_{DS}(\text{sat})$. This region of the i_D versus v_{DS} characteristic is referred to as the **saturation region**,[3] which is shown in Figure 5.9(d).

---

[3]The term saturation for MOSFETs is not to be confused with saturation for BJTs. We commonly say that a BJT is driven into *saturation*, which means that the collector current no longer increases with increasing base current and the collector–emitter voltage has reached a minimum value. When a MOSFET is biased in the *saturation region*, the drain current is essentially independent of drain voltage for a constant gate-to-source voltage.

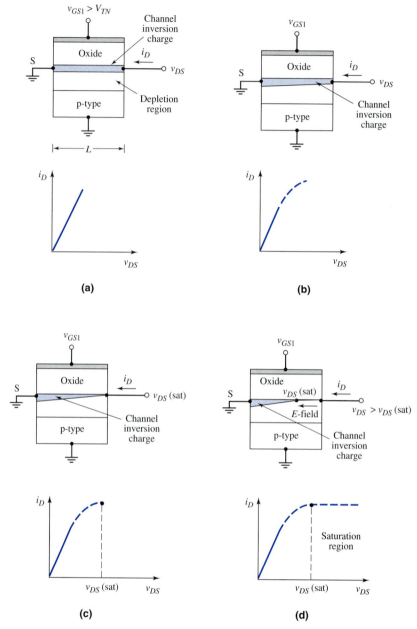

Figure 5.9 Cross section and i_D versus v_{DS} curve for an n-channel enhancement-mode MOSFET when $v_{GS} > V_{TN}$ for: (a) a small v_{DS} value, (b) a larger v_{DS} value, (c) $v_{DS} = v_{DS}(\text{sat})$, and (d) $v_{DS} > v_{DS}(\text{sat})$

As the applied gate-to-source voltage changes, the i_D versus v_{DS} curve changes. In Figure 5.8, we saw that the initial slope of i_D versus v_{DS} increases as v_{GS} increases. Also, Equation (5.1(b)) shows that $v_{DS}(\text{sat})$ is a function of v_{GS}. Therefore, we can generate the family of curves for this n-channel enhancement mode MOSFET as shown in Figure 5.10.

Although the derivation of the current–voltage characteristics of the MOSFET is beyond the scope of this text, we can define the relationships.

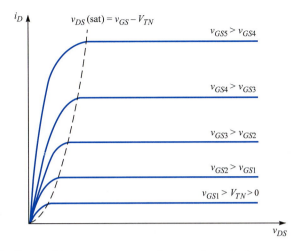

Figure 5.10   Family of i_D versus v_{DS} curves for an n-channel enhancement mode MOSFET

The region for which $v_{DS} < v_{DS}(\text{sat})$ is known as the **nonsaturation** or **triode region**. The ideal current–voltage characteristics in this region are described by the equation

$$i_D = K_n[2(v_{GS} - V_{TN})v_{DS} - v_{DS}^2]$$ **(5.2(a))**

In the saturation region, the ideal current–voltage characteristics for $v_{GS} > V_{TN}$ are described by the equation

$$i_D = K_n(v_{GS} - V_{TN})^2$$ **(5.2(b))**

In the saturation region, since the ideal drain current is independent of the drain-to-source voltage, the incremental or small-signal resistance is infinite. We see that

$$r_o = \Delta v_{DS}/\Delta i_D|_{v_{GS}=\text{const.}} = \infty$$

The parameter K_n is called the **conduction parameter** for the n-channel device and is given by

$$K_n = \frac{W\mu_n C_{ox}}{2L}$$ **(5.3(a))**

where C_{ox} is the oxide capacitance per unit area. The capacitance is given by

$$C_{ox} = \epsilon_{ox}/t_{ox}$$

where t_{ox} is the oxide thickness and ϵ_{ox} is the oxide permittivity. For silicon devices, $\epsilon_{ox} = (3.9)(8.85 \times 10^{-14})$ F/cm. The parameter μ_n is the mobility of the electrons in the inversion layer. The channel width W and channel length L were shown in Figure 5.5(a).

As Equation (5.3(a)) indicates, the conduction parameter is a function of both electrical and geometric parameters. The oxide capacitance and carrier mobility are essentially constants for a given fabrication technology. However, the geometry, or width-to-length ratio W/L, is a variable in the design of MOSFETs that is used to produce specific current–voltage characteristics in MOSFET circuits.

We can rewrite the conduction parameter in the form

$$K_n = \frac{k_n'}{2} \cdot \frac{W}{L} \tag{5.3(b)}$$

where $k_n' = \mu_n C_{ox}$. Normally, k_n' is considered to be a constant, so Equation (5.3(b)) emphasizes that the width-to-length ratio W/L is the transistor design variable.

Example 5.1   **Objective:** Calculate the current in an n-channel MOSFET.

Consider an n-channel enhancement mode MOSFET with the following parameters: $V_{TN} = 0.75\,\text{V}$, $W = 40\,\mu\text{m}$, $L = 4\,\mu\text{m}$, $\mu_n = 650\,\text{cm}^2/\text{V-s}$, $t_{ox} = 450\,\mathring{\text{A}}$, and $\epsilon_{ox} = (3.9)(8.85 \times 10^{-14})\,\text{F/cm}$. Determine the current when $V_{GS} = 2V_{TN}$, for the transistor biased in the saturation region.

Solution: The conduction parameter is determined by Equation (5.3(a)). First, consider the units involved in this equation, as follows:

$$K_n = \frac{W(\text{cm}) \cdot \mu_n \left(\dfrac{\text{cm}^2}{\text{V-s}}\right) \epsilon_{ox}\left(\dfrac{\text{F}}{\text{cm}}\right)}{2L(\text{cm}) \cdot t_{ox}(\text{cm})} = \frac{\text{F}}{\text{V-s}} = \frac{(\text{C/V})}{\text{V-s}} = \frac{\text{A}}{\text{V}^2}$$

The value of the conduction parameter is therefore

$$K_n = \frac{W\mu_n\epsilon_{ox}}{2Lt_{ox}} = \frac{(40 \times 10^{-4})(650)(3.9)(8.85 \times 10^{-14})}{2(4 \times 10^{-4})(450 \times 10^{-8})}$$

or

$$K_n = 0.249\,\text{mA/V}^2$$

From Equation (5.2(b)) for $v_{GS} = 2V_{TN}$, we find

$$i_D = K_n(v_{GS} - V_{TN})^2 = (0.249)(1.5 - 0.75)^2 = 0.140\,\text{mA}$$

Comment: The current capability of a transistor can be increased by increasing the conduction parameter. For a given fabrication technology, K_n is adjusted by varying the transistor width W.

Test Your Understanding

5.1   (a) An n-channel enhancement-mode MOSFET has a threshold voltage of $V_{TN} = 1.2\,\text{V}$ and an applied gate-to-source voltage of $v_{GS} = 2\,\text{V}$. Determine the region of operation when: (i) $v_{DS} = 0.4\,\text{V}$; (ii) $v_{DS} = 1\,\text{V}$; and (iii) $v_{DS} = 5\,\text{V}$. (b) Repeat part (a) for an n-channel depletion-mode MOSFET with a threshold voltage of $V_{TN} = -1.2\,\text{V}$. (Ans. (a) (i) nonsaturation, (ii) saturation, (iii) saturation; (b) (i) nonsaturation, (ii) nonsaturation, (iii) saturation)

5.2   The NMOS devices described in Exercise 5.1 have parameters $W = 100\,\mu\text{m}$, $L = 7\,\mu\text{m}$, $t_{ox} = 450\,\mathring{\text{A}}$, $\mu_n = 500\,\text{cm}^2/\text{V-s}$, and $\lambda = 0$. (a) Calculate the conduction parameter K_n for each device. (b) Calculate the drain current for each bias condition. (Ans. (a) $K_n = 0.274\,\text{mA/V}^2$ (b) $i_D = 0.132, 0.175$, and $0.175\,\text{mA}$; $i_D = 0.658, 1.48$, and $2.81\,\text{mA}$)

5.3 An NMOS transistor with $V_{TN} = 1\,V$ has a drain current $i_D = 0.8\,mA$ when $v_{GS} = 3\,V$ and $v_{DS} = 4.5\,V$. Calculate the drain current when: (a) $v_{GS} = 2\,V$, $v_{DS} = 4.5\,V$; and (b) $v_{GS} = 3\,V$, $v_{DS} = 1\,V$. (Ans. (a) $0.2\,mA$ (b) $0.6\,mA$)

5.1.4   Circuit Symbols and Conventions

The conventional circuit symbol for the n-channel enhancement-mode MOSFET is shown in Figure 5.11(a). The vertical solid line denotes the gate electrode, the vertical broken line denotes the channel (the broken line indicates the device is enhancement mode), and the separation between the gate line and channel line denotes the oxide that insulates the gate from the channel. The polarity of the pn junction between the substrate and the channel is indicated by the arrowhead on the body or substrate terminal. The direction of the arrowhead indicates the type of transistor, which in this case is an n-channel device.

In most applications in this text, we will implicitly assume that the source and substrate terminals are connected together. Explicitly drawing the substrate terminal for each transistor in a circuit becomes redundant and makes the circuits appear more complex. Instead, we will use the simplified circuit symbol for the n-channel MOSFET shown in Figure 5.11(b). In this symbol, the arrowhead is on the source terminal and it indicates the direction of current, which for the n-channel device is out of the source. By including the arrowhead in the symbol, we do not need to explicitly indicate the source and drain terminals. We will use the simplified circuit symbol throughout the text except in specific applications.

5.1.5   Additional MOSFET Structures and Circuit Symbols

Before we start analyzing MOSFET circuits, there are a number of other MOSFET structures in addition to the n-channel enhancement-mode device that need to be considered.

n-Channel Depletion-Mode MOSFET

Figure 5.12(a) shows the cross section of an n-channel **depletion-mode** MOSFET. When zero volts are applied to the gate, an n-channel region or inversion layer exists under the oxide as a result, for example, of impurities introduced during device fabrication. Since an n-region connects the n-source and n-drain, a drain-to-source current may be generated in the channel even with zero gate voltage. The term **depletion mode** means that a channel exists even at zero gate voltage. A negative gate voltage must be applied to the n-channel depletion-mode MOSFET to turn the device off.

Figure 5.12(b) shows the n-channel depletion mode MOSFET with a negative applied gate-to-source voltage. A negative gate voltage induces a space-charge region under the oxide, thereby reducing the thickness of the n-channel region. The reduced thickness decreases the channel conductance, which in turn reduces the drain current. When the gate voltage is equal to the threshold voltage, which is negative for this device, the induced space-charge region

(a)

(b)

Figure 5.11   The n-channel enhancement-mode MOSFET: (a) conventional circuit symbol and (b) simplified circuit symbol

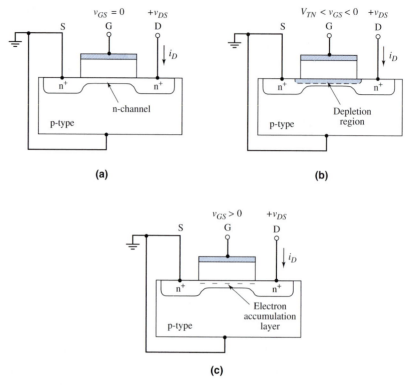

Figure 5.12 Cross section of an n-channel depletion mode MOSFET for: (a) $v_{GS} = 0$, (b) $v_{GS} < 0$, and (c) $v_{GS} > 0$

extends completely through the n-channel region, and the current goes to zero. A positive gate voltage creates an electron accumulation layer, as shown in Figure 5.12(c) which increases the drain current. The general i_D versus v_{DS} family of curves for the n-channel depletion-mode MOSFET is shown in Figure 5.13.

The current–voltage characteristics defined by Equations (5.2(a)) and (5.2(b)) apply to both enhancement- and depletion-mode n-channel devices. The only difference is that the threshold voltage V_{TN} is positive for the enhancement-mode MOSFET and negative for the depletion-mode MOSFET. Even though the current–voltage characteristics of enhancement- and depletion-mode devices are described by the same equations, different circuit symbols are used, simply for purposes of clarity.

The conventional circuit symbol for the n-channel depletion-mode MOSFET is shown in Figure 5.14(a). The vertical solid line denoting the channel indicates the device is depletion mode. A comparison of Figures 5.11(a) and 5.14(a) shows that the only difference between the enhancement- and depletion-mode symbols is the broken versus the solid line representing the channel.

A simplified symbol for the n-channel depletion-mode MOSFET is shown in Figure 5.14(b). The arrowhead is again on the source terminal and indicates the direction of current, which for the n-channel device is out of the source. The heavy solid line represents the depletion-mode channel region. Again,

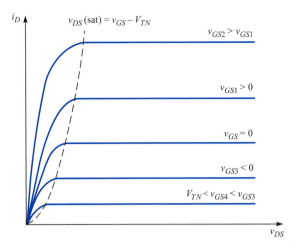

Figure 5.13   Family of i_D vesus v_{DS} curves for an n-channel depletion mode MOSFET

(a)

(b)

Figure 5.14   The n-channel depletion-mode MOSFET: (a) conventional circuit symbol and (b) simplified circuit symbol

using a different circuit symbol for the depletion-mode device compared to the enhancement-mode device is simply for clarity in a circuit diagram.

p-Channel MOSFETs

Figures 5.15(a) and 5.15(b) show cross sections of a p-channel enhancement-mode and a p-channel depletion-mode MOSFET, as well as the biasing configurations and currrent directions. The types of impurity doping in the source, drain, and substrate regions of the **p-channel MOSFET**, or **PMOS transistor**, are reversed compared to the n-channel device. In the p-channel enhancement-mode device, a negative gate-to-source voltage must be applied to create the inversion layer, or channel region, of holes that "connects" the source and drain regions. The threshold voltage, denoted as V_{TP} for the p-channel device,[4] is negative for an enhancement-mode device. The threshold voltage V_{TP} is positive for a p-channel depletion-mode device. Because holes flow from the source to the drain, the conventional current enters the source and leaves the drain. A p-channel region exists in the depletion-mode device with zero gate voltage.

The operation of the p-channel device is the same as that of the n-channel device, except that the hole is the charge carrier, rather than the electron, and the conventional current direction and voltage polarities are reversed. For the p-channel device biased in the nonsaturation region, the current is given by

$$i_D = K_p[2(v_{SG} + V_{TP})v_{SD} - v_{SD}^2] \qquad\qquad \textbf{(5.4(a))}$$

In the saturation region, the current is given by

$$i_D = K_p(v_{SG} + V_{TP})^2 \qquad\qquad \textbf{(5.4(b))}$$

---

[4]Using a different threshold voltage parameter for a PMOS device compared to that for the NMOS device is simply for clarity in particular applications.

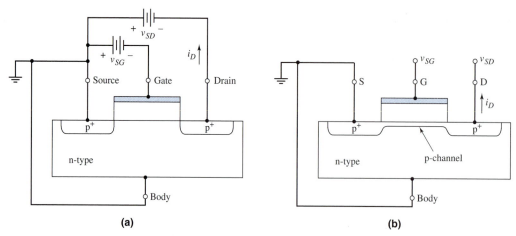

Figure 5.15   Cross section of p-channel MOSFETs: (a) enhancement-mode and (b) depletion-mode

and the drain current exits the drain terminal. The parameter K_p is the conduction parameter for the p-channel device and is given by

$$K_p = \frac{W \mu_p C_{ox}}{2L} \tag{5.5(a)}$$

where W, L, and C_{ox} are the channel width, length, and oxide capacitance per unit area, as previously defined. The parameter μ_p is the mobility of holes in the hole inversion layer. In general, the hole inversion layer mobility is less than the electron inversion layer mobility.

We can also rewrite Equation (5.5(a)) in the form

$$K_p = \frac{k_p'}{2} \cdot \frac{W}{L} \tag{5.5(b)}$$

where $k_p' = \mu_p C_{ox}$.

For a p-channel MOSFET biased in the saturation region, we have

$$v_{SD} > v_{SD}(\text{sat}) = v_{SG} + V_{TP} \tag{5.6}$$

Example 5.2   **Objective:**   Determine the source-to-drain voltage required to bias a p-channel depletion-mode MOSFET in the saturation region.

Consider a depletion-mode p-channel MOSFET for which $K_p = 0.2 \, \text{mA/V}^2$, $V_{TP} = +0.50 \, \text{V}$, and $i_D = 0.50 \, \text{mA}$.

Solution: In the saturation region, the drain current is given by

$$i_D = K_p (v_{SG} + V_{TP})^2$$

or

$$0.50 = 0.2(v_{SG} + 0.50)^2$$

which yields

$$v_{SG} = 1.08 \, \text{V}$$

To bias this p-channel MOSFET in the saturation region, the following must apply:

$$v_{SD} > v_{SD}(\text{sat}) = v_{SG} + V_{TP} = 1.08 + 0.5 = 1.58\,\text{V}$$

Comment: Biasing a transistor in either the saturation or the nonsaturation region depends on both the gate-to-source voltage and the drain-to-source voltage.

Test Your Understanding

5.4 (a) For a PMOS device, the threshold voltage is $V_{TP} = -2\,\text{V}$ and the applied source-to-gate voltage is $v_{SG} = 3\,\text{V}$. Determine the region of operation when: (i) $v_{SD} = 0.5\,\text{V}$; (ii) $v_{SD} = 2\,\text{V}$; and (iii) $v_{SD} = 5\,\text{V}$. (b) Repeat part (a) for a depletion-mode PMOS device with $V_{TP} = +0.5\,\text{V}$. (Ans. (a) (i) nonsaturation, (ii) saturation, (iii) saturation; (b) nonsaturation, (ii) nonsaturation, (iii) saturation)

The conventional circuit symbol for the p-channel enhancement-mode MOSFET appears in Figure 5.16(a). Note that the arrowhead direction on the substrate terminal is reversed from that in the n-channel enhancement-mode device. The simplified circuit symbol we will use is shown in Figure 5.16(b). The arrowhead is on the source terminal, indicating the direction of current, which for the p-channel device is into the source.

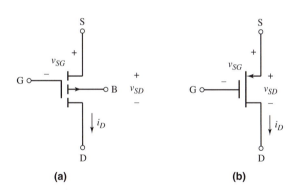

(a)   (b)

Figure 5.16 The p-channel enhancement-mode MOSFET: (a) conventional circuit symbol and (b) simplified circuit symbol

The conventional and simplified circuit symbols for the p-channel depletion-mode device are shown in Figure 5.17. The heavy solid line in the simplified symbol represents the channel region and denotes the depletion-mode device. The arrowhead is again on the source terminal and it indicates the current direction.

Complementary MOSFETs

Complementary MOS (CMOS) technology uses both n-channel and p-channel devices in the same circuit. Figure 5.18 shows the cross section of n-channel and p-channel devices fabricated on the same chip. CMOS circuits, in general,

(a)

(b)

Figure 5.17 The p-channel depletion mode MOSFET: (a) conventional circuit symbol and (b) simplified circuit symbol

Figure 5.18   Cross sections of n-channel and p-channel transistors fabricated with a p-well CMOS technology

are more complicated to fabricate than circuits using entirely NMOS or PMOS devices. Yet, as we will see in later chapters, CMOS circuits have great advantages over just NMOS or PMOS circuits.

In order to fabricate n-channel and p-channel devices that are electrically equivalent, the magnitude of the threshold voltages must be equal, and the n-channel and p-channel conduction parameters must be equal. Since, in general, μ_n and μ_p are not equal, the design of equivalent transistors involves adjusting the width-to-length ratios of the transistors.

5.1.6   Summary of Transistor Operation

We have presented a first-order model of the operation of the MOS transistor. For an n-channel enhancement-mode MOSFET, a positive gate-to-source voltage, greater than the threshold voltage V_{TN}, must be applied to induce an electron inversion layer. For $v_{GS} > V_{TN}$, the device is turned on. For an n-channel depletion-mode device, a channel between the source and drain exists even for $v_{GS} = 0$. The threshold voltage is negative, so that a negative value of v_{GS} is required to turn the device off.

For a p-channel device, all voltage polarities and current directions are reversed compared to the NMOS device. For the p-channel enhancement-mode transistor, $V_{TP} < 0$ and for the depletion-mode PMOS transistor, $V_{TP} > 0$.

Table 5.1 lists the first-order equations that describe the i–v relationships in MOS devices. We note that K_n and K_p are positive values and that the drain current i_D is positive into the drain for the NMOS device and positive out of the drain for the PMOS device.

Table 5.1   Summary of the MOSFET current–voltage relationships

NMOS	PMOS
Nonsaturation region ($v_{DS} < v_{DS}(\text{sat})$)	Nonsaturation region ($v_{SD} < v_{SD}(\text{sat})$)
$i_D = K_n[2(v_{GS} - V_{TN})v_{DS} - v_{DS}^2]$	$i_D = K_p[2(v_{SG} + V_{TP})v_{SD} - v_{SD}^2]$
Saturation region ($v_{DS} > v_{DS}(\text{sat})$)	Saturation region ($v_{SD} > v_{SD}(\text{sat})$)
$i_D = K_n(v_{GS} - V_{TN})^2$	$i_D = K_p(v_{SG} + V_{TP})^2$
Transition point	Transition point
$v_{DS}(\text{sat}) = v_{GS} - V_{TN}$	$v_{SD}(\text{sat}) = v_{SG} + V_{TP}$
Enhancement mode	Enhancement mode
$V_{TN} > 0$	$V_{TP} < 0$
Depletion mode	Depletion mode
$V_{TN} < 0$	$V_{TP} > 0$

5.1.7   Nonideal Current–Voltage Characteristics

The five nonideal effects in the current–voltage characteristics of MOS transistors are: the finite output resistance in the saturation region, the body effect, subthreshold conduction, breakdown effects, and temperature effects. This section will examine each of these effects.

Finite Output Resistance

In the ideal case, when a MOSFET is biased in the saturation region, the drain current i_D is independent of drain-to-source voltage v_{DS}. However, in actual MOSFET i_D versus v_{DS} characteristics, a nonzero slope does exist beyond the saturation point. For $v_{DS} > v_{DS}(\text{sat})$, the actual point in the channel at which the inversion charge goes to zero moves away from the drain terminal (see Figure 5.9(d)). The effective channel length decreases, producing the phenomenon called **channel length modulation**.

   An exaggerated view of the current–voltage characteristics is shown in Figure 5.19. The curves can be extrapolated so that they intercept the voltage axis at a point $v_{DS} = -V_A$. The voltage V_A is usually defined as a positive quantity and is similar to the Early voltage of a bipolar transistor (see Chapter 3). The slope of the curve in the saturation region can be described by expressing the i_D versus v_{DS} characteristic in the form, for an n-channel device,

$$i_D = K_n[(v_{GS} - V_{TN})^2(1 + \lambda v_{DS})] \tag{5.7}$$

where λ is a positive quantity called the channel-length modulation parameter.

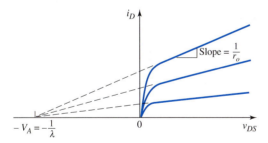

Figure 5.19   Effect of channel length modulation, resulting in a finite output resistance

   The parameters λ and V_A are related. From Equation (5.7), we have $(1 + \lambda v_{DS}) = 0$ at the extrapolated point where $i_D = 0$. At this point, $v_{DS} = -V_A$, which means that $V_A = 1/\lambda$.

   The output resistance due to the channel length modulation is defined as

$$r_o = \left(\frac{\partial i_D}{\partial v_{DS}}\right)^{-1}\bigg|_{v_{GS}=\text{const.}} \tag{5.8}$$

From Equation (5.7), the output resistance, evaluated at the Q-point, is

$$r_o = [\lambda K_n(V_{GSQ} - V_{TN})^2]^{-1} \tag{5.9(a)}$$

or

$$r_o \cong [\lambda I_{DQ}]^{-1} = \frac{1}{\lambda I_{DQ}} = \frac{V_A}{I_{DQ}}$$ (5.9(b))

The output resistance r_o is also a factor in the small-signal equivalent circuit of the MOSFET, which is discussed in the next chapter.

Test Your Understanding

5.5 For an NMOS enhancement-mode device, the parameters are: $V_{TN} = 0.8$ V and $K_n = 0.1$ mA/V^2. The device is biased at $v_{GS} = 2.5$ V. Calculate the drain current when $v_{DS} = 2$ V and $v_{DS} = 10$ V for: (a) $\lambda = 0$ and (b) $\lambda = 0.02$ V^{-1}. (c) Calculate the output resistance r_o for parts (a) and (b). (Ans. (a) $i_D = 0.289$ mA for both 2 and 10 V; (b) $i_D = 0.30$ mA (2 V), $i_D = 0.347$ mA (10 V); (c) $r_o = \infty$ (a), $r_o = 173$ kΩ (b))

Body Effect

Up to this point, we have assumed that the substrate, or body, is connected to the source. For this bias condition, the threshold voltage is a constant.

In integrated circuits, however, the substrates of all n-channel MOSFETs are usually common and are tied to the most negative potential in the circuit. An example of two n-channel MOSFETs in series is shown in Figure 5.20. The p-type substrate is common to the two transistors, and the drain of M_1 is common to the source of M_2. When the two transistors are conducting, there is a nonzero drain-to-source voltage on M_1, which means that the source of M_2 is not at the same potential as the substrate. These bias conditions mean that a zero or reverse-bias voltage exists across the source–substrate pn junction, and a change in the source–substrate junction voltage changes the threshold voltage. This is called the **body effect**. The same situation exists in p-channel devices.

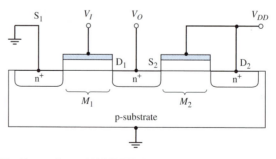

Figure 5.20   Two n-channel MOSFETs fabricated in series in the same substrate

Figure 5.21   An n-channel enhancement-mode MOSFET with a substrate voltage

For example, consider the n-channel device shown in Figure 5.21. To maintain a zero- or reverse-biased source–substrate pn junction, we must have $v_{SB} \geq 0$. The threshold voltage for this condition is given by

$$V_{TN} = V_{TNO} + \gamma\left[\sqrt{2\phi_f + v_{SB}} - \sqrt{2\phi_f}\right] \qquad \textbf{(5.10)}$$

where V_{TNO} is the threshold voltage for $v_{SB} = 0$; γ, called the bulk threshold or **body-effect parameter**, is related to device properties, and is typically on the order of $0.5\,\mathrm{V}^{1/2}$; and ϕ_f is a semiconductor parameter, typically on the order of $0.35\,\mathrm{V}$, and is a function of the semiconductor doping. We see from Equation (5.10) that the threshold voltage in n-channel devices increases due to this body effect.

The body effect can cause a degradation in circuit performance because of the changing threshold voltage. However, we will generally neglect the body effect in our circuit analyses, for simplicity.

Test Your Understanding

5.6 An NMOS transistor has parameters $V_{TNO} = 1\,\mathrm{V}$, $\gamma = 0.35\,\mathrm{V}^{1/2}$, and $\phi_f = 0.35\,\mathrm{V}$. Calculate the threshold voltage when: (a) $v_{SB} = 0$, (b) $v_{SB} = 1\,\mathrm{V}$, and (c) $v_{SB} = 4\,\mathrm{V}$. (Ans. (a) 1 V, (b) 1.16 V, (c) 1.47 V)

Subthreshold Conduction

If we consider the ideal current-voltage relationship for the n-channel MOSFET biased in the saturation region, we have, from Equation (5.2(b)),

$$i_D = K_n(v_{GS} - V_{TN})^2$$

Taking the square root of both sides of the equation, we obtain

$$\sqrt{i_D} = \sqrt{K_n}(v_{GS} - V_{TN}) \qquad \textbf{(5.11)}$$

From Equation (5.11), we see that $\sqrt{i_d}$ is a linear function of v_{GS}. Figure 5.22 shows a plot of this ideal relationship.

Also plotted in Figure 5.22 are experimental results, which show that when v_{GS} is slightly less than V_{TN}, the drain current is not zero, as previously assumed. This current is called the **subthreshold current**. The effect may not be significant for a single device, but if hundreds or thousands of devices on an integrated circuit are biased just slightly below the threshold voltage, the power supply current will not be zero but may contribute to significant power dissipation in the integrated circuit. One example of this is a dynamic random access memory (DRAM), as we will see in Chapter 16.

In this text, for simplicity we will not specifically consider the subthreshold current. However, when a MOSFET in a circuit is to be turned off, the "proper" design of the circuit must involve biasing the device at least a few tenths of a volt below the threshold voltage to achieve "true" cutoff.

Figure 5.22 Plot of $\sqrt{i_D}$ versus v_{GS} characteristic showing subthreshold conduction

Breakdown Effects

Several possible breakdown effects may occur in a MOSFET. The drain-to-substrate pn junction may break down if the applied drain voltage is too high and avalanche multiplication occurs. This breakdown is the same reverse-biased pn junction breakdown discussed in Chapter 1 in Section 1.2.5.

As the size of the device becomes smaller, another breakdown mechanism, called *punch-through*, may become significant. **Punch-through** occurs when the drain voltage is large enough for the depletion region around the drain to extend completely through the channel to the source terminal. This effect also causes the drain current to increase rapidly with only a small increase in drain voltage.

A third breakdown mechanism is called **near-avalanche** or **snapback breakdown**. This breakdown process is due to second-order effects within the MOSFET. The source-substrate-drain structure is equivalent to that of a bipolar transistor. As the device size shrinks, we may begin to see a parasitic bipolar transistor action with increases in the drain voltage. This parasitic action enhances the breakdown effect.

If the electric field in the oxide becomes large enough, breakdown can also occur in the oxide, which can lead to catastrophic failure. In silicon dioxide, the electric field at breakdown is on the order of 6×10^6 V/cm, which, to a first approximation, is given by $E_{ox} = V_G / t_{ox}$. A gate voltage of approximately 30 V would produce breakdown in an oxide with a thickness of $t_{ox} = 500$ Å. However, a safety margin of a factor of 3 is common, which means that the maximum safe gate voltage for $t_{ox} = 500$ Å would be 10 V. A safety margin is necessary since there may be defects in the oxide that lower the breakdown field. We must also keep in mind that the input impedance at the gate is very high, and a small amount of static charge accumulating on the gate can cause the breakdown voltage to be exceeded. To prevent the accumulation of static charge on the gate capacitance of a MOSFET, a gate protection device, such as a reverse-biased diode, is usually included at the input of a MOS integrated circuit.

Temperature Effects

Both the threshold voltage V_{TN} and conduction parameter K_n are functions of temperature. The magnitude of the threshold voltage decreases with temperature, which means that the drain current increases with temperature at a given V_{GS}. However, the conduction parameter is a direct function of the inversion carrier mobility, which decreases as the temperature increases. Since the temperature dependence of mobility is larger than that of the threshold voltage, the net effect of increasing temperature is a decrease in drain current at a given V_{GS}. This particular result provides a negative feedback condition in power MOSFETs. A decreasing value of K_n inherently limits the channel current and provides stability for a power MOSFET.

5.2   MOSFET DC CIRCUIT ANALYSIS

In the last section, we considered the basic MOSFET characteristics and properties. We now start analyzing and designing the dc biasing of MOS transistor circuits. A primary purpose of the rest of the chapter is to continue to become familiar and comfortable with the MOS transistor and MOSFET circuits. As with bipolars, the dc biasing of MOSFETs, the focus of this chapter, is an important part of the design of amplifiers. MOSFET amplifier design is the focus of the next chapter.

In most of the circuits presented in this chapter, resistors are used in conjunction with the MOS transistors. In a real MOSFET integrated circuit, however, the resistors are generally replaced by other MOSFETs, so the circuit is composed entirely of MOS devices. As we go through the chapter, we will begin to see how this is accomplished and as we finish the text, we will indeed analyze and design circuits containing only MOSFETs.

In the dc analysis of MOSFET circuits, we can use the ideal current–voltage equations listed in Table 5.1 in Section 5.1.

5.2.1   Common-Source Circuit

One of the basic MOSFET circuit configurations is called the **common-source circuit**. Figure 5.23 shows one example of this type of circuit using an n-channel enhancement-mode MOSFET. The source terminal is at ground potential and is common to both the input and output portions of the circuit. The coupling capacitor C_C acts as an open circuit to dc but it allows the signal voltage to be coupled to the gate of the MOSFET.

The dc equivalent circuit is shown in Figure 5.24(a). In the following dc analyses, we again use the notation for dc currents and voltages. Since the gate current into the transistor is zero, the voltage at the gate is given by a voltage divider, which can be written as

$$V_G = V_{GS} = \left(\frac{R_2}{R_1 + R_2}\right) V_{DD} \qquad (5.12)$$

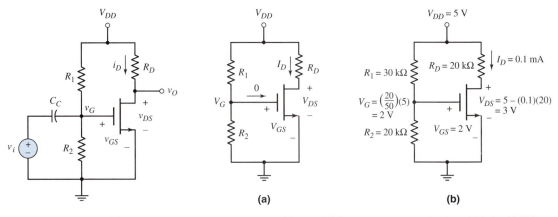

Figure 5.23   An NMOS common-source circuit

Figure 5.24   (a) An NMOS common-source circuit and (b) the NMOS circuit for Example 5.3

Assuming that the gate-to-source voltage given by Equation (5.12) is greater than V_{TN}, and that the transistor is biased in the saturation region, the drain current is

$$I_D = K_n(V_{GS} - V_{TN})^2 \qquad (5.13)$$

The drain-to-source voltage is

$$V_{DS} = V_{DD} - I_D R_D \qquad\qquad (5.14)$$

If $V_{DS} > V_{DS}(\text{sat}) = V_{GS} - V_{TN}$, then the transistor is biased in the saturation region, as we initially assumed, and our analysis is correct. If $V_{DS} < V_{DS}(\text{sat})$, then the transistor is biased in the nonsaturation region, and the drain current is given by Equation (5.2(a)).

Example 5.3   **Objective:**   Calculate the drain current and drain-to-source voltage of a common-source circuit with an n-channel enhancement-mode MOSFET.

For the circuit shown in Figure 5.24(a), assume that $R_1 = 30\,\text{k}\Omega$, $R_2 = 20\,\text{k}\Omega$, $R_D = 20\,\text{k}\Omega$, $V_{DD} = 5\,\text{V}$, $V_{TN} = 1\,\text{V}$, and $K_n = 0.1\,\text{mA/V}^2$.

Solution: From the circuit shown in Figure 5.24(b) and Equation (5.12), we have

$$V_G = V_{GS} = \left(\frac{R_2}{R_1 + R_2}\right)V_{DD} = \left(\frac{20}{20 + 30}\right)(5) = 2\,\text{V}$$

Assuming the transistor is biased in the saturation region, the drain current is

$$I_D = K_n(V_{GS} - V_{TN})^2 = (0.1)(2 - 1)^2 = 0.1\,\text{mA}$$

and the drain-to-source voltage is

$$V_{DS} = V_{DD} - I_D R_D = 5 - (0.1)(20) = 3\,\text{V}$$

Comment: Because $V_{DS} = 3\,\text{V} > V_{DS}(\text{sat}) = V_{GS} - V_{TN} = 2 - 1 = 1\,\text{V}$, the transistor is indeed biased in the saturation region and our analysis is valid.

Figure 5.25(a) shows a common-source circuit with a p-channel enhancement-mode MOSFET. The source terminal is tied to $+V_{DD}$, which becomes signal ground in the ac equivalent circuit. Thus the terminology common-source applies to this circuit.

Figure 5.25   (a) A PMOS common-source circuit, (b) results when saturation-region bias assumption is incorrect, and (c) results when nonsaturation-region bias assumption is correct

The dc analysis is essentially the same as for the n-channel MOSFET circuit. The gate voltage is

$$V_G = \left(\frac{R_2}{R_1 + R_2}\right)(V_{DD}) \tag{5.15(a)}$$

and the source-to-gate voltage is

$$V_{SG} = V_{DD} - V_G \tag{5.15(b)}$$

Assuming that $V_{GS} < V_{TP}$, or $V_{SG} > |V_{TP}|$, and that the device is biased in the saturation region, the drain current is given by

$$I_D = K_p(V_{SG} + V_{TP})^2 \tag{5.16}$$

and the source-to-drain voltage is

$$V_{SD} = V_{DD} - I_D R_D \tag{5.17}$$

If $V_{SD} > V_{SD}(\text{sat}) = V_{SG} + V_{TP}$, then the transistor is indeed biased in the saturation region, as we have assumed. However, if $V_{SD} < V_{SD}(\text{sat})$, the transistor is biased in the nonsaturation region.

Example 5.4 **Objective:** Calculate the drain current and source-to-drain voltage of a common-source circuit with a p-channel enhancement-mode MOSFET.

Consider the circuit shown in Figure 5.25(a). Assume that $R_1 = R_2 = 50\,\text{k}\Omega$, $V_{DD} = 5\,\text{V}$, $R_D = 7.5\,\text{k}\Omega$, $V_{TP} = -0.8\,\text{V}$, and $K_p = 0.2\,\text{mA/V}^2$.

Solution: From the circuit shown in Figure 5.25(b) and Equation (5.15(a)), we have

$$V_G = \left(\frac{R_2}{R_1 + R_2}\right)(V_{DD}) = \left(\frac{50}{50 + 50}\right)(5) = 2.5\,\text{V}$$

The source-to-gate voltage is therefore

$$V_{SG} = V_{DD} - V_G = 5 - 2.5 = 2.5\,\text{V}$$

Assuming the transistor is biased in the saturation region, the drain current is

$$I_D = K_p(V_{SG} + V_{TP})^2 = (0.2)(2.5 - 0.8)^2 = 0.578\,\text{mA}$$

and the source-to-drain voltage is

$$V_{SD} = V_{DD} - I_D R_D = 5 - (0.578)(7.5) = 0.665\,\text{V}$$

Since $V_{SD} = 0.665\,\text{V}$ is not greater than $V_{SD}(\text{sat}) = V_{SG} + V_{TP} = 2.5 - 0.8 = 1.7\,\text{V}$, the p-channel MOSFET is not biased in the saturation region, as we initially assumed.

In the nonsaturation region, the drain current is given by

$$I_D = K_p[2(V_{SG} + V_{TP})V_{SD} - V_{SD}^2]$$

and the source-to-drain voltage is

$$V_{SD} = V_{DD} - I_D R_D$$

Combining these two equations, we obtain

$$I_D = K_p[2(V_{SG} + V_{TP})(V_{DD} - I_D R_D) - (V_{DD} - I_D R_D)^2]$$

or

$$I_D = (0.2)[2(2.5 - 0.8)(5 - I_D(7.5)) - (5 - I_D(7.5))^2]$$

Solving this quadratic equation for I_D, we find

$$I_D = 0.515 \, \text{mA}$$

We also find that

$$V_{SD} = 1.14 \, \text{V}$$

Therefore, $V_{SD} < V_{SD}(\text{sat})$, which verifies that the transistor is biased in the nonsaturation region.

Comment: In solving the quadratic equation for I_D, we find a second solution that yields $V_{SD} = 2.93 \, \text{V}$. However, this value of V_{SD} is greater than $V_{SD}(\text{sat})$, so it is not a valid solution since we assumed the transistor to be biased in the nonsaturation region.

As Example 5.4 illustrated, we may not know initially whether a transistor is biased in the saturation or nonsaturation region. The approach involves making an educated guess and then verifying that assumption. If the assumption proves incorrect, we must then change it and reanalyze the circuit.

In linear amplifiers containing MOSFETs, the transistors are biased in the saturation region.

Test Your Understanding

*5.7 For the transistor in the circuit in Figure 5.26, the parameters are $V_{TN} = 1 \, \text{V}$ and $K_n = 0.5 \, \text{mA/V}^2$. Determine V_{GS}, I_D, and V_{DS}. (Ans. $V_{GS} = 2.65 \, \text{V}$, $I_D = 1.36 \, \text{mA}$, $V_{DS} = 5.92 \, \text{V}$)

*5.8 Consider the circuit shown in Figure 5.27. The transistor parameters are $V_{TP} = -1 \, \text{V}$ and $K_p = 0.25 \, \text{mA/V}^2$. Calculate V_{SG}, I_D, and V_{SD}. (Ans. $V_{SG} = 3.04 \, \text{V}$, $I_D = 1.04 \, \text{mA}$, $V_{SD} = 4.59 \, \text{V}$)

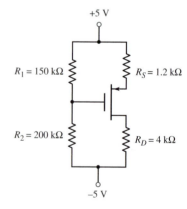

Figure 5.26   Circuit for Exercise 5.7   **Figure 5.27**   Circuit for Exercise 5.8

RD5.9 The transistor in the circuit shown in Figure 5.24(a) has parameters $V_{TN} = 0.8 \, \text{V}$ and $K_n = 0.25 \, \text{mA/V}^2$. The circuit is biased with $V_{DD} = 7.5 \, \text{V}$. Let $R_1 + R_2 = 250 \, \text{k}\Omega$. Redesign the circuit such that $I_D = 0.40 \, \text{mA}$ and $V_{DS} = 4 \, \text{V}$ (Ans. $R_2 = 68.7 \, \text{k}\Omega$, $R_1 = 181.3 \, \text{k}\Omega$, $R_D = 8.75 \, \text{k}\Omega$)

5.2.2   Load Line and Modes of Operation

The load line is helpful in visualizing the region in which the MOSFET is biased. Consider again the common-source circuit shown in Figure 5.24(b). Writing a Kirchhoff's voltage law equation around the drain–source loop results in Equation (5.14), which is the load line equation, showing a linear relationship between the drain current and drain-to-source voltage.

Figure 5.28 shows the $v_{DS}(\text{sat})$ characteristic for the transistor described in Example 5.3. The load line is given by

$$V_{DS} = V_{DD} - I_D R_D = 5 - I_D(20) \tag{5.18(a)}$$

or

$$I_D = \frac{5}{20} - \frac{V_{DS}}{20}\ (\text{mA}) \tag{5.18(b)}$$

and is also plotted in the figure. The two end points of the load line are determined in the usual manner. If $I_D = 0$, then $V_{DS} = 5\,\text{V}$; if $V_{DS} = 0$, then $I_D = 5/20 = 0.25\,\text{mA}$. The Q-point of the transistor is given by the dc drain current and drain-to-source voltage, and it is always on the load line, as shown in the figure. A few transistor characteristics are also shown on the figure.

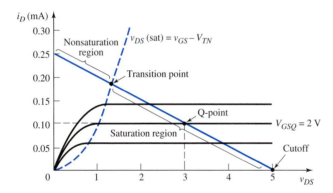

Figure 5.28   Transistor characteristics, $v_{DS}(\text{sat})$ curve, load line, and Q-point for the NMOS common-source circuit in Figure 5.24(b)

If the gate-to-source voltage is less than V_{TN}, the drain current is zero and the transistor is in cutoff. As the gate-to-source voltage becomes just greater than V_{TN}, the transistor turns on and is biased in the saturation region. As V_{GS} increases, the Q-point moves up the load line. The **transition point** is the boundary between the saturation and nonsaturation regions and is defined as the point where $V_{DS} = V_{DS}(\text{sat}) = V_{GS} - V_{TN}$. As V_{GS} increase above the transition point value, the transistor becomes biased in the nonsaturation region.

Example   5.5   **Objective:**   Determine the transition point parameters for a common-source circuit.

Consider the circuit shown in Figure 5.24(b). Assume transistor parameters of $V_{TN} = 1\,\text{V}$ and $K_n = 0.1\,\text{mA/V}^2$.

Solution: At the transition point,

$$V_{DS} = V_{DS}(\text{sat}) = V_{GS} - V_{TN} = V_{DD} - I_D R_D$$

The drain current is still

$$I_D = K_n(V_{GS} - V_{TN})^2$$

Combining the last two equations, we obtain

$$V_{GS} - V_{TN} = V_{DD} - K_n R_D(V_{GS} - V_{TN})^2$$

Rearranging this equation produces

$$K_n R_D(V_{GS} - V_{TN})^2 + (V_{GS} - V_{TN}) - V_{DD} = 0$$

or

$$(0.1)(20)(V_{GS} - V_{TN})^2 + (V_{GS} - V_{TN}) - 5 = 0$$

Solving the quadratic equation, we find that

$$V_{GS} - V_{TN} = 1.35\,\text{V} = V_{DS}$$

Therefore,

$$V_{GS} = 2.35\,\text{V}$$

and

$$I_D = (0.1)(2.35 - 1)^2 = 0.182\,\text{mA}$$

Comment: For $V_{GS} < 2.35\,\text{V}$, the transistor is biased in the saturation region; for $V_{GS} > 2.35\,\text{V}$, the transistor is biased in the nonsaturation region.

Problem-Solving Technique: MOSFET DC Analysis

Analyzing the dc response of a MOSFET circuit requires knowing the bias condition (saturation or nonsaturation) of the transistor. In some cases, the bias condition may not be obvious, which means that we have to guess the bias condition, then analyze the circuit to determine if we have a solution consistent with our initial guess. To do this, we can:

1. Assume that the transistor is biased in the saturation region, in which case $V_{GS} > V_{TN}$, $I_D > 0$, and $V_{DS} \geq V_{DS}(\text{sat})$.
2. Analyze the circuit using the saturation current-voltage relations.
3. Evaluate the resulting bias condition of the transistor. If the assumed parameter values in step 1 are valid, then the initial assumption is correct. If $V_{GS} < V_{TN}$, then the transistor is probably cutoff, and if $V_{DS} < V_{DS}(\text{sat})$, the transistor is likely biased in the nonsaturation region.
4. If the initial assumption is proved incorrect, then a new assumption must be made and the circuit reanalyzed. Step 3 must then be repeated.

5.2.3   Common MOSFET Configurations: DC Analysis

There are various other MOSFET circuit configurations, in addition to the basic common-source circuit. Several examples are discussed in this section. We continue the dc analysis and design of MOSFET circuits to increase our proficiency and to become more adept in the analysis of these types of circuits.

Design Example 5.6   **Objective:**   Design the dc bias of a MOSFET circuit to produce a specified drain current.

For the dc circuit in Figure 5.29, assume the MOSFET parameters are $V_{TN} = 2\,\text{V}$, $k_n' = 80\,\mu\text{A/V}^2$, and $W/L = 4$. Choose R_1 and R_2 such that the current in the bias resistors is approximately one-tenth of I_D. Design the circuit such that $I_D = 0.5\,\text{mA}$. In the final design, standard resistor values are to be used.

Solution: Assuming the transistor is biased in the saturation region, we have

$$I_D = \frac{k_n'}{2} \cdot \frac{W}{L} (V_{GS} - V_{TN})^2$$

or

$$0.5 = \left(\frac{0.080}{2}\right) \cdot (4) \cdot (V_{GS} - 2)^2$$

which yields

$$V_{GS} = 3.77\,\text{V}$$

The current through the bias resistors should be approximately 0.05 mA so that

$$R_1 + R_2 = \frac{10}{0.05} = 200\,\text{k}\Omega$$

We can write

$$V_{GS} = V_G - V_S = \left[\left(\frac{R_2}{R_1 + R_2}\right)(10) - 5\right] - [I_D R_S - 5]$$

so that

$$3.77 = \left(\frac{R_2}{200}\right)(10) - (0.5)(2)$$

We find that

$$R_2 = 95.4\,\text{k}\Omega \quad \text{and} \quad R_1 = 104.6\,\text{k}\Omega$$

The closest standard resistor values are $R_2 = 100\,\text{k}\Omega$ and $R_1 = 110\,\text{k}\Omega$.

Comment: Since $V_{DS} = 4\,\text{V}$, then $V_{DS} > V_{DS}(\text{sat}) = V_{GS} - V_{TN} = 3.77 - 2 = 1.77\,\text{V}$. Therefore, the transistor is biased in the saturation region, as initially assumed.

Design Pointer: We must keep in mind that there are certain tolerances in resistor values as well as tolerances in transistor conduction parameter and threshold voltage values. These tolerances lead to variations in the Q-point values. One implication of this variation is that we should not design the Q-point too close to the transition point, or the parameter variations may push the Q-point into the nonsaturation region. This is one situation in which a computer simulation may cut down the number of tedious calculations.

Figure 5.29   NMOS common-source circuit with source resistor

In bipolar circuits, we observed that the Q-point tended to be stabilized when an emitter resistor was included in the circuit. In a similar way, the Q-point of MOSFET circuits will tend to be stabilized against variations in transistor parameters by including a source resistor. The transistor conduction parameter may vary from one device to another because of fabrication tolerances in channel length, channel width, oxide thickness, or carrier mobility. The threshold voltage may also vary from one device to another. Variations in these device parameters will change the Q-point in a given circuit, but the change can be lessened by including a source resistor. Further, in many MOSFET circuits today, the source resistor is replaced by a constant-current source, which biases the transistor with a constant current that is independent of the transistor parameters, thereby stabilizing the Q-point.

Design Example 5.7 **Objective:** Design a MOSFET circuit biased with a constant-current source.

The parameters of the transistor in the circuit shown in Figure 5.30(a) are $V_{TN} = 0.8$ V, $k_n' = 80\,\mu\text{A/V}^2$, and $W/L = 3$. Design the circuit such that the quiescent values are $I_D = 250\,\mu\text{A}$ and $V_D = 2.5$ V.

Solution: The dc equivalent circuit is shown in Figure 5.30(b). Since $v_i = 0$, the gate is at ground potential and there is no gate current through R_G.

Figure 5.30 (a) NMOS common-source circuit biased with a constant-current source and (b) equivalent dc circuit

Assuming the transistor is biased in the saturation region, we have

$$I_D = \frac{k_n'}{2} \cdot \frac{W}{L}(V_{GS} - V_{TN})^2$$

or

$$250 = \left(\frac{80}{2}\right) \cdot (3)(V_{GS} - 0.8)^2$$

which yields

$$V_{GS} = 2.24\,\text{V}$$

The voltage at the source terminal is $V_S = -V_{GS} = -2.24$ V.

The drain current can also be written as

$$I_D = \frac{5 - V_D}{R_D}$$

For $V_D = 2.5\,\text{V}$, we have

$$R_D = \frac{5 - 2.5}{0.25} = 10\,\text{k}\Omega$$

The drain-to-source voltage is

$$V_{DS} = V_D - V_S = 2.5 - (-2.24) = 4.74\,\text{V}$$

Since $V_{DS} = 4.74\,\text{V} > V_{DS}(\text{sat}) = V_{GS} - V_{TN} = 2.24 - 0.8 = 1.44\,\text{V}$, the transistor is biased in the saturation region, as initially assumed.

Comment: MOSFET circuits can be biased by using constant-current sources, which in turn are designed by using other MOS transistors.

Test Your Understanding

RD5.10 For the circuit shown in Figure 5.30(b), the transistor parameters are $V_{TN} = 1.2\,\text{V}$ and $K_n = 0.080\,\text{mA/V}^2$. Redesign the circuit by replacing the current source with a source resistor such that $I_D = 100\,\mu\text{A}$ and $V_{DS} = 4.5\,\text{V}$. (Ans. $R_S = 26.8\,\text{k}\Omega$ and $R_D = 28.2\,\text{k}\Omega$)

D5.11 Consider the circuit shown in Figure 5.31. The transistor parameters are $V_{TP} = -0.8\,\text{V}$ and $K_p = 0.050\,\text{mA/V}^2$. Design the circuit such that $I_D = 120\,\mu\text{A}$ and $V_{SD} = 8\,\text{V}$. (Ans. $R_S = 63.75\,\text{k}\Omega$, $R_D = 36.25\,\text{k}\Omega$)

Figure 5.31 Circuit for Exercise 5.11

An enhancement-mode MOSFET connected in a configuration such as that shown in Figure 5.32 can be used as a nonlinear resistor. A transistor with this connection is called an **enhancement load device**. Since the transistor is an enhancement mode device, $V_{TN} > 0$. Also, for this circuit, $v_{DS} = v_{GS} > v_{DS}(\text{sat}) = v_{GS} - V_{TN}$, which means that the transistor is always biased in the saturation region. The general i_D versus v_{DS} characteristics can then be written as

$$i_D = K_n(v_{GS} - V_{TN})^2 = K_n(v_{DS} - V_{TN})^2 \qquad (5.19)$$

Figure 5.33 shows a plot of Equation (5.19) for the case when $K_n = 1\,\text{mA/V}^2$ and $V_{TN} = 1\,\text{V}$.

In the next chapter, we will see how this transistor is used in place of a load resistor in an amplifier circuit. In later chapters, we will show how this transistor is used in digital logic circuits.

Figure 5.32 Enhancement-mode NMOS device with the gate connected to the drain

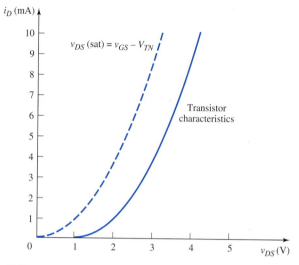

Figure 5.33 Current–voltage characteristic of an enhancement load device

Figure 5.34 Circuit containing an enhancement load device

Example 5.8 **Objective:** Calculate the characteristics of a circuit containing an enhancement load device.

Consider the circuit shown in Figure 5.34 with transistor parameters $V_{TN} = 0.8$ V and $K_n = 0.05$ mA/V^2.

Solution: Since the transistor is biased in the saturation region, the dc drain current is given by

$$I_D = K_n(V_{GS} - V_{TN})^2$$

and the dc drain-to-source voltage is

$$V_{DS} = V_{GS} = 5 - I_D R_S$$

Combining these two equations, we obtain

$$V_{GS} = 5 - K_n R_S(V_{GS} - V_{TN})^2$$

Substituting parameter values, we obtain

$$V_{GS} = 5 - (0.05)(10)(V_{GS} - 0.8)^2$$

which can be written as

$$0.5V_{GS}^2 + 0.2V_{GS} - 4.68 = 0$$

The two possible solutions are

$$V_{GS} = -3.27 \text{ V} \quad \text{and} \quad V_{GS} = +2.87 \text{ V}$$

Since we are assuming the transistor is conducting, the gate-to-source voltage must be greater than the threshold voltage. We therefore have the following solution:

$$V_{GS} = V_{DS} = 2.87 \text{ V} \quad \text{and} \quad I_D = 0.213 \text{ mA}$$

Comment: This particular circuit is obviously not an amplifier. However, the transistor connected in this configuration is extremely useful as an effective load resistor.

Test Your Understanding

5.12   The parameters for the circuit shown in Figure 5.34 are changed to $V_{DD} = 10\,\text{V}$ and $R_S = 10\,\text{k}\Omega$, and the transistor parameters are $V_{TN} = 2\,\text{V}$ and $K_n = 0.20\,\text{mA/V}^2$. Calculate I_D, V_{DS}, and the power dissipated in the transistor. (Ans. $V_{GS} = V_{DS} = 3.77\,\text{V}$, $I_D = 0.623\,\text{mA}$, $P = 2.35\,\text{mW}$)

D5.13   The parameters for the circuit shown in Figure 5.34 are $V_{DD} = 5\,\text{V}$ and $R_S = 5\,\text{k}\Omega$. The transistor threshold voltage is $V_{TN} = 1\,\text{V}$. If $k'_n = 40\,\mu\text{A/V}^2$, design the transistor width-to-length ratio such that $V_{DS} = 2.2\,\text{V}$. (Ans. $W/L = 19.5$)

If an enhancement load device is connected in a circuit with another MOSFET in the configuration in Figure 5.35, the circuit can be used as an amplifier or as an inverter in a digital logic circuit. The load device, M_L, is always biased in the saturation region, and the transistor M_D, called the **driver transistor**, can be biased in either the saturation or nonsaturation region, depending on the value of the input voltage. The next example addresses the dc analysis of this circuit for dc input voltages to the gate of M_D.

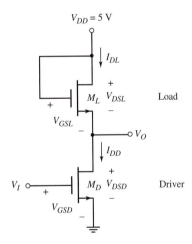

Figure 5.35   Circuit with enhancement load device and NMOS driver

Example   5.9   **Objective:** Determine the dc transistor currents and voltages in a circuit containing an enhancement load device.

The transistors in the circuit shown in Figure 5.35 have parameters $V_{TND} = V_{TNL} = 1\,\text{V}$, $K_{nD} = 50\,\mu\text{A/V}^2$, and $K_{nL} = 10\,\mu\text{A/V}^2$. (The subscript D applies to the driver transistor and the subscript L applies to the load transistor.) Determine V_O for $V_I = 5\,\text{V}$ and $V_I = 1.5\,\text{V}$.

Solution: ($V_I = 5\,\text{V}$) Assume that the driver transistor M_D is biased in the nonsaturation region. The drain current in the load device is equal to the drain current in the driver transistor. Writing these currents in generic form, we have

$$I_{DD} = I_{DL}$$

or

$$K_{nD}[2(V_{GSD} - V_{TND})V_{DSD} - V_{DSD}^2] = K_{nL}[V_{GSL} - V_{TNL}]^2$$

Since $V_{GSD} = V_I$, $V_{DSD} = V_O$, and $V_{GSL} = V_{DSL} = V_{DD} - V_O$, then

$$K_{nD}[2(V_I - V_{TND})V_O - V_O^2] = K_{nL}[V_{DD} - V_O - V_{TNL}]^2$$

Substituting numbers, we find

$$(50)[2(5 - 1)V_O - V_O^2] = (10)[5 - V_O - 1]^2$$

Rearranging the terms provides

$$3V_O^2 - 24V_O + 8 = 0$$

Using the quadratic formula, we obtain two possible solutions:

$$V_O = 7.65\,\text{V} \quad \text{or} \quad V_O = 0.349\,\text{V}$$

Since the output voltage cannot be greater than the supply voltage $V_{DD} = 5\,\text{V}$, the valid solution is $V_O = 0.349\,\text{V}$.

Also, since $V_{DSD} = V_O = 0.349\,\text{V} < V_{GSD} - V_{TND} = 5 - 1 = 4\,\text{V}$, the driver M_D is biased in the nonsaturation region, as initially assumed.

The current can be determined from

$$I_D = K_{nL}(V_{GSL} - V_{TNL})^2 = K_{nL}(V_{DD} - V_O - V_{TNL})^2$$

or

$$I_D = (10)(5 - 0.349 - 1)^2 = 133\,\mu\text{A}$$

Solution: ($V_I = 1.5\,\text{V}$) Assume that the driver transistor M_D is biased in the saturation region. Equating the currents in the two transistors and writing the current equations in generic form, we have

$$I_{DD} = I_{DL}$$

or

$$K_{nD}[V_{GSD} - V_{TND}]^2 = K_{nL}[V_{GSL} - V_{TNL}]^2$$

Again, since $V_{GSD} = V_I$ and $V_{GSL} = V_{DSL} = V_{DD} - V_O$, then

$$K_{nD}[V_I - V_{TND}]^2 = K_{nL}[V_{DD} - V_O - V_{TNL}]^2$$

Substituting numbers and taking the square root, we find

$$\sqrt{50}[1.5 - 1] = \sqrt{10}[5 - V_O - 1]$$

which yields $V_O = 2.88\,\text{V}$.

Since $V_{DSD} = V_O = 2.88\,\text{V} > V_{GSD} - V_{TND} = 1.5 - 1 = 0.5\,\text{V}$, the driver transistor M_D is biased in the saturation region, as initially assumed.

The current is

$$I_D = K_{nD}(V_{GSD} - V_{TND})^2 = (50)(1.5 - 1)^2 = 12.5\,\mu\text{A}$$

Comment: For this example, we made an initial guess as to whether the driver transistor was biased in the saturation or nonsaturation region. A more analytical approach is shown following this example.

Computer Simulation: The voltage transfer characteristics of the NMOS inverter circuit with the enhancement load shown in Figure 5.35 were obtained by a PSpice analysis. These results are shown in Figure 5.36. For an input voltage less than 1 V, the driver is cut off and the output voltage is $V_O = V_{DD} - V_{TNL} = 5 - 1 = 4\,\text{V}$. As the

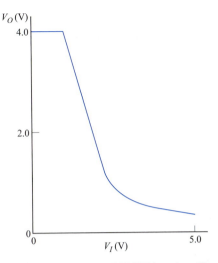

Figure 5.36   Voltage transfer characteristics of NMOS inverter with enhancement load device

input voltage decreases, the output voltage increases, charging and discharging capacitances in the transistors. When the current goes to zero at $V_I = 1$ V and $V_O = 4$ V, the capacitances cease charging and discharging so the output voltage cannot get to the full $V_{DD} = 5$ V value.

When the input voltage is just greater than 1 V, both transistors are biased in the saturation region as the previous analsyis for $V_I = 1.5$ V showed. The output voltage is a linear function of input voltage.

For an input voltage greater than approximately 2.25 V, the driver transistor is biased in the nonsaturation region and the output voltage is a nonlinear function of the input voltage.

In the circuit shown in Figure 5.35, we can determine the transition point for the driver transistor that separates the saturation and nonsaturation regions. The transition point is determined by the equation

$$V_{DSD}(\text{sat}) = V_{GSD} - V_{TND} \tag{5.20}$$

Again, the drain currents in the two transistors are equal. Using the saturation drain current relationship for the driver transistor, we have

$$I_{DD} = I_{DL} \tag{5.21(a)}$$

or

$$K_{nD}[V_{GSD} - V_{TND}]^2 = K_{nL}[V_{GSL} - V_{TNL}]^2 \tag{5.21(b)}$$

Again, noting that $V_{GSD} = V_I$ and $V_{GSL} = V_{DSL} = V_{DD} - V_O$, and taking the square root, we have

$$\sqrt{\frac{K_{nD}}{K_{nL}}}(V_I - V_{TND}) = (V_{DD} - V_O - V_{TNL}) \tag{5.22}$$

At the transition point, we can define the input voltage as $V_I = V_{It}$ and the output voltage as $V_{Ot} = V_{DSD}(\text{sat}) = V_{It} - V_{TND}$. Then, from Equation (5.22), the input voltage at the transition point is

$$V_{It} = \frac{V_{DD} - V_{TNL} + V_{TND}(1 + \sqrt{K_{nD}/K_{NL}})}{1 + \sqrt{K_{nD}/K_{nL}}} \qquad (5.23)$$

If we apply Equation (5.23) to the previous example, we can show that our initial assumptions were correct.

Test Your Understanding

5.14 For the circuit shown in Figure 5.35, use the transistor parameters given in Example 5.9. (a) Determine V_I and V_O at the transition point for the driver transistor. (b) Calculate the transistor currents at the transition point. (Ans. (a) $V_{It} = 2.24\,\text{V}$, $V_{Ot} = 1.24\,\text{V}$; (b) $I_D = 76.9\,\mu\text{A}$)

D5.15 Consider the circuit shown in Figure 5.35 with transistor parameters $V_{TND} = V_{TNL} = 1\,\text{V}$. (a) Design the ratio K_{nD}/K_{nL} that will produce a transition point at $V_I = 2.5\,\text{V}$. (b) Using the results of part (a), find V_O for $V_I = 5\,\text{V}$. (Ans. (a) $K_{nD}/K_{nL} = 2.78$ (b) $V_O = 0.57\,\text{V}$)

Up to this point, we have only considered the n-channel enhancement-mode MOSFET as a load device. An n-channel depletion-mode MOSFET can also be used. Consider the depletion-mode MOSFET with the gate and source connected together shown in Figure 5.37(a). The current–voltage characteristics are shown in Figure 5.37(b). The transistor may be biased in either the saturation or nonsaturation regions. The transition point is also shown on the plot. The threshold voltage of the n-channel depletion-mode MOSFET is negative so that $v_{DS}(\text{sat})$ is positive.

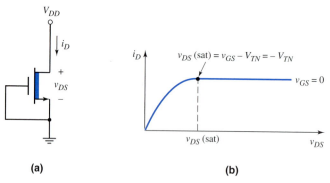

| (a) | (b) |

Figure 5.37   (a) Depletion-mode NMOS device with the gate connected to the source and (b) current–voltage characteristics

Figure 5.38 Circuit containing a depletion load device

Consider the circuit shown in Figure 5.38 in which the transistor is being used as a **depletion load device**. It may be biased in the saturation or nonsatura-tion region, depending on the values of the transistor parameters and V_{DD} and R_S.

Example 5.10 **Objective:** Calculate the characteristics of a circuit containing a depletion load device.

For the circuit shown in Figure 5.38 the transistor parameters are $V_{TN} = -2\,\text{V}$ and $K_n = 0.1\,\text{mA/V}^2$. Assume that $V_{DD} = 5\,\text{V}$ and $R_S = 5\,\text{k}\Omega$.

Solution: If we assume that the transistor is biased in the saturation region, then the dc drain current is

$$I_D = K_n(V_{GS} - V_{TN})^2 = K_n(-V_{TN})^2 = (0.1)(-(-2))^2 = 0.4\,\text{mA}$$

In this case, the transistor is acting as a constant-current source. The dc drain-to-source voltage is

$$V_{DS} = V_{DD} - I_D R_S = 5 - (0.4)(5) = 3\,\text{V}$$

Since

$$V_{DS} = 3\,\text{V} > V_{DS}(\text{sat}) = V_{GS} - V_{TN} = 0 - (-2) = 2\,\text{V}$$

the transistor is biased in the saturation region.

Comment: Although this circuit is also not an amplifier, this transistor configuration is useful as an effective load resistor in both analog and digital circuits.

Test Your Understanding

5.16 For the circuit shown in Figure 5.38, the circuit parameters are $V_{DD} = 10\,\text{V}$ and $R_S = 4\,\text{k}\Omega$, and the transistor parameters are $V_{TN} = -2.5\,\text{V}$ and $K_n = 0.25\,\text{mA/V}^2$. Calculate I_D, V_{DS}, and the power dissipated in the transistor. Is the transistor biased in the saturation or nonsaturation region? (Ans. $I_D = 1.56\,\text{mA}$, $V_{DS} = 3.76\,\text{V}$, $P = 5.87\,\text{mW}$, saturation region)

D5.17 The parameters for the circuit shown in Figure 5.38 are $V_{DD} = 5\,\text{V}$ and $R_S = 8\,\text{k}\Omega$. The transistor threshold voltage is $V_{TN} = -1.8\,\text{V}$. If $k_n' = 35\,\mu\text{A/V}^2$, design the transistor width-to-length ratio such that $V_{DS} = 1.2\,\text{V}$. Is the transistor biased in the saturation or nonsaturation region? (Ans. $W/L = 9.43$, nonsaturation)

A depletion load device can be used in conjunction with another MOSFET, as shown in Figure 5.39, to create a circuit that can be used as an amplifier or as an inverter in a digital logic circuit. Both the load device M_L and driver transistor M_D may be biased in either the saturation or nonsaturation region, depending on the value of the input voltage. We will perform the dc analysis of this circuit for a particular dc input voltage to the gate of the driver transistor.

Example 5.11 **Objective:** Determine the dc transistor currents and voltages in a circuit containing a depletion load device.

Consider the circuit shown in Figure 5.39 with transistor parameters: $V_{TND} = 1\,\text{V}$, $V_{TNL} = -2\,\text{V}$, $K_{nD} = 50\,\mu\text{A/V}^2$, and $K_{nL} = 10\,\mu\text{A/V}^2$. Determine V_O for $V_I = 5\,\text{V}$.

Figure 5.39   Circuit with depletion load device and NMOS driver

Solution: Assume the driver transistor M_D is biased in the nonsaturation region and the load transistor M_L is biased in the saturation region. The drain currents in the two transistors are equal. In generic form, these currents are

$$I_{DD} = I_{DL}$$

or

$$K_{nD}[2(V_{GSD} - V_{TND})V_{DSD} - V_{DSD}^2] = K_{nL}[V_{GSL} - V_{TNL}]^2$$

Since $V_{GSD} = V_I$, $V_{DSD} = V_O$, and $V_{GSL} = 0$, then

$$K_{nD}[2(V_I - V_{TND})V_O - V_O^2] = K_{nL}[-V_{TNL}]^2$$

Substituting numbers, we find

$$(50)[2(5 - 1)V_O - V_O^2] = (10)[-(-2)]^2$$

Rearranging the terms produces

$$5V_O^2 - 40V + 4 = 0$$

Using the quadratic formula, we obtain two possible solutions:

$$V_O = 7.90 \, \text{V} \quad \text{or} \quad V_O = 0.10 \, \text{V}$$

Since the output voltage cannot be greater than the supply voltage $V_{DD} = 5$ V, the valid solution is $V_O = 0.10$ V.

The current is

$$I_D = K_{nL}(-V_{TNL})^2 = (10)[-(-2)]^2 = 40 \, \mu\text{A}$$

Comment: Since $V_{DSD} = V_O = 0.10 \, \text{V} < V_{GSD} - V_{TND} = 5 - 1 = 4 \, \text{V}$, M_D is biased in the nonsaturation region, as assumed. Similarly, since $V_{DSL} = V_{DD} - V_O = 4.9 \, \text{V} > V_{GSL} - V_{TNL} = 0 - (-2) = 2 \, \text{V}$, M_L is biased in the saturation region, as originally assumed.

Computer Simulation: The voltage transfer characteristics of the NMOS inverter circuit with depletion load in Figure 5.39 were obtained using a PSpice analysis. These results are shown in Figure 5.40. For an input voltage less than 1 V, the driver is cut off and the output voltage is $V_O = V_{DD} = 5$ V.

When the input voltage is just greater than 1 V, the driver transistor is biased in the saturation region and the load device in the nonsaturation region. When the input voltage is approximately 1.9 V, both transistors are biased in the saturation region. If

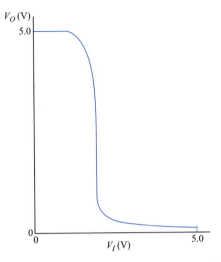

Figure 5.40 Voltage transfer characteristics of NMOS inverter with depletion load device

the channel length modulation parameter λ is assumed to be zero as in this example, there is no change in the input voltage during this transition region. As the input voltage becomes larger than 1.9 V, the driver is biased in the nonsaturation region and the load in the saturation region.

Test Your Understanding

5.18 For the circuit shown in Figure 5.39, use the transistor parameters given in Example 5.11. (a) Determine V_I and V_O at the transition point for the load transistor. (b) Determine V_I and V_O at the transition point for the driver transistor. (Ans. (a) $V_{It} = 1.89$ V, $V_{Ot} = 3$ V; (b) $V_{It} = 1.89$ V, $V_{Ot} = 0.89$ V)

D5.19 Consider the circuit shown in Figure 5.39 with transistor parameters $V_{TND} = 1$ V and $V_{TNL} = -2$ V. (a) Design the ratio K_{nD}/K_{nL} that will produce an output voltage of $V_O = 0.25$ V at $V_I = 5$ V. (b) Find K_{nD} and K_{nL} if the transistor currents are 0.2 mA when $V_I = 5$ V. (Ans. (a) $K_{nD}/K_{nL} = 2.06$ (b) $K_{nL} = 50\ \mu A/V^2$, $K_{nD} = 103\ \mu A/V^2$)

A p-channel enhancement-mode transistor can also be used as a load device to form a **complementary MOS (CMOS)** inverter. The term complementary implies that both n-channel and p-channel transistors are used in the same circuit. The CMOS technology is used extensively in both analog and digital electronic circuits.

Figure 5.41 shows one example of a CMOS inverter. The NMOS transistor is used as the amplifying device, or the driver, and the PMOS device is the load, which is referred to as an active load. This configuration is typically used in analog applications. In another configuration, the two gates are tied together and forms the input. This configuration will be discussed in detail in Chapter 16.

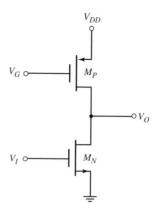

Figure 5.41   Example of CMOS inverter

As with the previous two NMOS inverters, the two transistors shown in Figure 5.41 may be biased in either the saturation or nonsaturation region, depending on the value of the input voltage. The voltage transfer characteristic is most easily determined from a PSpice analysis.

Example   5.12   **Objective:** Determine the voltage transfer characteristic of the CMOS inverter using a PSpice analysis.

For the circuit shown in Figure 5.41, assume transistor parameters of $V_{TN} = 1 \, \text{V}$, $V_{TP} = -1 \, \text{V}$, and $K_n = K_p$. Also assume $V_{DD} = 5 \, \text{V}$ and $V_G = 3.25 \, \text{V}$.

Solution: The voltage transfer characteristics are shown in Figure 5.42. In this case, there is a region, as was the case for an NMOS inverter with depletion load, in which both transistors are biased in the saturation region, and the input voltage is a constant over this transition region for the assumption that the channel length modulation parameter λ is zero.

Figure 5.42   Voltage transfer characteristics of CMOS inverter in Figure 5.41

Comment: In this example, the source-to-gate voltage of the PMOS device is only $V_{SG} = 0.75$ V. The effective resistance looking into the drain of the PMOS device is then relatively large. This is a desirable characteristic for an amplifier, as we will see in the next chapter.

5.2.4 Constant-Current Source Biasing

As was shown in Figure 5.30, a MOSFET can be biased by using a constant-current source I_Q. The advantage of this circuit is that the drain current is independent of the transistor parameters.

The constant-current source can be implemented by using MOSFETs as shown in Figure 5.43. The transistors M_2, M_3, and M_4 form the current source. Transistors M_3 and M_4 are each connected in a diode-type configuration, and they establish a reference current. We noted in the last section that this diode-type connection implies the transistor is always biased in the saturation region. Transistors M_3 and M_4 are therefore biased in the saturation region, and M_2 is assumed to be biased in the saturation region. The resulting gate-to-source voltage on M_3 is applied to M_2, and this establishes the bias current I_Q.

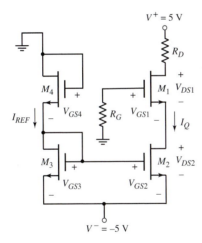

Figure 5.43   Implementation of a MOSFET constant-current source

Since the reference current is the same in transistors M_3 and M_4, we can write

$$K_{n3}(V_{GS3} - V_{TN3})^2 = K_{n4}(V_{GS4} - V_{TN4})^2 \tag{5.24}$$

We also know that

$$V_{GS4} + V_{GS3} = (-V^-) \tag{5.25}$$

Solving Equation (5.25) for V_{GS4} and substituting the result into Equation (5.24) yields

$$V_{GS3} = \frac{\sqrt{\dfrac{K_{n4}}{K_{n3}}[(-V^-) - V_{TN4}] + V_{TN3}}}{1 + \sqrt{\dfrac{K_{n4}}{K_{n3}}}}$$

(5.26)

Since $V_{GS3} = V_{GS2}$, the bias current is

$$I_Q = K_{n2}(V_{GS3} - V_{TN2})^2$$

(5.27)

Example 5.13 **Objective:** Determine the currents and voltages in a MOSFET constant-current source.

For the circuit shown in Figure 5.43, the transistor parameters are: $K_{n1} = 0.2\,\text{mA/V}^2$, $K_{n2} = K_{n3} = K_{n4} = 0.1\,\text{mA/V}^2$, and $V_{TN1} = V_{TN2} = V_{TN3} = V_{TN4} = 1\,\text{V}$.

Solution: From Equation (5.26), we can determine V_{GS3}, as follows:

$$V_{GS3} = \frac{\sqrt{\dfrac{0.1}{0.1}}[5 - 1] + 1}{1 + \sqrt{\dfrac{0.1}{0.1}}} = 2.5\,\text{V}$$

Since M_3 and M_4 are identical transistors, V_{GS3} should be one-half of the bias voltage. The bias current I_Q is then

$$I_Q = (0.1) \cdot (2.5 - 1)^2 = 0.225\,\text{mA}$$

The gate-to-source voltage on M_1 is found from

$$I_Q = K_{n1}(V_{GS1} - V_{TN1})^2$$

or

$$0.225 = (0.2) \cdot (V_{GS1} - 1)^2$$

which yields

$$V_{GS1} = 2.06\,\text{V}$$

The drain-to-source voltage on M_2 is

$$V_{DS2} = (-V^-) - V_{GS1} = 5 - 2.06 = 2.94\,\text{V}$$

Since $V_{DS2} = 2.94\,\text{V} > V_{DS}(\text{sat}) = V_{GS2} - V_{TN2} = 2.5 - 1 = 1.5\,\text{V}$, M_2 is biased in the saturation region.

Design Consideration: Since in this example M_2 and M_3 are identical transistors, the reference current I_{REF} and bias current I_Q are equal. By redesigning the width-to-length ratios of M_2, M_3, and M_4, we can obtain a specific bias current I_Q. If M_2 and M_3 are not identical, then I_Q and I_{REF} will not be equal. A variety of design options are possible with such a circuit configuration.

Test Your Understanding

D5.20 Consider the constant-current source shown in Figure 5.43. Assume that the threshold voltage of each transistor is $V_{TN} = 1\,\text{V}$. (a) Design the ratio of K_{n4}/K_{n3} such that $V_{GS3} = 2\,\text{V}$. (b) Determine K_{n2} such that $I_Q = 100\,\mu\text{A}$. (c) Find K_{n3} and K_{n4} such that $I_{REF} = 200\,\mu\text{A}$. (Ans. (a) $K_{n4}/K_{n3} = \frac{1}{4}$ (b) $K_{n2} = 0.1\,\text{mA/V}^2$ (c) $K_{n3} = 0.2\,\text{mA/V}^2$, $K_{n4} = 0.05\,\text{mA/V}^2$)

5.3 BASIC MOSFET APPLICATIONS: SWITCH, DIGITAL LOGIC GATE, AND AMPLIFIER

MOSFETs may be used to: switch currents, voltages, and power; perform digital logic functions; and amplify small time-varying signals. In this section, we will examine the switching properties of an NMOS transistor, analyze a simple NMOS transistor digital logic circuit, and discuss how the MOSFET can be used to amplify small signals.

5.3.1 NMOS Inverter

The MOSFET can be used as a switch in a wide variety of electronic applications. The transistor switch provides an advantage over mechanical switches in both speed and reliability. The transistor switch considered in this section is also called an inverter. Two other switch configurations, the NMOS transmission gate and the CMOS transmission gate, are discussed in Chapter 16.

Figure 5.44 shows the n-channel enhancement-mode MOSFET inverter circuit. If $v_I < V_{TN}$, the transistor is in cutoff and $i_D = 0$. There is no voltage drop across R_D, and the output voltage is $v_O = V_{DD}$. Also, since $i_D = 0$, no power is dissipated in the transistor.

If $v_I > V_{TN}$, the transistor is on and initially is biased in the saturation region, since $v_{DS} > v_{GS} - V_{TN}$. As the input voltage increases, the drain-to-source voltage decreases, and the transistor eventually becomes biased in the nonsaturation region. When $v_I = V_{DD}$, the transistor is biased in the non-saturation region, v_O reaches a minimum value, and the drain current reaches a maximum value. The current and voltage are given by

$$i_D = K_n[2(v_I - V_{TN})v_O - v_O^2] \qquad \textbf{(5.28)}$$

and

$$v_O = v_{DD} - i_D R_D \qquad \textbf{(5.29)}$$

where $v_O = v_{DS}$ and $v_I = v_{GS}$.

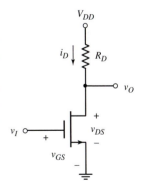

Figure 5.44 NMOS inverter circuit

Design Example 5.14 Objective: Design the size of a power MOSFET to meet the specification of a particular switch application.

The load in the inverter circuit in Figure 5.44 is a coil of an electromagnet that requires a current of 0.5 A when turned on. The effective load resistance varies between 8 and 10 Ω, depending on temperature and other variables. A 10 V power supply is available. The transistor parameters are $k'_n = 80\,\mu\text{A/V}^2$ and $V_{TN} = 1\,\text{V}$.

Solution: One solution is to bias the transistor in the saturation region so that the current is constant, independent of the load resistance.

The minimum V_{DS} value is 5 V. We need $V_{DS} > V_{DS}(\text{sat}) = V_{GS} - V_{TN}$. If we bias the transistor at $V_{GS} = 5\,\text{V}$, then the transistor will always be biased in the saturation region. We can then write

$$I_D = \frac{k'_n}{2} \cdot \frac{W}{L}(V_{GS} - V_{TN})^2$$

or

$$0.5 = \frac{80 \times 10^{-6}}{2}\left(\frac{W}{L}\right) \cdot (5-1)^2$$

which yields $W/L = 781$.

The maximum power dissipated in the transistor is

$$P(\text{max}) = V_{DS}(\text{max}) \cdot I_D = (6) \cdot (0.5) = 3\,\text{W}$$

Comment: We see that we can switch a relatively large drain current with essentially no input current to the transistor. The size of the transistor required is fairly large, which implies a power transistor is necessary. If a transistor with a slightly different width-to-length ratio is available, the applied V_{GS} can be changed to meet the specification.

Test Your Understanding

Figure 5.45

5.21 For the MOS inverter circuit shown in Figure 5.44, assume the circuit values are: $V_{DD} = 5\,\text{V}$ and $R_D = 10\,\text{k}\Omega$. The threshold voltage of the transistor is $V_{TN} = 1\,\text{V}$. Determine the value of the conduction parameter K_n such that $v_O = 1\,\text{V}$ when $v_I = 5\,\text{V}$. What is the power dissipated in the transistor? (Ans. $K_n = 0.057\,\text{mA/V}^2$, $P = 0.4\,\text{mW}$)

D5.22 The circuit shown in Figure 5.44 is biased with $V_{DD} = 10\,\text{V}$ and the transistor has parameters $V_{TN} = 0.70\,\text{V}$ and $K_n = 0.050\,\text{mA/V}^2$. Design the value of R_D for which the output voltage will be $v_O = 0.35\,\text{V}$, when $v_I = 10\,\text{V}$. (Ans. $R_D = 30.3\,\text{k}\Omega$)

D5.23 The transistor in the circuit shown in Figure 5.45 has parameters $K_n = 4\,\text{mA/V}^2$ and $V_{TN} = 0.8\,\text{V}$, and is used to switch the LED on and off. The LED cutin voltage is $V_\gamma = 1.5\,\text{V}$. The LED is turned on by applying an input voltage of $v_I = 5\,\text{V}$. (a) Determine the value of R such that the diode current is 12 mA. (b) From the results of part (a), what is the value of v_{DS}? (Ans. (a) $R = 261\,\Omega$, (b) $v_{DS} = 0.374\,\text{V}$)

5.3.2 Digital Logic Gate

For the transistor inverter circuit in Figure 5.44, when the input is low and approximately zero volts, the transistor is cut off, and the output is high and equal to V_{DD}. When the input is high and equal to V_{DD}, the transistor is biased in the nonsaturation region and the output reaches a low value. Since the input voltages will be either high or low, we can analyze the circuit in terms of dc parameters.

Now consider the case when a second transistor is connected in parallel, as shown in Figure 5.46. If the two inputs are zero, both M_1 and M_2 are cut off, and $V_O = 5\,V$. When $V_1 = 5\,V$ and $V_2 = 0$, the transistor M_1 turns on and M_2 is still cut off. Transistor M_1 is biased in the nonsaturation region, and V_O reaches a low value. If we reverse the input voltages such that $V_1 = 0$ and $V_2 = 5\,V$, then M_1 is cut off and M_2 is biased in the nonsaturation region. Again, V_O is at a low value. If both inputs are high, at $V_1 = V_2 = 5\,V$, then both transistors are biased in the nonsaturation region and V_O is low.

Figure 5.46 A two-input NMOS NOR logic gate

Table 5.2 shows these various conditions for the circuit in Figure 5.46. In a positive logic system, these results indicate that this circuit performs the NOR logic function, and, it is therefore called a two-input NOR logic circuit. In actual NMOS logic circuits, the resistor R_D is replaced by another NMOS transistor.

Table 5.2 NMOS NOR logic circuit response

V_1 (V)	V_2 (V)	V_O (V)
0	0	High
5	0	Low
0	5	Low
5	5	Low

Example 5.15   Objective: Determine the currents and voltages in a digital logic gate, for various input conditions.

Consider the circuit shown in Figure 5.46 with circuit and transistor parameters $R_D = 20\,\text{k}\Omega$, $K_n = 0.1\,\text{mA/V}^2$, and $V_{TN} = 0.8\,\text{V}$.

Solution: For $V_1 = V_2 = 0$, both M_1 and M_2 are cut off and $V_O = V_{DD} = 5\,\text{V}$. For $V_1 = 5\,\text{V}$ and $V_2 = 0$, the transistor M_1 is biased in the nonsaturation region, and we can write

$$I_R = I_{D1} = \frac{5 - V_O}{R_D} = K_n[2(V_1 - V_{TN})V_O - V_O^2]$$

Solving for the output voltage V_O, we obtain $V_O = 0.29\,\text{V}$.

The currents are

$$I_R = I_{D1} = \frac{5 - 0.29}{20} = 0.236\,\text{mA}$$

For $V_1 = 0$ and $V_2 = 5\,\text{V}$, we have $V_O = 0.29\,\text{V}$ and $I_R = I_{D2} = 0.236\,\text{mA}$. When both inputs go high to $V_1 = V_2 = 5\,\text{V}$, we have $I_R = I_{D1} + I_{D2}$, or

$$\frac{5 - V_O}{R_D} = K_n[2(V_1 - V_{TN})V_O - V_O^2] + K_n[2(V_2 - V_{TN})V_O - V_O^2]$$

which can be solved for V_O to yield $V_O = 0.147\,\text{V}$.

The currents are

$$I_R = \frac{5 - 0.147}{20} = 0.243\,\text{mA}$$

and

$$I_{D1} = I_{D2} = \frac{I_R}{2} = 0.121\,\text{mA}$$

Comment: When either transistor is biased on, it is biased in the nonsaturation region, since $V_{DS} < V_{DS}(\text{sat})$, and the output voltage reaches a low state.

This example and discussion illustrates that MOS transistors can be configured in a circuit to perform logic functions. A more detailed analysis and design of MOSFET logic gates and circuits is presented in Chapter 16.

Test Your Understanding

5.24   For the circuit in Figure 5.46, assume the circuit and transistor parameters are: $R_D = 30\,\text{k}\Omega$, $V_{TN} = 1\,\text{V}$, and $K_n = 50\,\mu\text{A/V}^2$. Determine V_O, I_R, I_{D1}, and I_{D2} for: (a) $V_1 = 5\,\text{V}$, $V_2 = 0$; and (b) $V_1 = V_2 = 5\,\text{V}$. (Ans. (a) $V_O = 0.40\,\text{V}$, $I_R = I_{D1} = 0.153\,\text{mA}$, $I_{D2} = 0$ (b) $V_O = 0.205\,\text{V}$, $I_R = 0.16\,\text{mA}$, $I_{D1} = I_{D2} = 0.080\,\text{mA}$)

5.25   In the circuit in Figure 5.46, let $R_D = 25\,\text{k}\Omega$ and $V_{TN} = 1\,\text{V}$. (a) Determine the value of the conduction parameter K_n required such that $V_O = 0.10\,\text{V}$ when $V_1 = 0$ and $V_2 = 5\,\text{V}$. (b) Using the results of part (a), find the value of V_O when $V_1 = V_2 = 5\,\text{V}$. (Ans. (a) $K_n = 0.248\,\text{mA/V}^2$, (b) $V_O = 0.0502\,\text{V}$)

5.3.3   MOSFET Small-Signal Amplifier

The MOSFET, in conjunction with other circuit elements, can amplify small time-varying signals. Figure 5.47(a) shows the MOSFET small-signal amplifier, which is a common-source circuit in which a time-varying signal is coupled to the gate through a coupling capacitor. Figure 5.47(b) shows the transistor characteristics and the load line. The load line is determined for $v_i = 0$.

We can establish a particular Q-point on the load line by designing the ratio of the bias resistors R_1 and R_2. If we assume that $v_i = V_i \sin \omega t$, the gate-to-source voltage will have a sinusoidal signal superimposed on the dc quiescent value. As the gate-to-source voltage changes over time, the Q-point will move up and down the line, as indicated in the figure.

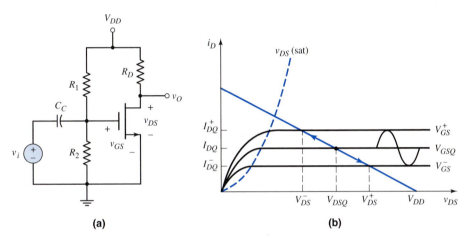

(a)	(b)

Figure 5.47   (a) An NMOS common-source circuit with a time-varying signal coupled to the gate and (b) transistor characteristics, load line, and superimposed sinusoidal signals

Moving up and down the load line translates into a sinusoidal variation in the drain current and in the drain-to-source voltage. The variation in output voltage can be larger than the input signal voltage, which means the input signal is amplified. The actual signal gain depends on both the transistor parameters and the circuit element values.

In the next chapter, we will develop an equivalent circuit for the transistor used to determine the time-varying small-signal gain and other characteristics of the circuit.

5.4   JUNCTION FIELD-EFFECT TRANSISTOR

The two general categories of **junction field-effect transistor (JFET)** are the pn junction FET, or **pn JFET**, and the **metal-semiconductor field-effect transistor (MESFET)**, which is fabricated with a Schottky barrier junction.

The current in a JFET is through a semiconductor region known as the channel, with ohmic contacts at each end. The basic transistor action is the modulation of the channel conductance by an electric field perpendicular to the channel. Since the modulating electric field is induced in the space-charge region of a reverse-biased pn junction or Schottky barrier junction, the field is a function of the gate voltage. Modulation of the channel conductance by the gate voltage modulates the channel current.

JFETs were developed before MOSFETs, but the applications and uses of the MOSFET have far surpassed those of the JFET. One reason is that the voltages applied to the gate and drain of a MOSFET are the same polarity (both positive or both negative), whereas the voltages applied to the gate and drain of most JFETs must have opposite polarities. Since the JFET is used only in specialized applications, our discussion will be brief.

5.4.1   pn JFET and MESFET Operation

pn JFET

A simplified cross section of a symmetrical pn JFET is shown in Figure 5.48. In the n-region channel between the two p-regions, majority carrier electrons flow from the source to the drain terminal; thus, the JFET is called a majority-carrier device. The two gate terminals shown in Figure 5.48 are connected to form a single gate.

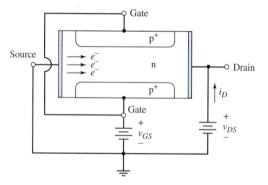

Figure 5.48   Cross section of a symmetrical n-channel pn junction field-effect transistor

In a p-channel JFET, the p- and n-regions are reversed from those of the n-channel device, and holes flow in the channel from the source to the drain. The current direction and voltage polarities in the p-channel JFET are reversed from those in the n-channel device. Also, the p-channel JFET is generally a lower-frequency device than the n-channel JFET, because hole mobility is lower than electron mobility.

Figure 5.49(a) shows an n-channel JFET with zero volts applied to the gate. If the source is at ground potential, and if a small positive drain voltage is applied, a drain current i_D is produced between the source and drain terminals. Since the n-channel acts essentially as a resistance, the i_D versus v_{DS} characteristic for small v_{DS} values is approximately linear, as shown in the figure.

If a voltage is applied to the gate of a pn JFET, the channel conductance changes. If a negative gate voltage is applied to the n-channel pn JFET in Figure 5.49, the gate-to-channel pn junction becomes reverse biased. The space-charge region widens, the channel region narrows, the resistance of the

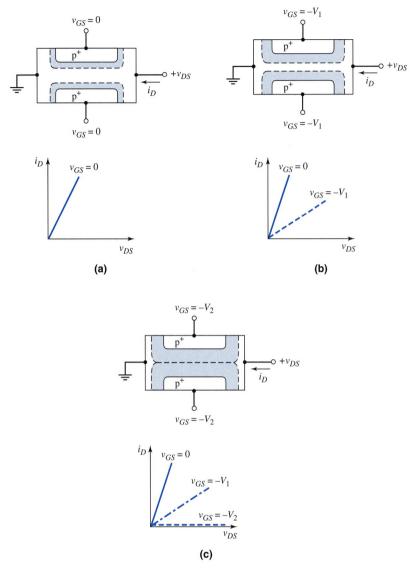

Figure 5.49 Gate-to-channel space-charge regions and current–voltage characteristics for small drain-to-source voltages and for: (a) zero gate voltage, (b) small reverse-biased gate voltage, and (c) a gate voltage that achieves pinchoff

n-channel increases, and the slope of the i_D versus v_{DS} curve, for small v_{DS}, decreases. These effects are shown in Figure 5.49(b). If a larger negative gate voltage is applied, the condition shown in Figure 5.49(c) can be achieved. The reverse-biased gate-to-channel space-charge region completely fills the channel region. This condition is known as **pinchoff**. Since the depletion region isolates the source and drain terminals, the drain current at pinchoff is essentially zero. The i_D versus v_{DS} curves are shown in Figure 5.49(c). The current in the channel is controlled by the gate voltage. The control of the current in one part of the device by a voltage in another part of the device is the basic transistor action. The pn JFET is a "normally on," or depletion mode, device; that is, a voltage must be applied to the gate terminal to turn the device off.

Consider the situation in which the gate voltage is zero, $v_{GS} = 0$, and the drain voltage changes, as shown in Figure 5.50(a). As the drain voltage increases (positive), the gate-to-channel pn junction becomes reverse biased near the drain terminal, and the space-charge region widens, extending farther into the channel. The channel acts essentially as a resistor, and the effective channel resistance increases as the space-charge region widens; therefore, the slope of the i_D versus v_{DS} characteristic decreases, as shown in Figure 5.50(b). The effective channel resistance now varies along the channel length, and, since the channel current must be constant, the voltage drop through the channel becomes dependent on position.

If the drain voltage increases further, the condition shown in Figure 5.50(c) can result. The channel is pinched off at the drain terminal. Any

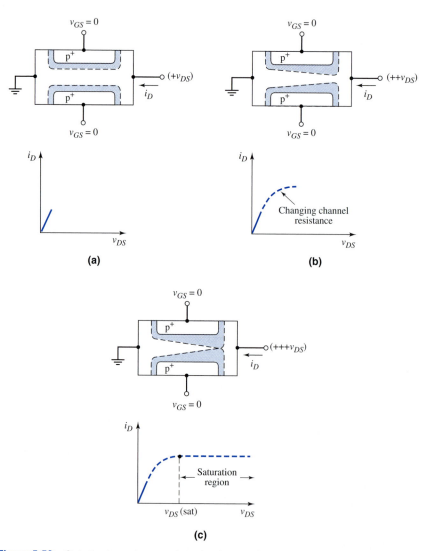

Figure 5.50 Gate-to-channel space-charge regions and current–voltage characteristics for zero gate voltage and for: (a) a small drain voltage, (b) a larger drain voltage, and (c) a drain voltage that achieves pinchoff at the drain terminal

further increase in drain voltage will not increase the drain current. The i_D–v_{DS} characteristic for this condition is also shown in the figure. The drain voltage at pinchoff is $v_{DS}(\text{sat})$. Therefore, for $v_{DS} > v_{DS}(\text{sat})$, the transistor is biased in the saturation region, and the drain current for this ideal case is independent of v_{DS}.

MESFET

In the MESFET, the gate junction is a Schottky barrier junction, instead of a pn junction. Although MESFETs can be fabricated in silicon, they are usually associated with gallium arsenide or other compound-semiconductor materials.

A simplified cross section of a GaAs MESFET is shown in Figure 5.51. A thin, epitaxial layer of GaAs is used for the active region; the substrate is a very high resistivity GaAs material, referred to as a semi-insulating substrate. The advantages of these devices include: higher electron mobility in GaAs, hence smaller transit time and faster response; and decreased parasitic capacitance and a simplified fabrication process, resulting from the semi-insulating GaAs substrate.

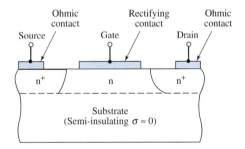

Figure 5.51   Cross section of an n-channel MESFET with a semi-insulating substrate

In the MESFET in Figure 5.51, a reverse-bias gate-to-source voltage induces a space-charge region under the metal gate, which modulates the channel conductance, as in the case of the pn JFET. If a negative applied gate voltage is sufficiently large, the space-charge region will eventually reach the substrate. Again, pinchoff will occur. Also, the device shown in the figure is a depletion mode device, since a gate voltage must be applied to pinch off the channel, that is, to turn the device off.

In another type of MESFET, the channel is pinched off even at $v_{GS} = 0$, as shown in Figure 5.52(a). For this MESFET, the channel thickness is smaller than the zero-biased space-charge width. To open a channel, the depletion region must be reduced; that is, a forward-biased voltage must be applied to the gate–semiconductor junction. When a slightly forward-bias voltage is applied, the depletion region extends just to the width of the channel as shown in Figure 5.52(b). The threshold voltage is the gate-to-source voltage required to create the pinchoff condition. The threshold voltage for this n-channel MESFET is positive, in contrast to the negative threshold voltage of the n-channel depletion-mode device. If a larger forward-bias voltage is applied, the channel region opens, as shown in Figure 5.52(c). The applied

Figure 5.52   Channel space-charge region of an enhancement-mode MESFET for:
(a) $v_{GS} = 0$, (b) $v_{GS} = V_{TN}$, and (c) $v_{GS} > V_{TN}$

forward-bias gate voltage is limited to a few tenths of a volt before a significant gate current occurs.

This device is an **n-channel enhancement-mode MESFET**. Enhancement-mode p-channel MESFETs and enhancement-mode pn JFETs have also been fabricated. The advantage of enhancement-mode MESFETs is that circuits can be designed in which the voltage polarities on the gate and drain are the same. However, the output voltage swing of these devices is quite small.

5.4.2   Current–Voltage Characteristics

The circuit symbols for the n-channel and p-channel JFETs are shown in Figure 5.53, along with the gate-to-source voltages and current directions. The ideal current–voltage characteristics, when the transistor is biased in the saturation region, are described by

$$i_D = I_{DSS}\left(1 - \frac{v_{GS}}{V_P}\right)^2 \tag{5.30}$$

where I_{DSS} is the saturation current when $v_{GS} = 0$, and V_P is the **pinchoff voltage**.

The current–voltage characteristics for n-channel and p-channel JFETs are shown in Figures 5.54(a) and 5.54(b), respectively. Note that the pinchoff voltage V_P for the n-channel JFET is negative and the gate-to-source voltage v_{GS} is usually negative; therefore, the ratio v_{GS}/V_P is positive. Similarly, the pinchoff voltage V_P for the p-channel JFET is positive and the gate-to-source voltage v_{GS} must be positive, and therefore the ratio v_{GS}/V_P is positive.

For the n-channel device, the saturation region occurs when $v_{DS} \geq v_{DS}(\text{sat})$ where

$$v_{DS}(\text{sat}) = v_{GS} - V_P \tag{5.31}$$

For the p-channel device, the saturation region occurs when $v_{SD} \geq v_{SD}(\text{sat})$ where

$$v_{SD}(\text{sat}) = V_P - v_{GS} \tag{5.32}$$

Figure 5.53   Circuit symbols for: (a) n-channel JFET and (b) p-channel JFET

The voltage transfer characteristics of i_D versus v_{GS}, when the transistor is biased in the saturation region, are shown in Figure 5.55, for both the n-channel and p-channel JFET.

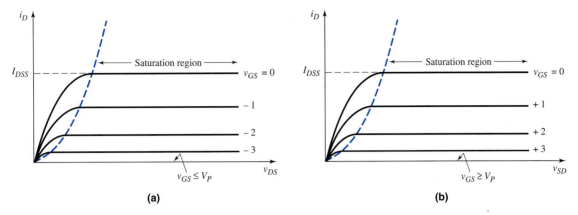

Figure 5.54 Current–voltage characteristics for: (a) n-channel JFET and (b) p-channel JFET

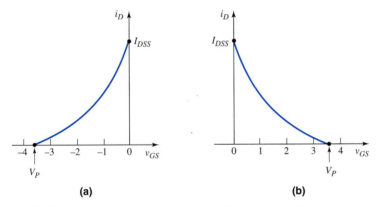

Figure 5.55 Drain current versus gate-to-source voltage characteristics for the transistor biased in the saturation region (a) n-channel JFET and (b) p-channel JFET.

Example 5.16   **Objective:** Calculate i_D and $v_{DS}(\text{sat})$ in an n-channel pn JFET.

Assume the saturation current is $I_{DSS} = 2\,\text{mA}$ and the pinchoff voltage is $V_P = -3.5\,\text{V}$. Calculate i_D and $v_{DS}(\text{sat})$ for $v_{GS} = 0$, $V_P/4$, and $V_P/2$.

Solution: From Equation (5.30), we have

$$i_D = I_{DSS}\left(1 - \frac{v_{GS}}{V_P}\right)^2 = (2)\left(1 - \frac{v_{GS}}{(-3.5)}\right)^2$$

Therefore, for $v_{GS} = 0$, $V_P/4$, and $V_P/2$, we obtain

$$i_D = 2,\,1.13,\ \text{and}\ 0.5\,\text{mA}$$

From Equation (5.31), we have

$$v_{DS}(\text{sat}) = v_{GS} - V_P = v_{GS} - (-3.5)$$

Therefore, for $v_{GS} = 0$, $V_P/4$, and $V_P/2$, we obtain

$$v_{DS}(\text{sat}) = 3.5,\,2.63,\ \text{and}\ 1.75\,\text{V}$$

Comment: The current capability of a JFET can be increased by increasing the value of I_{DSS}, which is a function of the transistor width.

As for the MOSFET, the i_D versus v_{DS} characteristic may have a nonzero slope beyond the saturation point. This nonzero slope can be described through the following equation:

$$i_D = I_{DSS}\left(1 - \frac{v_{GS}}{V_P}\right)^2(1 + \lambda v_{DS}) \tag{5.33}$$

where λ^{-1} is analogous to the Early voltage in bipolar transistors.

The output resistance r_o is defined as

$$r_o = \left(\frac{\partial i_D}{\partial v_{DS}}\right)^{-1}\Bigg|_{v_{GS}=\text{const.}} \tag{5.34}$$

Using Equation (5.33), we find that

$$r_o = \left[\lambda I_{DSS}\left(1 - \frac{V_{GSQ}}{V_P}\right)^2\right]^{-1} \tag{5.35(a)}$$

or

$$r_o \cong [\lambda I_{DQ}]^{-1} = \frac{1}{\lambda I_{DQ}} \tag{5.35(b)}$$

The output resistance will be considered again when we discuss the small-signal equivalent circuit of the JFET in the next chapter.

Enhancement-mode GaAs MESFETs can be fabricated with current–voltage characteristics much like those of the enhancement-mode MOSFET. Therefore, for the ideal enhancement-mode MESFET biased in the saturation region, we can write

$$i_D = K_n(v_{GS} - V_{TN})^2 \tag{5.36(a)}$$

For the ideal enhancement-mode MESFET biased in the nonsaturation region,

$$i_D = K_n[2(v_{GS} - V_{TN})v_{DS} - v_{DS}^2] \tag{5.36(b)}$$

where K_n is the conduction parameter and V_{TN} is the threshold voltage, which in this case is equivalent to the pinchoff voltage. For an n-channel enhancement-mode MESFET, the threshold voltage is positive.

Test Your Understanding

5.26 The parameters of an n-channel JFET are $I_{DSS} = 12\,\text{mA}$, $V_P = -4.5\,\text{V}$, and $\lambda = 0$. Determine $V_{DS}(\text{sat})$ for $V_{GS} = -1.2\,\text{V}$, and calculate I_D for $V_{DS} > V_{DS}(\text{sat})$. (Ans. $V_{DS}(\text{sat}) = 3.3\,\text{V}$, $I_D = 6.45\,\text{mA}$)

5.27 For an n-channel JFET, the parameters are: $I_{DSS} = 2\,\text{mA}$, $V_P = -2.5\,\text{V}$, and $\lambda = 0$. What is the value of V_{GS} when $I_D = 1.2\,\text{mA}$ and the transistor is biased in the saturation region? (Ans. $V_{GS} = -0.564\,\text{V}$)

5.28   A p-channel JFET has a pinchoff voltage of $V_P = 3.8\,\text{V}$. The drain current in the saturation region is $I_D = 3\,\text{mA}$ when $V_{GS} = 0.8\,\text{V}$. Determine I_{DSS} and $V_{SD}(\text{sat})$. (Ans. $I_{DSS} = 4.81\,\text{mA}$, $V_{SD}(\text{sat}) = 3.0\,\text{V}$)

5.29   An n-channel GaAs MESFET has parameters $K_n = 25\,\mu\text{A/V}^2$ and $V_{TN} = 0.25\,\text{V}$. Determine I_D in the saturation region for: (a) $V_{GS} = 0.35\,\text{V}$, and (b) $V_{GS} = 0.50\,\text{V}$. (Ans. (a) $I_D = 0.25\,\mu\text{A}$, (b) $I_D = 1.56\,\mu\text{A}$)

5.4.3   Common JFET Configurations: DC Analysis

There are several common JFET circuit configurations. We will look at a few of these, using examples, and will illustrate the dc analysis and design of such circuits.

Design Example   5.17   Objective: Design the dc bias of a JFET circuit with an n-channel depletion-mode JFET.

For the circuit in Figure 5.56(a), the transistor parameters are: $I_{DSS} = 5\,\text{mA}$, $V_P = -4\,\text{V}$, and $\lambda = 0$. Design the circuit such that $I_D = 2\,\text{mA}$ and $V_{DS} = 6\,\text{V}$.

(a)   **(b)**

Figure 5.56   (a) An n-channel JFET circuit with a self-biasing source resistor and (b) circuit for Example 5.16

Solution: Assume the transistor is biased in the saturation region. The dc drain current is then given by

$$I_D = I_{DSS}\left(1 - \frac{V_{GS}}{V_P}\right)^2$$

or

$$2 = 5\left(1 - \frac{V_{GS}}{(-4)}\right)^2$$

Therefore,

$$V_{GS} = -1.47\,\text{V}$$

From Figure 5.56(b) we see that the current through the source resistor can be written as

$$I_D = \frac{-V_{GS}}{R_S}$$

Therefore,

$$R_S = \frac{-V_{GS}}{I_D} = \frac{-(-1.47)}{2} = 0.735\,\text{k}\Omega$$

The drain-to-source voltage is

$$V_{DS} = V_{DD} - I_D R_D - I_D R_S$$

Therefore,

$$R_D = \frac{V_{DD} - V_{DS} - I_D R_S}{I_D} = \frac{10 - 6 - (2)(0.735)}{2} = 1.27\,\text{k}\Omega$$

We also see that

$$V_{DS} = 6\,\text{V} > V_{GS} - V_P = -1.47 - (-4) = 2.53\,\text{V}$$

which shows that the JFET is indeed biased in the saturation region, as initially assumed.

 Comment: Since the source terminal must be positive with respect to the gate in order to bias the transistor on, the source resistor self-biases the JFET, even though the gate and the "bottom" of R_S are at ground potential.

Design Example 5.18   Objective: Design a JFET circuit with a voltage divider biasing circuit.

Consider the circuit shown in Figure 5.57(a) with transistor parameters $I_{DSS} = 12\,\text{mA}$, $V_P = -3.5\,\text{V}$, and $\lambda = 0$. Let $R_1 + R_2 = 100\,\text{k}\Omega$. Design the circuit such that the dc drain current is $I_D = 5\,\text{mA}$ and the dc drain-to-source voltage is $V_{DS} = 5\,\text{V}$.

Solution: Assume the transistor is biased in the saturation region. The dc drain current is then given by

$$I_D = I_{DSS}\left(1 - \frac{V_{GS}}{V_P}\right)^2$$

Therefore,

$$5 = 12\left(1 - \frac{V_{GS}}{(-3.5)}\right)^2$$

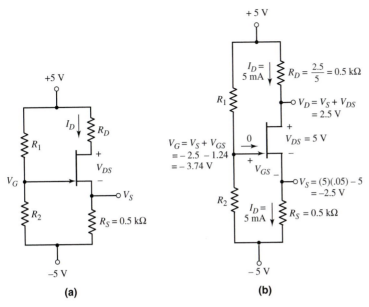

Figure 5.57 (a) An n-channel JFET circuit with voltage divider biasing and (b) the n-channel JFET circuit for Example 5.18

which yields

$$V_{GS} = -1.24\,\text{V}$$

From Figure 5.57(b), the voltage at the source terminal is

$$V_S = I_D R_S - 5 = (5)(0.5) - 5 = -2.5\,\text{V}$$

which means that the gate voltage is

$$V_G = V_{GS} + V_S = -1.24 - 2.5 = -3.74\,\text{V}$$

We can also write the gate voltage as

$$V_G = \left(\frac{R_2}{R_1 + R_2}\right)(10) - 5$$

or

$$-3.74 = \frac{R_2}{100}(10) - 5 \cdot$$

Therefore,

$$R_2 = 12.6\,\text{k}\Omega$$

and

$$R_1 = 87.4\,\text{k}\Omega$$

The drain-to-source voltage is

$$V_{DS} = 5 - I_D R_D - I_D R_S - (-5)$$

Therefore,

$$R_D = \frac{10 - V_{DS} - I_D R_S}{I_D} = \frac{10 - 5 - (5)(0.5)}{5} = 0.5\,\text{k}\Omega$$

We also see that

$$V_{DS} = 5\,\text{V} > V_{GS} - V_P = -1.24 - (-3.5) = 2.26\,\text{V}$$

which shows that the JFET is indeed biased in the saturation region, as initially assumed.

Comment: The dc analysis of the JFET circuit is essentially the same as that of the MOSFET circuit, since the gate current is assumed to be zero.

Example 5.19 Objective: Calculate the quiescent current and voltage values in a p-channel JFET circuit.

The parameters of the transistor in the circuit shown in Figure 5.58 are: $I_{DSS} = 2.5\,\text{mA}$, $V_P = +2.5\,\text{V}$, and $\lambda = 0$. The transistor is biased with a constant-current source.

Solution: From Figure 5.58, we can write the dc drain current as

$$I_D = I_Q = 0.8\,\text{mA} = \frac{V_D - (-9)}{R_D}$$

which yields

$$V_D = (0.8)(4) - 9 = -5.8\,\text{V}$$

Now, assume the transistor is biased in the saturation region. We then have

$$I_D = I_{DSS}\left(1 - \frac{V_{GS}}{V_P}\right)^2$$

or

$$0.8 = 2.5\left(1 - \frac{V_{GS}}{2.5}\right)^2$$

which yields

$$V_{GS} = 1.086\,\text{V}$$

Then

$$V_S = 1 - V_{GS} = 1 - 1.086 = -0.086\,\text{V}$$

and

$$V_{SD} = V_S - V_D = -0.086 - (-5.8) = 5.71\,\text{V}$$

Again, we see that

$$V_{SD} = 5.71\,\text{V} > V_P - V_{GS} = 2.5 - 1.086 = 1.41\,\text{V}$$

which verifies that the transistor is biased in the saturation region, as assumed.

Figure 5.58 A p-channel JFET circuit biased with a constant-current source

Comment: In the same way as bipolar or MOS transistors, junction field-effect transistors can be biased using constant-current sources.

Design Example 5.20 Objective: Design a circuit with an enhancement-mode MESFET.

Consider the circuit shown in Figure 5.59(a). The transistor parameters are: $V_{TN} = 0.24\,\text{V}$, $K_n = 1.1\,\text{mA/V}^2$, and $\lambda = 0$. Let $R_1 + R_2 = 50\,\text{k}\Omega$. Design the circuit such that $V_{GS} = 0.50\,\text{V}$ and $V_{DS} = 2.5\,\text{V}$.

Figure 5.59 (a) An n-channel enhancement-mode MESFET circuit and (b) the n-channel MESFET circuit for Example 5.20

Solution: From Equation (5.36(a)) the drain current is

$$I_D = K_n(V_{GS} - V_{TN})^2 = (1.1)(0.5 - 0.24)^2 = 74.4\,\mu\text{A}$$

From Figure 5.59(b), the voltage at the drain is

$$V_D = V_{DD} - I_D R_D = 4 - (0.0744)(6.7) = 3.5\,\text{V}$$

Therefore, the voltage at the source is

$$V_S = V_D - V_{DS} = 3.5 - 2.5 = 1\,\text{V}$$

The source resistance is then

$$R_S = \frac{V_S}{I_D} = \frac{1}{0.0744} = 13.4\,\text{k}\Omega$$

The voltage at the gate is

$$V_G = V_{GS} + V_S = 0.5 + 1 = 1.5\,\text{V}$$

Since the gate current is zero, the gate voltage is also given by a voltage divider equation, as follows:

$$V_G = \left(\frac{R_2}{R_1 + R_2}\right)(V_{DD})$$

or

$$1.5 = \left(\frac{R_2}{50}\right)(4)$$

which yields

$$R_2 = 18.75 \, \text{k}\Omega$$

and

$$R_1 = 31.25 \, \text{k}\Omega$$

Again, we see that

$$V_{DS} = 2.5 \, \text{V} > V_{GS} - V_{TN} = 0.5 - 0.24 = 0.26 \, \text{V}$$

which confirms that the transistor is biased in the saturation region, as initially assumed.

 Comment: The dc analysis and design of an enhancement-mode MESFET circuit is similar to that of MOSFET circuits, except that the gate-to-source voltage of the MESFET must be held to no more than a few tenths of a volt.

Test Your Understanding

5.30 For the circuit in Figure 5.60, the transistor parameters are: $V_P = -3.5 \, \text{V}$, $I_{DSS} = 18 \, \text{mA}$, and $\lambda = 0$. Calculate V_{GS} and V_{DS}. Is the transistor biased in the saturation or nonsaturation region? (Ans. $V_{GS} = -1.17 \, \text{V}$, $V_{DS} = 7.43 \, \text{V}$, saturation region)

5.31 The transistor in the circuit in Figure 5.61 has parameters $I_{DSS} = 6 \, \text{mA}$, $V_P = -4 \, \text{V}$, and $\lambda = 0$. Design the circuit such that $I_{DQ} = 2.5 \, \text{mA}$ and $V_{DS} = 6 \, \text{V}$, and the total power dissipated in R_1 and R_2 is 2 mW. (Ans. $R_D = 1.35 \, \text{k}\Omega$, $R_1 = 158 \, \text{k}\Omega$, $R_2 = 42 \, \text{k}\Omega$)

*5.32 For the p-channel transistor in the circuit in Figure 5.62, the parameters are: $I_{DSS} = 6 \, \text{mA}$, $V_P = 4 \, \text{V}$, and $\lambda = 0$. Calculate the quiescent values of I_D, V_{GS}, and V_{SD}. Is the transistor biased in the saturation or nonsaturation region? (Ans. $V_{GS} = 1.81 \, \text{V}$, $I_D = 1.81 \, \text{mA}$, $V_{SD} = 2.47 \, \text{V}$, saturation region)

*D5.33 Consider the circuit shown in Figure 5.63 with transistor parameters $I_{DSS} = 8 \, \text{mA}$, $V_P = 4 \, \text{V}$, and $\lambda = 0$. Design the circuit such that $R_{in} = 100 \, \text{k}\Omega$, $I_{DQ} = 5 \, \text{mA}$, and $V_{SDQ} = 12 \, \text{V}$. (Ans. $R_D = 0.4 \, \text{k}\Omega$, $R_1 = 387 \, \text{k}\Omega$, $R_2 = 135 \, \text{k}\Omega$)

Figure 5.60 Circuit for Exercise 5.30

Figure 5.61 Circuit for Exercise 5.31

Figure 5.62 Circuit for Exercise 5.32

5.34   The n-channel enhancement-mode MESFET in the circuit shown in Figure 5.64 has parameters $K_n = 50\,\mu\text{A}/\text{V}^2$ and $V_{TN} = 0.15\,\text{V}$. Find the value of V_{GG} so that $I_{DQ} = 5\,\mu\text{A}$. What are the values of V_{GS} and V_{DS}? (Ans. $V_{GG} = 0.516\,\text{V}$, $V_{GS} = 0.466\,\text{V}$, $V_{DS} = 4.45\,\text{V}$)

Figure 5.63   Circuit for Exercise 5.33

Figure 5.64   Circuit for Exercise 5.34

5.35   For the inverter circuit shown in Figure 5.65, the n-channel enhancement-mode MESFET parameters are $K_n = 100\,\mu\text{A}/\text{V}^2$ and $V_{TN} = 0.2\,\text{V}$. Determine the value of R_D required to produce $V_O = 0.10\,\text{V}$ when $V_I = 0.7\,\text{V}$. (Ans. $R_D = 267\,\text{k}\Omega$)

Figure 5.65   Circuit for Exercise 5.35

5.5   SUMMARY

- In this chapter, we have emphasized the structure and dc characteristics of the metal-oxide-semiconductor field-effect transistor (MOSFET). This device, because of its small size, has made possible the microprocessor and other high-density VLSI circuits, so this device is extremely important in integrated circuit technology.

- The current in the MOSFET is controlled by an electric field perpendicular to the surface of the semiconductor. This electric field is a function of the gate voltage. In the nonsaturation bias region of operation, the drain current is a function of the drain voltage, whereas in the saturation bias region of operation, the drain current is essentially independent of the drain voltage. The drain current is directly proportional to the width-to-length ratio of the transistor, so this parameter becomes the primary design variable in MOSFET circuit design.

- The dc analysis and the design of dc biasing of MOSFET circuits were emphasized in this chapter. Several circuit configurations were analyzed and designed by using the ideal current-voltage relationships. The use of MOSFETs, both enhancement-mode and depletion-mode devices, in place of resistors was developed. This leads to the design of all-MOSFET circuits.
- Basic applications of the MOSFET were discussed. These include switching currents and voltages, performing digital logic functions, and amplifying time-varying signals. The amplifying characteristics will be considered in the next chapter and the important digital applications will be considered in Chapter 16.

CHECKPOINT

After studying this chapter, the reader should have the ability to:

✔ Understand and describe the general operation of n-channel and p-channel enhancement-mode and depletion-mode MOSFETs. (Section 5.1)

✔ Understand the meaning of the various transistor parameters, including threshold voltage, width-to-length ratio, and drain-to-source saturation voltage. (Section 5.1)

✔ Apply the ideal current–voltage relations in the dc analysis and design of various MOSFET circuits using any of the four basic MOSFETs. (Section 5.2)

✔ Understand how MOSFETs can be used in place of resistor load devices to create all-MOSFET circuits. (Section 5.2)

✔ Qualitatively understand how MOSFETs can be used to switch currents and voltages, to perform digital logic functions, and to amplify time-varying signals. (Section 5.3)

✔ Understand the general operation and characteristics of junction FETs. (Section 5.4)

REVIEW QUESTIONS

1. Describe the basic operation of a MOSFET. Define enhancement mode and depletion mode.
2. Describe the general current–voltage characteristics for both enhancement-mode and depletion-mode MOSFETs.
3. Describe what is meant by threshold voltage, width-to-length ratio, and drain-to-source saturation voltage.
4. Define the saturation and nonsaturation bias regions.
5. Describe the channel length modulation effect and define the parameter λ. Describe the body effect and define the parameter γ,
6. Describe a simple common-source MOSFET circuit with an n-channel enhancement-mode device and discuss the relation between the drain-to-source voltage and gate-to-source voltage.
7. What are the steps in the dc analysis of a MOSFET circuit?
8. How do you prove that a MOSFET is biased in the saturation region?
9. In the dc analysis of some MOSFET circuits, quadratic equations in gate-to-source voltage are developed. How do you determine which of the two possible solutions is the correct one?
10. How can the Q-point be stabilized against variations in transistor parameters?

11. Describe the current–voltage relation of an n-channel enhancement-mode MOSFET with the gate connected to the drain.

12. Describe the current–voltage relation of an n-channel depletion-mode MOSFET with the gate connected to the source.

13. What is the principal difference between biasing techniques used in discrete transistor circuits and integrated circuits?

14. Describe how an n-channel enhancement-mode MOSFET can be used to switch a motor on and off.

15. Describe a MOSFET NOR logic circuit.

16. Describe how a MOSFET can be used to amplify a time-varying voltage.

17. Describe the basic operation of a junction FET.

18. What is the difference between a MESFET and a pn junction FET?

PROBLEMS

[Note: In all problems, assume the transistor parameter $\lambda = 0$, unless otherwise stated.]

Section 5.1   MOS Field-Effect Transistor

5.1 Consider an n-channel enhancement-mode MOSFET with parameters $V_{TN} = 1.5\,\text{V}$ and $K_n = 0.25\,\text{mA/V}^2$. Determine I_D for: (a) $V_{GS} = 5\,\text{V}$, $V_{DS} = 6\,\text{V}$; and (b) $V_{GS} = 5\,\text{V}$, $V_{DS} = 2.5\,\text{V}$.

5.2 The parameters of an n-channel enhancement-mode MOSFET are $V_{TN} = 0.8\,\text{V}$, $k_n' = 80\,\mu\text{A/V}^2$, and $W/L = 5$. (a) Assume the transistor is biased in the saturation region and $I_D = 0.5\,\text{mA}$. Determine $V_{DS}(\text{sat})$ and the required V_{GS}. (b) Repeat part (a) for $I_D = 1.5\,\text{mA}$.

5.3 For an n-channel depletion-mode MOSFET, the parameters are $V_{TN} = -2.5\,\text{V}$ and $K_n = 1.1\,\text{mA/V}^2$. (a) Determine I_D for $V_{GS} = 0$; and: (i) $V_{DS} = 0.5\,\text{V}$, (ii) $V_{DS} = 2.5\,\text{V}$, and (iii) $V_{DS} = 5\,\text{V}$. (b) Repeat part (a) for $V_{GS} = 2\,\text{V}$.

5.4 Consider an n-channel depletion-mode MOSFET with parameters $V_{TN} = -2\,\text{V}$ and $k_n' = 80\,\mu\text{A/V}^2$. The drain current is $I_D = 1.5\,\text{mA}$ at $V_{GS} = 0$ and $V_{DS} = 3\,\text{V}$. Determine the W/L ratio.

5.5 An n-channel enhancement-mode MOSFET has parameters $V_{TN} = 0.8\,\text{V}$, $W = 64\,\mu\text{m}$, $L = 4\,\mu\text{m}$, $t_{ox} = 450\,\text{Å}$, and $\mu_n = 650\,\text{cm}^2/\text{V–s}$. (a) Calculate the conduction parameter K_n. (b) Determine the drain current when $V_{GS} = V_{DS} = 3\,\text{V}$.

5.6 For a depletion NMOS device, the parameters are: $V_{TN} = -2\,\text{V}$, $W = 100\,\mu\text{m}$, $L = 5\,\mu\text{m}$, $t_{ox} = 600\,\text{Å}$, and $\mu_n = 500\,\text{cm}^2/\text{V–s}$. (a) Calculate the conduction parameter K_n. (b) Determine the drain current when: (i) $V_{GS} = 0$, $V_{DS} = 5\,\text{V}$, and (ii) $V_{GS} = 2\,\text{V}$, $V_{DS} = 1\,\text{V}$.

5.7 A particular NMOS device has parameters $V_{TN} = 1\,\text{V}$, $L = 2.5\,\mu\text{m}$, $t_{ox} = 400\,\text{Å}$, and $\mu_n = 600\,\text{cm}^2/\text{V–s}$. A drain current of $I_D = 1.2\,\text{mA}$ is requried when the device is biased in the saturation region at $V_{GS} = 5\,\text{V}$. Determine the necessary channel width of the device.

5.8 For a p-channel enhancement-mode MOSFET, $k_p' = 40\,\mu\text{A/V}^2$. The device has drain currents of $I_D = 0.225\,\text{mA}$ at $V_{SG} = V_{SD} = 3\,\text{V}$ and $I_D = 1.40\,\text{mA}$ at $V_{SG} = V_{SD} = 4\,\text{V}$. Determine the W/L ratio and the value of V_{TP}.

5.9 For a p-channel enhancement-mode MOSFET, the parameters are $K_P = 2\,\text{mA/V}^2$ and $V_{TP} = -0.5\,\text{V}$. The gate is at ground potential, and the source and substrate terminals are at $+5\,\text{V}$. Determine I_D when the drain terminal voltage is: (a) $V_D = 0\,\text{V}$, (b) $V_D = 2\,\text{V}$, (c) $V_D = 4\,\text{V}$, and (d) $V_D = 5\,\text{V}$.

5.10   A p-channel depletion-mode MOSFET has parameters $V_{TP} = +2\,\text{V}$, $k_p' = 40\,\mu\text{A}/\text{V}^2$, and $W/L = 6$. Determine $V_{SD}(\text{sat})$ for: (a) $V_{SG} = -1\,\text{V}$, (b) $V_{SG} = 0$, and (c) $V_{SG} = +1\,\text{V}$. If the transistor is biased in the saturation region, calculate the drain current for each value of V_{SG}.

5.11   Consider a p-channel depletion-mode MOSFET with parameters $K_P = 0.5\,\text{mA}/\text{V}^2$ and $V_{TP} = +2\,\text{V}$. If $V_{SG} = 0$, determine I_D for: (a) $V_{SD} = 1\,\text{V}$, (b) $V_{SD} = 2\,\text{V}$, and (c) $V_{SD} = 3\,\text{V}$.

D5.12   Enhancement-mode NMOS and PMOS devices both have parameters $L = 4\,\mu\text{m}$ and $t_{ox} = 500\,\text{Å}$. For the NMOS transistor, $V_{TN} = +0.6\,\text{V}$, $\mu_n = 675\,\text{cm}^2/\text{V–s}$, and the channel width is W_n; for the PMOS transistor, $V_{TP} = -0.6\,\text{V}$, $\mu_p = 375\,\text{cm}^2/\text{V–s}$, and the channel width is W_p. Design the widths of the two transistors such that they are electrically equivalent and the drain current in the PMOS transistor is $I_D = 0.8\,\text{mA}$ when it is biased in the saturation region at $V_{SG} = 5\,\text{V}$. What are the values of K_n, K_p, W_n, and W_p?

5.13   For an NMOS enhancement-mode transistor, the parameters are: $V_{TN} = 1.2\,\text{V}$, $K_n = 0.20\,\text{mA}/\text{V}^2$, and $\lambda = 0.01\,\text{V}^{-1}$. Calculate the output resistance r_o for $V_{GS} = 2.0\,\text{V}$ and for $V_{GS} = 4.0\,\text{V}$. What is the value of V_A?

5.14   The parameters of an n-channel enhancement-mode MOSFET are $V_{TN} = 0.8\,\text{V}$, $k_n' = 80\,\mu\text{A}/\text{V}^2$, and $W/L = 4$. What is the maximum value of λ and the minimum value of V_A such that for $V_{GS} = 3\,\text{V}$, $r_o \geq 200\,\text{k}\Omega$?

5.15   An enhancement-mode NMOS transistor has parameters $V_{TNO} = 0.8\,\text{V}$, $\gamma = 0.8\,\text{V}^{1/2}$, and $\phi_f = 0.35\,\text{V}$. At what value of V_{SB} will the threshold voltage change by $2\,\text{V}$ due to the body effect?

5.16   Consider an NMOS device with parameters $V_{TNO} = 1\,\text{V}$ and $\phi_f = 0.37\,\text{V}$. Determine the maximum value of γ such that the shift in threshold voltage between $V_{SB} = 0$ and $V_{SB} = 10\,\text{V}$ is no more than $1.2\,\text{V}$.

5.17   The silicon dioxide gate insulator of an MOS transistor has a thickness of $t_{ox} = 275\,\text{Å}$. (a) Calculate the ideal oxide breakdown voltage. (b) If a safety factor of three is required, determine the maximum safe gate voltage that may be applied.

5.18   In a power MOS transistor, the maximum applied gate voltage is $24\,\text{V}$. If a safety factor of three is specified, determine the minimum thickness necessary for the silicon dioxide gate insulator.

Section 5.2   Transistor DC Analysis

5.19   In the circuit in Figure P5.19, the transistor parameters are $V_{TN} = 0.8\,\text{V}$ and $K_n = 0.5\,\text{mA}/\text{V}^2$. Calculate V_{GS}, I_D, and V_{DS}.

5.20   For the transistor in the circuit in Figure P5.20, the parameters are $V_{TN} = 2\,\text{V}$, $k_n' = 60\,\mu\text{A}/\text{V}^2$, and $W/L = 60$. Determine V_{GS}, I_D, and V_{DS}.

5.21   Consider the circuit in Figure P5.21. The transistor parameters are $V_{TP} = -2\,\text{V}$ and $K_p = 1\,\text{mA}/\text{V}^2$. Determine I_D, V_{SG}, and V_{SD}.

5.22   For the circuit in Figure P5.22, the transistor parameters are $V_{TP} = -0.8\,\text{V}$ and $K_p = 200\,\mu\text{A}/\text{V}^2$. Determine V_S and V_{SD}.

*D5.23   Design a MOSFET circuit in the configuration shown in Figure P5.19. The transistor parameters are $V_{TN} = 1.2\,\text{V}$, $k_n' = 60\,\mu\text{A}/\text{V}^2$, and $\lambda = 0$. The circuit parameters are $V_{DD} = 10\,\text{V}$ and $R_D = 5\,\text{k}\Omega$. Design the circuit so that $V_{DSQ} \cong 5\,\text{V}$, the voltage across

Figure P5.19   Figure P5.20   Figure P5.21

Figure P5.22   Figure P5.24

R_S is approximately equal to V_{GS}, and the current through the bias resistors is approximately 5 percent of the drain current.

5.24 The parameters of the transistor in the circuit in Figure P5.24 are $V_{TN} = -1\,\text{V}$ and $K_n = 0.5\,\text{mA/V}^2$. Determine V_{GS}, I_D, and V_{DS}.

*D5.25 Design a MOSFET circuit with the configuration shown in Figure P5.21. The transistor parameters are $V_{TP} = -2\,\text{V}$, $k'_p = 40\,\mu\text{A/V}^2$, and $\lambda = 0$. The circuit bias is $\pm 10\,\text{V}$, the drain current is to be $0.8\,\text{mA}$, the drain-to-source voltage is to be approximately $10\,\text{V}$, and the voltage across R_S is to be approximately equal to V_{GS}. In addition, the current through the bias resistors is to be no more than 10 percent of the drain current.

5.26 The parameters of the transistors in Figures P5.26(a) and (b) are $K_n = 0.5\,\text{mA/V}^2$, $V_{TN} = 1.2\,\text{V}$, and $\lambda = 0$. Determine v_{GS} and v_{DS} for each transistor when (i) $I_Q = 50\,\mu\text{A}$ and (ii) $I_Q = 1\,\text{mA}$.

5.27 For the circuit in Figure P5.27, the transistor parameters are $V_{TN} = 0.6\,\text{V}$ and $K_n = 200\,\mu\text{A/V}^2$. Determine V_S and V_D.

D5.28 Design the circuit in Figure P5.28, such that $I_D = 0.8\,\text{mA}$ and $V_D = 1\,\text{V}$. The transistor parameters are $K_n = 400\,\mu\text{A/V}^2$ and $V_{TN} = 1.7\,\text{V}$.

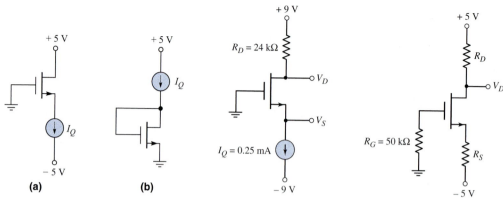

Figure P5.26   **Figure P5.27**   **Figure P5.28**

D5.29 The PMOS transistor in Figure P5.29 has parameters $V_{TP} = -1.5\,\text{V}$, $k_p' = 25\,\mu\text{A/V}^2$, $L = 4\,\mu\text{m}$, and $\lambda = 0$. Determine the values of W and R such that $I_D = 0.1\,\text{mA}$ and $V_{SD} = 2.5\,\text{V}$.

D5.30 Design the circuit in Figure P5.30 so that $V_{SD} = 2.5\,\text{V}$. The current in the bias resistors should be no more than 10 percent of the drain current. The transistor parameters are $V_{TP} = +1.5\,\text{V}$ and $K_p = 0.5\,\text{mA/V}^2$.

Figure P5.29   **Figure P5.30**   **Figure P5.31**

D5.31 Design the circuit in Figure P5.31 such that $I_D = 0.5\,\text{mA}$ and $V_D = -3\,\text{V}$. The transistor parameters are $k_p' = 30\,\mu\text{A/V}^2$, $W/L = 20$, and $V_{TP} = -1.2\,\text{V}$.

D5.32 The parameters of the transistor in the circuit in Figure P5.32 are $V_{TP} = -1.75\,\text{V}$ and $K_p = 3\,\text{mA/V}^2$. Design the circuit such that $I_D = 5\,\text{mA}$, $V_{SD} = 6\,\text{V}$, and $R_{\text{in}} = 80\,\text{k}\Omega$.

*5.33 For each transistor in the circuit in Figure P5.33, $k_n' = 60\,\mu\text{A/V}^2$. Also for M_1, $W/L = 4$ and $V_{TN} = +0.8\,\text{V}$, and for M_2, $W/L = 1$ and $V_{TN} = -1.8\,\text{V}$. Determine the region of operation of each transistor and the output voltage v_O for: (a) $v_I = 1\,\text{V}$, (b) $v_I = 3\,\text{V}$, and (c) $v_I = 5\,\text{V}$.

*D5.34 Consider the circuit in Figure P5.33. The transistor parameters are for M_1, $V_{TN} = +0.8\,\text{V}$ and $k_n' = 40\,\mu\text{A/V}^2$, and for M_2, $V_{TN} = -2\,\text{V}$, $k_n' = 40\,\mu\text{A/V}^2$, and $W/L = 1$. Determine the W/L ratio for M_1 such that $v_O = 0.15\,\text{V}$ when $v_I = 5\,\text{V}$.

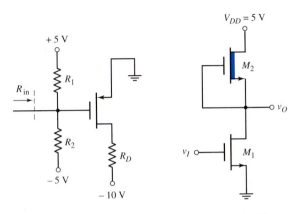

| Figure P5.32 | Figure P5.33 |

*5.35 The transistors in the circuit in Figure P5.35 both have parameters $V_{TN} = 0.8\,\text{V}$ and $k_n' = 30\,\mu\text{A/V}^2$. (a) If the width-to-length ratios of M_1 and M_2 are $(W/L)_1 = (W/L)_2 = 40$, determine V_{GS1}, V_{GS2}, V_O, and I_D. (b) Repeat part (a) if the width-to-length ratios are changed to $(W/L)_1 = 40$ and $(W/L)_2 = 15$.

D5.36 Consider the circuit in Figure P5.36. The transistor parameters are $V_{TN} = 1\,\text{V}$ and $k_n' = 36\,\mu\text{A/V}^2$. Design the width-to-length ratio required in each transistor such that $I_D = 0.5\,\text{mA}$, $V_1 = 2\,\text{V}$, and $V_2 = 5\,\text{V}$.

D5.37 The transistors in the circuit in Figure 5.35 in the text have parameters $V_{TN} = 0.8\,\text{V}$, $k_n' = 40\,\mu\text{A/V}^2$, and $\lambda = 0$. The width-to-length ratio of M_L is $(W/L)_L = 1$. Design the width-to-length ratio of the driver transistor such that $V_O = 0.10\,\text{V}$ when $V_I = 5\,\text{V}$.

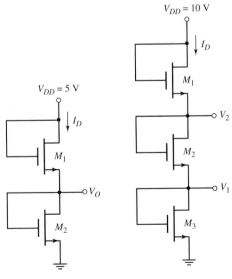

| Figure P5.35 | Figure P5.36 |

D5.38 For the circuit in Figure 5.39 in the text, the transistor parameters are: $V_{TND} = 0.8\,\mathrm{V}$, $V_{TNL} = -1.8\,\mathrm{V}$, $k_n' = 40\,\mu\mathrm{A/V^2}$, and $\lambda = 0$. Let $V_{DD} = 5\,\mathrm{V}$. The width-to-length ratio of M_L is $(W/L)_L = 1$. Design the width-to-length ratio of the driver transistor such that $V_O = 0.05\,\mathrm{V}$ when $V_I = 5\,\mathrm{V}$.

5.39 All transistors in the circuit in Figure 5.43 in the text have parameters $V_{TN} = 1\,\mathrm{V}$ and $\lambda = 0$. Assume the conduction parameters are $K_{n1} = 80\,\mu\mathrm{A/V^2}$, $K_{n2} = 100\,\mu\mathrm{A/V^2}$, $K_{n3} = 200\,\mu\mathrm{A/V^2}$, and $K_{n4} = 400\,\mu\mathrm{A/V^2}$. Determine I_{REF}, I_Q, and each gate-to-source voltage.

Section 5.3   MOSFET Switch and Amplifier

5.40 Consider the circuit in Figure P5.40. The transistor parameters are $V_{TN} = 0.8\,\mathrm{V}$ and $k_n' = 30\,\mu\mathrm{A/V^2}$. The resistor is $R_D = 10\,\mathrm{k\Omega}$. Determine the transistor width-to-length ratio (W/L) such that $V_O = 0.1\,\mathrm{V}$ when $V_I = 4.2\,\mathrm{V}$.

D5.41 The transistor in the circuit in Figure P5.41 is used to turn the LED on and off. The transistor parameters are $V_{TN} = 0.8\,\mathrm{V}$, $k_n' = 40\,\mu\mathrm{A/V^2}$, and $\lambda = 0$. The diode cut-in voltage is $V_\gamma = 1.6\,\mathrm{V}$. Design R_D and the transistor W/L ratio such that $I_D = 12\,\mathrm{mA}$ for $V_I = 5\,\mathrm{V}$ and $V_{DS} = 0.2\,\mathrm{V}$.

D5.42 The circuit in Figure P5.42 is another configuration used to switch an LED on and off. The transistor parameters are $V_{TP} = -0.8\,\mathrm{V}$, $k_p' = 20\,\mu\mathrm{A/V^2}$, and $\lambda = 0$. The diode cut-in voltage is $V_\gamma = 1.6\,\mathrm{V}$. Design R_D and the transistor W/L ratio such that $I_D = 15\,\mathrm{mA}$ for $V_I = 0\,\mathrm{V}$ and $V_{SD} = 0.15\,\mathrm{V}$.

Figure P5.40   **Figure P5.41**   **Figure P5.42**

Section 5.4   Junction Field-Effect Transistor

5.43 The gate and source of an n-channel depletion-mode JFET are connected together. What value of V_{DS} will ensure that this two-terminal device is biased in the saturation region. What is the drain current for this bias condition?

5.44 For an n-channel JFET, the parameters are $I_{DSS} = 6\,\mathrm{mA}$ and $V_P = -3\,\mathrm{V}$. Calculate $V_{DS}(\mathrm{sat})$. If $V_{DS} > V_{DS}(\mathrm{sat})$, determine I_D for: (a) $V_{GS} = 0$, (b) $V_{GS} = -1\,\mathrm{V}$, (c) $V_{GS} = -2\,\mathrm{V}$, and (d) $V_{GS} = -3\,\mathrm{V}$.

5.45 A p-channel JFET biased in the saturation region with $V_{SD} = 5\,\mathrm{V}$ has a drain current of $I_D = 2.8\,\mathrm{mA}$ at $V_{GS} = 1\,\mathrm{V}$ and $I_D = 0.30\,\mathrm{mA}$ at $V_{GS} = 3\,\mathrm{V}$. Determine I_{DSS} and V_P.

5.46 Consider the p-channel JFET in Figure P5.46. Determine the range of V_{DD} that will bias the transistor in the saturation region. If $I_{DSS} = 6\,\text{mA}$ and $V_P = 2.5\,\text{V}$, find V_S.

5.47 Consider a GaAs MESFET. When the device is biased in the saturation region, we find that $I_D = 18.5\,\mu\text{A}$ at $V_{GS} = 0.35\,\text{V}$ and $I_D = 86.2\,\mu\text{A}$ at $V_{GS} = 0.50\,\text{V}$. Determine the conduction parameter k and the threshold voltage V_{TN}.

5.48 The threshold voltage of a GaAs MESFET is $V_{TN} = 0.24\,\text{V}$. The maximum allowable gate-to-source voltage is $V_{GS} = 0.75\,\text{V}$. When the transistor is biased in the saturation region, the maximum drain current is $I_D = 250\,\mu\text{A}$. What is the value of the conduction parameter k?

***5.49** For the transistor in the circuit in Figure P5.49, the parameters are: $I_{DSS} = 10\,\text{mA}$ and $V_P = -5\,\text{V}$. Determine I_{DQ}, V_{GSQ}, and V_{DSQ}.

D5.50 Consider the source follower with the n-channel JFET in Figure P5.50. The input resistance is to be $R_{\text{in}} = 500\,\text{k}\Omega$. We wish to have $I_{DQ} = 5\,\text{mA}$, $V_{DSQ} = 8\,\text{V}$, and $V_{GSQ} = -1\,\text{V}$. Determine R_S, R_1, and R_2, and the required transistor values of I_{DSS} and V_P.

Figure P5.46   Figure P5.49   Figure P5.50

D5.51 The transistor in the circuit in Figure P5.51 has parameters $I_{DSS} = 8\,\text{mA}$ and $V_P = 4\,\text{V}$. Design the circuit such that $I_D = 5\,\text{mA}$. Assume $R_{\text{in}} = 100\,\text{k}\Omega$. Determine V_{GS} and V_{SD}.

D5.52 For the circuit in Figure P5.52, the transistor parameters are $I_{DSS} = 7\,\text{mA}$ and $V_P = 3\,\text{V}$. Let $R_1 + R_2 = 100\,\text{k}\Omega$. Design the circuit such that $I_{DQ} = 5.0\,\text{mA}$ and $V_{SDQ} = 6\,\text{V}$.

5.53 The transistor in the circuit in Figure P5.53 has parameters $I_{DSS} = 8\,\text{mA}$ and $V_P = -4\,\text{V}$. Determine V_G, I_{DQ}, V_{GSQ}, and V_{DSQ}.

5.54 Consider the circuit in Figure P5.54. The quiescent value of V_{DS} is found to be $V_{DSQ} = 5\,\text{V}$. If $I_{DSS} = 10\,\text{mA}$, determine I_{DQ}, V_{GSQ}, and V_P.

D5.55 For the circuit in Figure P5.55, the transistor parameters are $I_{DSS} = 4\,\text{mA}$ and $V_P = -3\,\text{V}$. Design R_D such that $V_{DS} = |V_P|$. What is the value of I_D?

D5.56 Consider the source-follower circuit in Figure P5.56. The transistor parameters are $I_{DSS} = 2\,\text{mA}$ and $V_P = 2\,\text{V}$. Design the circuit such that $I_{DQ} = 1\,\text{mA}$, $V_{SDQ} = 10\,\text{V}$, and the current through R_1 and R_2 is 0.1 mA.

Figure P5.51

Figure P5.52

Figure P5.53

Figure P5.54

Figure P5.55

Figure P5.56

D5.57 The GaAs MESFET in the circuit in Figure P5.57 has parameters $k = 250\,\mu A/V^2$ and $V_{TN} = 0.20\,V$. Let $R_1 + R_2 = 150\,k\Omega$. Design the circuit such that $I_D = 40\,\mu A$ and $V_{DS} = 2\,V$.

5.58 For the circuit in Figure P5.58, the GaAs MESFET threshold voltage is $V_{TN} = 0.15\,V$. Let $R_D = 50\,k\Omega$. Determine the value of the conduction parameter required so that $V_O = 0.70\,V$ when $V_I = 0.75\,V$.

Figure P5.57

Figure P5.58

COMPUTER SIMULATION PROBLEMS

5.59   Generate the i_D versus v_{DS} characteristics for an n-channel enhancement-mode silicon MOSFET at $T = 300\,°K$. Limit the characteristics to $v_{DS}(max) = 10\,V$ and $v_{GS}(max) = 10\,V$. Plot curves for: (a) $W/L = 4$, $\lambda = 0$; (b) $W/L = 40$, $\lambda = 0$; and (c) $W/L = 4$, $\lambda = 0.02\,V^{-1}$.

5.60   Consider the NMOS circuit with enhancement load shown in Figure 5.35. Assume a width-to-length ratio of $W/L = 1$ for M_L. From a computer analysis, plot the dc voltage transfer characteristics V_O versus V_I for M_D width-to-length ratios of: (a) $W/L = 2$, (b) $W/L = 9$, (c) $W/L = 16$, and (d) $W/L = 100$. Consider the case when the body effect is neglected, and then when the body effect is included.

5.61   Consider the NMOS circuit with depletion load shown in Figure 5.39. Use a computer analysis to plot the dc voltage transfer characteristics V_O versus V_I for the same parameters listed in Problem 5.60. Consider the case when the body effect is neglected, and then when the body effect is included.

5.62   (a) Correlate the results of Example 5.13 with a computer analysis. (b) Repeat the analysis if the width-to-length ratio of M_3 is doubled.

5.63   Correlate the JFET design in Example 5.18 with a computer analysis.

DESIGN PROBLEMS

[Note: All design should be correlated with a computer simulation.]

*D5.64   Consider a discrete common-source circuit with the configuration shown in Figure 5.29. The circuit and transistor parameters are: $V_{DD} = 10\,V$, $R_S = 0.5\,k\Omega$, $R_D = 4\,k\Omega$, and $V_{TN} = 2\,V$. Design the circuit such that the nominal Q-point is midway between the transition point and cutoff, and determine the conduction parameter. The dc currents in R_1 and R_2 should be approximately a factor of ten smaller than the quiescent drain current.

*D5.65   For the circuit shown in Figure 5.43, the threshold voltage of each transistor is $V_{TN} = 1\,V$, and the parameter value $k_n' = 40\,\mu A/V^2$ is the same for all devices. If $R_D = 4\,k\Omega$, design the cicuit such that the quiescent drain-to-source voltage of M_1 is $4\,V$ and $I_Q = \left(\frac{1}{2}\right)I_{REF}$.

*D5.66   The NMOS circuit with depletion load shown in Figure 5.39 is biased at $V_{DD} = 5\,V$. The threshold voltage of M_D is $V_{TND} = 0.8\,V$ and that of M_L is $V_{TNL} = -2\,V$. For each transistor, $k_n' = 40\,\mu A/V^2$. Design the transistors such that $V_O = 0.1\,V$ when $V_I = 5\,V$ and the maximum power dissipated in the circuit is $1.0\,mW$.

*D5.67   The threshold voltage of the load transistor M_L in Figure 5.35 is $V_{TNL} = 0.8\,V$. All other parameters are the same as given in Problem 5.66. Design the transistors to meet the same specifications given in Problem 5.66.

*D5.68   Consider the JFET common-source circuit shown in Figure 5.57(a). The transistor pinchoff voltage is $V_P = -4\,V$ and the saturation current is in the range $1 \leq I_{DSS} \leq 2\,mA$. Design the circuit such that the nominal Q-point is in the center of the load line and the Q-point parameters do not deviate from the nominal value by more than 10 percent. The value of R_S may be changed, the current in R_1 and R_2 should be approximately a factor of ten less than the quiescent drain current, and the standard tolerance resistance value of 5 percent should be used.

6

Basic FET Amplifiers

6.0 PREVIEW

In the last chapter, we described the operation of the FET, in particular the MOSFET, and analyzed and designed the dc response of circuits containing these devices. In this chapter, we emphasize the use of FETs in linear amplifier applications. Although a major use of MOSFETs is in digital applications, they are also used in linear amplifier circuits.

There are three basic configurations of single-stage or single-transistor FET amplifiers. These are the common-source, source-follower, and common-gate configurations. We investigate the characteristics of each configuration and show how these properties are used in various applications. Since MOSFET integrated circuit amplifiers normally use MOSFETs as load devices instead of resistors because of their small size, we introduce the technique of using MOSFET enhancement or depletion devices as loads. These three configurations form the building blocks for more complex amplifiers, so gaining a good understanding of these three amplifier circuits is an important goal of this chapter.

In integrated circuit systems, amplifiers are usually connected in series or cascade, forming a multistage configuration, to increase the overall voltage gain, or to provide a particular combination of voltage gain and output resistance. We consider a few of the many possible multistage configurations, to introduce the analysis methods required for such circuits, as well as their properties.

JFET amplifiers are also considered. These circuits, again, tend to be specialized, so the JFET discussion is brief.

6.1 THE MOSFET AMPLIFIER

In Chapter 4, we discussed the reasons linear amplifiers are necessary in analog electronic systems. In this chapter, we continue the analysis and design of linear amplifiers that use field-effect transistors as the amplifying device. The term **small signal** means that we can linearize the ac equivalent circuit. We will define what is meant by small signal in the case of MOSFET circuits. The term linear amplifiers means that we can use superposition so that the dc analysis and ac analysis of the circuits can be performed separately and the total response is the sum of the two individual responses.

The mechanism with which MOSFET circuits amplify small time-varying signals was introduced in the last chapter. In this section, we will expand that discussion using the graphical technique, dc load line, and ac load line. In the process, we will develop the various small-signal parameters of linear circuits and the corresponding equivalent circuits.

There are four possible equivalent circuits that can be used. These are listed in Table 4.3 of Chapter 4. The most common equivalent circuit that is used for the FET amplifiers is the transconductance amplifier, in which the input signal is a voltage and the output signal is a current. The small-signal parameters associated with this equivalent circuit are developed in the following section.

6.1.1   Graphical Analysis, Load Lines, and Small-Signal Parameters

Figure 6.1 shows an NMOS common-source circuit with a time-varying voltage source in series with the dc source. We assume the time-varying input signal is sinusoidal. Figure 6.2 shows the transistor characteristics, dc load line, and Q-point, where the dc load line and Q-point are functions of v_{GS}, V_{DD}, R_D, and the transistor parameters. For the output voltage to be a linear function of the input voltage, the transistor must be biased in the saturation region. (Note that, although we primarily use n-channel, enhancement-mode MOSFETs in our discussions, the same results apply to the other MOSFETs.)

Figure 6.1   NMOS common-source circuit with time-varying signal source in series with gate dc source

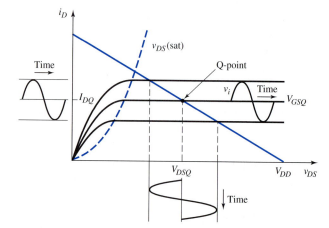

Figure 6.2   Common-source transistor characteristics, dc load line, and sinusoidal variation in gate-to-source voltage, drain current, and drain-to-source voltage

Also shown in Figure 6.2 are the sinusoidal variations in the gate-to-source voltage, drain current, and drain-to-source voltage, as a result of the sinusoidal source v_i. The total gate-to-source voltage is the sum of V_{GSQ} and v_i. As v_i increases, the instantaneous value of v_{GS} increases, and the bias point moves up the load line. A larger value of v_{GS} means a larger drain current and a smaller value of v_{DS}. For a negative v_i (the negative portion of the sine wave), the instantaneous value of v_{GS} decreases below the quiescent value, and the bias point moves down the load line. A smaller v_{GS} value means a smaller drain current and increased value of v_{DS}. Once the Q-point is established, we can develop a mathematical model for the sinusoidal, or small-signal, variations in gate-to-source voltage, drain-to-source voltage, and drain current.

The time-varying signal source v_i in Figure 6.1 generates a time-varying component of the gate-to-source voltage. In this case, $v_{gs} = v_i$, where v_{gs} is the time-varying component of the gate-to-source voltage. For the FET to operate as a linear amplifier, the transistor must be biased in the saturation region, and the instantaneous drain current and drain-to-source voltage must also be confined to the saturation region.

Transistor Parameters

The instantaneous gate-to-source voltage is

$$v_{GS} = V_{GSQ} + v_i = V_{GSQ} + v_{gs} \tag{6.1}$$

where V_{GSQ} is the dc component and v_{gs} is the ac component. The instantaneous drain current is

$$i_D = K_n(v_{GS} - V_{TN})^2 \tag{6.2}$$

Substituting Equation (6.1) into (6.2) produces

$$i_D = K_n[V_{GSQ} + v_{gs} - V_{TN}]^2 = K_n[(V_{GSQ} - V_{TN}) + v_{gs}]^2 \tag{6.3(a)}$$

or

$$i_D = K_n(V_{GSQ} - V_{TN})^2 + 2K_n(V_{GSQ} - V_{TN})v_{gs} + K_n v_{gs}^2 \tag{6.3(b)}$$

The first term in Equation (6.3(b)) is the dc or quiescent drain current I_{DQ}, the second term is the time-varying drain current component that is linearly related to the signal v_{gs}, and the third term is proportional to the square of the signal voltage. For a sinusoidal input signal, the squared term produces undesirable harmonics, or nonlinear distortion, in the output voltage. To minimize these harmonics, we require

$$v_{gs} \ll 2(V_{GSQ} - V_{TN}) \tag{6.4}$$

which means that the third term in Equation (6.3(b)) will be much smaller than the second term. *Equation (6.4) represents the small-signal condition that must be satisfied for linear amplifiers.*

Neglecting the v_{gs}^2 term, we can write Equation (6.3(b))

$$i_D = I_{DQ} + i_d \tag{6.5}$$

Again, small-signal implies linearity so that the total current can be separated into a dc component and an ac component. The ac component of the drain current is given by

$$i_d = 2K_n(V_{GSQ} - V_{TN})v_{gs} \tag{6.6}$$

The small-signal drain current is related to the small-signal gate-to-source voltage by the transconductance g_m. The relationship is

$$g_m = \frac{i_d}{v_{gs}} = 2K_n(V_{GSQ} - V_{TN}) \tag{6.7}$$

The transconductance is a transfer coefficient relating output current to input voltage and can be thought of as representing the gain of the transistor.

The transconductance can also be obtained from the derivative

$$g_m = \left. \frac{\partial i_D}{\partial v_{GS}} \right|_{v_{GS}=V_{GSQ}=\text{const.}} = 2K_n(V_{GSQ} - V_{TN}) \tag{6.8(a)}$$

which can be written

$$g_m = 2\sqrt{K_n I_{DQ}} \tag{6.8(b)}$$

The drain current versus gate-to-source voltage for the transistor biased in the saturation region is given in Equation (6.2) and is shown in Figure 6.3. The transconductance g_m is the slope of the curve. If the time-varying signal v_{gs} is sufficiently small, the transconductance g_m is a constant. With the Q-point in the saturation region, the transistor operates as a current source that is linearly controlled by v_{gs}. If the Q-point moves into the nonsaturation region, the transistor no longer operates as a linearly controlled current source.

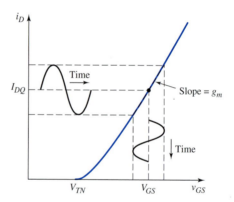

Figure 6.3 Drain current versus gate-to-source voltage characteristics, with superimposed sinusoidal signals

As shown in Equation (6.8(a)), the transconductance is directly proportional to the conduction parameter K_n, which in turn is a function of the width-to-length ratio. Therefore, increasing the width of the transistor increases the transconductance, or gain, of the transistor.

Example 6.1   **Objective:** Calculate the transconductance of an n-channel MOSFET.

Consider an n-channel MOSFET with parameters $V_{TN} = 1\,\text{V}$, $\left(\frac{1}{2}\right)\mu_n C_{ox} = 20\,\mu\text{A/V}^2$, and $W/L = 40$. Assume the drain current is $I_D = 1\,\text{mA}$.

Solution: The conduction parameter is

$$K_n = \left(\frac{1}{2}\mu_n C_{ox}\right)\left(\frac{W}{L}\right) = (20)(40)\,\mu\text{A/V}^2 \Rightarrow 0.80\,\text{mA/V}^2$$

Assuming the transistor is biased in the saturation region, the transconductance is determined from Equation (6.8(b)),

$$g_m = 2\sqrt{K_n I_{DQ}} = 2\sqrt{(0.8)(1)} = 1.79 \, \text{mA/V}$$

Comment: The transconductance of a bipolar transistor is $g_m = (I_{CQ}/V_T)$, which is 38.5 mA/V for a collector current of 1 mA. The transconductance values of MOSFETs tend to be small compared to those of BJTs. However, the advantages of MOSFETs include high input impedance, small size, and low power dissipation.

AC Equivalent Circuit

From Figure 6.1, we see that the output voltage is

$$v_{DS} = v_O = V_{DD} - i_D R_D \tag{6.9}$$

Using Equation (6.5), we obtain

$$v_O = V_{DD} - (I_{DQ} + i_d)R_D = (V_{DD} - I_{DQ}R_D) - i_d R_D \tag{6.10}$$

The output voltage is also a combination of dc and ac values. The time-varying output signal is the time-varying drain-to-source voltage, or

$$v_o = v_{ds} = -i_d R_D \tag{6.11}$$

Also, from Equations (6.6) and (6.7), we have

$$i_d = g_m v_{gs} \tag{6.12}$$

In summary, the following relationships exist between the time-varying signals for the circuit in Figure 6.1. The equations are given in terms of the instantaneous ac values, as well as the phasors. We have

$$v_{gs} = v_i \tag{6.13(a)}$$

or

$$V_{gs} = V_i \tag{6.13(b)}$$

and

$$i_d = g_m v_{gs} \tag{6.14(a)}$$

or

$$I_d = g_m V_{gs} \tag{6.14(b)}$$

Also,

$$v_{ds} = -i_d R_D \tag{6.15(a)}$$

or

$$V_{ds} = -I_d R_D \tag{6.15(b)}$$

The ac equivalent circuit in Figure 6.4 is developed by setting the dc sources in Figure 6.1 equal to zero. The small-signal relationships are given in Equations (6.13), (6.14), and (6.15). As shown in Figure 6.1, the drain current, which is composed of ac signals superimposed on the quiescent value, flows through the voltage source V_{DD}. Since the voltage across this

Figure 6.4 AC equivalent circuit of common-source amplifier with NMOS transistor

source is assumed to be constant, the sinusoidal current produces no sinusoidal voltage component across this element. The equivalent ac impedance is therefore zero, or a short circuit. Consequently, in the ac equivalent circuit, the dc voltage sources are equal to zero. We say that the node connecting R_D and V_{DD} is at signal ground.

6.1.2   Small-Signal Equivalent Circuit

Now that we have the ac equivalent circuit for the NMOS amplifier circuit, (Figure 6.4), we must develop a small-signal equivalent circuit for the transistor.

Initially, we assume that the signal frequency is sufficiently low so that any capacitance at the gate terminal can be neglected. The input to the gate thus appears as an open circuit, or an infinite resistance. Equation (6.14) relates the small-signal drain current to the small-signal input voltage, and Equation (6.7) shows that the transconductance g_m is a function of the Q-point. The resulting simplified small-signal equivalent circuit for the NMOS device is shown in Figure 6.5. (The phasor components are in parentheses.)

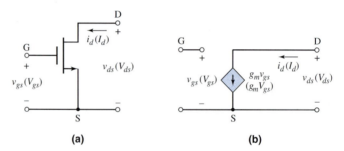

(a)   **(b)**

Figure 6.5   (a) Common-source NMOS transistor with small-signal parameters and (b) simplified small-signal equivalent circuit for NMOS transistor

This small-signal equivalent circuit can also be expanded to take into account the finite output resistance of a MOSFET biased in the saturation region. This effect, discussed in the last chapter, is a result of the nonzero slope in the i_D versus v_{DS} curve.

We know that

$$i_D = K_n[(v_{GS} - V_{TN})^2(1 + \lambda v_{DS})] \tag{6.16}$$

where λ is the channel-length modulation parameter and is a positive quantity. The small-signal output resistance, as previously defined, is

$$r_o = \left(\frac{\partial i_D}{\partial v_{DS}}\right)^{-1}\Bigg|_{v_{GS}=V_{GSQ}=\text{const.}} \tag{6.17}$$

or

$$r_o = [\lambda K_n(V_{GSQ} - V_{TN})^2]^{-1} \cong [\lambda I_{DQ}]^{-1} \tag{6.18}$$

This small-signal output resistance is also a function of the Q-point parameters.

The expanded small-signal equivalent circuit of the n-channel MOSFET is shown in Figure 6.6 in phasor notation. Note that this equivalent circuit is a transconductance amplifier (see Table 4.3) in that the input signal is a voltage and the output signal is a current. This equivalent circuit can now be inserted into the amplifier ac equivalent circuit in Figure 6.4 to produce the circuit in Figure 6.7. We may note that the small-signal equivalent circuit for the MOSFET circuit is very similar to that of the BJT circuits considered in Chapter 4.

Figure 6.6 Expanded small-signal equivalent circuit, including output resistance, for NMOS transistor

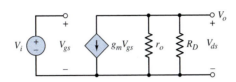

Figure 6.7 Small-signal equivalent circuit of common-source circuit with NMOS transistor model

Example 6.2 **Objective:** Determine the small-signal voltage gain of a MOSFET circuit.

For the circuit in Figure 6.1, assume parameters are: $V_{GSQ} = 2.12\,\text{V}$, $V_{DD} = 5\,\text{V}$, and $R_D = 2.5\,\text{k}\Omega$. Assume transistor parameters are: $V_{TN} = 1\,\text{V}$, $K_n = 0.80\,\text{mA/V}^2$, and $\lambda = 0.02\,\text{V}^{-1}$. Assume the transistor is biased in the saturation region.

Solution: The quiescent values are

$$I_{DQ} \cong K_n(V_{GSQ} - V_{TN})^2 = (0.8)(2.12 - 1)^2 = 1.0\,\text{mA}$$

and

$$V_{DSQ} = V_{DD} - I_{DQ}R_D = 5 - (1)(2.5) = 2.5\,\text{V}$$

Therefore,

$$V_{DSQ} = 2.5\,\text{V} > V_{DS}(\text{sat}) = V_{GS} - V_{TN} = 1.82 - 1 = 0.82\,\text{V}$$

which means that the transistor is biased in the saturation region, as initially assumed, and as required for a linear amplifier. The transconductance is

$$g_m = 2K_n(V_{GSQ} - V_{TN}) = 2(0.8)(2.12 - 1) = 1.79\,\text{mA/V}$$

and the output resistance is

$$r_o = [\lambda I_{DQ}]^{-1} = [(0.02)(1)]^{-1} = 50\,\text{k}\Omega$$

From Figure 6.7, the output voltage is

$$V_o = -g_m V_{gs}(r_o \| R_D)$$

Since $V_{gs} = V_i$, the small-signal voltage gain is

$$A_v = \frac{V_o}{V_i} = -g_m(r_o \| R_d) = -(1.79)(50\|2.5) = -4.26$$

Comment: Because of the relatively low value of transconductance, MOSFET circuits tend to have a lower small-signal voltage gain than comparable bipolar circuits. Also, the small-signal voltage gain contains a minus sign, which means that the sinusoidal output voltage is 180 degrees out of phase with respect to the input sinusoidal signal.

Problem-Solving Technique: MOSFET AC Analysis

Since we are dealing with linear amplifiers, superposition applies, which means that we can perform the dc and ac analyses separately. The analysis of the MOSFET amplifier proceeds as follows:

1. Analyze the circuit with only the dc sources present. This solution is the dc or quiescent solution. The transistor must be biased in the saturation region in order to produce a linear amplifier.
2. Replace each element in the circuit with its small-signal model, which means replacing the transistor by its small-signal equivalent cicuit.
3. Analyze the small-signal equivalent circuit, setting the dc source components equal to zero, to produce the response of the circuit to the time-varying input signals only.

Test Your Understanding

6.1 For an n-channel MOSFET biased in the saturation region, the parameters are $K_n = 0.5 \, \text{mA/V}^2$, $V_{TN} = 0.8 \, \text{V}$, and $\lambda = 0.01 \, \text{V}^{-1}$, and $I_{DQ} = 0.75 \, \text{mA}$. Determine g_m and r_o. (Ans. $g_m = 1.22 \, \text{mA/V}$, $r_o = 1.33 \, \text{k}\Omega$)

6.2 The parameters of an n-channel MOSFET are: $V_{TN} = 1 \, \text{V}$, $\frac{1}{2}\mu_n C_{ox} = 18 \, \mu\text{A/V}^2$, and $\lambda = 0.015 \, \text{V}^{-1}$. The transistor is to be biased in the saturation region with $I_{DQ} = 2 \, \text{mA}$. Design the width-to-length ratio such that the transconductance is $g_m = 3.4 \, \text{mA/V}$. Calculate r_o for this condition. (Ans. $W/L = 80.6$, $r_o = 33.3 \, \text{k}\Omega$)

6.3 For the circuit shown in Figure 6.1, $V_{DD} = 10 \, \text{V}$ and $R_D = 10 \, \text{k}\Omega$. The transistor parameters are: $V_{TN} = 2 \, \text{V}$, $K_n = 0.5 \, \text{mA/V}^2$, and $\lambda = 0$. (a) Determine V_{GSQ} such that $I_{DQ} = 0.4 \, \text{mA}$. Calculate V_{DSQ}. (b) Calculate g_m and r_o, and determine the small-signal voltage gain. (c) If $v_i = 0.4 \sin \omega t$, find v_{ds}. Does the transistor remain in the saturation region? (Ans. (a) $V_{GSQ} = 2.89 \, \text{V}$, $V_{DSQ} = 6 \, \text{V}$; (b) $g_m = 0.89 \, \text{mA/V}$, $r_o = \infty$, $A_v = -8.9$; (c) $v_{ds} = -3.56 \sin \omega t$, yes)

The previous discussion was for an n-channel MOSFET amplifier. The same basic analysis and equivalent circuit also applies to the p-channel transistor. Figure 6.8(a) shows a circuit containing a p-channel MOSFET. Note that the power supply voltage V_{DD} is connected to the source. (The subscript DD can be used to indicate that the supply is connected to the drain terminal. Here, however, V_{DD} is simply the usual notation for the power supply voltage in MOSFET circuits.) Also note the change in current directions and voltage polarities compared to the circuit containing the NMOS transistor. Figure 6.8(b) shows the ac equivalent circuit, with the dc voltage sources replaced

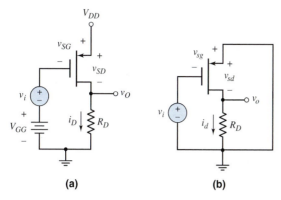

(a) **(b)**

Figure 6.8 (a) Common-source circuit with PMOS transistor and (b) corresponding ac equivalent circuit

by ac short circuits, and all currents and voltages shown are the time-varying components.

In the circuit of Figure 6.8(b), the transistor can be replaced by the equivalent circuit in Figure 6.9. The equivalent circuit of the p-channel MOSFET is the same as that of the n-channel device, except that all current directions and voltage polarities are reversed.

The final small-signal equivalent circuit of the p-channel MOSFET amplifier is shown in Figure 6.10. The output voltage is

$$V_o = g_m V_{sg}(r_o \| R_D) \tag{6.19}$$

Figure 6.9 Small-signal equivalent circuit of PMOS transistor

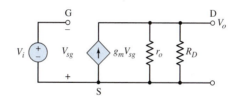

Figure 6.10 Small-signal equivalent circuit of common-source amplifier with PMOS transistor model

The control voltage V_{sg}, given in terms of the input signal voltage, is

$$V_{sg} = -V_i \tag{6.20}$$

and the small-signal voltage gain is

$$A_v = \frac{V_o}{V_i} = -g_m(r_o \| R_D) \tag{6.21}$$

This expression for the small-signal voltage gain of the p-channel MOSFET amplifier is exactly the same as that for the n-channel MOSFET amplifier. The negative sign indicates that a 180-degree phase reversal exists between the ouput and input signals, for both the PMOS and the NMOS circuit.

We may again note that if the polarity of the small-signal gate-to-source voltage is reversed, then the small-signal drain current direction is reversed and the small-signal equivalent circuit of the PMOS device is exactly identical to that of the NMOS device. However, the author prefers to use the small-signal equivalent circuit in Figure 6.9 to be consistent with the voltage polarities and current directions of the PMOS transistor.

6.1.3   Modeling the Body Effect

As mentioned in Section 5.1.7, Chapter 5, the body effect occurs in a MOSFET in which the substrate, or body, is not connected to the source. For an NMOS device, the body is connected to the most negative potential in the circuit and will be at signal ground. Figure 6.11(a) shows the four-terminal MOSFET with dc voltages and Figure 6.11(b) shows the device with ac voltages. Keep in mind that v_{SB} must be greater than or equal to zero. The simplified current-voltage relation is

$$i_D = K_n(v_{GS} - V_{TN})^2 \tag{6.22}$$

and the threshold voltage is given by

$$V_{TN} = V_{TNO} + \gamma\left[\sqrt{2\phi_f + v_{SB}} - \sqrt{2\phi_f}\right] \tag{6.23}$$

Figure 6.11   The four-terminal NMOS device with (a) dc voltages and (b) ac voltages

If an ac component exists in the source-to-body voltage, v_{SB}, there will be an ac component induced in the threshold voltage, which causes an ac component in the drain current. Thus, a back-gate transconductance can be defined as

$$g_{mb} = \left.\frac{\partial i_D}{\partial v_{BS}}\right|_{Q\text{-}pt} = \left.\frac{-\partial i_D}{\partial v_{SB}}\right|_{Q\text{-}pt} = -\left(\frac{\partial i_D}{\partial V_{TN}}\right) \cdot \left.\left(\frac{\partial V_{TN}}{\partial v_{SB}}\right)\right|_{Q\text{-}pt} \tag{6.24}$$

Using Equation (6.22), we find

$$\frac{\partial i_D}{\partial V_{TN}} = -2K_n(v_{GS} - V_{TN}) = -g_m \tag{6.25(a)}$$

and using Equation (6.23), we find

$$\frac{\partial V_{TN}}{\partial v_{SB}} = \frac{\gamma}{2\sqrt{2\phi_f + v_{SB}}} \equiv \eta \tag{6.25(b)}$$

The back-gate transconductance is then

$$g_{mb} = -(-g_m) \cdot (\eta) = g_m \eta \qquad (6.26)$$

Including the body effect, the small-signal equivalent circuit of the MOSFET is shown in Figure 6.12. We note the direction of the current and the polarity of the small-signal source-to-body voltage. If $v_{bs} > 0$, then v_{SB} decreases, V_{TN} decreases, and i_D increases. The current direction and voltage polarity are thus consistent.

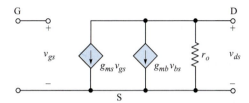

Figure 6.12 Small-signal equivalent circuit of NMOS device including body effect

For $\phi_f = 0.35\,\text{V}$ and $\gamma = 0.35\,\text{V}^{1/2}$, the value of η from Equation (6.25(b)) is $\eta \cong 0.23$. Therefore, η will be in the range $0 \leq \eta \leq 0.23$. The value of v_{bs} will depend on the particular circuit.

In general, we will neglect g_{mb} in our hand analyses and designs, but will investigate the body effect in PSpice analyses.

Test Your Understanding

6.4 The parameters for the circuit in Figure 6.8 are $V_{DD} = 12\,\text{V}$ and $R_D = 6\,\text{k}\Omega$. The transistor parameters are: $V_{TP} = -1\,\text{V}$, $K_p = 2\,\text{mA/V}^2$, and $\lambda = 0$. (a) Determine V_{SG} such that $V_{SDQ} = 7\,\text{V}$. (b) Determine g_m and r_o, and calculate the small-signal voltage gain. (Ans. (a) $V_{SG} = 1.65\,\text{V}$; (b) $g_m = 2.6\,\text{mA/V}$, $r_o = \infty$, $A_v = -15.6$)

6.5 Show that, for an NMOS transistor biased in the saturation region, with a drain current of I_{DQ}, the transconductance can be expressed as given in Equation (6.8(b)), that is

$$g_m = 2\sqrt{K_n I_{DQ}}$$

6.6 A transistor has the same parameters as those given in Exercise 6.1. In addition, the body effect coefficient is $\gamma = 0.40\,\text{V}^{1/2}$ and $\phi_f = 0.35\,\text{V}$. Determine the value of η and the back-gate transconductance g_{mb} for (a) $v_{SB} = 1\,\text{V}$ and (b) $v_{SB} = 3\,\text{V}$.

6.2 BASIC TRANSISTOR AMPLIFIER CONFIGURATIONS

As we have seen, the MOSFET is a three-terminal device. Three basic single-transistor amplifier configurations can be formed, depending on which of the three transistor terminals is used as signal ground. These three basic configurations are appropriately called **common source**, **common drain (source follower)**, and **common gate**. These three circuit configurations correspond to the common-emitter, emitter-follower, and common-base configurations using

BJTs. The similarities and differences between the FET and BJT circuits will be discussed.

The input and output resistance characteristics of amplifiers are important in determining loading effects. These parameters, as well as voltage gain, for the three basic MOSFET circuit configurations will be determined in the following sections. The characteristics of the three types of amplifiers will then allow us to understand under what condition each amplifier is most useful.

Initially, we will consider MOSFET amplifier circuits that emphasize discrete designs, in that resistor biasing will be used. The purpose is to become familiar with basic MOSFET amplifier designs and their characteristics, using biasing techniques similar to those used in BJT amplifiers in previous chapters. In Section 6.7, we will begin to consider integrated circuit MOSFET designs that involve all-transistor circuits and current source biasing. These initial designs provide an introduction to more advanced MOS amplifier designs that will be considered in Part II of the text.

6.3   THE COMMON-SOURCE AMPLIFIER

In this section, we consider the first of the three basic circuits—the common-source amplifier. We will analyze several basic common-source circuits, and will determine small-signal voltage gain and input and output impedances.

6.3.1   A Basic Common-Source Configuration

For the circuit shown in Figure 6.13, assume that the transistor is biased in the saturation region by resistors R_1 and R_2, and that the signal frequency is sufficiently large for the coupling capacitor to act essentially as a short circuit. The signal source is represented by a Thevenin equivalent circuit, in which the signal voltage source v_i is in series with an equivalent source resistance R_{Si}. As we will see, R_{Si} should be much less than the amplifier input resistance, $R_i = R_1 \| R_2$, in order to minimize loading effects.

Figure 6.14 shows the resulting small-signal equivalent circuit. The small-signal variables, such as the input signal voltage V_i, are given in phasor form.

Figure 6.13   Common-source circuit with voltage divider biasing and coupling capacitor

Figure 6.14 Small-signal equivalent circuit, assuming coupling capacitor acts as a short circuit

Since the source is at ground potential, there is no body effect. The output voltage is

$$V_o = -g_m V_{gs}(r_o \| R_D) \qquad (6.27)$$

The input gate-to-source voltage is

$$V_{gs} = \left(\frac{R_i}{R_i + R_{Si}}\right) \cdot V_i \qquad (6.28)$$

so the small-signal voltage gain is

$$A_v = \frac{V_o}{V_i} = -g_m(r_o \| R_D) \cdot \left(\frac{R_i}{R_i + R_{Si}}\right) \qquad (6.29)$$

We can also relate the ac drain current to the ac drain-to-source voltage, as $V_{ds} = -I_d(R_D)$.

Figure 6.15 shows the dc load line, the transition point, and the Q-point, which is in the saturation region. As previously stated, in order to provide the maximum symmetrical output voltage swing and keep the transistor biased in the saturation region, the Q-point must be near the middle of the saturation region. At the same time, the input signal must be small enough for the amplifier to remain linear.

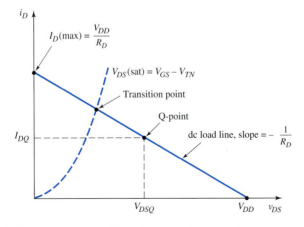

Figure 6.15 DC load line and transition point separating saturation and nonsaturation regions

The input and output resistances of the amplifier can be determined from Figure 6.14. The input resistance to the amplifier is $R_{is} = R_1 \| R_2$. Since the low-frequency input resistance looking into the gate of the MOSFET is essentially infinite, the input resistance is only a function of the bias resistors. The output resistance looking back into the output terminals is found by setting the independent input source V_i equal to zero, which means that $V_{gs} = 0$. The output resistance is therefore $R_o = R_D \| r_o$.

Example 6.3   Objective: Determine the small-signal voltage gain and input and output resistances of a common-source amplifier.

For the circuit shown in Figure 6.13, the parameters are: $V_{DD} = 10\,\text{V}$, $R_1 = 70.9\,\text{k}\Omega$, $R_2 = 29.1\,\text{k}\Omega$, and $R_D = 5\,\text{k}\Omega$. The transistor parameters are: $V_{TN} = 1.5\,\text{V}$, $K_n = 0.5\,\text{mA/V}^2$, and $\lambda = 0.01\,\text{V}^{-1}$. Assume $R_{Si} = 4\,\text{k}\Omega$.

Solution: DC Calculations: The dc or quiescent gate-to-source voltage is

$$V_{GSQ} = \left(\frac{R_2}{R_1 + R_2}\right)(V_{DD}) = \left(\frac{29.1}{70.9 + 29.1}\right)(10) = 2.91\,\text{V}$$

The quiescent drain current is

$$I_{DQ} = K_n(V_{GSQ} - V_{TN})^2 = (0.5)(2.91 - 1.5)^2 = 1\,\text{mA}$$

and the quiescent drain-to-source voltage is

$$V_{DSQ} = V_{DD} - I_{DQ}R_D = 10 - (1)(5) = 5\,\text{V}$$

Since $V_{DSQ} > V_{GSQ} - V_{TN}$, the transistor is biased in the saturation region.

Small-signal Voltage Gain: The small-signal transconductance g_m is then

$$g_m = 2K_n(V_{GSQ} - V_{TN}) = 2(0.5)(2.91 - 1.5) = 1.41\,\text{mA/V}$$

and the small-signal output resistance r_o is

$$r_o \cong [\lambda I_{DQ}]^{-1} = [(0.01)(1)]^{-1} = 100\,\text{k}\Omega$$

The amplifier input resistance is

$$R_i = R_1 \| R_2 = 70.9 \| 29.1 = 20.6\,\text{k}\Omega$$

From Figure 6.14 and Equation (6.29), the small-signal voltage gain is

$$A_v = -g_m(r_o \| R_D) \cdot \left(\frac{R_i}{R_i + R_{Si}}\right) = -(1.41)(100 \| 5)\left(\frac{20.6}{20.6 + 4}\right)$$

or

$$A_v = -5.62$$

Input and Output Resistances: As already calculated, the amplifier input resistance is

$$R_i = R_1 \| R_2 = 70.9 \| 29.1 = 20.6\,\text{k}\Omega$$

and the amplifier output resistance is

$$R_o = R_D \| r_o = 5 \| 100 = 4.76\,\text{k}\Omega$$

Comment: The resulting Q-point is in the center of the load line but not in the center of the saturation region. Therefore, this circuit does not achieve the maximum symmetrical output voltage swing in this case.

Discussion: The small-signal input gate-to-source voltage is

$$V_{gs} = \left(\frac{R_i}{R_i + R_{Si}}\right) \cdot V_i = \left(\frac{20.6}{20.6 + 4}\right) \cdot V_i = (0.837) \cdot V_i$$

Since R_{Si} is not zero, the amplifier input signal V_{gs} is approximately 84 percent of the signal voltage. This is again called a loading effect. Even though the input resistance to the gate of the transistor is essentially infinite, the bias resistors greatly influence the amplifier input resistance and loading effect.

Test Your Understanding

6.7 Consider the circuit in Figure 6.1 with circuit parameters $V_{DD} = 5\,\text{V}$, $R_D = 5\,\text{k}\Omega$, $V_{GSQ} = 2\,\text{V}$, and with transistor paramaeters $K_n = 0.25\,\text{mA/V}^2$, $V_{TN} = 0.8\,\text{V}$, and $\lambda = 0$. (a) Calculate the quiescent values I_{DQ} and V_{DSQ}. (b) Calculate the transconductance g_m. (c) Determine the small-signal voltage gain $A_v = v_o/v_i$. (Ans. (a) $I_{DQ} = 0.36\,\text{mA}$, $V_{DSQ} = 3.2\,\text{V}$; (b) $g_m = 0.6\,\text{mA/V}$, $r_o = \infty$; (c) $A_v = -3.0$)

6.8 For the circuit in Figure 6.1, the circuit and transistor parameters are given in Exercise 6.7. If $v_i = 0.1 \sin \omega t\,\text{V}$, determine i_D and v_{DS}. (Ans. $i_D = (0.36 + 0.06 \sin \omega t)\,\text{mA}$, $v_{DS} = (3.2 - 0.3 \sin \omega t)\,\text{V}$)

Design Example 6.4

Objective: Design the bias of a MOSFET such that the Q-point is in the middle of the saturation region.

Consider the circuit in Figure 6.16 with transistor parameters $V_{TN} = 1\,\text{V}$, $K_n = 1\,\text{mA/V}^2$, and $\lambda = 0.015\,\text{V}^{-1}$. Let $R_i = R_1 \| R_2 = 100\,\text{k}\Omega$. Design the circuit such that $I_{DQ} = 2\,\text{mA}$ and the Q-point is in the middle of the saturation region.

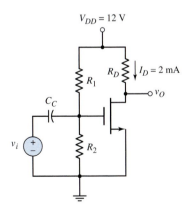

Figure 6.16 Common-source NMOS transistor circuit

Solution: The load line and the desired Q-point are given in Figure 6.17. If the Q-point is to be in the middle of the saturation region, the current at the transition point must be 4 mA.

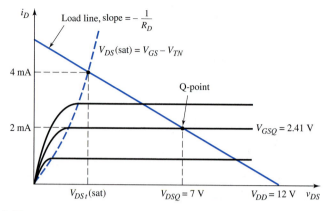

Figure 6.17   DC load line and transition point for NMOS circuit shown in Figure 6.16

We can now calculate $V_{DS}(\text{sat})$ at the transition point. The subscript t indicates transition point values. To determine V_{GSt}, we use

$$I_{Dt} = 4 = K_n(V_{GSt} - V_{TN})^2 = 1(V_{GSt} - 1)^2$$

which yields

$$V_{GSt} = 3\,\text{V}$$

Therefore

$$V_{DSt} = V_{GSt} - V_{TN} = 3 - 1 = 2\,\text{V}$$

If the Q-point is in the middle of the saturation region, then $V_{DSQ} = 7\,\text{V}$, which would yield a 10 V peak-to-peak symmetrical output voltage. From Figure 6.16, we can write

$$V_{DSQ} = V_{DD} - I_{DQ}R_D$$

or

$$R_D = \frac{V_{DD} - V_{DSQ}}{I_{DQ}} = \frac{12 - 7}{2} = 2.5\,\text{k}\Omega$$

We can determine the required quiescent gate-to-source voltage from the current equation, as follows:

$$I_{DQ} = 2 = K_n(V_{GSQ} - V_{TN})^2 = (1)(V_{GSQ} - 1)^2$$

or

$$V_{GSQ} = 2.41\,\text{V}$$

Then

$$V_{GSQ} = 2.41 = \left(\frac{R_2}{R_1 + R_2}\right)(V_{DD}) = \left(\frac{1}{R_1}\right)\left(\frac{R_1 R_2}{R_1 + R_2}\right)(V_{DD})$$
$$= \frac{R_i}{R_1} \cdot V_{DD} = \frac{(100)(12)}{R_1}$$

which yields

$$R_1 = 498\,\text{k}\Omega \quad \text{and} \quad R_2 = 125\,\text{k}\Omega$$

We can then determine the small-signal equivalent circuit parameters from the Q-point values. The transconductance is $g_m = 2.82\,\text{mA/V}$, the transistor output resistance is $r_o = 33.3\,\text{k}\Omega$, and the small-signal voltage gain, assuming an ideal signal source, is

$$A_v = \frac{V_o}{V_i} = -g_m(r_o\|R_D) = -(2.82)(33.3\|2.5) = -6.56$$

Comment: Establishing the Q-point in the middle of the saturation region allows the maximum symmetrical swing in the output voltage, while keeping the transistor biased in the saturation region.

Design Pointer: If the circuit were to contain bypass or load capacitors, then an ac load line would be superimposed on the figure at the Q-point. Establishing the Q-point in the middle of the saturation region, then, may not be optimal in terms of obtaining the maximum symmetrical swing.

6.3.2 Common-Source Amplifier with Source Resistor

A source resistor R_S tends to stabilize the Q-point against variations in transistor parameters (Figure 6.18). If, for example, the value of the conduction parameter varies from one transistor to another, the Q-point will not vary as much if a source resistor is included in the circuit. However, as shown in the following example, a source resistor also reduces the signal gain. This same effect was observed in BJT circuits when an emitter resistor was included.

The circuit in Figure 6.18 is an example of a situation in which the body effect should be taken into account. The substrate (not shown) would normally be connected to the -5 V supply, so that the body and substrate terminals are not at the same potential. However, in the following example, we will neglect this effect.

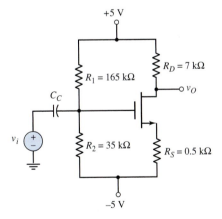

Figure 6.18 Common-source circuit with source resistor and positive and negative supply voltages

Example 6.5 **Objective:** Determine the small-signal voltage gain of a common-source circuit containing a source resistor.

Consider the circuit in Figure 6.18. The transistor parameters are $V_{TN} = 0.8$ V, $K_n = 1$ mA/V^2, and $\lambda = 0$.

Solution: From the dc analysis of the circuit, we find that $V_{GSQ} = 1.50\,\text{V}$, $I_{DQ} = 0.50\,\text{mA}$, and $V_{DSQ} = 6.25\,\text{V}$. The small-signal transconductance is

$$g_m = 2K_n(V_{GS} - V_{TN}) = 2(1)(1.50 - 0.8) = 1.4\,\text{mA/V}$$

and the small-signal resistance is

$$r_o \cong [\lambda I_{DQ}]^{-1} = \infty$$

Figure 6.19 shows the resulting small-signal equivalent circuit.

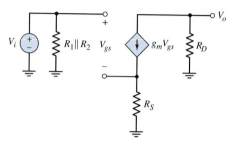

Figure 6.19   Small-signal equivalent circuit of NMOS common-source amplifier with source resistor

The output voltage is

$$V_o = -g_m V_{gs} R_D$$

Writing a KVL equation from the input around the gate–source loop, we find

$$V_i = V_{gs} + (g_m V_{gs})R_S = V_{gs}(1 + g_m R_S)$$

or

$$V_{gs} = \frac{V_i}{1 + g_m R_S}$$

The small-signal voltage gain is

$$A_v = \frac{V_o}{V_i} = \frac{-g_m R_D}{1 + g_m R_S}$$

We may note that if g_m were large, then the small-signal voltage gain would be approximately

$$A_v \cong \frac{-R_D}{R_S}$$

Substituting the appropriate parameters into the actual voltage gain expression, we find

$$A_v = \frac{-(1.4)(7)}{1 + (1.4)(0.5)} = -5.76$$

Comment: A source resistor reduces the small-signal voltage gain. However, as discussed in the last chapter, the Q-point is more stabilized against variations in the transistor parameters. We may note that the approximate voltage gain gives $A_v \cong -R_D/R_S = -14$. Since the transconductance of MOSFETs is generally low, the approximate gain expression is a poor one at best.

Discussion: We mentioned that including a source resistor tends to stabilize the circuit characteristics against any changes in transistor parameters. If, for example, the conduction parameter K_n varies by ± 20 percent, we find the following results.

K_n (mA/V^2)	g_m (mA/V)	A_v
0.8	1.17	-5.17
1.0	1.40	-5.76
1.2	1.62	-6.27

The change in K_n produces a fairly large change in g_m. The resulting change in the voltage gain is approximately ± 9.5 percent. This change is larger than might be expected because the initial value of g_m is smaller than that of the bipolar circuit.

Test Your Understanding

6.9 For the circuit shown in Figure 6.20, the transistor parameters are $K_p = 1$ mA/V^2, $V_{TP} = -1$ V, and $\lambda = 0$. The source-to-drain voltage is $v_{SD} = 3 + 0.46 \sin \omega t$ V, and the quiescent drain current is $I_{DQ} = 0.5$ mA. Determine R_D, V_{GG}, v_i, and the small-signal voltage gain. (Ans. $R_D = 4$ kΩ, $V_{GG} = 3.29$ V, $A_v = -5.64$, $v_i = +0.0816 \sin \omega t$ V)

6.10 The common-source amplifier in Figure 6.21 has transistor parameters $K_p = 2$ mA/V^2, $V_{TP} = -2$ V, and $\lambda = 0.01$ V^{-1}. (a) Determine I_{DQ} and V_{SDQ}. (b) Calculate the small-signal voltage gain. (Ans. (a) $I_{DQ} = 4.56$ mA, $V_{SDQ} = 7.97$ V; (b) $A_v = -6.04$)

Figure 6.20 Figure for Exercise 6.9

Figure 6.21 Figure for Exercise 6.10

6.3.3 Common-Source Circuit with Source Bypass Capacitor

A source bypass capacitor added to the common-source circuit with a source resistor will minimize the loss in the small-signal voltage gain, while maintaining the Q-point stability. The Q-point stability can be further increased by replacing the source resistor with a constant-current source. The resulting circuit is shown in Figure 6.22, assuming an ideal signal source. If the signal frequency is sufficiently large so that the bypass capacitor acts essentially as an ac short-circuit, the source will be held at signal ground.

Figure 6.22   NMOS common-source circuit with source bypass capacitor

Example 6.6   **Objective:** Determine the small-signal voltage gain of a circuit biased with a constant-current source and incorporating a source bypass capacitor.

For the circuit shown in Figure 6.22, the transistor parameters are: $V_{TN} = 0.8\,\text{V}$, $K_n = 1\,\text{mA/V}^2$, and $\lambda = 0$.

Solution: Since the dc gate current is zero, the dc voltage at the source terminal is $V_S = -V_{GSQ}$, and the gate-to-source voltage is determined from

$$I_{DQ} = I_Q = K_n(V_{GSQ} - V_{TN})^2$$

or

$$0.5 = (1)(V_{GSQ} - 0.8)^2$$

which yields

$$V_{GSQ} = -V_S = 1.51\,\text{V}$$

The quiescent drain-to-source voltage is

$$V_{DSQ} = V_{DD} - I_{DQ}R_D - V_S = 5 - (0.5)(7) - (-1.51) = 3.01\,\text{V}$$

The transistor is therefore biased in the saturation region.

The small-signal equivalent circuit is shown in Figure 6.23. The output voltage is

Figure 6.23   Small-signal equivalent circuit, assuming the source bypass capacitor acts as a short circuit

$$V_o = -g_m V_{gs} R_D$$

Since $V_{gs} = V_i$, the small-signal voltage gain is

$$A_v = \frac{V_o}{V_i} = -g_m R_D = -(1.4)(7) = -9.8$$

Comment: Comparing the small-signal voltage gain of 9.8 in this example to the 5.76 calculated in Example 6.5, we see that the magnitude of the gain increases when a source bypass capacitor is included.

Test Your Understanding

*D6.11 The common-source amplifier in Figure 6.24 has transistor parameters $V_{TN} = 1.5\,\text{V}$, $\frac{1}{2}\mu_n C_{ox} = 20\,\mu\text{A/V}^2$, and $\lambda = 0$. Design the circuit such that $I_{DQ} = 0.5\,\text{mA}$ and the small-signal voltage gain is $A_v = -4.0$. (Ans. For example: For $V_{GS} = 2.5\,\text{V}$, then $W/L = 25$, $R_D = 4.0\,\text{k}\Omega$)

6.12 Consider the common-source amplifier in Figure 6.25 with transistor parameters $V_{TN} = 1.8\,\text{V}$, $K_n = 0.15\,\text{mA/V}^2$, and $\lambda = 0$. (a) Calculate I_{DQ} and V_{DSQ}. (b) Determine the small-signal voltage gain. (c) Discuss the purpose of R_G and its effect on the small-signal operation of the amplifier. (Ans. (a) $I_{DQ} = 1.05\,\text{mA}$, $V_{DSQ} = 4.45\,\text{V}$; (b) $A_v = -2.65$)

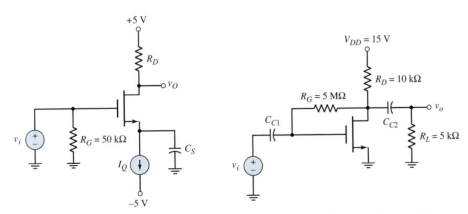

Figure 6.24 Figure for Exercise 6.11

Figure 6.25 Figure for Exercise 6.12

*6.13 For the circuit in Figure 6.26, the n-channel depletion-mode transistor parameters are: $K_n = 0.8\,\text{mA/V}^2$, $V_{TN} = -2\,\text{V}$, and $\lambda = 0$. (a) Calculate I_{DQ}. (b) Find R_D such that $V_{DSQ} = 6\,\text{V}$. (c) Determine the small-signal voltage gain. (Ans. (a) $I_{DQ} = 0.338\,\text{mA}$; (b) $R_D = 7.83\,\text{k}\Omega$; (c) $A_v = -1.58$)

6.14 The parameters of the transistor shown in Figure 6.27 are: $V_{TP} = +0.8\,\text{V}$, $K_p = 0.5\,\text{mA/V}^2$, and $\lambda = 0.02\,\text{V}^{-1}$. (a) Determine R_S and R_D such that $I_{DQ} = 0.8\,\text{mA}$ and $V_{SDQ} = 3\,\text{V}$. (b) Find the small-signal voltage gain. (Ans. (a) $R_S = 5.67\,\text{k}\Omega$, $R_D = 3.08\,\text{k}\Omega$; (b) $A_v = -3.73$)

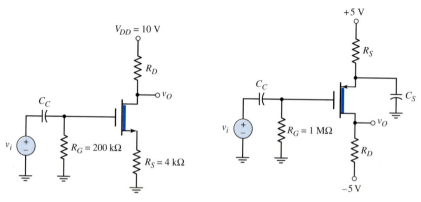

Figure 6.26   Figure for Exercise 6.13   **Figure 6.27**   Figure for Exercise 6.14

6.4   THE SOURCE-FOLLOWER AMPLIFIER

The second type of MOSFET amplifier to be considered is the **common-drain circuit**. An example of this circuit configuration is shown in Figure 6.28. As seen in the figure, the output signal is taken off the source with respect to ground and the drain is connected directly to V_{DD}. Since V_{DD} becomes signal ground in the ac equivalent circuit, we have the name common drain. The more common name is **source follower**. The reason for this name will become apparent as we proceed through the analysis.

Figure 6.28   NMOS source-follower or common-drain amplifier

6.4.1   Small-Signal Voltage Gain

The dc analysis of the circuit is exactly the same as we have already seen, so we will concentrate on the small-signal analysis. The small-signal equivalent circuit, assuming the coupling capacitor acts as a short circuit, is shown in Figure 6.29(a). The drain is at signal ground, and the small-signal resistance r_o of the transistor is in parallel with the dependent current source. Figure 6.29(b) is the same equivalent circuit, but with all signal grounds at a common point. We are again neglecting the body effect. The output voltage is

Figure 6.29 (a) Small-signal equivalent circuit of NMOS source follower and (b) small-signal equivalent circuit of NMOS source follower with all signal grounds at a common point

$$V_o = (g_m V_{gs})(R_s \| r_o) \qquad (6.30)$$

Writing a KVL equation from input to output results in the following:

$$V_{in} = V_{gs} + V_o = V_{gs} + g_m V_{gs}(R_S \| r_o) \qquad (6.31(a))$$

Therefore, the gate-to-source voltage is

$$V_{gs} = \frac{V_{in}}{1 + g_m(R_S \| r_o)} = \left[\frac{\dfrac{1}{g_m}}{\dfrac{1}{g_m} + (R_S \| r_o)} \right] \cdot V_{in} \qquad (6.31(b))$$

Equation (6.31(b)) is written in the form of a voltage-divider equation, in which the gate-to-source of the NMOS device looks like a resistance with a value of $1/g_m$. More accurately, the effective resistance looking into the source terminal (ignoring r_o) is $1/g_m$. The voltage V_{in} is related to the source input voltage V_i by

$$V_{in} = \left(\frac{R_i}{R_i + R_{Si}} \right) \cdot V_i \qquad (6.32)$$

where $R_i = R_1 \| R_2$ is the input resistance to the amplifier.

Substituting Equations (6.31(b)) and (6.32) into (6.30), we have the small-signal voltage gain:

$$A_v = \frac{V_o}{V_i} = \frac{g_m(R_S \| r_o)}{1 + g_m(R_S \| r_o)} \cdot \left(\frac{R_i}{R_i + R_{Si}} \right) \qquad (6.33(a))$$

or

$$A_v = \frac{R_S \| r_o}{\dfrac{1}{g_m} + R_S \| r_o} \cdot \left(\frac{R_i}{R_i + R_{Si}} \right) \qquad (6.33(b))$$

which again is written in the form of a voltage-divider equation. An inspection of Equation 6.33(b) shows that the magnitude of the voltage gain is always less than unity. This result is consistent with the results of the BJT emitter-follower circuit.

Example 6.7 **Objective:** Calculate the small-signal voltage gain of the source-follower circuit in Figure 6.28.

Assume the circuit parameters are $V_{DD} = 12\,\text{V}$, $R_1 = 162\,\text{k}\Omega$, $R_2 = 463\,\text{k}\Omega$, and $R_S = 0.75\,\text{k}\Omega$, and the transistor parameters are $V_{TN} = 1.5\,\text{V}$, $K_n = 4\,\text{mA/V}^2$, and $\lambda = 0.01\,\text{V}^{-1}$. Also assume $R_{Si} = 4\,\text{k}\Omega$.

Solution: The dc analysis results are $I_{DQ} = 7.97\,\text{mA}$ and $V_{GSQ} = 2.91\,\text{V}$. The small-signal transconductance is therefore

$$g_m = 2K_n(V_{GSQ} - V_{TN}) = 2(4)(2.91 - 1.5) = 11.3\,\text{mA/V}$$

and the small-signal transistor resistance is

$$r_o \cong [\lambda I_{DQ}]^{-1} = [(0.01)(7.97)]^{-1} = 12.5\,\text{k}\Omega$$

The amplifier input resistance is

$$R_i = R_1 \| R_2 = 162 \| 463 = 120\,\text{k}\Omega$$

The small-signal voltage gain then becomes

$$A_v = \frac{g_m(R_S \| r_o)}{1 + g_m(R_S \| r_o)} \cdot \frac{R_i}{R_i + R_{Si}} = \frac{(11.3)(0.75\|12.5)}{1 + (11.3)(0.75\|12.5)} \cdot \frac{120}{120 + 4} = +0.860$$

Comment: The magnitude of the small-signal voltage gain is less than 1. An examination of Equation (6.33(b)) shows that this is always true. Also, the voltage gain is positive, which means that the output signal voltage is in phase with the input signal voltage. Since the output signal is essentially equal to the input signal, the circuit is called a source follower.

Discussion: The expression for the voltage gain of the source follower is essentially identical to that of the bipolar emitter follower. Since the transconductance of the BJT is, in general, larger than that of the MOSFET, the voltage gain of the emitter follower will be closer to unity than that of the MOSFET source follower.

Although the voltage gain is slightly less than 1, the source follower is an extremely useful circuit because the output resistance is less than that of a common-source circuit. A small output resistance is desirable when the circuit is to act as an ideal voltage source and drive a load circuit without suffering any loading effects.

Test Your Understanding

6.15 For an NMOS source-follower circuit, the parameters are $g_m = 4\,\text{mA/V}$ and $r_o = 50\,\text{k}\Omega$. (a) Find the no load ($R_S = \infty$) small-signal voltage gain and the output resistance. (b) Determine the small-signal voltage gain when a $4\,\text{k}\Omega$ load is connected to the output. (Ans. (a) $A_v = 0.995$, $R_o \cong 0.25\,\text{k}\Omega$; (b) $A_v = 0.937$)

D6.16 The source-follower circuit in Figure 6.28 has transistor parameters $V_{TN} = +0.8\,\text{V}$, $K_n = 1\,\text{mA/V}^2$, and $\lambda = 0.015\,\text{V}^{-1}$. Let $V_{DD} = 10\,\text{V}$, $R_{Si} = 200\,\Omega$, and $R_1 + R_2 = 400\,\text{k}\Omega$. Design the circuit such that $I_{DQ} = 1.5\,\text{mA}$ and $V_{DSQ} = 5\,\text{V}$. Determine the small-signal voltage gain and the output resistance. (Ans. $R_S = 3.33\,\text{k}\Omega$, $R_1 = 119\,\text{k}\Omega$, $R_2 = 281$, $\text{k}\Omega$, $A_v = 0.884$, and $R_o = 0.36\,\text{k}\Omega$)

Design Example 6.8 **Objective:** Design a specific source follower with a p-channel enhancement-mode MOSFET.

For the circuit in Figure 6.30, the transistor parameters are $V_{TP} = -2\,\text{V}$, $k'_p = 40\,\mu\text{A/V}^2$, and $\lambda = 0$. The circuit parameters are $V_{DD} = 20\,\text{V}$ and $R_{Si} = 4\,\text{k}\Omega$.

The circuit is to be designed such that $V_{SDQ} = 10\,\text{V}$, $I_{DQ} = 2.5\,\text{mA}$, and $R_i = 50\,\text{k}\Omega$, and the transistor width-to-length ratio is to be designed such that the small-signal voltage gain is $A_v = 0.90$.

Figure 6.30   PMOS source follower

Solution: From the dc analysis, we have

$$V_{DD} = V_{SDQ} + I_{DQ}R_S$$

or

$$20 = 10 + 2.5R_S$$

The required source resistance value is therefore

$$R_S = 4\,\text{k}\Omega$$

The small-signal voltage gain of this circuit is the same as that of a source follower using an NMOS device. From Equation (6.33(a)), we have

$$A_v = \frac{V_o}{V_i} = \frac{g_m R_S}{1 + g_m R_S} \cdot \left(\frac{R_i}{R_i + R_{Si}}\right)$$

or

$$0.90 = \frac{g_m(4)}{1 + g_m(4)} \cdot \left(\frac{50}{50 + 4}\right)$$

which yields

$$0.972 = \frac{g_m(4)}{1 + g_m(4)}$$

Therefore, the required small-signal transconductance is

$$g_m = 8.68\,\text{mA/V}$$

Since the transconductance can be written as

$$g_m = 2\sqrt{K_p I_{DQ}}$$

we have

$$8.68 \times 10^{-3} = 2\sqrt{K_p(2.5 \times 10^{-3})}$$

which yields

$$K_p = 7.53 \times 10^{-3} \, \text{A/V}^2$$

The conduction parameter, which is a function of the width-to-length ratio, is

$$K_p = 7.53 \times 10^{-3} = \left(\frac{W}{L}\right)\left(\frac{1}{2}k_p'\right) = \left(\frac{W}{L}\right)\left(\frac{40 \times 10^{-6}}{2}\right)$$

which means that the width-to-length ratio is

$$\frac{W}{L} = 377$$

This is a relatively large p-channel transistor.

Completing the dc analysis, we have

$$I_{DQ} = K_p(V_{SGQ} + V_{TP})^2$$

or

$$2.5 = 7.53(V_{SGQ} - 2)^2$$

which yields a quiescent source-to-gate voltage of

$$V_{SGQ} = 2.58 \, \text{V}$$

The quiescent source-to-gate voltage can also be written as

$$V_{SGQ} = (V_{DD} - I_{DQ}R_S) - \left(\frac{R_2}{R_1 + R_2}\right)(V_{DD})$$

Since

$$\left(\frac{R_2}{R_1 + R_2}\right) = \left(\frac{1}{R_1}\right)\left(\frac{R_1 R_2}{R_1 + R_2}\right) = \left(\frac{1}{R_1}\right)R_i$$

we have

$$2.58 = [20 - (2.5)(4)] - \left(\frac{1}{R_1}\right)(50)(20)$$

the bias resistor R_1 is then found to be

$$R_1 = 135 \, \text{k}\Omega$$

Since

$$R_i = R_1 \| R_2 = 50 \, \text{k}\Omega$$

then

$$R_2 = 79.4 \, \text{k}\Omega$$

 Comment: In order to achieve the desired specifications, a relatively large transconductance is required, which means that a large transistor is needed. If the load effect were reduced, that is, if R_i were made larger, the required size of the transistor could be reduced.

6.4.2   Input and Output Impedance

The input resistance R_i as defined in Figure 6.29(b), for example, is the Thevenin equivalent resistance of the bias resistors. Even though the input resistance to the gate of the MOSFET is essentially infinite, the input bias resistances do provide a loading effect. This same effect was seen in the common-source circuits.

To calculate the output resistance, we set all independent small-signal sources equal to zero, apply a test voltage to the output terminals, and measure a test current. Figure 6.31 shows the circuit we will use to determine the output resistance of the source follower shown in Figure 6.28. We set $V_i = 0$ and apply a test voltage V_x. Since there are no capacitances in the circuit, the output impedance is simply an output resistance, which is defined as

$$R_o = \frac{V_x}{I_x} \tag{6.34}$$

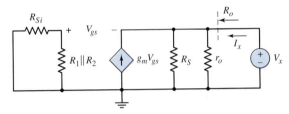

Figure 6.31   Equivalent circuit of NMOS source follower, for determining output resistance

Writing a KCL equation at the output source terminal produces

$$I_x + g_m V_{gs} = \frac{V_x}{R_S} + \frac{V_x}{r_o} \tag{6.35}$$

Since there is no current in the input portion of the circuit, we see that $V_{gs} = -V_x$. Therefore, Equation (6.35) becomes

$$I_x = V_x \left(g_m + \frac{1}{R_S} + \frac{1}{r_o} \right) \tag{6.36(a)}$$

or

$$\frac{I_x}{V_x} = \frac{1}{R_o} = g_m + \frac{1}{R_S} + \frac{1}{r_o} \tag{6.36(b)}$$

The output resistance is then

$$R_o = \frac{1}{g_m} \| R_S \| r_o \tag{6.37}$$

From Figure 6.31, we see that the voltage V_{gs} is directly across the current source $g_m V_{gs}$. This means that the effective resistance of the device is $1/g_m$. The output resistance given by Equation (6.37) can therefore be written directly. This result also means that the resistance looking into the source terminal (ignoring r_o) is $1/g_m$, as previously noted.

340 Part I Semiconductor Devices and Basic Applications

Example 6.9 Objective: Calculate the output resistance of a source-follower circuit.

Consider the circuit shown in Figure 6.28 with circuit and transistor parameters given in Example 6.7.

Solution: The results of Example 6.7 are: $R_S = 0.75\,\text{k}\Omega$, $r_o = 12.5\,\text{k}\Omega$, and $g_m = 11.3\,\text{mA/V}$. Using Figure 6.31 and Equation (6.37), we find

$$R_o = \frac{1}{g_m}\|R_S\|r_o = \frac{1}{11.3}\|0.75\|12.5$$

or

$$R_o = 0.0787\,\text{k}\Omega = 78.7\,\Omega$$

Comment: The output resistance of a source-follower circuit is dominated by the transconductance parameter. Also, because the output resistance is very low, the source follower tends to act like an ideal voltage source, which means that the output can drive another circuit without significant loading effects.

Test Your Understanding

*6.17 Consider the circuit shown in Figure 6.30 with circuit parameters $V_{DD} = 5\,\text{V}$, $R_S = 5\,\text{k}\Omega$, $R_1 = 70.7\,\text{k}\Omega$, $R_2 = 9.3\,\text{k}\Omega$, and $R_{Si} = 500\,\Omega$. The transistor parameters are: $V_{TP} = -0.8\,\text{V}$, $K_p = 0.4\,\text{mA/V}^2$, and $\lambda = 0$. Calculate the small-signal voltage gain $A_v = v_o/v_i$ and the output resistance R_o seen looking back into the circuit. (Ans. $A_v = 0.817$, $R_o = 0.915\,\text{k}\Omega$)

D6.18 The transistor in the source-follower circuit shown in Figure 6.32 is biased with a constant current source. The transistor parameters are: $V_{TN} = 2\,\text{V}$, $k_n' = 40\,\mu\text{A/V}^2$, and $\lambda = 0.01\,\text{V}^{-1}$. The load resistor is $R_L = 4\,\text{k}\Omega$. (a) Design the transistor width-to-length ratio such that $g_m = 2\,\text{mA/V}$ when $I = 0.8\,\text{mA}$. What is the corresponding value for V_{GS}? (b) Determine the small-signal voltage gain and the output resistance R_o. (Ans. (a) $W/L = 62.5$, $V_{GS} = 2.8\,\text{V}$; (b) $A_v = 0.886$, $R_o \cong 0.5\,\text{k}\Omega$)

Figure 6.32 Figure for Exercise 6.18

*D6.19 The parameters of the transistor in the source-follower circuit shown in Figure 6.33 are: $V_{TP} = -2\,\text{V}$, $K_p = 2\,\text{mA/V}^2$, and $\lambda = 0.02\,\text{V}^{-1}$. Design the circuit such that $I_{DQ} = 3\,\text{mA}$. Determine the open-circuit ($R_L = \infty$) small-signal voltage gain. What

Figure 6.33   Figure for Exercise 6.19

value of R_L will result in a 10 percent reduction in the gain? (Ans. $R_S = 0.593 \, \text{k}\Omega$, $A_v = 0.737$, $R_L = 1.35 \, \text{k}\Omega$)

6.5   THE COMMON-GATE CONFIGURATION

The third amplifier configuration is the **common-gate circuit**. To determine the small-signal voltage and current gains, and the input and output impedances, we will use the same small-signal equivalent circuit for the transistor that was used previously. The dc analysis of the common-gate circuit is the same as that of previous MOSFET circuits.

6.5.1   Small-Signal Voltage and Current Gains

In the common-gate configuration, the input signal is applied to the source terminal and the gate is at signal ground. The common-gate configuration shown in Figure 6.34 is biased with a constant-current source I_Q. The gate resistor R_G prevents the buildup of static charge on the gate terminal, and the capacitor C_G ensures that the gate is at signal ground. The coupling capacitor C_{C1} couples the signal to the source, and coupling capacitor C_{C2} couples the output voltage to load resistance R_L.

The small-signal equivalent circuit is shown in Figure 6.35. The small-signal transistor resistance r_o is assumed to be infinite. The output voltage is

Figure 6.34   Common-gate circuit

Figure 6.35   Small-signal equivalent circuit of common-gate amplifier

$$V_o = -(g_m V_{gs})(R_D \| R_L) \tag{6.38}$$

Writing the KVL equation around the input, we find

$$V_i = I_i R_{Si} - V_{gs} \tag{6.39}$$

where $I_i = -g_m V_{gs}$. The gate-to-source voltage can then be written as

$$V_{gs} = \frac{-V_i}{1 + g_m R_{Si}} \tag{6.40}$$

The small-signal voltage gain is found to be

$$A_v = \frac{V_o}{V_i} = \frac{g_m(R_D \| R_L)}{1 + g_m R_{Si}} \tag{6.41}$$

Also, since the voltage gain is positive, the output and input signals are in phase.

In many cases, the signal input to a common-gate circuit is a current. Figure 6.36 shows the small-signal equivalent common-gate circuit with a Norton equivalent circuit as the signal source. We can calculate a current gain. The output current I_o can be written

$$I_o = \left(\frac{R_D}{R_D + R_L}\right)(-g_m V_{gs}) \tag{6.42}$$

At the input we have

$$I_i + g_m V_{gs} + \frac{V_{gs}}{R_{Si}} = 0 \tag{6.43}$$

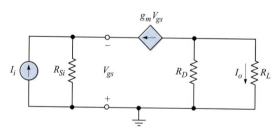

Figure 6.36   Small-signal equivalent circuit of common-gate amplifier with a Norton equivalent signal source

or

$$V_{gs} = -I_i \left(\frac{R_{Si}}{1 + g_m R_{Si}} \right)$$ (6.44)

The small-signal current gain is then

$$A_i = \frac{I_o}{I_i} = \left(\frac{R_D}{R_D + R_L} \right) \cdot \left(\frac{g_m R_{Si}}{1 + g_m R_{Si}} \right)$$ (6.45)

We may note that if $R_D \gg R_L$ and $g_m R_{Si} \gg 1$, then the current gain is essentially unity as it is for an ideal BJT common-base circuit.

6.5.2 Input and Output Impedance

In contrast to the common-source and source-follower amplifiers, the common-gate circuit has a low input resistance because of the transistor. However, if the input signal is a current, a low input resistance is an advantage. The input resistance is defined as

$$R_i = \frac{-V_{gs}}{I_i}$$ (6.46)

Since $I_i = -g_m V_{gs}$, the input resistance is

$$R_i = \frac{1}{g_m}$$ (6.47)

This result has been obtained previously.

We can find the output resistance by setting the input signal voltage equal to zero. From Figure 6.35, we see that $V_{gs} = -g_m V_{gs} R_{Si}$, which means that $V_{gs} = 0$. Consequently, $g_m V_{gs} = 0$. The output resistance, looking back from the load resistance, is therefore

$$R_o = R_D$$ (6.48)

Example 6.10 **Objective:** For the common-gate circuit, determine the output voltage for a given input current.

For the circuits shown in Figures 6.34 and 6.36, the circuit parameters are: $I_Q = 1\,\text{mA}$, $V^+ = 5\,\text{V}$, $V^- = -5\,\text{V}$, $R_G = 100\,\text{k}\Omega$, $R_D = 4\,\text{k}\Omega$, and $R_L = 10\,\text{k}\Omega$. The transistor parameters are: $V_{TN} = 1\,\text{V}$, $K_n = 1\,\text{mA/V}^2$, and $\lambda = 0$. Assume the input current is $100 \sin \omega t\,\mu\text{A}$ and assume $R_{Si} = 50\,\text{k}\Omega$.

Solution: The quiescent gate-to-source voltage is determined from

$$I_Q = I_{DQ} = K_n (V_{GSQ} - V_{TN})^2$$

or

$$1 = 1(V_{GSQ} - 1)^2$$

which yields

$$V_{GSQ} = 2\,\text{V}$$

The small-signal transconductance is

$$g_m = 2K_n(V_{GSQ} - V_{TN}) = 2(1)(2 - 1) = 2 \, \text{mA/V}$$

From Equation (6.45), we can write the output current as

$$I_o = I_i \left(\frac{R_D}{R_D + R_L} \right) \cdot \left(\frac{g_m R_{Si}}{1 + g_m R_{Si}} \right)$$

The output voltage is $V_o = I_o R_L$, so we find

$$V_o = I_i \left(\frac{R_L R_D}{R_D + R_L} \right) \cdot \left(\frac{g_m R_{Si}}{1 + g_m R_{Si}} \right)$$

$$= \left[\frac{(10)(4)}{4 + 10} \right] \cdot \left[\frac{(2)(50)}{1 + (2)(50)} \right] \cdot (0.1) \sin \omega t$$

or

$$V_o = 0.283 \sin \omega t \, \text{V}$$

Comment: As with the BJT common-base circuit, the MOSFET common-gate amplifier is useful if the input signal is a current.

Test Your Understanding

RD6.20   For the circuit shown in Figure 6.34, the circuit parameters are: $V^+ = 5 \, \text{V}$, $V^- = -5 \, \text{V}$, $R_G = 100 \, \text{k}\Omega$, $R_L = 4 \, \text{k}\Omega$, and $I_Q = 0.5 \, \text{mA}$. The transistor parameters are $V_{TN} = 1 \, \text{V}$ and $\lambda = 0$. The circuit is driven by a signal current source I_i. Redesign R_D and g_m such that the transfer function V_o/I_i is $2.4 \, \text{k}\Omega$ and the output resistance is $R_i = 350 \, \Omega$. Determine V_{GSQ} and show that the transistor is biased in the saturation region. (Ans. $g_m = 2.86 \, \text{mA/V}$, $R_D = 6 \, \text{k}\Omega$, $V_{GSQ} = 1.35 \, \text{V}$)

6.21   Consider the circuit shown in Figure 6.37 with circuit parameters $V^+ = 5 \, \text{V}$, $V^- = -5 \, \text{V}$, $R_S = 4 \, \text{k}\Omega$, $R_D = 2 \, \text{k}\Omega$, $R_L = 4 \, \text{k}\Omega$ and $R_G = 50 \, \text{k}\Omega$. The transistor parameters are: $K_p = 1 \, \text{mA/V}^2$, $V_{TP} = -0.8 \, \text{V}$, and $\lambda = 0$. Draw the small-signal equivalent circuit, determine the small-signal voltage gain $A_v = V_o/V_i$, and find the input resistance R_i. (Ans. $A_v = 2.41$, $R_i = 0.485 \, \text{k}\Omega$)

Figure 6.37   Figure for Exercise 6.21

6.6 THE THREE BASIC AMPLIFIER CONFIGURATIONS: SUMMARY AND COMPARISON

Table 6.1 is a summary of the small-signal characteristics of the three amplifier configurations.

Table 6.1 Characteristics of the three MOSFET amplifier configurations

Configuration	Voltage gain	Current gain	Input resistance	Output resistance
Common source	$A_v > 1$	—	R_{TH}	Moderate to high
Source follower	$A_v \cong 1$	—	R_{TH}	Low
Common gate	$A_v > 1$	$A_i \cong 1$	Low	Moderate to high

The common-source amplifier voltage gain is generally greater than 1. The voltage gain of the source follower is slightly less than 1, and that of the common-gate circuit is generally greater than 1.

The input resistance looking directly into the gate of the common-source and source-follower circuits is essentially infinite at low to moderate signal frequencies. However, the input resistance of these discrete amplifiers is the Thevenin equivalent resistance R_{TH} of the bias resistors. In contrast, the input resistance to the common-gate circuit is generally in the range of only a few hundred ohms.

The output resistance of the source follower is generally in the range of a few hundred ohms or less. The output resistance of the common-source and common-gate configurations is dominated by the resistance R_D. In Chapters 10 and 11, we will see that the output resistance of these configurations is dominated by the resistance r_o when transistors are used as load devices in ICs.

The specific characteristics of these single-stage amplifiers are used in the design of multistage amplifiers.

6.7 SINGLE-STAGE INTEGRATED CIRCUIT MOSFET AMPLIFIERS

In the last chapter, we considered three all-MOSFET inverters and plotted the voltage transfer characteristics. All three inverters use an n-channel enhancement-mode driver transistor. The three types of load devices are an n-channel enhancement-mode device, an n-channel depletion-mode device, and a p-channel enhancement-mode device. The MOS transistor used as a load device is referred to as an **active load**. We mentioned that these three circuits can be used as amplifiers.

In this section, we revisit these three circuits and consider their amplifier characteristics. We will emphasize the small-signal equivalent circuits. This section serves as an introduction to more advanced MOS integrated circuit amplifier designs considered in Part II of the text.

6.7.1 NMOS Amplifiers with Enhancement Load

The characteristics of an n-channel enhancement load device were presented in the last chapter. Figure 6.38(a) shows an NMOS enhancement load transistor,

Figure 6.38 (a) NMOS enhancement-mode transistor with gate and drain connected in a load device configuration and (b) current–voltage characteristics of NMOS enhancement load transistor

and Figure 6.38(b) shows the current–voltage characteristics. The threshold voltage is V_{TNL}.

Figure 6.39(a) shows an NMOS amplifier with enhancement load. The driver transistor is M_D and the load transistor is M_L. The characteristics of transistor M_D and the load curve are shown in Figure 6.39(b). The load curve is essentially the mirror image of the i–v characteristic of the load device. Since the i–v characteristics of the load device are nonlinear, the load curve is also nonlinear. The load curve intersects the voltage axis at $V_{DD} - V_{TNL}$, which is the point where the current in the enhancement load device goes to zero. The transition point is also shown on the curve.

The voltage transfer characteristic is also useful in visualizing the operation of the amplifier. This curve is shown in Figure 6.39(c). When the enhancement-mode driver first begins to conduct, it is biased in the saturation region. For use as an amplifier, the circuit Q-point should be in this region, as shown in both Figures 6.39(b) and (c).

We can now apply the small-signal equivalent circuits to find the voltage gain. In the discussion of the source follower, we found that the equivalent resistance looking into the source terminal (with $R_S = \infty$) was $R_o = (1/g_m)\|r_o$. The small-signal equivalent circuit of the inverter is given in Figure 6.40, where the subscripts D and L refer to the driver and load transistors, respectively. We are again neglecting the body effect of the load transistor.

The small-signal voltage gain is then

$$A_v = \frac{V_o}{V_i} = -g_{mD}\left(r_{oD}\left\|\frac{1}{g_{mL}}\right\|r_{oL}\right) \tag{6.49}$$

Since, generally, $1/g_{mL} \ll r_{oL}$ and $1/g_{mD} \ll r_{oD}$, the voltage gain, to a good approximation is given by

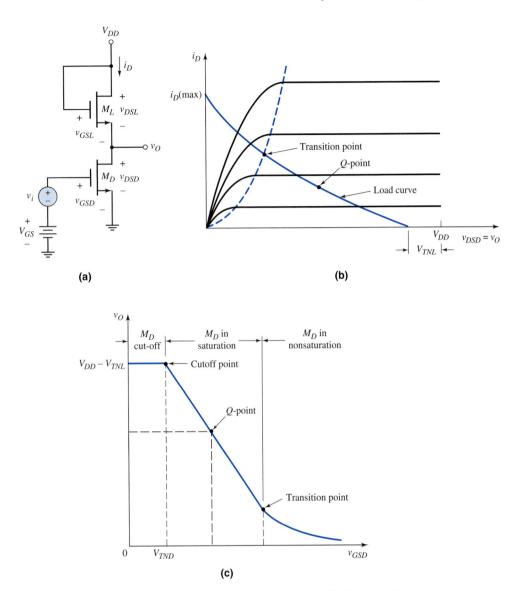

(a)

(b)

(c)

Figure 6.39 (a) NMOS amplifier with enhancement load device; (b) driver transistor characteristics and enhancement load curve with transition point; and (c) voltage transfer characteristics of NMOS amplifier with enhancement load device

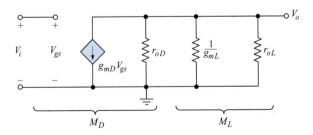

Figure 6.40 Small-signal equivalent circuit of NMOS inverter with enhancement load device

$$A_v = \frac{-g_{mD}}{g_{mL}} = -\sqrt{\frac{K_{nD}}{K_{nL}}} = -\sqrt{\frac{(W/L)_D}{(W/L)_L}} \qquad (6.50)$$

The voltage gain, then, is related to the size of the two transistors.

Design Example 6.11 **Objective:** Design the small-signal voltage gain of an NMOS amplifier with enhancement load, and establish the Q-point in the middle of the saturation region.

Consider the circuit shown in Figure 6.39(a) with transistor parameters $V_{TND} = V_{TNL} = 1\,\text{V}$, $k_n' = 30\,\mu\text{A/V}^2$, and $(W/L)_L = 1$. The circuit parameter is $V_{DD} = 5\,\text{V}$.

Design the circuit such that the voltage gain is $|A_v| = 10$.

Solution: From Equation (6.50), we have

$$|A_v| = 10 = \sqrt{\frac{(W/L)_D}{(W/L)_L}}$$

Therefore, the width-to-length ratio of the driver transistor must be

$$\left(\frac{W}{L}\right)_D = (10)^2 \left(\frac{W}{L}\right)_L = (100)(1) = 100$$

The conduction parameters are then

$$K_{nD} = \left(\frac{W}{L}\right)_D \left(\frac{1}{2}k_n'\right) = (100)(15) \Rightarrow 1.5\,\text{mA/V}^2$$

and

$$K_{nL} = \left(\frac{W}{L}\right)_L \left(\frac{1}{2}k_n'\right) = (1)(15) \Rightarrow 0.015\,\text{mA/V}^2$$

We can determine the transition point by setting

$$v_O = v_{GSD} - V_{TND}$$

Therefore,

$$v_{GSD} - V_{TND} = (V_{DD} - V_{TNL}) - \sqrt{\frac{K_{nD}}{K_{nL}}}\,(v_{GSD} - V_{TND})$$

or

$$v_{GSD} - 1 = (5 - 1) - \sqrt{\frac{1.5}{0.015}}\,(v_{GSD} - 1)$$

which yields transition point values of

$$v_{GSD} = 1.36\,\text{V} \quad \text{and} \quad v_{DSD} = 0.36\,\text{V}$$

Considering the resulting voltage transfer characteristics shown in Figure 6.41, the middle of the saturation region is halfway between the cutoff point ($v_{GSD} = V_{TND} = 1$ V) and the transition point ($v_{GSD} = 1.36$ V), or

$$V_{GSQ} = 1.18 \text{ V}$$

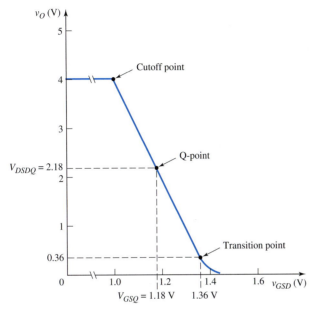

Figure 6.41 Voltage transfer characteristics and *Q*-point of NMOS amplifier with enhancement load, for Example 6.11

Comment: These results show that a very large difference is required in the sizes of the two transistors to produce a gain of 10. In fact, a gain of 10 is about the largest practical gain that can be produced by an enhancement load device. A larger small-signal gain can be obtained by using a depletion-mode MOSFET as a load device, as shown in the next section.

Design Pointer: The body effect of the load transistor was neglected in this analysis. The body effect will actually lower the small-signal voltage gain from that determined in the example.

Test Your Understanding

6.22 For the enhancement load amplifier shown in figure 6.39(a), the parameters are: $V_{TND} = V_{TNL} = 0.8$ V, $k'_n = 40 \, \mu\text{A/V}^2$, $(W/L)_D = 80$, $(W/L)_L = 1$, and $V_{DD} = 5$ V. Determine the small-signal voltage gain. Determine V_{GS} such that the *Q*-point is in the middle of the saturation region. (Ans. $A_v = -8.94$, $V_{GS} = 1.01$ V)

D6.23 For the enhancement load amplifier shown in Figure 6.39(a), the parameters are: $V_{TND} = V_{TNL} = 1$ V, $k'_n = 30 \, \mu\text{A/V}^2$, $(W/L)_L = 2$, and $V_{DD} = 10$ V. Design the circuit such that the small-signal voltage gain is $|A_v| = 6$ and the *Q*-point is in the center of the saturation region. (Ans. $(W/L)_D = 72$, $V_{GS} = 1.645$ V)

6.7.2   NMOS Amplifier with Depletion Load

Figure 6.42(a) shows the NMOS depletion-mode transistor connected as a load device and Figure 6.42(b) shows the current–voltage characteristics. The transition point is also indicated. The threshold voltage V_{TNL} of this device is negative, which means that the v_{DS} value at the transition point is positive. Also, the slope of the curve in the saturation region is not zero; therefore, a finite resistance r_o exists in this region.

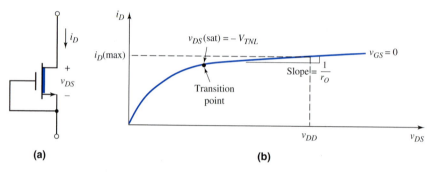

(a)   **(b)**

Figure 6.42   (a) NMOS depletion-mode transistor with gate and source connected in a load device configuration and (b) current–voltage characteristic of NMOS depletion load transistor

Figure 6.43(a) shows an **NMOS depletion load amplifier**. The transistor characteristics of M_D and the load curve for the circuit are shown in Figure 6.43(b). The load curve, again, is the mirror image of the i–v characteristic of the load device. Since the i–v characteristics of the load device are nonlinear, the load curve is also nonlinear. The transition points for both M_D and M_L are also indicated. Point A is the transition point for M_D, and point B is the transition point for M_L. The Q-point should be approximately midway between the two transition points.

The dc voltage V_{GSDQ} biases transistor M_D in the saturation region at the Q-point. The signal voltage v_i superimposes a sinuosoidal gate-to-source voltage on the dc value, and the bias point moves along the load curve about the Q-point. Again, both M_D and M_L must be biased in their saturation regions at all times.

The voltage transfer characteristic of this circuit is shown in Figure 6.43(c). Region III corresponds to the condition in which both transistors are biased in the saturation region. The desired Q-point is indicated.

We can again apply the small-signal equivalent circuit to find the small-signal voltage gain. Since the gate-to-source voltage of the depletion-load device is held at zero, the equivalent resistance looking into the source terminal is $R_o = r_o$. The small-signal equivalent circuit of the inverter is given in Figure 6.44, where the subscripts D and L refer to the driver and load transistors, respectively. We are again neglecting the body effect of the load device.

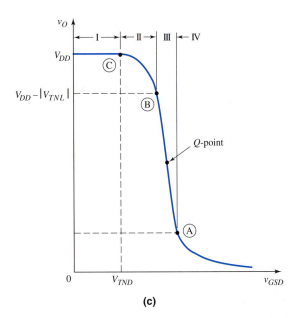

Figure 6.43 (a) NMOS amplifier with depletion load device; (b) driver transistor characteristics and depletion load curve, with transition points; and (c) voltage transfer characteristics

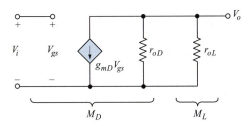

Figure 6.44 Small-signal equivalent circuit of NMOS inverter with depletion load device

The small-signal voltage gain is then

$$A_v = \frac{V_o}{V_i} = -g_{mD}(r_{oD} \| r_{oL}) \tag{6.51}$$

In this circuit, the voltage gain is directly proportional to the output resistances of the two transistors.

Example 6.12   **Objective:**   Determine the small-signal voltage gain of the NMOS amplifier with depletion load.

For the circuit shown in Figure 6.43(a), assume transistor parameters of $V_{TND} = +0.8$ V, $V_{TNL} = -1.5$ V, $K_{nD} = 1$ mA/V^2, $K_{nL} = 0.2$ mA/V^2, and $\lambda_D = \lambda_L = 0.01$ V^{-1}. Assume the transistors are biased at $I_{DQ} = 0.2$ mA.

Solution: The transconductance of the driver is

$$g_{mD} = 2\sqrt{K_{nD}I_{DQ}} = 2\sqrt{(1)(0.2)} = 0.894 \text{ mA/V}$$

Since $\lambda_D = \lambda_L$, the output resistances are

$$r_{oD} = r_{oL} = \frac{1}{\lambda I_{DQ}} = \frac{1}{(0.01)(0.2)} = 500 \text{ k}\Omega$$

The small-signal voltage gain is then

$$A_v = -g_{mD}(r_{oD} \| r_{oL}) = -(0.894)(500\|500) = -224$$

Comment: The voltage gain of the NMOS amplifier with depletion load is, in general, significantly larger than that with the enhancement load device. The body effect will lower the ideal gain factor.

Discussion: One aspect of this circuit design that we have not emphasized is the dc biasing. We mentioned that both transistors need to be biased in their saturation regions. From Figure 6.43(a), this dc biasing is accomplished with the dc source V_{GSDQ}. However, because of the steep slope of the transfer characteristics (Figure 6.43(c)), applying the "correct" voltage becomes difficult. As we will see in the next section, dc biasing is generally accomplished with current source biasing.

Test Your Understanding

*6.24   For the depletion load amplifier in Figure 6.43(a), the parameters are: $V_{TND} = 0.8$ V, $V_{TNL} = -1.2$ V, $K_{nD} = 250$ μA/V^2, $K_{nL} = 25$ μA/V^2, $\lambda_D = \lambda_L = 0.01$ V^{-1}, and $V_{DD} = 5$ V. (a) Determine V_{GS} such that the Q-point is in the middle of the saturation region. (b) Calculate the quiescent drain current. (c) Determine the small-signal voltage gain. (Ans. (a) $V_{GS} = 1.18$ V; (b) $I_{DQ} = 37$ μA; (c) $A_v = -257$)

6.7.3   NMOS Amplifier with PMOS Load

Common-Source Amplifier

An amplifier using an n-channel enhancement-mode driver and a p-channel enhancement mode active load is shown in Figure 6.45(a) in a common-source configuration. The p-channel active load transistor M_2 is biased from M_3 and I_{Bias}. This configuration is similar to the MOSFET current source shown in Figure 5.39 in Chapter 5. With both n- and p-channel transistors in the same circuit, this circuit is now referred to as a CMOS amplifier.

The i–v characteristic curve for M_2 is shown in Figure 6.45(b). The source-to-gate voltage is a constant and is established by M_3. The driver transistor characteristics and the load curve are shown in Figure 6.45(c). The transition points of both M_1 and M_2 are shown. Point A is the transition point for M_1 and point B is the transition point for M_2. The Q-point, to establish an amplifier, should be approximately halfway between points A and B, so that both transistors are biased in their saturation regions. The voltage transfer characteristics are shown in Figure 6.45(d). Shown on the curve are the same transition points A and B and the desired Q-point.

We again apply the small-signal equivalent circuits to find the small-signal voltage gain. With v_{SG2} held constant, the equivalent resistance looking into the

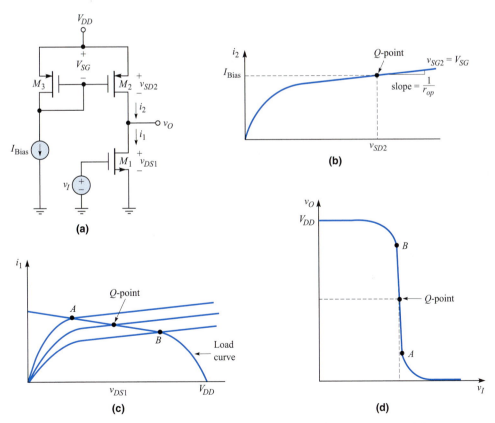

Figure 6.45   (a) CMOS common-source amplifier; (b) PMOS active load i–v characteristic, (c) driver transistor characteristics with load curve, (d) voltage transfer characteristics

drain of M_2 is just $R_o = r_{op}$. The small-signal equivalent circuit of the inverter is then as given in Figure 6.46. The subscripts n and p refer to the n-channel and p-channel transistors, respectively. We may note that the body terminal of M_1 will be tied to ground, which is the same as the source of M_1, and the body terminal of M_2 will be tied to V_{DD}, which is the same as the source of M_2. Hence, there is no body effect in this circuit.

The small-signal voltage gain is

$$A_v = \frac{V_o}{V_i} = -g_{mn}(r_{on} \| r_{op}) \tag{6.52}$$

Again for this circuit, the small-signal voltage gain is directly proportional to the output resistances of the two transistors.

Figure 6.46   Small-signal equivalent circuit of the CMOS common-source amplifier

Example 6.13   **Objective:** Determine the small-signal voltage gain of the CMOS amplifier.

For the circuit shown in Figure 6.45(a), assume transistor parameters of $V_{TN} = +0.8\,\text{V}$, $V_{TP} = -0.8\,\text{V}$, $k_n' = 80\,\mu\text{A/V}^2$, $k_p' = 40\,\mu\text{A/V}^2$, $(W/L)_n = 15$, $(W/L)_p = 30$, and $\lambda_n = \lambda_p = 0.01\,\text{V}^{-1}$. Also, assume $I_{\text{Bias}} = 0.2\,\text{mA}$.

Solution: The transconductance of the NMOS driver is

$$g_{mn} = 2\sqrt{K_n I_{DQ}} = 2\sqrt{\left(\frac{k_n'}{2}\right)\left(\frac{W}{L}\right)_n I_{\text{Bias}}}$$

$$= 2\sqrt{\left(\frac{0.08}{2}\right)(15)(0.2)} = 0.693\,\text{mA/V}$$

Since $\lambda_n = \lambda_p$, the output resistances are

$$r_{on} = r_{op} = \frac{1}{\lambda I_{DQ}} = \frac{1}{(0.01)(0.2)} = 500\,\text{k}\Omega$$

The small-signal voltage gain is then

$$A_v = -g_m(r_{on} \| r_{op}) = -(0.693)(500 \| 500) = -173$$

Comment: The voltage gain of the CMOS amplifier is on the same order of magnitude as the NMOS amplifier with depletion load. However, the CMOS amplifier does not suffer from the body effect.

Discussion: In the circuit configuration shown in Figure 6.45(a), we must again apply a dc voltage to the gate of M_1 to achieve the "proper" Q-point. We will show in later chapters using more sophisticated circuits how the Q-point is more easily established with current-source biasing. However, this circuit demonstrates the basic principles of the CMOS common-source amplifier.

CMOS Source-Follower and Common-Gate Amplifiers

The same basic CMOS circuit configuration can be used to form CMOS source-follower and common-gate configurations. Figure 6.47(a) and (b) show these circuits.

 We see that for the source-follower circuit, the active load (M_2) is an n-channel rather than a p-channel device. The input is applied to the gate of M_1 and the output is at the source of M_1. For the common-gate amplifier, the active load (M_2) is again a p-channel device. The input is applied to the source of M_1 and the output is at the drain of M_1.

 We may note that in both the source-follower and common-gate circuits, the body effect will need to be taken into account. In both circuits, the body terminal of the amplifying transistor M_1 will be connected to the most negative voltage, which is not the same as the source terminal. We will consider these types of circuits in detail in later chapters.

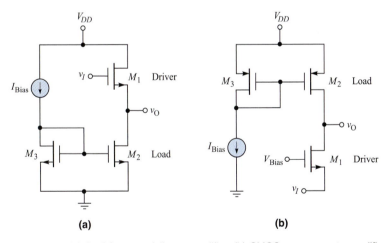

(a)   **(b)**

Figure 6.47   (a) CMOS source-follower amplifier; (b) CMOS common-gate amplifier

6.8   MULTISTAGE AMPLIFIERS

In most applications, a single-transistor amplifier will not be able to meet the combined specifications of a given amplification factor, input resistance, and output resistance. For example, the required voltage gain may exceed that which can be obtained in a single-transistor circuit.

Transistor amplifier circuits can be connected in series, or **cascaded**, as shown in Figure 6.48. This may be done either to increase the overall small-signal voltage gain, or provide an overall voltage gain greater than 1, with a very low output resistance. The overall voltage gain may not simply be the product of the individual amplification factors. Loading effects, in general, need to be taken into account.

There are many possible multistage configurations; we will examine a few here, in order to understand the type of analysis required.

Figure 6.48   Generalized two-stage amplifier

6.8.1   DC Analysis

The circuit shown in Figure 6.49 is a cascade of a common-source amplifier followed by a source-follower amplifier. As shown previously, the common-source amplifier provides a small-signal voltage gain and the source follower has a low output impedance.

Figure 6.49   Common-source amplifier in cascade with source follower

Design Example 6.14   **Objective:**   Design the biasing of a multistage MOSFET circuit to meet specific requirements.

Consider the circuit shown in Figure 6.49 with transistor parameters $K_{n1} = 500\,\mu\text{A/V}^2$, $K_{n2} = 200\,\mu\text{A/V}^2$, $V_{TN1} = V_{TN2} = 1.2\,\text{V}$, and $\lambda_1 = \lambda_2 = 0$. Design the circuit such that $I_{DQ1} = 0.2\,\text{mA}$, $I_{DQ2} = 0.5\,\text{mA}$, $V_{DSQ1} = V_{DSQ2} = 6\,\text{V}$, and $R_i = 100\,\text{k}\Omega$. Let $R_{Si} = 4\,\text{k}\Omega$.

Solution: For output transistor M_2, we have

$$V_{DSQ2} = 5 - (-5) - I_{DQ2}R_{S2}$$

or

$$6 = 10 - (0.5)R_{S2}$$

which yields $R_{S2} = 8\,\text{k}\Omega$. Also,

$$I_{DQ2} = K_{n2}(V_{GS2} - V_{TN2})^2$$

or

$$0.5 = 0.2(V_{GS2} - 1.2)^2$$

which yields

$$V_{GS2} = 2.78\,\text{V}$$

Since $V_{DSQ2} = 6\,\text{V}$, the source voltage of M_2 is $V_{S2} = -1\,\text{V}$. With $V_{GS2} = 2.78\,\text{V}$, the gate voltage on M_2 must be

$$V_{G2} = -1 + 2.78 = 1.78\,\text{V}$$

The resistor R_{D1} is then

$$R_{D1} = \frac{5 - 1.78}{0.2} = 16.1\,\text{k}\Omega$$

For $V_{DSQ1} = 6\,\text{V}$, the source voltage of M_1 is

$$V_{S1} = 1.78 - 6 = -4.22\,\text{V}$$

The resistor R_{S1} is then

$$R_{S1} = \frac{-4.22 - (-5)}{0.2} = 3.9\,\text{k}\Omega$$

For transistor M_1, we have

$$I_{DQ1} = K_{n1}(V_{GS1} - V_{TN1})^2$$

or

$$0.2 = 0.50(V_{GS1} - 1.2)^2$$

which yields

$$V_{GS1} = 1.83\,\text{V}$$

To find R_1 and R_2, we can write

$$V_{GS1} = \left(\frac{R_2}{R_1 + R_2}\right)(10) - I_{DQ1}R_{S1}$$

Since

$$\frac{R_2}{R_1 + R_2} = \frac{1}{R_1} \cdot \left(\frac{R_1 R_2}{R_1 + R_2} \right) = \frac{1}{R_1} \cdot R_i$$

then

$$1.83 = \frac{1}{R_1}(100)(10) - (0.2)(3.9)$$

which yields $R_1 = 383\,\text{k}\Omega$. From $R_i = 100\,\text{k}\Omega$, we find that $R_2 = 135\,\text{k}\Omega$.

 Comment: Both transistors are biased in the saturation region, which is desired for linear amplifiers.

Figure 6.50 shows a cascode circuit with n-channel MOSFETs. Transistor M_1 is connected in a common-source configuration and M_2 is connected in a common-gate configuration. The advantage of this type of circuit is a higher frequency response, which is discussed in a later chapter.

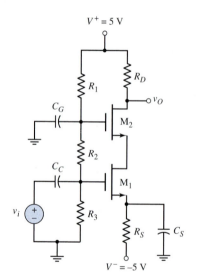

Figure 6.50   NMOS cascode circuit

Design Example 6.15   **Objective:**   Design the biasing of the cascode circuit to meet specific requirements.

For the circuit shown in Figure 6.50, the transistor parameters are: $V_{TN1} = V_{TN2} = 1.2\,\text{V}$, $K_{n1} = K_{n2} = 0.8\,\text{mA/V}^2$, and $\lambda_1 = \lambda_2 = 0$. Let $R_1 + R_2 + R_3 = 300\,\text{k}\Omega$ and $R_S = 10\,\text{k}\Omega$. Design the circuit such that $I_{DQ} = 0.4\,\text{mA}$ and $V_{DSQ1} = V_{DSQ2} = 2.5\,\text{V}$.

Solution: The dc voltage at the source of M_1 is

$$V_{S1} = I_{DQ}R_S - 5 = (0.4)(10) - 5 = -1\,\text{V}$$

Since M_1 and M_2 are identical transistors, and since the same current exists in the two transistors, the gate-to-source voltage is the same for both devices. We have

$$I_D = K_n(V_{GS} - V_{TN})^2$$

or

$$0.4 = 0.8(V_{GS} - 1.2)^2$$

which yields

$$V_{GS} = 1.91 \text{ V}$$

Then,

$$V_{G1} = \left(\frac{R_3}{R_1 + R_2 + R_3} \right)(5) = V_{GS} + V_{S1}$$

or

$$\left(\frac{R_3}{300} \right)(5) = 1.91 - 1 = 0.91$$

which yields

$$R_3 = 54.6 \text{ k}\Omega$$

The voltage at the source of M_2 is

$$V_{S2} = V_{DSQ2} + V_{S2} = 2.5 - 1 = 1.5 \text{ V}$$

Then,

$$V_{G2} = \left(\frac{R_2 + R_3}{R_1 + R_2 + R_3} \right)(5) = V_{GS} + V_{S2}$$

or

$$\left(\frac{R_2 + R_3}{300} \right)(5) = 1.91 + 1.5 = 3.41 \text{ V}$$

which yields

$$R_2 + R_3 = 204.6 \text{ k}\Omega$$

and

$$R_2 = 150 \text{ k}\Omega$$

Therefore

$$R_1 = 95.4 \text{ k}\Omega$$

The voltage at the drain of M_2 is

$$V_{D2} = V_{DSQ2} + V_{S2} = 2.5 + 1.5 = 4 \text{ V}$$

The drain resistor is therefore

$$R_D = \frac{5 - V_{D2}}{I_{DQ}} = \frac{5 - 4}{0.4} = 2.5 \text{ k}\Omega$$

Comment: Since $V_{DS} = 2.5 \text{ V} > V_{GS} - V_{TN} = 1.91 - 1.2 = 0.71 \text{ V}$, each transistor is biased in the saturation region.

6.8.2   Small-Signal Analysis

The midband small-signal voltage gain of multistage amplifiers is determined by assuming that all external coupling capacitors act as short circuits and inserting the small-signal equivalent circuits for the transistors.

Example 6.16   Objective: Determine the small-signal voltage gain of a multistage cascade circuit.

Consider the circuit shown in Figure 6.49 with transistor and circuit parameters given in Example 6.14.

Solution: The small-signal transconductance parameters are

$$g_{m1} = 2K_{n1}(V_{GS1} - V_{TN1}) = 2(0.50)(1.83 - 1.2) = 0.63 \, \text{mA/V}$$

and

$$g_{m2} = 2K_{n2}(V_{GS2} - V_{TN2}) = 2(0.2)(2.78 - 1.2) = 0.632 \, \text{mA/V}$$

The small-signal equivalent circuit is shown in Figure 6.51.

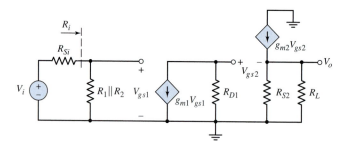

Figure 6.51   Small-signal equivalent circuit of NMOS cascade circuit

The output voltage is

$$V_o = g_{m2}V_{gs2}(R_{S2} \| R_L)$$

Also,

$$V_{gs2} + V_o = -g_{m1}V_{gs1}R_{D1}$$

where

$$V_{gs1} = \left(\frac{R_i}{R_i + R_{Si}}\right) \cdot V_i$$

Then

$$V_{gs2} = -g_{m1}R_{D1}\left(\frac{R_i}{R_i + R_{Si}}\right) \cdot V_i - V_o$$

Therefore

$$V_o = g_{m2}\left[-g_{m1}R_{D1}\left(\frac{R_i}{R_i + R_{Si}}\right) \cdot V_i - V_o\right](R_{S2} \| R_L)$$

The small-signal voltage gain is then

$$A_v = \frac{V_o}{V_i} = \frac{-g_{m1}g_{m2}R_{D1}(R_{S2}\|R_L)}{1 + g_{m2}(R_{S2}\|R_L)} \cdot \left(\frac{R_i}{R_i + R_{Si}}\right)$$

or

$$A_v = \frac{-(0.63)(0.632)(16.1)(8\|4)}{1 + (0.632)(8\|4)} \cdot \left(\frac{100}{100 + 4}\right) = -6.13$$

Comment: Since the small-signal voltage gain of the source follower is slightly less than 1, the overall gain is due essentially to the common-source input stage. Also, as shown previously, the output resistance of the source follower is small, which is desirable in many applications.

Example 6.17 **Objective:** Determine the small-signal voltage gain of a cascode circuit.

For the circuit shown in Figure 6.50, the transistor and circuit parameters are as given in Example 6.15. The input signal to the circuit is an ideal voltage source.

Solution: Since the transistors are identical, the small-signal transconductance parameters of the two transistors are equal. Therefore,

$$g_{m1} = g_{m2} = 2K_n(V_{GS} - V_{TN}) = 2(0.8)(1.91 - 1.2) = 1.14\,\text{mA/V}$$

The small-signal equivalent circuit is shown in Figure 6.52. Transistor M_1 supplies the source current of M_2 with the signal current $(g_{m1}V_i)$. Transistor M_2 acts as a current follower and passes this current on to its drain terminal. The output voltage is therefore

$$V_o = -g_{m1}V_{gs1}R_D$$

Since $V_{gs1} = V_i$, the small-signal voltage gain is

$$A_v = \frac{V_o}{V_i} = -g_{m1}R_D$$

or

$$A_v = -(1.14)(2.5) = -2.85$$

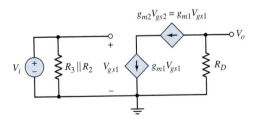

Figure 6.52 Small-signal equivalent circuit of NMOS cascode circuit

Comment: The small-signal voltage gain is essentially the same as that of a single common-source amplifier stage. The addition of a common-gate transistor will increase the frequency bandwidth, as we will see in a later chapter.

Test Your Understanding

6.25 For the cascade circuit shown in Figure 6.49, the transistor and circuit parameters are given in Example 6.15. Calculate the small-signal output resistance R_o. (The small-signal equvialent circuit is shown in Figure 6.51.) (Ans. $R_o = 1.32\,\text{k}\Omega$)

*D6.26 The supply voltages to the cascade circuit shown in Figure 6.49 are changed to $V^+ = 10\,\text{V}$ and $V^- = -10\,\text{V}$. The transistor parameters are: $K_{n1} = K_{n2} = 1\,\text{mA/V}^2$, $V_{TN1} = V_{TN2} = 2\,\text{V}$, and $\lambda_1 = \lambda_2 = 0.01\,\text{V}^{-1}$. (a) Let $R_L = 4\,\text{k}\Omega$, and design the circuit such that $I_{DQ1} = I_{DQ2} = 2\,\text{mA}$, $V_{DSQ1} = V_{DSQ2} = 10\,\text{V}$, and $R_{is} = 200\,\text{k}\Omega$. (b) Calculate the small-signal voltage gain and the output resistance R_o. (Ans. (a) $R_{S2} = 5\,\text{k}\Omega$, $R_{D1} = 3.3\,\text{k}\Omega$, $R_{S1} = 1.71\,\text{k}\Omega$, $R_1 = 586\,\text{k}\Omega$, $R_2 = 304\,\text{k}\Omega$; (b) $A_v = -8.06$, $R_o = 0.330\,\text{k}\Omega$)

D6.27 The supply voltages to the cascode circuit shown in Figure 6.50 are changed to $V^+ = 10\,\text{V}$ and $V^- = -10\,\text{V}$. The transistor parameters are: $K_{n1} = K_{n2} = 1.2\,\text{mA/V}^2$, $V_{TN1} = V_{TN2} = 2\,\text{V}$, and $\lambda_1 = \lambda_2 = 0$. (a) Let $R_1 + R_2 + R_3 = 500\,\text{k}\Omega$, and $R_S = 10\,\text{k}\Omega$. Design the circuit such that $I_{DQ} = 1\,\text{mA}$ and $V_{DSQ1} = V_{DSQ2} = 3.5\,\text{V}$. (b) Determine the small-signal voltage gain. (Ans. (a) $R_3 = 145.5\,\text{k}\Omega$, $R_2 = 175\,\text{k}\Omega$, $R_1 = 179.5\,\text{k}\Omega$, $R_D = 3\,\text{k}\Omega$; (b) $A_v = -6.57$)

6.9   BASIC JFET AMPLIFIERS

Like MOSFETs, JFETs can be used to amplify small time-varying signals. Initially, we will develop the small-signal model and equivalent circuit of the JFET. We will then use the model in the analysis of JFET amplifiers.

6.9.1   Small-Signal Equivalent Circuit

Figure 6.53 JFET common-source circuit with time-varying signal source in series with gate dc source

Figure 6.53 shows a JFET circuit with a time-varying signal applied to the gate. The instantaneous gate-to-source voltage is

$$v_{GS} = V_{GS} + v_i = V_{GS} + v_{gs} \tag{6.53}$$

where v_{gs} is the small-signal gate-to-source voltage. Assuming the transistor is biased in the saturation region, the instantaneous drain current is

$$i_D = I_{DSS}\left(1 - \frac{v_{GS}}{V_P}\right)^2 \tag{6.54}$$

where I_{DSS} is the saturation current and V_P is the pinchoff voltage. Substituting Equation (6.53) into (6.54), we obtain

$$i_D = I_{DSS}\left[\left(1 - \frac{V_{GS}}{V_P}\right) - \left(\frac{v_{gs}}{V_P}\right)\right]^2 \tag{6.55}$$

If we expand the squared term, we have

$$i_D = I_{DSS}\left(1 - \frac{V_{GS}}{V_P}\right)^2 - 2I_{DSS}\left(1 - \frac{V_{GS}}{V_P}\right)\left(\frac{v_{gs}}{V_P}\right) + I_{DSS}\left(\frac{v_{gs}}{V_P}\right)^2 \tag{6.56}$$

The first term in Equation (6.56) is the dc or quiescent drain current I_{DQ}, the second term is the time-varying drain current component, which is linearly related to the signal voltage v_{gs}, and the third term is proportional to the square of the signal voltage. As in the case of the MOSFET, the third term produces a nonlinear distortion in the output current. To minimize this distortion, we will usually impose the following condition:

$$\left|\frac{v_{gs}}{V_P}\right| \ll 2\left(1 - \frac{V_{GS}}{V_P}\right) \tag{6.57}$$

Equation (6.57) represents the small-signal condition that must be satisfied for JFET amplifiers to be linear.

Neglecting the term v_{gs}^2 in Equation (6.56), we can write

$$i_D = I_{DQ} + i_d \tag{6.58}$$

where the time-varying signal current is

$$i_d = +\frac{2I_{DSS}}{(-V_P)}\left(1 - \frac{V_{GS}}{V_P}\right)v_{gs} \tag{6.59}$$

The constant relating the small-signal drain current and small-signal gate-to-source voltage is the transconductance g_m. We can write

$$i_d = g_m v_{gs} \tag{6.60}$$

where

$$g_m = +\frac{2I_{DSS}}{(-V_P)}\left(1 - \frac{V_{GS}}{V_P}\right) \tag{6.61}$$

Since V_P is negative for n-channel JFETs, the transconductance is positive. A relationship that applies to both n-channel and p-channel JFETs is

$$g_m = \frac{2I_{DSS}}{|V_P|}\left(1 - \frac{V_{GS}}{V_P}\right) \tag{6.62}$$

We can also obtain the transconductance from

$$g_m = \left.\frac{\partial i_D}{\partial v_{GS}}\right|_{v_{GS}=V_{GSQ}} \tag{6.63}$$

Since the transconductance is directly proportional to the saturation current I_{DSS}, the transconductance is also a function of the width-to-length ratio of the transistor.

Since we are looking into a reverse-biased pn junction, we assume that the input gate current i_g is zero, which means that the small-signal input resistance is infinite. Equation (6.54) can be expanded to take into account the finite output resistance of a JFET biased in the saturation region. The equation becomes

$$i_D = I_{DSS}\left(1 - \frac{v_{GS}}{V_P}\right)^2 (1 + \lambda v_{DS}) \tag{6.64}$$

The small-signal output resistance is

$$r_o = \left.\left(\frac{\partial i_D}{\partial v_{DS}}\right)^{-1}\right|_{v_{GS}=\text{const.}} \tag{6.65}$$

Using Equation (6.64), we obtain

$$r_o = \left[\lambda I_{DSS} \left(1 - \frac{V_{GS}}{V_P} \right)^2 \right]^{-1} \tag{6.66(a)}$$

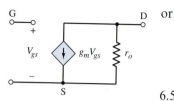

Figure 6.54   Small-signal equivalent circuit of n-channel JFET

or

$$r_o \cong [\lambda I_{DQ}]^{-1} = \frac{1}{\lambda I_{DQ}} \tag{6.66(b)}$$

The small-signal equivalent circuit of the n-channel JFET, shown in Figure 6.54, is exactly the same as that of the n-channel MOSFET. The small-signal equivalent circuit of the p-channel JFET is also the same as that of the p-channel MOSFET. However, the polarity of the controlling gate-to-source voltage and the direction of the dependent current source are reversed from those of the n-channel device.

6.9.2   Small-Signal Analysis

Since the small-signal equivalent circuit of the JFET is the same as that of the MOSFET, the small-signal analyses of the two types of circuits are identical. For illustration purposes, we will analyze two JFET circuits.

Example   6.18   Objective: Determine the small-signal voltage gain of a JFET amplifier.

Consider the circuit shown in Figure 6.55 with transistor parameters $I_{DSS} = 12\,\text{mA}$, $V_P = -4\,\text{V}$, and $\lambda = 0.008\,\text{V}^{-1}$. Determine the small-signal voltage gain $A_v = v_o/v_i$.

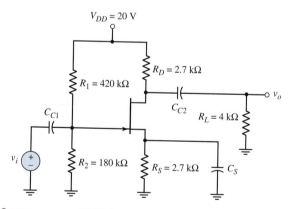

Figure 6.55   Common-source JFET circuit with source resistor and source bypass capacitor

Solution: The dc quiescent gate-to-source voltage is determined from

$$V_{GSQ} = \left(\frac{R_2}{R_1 + R_2} \right) V_{DD} - I_{DQ} R_S$$

where

$$I_{DQ} = I_{DSS}\left(1 - \frac{V_{GSQ}}{V_P}\right)^2$$

Combining these two equations produces

$$V_{GSQ} = \left(\frac{180}{180 + 420}\right)(20) - (12)(2.7)\left(1 - \frac{V_{GSQ}}{(-4)}\right)^2$$

which reduces to

$$2.025V_{GSQ}^2 + 17.2V_{GSQ} + 26.4 = 0$$

The appropriate solution is

$$V_{GSQ} = -2.01 \text{ V}$$

The quiescent drain current is

$$I_{DQ} = I_{DSS}\left(1 - \frac{V_{GSQ}}{V_P}\right)^2 = (12)\left(1 - \frac{(-2.01)}{(-4)}\right)^2 = 2.97\,\text{mA}$$

The small-signal parameters are then

$$g_m = \frac{2I_{DSS}}{(-V_P)}\left(1 - \frac{V_{GS}}{V_P}\right) = \frac{2(12)}{(4)}\left(1 - \frac{(-2.01)}{(-4)}\right) = 2.98\,\text{mA/V}$$

and

$$r_o = \frac{1}{\lambda I_{DQ}} = \frac{1}{(0.008)(2.97)} = 42.1\,\text{k}\Omega$$

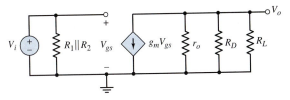

Figure 6.56 Small-signal equivalent circuit of common-source JFET, assuming bypass capacitor acts as a short circuit

The small-signal equivalent circuit is shown in Figure 6.56.
Since $V_{gs} = V_i$, the small-signal voltage gain is

$$A_v = \frac{V_o}{V_i} = -g_m(r_o\|R_D\|R_L)$$

or

$$A_v = -(2.98)(42.1\|2.7\|4) = -4.62$$

Comment: The voltage gain of JFET amplifiers is the same order of magnitude as that of MOSFET amplifiers.

Design Example 6.19   **Objective:**   Design a JFET source-follower circuit with a specified small-signal voltage gain.

For the source-follower circuit shown in Figure 6.57, the transistor parameters are: $I_{DSS} = 12\,\text{mA}$, $V_P = -4\,\text{V}$, and $\lambda = 0.01\,\text{V}^{-1}$. Determine R_S and I_{DQ} such that the small-signal voltage gain is at least $A_v = v_o/v_i = 0.90$.

Figure 6.57   JFET source-follower circuit

Solution: The small-signal equivalent circuit is shown in Figure 6.58. The output voltage is

$$V_o = g_m V_{gs}(R_S \| R_L \| r_o)$$

Also

$$V_i = V_{gs} + V_o$$

or

$$V_{gs} = V_i - V_o$$

Therefore, the output voltage is

$$V_o = g_m(V_i - V_o)(R_S \| R_L \| r_o)$$

The small-signal voltage gain becomes

$$A_v = \frac{V_o}{V_i} = \frac{g_m(R_S \| R_L \| r_o)}{1 + g_m(R_S \| R_L \| r_o)}$$

As a first approximation, assume r_o is sufficiently large for the effect of r_o to be neglected.

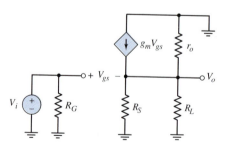

Figure 6.58   Small-signal equivalent circuit of JFET source-follower circuit

The transconductance is

$$g_m = \frac{2I_{DSS}}{(-V_P)}\left(1 - \frac{V_{GS}}{V_P}\right) = \frac{2(12)}{4}\left(1 - \frac{V_{GS}}{(-4)}\right)$$

If we pick a nominal transconductance value of $g_m = 2\,\text{mA/V}$, then $V_{GS} = -2.67\,\text{V}$ and the quiescent drain current is

$$I_{DQ} = I_{DSS}\left(1 - \frac{V_{GS}}{V_P}\right)^2 = (12)\left(1 - \frac{(-2.67)}{(-4)}\right)^2 = 1.33\,\text{mA}$$

The value of R_S is then determined from

$$R_S = \frac{-V_{GS} - (-10)}{I_{DQ}} = \frac{2.67 + 10}{1.33} = 9.53\,\text{k}\Omega$$

Also, the value of r_o is

$$r_o = \frac{1}{\lambda I_{DQ}} = \frac{1}{(0.01)(1.33)} = 75.2\,\text{k}\Omega$$

The small-signal voltage gain, including the effect of r_o, is

$$A_v = \frac{g_m(R_S\|R_L\|r_o)}{1 + g_m(R_S\|R_L\|r_o)} = \frac{(2)(9.53\|10\|75.2)}{1 + (2)(9.53\|10\|75.2)} = 0.902$$

Comment: This particular design meets the design criteria, but the solution is not unique.

In the last example, we chose a value of transconductance and continued through the design. A more detailed examination shows that both g_m and R_S depend upon the drain current I_{DQ} in such a way that the product $g_m R_S$ is approximately a constant. This means the small-signal voltage gain is insensitive to the initial value of the transconductance.

Test Your Understanding

6.28 Reconsider the JFET amplifier shown in Figure 6.55 with transistor parameters given in Example 6.19. Determine the small-signal voltage gain if a 20 kΩ resistor is in series with the signal source v_i. (Ans. $A_v = -3.98$)

RD6.29 For the JFET amplifier shown in Figure 6.55, the transistor parameters are: $I_{DSS} = 4\,\text{mA}$, $V_P = -3\,\text{V}$, and $\lambda = 0.005\,\text{V}^{-1}$. Let $R_L = 4\,\text{k}\Omega$, $R_S = 2.7\,\text{k}\Omega$, and $R_1 + R_2 = 500\,\text{k}\Omega$. Redesign the circuit such that $I_{DQ} = 1.2\,\text{mA}$ and $V_{DSQ} = 12\,\text{V}$. Calculate the small-signal voltage gain. (Ans. $R_D = 3.97\,\text{k}\Omega$, $R_1 = 453\,\text{k}\Omega$, $R_2 = 47\,\text{k}\Omega$, $A_v = -2.87$)

*6.30 For the circuit shown in Figure 6.59, the transistor parameters are: $I_{DSS} = 6\,\text{mA}$, $|V_P| = 2\,\text{V}$, and $\lambda = 0$. (a) Calculate the quiescent drain current and drain-to-

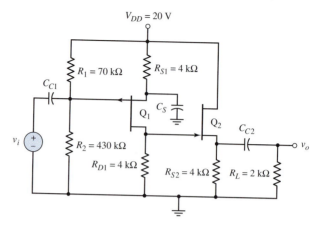

Figure 6.59   Figure for Exercise 6.30

source voltage of each transistor. (b) Determine the overall small-signal voltage gain $A_v = v_o/v_i$. (Ans. (a) $I_{DQ1} = 1\,\text{mA}$, $V_{SDQ1} = 12\,\text{V}$, $I_{DQ2} = 1.27\,\text{mA}$, $V_{DSQ2} = 14.9\,\text{V}$; (b) $A_v = -2.05$)

6.31   Reconsider the source-follower circuit shown in Figure 6.57 with transistor parameters $I_{DSS} = 8\,\text{mA}$, $V_P = -3.5\,\text{V}$, and $\lambda = 0.01\,\text{V}^{-1}$. (a) Design the circuit such that $I_{DQ} = 2\,\text{mA}$. (b) Calculate the small-signal voltage gain if R_L approaches infinity. (c) Determine the value of R_L at which the small-signal gain is reduced by 20 percent from its value for (b). (Ans. (a) $R_S = 5.88\,\text{k}\Omega$, (b) $A_v = 0.923$, $R_L = 1.64\,\text{k}\Omega$)

6.10   SUMMARY

- The application of MOSFET transistors in linear amplifier circuits was emphasized in this chapter. A small-signal equivalent circuit for the transistor was developed, which is used in the analysis and design of linear amplifiers.

- Three basic circuit configurations were considered: the common source, source follower, and common gate. These three configurations form the basic building blocks for complex integrated circuits. The small-signal voltage gains and output resistances for these circuits were analyzed. The circuit characteristics of the three circuits were compared in Table 6.1.

- The ac analysis of circuits with enhancement load devices, with depletion load devices, and complementary (CMOS) devices were analyzed. These circuits are examples of all MOSFET circuits and act as an introduction to more complex all-MOSFET integrated circuits considered later in the text.

- The small-signal equivalent circuit of a JFET was developed and used in the analysis of several configurations of JFET amplifiers.

CHECKPOINT

After studying this chapter, the reader should have the ability to:

✔ Explain graphically the amplification process in a simple MOSFET amplifier circuit. (Section 6.1)

✔ Describe the small-signal equivalent circuit of the MOSFET and to determine the values of the small-signal parameters. (Section 6.1)

✔ Apply the small-signal equivalent circuit to various MOSFET amplifier circuits to obtain the time-varying circuit characteristics. (Section 6.1)

✔ Characterize the small-signal voltage gain and output resistance of a common-source amplifier. (Section 6.3)

✔ Characterize the small-signal voltage gain and output resistance of a source-follower amplifier. (Section 6.4)

✔ Characterize the small-signal voltage gain and output resistance of a common-gate amplifier. (Section 6.5)

✔ Describe the operation of an NMOS amplifier with either an enhancement load, a depletion load, or a PMOS load. (Section 6.7)

✔ Apply the MOSFET small-signal equivalent circuit in the analysis of multistage amplifier circuits. (Section 6.8)

✔ Describe the operation and analyze basic JFET amplifier circuits. (Section 6.9)

REVIEW QUESTIONS

1. Discuss, using the concept of a load line superimposed on the transistor characteristics, how a simple common-source circuit can amplify a time-varying signal.

2. How does a transistor width-to-length ratio affect the small-signal voltage gain of a common-source amplifier?

3. Discuss the physical meaning of the small-signal circuit parameter r_o.

4. How does the body effect change the small-signal equivalent circuit of the MOSFET?

5. Sketch a simple common-source amplifier circuit and discuss the general ac circuit characteristics (voltage gain and output resistance).

6. Discuss the general conditions under which a common-source amplifier would be used.

7. Why, in general, is the magnitude of the voltage gain of a common-source amplifier smaller than that of a bipolar common-emitter amplifier?

8. What are the changes in the ac characteristics of a common-source amplifier when a source resistor and a source bypass capacitor are incorporated in the design?

9. Sketch a simple source-follower amplifier circuit and discuss the general ac circuit characteristics (voltage gain and output resistance).

10. Discuss the general conditions under which a source-follower amplifier would be used.

11. Sketch a simple common-gate amplifier circuit and discuss the general ac circuit characteristics (voltage gain and output resistance).

12. Discuss the general conditions under which a common-gate amplifier would be used.

13. Compare the ac circuit characteristics of the common-source, source-follower, and common-gate circuits.

14. State the general advantage of using transistors in place of resistors in integrated circuits.

15. State at least two reasons why a multistage amplifier circuit would be required in a design compared to using a single-stage circuit.

16. Give one reason why a JFET might be used as an input device in a circuit as opposed to a MOSFET.

PROBLEMS

Section 6.1   The MOSFET Amplifier

6.1 An NMOS transistor has parameters $V_{TN} = 0.8$ V, $k_n' = 40\,\mu\text{A/V}^2$, and $\lambda = 0$. (a) Determine the width-to-length ratio (W/L) such that $g_m = 0.5\,\text{mA/V}$ at $I_D = 0.5\,\text{mA}$ when biased in the saturation region. (b) Calculate the required value of V_{GS}.

6.2 A PMOS transistor has parameters $V_{TP} = -1.2$ V, $k_p' = 20\,\mu\text{A/V}$, and $\lambda = 0$. (a) Determine the width-to-length ratio (W/L) such that $g_m = 50\,\mu\text{A/V}$ at $I_D = 0.1\,\text{mA}$ when biased in the saturation region. (b) Calculate the required value of V_{SG}.

6.3 An NMOS transistor is biased in the saturation region at a constant V_{GS}. The drain current is $I_D = 3$ mA at $V_{DS} = 5$ V and $I_D = 3.4\,\text{mA}$ at $V_{DS} = 10$ V. Determine λ and r_o.

6.4 The minimum value of small-signal resistance of a PMOS transistor is to be $r_o = 100\,\text{k}\Omega$. If $\lambda = 0.012\,\text{V}^{-1}$, calculate the maximum allowed value of I_D.

6.5 Calculate the small-signal voltage gain of the circuit shown in Figure 6.1, for $g_m = 1\,\text{mA/V}$, $r_o = 50\,\text{k}\Omega$, and $R_D = 10\,\text{k}\Omega$.

***D6.6** For the circuit shown in Figure 6.1, the transistor parameters are: $V_{TN} = +0.8$ V, $\lambda = 0.015\,\text{V}^{-1}$, and $k_n' = 60\,\mu\text{A/V}^2$. Let $V_{DD} = 10$ V. (a) Design the transistor width-to-length ratio (W/L) and the resistance R_D such that $I_{DQ} = 0.5\,\text{mA}$ and $V_{DSQ} = 6$ V. (b) Calculate g_m and r_o. (c) What is the small-signal voltage gain $A_v = v_o/v_i$?

***6.7** In our analyses, we assumed the small-signal condition given by Equation (6.4). Now consider Equation (6.3(b)) and let $v_{gs} = V_{gs} \sin \omega t$. Show that the ratio of the signal at frequency 2ω to the signal at frequency ω is given by $V_{gs}/[4(V_{GS} - V_{TN})]$. This ratio, expressed in a percentage, is called the **second-harmonic distortion**. [Hint: Use the trigonometric identity $\sin^2 \theta = \frac{1}{2} - \frac{1}{2} \cos 2\theta$.]

6.8 Using the results of Problem 6.7, find the peak amplitude V_{gs} that produces a second-harmonic distortion of 1 percent if $V_{GS} = 3$ V and $V_{TN} = 1$ V.

Section 6.3   The Common-Source Amplifier

6.9 Calculate the small-signal voltage gain of a common-source amplifier, such as that shown in Figure 6.13, assuming $g_m = 1\,\text{mA/V}$, $r_o = 50\,\text{k}\Omega$, and $R_D = 10\,\text{k}\Omega$. Also assume $R_{Si} = 2\,\text{k}\Omega$ and $R_1 \| R_2 = 50\,\text{k}\Omega$.

6.10 A common-source amplifier, such as shown in Figure 6.13 in the text, has parameters $r_o = 100\,\text{k}\Omega$ and $R_D = 5\,\text{k}\Omega$. Determine the transconductance of the transistor if the small-signal voltage gain is $A_v = -10$. Assume $R_{Si} = 0$.

6.11 For the NMOS common-source amplifier in Figure P6.11, the transistor parameters are: $V_{TN} = 2$ V, $K_n = 1\,\text{mA/V}^2$, and $\lambda = 0$. The circuit parameters are: $V_{DD} = 12$ V, $R_S = 2\,\text{k}\Omega$, $R_D = 3\,\text{k}\Omega$, $R_1 = 300\,\text{k}\Omega$, and $R_2 = 200\,\text{k}\Omega$. Assume $R_{Si} = 2\,\text{k}\Omega$ and assume a load resistance $R_L = 3\,\text{k}\Omega$ is capacitively coupled to the output. (a) Determine the quiescent values of I_D and V_{DS}. (b) Find the small-signal voltage gain. (c) Determine the maximum symmetrical swing in the output voltage.

6.12 In the circuit in Figure P6.11, $V_{DD} = 15$ V, $R_D = 2\,\text{k}\Omega$, $R_L = 5\,\text{k}\Omega$, $R_S = 0.5\,\text{k}\Omega$, and $R_{in} = 200\,\text{k}\Omega$. (a) Find R_1 and R_2 such that $I_{DQ} = 3$ mA for $V_{TN} = 2$ V, $K_n = 2\,\text{mA/V}^2$, and $\lambda = 0$. (b) Determine the small-signal voltage gain.

6.13 Repeat Problem 6.11 if the source resistor is bypassed by a source capacitor C_S.

Figure P6.11

*6.14 The transistor in the common-source amplifier in Figure P6.14 has parameters $V_{TN} = 1\,\mathrm{V}$, $K_n = 0.5\,\mathrm{mA/V^2}$, and $\lambda = 0.01\,\mathrm{V^{-1}}$. The circuit parameters are: $V^+ = 5\,\mathrm{V}$, $V^- = -5\,\mathrm{V}$, and $R_D = R_L = 10\,\mathrm{k\Omega}$. (a) Determine I_{DQ} to achieve the maximum symmetrical swing in the output voltage. (b) Find the small-signal voltage gain.

D6.15 For the common-source amplifier in Figure P6.15, the transistor parameters are: $V_{TN} = -1\,\mathrm{V}$, $K_n = 4\,\mathrm{mA/V^2}$, and $\lambda = 0$. The circuit parameters are $V_{DD} = 10\,\mathrm{V}$ and $R_L = 2\,\mathrm{k\Omega}$. (a) Design the circuit such that $I_{DQ} = 2\,\mathrm{mA}$ and $V_{DSQ} = 6\,\mathrm{V}$. (b) Determine the small-signal voltage gain. (c) If $v_i = V_i \sin \omega t$, determine the maximum value of V_i such that v_o is an undistorted sine wave.

Figure P6.14

Figure P6.15

*6.16 The transistor in the common-source circuit in Figure P6.15 has the same parameters as given in Problem 6.15. The circuit parameters are $V_{DD} = 5\,\mathrm{V}$ and $R_D = R_L = 2\,\mathrm{k\Omega}$. (a) Find R_S and $V_{DSQ} = 2.5\,\mathrm{V}$. (b) Determine the small-signal voltage gain.

*6.17 Consider the PMOS common-source circuit in Figure P6.17 with transistor parameters $V_{TP} = -2\,\mathrm{V}$ and $\lambda = 0$, and circuit parameters $R_D = R_L = 10\,\mathrm{k\Omega}$. (a) Determine the values of K_p and R_S such that $V_{SDQ} = 6\,\mathrm{V}$. (b) Determine the resulting value of I_{DQ} and the small-signal voltage gain. (c) Can the values of K_p and R_S from

Figure P6.17

part (a) be changed to achieve a larger voltage gain, while still meeting the requirements of part (a)?

D6.18 For the common-source circuit in Figure P6.17, the PMOS transistor parameters are: $V_{TP} = -1.5\,\text{V}$, $K_p = 5\,\text{mA/V}^2$, and $\lambda = 0$. The load resistor is $R_L = 2\,\text{k}\Omega$. (a) Design the circuit such that $I_{DQ} = 1\,\text{mA}$ and $V_{SDQ} = 5\,\text{V}$. (b) Determine the small-signal voltage gain $A_v = v_o/v_i$. (c) What is the maximum symmetrical swing in the output voltage?

*D6.19 Design the common-source circuit in Figure P6.19 using an n-channel MOSFET with $\lambda = 0$. The quiescent values are to be $I_{DQ} = 6\,\text{mA}$, $V_{GSQ} = 2.8\,\text{V}$, and $V_{DSQ} = 10\,\text{V}$. The transconductance is $g_m = 2.2\,\text{mA/V}$. Let $R_L = 1\,\text{k}\Omega$, $A_v = -1$, and $R_{\text{in}} = 100\,\text{k}\Omega$. Find R_1, R_2, R_S, R_D, K_n, and V_{TN}.

*6.20 For the common-source amplifier in Figure P6.20, the transistor parameters are: $V_{TP} = -1.5\,\text{V}$, $K_p = 2\,\text{mA/V}^2$, and $\lambda = 0.01\,\text{V}^{-1}$. The circuit is to drive a load resistance of $R_L = 20\,\text{k}\Omega$. To minimize loading effects, the drain resistance should be $R_D \leq 0.1 R_L$. (a) Determine I_Q such that the Q-point is in the center of the saturation region. (b) Determine the open-circuit ($R_L = \infty$) small-signal voltage gain. (c) By what percentage does the small-signal voltage gain decrease when R_L is connected?

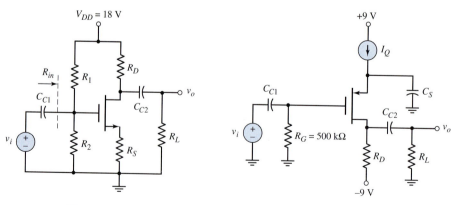

Figure P6.19　　　　　　　　　　　**Figure P6.20**

D6.21 For the circuit shown in Figure P6.21, the transistor parameters are: $V_{TP} = 2\,\text{V}$, $K_p = 0.5\,\text{mA/V}^2$, and $\lambda = 0$. (a) Design the circuit such that $I_{DQ} = 2\,\text{mA}$ and $V_{SDQ} = 6\,\text{V}$. (b) Determine the small-signal voltage gain $A_v = v_o/v_i$.

*D6.22 Design a common-source amplifier, such as that in Figure P6.22, to achieve a small-signal voltage gain of at least $A_v = v_o/v_i = -10$ for $R_L = 20\,\text{k}\Omega$ and $R_{\text{in}} = 200\,\text{k}\Omega$. Assume the Q-point is chosen at $I_{DQ} = 1\,\text{mA}$ and $V_{DSQ} = 10\,\text{V}$. Let $V_{TN} = 2\,\text{V}$, and $\lambda = 0$.

Figure P6.21   Figure P6.22

Section 6.4   The Source-Follower Amplifier

6.23 For an enhancement-mode MOSFET source follower, $g_m = 4\,\text{mA/V}$ and $r_o = 50\,\text{k}\Omega$. Determine the no-load voltage gain and the output resistance. Calculate the small-signal voltage gain when a load resistance $R_S = 2.5\,\text{k}\Omega$ is connected.

6.24 The transistor in the source-follower circuit in Figure P6.24 has parameters $K_p = 2\,\text{mA/V}^2$, $V_{TP} = -2\,\text{V}$, and $\lambda = 0.02\,\text{V}^{-1}$. The circuit parameters are: $R_L = 4\,\text{k}\Omega$, $R_S = 4\,\text{k}\Omega$, $R_1 = 1.24\,\text{M}\Omega$, and $R_2 = 396\,\text{k}\Omega$. (a) Calculate I_{DQ} and V_{SDQ}. (b) Determine the small-signal gains $A_v = v_o/v_i$ and $A_i = i_o/i_i$, and the output resistance R_o.

Figure P6.24

6.25 Consider the source-follower circuit in Figure P6.25 with transistor parameters $V_{TN} = 1.2\,\text{V}$, $K_n = 1\,\text{mA/V}^2$, and $\lambda = 0.01\,\text{V}^{-1}$. If $I_Q = 1\,\text{mA}$, determine the small-signal voltage gain $A_v = v_o/v_i$ and the output resistance R_o.

Figure P6.25

6.26 For the source-follower circuit shown in Figure P6.25, the transistor parameters are: $V_{TN} = 1\,\text{V}$, $k'_n = 60\,\mu\text{A/V}^2$, and $\lambda = 0$. The small-signal voltage gain is to be $A_v = v_o/v_i = 0.95$. (a) Determine the required width-to-length ratio (W/L) for $I_Q = 4\,\text{mA}$. (b) Determine the required I_Q if $(W/L) = 60$.

***D6.27** In the source-follower circuit in Figure P6.27 with a depletion NMOS transistor, the device parameters are: $V_{TN} = -2\,\text{V}$, $K_n = 5\,\text{mA/V}^2$, and $\lambda = 0.01\,\text{V}^{-1}$. Design the circuit such that $I_{DQ} = 5\,\text{mA}$. Find the small-signal voltage gain $A_v = v_o/v_i$ and the output resistance R_o.

Figure P6.27

6.28 Consider the circuit in Figure P6.27. Let $R_S = 10\,\text{k}\Omega$ and $\lambda = 0$. The open-circuit voltage gain $(R_L = \infty)$ is $A_v = v_o/v_i = 0.90$. Determine g_m and R_o. Determine the value of the voltage gain if a load resistor $R_L = 2\,\text{k}\Omega$ is connected.

D6.29 For the source-follower circuit in Figure P6.27, the transistor parameters are: $V_{TN} = -2\,\text{V}$, $K_n = 4\,\text{mA/V}^2$, and $\lambda = 0$. Design the circuit such that $R_o \le 200\,\Omega$. Determine the resulting small-signal voltage gain.

6.30 The current source in the source-follower circuit in Figure P6.30 is $I_Q = 5\,\text{mA}$ and the transistor parameters are: $V_{TP} = -2\,\text{V}$, $K_p = 5\,\text{mA/V}^2$, and $\lambda = 0$. (a) Determine the output resistance R_o. (b) Determine the value of R_L that reduces the small-signal voltage gain to one-half the open-circuit $(R_L = \infty)$ value.

Figure P6.30

6.31 Consider the source-follower circuit shown in Figure P6.31. The most negative output signal voltage occurs when the transistor just cuts off. Show that this output voltage $v_o(\text{min})$ is given by

$$v_o(\text{min}) = \frac{-I_{DQ}R_S}{1 + \dfrac{R_S}{R_L}}$$

Show that the corresponding input voltage is given by

$$v_i(\text{min}) = -\frac{I_{DQ}}{g_m}(1 + g_m(R_S \| R_L))$$

D6.32 The transistor in the circuit in Figure P6.32 has parameters $V_{TN} = 1\,\text{V}$, $K_n = 1\,\text{mA/V}^2$, and $\lambda = 0$. The circuit parameters are $V_{DD} = 5\,\text{V}$ and $R_i = 300\,\text{k}\Omega$. (a) Design the circuit such that $I_{DQ} = 1.7\,\text{mA}$ and $V_{DSQ} = 3\,\text{V}$. (b) Determine the small-signal voltage gain $A_v = v_o/v_i$ and the output resistance R_o.

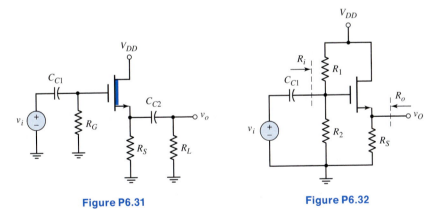

Figure P6.31   Figure P6.32

Section 6.5   The Common-Gate Configuration

6.33 For the common-gate circuit in Figure P6.33, the NMOS transistor parameters are: $V_{TN} = 1\,\text{V}$, $K_n = 3\,\text{mA/V}^2$, and $\lambda = 0$. (a) Determine I_{DQ} and V_{DSQ}. (b) Calculate g_m and r_o. (c) Find the small-signal voltage gain $A_v = v_o/v_i$.

6.34 Consider the PMOS common-gate circuit in Figure P6.34. The transistor parameters are: $V_{TP} = -1\,\text{V}$, $K_p = 0.5\,\text{mA/V}^2$, and $\lambda = 0$. (a) Determine R_S and R_D such that $I_{DQ} = 0.75\,\text{mA}$ and $V_{SDQ} = 6\,\text{V}$. (b) Determine the input impedance R_i and

Figure P6.33   **Figure P6.34**

the output impedance R_o. (c) Determine the load current i_o and the output voltage v_o, if $i_i = 5 \sin \omega t \, \mu A$.

6.35 The parameters of the transistor in the circuit in Figure 6.34 in the text are: $V_{TN} = 2\,V$, $K_n = 4\,mA/V^2$, and $\lambda = 0$. The circuit parameters are: $V^+ = 10\,V$, $V^- = -10\,V$, $R_G = 100\,k\Omega$, $R_L = 2\,k\Omega$, and $I_Q = 5\,mA$. (a) Find R_D such that $V_{DSQ} = 12\,V$. (b) Calculate g_m and R_i. (c) Determine the small-signal voltage gain $A_v = v_o/v_i$.

6.36 For the common-gate amplifier in Figure 6.37 in the text, the PMOS transistor parameters are: $V_{TP} = -2\,V$, $K_p = 2\,mA/V^2$, and $\lambda = 0$. The circuit parameters are: $V^+ = 10\,V$, $V^- = -10\,V$, $R_G = 200\,k\Omega$, and $R_L = 10\,k\Omega$. (a) Determine R_S and R_D such that $I_{DQ} = 3\,mA$ and $V_{SDQ} = 10\,V$. (b) Determine the small-signal voltage gain $A_v = v_o/v_i$.

Section 6.7   Amplifiers with MOSFET Load Devices

D6.37 Consider the NMOS amplifier with saturated load in Figure 6.39(a). The transistor parameters are: $V_{TND} = V_{TNL} = 2\,V$, $k'_n = 60\,\mu A/V^2$, $\lambda = 0$, and $(W/L)_L = 0.5$. Let $V_{DD} = 10\,V$. (a) Design the circuit such that the small-signal voltage gain is $|A_v| = 5$ and the Q-point is in the center of the saturation region. (b) Determine I_{DQ} and the dc value of v_o.

*6.38 For the NMOS amplifier with depletion load in Figure 6.43(a), the transistor parameters are: $V_{TND} = 1.2\,V$, $V_{TNL} = -2\,V$, $K_{nD} = 0.5\,mA/V^2$, $K_{nL} = 0.1\,mA/V^2$, and $\lambda_D = \lambda_L = 0.02\,V^{-1}$. Let $V_{DD} = 10\,V$. (a) Determine V_{GS} such that the Q-point is in the middle of the saturation region. (b) Calculate I_{DQ} and the dc value of v_o. (c) Determine the small-signal voltage gain.

6.39 Consider a saturated load device in which the gate and drain of an enhancement mode MOSFET are connected together. The transistor drain current becomes zero when $V_{DS} = 1.5\,V$. When $V_{DS} = 3\,V$, the drain current is $0.8\,mA$. Determine the small-signal resistance at this operating point.

6.40 The parameters of the transistors in the circuit in Figure P6.40 are $V_{TND} = -1\,V$, $K_{nD} = 0.5\,mA/V^2$ for transistor M_D, and $V_{TNL} = +1\,V$, $K_{nL} = 30\,\mu A/V^2$ for transistor M_L. Assume $\lambda = 0$ for both transistors. (a) Calculate the quiescent drain current I_{DQ} and the dc value of the output voltage. (b) Determine the small-signal voltage gain $A_v = v_o/v_i$ about the Q-point.

6.41 A source-follower circuit with a saturated load is shown in Figure P6.41. The transistor parameters are $V_{TND} = 1\,V$, $K_{nD} = 1\,mA/V^2$ for M_D, and $V_{TNL} = 1\,V$, $K_{nL} = 0.1\,mA/V^2$ for M_L. Assume $\lambda = 0$ for both transistors. Let $V_{DD} = 9\,V$. (a) Determine

Figure P6.40

Figure P6.41

V_{GG} such that the quiescent value of v_{DSL} is 4 V. (b) Show that the small-signal open-circuit ($R_L = \infty$) voltage gain about this Q-point is given by $A_v = 1/[1 + \sqrt{K_{nL}/K_{nD}}]$. (c) Calculate the small-signal voltage gain for $R_L = 4 \, k\Omega$.

6.42 For the source-follower circuit with a saturated load, as shown in Figure P6.41, assume the same transistor parameters as given in Problem 6.41. (a) Determine the small-signal voltage gain if $R_L = 10 \, k\Omega$. (b) Determine the small-signal output resistance R_o.

Section 6.8   Multistage Amplifiers

*D6.43 The transistor parameters in the circuit in Figure P6.43 are: $K_{n1} = 0.1 \, mA/V^2$, $K_{p2} = 1.0 \, mA/V^2$, $V_{TN1} = +2 \, V$, $V_{TD2} = -2 \, V$, and $\lambda_1 = \lambda_2 = 0$. The circuit parameters are: $V_{DD} = 10 \, V$, $R_{S1} = 4 \, k\Omega$, and $R_{in} = 200 \, k\Omega$. (a) Design the circuit such that $I_{DQ1} = 0.4 \, mA$, $I_{DQ2} = 2 \, mA$, $V_{DSQ1} = 4 \, V$, and $V_{SDQ2} = 5 \, V$. (b) Calculate the small-signal voltage gain $A_v = v_o/v_i$. (c) Determine the maximum symmetrical swing in the output voltage.

Figure P6.43

D6.44 The transistor parameters in the circuit in Figure P6.43 are the same as those given in Problem 6.43. The circuit parameters are: $V_{DD} = 10 \, V$, $R_{S1} = 1 \, k\Omega$, $R_{in} = 200 \, k\Omega$, $R_{D2} = 2 \, k\Omega$, and $R_{S2} = 0.5 \, k\Omega$. (a) Design the circuit such that the Q-point of M_2 is in the center of the saturation region and $I_{DQ1} = 0.4 \, mA$. (b) Determine the resulting values of I_{DQ2}, V_{SDQ2}, and V_{DSQ1}. (c) Determine the resulting small-signal voltage gain.

D6.45 Consider the circuit in Figure P6.45 with transistor parameters $K_{n1} = K_{n2} = 200\,\mu\text{A/V}^2$, $V_{TN1} = V_{TN2} = 0.8\,\text{V}$, and $\lambda_1 = \lambda_2 = 0$. (a) Design the circuit such that $V_{DSQ2} = 7\,\text{V}$ and $R_{\text{in}} = 400\,\text{k}\Omega$. (b) Determine the resulting values of I_{DQ1}, I_{DQ2}, and V_{DSQ1}. (c) Calculate the resulting small-signal voltage gain $A_v = v_o/v_i$ and the output resistance R_o.

6.46 For the circuit in Figure P6.46, the transistor parameters are: $K_{n1} = K_{n2} = 4\,\text{mA/V}^2$, $V_{TN1} = V_{TN2} = 2\,\text{V}$, and $\lambda_1 = \lambda_2 = 0$. (a) Determine I_{DQ1}, I_{DQ2}, V_{DSQ1}, and V_{DSQ2}. (b) Determine g_{m1} and g_{m2}. (c) Determine the overall small-signal voltage gain $A_v = v_o/v_i$.

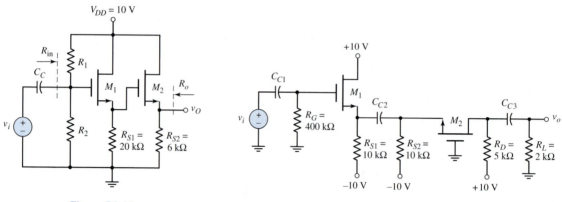

Figure P6.45   **Figure P6.46**

D6.47 For the cascode circuit in Figure 6.50 in the text, the transistor parameters are: $V_{TN1} = V_{TN2} = 1\,\text{V}$, $K_{n1} = K_{n2} = 2\,\text{mA/V}^2$, and $\lambda_1 = \lambda_2 = 0$. (a) Let $R_S = 1.2\,\text{k}\Omega$ and $R_1 + R_2 + R_3 = 500\,\text{k}\Omega$. Design the circuit such that $I_{DQ} = 3\,\text{mA}$ and $V_{DSQ1} = V_{DSQ2} = 2.5\,\text{V}$. (b) Determine the small-signal voltage gain $A_v = v_o/v_i$.

D6.48 The supply voltages to the cascode circuit in Figure 6.50 in the text are changed to $V^+ = 10\,\text{V}$ and $V^- = -10\,\text{V}$. The transistor parameters are: $K_{n1} = K_{n2} = 4\,\text{mA/V}^2$, $V_{TN1} = V_{TN2} = 1.5\,\text{V}$, and $\lambda_1 = \lambda_2 = 0$. (a) Let $R_S = 2\,\text{k}\Omega$, and assume the current in the bias resistors is 0.1 mA. Design the circuit such that $I_{DQ} = 5\,\text{mA}$ and $V_{DSQ1} = V_{DSQ2} = 3.5\,\text{V}$. (b) Determine the resulting small-signal voltage gain.

Section 6.9   Basic JFET Amplifiers

6.49 Consider the JFET amplifier in Figure 6.53 with transistor parameters $I_{DSS} = 6\,\text{mA}$, $V_P = -3\,\text{V}$, and $\lambda = 0.01\,\text{V}^{-1}$. Let $V_{DD} = 10\,\text{V}$. (a) Determine R_D and V_{GS} such that $I_{DQ} = 4\,\text{mA}$ and $V_{DSQ} = 6\,\text{V}$. (b) Determine g_m and r_o at the Q-point. (c) Determine the small-signal voltage gain $A_v = v_o/v_i$ where v_o is the time-varying portion of the output voltage v_O.

6.50 For the JFET amplifier in Figure P6.50, the transistor parameters are: $I_{DSS} = 2\,\text{mA}$, $V_P = -2\,\text{V}$, and $\lambda = 0$. Determine g_m, $A_v = v_o/v_i$, and $A_i = i_o/i_i$.

D6.51 The parameters of the transistor in the JFET common-source amplifier shown in Figure P6.51 are: $I_{DSS} = 8\,\text{mA}$, $V_P = -4.2\,\text{V}$, and $\lambda = 0$. Let $V_{DD} = 20\,\text{V}$ and $R_L = 16\,\text{k}\Omega$. Design the circuit such that $V_S = 2\,\text{V}$, $R_1 + R_2 = 100\,\text{k}\Omega$, and the Q-point is at $I_{DQ} = I_{DSS}/2$ and $V_{DSQ} = V_{DD}/2$.

Figure P6.50

Figure P6.51

*D6.52 Consider the source-follower JFET amplifier in Figure P6.52 with transistor parameters $I_{DSS} = 10\,\text{mA}$, $V_P = -5\,\text{V}$, and $\lambda = 0.01\,\text{V}^{-1}$. Let $V_{DD} = 12\,\text{V}$ and $R_L = 0.5\,\text{k}\Omega$. (a) Design the circuit such that $R_{\text{in}} = 100\,\text{k}\Omega$, and the Q-point is at $I_{DQ} = I_{DSS}/2$ and $V_{DSQ} = V_{DD}/2$. (b) Determine the resulting small-signal voltage gain $A_v = v_o/v_i$ and the output resistance R_o.

6.53 For the p-channel JFET source-follower circuit in Figure P6.53, the transistor parameters are: $I_{DSS} = 2\,\text{mA}$, $V_P = +1.75\,\text{V}$, and $\lambda = 0$. (a) Determine I_{DQ} and V_{SDQ}. (b) Determine the small-signal gains $A_v = v_o/v_i$ and $A_i = i_o/i_i$. (c) Determine the maximum symmetrical swing in the output voltage.

D6.54 The p-channel JFET common-source amplifier in Figure P6.54 has transistor parameters $I_{DSS} = 8\,\text{mA}$, $V_P = 4\,\text{V}$, and $\lambda = 0$. Design the circuit such that $I_{DQ} = 4\,\text{mA}$, $V_{SDQ} = 7.5\,\text{V}$, $A_v = v_o/v_i = -3$, and $R_1 + R_2 = 400\,\text{k}\Omega$.

Figure P6.52

Figure P6.53

Figure P6.54

COMPUTER SIMULATION PROBLEMS

6.55   Consider the circuit in Figure 6.22 with transistor parameters given in Example 6.6. Using a computer analysis, investigate the effect of the channel-length modulation parameter λ and the body-effect parameter γ on the small-signal voltage gain.

6.56   Using a computer analysis, investigate the effect of the transistor parameters λ and γ on the small-signal voltage gain and output resistance of the source-follower circuit in Figure 6.28. The circuit and transistor parameters are given in Example 6.7.

6.57   For the common-gate circuit in Figure 6.34 the circuit and transistor parameters are as given in Example 6.10. Using a computer analysis, determine the small-signal voltage gain, current gain, input resistance R_i, and output resistance (looking into the drain of the transistor). As part of the analysis, investigate the effect of the transistor parameters λ and γ on the circuit characteristics.

6.58   Perform a computer analysis of Exercise 6.22, including the body effect. Determine the change in the small-signal voltage gain when the body effect is included. If the dc output voltage is approximately 2.5 V, determine the required change in the dc bias on the driver transistor when the body effect is included.

6.59   Repeat Problem 6.58 for Exercise 6.24.

DESIGN PROBLEMS

[Note: Each design should be correlated with a computer analysis.]

*D6.60   A discrete common-source circuit with the configuration shown in Figure 6.16 is to be designed to provide a voltage gain of 20 and a symmetrical output voltage swing. The power supply voltage is $V_{DD} = 5\,\text{V}$, the output resistance of the signal source is $1\,\text{k}\Omega$, and the transistor parameters are: $V_{TN} = 0.8\,\text{V}$, $k_n' = 40\,\mu\text{A/V}^2$, and $\lambda = 0.01\,\text{V}^{-1}$. Plot W/L and R_D versus quiescent drain current. Determine W/L and R_D for $I_{DQ} = 0.1\,\text{mA}$.

*D6.61   For a common-gate amplifier in Figure 6.37 the available power supplies are $\pm 10\,\text{V}$, the output resistance of the signal source is $200\,\Omega$, and the input resistance of the amplifier is to be $200\,\Omega$. The transistor parameters are: $k_p' = 30\,\mu\text{A/V}^2$, $V_{TP} = -2\,\text{V}$, and $\lambda = 0$. The output load resistance is $R_L = 5\,\text{k}\Omega$. Design the circuit such that the output voltage has a peak-to-peak symmetrical swing of at least 5 V.

*D6.62   A source-follower amplifier with the general configuration shown in Figure 6.32 is to be designed. The available power supplies are $\pm 12\,\text{V}$, and the transistor parameters are: $V_{TN} = 1.5\,\text{V}$, $k_n' = 40\,\mu\text{A/V}^2$, and $\lambda = 0$. The load resistance is $R_L = 100\,\Omega$. Design the circuit such that $200\,\text{mW}$ of signal power is delivered to the load. As part of the design, a constant-current source circuit is also to be designed.

*D6.63   For an NMOS amplifier with a depletion load, such as shown in Figure 6.43(a), the available power supplies are $\pm 5\,\text{V}$, and the transistor parameters are: $V_{TN}(M_D) = +1\,\text{V}$, $V_{TN}(M_L) = -2\,\text{V}$, $k_n' = 40\,\mu\text{A/V}^2$, $\lambda = 0.01\,\text{V}^{-1}$, and $\gamma = 0.35\,\text{V}^{1/2}$. Design the circuit such that the small-signal voltage gain is at least $|A_v| = 200$ when the output is an open circuit. Use a constant-current source to establish the quiescent Q-point, and couple the signal source v_i directly to the gate of M_D.

*D6.64   For the cascode circuit shown in Figure 6.50, the transistor parameters are: $V_{TN} = 1\,\text{V}$, $k_n' = 40\,\mu\text{A/V}^2$, and $\lambda = 0$. Design the circuit such that the minimum open-circuit voltage gain is 10. Determine the maximum symmetrical swing in the output voltage.

7

Frequency Response

7.0 PREVIEW

Thus far in our linear amplifier analyses, we have assumed that coupling capacitors and bypass capacitors act as short circuits to the signal voltages and open circuits to dc voltages. However, capacitors do not change instantaneously from a short circuit to an open circuit as the frequency approaches zero. We have also assumed that transistors are ideal in that output signals respond instantaneously to input signals. However, there are internal capacitances in both the bipolar transistor and field-effect transistor that affect the frequency response. The major goal of this chapter is to determine the frequency response of amplifier circuits due to circuit capacitors and transistor capacitances.

Initially, we derive transfer functions, using the complex frequency s, of several passive circuits as a basic review of frequency response. We introduce Bode plots of the transfer function magnitude and phase, and a time constant technique for determining the corner, or 3 dB, frequencies of the circuit response. The goal of this analysis is to help the reader become comfortable with basic frequency response analysis and sketching Bode plots. When there is more than one capacitor in a circuit, computer simulation becomes an attractive analysis tool for determining the frequency response.

The frequency response of electronic amplifiers is analyzed, taking into account various circuit capacitors, to determine the bandwidth of the circuit. The parameters that affect low-frequency cutoff and high-frequency cutoff are determined. These parameters become important in the design of amplifiers with specified frequency response characteristics. By knowing the position or function of a capacitor in a circuit, the reader should be able to easily identify whether the low-frequency or high-frequency characteristics will be affected.

The capacitances in the transistor structure are identified and the resulting frequency response of the transistor itself is analyzed. The combination of circuit capacitors and transistor effects results in amplifiers with specified bandwidths, or frequency ranges over which the amplifier can produce output signals. Much of the chapter deals with analysis. However, the results of the analyses will aid in the design of amplifiers with specific bandwidth requirements.

7.1   AMPLIFIER FREQUENCY RESPONSE

All amplifier gain factors are functions of signal frequency. These gain factors include voltage, current, transconductance, and transresistance. Up to this point, we have assumed that the signal frequency is high enough that coupling and bypass capacitors can be treated as short circuits and, at the same time, we have assumed that the signal frequency is low enough that parasitic, load, and transistor capacitances can be treated as open circuits. In this chapter, we consider the amplifier response over the entire frequency range.

In general, an amplifier gain factor versus frequency will resemble that shown in Figure 7.1.[1] Both the gain factor and frequency are plotted on logarithmic scales (the gain factor in terms of decibels). Three frequency ranges, low, midband, and high, are indicated. In the **low-frequency range**, $f < f_L$, the gain decreases as the frequency decreases because of coupling and bypass capacitor effects. In the **high-frequency range**, $f > f_H$, stray capacitance and transistor capacitance effects cause the gain to decrease as the frequency increases. The **midband range** is the region where coupling and bypass capacitors act as short circuits, and stray and transistor capacitances act as open circuits. In this region, the gain is almost a constant. As we will show, the gain at $f = f_L$ and at $f = f_H$ is 3 dB less than the maximum midband gain. The bandwidth of the amplifier (in Hz) is defined as $f_{BW} = f_H - f_L$.

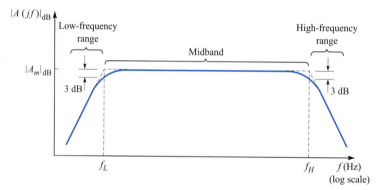

Figure 7.1   Amplifier gain versus frequency

For an audio amplifier, for example, signal frequencies in the range of $20\,\text{Hz} < f < 20\,\text{kHz}$ need to be amplified equally so as to reproduce the sound as accurately as possible. Therefore, in the design of a good audio amplifier, the frequency f_L must be designed to be less than 20 Hz and f_H must be designed to be greater than 20 kHz.

7.1.1   Equivalent Circuits

Each capacitor in a circuit is important to only one end of the frequency spectrum. For this reason, we can develop specific equivalent circuits that

---

[1]In many references, the gain is plotted as a function of the radian frequency ω. All curves in this chapter, for consistency, will be plotted as a function of cyclical frequency f (Hz). We note that $\omega = 2\pi f$. The amplifier gain is also plotted in terms of decibels (dB), where $|A|_{dB} = 20\log_{10}|A|$.

apply to the low-frequency range, to midband, and to the high-frequency range.

Midband Range

The equivalent circuits used for calculations in the midband range are the same as those considered up to this point in the text. As already mentioned, the coupling and bypass capacitors in this region are treated as short circuits. The stray and transistor capacitances are treated as open circuits. In this frequency range, there are no capacitances in the equivalent circuit. These circuits are referred to as midband equivalent circuits.

Low-Frequency Range

In this frequency range, we use a low-frequency equivalent circuit. In this range, coupling and bypass capacitors must be included in the equivalent circuit and in the amplification factor equations. The stray and transistor capacitances are treated as open circuits. The mathematical expressions obtained for the amplification factor in this frequency range must approach the midband results as f approaches the midband frequency range, since in this limit the capacitors approach short-circuit conditions.

High-Frequency Range

In the high-frequency range, we use a high-frequency equivalent circuit. In this region, coupling and bypass capacitors are treated as short circuits. The transistor and any parasitic or load capacitances must be taken into account in this equivalent circuit. The mathematical expressions obtained for the amplification factor in this frequency range must approach the midband results as f approaches the midband frequency range, since in this limit the capacitors approach open-circuit conditions.

7.1.2   Frequency Response Analysis

Using the three equivalent circuits just considered rather than a complete circuit is an approximation technique that produces useful hand-analysis results while avoiding complex transfer functions. This technique is valid if there is a large separation between f_L and f_H, that is $f_H \gg f_L$. This condition is satisfied in many electronic circuits that we will consider.

Computer simulations, such as PSpice, can take into account all capacitances and can produce frequency response curves that are more accurate than the hand-analysis results. However, the computer results do not provide any physical insight into a particular result and hence do not provide any suggestions as to design changes that can be made to improve a particular frequency response. A hand analysis can provide insight into the "whys and wherefores" of a particular response. This basic understanding can then lead to a better circuit design.

In the next section, we introduce two simple circuits to begin our frequency analysis study. We initially derive the mathematical expressions relating output voltage to input voltage (transfer function) as a function of signal frequency. From these functions, we can develop the response curves. The two frequency response curves give the magnitude of the transfer function versus frequency and the phase of the transfer function versus frequency. The phase response relates the phase of the output signal to the phase of the input signal.

We will then develop a technique by which we can easily sketch the frequency response curves without resorting to a full analysis of the transfer function. This simplified approach will lead to a general understanding of the frequency response of electronic circuits. We will then rely on a computer simulation to provide more detailed calculations when required.

7.2   SYSTEM TRANSFER FUNCTIONS

The frequency response of a circuit is usually determined by using the **complex frequency** s. Each capacitor is represented by its complex impedance, $1/sC$, and each inductor is represented by its complex impedance, sL. The circuit equations are then formulated in the usual way. Using the complex frequency, the mathematical expressions obtained for voltage gain, current gain, input impedance, or output impedance are ratios of polynomials in s.

We will be concerned in many cases with system transfer functions. These will be in the form of ratios of, for example, output voltage to input voltage (voltage transfer function) or output current to input voltage (transconductance function). The four general transfer functions are listed in Table 7.1.

Once a transfer function is found, we can find the result of a steady-state sinusoidal excitation by setting $s = j\omega = j2\pi f$. The ratio of polynomials in s then reduces to a complex number for each frequency f. The complex number can be reduced to a magnitude and a phase.

Table 7.1   Transfer functions of the complex frequency s

Name of function	Expression
Voltage transfer function	$T(s) = V_o(s)/V_i(s)$
Current transfer function	$I_o(s)/I_i(s)$
Transresistance function	$V_o(s)/I_i(s)$
Transconductance function	$I_o(s)/V_i(s)$

7.2.1   *s*-Domain Analysis

In general, a transfer function in the s-domain can be expressed in the form

$$T(s) = K\frac{(s - z_1)(s - z_2)\cdots(s - z_m)}{(s - p_1)(s - p_2)\ldots(s - p_n)} \tag{7.1}$$

where K is a constant, $z_1, z_2, \ldots, z_m$ are the transfer function "zeros," and $p_1, p_2, \ldots, p_n$ are the transfer function "poles." When the complex frequency is equal to a zero, $s = z_i$, the transfer function is zero; when the complex frequency is equal to a pole, $s = p_i$, the transfer function diverges and becomes infinite. The transfer function can be evaluated for physical frequencies by replacing s with $j\omega$. In general, the resulting transfer function $T(j\omega)$ is a complex function, that is, its magnitude and phase are both functions of frequency. These topics are usually discussed in a basic circuit analysis course.

For a simple transfer function of the form

$$T(s) = \frac{K}{s + \omega_o} \tag{7.2(a)}$$

we can rearrange the terms and write the function as

$$T(s) = K_1 \left(\frac{1}{1 + s\tau_1} \right) \tag{7.2(b)}$$

where τ_1 is a **time constant**. Other transfer functions may be written as

$$T(s) = K_2 \left(\frac{s\tau_2}{1 + s\tau_2} \right) \tag{7.2(c)}$$

where τ_2 is also a time constant. In most cases, we will write the transfer functions in terms of the time constants.

To introduce the frequency response analysis of transistor circuits, we will examine the circuits shown in Figures 7.2 and 7.3. The voltage transfer function for the circuit in Figure 7.2 can be expressed in a voltage divider format, as follows:

Figure 7.2 Series coupling capacitor circuit

$$\frac{V_o(s)}{V_i(s)} = \frac{R_P}{R_S + R_P + \dfrac{1}{sC_S}} \tag{7.3}$$

The elements R_S and C_S are in series between the input and output signals, and the element R_P is in parallel with the output signal. Equation (7.3) can be written in the form

$$\frac{V_o(s)}{V_i(s)} = \frac{sR_P C_S}{1 + s(R_S + R_P)C_S} \tag{7.4}$$

Figure 7.3 Parallel load capacitor circuit

which can be rearranged and written as

$$\frac{V_o(s)}{V_i(s)} = \left(\frac{R_P}{R_S + R_P} \right) \left[\frac{s(R_S + R_P)C_S}{1 + s(R_S + R_P)C_S} \right] = K_2 \left(\frac{s\tau}{1 + s\tau} \right) \tag{7.5}$$

In this equation, the time constant is

$$\tau = (R_S + R_P)C_S$$

Writing a Kirchhoff current law (KCL) equation at the output node, we can determine the voltage transfer function for the circuit shown in Figure 7.3, as follows:

$$\frac{V_o - V_i}{R_S} + \frac{V_o}{R_P} + \frac{V_o}{(1/sC_P)} = 0 \tag{7.6}$$

In this case, the element R_S is in series between the input and output signals, and the elements R_P and C_P are in parallel with the output signal. Rearranging the terms in Equation (7.6) produces

$$\frac{V_o(s)}{V_i(s)} = \left(\frac{R_P}{R_S + R_P} \right) \left[\frac{1}{1 + s\left(\dfrac{R_S R_P}{R_S + R_P} \right) C_P} \right] \tag{7.7(a)}$$

or

$$\frac{V_o(s)}{V_i(s)} = \left(\frac{R_P}{R_S + R_P}\right)\left[\frac{1}{1 + s(R_S \| R_P)C_P}\right] \tag{7.7(b)}$$

In Equation (7.7(b)), the time constant is

$$\tau = (R_S \| R_P)C_P$$

7.2.2   First-Order Functions

In our hand analysis of transistor circuits in this chapter, we will, in general, limit ourselves to the consideration of only one capacitance at a time. We will therefore be dealing with **first-order transfer functions** that, in most cases, will have the general form of either Equation (7.5) or (7.7(b)). This simplified analysis will allow us to present the frequency responses of specific capacitances and of the transistors themselves. We will then compare our hand analysis results with more rigorous solutions, using a computer simulation.

7.2.3   Bode Plots

A simplified technique for obtaining approximate plots of the magnitude and phase of a transfer function, given the poles and zeros or the equivalent time constants, was developed by H. Bode, and the resulting diagrams are called **Bode plots**.

Bode Plot for Figure 7.2

For the transfer function in Equation (7.5), for example, if we replace s by $j\omega$ and define a time constant τ_s as

$$\tau_s = (R_S + R_P)C_S$$

we have

$$T(j\omega) = \frac{V_o(j\omega)}{V_i(j\omega)} = \left(\frac{R_P}{R_S + R_P}\right)\left[\frac{j\omega\tau_S}{1 + j\omega\tau_S}\right] \tag{7.8}$$

The magnitude of Equation (7.8) is

$$|T(j\omega)| = \left(\frac{R_P}{R_S + R_P}\right)\left[\frac{\omega\tau_S}{\sqrt{1 + \omega^2\tau_S^2}}\right] \tag{7.9(a)}$$

or

$$|T(jf)| = \left(\frac{R_P}{R_S + R_P}\right)\left[\frac{2\pi f\tau_S}{\sqrt{1 + (2\pi f\tau_S)^2}}\right] \tag{7.9(b)}$$

We can develop the Bode plot of the gain magnitude versus frequency. We may note that $|T(jf)|_{dB} = 20\log_{10}|T(jf)|$. From Equation (7.9(b)), we can write

$$|T(jf)|_{\text{dB}} = 20 \log_{10}\left[\left(\frac{R_P}{R_S + R_P}\right) \cdot \frac{2\pi f \tau_S}{\sqrt{1 + (2\pi f \tau_S)^2}}\right] \qquad \text{(7.10(a))}$$

or

$$|T(jf)|_{\text{dB}} = 20 \log_{10}\left(\frac{R_P}{R_S + R_P}\right) + 20 \log_{10}(2\pi f \tau_S)$$
$$- 20 \log_{10}\sqrt{1 + (2\pi f \tau_S)^2} \qquad \text{(7.10(b))}$$

We can plot each term of Equation (7.10(b)) and then combine the three plots to form the final Bode plot of the gain magnitude.

Figure 7.4(a) is the plot of the first term of equation (7.10(b)), which is just a constant independent of frequency. We may note that, since $[R_P/(R_S + R_P)]$ is less than unity, the dB value is less than zero.

Figure 7.4(b) is the plot of the second term of Equation (7.10(b)). When $f = 1/2\pi\tau_S$, we have $20 \log_{10}(1) = 0$. The slopes in Bode plot magnitudes are described in units of either dB/octave or dB/decade. An **octave** means that frequency is increased by a factor of two, and a **decade** implies that the

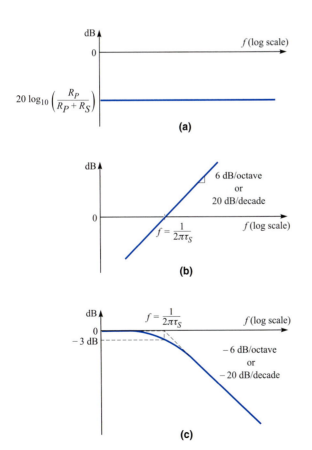

Figure 7.4 Plots of (a) the first term, (b) the second term, and (c) the third term of Equation (7.10(b))

frequency is increased by a factor of 10. The value of the function $20 \log_{10}(2\pi f\tau_S)$ increases by a factor of $6.02 \cong 6\,dB$ for every factor of 2 increase in frequency, and the value of the function increases by a factor of $20\,dB$ for every factor of 10 increase in frequency. Hence, we can consider a slope of $6\,dB$/octave or $20\,dB$/decade.

Figure 7.4(c) is the plot of the third term in Equation (7.10(b)). For $f \ll 1/2\pi\tau_S$, the value of the function is essentially $0\,dB$ and when $f = 1/2\pi\tau_S$, the value is $-3\,dB$. For $f \gg 1/2\pi\tau_S$, the function becomes $-20 \log_{10}(2\pi f\tau_S)$, so the slope becomes $-6\,dB$/octave or $-20\,dB$/decade. A straight-line projection of this slope passes through $0\,dB$ at $f = 1/2\pi\tau_S$. We can then approximate the Bode plot for this term by two straight line asymptotes intersecting at $0\,dB$ and $f = 1/2\pi\tau_S$. This particular frequency is known as a **breakpoint frequency**, **corner frequency**, or **−3 dB frequency**.

The complete Bode plot of Equation (7.10(b)) is shown in Figure 7.5. For $f \gg 1/2\pi\tau_S$, the second and third terms of Equation (7.10(b)) cancel, and for $f \ll 1/2\pi\tau_S$, the large negative dB value from Figure 7.4(b) dominates.

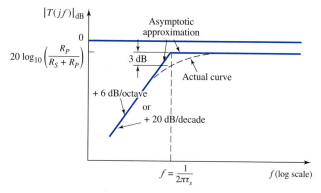

Figure 7.5   Bode plot of the voltage transfer function magnitude for the circuit in Figure 7.2

The transfer function given by Equation (7.9) is for the circuit shown in Figure 7.2. The series capacitor C_S is a coupling capacitor between the input and output signals. At a high enough frequency, capacitor C_S acts as a short circuit, and the output voltage, from a voltage divider, is

$$V_o = [R_P/(R_S + R_P)]V_i$$

For very low frequencies, the impedance of C_S approaches that of an open circuit, and the output voltage approaches zero. This circuit is called a **high-pass network** since the high-frequency signals are passed through to the output. We can now understand the form of the Bode plot shown in Figure 7.5.

The Bode plot of the phase function can be easily developed by recalling the relation between the rectangular and polar form of a complex number. We can write $A + jB = Ke^{j\theta}$, where $K = \sqrt{A^2 + B^2}$ and $\theta = \tan^{-1}(B/A)$. This relationship is shown in Figure 7.6.

For the function given in Equation (7.8), we can write the function in the form

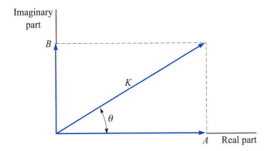

Figure 7.6   Relation between rectangular and polar forms of a complex number

$$T(jf) = \left(\frac{R_P}{R_S + R_P}\right) \cdot \left[\frac{j2\pi f \tau_S}{1 + j2\pi f \tau_S}\right]$$

$$= \left[\left|\frac{R_P}{R_S + R_P}\right| e^{j\theta_1}\right] \frac{[|j2\pi f \tau_S| e^{j\theta_2}]}{[|1 + j2\pi f \tau_S| e^{j\theta_3}]}$$ **(7.11(a))**

or

$$T(jf) = [K_1 e^{j\theta_1}]\frac{[K_2 e^{j\theta_2}]}{[K_3 e^{j\theta_3}]} = \frac{K_1 K_2}{K_3} e^{j(\theta_1 + \theta_2 - \theta_3)}$$ **(7.11(b))**

The net phase of the function $T(jf)$ is then $\theta = \theta_1 + \theta_2 - \theta_3$.

Since the first term, $[R_P/(R_S + R_P)]$, is a positive real quantity, the phase is $\theta_1 = 0$. The second term, $(j2\pi f \tau_S)$, is purely imaginary so that the phase is $\theta_2 = 90°$. The third term is complex so that its phase is $\theta_3 = \tan^{-1}(2\pi f \tau_S)$. The net phase of the function is now

$$\theta = 90 - \tan^{-1}(2\pi f \tau_S)$$ **(7.12)**

For the limiting case of $f \to 0$, we have $\tan^{-1}(0) = 0$, and for $f \to \infty$, we have $\tan^{-1}(\infty) = 90°$. At the corner frequency of $f = 1/(2\pi\tau_S)$, the phase is $\tan^{-1}(1) = 45°$. The Bode plot of the phase of the function given in Equation (7.11(a)) is given in Figure 7.7. The actual plot as well as an asymptotic approximation is shown. The phase is especially important in feedback circuits since this can influence stability. We will see this effect in Chapter 12.

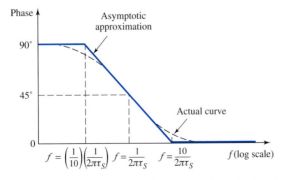

Figure 7.7   Bode plot of the voltage transfer function phase for the circuit in Figure 7.2

Bode Plot for Figure 7.3

The transfer function given by Equation (7.7(b)) is for the circuit that was shown in Figure 7.3. If we replace s by $s = j\omega = j2\pi f$ and define a time constant τ_P as $\tau_P = (R_S \| R_P)C_P$, then the transfer function is

$$T(jf) = \left(\frac{R_P}{R_S + R_P} \right) \left[\frac{1}{1 + j2\pi f \tau_P} \right] \qquad (7.13)$$

The magnitude of Equation (7.13) is

$$|T(jf)| = \left(\frac{R_P}{R_S + R_P} \right) \cdot \left[\frac{1}{\sqrt{1 + (2\pi f \tau_P)^2}} \right] \qquad (7.14)$$

A Bode plot of this magnitude expression is shown in Figure 7.8. The low-frequency asymptote is a horizontal line, and the high-frequency asymptote is a straight line with a slope of $-20\,\mathrm{dB/decade}$, or $-6\,\mathrm{dB/octave}$. The two asymptotes meet at the frequency $f = 1/2\pi\tau_P$, which is the corner, or 3 dB, frequency for this circuit. Again, the actual magnitude of the transfer function at the corner frequency differs from the maximum asymptotic value by 3 dB.

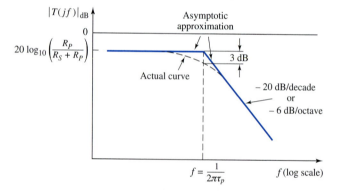

Figure 7.8   Bode plot of the voltage transfer function magnitude for the circuit in Figure 7.3

Again, the magnitude of the transfer function given by Equation (7.13) is for the circuit shown in Figure 7.3. The parallel capacitor C_P is a load, or parasitic, capacitance. At low frequencies, C_P acts as an open circuit, and the output voltage, from a voltage divider, is

$$V_o = [R_P/(R_S + R_P)]V_i$$

As the frequency increases, the magnitude of the impedance of C_P decreases and approaches that of a short circuit, and the output voltage approaches zero. This circuit is called a **low-pass network**, since the low-frequency signals are passed through to the output.

The phase of the transfer function given by Equation (7.13) is

$$\text{Phase} = -\angle \tan^{-1}(2\pi f \tau_P) \qquad (7.15)$$

The Bode plot of the phase is shown in Figure 7.9. The phase is −45 degrees at the corner frequency and 0 degrees at the low-frequency asymptote, where C_P is effectively out of the circuit.

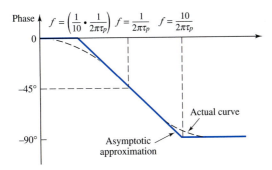

Figure 7.9 Bode plot of the voltage transfer function phase for the circuit in Figure 7.3

Example 7.1 **Objective:** Determine the corner frequencies and maximum-magnitude asymptotes of the Bode plots for a specified circuit.

For the circuits in Figures 7.2 and 7.3, the parameters are: $R_S = 1\,k\Omega$, $R_P = 10\,k\Omega$, $C_S = 1\,\mu F$, and $C_P = 3\,pF$.

Solution: (Figure 7.2) The time constant is

$$\tau_S = (R_S + R_P)C_S = (10^3 + 10 \times 10^3)(10^{-6}) = 1.1 \times 10^{-2}\,s$$

or

$$\tau_S = 11\,ms$$

The corner frequency of the Bode plot shown in Figure 7.5 is then

$$f = \frac{1}{2\pi\tau_S} = \frac{1}{(2\pi)(11 \times 10^{-3})} = 14.5\,Hz$$

The maximum magnitude is

$$\frac{R_P}{R_S + R_P} = \frac{10}{1 + 10} = 0.909$$

or

$$20\log_{10}\left(\frac{R_P}{R_S + R_P}\right) = -0.828\,dB$$

Solution: (Figure 7.3) The time constant is

$$\tau_P = (R_S \| R_P)C_P = (10^3 \| (10 \times 10^3))(3 \times 10^{-12}) = 2.73 \times 10^{-9}\,s$$

or

$$\tau_P = 2.73\,ns$$

The corner frequency of the Bode plot in Figure 7.8 is then

$$f = \frac{1}{2\pi\tau_P} = \frac{1}{(2\pi)(2.73 \times 10^{-9})} \Rightarrow 58.3\,MHz$$

The maximum magnitude is the same as just calculated: 0.909 or −0.828 dB.

Comment: Since the two capacitance values are substantially different, the two time constants differ by orders of magnitude, which means that the two corner frequencies also differ by orders of magnitude. Later in this text, we will take advantage of these differences in our analysis of transistor circuits.

Test Your Understanding

7.1   For the circuit shown in Figure 7.2, the parameters are $R_S = R_P = 4\,\text{k}\Omega$. (a) If the corner frequency is $f = 20\,\text{Hz}$, determine the value of C_S. (b) Find the magnitude of the transfer function at $f = 40\,\text{Hz}$, $80\,\text{Hz}$, and $200\,\text{Hz}$. (Ans. (a) $C_S = 0.995\,\mu\text{F}$ (b) $|T(j\omega)| = 0.447, 0.485,$ and 0.498)

7.2   Consider the circuit shown in Figure 7.3 with parameters $R_S = R_P = 10\,\text{k}\Omega$. If the corner frequency is $f = 500\,\text{kHz}$, determine the value of C_P. (Ans. $C_P = 63.7\,\text{pF}$)

7.2.4   Short-Circuit and Open-Circuit Time Constants

The two circuits shown in Figures 7.2 and 7.3 each contain only one capacitor. The circuit in Figure 7.10 is the same basic configuration but contains both capacitors. Capacitor C_S is the coupling capacitor and is in series with the input and output; capacitor C_P is the load capacitor and is in parallel with the output and ground.

Figure 7.10   Circuit with both a series coupling and a parallel load capacitor

We can determine the voltage transfer function of this circuit by writing a KCL equation at the output node. The result is

$$\frac{V_o(s)}{V_i(s)} = \left(\frac{R_P}{R_S + R_P}\right) \times \frac{1}{\left[1 + \left(\dfrac{R_P}{R_S + R_P}\right)\left(\dfrac{C_P}{C_S}\right) + \dfrac{1}{s\tau_S} + s\tau_P\right]} \quad (7.16)$$

where τ_S and τ_P are the same time constants as previously defined.

Although Equation (7.16) is the exact transfer function, it is awkward to deal with in this form.

We have seen in the previous analysis, however, that C_S affects the low-frequency response and C_P affects the high-frequency response. Further, if $C_P \ll C_S$ and if R_S and R_P are of the same order of magnitude, then the corner frequencies of the Bode plots created by C_S and C_P will differ by orders of magnitude. (We actually encounter this situation in real circuits.) Consequently, when a circuit contains both coupling and load capacitors, and when the values of the capacitors differ by orders of magnitude, then we can determine the effect of each capacitor individually.

At low frequencies, we can treat the load capacitor C_P as an open circuit. To find the equivalent resistance seen by a capacitor, set all independent sources equal to zero. Therefore, the effective resistance seen by C_S is the series combination of R_S and R_P. The time constant associated with C_S is

$$\tau_S = (R_S + R_P)C_S \qquad (7.17)$$

Since C_P was made an open circuit, τ_S is called an **open-circuit time constant**. The subscript S is associated with the coupling capacitor, or the capacitor in series with the input and output signals.

At high frequencies, we can treat the coupling capacitor C_S as a short circuit. The effective resistance seen by C_P is the parallel combination of R_S and R_P, and the associated time constant is

$$\tau_P = (R_S \| R_P)C_P \qquad (7.18)$$

which is called the **short-circuit time constant**. The subscript P is associated with the load capacitor, or the capacitor in parallel with the output and ground.

We can now define the corner frequencies of the Bode plot. The **lower corner**, or 3 dB frequency, which is at the low end of the frequency scale, is a function of the open-circuit time constant and is defined as

$$f_L = \frac{1}{2\pi\tau_S} \qquad (7.19(a))$$

The **upper corner**, or 3 dB, frequency, which is at the high end of the frequency scale, is a function of the short-circuit time constant and is defined as

$$f_H = \frac{1}{2\pi\tau_P} \qquad (7.19(b))$$

The resulting Bode plot of the magnitude of the voltage transfer function for the circuit in Figure 7.9 is shown in Figure 7.11.

This Bode plot is for a passive circuit; the Bode plots for transistor amplifiers are similar. The amplifier gain is constant over a wide frequency range, called the **midband**. In this frequency range, all capacitance effects are negligible and can be neglected in the gain calculations. At the high end of the frequency spectrum, the gain drops as a result of the load capacitance and, as we will see later, the transistor effects. At the low end of the frequency spectrum, the gain decreases because coupling capacitors and bypass capacitors do not act as perfect short circuits.

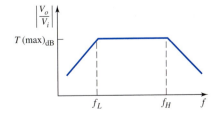

Figure 7.11 Bode plot of the voltage transfer function magnitude for the circuit in Figure 7.10

The midband range, or **bandwidth**, is defined by the corner frequencies f_L and f_H, as follows:

$$f_{BW} = f_H - f_L \tag{7.20}$$

Since $f_H \gg f_L$, as we have seen in our examples, the bandwidth is essentially given by

$$f_{BW} \cong f_H \tag{7.21}$$

Example 7.2   Objective: Determine the corner frequencies and bandwidth of a passive circuit containing two capacitors.

Consider the circuit shown in Figure 7.10 with parameters $R_S = 1\,k\Omega$, $R_P = 10\,k\Omega$, $C_S = 1\,\mu F$, and $C_P = 3\,pF$.

Solution: Since C_P is less than C_S by approximately six orders of magnitude, we can treat the effect of each capacitor separately. The open-circuit time constant is

$$\tau_S = (R_S + R_P)C_S = (10^3 + 10 \times 10^3)(10^{-6}) = 1.1 \times 10^{-2}\,s$$

and the short-circuit time constant is

$$\tau_P = (R_S \| R_P)C_P = [10^3 \| (10 \times 10^3)](3 \times 10^{-12}) = 2.73 \times 10^{-9}\,s$$

The corner frequencies are then

$$f_L = \frac{1}{2\pi\tau_S} = \frac{1}{2\pi(1.1 \times 10^{-2})} = 14.5\,Hz$$

and

$$f_H = \frac{1}{2\pi\tau_P} = \frac{1}{2\pi(2.73 \times 10^{-9})} \Rightarrow 58.3\,MHz$$

Finally, the bandwidth is

$$f_{BW} = f_H - f_L = 58.3\,MHz - 14.5\,Hz \cong 58.3\,MHz$$

Comment: The corner frequencies in this example are exactly the same as those determined in Example 7.1. This occurred because the two corner frequencies are far apart. The maximum magnitude of the voltage transfer function is again

$$\frac{R_P}{R_S + R_P} = \frac{10}{1 + 10} = 0.909 \Rightarrow -0.828\,dB$$

The Bode plot of the magnitude of the voltage transfer function is shown in Figure 7.12.

We will continue, in the following sections of the chapter, to use the concept of open-circuit and short-circuit time constants to determine the corner frequencies of the Bode plots of transistor circuits. An implicit assumption in this technique is that coupling and load capacitance values differ by many orders of magnitude.

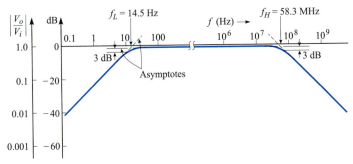

Figure 7.12 Bode plot of the magnitude of the voltage transfer function for the circuit in Figure 7.10

Test Your Understanding

7.3 For the equivalent circuit shown in Figure 7.13, the parameters are: $R_S = 1\,\text{k}\Omega$, $r_\pi = 2\,\text{k}\Omega$, $R_L = 4\,\text{k}\Omega$, $g_m = 50\,\text{mA/V}$, and $C_C = 1\,\mu\text{F}$. (a) Determine the expression for the circuit time constant. (b) Calculate the 3 dB frequency and maximum gain asymptote. (c) Sketch the Bode plot of the transfer function magnitude. (Ans. (a) $\tau = (r_\pi + R_S)C_C$, (b) $f_{3\,\text{dB}} = 53.1\,\text{Hz}$, $|T(j\omega)|_{\text{max}} = 133$)

Figure 7.13 Figure for Exercise 7.3

Figure 7.14 Figure for Exercise 7.4

7.4 The equivalent circuit in Figure 7.14 has circuit parameters $R_S = 0.5\,\text{k}\Omega$, $r_\pi = 1.5\,\text{k}\Omega$, $g_m = 75\,\text{mA/V}$, $R_L = 5\,\text{k}\Omega$, and $C_L = 10\,\text{pF}$. (a) Determine the expression for the circuit time constant. (b) Calculate the 3 dB frequency and maximum gain asymptote. (c) Sketch the Bode plot of the transfer function magnitude. (Ans. (a) $\tau = R_L C_L$, (b) $f_{3\,\text{dB}} = 3.18\,\text{MHz}$, $|T(j\omega)|_{\text{max}} = 281$)

7.5 The value of R_S in the circuit in Figure 7.10 is $R_S = 1\,\text{k}\Omega$. The midband gain is $-1\,\text{dB}$, and the corner frequencies are $f_L = 100\,\text{Hz}$ and $f_H = 1\,\text{MHz}$. (a) Determine R_P, C_S, and C_P. (b) Determine the open-circuit and short-circuit time constants. (Ans. (a) $R_P = 8.17\,\text{k}\Omega$, $C_S = 0.174\,\mu\text{F}$, $C_P = 179\,\text{pF}$, (b) $\tau_S = 1.60\,\text{ms}$, $\tau_P = 0.160\,\mu\text{s}$)

7.6 The parameters in the circuit in Figure 7.15 are $R_S = 0.25\,\text{k}\Omega$, $r_\pi = 2\,\text{k}\Omega$, $R_L = 4\,\text{k}\Omega$, $g_m = 65\,\text{mA/V}$, $C_C = 2\,\mu\text{F}$, and $C_L = 50\,\text{pF}$. (a) Find the open-circuit and short-

Figure 7.15 Figure for Exercise 7.6

circuit time constants. (b) Calculate the midband voltage gain. (c) Determine the upper and lower 3 dB frequencies. (d) Verify the results with a PSpice analysis. (Ans. (a) $\tau_S = 4.5\,\text{ms}$, $\tau_P = 0.2\,\mu\text{s}$, (b) $A_v = -231$, (c) $f_L = 35.4\,\text{Hz}$, $f_H = 0.796\,\text{MHz}$)

7.3 FREQUENCY RESPONSE: TRANSISTOR AMPLIFIERS WITH CIRCUIT CAPACITORS

In this section, we will analyze the basic single-stage amplifier that includes circuit capacitors. Three types of capacitors will be considered: coupling capacitor, load capacitor, and bypass capacitor. In our hand analysis, we will consider each type of capacitor individually and determine its frequency response. In the last part of this section, we will consider the effect of multiple capacitors using a PSpice analysis.

The frequency response of multistage circuits will be considered in Chapter 12 when the stability of amplifiers is considered.

7.3.1 Coupling Capacitor Effects

Input Coupling Capacitor: Common-Emitter Circuit

Figure 7.16(a) shows a bipolar common-emitter circuit with a coupling capacitor. Figure 7.16(b) shows the corresponding small-signal equivalent circuit, with the transistor small-signal output resistance r_o assumed to be infinite. This assumption is valid since $r_o \gg R_C$ and $r_o \gg R_E$ in most cases. Initially, we will use a current–voltage analysis to determine the frequency response of the circuit. Then, we will use the equivalent time constant technique.

Figure 7.16 (a) Common-emitter circuit with coupling capacitor and (b) small-signal equivalent circuit

From the analysis in the previous section, we note that this circuit is a high-pass network. At high frequencies, the capacitor C_C acts as a short circuit, and the input signal is coupled through the transistor to the output. At low frequencies, the impedance of C_C becomes large and the output approaches zero.

Current–Voltage Analysis The input current can be written as

$$I_i = \frac{V_i}{R_{Si} + \dfrac{1}{sC_C} + R_i} \tag{7.22}$$

where the input resistance R_i is given by

$$R_i = R_B \| [r_\pi + (1+\beta)R_E] = R_B \| R_{ib} \tag{7.23}$$

In writing Equation (7.23), we used the resistance reflection rule given in Chapter 3. To determine the input resistance to the base of the transistor, we multiplied the emitter resistance by the factor $(1+\beta)$.

Using a current divider, we determine the base current to be

$$I_b = \left(\frac{R_B}{R_B + R_{ib}} \right) I_i \tag{7.24}$$

and then

$$V_\pi = I_b r_\pi \tag{7.25}$$

The output voltage is given by

$$V_o = -g_m V_\pi R_C \tag{7.26}$$

Combining Equations (7.22) through (7.26) produces

$$V_o = -g_m R_C (I_b r_\pi) = -g_m r_\pi R_C \left(\frac{R_B}{R_B + R_{ib}} \right) I_i$$

$$= -g_m r_\pi R_C \left(\frac{R_B}{R_B + R_{ib}} \right) \left(\frac{V_i}{R_{Si} + \dfrac{1}{sC_C} + R_i} \right) \tag{7.27}$$

Therefore, the small-signal voltage gain is

$$A_v(s) = \frac{V_o(s)}{V_i(s)} = -g_m r_\pi R_C \left(\frac{R_B}{R_B + R_{ib}} \right) \left(\frac{sC_C}{1 + s(R_{Si} + R_i)C_C} \right) \tag{7.28}$$

which can be written in the form

$$A_v(s) = \frac{V_o(s)}{V_i(s)} = \frac{-g_m r_\pi R_C}{(R_{Si} + R_i)} \left(\frac{R_B}{R_B + R_{ib}} \right) \left(\frac{s\tau_S}{1 + s\tau_S} \right) \tag{7.29}$$

where the time constant is

$$\tau_S = (R_{Si} + R_i)C_C \tag{7.30}$$

The form of the voltage transfer function as given in Equation (7.29) is the same as that of Equation (7.5), for the coupling capacitor circuit in Figure 7.2. The Bode plot is therefore similar to that shown in Figure 7.5. The corner frequency is

$$f_L = \frac{1}{2\pi\tau_S} = \frac{1}{2\pi(R_{Si} + R_i)C_C} \tag{7.31}$$

and the maximum magnitude, in decibels, is

$$|A_v(\text{max})|_{dB} = 20 \log_{10} \left(\frac{g_m r_\pi R_C}{R_{Si} + R_i} \right) \left(\frac{R_B}{R_B + R_{ib}} \right) \tag{7.32}$$

Example 7.3 **Objective:** Calculate the corner frequency and maximum gain of a bipolar common-emitter circuit with a coupling capacitor.

For the circuit shown in Figure 7.16, the parameters are: $R_1 = 51.2\,\text{k}\Omega$, $R_2 = 9.6\,\text{k}\Omega$, $R_C = 2\,\text{k}\Omega$, $R_E = 0.4\,\text{k}\Omega$, $R_{Si} = 0.1\,\text{k}\Omega$, $C_C = 1\,\mu\text{F}$, and $V_{CC} = 10\,\text{V}$. The transistor parameters are: $V_{BE}(\text{on}) = 0.7\,\text{V}$, $\beta = 100$, and $V_A = \infty$.

Solution: From a dc analysis, the quiescent collector current is $I_{CQ} = 1.81\,\text{mA}$. The transconductance is therefore

$$g_m = \frac{I_{CQ}}{V_T} = \frac{1.81}{0.026} = 69.6\,\text{mA/V}$$

and the diffusion resistance is

$$r_\pi = \frac{\beta V_T}{I_{CQ}} = \frac{(100)(0.026)}{1.81} = 1.44\,\text{k}\Omega$$

The input resistance is

$$R_i = R_1 \| R_2 \| [r_\pi + (1 + \beta)R_E]$$
$$= 51.2 \| 9.6 \| [1.44 + (101)(0.4)] = 6.77\,\text{k}\Omega$$

and the time constant is therefore

$$\tau_S = (R_{Si} + R_i)C_C = (0.1 \times 10^3 + 6.77 \times 10^3)(1 \times 10^{-6}) = 6.87 \times 10^{-3}\,\text{s}$$

or

$$\tau_S = 6.87\,\text{ms}$$

The corner frequency is

$$f_L = \frac{1}{2\pi\tau_S} = \frac{1}{2\pi(6.87 \times 10^{-3})} = 23.2\,\text{Hz}$$

Finally, the maximum voltage gain magnitude is

$$|A_v|_{\max} = \frac{g_m r_\pi R_C}{(R_{Si} + R_i)}\left(\frac{R_B}{R_B + R_{ib}}\right)$$

where

$$R_{ib} = r_\pi + (1 + \beta)R_E = 1.44 + (101)(0.4) = 41.8\,\text{k}\Omega$$

Therefore,

$$|A_v|_{\max} = \frac{(69.6)(1.44)}{(0.1 + 6.77)}\left(\frac{8.08}{8.08 + 41.8}\right) = 4.73$$

Comment: The coupling capacitor produces a high-pass network. In this circuit, if the signal frequency is approximately two octaves above the corner frequency, the coupling capacitor acts as a short circuit.

Time Constant Technique In general, we do not need to derive the complete circuit transfer function including capacitance effects in order to complete the Bode plot and determine the frequency response. By looking at a circuit with, initially, only one capacitor, we can determine if the amplifier is a low-pass or high-pass circuit. We can then specify the Bode plot if we know the time constant and the maximum midband gain. The time constant determines the corner frequency. The midband gain is found in the usual way when capacitances are eliminated from the circuit.

This time constant technique yields good results when all poles are real, as will be the case in this chapter. In addition, this technique does not determine the corner frequencies due to system zeros. The major benefit of using the time constant approach is that it gives infomation about which circuit elements affect the $-3\,\text{dB}$ frequency of the circuit. A coupling capacitor produces a high-pass network, so the form of the Bode plot will be the same as that shown in Figure 7.5. Also, the maximum gain is determined when the coupling capacitor acts as a short circuit, as was assumed in Chapters 4 and 6.

The time constant for the circuit is a function of the equivalent resistance seen by the capacitor. The small-signal equivalent circuit is shown in Figure 7.16(b). If we set the independent voltage source equal to zero, the equivalent resistance seen by the coupling capacitor C_C is $(R_{Si} + R_i)$. The time constant is then

$$\tau_S = (R_{Si} + R_i)C_C \tag{7.33}$$

This is the same as Equation (7.30), which was determined by using a current–voltage analysis.

Output Coupling Capacitor: Common-Source Circuit

Figure 7.17(a) shows a common-source MOSFET amplifier. We assume that the resistance of the signal generator is much less than R_G and can therefore be neglected. In this case, the output signal is connected to the load through a coupling capacitor.

Figure 7.17 (a) Common-source circuit with output coupling capacitor and (b) small-signal equivalent circuit

The small-signal equivalent circuit, assuming r_o is infinite, is shown in Figure 7.17(b). The maximum output voltage, assuming C_C is a short circuit, is

$$|V_o|_{\max} = g_m V_{gs}(R_D\|R_L) \tag{7.34}$$

and the input voltage can be written as

$$V_i = V_{gs} + g_m R_S V_{gs} \tag{7.35}$$

Therefore, the maximum small-signal gain is

$$|A_v|_{\max} = \frac{g_m(R_D \| R_L)}{1 + g_m R_S} \tag{7.36}$$

Even though the coupling capacitor is in the output portion of the circuit, the Bode plot will still be that of a high-pass network, as shown in Figure 7.5. Using the time constant technique to determine the corner frequency will substantially simplify the circuit analysis, since we do not specifically need to determine the transfer function for the frequency response.

The time constant is a function of the effective resistance seen by capacitor C_C, which is determined by setting all independent sources equal to zero. Since $V_i = 0$, then $V_{gs} = 0$ and $g_m V_{gs} = 0$, and the effective resistance seen by C_C is $(R_D + R_L)$. The time constant is then

$$\tau_S = (R_D + R_L)C_C \tag{7.37}$$

and the corner frequency is $f_L = 1/2\pi\tau_S$.

Design Example 7.4   Objective: The circuit in Figure 7.17(a) is to be used as a simple audio amplifier. Design the circuit such that the lower corner frequency is $f_L = 20\,\text{Hz}$.

Solution: The corner frequency can be written in terms of the time constant, as follows:

$$f_L = \frac{1}{2\pi\tau_S}$$

The time constant is then

$$\tau_S = \frac{1}{2\pi f} = \frac{1}{2\pi(20)} \Rightarrow 7.96\,\text{ms}$$

Therefore, from Equation (7.37) the coupling capacitance is

$$C_C = \frac{\tau_S}{R_D + R_L} = \frac{7.96 \times 10^{-3}}{6.7 \times 10^3 + 10 \times 10^3} = 4.77 \times 10^{-7}\,\text{F}$$

or

$$C_C = 0.477\,\mu\text{F}$$

 Comment: Using the time constant technique to find the corner frequency is substantially easier than using the circuit analysis approach.

Test Your Understanding

7.7   For the circuit in Figure 7.16(a), the parameters are: $R_{Si} = 0.1\,\text{k}\Omega$, $R_1 = 20\,\text{k}\Omega$, $R_2 = 2.2\,\text{k}\Omega$, $R_E = 0.1\,\text{k}\Omega$, $R_C = 2\,\text{k}\Omega$, $C_C = 47\,\mu\text{F}$, and $V_{CC} = 10\,\text{V}$. The transistor parameters are: $V_{BE}(\text{on}) = 0.7\,\text{V}$, $\beta = 200$, and $V_A = \infty$. (a) Determine the expression for the time constant τ_S. (b) Determine the corner frequency and midband voltage gain. (Ans. (a) $\tau_S = (R_i + R_{Si})C_C$, (b) $f = 1.76\,\text{Hz}$, $A_v = -17.2$)

D7.8   Consider the circuit shown in Figure 7.17(a) with transistor parameters $V_{TN} = 2\,\text{V}$, $K_n = 0.5\,\text{mA/V}^2$, and $\lambda = 0$. The circuit parameters are $R_G = 50\,\text{k}\Omega$ and $R_L = 10\,\text{k}\Omega$. (a) Determine new values of R_S and R_D such that $I_{DQ} = 0.8\,\text{mA}$ and the quiescent drain voltage is $V_{DQ} = 0$. (b) Find the required value of C_C for a corner frequency of $f = 20\,\text{Hz}$. (Ans. (a) $R_S = 2.18\,\text{k}\Omega$, $R_D = 6.25\,\text{k}\Omega$, (b) $C_C = 0.49\,\mu\text{F}$)

Output Coupling Capacitor: Emitter-Follower Circuit   An emitter follower with a coupling capacitor in the output portion of the circuit is shown in Figure 7.18(a). We assume that coupling capacitor C_{C1}, which is part of the original emitter follower, is very large, and that it acts as a short circuit to the input signal.

(a)   (b)

Figure 7.18   (a) Emitter-follower circuit with output coupling capacitor and (b) small-signal equivalent circuit

The small-signal equivalent circuit, including the small-signal transistor resistance r_o, is shown in Figure 7.18(b). The equivalent resistance seen by the coupling capacitor C_{C2} is $[R_o + R_L]$, and the time constant is

$$\tau_S = [R_o + R_L]C_{C2} \tag{7.38}$$

where R_o is the output resistance as defined in Figure 7.18(b). As shown in Chapter 4, the output resistance is

$$R_o = R_E \| r_o \| \left\{ \frac{[r_\pi + (R_S \| R_B)]}{1 + \beta} \right\} \tag{7.39}$$

If we combine Equations (7.39) and (7.38), the time constant expression becomes fairly complicated. However, the current–voltage analysis of this circuit including C_{C2} is even more cumbersome. The time constant technique again simplifies the analysis substantially.

Example   7.5   **Objective:** Determine the 3 dB frequency of an emitter-follower amplifier circuit with an output coupling capacitor.

Consider the circuit shown in Figure 7.18(a) with transistor parameters $\beta = 100$, $V_{BE}(\text{on}) = 0.7\,\text{V}$, and $V_A = 120\,\text{V}$. The output coupling capacitance is $C_{C2} = 1\,\mu\text{F}$.

Solution: A dc analysis shows that $I_{CQ} = 0.838\,\text{mA}$. Therefore, the small-signal parameters are: $r_\pi = 3.10\,\text{k}\Omega$, $g_m = 32.2\,\text{mA/V}$, and $r_o = 143\,\text{k}\Omega$.

From Equation (7.39), the output resistance R_o of the emitter follower is

$$R_o = R_E \| r_o \| \left\{ \frac{[r_\pi + (R_S \| R_B)]}{1 + \beta} \right\}$$

$$= 10 \| 143 \| \left\{ \frac{[3.10 + (0.5 \| 100)]}{101} \right\} = 10 \| 143 \| 0.0356\,\text{k}\Omega$$

or

$$R_o \cong 35.5\,\Omega$$

From Equation (7.38), the time constant is

$$\tau_S = [R_o + R_L]C_{C2} = [35.5 + 10^4](10^{-6}) \cong 1 \times 10^{-2}\,\text{s}$$

The 3 dB frequency is then

$$f_L = \frac{1}{2\pi\tau_S} = \frac{1}{2\pi(10^{-2})} = 15.9\,\text{Hz}$$

Comment: Determining the 3 dB or corner frequency is very direct with the time constant technique.

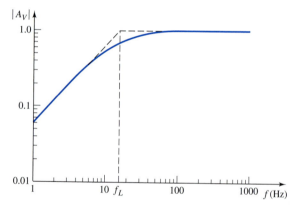

Figure 7.19   PSpice analysis results for the emitter-follower circuit in Figure 7.18(a)

Computer Verification: Based on a PSpice analysis, Figure 7.19 is a Bode plot of the voltage gain magnitude of the emitter-follower circuit shown in Figure 7.18(a). The corner frequency is essentially identical to that obtained by the time constant technique. Also, the asymptotic value of the small-signal voltage gain is $A_v = 0.988$, as expected for an emitter-follower circuit.

Problem-Solving Technique: Bode Plot of Gain Magnitude

1. For a particular capacitor in a circuit, determine whether the capacitor is producing a low-pass or high-pass circuit. From this, sketch the general shape of the Bode plot.

2. The corner frequency is found from $f = 1/2\pi\tau$ where the time constant is $\tau = R_{eq}C$. The equivalent resistance R_{eq} is the equivalent resistance seen by the capacitor.
3. The maximum gain magnitude is the midband gain. Coupling and bypass capacitors act as short circuits and load capacitors act as open circuits.

Test Your Understanding

7.9 For the emitter-follower circuit shown in Figure 7.18(a), determine the required value of C_{C2} to yield a corner frequency of 10 Hz. (Ans. $C_{C2} = 1.59\,\mu F$)

7.3.2 Load Capacitor Effects

An amplifier output may be connected to a load or to the input or another amplifier. The model of the load circuit input impedance is generally a capacitance in parallel with a resistance. In addition, there is a parasitic capacitance between ground and the line that connects the amplifier output to the load circuit.

Figure 7.20(a) shows a MOSFET common-source amplifier with a load resistance R_L and a load capacitance C_L connected to the output, and Figure 7.20(b) shows the small-signal equivalent circuit. The transistor small-signal output resistance r_o is assumed to be infinite. This circuit configuration is essentially the same as that in Figure 7.3, which is a low-pass network. At high frequencies, the impedance of C_L decreases and acts as a shunt between the output and ground, and the output voltage tends toward zero. The Bode plot is similar to that shown in Figure 7.8, with an upper corner frequency and a maximum gain asymptote.

The equivalent resistance seen by the load capacitor C_L is $R_D \| R_L$. Since we set $V_i = 0$, then $g_m V_{sg} = 0$, which means that the dependent current source does not affect the equivalent resistance.

The time constant for this circuit is

$$\tau_P = (R_D \| R_L)C_L \tag{7.40}$$

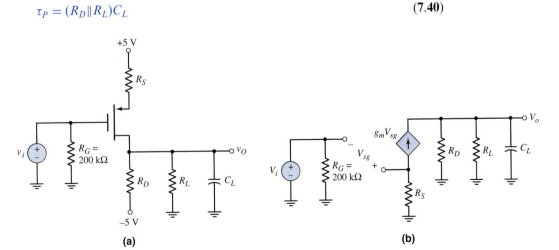

Figure 7.20 (a) MOSFET common-source circuit with a load capacitor and (b) small-signal equivalent circuit

The maximum gain asymptote, which is found by assuming C_L is an open circuit, is

$$|A_v|_{\max} = \frac{g_m(R_D \| R_L)}{1 + g_m R_S} \tag{7.41}$$

Example 7.6   Objective: Determine the corner frequency and maximum gain asymptote of a MOSFET amplifier.

For the circuit in Figure 7.20(a), the parameters are: $R_S = 3.2\,\text{k}\Omega$, $R_D = 10\,\text{k}\Omega$, $R_L = 20\,\text{k}\Omega$, and $C_L = 10\,\text{pF}$. The transistor parameters are: $V_{TP} = -2\,\text{V}$, $K_p = 0.25\,\text{mA/V}^2$, and $\lambda = 0$.

Solution: From the dc analysis, we find that $I_{DQ} = 0.5\,\text{mA}$, $V_{SGQ} = 3.41\,\text{V}$, and $V_{SDQ} = 3.41\,\text{V}$. The transconductance is therefore

$$g_m = 2K_p(V_{SG} + V_{TP}) = 2(0.25)(3.41 - 2) = 0.705\,\text{mA/V}$$

From Equation (7.40), the time constant is

$$\tau_P = (R_D \| R_L)C_L = ((10 \times 10^3)\|(20 \times 10^3))(10 \times 10^{-12}) = 6.67 \times 10^{-8}\,\text{s}$$

or

$$\tau_P = 66.7\,\text{ns}$$

Therefore, the corner frequency is

$$f_H = \frac{1}{2\pi\tau_P} = \frac{1}{2\pi(66.7 \times 10^{-9})} \Rightarrow 2.39\,\text{MHz}$$

Finally, from Equation (7.41), the maximum gain asymptote is

$$|A_v|_{\max} = \frac{g_m(R_D \| R_L)}{1 + g_m R_S} = \frac{(0.705)(10\|20)}{1 + (0.705)(3.2)} = 1.44$$

Comment: The Bode plot for this circuit is similar to that in Figure 7.8, and it represents a low-pass network. The relatively large value of R_S results in a low voltage gain.

Computer Simulation: Figure 7.21 shows the results of a PSpice analysis of the circuit given in Figure 7.20(a). Figure 7.21(a) is a Bode plot of the voltage gain magnitude. The

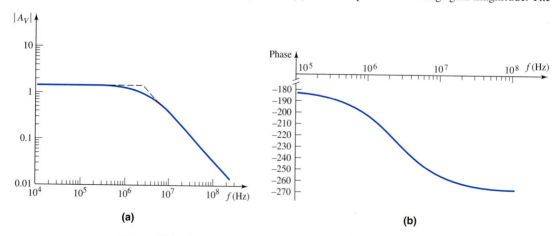

(a)   **(b)**

Figure 7.21   PSpice analysis results for the circuit in Figure 7.20(a): (a) voltage gain magnitude response, and (b) phase response midband gain is 1.44 and the corner frequency is 2.4 MHz, which agrees extremely well with the hand analysis results. The phase of the voltage gain is shown in Figure 7.21(b). The midband phase is -180 degrees, as expected. Also, as the frequency increases, the phase approaches -270 degrees. As was shown in Figure 7.9, a phase change of -90 degrees is expected for a load capacitor.

Test Your Understanding

D7.10   The PMOS common-source circuit shown in Figure 7.20(a) has a load resistance $R_L = 10\,\text{k}\Omega$. The transistor parameters are: $V_{TN} = -2\,\text{V}$, $K_p = 0.5\,\text{mA/V}^2$, and $\lambda = 0$. (a) Design the circuit such that $I_{DQ} = 1\,\text{mA}$ and $V_{SDQ} = V_{SGQ}$. (b) Determine the value of C_L such that the corner frequency is $f = 1\,\text{MHz}$. (Ans. (a) $R_S = 1.59\,\text{k}\Omega$, $R_D = 5\,\text{k}\Omega$, (b) $C_L = 47.7\,\text{pF}$)

7.3.3   Coupling and Load Capacitors

A circuit with both a coupling capacitor and a load capacitor is shown in Figure 7.22(a). Since the values of the coupling capacitance and load capacitance differ by orders of magnitude, the corner frequencies are far apart and can be treated separately as discussed previously. The small-signal equivalent circuit is shown in Figure 7.22(b), assuming the transistor small-signal resistance r_o is infinite.

The Bode plot of the voltage gain magnitude is similar to that shown in Figure 7.11. The lower corner frequency f_L is given by

$$f_L = \frac{1}{2\pi\tau_S} \tag{7.42}$$

where τ_S is the time constant associated with the coupling capacitor C_C, and the upper corner frequency f_H is given by

$$f_H = \frac{1}{2\pi\tau_P} \tag{7.43}$$

where τ_P is the time constant associated with the load capacitor C_L. It should be emphasized that Equations (7.42) and (7.43) are valid only as long as the two corner frequencies are far apart.

Using the small-signal equivalent circuit in Figure 7.22(b), we set the signal source equal to zero to find the equivalent resistance associated with the coupling capacitor. The related time constant is

$$\tau_S = [R_S + (R_1\|R_2\|R_i)]C_C \tag{7.44}$$

where

$$R_i = r_\pi + (1+\beta)R_E \tag{7.45}$$

Similarly, the time constant related to C_L is

$$\tau_P = (R_C\|R_L)C_L \tag{7.46}$$

(a)

(b)

Figure 7.22   (c) Circuit with both a coupling and a load capacitor and (b) small-signal equivalent circuit

Since the two corner frequencies are far apart, the gain will reach a maximum value in the frequency range between f_L and f_H, which is the midband. We can calculate the midband gain by assuming that the coupling capacitor is a short circuit and the load capacitor is an open circuit.

From Figure 7.22(b), we see that in midband we have

$$I_i = \frac{V_i}{R_S + R_1 \| R_2 \| R_i} \tag{7.47}$$

and

$$I_b = \left(\frac{R_1 \| R_2}{(R_1 \| R_2) + R_i} \right) I_i \tag{7.48}$$

Also,

$$V_\pi = I_b r_\pi \tag{7.49}$$

and the output voltage is

$$V_o = -g_m V_\pi (R_C \| R_L) \tag{7.50}$$

Finally, combining Equations (7.47) through (7.50), we find the magnitude of the midband gain, as follows:

$$
\begin{aligned}
|A_v| &= \left|\frac{V_o}{V_i}\right| \\
&= g_m r_\pi (R_C \| R_L)\left(\frac{R_1 \| R_2}{(R_1 \| R_2) + R_i}\right)\left(\frac{1}{[R_S + (R_1 \| R_2 \| R_i)]}\right)
\end{aligned}
\tag{7.51}
$$

Example 7.7 **Objective:** Determine the midband gain, corner frequencies, and bandwidth of a circuit containing both a coupling capacitor and a load capacitor.

Consider the circuit shown in Figure 7.22(a) with transistor parameters $V_{BE}(\text{on}) = 0.7\,\text{V}$, $\beta = 100$, and $V_A = \infty$.

Solution: The dc analysis of this circuit yields a quiescent collector current of $I_{CQ} = 0.99\,\text{mA}$. The transconductance is

$$
g_m = \frac{I_{CQ}}{V_T} = \frac{0.99}{0.026} = 38.1\,\text{mA/V}
$$

and the base diffusion resistance is

$$
r_\pi = \frac{\beta V_T}{I_{CQ}} = \frac{(100)(0.026)}{0.99} = 2.63\,\text{k}\Omega
$$

The input resistance R_i is therefore

$$
R_i = r_\pi + (1 + \beta)R_E = 2.63 + (101)(0.5) = 53.1\,\text{k}\Omega
$$

From Equation (7.51), the midband gain is

$$
\begin{aligned}
|A_v|_{\text{max}} &= \left|\frac{V_o}{V_i}\right|_{\text{max}} = g_m r_\pi (R_C \| R_L)\left(\frac{R_1 \| R_2}{(R_1 \| R_2) + R_i}\right)\left(\frac{1}{[R_S + (R_1 \| R_2 \| R_i)]}\right) \\
&= (38.1)(2.63)(5\|10)\left(\frac{40\|5.7}{(40\|5.7) + 53.1}\right)\left(\frac{1}{[0.1 + (40\|5.7\|53.1)]}\right)
\end{aligned}
$$

or

$$
|A_v|_{\text{max}} = 6.16
$$

The time constant τ_S is

$$
\begin{aligned}
\tau_S &= (R_S + R_1 \| R_2 \| R_i)C_C \\
&= (0.1 \times 10^3 + (5.7 \times 10^3)\|(40 \times 10^3)\|(53.1 \times 10^3))(10 \times 10^{-6}) = 4.66 \times 10^{-2}\,\text{s}
\end{aligned}
$$

or

$$
\tau_S = 46.6\,\text{ms}
$$

and the time constant τ_P is

$$
\tau_P = (R_C \| R_L)C_L = ((5 \times 10^3)\|(10 \times 10^3))(15 \times 10^{-12}) = 5 \times 10^{-8}\,\text{s}
$$

or

$$
\tau_P = 50\,\text{ns}
$$

The lower corner frequency is

$$
f_L = \frac{1}{2\pi \tau_S} = \frac{1}{2\pi(46.6 \times 10^{-3})} = 3.42\,\text{Hz}
$$

and the upper corner frequency is

$$f_H = \frac{1}{2\pi\tau_P} = \frac{1}{2\pi(50 \times 10^{-9})} \Rightarrow 3.18 \, \text{MHz}$$

Finally, the bandwidth is

$$f_{BW} = f_H - f_L = 3.18 \, \text{MHz} - 3.4 \, \text{Hz} \cong 3.18 \, \text{MHz}$$

Comment: The two corner frequencies differ by approximately six orders of magnitude; therefore, considering one capacitor at a time is a valid approach.

A figure of merit for an amplifier is the **gain–bandwidth product**. Assuming the corner frequencies are far apart, the bandwidth is

$$f_{BW} = f_H - f_L \cong f_H \tag{7.52}$$

and the maximum gain is $|A_v|_{\text{max}}$. The gain–bandwidth product is therefore

$$GB = |A_v|_{\text{max}} \cdot f_H \tag{7.53}$$

Later we will show that, for a given load capacitance, this product is essentially a constant. We will also describe how trade-offs must be made between gain and bandwidth in amplifier design.

7.3.4   Bypass Capacitor Effects

In bipolar and FET discrete amplifiers, emitter and source bypass capacitors are often included so that emitter and source resistors can be used to stabilize the Q-point without sacrificing the small-signal gain. The bypass capacitors are assumed to act as short circuits at the signal frequency. However, to guide us in choosing a bypass capacitor, we must determine the circuit response in the frequency range where these capacitors are neither open nor short circuits.

Figure 7.23(a) shows a common-emitter circuit with an emitter bypass capacitor. The small-signal equivalent circuit is shown in Figure 7.23(b). We

(a)   **(b)**

Figure 7.23   (a) Circuit with emitter bypass capacitor and (b) small-signal equivalent circuit can find the small-signal voltage gain as a function of frequency. Using the impedance reflection rule, the small-signal input current is

$$I_b = \frac{V_i}{R_S + r_\pi + (1 + \beta)\left(R_E \left\| \dfrac{1}{sC_E}\right.\right)} \tag{7.54}$$

The total impedance in the emitter is multiplied by the factor $(1 + \beta)$. The control voltage is

$$V_\pi = I_b r_\pi \tag{7.55}$$

and the output voltage is

$$V_o = -g_m V_\pi R_C \tag{7.56}$$

Combining equations produces the small-signal voltage gain, as follows:

$$A_v(s) = \frac{V_o(s)}{V_i(s)} = \frac{-g_m r_\pi R_C}{R_S + r_\pi + (1 + \beta)\left(R_E \left\| \dfrac{1}{sC_E}\right.\right)} \tag{7.57}$$

Expanding the parallel combination of R_E and $1/sC_E$ and rearranging terms, we find

$$A_v = \frac{-g_m r_\pi R_C}{[R_S + r_\pi + (1 + \beta)R_E]} \times \frac{(1 + sR_E C_E)}{\left\{1 + \dfrac{sR_E(R_S + r_\pi)C_E}{[R_S + r_\pi + (1 + \beta)R_E]}\right\}} \tag{7.58}$$

Equation (7.58) can be written in terms of time constants as

$$A_v = \frac{-g_m r_\pi R_C}{[R_S + r_\pi + (1 + \beta)R_E]}\left\{\frac{1 + s\tau_A}{1 + s\tau_B}\right\} \tag{7.59}$$

The form of this transfer function is somewhat different from what we have previously encountered in that we have both a zero and a pole.

The Bode plot of the voltage gain magnitude has two limiting horizontal asymptotes. If we set $s = j\omega$, we can then consider the limit as $\omega \to 0$ and the limit as $\omega \to \infty$. For $\omega \to 0$, C_E acts as an open circuit; for $\omega \to \infty$, C_E acts as a short circuit. From Equation (7.58), we have

$$|A_v|_{\omega \to 0} = \frac{g_m r_\pi R_C}{[R_S + r_\pi + (1 + \beta)R_E]} \tag{7.60(a)}$$

and

$$|A_v|_{\omega \to \infty} = \frac{g_m r_\pi R_C}{R_S + r_\pi} \tag{7.60(b)}$$

From these results, we see that for $\omega \to 0$, R_E is included in the gain expression, and for $\omega \to \infty$, R_E is not part of the gain expression, since it has been effectively shorted out by C_E.

If we assume that the time constants τ_A and τ_B in Equation (7.59) differ substantially in magnitude, then the corner frequency due to τ_B is

$$f_B = \frac{1}{2\pi\tau_B} \tag{7.61(a)}$$

and the corner frequency due to τ_A is

$$f_A = \frac{1}{2\pi\tau_A} \tag{7.61(b)}$$

The resulting Bode plot of the voltage gain magnitude is shown in Figure 7.24.

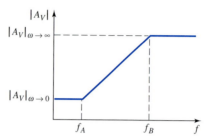

Figure 7.24 Bode plot of the voltage gain magnitude for the circuit with an emitter bypass capacitor

Example   7.8   Objective: Determine the corner frequencies and limiting horizontal asymptotes of a common-emitter circuit with an emitter bypass capacitor.

Consider the circuit in Figure 7.23(a) with parameters $R_E = 4\,\text{k}\Omega$, $R_C = 2\,\text{k}\Omega$, $R_S = 0.5\,\text{k}\Omega$, $C_E = 1\,\mu\text{F}$, $V^+ = 5\,\text{V}$, and $V^- = -5\,\text{V}$. The transistor parameters are: $\beta = 100$, $V_{BE}(\text{on}) = 0.7\,\text{V}$, and $r_o = \infty$.

Solution: From the dc analysis, we find the quiescent collector current as $I_{CQ} = 1.06\,\text{mA}$. The transconductance is

$$g_m = \frac{I_{CQ}}{V_T} = \frac{1.06}{0.026} = 40.8\,\text{mA/V}$$

and the input base resistance is

$$r_\pi = \frac{\beta V_T}{I_{CQ}} = \frac{(100)(0.026)}{1.06} = 2.45\,\text{k}\Omega$$

The time constant τ_A is

$$\tau_A = R_E C_E = (4 \times 10^3)(1 \times 10^{-6}) = 4 \times 10^{-3}\,\text{s}$$

and the time constant τ_B is

$$\tau_B = \frac{R_E(R_S + r_\pi)C_E}{[R_S + r_\pi + (1+\beta)R_E]}$$
$$= \frac{(4 \times 10^3)(0.5 \times 10^3 + 2.45 \times 10^3)(1 \times 10^{-6})}{[0.5 \times 10^3 + 2.45 \times 10^3 + (101)(4 \times 10^3)]}$$

or

$$\tau_B = 2.90 \times 10^{-5}\,\text{s}$$

The corner frequencies are then

$$f_A = \frac{1}{2\pi\tau_A} = \frac{1}{2\pi(4 \times 10^{-3})} = 39.8\,\text{Hz}$$

and

$$f_B = \frac{1}{2\pi\tau_B} = \frac{1}{2\pi(2.9 \times 10^{-5})} \Rightarrow 5.49\,\text{kHz}$$

The limiting low-frequency horizontal asymptote, given by Equation (7.60(a)) is

$$|A_v|_{\omega\to 0} = \frac{g_m r_\pi R_C}{[R_S + r_\pi + (1 + \beta)R_E]} = \frac{(40.8)(2.45)(2)}{[0.5 + 2.45 + (101)(4)]}$$

or

$$|A_v|_{\omega\to 0} = 0.49$$

The limiting high-frequency horizontal asymptote, given by Equation (7.60(b)) is

$$|A_v|_{\omega\to\infty} = \frac{g_m r_\pi R_C}{R_S + r_\pi} = \frac{(40.8)(2.45)(2)}{0.5 + 2.45} = 67.8$$

Comment: Comparing the two limiting values of voltage gain, we see that including a bypass capacitor produces a large high-frequency gain.

Computer Verification: The results of a PSpice analysis are given in Figure 7.25. The magnitude of the small-signal voltage gain is shown in Figure 7.25(a). The two corner frequencies are approximately 39 Hz and 5600 Hz, which agree very well with the results

(a)

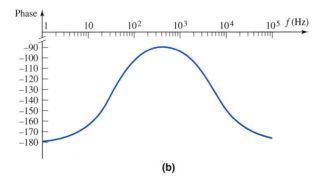

(b)

Figure 7.25 PSpice analysis results for the circuit with an emitter bypass capacitor: (a) voltage gain magnitude response and (b) phase response from the time constant analysis. The two limiting magnitudes of 0.49 and 68 also correlate extremely well with the hand analysis results.

Figure 7.25(b) is a plot of the phase response versus frequency. At very low and very high frequencies, where the capacitor acts as either an open circuit or short circuit, the phase is −180 degrees, as expected for a common-emitter circuit. Between the two corner frequencies, the phase changes substantially, approaching −90 degrees.

The analysis of an FET amplifier with a source bypass capacitor is essentially the same as for the bipolar circuit. The general form of the voltage gain expression is the same as Equation (7.59), and the Bode plot of the gain is essentially the same as that shown in Figure 7.24.

Test Your Understanding

∗7.11   The circuit shown in Figure 7.23(a) has parameters $V^+ = 10\,\text{V}$, $V^- = -10\,\text{V}$, $R_S = 0.5\,\text{k}\Omega$, $R_E = 4\,\text{k}\Omega$, and $R_C = 2\,\text{k}\Omega$. The transistor parameters are: $V_{BE}(\text{on}) = 0.7\,\text{V}$, $V_A = \infty$, and $\beta = 100$. (a) Determine the value of C_E such that the low-frequency 3 dB point is $f_B = 200\,\text{Hz}$. (b) Using the results from part (a), determine f_A. (Ans. (a) $C_E = 49.5\,\mu\text{F}$, (b) $f_A = 0.80\,\text{Hz}$)

7.3.5   Combined Effects: Coupling and Bypass Capacitors

When a circuit contains multiple capacitors, the frequency response analysis becomes more complex. In many amplifier applications, the circuit is to amplify an input signal whose frequency is confined to the midband range. In this case, the actual frequency response outside the midband range is not of interest. The end points of the midband range are defined to be those frequencies at which the gain decreases by 3 dB from the maximum midband value. These endpoint frequencies are a function of the high- and low-frequency capacitors. These capacitors introduce a pole to the amplifier transfer function.

If multiple coupling capacitors, for example, exist in a circuit, one capacitor may introduce the pole that produces the 3 dB reduction in the maximum gain at the low frequency. This pole is referred to as the **dominant pole**. A more detailed discussion of dominant poles is given in Chapter 12. At this point in the text, we will determine the frequency response of circuits containing multiple capacitors with a computer simulation.

As an example, Figure 7.26 shows a circuit with two coupling capacitors and an emitter bypass capacitor, all of which affect the circuit response at low frequencies. We could develop a transfer function that includes all the capacitors, but the analysis of such circuits is most easily handled by computer.

Figure 7.27 is the Bode plot of the voltage gain magnitude for the example circuit, taking into account the effects of the two coupling capacitors. In this case, the bypass capacitor is assumed to be a short circuit. The plots consider C_1 and C_2 individually, as well as together. As expected, with two capacitors both acting at the same time, the slope is 40 dB/decade or 12 dB/octave. Since the poles are not far apart, in the actual circuit, we cannot consider the effect of each capacitor individually.

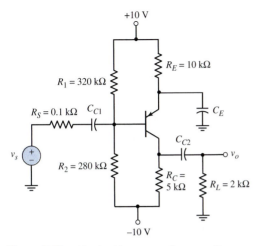

Figure 7.26   Circuit with two coupling capacitors and an emitter bypass capacitor

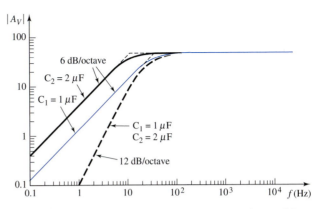

Figure 7.27   PSpice results for each coupling capacitor, and the combined effect for the circuit in Figure 7.26 ($C_E \rightarrow \infty$)

Figure 7.28 is the Bode plot of the voltage gain magnitude, taking into account the emitter bypass capacitor and the two coupling capacitors. The plot shows the effect of the bypass capacitor, the effect of the two coupling capacitors, and the net effect of the three capacitors together. When all three capacitors are taken into account, the slope is continually changing; there is no definitive corner frequency. However, at approximately $f = 150$ Hz, the curve is 3 dB below the maximum asymptotic value, and this frequency is defined as the **lower corner frequency**, or **lower cutoff frequency**.

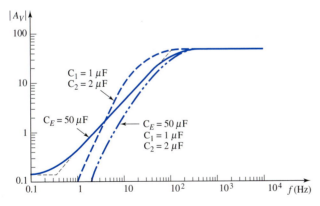

Figure 7.28   PSpice results for the two coupling capacitors, the bypass capacitor, and the combined effects

Test Your Understanding

*7.12   Consider the common-base circuit shown in Figure 7.29. Can the two coupling capacitors be treated separately? (a) From a computer analysis, determine the cutoff frequency. Assume the parameter values are $\beta = 100$ and $I_S = 2 \times 10^{-15}$ A. (b) Determine the midband small-signal voltage gain. (Ans. (a) $f_{3\,\text{dB}} = 1.2$ kHz, (b) $A_v = 118$)

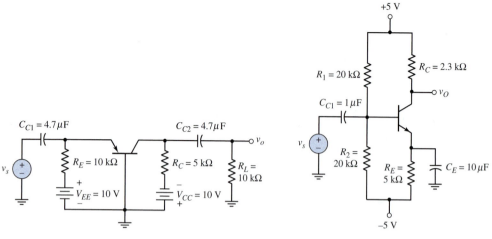

Figure 7.29   Figure for Exercise 7.12   **Figure 7.30**   Figure for Exercise 7.13

*7.13   The common-emitter circuit shown in Figure 7.30 contains both a coupling capacitor and an emitter bypass capacitor. (a) From a computer analysis, determine the 3 dB frequency. Assume the parameter values are $\beta = 100$ and $I_S = 2 \times 10^{-15}$ A. (b) Determine the midband small-signal voltage gain. (Ans. (a) $f_{3\,\text{dB}} \approx 575$ Hz, (b) $|A_v|_{\text{max}} = 74.4$)

7.4   FREQUENCY RESPONSE: BIPOLAR TRANSISTOR

Thus far, we have considered the frequency response of circuits as a function of external resistors and capacitors, and we have assumed the transistor to be ideal. However, both bipolar transistors and FETs have internal capacitances that influence the high-frequency response of circuits. In this section, we will first develop an expanded small-signal hybrid-π model of the bipolar transistor, taking these capacitances into account. We will then use this model to analyze the frequency characteristics of the bipolar transistor.

7.4.1   Expanded Hybrid-π Equivalent Circuit

When a bipolar transistor is used in a linear amplifier circuit, the transistor is biased in the forward-active region, and small sinusoidal voltages and currents are superimposed on the dc voltages and currents. Figure 7.31(a) shows an npn bipolar transistor in a common-emitter configuration, along with the small-signal voltages and currents. Figure 7.31(b) is a cross section of the npn transistor in a classic integrated circuit configuration. The C, B, and E terminals are the external connections to the transistor, and the C', B', and E' points are the idealized internal collector, base, and emitter regions.

To construct the equivalent circuit of the transistor, we will first consider various pairs of terminals. Figure 7.32(a) shows the equivalent circuit for the connection between the external base input terminal and the external emitter terminal. Resistance r_b is the base series resistance between the external base

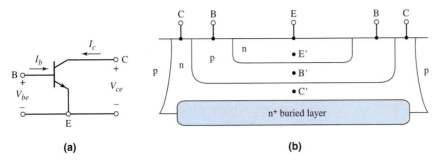

Figure 7.31 (a) Common-emitter npn bipolar transistor with small-signal currents and voltages and (b) cross section of an npn bipolar transistor, for the hybrid-π model

Figure 7.32 Components of the hybrid-π equivalent circuit: (a) base to emitter, (b) collector to emitter, and (c) base to collector

terminal B and the internal base region B′. the B′–E′ junction is forward biased; therefore, C_π is the forward-biased junction capacitance and r_π is the forward-biased junction diffusion resistance. Both parameters are functions of the junction current. Finally, r_{ex} is the emitter series resistance between the external emitter terminal and the internal emitter region. This resistance is usually very small, on the order of 1 to 2 Ω.

Figure 7.32(b) shows the equivalent circuit looking into the collector terminal. Resistance r_c is the collector series resistance between the external and internal collector connections, and capacitance C_s is the junction capacitance of the reverse-biased collector–substrate junction. The dependent current source, $g_m V_\pi$, is the transistor collector current controlled by the internal base–emitter voltage. Resistance r_o is the inverse of the output conductance g_o and is due primarily to the Early effect.

Finally, Figure 7.32(c) shows the equivalent circuit of the reverse-biased B′–C′ junction. Capacitance C_μ is the reverse-biased junction capacitance, and r_μ is the reverse-biased diffusion resistance. Normally, r_μ is on the order of megohms and can be neglected. The value of C_μ is usually much smaller than C_π; however, because of a phenomenon known as the Miller effect, C_μ usually cannot be neglected. (We will consider the Miller effect later in this chapter.)

The complete hybrid-π equivalent circuit for the bipolar transistor is shown in Figure 7.33. The capacitances lead to frequency effects in the transistor. One

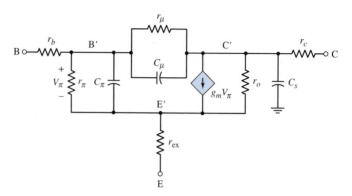

Figure 7.33   Hybrid-π equivalent circuit

result is that the gain is a function of the input signal frequency. Because of the large number of elements, a computer simulation of this complete model is easier than a hand analysis. However, we can make some simplifications in order to evaluate some fundamental frequency effects of bipolar transistors.

7.4.2   Short-Circuit Current Gain

We can begin to understand the frequency effects of the bipolar transistor by first determining the **short-circuit current gain**, after simplifying the hybrid-π model. Figure 7.34 shows a simplified equivalent circuit for the transistor, in which we neglect the parasitic resistances r_b, r_c, and r_{ex}, the B–C diffusion resistance r_μ, and the substrate capacitance C_s. Also, the collector is connected to signal ground. Keep in mind that the transistor must still be biased in the forward-active region. We will determine the small-signal current gain $A_i = I_c/I_b$.

Writing a KCL equation at the input node, we find that

$$I_b = \frac{V_\pi}{r_\pi} + \frac{V_\pi}{\dfrac{1}{j\omega C_\pi}} + \frac{V_\pi}{\dfrac{1}{j\omega C_\mu}} = V_\pi\left[\frac{1}{r_\pi} + j\omega(C_\pi + C_\mu)\right] \qquad (7.62)$$

We see that V_π is no longer equal to $I_b r_\pi$, since a portion of I_b is now shunted through C_π and C_μ.

From a KCL equation at the output node, we obtain

$$\frac{V_\pi}{\dfrac{1}{j\omega C_\mu}} + I_c = g_m V_\pi \qquad (7.63(a))$$

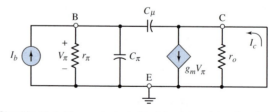

Figure 7.34   Simplified hybrid-π equivalent circuit for determining the short-circuit current gain

or

$$I_c = V_\pi(g_m - j\omega C_\mu) \tag{7.63(b)}$$

The input voltage V_π can then be written as

$$V_\pi = \frac{I_c}{(g_m - j\omega C_\mu)} \tag{7.63(c)}$$

Substituting this expression for V_π into Equation (7.62) yields

$$I_b = I_c \cdot \frac{\left[\dfrac{1}{r_\pi} + j\omega(C_\pi + C_\mu)\right]}{(g_m - j\omega C_\mu)} \tag{7.64}$$

The small-signal current gain usually designated as h_{fe}, becomes

$$A_i = \frac{I_c}{I_b} = h_{fe} = \frac{(g_m - j\omega C_\mu)}{\left[\dfrac{1}{r_\pi} + j\omega(C_\pi + C_\mu)\right]} \tag{7.65}$$

If we assume typical circuit parameter values of $C_\mu = 0.2\,\text{pF}$, $g_m = 50\,\text{mA/V}$, and a maximum frequency of $f = 100\,\text{MHz}$, then we see that $\omega C_\mu \ll g_m$. Therefore, to a good approximation, the small-signal current gain is

$$h_{fe} \cong \frac{g_m}{\left[\dfrac{1}{r_\pi} + j\omega(c_\pi + C_\mu)\right]} = \frac{g_m r_\pi}{1 + j\omega r_\pi(C_\pi + C_\mu)} \tag{7.66}$$

Since $g_m r_\pi = \beta$, then the low frequency current gain is just β, as we previously assumed. Equation (7.66) shows that, in a bipolar transistor, the magnitude and phase of the current gain are both functions of the frequency.

Figure 7.35(a) is a Bode plot of the short-circuit current gain magnitude. The corner frequency, which is also the **beta cutoff frequency** f_β in this case, is given by

$$f_\beta = \frac{1}{2\pi r_\pi(C_\pi + C_\mu)} \tag{7.67}$$

Figure 7.35(b) shows the phase of the current gain. As the frequency increases, the small-signal collector current is no longer in phase with the

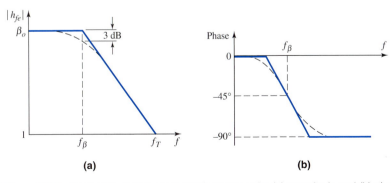

Figure 7.35 Bode plots for the short-circuit current gain: (a) magnitude and (b) phase

small-signal base current. At high frequencies, the collector current lags the input current by 90 degrees.

Example 7.9   **Objective:**   Determine the 3 dB frequency of the short-circuit current gain of a bipolar transistor.

Consider a bipolar transistor with parameters $r_\pi = 2.6 \, \text{k}\Omega$, $C_\pi = 2 \, \text{pF}$, and $C_\mu = 0.1 \, \text{pF}$.

Solution: From Equation (7.67), we have

$$f_\beta = \frac{1}{2\pi r_\pi (C_\pi + C_\mu)} = \frac{1}{2\pi (2.6 \times 10^3)(2 + 0.1)(10^{-12})}$$

or

$$f_\beta = 29.1 \, \text{MHz}$$

Comment: High-frequency transistors must have small capacitances; therefore, small devices must be used.

Test Your Understanding

7.14   A bipolar transistor has parameters $\beta_o = 150$, $C_\pi = 2 \, \text{pF}$, and $C_\mu = 0.3 \, \text{pF}$, and is biased at $I_{CQ} = 0.5 \, \text{mA}$. Determine the beta cutoff frequency. (Ans. $f_\beta = 8.87 \, \text{MHz}$)

7.15   A BJT is biased at $I_{CQ} = 0.25 \, \text{mA}$, and its parameters are $\beta_o = 100$ and $C_\mu = 0.1 \, \text{pF}$. The beta cutoff frequency is $f_\beta = 11.5 \, \text{MHz}$. Determine the capacitance C_π. (Ans. $C_\pi = 1.23 \, \text{pF}$)

7.4.3   Cutoff Frequency

Figure 7.35(a) shows that the magnitude of the small-signal current gain decreases with increasing frequency. At frequency f_T, which is the **cutoff frequency**, this gain goes to 1. The cutoff frequency is a figure of merit for transistors.

From Equation (7.66), we can write the small-signal current gain in the form

$$h_{fe} = \frac{\beta_o}{1 + j\left(\dfrac{f}{f_\beta}\right)} \tag{7.68}$$

where f_β is the beta cutoff frequency defined by Equation (7.67). The magnitude of h_{fe} is

$$|h_{fe}| = \frac{\beta_o}{\sqrt{1 + \left(\dfrac{f}{f_\beta}\right)^2}} \tag{7.69}$$

At the cutoff frequency f_T, $|h_{fe}| = 1$, and Equation (7.69) becomes

$$|h_{fe}| = 1 = \frac{\beta_o}{\sqrt{1 + \left(\dfrac{f_T}{f_\beta}\right)^2}} \qquad (7.70)$$

Normally, $\beta_o \gg 1$, which implies that $f_T \gg f_\beta$. Then Equation (7.70) can be written as

$$1 \cong \frac{\beta_o}{\sqrt{\left(\dfrac{f_T}{f_\beta}\right)^2}} = \frac{\beta_o f_\beta}{f_T} \qquad (7.71(a))$$

or

$$f_T = \beta_o f_\beta \qquad (7.71(b))$$

Frequency f_β is also called the bandwidth of the transistor. Therefore, from Equation (7.71(b)), the cutoff frequency f_T is the gain–bandwidth product of the transistor, or more commonly the **unity-gain bandwidth**. From Equation (7.67), the unity-gain bandwidth is

$$f_T = \beta_o \left[\frac{1}{2\pi r_\pi (C_\pi + C_\mu)} \right] = \frac{g_m}{2\pi (C_\pi + C_\mu)} \qquad (7.72)$$

Since the capacitances are a function of the size of the transistor, we see again that high frequency transistors imply small device sizes.

The cutoff frequency f_T is also a function of the dc collector current I_C, and the general characteristic of f_T versus I_C is shown in Figure 7.36. The transconductance g_m is directly proportional to I_C, but only a portion of C_π is related to I_C. The cutoff frequency is therefore lower at low collector current levels. However, the cutoff frequency also decreases at high current levels, in the same way that β decreases at large currents.

The cutoff frequency or unity-gain bandwidth of a transistor is usually specified on the device data sheets. Since the low-frequency current gain is also given, the beta cutoff frequency, or bandwidth, of the transistor can be determined from

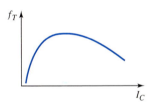

Figure 7.36 Cutoff frequency versus collector current

$$f_\beta = \frac{f_T}{\beta_o} \qquad (7.73)$$

The cutoff frequency of the general-purpose 2N2222A discrete bipolar transistor is $f_T = 300\,\text{MHz}$. For the MSC3130 discrete bipolar transistor, which has a special surface mount package, the cutoff frequency is $f_T = 1.4\,\text{GHz}$. This tells us that very small transistors fabricated in integrated circuits can have cutoff frequencies in the low GHz range.

Example 7.10   **Objective:** Calculate the bandwidth f_β and capacitance C_π of a bipolar transistor.

Consider a transistor that has parameters $f_T = 500\,\text{MHz}$ at $I_C = 1\,\text{mA}$, $\beta_o = 100$, and $C_\mu = 0.3\,\text{pF}$.

Solution: From Equation (7.73), the bandwidth is

$$f_\beta = \frac{f_T}{\beta_o} = \frac{500}{100} = 5\,\text{MHz}$$

The transconductance is

$$g_m = \frac{I_C}{V_T} = \frac{1}{0.026} = 38.5\,\text{mA/V}$$

The C_π capacitance is determined from Equation (7.72). We have

$$f_T = \frac{g_m}{2\pi(C_\pi + C_\mu)}$$

or

$$500 \times 10^6 = \frac{38.5 \times 10^{-3}}{2\pi(C_\pi + 0.3 \times 10^{-12})}$$

which yields $C_\pi = 12.0\,\text{pF}$.

Comment: Although the value of C_π may be much larger than that of C_μ, C_μ cannot be neglected in circuit applications as we will see in the next section.

The hybrid-π equivalent circuit for the bipolar transistor uses discrete or lumped elements. However, when cutoff frequencies are on the order of $f_T \cong 1\,\text{GHz}$ and the transistor is operated at microwave frequencies, other parasitic elements and distributed parameters must be included in the transistor model. For simplicity, we will assume in this text that the hybrid-π model is sufficient to model the transistor characteristics up through the beta cutoff frequency.

Test Your Understanding

7.16 For the transistor described in Example 7.10 and biased at the same Q-point, determine $|h_{fe}|$ and the phase at $f = 50\,\text{MHz}$. (Ans. $|h_{fe}| = 9.95$, Phase $= -84.3°$)

7.17 The parameters of a transistor are: $\beta_o = 120$, $f_T = 500\,\text{MHz}$, $r_\pi = 5\,\text{k}\Omega$, and $C_\mu = 0.2\,\text{pF}$. Determine C_π and f_β. (Ans. $f_\beta = 4.17\,\text{MHz}$, $C_\pi = 7.43\,\text{pF}$)

7.18 A BJT is biased at $I_C = 1\,\text{mA}$, and its parameters are: $\beta_o = 150$, $C_\pi = 4\,\text{pF}$, and $C_\mu = 0.5\,\text{pF}$. Determine f_β and f_T. (Ans. $f_\beta = 9.07\,\text{MHz}$, $f_T = 1.36\,\text{GHz}$)

7.4.4   Miller Effect and Miller Capacitance

As previously mentioned, the C_μ capacitance cannot in reality be ignored. The **Miller effect**, or feedback effect, is a multiplication effect of C_μ in circuit appliations.

Figure 7.37(a) is a common-emitter circuit with a signal current source at the input. We will determine the small-signal current gain $A_i = i_o/i_s$ of the circuit. Figure 7.37(b) is the small-signal equivalent circuit, assuming the frequency is sufficiently high for the coupling and bypass capacitors to act as short circuits. The transistor model is the simplified hybrid-π circuit in

(a)   **(b)**

Figure 7.37 (a) Common-emitter circuit with current source input; (b) small-signal equivalent circuit with simplified hybrid-π model

Figure 7.34 (assuming $r_o = \infty$). Capacitor C_μ is a feedback element that connects the output back to the input. The output voltage and current will therefore influence the input characteristics.

We can determine the effect of C_μ on the input characteristics by finding an equivalent impedance Z_A across the plane A-A in Figure 7.38(a), producing the equivalent circuit shown in Figure 7.38(b). The current I_1 from Figure 7.38(a) can be written as

$$I_1 = \frac{V_\pi - V_o}{(1/j\omega C_\mu)} = (V_\pi - V_o)(j\omega C_\mu) \tag{7.74}$$

Summing the currents at the output gives

$$I_1 = (V_\pi - V_o)(j\omega C_\mu) = g_m V_\pi + \frac{V_o}{R_C \| R_L} \tag{7.75(a)}$$

or combining terms, we have

$$V_\pi(j\omega C_\mu - g_m) = V_o\left(\frac{1}{R_C \| R_L} + j\omega C_\mu\right) \tag{7.75(b)}$$

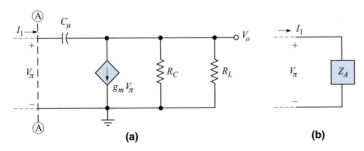

(a)   **(b)**

Figure 7.38 (a) Output portion of small-signal equivalent circuit; (b) equivalent impedance of this portion of the circuit

From our previous discussion of Equation (7.65), we noted that $|j\omega C_\mu| \ll g_m$ for typical transistor parameters, so the left side of Equation (7.75(b)) is just $-g_m V_\pi$. If, for example, $R_C = R_L = 4\,\text{k}\Omega$ and $C_\mu = 0.2\,\text{pF}$, then $|j\omega C_\mu| \ll 1/(R_C\|R_L)$ for $f \ll 400\,\text{MHz}$. If this condition is valid, then Equation (7.75(b)) becomes

$$-g_m V_\pi \cong V_o\left(\frac{1}{R_C\|R_L}\right) \tag{7.76(a)}$$

or

$$V_o = -g_m(R_C\|R_L)V_\pi \tag{7.76(b)}$$

Substituting this expression for V_o into Equation (7.74) yields

$$I_1 = \{V_\pi - [-g_m(R_C\|R_L)V_\pi]\}j\omega C_\mu \tag{7.77(a)}$$

or

$$I_1 = V_\pi \cdot j\omega C_\mu[1 + g_m(R_C R_L)] \tag{7.77(b)}$$

From Equation (7.77(b)), we see that the equivalent impedance Z_A in Figure 7.38(b) then corresponds to a capacitance whose value is

$$C_M = C_\mu[1 + g_m(R_C\|R_L)] \tag{7.78}$$

The small-signal equivalent circuit in Figure 7.37(b) can be redrawn as shown in Figure 7.39. Capacitance C_M is called the **Miller capacitance** and the multiplication effect of C_μ is called the **Miller effect**. [Note: If the condition $|j\omega C_\mu| \ll 1/(R_C\|R_L)$ is not valid, then the C_μ capacitance is also reflected in the output circuit.

Figure 7.39 Small-signal equivalent circuit, including the equivalent Miller capacitance

From the equivalent circuit in Figure 7.39, the input capacitance is now $C_\pi + C_M$, rather than just C_π if C_μ had been ignored.

Example 7.11   **Objective:** Determine the 3 dB frequency of the current gain for the circuit shown in Figure 7.39, both with and without the effect of C_M.

The circuit parameters are: $R_C = R_L = 4\,\text{k}\Omega$, $r_\pi = 2.6\,\text{k}\Omega$, $R_B = 200\,\text{k}\Omega$, $C_\pi = 4\,\text{pF}$, $C_\mu = 0.2\,\text{pF}$, and $g_m = 38.5\,\text{mA/V}$.

Solution: The output current can be written as

$$I_o = -(g_m V_\pi)\left(\frac{R_C}{R_C + R_L}\right)$$

Also, the input voltage is

$$V_\pi = I_s\left[R_B\|r_\pi\left\|\frac{1}{j\omega C_\pi}\right\|\frac{1}{j\omega C_M}\right]$$
$$= I_s\left[\frac{R_B\|r_\pi}{1 + j\omega(R_B\|r_\pi)(C_\pi + C_M)}\right]$$

Therefore, the current gain is

$$A_i = \frac{I_o}{I_s} = -g_m\left(\frac{R_C}{R_C + R_L}\right)\left[\frac{R_B\|r_\pi}{1 + j\omega(R_B\|r_\pi)(C_\pi + C_M)}\right]$$

The 3 dB frequency is

$$f_{3\,\text{dB}} = \frac{1}{2\pi(R_B\|r_\pi)(C_\pi + C_M)}$$

Neglecting the effect of $C_\mu(C_M = 0)$, we find that

$$f_{3\,\text{dB}} = \frac{1}{2\pi[(200 \times 10^3)\|(2.6 \times 10^3)](4 \times 10^{-12})} \Rightarrow 15.5\,\text{MHz}$$

The Miller capacitance is

$$C_M = C_\mu[1 + g_m(R_C\|R_L)] = (0.2)[1 + (38.5)(4\|4)] = 15.6\,\text{pF}$$

Taking into account the Miller capacitance, the 3 dB frequency is

$$f_{3\,\text{dB}} = \frac{1}{2\pi(R_B\|r_\pi)(C_\pi + C_M)}$$
$$= \frac{1}{2\pi[(200 \times 10^3)\|(2.6 \times 10^3)][4 + 15.6](10^{-12})}$$

or

$$f_{3\,\text{dB}} = 3.16\,\text{MHz}$$

Comment: The Miller effect, or multiplication factor of C_μ, is 78, giving a Miller capacitance of $C_M = 15.6\,\text{pF}$. The Miller capacitance, in this case, is approximately a factor of four larger than C_π. This means that the actual transistor bandwidth is approximately five times less than the bandwidth expected if C_μ is neglected.

The Miller capacitance, from Equation (7.78), can be written in the form

$$C_M = C_\mu(1 + |A_v|) \tag{7.79}$$

where A_v is the internal base-to-collector voltage gain. The physical origin of the Miller effect is in the voltage gain factor appearing across the feedback element C_μ. A small input voltage V_π produces a large output voltage $V_o = -|A_v| \cdot V_\pi$ of the opposite polarity at the output of C_μ. Thus the voltage across C_μ is $(1 + |A_v|)V_\pi$, which induces a large current through C_μ. For this reason, the effect of C_μ on the input portion of the circuit is significant.

We can now see one of the trade-offs that can be made in an amplifier design. The tradeoff is between amplifier gain and bandwidth. If the gain is reduced, then the Miller capacitance will be reduced and the bandwidth will be increased. We will consider this tradeoff again when we consider the cascode amplifier later in the chapter.

Discussion: In Equation (7.75(b)), we assumed that $|j\omega C_\mu| \ll g_m$, which is valid even for frequencies in the 100 MHz range. Equation (7.75(b)) can then be written as

$$-g_m V_\pi = V_o\left(\frac{1}{R_C\|R_L} + j\omega C_\mu\right) \tag{7.80}$$

The right side of Equation (7.80) implies that a capacitance C_μ should be in parallel with R_C and R_L in the output portion of the equivalent circuit in Figure 7.39. For $R_C = R_L = 4\,\text{k}\Omega$ and $C_\mu = 0.2\,\text{pF}$, we indicated that this capacitance is negligible for $f \ll 400\,\text{MHz}$ [yielding Equation (7.76(a))]. However, in special circuits involving, for example, active loads, the equivalent R_C and R_L resistances may be on the order of $100\,\text{k}\Omega$. This means that the C_μ capacitance in the output part of the circuit is not negligible for frequencies even in the low-megahertz range. We will consider a few special cases in which C_μ in the output circuit is not negligible.

Test Your Understanding

*7.19   For the circuit in Figure 7.37(a), the parameters are: $R_1 = 200\,\text{k}\Omega$, $R_2 = 220\,\text{k}\Omega$, $R_C = 2.2\,\text{k}\Omega$, $R_L = 4.7\,\text{k}\Omega$, $R_E = 1\,\text{k}\Omega$, $r_s = 100\,\text{k}\Omega$, and $V_{CC} = 5\,\text{V}$. The transistor parameters are: $\beta_o = 100$, $V_{BE}(\text{on}) = 0.7\,\text{V}$, $V_A = \infty$, $C_\pi = 10\,\text{pF}$, and $C_\mu = 2\,\text{pF}$. Using the simplified hybrid-π model shown in Figure 7.34, calculate: (a) the Miller capacitance, and (b) the 3 dB frequency. (Ans. (a) $C_M = 109\,\text{pF}$, (b) $f_{3\,\text{dB}} = 0.505\,\text{MHz}$)

7.5   FREQUENCY RESPONSE: THE FET

We have considered the expanded hybrid-π equivalent circuit of the bipolar transistor that models the high-frequency response of the transistor. We will now develop the high-frequency equivalent circuit of the FET that takes into account various capacitances in the device. We will develop the model for a MOSFET, but it also applies to JFETs and MESFETs.

7.5.1   High-Frequency Equivalent Circuit

We will construct the small-signal equivalent circuit of a MOSFET from the basic MOSFET geometry, as described in Chapter 5. Figure 7.40 shows a model based on the inherent capacitances and resistances in an n-channel MOSFET, as well as the elements representing the basic device equations.

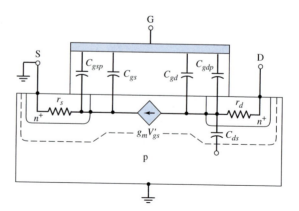

Figure 7.40   Inherent resistances and capacitances in the n-channel MOSFET structure

We make one simplifying assumption in the equivalent circuit: The source and substrate are both tied to ground.

Two capacitances connected to the gate are inherent in the transistor. These capacitances, C_{gs} and C_{gd}, represent the interaction between the gate and the channel inversion charge near the source and drain terminals, respectively. If the device is biased in the nonsaturation region and v_{DS} is small, the channel inversion charge is approximately uniform, which means that

$$C_{gs} \cong C_{gd} \cong \left(\tfrac{1}{2}\right) WLC_{ox}$$

where $C_{ox}(\mathrm{F/cm}^2) = \epsilon_{ox}/t_{ox}$. The parameter ϵ_{ox} is the oxide permittivity, which for silicon MOSFETs is $\epsilon_{ox} = 3.9\,\epsilon_o$, where $\epsilon_o = 8.85 \times 10^{-14}$ F/cm is the permittivity of free space. The parameter t_{ox} is the oxide thickness in cm.

However, when the transistor is biased in the saturation region, the channel is pinched off at the drain and the inversion charge is no longer uniform. The value of C_{gd} essentially goes to zero, and C_{gs} approximately equals $(2/3)WLC_{ox}$. As an example, if a device has an oxide thickness of 500 Å, a channel length of $L = 5\,\mu$m, and a channel width of $W = 50\,\mu$m, the value of C_{gs} is $C_{gs} \cong 0.12\,\mathrm{pF}$. The value of C_{gs} changes as the device size changes, but typical values are in the tenths of picofarad range.

The remaining two gate capacitances, C_{gsp} and C_{gdp}, are parasitic or **overlap capacitances**, so called because, in actual devices, the gate oxide overlaps the source and drain contacts, because of tolerances or other fabrication factors. As we will see, the drain overlap capacitance C_{gdp} lowers the bandwidth of the FET. The parameter C_{ds} is the drain-to-substrate pn junction capacitance, and r_s and r_d are the series resistances of the source and drain terminals. The internal gate-to-source voltage controls the small-signal channel current through the transconductance.

The small-signal equivalent circuit for the n-channel common-source MOSFET is shown in Figure 7.41. Voltage V'_{gs} is the internal gate-to-source voltage that controls the channel current. We will assume that the gate-to-source and gate-to-drain capacitances, C_{gs} and C_{gd}, contain the parasitic overlap capacitances. One parameter, r_o, shown in Figure 7.41 is not shown in Figure 7.40. This resistance is associated with the slope of I_D versus V_{DS}. In the ideal MOSFET biased in the saturation region, I_D is independent of V_{DS}, which means that r_o is infinite. However, r_o is finite in short-channel-length

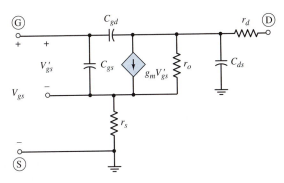

Figure 7.41 Equivalent circuit of the n-channel common-source MOSFET

devices, because of channel-length modulation, and is therefore included in the equivalent circuit.

Source resistance r_s can have a significant effect on the transistor characteristics. To illustrate, Figure 7.42 shows a simplified low-frequency equivalent circuit including r_s but not r_o. For this circuit, the drain current is

$$I_d = g_m V'_{gs} \tag{7.81}$$

and the relationship between V_{gs} and V'_{gs} is

$$V_{gs} = V'_{gs} + (g_m V'_{gs})r_s = (1 + g_m r_s)V'_{gs} \tag{7.82}$$

From Equation (7.81), the drain current can now be written as

$$I_d = \left(\frac{g_m}{1 + g_m r_s} \right) V_{gs} = g'_m V_{gs} \tag{7.83}$$

Figure 7.42 Simplified low-frequency equivalent circuit of the n-channel common-source MOSFET including source resistance r_s but not resistance r_o

Equation (7.83) shows that the source resistance reduces the effective transconductance, or the transistor gain.

The equivalent circuit of a p-channel MOSFET is exactly the same as that of an n-channel device, except that all voltage polarities and current directions are reversed. The capacitances and resistances are the same for both models.

Test Your Understanding

7.20 An n-channel MOSFET has parameters $K_n = 0.4\,\text{mA/V}^2$, $V_{TN} = 1\,\text{V}$, and $\lambda = 0$. (a) Determine the maximum source resistance such that the transconductance is reduced by no more than 20 percent from its ideal value when $V_{GS} = 3\,\text{V}$. (b) Using the value of r_s calculated in part (a), determine how much g_m is reduced from its ideal value when $V_{GS} = 5\,\text{V}$. (Ans. (a) $r_s = 156\,\Omega$, (b) 33.4%)

7.5.2   Unity-Gain Bandwidth

As for the bipolar transistor, the unity-gain frequency or bandwidth is a figure of merit for the FETs. If we neglect r_s, r_d, r_o, and C_{ds}, and connect the drain to signal ground, the resulting equivalent small-signal circuit is shown in Figure 7.43. Since the input gate impedance is no longer infinite at high frequency, we

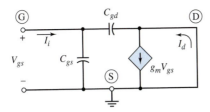

Figure 7.43 Equivalent high-frequency small-signal circuit of a MOSFET, for calculating short-circuit current gain

can define the short-circuit current gain. From that we can define and calculate the unity-gain bandwidth.

Writing a KCL equation at the input node, we find that

$$I_i = \frac{V_{gs}}{\dfrac{1}{j\omega C_{gs}}} + \frac{V_{gs}}{\dfrac{1}{j\omega C_{gd}}} = V_{gs}[j\omega(C_{gs} + C_{gd})] \tag{7.84}$$

From a KCL equation at the output node, we obtain

$$\frac{V_{gs}}{\dfrac{1}{j\omega C_{gd}}} + I_d = g_m V_{gs} \tag{7.85(a)}$$

or

$$I_d = V_{gs}(g_m - j\omega C_{gd}) \tag{7.85(b)}$$

Solving Equation (7.85(b)) for V_{gs} produces

$$V_{gs} = \frac{I_d}{(g_m - j\omega C_{gd})} \tag{7.86}$$

Substituting Equation (7.86) into (7.84) yields

$$I_i = I_d \cdot \frac{[j\omega(C_{gs} + C_{gd})]}{(g_m - j\omega C_{gd})} \tag{7.87}$$

Therefore, the small-signal current gain is

$$A_i = \frac{I_d}{I_i} = \frac{g_m - j\omega C_{gd}}{j\omega(C_{gs} + C_{gd})} \tag{7.88}$$

If we assume typical values of $C_{gd} = 0.05\,\text{pF}$ and $g_m = 1\,\text{mA/V}$, and a maximum frequency of $f = 100\,\text{MHz}$, we find that $\omega C_{gd} \ll g_m$. The small-signal current gain, to a good approximation, is then

$$A_i = \frac{I_d}{I_i} \cong \frac{g_m}{j\omega(C_{gs} + C_{gd})} \tag{7.89}$$

The unity-gain frequency f_T is defined as the frequency at which the magnitude of the short-circuit current gain goes to 1. From Equation (7.89) we find that

$$f_T = \frac{g_m}{2\pi(C_{gs} + C_{gd})} \tag{7.90}$$

The unity-gain frequency or bandwidth is a parameter of the transistor and is independent of the circuit.

Example 7.12 **Objective:** Determine the unity-gain bandwidth of an FET.

Consider an n-channel MOSFET with parameters $K_n = 0.25\,\text{mA/V}^2$, $V_{TN} = 1\,\text{V}$, $\lambda = 0$, $C_{gd} = 0.04\,\text{pF}$, and $C_{gs} = 0.2\,\text{pF}$. Assume the transistor is biased at $V_{GS} = 3\,\text{V}$.

Solution: The transconductance is

$$g_m = 2K_n(V_{GS} - V_{TN}) = 2(0.25)(3-1) = 1\,\text{mA/V}$$

From Equation (7.90), the unity-gain bandwidth, or frequency, is

$$f_T = \frac{g_m}{2\pi(C_{gs} + C_{gd})} = \frac{10^{-3}}{2\pi(0.2 + 0.04)\times 10^{-12}} = 6.63 \times 10^8\,\text{Hz}$$

or

$$f_T = 663\,\text{MHz}$$

Comment: As with bipolar transistors, high-frequency FETs require small capacitances and a small device size.

Typically, values of C_{gs} for MOSFETs are in the range of 0.1 to 0.5 pF and values of C_{gd} are typically from 0.01 to 0.04 pF.

As previously stated, the equivalent circuit is the same for MOSFETs, JFETS, and MESFETs. For JFETs, and MESFETS, capacitances C_{gs} and C_{gd} are depletion capacitances rather than oxide capacitances. Typically, for JFETs, C_{gs} and C_{gd} are larger than for MOSFETs, while the values for MESFETs are smaller. Also, for MESFETs fabricated in gallium arsenide, the unity-gain bandwidths may be in the range of tens of GHz. For this reason, gallium arsenide MESFETs are often used in microwave amplifiers.

Test Your Understanding

7.21 For an n-channel MOSFET, the parameters are: $K_n = 0.2\,\text{mA/V}^2$, $V_{TN} = 1\,\text{V}$, $\lambda = 0$, $C_{gd} = 0.02\,\text{pF}$, $C_{gs} = 0.25\,\text{pF}$. The device is biased at $I_{DQ} = 0.4\,\text{mA}$. Determine the unity-gain frequency. (Ans. $f_T = 332\,\text{MHz}$)

7.22 A MOSFET has a unity-gain bandwidth of $f_T = 500\,\text{MHz}$. Assume overlap capacitances of $C_{gsp} = C_{gdp} = 0.01\,\text{pF}$. If the transistor is biased such that $g_m = 0.5$ mA/V, determine C_{gs}. (Assume C_{gd} is equal to the overlap capacitance.) (Ans. $C_{gs} = 0.139\,\text{pF}$)

7.23 For a MOSFET, assume that $g_m = 1\,\text{mA/V}$. The basic gate capacitances are $C_{gs} = 0.4\,\text{pF}$, $C_{gd} = 0$, and the overlap capacitances are $C_{gsp} = C_{gdp}$. Determine the maximum overlap capacitance for a unity-gain bandwidth of 350 MHz. (Ans. $C_{gsp} = C_{gdp} = 0.0274\,\text{pF}$)

7.5.3 Miller Effect and Miller Capacitance

As for the bipolar transistor, the Miller effect and Miller capacitance are factors in the high-frequency characteristics of FET circuits. Figure 7.44 is a simplified high-frequency transistor model, with a load resistor R_L connected to the output. We will determine the current gain in order to demonstrate the impact of the Miller effect.

Figure 7.44 Equivalent high-frequency small-signal circuit of a MOSFET with a load resistance R_L

Writing a Kirchhoff current law (KCL) equation at the input gate node, we have

$$I_i = j\omega C_{gs} V_{gs} + j\omega C_{gd}(V_{gs} - V_{ds}) \qquad (7.91)$$

where I_i is the input current. Likewise, summing currents at the output drain node, we have

$$\frac{V_{ds}}{R_L} + g_m V_{gs} + j\omega C_{gd}(V_{ds} - V_{gs}) = 0 \qquad (7.92)$$

We can combine Equations (7.91) and (7.92) to eliminate voltage V_{ds}. The input current is then

$$I_i = j\omega \left\{ C_{gs} + C_{gd}\left[\frac{1 + g_m R_L}{1 + j\omega R_L C_{gd}} \right] \right\} V_{gs} \qquad (7.93)$$

Normally, $(\omega R_L C_{gd})$ is much less than 1; therefore, we can neglect $(j\omega R_L C_{gd})$ and Equation (7.93) becomes

$$I_i = j\omega[C_{gs} + C_{gd}(1 + g_m R_L)]V_{gs} \qquad (7.94)$$

Figure 7.45 shows the equivalent circuit described by Equation (7.94). The parameter C_M is the Miller capacitance and is given by

$$C_M = C_{gd}(1 + g_m R_L) \qquad (7.95)$$

Equation (7.95) clearly shows the effect of the parasitic drain overlap capacitance. When the transistor is biased in the saturation region, as in an amplifier circuit, the major contribution to the total gate-to-drain capacitance C_{gd} is the overlap capacitance. This overlap capacitance is multiplied because of the Miller effect and may become a significant factor in the bandwidth of an amplifier. Minimizing the overlap capacitance is one of the challenges of fabrication technology.

Figure 7.45 MOSFET high-frequency circuit, including the equivalent Miller capacitance

The cutoff frequency f_T of a MOSFET is defined as the frequency at which the current gain magnitude is 1, or the magnitude of the input current I_i is equal to the ideal load current I_d. From Figure 7.45, we see that

$$I_i = j\omega(C_{gs} + C_M)V_{gs} \tag{7.96}$$

and the ideal load current is

$$I_d = g_m V_{gs} \tag{7.97}$$

The magnitude of the current gain is therefore

$$|A_i| = \left|\frac{I_d}{I_i}\right| = \frac{g_m}{2\pi f(C_{gs} + C_M)} \tag{7.98}$$

Setting Equation (7.98) equal to 1, we find the cutoff frequency

$$f_T = \frac{g_m}{2\pi(C_{gs} + C_M)} = \frac{g_m}{2\pi C_G} \tag{7.99}$$

where C_G is the equivalent input gate capacitance.

Example 7.13 **Objective:** Determine the Miller capacitance and cutoff frequency of an FET circuit.

The n-channel MOSFET described in Example 7.12 is biased at the same current, and a $10\,\text{k}\Omega$ load is connected to the output.

Solution: From Example 7.12, the transconductance is $g_m = 1\,\text{mA/V}$. The Miller capacitance is therefore

$$C_M = C_{gd}(1 + g_m R_L) = (0.04)[1 + (1)(10)] = 0.44\,\text{pF}$$

From Equation (7.99), the cutoff frequency is

$$f_T = \frac{g_m}{2\pi(C_{gs} + C_M)} = \frac{10^{-3}}{2\pi(0.2 + 0.44) \times 10^{-12}} = 2.49 \times 10^8\,\text{Hz}$$

or

$$f_T = 249\,\text{MHz}$$

Comment: The Miller effect and equivalent Miller capacitance reduce the cutoff frequency of an FET circuit, just as they do in a bipolar circuit.

Test Your Understanding

*7.24 For the circuit in Figure 7.46, the transistor parameters are: $K_n = 0.5 \, \text{mA/V}^2$, $V_{TN} = 2 \, \text{V}$, $\lambda = 0$, $C_{gd} = 0.1 \, \text{pF}$, and $C_{gs} = 1 \, \text{pF}$. Calculate: (a) the Miller capacitance, and (b) the 3 dB frequency of the small-signal voltage gain. (Ans. (a) $C_M = 0.617 \, \text{pF}$, (b) $f_H = 10.9 \, \text{MHz}$)

Figure 7.46 Figure for Exercise 7.24

7.6 HIGH-FREQUENCY RESPONSE OF TRANSISTOR CIRCUITS

In the last sections, we developed the high-frequency equivalent circuits for the bipolar and field-effect transistors. We also discussed the Miller effect, which occurs when transistors are operating in a circuit configuration. In this section, we will expand our analysis of the high-frequency characteristics of transistor circuits.

Initially, we will look at the high-frequency response of the common-emitter and common-source configurations. We will then examine common-base and common-gate circuits, and a cascode circuit that is a combination of the common-emitter and common-base circuits. Finally, we will analyze the high-frequency characteristics of emitter-follower and source-follower circuits. In the following examples, we will use the same basic bipolar transistor circuit so that a good comparison can be made between the three circuit configurations.

7.6.1 Common-Emitter and Common-Source Circuits

The transistor capacitances and the load capacitance in the common-emitter amplifier shown in Figure 7.47 affect the high-frequency response of the circuit. Initially, we will use a hand analysis to determine the effects of the transistor on the high-frequency response. In this analysis, we will assume that C_C and C_E

Figure 7.47   Common-emitter amplifier

are short circuits, and C_L is an open circuit. A computer analysis will then be used to determine the effect of both the transistor and load capacitances.

The high-frequency small-signal equivalent circuit of the common-emitter circuit is shown in Figure 7.48(a) in which C_L is assumed to be an open circuit. We replace the capacitor C_μ with the equivalent Miller capacitance C_M as shown in Figure 7.48(b). From our previous analysis of the Miller capacitance, we can write

$$C_M = C_\mu(1 + g_m R_L')$$ (7.100)

where the output resistance R_l' is $r_o \| R_C \| R_L$.

The upper 3 dB frequency can be determined by using the time constant technique. We can write

$$f_H = \frac{1}{2\pi\tau_P}$$ (7.101)

where $\tau_P = R_{eq}C_{eq}$. In this case, the equivalent capacitance is $C_{eq} = C_\pi + C_M$, and the equivalent resistance is the effective resistance seen by the capacitance, $R_{eq} = r_\pi \| R_B \| R_S$. The upper corner frequency is therefore

$$f_H = \frac{1}{2\pi[r_\pi \| R_B \| R_S](C_\pi + C_M)}$$ (7.102)

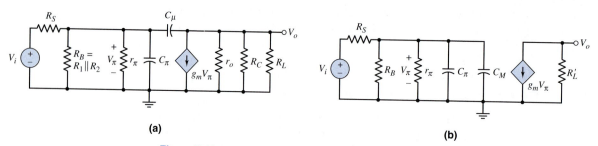

(a)

(b)

Figure 7.48   (a) High-frequency equivalent circuit of common-emitter amplifier; (b) high-frequency equivalent circuit of common-emitter amplifier, including the Miller capacitance

We determine the midband voltage gain magnitude by assuming C_π and C_M are open circuits. We find that

$$|A_v|_M = \left|\frac{V_o}{V_i}\right|_M = g_m R_L' \left[\frac{R_B \| r_\pi}{R_B \| r_\pi + R_S}\right] \qquad (7.103)$$

The Bode plot of the high-frequency voltage gain magnitude is shown in Figure 7.49.

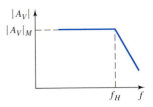

Figure 7.49 Bode plot of the high-frequency voltage gain magnitude for the common-emitter amplifier

Example 7.14 **Objective:** Determine the upper corner frequency and midband gain of a common-emitter circuit.

For the circuit in Figure 7.47, the parameters are: $V^+ = 5\,\text{V}$, $V^- = -5\,\text{V}$, $R_S = 0.1\,\text{k}\Omega$, $R_1 = 40\,\text{k}\Omega$, $R_2 = 5.72\,\text{k}\Omega$, $R_E = 0.5\,\text{k}\Omega$, $R_C = 5\,\text{k}\Omega$, and $R_L = 10\,\text{k}\Omega$. The transistor parameters are: $\beta = 150$, $V_{BE}(\text{on}) = 0.7\,\text{V}$, $V_A = \infty$, $C_\pi = 35\,\text{pF}$, and $C_\mu = 4\,\text{pF}$.

Solution: From a dc analysis, we find that $I_{CQ} = 1.02\,\text{mA}$. The small-signal parameters are therefore

$$r_\pi = \frac{\beta V_T}{I_{CQ}} = \frac{(150)(0.026)}{1.02} = 3.82\,\text{k}\Omega$$

and

$$g_m = \frac{I_{CQ}}{V_T} = \frac{1.02}{0.026} = 39.2\,\text{mA/V}$$

The Miller capacitance is then

$$C_M = C_\mu (1 + g_m R_L') = C_\mu [1 + g_m (R_C \| R_L)]$$

or

$$C_M = (4)[1 + (39.2)(5\|10)] = 527\,\text{pF}$$

and the upper 3 dB frequency is therefore

$$f_H = \frac{1}{2\pi [r_\pi \| R_B \| R_S](C_\pi + C_M)}$$

$$= \frac{1}{2\pi [3.82\|40\|5.72\|0.1](10^3)(35 + 527)(10^{-12})}$$

or

$$f_H = 2.96\,\text{MHz}$$

Finally, the midband gain is

$$|A_v|_M = g_m R_L' \left[\frac{R_B \| r_\pi}{R_B \| r_\pi + R_S}\right]$$

$$= (39.2)(5\|10)\left[\frac{40\|5.72\|3.82}{40\|5.72\|3.82 + 0.1}\right]$$

or

$$|A_v|_M = 125$$

Comments: This example demonstrates the importance of the Miller effect. The feedback capacitance C_μ is multiplied by a factor of 132 (from 4 pF to 527 pF), and the resulting Miller capacitance C_M is approximately 15 times larger than C_π. The actual corner frequency is therefore approximately 15 times smaller than it would be if C_μ were neglected.

PSpice Verification: Figure 7.50 shows the results of a PSpice analysis of this common-emitter circuit. The computer values are: $C_\pi = 35.5\,\text{pF}$ and $C_\mu = 3.89\,\text{pF}$. The curve marked "C_π only" is the circuit frequency response if C_μ is neglected; the curve marked "C_π and C_μ only" is the response due to C_π, C_μ, and the Miller effect. These curves illustrate that the bandwidth of this circuit is drastically reduced by the Miller effect.

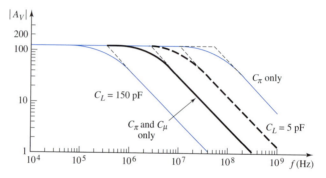

Figure 7.50   PSpice analysis results for common-emitter amplifier

The corner frequency is approximately 2.5 MHz and the midband gain is 125, which agree very well with the hand analysis results.

The curves marked "$C_L = 5\,\text{pF}$" and "$C_L = 150\,\text{pF}$" show the circuit response if the transistor is ideal, with zero C_π and C_μ capacitances and a load capacitance connected to the output. These results show that, for $C_L = 5\,\text{pF}$, the circuit response is dominated by the C_π and C_μ capacitances of the transistor. However, if a large load capacitance, such as $C_L = 150\,\text{pF}$, is connected to the output, the circuit response is dominated by the C_L capacitance.

The high-frequency response of the common-source circuit is similar to that of the common-emitter circuit, and the discussion and conclusions are the same. Capacitance C_π is replaced by C_{gs}, and C_μ is replaced by C_{gd}. The high-frequency small-signal equivalent circuit of the FET is then essentially identical to that of the bipolar transistor.

7.6.2   Common-Base, Common-Gate, and Cascode Circuits

As we have just seen, the bandwidth of the common-emitter and common-source circuits is reduced by the Miller effect. To increase the bandwidth, the Miller effect, or the C_μ multiplication factor, must be minimized or eliminated. The common-base and common-gate amplifier configurations achieve this result. We will analyze a common-base circuit; the analysis is the same for the common-gate circuit.

Common-Base Circuit

Figure 7.51 shows a common-base circuit. The circuit configuration is the same as the common-emitter circuit considered previously, except a bypass capacitor is added to the base and the input is capacitively coupled to the emitter.

Figure 7.51   Common-base amplifier

Figure 7.52(a) shows the high-frequency equivalent circuit, with the coupling and bypass capacitors replaced by short circuits. Resistors R_1 and R_2 are then effectively short circuited. Also, resistance r_o is assumed to be infinite. Capacitance C_μ, which led to the multiplication effect, is no longer between the input and output terminals. One side of capacitor C_μ is tied to signal ground.

Writing a KCL equation at the emitter, we find that

$$I_e + g_m V_\pi + \frac{V_\pi}{(1/sC_\pi)} + \frac{V_\pi}{r_\pi} = 0 \tag{7.104}$$

Since $V_\pi = -V_e$, Equation (7.104) becomes

$$\frac{I_e}{V_e} = \frac{1}{Z_i} = \frac{1}{r_\pi} + g_m + sC_\pi \tag{7.105}$$

where Z_i is the impedance looking into the emitter. Rearranging terms, we have

$$\frac{1}{Z_i} = \frac{1 + r_\pi g_m}{r_\pi} + sC_\pi = \frac{1+\beta}{r_\pi} + sC_\pi \tag{7.106}$$

The equivalent input portion of the circuit is shown in Figure 7.52(b).

Figure 7.52(c) shows the equivalent output portion of the circuit. Again, one side of C_μ is tied to ground, which eliminates the feedback or Miller multiplication effect. We then expect the upper 3 dB frequency to be larger than that observed in the common-emitter configuration.

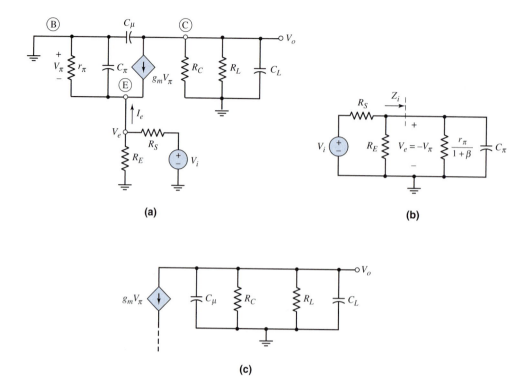

Figure 7.52 (a) High-frequency common-base equivalent circuit, (b) equivalent input circuit, and (c) equivalent output circuit

For the input portion of the circuit, the upper 3 dB frequency is given by

$$f_{H\pi} = \frac{1}{2\pi \tau_{p\pi}} \tag{7.107(a)}$$

where the time constant is

$$\tau_{P\pi} = \left[\left(\frac{r_\pi}{1+\beta}\right) \| R_E \| R_S\right] \cdot C_\pi \tag{7.107(b)}$$

In the hand analysis, we assume that C_L is an open circuit. Capacitance C_μ will also produce an upper 3 dB frequency, given by

$$f_{H\mu} = \frac{1}{2\pi \tau_{P\mu}} \tag{7.108(a)}$$

where the time constant is

$$\tau_{P\mu} = [R_C \| R_L] \cdot C_\mu \tag{7.108(b)}$$

If C_μ is much smaller than C_π, we would expect the 3 dB frequency $f_{H\pi}$ due to C_π to dominate the high-frequency response. However, the factor $r_\pi/(1+\beta)$ in the time constant $\tau_{P\pi}$ is small; therefore, the two time constants may be the same order of magnitude.

Example 7.15 **Objective:** Determine the upper corner frequencies and midband gain of a common-base circuit.

Consider the circuit shown in Figure 7.51 with circuit parameters $V^+ = 5\,\text{V}$, $V^- = -5\,\text{V}$, $R_S = 0.1\,\text{k}\Omega$, $R_1 = 40\,\text{k}\Omega$, $R_2 = 5.72\,\text{k}\Omega$, $R_E = 0.5\,\text{k}\Omega$, $R_C = 5\,\text{k}\Omega$, and $R_L = 10\,\text{k}\Omega$. (These are the same values as those used for the common-emitter circuit in Example 7.14.) The transistor parameters are: $\beta = 150$, $V_{BE}(\text{on}) = 0.7\,\text{V}$, $V_A = \infty$, $C_\pi = 35\,\text{pF}$, and $C_\mu = 4\,\text{pF}$.

Solution: The dc analysis is the same as in Example 7.14; therefore, $I_{CQ} = 1.02\,\text{mA}$, $g_m = 39.2\,\text{mA/V}$, and $r_\pi = 3.82\,\text{k}\Omega$. The time constant associated with C_π is

$$
\tau_{P\pi} = \left[\left(\frac{r_\pi}{1+\beta}\right)\|R_E\|R_S\right] \cdot C_\pi
$$

$$
= \left[\left(\frac{3.82}{151}\right)\|(0.5)\|(0.1)\right] \times 10^3 (35 \times 10^{-12})
$$

or

$$
\tau_{P\pi} = 0.679\,\text{ns}
$$

The upper 3 dB frequency corresponding to C_π is therefore

$$
f_{H\pi} = \frac{1}{2\pi\tau_{P\pi}} = \frac{1}{2\pi(0.679 \times 10^{-9})} \Rightarrow 234\,\text{MHz}
$$

The time constant associated with C_μ in the output portion of the circuit is

$$
\tau_{P\mu} = [R_C\|R_L] \cdot C_\mu = [5\|10] \times 10^3 (4 \times 10^{-12}) \Rightarrow 13.3\,\text{ns}
$$

The upper 3 dB frequency corresponding to C_μ is therefore

$$
f_{H\mu} = \frac{1}{2\pi\tau_{P\mu}} = \frac{1}{2\pi(13.3 \times 10^{-9})} \Rightarrow 12.0\,\text{MHz}
$$

So in this case, $f_{H\mu}$ is the dominant pole frequency.

The magnitude of the midband voltage gain is

$$
|A_v|_M = g_m(R_C\|R_L)\left[\frac{R_E\|\left(\frac{r_\pi}{1+\beta}\right)}{R_E\|\left(\frac{r_\pi}{1+\beta}\right) + R_S}\right]
$$

$$
= (39.2)(5\|10)\left[\frac{0.5\|\left(\frac{3.82}{151}\right)}{0.5\|\left(\frac{3.82}{151}\right) + 0.1}\right] = 25.3
$$

Comment: The results of this example show that the bandwidth of the common-base circuit is limited by the capacitance C_μ in the output portion of the circuit. The bandwidth of this particular circuit is 12 MHz, which is approximately a factor of four greater than the bandwidth of the common-emitter circuit in Example 7.14.

Computer Verification: Figure 7.53 shows the results of a PSpice analysis of the common-base circuit. The computer values are $C_\pi = 35.5\,\text{pF}$ and $C_\mu = 3.89\,\text{pF}$, which are the same as those in Example 7.14. The curve marked "C_π only" is the circuit frequency response if C_μ is neglected. The curve marked "C_π and C_μ only" includes the effect of both C_π and C_μ. As the hand analysis predicted, C_μ dominates the circuit high-frequency response.

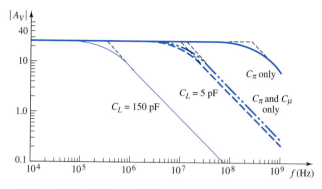

Figure 7.53 PSpice analysis results for common-base circuit

The corner frequency is approximately 13.5 MHz and the midband gain is 25.5, both of which agree very well with the hand analysis results.

The curves marked "$C_L = 5\,\text{pF}$" and "$C_L = 150\,\text{pF}$" are the circuit response if the transistor is ideal and only a load capacitance is included. These results again show that if a load capacitance of $C_L = 150\,\text{pF}$ were connected to the output, the circuit response would be dominated by this capacitance. However, if a 5 pF load capacitor were connected to the output, the circuit response would be a function of both the C_L and C_μ capacitances, since the two response characteristics are almost identical.

Cascode Circuit

The cascode circuit, as shown in Figure 7.54, combines the advantages of the common-emitter and common-base circuits. The input signal is applied to the common-emitter circuit (Q_1), and the output signal from the common emitter

Figure 7.54 Cascode circuit

is fed into the common-base circuit (Q_2). The input impedance to the common-emitter circuit (Q_1) is relatively large, and the load resistance seen by Q_1 is the input impedance to the emitter of Q_2 and is fairly small. The low output resistance seen by Q_1 reduces the Miller multiplication factor on $C_{\mu 1}$ and therefore extends the bandwidth of the circuit.

Figure 7.55(a) shows the high-frequency small-signal equivalent circuit. The coupling and bypass capacitors are equivalent to short circuits, and resistance r_o for Q_2 is assumed to be infinite.

The input impedance to the emitter of Q_2 is Z_{ie2}. From Equation (7.106) in our previous analysis, we have

$$Z_{ie2} = \left(\frac{r_{\pi 2}}{1+\beta}\right) \left\| \left(\frac{1}{sC_{\pi 2}}\right)\right. \tag{7.109}$$

The input portion of the small-signal equivalent circuit can be transformed to that shown in Figure 7.55(b). The input impedance Z_{ie2} is again shown.

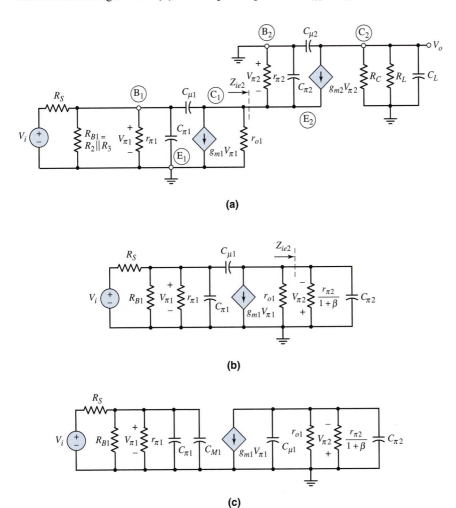

(a)

(b)

(c)

Figure 7.55 (a) High-frequency equivalent circuit of cascode configuration, (b) rearranged high-frequency equivalent circuit, and (c) variation of the high-frequency circuit, including the Miller capacitance

The input portion of the circuit shown in Figure 7.55(b) can be transformed to that given in Figure 7.55(c), which shows the Miller capacitance. The Miller capacitance C_{M1} is included in the input, and capacitance $C_{\mu1}$ is included in the output portion of the Q_1 model. The possibility of including C_{μ} in the output circuit was discussed previously.

In the center of this equivalent circuit, r_{o1} is in parallel with $r_{\pi2}/(1+\beta)$. Since r_{o1} is usually large, it can be approximated as an open circuit. The Miller capacitance is then

$$C_{M1} = C_{\mu1}\left[1 + g_{m1}\left(\frac{r_{\pi2}}{1+\beta}\right)\right] \tag{7.110}$$

Transistors Q_1 and Q_2 are biased with essentially the same current; therefore,

$$r_{\pi1} \cong r_{\pi2} \quad \text{and} \quad g_{m1} \cong g_{m2}$$

Then

$$g_{m1}r_{\pi2} = \beta$$

which yields

$$C_{M1} \cong 2\,C_{\mu1} \tag{7.111}$$

Equation (7.111) shows that this cascode circuit greatly reduces the Miller multiplication factor.

The time constant related to $C_{\pi2}$ involves resistance $r_{\pi2}/(1+\beta)$. Since this resistance is small, the time constant is small, and the corner frequency related to $C_{\pi2}$ is very large. We can therefore neglect the effects of $C_{\mu1}$ and $C_{\pi2}$ in the center portion of the circuit.

The time constant for the input portion of the circuit is

$$\tau_{P\pi} = [R_S \| R_{B1} \| r_{\pi1}](C_{\pi1} + C_{M1}) \tag{7.112(a)}$$

where $C_{M1} = 2C_{\mu1}$. The corresponding 3 dB frequency is

$$f_{H\pi} = \frac{1}{2\pi\tau_{P\pi}} \tag{7.112(b)}$$

Assuming C_L acts as an open circuit, the time constant of the output portion of the circuit, from Figure 7.55, is

$$\tau_{P\mu} = [R_C \| R_L](C_{\mu2}) \tag{7.113(a)}$$

and the corresponding corner frequency is

$$f_{H\mu} = \frac{1}{2\pi\tau_{P\mu}} \tag{7.113(b)}$$

To determine the midband voltage gain we assume that all capacitances in the circuit in Figure 7.55(c) are open circuits. The output voltage is then

$$V_o = -g_{m2}V_{\pi2}(R_C \| R_L) \tag{7.114}$$

and

$$V_{\pi2} = g_{m1}V_{\pi1}\left[r_{o1}\left\|\left(\frac{r_{\pi2}}{1+\beta}\right)\right.\right] \tag{7.115}$$

We can neglect the effect of r_{o1} compared to $r_{\pi2}/(1+\beta)$. Also, since $g_{m1}r_{\pi2} = \beta$, Equation (7.115) becomes

$$V_{\pi2} \cong V_{\pi1} \tag{7.116}$$

and, from the input portion of the circuit,

$$V_{\pi 1} = \frac{R_{B1} \| r_{\pi 1}}{R_{B1} \| r_{\pi 1} + R_S} \times V_i \tag{7.117}$$

Finally, combining equations, we find the midband voltage gain is

$$A_{vM} = \frac{V_o}{V_i} = -g_{m2}(R_C \| R_L) \left[\frac{R_{B1} \| r_{\pi 1}}{R_{B1} \| r_{\pi 1} + R_S} \right] \tag{7.118}$$

If we compare Equation (7.118) to Equation (7.103) for the common-emitter circuit, we see that the expression for the midband gain of the cascode circuit is identical to that of the common-emitter circuit. The cascode circuit achieves a relatively large voltage gain, while extending the bandwidth.

Example 7.16 **Objective:** Determine the 3 dB frequencies and midband gain of a cascode circuit.

For the circuit in Figure 7.54, the parameters are: $V^+ = 10\,\text{V}$, $V^- = -10\,\text{V}$, $R_S = 0.1\,\text{k}\Omega$, $R_1 = 42.5\,\text{k}\Omega$, $R_2 = 20.5\,\text{k}\Omega$, $R_3 = 28.3\,\text{k}\Omega$, $R_E = 5.4\,\text{k}\Omega$, $R_C = 5\,\text{k}\Omega$, $R_L = 10\,\text{k}\Omega$, and $C_L = 0$. The transistor parameters are: $\beta = 150$, $V_{BE}(\text{on}) = 0.7\,\text{V}$, $V_A = \infty$, $C_\pi = 35\,\text{pF}$, and $C_\mu = 4\,\text{pF}$.

Solution: Since β is large for each transistor, the quiescent collector current is essentially the same in each transistor and is $I_{CQ} = 1.02\,\text{mA}$. The small-signal parameters are: $r_{\pi 1} = r_{\pi 2} \equiv r_\pi = 3.82\,\text{k}\Omega$ and $g_{m1} = g_{m2} \equiv g_m = 39.2\,\text{mA/V}$.

From Equation (7.112(a)), the time consant related to the input portion of the circuit is

$$\tau_{P\pi} = [R_S \| R_{B1} \| r_{\pi 1}](C_{\pi 1} + C_{M1})$$

Since $R_{B1} = R_2 \| R_3$ and $C_{M1} = 2\,C_{\mu 1}$, then

$$\tau_{P\pi} = [(0.1)\|20.5\|28.3\|3.82] \times 10^3 [35 + 2(4)] \times 10^{-12} \Rightarrow 4.16\,\text{ns}$$

The corresponding 3 dB frequency is

$$f_{H\pi} = \frac{1}{2\pi\tau_{P\pi}} = \frac{1}{2\pi(4.16 \times 10^{-9})} \Rightarrow 38.3\,\text{MHz}$$

From Equation (7.113(a)), the time constant of the output portion of the circuit is

$$\tau_{P\mu} = [R_C \| R_L]C_{\mu 2} = [5\|10] \times 10^3 (4 \times 10^{-12}) \Rightarrow 13.3\,\text{ns}$$

and the corresponding 3 dB frequency is

$$f_{H\mu} = \frac{1}{2\pi\tau_{P\mu}} = \frac{1}{2\pi(13.3 \times 10^{-9})} \Rightarrow 12\,\text{MHz}$$

From Equation (7.118), the midband voltage gain is

$$|A_v|_M = g_{m2}(R_C \| R_L) \left[\frac{R_{B1} \| r_{\pi 1}}{R_{B1} \| r_{\pi 1} + R_S} \right]$$

$$= (39.2)(5\|10) \left[\frac{(20.5\|28.3\|3.82)}{(20.5\|28.3\|3.82) + (0.1)} \right] = 126$$

Comment: As was the case for the common-base circuit, the 3 dB frequency for the cascode circuit is determined by capacitance C_μ in the output stage. The bandwidth of the cascode circuit is 12 Mz, compared to approximately 3 MHz for the common-emitter circuit. The midband voltage gains for the two circuits are essentially the same.

Computer Verification: Figure 7.56 shows the results of a PSpice analysis of the cascode circuit. From the hand analysis, the two corner frequencies are 12 Mz and 38.3 MHz. Since these frequencies are fairly close, we expect the actual response to show the effects of both capacitances. This hypothesis is verified and demonstrated in the computer analysis results. The curves marked "C_π only" and "C_π and C_μ only" are fairly close together, and their slopes are steeper than -6 dB/octave, which shows that more than one capacitor is involved in the response. At a frequency of 12 MHz, the response curve is 3 dB below the maximum asymptotic gain, and the midband gain is 120. These values closely agree with the hand analysis results.

The curves marked "$C_L = 5$ pF" and "$C_L = 150$ pF" show the circuit response if the transistor is ideal and only a load capacitance is included.

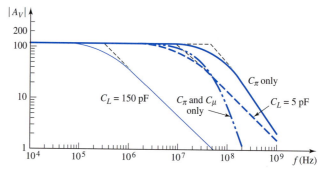

Figure 7.56   PSpice analysis results for cascode circuit

7.6.3   Emitter- and Source-Follower Circuits

In this section, we analyze the high-frequency response of the emitter follower. We will analyze the same basic circuit configuration that we have considered previously. The results and discussions also apply to the source follower.

Figure 7.57 shows an emitter-follower circuit with the output signal at the emitter capacitively coupled to a load. Figure 7.58(a) shows the high-frequency

Figure 7.57   Emitter-follower circuit

small-signal equivalent circuit, with the coupling capacitors acting effectively as short circuits.

We will rearrange the circuit so that we can gain a better insight into the circuit behavior. We see that C_μ is tied to ground potential and also that r_o is in parallel with R_E and R_L. We may define

$$R_L' = R_E \| R_L \| r_o$$

In this analysis we neglect the effect of C_L. Figure 7.58(b) shows a rearrangement of the circuit.

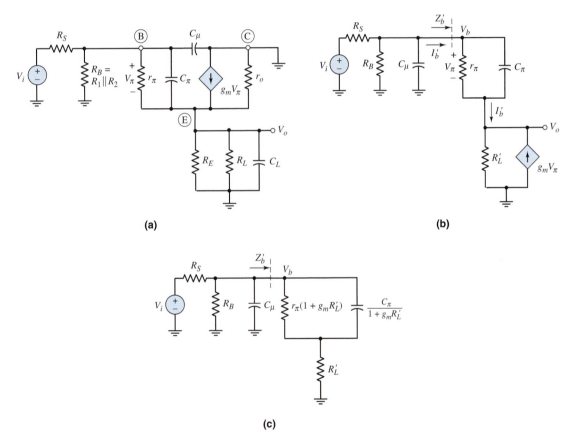

(a)

(b)

(c)

Figure 7.58 (a) High-frequency equivalent circuit of emitter follower, (b) rearranged high-frequency equivalent circuit, and (c) high-frequency equivalent circuit with effective input base impedance

We can find the impedance Z_b' looking into the base without capacitance C_μ. The current I_b' entering the parallel combination of r_π and C_π is the same as that coming out of the combination. The output voltage is then

$$V_o = (I_b' + g_m V_\pi)R_L' \tag{7.119}$$

Voltage V_π is given by

$$V_\pi = \frac{I_b'}{y_\pi} \tag{7.120}$$

where

$$y_\pi = (1/r_\pi) + sC_\pi$$

Voltage V_b is

$$V_b = V_\pi + V_o$$

Therefore,

$$Z_b' = \frac{V_b}{I_b'} = \frac{V_\pi + V_o}{I_b'} \tag{7.121}$$

Combining Equations (7.119), (7.120), and (7.121), we obtain

$$Z_b' = \frac{1}{y_\pi} + R_L' + \frac{g_m R_L'}{y_\pi} \tag{7.122(a)}$$

or

$$Z_b' = \frac{1}{y_\pi}(1 + g_m R_L') + R_L' \tag{7.122(b)}$$

Substituting the expression for y_π, we find

$$Z_b' = \frac{1}{\dfrac{1}{r_\pi} + sC_\pi} \times (1 + g_m R_L') + R_L' \tag{7.123(a)}$$

This can then be written as

$$Z_b' = \frac{1}{\dfrac{1}{r_\pi(1 + g_m R_L')} + \dfrac{sC_\pi}{(1 + g_m R_L')}} + R_L' \tag{7.123(b)}$$

Impedance Z_b' is shown in the equivalent circuit in Figure 7.58(c). Equation (7.123(b)) shows that the effect of capacitance C_π is reduced in the emitter-follower configuration.

Since the emitter-follower circuit has a zero and two poles, a detailed analysis of the circuit is very tedious. From Equations (7.119) and (7.120), we have

$$V_o = V_\pi(y_\pi + g_m)R_L' \tag{7.124}$$

which yields a zero when $y_\pi + g_m = 0$. Using the definition of y_π, the zero occurs at

$$f_o = \frac{1}{2\pi C_\pi\left(\dfrac{r_\pi}{1 + \beta}\right)} \tag{7.125}$$

Since $r_\pi/(1 + \beta)$ is small, frequency f_o is usually very high.

If we make a simplifying assumption, we can determine an approximate value of one pole. In many applications, the impedance of $r_\pi(1 + g_m R_L')$ in parallel with $C_\pi/(1 + g_m R'_L)$ is large compared to R'_L. If we neglect R'_L, then the time constant is

$$\tau_P = [R_S \| R_B \| (1 + g_m R'_L) r_\pi] \left(C_\mu + \frac{C_\pi}{1 + g_m R'_L} \right) \qquad (7.126(a))$$

and the 3 dB frequency (or pole) is

$$f_H = \frac{1}{2\pi\tau_P} \qquad (7.126(b))$$

Example 7.17 **Objective:** Determine the frequency of a zero and a pole in the high-frequency response of an emitter follower.

Consider the emitter-follower circuit in Figure 7.57 with parameters $V^+ = 5\,\text{V}$, $V^- = -5\,\text{V}$, $R_S = 0.1\,\text{k}\Omega$, $R_1 = 40\,\text{k}\Omega$, $R_2 = 5.72\,\text{k}\Omega$, $R_E = 0.5\,\text{k}\Omega$, and $R_L = 10\,\text{k}\Omega$. The transistor parameters are: $\beta = 150$, $V_{BE}(\text{on}) = 0.7\,\text{V}$, $V_A = \infty$, $C_\pi = 35\,\text{pF}$, and $C_\mu = 4\,\text{pF}$.

Solution: As in previous examples, the dc analysis yields $I_{CQ} = 1.02\,\text{mA}$. Therefore, $g_m = 39.2\,\text{mA/V}$ and $r_\pi = 3.82\,\text{k}\Omega$.

From Equation (7.125), the zero occurs at

$$f_o = \frac{1}{2\pi C_\pi \left(\dfrac{r_\pi}{1 + \beta} \right)} = \frac{1}{2\pi(35 \times 10^{-12}) \left(\dfrac{3.82 \times 10^3}{151} \right)} \Rightarrow 180\,\text{MHz}$$

To determine the time constant for the high-frequency pole calculation, we know that

$$1 + g_m R'_L = 1 + g_m(R_E \| R_L) = 1 + (39.2)(0.5 \| 10) = 19.7$$

and

$$R_B = R_1 \| R_2 = 40 \| 5.72 = 5\,\text{k}\Omega$$

The time constant is therefore

$$\tau_P = [R_S \| R_B \| (1 + g_m R'_L) r_\pi] \left(C_\mu + \frac{C_\pi}{1 + g_m R'_L} \right)$$

$$= [(0.1) \| 5 \| (19.7)(3.82)] \times 10^3 \left(4 + \frac{35}{19.7} \right) \times 10^{-12} \Rightarrow 0.566\,\text{ns}$$

The 3 dB frequency (or pole) is then

$$f_H = \frac{1}{2\pi\tau_P} = \frac{1}{2\pi(0.566 \times 10^{-9})} \Rightarrow 281\,\text{MHz}$$

Comment: The frequencies for the zero and the pole are very high and are not far apart. This makes the calculations suspect. However, since the frequencies are high, the emitter follower is a wide-bandwidth circuit.

Computer Verification: Figure 7.59 shows the results of a PSpice analysis of the emitter follower. From the hand analysis, the 3 dB frequency is on the order of 281 MHz. However, the computer results show the 3 dB frequency to be approximately 400 MHz. We must keep in mind that at these high frequencies, distributed parameter

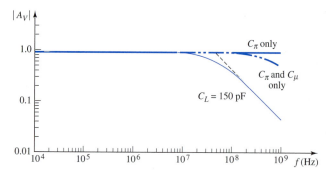

Figure 7.59   PSpice analysis results for emitter follower

effects may need to be considered in the transistor to more accurately predict the frequency response.

Also shown in the figure is the frequency response due to a 150 pF load capacitance. Comparing this result to the common-emitter circuit, for example, we see that the bandwidth of the emitter-follower circuit is approximately two orders of magnitude larger.

7.6.4   High-Frequency Amplifier Design

Our analysis shows that the frequency response, or the high-frequency cutoff point of an amplifier, depends on the transistor used, the circuit parameters, and the amplifier configuration.

We also saw that a computer simulation is easier than a hand analysis, particularly for the emitter-follower circuit. However, the parameters of the actual transistor used in the circuit must be used in the simulation if it is to predict the circuit frequency response accurately. Also, at high frequencies, additional parasitic capacitances, such as the collector–substrate capacitance, may need to be included. This was not done in our examples. Finally, in high-frequency amplifiers, the parasitic capacitances of the interconnect lines between the devices in an IC may also be a factor in the overall circuit response.

Test Your Understanding

*7.25   The transistor in the circuit in Figure 7.60 has parameters $\beta = 125$, $V_{BE}(\text{on}) = 0.7\,\text{V}$, $V_A = 200\,\text{V}$, $C_\pi = 24\,\text{pF}$, and $C_\mu = 3\,\text{pF}$. (a) Calculate the Miller capacitance. (b) Determine the upper 3 dB frequency. (c) Determine the small-signal midband voltage gain. (Ans. (a) $C_M = 155\,\text{pF}$, (b) $f_H = 1.21\,\text{MHz}$, (c) $|A_v| = 37.3$)

*7.26   For the circuit in Figure 7.61, the transistor parameters are: $K_n = 1\,\text{mA/V}^2$, $V_{TN} = 0.8\,\text{V}$, $\lambda = 0$, $C_{gs} = 2\,\text{pF}$, and $C_{gd} = 0.2\,\text{pF}$. Determine: (a) the Miller capacitance, (b) the upper 3 dB frequency, and (c) the midband voltage gain. (d) Correlate the results from parts (b) and (c) with a computer analysis. (Ans. (a) $C_M = 1.62\,\text{pF}$, (b) $f_H = 3.38\,\text{MHz}$, (c) $|A_v| = 4.63$)

Figure 7.60 Figure for Exercise 7.25   **Figure 7.61** Figure for Exercise 7.26

Figure 7.62 Figure for Exercise 7.27

*7.27 Consider the common-base circuit in Figure 7.62. The transistor parameters are $\beta = 100$, $V_{BE}(\text{on}) = 0.7\,\text{V}$, $V_A = \infty$, $C_\pi = 24\,\text{pF}$, and $C_\mu = 3\,\text{pF}$. (a) Determine the upper 3 dB frequencies corresponding to the input and output portions of the equivalent circuit. (b) Calculate the small-signal midband voltage gain. (Ans. (a) $f_{H\pi} = 223\,\text{MHz}$, $f_{H\mu} = 58.3\,\text{MHz}$, (b) $A_v = 0.869$)

*7.28 For the circuit in Figure 7.63, the transistor parameters are: $V_{TN} = 1\,\text{V}$, $K_n = 1\,\text{mA/V}^2$, $\lambda = 0$, $C_{gd} = 0.4\,\text{pF}$, and $C_{gs} = 5\,\text{pF}$. Perform a computer simulation to determine the upper 3 dB frequency and the midband small-signal voltage gain. (Ans. $f_H = 64.5\,\text{MHz}$, $|A_v| = 0.127$)

Figure 7.63 Figure for Exercise 7.28

*7.29   The cascode circuit in Figure 7.54 has parameters $V^+ = 12\,V$, $V^- = 0$, $R_1 = 58.8\,k\Omega$, $R_2 = 33.3\,k\Omega$, $R_3 = 7.92\,k\Omega$, $R_C = 7.5\,k\Omega$, $R_S = 1\,k\Omega$, $R_E = 0.5\,k\Omega$, and $R_L = 2\,k\Omega$. The transistor parameters are: $\beta = 100$, $V_{BE}(on) = 0.7\,V$, $V_A = \infty$, $C_\pi = 24\,pF$, and $C_\mu = 3\,pF$. Let C_L be an open circuit. (a) Determine the 3 dB frequencies corresponding to the input and output portions of the equivalent circuit. (b) Calculate the small-signal midband voltage gain. (c) Correlate the results from parts (a) and (b) with a computer analysis. (Ans. (a) $f_{H\pi} = 7.17\,MHz$, $f_{H\mu} = 33.6\,MHz$, (b) $|A_v| = 22.5$)

7.7   SUMMARY

- In this chapter, we studied the frequency response of transistor circuits. We determined the effects due to circuit capacitors, including coupling, bypass, and load capacitors, and also analyzed the expanded equivalent circuits of BJTs and FETs to determine the frequency response of the transistors.

- A time constant technique was developed so that Bode plots can be constructed without the need of deriving complex transfer functions. The high and low corner frequencies or 3 dB frequencies can be determined directly from the time constants.

- Coupling and bypass capacitors affect the low-frequency characteristics of a circuit. In general, capacitance values in the microfarad range typically result in cutoff frequencies in the hertz or tens of hertz range. A load capacitor affects the high-frequency characteristics of a circuit. Load capacitances in the picofarad range typically result in cutoff frequencies in the vicinity of a megahertz or higher.

- An expanded hybrid-π model for the bipolar transistor and a high-frequency model for the field-effect transistor were developed. The capacitances included in these models result in reduced transistor gain at high frequencies. The cutoff frequency is a figure of merit for the transistor and is defined as the frequency at which the magnitude of the current gain is unity.

- The Miller effect is a multiplication of the base-collector or gate-drain capacitance due to feedback between the output and input of the transistor circuit. The bandwidth of the amplifier is reduced by this effect.

- The common-emitter (common-source) amplifier, in general, shows the greatest effect of the Miller multiplication factor, so the bandwidth of this circuit is the smallest of the three basic types of amplifiers. The common-base (common-gate) amplifier has a larger bandwidth because of a smaller Miller multiplication factor. The cascode configuration, a combination of common-emitter and common-base stages, combines the advantages of high gain and wide bandwidth. The emitter-follower (source-follower) amplifier generally has the largest bandwidth of the three basic amplifier configurations.

CHECKPOINT

After studying this chapter, the reader should have the ability to:

✔ Construct the Bode plots of the gain magnitude and phase from a transfer function written in terms of the complex frequency s. (Section 7.2)

✔ Construct the Bode plots of the gain magnitude and phase of electronic amplifier circuits, taking into account circuit capacitors, using the time constant technique. (Section 7.3)

✔ Determine the short-circuit current gain versus frequency of a BJT and determine the Miller capacitance of a BJT circuit using the expanded hybrid-π equivalent circuit. (Section 7.4)

✔ Determine the unity-gain bandwidth of an FET and determine the Miller capacitance of an FET circuit using the expanded small-signal equivalent circuit. (Section 7.5)

✔ Describe the relative frequency responses of the three basic amplifier configurations and the cascode amplifier. (Section 7.6)

REVIEW QUESTIONS

1. Describe the general frequency response of an amplifier and define the low-frequency, midband, and high-frequency ranges.
2. Describe the general characteristics of the equivalent circuits that apply to the low-frequency, midband, and high-frequency ranges.
3. Describe what is meant by a system transfer function in the s-domain.
4. What is the criterion that defines a corner, or 3 dB, frequency?
5. Define octave and decade.
6. Describe what is meant by the phase of the transfer function.
7. Describe the time constant technique for determining the corner frequencies.
8. Describe the general frequency response of a coupling capacitor.
9. Describe the general frequency response of a bypass capacitor.
10. Describe the general frequency response of a load capacitor.
11. Sketch the expanded hybrid-π model of the BJT.
12. Describe the short-circuit current gain versus frequency characteristics of the BJT.
13. Define the cutoff frequency for a BJT.
14. Describe the Miller effect and the Miller capacitance.
15. What effect does the Miller capacitance have on the amplifier bandwidth?
16. Sketch the expanded small-signal equivalent circuit of a MOSFET.
17. Define the cutoff frequency for a MOSFET.
18. What is the major contribution to the Miller capacitance in a MOSFET?
19. Why is there not a Miller effect in a common-base circuit?
20. Describe the configuration of a cascode amplifier.
21. Why is the bandwidth of a cascode amplifier larger, in general, than that of a simple common-emitter amplifier?
22. Why is the bandwidth of the emitter-follower amplifier the largest of the three basic BJT amplifiers?

PROBLEMS

Section 7.2   System Transfer Functions

7.1 (a) Determine the voltage transfer function $T(s) = V_o(s)/V_i(s)$ for the circuit shown in Figure P7.1. (b) Sketch the Bode magnitude plot and determine the corner frequency. (c) Determine the time response of the circuit to an input step function of magnitude V_{Io}.

7.2 Repeat Problem 7.1 for the circuit in Figure P7.2.

*7.3 (a) Derive the voltage transfer function $T(s) = V_o(s)/V_i(s)$ for the circuit shown in Figure 7.10, taking both capacitors into account. (b) Let $R_S = R_P = 10\,\text{k}\Omega$, $C_S = 1\,\mu\text{F}$, and $C_P = 10\,\text{pF}$. Calculate the actual magnitude of the transfer function at $f_L = 1/[(2\pi)(R_S + R_P)C_S]$ and at $f_H = 1/[(2\pi)(R_S\|R_P)C_P]$. How do these magnitudes

Figure P7.1

$C_2 = 10\,\mu F$

V_i o———| |———o V_o

$R_2 = 10\,k\Omega$

Figure P7.2

compare to the maximum magnitude of $R_P/(R_S + R_P)$? (c) Repeat part (b) for $R_S = R_P = 10\,k\Omega$ and $C_S = C_P = 0.1\,\mu F$.

7.4 (a) For the two circuits in Figure P7.4, sketch the Bode magnitude plot and Bode phase plot of the voltage transfer function. (b) Verify the results of part (a) with a computer simulation.

$R_1 = 10\,k\Omega$

$C_1 = 10\,\mu F$

$R_2 = 20\,k\Omega$

(a)

$R_1 = 10\,k\Omega$

$R_2 = 20\,k\Omega$

$C_2 = 10\,\mu F$

(b)

Figure P7.4

C_S

I_i R_i R_P C_P o V_o

Figure P7.5

7.5 Consider the circuit in Figure P7.5 with a signal current source. The circuit parameters are $R_i = 30\,k\Omega$, $R_P = 10\,k\Omega$, $C_S = 10\,\mu F$, and $C_P = 50\,pF$. (a) Determine the open-circuit time constant associated with C_S and the short-circuit time constant associated with C_P. (b) Determine the corner frequencies and the magnitude of the transfer function $T(s) = V_o(s)/I_i(s)$ at midband. (c) Sketch the Bode magnitude plot.

7.6 A voltage transfer function is given by $T(jf) = 1/(1 + j2\pi f \tau)^2$. (a) Show that the actual response at $f = 1/(2\pi\tau)$ is approximately $-6\,dB$ below the maximum value. What is the phase angle at this frequency? (b) What is the slope of the magnitude plot for $f \gg 1/(2\pi\tau)$? What is the phase angle in this frequency range?

7.7 Sketch the Bode magnitude plots for the following functions:

(a) $T(s) = \dfrac{-10s}{(s+20)(s+2000)}$ (b) $T(s) = \dfrac{10(s+10)}{(s+1000)}$

7.8 Consider the circuit shown in Figure 7.15 with parameters $R_S = 0.5\,k\Omega$, $r_\pi = 5.2\,k\Omega$, $g_m = 29\,mA/V$, and $R_L = 6\,k\Omega$. The corner frequencies are $f_L = 30\,Hz$ and $f_H = 480\,kHz$. (a) Calculate the midband voltage gain. (b) What are the open-circuit and short-circuit time constants? (c) Determine C_C and C_L.

7.9 For the circuit shown in Figure P7.9, the parameters are: $R_1 = 10\,k\Omega$, $R_2 = 10\,k\Omega$, $R_3 = 40\,k\Omega$, and $C = 10\,\mu F$. Using a computer simulation, plot the magnitude and phase of the voltage transfer function. From the computer analysis, determine the

R_1

R_2

V_i R_3 C V_o

Figure P7.9

frequency at which the magnitude of the voltage transfer function is 3 dB below the maximum asymptotic value.

7.10 The circuit shown in Figure 7.10 has parameters $R_S = 1 \, k\Omega$, $R_P = 10 \, k\Omega$, and $C_S = C_P = 0.01 \, \mu F$. Using PSpice, plot the magnitude and phase of the voltage transfer function. Determine the maximum value of the voltage transfer function. Determine the frequencies at which the magnitude is $1/\sqrt{2}$ of the peak value.

Section 7.3   Frequency Response: Transistor Circuits

7.11 For the common-emitter circuit in Figure P7.11, the transistor parameters are: $\beta = 100$, $V_{BE}(\text{on}) = 0.7 \, V$, and $V_A = \infty$. (a) Calculate the lower corner frequency. (b) Determine the midband voltage gain. (c) Sketch the Bode plot of the voltage gain magnitude.

| Figure P7.11 | Figure P7.12 |

D7.12 Design the circuit shown in Figure P7.12 such that $I_{DQ} = 0.5 \, mA$, $V_{DSQ} = 4.5 \, V$, $R_{in} = 200 \, k\Omega$, and the lower corner frequency is $f_L = 20 \, Hz$. The transistor parameters are: $K_n = 0.2 \, mA/V^2$, $V_{TN} = 1.5 \, V$, and $\lambda = 0$. Sketch the Bode plot of the voltage magnitude and phase.

D7.13 The transistor in the circuit in Figure P7.13 has parameters $K_n = 0.5 \, mA/V^2$, $V_{TN} = 1 \, V$, and $\lambda = 0$. (a) Design the circuit such that $I_{DQ} = 1 \, mA$ and $V_{DSQ} = 3 \, V$.

Figure P7.13

(b) Derive the expression for the transfer function $T(s) = I_o(s)/V_i(s)$. What is the expression for the circuit time constant? (c) Determine C_C such that the lower 3 dB frequency is 10 Hz. (d) Verify the results of parts (a) and (c) with a computer simulation.

*D7.14 The transistor in the circuit in Figure P7.14 has parameters $K_p = 0.5\,\text{mA/V}^2$, $V_{TP} = -2\,\text{V}$, and $\lambda = 0$. (a) Determine R_o. (b) What is the expression for the circuit time constant? (c) Determine C_C such that the lower 3 dB frequency is 20 Hz.

Figure P7.14   **Figure P7.15**

*D7.15 For the circuit in Figure P7.15, the transistor parameters are: $\beta = 120$, $V_{BE}(\text{on}) = 0.7\,\text{V}$, and $V_A = 80\,\text{V}$. (a) Design a bias-stable circuit such that $I_{CQ} = 1\,\text{mA}$. (b) Determine the output resistance R_o. (c) What is the lower 3 dB corner frequency?

7.16 The parameters of the transistor in the circuit in Figure P7.16 are $K_p = 1\,\text{mA/V}^2$, $V_{TP} = -1.5\,\text{V}$, and $\lambda = 0$. (a) Determine the quiescent and small-signal parameters of the transistor. (b) Find the time constants associated with C_{C1} and C_{C2}. (c) Is there a dominant pole frequency? Estimate the $-3\,\text{dB}$ frequency.

Figure P7.16

*D7.17 A MOSFET amplifier with the configuration in Figure P7.17 is to be designed for use in a telephone circuit. The magnitude of the voltage gain should be 10 in the midband range, and the midband frequency range should extend from 200 Hz to 3 kHz. [Note: A telephone's frequency range does not correspond to a high-fidelity system's.] All resistor, capacitor, and MOSFET parameters should be specified.

Figure P7.17   **Figure P7.18**

*7.18 Consider the circuit in Figure P7.18. (a) Derive the expression for the voltage transfer function $T(s) = V_o(s)/V_i(s)$. Arrange the terms in the form $T(s) \propto (1 + s\tau_A)/(1 + s\tau_B)$. (b) Sketch the Bode magnitude plot. (c) Determine the time constants and corner frequencies.

7.19 The circuit in Figure P7.19 is a simple output stage of an audio amplifier. The transistor parameters are $\beta = 200$, $V_{BE}(\text{on}) = 0.7\,\text{V}$, and $V_A = \infty$. Determine C_C such that the lower $-3\,\text{dB}$ frequency is 15 Hz.

Figure P7.19   **Figure P7.20**

D7.20 The parameters of the transistor in the circuit in Figure P7.20 are $\beta = 100$, $V_{BE}(\text{on}) = 0.7\,\text{V}$, and $V_A = \infty$. The time constant associated with C_{C1} is a factor of 100 larger than the time constant associated with C_{C2}. (a) Determine C_{C2} such that the $-3\,\text{dB}$ frequency associated with this capacitor is 25 Hz. (b) Determine C_{C1}.

*D7.21 For the transistor in the circuit in Figure P7.21, the parameters are: $K_n = 0.5\,\text{mA/V}^2$, $V_{TN} = 0.8\,\text{V}$, and $\lambda = 0$. (a) Design the circuit such that $I_{DQ} = 0.5\,\text{mA}$ and $V_{DSQ} = 4\,\text{V}$. (b) Determine the 3 dB frequencies. (c) If the R_S resistor is replaced by a constant-current source producing the same I_{DQ} quiescent current, determine the 3 dB corner frequencies.

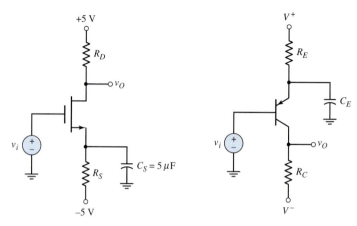

Figure P7.21   Figure P7.23

*7.22 For the circuit in Figure 7.23(a) in the text, the parameters are: $V^+ = 10\,V$, $V^- = -10\,V$, $R_S = 0$, $R_E = 5\,k\Omega$, and $R_C = 1.5\,k\Omega$. The transistor parameters are $V_{BE}(\text{on}) = 0.7\,V$ and $V_A = \infty$, and the transistor current gain β is in the range $75 \leq \beta \leq 125$. (a) Determine the value of C_E such that the low-frequency 3 dB point is $f_B \leq 200\,Hz$. (b) From the results of part (a), determine the range in frequencies f_B and f_A.

*7.23 The common-emitter circuit in Figure P7.23 has an emitter bypass capacitor. (a) Derive the expression for the small-signal voltage gain $A_v(s) = V_o(s)/V_i(s)$. Write the expression in a form similar to that of Equation (7.59). (b) What are the expressions for the time constants τ_A and τ_B?

7.24 In the common-base circuit in Figure 7.29 in the text, the transistor parameters are: $\beta = 100$, $V_{EB}(\text{on}) = 0.7\,V$, and $V_A = \infty$. A load capacitance $C_L = 15\,pF$ is connected in parallel with R_L. Determine the upper 3 dB frequency and the small-signal midband voltage gain.

7.25 For the circuit in Figure P7.25, the transistor parameters are: $K_n = 0.5\,mA/V^2$, $V_{TN} = 2\,V$, and $\lambda = 0$. Determine the maximum value of C_L such that the bandwidth is at least $BW = 5\,MHz$. State any approximations or assumptions that you make. What is the magnitude of the small-signal midband voltage gain? Verify the results with a computer simulation.

Figure P7.25

7.26 The parameters of the transistor in the circuit in Figure P7.26 are $\beta = 100$, $V_{BE}(\text{on}) = 0.7\,\text{V}$, and $V_A = \infty$. Neglect the capacitance effects of the transistor. (a) Draw the three equivalent circuits that represent the amplifier in the low-frequency range, midband range, and the high frequency range. (b) Sketch the Bode magnitude plot. (c) Determine the values of $|A_m|_{\text{dB}}$, f_L, and f_H.

Figure P7.26

7.27 In the common-source amplifier in Figure 7.20(a) in the text, a source bypass capacitor is to be added between the source terminal and ground potential. The circuit and transistor parameters are as described in Example 7.6. (a) Derive the small-signal voltage gain expression, as a function of s, that describes the circuit behavior in the high-frequency range. (b) What is the expression for the time constant associated with the upper 3 dB frequency? (c) Determine the time constant, upper 3 dB frequency, and small-signal midband voltage gain.

*7.28 Consider the common-base circuit in Figure P7.28. Choose appropriate transistor parameters. (a) Using a computer analysis, generate the Bode plot of the voltage gain magnitude from a very low frequency to the midband frequency range. At what frequency is the voltage gain magnitude 3 dB below the maximum value? What is the slope of the curve at very low frequencies? (b) Using the PSpice analysis, determine the voltage gain magnitude, input resistance R_i, and output resistance R_o at midband.

Figure P7.28   **Figure P7.29**

*7.29 For the common-emitter circuit in Figure P7.29, choose appropriate transistor parameters and perform a computer analysis. Generate the Bode plot of the voltage gain magnitude from a very low frequency to the midband frequency range. At what frequency is the voltage gain magnitude 3 dB below the maximum value? Does one capacitor dominate this 3 dB frequency? If so, which one?

*7.30 For the multitransistor amplifier in Figure P7.30, choose appropriate transistor parameters. The lower 3 dB frequency is to be less than or equal to 20 Hz. Assume that all three coupling capacitors are equal. Let $C_B \rightarrow \infty$. Using a computer analysis, determine the maximum values of the coupling capacitors. Determine the slope of the Bode plot of the voltage gain magnitude at very low frequencies.

Figure P7.30

Section 7.4   Frequency Response: Bipolar Transistor

7.31 A bipolar transistor is biased at $I_{CQ} = 1\,\text{mA}$ and has parameters $C_\pi = 10\,\text{pF}$, $C_\mu = 2\,\text{pF}$, and $\beta_o = 120$. Determine f_β and f_T.

7.32 A high-frequency bipolar transistor is biased at $I_{CQ} = 0.5\,\text{mA}$ and has parameters $C_\mu = 0.15\,\text{pF}$, $f_T = 5\,\text{GHz}$, and $\beta_o = 150$. Determine C_π and f_β.

7.33 For a bipolar transistor, the unity-gain bandwidth is $f_T = 2\,\text{GHz}$ and the low-frequency current gain is $\beta_o = 150$. (a) Determine f_β. (b) Find the frequency at which the magnitude of h_{fe} is 10.

7.34 The circuit in Figure P7.34 is a hybrid-π equivalent circuit including the resistance r_b. (a) Derive the expression for the voltage gain transfer function $A_v(s) = V_o(s)/V_i(s)$. (b) If the transistor is biased at $I_{CQ} = 1\,\text{mA}$, and if $R_L = 4\,\text{k}\Omega$ and $\beta_o = 100$, determine the midband voltage gain for (i) $r_b = 100\,\Omega$ and (ii) $r_b = 500\,\Omega$. (c) For $C_1 = 2.2\,\text{pF}$, determine the $-3\,\text{dB}$ frequency for (i) $r_b = 100\,\Omega$ and (ii) $r_b = 500\,\Omega$.

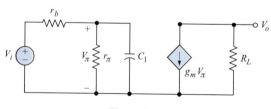

Figure P7.34

*7.35 A common-emitter equivalent circuit is shown in Figure P7.35. (a) What is the expression for the Miller capacitance? (b) Derive the expression for the voltage gain $A_v(s) = V_o(s)/V_i(s)$ in terms of the Miller capacitance and other circuit parameters. (c) What is the expression for the upper 3 dB frequency?

Figure P7.35

7.36 For the common-emitter circuit in Figure 7.37(a) in the text, assume that $r_s = \infty$, $R_1 \| R_2 = 5 \text{ k}\Omega$, and $R_C = R_L = 1 \text{ k}\Omega$. The transistor is biased at $I_{CQ} = 5 \text{ mA}$ and the parameters are: $\beta_o = 200$, $V_A = \infty$, $C_\mu = 5 \text{ pF}$, and $f_T = 250 \text{ MHz}$. Determine the upper 3 dB frequency for the small-signal current gain.

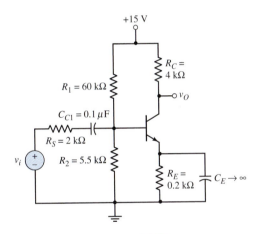

Figure P7.37

*7.37 For the common-emitter circuit in Figure P7.37, assume the emitter bypass capacitor C_E is very large, and the transistor parameters are: $\beta_o = 100$, $V_{BE}(\text{on}) = 0.7 \text{ V}$, $V_A = \infty$, $C_\mu = 2 \text{ pF}$, and $f_T = 400 \text{ MHz}$. Determine the lower and upper 3 dB frequencies for the small-signal voltage gain. Use the simplified hybrid-π model for the transistor.

Section 7.5   Frequency Response: The FET

7.38 A MOSFET is biased at $I_{DQ} = 100 \text{ μA}$, and the parameters are: $\left(\frac{1}{2}\right)\mu_n C_{ox} = 15 \text{ μA/V}^2$, $W = 40 \text{ μm}$, $L = 10 \text{ μm}$, $C_{gs} = 0.5 \text{ pF}$, and $C_{gd} = 0.05 \text{ pF}$. Determine f_T.

7.39 A common-source equivalent circuit is shown in Figure P7.39. The transistor transconductance is $g_m = 3 \text{ mA/V}$. (a) Calculate the equivalent Miller capacitance. (b) Determine the uper 3 dB frequency for the small-signal voltage gain.

Figure P7.39

7.40 Starting with the definition of unity-gain frequency, as given by Equation (7.90), neglect the overlap capacitance, assume $C_{gd} \cong 0$ and $C_{gs} \cong \left(\frac{2}{3}\right)WLC_{ox}$, and show that

$$f_T = \frac{3}{2\pi L} \cdot \sqrt{\frac{\mu_n I_D}{2C_{ox}WL}}$$

Since I_D is proportional to W, this relationship indicates that to increase f_T, the channel length L must be small.

7.41 For an ideal n-channel MOSFET, $(W/L) = 10$, $\mu_n = 400 \, \text{cm}^2/\text{V–s}$, $C_{ox} = 7.25 \times 10^{-8} \, \text{F/cm}^2$, and $V_{TN} = 0.65 \, \text{V}$. (a) Determine the maximum source resistance such that the transconductance g_m is reduced by no more than 20 percent from its ideal value when $V_{GS} = 5 \, \text{V}$. (b) Using the value of r_s calculated in part (a), determine how much g_m is reduced from its ideal value when $V_{GS} = 3 \, \text{V}$.

*7.42 Figure P7.42 shows the high-frequency equivalent circuit of an FET, including a source resistance r_s. (a) Derive an expression for the low-frequency current gain $A_i = I_o/I_i$. (b) Assuming R_i is very large, derive an expression for the current gain transfer function $A_i(s) = I_o(s)/I_i(s)$. (c) How does the magnitude of the current gain behave as r_s increases?

Figure P7.42   **Figure P7.43**

7.43 For the FET circuit in Figure P7.43, the transistor parameters are: $K_n = 1 \, \text{mA/V}^2$, $V_{TN} = 2 \, \text{V}$, $\lambda = 0$, $C_{gs} = 5 \, \text{pF}$, and $C_{gd} = 1 \, \text{pF}$. (a) Draw the simplified high-frequency equivalent circuit. (b) Calculate the equivalent Miller capacitance. (c) Determine the upper 3 dB frequency for the small-signal voltage gain and find the midband voltage gain.

Section 7.6   High-Frequency Response of Transistor Circuits

7.44 In the circuit in Figure P7.44, the transistor parameters are: $\beta = 120$, $V_{BE}(\text{on}) = 0.7\,\text{V}$, $V_A = 100\,\text{V}$, $C_\mu = 1\,\text{pF}$, and $f_T = 600\,\text{MHz}$. (a) Determine C_π and the equivalent Miller capacitance C_M. State any approximations or assumptions that you make. (b) Find the upper 3 dB frequency and the midband voltage gain.

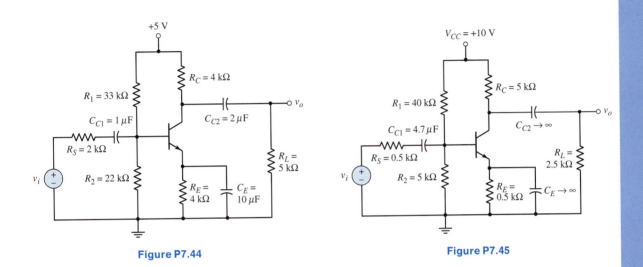

Figure P7.44   **Figure P7.45**

*7.45 In the circuit in Figure P7.45, the transistor parameters are: $\beta = 120$, $V_{BE}(\text{on}) = 0.7\,\text{V}$, $V_A = \infty$, $C_\mu = 3\,\text{pF}$, and $f_T = 250\,\text{MHz}$. Assume the emitter bypass capacitor C_E and the coupling capacitor C_{C2} are very large. (a) Determine the lower and upper 3 dB frequencies. Use the simplified hybrid-π model for the transistor. (b) Sketch the Bode plot of the voltage gain magnitude.

7.46 The parameters of the transistor in the common-source circuit in Figure P7.46 are: $K_p = 2\,\text{mA/V}^2$, $V_{TP} = -2\,\text{V}$, $\lambda = 0.01\,\text{V}^{-1}$, $C_{gs} = 10\,\text{pF}$, and $C_{gd} = 1\,\text{pF}$. (a) Determine the equivalent Miller capacitance C_M. (b) Find the upper 3 dB frequency and midband voltage gain.

Figure P7.46

7.47 For the PMOS common-source circuit shown in Figure P7.47, the transistor parameters are: $V_{TP} = -2\,\text{V}$, $K_p = 1\,\text{mA/V}^2$, $\lambda = 0$, $C_{gs} = 15\,\text{pF}$, and $C_{gd} = 3\,\text{pF}$. (a) Determine the upper 3 dB frequency. (b) What is the equivalent Miller capacitance? State any assumptions or approximations that you make. (c) Find the midband voltage gain.

Figure P7.47   Figure P7.48

*7.48 In the common-base circuit shown in Figure P7.48, the transistor parameters are: $\beta = 100$, $V_{BE}(\text{on}) = 0.7\,\text{V}$, $V_A = \infty$, $C_\pi = 10\,\text{pF}$, and $C_\mu = 1\,\text{pF}$. (a) Determine the upper 3 dB frequencies corresponding to the input and output portions of the equivalent circuit. (b) Calculate the small-signal midband voltage gain. (c) If a load capacitor $C_L = 15\,\text{pF}$ is connected between the output and ground, determine if the upper 3 dB frequency will be dominated by the C_L load capacitor or by the transistor characteristics.

*7.49 Repeat Problem 7.48 for the common-base circuit in Figure P7.49. Assume V_{EB} (on) $= 0.7$ for the pnp transistor. The remaining transistor parameters are the same as given in Problem 7.48.

Figure P7.49   Figure P7.50

*7.50 In the common-gate circuit in Figure P7.50, the transistor parameters are: $V_{TN} = 1\,\text{V}$, $K_n = 3\,\text{mA/V}^2$, $\lambda = 0$, $C_{gs} = 15\,\text{pF}$, and $C_{gd} = 4\,\text{pF}$. Determine the upper 3 dB frequency and midband voltage gain.

Figure P7.51

7.51 Consider the common-gate circuit in Figure P7.51 with parameters $V^+ = 5\,\text{V}$, $V^- = -5\,\text{V}$, $R_S = 4\,\text{k}\Omega$, $R_D = 2\,\text{k}\Omega$, $R_L = 4\,\text{k}\Omega$, $R_G = 50\,\text{k}\Omega$, and $R_i = 0.5\,\text{k}\Omega$. The transistor parameters are: $K_p = 1\,\text{mA/V}^2$, $V_{TP} = -0.8\,\text{V}$, $\lambda = 0$, $C_{gs} = 4\,\text{pF}$, and $C_{gd} = 1\,\text{pF}$. Determine the upper 3 dB frequency and midband voltage gain.

*7.52 For the cascode circuit in Figure 7.54 in the text, circuit parameters are the same as described in Example 7.16. The transistor parameters are: $\beta_o = 120$, $V_A = \infty$, $V_{BE}(\text{on}) = 0.7\,\text{V}$, $C_\pi = 12\,\text{pF}$, and $C_\mu = 2\,\text{pF}$. (a) If C_L is an open circuit, determine the 3 dB frequencies corresponding to the input and output portions of the equivalent circuit. (b) Determine the midband voltage gain. (c) If a load capacitance $C_L = 15\,\text{pF}$ is connected to the output, determine if the upper 3 dB frequency is dominated by the load capacitance or by the transistor characteristics.

COMPUTER SIMULATION PROBLEMS

*7.53 An emitter-follower circuit is shown in Figure P7.53. Assume the transistor parameters are: $\beta_0 = 100$, $V_A = \infty$, $C_\pi = 35\,\text{pF}$, and $C_\mu = 4\,\text{pF}$. From a PSpice analysis, determine the upper 3 dB frequency and midband voltage gain for: (a) $R_L = 0.2\,\text{k}\Omega$, (b) $R_L = 2\,\text{k}\Omega$, and (c) $R_L = 20\,\text{k}\Omega$. Explain any differences between the results.

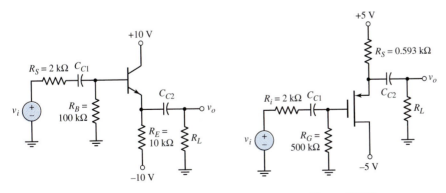

Figure P7.53   **Figure P7.54**

*7.54 For the source-follower circuit shown in Figure P7.54, assume the transistor parameters are: $V_{TP} = -2\,\text{V}$, $K_p = 2\,\text{mA/V}^2$, $\lambda = 0.02\,\text{V}^{-1}$, $C_{gs} = 5\,\text{pF}$, and $C_{gd} = 0.8\,\text{pF}$. From a PSpice analysis, determine the upper 3 dB frequency and midband voltage gain for: (a) $R_L = 0.2\,\text{k}\Omega$, (b) $R_L = 2\,\text{k}\Omega$, and (c) $R_L = 20\,\text{k}\Omega$. Explain any differences between the results.

*7.55 The emitter-follower is a wide bandwidth circuit, but the voltage gain is slightly less than unity. Figure P7.55 shows a cascade configuration of an emitter follower and a common emitter. Investigate the possibility of obtaining the properties of wide bandwidth from the emitter follower and a large voltage gain from the common emitter in a single circuit. Assume the transistor parameters are identical and are: $\beta_o = 150$, $V_A = \infty$, $C_\pi = 24\,\text{pF}$, and $C_\mu = 4\,\text{pF}$. Determine the upper 3 dB frequency and midband gain. How do these results compare to those of a cascode circuit?

Figure P7.55

*7.56 The transistor circuit in Figure P7.56 is a Darlington pair configuration. Assume $\beta_o = 100$ and $V_A = \infty$ for each transistor, the capacitance values of Q_1 and Q_2 are identical and given by $C_\pi = 24\,\text{pF}$ and $C_\mu = 4\,\text{pF}$. From a PSpice analysis, determine the upper 3 dB frequency and midband voltage gain for: (a) $R_{E1} = 10\,\text{k}\Omega$, (b) $R_{E1} = 40\,\text{k}\Omega$, and (c) $R_{E1} = \infty$. Explain any differences between the results.

Figure P7.56

*7.57 For the common-source circuit in Figure P7.57(a) and the NMOS cascode circuit in Figure P7.57(b), all transistors have the following identical parameters: $K_n = 1.2\,\text{mA/V}^2$, $V_{TN} = 2\,\text{V}$, $\lambda = 0$, $C_{gs} = 5\,\text{pF}$, and $C_{gd} = 0.8\,\text{pF}$. From a PSpice simulation of each circuit, determine the upper 3 dB frequencies and the midband voltage gains. Compare the 3 dB frequencies and midband voltage gains.

(a)

Figure P7.57(a)

(b)

Figure P7.57(b)

*7.58 For the circuit in Figure P7.58, the transistors are identical, and the parameters are: $K_n = 4\,\text{mA/V}^2$, $V_{TN} = 2\,\text{V}$, $\lambda = 0$, $C_{gs} = 10\,\text{pF}$, and $C_{gd} = 2\,\text{pF}$. The coupling capacitors are all equal at $C_C = 4.7\,\mu\text{F}$. Using a PSpice analysis, determine the lower and upper 3 dB frequencies. What is the bandwidth and midband voltage gain? What value of load capacitance will change the bandwidth by a factor of two?

Figure P7.58

DESIGN PROBLEMS

[Note: Each design should be verified with a computer analysis.]

*D7.59   A simplified high-frequency equivalent circuit of an FET amplifier with a source resistor R_S is shown in Figure P7.59. Including the source resistor decreases the small-signal voltage gain. Investigate the amplifier bandwidth as a function of the source resistance to determine the trade-offs required between gain and bandwidth in amplifier designs. (a) Derive an approximate single-pole expression for the voltage gain $A_v(s) = V_o(s)/V_i(s)$, the midband gain, and the upper 3 dB frequency. (b) Assume the circuit parameters are: $R = 1\,k\Omega$, $R_L = 4\,k\Omega$, $C_{gs} = 5\,pF$, $C_{gd} = 1\,pF$, and $g_m = 2\,mA/V$. Determine the magnitude of the midband gain and upper 3 dB frequency for $R_S = 0$, 100, 250, and 500 Ω. (c) Plot the gain–bandwidth versus source resistance.

Figure P7.59

*D7.60   (a) Design a common-emitter amplifier using a 2N2222A transistor biased at $I_{CQ} = 1\,mA$ and $V_{CEQ} = 10\,V$. The available power supplies are $\pm15\,V$, the load resistance is $R_L = 20\,k\Omega$, the source resistance is $R_S = 0.5\,k\Omega$, the input and output are ac coupled to the amplifier, and the lower 3 dB frequency is to be less than 10 Hz. Design the circuit to maximize the midband gain. What is the upper 3 dB frequency? (b) Repeat the design for $I_{CQ} = 50\,\mu A$. Assume f_T is the same as the case when $I_{CQ} = 1\,mA$. Compare the midband gain and bandwidth of the two designs.

*D7.61   Design a bipolar amplifier with a midband gain of $|A_v| = 50$ and a lower 3 dB frequency of 10 Hz. The available transistors are 2N2222A, and the available power supplies are $\pm10\,V$. All transistors in the circuit should be biased at approximately 0.5 mA. The load resistance is $R_L = 5\,k\Omega$, the source resistance is $R_S = 0.1\,k\Omega$, and the input and output are ac coupled to the amplifier. Compare the bandwidth of a single-stage design to that of a cascode design.

*D7.62   A common-emitter amplifier is designed to provide a particular midband gain and a particular bandwidth, using device A from Table P7.62. Assume $I_{CQ} = 1\,mA$. Investigate the effect on midband gain and bandwidth if devices B and C are inserted into the circuit. Which device provides the largest bandwidth? What is the gain–bandwidth product in each case?

Table P7.62   Device specifications for Problem 7.62

Device	f_T (MHz)	C_μ (pF)	β	r_b (Ω)
A	350	2	100	15
B	400	5	100	10
C	500	2	50	5

*D7.63 A simplified high-frequency equivalent circuit of a common-emitter amplifier is shown in Figure P7.63. The input signal is coupled into the amplifier through C_{C1}, the output signal is coupled to the load through C_{C2}, and the amplifier provides a midband gain of $|A_m|$ and an upper 3 dB frequency of f_H. Compare this single-stage amplifier design to one in which three amplifier stages are used between the signal and load. In the three-stage amplifier, assume all parameters are the same, except g_m for each stage is one-third that of the single-stage amplifier. Compare the midband gains and the bandwidths.

Figure P7.63

8

Output Stages and Power Amplifiers

8.0 PREVIEW

In previous chapters, we dealt mainly with small-signal voltage gains, current gains, and impedance characteristics. In this chapter, we analyze and design circuits that deliver a specified power to a load. We will, therefore, be concerned with power dissipation in transistors, especially in the output stage since the output stage must deliver the signal power. Linearity in the output signal is still a priority, however. A figure of merit for the output stage linearity characteristic is the total harmonic distortion that is present.

Initially, we will look at the characteristics of power BJTs and MOSFETs. Such characteristics include the current, voltage, and power ratings of these devices, as well as the safe operating area. The heat generated in these transistors from power dissipation must be removed in order to limit the device temperature to a specified maximum rated value. This maximum device temperature is a function of the thermal resistance between the transistor and the ambient and determines the maximum safe operating power of the transistor.

One important aspect in the design of power amplifiers is that it delivers the specified power to the load efficiently. Power amplifiers are classified according to the percent of time the output transistors are conducting. Three principal classes of power amplifiers are analyzed. In our discussion, we will determine the maximum possible conversion efficiency for each type of power amplifier.

An often-used output stage for power amplifiers, called a class-AB circuit, uses complementary pairs of transistors. We will analyze several configurations of this type of output stage, in both BJT and MOSFET configurations. One principal goal of this chapter is that the reader will be able to understand the characteristics of a class AB output stage and design one to meet particular specifications.

8.1 POWER AMPLIFIERS

A multistage amplifier may be required to deliver a large amount of power to a passive load. This power may be in the form of a large current delivered to a relatively small load resistance such as an audio speaker, or may be in the form of a large voltage delivered to a relatively large load resistance such as in a switching power supply. The output stage of the power amplifier must be designed to meet the power requirements. In this chapter, we are interested only in power amplifiers using BJTs or MOSFETS, and will not consider other types of power electronics that, for example, use thyristors.

Two important functions of the output stage are to provide a low output resistance so that it can deliver the signal power to the load without loss of gain and to maintain linearity in the output signal. A low output resistance implies the use of emitter-follower or source-follower circuit configurations. A measure of the linearity of the output signal is the **total harmonic distortion (THD)**. This figure of merit is the rms value of the harmonic components of the output signal, excluding the fundamental, expressed as a percentage of the fundamental.

A particular concern in the design of the output stage is to deliver the required signal power to the load efficiently. This specification implies that the power dissipated in the transistors of the output stage should be as small as possible. The output transistors must be capable of delivering the required current to the load, and must be capable of sustaining the required output voltage.

We will initially discuss power transistors and will then consider several output stages of power amplifiers.

8.2   POWER TRANSISTORS

In our previous discussions, we have ignored any physical transistor limitations in terms of maximum current, voltage, and power. We implicitly assumed that the transistors were capable of handling the current and voltage, and could handle the power dissipated within the transistor without suffering any damage.

However, since we are now discussing power amplifiers, we must be concerned with transistor limitations. The limitations involve: maximum rated current (on the order of amperes), maximum rated voltage (on the order of 100 V), and maximum rated power (on the order of watts or tens of watts).[1] We will consider these effects in the BJT and then in the MOSFET. The maximum power limitation is related to the maximum allowed temperature of the transistor, which in turn is a function of the rate at which heat is removed. We will therefore briefly consider heat sinks and heat flow.

8.2.1   Power BJTs

Power transistors are large-area devices. Because of differences in geometry and doping concentrations, their properties tend to vary from those of the small-signal devices. Table 8.1 compares the parameters of a general-purpose small-signal BJT to those of two power BJTs. The current gain is generally smaller in the power transistors, typically in the range of 20 to 100, and may be a strong function of collector current and temperature. Figure 8.1 shows typical current gain versus collector current characteristics for the 2N3055 power

---

[1]We must note that, in general, the maximum rated current and maximum rated voltage cannot occur at the same time.

Table 8.1 Comparison of the characteristics and maximum ratings of a small-signal and power BJT

Parameter	Small-Signal BJT (2N2222A)	Power BJT (2N3055)	Power BJT (2N6078)
V_{CE}(max) (V)	40	60	250
I_C(max) (A)	0.8	15	7
P_D(max) (W) (at $T = 25\,°C$)	1.2	115	45
β	35–100	5–20	12–70
f_T (MHz)	300	0.8	1

Figure 8.1 Typical dc beta characteristics (h_{FE} versus I_C) for 2N3055

BJT at various temperatures. At high current levels, the current gain tends to drop off significantly, and parasitic resistances in the base and collector regions may become significant, affecting the transistor terminal characteristics.

The **maximum rated collector current** $I_{C,\text{rated}}$ may be related to: the maximum current that the wires connecting the semiconductor to the external terminals can handle; the collector current at which the current gain falls below a minimum specified value; or the current that leads to the maximum power dissipation when the transistor is in saturation.

The maximum voltage limitation in a BJT is generally associated with avalanche breakdown in the reverse-biased base–collector junction. In the common-emitter configuration, the breakdown voltage mechanism also involves the transistor gain, as well as the breakdown phenomenon on the pn junction. Typical I_C versus V_{CE} characteristics are shown in Figure 8.2. The breakdown voltage when the base terminal is open circuited ($I_B = 0$) is V_{CEO}. From the data in Figure 8.2, this value is approximately 130 V.

When the transistor is biased in the active region, the collector current begins to increase significantly before breakdown voltage V_{CEO} is reached, and all the curves tend to merge to the same collector–emitter voltage once breakdown has occurred. The voltage at which these curves merge is denoted $V_{CE(\text{sus})}$ and is the minimum voltage necessary to sustain the transistor in breakdown. From the data in Figure 8.2, the value of $V_{CE(\text{sus})}$ is approximately 115 V.

Figure 8.2   Typical collector current versus collector–emitter voltage characteristics of a bipolar transistor, showing breakdown effects

Another breakdown effect is called **second breakdown**, which occurs in a BJT operating at high voltage and a fairly high current. Slight nonuniformities in current density produce local regions of increased heating that decreases the resistance of the semiconductor material, which in turn increases the current in those regions. This effect results in positive feedback, and the current continues to increase, producing a further increase in temperature, until the semiconductor material may actually melt, creating a short circuit between the collector and emitter and producing a permanent failure.

The instantaneous power dissipation in a BJT is given by

$$p_Q = v_{CE}i_C + v_{BE}i_B \tag{8.1}$$

The base current is generally much smaller than the collector current; therefore, to a good approximation, the instantaneous power dissipation is

$$p_Q \cong v_{CE}i_C \tag{8.2}$$

The average power, which is found by integrating Equation (8.2) over one cycle of the signal, is

$$\overline{P}_Q = \frac{1}{T}\int_0^T v_{CE}i_C dt \tag{8.3}$$

The average power dissipated in a BJT must be kept below a specified maximum value, to ensure that the temperature of the device remains below a maximum value. If we assume that the collector current and collector–emitter voltage are dc quantities, then at the **maximum rated power** P_T for the transistor, we can write

$$P_T = V_{CE}I_C \tag{8.4}$$

The maximum current, voltage, and power limitations can be illustrated on the I_C versus V_{CE} characteristics, as shown in Figure 8.3. The average power limitation P_T is a hyperbola described by Equation (8.4). The region where the transistor can be operated safely is known as the **safe operating area (SOA)** and is bounded by $I_{C,max}$, $V_{CE(sus)}$, P_T, and the transistor's second breakdown

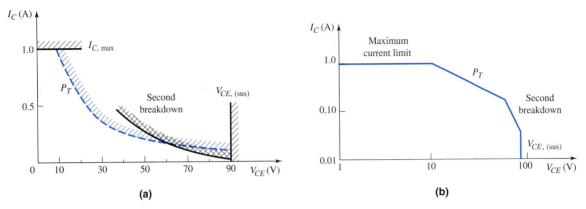

Figure 8.3 The safe operating area of a bipolar transistor plotted on: (a) linear scales and (b) logarithmic scales

characteristics curve. Figure 8.3(a) shows the safe operating area, using linear current and voltage scales; Figure 8.3(b) shows the same characteristics using logarithmic scales.

The i_C–v_{CE} operating point may move momentarily outside the safe operating area without damaging the transistor, but this depends on how far the Q-point moves outside the area and for how long. For our purposes, we will assume that the device must remain within the safe operating area at all times.

Example 8.1 **Objective:** Determine the required current, voltage, and power ratings of a power BJT.

Consider the common-emitter circuit in Figure 8.4. The parameters are $R_L = 8\,\Omega$ and $V_{CC} = 24\,\text{V}$.

Solution: For $V_{CE} \approx 0$, the maximum collector current is

$$I_C(\text{max}) = \frac{V_{CC}}{R_L} = \frac{24}{8} = 3\,\text{A}$$

For $I_C = 0$, the maximum collector–emitter voltage is

$$V_{CE}(\text{max}) = V_{CC} = 24\,\text{V}$$

The load line is given by

$$V_{CE} = V_{CC} - I_C R_L$$

and must remain within the safe operating area, as shown in Figure 8.5.

The transistor power dissipation is therefore

$$P_T = V_{CE} I_C = (V_{CC} - I_C R_L) I_C = V_{CC} I_C - I_C^2 R_L$$

The current at which the maximum power occurs is found by setting the derivative of this equation equal to zero as follows:

$$\frac{dP_T}{dI_C} = 0 = V_{CC} - 2 I_C R_L$$

Figure 8.4 Figure for Example 8.1

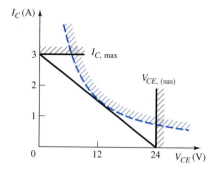

Figure 8.5 DC load line within the safe operating area

which yields

$$I_C = \frac{V_{CC}}{2R_L} = \frac{24}{2(8)} = 1.5 \, \text{A}$$

The C–E voltage at the maximum power point is

$$V_{CE} = V_{CC} - I_C R_L = 24 - (1.5)(8) = 12 \, \text{V}$$

The maximum power dissipation in the transistor occurs at the center of the load line. The maximum transistor power dissipation is therefore

$$P_T = V_{CE} I_C = 12(1.5) = 18 \, \text{W}$$

Comment: To find a transistor for a given application, safety factors are normally used. For this example, a transistor with a current rating greater than 3 A, a voltage rating greater than 24 V, and a power rating greater than 18 W would be required.

Power transistors, which are designed to handle large currents, require large emitter areas to maintain reasonable current densities. These transistors are usually designed with narrow emitter widths to minimize the parasitic base resistance, and may be fabricated as an **interdigitated structure**, as shown in Figure 8.6. Also, emitter ballast resistors, which are small resistors in each emitter leg, are usually incorporated in the design. These resistors help maintain equal currents in each B–E junction.

8.2.2 Power MOSFETs

Table 8.2 lists the basic parameters of two n-channel power MOSFETs. The drain currents are in the ampere range and the breakdown voltages are in the hundreds of volts range. These transistors must also operate within a safe operating area as discussed for the BJTs.

Table 8.2 Characteristics of two power MOSFETs

Parameter	2N6757	2N6792
$V_{DS}(\text{max})$ (V)	150	400
$I_D(\text{max})$ (at $T = 25\,^\circ\text{C}$)	8	2
P_D (W)	75	20

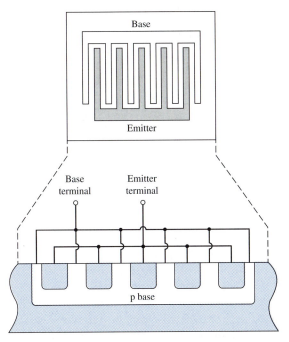

Figure 8.6 An interdigitated bipolar transistor structure showing the top view and cross-sectional view

Power MOSFETs differ from bipolar power transistors both in operating principles and performance. The superior performance characteristics of power MOSFETs are: faster switching times, no second breakdown, and stable gain and response time over a wide temperature range. Figure 8.7(a) shows the transconductance of the 2N6757 versus temperature. The variation with temperature of the MOSFET transconductance is less than the variation in the BJT current gain shown in Figure 8.1.

Figure 8.7 Typical characteristics for high-power MOSFETs: (a) transconductance versus drain current; (b) transfer characteristics

Also, since a MOSFET is a high input impedance, voltage-controlled device, the drive circuitry is simpler. The gate of a 10 A power MOSFET may be driven by the output of a standard logic circuit. In contrast, if the current gain of a 10 A BJT is $\beta = 10$, then a base current of 1 A is required for a collector current of 10 A. However, this required input current is much larger than the output drive capability of most logic circuits, which means that the drive circuitry for power BJTs is more complicated.

The MOSFET is a majority carrier device. Majority carrier mobility decreases with increasing temperature, which makes the semiconductor more resistive. This means that MOSFETs are more immune to the thermal runaway effects and second breakdown phenomena experienced in bipolars. Figure 8.7(b) shows typical I_D versus V_{GS} characteristics at several temperatures, clearly demonstrating that at high current levels, the current actually decreases with increasing temperature, for a given gate-to-source voltage.

Power MOSFETs are often manufactured by a vertical or double-diffused process, called VMOS or DMOS, respectively. The cross section of a VMOS device is shown in Figure 8.8(a) and the cross section of the DMOS device is shown in Figure 8.8(b). The DMOS process can be used to produce a large number of closely packed hexagonal cells on a single silicon chip, as shown in Figure 8.8(c). Also, such MOSFETs can be paralleled to form large-area devices, without the need of an equivalent emitter ballast resistance to equalize the current density. A single power MOSFET chip may contain as many as 25,000 paralleled cells.

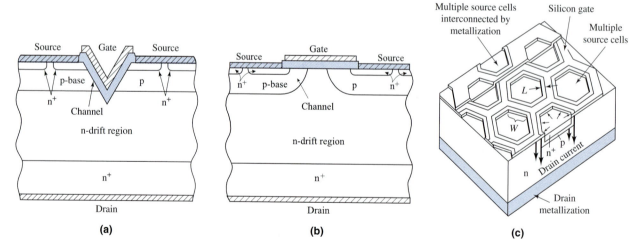

Figure 8.8   (a) Cross section of a VMOS device; (b) cross section of DMOS device; (c) HEXFET structure

Since the path between the drain and the source is essentially resistive, the **on resistance** $r_{ds}(\text{on})$ is an important parameter in the power capability of a MOSFET. Figure 8.9 shows a typical $r_{ds}(\text{on})$ characteristic as a function of drain current. Values in the tens of milliohm range have been obtained.

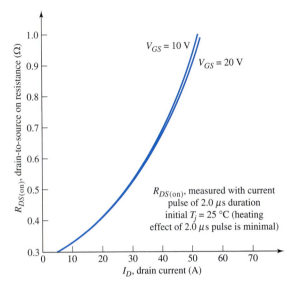

Figure 8.9 Typical drain-to-source resistance versus drain current characteristics of a MOSFET

8.2.3 Heat Sinks

The power dissipated in a transistor increases its internal temperature above the ambient temperature. If the device or junction temperature T_j becomes too high, the transistor may suffer permanent damage. Special precautions must be taken in packaging power transistors and in providing heat sinks so that heat can be conducted from the transistor. Figures 8.10(a) and (b) show two packaging schemes, and Figure 8.10(c) shows a typical heat sink.

(a)   **(b)**   **(c)**

Figure 8.10 Two packaging schemes: (a) and (b) for power transistors and (c) typical heat sink

To design a heat sink for a power transistor, we must first consider the concept of **thermal resistance** θ, which has units of °C/W. The temperature difference, $T_2 - T_1$, across an element with a thermal resistance θ is

$$T_2 - T_1 = P\theta \tag{8.5}$$

where P is the thermal power through the element. Temperature difference is the electrical analog of voltage, and power or heat flow is the electrical analog of current.

Manufacturers' data sheets for power devices generally give the maximum operating junction or device temperature $T_{j,max}$ and the thermal resistance from the junction to the case $\theta_{jc} = \theta_{dev-case}$.[2] By definition, the thermal resistance between the case and heat sink is $\theta_{case-snk}$, and between the heat sink and ambient is $\theta_{snk-amb}$.

The temperature difference between the device and the ambient can now be written as follows, when a heat sink is used:

$$T_{dev} - T_{amb} = P_D(\theta_{dev-case} + \theta_{case-snk} + \theta_{snk-amb}) \tag{8.6}$$

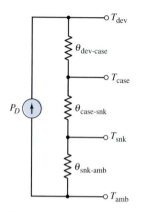

Figure 8.11 Electrical equivalent circuit for heat flow from the device to the ambient

where P_D is the power dissipated in the device. Equation (8.6) may also be modeled by its equivalent electrical elements, as shown in Figure 8.11. The temperature difference across the elements, such as the case and heat sink, is the dissipated power P_D multiplied by the applicable thermal resistance, which is $\theta_{case-snk}$ for this example.

If a heat sink is not used, the temperature difference between the device and ambient is written as

$$T_{dev} - T_{amb} = P_D(\theta_{dev-case} + \theta_{case-amb}) \tag{8.7}$$

where $\theta_{case-amb}$ is the thermal resistance between the case and ambient.

Example 8.2   Objective: Determine the maximum power dissipation in a transistor. Consider a power MOSFET for which the thermal resistance parameters are:

$$\theta_{dev-case} = 1.75\,°C/W \qquad \theta_{case-snk} = 1\,°C/W$$

$$\theta_{snk-amb} = 5\,°C/W \qquad \theta_{case-amb} = 50\,°C/W$$

The ambient temperature is $T_{amb} = 30\,°C$, and the maximum junction or device temperature is $T_{j,max} = T_{dev} = 150\,°C$.

Solution: When no heat sink is used, the maximum device power dissipation is found from Equation (8.7) as

$$P_{D,max} = \frac{T_{j,max} - T_{amb}}{\theta_{dev-case} + \theta_{case-amb}} = \frac{150 - 30}{1.75 + 50} = 2.32\,W$$

When a heat sink is used, the maximum device power dissipation is found from Equation (8.6) as

$$P_{D,max} = \frac{T_{j,max} - T_{amb}}{\theta_{dev-case} + \theta_{case-snk} + \theta_{snk-amb}}$$

$$= \frac{150 - 30}{1.75 + 1 + 5} = 15.5\,W$$

Comment: These results illustrate that the use of a heat sink allows more power to be dissipated in the device, while keeping the device temperature at or below its maximum limit.

---

[2]In this short discussion, we use a more descriptive subscript notation to help clarify the discussion.

The maximum safe power dissipation in a device is a function of: (1) the temperature difference between the junction and case, and (2) the thermal resistance between the device and the case $\theta_{\text{dev}-\text{case}}$, or

$$P_{D,\text{max}} = \frac{T_{j,\text{max}} - T_{\text{case}}}{\theta_{\text{dev}-\text{case}}} \qquad \qquad \textbf{(8.8)}$$

A plot of $P_{D,\text{max}}$ versus T_{case}, called the **power derating curve** of the transistor, is shown in Figure 8.12. The temperature at which the power derating curve crosses the horizontal axis corresponds to T_{jmax}. At this temperature, no additional temperature rise in the device can be tolerated; therefore, the allowed power dissipation must be zero, which implies a zero input signal.

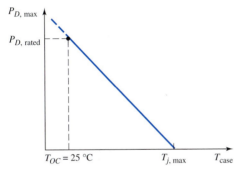

Figure 8.12 A power derating curve

The rated power of a device is generally defined as the power at which the device reaches its maximum temperature, while the case temperature remains at room or ambient temperature, that is, $T_{\text{case}} = 25\,°\text{C}$. Maintaining the case at ambient temperature implies that the thermal resistance between the case and ambient is zero, or that an infinite heat sink is used. However, with nonzero values of $\theta_{\text{case}-\text{snk}}$ and $\theta_{\text{snk}-\text{amb}}$, the case temperature rises above the ambient, and the maximum rated power of the device cannot be achieved. This effect can be seen by examining the equivalent circuit model in Figure 8.11. If the device temperature is at its maximum allowed value of $T_{\text{dev}} = T_{j,\text{max}}$, then as T_{case} increases, the temperature difference across $\theta_{\text{dev}-\text{case}}$ decreases, which means that the power through the element must decrease.

Example 8.3 **Objective:** Determine the maximum safe power dissipation in a transistor.

Consider a BJT with a rated power of 20 W and a maximum junction temperature of $T_{j,\text{max}} = 175\,°\text{C}$. The transistor is mounted on a heat sink with parameters $\theta_{\text{case}-\text{snk}} = 1\,°\text{C/W}$ and $\theta_{\text{snk}-\text{amb}} = 5\,°\text{C/W}$.

Solution: From Equation (8.8), the device-to-case thermal resistance is

$$\theta_{\text{dev}-\text{case}} = \frac{T_{j,\text{max}} - T_{OC}}{P_{D,\text{rated}}} = \frac{175 - 25}{20} = 7.5\,°\text{C/W}$$

From Equation (8.6), the maximum power dissipation is

$$P_{D,\max} = \frac{T_{j,\max} - T_{amb}}{\theta_{dev-case} + \theta_{case-snk} + \theta_{snk-amb}}$$

$$= \frac{175 - 25}{7.5 + 1 + 5} = 11.1\,\text{W}$$

Comment: The actual maximum safe power dissipation in a device may be less than the rated value. This occurs when the case temperature cannot be held at the ambient temperature, because of the nonzero thermal resistance factors between the case and ambient.

Test Your Understanding

Figure 8.13 Figure for Exercise 8.1 and Example 8.4

8.1 Consider the common-source circuit shown in Figure 8.13. The parameters are $R_D = 20\,\Omega$ and $V_{DD} = 24\,\text{V}$. Determine the required current, voltage, and power ratings of the MOSFET. (Ans. $I_D(\max) = 1.2\,\text{A}$, $V_{DS}(\max) = 24\,\text{V}$, $P_D(\max) = 7.2\,\text{W}$)

8.2 Assume that the BJT in the common-emitter circuit shown in Figure 8.4 has limiting factors of: $I_{C,\max} = 2\,\text{A}$, $V_{CE(sus)} = 50\,\text{V}$, and $P_T = 10\,\text{W}$. Neglecting second breakdown effects, determine the minimum value of R_L such that the Q-point of the transistor always stays within the safe operating area for: (a) $V_{CC} = 30\,\text{V}$, and (b) $V_{CC} = 15\,\text{V}$. In each case, determine the maximum collector current and maximum transistor power dissipation. (Ans. (a) $R_L = 22.5\,\Omega$, $I_{c,\max} = 1.33\,\text{A}$, $P_{Q,\max} = 10\,\text{W}$ (b) $R_L = 7.5\,\Omega$, $I_{C,\max} = 2\,\text{A}$, $P_{Q,\max} = 7.5\,\text{W}$)

8.3 For the emitter-follower circuit in Figure 8.14, the parameters are $V_{CC} = 10\,\text{V}$ and $R_E = 200\,\Omega$. The transistor current gain is $\beta = 150$, and the current and voltage limitations are $I_{C,\max} = 200\,\text{mA}$ and $V_{CE(sus)} = 50\,\text{V}$. Determine the minimum transistor power rating such that the transistor Q-point is always inside the safe operating area. (Ans. $P_{\max} = 0.5\,\text{W}$)

Figure 8.14 Figure for Exercise 8.3

8.4 A power MOSFET with $\theta_{dev-case} = 3\,°\text{C/W}$ is operating with an average drain current of $\bar{I}_D = 1\,\text{A}$ and an average drain-source voltage of $\bar{V}_{DS} = 12\,\text{V}$. The device is mounted on a heat sink with parameters $\theta_{snk-amb} = 4\,°\text{C/W}$ and $\theta_{case-snk} = 1\,°\text{C/W}$. If the ambient temperature is $T_{amb} = 25\,°\text{C}$, determine the temperature of the: (a) device, (b) case, and (c) heat sink. (Ans. (a) $121\,°\text{C}$ (b) $85\,°\text{C}$ (c) $73\,°\text{C}$)

8.5 The rated power of a power BJT is $P_{D,rated} = 50\,\text{W}$, the maximum allowed junction temperature is $T_{j,\max} = 200\,°\text{C}$, and the ambient temperature is $T_{amb} = 25\,°\text{C}$. The thermal resistance between the heat sink and air is $\theta_{snk-amb} = 2\,°\text{C/W}$, and that between the case and heat sink is $\theta_{case-snk} = 0.5\,°\text{C/W}$. Find the maximum safe power dissipation and the temperature of the case. (Ans. $P_{D,\max} = 29.2\,\text{W}$, $T_{case} = 98\,°\text{C}$)

8.3   CLASSES OF AMPLIFIERS

Power amplifiers are classified according to the percent of time the output transistors are conducting, or "turned on." The four principal classifications are: class A, class B, class AB, and class C. These classifications are illustrated in Figure 8.15, for a sinusoidal input signal. In **class-A operation**, an output transistor is biased at a quiescent current I_Q and conducts for the entire cycle of the input signal. For **class-B operation**, an output transistor conducts for only

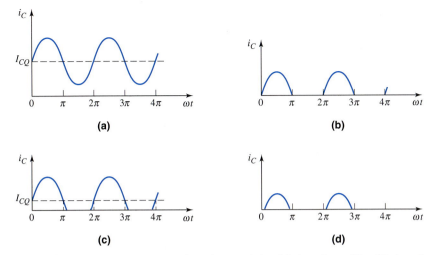

Figure 8.15 Collector current versus time characteristics: (a) class-A amplifier, (b) class-B amplifier, (c) class-AB amplifier, and (d) class-C amplifier

one-half of each sine wave input cycle. In **class-AB operation**, an output transistor is biased at a small quiescent current I_Q and conducts for slightly more than half a cycle. In contrast, in **class-C operation** an output transistor conducts for less than half a cycle. We will analyze the biasing, load lines, and efficiency of each class of power amplifier.

8.3.1 Class-A Operation

The small-signal amplifiers considered in Chapters 4 and 6 were all biased for class-A operation. A basic common-emitter configuration is shown in Figure 8.16(a). The bias circuitry has been omitted, for convenience. Also, in this **standard class-A amplifier** configuration, no inductors or transformers are used.

The dc load line is shown in Figure 8.16(b). The Q-point is assumed to be in the center of the load line, so that $V_{CEQ} = V_{CC}/2$. If a sinusoidal input signal

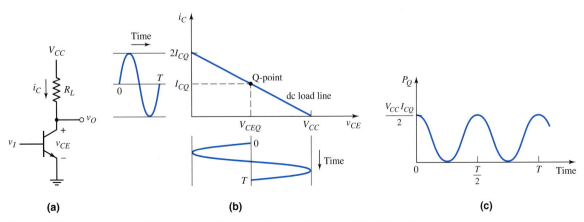

Figure 8.16 (a) Common-emitter amplifier, (b) dc load line, and (c) instantaneous power dissipation versus time in the transistor

is applied, sinusoidal variations are induced in the collector current and collector–emitter voltage. The absolute possible variations are shown in the figure, although values of $v_{CE} = 0$ and $i_C = 2I_{CQ}$ cannot actually be attained.

The instantaneous power dissipation in the transistor, neglecting the base current, is

$$p_Q = v_{CE}i_C \qquad \qquad (8.9)$$

For a sinusoidal input signal, the collector current and collector-emitter voltage can be written

$$i_C = I_{CQ} + I_p \sin \omega t \qquad \qquad (8.10(a))$$

and

$$v_{CE} = \frac{V_{CC}}{2} - V_p \sin \omega t \qquad \qquad (8.10(b))$$

If we consider the absolute possible variations, then $I_p = I_{CQ}$ and $V_p = V_{CC}/2$. Therefore, the instantaneous power dissipation in the transistor, from Equation (8.9), is

$$p_Q = \frac{V_{CC}I_{CQ}}{2}(1 - \sin^2 \omega t) \qquad \qquad (8.11)$$

Figure 8.16(c) is a plot of the instantaneous transistor power dissipation. Since the maximum power dissipation corresponds to the quiescent value (see Figure 8.5), the transistor must be capable of handling a continuous power dissipation of $V_{CC}I_{CQ}/2$ when the input signal is zero.

The **power conversion efficiency** is defined as

$$\eta = \frac{\text{signal load power } (\overline{P}_L)}{\text{supply power}(\overline{P}_S)} \qquad \qquad (8.12)$$

where $\overline{P}_L$ is the average ac power delivered to the load and $\overline{P}_S$ is the average power supplied by the V_{CC} power source(s). For the standard class-A amplifier and sinusoidal input signals, the average ac power delivered to the load is $\left(\frac{1}{2}\right)V_pI_p$. Using the absolute possible variations, we have

$$\overline{P}_L(\text{max}) = \left(\frac{1}{2}\right)\left(\frac{V_{CC}}{2}\right)(I_{CQ}) = \frac{V_{CC}I_{CQ}}{4} \qquad \qquad (8.13)$$

The average power supplied by the V_{CC} source is

$$\overline{P}_S = V_{CC}I_{CQ} \qquad \qquad (8.14)$$

The maximum attainable conversion efficiency is therefore

$$\eta(\text{max}) = \frac{\frac{1}{4}V_{CC}I_{CQ}}{V_{CC}I_{CQ}} \Rightarrow 25\% \qquad \qquad (8.15)$$

We must keep in mind that the maximum possible conversion efficiency may change when a load is connected to the output of the amplifier. This efficiency is relatively low; therefore, standard class-A amplifiers are normally not used when signal powers greater than approximately 1 W are required.

Design Pointer: We must emphasize that in practice, a maximum signal voltage of $V_{CC}/2$ and a maximum signal current of I_{CQ} are not possible. The output signal voltage must be limited to smaller values in order to avoid transistor saturation and cutoff, and the resulting nonlinear distortion. The calculation for the maximum possible efficiency also neglects power dissipation in the bias circuitry. Consequently, the realistic maximum conversion efficiency in a standard class-A amplifier is on the order of 20 percent or less.

Example   8.4   **Objective:**   Calculate the actual efficiency of a class-A output stage.

Consider the common-source circuit in Figure 8.13. The circuit parameters are $V_{DD} = 10\,\text{V}$ and $R_D = 5\,\text{k}\Omega$, and the transistor parameters are: $K_n = 1\,\text{mA/V}^2$, $V_{TN} = 1\,\text{V}$, and $\lambda = 0$. Assume the output voltage swing is limited to the range between the transition point and $v_{DS} = 9\,\text{V}$, to minimize nonlinear distortion.

Solution: The load line is given by

$$V_{DS} = V_{DD} - I_D R_D$$

At the transition point, we have

$$V_{DS}(\text{sat}) = V_{GS} - V_{TN}$$

and

$$I_D = K_n(V_{GS} - V_{TN})^2$$

Combining these expressions, the transition point is determined from

$$V_{DS}(\text{sat}) = V_{DD} - K_n R_D V_{DS}^2(\text{sat})$$

or

$$(1)(5)V_{DS}^2(\text{sat}) + V_{DS}(\text{sat}) - 10 = 0$$

which yields

$$V_{DS}(\text{sat}) = 1.32\,\text{V}$$

To obtain the maximum symmetrical swing under the conditions specified, we want the Q-point midway between $V_{DS} = 1.32\,\text{V}$ and $V_{DS} = 9\,\text{V}$, or

$$V_{DSQ} = 5.16\,\text{V}$$

The maximum ac component of voltage across the load resistor is then

$$v_r = 3.84 \sin \omega t$$

and the average power delivered to the load is

$$\overline{P}_L = \frac{1}{2} \cdot \frac{(3.84)^2}{5} = 1.47\,\text{mW}$$

The quiescent drain current is found to be

$$I_{DQ} = \frac{10 - 5.16}{5} = 0.968\,\text{mA}$$

The average power supplied by the V_{DD} source is

$$\overline{P}_S = V_{DD} I_{DQ} = (10)(0.968) = 9.68\,\text{mW}$$

and the power conversion efficiency, from Equation (8.12), is

$$\eta = \frac{\overline{P}_L}{\overline{P}_S} = \frac{1.47}{9.68} \Rightarrow 15.2\%$$

Comment: By limiting the swing in the drain–source voltage, to avoid nonsaturation and cutoff and the resulting nonlinear distortion, we reduce the output stage power conversion efficiency considerably, compared to the theoretical maximum possible value of 25 percent for the standard class-A amplifier.

Class-A operation also applies to the emitter-follower, common-base, source-follower, and common-gate configurations. As previously stated, the circuits considered in Figures 8.13 and 8.16(a) are standard class-A amplifiers in that no inductors or transformers are used. Later in this chapter, we will analyze inductively coupled and transformer-coupled power amplifiers that also operate in the class-A mode. We will show that, for these circuits, the maximum conversion efficiency is 50 percent.

Test Your Understanding

8.6 Consider the common-emitter output stage shown in Figure 8.16(a). Let $V_{CC} = 15\,\text{V}$ and $R_L = 1\,\text{k}\Omega$, and assume the Q-point is in the center of the load line. (a) Find the quiescent power dissipated in the transistor. (b) If the sinusoidal output signal is limited to a 13 V peak-to-peak value, determine: the average signal power delivered to the load, the power conversion efficiency, and the average power dissipated in the transistor. (Ans. (a) $P_Q = 56.3\,\text{mW}$ (b) $\overline{P}_L = 21.1\,\text{mW}$, $\eta = 18.7\%$, $\overline{P}_Q = 35.2\,\text{mW}$)

Figure 8.17   Figure for Exercise 8.7

*8.7   For the common-source circuit shown in Figure 8.17, the Q-point is $V_{DSQ} = 4\,\text{V}$. (a) Find I_{DQ}. (b) The minimum value of the instantaneous drain current must be no less than $\left(\frac{1}{10}\right)I_{DQ}$, and the minimum value of the instantaneous drain–source voltage must be no less than $v_{DS} = 1.5\,\text{V}$. Determine the maximum peak-to-peak amplitude of a symmetrical sinusoidal output voltage. (c) For the conditions of part (b), calculate the power conversion efficiency, where the signal power is the power delivered to R_L. (Ans. (a) $I_{DQ} = 60\,\text{mA}$ (b) $V_{p-p} = 5.0\,\text{V}$ (c) $\overline{P}_L = 31.25\,\text{mW}$, $\eta = 5.2\%$)

8.3.2   Class-B Operation

Figure 8.18 shows an output stage that consists of a complementary pair of bipolar transistors. When the input voltage is $v_I = 0$, both transistors are cut off and the output voltage is $v_O = 0$. If we assume a B–E cut-in voltage of 0.6 V, then the output voltage v_O remains zero as long as the input voltage is in the range $-0.6 \leq v_I \leq +0.6\,\text{V}$.

If v_I becomes positive and is greater than 0.6 V, then Q_n turns on and operates as an emitter follower. The load current i_L is positive and is supplied through Q_n, and the B–E junction of Q_p is reverse biased. If v_I becomes negative by more than 0.6 V, then Q_p turns on and operates as an emitter follower. Transistor Q_p is a sink for the load current, which means that i_L is negative.

This circuit is called a **complementary push–pull** output stage. Transistor Q_n conducts during the positive half of the input cycle, and Q_p conducts during the negative half-cycle. The transistors do not both conduct at the same time.

Figure 8.18 Basic complementary push–pull output stage

Crossover Distortion

Figure 8.19 shows the voltage transfer characteristics for this circuit. When either transistor is conducting, the voltage gain, which is the slope of the curve, is essentially unity as a result of the emitter follower. Also, there is a range of input voltage around zero volts where both transistors are cut off and v_O is zero. This portion of the curve is called the **dead band**, and it produces a **crossover distortion**, as illustrated in Figure 8.20, for a sinuosidal input signal. (Crossover distortion can be virtually eliminated by biasing both Q_n and Q_p with a small quiescent collector current when v_I is zero. This techique is discussed in the next section.)

Figure 8.19   Voltage transfer characteristics of basic complementary push–pull output stage

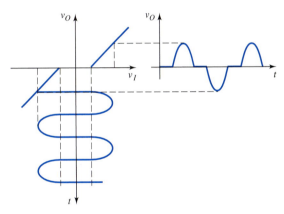

Figure 8.20   Crossover distortion of basic complementary push–pull output stage

---

Example   8.5   **Objective:** Determine the total harmonic distortion (THD) of the class B complementary push–pull output stage in Figure 8.18.

A PSpice analysis was performed, which yielded the harmonic content of the output signal.

Solution: A 1 kHz sinusoidal signal with an amplitude of 2 V was applied to the input of the circuit shown in Figure 8.18. The circuit was biased at ±10 V. The transistors used in the circuit were 2N3904 npn and 2N3906 pnp devices. A 1 kΩ load was connected to the output.

The harmonic content for the first nine harmonics is shown in Table 8.3. We see that the output is rich in odd harmonics with the 3 kHz third harmonic being 18 percent as large as the 1 kHz principal output signal. The total harmonic distortion is 19.7 percent, which is large.

Table 8.3 Harmonic content for Example 8.5

Frequency (Hz)	Fourier component	Normalized component	Phase (degrees)
1.000E+03	1.151E+00	1.000E+00	-1.626E-01
2.000E+03	6.313E-03	5.485E-03	-9.322E+01
3.000E+03	2.103E-01	1.827E-01	-1.793E+02
4.000E+03	4.984E-03	4.331E-03	-9.728E+01
5.000E+03	8.064E-02	7.006E-02	-1.792E+02
6.000E+03	3.456E-03	3.003E-03	-9.702E+01
7.000E+03	2.835E-02	2.464E-02	1.770E+02
8.000E+03	2.019E-03	1.754E-03	-8.029E+01
9.000E+03	6.679E-03	5.803E-03	1.472E+02

TOTAL HARMONIC DISTORTION = 1.974899E+01 PERCENT

Comment: These results show the obvious effects of the dead band region. If the input signal amplitude increases, the total harmonic distortion decreases, but if the amplitude decreases, the total harmonic distortion will increase above the 19 percent value.

Power Efficiency

If we consider an idealized version of the circuit in Figure 8.18 in which the base–emitter turn-on voltages are zero, then each transistor would conduct for exactly one-half cycle of the sinusoidal input signal. This circuit would be a class-B output stage, and the output voltage and load current would be replicas of the input signal. The collector–emitter voltages would also show the same sinusoidal variation.

Figure 8.21 illustrates the applicable dc load line. The Q-point is at zero collector current, or at cutoff for both transistors. The quiescent power dissipation in each transistor is then zero.

The output voltage for this idealized class-B output stage can be written

$$v_O = V_p \sin \omega t \tag{8.16}$$

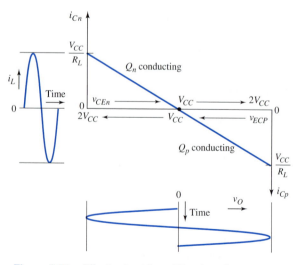

Figure 8.21   Effective load line of the class-B output stage

where the maximum possible value of V_p is V_{CC}.

The instantaneous power dissipation in Q_n is

$$p_{Qn} = v_{CEn} i_{Cn} \tag{8.17}$$

and the collector current is

$$i_{Cn} = \frac{V_p}{R_L} \sin \omega t \tag{8.18(a)}$$

for $0 \le \omega t \le \pi$, and

$$i_{Cn} = 0 \tag{8.18(b)}$$

for $\pi \le \omega t \le 2\pi$, where V_p is the peak output voltage.

From Figure 8.21, we see that the collector–emitter voltage can be written as

$$v_{CEn} = V_{CC} - V_p \sin \omega t \tag{8.19}$$

Therefore, the total instantaneous power dissipation in Q_n is

$$p_{Qn} = (V_{CC} - V_p \sin \omega t)\left(\frac{V_p}{R_L} \sin \omega t\right) \tag{8.20}$$

for $0 \le \omega t \le \pi$, and

$$p_{Qn} = 0$$

for $\pi \le \omega t \le 2\pi$. The average power dissipation is therefore

$$\overline{P}_{Qn} = \frac{V_{CC} V_p}{\pi R_L} - \frac{V_p^2}{4 R_L} \tag{8.21}$$

The average power dissipation in transistor Q_p is exactly the same as that for Q_n, because of symmetry.

A plot of the average power dissipation in each transistor, as a function of V_p, is shown in Figure 8.22. The power dissipation first increases with increasing output voltage, reaches a maximum, and finally decreases with increasing V_p. We determine the maximum average power dissipation by setting the derivative of $\overline{P}_{Qn}$ with respect to V_p equal to zero, producing

$$\overline{P}_{Qn}(\text{max}) = \frac{V_{CC}^2}{\pi^2 R_L} \tag{8.22}$$

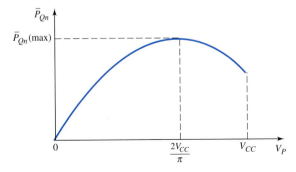

Figure 8.22 Average power dissipation in each transistor versus peak output voltage for class-B output stage

which occurs when

$$V_p\big|_{\overline{P}_{Q_n}(\max)} = \frac{2V_{CC}}{\pi} \tag{8.23}$$

The average power delivered to the load is

$$\overline{P}_L = \frac{1}{2} \cdot \frac{V_p^2}{R_L} \tag{8.24}$$

Since the current supplied by each power supply is half a sine wave, the average current is $V_p/(\pi R_L)$. The average power supplied by each source is therefore

$$\overline{P}_{S+} = \overline{P}_{S-} = V_{CC}\left(\frac{V_p}{\pi R_L}\right) \tag{8.25}$$

and the total average power supplied by the two sources is

$$\overline{P}_S = 2V_{CC}\left(\frac{V_p}{\pi R_L}\right) \tag{8.26}$$

From Equation (8.12), the conversion efficiency is

$$\eta = \frac{\dfrac{1}{2} \cdot \dfrac{V_p^2}{R_L}}{2V_{CC}\left(\dfrac{V_p}{\pi R_L}\right)} = \frac{\pi}{4} \cdot \frac{V_p}{V_{CC}} \tag{8.27}$$

The maximum possible efficiency, which occurs when $V_p = V_{CC}$, is

$$\eta(\max) = \frac{\pi}{4} \Rightarrow 78.5\% \tag{8.28}$$

This maximum efficiency value is substantially larger than that of the standard class-A amplifier.

From Equation (8.24), we find the maximum possible average power that can be delivered to the load, as follows:

$$\overline{P}_L(\max) = \frac{1}{2} \cdot \frac{V_{CC}^2}{R_L} \tag{8.29}$$

The actual conversion efficiency obtained in practice is less than the maximum value because of other circuit losses, and because the peak output voltage must remain less than V_{CC} to avoid transistor saturation. As the output voltage amplitude increases, output signal distortion also increases. To limit this distortion to an acceptable level, the peak output voltage is usually limited to several volts below V_{CC}. From Figure 8.22 and Equation (8.23), we see that the maximum transistor power dissipation occurs when $V_p = 2V_{CC}/\pi$. At this peak output voltage, the conversion efficiency of the class-B amplifier is, from Equation (8.27),

$$\eta = \frac{\pi}{4V_{CC}} \cdot V_p = \left(\frac{\pi}{4V_{CC}}\right) \cdot \left(\frac{2V_{CC}}{\pi}\right) = \frac{1}{2} \Rightarrow 50\% \tag{8.30}$$

Test Your Understanding

8.8 Design an idealized class-B output stage, as shown in Figure 8.18, to deliver an average of 25 W to an 8 Ω speaker. The peak output voltage must be no larger than 80 percent of supply voltages V_{CC}. Determine: (a) the required value of V_{CC}, (b) the peak current in each transistor, (c) the average power dissipated in each transistor, and (d) the power conversion efficiency. (Ans. (a) $V_{CC} = 25$ V (b) $I_p = 2.5$ A (c) $\overline{P}_Q = 7.4$ W (d) $\eta = 62.8\%$)

8.9 For the idealized class-B output stage shown in Figure 8.18, the parameters are $V_{CC} = 5$ V and $R_L = 100$ Ω. The measured output signal is $v_o = 4 \sin \omega t$ (V). Determine: (a) the average signal load power, (b) the peak current in each transistor, (c) the average power dissipated in each transistor, and (d) the power conversion efficiency. (Ans. (a) $\overline{P}_L = 80$ mW (b) $I_p = 40$ mA (c) $\overline{P}_Q = 23.7$ mW (d) $\eta = 62.8\%$)

8.3.3 Class-AB Operation

Crossover distortion can be virtually eliminated by applying a small quiescent bias on each output transistor, for a zero input signal. This is called a class-AB output stage and is shown schematically in the circuit in Figure 8.23. If Q_n and Q_p are matched, then for $v_I = 0$, $V_{BB}/2$ is applied to the B–E junction of Q_n, $V_{BB}/2$ is applied to the E–B junction of Q_p, and $v_O = 0$. The quiescent collector currents in each transistor are given by

$$i_{Cn} = i_{Cp} = I_S e^{V_{BB}/2V_T} \tag{8.31}$$

As v_I increases, the voltage at the base of Q_n increases and v_O increases. Transistor Q_n operates as an emitter follower, supplying the load current to R_L. The output voltage is given by

$$v_O = v_I + \frac{V_{BB}}{2} - v_{BEn} \tag{8.32}$$

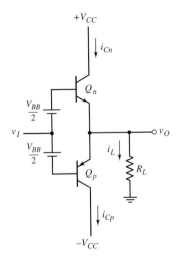

Figure 8.23 Bipolar class-AB output stage

and the collector current of Q_n (neglecting base currents) is

$$i_{Cn} = i_L + i_{Cp} \qquad\qquad (8.33)$$

Since i_{Cn} must increase to supply the load current, v_{BEn} increases. Assuming V_{BB} remains constant, as v_{BEn} increases, v_{EBp} decreases resulting in a decrease in i_{Cp}.

As v_I goes negative, the voltage at the base of Q_p decreases and v_O decreases. Transistor Q_p operates as an emitter follower, sinking current from the load. As i_{Cp} increases, v_{EBp} increases, causing a decrease in v_{BEn} and i_{Cn}.

Figure 8.24(a) shows the voltage transfer characteristics for this class-AB output stage. If v_{BEn} and v_{EBp} do not change significantly, then the voltage gain, or the slope of the transfer curve, is essentially unity. A sinusoidal input signal voltage and the resulting collector currents and load current are shown in Figures 8.24(b), (c), and (d). Each transistor conducts for more than one-half cycle, which is the definition of class-AB operation.

There is a relationship between i_{Cn} and i_{Cp}. We know that

$$v_{BEn} + v_{EBp} = V_{BB} \qquad\qquad (8.34(a))$$

which can be written

$$V_T \ln\left(\frac{i_{Cn}}{I_S}\right) + V_T \ln\left(\frac{i_{Cp}}{I_S}\right) = 2V_T \ln\left(\frac{I_{CQ}}{I_S}\right) \qquad\qquad (8.34(b))$$

Combining terms in Equation (8.34(b)), we find

$$i_{Cn} i_{Cp} = I_{CQ}^2 \qquad\qquad (8.35)$$

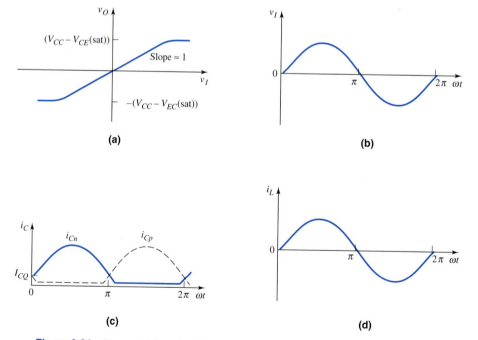

(a)

(b)

(c)

(d)

Figure 8.24   Characteristics of a class-AB output stage: (a) voltage transfer curve, (b) sinusoidal input signal, (c) collector currents, and (d) output current

The product of i_{Cn} and i_{Cp} is a cosntant; therefore, if i_{Cn} increases, i_{Cp} decreases, but does not go to zero.

Since, for a zero input signal, quiescent collector currents exist in the output transistors, the average power supplied by each source and the average power dissipated in each transistor are larger than for a class-B configuration. This means that the power conversion efficiency for a class-AB output stage is less than that for an idealized class-B circuit. In addition, the required power handling capability of the transistors in a class-AB circuit must be slightly larger than in a class-B circuit. However, since the quiescent collector currents I_{CQ} are usually small compared to the peak current, this increase in power dissipation is not great. The advantage of eliminating crossover distortion in the class-AB output stage greatly outweighs the slight disadvantage of reduced conversion efficiency and increased power dissipation.

Example 8.6   **Objective:** Determine the total harmonic distortion (THD) of the class AB complementary push–pull output stage shown in Figure 8.23.

A PSpice analysis was performed, which yielded the harmonic content of the output signal.

Solution: A 1 kHz sinusoidal signal with an amplitude of 2 V was applied to the input of the circuit. The bias voltages $V_{BB}/2$ were varied. The circuit was biased at ± 10 V and a 1 kΩ load was connected to the output. Shown in Table 8.4 are the $V_{BB}/2$ bias voltages applied, the quiescent transistor currents, and the total harmonic distortion (THD).

Table 8.4 Quiescent collector currents and total harmonic distortion of class-AB circuit

$V_{BB}/2$ (V)	I_{CQ} (mA)	THD (%)
0.60	0.048	1.22
0.65	0.33	0.244
0.70	2.20	0.0068
0.75	13.3	0.0028

Discussion: With a peak input voltage of 2 V and a 1 kΩ load, the peak load current is on the order of 2 mA. From the results shown in Table 8.4, the THD decreases as the ratio of quiescent transistor current to peak load current increases. In other words, for a given input voltage, the smaller the variation in collector current when the signal is applied compared to the quiescent collector current, the smaller the distortion. However, there is a trade-off. As the quiescent transistor current increases, the power efficiency is reduced. The circuit should be designed such that the transistor quiescent current is the smallest value while meeting the maximum total harmonic distortion specification.

Comment: We see that the class-AB output stage results in a much smaller THD value than the class-B circuit, but as with most circuits, there are no uniquely specified bias voltages.

A class-AB output stage using enhancement-mode MOSFETs is shown in Figure 8.25. If M_n and M_p are matched, and if $v_I = 0$, then $V_{BB}/2$ is applied across the gate–source terminals of M_n and the source–gate terminals of M_p. The quiescent drain currents established in each transistor are given by

$$i_{Dn} = i_{Dp} = I_{DQ} = K\left(\frac{V_{BB}}{2} - |V_T|\right)^2 \tag{8.36}$$

As v_I increases, the voltage at the gate of M_n increases and v_O increases. Transistor M_n operates as a source follower, supplying the load current to R_L. Since i_{Dn} must increase to supply the load current, v_{GSn} must also increase. Assuming V_{BB} remains constant, an increase in v_{GSn} implies a decrease in v_{SGp} and a resulting decrease in i_{Dp}. As v_I goes negative, the voltage at the base of M_p decreases and v_O decreases. Transistor M_p then operates as a source follower, sinking current from the load.

Example　8.7　**Objective:**　Determine the required biasing in a MOSFET class-AB output stage.

The circuit is shown in Figure 8.25. The parameters are $V_{DD} = 10\,\text{V}$ and $R_L = 20\,\Omega$. The transistors are matched, and the parameters are $K = 0.20\,\text{A/V}^2$ and $|V_T| = 1\,\text{V}$. The quiescent drain current is to be 20 percent of the load current when $v_O = 5\,\text{V}$.

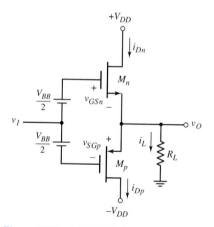

Figure 8.25　MOSFET class-AB output stage

Solution:　For $v_O = 5\,\text{V}$,

$$i_L = 5/20 = 0.25\,\text{A}$$

Then, for $I_Q = 0.05\,\text{A}$ when $v_O = 0$, we have

$$I_{DQ} = 0.05 = K\left(\frac{V_{BB}}{2} - |V_T|\right)^2 = (0.20)\left(\frac{V_{BB}}{2} - 1\right)^2$$

which yields

$$V_{BB}/2 = 1.50\,\text{V}$$

The input voltage for v_O positive is

$$v_I = v_O + v_{GSn} - \frac{V_{BB}}{2}$$

For $v_O = 5\,\text{V}$ and $i_{Dn} \cong i_L = 0.25\,\text{A}$, we have

$$v_{GSn} = \sqrt{\frac{i_{Dn}}{K}} + |V_T| = \sqrt{\frac{0.25}{0.20}} + 1 = 2.12\,\text{V}$$

The source-to-gate voltage of M_p is

$$v_{SGp} = V_{BB} - V_{GSn} = 3 - 2.12 = 0.88\,\text{V}$$

which means that M_p is cut off and $i_{Dn} = i_L$. Finally, the input voltage is

$$v_I = 5 + 2.12 - 1.5 = 5.62\,\text{V}$$

Comment: Since $v_I > v_O$, the voltage gain of this output stage is less than unity, as expected.

Voltage V_{BB} can be established in a MOSFET class-AB circuit by using additional enhancement-mode MOSFETs and a constant current I_{Bias}. This will be considered in a problem at the end of the chapter.

Test Your Understanding

*8.10   Consider the MOSFET class-AB output stage shown in Figure 8.25, with the circuit and transistor parameters as given in Example 8.7. Let $V_{BB} = 3.0\,\text{V}$. Determine the small-signal voltage gain $A_v = dv_O/dv_I$ evaluated at: (a) $v_O = 0$, and (b) $v_O = 5.0\,\text{V}$. (Ans. (a) $A_v = 0.889$ (b) $A_v = 0.899$)

8.3.4   Class-C Operation

The transistor circuit ac load line, including an extension beyond cutoff, is shown in Figure 8.26. For class-C operation, the transistor has a reverse-biased B–E voltage at the Q-point. This effect is illustrated in Figure 8.26. Note that

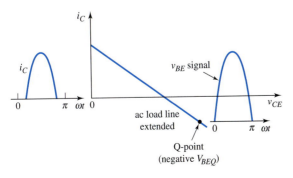

Figure 8.26   Effective ac load line of a class-C amplifier

the collector current is not negative, but is zero at the quiescent point. The transistor conducts only when the input signal becomes sufficiently positive during its positive half-cycle. The transistor therefore conducts for less than a half-cycle, which defines class-C operation.

Class-C amplifiers are capable of providing large amounts of power, with conversion efficiencies larger than 78.5 percent. These amplifiers are normally used for radio-frequency (RF) circuits, with tuned RLC loads that are commonly used in radio and television transmitters. The RLC circuits convert drive current pulses into sinusoidal signals. Since this is a specialized area, we will not analyze these circuits here.

8.4   CLASS-A POWER AMPLIFIERS

The standard class-A amplifier was analyzed previously, and the maximum possible power conversion efficiency was found to be 25 percent. This conversion efficiency can be increased with the use of inductors and transformers.

8.4.1   Inductively Coupled Amplifier

Delivering a large power to a load generally requires both a large voltage and a high current. In a common-emitter circuit, this requirement can be met by replacing the collector resistor with an inductor, as shown in Figure 8.27(a). The inductor is a short circuit to a dc current, but acts as an open circuit to an ac signal operating at a sufficiently high frequency. The entire ac current is therefore coupled to the load. We assume that $\omega L \gg R_L$ at the lowest signal frequency.

The dc and ac load lines are shown in Figure 8.27(b). We assume that the resistance of the inductor is negligible, and that the emitter resistor value is small. The quiescent collector–emitter voltage is then approximately $V_{CEQ} \cong V_{CC}$. The ac collector current is

$$i_c = \frac{-v_{ce}}{R_L} \tag{8.37}$$

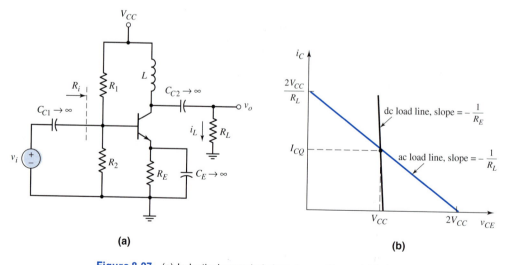

(a)   **(b)**

Figure 8.27   (a) Inductively coupled class-A amplifier and (b) dc and ac load lines

To obtain the maximum symmetrical output-signal swing, which will in turn produce the maximum power, we want

$$I_{CQ} \cong \frac{V_{CC}}{R_L} \tag{8.38}$$

For this condition, the ac load line intersects the v_{CE} axis at $2V_{CC}$.

The use of an inductor or storage device results in an output ac voltage swing that is larger than V_{CC}. The polarity of the induced voltage across the inductor may be such that the voltage adds to V_{CC}, producing an output voltage that is larger than V_{CC}.

The absolute maximum amplitude of the signal current in the load is I_{CQ}; therefore, the maximum possible average signal power delivered to the load is

$$\overline{P}_L(\text{max}) = \frac{1}{2}I_{CQ}^2 R_L = \frac{1}{2} \cdot \frac{V_{CC}^2}{R_L} \tag{8.39}$$

If we neglect the power dissipation in the bias resistors R_1 and R_2, the average power supplied by the V_{CC} source is

$$\overline{P}_S = V_{CC}I_{CQ} = \frac{V_{CC}^2}{R_L} \tag{8.40}$$

The maximum possible power conversion efficiency is then

$$\eta(\text{max}) = \frac{\overline{P}_L(\text{max})}{\overline{P}_S} = \frac{\frac{1}{2} \cdot \frac{V_{CC}^2}{R_L}}{\frac{V_{CC}^2}{R_L}} = \frac{1}{2} \Rightarrow 50\% \tag{8.41}$$

This demonstrates that, in a standard class-A amplifier, replacing the collector resistor with an inductor doubles the maximum possible power conversion efficiency.

8.4.2 Transformer-Coupled Common-Emitter Amplifier

The design of an inductively coupled amplifier to achieve high power conversion efficiency may be difficult, depending on the relationship between the supply voltage V_{CC} and the load resistance R_L. The effective load resistance can be optimized by using a transformer with the proper turns ratio.

Figure 8.28(a) shows a common-emitter amplifier with a transformer-coupled load in the collector circuit.

The dc and ac load lines are shown in Figure 8.28(b). If we neglect any resistance in the transformer and assume that R_E is small, the quiescent collector–emitter voltage is

$$V_{CEQ} \cong V_{CC}$$

The transformed load resistance is

$$R_L' = a^2 R_L \tag{8.42}$$

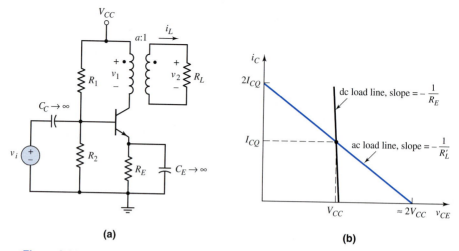

Figure 8.28 (a) Transformer-coupled common-emitter amplifier and (b) dc and ac load lines

where a is the ratio of primary to secondary turns, or simply the turns ratio. The turns ratio is designed to produce the maximum symmetrical swing in the output current and voltage; therefore,

$$R'_L = \frac{2V_{CC}}{2I_{CQ}} = \frac{V_{CC}}{I_{CQ}} = a^2 R_L \qquad (8.43)$$

The maximum average power delivered to the load is equal to the maximum average power delivered to the primary of the ideal transformer, as follows:

$$\overline{P}_L(\text{max}) = \frac{1}{2} V_{CC} I_{CQ} \qquad (8.44)$$

where V_{CC} and I_{CQ} are the maximum possible amplitudes of the sinusoidal signals. If we neglect the power dissipation in the bias resistors R_1 and R_2, the average power supplied by the V_{CC} source is

$$\overline{P}_S = V_{CC} I_{CQ}$$

and the maximum possible power conversion efficiency is again

$$\eta(\text{max}) = 50\%$$

Test Your Understanding

*D8.11   For the inductively coupled amplifier shown in Figure 8.27(a), the parameters are: $V_{CC} = 12\,\text{V}$, $V_{BE}(\text{on}) = 0.7\,\text{V}$, $R_E = 0.1\,\text{k}\Omega$, $R_L = 1.5\,\text{k}\Omega$, and $\beta = 75$. (a) Design R_1 and R_2 for maximum symmetrical swing in the output current and voltage. (Let $R_{TH} = (1 + \beta)R_E$.) (b) If the peak output voltage amplitude is limited to $0.9V_{CC}$, and the peak output current amplitude is limited to $0.9I_{CQ}$, determine the average power delivered to the load, the average power dissipated in the transistor, and the power conversion efficiency. (Ans. (a) $R_1 = 39.1\,\text{k}\Omega$, $R_2 = 9.43\,\text{k}\Omega$ (b) $\overline{P}_L = 38.9\,\text{mW}$, $\overline{P}_Q = 57.1\,\text{mW}$, $\eta = 40.5\%$)

8.4.3 Transformer-Coupled Emitter-Follower Amplifier

Since the emitter follower has a low output impedance, it is often used as the output stage of an amplifier. A transformer-coupled emitter follower is shown in Figure 8.29(a). The dc and ac load lines are shown in Figure 8.29(b). As before, the resistance of the transformer is assumed to be negligible.

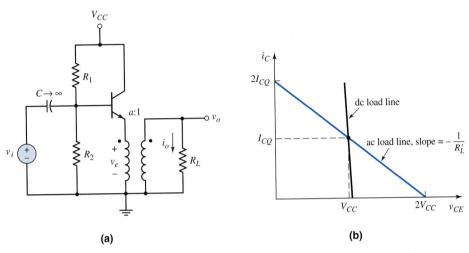

(a)   **(b)**

Figure 8.29 (a) Transformer-coupled emitter-follower amplifier and (b) dc and ac load lines

The transformed load resistance is again $R_L' = a^2 R_L$. By correctly designing the turns ratio, we can achieve the maximum symmetrical swing in the output voltage and current.

The average power delivered to the load is

$$\overline{P}_L = \frac{1}{2} \cdot \frac{V_p^2}{R_L} \tag{8.45}$$

where V_p is the peak amplitude of the sinusoidal output voltage. The maximum peak amplitude of the emitter voltage is V_{CC}, so that the maximum peak amplitude of the output signal is

$$V_p(\text{max}) = V_{CC}/a$$

The maximum average output signal power is therefore

$$\overline{P}_L(\text{max}) = \frac{1}{2} \cdot \frac{[V_p(\text{max})]^2}{R_L} = \frac{V_{CC}^2}{2a^2 R_L} \tag{8.46}$$

The maximum power conversion efficiency for this circuit is also 50 percent.

Design Example 8.8 **Objective:** Design a transformer-coupled emitter-follower amplifier to deliver a specified signal power.

Consider the circuit shown in Figure 8.29(a), with parameters $V_{CC} = 24$ V and $R_L = 8\ \Omega$. The average power delivered to the load is to be 5 W, the peak amplitude of the signal emitter current is to be no more than $0.9I_{CQ}$, and that of the signal emitter voltage is to be no more than $0.9V_{CC}$. Let $\beta = 100$.

Solution: The average power delivered to the load is given by Equation (8.45). The peak output voltage must therefore be

$$V_p = \sqrt{2R_L \overline{P}_L} = \sqrt{2(8)(5)} = 8.94\ \text{V}$$

and the peak output current is

$$I_p = \frac{V_p}{R_L} = \frac{8.94}{8} = 1.12\ \text{A}$$

Since

$$V_e = 0.9V_{CC} = aV_p$$

then

$$a = \frac{0.9V_{CC}}{V_p} = \frac{(0.9)(24)}{8.94} = 2.42$$

Also, since

$$I_e = 0.9I_{CQ} = I_p/a$$

then

$$I_{CQ} = \frac{1}{0.9} \cdot \frac{I_p}{a} = \frac{1.12}{(0.9)(2.42)} = 0.514\ \text{A}$$

The maximum power dissipated in the transistor, for this class-A operation, is

$$P_Q = V_{CC}I_{CQ} = (24)(0.514) = 12.3\ \text{W}$$

so the transistor must be capable of handling this power.

Bias resistors R_1 and R_2 are found from a dc analysis. The Thevenin equivalent voltage is

$$V_{TH} = I_{BQ}R_{TH} + V_{BE}(\text{on})$$

where

$$R_{TH} = R_1 \| R_2 \qquad \text{and} \qquad V_{TH} = [R_2/(R_1 + R_2)] \cdot V_{CC}$$

We also have

$$I_{BQ} = \frac{I_{CQ}}{\beta} = \frac{0.514}{100} \Rightarrow 5.14\ \text{mA}$$

Since $V_{TH} < V_{CC}$ and $I_{BQ} \approx 5$ mA, then R_{TH} cannot be unduly large. However, if R_{TH} is small, then the power dissipation in R_1 and R_2 becomes unacceptably high. We choose $R_{TH} = 2.5$ kΩ, so that

$$V_{TH} = \frac{1}{R_1}(R_{TH})V_{CC} = \frac{1}{R_1}(2.5)(24) = (5.14)(2.5) + 0.7$$

Therefore, $R_1 = 4.43$ kΩ and $R_2 = 5.74$ kΩ.

Comment: The average power delivered by V_{CC} (neglecting bias resistor effects) is $\overline{P}_S = V_{CC}I_{CQ} = 12.3$ W, which means that the power conversion efficiency is $\eta = 5/12.3 \Rightarrow 40.7\%$. The efficiency will always be less than the 50% maximum value, if transistor saturation and distortion are to be minimized.

Test Your Understanding

*D8.12 A transformer-coupled emitter-follower amplifier is shown in Figure 8.29(a). The parameters are: $V_{CC} = 18\,\text{V}$, $V_{BE}(\text{on}) = 0.7\,\text{V}$, $\beta = 100$, $a = 10$, and $R_L = 8\,\Omega$. (a) Design R_1 and R_2 to deliver the maximum power to the load. The input resistance seen by the v_i source is to be $1.5\,\text{k}\Omega$. (b) If the peak amplitude of the emitter voltage v_E is limited to $0.9V_{CC}$, and the peak amplitude of the emitter current i_E is limited to $0.9I_{CQ}$, determine the maximum amplitude of the output signal voltage, and the average power delivered to the load. (Ans. (a) $R_1 = 26.4\,\text{k}\Omega$, $R_2 = 1.62\,\text{k}\Omega$ (b) $V_p = 1.62\,\text{V}$, $I_p = 203\,\text{mA}$, $\overline{P}_L = 0.164\,\text{W}$)

8.5 CLASS-AB PUSH–PULL COMPLEMENTARY OUTPUT STAGES

A class-AB output stage eliminates the crossover distortion that occurs in a class-B circuit. In this section, we will analyze several circuits that provide a small quiescent bias to the output transistors. Such circuits are used as the output stage of power amplifiers, as well as the output stage of operational amplifiers, and will be discussed in Chapter 13.

8.5.1 Class-AB Output Stage with Diode Biasing

In a class-AB circuit, the V_{BB} voltage that provides the quiescent bias for the output transistors can be established by voltage drops across diodes, as shown in Figure 8.30. A constant current I_{Bias} is used to establish the required voltage across the pair of diodes, or the diode-connected transistors, D_1 and D_2. Since D_1 and D_2 are not necessarily matched with Q_n and Q_p, the quiescent transistor currents may not be equal to I_{Bias}.

As the input voltage increases, the output voltage increases, causing an increase in i_{Cn}. This in turn produces an increase in the base current i_{Bn}. Since

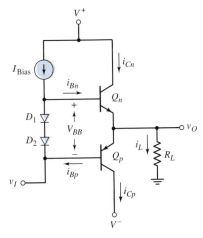

Figure 8.30 Class-AB output stage with quiescent bias established by diodes

the increase base current is supplied by I_{Bias}, the current through D_1 and D_2, and hence the voltage V_{BB}, decreases slightly. Since voltage V_{BB} does not remain constant in this circuit, the relationship between i_{Cn} and i_{Cp}, as given by Equation (8.35), is not precisely valid for this situation. The analysis in the previous section must therefore be modified slightly, but the basic operation of this class-AB circuit is the same.

Design Example 8.9   **Objective:**   Design the class-AB output stage in Figure 8.30 to meet specific design criteria.

     Assume $I_{SD} = 3 \times 10^{-14}$ A for D_1 and D_2, $I_{SQ} = 10^{-13}$ A for Q_n and Q_p, and $\beta_n = \beta_p = 75$. Let $R_L = 8\,\Omega$. The average power delivered to the load is to be 5 W. The peak output voltage is to be no more than 80 percent of V_{CC}, and the minimum value of diode current I_D is to be no less than 5 mA.

Solution: The average power delivered to the load, from Equation (8.24), is

$$\overline{P}_L = \frac{1}{2} \cdot \frac{V_p^2}{R_L}$$

Therefore,

$$V_p = \sqrt{2 R_L \overline{P}_L} = \sqrt{2(8)(5)} = 8.94 \text{ V}$$

The supply voltages must then be

$$V_{CC} = \frac{V_p}{0.8} = \frac{8.94}{0.8} = 11.2 \text{ V}$$

At this peak output voltage, the emitter current of Q_n is approximately equal to the load current, or

$$i_{En} \cong i_L(\text{max}) = \frac{V_p(\text{max})}{R_L} = \frac{8.94}{8} = 1.12 \text{ A}$$

and the base current is

$$i_{Bn} = \frac{i_{En}}{1 + \beta_n} = \frac{1.12}{76} \Rightarrow 14.7 \text{ mA}$$

     For a minimum $I_D = 5$ mA, we can choose $I_{\text{Bias}} = 20$ mA. For a zero input signal, neglecting base currents, we find that

$$V_{BB} = 2 V_T \ln\left(\frac{I_D}{I_{SD}}\right) = 2(0.026) \ln\left(\frac{20 \times 10^{-3}}{3 \times 10^{-14}}\right) = 1.416 \text{ V}$$

The quiescent collector currents are then

$$I_{CQ} = I_{SQ} e^{(V_{BB}/2)/V_T} = 10^{-13} e^{1.416/2(0.026)} \Rightarrow 67.0 \text{ mA}$$

For $v_O = 8.94$ V and $i_L = 1.12$ A, the base current is $i_{Bn} = 14.7$ mA, and

$$I_D = I_{\text{Bias}} - i_{Bn} = 5.3 \text{ mA}$$

The new value of V_{BB} is then

$$V'_{BB} = 2 V_T \ln\left(\frac{I_D}{I_{SD}}\right) = 2(0.026) \ln\left(\frac{5.3 \times 10^{-3}}{3 \times 10^{-14}}\right) = 1.347 \text{ V}$$

The B–E voltage of Q_n is

$$v_{BEn} = V_T \ln\left(\frac{i_{Cn}}{I_{SQ}}\right) = (0.026)\ln\left(\frac{1.12}{10^{-13}}\right) = 0.781\ \text{V}$$

The emitter–base voltage of Q_p is then

$$v_{EBp} = V'_{BB} - v_{BEn} = 1.347 - 0.781 = 0.566\ \text{V}$$

and

$$i_{Cp} = I_{SQ}e^{v_{EBp}/V_T} = (10^{-13})e^{0.566/0.026} \Rightarrow 0.285\ \text{mA}$$

Comment: When the output goes positive, the current in Q_p decreases significantly, as expected, but it does not go to zero. There is a factor of approximately 10^3 difference in the currents between Q_n and Q_p.

Design Pointer: If the output signal currents are large, the base currents in the output transistors may become significant compared to the bias current through the diodes D_1 and D_2. The change in the diode bias current should be minimized in order to keep the small-signal voltage gain of the output stage close to unity.

Test Your Understanding

*8.13 Consider the class-AB output stage in Figure 8.30. The circuit is biased with $V^+ = 12\ \text{V}$, $V^- = -12\ \text{V}$, and the load resistance is $R_L = 75\ \Omega$. The device parameters are: $I_{SD} = 5 \times 10^{-13}\ \text{A}$ for D_1 and D_2, and $I_{SQ} = 2 \times 10^{-13}\ \text{A}$ for Q_n and Q_p. (a) Neglecting base currents, determine the required value of I_{Bias} such that the quiescent currents in Q_n and Q_p are $I_{CQ} = 5\ \text{mA}$. (b) Assuming $\beta_n = \beta_p = 60$, determine i_{Cn}, i_{Cp}, v_{BEn}, v_{EBp}, and I_D when $v_O = 2\ \text{V}$. (c) Repeat part (b) for $v_O = 10\ \text{V}$. (Ans. (a) $I_{Bias} = 12.5\ \text{mA}$ (b) $i_{Cn} = 27.1\ \text{mA}$, $I_D = 12.05\ \text{mA}$, $v_{BEn} = 0.6664\ \text{V}$, $v_{EBp} = 0.5766\ \text{V}$, $i_{Cp} = 0.856\ \text{mA}$ (c) $i_{Cn} = 131\ \text{mA}$, $I_D = 10.3\ \text{mA}$, $v_{BEn} = 0.7074\ \text{V}$, $v_{EBp} = 0.5276\ \text{V}$, $i_{Cp} = 0.130\ \text{mA}$)

8.5.2   Class-AB Biasing Using the V_{BE} Multiplier

An alternative biasing scheme, which provides more flexibility in the design of the output stage, is shown in Figure 8.31. The bias circuit that provides voltage V_{BB} consists of transistor Q_1 and resistors R_1 and R_2, biased by a constant-current source I_{Bias}.

If we neglect the base current in Q_1, then

$$I_R = \frac{V_{BE1}}{R_2} \tag{8.47}$$

and voltage V_{BB} is

$$V_{BB} = I_R(R_1 + R_2) = V_{BE1}\left(1 + \frac{R_1}{R_2}\right) \tag{8.48}$$

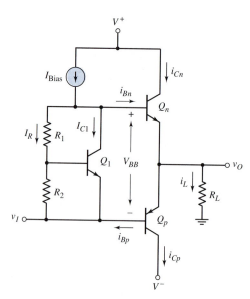

Figure 8.31   Class-AB output stage with V_{BE} multiplier bias circuit

Since voltage V_{BB} is a multiplication of the junction voltage V_{BE1}, the circuit is called a **V_{BE} multiplier**. The multiplication factor can be designed to yield the required value of V_{BB}.

A fraction of the constant current I_{Bias} flows through Q_1, so that

$$V_{BE1} = V_T \ln\left(\frac{I_{C1}}{I_{S1}}\right) \tag{8.49}$$

Also, the quiescent bias currents i_{Cn} and i_{Cp} are normally small; therefore, we can neglect i_{Bn} and i_{Bp}. Current I_{Bias} divides between I_R and I_{C1}, satisfying both Equations (8.47) and (8.49).

As v_I increases, v_O becomes positive, and i_{Cn} and i_{Bn} increase, which reduces the collector current in Q_1. However, the logarithmic dependence of I_{C1}, shown in Equation (8.49), means that V_{BE1} and, in turn V_{BB} remain essentially constant as the output voltage changes.

Design Example   8.10   Objective: Design a Class-AB output stage using the V_{BE} multiplier circuit to meet a specified total harmonic distortion.

Assume the circuit in Figure 8.31, biased at $V^+ = 15$ V and $V^- = -15$ V, is the output stage of an audio amplifier that is to drive another power amplifier whose input resistance is $1\,\text{k}\Omega$. The maximum peak sinusoidal output voltage is to be 10 V and the total harmonic distortion is to be less than 0.1 percent.

Solution: Standard 2N3904 and 2N3906 transistors are to be used. From the results of Example 8.6, the THD is a function of the output transistor quiescent currents. For the basic circuit in Figure 8.23, the THD is found to be 0.097 percent for $V_{BB} = 1.346$ V, quiescent collector currents of 0.88 mA, and a peak sinusoidal output voltage of 10 V.

Figure 8.32 is the PSpice circuit schematic. For a peak output voltage of 10 V, the peak load current is 10 mA. Assuming $\beta \cong 100$, the peak base current is 0.1 mA. A bias current of 1 mA is chosen to bias the V_{BE} multiplier. The peak 0.1 mA base current, then, will not greatly disturb the current through the multiplier circuit.

Figure 8.32 PSpice circuit schematic for Example 8.10

We may select $I_R = 0.2$ mA (current through R_1 and R_2) and $I_{C3} = 0.8$ mA. We then have

$$R_1 + R_2 = \frac{V_{BB}}{I_R} = \frac{1.346}{0.2} = 6.73 \text{ k}\Omega$$

For the 2N3904, we find that $V_{BE} \cong 0.65$ V for a quiescent collector current of approximately 0.8 mA. Therefore

$$R_2 = \frac{V_{BE3}}{I_R} = \frac{0.65}{0.2} = 3.25 \text{ k}\Omega$$

so that $R_1 = 3.48$ kΩ.

From the PSpice results, we find that the voltage at the base of Q_1 to be 0.6895 V and the voltage at the base of Q_2 to be -0.6961 V, which means that $V_{BB} = 1.3856$ V. This voltage is slightly greater than the design value of $V_{BB} = 1.346$ V. Listed below are the quiescent transistor parameters. The quiescent collector currents of the output transistors are 1.88 mA, approximately twice the design value of 0.88 mA. The total harmonic distortion is 0.0356 percent, which is well within the design specification.

NAME	Q_Q1	Q_Q2	Q_Q3	Q_Q4
MODEL	Q2N3904	Q2N3906	Q2N3904	Q2N3906
IB	1.12E-05	-5.96E-06	6.01E-06	-3.20E-06
IC	1.88E-03	-1.88E-03	7.80E-04	-9.92E-04
VBE	6.78E-01	-7.08E-01	6.59E-01	-6.92E-01
VBC	-1.43E+01	1.43E+01	-7.27E-01	1.36E+01
VCE	1.50E+01	-1.50E+01	1.39E+00	-1.43E+01
BETADC	1.67E+02	3.15E+02	1.30E+02	3.10E+02
GM	7.11E-02	7.15E-02	2.98E-02	3.80E-02
RPI	2.66E+03	4.34E+03	5.01E+03	8.09E+03

Comment: Since the resulting V_{BB} voltage is slightly larger than the design value, the quiescent output transistor currents are approximately double the design value. Although the THD specification is met, the larger collector currents mean a larger quiescent power dissipation. For this reason, the circuit may need to be redesigned slightly to lower the quiescent currents.

8.5.3   Class-AB Output Stage with Input Buffer Transistors

The output stage in Figure 8.33 is a class-AB configuration composed of the complementary transistor pair Q_3 and Q_4. Resistors R_1 and R_2 and the emitter-follower transistors Q_1 and Q_2 establish the quiescent bias required in this configuration. Resistors R_3 and R_4, used in conjunction with short-circuit protection devices not shown in the figure, also provide thermal stability for the output transistors. The input signal v_I may be the output of a low-power

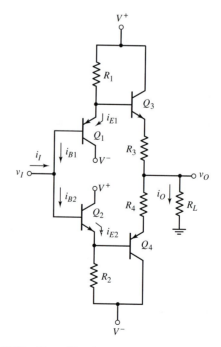

Figure 8.33   Class-AB output stage with input buffer transistors

amplifier. Also, since this is an emitter follower, the output voltage is approximately equal to the input voltage.

When the input voltage v_I increases from zero, the base voltage of Q_3 increases, and the output voltage v_O increases. The load current i_O is positive, and the emitter current in Q_3 increases to supply the load current, which causes an increase in the base current into Q_3. Since the base voltage of Q_3 increases, the voltage drop across R_1 decreases, resulting in a smaller current in R_1. This means that i_{E1} and i_{B1} also decrease. As v_I increases, the voltage across R_2 increases, and i_{E2} and i_{B2} increase. A net input current i_I is then produced, to account for the reduction in i_{B1} and the increase in i_{B2}.

The net input current is

$$i_I = i_{B2} - i_{B1} \tag{8.50}$$

Neglecting the voltage drops across R_3 and R_4 and the base currents in Q_3 and Q_4, we have

$$i_{B2} = \frac{(v_I - V_{BE}) - V^-}{(1 + \beta_n)R_2} \tag{8.51(a)}$$

and

$$i_{B1} = \frac{V^+ - (v_I + V_{EB})}{(1 + \beta_p)R_1} \tag{8.51(b)}$$

where β_n and β_p are the current gains of the npn and pnp transistors, respectively. If $V^+ = -V^-$, $V_{BE} = V_{EB}$, $R_1 = R_2 \equiv R$, and $\beta_n = \beta_p \equiv \beta$, then combining Equations (8.51(a)), (8.51(b)), and (8.50) produces

$$i_I = \frac{(v_I - V_{BE} - V^-)}{(1 + \beta)R} - \frac{(V^+ - v_I - V_{EB})}{(1 + \beta)R} = \frac{2v_I}{(1 + \beta)R} \tag{8.52}$$

Since the voltage gain of this output stage is approximately unity, the output current is

$$i_O = \frac{v_O}{R_L} \cong \frac{v_I}{R_L} \tag{8.53}$$

Using Equations (8.52) and (8.53), we find the current gain of this output stage to be

$$A_i = \frac{i_O}{i_I} = \frac{(1 + \beta)R}{2R_L} \tag{8.54}$$

With β in the numerator, this current gain should be substantial. A large current gain is desirable, since the output stage of power amplifiers must provide the current necessary to meet the power requirements.

Example 8.11 **Objective:** Determine the currents and the current gain for the output stage with input buffer transistors.

For the circuit in Figure 8.33, the parameters are: $R_1 = R_2 = 2\,\text{k}\Omega$, $R_L = 100\,\Omega$, $R_3 = R_4 = 0$, and $V^+ = -V^- = 15\,\text{V}$. Assume all transistors are matched, with $\beta = 60$ and $V_{BE}(\text{npn}) = V_{EB}(\text{pnp}) = 0.6\,\text{V}$.

Solution: For $v_I = 0$,

$$i_{R1} = i_{R2} \cong i_{E1} = i_{E2} = \frac{15 - 0.6}{2} = 7.2\,\text{mA}$$

Assuming all transistors are matched, the bias currents in Q_3 and Q_4 are also approximately 7.2 mA, since the base–emitter voltages of Q_1 and Q_3 are equal and those of Q_2 and Q_4 are equal.

Solution: For $v_I = 10\,\text{V}$, the output current is approximately

$$i_O = \frac{v_O}{R_L} \cong \frac{10}{0.1} = 100\,\text{mA}$$

The emitter current in Q_3 is essentially equal to the load current, which means that the base current in Q_3 is approximately

$$i_{B3} = 100/61 = 1.64\,\text{mA}$$

The current in R_1 is

$$i_{R1} = \frac{15 - (10 + 0.6)}{2} = 2.2\,\text{mA}$$

which means that

$$i_{E1} = i_{R1} - i_{B3} = 0.56\,\text{mA}$$

and

$$i_{B1} = i_{E1}/(1 + \beta) = 0.56/61 \Rightarrow 9.18\,\mu\text{A}$$

Since Q_4 tends to turn off when v_O increases, we have

$$i_{E2} \cong i_{R2} = \frac{10 - 0.6 - (-15)}{2} = 12.2\,\text{mA}$$

and

$$i_{B2} = i_{E2}/(1 + \beta) = 12.2/61 \Rightarrow 200\,\mu\text{A}$$

The input current is then

$$i_I = i_{B2} - i_{B1} = 200 - 9.18 \cong 191\,\mu\text{A}$$

The current gain is then

$$A_i = \frac{i_O}{i_I} = \frac{100}{0.191} = 524$$

From Equation (8.54), the predicted current gain is

$$A_i = \frac{i_O}{i_I} = \frac{(1 + \beta)R}{2R_L} = \frac{(61)(2)}{2(0.1)} = 610$$

Comment: Since the current gain determined from Equation (8.54) neglects base currents in Q_3 and Q_4, the actual current gain is less than the predicted value, as expected. The input current of 191 μA can easily be supplied by a low-power amplifier.

Test Your Understanding

*8.14   Consider the class-AB output stage in Figure 8.33. The parameters are: $V^+ = -V^- = 12\,\text{V}$, $R_1 = R_2 = 250\,\Omega$, $R_L = 8\,\Omega$, and $R_3 = R_4 = 0$. Assume all transistors are matched, with $\beta = 40$ and $V_{BE}(\text{npn}) = V_{EB}(\text{pnp}) = 0.7\,\text{V}$. (a) For $v_I = 0$, determine i_{E1}, i_{E2}, i_{B1}, and i_{B2}. (b) For $v_I = 5\,\text{V}$, find i_O, i_{E1}, i_{E2}, i_{B1}, i_{B2}, and i_I. (c) Using the results of part (b), determine the current gain of the output stage. Compare this value to that found using Equation (8.54). (Ans. (a) $i_{E1} = i_{E2} = 44.1\,\text{mA}$, $i_{B1} = i_{B2} = 1.08\,\text{mA}$ (b) $i_O = 0.625\,\text{A}$, $i_{E1} = 10.0\,\text{mA}$, $i_{B1} = 0.244\,\text{mA}$, $i_{E2} = 65.2\,\text{mA}$, $i_{B2} = 1.59\,\text{mA}$, $i_I = 1.35\,\text{mA}$ (c) $A_i = 463$, from Equation (8.54) $A_i = 641$)

8.5.4 Class-AB Output Stage Utilizing the Darlington Configuration

The complementary push–pull output stage uses npn and pnp bipolar transistors. Usually in IC design, the pnp transistors are fabricated as lateral devices with low β values that are typically in the range of 5 to 10, and the npn transistors are fabricated as vertical devices with β values on the order of 200. This means that the npn and pnp transistors are not well matched, as we have assumed in our analyses.

Consider the two-transistor configuration shown in Figure 8.34(a). Assume the transistor current gains are β_n and β_p for the npn and pnp transistors, respectively. We can write

$$i_{Cp} = i_{Bn} = \beta_p i_{Bp} \tag{8.55}$$

and

$$i_2 = (1 + \beta_n)i_{Bn} = (1 + \beta_n)\beta_p i_{Bp} \cong \beta_n \beta_p i_{Bp} \tag{8.56}$$

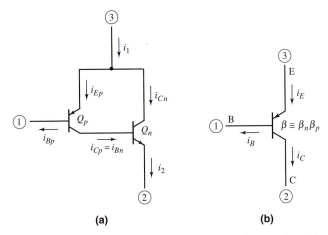

(a)   **(b)**

Figure 8.34   (a) A two-transistor configuration of an equivalent pnp transistor; (b) the equivalent pnp transistor

Terminal 1 acts as the base of the composite three-terminal device, terminal 2 acts as the collector, and terminal 3 is the emitter. The current gain of the device is then approximately $\beta_n \beta_p$. The equivalent circuit is shown in Figure 8.34(b). We can use the two-transistor configuration in Figure 8.34(a) as a single equivalent pnp transistor with a current gain on the same order of magnitude as that of an npn device.

In Figure 8.35, the output stage uses Darlington pairs to provide the necessary current gain. Transistors Q_1 and Q_2 constitute the npn Darlington

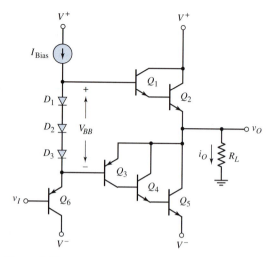

Figure 8.35   Class-AB output stage with Darlington pairs

emitter-follower that sources current to the load. Transistors Q_3, Q_4, and Q_5 constitute a composite pnp Darlington emitter follower that sinks current from the load. The three diodes D_1, D_2, and D_3 establish the quiescent bias for the output transistors.

The effective current gain of the three-transistor configuration Q_3–Q_4–Q_5 is essentially the product of the three individual gains. With the low current gain of the pnp device Q_3, the overall current gain of the Q_3–Q_4–Q_5 configuration is similar to that of the Q_1–Q_2 pair.

Test Your Understanding

8.15   From Figure 8.35, show that the overall current gain of the three-transistor configuration composed of Q_3, Q_4, and Q_5 is approximately $\beta = \beta_3\beta_4\beta_5$.

8.6   SUMMARY

- In this chapter, we analyzed and designed amplifiers and output stages capable of delivering a substantial amount of power to a load.
- The current, voltage, and power ratings of BJTs and MOSFETs were considered, and the safe operating area for the transistors was defined in terms of these limiting parameters. The maximum power rating of a transistor is related to the maximum allowed device temperature at which the device can operate without being damaged.
- In a class-A amplifier, the output transistor conducts 100 percent of the time. The theoretical maximum power conversion efficiency for a standard class-A amplifier is 25 percent. This efficiency can be theoretically increased to 50 percent by incorporating inductors or transformers in the class-A circuit.
- Class-B output stages are composed of complementary pairs of transistors operating in a push–pull manner. In an ideal class-B operation, each output transistor conducts 50 percent of the time. For an idealized Class-B output stage, the theoretical max-

imum power conversion efficiency is 78.5 percent. However, practical class-B output stages tend to suffer from crossover distortion effects when the output is in the vicinity of zero volts.

- The class-AB output stage is similar to the class-B circuit, except that each output transistor is provided with a small quiescent bias and conducts more than 50 percent of the time. The power conversion efficiency of a class-AB output stage is less than that of the ideal class-B circuit, but is substantially larger than that of the class-A circuit.

CHECKPOINT

After studying this chapter, the reader should have the ability to:

✔ Describe what factors are related to the maximum transistor current and maximum transistor voltage. (Section 8.2)

✔ Define the safe operating area of a transistor and define the power derating curve. (Section 8.2)

✔ Define the power conversion efficiency of an output stage. (Section 8.3)

✔ Describe the operation of a class-A output stage. (Section 8.3)

✔ Describe the operation of an ideal class-B output stage and discuss the concept of crossover distortion. (Section 8.3)

✔ Describe and design a class-AB output stage and discuss why crossover distortion is essentially eliminated. (Sections 8.3 and 8.5)

REVIEW QUESTIONS

1. Discuss the limiting factors for the maximum rated current in a BJT and MOSFET.
2. Discuss the limiting factors for the maximum rated voltage in a BJT and MOSFET.
3. Discuss the safe operating area of a transistor.
4. Why is an interdigitated structure typically used in a high-power BJT design?
5. Discuss the role of thermal resistance between various junctions in a high-power transistor structure.
6. Define and describe the power derating curve for a transistor.
7. Define class-A, class-B, and class-AB operation.
8. Define power conversion efficiency for an output stage.
9. Describe the operation of a class-A output stage.
10. Describe the operation of a class-B output stage.
11. Discuss crossover distortion.
12. What is meant by harmonic distortion?
13. Describe the operation of a class-AB output stage and why a class-AB output stage is important.
14. Describe the operation of a transformer-coupled class-A common-emitter amplifier.
15. Sketch a class-AB complementary BJT push–pull output stage using a V_{BE} multiplier circuit.
16. What are the advantages of a Darlington pair configuration?
17. Sketch a two-transistor configuration using npn and pnp BJTs that are equivalent to a single pnp BJT.

PROBLEMS

Section 8.2   Power Transistors

V_{DD}

R_D

V_{GG}

Figure P8.1

8.1 The maximum current, voltage, and power ratings of a power transistor are: 5 A, 80 V, and 25 W, respectively. (a) Sketch and label the safe operating area for this transistor, using linear current and voltage scales. (b) For the common-source circuit in Figure P8.1, determine the value of R_D, and sketch the load line that produces a maximum power in the transistor for: (i) $V_{DD} = 80$ V and (ii) $V_{DD} = 50$ V.

8.2 The common-emitter circuit in Figure P8.2 is biased at $V_{CC} = 24$ V. The maximum transistor power is $P_{D,\max} = 20$ W and the current gain is $\beta = 80$. (a) Determine R_L and R_B such that the maximum power is delivered to the load R_L. (b) Find the value of V_p for the input signal that delivers the maximum power. State any assumptions.

D8.3 For the transistor in the common-emitter circuit in Figure P8.2, the parameters are: $\beta = 100$, $P_{D,\max} = 2.5$ W, $V_{CE(\text{sus})} = 25$ V, and $I_{C,\max} = 500$ mA. Let $R_L = 100\,\Omega$. (a) Design V_{CC} and R_B to deliver the maximum power to the load. (b) Using the results of part (a), calculate the maximum undistorted ac power that can be delivered to R_L.

V_{CC}

R_B R_L

C

$v_S = V_p \sin\omega t$
(V)

Figure P8.2

8.4 Sketch the safe operating region for a MOSFET. Label three arbitrary points on the maximum hyperbola. Assume each of the labeled points is a Q-point and draw a tangent load line through each point. Discuss the advantages or disadvantages of each point relative to the maximum possible signal amplitude.

8.5 A power MOSFET is connected in a common-source configuration as shown in Figure P8.1. The parameters are: $I_{D,\max} = 4$ A, $V_{DS,\max} = 50$ V, $P_{D,\max} = 35$ W, $V_{TN} = 4$ V, and $K_n = 0.25$ A/V^2. The circuit parameters are $V_{DD} = 40$ V and $R_L = 10\,\Omega$. (a) Sketch and label the safe operating area for this transistor, using linear current and voltage scales. Also sketch the load line on the same graph. (b) Calculate the power dissipated in the transistor for $V_{GG} = 5, 6, 7, 8$, and 9 V. (c) Is there a possibility of damaging the transistor? Explain.

D8.6 Consider the common-source circuit shown in Figure P8.6. The transistor parameters are $V_{TN} = 4$ V and $K_n = 0.2$ A/V^2. (a) Design the bias circuit such that the Q-point is in the center of the load line. (b) What is the power dissipated in the transistor at the Q-point? (c) Determine the minimum rated $I_{D,\max}$, $V_{DS,\max}$, and $P_{D,\max}$ values. (d) If $v_i = 0.5 \sin\omega t$ V, calculate the ac power delivered to R_L, and determine the average power dissipated in the transistor.

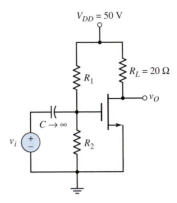

$V_{DD} = 50$ V

R_1

$R_L = 20 \, \Omega$

v_O

$C \to \infty$

v_i

R_2

Figure P8.6

8.7 A particular transistor is rated for a maximum power dissipation of 60 W if the case temperature is at 25 °C. Above 25 °C, the allowed power dissipation is reduced by 0.5 W/°C. (a) Sketch the power derating curve. (b) What is the maximum allowed junction temperature? (c) What is the value of $\theta_{dev-case}$?

8.8 A MOSFET has a rated power of 50 W and a maximum specified junction temperature of 150 °C. The ambient is $T_{amb} = 25$ °C. Find the relationship between the actual operating power and $\theta_{case-amb}$.

8.9 For a power MOSFET, $\theta_{dev-case} = 1.75$ °C/W, the drain current is $I_D = 4$ A, and the average drain-to-source voltage is 5 V. The device is mounted on a heat sink with parameters $\theta_{snk-amb} = 3$ °C/W and $\theta_{case-snk} = 0.8$ °C/W. If the ambient temperature is $T_{amb} = 25$ °C, determine the temperature of: (a) device, (b) case, and (c) heat sink.

8.10 A BJT must dissipate 25 W of power. The maximum junction temperature is $T_{j,max} = 200$ °C, the ambient temperature is 25 °C, and the device-to-case thermal resistance is 3 °C/W. Determine the maximum permissible thermal resistance between the case and ambient.

8.11 A BJT has a rated power of 15 W and a maximum junction temperature of 175 °C. The ambient temperature is 25 °C, and the thermal resistance parameters are: $\theta_{snk-amb} = 4$ °C/W and $\theta_{case-snk} = 1$ °C/W. Determine the actual power that can be safely dissipated in the transistor.

Section 8.3   Classes of Amplifiers

8.12 For the class-A amplifier shown in Figure 8.16(a), show that the maximum theoretical conversion efficiency for a symmetrical square-wave input signal is 50 percent.

*D8.13 A class-A emitter follower biased with a constant-current source is shown in Figure P8.13. Assume the circuit parameters are: $V^+ = 10$ V, $V^- = -10$ V, and $R_L = 1$ kΩ. The transistor parameters are: $\beta = 200$, $V_{BE} = 0.7$ V, and $V_{CE}(sat) = 0.2$ V. (a) Determine the value of R that will produce the maximum possible output signal swing. What is the value of I_Q, and the maximum and minimum values of i_{E1} and i_L? (b) Using the results of part (a), calculate the conversion efficiency.

8.14 The circuit parameters for the class-A emitter follower shown in Figure P8.13 are: $V^+ = 12$ V, $V^- = -12$ V, and $R_L = 100$ Ω. The transistor parameters are: $\beta = 200$, $V_{BE} = 0.7$ V, and $V_{CE}(sat) = 0.2$ V. The output voltage is to vary between +10 V and −10 V. (a) Find the minimum required I_Q and the value of R. (b) For $v_O = 0$, find the

Figure P8.13

power dissipated in the transistor Q_1, and the power dissipated in the current source (Q_2, Q_3, and R). (c) Determine the conversion efficiency for a symmetrical sine-wave output voltage with a peak value of 10 V.

8.15 For the idealized class-B output stage in Figure 8.18, show that the maximum theoretical conversion efficiency for a symmetrical square-wave input signal is 100 percent.

8.16 Consider the class-B output stage with complementary MOSFETs shown in Figure P8.16. The transistor parameters are $V_{TN} = V_{TP} = 0$ and $K_n = K_p = 0.4\,\text{mA/V}^2$. Let $R_L = 5\,\text{k}\Omega$. (a) Find the maximum output voltage such that M_n remains biased in the saturation region. What are the corresponding values of i_L and v_I for this condition? (b) Determine the conversion efficiency for a symmetrical sine-wave output signal with the peak value found in part (a).

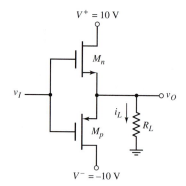

Figure P8.16

8.17 Using the same transistor parameters listed in Problem 8.16 for the circuit shown in Figure P8.16, plot v_O versus v_I for $-10 \leq v_I \leq +10\,\text{V}$. What is the voltage gain (slope of the curve) at $v_I = 0$ and at $v_I = 10\,\text{V}$?

*8.18 A simplified class-AB output stage with BJTs is shown in Figure 8.23. The circuit parameters are $V_{CC} = 10\,\text{V}$ and $R_L = 100\,\Omega$. The parameter I_S for each transistor is $I_S = 5 \times 10^{-13}$ A. (a) Determine the value of V_{BB} such that $i_{Cn} = i_{Cp} = 5$ mA when $v_I = 0$. What is the power dissipated in each transistor? (b) For $v_O = -8$ V, determine i_L, i_{Cn}, i_{Cp}, and v_I. What is the power dissipated in Q_n, Q_p, and R_L?

*8.19 A simplified class-AB output stage with enhancement-mode MOSFETs is shown in Figure 8.25. The circuit parameters are $V_{DD} = 10$ V and $R_L = 1$ kΩ. The transistor parameters are $V_{TN} = -V_{TP} = 2$ V and $K_n = K_p = 2$ mA/V^2. (a) Determine the value of V_{BB} such that $i_{Dn} = i_{Dp} = 0.5$ mA when $v_I = 0$. What is the power dissipated in each transistor? (b) Determine the maximum output voltage such that M_n remains biased in the saturation region. What are the values of i_{Dn}, i_{Dp}, i_L, and v_I for this case? Calculate the power dissipated in M_n, M_p, and R_L.

8.20 Consider the class-AB output stage in Figure P8.20. The diodes and transistors are matched, with parameters $I_S = 6 \times 10^{-12}$ A, and $\beta = 40$. (a) Determine R_1 such that the minimum current in the diodes is 25 mA when $v_O = 24$ V. Find i_N and i_P for this condition. (b) Using the results of part (a), determine the diode and transistor currents when $v_O = 0$.

Figure P8.20   **Figure P8.21**

*8.21 An enhancement-mode MOSFET class-AB output stage is shown in Figure P8.21. The threshold voltage of each transistor is $V_{TN} = -V_{TP} = 1$V and the conduction parameters of the output transistors are $K_{n1} = K_{p2} = 5$ mA/V^2. Let $I_{Bias} = 200$ μA. (a) Determine $K_{n3} = K_{p4}$ such that the quiescent drain currents in M_1 and M_2 are 5 mA. (b) Using the results of part (a), find the small-signal voltage gain $A_v = dv_O/dv_I$ evaluated at: (i) $v_O = 0$, and (ii) $v_O = 5$ V.

D8.22 Consider the MOSFET class-AB output stage in Figure 8.25. The parameters are: $V_{DD} = 10$ V and $R_L = 100$ Ω. For transistors M_n and M_p, $V_{TN} = -V_{TP} = 1$ V. The peak amplitude of the output voltage is limited to 5 V. Design the circuit such that the small-signal voltage gain is $A_v = dv_O/dv_I = 0.95$ when $v_O = 0$.

Section 8.4   Class-A Power Amplifiers

D8.23   Design an inductively coupled common-emitter amplifier, such as that in Figure 8.27(a), to provide a small-signal voltage gain of $A_v = -12$. The circuit and transistor parameters are: $R_i = 6\,\text{k}\Omega$, $R_L = 2\,\text{k}\Omega$, $V_{CC} = 10\,\text{V}$, $\beta = 180$, and $V_{BE} = 0.7\,\text{V}$. Determine the maximum power that can be delivered to the load, and the conversion efficiency.

D8.24   For the inductively coupled amplifier in Figure 8.27(a), the parameters are: $V_{CC} = 15\,\text{V}$, $R_E = 0.1\,\text{k}\Omega$, and $R_L = 1\,\text{k}\Omega$. The transistor parameters are $\beta = 100$ and $V_{BE} = 0.7\,\text{V}$. Design R_1 and R_2 to deliver the maximum power to the load. What is the maximum power that can be delivered to the load?

8.25   Consider the transformer-coupled common-emitter circuit shown in Figure P8.25. The parameters are: $V_{CC} = 10\,\text{V}$, $R_L = 8\,\Omega$, $n_1 : n_2 = 3 : 1$, $R_1 = 0.73\,\text{k}\Omega$, $R_2 = 1.55\,\text{k}\Omega$, and $R_E = 20\,\Omega$. The transistor parameters are $\beta = 25$ and $V_{BE}(\text{on}) = 0.7\,\text{V}$. The amplitude of the sinusoidal input voltage is $17\,\text{mV}$. Determine the ac power delivered to the load, and the conversion efficiency.

Figure P8.25

8.26   The parameters for the transformer-coupled common-emitter circuit in Figure P8.25 are $V_{CC} = 36\,\text{V}$ and $n_1 : n_2 = 4 : 1$. The signal power delivered to the load is $2\,\text{W}$. Determine: (a) the rms voltage across the load; (b) the rms voltage across the transformer primary; and (c) the primary and secondary currents. (d) If $I_{CQ} = 150\,\text{mA}$, what is the conversion efficiency?

8.27   A BJT emitter follower is coupled to a load with an ideal transformer, as shown in Figure P8.27. The bias circuit is not shown. The transistor current gain is $\beta = 49$, and the transistor is biased such that $I_{CQ} = 100\,\text{mA}$. (a) Derive the expressions for the voltage transfer functions v_e/v_i and v_o/v_i. (b) Find $n_1 : n_2$ for maximum ac power transfer to R_L. (c) Determine the small-signal output resistance looking back into the emitter.

D8.28   Consider the transformer-coupled emitter follower in Figure P8.28. Assume an ideal transformer. The transistor parameters are $\beta = 100$ and $V_{BE} = 0.7\,\text{V}$. (a) Design the circuit to provide a current gain at $A_i = i_o/i_i = 80$. (b) If the magnitude of the signal emitter current is limited to $0.9\,I_{CQ}$ to prevent distortion, determine the power delivered to the load, and the conversion efficiency.

Figure P8.27   **Figure P8.28**

D8.29 A class-A transformer-coupled emitter follower must deliver 2 W to an 8 Ω speaker. Let $V_{CC} = 18$ V, $\beta = 100$, and $V_{BE} = 0.7$ V. (a) Determine the required transformer ratio $n_1 : n_2$. (b) Determine the minimum transistor power rating.

D8.30 Repeat Problem 8.28 if the primary side of the transformer has a resistance of 100 Ω.

Section 8.5   Class-AB Push–Pull Complementary Output Stages

8.31 The transistors in the output stage in Figure 8.33 are all matched. Their parameters are $\beta = 60$ and $I_S = 5 \times 10^{-13}$ A. Resistors R_1 and R_2 are replaced by 3 mA ideal current sources, and $R_3 = R_4 = 0$. Let $V^+ = 10$ V and $V^- = -10$ V. (a) Determine the quiescent collector currents in the four transistors for $v_I = v_O = 0$. (b) For a load resistance of $R_L = 20\,\Omega$ and a peak output voltage of 6 V, determine the current gain and voltage gain of the circuit.

*8.32 Consider the circuit in Figure 8.33. The supply voltages are $V^+ = 10$ V and $V^- = -10$ V, and the R_3 and R_4 resistor values are zero. The transistor parameters are: $\beta_1 = \beta_2 = 120$, $\beta_3 = \beta_4 = 50$, $I_{S1} = I_{S2} = 2 \times 10^{-13}$ A, and $I_{S3} = I_{S4} = 2 \times 10^{-12}$ A. (a) The range in output current is $-1 \le i_O \le +1$ A. Determine the values of R_1 and R_2 such that the currents in Q_1 and Q_2 do not vary by more than 2 : 1. (b) Using the results of part (a), determine the quiescent collector currents in the four transistors for $v_I = v_O = 0$. (c) Calculate the output resistance, excluding R_L, for a quiescent output voltage of zero. Assume the source resistance of v_I is zero.

8.33 Using the parameters given in Example 8.11 for the circuit in Figure 8.33, calculate the input resistance when the quiescent output voltage is zero.

D8.34 (a) Redesign the class-AB output stage in Figure 8.33 using enhancement-mode MOSFETs. Let $R_3 = R_4 = 0$. (b) Assume the MOSFETs are all matched, and their parameters are $K = 10\,\text{mA/V}^2$ and $V_{TN} = -V_{TP} = 2$ V. Let $V^+ = 10$ V and $V^- = -10$ V. Find R_1 and R_2 such that the quiescent current in each transistor is 5 mA. (c) If $R_L = 100\,\Omega$, determine the current in each transistor; determine the power delivered to the load if $v_O = 5$ V.

8.35 Figure P8.35 shows a composite pnp Darlington emitter follower that sinks current from a load. Parameter I_Q is the equivalent bias current and Z is the equivalent impedance in the base of Q_1. Assume the transistor parameters are: $\beta(\text{pnp}) = 10$, $\beta(\text{npn}) = 50$, $V_{AP} = 50$ V, and $V_{AN} = 100$ V, where V_{AP} and V_{AN} are the Early voltages of the pnp and npn devices, respectively. Calculate the output resistance R_o.

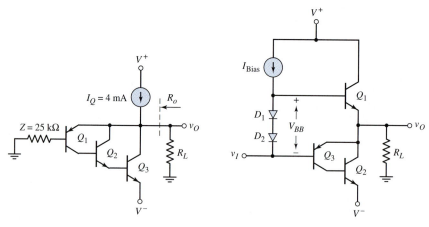

Figure P8.35   **Figure P8.36**

*8.36   Consider the class-AB output stage in Figure P8.36. The parameters are: $V^+ = 12\,\text{V}$, $V^- = -12\,\text{V}$, $R_L = 100\,\Omega$, and $I_{\text{Bias}} = 5\,\text{mA}$. The transistor and diode parameters are $I_S = 10^{-13}\,\text{A}$. The transistor current gains are $\beta_n = 100$ and $\beta_p = 20$ for the npn and pnp devices, respectively. (a) For $v_O = 0$, determine V_{BB}, and the quiescent collector current and base–emitter voltage for each transistor. (b) Repeat part (a) for $v_O = 10\,\text{V}$. What is the power delivered to the load and what is the power dissipated in each transistor?

*8.37   For the class-AB output stage in Figure 8.35, the parameters are: $V^+ = 24\,\text{V}$, $V^- = -24\,\text{V}$, $R_L = 20\,\Omega$, and $I_{\text{Bias}} = 10\,\text{mA}$. The diode and transistor parameters are $I_S = 2 \times 10^{-12}\,\text{A}$. The transistor current gains are $\beta_n = 20$ and $\beta_p = 5$ for the npn and pnp devices, respectively. (a) For $v_O = 0$, determine V_{BB}, and the quiescent collector and base–emitter voltage for each transistor. (b) An average power of 10 watts is to be delivered to the load. Determine the quiescent collector current in each transistor and the instantaneous power dissipated in Q_2, Q_5, and R_L when the output voltage is at its peak negative amplitude.

COMPUTER SIMULATION PROBLEMS

8.38   (a) Simulate the class-B output stage in Figure 8.18 and plot the voltage transfer function to demonstrate the crossover distortion region. (b) Repeat part (a) for the class-AB output stage shown in Figure 8.30. Use diode-connected transistors for D_1 and D_2 and assume all devices are matched. Has the crossover distortion region been eliminated?

8.39   Verify the design of the class-AB output stage in Example 8.9 with a computer analysis.

8.40   (a) Simulate the class-AB output stage shown in Figure 8.33, using parameters $V^+ = -V^- = 15\,\text{V}$, $R_1 = R_2 = 2\,\text{k}\Omega$, $R_L = 100\,\Omega$, and $R_3 = R_4 = 0$. Assume all transistors are matched, with parameters $I_S = 10^{-13}\,\text{A}$ and $\beta = 60$. Plot v_O versus v_I and i_O versus i_I for $-10 \leq v_I \leq +10\,\text{V}$. Determine the voltage and current gains. (b) Repeat part (a) for $R_3 = R_4 = 20\,\Omega$.

8.41   Consider the class-AB output stage shown in Figure 8.33. The parameters are: $V^+ = -V^- = 15\,\text{V}$, $R_1 = R_2 = 2\,\text{k}\Omega$, $R_L = 8\,\Omega$, and $R_3 = R_4 = 0$. Assume all transistors are matched, with $I_S = 10^{-13}$ A and $\beta = 60$. Assume the input voltage is given by $v_I = V_p \sin \omega t$. Determine the average power delivered to the load, and the average power dissipated in Q_3 and Q_4, for $0 \leq V_p \leq 13$ V.

DESIGN PROBLEMS

[Note: Each design should be correlated with a computer analysis.]

*D8.42 Design an audio amplifier to deliver an average of 60 W to an 8 Ω speaker. The bandwidth is to cover the range from 10 Hz to 15 kHz. Specify minimum current gains, and current, voltage, and power ratings of all transistors.

*D8.43 Design a class-A transformer-coupled emitter-follower amplifier to deliver an average power of 20 W to an 8 Ω speaker. The ambient temperature is 25 °C, and the maximum junction temperature is $T_{j,\text{max}} = 125$ °C. Assume the thermal resistance values are: $\theta_{\text{dev-case}} = 3.6$ °C/W, $\theta_{\text{case-snk}} = 0.5$ °C/W, and $\theta_{\text{snk-amb}} = 4.5$ °C/W. Specify the power supply voltage, transformer turns ratio, bias resistor values, and transistor current, voltage, and power ratings.

*D8.44 Design the class-AB output stage with the V_{BE} multiplier shown in Figure 8.31 to deliver an average of 5 W to an 8 Ω load. The peak output voltage must be no more than 80 percent of V^+. Let $V^- = -V^+$. Specify the circuit and transistor parameters.

Industry Insight

LINLEY GUMM
Principal Engineer
Tektronix, Inc.

''Virtually every electronics designer uses the principles in this part. These are core principles; they are used at some level in every design, analog or digital. They must be mastered before the design of specialized electronics used by advanced professionals can be understood or attempted.

Of course, analog circuit designers use the contents of this part daily. As an example, I recently designed the baseband analog subsystem for a digitally based cellular telephone test system. It consisted of a digital-to-analog converter, followed by a multistage active and passive filter. The design used operational amplifiers in many configurations. Designing the system required the basic operational amplifier principles illustrated in Chapter 9, a knowledge of how they are internally configured, as given in Chapter 13, and the nonideal effects described in Chapter 14. The active filter uses several combinations of the active filters shown in Chapter 15. Feedback was used around every individual stage, plus around the whole filter. Each of these paths required analysis using the principles shown in Chapter 12. The accuracy of the system was based on a voltage reference IC that is similar to the voltage regulators shown in Chapter 13. This is not to say that the rest of the circuits shown in this part are not useful. They are also very important and are frequently used in my other designs.

Many have concluded that analog circuits are obsolete and are not worth studying. It must be remembered that, at base, every electronic circuit is analog! The logic gate, the basic element of digital design, is an analog circuit using the circuit elements and principles described here. Beyond that, every digital system in the world uses analog subsystems for conditioning their input and output signals. Far from being obsolete, analog circuits are experiencing a resurgence of interest, as they are used to provide the physical layers of the digital communications systems that are transforming the world in which we live. These principles are important and will continue to be. Learn them well, because they provide the basis for understanding everything electronic.''

ANALOG ELECTRONICS

Part I of this text dealt with the basic electronic devices, and the fundamental circuit configurations and characteristics. Part II deals with more complex analog circuits, of which amplifiers are a very significant category.

Chapter 9 introduces the ideal op-amp and related circuits. The op-amp is one of the most common analog integrated circuits. IC biasing techniques, which primarily use constant-current sources, are described in Chapter 10. One of the most widely used amplifier configurations is the differential amplifier, which is analyzed in Chapter 11. Chapter 12 covers the fundamentals of feedback, which is used extensively in analog circuits to set or control gain values more precisely, and to alter, in a favorable way, input and output impedances.

More complex analog integrated circuits, including circuits that form operational amplifiers, are discussed in Chapter 13. These circuits are composed of fundamental configurations, such as the diff-amp, constant-current source, active load, and output stage, all of which have been previously analyzed. Then, Chapter 14 considers the non-ideal effects in these operational amplifiers, and discusses the impacts of such effects on op-amp applications. Integrated circuit applications and designs are considered in Chapter 15. Such applications include: active filters, tuned amplifiers, and oscillators.

9

The Ideal Operational Amplifier

9.0   PREVIEW

An operational amplifier (op-amp) is an integrated circuit that amplifies the difference between two input voltages and produces a single output. The op-amp is prevalent in analog electronics, and can be thought of as another electronic device, in much the same way as the bipolar or field-effect transistor.

The term operational amplifier comes from the original applications of the device in the early 1960s. Op-amps, in conjunction with resistors and capacitors, were used in analog computers to perform mathematical operations to solve differential and integral equations. The application of op-amps has expanded significantly since those early days.

The main reason for postponing the discussion of op-amp circuits until now is that we can use a relatively simple transistor circuit to develop the ideal characteristics of the op-amp, instead of simply stating the ideal parameters as postulates. Once the ideal properties have been developed, the reader can then be more comfortable applying these ideal characteristics in the design of op-amp circuits. Just as we developed equivalent circuits of transistors that include dependent sources representing gain factors, we will develop a basic op-amp equivalent circuit with a dependent source that represents the device gain that can be used to determine some of the nonideal properties of op-amp circuits.

Op-amps are used extensively in the design of electronic circuits and systems because of their relatively low cost and versatility, and because integrated circuit op-amp characteristics approach the ideal. We will consider, in this chapter, a few of the many applications of op-amps. By the end of the chapter, the reader should be able to design fairly sophisticated circuits using op-amps. (See the Overview of Electronic Design section in this chapter.)

9.1   THE OPERATIONAL AMPLIFIER

The integrated circuit operational amplifier evolved soon after development of the first bipolar integrated circuit. The μA-709 was introduced by Fairchild Semiconductor in 1965 and was one of the first widely used general purpose op-amps. The now classic μA-741, also by Fairchild, was introduced in the late 1960s. Since then, a vast array of op-amps with improved characteristics, using both bipolar and MOS technologies, have been designed. Most op-amps are very inexpensive (less than a dollar) and are available from a wide range of suppliers.

From a signal point of view, the op-amp has two input terminals and one output terminal, as shown in the small-signal circuit symbol in Figure 9.1(a). The op-amp also requires dc power, as do all transistor circuits, so that the transistors are biased in the active region. Also, most op-amps are biased with both a positive and a negative voltage supply, as indicated in Figure 9.1(b). As

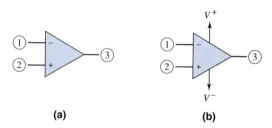

(a)   (b)

Figure 9.1   (a) Small-signal circuit symbol of the op-amp; (b) op-amp with positive and negative supply voltages

before, the positive voltage is indicated by V^+ and the negative voltage by V^-.

There are normally 20 to 30 transistors that make up an op-amp circuit. The typical IC op-amp has parameters that approach the ideal characteristics. For this reason, then, we can treat the op-amp as a "simple" electronic device, which means that it is quite easy to design a wide range of circuits using the IC op-amp.

In this chapter, we develop the ideal set of op-amp parameters and then consider the analysis and design of a wide variety of op-amp circuits, which will aid in our understanding of the design process of electronic circuits. We generally assume, in this chapter, that the op-amp is ideal. In the following chapters, we consider the differential amplifier, current-source biasing, and feedback, which leads to the development of the actual operational amplifier circuit in Chapter 13. Once the actual op-amp circuit is studied, then the source of nonideal characteristics can be understood. The effect of nonideal op-amp parameters is then considered in Chapter 14. Additional op-amp applications are given in Chapter 15.

9.1.1   Ideal Parameters

The ideal op-amp senses the difference between two input signals and amplifies this difference to produce an output signal. The terminal voltage is the voltage at a terminal measured with respect to ground. The ideal op-amp equivalent circuit is shown in Figure 9.2(a).

Ideally, the input impedance is infinite, which means that the input current is zero. The output terminal of the ideal op-amp acts as the output of an ideal voltage source, meaning that the small-signal output impedance is zero.

The parameter A_{od} shown in the equivalent circuit is the **differential gain** of the op-amp. The output is out of phase with respect to v_1 and in phase with respect to v_2. Terminal (1) then is the **inverting input terminal**, designated by the

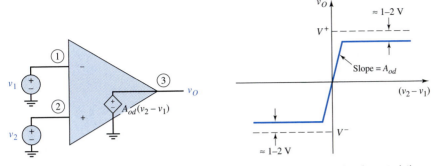

Figure 9.2   (a) Ideal op-amp equivalent circuit and (b) op-amp transfer characteristics

"−" notation, and terminal (2) is the **noninverting input terminal**, designated by the "+" notation.

Since the ideal op-amp responds only to the difference between the two input signals v_1 and v_2, the ideal op-amp maintains a zero output signal for $v_1 = v_2$. When $v_1 = v_2 \neq 0$, there is what is called a **common-mode input signal**. For the ideal op-amp, the common-mode output signal is zero. This characteristic is referred to as **common-mode rejection**.

Because the device is biased with both positive and negative power supplies, most op-amps are direct-coupled devices (i.e., no coupling capacitors are used on the input). Therefore, the input voltages v_1 and v_2 shown in Figure 9.2(a) can be dc voltages, which will produce a dc output voltage v_O.

Since the op-amp is composed of transistors biased in the active region by the dc input voltages V^+ and V^-, the output voltage is limited. When v_O approaches V^+, it will saturate, or be limited to a value nearly equal to V^+, since it cannot go above the positive bias voltage. Similarly, when the output voltage approaches V^-, it will saturate at a value nearly equal to V^-. The actual saturation voltages vary from one op-amp to another, but in general the output volage is limited to $V^- + \Delta V < v_O < V^+ - \Delta V$, where ΔV is generally between 1 and 2 V. Figure 9.2(b) is a simplified voltage transfer characteristic for the op-amp, showing this saturation effect.

The ideal op-amp is being considered in this chapter, in order to gain an appreciation of the properties and characteristics of op-amp circuits. Two important op-amp parameters are: the differential gain A_{od} and the bandwidth, or frequency response. The differential gain A_{od} is very large, ideally infinite. We will see how this property produces ideal op-amp circuit characteristics. The bandwidth or frequency response of op-amps is discussed in Chapter 13.

9.1.2   Development of the Ideal Parameters

To develop the ideal op-amp parameters, we start with the MOSFET small-signal equivalent circuit and apply this model to a particular circuit. Figure 9.3(a) shows an n-channel enhancement-mode MOSFET, and Figure 9.3(b) is the simplified low-frequency small-signal equivalent circuit. In our analysis, the transistor small-signal output resistance r_o is assumed to be infinite.

Figure 9.4 shows the MOSFET equivalent circuit with two external circuit resistors, R_I and R_F, and an input voltage v_I. Resistor R_F is a **feedback resistor**

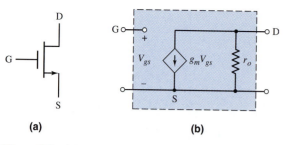

Figure 9.3 (a) n-channel enhancement-mode MOSFET and (b) small-signal equivalent circuit

Figure 9.4 Simplified small-signal equivalent circuit of a MOSFET with input and feedback resistors

that connects the output back to the input of the transistor. This circuit is therefore called a feedback circuit. In this example, we use a single transistor as the basic amplifier of the feedback circuit.

Writing a KCL equation at the gate terminal, we obtain

$$\frac{v_I - V_{gs}}{R_I} = \frac{V_{gs} - v_O}{R_F} \tag{9.1(a)}$$

which can be arranged as

$$\frac{v_I}{R_I} + \frac{v_O}{R_F} = V_{gs}\left(\frac{1}{R_I} + \frac{1}{R_F}\right) \tag{9.1(b)}$$

Since the input impedance to the transistor is infinite, the current into the device is zero.

A KCL equation at the output node yields

$$\frac{V_{gs} - v_O}{R_F} = g_m V_{gs} \tag{9.2(a)}$$

which can be solved for V_{gs}, as follows:

$$V_{gs} = \frac{v_O}{R_F} \cdot \frac{1}{\left(\dfrac{1}{R_F} - g_m\right)} \tag{9.2(b)}$$

Substituting Equation (9.2(b)) into (9.1(b)) results in the overall voltage gain of the circuit

$$\frac{v_O}{v_I} = -\frac{R_F}{R_I} \cdot \frac{\left(1 - \dfrac{1}{g_m R_F}\right)}{\left(1 + \dfrac{1}{g_m R_F}\right)} \tag{9.3}$$

If we let the gain g_m of the basic amplifier (i.e., the transistor) go to infinity, then the overall voltage gain becomes

$$\frac{v_O}{v_I} = -\frac{R_F}{R_I} \tag{9.4}$$

Equation (9.4) shows that the overall voltage gain is the ratio of two external circuit resistors, which is one result of using an ideal op-amp. The negative sign indicates a 180 degree phase shift between the input and the output, which means that the input to the transistor corresponds to the inverting terminal of an op-amp. The voltage gain given by Equations (9.3) and (9.4) is called a **closed-loop voltage gain**, since feedback is incorporated into the circuit. Conversely, the voltage gain A_{od} is an **open-loop gain**.

Voltage V_{gs} at the input of the basic amplifier (transistor) is given by Equation (9.2(b)). Again, if we let the gain g_m go to infinity, then $V_{gs} \cong 0$; that is, the voltage at the input terminal to the basic amplifier is almost at ground potential. This terminal is said to be at **virtual ground**, which is another characteristic that we will observe in ideal op-amp circuits.

The output resistance of this circuit can be determined from the equivalent circuit shown in Figure 9.5. The input signal source is set at zero. A KCL equation at the output node, written in phasor notation, is

$$I_x = g_m V_{gs} + \frac{V_x}{R_I + R_F} \tag{9.5}$$

Voltage V_{gs} can be written in terms of the test voltage V_x, as

$$V_{gs} = V_x \left(\frac{R_I}{R_I + R_F} \right) \tag{9.6}$$

Substituting Equation (9.6) into (9.5), we find that

$$\frac{I_x}{V_x} = \frac{1}{R_o} = \frac{1 + g_m R_I}{R_I + R_F} \tag{9.7}$$

If the gain g_m goes to infinity, then $1/R_o \to \infty$, or $R_o \to 0$. The output resistance of the circuit with negative feedback included goes to zero. This is also a property of an ideal op-amp circuit.

A simplified MOSFET model with a large gain has thus provided the properties of an ideal op-amp.

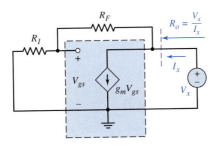

Figure 9.5   Equivalent circuit determining output resistance

9.1.3   Analysis Method

Usually, an op-amp is not used in the open-loop configuration shown in Figure 9.2(a). Instead, feedback is added to close the loop between the output and the input. In this chapter, we will limit our discussion to **negative feedback**, in which the connection from the output goes to the inverting terminal, or terminal (1). As we will see later, this configuration produces stable circuits; positive feedback, in which the output is connected to the noninverting terminal, can be used to produce oscillators.

The ideal op-amp characteristics resulting from our negative feedback analysis are shown in Figure 9.6 and summarized below.

1. The internal differential gain A_{od} is considered to be infinite.
2. The differential input voltage $(v_2 - v_1)$ is assumed to be zero. If A_{od} is very large and if the output voltage v_O is finite, then the two input voltages must be nearly equal.
3. The effective input resistance to the op-amp is assumed to be infinite, so the two input currents, i_1 and i_2, are essentially zero.
4. The output resistance R_o is assumed to be zero in the ideal case, so the output voltage is connected directly to the dependent voltage source, and the output voltage is independent of any load connected to the output.

We use these ideal characteristics in the analysis and design of op-amp circuits.

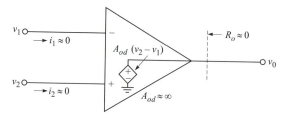

Figure 9.6   Parameters of the ideal op-amp

9.1.4   PSpice Modeling

Three general purpose op-amps are included in the PSpice library. The PSpice circuit simulation uses a macromodel, which is a simpified version of the op-amp, to model the op-amp characteristics. For example, the μA-741 op-amp has parameters $R_i = 2\,\text{M}\Omega$, $R_o = 75\,\Omega$, $A_{od} = 2 \times 10^5$, and a unity-gain bandwidth of $f_{BW} = 1\,\text{MHz}$. This device is also capable of producing output voltages of $\pm 14\,\text{V}$ with dc power supply voltages of $\pm 15\,\text{V}$. We will see in several examples as to whether these nonideal parameters affect actual circuit properties.

9.2   INVERTING AMPLIFIER

One of the most widely used op-amp circuits is the **inverting amplifier**. Figure 9.7 shows the closed-loop configuration of this circuit. We must keep in mind that the op-amp is biased with dc voltages, although those connections are seldom explicitly shown.

Figure 9.7   Inverting op-amp circuit

9.2.1 Basic Amplifier

We analyze the circuit in Figure 9.7 by considering the ideal equivalent circuit shown in Figure 9.8. The **closed-loop voltage gain**, or simply the voltage gain, is defined as

$$A_v = \frac{v_O}{v_I} \tag{9.8}$$

We stated that if the open-loop gain A_{od} is very large, then the two inputs v_1 and v_2 must be nearly equal. Since v_2 is at ground potential, voltage v_1 must also be approximately zero volts. We must point out, however, that having v_1 be essentially at ground potential does not imply that terminal (1) is grounded. Rather, terminal (1) is said to be at **virtual ground**; that is, it is essentially zero volts, but it does not provide a current path to ground.

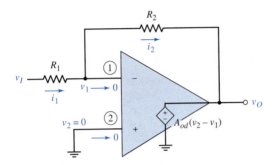

Figure 9.8 Inverting op-amp equivalent circuit

From Figure 9.8, we can write

$$i_1 = \frac{v_I - v_1}{R_1} = \frac{v_I}{R_1} \tag{9.9}$$

Since the current into the op-amp is assumed to be zero, current i_1 must flow through resistor R_2 to the output terminal, which means that $i_1 = i_2$.

The output voltage is given by

$$v_O = v_1 - i_2 R_2 = 0 - \left(\frac{v_I}{R_1}\right) R_2 \tag{9.10}$$

Therefore, the closed-loop voltage gain is

$$A_v = \frac{v_O}{v_I} = -\frac{R_2}{R_1} \tag{9.11}$$

For the ideal op-amp, the closed-loop voltage gain is a function of the ratio of two resistors; it is not a function of the transistor parameters within the op-amp circuit. Again, the minus sign implies a phase reversal. If the input voltage v_I is positive, then, because v_1 is essentially at ground potential, the output voltage v_O must be negative, or below ground potential. Also note that if the output terminal is open-circuited, current i_2 must flow back into the op-amp. However, since the output impedance for the ideal case is zero, the output voltage is not a function of this current that flows back into the op-amp and is not dependent on the load.

We can also determine the input resistance seen by the voltage source v_I. Because of the virtual ground, we have, from Equation (9.9)

$$i_1 = v_I/R_1$$

The **input resistance** is then defined as

$$R_i = \frac{v_I}{i_1} = R_1 \tag{9.12}$$

This shows that the input resistance seen by the source is a function of R_1 only, and is a result of the "virtual ground" concept. Figure 9.9 summarizes our analysis of the inverting amplifier circuit.

Figure 9.9   Currents and voltages in the inverting op-amp

Design Example 9.1   Objective: Design an inverting amplifier with a closed-loop voltage gain of $A_v = -5$.

Assume the op-amp is driven by a sinusoidal source, $v_S = 0.1 \sin \omega t$ volts, which has a source resistance of $R_S = 1\,k\Omega$ and which can supply a maximum current of $5\,\mu A$. Assume that frequency ω is low, which means that any frequency effects can be neglected.

Solution: The signal source resistance is in series with the input resistor R_1 (Figure 9.9), so

$$i_1 = \frac{v_S}{R_S + R_1}$$

If $i_1(\text{max}) = 5\,\mu A$, then we can write

$$R_1(\text{min}) + R_S = \frac{v_S(\text{max})}{i_1(\text{max})} = \frac{0.1}{5 \times 10^{-6}} \Rightarrow 20\,k\Omega$$

The resistance R_1, then, should be $19\,k\Omega$. The closed-loop gain is given by

$$A_v = \frac{-R_2}{R_S + R_1} = -5$$

where again, the source resistance must be taken into account. We then have

$$R_2 = 5(R_S + R_1) = 5(20) = 100\,k\Omega$$

Comment: An inverting op-amp with $R_1 = 19\,k\Omega$ and $R_2 = 100\,k\Omega$ will satisfy the specified requirements. We note that the resistance of the source must be taken into account.

Computer Verification: Figure 9.10(a) shows the PSpice circuit schematic used in the computer simulation and Figure 9.10(b) shows the 100 mV sinusoidal input signal. Figure 9.10(c) is the output signal which shows that a gain of 5 (magnitude) has been achieved and also shows that the output signal is 180 degrees out of phase with respect to the input signal. Finally, the input current is shown in Figure 9.10(d) with a maximum value of 5 μA. The actual circuit characteristics are not influenced to any great extent by the nonideal parameters of the μA-741 op-amp used in the circuit simulation.

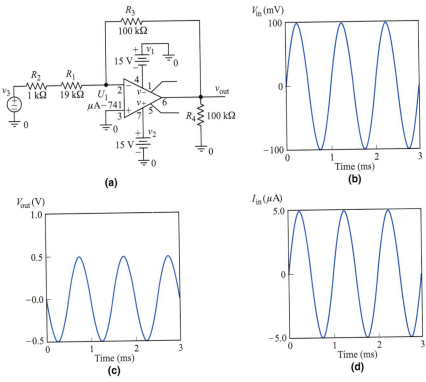

Figure 9.10 (a) PSpice circuit schematic, (b) input signal, (c) output signal, and (d) input current signal for Example 9.1

Problem-Solving Technique: Ideal Op-Amp Circuits

1. If the noninverting terminal of the op-amp is at ground potential, then the inverting terminal is at virtual ground. Sum currents at this node, assuming zero current enters the op-amp itself.

2. If the noninverting terminal of the op-amp is not at ground potential, then the inverting terminal voltage is equal to that at the noninverting terminal. Sum currents at the inverting terminal node, assuming zero current enters the op-amp itself.

3. For the ideal op-amp circuit, the output voltage is determined from either step 1 or step 2 above and is independent of any load connected to the output terminal.

Test Your Understanding

9.1   In the ideal inverting op-amp in Figure 9.9, let $R_1 = 10\,k\Omega$ and $R_2 = 100\,k\Omega$. Determine A_v, v_O, i_1, i_2, and the input resistance when $v_I = 0.25\,V$. (Ans. $A_v = -10$, $v_O = -2.5\,V$, $i_1 = i_2 = 25\,\mu A$, and $R_i = 10\,k\Omega$)

RD9.2   Redesign the ideal inverting op-amp such that the closed-loop voltage gain is $A_v = -15$ and the input resistance is $R_i = 20\,k\Omega$. (Ans. $R_1 = 20\,k\Omega$, $R_2 = 300\,k\Omega$)

9.3   Consider Example 9.1. Suppose the source resistance is not a constant, but varies within the range $0.7k\Omega \leq R_S \leq 1.3\,k\Omega$. Using the results of Example 9.1, what is the range in (a) the voltage gain A_v and (b) the input current i_1. (c) Is the specified maximum input current still maintained? (Ans. (a) $4.926 \leq A_v \leq 5.076$, (b) $4.926 \leq i_1 \leq 5.076\,\mu A$)

9.2.2   Amplifier with a T-Network

Assume that an inverting amplifier is to be designed having a closed-loop voltage gain of $A_v = -100$ and an input resistance of $R_i = R_1 = 50\,k\Omega$. The feedback resistor R_2 would then have to be $5\,M\Omega$. However this resistance value is too large for most practical circuits.

Consider the op-amp circuit shown in Figure 9.11 with a T-network in the feedback loop. The analysis of this circuit is similar to that of the inverting op-amp circuit of Figure 9.9. At the input, we have

$$i_1 = \frac{v_I}{R_1} = i_2 \tag{9.13}$$

Figure 9.11   Inverting op-amp with T-network

We can also write that

$$v_X = 0 - i_2 R_2 = -v_I\left(\frac{R_2}{R_1}\right) \tag{9.14}$$

If we sum the currents at the node v_X, we have

$$i_2 + i_4 = i_3$$

which can be written

$$-\frac{v_X}{R_2} - \frac{v_X}{R_4} = \frac{v_X - v_O}{R_3} \tag{9.15}$$

or

$$v_X \left(\frac{1}{R_2} + \frac{1}{R_4} + \frac{1}{R_3} \right) = \frac{v_O}{R_3} \tag{9.16}$$

Substituting the expression for v_X from Equation (9.14), we obtain

$$-v_I \left(\frac{R_2}{R_1} \right) \left(\frac{1}{R_2} + \frac{1}{R_4} + \frac{1}{R_3} \right) = \frac{v_O}{R_3} \tag{9.17}$$

The closed-loop voltage gain is therefore

$$A_v = \frac{v_O}{v_I} = -\frac{R_2}{R_1} \left(1 + \frac{R_3}{R_4} + \frac{R_3}{R_2} \right) \tag{9.18}$$

The advantage of using a T-network is demonstrated in the following example.

Design Example 9.2   Objective: An op-amp with a T-network is to be used as a preamplifier for a microphone. The maximum microphone output voltage is 12 mV (rms) and the microphone has an output resistance of 1 kΩ.

The op-amp circuit is to be designed such that the maximum output voltage is 1.2 V(rms). The input amplifier resistance should be fairly large, but all resistance values should be less than 500 kΩ.

Solution: We need a voltage gain of

$$|A_v| = \frac{1.2}{0.012} = 100$$

Equation (9.18) can be written in the form

$$A_v = -\frac{R_2}{R_1} \left(1 + \frac{R_3}{R_4} \right) - \frac{R_3}{R_1}$$

If, for example, we arbitrarily choose $\frac{R_2}{R_1} = \frac{R_3}{R_1} = 8$, then

$$-100 = -8 \left(1 + \frac{R_3}{R_4} \right) - 8$$

which yields

$$\frac{R_3}{R_4} = 10.5$$

The effective R_1 must include the R_S resistance of the microphone. If we set $R_1 = 49\,\text{k}\Omega$ so that $R_{1,\text{eff}} = 50\,\text{k}\Omega$, then

$$R_2 = R_3 = 400\,\text{k}\Omega$$

and

$$R_4 = 38.1\,\text{k}\Omega$$

Comment: As required, all resistor values are less than 500 kΩ. Also the resistance ratios in the closed-loop gain equation are approximately equal. As with most design problems, there is no unique solution.

Design Pointer: If we need to use standard resistance values in our design, then, using Appendix D, we can choose $R_1 = 51\,\text{k}\Omega$ so that $R_{1,\text{eff}} = 52\,\text{k}\Omega$, and we can choose $R_2 = R_3 = 390\,\text{k}\Omega$. Then, using Equation (9.18), we have

$$A_v = -100 = \frac{-R_2}{R_{1,\text{eff}}}\left(1 + \frac{R_3}{R_4}\right) - \frac{R_3}{R_{1,\text{eff}}} = \frac{-390}{52}\left(1 + \frac{390}{R_4}\right) - \frac{390}{52}$$

which yields $R_4 = 34.4\,\text{k}\Omega$. We could, for example, use a $50\,\text{k}\Omega$ potentiometer for R_4 and set the value at $34.4\,\text{k}\Omega$ to produce a gain of 100. The use of a potentiometer will also allow us to adjust the voltage gain of the circuit to take into account tolerance variations in resistor values.

The amplifier with a T-network allows us to obtain a large gain using reasonably sized resistors.

Test Your Understanding

D9.4   Design an ideal inverting op-amp with a T-network that has a closed-loop voltage gain of $A_V = -50$ and an input resistance of $10\,\text{k}\Omega$. All resistors must be no larger than $50\,\text{k}\Omega$. Verify your design with a PSpice analysis. (Ans. For example: $R_1 = 10\,\text{k}\Omega$, $R_2 = R_3 = 50\,\text{k}\Omega$, and $R_4 = 6.25\,\text{k}\Omega$)

9.2.3   Effect of Finite Gain

A **finite open-loop gain** A_{od}, also called the finite differential-mode gain, affects the closed-loop gain of an inverting amplifier. We will consider nonideal effects in op-amps in a later chapter; here, we will determine the magnitude of A_{od} required to approach the ideal case.

Consider the inverting op-amps shown in Figure 9.12. As before, we assume an infinite input resistance at terminals (1) and (2), which means the input currents to the op-amp are zero.

The current through R_1 can be written as

$$i_1 = \frac{v_I - v_1}{R_1}$$

(9.19)

Figure 9.12   Equivalent circuit of the inverting op-amp with a finite differential-mode gain

and the current through R_2 is

$$i_2 = \frac{v_1 - v_O}{R_2} \qquad\qquad \textbf{(9.20)}$$

The output voltage is now given by

$$v_O = -A_{od}v_1$$

so that the terminal (1) voltage can be written as

$$v_1 = -\frac{v_O}{A_{od}} \qquad\qquad \textbf{(9.21)}$$

Combining Equations (9.21), (9.19), and (9.20), and setting $i_1 = i_2$, we obtain

$$i_1 = \frac{v_I + \dfrac{v_O}{A_{od}}}{R_1} = i_2 = \frac{-\dfrac{v_O}{A_{od}} - v_O}{R_2} \qquad\qquad \textbf{(9.22)}$$

Solving for the closed-loop voltage gain, we find that

$$A_v = \frac{v_O}{v_I} = -\frac{R_2}{R_1} \frac{1}{\left[1 + \dfrac{1}{A_{od}}\left(1 + \dfrac{R_2}{R_1}\right)\right]} \qquad\qquad \textbf{(9.23)}$$

Equation (9.23) shows that if $A_{od} \to \infty$, the ideal closed-loop voltage gain reduces to that given by Equation (9.11).

Example 9.3   Objective: Determine the deviation from the ideal due to a finite differential gain.

Consider an inverting op-amp with $R_1 = 10\,\text{k}\Omega$ and $R_2 = 100\,\text{k}\Omega$. Determine the closed-looop gain for: $A_{od} = 10^2, 10^3, 10^4, 10^5,$ and 10^6. Calculate the percent deviation from the ideal gain.

Solution: The ideal closed-loop gain is

$$A_v = -\frac{R_2}{R_1} = -\frac{100}{10} = -10$$

If $A_{od} = 10^2$, we have, from Equation (9.3),

$$A_v = -\frac{100}{10} \cdot \frac{1}{\left[1 + \dfrac{1}{10^2}\left(1 + \dfrac{100}{10}\right)\right]} = \frac{-10}{(1 + 0.11)} = -9.01$$

which is a 9.9 percent deviation from the ideal. For the other differential gain values we have the following results:

A_{od}	A_v	Deviation (%)
10^2	-9.01	9.9
10^3	-9.89	1.1
10^4	-9.989	0.11
10^5	-9.999	0.01
10^6	-9.9999	0.001

Comment: For this case, the open-loop gain must be on the order of at least 10^3 in order to be within 1 percent of the ideal gain. If the ideal closed-loop gain changes, a

new value of open-loop gain must be determined in order to meet the specified requirements. As we will see in Chapter 14, at low frequencies, most op-amp circuits have gains
on the order of 10^5, so achieving the required accuracy is not difficult.

Test Your Understanding

9.5 An op-amp is ideal, except that its open-loop differential voltage gain is limited
to $A_{od} = 10^3$. The voltages at two of the three signal terminals are measured. Determine
the voltage at the third signal terminal for: (a) $v_2 = 0$ V and $v_O = 5$ V; (b) $v_1 = 5$ V and
$v_O = -10$ V; (c) $v_1 = 0.001$ V and $v_2 = -0.001$ V; (d) $v_2 = 3$ V and $v_O = 3$ V. (Ans.
(a) $v_1 = -5$ mV, (b) $v_2 = 4.99$ V, (c) $v_O = -2$ V, and (d) $v_1 = 2.997$ V)

D9.6 An inverting op-amp is ideal, except that the differential voltage gain is finite
and is $A_{od} = 5 \times 10^3$. Design the circuit such that the closed-loop voltage gain is $A_v =
-12.0$ and the input resistor is $R_1 = 25$ kΩ. Determine the required value of R_2. (Ans.
$R_2 = 300.78$ kΩ)

9.3   SUMMING AMPLIFIER

To analyze the op-amp circuit shown in Figure 9.13(a), we will use the superposition theorem and the concept of virtual ground. Using the superposition
theorem, we will determine the output voltage due to each input acting alone.

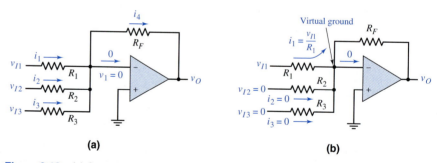

Figure 9.13   (a) Summing op-amp amplifier circuit and (b) currents and voltages in the
summing amplifier

We will then algebraically sum these terms to determine the total output.
If we set $v_{I2} = v_{I3} = 0$, the current i_1 is

$$i_1 = \frac{v_{I1}}{R_1} \tag{9.24}$$

Since $v_{I2} = v_{I3} = 0$ and the inverting terminal is at virtual ground, the currents
i_2 and i_3 must both be zero. Current i_1 does not flow through either R_2 or R_3,
but the entire current must flow through the feedback resistor R_F, as indicated
in Figure 9.13(b). The output voltage due to v_{I1} acting alone is

$$v_O(v_{I1}) = -i_1 R_F = -\left(\frac{R_F}{R_1}\right)v_{I1} \tag{9.25}$$

Similarly, the output voltages due to v_{I2} and v_{I3} acting individually are

$$v_O(v_{I2}) = -i_2 R_F = -\left(\frac{R_F}{R_2}\right)v_{I2} \qquad\qquad (9.26)$$

and

$$v_O(v_{I3}) = -i_3 R_F = -\left(\frac{R_F}{R_3}\right)v_{I3} \qquad\qquad (9.27)$$

The total output voltage is the algebraic sum of the individual output voltages, or

$$v_O = v_O(v_{I1}) + v_O(v_{I2}) + v_O(v_{I3}) \qquad\qquad (9.28)$$

which becomes

$$v_O = -\left(\frac{R_F}{R_1}v_{I1} + \frac{R_F}{R_2}v_{I2} + \frac{R_F}{R_3}v_{I3}\right) \qquad\qquad (9.29)$$

The output voltage is the sum of the three input voltages, with different weighting factors. This circuit is therefore called the **inverting summing amplifier**. The number of input terminals and input resistors can be changed to add more or fewer voltages.

A special case occurs when the three input resistances are equal. When $R_1 = R_2 = R_3 \equiv R$, then

$$v_O = -\frac{R_F}{R}(v_{I1} + v_{I2} + v_{I3}) \qquad\qquad (9.30)$$

This means that the output voltage is the sum of the input voltages, with a single amplification factor.

Up to this point, we have seen that op-amps can be used to multiply a signal by a constant and sum a number of signals with prescribed weights. These are mathematical operations. Later in the chapter, we will see that op-amps can also be used to integrate and differentiate. These circuits are the building blocks needed to perform analog computations—hence the original name of operational amplifier. Op-amps, however, are versatile and can do much more than just perform mathematical operations, as we will continue to observe through the remainder of the chapter.

Design Example 9.4 **Objective:** Design the summing amplifier to produce a specified output signal.

The output signal generated from a BJT emitter-follower amplifier is $v_{O1} = (5 - 0.5\sin\omega t)$ V and the effective output resistance is $R_o = 50\,\Omega$. Design a summing amplifier such that its output signal is $v_O = 2\sin\omega$ V. One input to the summing amplifier is to be the output of the BJT emitter follower. (Assume that the frequency ω goes very low, making the use of a coupling capacitor impractical.)

Solution: In this case, we only need two inputs to a summing amplifier, as shown in Figure 9.13. We can apply -5 V to one input of the summing amplifier to cancel the $+5$ V from the emitter-follower signal. Assume the input signal v_{I1} is from the emitter follower, the input v_{I2} is -5 V, and $v_{I3} = 0$. If we make $R_1 = 5\,\text{k}\Omega$, then the effect of the output resistance, $R_o = 50\,\Omega$, can be neglected (to within 1 percent of the ideal). The gain needs to be

$$A_v = \frac{-R_F}{R_1} = \frac{2}{-0.5} = -4$$

The feedback resistor must then be $R_F = 20\,\text{k}\Omega$. We set the input signal $v_{I2} = -5\,\text{V}$ with $R_2 = 5\,\text{k}\Omega$.

Comment: In this example, we have used a summing amplifier to amplify a time-varying signal and eliminate a dc offset voltage.

Test Your Understanding

9.7   Consider an ideal summing amplifier as shown in Figure 9.13(a), with $R_1 = 10\,\text{k}\Omega$, $R_2 = 20\,\text{k}\Omega$, $R_3 = 30\,\text{k}\Omega$, and $R_F = 40\,\text{k}\Omega$. Determine the output voltage v_O if $v_{I1} = 250\,\mu\text{V}$, $v_{I2} = 200\,\mu\text{V}$, and $v_{I3} = 75\,\mu\text{V}$. (Ans. $v_O = -1.5\,\text{mV}$)

D9.8   Design a summing amplifier that will produce an output voltage of $v_O = -(7v_{I1} + 14v_{I2} + 3.5v_{I3} + 10v_{I4})$. The maximum allowable resistance value is $280\,\text{k}\Omega$. (Ans. For example: $R_F = 280\,\text{k}\Omega$, $R_1 = 40\,\text{k}\Omega$, $R_2 = 20\,\text{k}\Omega$, $R_3 = 80\,\text{k}\Omega$, and $R_4 = 28\,\text{k}\Omega$)

D9.9   Design the summing amplifier in Figure 9.13 to produce the average (magnitude) of three input voltages, i.e., $v_O = (v_{I1} + v_{I2} + v_{I3})/3$. The amplifier is to be designed such that each input signal sees the maximum possible input resistance under the condition that the maximum allowed resistance in the circuit is 1 M. (Ans. $R_1 = R_2 = R_3 = 1\,\text{M}\Omega$, $R_F = 333\,\text{k}\Omega$)

9.4   NONINVERTING AMPLIFIER

In our previous discussions, the feedback element was connected between the output and the inverting terminal. However, a signal can be applied to the noninverting terminal while still maintaining negative feedback.

9.4.1   Basic Amplifier

Figure 9.14 shows the basic **noninverting amplifier**. The input signal v_I is applied directly to the noninverting terminal, while one side of resistor R_1 is connected to the inverting terminal and the other side is at ground.

Previously, when v_2 was at ground potential, we argued that v_1 was also essentially at ground potential, and we stated that terminal (1) was at virtual ground. The same principle applies to the circuit in Figure 9.14, with slightly different terminology. The negative feedback connection forces the terminal voltages v_1 and v_2 to be essentially equal. Such a condition is referred to as a **virtual short**. This condition exists since a change in v_2 will cause the output voltage v_O to change in such a way that v_1 is forced to track v_2. The virtual short means that the voltage difference between v_1 and v_2 is, for all practical purposes, zero. However, unlike a true short circuit, there is no current flow

Figure 9.14   Noninverting op-amp circuit

directly from one terminal to the other. We use the virtual short concept, i.e. $v_1 = v_2$, as an ideal op-amp characteristic and use this property in our circuit analysis.

The analysis of the noninverting amplifier is essentially the same as for the inverting amplifier. We assume that no current enters the input terminals. Since $v_1 = v_2$, then $v_1 = v_I$, and current i_1 is given by

$$i_1 = -\frac{v_1}{R_1} = -\frac{v_I}{R_1} \qquad (9.31)$$

Current i_2 is given by

$$i_2 = \frac{v_1 - v_O}{R_2} = \frac{v_I - v_O}{R_2} \qquad (9.32)$$

As before, $i_1 = i_2$, so that

$$-\frac{v_I}{R_1} = \frac{v_I - v_O}{R_2} \qquad (9.33)$$

Solving for the closed-loop voltage gain, we find

$$A_v = \frac{v_O}{v_I} = 1 + \frac{R_2}{R_1} \qquad (9.34)$$

From this equation, we see that the output is in phase with the input, as expected. Also note that the gain is always greater than unity.

The input signal v_I is connected directly to the noninverting terminal; therefore, since the input current is essentially zero, the input impedance seen by the source is very large, ideally infinite. The ideal equivalent circuit of the noninverting op-amp is shown in Figure 9.15.

9.4.2   Voltage Follower

An interesting property of the noninverting op-amp occurs when $R_1 = \infty$, an open circuit. The closed-loop gain then becomes

$$A_v = \frac{v_O}{v_I} = 1 \qquad (9.35)$$

Since the output voltage follows the input, this op-amp circuit is called a **voltage follower**. The closed-loop gain is independent of resistor R_2 (except when $R_2 = \infty$), so we can set $R_2 = 0$ to create a short circuit.

Figure 9.15   Equivalent circuit of ideal noninverting op-amp

Figure 9.16 Voltage-follower op-amp

The voltage-follower op-amp circuit is shown in Figure 9.16. At first glance, it might seem that this circuit, with unity voltage gain, would be of little value. However, other terms used for the voltage follower are **impedance transformer** or **buffer**. The input impedance is essentially infinite, and the output impedance is essentially zero. If, for example, the output impedance of a signal source is large, a voltage follower inserted between the source and a load will prevent loading effects, that is, it will act as a buffer between the source and the load.

Consider the case of a voltage source with a $100\,\text{k}\Omega$ output impedance driving a $1\,\text{k}\Omega$ load impedance, as shown in Figure 9.17(a). This situation may occur if the source is a transducer. (We will see an example of this later in the chapter when we consider a temperature-sensitive resistor, or thermistor, in a bridge circuit.) The ratio of output voltage to input voltage is

$$\frac{v_O}{v_I} = \frac{R_L}{R_L + R_S} = \frac{1}{1 + 100} \cong 0.01$$

This equation indicates that, for this case, there is a severe loading effect, or **attenuation**, in the signal voltage.

Figure 9.17(b) shows a voltage follower inserted between the source and the load. Since the input impedance to the noninverting terminal is usually much greater than $100\,\text{k}\Omega$, then $v_O \cong v_I$ and the loading effect is eliminated.

(a)   (b)

Figure 9.17   (a) Source with a $100\,\text{k}\Omega$ output resistance driving a $1\,\text{k}\Omega$ load and (b) source with a $100\,\text{k}\Omega$ output resistance, voltage follower, and $1\,\text{k}\Omega$ load

Test Your Understanding

D9.10   Design a noninverting amplifier with a closed-loop gain of $A_v = 5$. The output voltage is limited to $-10\,\text{V} \le v_O \le +10\,\text{V}$, and the maximum current in any resistor is limited to $50\,\mu\text{A}$. (Ans. $R_1 = 40\,\text{k}\Omega$, $R_2 = 160\,\text{k}\Omega$)

*9.11   The noninverting op-amp in Figure 9.14 has a finite differential gain of A_{od}. Show that the closed-loop gain is

$$A_v = \frac{v_O}{v_I} = \frac{\left(1 + \dfrac{R_2}{R_1}\right)}{\left[1 + \dfrac{1}{A_{od}}\left(1 + \dfrac{R_2}{R_1}\right)\right]}$$

9.12   Use superposition to determine the output voltage v_O in the ideal op-amp circuit in Figure 9.18. (Ans. $v_O = 10v_{I1} + 5v_{I2}$)

Figure 9.18   Figure for Exercise 9.12

9.5   OP-AMP APPLICATIONS

The summing amplifier is one example of special functional capabilities that can be provided by the op-amp. In this section, we will look at other examples of op-amp versatility.

9.5.1   Current-to-Voltage Converter

In some situations, the output of a device or circuit is a current. An example is the output of a photodiode or photodetector. We may need to convert this output current to an output voltage.

Consider the circuit in Figure 9.19. The input resistance R_i at the virtual ground node is

$$R_i = \frac{v_1}{i_1} \approx 0 \tag{9.36}$$

In most cases, we can assume that $R_S \gg R_i$; therefore, current i_1 is essentially equal to the signal current i_S. Then,

$$i_2 = i_1 = i_S \tag{9.37}$$

and

$$v_O = -i_2 R_F = -i_S R_F \tag{9.38}$$

Figure 9.19   Current-to-voltage converter

The output voltage is directly proportional to the signal current, and the feedback resistance R_F is the magnitude of the ratio of the output voltage to the signal current.

Test Your Understanding

D9.13 A current source has an output impedance of $R_S = 100\,\text{k}\Omega$. Design a current-to-voltage converter with an output voltage of $v_O = -10\,\text{V}$ when the signal current is $i_S = 100\,\mu\text{A}$. (Ans. Figure 9.19 with $R_F = 100\,\text{k}\Omega$)

9.5.2 Voltage-to-Current Converter

The complement of the current-to-voltage converter is the voltage-to-current converter. For example, we may want to drive a coil in a magnetic circuit with a given current, using a voltage source. We could use the inverting op-amp shown in Figure 9.20. For this circuit,

$$i_2 = i_1 = \frac{v_I}{R_1} \tag{9.39}$$

which means that current i_2 is directly proportional to input voltage v_I and is independent of the load impedance or resistance R_2. However, one side of the load device might need to be at ground potential, so the circuit in Figure 9.20 would not be practical for such applications.

Figure 9.20 Simple voltage-to-current converter

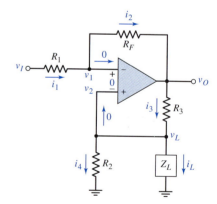

Figure 9.21 Voltage-to-current converter

Consider the circuit in Figure 9.21. In this case, one terminal of the load device, which has an impedance of Z_L, is at ground potential. The inverting terminal (1) is not at virtual ground. From the virtual short concept, $v_1 = v_2$. We also note that $v_1 = v_2 = v_L = i_L Z_L$. Equating the currents i_1 and i_2, we have

$$\frac{v_I - i_L Z_L}{R_1} = \frac{i_L Z_L - v_O}{R_F} \tag{9.40}$$

Summing the currents at the noninverting terminal gives

$$\frac{v_O - i_L Z_L}{R_3} = i_L + \frac{i_L Z_L}{R_2} \tag{9.41}$$

Solving for $(v_O - i_L Z_L)$ from Equation (9.40) and substituting into Equation (9.41) produces

$$\frac{R_F}{R_1} \cdot \frac{(i_L Z_L - v_I)}{R_3} = i_L + \frac{i_L Z_L}{R_2} \tag{9.42}$$

Combining terms in i_L, we obtain

$$i_L \left(\frac{R_F Z_L}{R_1 R_3} - 1 - \frac{Z_L}{R_2} \right) = v_I \left(\frac{R_F}{R_1 R_3} \right) \tag{9.43}$$

In order to make i_L independent of Z_L, we can design the circuit such that the coefficient of Z_L is zero, or

$$\frac{R_F}{R_1 R_3} = \frac{1}{R_2} \tag{9.44}$$

Equation (9.43) then reduces to

$$i_L = -v_I \left(\frac{R_F}{R_1 R_3} \right) = \frac{-v_I}{R_2} \tag{9.45}$$

which means that the load current is proportional to the input voltage and is independent of the load impedance Z_L, as long as the output voltage remains between allowed limits.

We may note that the input resistance seen by the source v_I is finite, and is actually a function of the load impedance Z_L. For a constant i_L, a change in Z_L produces a change in $v_L = v_2 = v_1$, which causes a change in i_1. A voltage follower may be inserted between the voltage source v_I and the resistor R_1 to eliminate any loading effects due to a variable input resistance.

Example 9.5 **Objective:** Determine a load current in a voltage-to-current converter.

Consider the circuit in Figure 9.21. Let $Z_L = 100\,\Omega$, $R_1 = 10\,\text{k}\Omega$, $R_2 = 1\,\text{k}\Omega$, $R_3 = 1\,\text{k}\Omega$, and $R_F = 10\,\text{k}\Omega$. If $v_I = -5\,\text{V}$, determine the load current i_L and the output voltage v_O.

Solution: We note first that the condition expressed by Equation (9.44) is satisfied; that is,

$$\frac{1}{R_2} = \frac{R_F}{R_1 R_3} = \frac{10}{(10)(1)} \rightarrow \frac{1}{1}$$

The load current is

$$i_L = \frac{-v_I}{R_2} = \frac{-(-5)}{1\,\text{k}\Omega} = 5\,\text{mA}$$

and the voltage across the load is

$$v_L = i_L Z_L = (5 \times 10^{-3})(100) = 0.5\,\text{V}$$

Currents i_4 and i_3 are

$$i_4 = \frac{v_L}{R_2} = \frac{0.5}{1} = 0.5\,\text{mA}$$

and

$$i_3 = i_4 + i_L = 0.5 + 5 = 5.5 \, \text{mA}$$

The output voltage is then

$$v_O = i_3 R_3 + v_L = (5.5 \times 10^{-3})(10^3) + 0.5 = 6 \, \text{V}$$

We could also calculate i_1 and i_2 as

$$i_1 = i_2 = -0.55 \, \text{mA}$$

Comment: In this example, we implicitly assume that the op-amp is not in saturation, which means that the applied dc bias voltage must be greater than 6 V. In addition, since currents i_2 (which is negative) and i_3 must be supplied by the op-amp, we are assuming that the op-amp is capable of supplying 6.05 mA.

Computer Verification: The PSpice circuit schematic of the voltage-to-current converter is shown in Figure 9.22(a). The input voltage was varied between 0 and −10 V. Figure 9.22(b) shows the current through the 100 Ω load and Figure 9.22(c) shows the

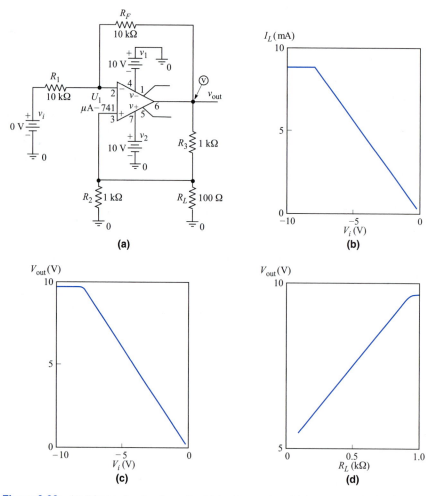

Figure 9.22 (a) PSpice circuit schematic; (b) load current and (c) op-amp output voltage versus input voltage; (d) op-amp output voltage versus load resistance for $v_I = -5$ V

op-amp output voltage as a function of the input voltage. At approximately $v_I = -7.5\,\text{V}$, the op-amp saturates, so the load current and output voltage no longer increase with input voltage. This result demonstrates that the ideal voltage-to-current conversion is valid only if the op-amp is operating in its linear region. Figure 9.22(d) shows the output voltage as a function of load resistance for an input voltage of $v_I = -5\,\text{V}$. At a load resistance greater than approximately $900\,\Omega$, the op-amp saturates. The range over which the op-amp remains linear could be increased by increasing the bias to $\pm15\,\text{V}$, for example.

Test Your Understanding

9.14 Consider the voltage-to-current converter shown in Figure 9.21. The load impedance is $Z_L = 200\,\Omega$ and the input voltage is $v_I = -3\,\text{V}$. Determine i_L and v_O if $R_1 = 10\,\text{k}\Omega$, $R_2 = 1.5\,\text{k}\Omega$, $R_3 = 3\,\text{k}\Omega$, and $R_F = 20\,\text{k}\Omega$. (Ans. $i_L = 2\,\text{mA}$, $v_O = 7.2\,\text{V}$)

D9.15 Design the voltage-to-current converter shown in Figure 9.21 such that the load current in a $500\,\Omega$ load can be varied between 0 and $1\,\text{mA}$ with an input voltage between 0 and $-5\,\text{V}$. Assume the op-amp is biased at $\pm10\,\text{V}$. (Ans. $R_2 = 5\,\text{k}\Omega$; for example, let $R_3 = 7\,\text{k}\Omega$, $R_1 = 10\,\text{k}\Omega$, $R_F = 14\,\text{k}\Omega$)

9.5.3 Difference Amplifier

An ideal difference amplifier amplifies only the difference between two signals; it rejects any common signals to the two input terminals. For example, a microphone system amplifies an audio signal applied to one terminal of a difference amplifier, and rejects any 60 Hz noise signal or "hum" existing on both terminals. The basic op-amp also amplifies the difference between two input signals. However, we would like to make a difference amplifier, in which the output is a function of the ratio of resistors, as we had for the inverting and noninverting amplifiers.

Consider the circuit shown in Figure 9.23(a), with inputs v_{I1} and v_{I2}. To analyze the circuit, we will use superposition and the virtual short concept. Figure 9.23(b) shows the circuit with input $v_{I2} = 0$. There are no currents in R_3 and R_4; therefore, $v_{2a} = 0$. The resulting circuit is the inverting amplifier previously considered, for which

$$v_{O1} = -\frac{R_2}{R_1} v_{I1} \tag{9.46}$$

Figure 9.23(c) shows the circuit with $v_{I1} = 0$. Since the current into the op-amp is zero, R_3 and R_4 form a voltage divider. Therefore,

$$v_{2b} = \frac{R_4}{R_3 + R_4} v_{I2} \tag{9.47}$$

From the virtual short concept, $v_{1b} = v_{2b}$ and the circuit becomes a noninverting amplifier, for which

$$v_{O2} = \left(1 + \frac{R_2}{R_1}\right) v_{1b} = \left(1 + \frac{R_2}{R_1}\right) v_{2b} \tag{9.48}$$

(a)

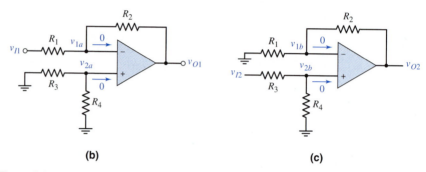

(b)   (c)

Figure 9.23   (a) Op-amp difference amplifier, (b) difference amplifier with $v_{I2} = 0$ and (c) difference amplifier with $v_{I1} = 0$

Substituting Equation (9.47) into (9.48), we obtain

$$v_{O2} = \left(1 + \frac{R_2}{R_1}\right)\left(\frac{R_4}{R_3 + R_4}\right)v_{I2}$$

(**9.49(a)**)

which can be rearranged as follows:

$$v_{O2} = (1 + R_2/R_1)\left(\frac{R_4/R_3}{1 + R_4/R_3}\right)v_{I2}$$

(**9.49(b)**)

Since the net output voltage is the sum of the individual terms, we have

$$v_O = v_{O1} + v_{O2}$$

(**9.50(a)**)

or

$$v_O = \left(1 + \frac{R_2}{R_1}\right)\left(\frac{\dfrac{R_4}{R_3}}{1 + \dfrac{R_4}{R_3}}\right)v_{I2} - \left(\frac{R_2}{R_1}\right)v_{I1}$$

(**9.50(b)**)

A property of the ideal difference amplifier is that the output voltage is zero when $v_{I1} = v_{I2}$. An inspection of Equation (9.50(b)) shows that this condition is met if

$$\frac{R_4}{R_3} = \frac{R_2}{R_1}$$

(**9.51**)

The output voltage is then

$$v_O = \frac{R_2}{R_1}(v_{I2} - v_{I1})$$

(**9.52**)

OVERVIEW OF ELECTRONICS DESIGN

PREVIEW

In Part I, we dealt with discrete electronic circuits; that is, circuits containing discrete resistors, capacitors, and transistors. The analysis of these fundamental circuits provided a basic understanding of circuit operation and characteristics. Some design discussions were also included to introduce the concept of electronic circuit design.

In Part II, we will develop, analyze, and design more complex analog electronic circuits. We will combine and expand the basic circuits considered in Part I, to form these more complex circuits. Although, for the most part, we will continue to analyze and design discrete circuits, these circuits are usually fabricated as integrated circuits. In this short overview, we discuss the basic aspects of the electronic design process.

DESIGN APPROACH

The design process can be viewed from two directions. The top-down design process begins with a proposed overall system design. The system is then divided into subsystems, one of which may include the electronics associated with the project. The electronics is then divided into its own set of subsystems.

The top-down approach usually relies on existing technologies and devices, which means that these electronic subsystems are usually designed with existing ICs. New or customized ICs may be designed and fabricated for a specialized application, although this may increase the cost of the system.

The design engineer must be able to evaluate existing design strategies and technologies to determine if they are able to meet the design or performance objectives. Insight into the operation and characteristics of basic circuits is essential for circuit design, and for being able to make appropriate choices in a top-down design process.

The bottom-up design process usually begins in a research laboratory, with the development of new and unique semiconductor materials. Silicon-based devices and circuits still dominate the electronics technology, but new compound semiconductors are gaining in importance. These new semiconductor materials are being used in the development and design of new discrete devices, such as high-performance JFETs and improved optoelectronic devices. These new devices may be incorporated into integrated circuits, which may eventually lead to the development of new systems based on the characteristics and properties of the new devices.

SYSTEM DESIGN

Consider a top-down approach in which the design of an electronic circuit or system begins with a proposed design for a large system, such as a new airplane for example. Designing and building such a system may involve hundreds or even thousands of engineers from the initial concept to the final working system. The concept begins with a set of specifications or performance objectives. The large system can be broken into subsystems.

The specifications for the electronics subsystem are usually dictated by the overall system specifications, which may include such things as size, weight, and power consumption. Design is an iterative process, and trade-offs are an integral part of the process, all the way from the overall system to each individual circuit. As the work progresses, the overall system or subsystem specifications may be refined or modified. During the design, issues may arise that were not anticipated, or design trade-offs may be required. For example, there may be tradeoffs between airplane performance and cost. High performance may require very expensive electronics and higher than expected costs in the development of high-performance engines. Once the system or subsystem requirements are finalized, design engineers then evaluate different approaches for meeting the design specifications. There are seldom unique solutions for a design, and engineering creativity is an integral part of this phase.

Once suitable approaches are selected for an overall subsystem, such as the electronics subsystem, it may then be broken down into smaller subsystems. For example, the initial electronics breakdown may include radar systems, voice communication systems, and aircraft performance monitoring systems. The specifications for each subsystem are developed from the overall set of specifications.

ELECTRONIC DESIGN

A flowchart of the general electronic design process is usually developed. This chart can apply to an entire subsystem, or to an individual circuit. A set of specifications is developed for each electronic system, and then each system is divided into many simpler circuits. For example, one relatively simple electronic system may be a temperature warning indicator. If the temperature of an engine or a particular engine part becomes greater than some predetermined value, a warning light would go on in the cockpit.

Initial design approaches are considered and a circuit configuration is proposed, based on the experience and creativity of the circuit design engineer. This is where experience in the analysis of many different types of electronic circuits becomes important. Knowledge of particular characteristics, such as input impedance, output impedance, gain characteristics, and bandwidth, for many types of circuits, is used to choose a particular circuit configuration.

A block diagram of a proposed circuit configuration can serve as the starting point for the design of the temperature indicator. For example, the output of a temperature transducer may be connected to an amplifier whose output, in turn, may be connected to a comparator. The output of the comparator is then connected to the warning light.

The choice of a temperature transducer is based in part on the required temperature range. The design of the amplifier begins with the anticipated output signal of the temperature transducer, which in turn implies a required level of input impedance and signal gain for the amplifier. The necessary stability of the amplifier gain may determine whether a simple amplifier design may be used, or a more complex amplifier using feedback may be necessary. In addition, the location of the amplifier on the aircraft may determine the temperature range over which the amplifier must function. The comparator compares the amplifier output, which corresponds to a particular temperature, to a preset value. If the amplifier output is greater than the preset value, the output of the comparator must then be able to activate the warning light. The voltage and current levels required to activate the warning light are determined and are incorporated into the design of the comparator.

In proposing the initial circuit configuration and component values, the design engineer may use an intuitive approach based to a large extent on experience. However, once this initial design is completed, the design must be verified by a mathematical analysis or computer simulation. The initial design process may also include calculations using simple models for the transistors and circuits. Normally, a more sophisticated analysis is required to take into account such things as temperature variations, tolerances in component values, and other parameter variations based on a particular application.

The circuit performance estimation or simulation is usually a very important phase of the design process. To validate the final design, it is necessary to simulate, as precisely as possible, the performance of the discrete devices and ICs used in the design. Simulation models are required for each circuit component in order to predict the operation and characteristics of the designed circuits. Based on these simulation models, trade-offs between technologies and devices may be evaluated to obtain the optimum performance. With improved simulation models, the breadboard development stage may be eliminated and the design process may proceed directly to the construction of a prototype circuit. Since the prototype circuit may involve the fabrication of specialized or customized integrated circuits, this phase of the design process may be expensive; therefore, costly mistakes in the design must be avoided. A good circuit simulation may identify potential problems that can be corrected before the prototype circuit is fabricated.

The prototype circuit is then tested and evaluated. At this point, a minor redesign may only involve selecting slightly different component values. A more extensive redesign may require selecting an entirely different circuit configuration in order to meet the system requirements. Finally, the entire system is constructed from the operating subsystems.

Design involves creativity, and it can be challenging and rewarding. Design is based on experience. The design process in Part II of this text is based on the experience gained in Part I. Our design experience should continue to grow as we proceed through the remainder of this book.

which indicates that this amplifier has a differential gain of $A_d = R_2/R_1$. This factor is a closed-loop differential gain, rather than the open-loop differential gain A_{od} of the op-amp itself.

As previously stated, another important characteristic of electronic circuits is the input resistance. The **differential input resistance** of the differential amplifier can be determined by using the circuit shown in Figure 9.24. In the figure, we have imposed the condition given in Equation (9.51) and have set $R_1 = R_3$ and $R_2 = R_4$. The input resistance is then defined as

$$R_i = \frac{V_I}{i} \tag{9.53}$$

Taking into account the virtual short concept, we can write a loop equation, as follows:

$$v_I = iR_1 + iR_1 = i(2R_1) \tag{9.54}$$

Therefore, the input resistance is

$$R_i = 2R_1 \tag{9.55}$$

Figure 9.24 Circuit for measuring differential input resistance of op-amp difference amplifier

---

Design Example 9.6 **Objective:** Design a difference amplifier with a specified gain and minimum differential input resistance.

Consider the difference amplifier in Figure 9.23(a). Design the circuit such that the differential gain is 30 and the minimum differential input resistance is $R_i = 50\,\text{k}\Omega$.

Solution: From Equation (9.55), the differential input resistance is

$$R_i = 2R_1 = 50\,\text{k}\Omega$$

Therefore,

$$R_1 = R_3 = 25\,\text{k}\Omega$$

Since the differential gain is

$$R_2/R_1 = 30$$

we must have

$$R_2 = R_4 = 750\,\text{k}\Omega$$

Comment: This example illustrates a disadvantage of this difference amplifier design. It cannot achieve both high gain and high input impedance without using extremely large resistance values.

In the ideal difference amplifier, the output v_O is zero when $v_{I1} = v_{I2}$. However, an inspection of Equation (9.50(b)) shows that this condition is not satisfied if $R_4/R_3 \neq R_2/R_1$. When $v_{I1} = v_{I2}$, the input is called a **common-mode input signal**. The common-mode input voltage is defined as

$$v_{cm} = (v_{I1} + v_{I2})/2 \tag{9.56}$$

The common-mode gain is then defined as

$$A_{cm} = \frac{v_O}{v_{cm}} \tag{9.57}$$

Ideally, when a common-mode signal is applied, $v_O = 0$ and $A_{cm} = 0$.

A nonzero common-mode gain may be generated in actual op-amp circuits. This is discussed in Chapter 14.

A figure of merit for a difference amplifier is the **common-mode rejection ratio (CMRR)**, which is defined as the magnitude of the ratio of differential gain to common-mode gain, or

$$CMRR = \left| \frac{A_d}{A_{cm}} \right| \tag{9.58}$$

Usually, the CMRR is expressed in decibels, as follows:

$$CMRR(dB) = 20 \log_{10} \left| \frac{A_d}{A_{cm}} \right| \tag{9.59}$$

Ideally, the common-mode rejection ratio is infinite. In an actual differential amplifier, we would like the common-mode rejection ratio to be as large as possible.

Example 9.7   Objective: Calculate the common-mode rejection ratio of a difference amplifier.

Consider the difference amplifier shown in Figure 9.23(a). Let $R_2/R_1 = 10$ and $R_4/R_3 = 11$. Determine CMRR(dB).

Solution: From Equation (9.50(b)), we have

$$v_O = (1 + 10) \left(\frac{11}{1 + 11} \right) v_{I2} - (10) v_{I1}$$

or

$$v_O = 10.0833 v_{I2} - 10 v_{I1} \tag{9.60}$$

The differential-mode input voltage is defined as

$$v_d = v_{I2} - v_{I1}$$

and the common-mode input voltage is defined as

$$v_{cm} = (v_{I1} + v_{I2})/2$$

Combining these two equations produces

$$v_{I1} = v_{cm} - \frac{v_d}{2} \tag{9.61(a)}$$

and

$$v_{I2} = v_{cm} + \frac{v_d}{2}$$ **(9.61(b))**

If we substitute Equations (9.61(a)) and (9.61(b)) in Equation (9.60), we obtain

$$v_O = (10.0833)\left(v_{cm} + \frac{v_d}{2}\right) - (10)\left(v_{cm} - \frac{v_d}{2}\right)$$

or

$$v_O = 10.042 v_d + 0.0833 v_{cm}$$ **(9.62)**

The output voltage is also

$$v_O = A_d v_d + A_{cm} v_{cm}$$ **(9.63)**

If we compare Equations (9.62) and (9.63), we see that

$$A_d = 10.042 \quad \text{and} \quad A_{cm} = 0.0833$$

Therefore, from Equation (9.59), the common-mode rejection ratio, is

$$\text{CMRR(dB)} = 20 \log_{10}\left(\frac{10.042}{0.0833}\right) = 41.6\,\text{dB}$$

Comment: For good differential amplifiers, typical CMRR values are in the range of 80–100 dB. This example shows how close the ratios R_2/R_1 and R_4/R_3 must be in order to achieve a CMRR value in that range.

Computer Verification: A PSpice analysis was performed on the differential amplifier in this example with a μA-741 op-amp. For input voltages of $v_{I1} = -50\,\text{mV}$ and $v_{I2} = +50\,\text{mV}$, the output voltage is $v_O = 1.0043\,\text{V}$, which gives a differential voltage gain of 10.043. For input voltages of $v_{I1} = v_{I2} = 5\,\text{V}$, the output voltage is $v_O = 0.4153\,\text{V}$, which gives a common-mode voltage gain of $A_{cm} = 0.4153/5 = 0.0831$. The common-mode rejection ratio is then $\text{CMRR} = 10.043/0.0831 = 120.9 \Rightarrow 41.6\,\text{dB}$, which agrees with the hand analysis. This result demonstrates that at this point, the nonideal characteristics of the μA-741 op-amp do not affect these results.

Test Your Understanding

D9.16 Design a difference amplifier with a differential input impedance of $R_i = 5\,\text{k}\Omega$, a differential voltage gain of 100, and a common-mode gain of zero. (Ans. $R_1 = 2.5\,\text{k}\Omega$, $R_2 = 250\,\text{k}\Omega$)

*9.17 In the difference amplifier shown in Figure 9.23(a), $R_1 = R_3 = 10\,\text{k}\Omega$, $R_2 = 20\,\text{k}\Omega$, and $R_4 = 21\,\text{k}\Omega$. Determine v_O when: (a) $v_{I1} = +1\,\text{V}$, $v_{I2} = -1\,\text{V}$; and (b) $v_{I1} = v_{I2} = +1\,\text{V}$. (c) Determine the common-mode gain. (d) Determine the CMRR(dB). (Ans. (a) $v_O = -4.032\,\text{V}$, (b) $v_O = 0.0323\,\text{V}$, (c) $A_{cm} = 0.0323$, (d) CMRR(db) = 35.9 dB)

9.5.4   Instrumentation Amplifier

We saw in the last section that it is difficult to obtain a high input impedance and a high gain in a difference amplifier with reasonable resistor values. One solution is to insert a voltage follower between each source and the corresponding input. However, a disadvantage of this design is that the gain of the amplifier cannot easily be changed. We would need to change two resistance values and still maintain equal ratios between R_2/R_1 and R_4/R_3. Optimally, we would like to be able to change the gain by changing only a single resistance value. The circuit in Figure 9.25, called an instrumentation amplifier, allows this flexibility. Note that two noninverting amplifiers, A_1 and A_2, are used as the input stage, and a difference amplifier, A_3 is the second, or amplifying, stage.

Figure 9.25   Instrumentation amplifier

We begin the analysis using the virtual short concept. The voltages at the inverting terminals of the voltage followers are equal to the input voltages. The currents and voltages in the amplifier are shown in Figure 9.26. The current in resistor R_1 is then

$$i_1 = \frac{v_{I1} - v_{I2}}{R_1}$$

(9.64)

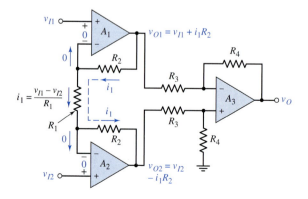

Figure 9.26   Voltages and currents in instrumentation amplifier

The current in resistors R_2 is also i_1, as shown in the figure, and the output voltages of op-amps A_1 and A_2 are, respectively,

$$v_{O1} = v_{I1} + i_1 R_2 = \left(1 + \frac{R_2}{R_1}\right)v_{I1} - \frac{R_2}{R_1}v_{I2} \qquad \text{(9.65(a))}$$

and

$$v_{O2} = v_{I2} - i_1 R_2 = \left(1 + \frac{R_2}{R_1}\right)v_{I2} - \frac{R_2}{R_1}v_{I1} \qquad \text{(9.65(b))}$$

From previous results, the output of the difference amplifier is given as

$$v_O = \frac{R_4}{R_3}(v_{O2} - v_{O1}) \qquad \text{(9.66)}$$

Substituting Equations (9.65(a)) and (9.65(b)) into Equation (9.66), we find the output voltage, as follows:

$$v_O = \frac{R_4}{R_3}\left(1 + \frac{2R_2}{R_1}\right)(v_{I2} - v_{I1}) \qquad \text{(9.67)}$$

Since the input signal voltages are applied directly to the noninverting terminals of A_1 and A_2, the input impedance is very large, ideally infinite, which is one desirable characteristic of the instrumentation amplifier. Also, the differential gain is a function of resistor R_1, which can easily be varied by using a potentiometer, thus providing a variable amplifier gain with the adjustment of only one resistance.

Example 9.8   Objective: Determine the range required for resistor R_1, to realize a differential gain adjustable from 5 to 500.

The instrumentation amplifier circuit is shown in Figure 9.25. Assume that $R_4 = 2R_3$, so that the difference amplifier gain is 2.

Solution: Assume that resistance R_1 is a combination of a fixed resistance R_{1f} and a variable resistance R_{1v}, as shown in Figure 9.27. The fixed resistance ensures that the gain is limited to a maximum value, even if the variable resistance is set equal to zero. Assume the variable resistance is a 100 kΩ potentiometer.

From Equation (9.67), the maximum differential gain is

$$500 = 2\left(1 + \frac{2R_2}{R_{1f}}\right)$$

and the minimum differential gain is

$$5 = 2\left(1 + \frac{2R_2}{R_{1f} + 100}\right)$$

From the maximum gain expression, we find that

$$2R_2 = 249 R_{1f}$$

Substituting this R_2 value into the minimum gain expression, we have

$$1.5 = \frac{2R_2}{R_{1f} + 100} = \frac{249 R_{1f}}{R_{1f} + 100}$$

The resulting value of R_{1f} is $R_{1f} = 0.606\,\text{k}\Omega$, which yields $R_2 = 75.5\,\text{k}\Omega$.

Figure 9.27 Equivalent resistance R_1 in instrumentation amplifier

Comment: We can select standard resistance values that are close to the values calculated, and the range of the gain will be approximately in the desired range.

Test Your Understanding

9.18   For the instrumentation amplifier in Figure 9.25, the parameters are: $R_3 = R_4 = 20\,\text{k}\Omega$, and $R_2 = 100\,\text{k}\Omega$. Resistance R_1 is a series combination of a fixed resistance of $1\,\text{k}\Omega$ and a $50\,\text{k}\Omega$ potentiometer. Determine the range of the differential voltage gain. (Ans. $4.92 \le A_d \le 201$)

*9.19   All parameters associated with the instrumentation amplifier in Figure 9.25 are as given in Exercise 9.18, except that resistor R_2 associated with the A_1 op-amp is $R_2 = 100\,\text{k}\Omega \pm 5\%$. (a) Determine the maximum and minimum possible values of the common-mode gain. (b) Determine the maximum and minimum possible values of the differential-mode gain. (c) Determine the minimum CMRR(dB). (Ans. (a) $A_{cm} = 0$, (b) $A_d(\text{min}) = 4.82$, $A_d(\text{max}) = 206$, (c) CMRR $= \infty$)

D9.20   Design the instrumentation amplifier in Figure 9.25 such that the variable differential voltage gain is in the range of 2 to 1000

9.5.5   Integrator and Differentiator

In the op-amp circuits previously considered, the elements exterior to the op-amp have been resistors. Other elements can be used, with differing results. Figure 9.28 shows a generalized inverting amplifier for which the voltage transfer function has the same general form as before, that is,

$$\frac{v_O}{v_I} = -\frac{Z_2}{Z_1} \tag{9.68}$$

where Z_1 and Z_2 are generalized impedances. Two special circuits can be developed from this generalized inverting amplifier.

Figure 9.28   Generalized inverting amplifier

In the first, Z_1 corresponds to a resistor and Z_2 to a capacitor. The impedances are then $Z_1 = R_1$ and $Z_2 = 1/sC_2$, where s again is the complex frequency. The output voltage is

$$v_O = -\frac{Z_2}{Z_1}v_I = \frac{-1}{sR_1C_2}v_I \tag{9.69}$$

Equation (9.69) represents integration in the time domain. If V_C is the voltage across the capacitor at $t = 0$, the output voltage is

$$v_O = V_C - \frac{1}{R_1 C_2} \int_0^t v_I(t')dt' \qquad (9.70)$$

where t' is the variable of integration. Figure 9.29 summarizes these results.

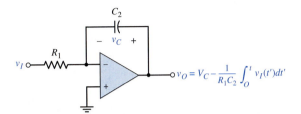

Figure 9.29   Op-amp integrator

Equation (9.70) is the output response of the integrator circuit, shown in Figure 9.29, for any input voltage v_I. Note that if $v_I(t)$ is a finite step function, output v_O will be a linear function of time. The output v_O will be a ramp function and will eventually saturate at a voltage near either the positive or negative supply voltage. We will use the integrator in filter circuits, which are covered in Chapter 15.

We will show in Chapter 14 that nonzero bias currents into the op-amp greatly influence the characteristics of this circuit. A dc current through the capacitor will cause the output voltage to linearly change with time until the positive or negative supply voltage is reached. In many applications, a transistor switch needs to be added in parallel with the capacitor to periodically set the capacitor voltage to zero.

The second generalized inverting op-amp uses a capacitor for Z_1 and a resistor for Z_2, as shown in Figure 9.30. The impedances are $Z_1 = 1/sC_1$ and $Z_2 = R_2$, and the voltage transfer function is

$$\frac{v_O}{v_I} = -\frac{Z_2}{Z_1} = -sR_2C_1 \qquad (9.71(a))$$

The output voltage is

$$v_O = -sR_2C_1v_I \qquad (9.71(b))$$

Figure 9.30   Op-amp differentiator

Equation (9.71(b)) represents differentiation in the time domain, as follows:

$$v_O(t) = -R_2 C_1 \frac{dv_I(t)}{dt} \tag{9.72}$$

The circuit in Figure 9.30 is therefore a differentiator.

Differentiator circuits are more susceptible to noise than are the integrator circuits. Input noise fluctuations of small amplitudes may have large derivatives. When differentiated, these noise fluctuations may generate large noise signals at the output, creating a poor output signal to noise ratio. This problem may be alleviated by placing a resistor in series with the input capacitor. This modified circuit then differentiates low-frequency signals but has a constant high-frequency gain.

Example 9.9   Objective: Determine the time constant required in an integrator.

Consider the integrator shown in Figure 9.29. Assume that voltage V_C across the capacitor is zero at $t = 0$. A step input voltage of $v_I = -1\,\text{V}$ is applied at $t = 0$. Determine the time constant required such that the output reaches $+10\,\text{V}$ at $t = 1\,\text{ms}$.

Solution: From Equation (9.70), we have

$$v_o = \frac{-1}{R_1 C_2} \int_0^t (-1)\,dt' = \frac{1}{R_1 C_2} t' \bigg|_0^t = \frac{t}{R_1 C_2}$$

At $t = 1\,\text{ms}$, we want $v_O = 10\,\text{V}$. Therefore,

$$10 = \frac{10^{-3}}{R_1 C_2}$$

which means the time constant is $R_1 C_2 = 0.1\,\text{ms}$.

Comment: As an example, for a time constant of 0.1 ms, we could have $R_1 = 10\,\text{k}\Omega$ and $C_2 = 0.01\,\mu\text{F}$, which are reasonable values of resistance and capacitance.

Test Your Understanding

9.21   An integrator with input and output voltages that are zero at $t = 0$ is driven by the input signal shown in Figure 9.31. The resistance and capacitance in the circuit are $R_1 = 10\,\text{k}\Omega$ and $C_2 = 0.1\,\mu\text{F}$. Sketch and label the resulting output waveform.

D9.22   An integrator is driven by the series of pulses shown in Figure 9.32. At the end of the tenth pulse, the output voltage is to be $v_O = -5\,\text{V}$. Assume $V_C = 0$ at $t = 0$.

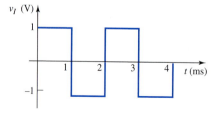

Figure 9.31   Figure for Exercise 9.21

Figure 9.32   Figure for Exercise 9.22

Determine the time constant and values of R_1 and C_2 that will meet these specifications. (Ans. $\tau = 20\,\mu s$; for example, let $C_2 = 0.01\,\mu F$, $R_1 = 2\,k\Omega$)

9.5.6 Nonlinear Circuit Applications

Up to this point in the chapter, we have used linear passive elements in conjunction with the op-amp. Many useful circuits can be fabricated if nonlinear elements, such as diodes or transistors, are used in the op-amp circuits. We will consider three simple examples to illustrate the types of nonlinear characteristics that can be generated and to illustrate the general analysis technique.

Precision Half-Wave Rectifier

An op-amp and diode are combined as shown in Figure 9.33 to form a precision half-wave rectifier. For $v_I > 0$, the circuit behaves as a voltage follower. The output voltage is $v_O = v_I$, the load current i_L is positive, and a positive diode current is induced such that $i_D = i_L$. The feedback loop is closed through the forward-biased diode. The output voltage of the op-amp, v_{O1}, adjusts itself to exactly absorb the forward voltage drop of the diode.

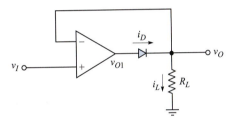

Figure 9.33   Precision half-wave rectifier circuit

For $v_I < 0$, the output voltage tends to go negative, which tends to produce negative load and diode currents. However, a negative diode current cannot exist, so the diode cuts off, the feedback loop is broken, and $v_O = 0$.

The voltage transfer characteristics are shown in Figure 9.34. The rectification is precise in that, even at small positive input voltages, $v_O = v_I$ and we do not observe a diode cut-in voltage.

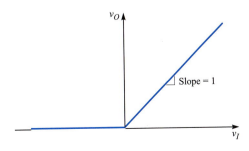

Figure 9.34   Voltage transfer characteristics of precision half-wave rectifier

A potential problem in this circuit exists for $v_I < 0$. The feedback loop is broken so that the op-amp output voltage v_{O1} will saturate near the negative supply voltage. When v_I switches positive, it will take time for the internal circuit to recover, so the response time of the output voltage may be relatively slow. In addition, for $v_I < 0$ and $v_O = 0$, there is now a voltage difference applied across the input terminals of the op-amp. Most op-amps provide input voltage protection so the op-amp will not be damaged in this case. However, if the op-amp does not have input protection, the op-amp may be damaged if the input voltage is larger than 5 or 6 V.

Log Amplifier

Consider the circuit in Figure 9.35. The diode is to be forward biased, so the input signal voltage is limited to positive values. The diode current is

$$i_D = I_S\left(e^{v_D/V_T} - 1\right) \tag{9.73(a)}$$

Figure 9.35 Simple log amplifier

If the diode is sufficiently forward biased, the (-1) term is negligible, and

$$i_D \cong I_S e^{v_D/V_T} \tag{9.73(b)}$$

The input current can be written

$$i_1 = \frac{v_I}{R_1} \tag{9.74}$$

and the output voltage, since v_1 is at virtual ground, is given by

$$v_O = -v_D \tag{9.75}$$

Noting that $i_1 = i_D$, we can write

$$i_1 = \frac{v_I}{R_1} = i_D = I_s e^{-v_O/V_T} \tag{9.76}$$

If we take the natural log of both sides of this equation, we obtain

$$\ln\left(\frac{v_I}{I_S R_1}\right) = -\frac{v_O}{V_T} \tag{9.77(a)}$$

or

$$v_O = -V_T \ln\left(\frac{v_I}{I_S R_1}\right) \tag{9.77(b)}$$

Equation (9.77(b)) indicates that, for this circuit, the output voltage is proportional to the log of the input voltage. One disadvantage of this circuit is that the reverse-saturation current I_S is a strong function of temperature, and it varies substantially from one diode to another. A more sophisticated circuit uses bipolar transistors to eliminate the I_S parameter in the log term. This circuit will not be considered here.

Antilog or Exponential Amplifier

The complement, or inverse function, of the log amplifier is the antilog, or exponential, amplifier. A simple example using a diode is shown in Figure 9.36. Since v_1 is at virtual ground, we can write for $v_I > 0$

$$i_D \cong I_S e^{v_I/V_T} \tag{9.78}$$

and

$$v_O = -i_2 R = -i_D R \tag{9.79(a)}$$

or

$$v_O = -I_S R \cdot e^{v_I/V_T} \tag{9.79(b)}$$

The output voltage is an exponential function of the input voltage. Again, there are more sophisticated circuits that perform this function, but they will not be considered here.

Figure 9.36   A simple antilog, or exponential, amplifier

9.6   OP-AMP CIRCUIT DESIGN

Up to this point, we have mainly been concerned with analyzing op-amp circuits. In this section, we will design three specific circuits.

9.6.1   Summing Op-Amp Circuit Design

In an inverting summing op-amp, each input is connected to the inverting terminal through a resistor. The summing op-amp can be designed such that the output is

$$v_O = -a_1 v_{I1} - a_2 v_{I2} + a_3 v_{I3} + a_4 v_{I4} \tag{9.80}$$

where the coefficients a_i are all positive. In one design, we could apply voltages v_{I3} and v_{I4} to inverter amplifiers and use the summing op-amp considered previously. This design would require three such op-amps. Alternatively, we could use the results of Exercise 9.12 to design a summing circuit that uses only one op-amp and is more versatile.

Consider the circuit shown in Figure 9.37. Resistor R_C provides more versatility in the design. When we consider nonideal effects, such as bias currents, in op-circuits, in Chapter 14, we will impose a design constraint on the relationship between the resistors connected to the inverting and noninverting terminals. In this section, we will continue to use the ideal op-amp.

Figure 9.37   Generalized op-amp summing amplifier

To determine the output voltage of our circuit, we use superposition. The inputs v_{I1} and v_{I2} produce the usual outputs, as follows:

$$v_O(v_{I1}) = -\frac{R_F}{R_1}v_{I1}$$

(9.81(a))

and

$$v_O(v_{I2}) = -\frac{R_F}{R_2}v_{I2}$$

(9.81(b))

We then determine the output due to v_{I3}, with all other inputs set equal to zero. We can write

$$v_2(v_{I3}) = \frac{R_B\|R_C}{R_A + R_B\|R_c}v_{I3} = v_1(v_{I3})$$

(9.82)

Since $v_{I1} = v_{I2} = 0$, the voltage $v_2(v_{I3})$ is the input to a noninverting op-amp with R_1 and R_2 in parallel.

Then,

$$v_O(v_{I3}) = (1 + \frac{R_F}{R_1\|R_2})v_1(v_{I3}) = \left(1 + \frac{R_F}{R_1\|R_2}\right)\left(\frac{R_B\|R_C}{R_A + R_B\|R_C}\right)v_{I3}$$

(9.83)

which can be rearranged as follows:

$$v_O(v_{I3}) = \left(1 + \frac{R_F}{R_N}\right)\left(\frac{R_P}{R_A}\right)v_{I3}$$

(9.84)

Here, we define

$$R_N = R_1\|R_2$$

(9.85(a))

and

$$R_P = R_A\|R_B\|R_C$$

(9.85(b))

The output voltage due to v_{I4} is similarly determined and is

$$v_O(v_{I4}) = \left(1 + \frac{R_F}{R_N}\right)\left(\frac{R_P}{R_B}\right)v_{I4} \qquad (9.86)$$

The total output voltage is then the sum of the individual terms, or

$$v_O = -\frac{R_F}{R_1}v_{I1} - \frac{R_F}{R_2}v_{I2} + \left(1 + \frac{R_F}{R_N}\right)\left[\frac{R_P}{R_A}v_{I3} + \frac{R_P}{R_B}v_{I4}\right] \qquad (9.87)$$

This form of the output voltage is the same as the desired output given by Equation (9.80).

Design Example 9.10 Objective: Design a summing op-amp to produce the output

$$v_O = -10v_{I1} - 4v_{I2} + 5v_{I3} + 2v_{I4}$$

The smallest resistor value allowable is 20 kΩ.

Solution: First we determine the values of resistors R_1, R_2, and R_F, and then we can determine the inverting terms. We know that

$$\frac{R_F}{R_1} = 10 \quad \text{and} \quad \frac{R_F}{R_2} = 4$$

Resistor R_1 will be the smallest value, so we can set $R_1 = 20\,\text{k}\Omega$. Then,

$$R_F = 200\,\text{k}\Omega \quad \text{and} \quad R_2 = 50\,\text{k}\Omega$$

The multiplying factor in the noninverting terms becomes

$$\left(1 + \frac{R_F}{R_1\|R_2}\right) = \left(1 + \frac{200}{20\|50}\right) = 15$$

We then need

$$(15)\left(\frac{R_P}{R_A}\right) = 5 \quad \text{and} \quad (15)\left(\frac{R_P}{R_B}\right) = 2$$

If we take the ratio of these two expressions, we have

$$\frac{R_B}{R_A} = \frac{5}{2}$$

If we choose $R_A = 80\,\text{k}\Omega$, then $R_B = 200\,\text{k}\Omega$, $R_P = 26.67\,\text{k}\Omega$, and R_C becomes $R_C = 50\,\text{k}\Omega$.

Comment: We could change the number of inputs to either the inverting or noninverting terminal, depending on the desired output versus input voltage response.

Test Your Understanding

*D9.23 Design a summing op-amp to produce the output $v_O = v_{I1} + 10v_{I2} - 25v_{I3} - 80v_{I4}$.

9.6.2   Reference Voltage Source Design

In Chapter 2, we discussed the use of Zener diodes to provide a constant or reference voltage source. A limitation, however, was that the reference voltage could never be greater than the Zener voltage. Now, we can combine a Zener diode with an op-amp to provide more flexibility in the design of reference voltage sources.

Consider the circuit shown in Figure 9.38. Voltage source V^+ and resistor R_S bias the Zener diode in the breakdown region. The op-amp is then used as a noninverting amplifier. The output voltage is

$$V_O = \left(1 + \frac{R_2}{R_1}\right)V_Z \tag{9.88}$$

The output current to the load circuit is supplied by the op-amp. A change in the load current will not produce a change in the Zener diode current; consequently, voltage regulation is much improved compared to the simple Zener diode voltage source previously considered.

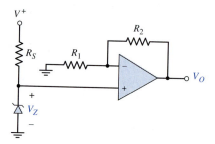

Figure 9.38   Simple op-amp voltage reference circuit

Since the incremental Zener resistance is not zero, the Zener diode voltage is a slight function of the diode current. The circuit shown in Figure 9.39 is less affected by variations in V_S, since V_S is used only to start up the circuit. The Zener diode begins to conduct when

$$\frac{R_4}{R_3 + R_4}V_S > V_Z + V_D \cong V_Z + 0.7 \tag{9.89}$$

At this specific voltage, we have

$$V_O = \left(1 + \frac{R_2}{R_1}\right)V_Z \tag{9.90}$$

Figure 9.39 Op-amp voltage reference circuit

and

$$I_F = \frac{V_O - V_Z}{R_F} = \frac{R_2 V_Z}{R_1 R_F} \qquad \text{(9.91)}$$

If V_S decreases and diode D_1 becomes reverse biased, the Zener diode continues to conduct; the Zener diode current is then constant. However, if diode D_1 is conducting, the circuit can be designed such that variations in Zener diode current will be small.

Design Example 9.11 Objective: Design a voltage reference source with an output of 10.0 V. Use a Zener diode with a breakdown voltage of 5.6 V. Assume the voltage regulation will be within specifications if the Zener diode is biased between 1–1.2 mA.

Solution: Consider the circuit shown in Figure 9.39. For this example, we need

$$\frac{V_O}{V_Z} = \left(1 + \frac{R_2}{R_1}\right) = \frac{10.0}{5.6}$$

Therefore,

$$\frac{R_2}{R_1} = 0.786$$

We know that

$$I_F = \frac{V_O - V_Z}{R_F}$$

If we set I_F equal to the minimum bias current, we have

$$1\,\text{mA} = \frac{10 - 5.6}{R_F}$$

which means that $R_F = 4.4\,\text{k}\Omega$. If we choose $R_2 = 30\,\text{k}\Omega$, then $R_1 = 38.17\,\text{k}\Omega$.

Resistors R_3 and R_4 can be determined from Figure 9.40. The maximum Zener current supplied by V_S, R_3, and R_4 should be no more than 0.2 mA. We set the current through D_1 equal to 0.2 mA, for $V_S = 10\,\text{V}$. We then have

$$V_2' = V_Z + 0.7 = 5.6 + 0.7 = 6.3\,\text{V}$$

Figure 9.40 Input circuit of the op-amp voltage reference circuit

Also,

$$I_4 = \frac{V_2'}{R_4} = \frac{6.3}{R_4}$$

and

$$I_3 = \frac{V_S - V_2'}{R_3} = \frac{10 - 6.3}{R_3} = \frac{3.7}{R_3}$$

If we set $I_4 = 0.2\,\text{mA}$, then

$$I_3 = 0.4\,\text{mA} \qquad R_3 = 9.25\,\text{k}\Omega \qquad R_4 = 31.5\,\text{k}\Omega$$

 Comment: Voltage V_S is used as a start-up source. Once the Zener diode is biased in breakdown, the output will be maintained at 10.0 V, even if V_S is reduced to zero.

Test Your Understanding

9.24 Consider the op-amp voltage reference circuit in Figure 9.39 with parameters given in Example 9.11. Initially set $V_S = 10\,\text{V}$ and then plot, using PSpice, v_O and I_F versus V_S as V_S decreases from 10 to 0 V. Bias the op-amp at $\pm 15\,\text{V}$.

9.6.3 Difference Amplifier and Bridge Circuit Design

A transducer is a device that transforms one form of energy into another form. One type of transducer uses nonelectrical inputs to produce electrical outputs. For example, a microphone converts acoustical energy into electrical energy. A pressure transducer is a device in which, for example, a resistance is a function of pressure, so that pressure can be converted to an electrical signal. Often, the output characteristics of these transducers are measured with a bridge circuit.

Figure 9.41 shows a bridge circuit. Resistance R_3 represents the transducer, and parameter δ is the deviation of R_3 from R_2 due to the input response of the transducer. The output voltage v_{O1} is a measure of δ. If v_{O1} is an open-circuit voltage, then

$$v_{O1} = \left[\frac{R_2(1+\delta)}{R_2(1+\delta)+R_1} - \frac{R_2}{R_1+R_2}\right]V^+ \tag{9.92}$$

which reduces to

$$v_{O1} = \delta\left(\frac{R_1 \| R_2}{R_1 + R_2}\right)V^+ \tag{9.93}$$

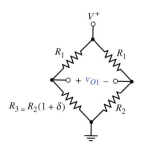

Figure 9.41 Bridge circuit

Since neither side of voltage v_{O1} is at ground potential, we must connect v_{O1} to an instrumentation amplifier. In addition, v_{O1} is directly proportional to supply voltage V^+; therefore, this bias should be a well-defined voltage reference.

Design Example 9.12   **Objective:**   Design an amplifier system that will produce an output voltage of ± 5 V when the resistance R_3 deviates by $\pm 1\%$ from the value of R_2. This would occur, for example, in a system where R_3 is a thermistor whose resistance is given by

$$R_3 = 200 \left[1 + \frac{(0.040)(T - 300)}{300} \right] \text{k}\Omega$$

where T is the absolute temperature. For R_3 to vary by $\pm 1\%$ means the temperature is in the range $225 \leq T \leq 375\,^\circ$K.

Consider biasing the bridge circuit at $V^+ = 7.5$ V using a 5.6 V Zener diode. Assume ± 10 V is available for biasing the op-amp and reference voltage source, and that $R_1 = R_2 = 200\,\text{k}\Omega$.

Solution: With $R_1 = R_2$, from Equation (9.93), we have

$$v_{O1} = \left(\frac{\delta}{4} \right) V^+$$

For $V^+ = 7.5$ V and $\delta = 0.01$, the maximum output of the bridge circuit is $v_{O1} = 0.01875$ V. If the output of the amplifier system is to be $+5$ V, the gain of the instrumentation amplifier must be $5/0.01875 = 266.7$. Consider the instrumentation amplifier shown in Figure 9.25. The output voltage is given by Equation (9.67), which can be written

Figure 9.42   Complete amplifier system

$$\frac{v_O}{v_{O1}} = \frac{R_4'}{R_3'}\left(1 + \frac{2R_2'}{R_1'}\right) = 266.7$$

We would like the ratios R_4'/R_3' and R_2'/R_1' to be the same order of magnitude. If we let $R_3' = 15.0\,k\Omega$ and $R_4' = 187.0\,k\Omega$, then $R_4'/R_3' = 12.467$ and $R_2'/R_1' = 10.195$. If we set $R_2' = 200.0\,k\Omega$, then $R_1' = 19.62\,k\Omega$.

Resistance R_1' can be a combination of a fixed resistance in series with a potentiometer, to permit adjustment of the gain.

Comment: The complete design of this instrumentation amplifier is shown in Figure 9.42. Correlation of the reference voltage source design is left as an exercise.

Design Pointer: The design of fairly sophisticated op-amp circuits is quite straightforward when the ideal op-amp parameters are used.

Test Your Understanding

D9.25 Consider the bridge circuit in Figure 9.43. The resistance is $R = 10\,k\Omega$ and the maximum ΔR is $50\,\Omega$. The bridge circuit is to be biased with $V^+ = 3.5\,V$. Design an amplifier system such that the output is $5.0\,V$ when $\Delta R = 50\,\Omega$. Use reasonable resistance values.

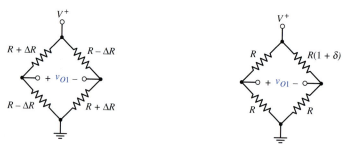

Figure 9.43 Figure for Exercise 9.25   **Figure 9.44** Figure for Exercise 9.26

D9.26 Consider the bridge circuit in Figure 9.44. The resistance is $R = 100\,k\Omega$, and the bridge circuit is to be biased with $V^+ = 5.0\,V$. Design an amplifier system such that the output varies from $+5\,V$ to $-5\,V$ as δ varies from $+0.01$ to -0.01. Use reasonable resistance values.

9.7   SUMMARY

- In this chapter, we considered the ideal operational amplifier (op-amp) and various op-amp applications. The op-amp is a three-terminal device (three signal terminals) that ideally amplifies only the difference between two input signals. The op-amp, then, is a high-gain differential amplifier.

- The ideal op-amp model has infinite input impedance (zero input bias currents, infinite differential voltage gain (zero voltage between the two input terminals), and zero output impedance.

- Two basic op-amp circuits are the inverting amplifier and the noninverting amplifier. In the ideal model of the op-amp, the voltage gain of these circuits is just a function of the ratio of resistors.

- Other amplifier configurations considered were the summing amplifier, voltage follower, current-to-voltage converter, and voltage-to-current converter.
- If a capacitor is included as a feedback element, the output voltage is the integral of the input voltage. If a capacitor is included as an input element, the output voltage is the derivative of the input voltage. Nonlinear feedback elements, such as a diode or transistor, produce nonlinear transfer functions such as a logarithmic function.

CHECKPOINT

After studying this chapter, the reader should have the ability to:

✔ Analyze various op-amp circuits using the ideal op-amp model. (Sections 9.2, 9.3, 9.4, and 9.5)

✔ Analyze various op-amp circuits, taking into account the finite gain of the op-amp. (Section 9.2)

✔ Understand and describe the characteristics and operation of various op-amp circuits, such as the difference amplifier and instrumentation amplifier. (Section 9.5)

✔ Design various op-amp circuits to perform specific functions using the ideal op-amp model. (Section 9.6)

REVIEW QUESTIONS

1. Describe the ideal op-amp model and describe the implications of this ideal model in terms of input currents and voltages.
2. Describe the op-amp model including the effect of a finite op-amp voltage gain.
3. Describe the operation and characteristics of the inverting amplifier.
4. What is the concept of virtual ground?
5. When a finite op-amp gain is taken into account, is the magnitude of the resulting amplifier voltage gain less than or greater than the ideal value?
6. What is the significance of a zero output resistance?
7. Describe the operation and characteristics of a summing amplifier.
8. Describe the operation and characteristics of a noninverting amplifier.
9. Describe a voltage follower.
10. What is the input resistance of an ideal current-to-voltage converter?
11. Describe the operation and characteristics of a difference amplifier.
12. Describe the operation and characteristics of an instrumentation amplifier.

PROBLEMS

Section 9.2   Inverting Amplifier

9.1   Assume the op-amps in Figure P9.1 are ideal. Find the voltage gain $A_v = v_O/v_I$ and the input resistance R_i of each circuit.

9.2   Consider an ideal inverting op-amp with $R_2 = 100\,\text{k}\Omega$ and $R_1 = 10\,\text{k}\Omega$. (a) Determine the ideal voltage gain and input resistance R_i. (b) Repeat part (a) for a second $100\,\text{k}\Omega$ resistor connected in parallel with R_2. (c) Repeat part (a) for a second $10\,\text{k}\Omega$ resistance connected in series with R_1.

D9.3   Design an inverting op-amp circuit with a voltage gain of $A_v = v_O/v_I = -12$ and an input resistance of $R_i = 25\,\text{k}\Omega$.

Figure P9.1

D9.4 Design an inverting op-amp circuit with a voltage gain of $A_v = v_O/v_I = -8$. When the input voltage is $v_I = -1\,\text{V}$, the maximum current in R_1 and R_2 must be no larger than $15\,\mu\text{A}$. Determine the minimum values of R_1 and R_2.

D9.5 Design an inverting op-amp circuit with a voltage gain of $A_v = v_O/v_I = -30$ and an input resistance that is the largest value possible but under the constraint that the largest resistance value is limited to $1\,\text{M}\Omega$.

9.6 (a) In an inverting op-amp circuit, the nominal resistance values are $R_2 = 300\,\text{k}\Omega$ and $R_1 = 15\,\text{k}\Omega$. The tolerance of each resistor is $\pm 5\%$, which means that each resistance can deviate from its nominal value by $\pm 5\%$. What is the maximum deviation in the voltage gain from its nominal value? (b) Repeat part (a) if the resistor tolerance is reduced to $\pm 1\%$.

9.7 The input to the circuit in Figure P9.7 is $v_I = 10 \sin \omega t\,\text{mV}$. (a) What is the output voltage v_O? (b) Determine the currents i_2, i_L, and i_O.

Figure P9.7

D9.8 Design an inverting amplifier to provide a nominal closed-loop voltage gain of $A_v = -30$. The maximum input voltage signal is 25 mV with a source resistance in the range $1\,k\Omega \le R_S \le 2\,k\Omega$. The variable source resistance should introduce no more than a 5 percent difference in the gain factor. What is the range in output voltage?

9.9 Consider two inverting op-amp circuits connected in cascade, as shown in Figure P9.9. Let $R_1 = 10\,k\Omega$, $R_2 = 50\,k\Omega$, $R_3 = 25\,k\Omega$, and $R_4 = 150\,k\Omega$. If $v_I = 0.15\,V$, calculate v_{O1}, v_O, i_1, i_2, i_3, and i_4. Discuss the current sum and the current directions at the node v_{O1}.

Figure P9.9   Figure P9.10

*9.10 Consider the circuit shown in Figure P9.10. (a) Determine the ideal voltage gain. (b) Find the actual voltage gain if the open-loop gain is $A_{od} = 150,000$. (c) Determine the required valued of A_{od} in order that the actual voltage gain be within 1 percent of the ideal value.

9.11 For the ideal noninverting op-amp with T-network, shown in Figure 9.11, the circuit parameters are $R_1 = R_3 = R_4 = 100\,k\Omega$. Determine R_2 such that: (a) $A_v = v_O/v_I = -10$, and (b) $A_v = v_O/v_I = -100$.

D9.12 Consider the ideal inverting op-amp circuit with T-network in Figure 9.11. (a) Design the circuit such that the input resistance is $500\,k\Omega$ and the gain is $A_v = -80$. Do not use resistor values greater than $500\,k\Omega$. (b) For the design in part (a), determine the current in each resistor if $v_I = -0.05\,V$.

9.13 For the op-amp circuit shown in Figure P9.13, determine the gain $A_v = v_O/v_I$. Compare this result to the gain of the circuit shown in Figure 9.11, assuming all resistor values are equal.

Figure P9.13

*9.14 The inverting op-amp circuit in Figure 9.8 has parameters $R_1 = 10\,k\Omega$, $R_2 = 50\,k\Omega$, and $A_{od} = 2 \times 10^5$. The input voltage is from an ideal voltage source whose value is $v_I = 100\,mV$. (a) Calculate the closed-loop voltage gain, (b) the output voltage, and (c) the error in the output voltage due to the finite open-loop gain.

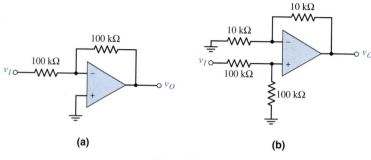

(a)   **(b)**

Figure P9.15

*9.15   Consider the two op-amp circuits in Figure P9.15. If the open-loop differential gain for each op-amp is $A_{od} = 10^3$, determine the output voltage v_O when $v_I = 2\,\text{V}$.

*9.16   The circuit in Figure P9.16 is similar to the inverting amplifier except the resistor R_3 has been added. (a) Derive the expression for v_O in terms of v_I and the resistors. (b) Derive the expression for i_3 in terms of v_I and the resistors.

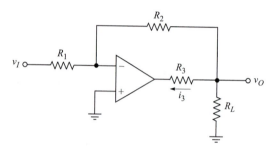

Figure P9.16

*D9.17   Design the amplifier in Figure P9.17 such that the output voltage varies between $\pm 10\,\text{V}$ as the wiper arm of the potentiometer changes from $-10\,\text{V}$ to $+10\,\text{V}$. What is the purpose of including R_3 and R_4 instead of connecting R_1 directly to the wiper arm?

Figure P9.17

9.18 Assume an op-amp is ideal, except for having a finite open-loop differential gain. Measurements were made with the op-amp in the open-loop mode. Determine the open-loop gain and complete the following table, which shows the results of those measurements.

v_1	v_2	v_O
$-1\,\text{mV}$	$+1\,\text{mV}$	$1\,\text{V}$
$+1\,\text{mV}$		$1\,\text{V}$
	$1\,\text{V}$	$5\,\text{V}$
$-1\,\text{V}$	$-1\,\text{V}$	
$-0.5\,\text{V}$		$-3\,\text{V}$

Section 9.3   Summing Amplifier

9.19 Consider the ideal inverting summing amplifier in Figure 9.13(a). Let $R_1 = 20\,\text{k}\Omega$, $R_2 = 40\,\text{k}\Omega$, $R_3 = 60\,\text{k}\Omega$, and $R_F = 80\,\text{k}\Omega$. (a) Determine v_O if $v_{I1} = 0.5\,\text{V}$, $v_{I2} = -1\,\text{V}$, and $v_{I3} = 2\,\text{V}$. (b) Determine v_{I3} if $v_{I1} = 1\,\text{V}$, $v_{I2} = 0.25\,\text{V}$, and $v_O = -5.2\,\text{V}$.

D9.20 Design an ideal inverting summing amplifier to produce an output voltage of $v_O = -2(4v_{I1} + v_{I2} + 2.5v_{I3})$. Design the circuit to produce the largest possible input resistance, assuming the largest usable resistance value is $500\,\text{k}\Omega$.

D9.21 Design an ideal inverting summing amplifier to produce an output voltage of $v_O = -4v_{I1} - 0.5v_{I2}$. The input voltages are limited to the range $-2 \le v_I \le 2\,\text{V}$, and the current in any resistor is limited to a maximum value of $100\,\mu\text{A}$.

9.22 Consider the summing amplifier in Figure 9.13 with $R_F = 10\,\text{k}\Omega$, $R_1 = 1\,\text{k}\Omega$, $R_2 = 5\,\text{k}\Omega$, and $R_3 = 10\,\text{k}\Omega$. If v_{I1} is a 1 kHz sine wave with an rms value of $50\,\text{mV}$, if v_{I2} is a 100 Hz square wave with an amplitude of $\pm1\,\text{V}$, and if $v_{I3} = 0$, sketch the output voltage v_O.

9.23 The parameters for the summing amplifier in Figure 9.13 are $R_F = 20\,\text{k}\Omega$, $R_1 = 10\,\text{k}\Omega$, $R_2 = 5\,\text{k}\Omega$, and $R_3 = 2\,\text{k}\Omega$. Determine the voltage v_{I1} to ensure the output voltage is symmetrical about 0 V for $v_{I2} = 2 + 100\sin\omega t\,\text{mV}$ and $v_{I3} = 0$.

9.24 A summing amplifier can be used as a digital-to-analog converter (DAC). An example of a 4-bit DAC is shown in Figure P9.24. When switch S_3 is connected to the $-5\,\text{V}$ supply, the most significant bit is $a_3 = 1$; when S_3 is connected to ground, the most

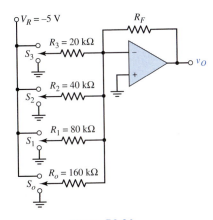

Figure P9.24

significant bit is $a_3 = 0$. The same condition applies to the other switches S_2, S_1, and S_o, corresponding to bits a_2, a_1, and a_o, where a_o is the least significant bit. (a) Show that the output voltage is given by

$$v_O = \frac{R_F}{10}\left[\frac{a_3}{2} + \frac{a_2}{4} + \frac{a_1}{8} + \frac{a_o}{16}\right] (5)$$

where R_F is in kΩ. (b) Find the value of R_F such that $v_O = 2.5$ V when the digital input is $a_3 a_2 a_1 a_o = 1000$. (c) Using the results of part (b), find v_o for: (i) $a_3 a_2 a_1 a_o = 0001$, and (ii) $a_3 a_2 a_1 a_o = 1111$.

9.25 For the circuit in Figure P9.25, (a) derive the expression for v_O in terms of v_{I1} and v_{I2}, and (b) find v_O if $v_{I1} = 1 + 2 \sin \omega t$ mV and $v_{I2} = -10$ mV.

Figure P9.25

*9.26 Consider the summing amplifier in Figure 9.13(a). Assume the op-amp has a finite open-loop differential gain A_{od}. Using the principle of superposition, show that the output voltage is given by

$$v_O = \frac{-1}{1 + \dfrac{(1 + R_F/R_P)}{A_{od}}}\left[\frac{R_F}{R_1}v_{I1} + \frac{R_F}{R_2}v_{I2} + \frac{R_F}{R_3}v_{I3}\right]$$

where $R_P = R_1 \| R_2 \| R_3$. Demonstrate how the expression will change if more or fewer inputs are included.

Section 9.4   Noninverting Amplifier

D9.27 Design an ideal noninverting op-amp with a closed-loop voltage gain of $A_v = 10$. When $v_I = 0.8$ V, the current in any resistor is limited to a maximum of 100 μA.

9.28 For the circuit in Figure P9.28, the input voltage is $v_I = 5$ V. (a) If $v_O = 2.5$ V, determine the finite open-loop differential gain of the op-amp. (b) If the open-loop differential gain of the op-amp is 5000, determine v_O.

Figure P9.28   **Figure P9.29**

9.29 Determine v_O as a function of v_{I1} and v_{I2} for the ideal noninverting op-amp circuit in Figure P9.29.

9.30 Consider the ideal noninverting op-amp in Figure P9.30. Determine v_O as a function of v_{I1} and v_{I2}.

Figure P9.30   **Figure P9.31**

9.31 Determine the gain $A_v = v_O/v_I$ for the ideal op-amp circuit in Figure P9.31.

9.32 For the amplifier in Figure P9.32, determine (a) the ideal closed-loop voltage gain, (b) the actual closed-loop voltage gain if the open-loop gain is $A_{od} = 150{,}000$, and (c) the open-loop gain such that the actual closed-loop gain is within 1 percent of the ideal.

9.33 For the voltage follower in Figure 9.16, determine the closed-loop gain if the open-loop differential gain is $A_{od} = 10^4$, 10^3, 10^2, and 10.

Figure P9.32   **Figure P9.34**

9.34 Consider the ideal op-amp circuit shown in Figure P9.34. Determine the voltage gains $A_{v1} = v_{O1}/v_I$ and $A_{v2} = v_{O2}/v_I$. What is the relationship between v_{O1} and v_{O2}?

9.35 (a) Assume the op-amp in the circuit in Figure P9.35 is ideal. Determine i_L as a function of v_I. (b) Let $R_1 = 9\,k\Omega$ and $R_L = 1\,k\Omega$. If the op-amp is biased at $\pm 10\,V$, determine the maximum value of v_I and i_L before the op-amp saturates.

9.36 The input voltage is $v_I = 6\,V$ for each ideal op-amp circuit shown in Figure P9.36. Determine each output voltage.

Figure P9.35

(a)

(b)

(c)

Figure P9.36

Section 9.5   Op-Amp Applications

*9.37 A current-to-voltage converter is shown in Figure P9.37. The current source has a finite output resistance R_S, and the op-amp has a finite open-loop differential gain A_{od}. (a) Show that the input resistance is given by

$$R_{in} = \frac{R_F}{1 + A_{od}}$$

(b) If $R_F = 10\,\text{k}\Omega$ and $A_{od} = 1000$, determine the range of R_S such that the output voltage deviates from its ideal value by less than 1 percent.

Figure P9.37

*D9.38 Figure P9.38 shows a phototransistor that converts light intensity into an output current. The transistor must be biased as shown. The transistor output versus input characteristics are shown. Design a current-to-voltage converter to produce an output voltage between 0 and 8 V for an input light intensity between 0 and 20 mW/cm². Power supplies of $+10$ V and -10 V are available.

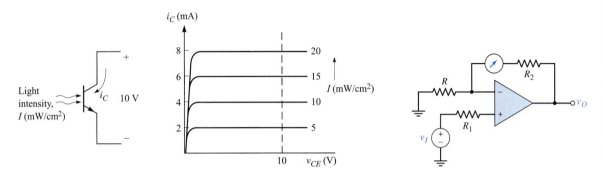

Figure P9.38   **Figure P9.39**

D9.39 The circuit in Figure P9.39 is an analog voltmeter in which the meter reading is directly proportional to the input voltage v_I. Design the circuit such that a 1 mA full-scale reading corresponds to $v_I = 10$ V. Resistance R_2 corresponds to the meter resistance, and R_1 corresponds to the source resistance. How do these resistances influence the design?

D9.40 Consider the voltage-to-current converter in Figure 9.21. Design the circuit such that the current in a 100 Ω load can be varied between 0 and 10 mA with an input voltage between 0 and -10 V. Assume the op-amp is biased at ±15 V.

D9.41 The circuit in Figure P9.41 is used to drive an LED with a voltage source. The circuit can also be thought of as a current amplifier in that, with the proper design, $i_D > i_1$. (a) Derive the expression for i_D in terms of i_1 and the resistors. (b) Design the circuit such that $i_D = 12$ mA and $i_1 = 1$ mA for $v_I = 5$ V.

Figure P9.41   **Figure P9.42**

*9.42 Figure P9.42 is used to calculate the resistance seen by the load in the voltage-to-current converter given in Figure 9.21. (a) Show that the output resistance is given by

$$R_o = \frac{R_1 R_2 R_3}{R_1 R_3 - R_2 R_F}$$

(b) Using the parameters given in Example 9.5, determine R_o. Is this result unexpected?
(c) Consider the design specification given by Equation (9.44). What is the expected value of R_o?

9.43   For the op-amp difference amplifier in Figure 9.23(a), let $R_1 = R_3$ and $R_2 = R_4$. A load resistor $R_L = 5\,\text{k}\Omega$ is connected between v_O and ground. The circuit has a differential voltage gain of $A_d = 5$, and the minimum resistance seen by the signal sources v_{I1} and v_{I2} is $25\,\text{k}\Omega$. If the load current is $i_L = 0.5\,\text{mA}$ when $v_{I1} = 2\,\text{V}$, determine v_{I2}.

*9.44   The circuit in Figure P9.44 is a representation of the common-mode and differential-mode input signals to a difference amplifier. The output voltage can be written as

$$v_O = A_d v_d + A_{cm} v_{cm}$$

where A_d is the differential-mode gain and A_{cm} is the common-mode gain. (a) Setting $v_d = 0$, show that the common-mode gain is given by

$$A_{cm} = \frac{\left(\dfrac{R_4}{R_3} - \dfrac{R_2}{R_1}\right)}{(1 + R_4/R_3)}$$

Figure P9.44

(b) If $R_1 = 10\,\text{k}\Omega \pm 5\%$, $R_3 = 10\,\text{k}\Omega \pm 5\%$, $R_2 = 50\,\text{k}\Omega \pm 5\%$, and $R_4 = 50\,\text{k}\Omega \pm 5\%$, determine the maximum value of $|A_{cm}|$.

*9.45   Consider the adjustable gain difference amplifier in Figure P9.45. Variable resistor R_V is used to vary the gain. Show that the output voltage v_O, as a function of v_{I1} and v_{I2}, is given by

$$v_O = \frac{2R_2}{R_1}\left(1 + \frac{R_2}{R_V}\right)(v_{I2} - v_{I1})$$

Figure P9.45

9.46   Consider the instrumentation amplifier in Figure 9.25. The circuit parameters are: $R_1 = 10\,\text{k}\Omega$, $R_2 = 50\,\text{k}\Omega$, $R_3 = 20\,\text{k}\Omega$, and $R_4 = 30\,\text{k}\Omega$. Let $v_{I2} = 25\sin\omega t\,\text{mV}$ and $v_{I1} = -25\sin\omega t\,\text{mV}$. Find v_{O1}, v_{O2}, v_O, and the current in each resistor.

9.47   The instrumentation amplifier in Figure 9.25 has the same circuit parameters and input voltages as given in Problem 9.46, except that R_1 is replaced by a fixed resistance R_{1f} in series with a potentiometer, as shown in Figure 9.27. Determine the values of R_{1f} and the potentiometer resistance if the magnitude of the output has a minimum value of $|v_O| = 0.1\,\text{V}$ and a maximum value of $|v_O| = 5\,\text{V}$.

D9.48   Design the instrumentation amplifier in Figure 9.25 to have a variable differential gain in the range 0.5–200. Use a $50\,\text{k}\Omega$ potentiometer.

9.49   All parameters associated with the instrumentation amplifier in Figure 9.25 are the same as given in Exercise 9.18, except that resistor R_3, which is connected to the inverting terminal of A3, is $R_3 = 20\,\text{k}\Omega \pm 5\%$. Determine the maximum common-mode gain.

9.50   For the integrator in Figure 9.29, the circuit parameters are $R_1 = 50\,\text{k}\Omega$ and $C_2 = 0.1\,\mu\text{F}$. The input signal is $v_I = 0.5\sin\omega t\,\text{V}$. (a) At what frequency will the input and output signals have equal amplitudes? At this frequency, what is the phase of the output signal with respect to the input? (b) At what frequency will the output signal amplitude be: (i) $|v_O| = 1\,\text{V}$, and (ii) $|v_O| = 0.1\,\text{V}$?

9.51   For the ideal integrator, the RC time constant is $RC = 200\,\text{ms}$. Assume that the capacitor is initially uncharged. (a) Determine the output voltage 2 seconds after applying a voltage of $0.5\,\text{V}$ to the input. (b) How long will it take to reach $-15\,\text{V}$?

9.52   The circuit in Figure P9.52 is a first-order low-pass active filter. (a) Derive the voltage transfer function $A_v = v_O/v_I$ as a function of frequency. (b) What is the voltage gain at dc ($\omega = 0$)? (c) At what frequency is the magnitude of the gain a factor of $\sqrt{2}$ less than the dc value? (This is the $-3\,\text{dB}$ frequency.)

Figure P9.52   **Figure P9.53**

9.53   The circuit shown in Figure P9.53 is a first-order high-pass active filter. (a) Derive the voltage transfer function $A_v = v_O/v_I$ as a function of frequency. (b) What is the voltage gain as the frequency becomes large? (c) At what frequency is the magnitude of the gain a factor of $\sqrt{2}$ less than the high-frequency limiting value?

9.54   Consider the voltage reference circuit shown in Figure P9.54. Determine v_O, i_2, and i_Z.

9.55   Consider the circuit in Figure 9.35. The diode parameter is $I_S = 10^{-14}\,\text{A}$ and the resistance is $R_1 = 10\,\text{k}\Omega$. Plot v_O versus v_I over the range $20\,\text{mV} \leq v_I \leq 2\,\text{V}$. (Plot v_I on a log scale.)

Figure P9.54

Figure P9.56

*9.56   In the circuit in Figure P9.56, assume that Q_1 and Q_2 are identical transistors. If $T = 300\,°\mathrm{K}$, show that the output voltage is

$$v_O = 1.0 \log_{10}\left(\frac{v_2 R_1}{v_1 R_2}\right)$$

9.57   Consider the circuit in Figure 9.36. The diode parameter is $I_S = 10^{-14}\,\mathrm{A}$ and the resistance is $R_1 = 10\,\mathrm{k\Omega}$. Plot v_O versus v_I for $0.30 \le v_I \le 0.60\,\mathrm{V}$. (Plot v_O on a log scale.)

Section 9.6   Op-Amp Circuit Design

*D9.58   Design an op-amp summer to produce the output voltage $v_O = 2v_{I1} - 10v_{I2} + 3v_{I3} - v_{I4}$. Assume the largest resistor value is $500\,\mathrm{k\Omega}$, and the input impedance seen by each source is the largest value possible.

*9.59   Design an op-amp summer to produce the output voltage $v_O = 6v_{I1} + 3v_{I2} + 5v_{I3} - v_{I4} - 2v_{I5}$. The largest resistor value is $250\,\mathrm{k\Omega}$.

*9.60   Design a voltage reference source as shown in Figure 9.39, with an output of 9.0 V, using a Zener diode with a breakdown voltage of 5.6 V. Assume the voltage regulation will be within specifications if the Zener diode is biased between 0.8 and 0.9 mA.

*D9.61   Consider the voltage reference circuit in Figure P9.61. Using a Zener diode with a breakdown voltage of 5.6 V, design the circuit to produce an output voltage of 10 V. Assume the input voltage is 12 V and the Zener diode current is $I_Z = 1\,\mathrm{mA}$.

*D9.62   Consider the bridge circuit in Figure P9.62. The resistor R_T is a thermistor with values of $10\,\mathrm{k\Omega}$ at $T = 300\,°\mathrm{K}$ and $12\,\mathrm{k\Omega}$ at $T = 250\,°\mathrm{K}$. Assume that the thermistor resistance is linear with temperature, and that the bridge is biased at $V^+ = 10\,\mathrm{V}$. Design an amplifier system with an output of 0 V at $T = 250\,°\mathrm{K}$ and 5 V at $T = 300\,°\mathrm{K}$.

*D9.63   Consider the bridge circuit in Figure 9.43. Resistance R is $R = 50\,\mathrm{k\Omega}$ and the bias is $V^+ = 10\,\mathrm{V}$. Design an amplifier system such that the output varies from 0 V to 5 V as δ varies from 0 to +0.02.

Figure P9.61   Figure P9.62

....

COMPUTER SIMULATION PROBLEMS

9.64 Assume the input signal to the op-amp integrator is a 500 Hz square wave with amplitudes of ± 0.5 V. Design the integrator such that the steady-state output signal is a triangular wave with peak values of 0 and -5 V. Verify the design with a computer analysis.

9.65 The parameters of the filter circuit in Figure P9.52 are $R_1 = 5\,\text{k}\Omega$, $R_2 = 50\,\text{k}\Omega$, and $C_2 = 0.03\,\mu\text{F}$. Using a computer simulation plot v_O versus frequency over the range $1\,\text{Hz} \leq f \leq 10\,\text{kHz}$. Determine the corner frequency.

9.66 The parameters of the filter circuit shown in Figure P9.53 are $R_1 = 50\,\text{k}\Omega$, $R_2 = 500\,\text{k}\Omega$, and $C_1 = 50\,\text{pF}$. Using a computer simulation, plot v_O versus frequency over the range $10\,\text{kHz} \leq f \leq 10\,\text{MHz}$. Determine the corner frequency.

9.67 Verify that the design given in Example 9.12 meets the specifications.

10

Integrated Circuit Biasing and Active Loads

10.0   PREVIEW

The biasing techniques in Chapters 3 through 6 for BJT and FET amplifiers for the most part used voltage-divider resistor networks. While this technique can be used for discrete circuits, it is not suitable for integrated circuits. Resistors require relatively large areas on an IC compared to transistors; therefore, a resistor-intensive circuit would necessitate a large chip area. Also, the resistor biasing technique uses coupling and bypass capacitors extensively. On an IC, it is almost impossible to fabricate capacitors in the microfarad range, as would be required for the coupling capacitors.

Biasing transistors and transistor circuits in ICs is considerably different from that in discrete transistor designs. Essentially, biasing integrated circuit amplifiers involves the use of constant-current sources. In this chapter, we will analyze and design both bipolar and FET circuits that form these constant-current sources. We will begin to see for the first time in this chapter the use of matched or identical transistor characteristics as a specific design parameter. Transistors can easily be fabricated in ICs with matched or identical parameters. A principal goal of this chapter is to help the reader understand how matched transistor characteristics are used in design and to be able to design BJT and MOSFET current source circuits.

Transistors are also used as load devices in amplifier circuits. These transistors, called active loads, replace the discrete collector and drain resistors in BJT and FET circuits. Using an active load eliminates resistors from the IC and achieves a higher small-signal voltage gain. The active load is essentially an "upside down" constant-current source, so an initial discussion of active loads is entirely appropriate in this chapter.

10.1   BIPOLAR TRANSISTOR CURRENT SOURCES

As we saw in previous chapters, when the bipolar transistor is used as a linear amplifying device, it must be biased in the forward-active mode. The bias may be a current source that establishes the quiescent collector current as shown in Figure 10.1. We now need to consider the types of circuits that

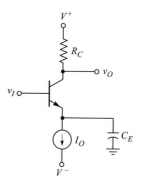

Figure 10.1   Bipolar circuit with current-source biasing

can be designed to establish the bias current I_O. We will discuss a simple two-transistor current-source circuit and then two improved versions of the constant-current source. We will then analyze another current-source circuit, known as the Widlar current source. Finally, we will discuss a multitransistor current source.

10.1.1   Two-Transistor Current Source

The **two-transistor current source**, also called a **current mirror**, is the basic building block in the design of integrated circuit current sources. Figure 10.2(a) shows the basic current-source circuit, which consists of two *matched* or *identical* transistors, Q_1 and Q_2, operating at the same temperature, with their base terminals and emitter terminals connected together. The B–E voltage is therefore the same in the two transistors. Transistor Q_1 is connected as a diode; consequently, when the supply voltages are applied, the B–E junction of Q_1 is forward biased and a reference current I_{REF} is established. Although there

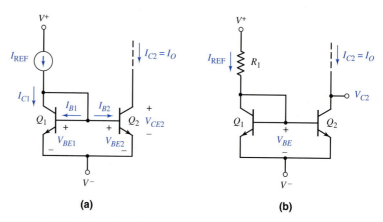

Figure 10.2   (a) Basic two-transistor current source; (b) two-transistor current source with reference resistor R_1

is a specific relationship between I_{REF} and V_{BE1}, we can think of V_{BE1} as being the result of I_{REF}. Once V_{BE1} is established, it is applied to the B–E junction of Q_2. The applied V_{BE2} turns Q_2 on and generates the load current I_O, which is used to bias a transistor or transistor circuit.

The reference current in the two-transistor current source can be established by connecting a resistor to the positive voltage source, as shown in Figure 10.2(b). The reference current is then

$$I_{\text{REF}} = \frac{V^+ - V_{BE} - V^-}{R_1} \qquad (10.1)$$

where V_{BE} is the B–E voltage corresponding to the collector current, which is essentially equal to I_{REF}.

Connecting the base and collector terminals of a bipolar transistor effectively produces a two-terminal device with I–V characteristics that are identical to the i_C versus v_{BE} characteristic of the BJT. For $v_{CB} = 0$, the transistor is still biased in the forward-active mode, and the base, collector, and emitter currents are related through the current gain β. In constant-current source circuits, β is a dc term that is the ratio of the dc collector current to the dc base current. However, as discussed in Chapter 4, we assume the dc leakage currents are negligible; therefore, the dc beta and ac beta are essentially the same. We do not distinguish between the two values.

Current Relationships

Figure 10.2(a) shows the currents in the two-transistor current source. Since V_{BE} is the same in both devices, and the transistors are identical, then $I_{B1} = I_{B2}$ and $I_{C1} = I_{C2}$. Transistor Q_2 is assumed to be biased in the forward-active region. If we sum the currents at the collector node of Q_1, we have

$$I_{\text{REF}} = I_{C1} + I_{B1} + I_{B2} = I_{C1} + 2I_{B2} \qquad (10.2)$$

Replacing I_{C1} by I_{C2} and noting that $I_{B2} = I_{C2}/\beta$, Equation (10.2) becomes

$$I_{\text{REF}} = I_{C2} + 2\frac{I_{C2}}{\beta} = I_{C2}\left(1 + \frac{2}{\beta}\right) \qquad (10.3)$$

The output current is then

$$I_{C2} = I_O = \frac{I_{\text{REF}}}{1 + \dfrac{2}{\beta}} \qquad (10.4)$$

Equation (10.4) gives the ideal output current of the two-transistor current source, taking into account the finite current gain of the transistors. Implicit in Equation (10.4) is that Q_2 is biased in the forward-active region (the base–collector junction is zero or reverse biased, meaning $V_{CE2} > V_{BE2}$[1]) and the Early voltage is infinite, or $V_A = \infty$. We will consider the effects of a finite Early voltage later in this chapter.

---

[1] In actual circuits, the collector–emitter voltage may decrease to values as low as 0.2 or 0.3 V, and the circuit will still behave as a constant-current source.

Design Example   10.1   Objective: Design a two-transistor current source to provide a specific output current.

Consider the circuit shown in Figure 10.2(b). The transistor parameters are: $V_{BE}(\text{on}) = 0.6\,\text{V}$, $\beta = 100$, and $V_A = \infty$. The output current is to be $200\,\mu\text{A}$ with $V^+ = 5\,\text{V}$ and $V^- = 0$.

Solution: The reference current can be written as

$$I_{\text{REF}} = I_O\left(1 + \frac{2}{\beta}\right) = (200)\left(1 + \frac{2}{100}\right) = 204\,\mu\text{A}$$

From Equation (10.1), resistor R_1 is found to be

$$R_1 = \frac{V^+ - V_{BE}}{I_{\text{REF}}} = \frac{5 - 0.6}{0.204} = 21.6\,\text{k}\Omega$$

Comment: In this example, we assumed a B–E voltage of 0.6 V. This approximation is satisfactory for many cases in which the B–E voltage is the same for all transistors in the current source. In other cases, we need to determine the B–E voltage corresponding to a particular collector current.

Design Pointer: We see in this example that, for $\beta = 100$, the reference and load currents are within 2 percent of each other in this two-transistor current source. In most circuit applications, we can use the approximation that $I_O \cong I_{\text{REF}}$.

Test Your Understanding

10.1   Consider the two-transistor current source shown in Figure 10.2(b). The circuit parameters are: $V^+ = 10\,\text{V}$, $V^- = 0$, and $R_1 = 15\,\text{k}\Omega$, and the transistor parameters are: $V_{BE}(\text{on}) = 0.7\,\text{V}$, $\beta = 75$, and $V_A = \infty$. Determine I_{REF} and I_O. (Ans. $I_{\text{REF}} = 0.62\,\text{mA}$, $I_O = 0.604\,\text{mA}$)

D10.2   For the current source shown in Figure 10.2(b), the circuit parameters are $V^+ = 5$ and $V^- = -5\,\text{V}$, and the transistor parameters are: $V_{BE}(\text{on}) = 0.7\,\text{V}$, $\beta = 100$, and $V_A = \infty$. Design the circuit such that $I_O = 0.75\,\text{mA}$. What is the value of I_{REF}? (Ans. $I_{\text{REF}} = 0.765\,\text{mA}$, $R_1 = 12.2\,\text{k}\Omega$)

Output Resistance

In our previous analysis, we assumed the Early voltage was infinite, so that $r_O = \infty$. In actual transistors, the Early voltage is finite, which means that the collector current is a function of the collector–emitter voltage. The stability of a load current generated in a constant-current source is a function of the output resistance looking back into the output transistor.

Figure 10.3 shows the dc equivalent circuit of a simple transistor circuit biased with a two-transistor current source. The voltage V_I applied to the base of Q_o is a dc voltage. If the value of V_I changes, the collector–emitter voltage V_{CE2} changes since the B–E voltage of Q_o is essentially a constant. A variation in V_{CE2} in turn changes the output current I_O, because of the Early

Figure 10.3   The dc equivalent circuit of simple amplifier biased with two-transistor current source

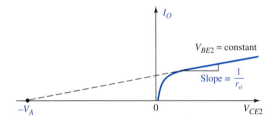

Figure 10.4   Output current versus collector–emitter voltage, showing the Early voltage

effect. Figure 10.4 shows that I_O versus V_{CE2} characteristic at a constant B–E voltage.

The ratio of load current to reference current, taking the Early effect into account, is

$$\frac{I_O}{I_{REF}} = \frac{1}{\left(1 + \dfrac{2}{\beta}\right)} \times \frac{\left(1 + \dfrac{V_{CE2}}{V_A}\right)}{\left(1 + \dfrac{V_{CE1}}{V_A}\right)} \tag{10.5}$$

where V_A is the Early voltage and the factor $(1 + 2/\beta)$ accounts for the finite gain. From the circuit configuration, we see that $V_{CE1} = V_{BE}$, which is essentially a constant.

From Figure 10.3, the collector–emitter voltage of Q_2 can be written

$$V_{CE2} = V_I - V_{BEo} - V^- \tag{10.6}$$

If the dc voltage V_I at the base of Q_o changes, then V_{CE2} changes. A change in the dc bias conditions in the load circuit affects the collector–emitter voltage of Q_2.

The differential change in I_O with respect to a change in V_{CE2}, is, from Equation (10.5),

$$\frac{dI_O}{dV_{CE2}} = \frac{I_{REF}}{\left(1 + \frac{2}{\beta}\right)} \times \frac{1}{V_A} \times \frac{1}{\left(1 + \frac{V_{BE}}{V_A}\right)} \tag{10.7}$$

If we assume $V_{BE} \ll V_A$, then Equation (10.7) becomes

$$\frac{dI_O}{dV_{CE2}} \cong \frac{I_O}{V_A} = \frac{1}{r_o} \tag{10.8}$$

where r_o is the small-signal output resistance looking into the collector of Q_2.

Example 10.2 **Objective:** Determine the change in load current produced by a change in collector–emitter voltage in a two-transistor current source.

Consider the circuit shown in Figure 10.3. The circuit parameters are: $V^+ = 5\,V$, $V^- = -5\,V$, and $R_1 = 9.3\,k\Omega$. Assume the transistor parameters are: $\beta = 50$, $V_{BE}(on) = 0.7\,V$, and $V_A = 80\,V$. Determine the change in I_O as V_{CE2} changes from $0.7\,V$ to $5\,V$.

Solution: The reference current is

$$I_{REF} = \frac{V^+ - V_{BE}(on) - V^-}{R_1} = \frac{5 - 0.7 - (-5)}{9.3} = 1.0\,mA$$

For $V_{CE2} = 0.7\,V$, transistors Q_1 and Q_2 are identically biased. From Equation (10.5), we then have

$$I_O = \frac{I_{REF}}{1 + \frac{2}{\beta}} = \frac{1.0}{1 + \frac{2}{50}} = 0.962\,mA$$

From Equation (10.8), the small-signal output resistance is

$$r_o = \frac{V_A}{I_O} = \frac{80}{0.962} = 83.2\,k\Omega$$

The change in load current is determined from

$$\frac{dI_O}{dV_{CE2}} = \frac{1}{r_o}$$

or

$$dI_O = \frac{1}{r_o}dV_{CE2} = \frac{1}{83.2}(5 - 0.7) = 0.052\,mA$$

The percent change in output current is therefore

$$\frac{dI_O}{I_O} = \frac{0.052}{0.962} = 0.054 \Rightarrow 5.4\%$$

Comment: Although in many circuits a 5 percent change in bias current is insignificant, there are cases, such as digital-to-analog converters, in which the bias current must be held to very tight tolerances. The stability of the load current can be significantly affected by a change in collector–emitter voltage. The stability is a function of the output impedance of the current source.

Test Your Understanding

10.3 Consider the circuit shown in Figure 10.3. The circuit parameters are: $V^+ = 5\,\text{V}$, $V^- = -5\,\text{V}$, and $R_1 = 12\,\text{k}\Omega$. The transistor parameters are $\beta = 75$ and $V_{BE}(\text{on}) = 0.7\,\text{V}$. The percentage change in load current $\Delta I_O/I_O$ must be no more than 2 percent for a change in V_{CE2} from 1 V to 5 V. Determine the minimum required value of Early voltage. (Ans. $V_A \cong 200\,\text{V}$)

Mismatched Transistors

In practice, transistors Q_1 and Q_2 may not be exactly identical. If $\beta \gg 1$, we can neglect base currents. The current–voltage relationship for the circuit in Figure 10.2(b) is then

$$I_{\text{REF}} \cong I_{C1} = I_{S1}e^{V_{BE}/V_T} \tag{10.9(a)}$$

and

$$I_O = I_{C2} = I_{S2}e^{V_{BE}/V_T} \tag{10.9(b)}$$

Here, we are neglecting the Early effect. The parameters I_{S1} and I_{S2} contain both the electrical and geometric parameters of Q_1 and Q_2. If Q_1 and Q_2 are not identical, then $I_{S1} \neq I_{S2}$.

Combining Equations (10.9(a)) and (10.9(b)), we obtain the relationship between the bias and reference currents, neglecting base currents, as follows:

$$I_O = I_{\text{REF}}\left(\frac{I_{S2}}{I_{S1}}\right) \tag{10.10}$$

Any deviation in bias current from the ideal, as a function of mismatch between Q_1 and Q_2, is directly related to the ratio of the reverse-saturation currents I_{S1} and I_{S2}. The parameter I_S is a strong function of temperature. The temperatures of Q_1 and Q_2 must be the same in order for the circuit to operate properly. Therefore, Q_1 and Q_2 must be close to one another on the semiconductor chip. If Q_1 and Q_2 are not maintained at the same temperature, then the relationship between I_O and I_{REF} is a function of temperature, which is undesirable.

Also, the parameters I_{S1} and I_{S2} are functions of the cross-sectional area of the B–E junctions. Therefore, we can use Equation (10.10) to our advantage. By using different sizes of transistors, we can design the circuit such that $I_O \neq I_{\text{REF}}$. This is discussed further later in this chapter.

10.1.2   Improved Current-Source Circuits

In many IC designs, critical current-source characteristics are the changes in bias current with variations in β and with changes in the output transistor collector voltage. In this section, we will look at two constant-current circuits that have improved load current stability against changes in β and changes in output collector voltage.

Basic Three-Transistor Current Source

A basic three-transistor current source is shown in Figure 10.5. We again assume that all transistors are identical; therefore, since the B–E voltage is the same for Q_1 and Q_2, $I_{B1} = I_{B2}$ and $I_{C1} = I_{C2}$. Transistor Q_3 supplies the base currents to Q_1 and Q_2, so these base currents should be less dependent on the reference current. Also, since the current in Q_3 is substantially smaller than that in either Q_1 or Q_2, we expect the current gain of Q_3 to be less than those of Q_1 and Q_2. We define the current gains of Q_1 and Q_2 as $\beta_1 = \beta_2 \equiv \beta$, and the current gain of Q_3 as β_3. Summing the currents at the collector node of Q_1, we obtain

$$I_{REF} = I_{C1} + I_{B3} \tag{10.11}$$

Since

$$I_{B1} = I_{B2} = 2I_{B2} = I_{E3} \tag{10.12}$$

and

$$I_{E3} = (1 + \beta_3)I_{B3} \tag{10.13}$$

then combining Equations (10.11), (10.12), and (10.13) produces

$$I_{REF} = I_{C1} + \frac{I_{E3}}{(1 + \beta_3)} = I_{C1} + \frac{2I_{B2}}{(1 + \beta_3)} \tag{10.14}$$

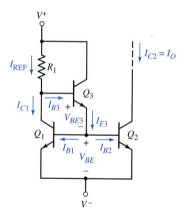

Figure 10.5   Basic three-transistor current source

Replacing I_{C1} by I_{C2} and noting that $I_{B2} = I_{C2}/\beta$, we can rewrite Equation (10.14) as

$$I_{REF} = I_{C2} + \frac{2I_{C2}}{\beta(1 + \beta_3)} = I_{C2}\left[1 + \frac{2}{\beta(1 + \beta_3)}\right] \tag{10.15}$$

The output or bias current is then

$$I_{C2} = I_O = \frac{I_{REF}}{\left[1 + \dfrac{2}{\beta(1 + \beta_3)}\right]} \tag{10.16}$$

The reference current is given by

$$I_{REF} = \frac{V^+ - V_{BE3} - V_{BE} - V^-}{R_1} \cong \frac{V^+ - 2V_{BE} - V^-}{R_1} \tag{10.17}$$

As a first approximation, we usually assume that the B–E voltage of Q_3 and Q_1 are equal, as indicated in Equation (10.17).

A comparison of Equation (10.16) for the three-transistor current source and Equation (10.4) for the two-transistor current source shows that the approximation of $I_O \cong I_{REF}$ is better for the three-transistor circuit. In addition, as we will see in the following example, the change in load current with a change in β is much smaller in the three-transistor current source.

Example 10.3   **Objective:** Compare the variation in bias current between the two- and three-transistor current-source circuits as a result of variations in β. A PSpice analysis is used.

Figure 10.6(a) shows the two-transistor PSpice circuit schematic and Figure 10.6(b) shows the three-transistor PSpice circuit schematic used in this analysis.

Solution: In both circuits, the current gain β of all transistors was assumed to be equal, but the actual value was varied between 20 and 200. Since the change in β is very large, we cannot use derivatives to determine the changes in bias currents. Standard 2N3904 transistors were used, which means that the Early voltage is $74\,V$, and not infinite as in the ideal circuit. The Early voltage will influence the actual value of bias current, but has very little effect in terms of the change in bias current with a change in current gain.

Figure 10.6(c) shows the bias current versus current gain for both the two-transistor and three-transistor current-source circuits.

Comment: There is a significant decrease in the variation in bias current for the three-transistor circuit compared to that of the two-transistor circuit. For values of β greater than approximately 50, there is no perceptible change in bias current for the three-transistor current mirror

---

The output resistance looking into the collector of the output transistor Q_2 of the basic three-transistor current source shown in Figure 10.5 is the same as that of the two-transistor current source; that is,

$$\frac{dI_O}{dV_{CE2}} = \frac{1}{r_{o2}} \tag{10.18}$$

This means that, in the three-transistor current source, the change in bias current I_O with a change in V_{CE2} is the same as that in the two-transistor current-source circuit. In addition, any mismatch between Q_1 and Q_2 produces a deviation in the bias current from the ideal, as given by Equation (10.10).

Figure 10.6   (a) Two-transistor current mirror; (b) three-transistor current mirror; (c) variation in bias currents with a change in β

Test Your Understanding

10.4   The parameters for the circuit shown in Figure 10.5 are: $V^+ = 9\,\text{V}$, $V^- = 0$, and $R_1 = 12\,\text{k}\Omega$. The transistor parameters, for all transistors, are: $V_{BE}(\text{on}) = 0.7\,\text{V}$, $\beta = 75$, and $V_A = \infty$. Calculate the value of each current shown in the figure. (Ans. $I_{REF} = 0.6333\,\text{mA}$, $I_O = 0.6331\,\text{mA} = I_{C1}$, $I_{B1} = I_{B2} = 8.44\,\mu\text{A}$, $I_{E3} = 16.88\,\mu\text{A}$, $I_{B3} = 0.222\,\mu\text{A}$)

10.5   The current source shown in Figure 10.5 utilizes BJTs, with parameters $\beta = 100$, $V_{BE}(\text{on}) = 0.7\,\text{V}$, and $V_A = 100\,\text{V}$. The circuit parameters are: $V^+ = 10\,\text{V}$, $V^- = 0$, and $R_1 = 12\,\text{k}\Omega$. Determine the output resistance and the change in load current if the collector voltage at Q_2 changes from 1 V to 5 V. (Ans. $r_o = 139\,\text{k}\Omega$, $\Delta I_O = 0.0288\,\text{mA}$)

Cascode Current Source

Current-source circuits can be designed such that the output resistance is much greater than that of the two-transistor circuit. One example is the cascode circuit shown in Figure 10.7(a). In this case, if the transistors are matched, then the load and reference currents are essentially equal.

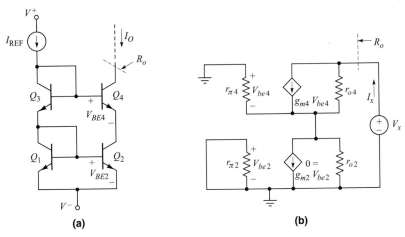

Figure 10.7 (a) Bipolar cascode current mirror; (b) small-signal equivalent circuit

We may calculate the output resistance R_o by considering the small-signal equivalent transistor circuits. For a constant reference current, the base voltages of Q_2 and Q_4 are constant, which implies these terminals are at signal ground. The equivalent circuit is then shown in Figure 10.7(b). Since $g_{m2}V_{be2} = 0$, then $V_{be4} = -I_x(r_{o2}\|r_{\pi4})$. Summing currents at the output node yields

$$I_x = g_{m4}V_{be4} + \left(\frac{V_x - I_x(r_{o2}\|r_{\pi4})}{r_{o4}}\right)$$

$$= -g_{m4}I_x(r_{o2}\|r_{\pi4}) + \left(\frac{V_x - I_x(r_{o2}\|r_{\pi4})}{r_{o4}}\right) \tag{10.19}$$

Combining terms, we find

$$R_o = \frac{V_x}{I_x} = r_{o4}(1 + \beta) + r_{\pi4} \cong \beta r_{o4} \tag{10.20}$$

The output resistance has increased by a factor of β compared to the two-transistor current source, which increases the stability of the current source with changes in output voltage.

Wilson Current Source

Another configuration of a three-transistor current source, called a **Wilson current source**, is shown in Figure 10.8. This circuit also has a large output resistance. Our analysis again assumes identical transistors, with $I_{B1} = I_{B2}$ and $I_{C1} = I_{C2}$. The current levels in all three transistors are nearly the same;

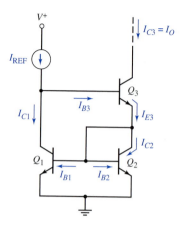

Figure 10.8 Wilson current source

therefore, we can assume that the current gains of the three transistors are equal. Nodal equations at the collector of Q_1 and the emitter of Q_3 yield

$$I_{REF} = I_{C1} + I_{B3} \qquad (10.21)$$

and

$$I_{E3} = I_{C2} + 2I_{B2} = I_{C2}\left(1 + \frac{2}{\beta}\right) \qquad (10.22)$$

Using the relationships between the base, collector, and emitter currents in Q_3, we can write the collector current I_{C2}, from Equation (10.22), as follows:

$$I_{C2} = \frac{I_{E3}}{\left(1 + \frac{2}{\beta}\right)} = \frac{1}{\left(1 + \frac{2}{\beta}\right)} \times \left(\frac{1 + \beta}{\beta}\right)I_{C3} = \left(\frac{1 + \beta}{2 + \beta}\right)I_{C3} \qquad (10.23)$$

If we replace I_{C1} by I_{C2} in Equation (10.21), the reference current becomes

$$I_{REF} = I_{C2} + I_{B3} = \left(\frac{1 + \beta}{2 + \beta}\right)I_{C3} + \frac{I_{C3}}{\beta} \qquad (10.24)$$

Rearranging terms, we can solve for the output current,

$$I_{C3} = I_O = I_{REF} \times \frac{1}{1 + \frac{2}{\beta(2 + \beta)}} \qquad (10.25)$$

This current relationship is essentially the same as that of the previous three-transistor current source.

The difference between the two three-transistor current-source circuits is the output resistance. In the Wilson current source, the output resistance looking into the collector of Q_3 is $R_o \cong \beta r_{o3}/2$, which is approximately a factor $\beta/2$ larger than that of either the two-transistor source or the basic three-transistor source. This means that, in the Wilson current source, the change in bias current I_O with a change in output collector voltage is much smaller.

Output Voltage Swing

If we consider the equivalent circuit in Figure 10.3, we see that the maximum possible swing in the output voltage is a function of the minimum possible collector–emitter voltage of Q_2. For the two-transistor current source in this figure, the minimum value of $V_{CE2} = V_{CE}(\text{sat})$, which may be on the order of 0.1 to 0.3 V.

For the cascode and Wilson current sources, the minimum output voltage is $V_{BE} + V_{CE}(\text{sat})$ above the negative power supply voltage, which may be on the order of 0.7 to 0.9 V. For circuits biased at ± 5 V, for example, this increased minimum voltage may not be a serious problem. However, as the voltages decrease in low-power circuits, this minimum voltage effect may become more serious.

Problem-Solving Technique: BJT Current Source Circuits

1. Sum currents at the various nodes in the circuit to find the relation between the reference current and the bias currrent.

2. To find the output resistance of the current source circuit, place a test voltage at the output node and analyze the small-signal equivalent circuit. Keep in mind that the reference current is a constant, which may make some of the base voltages constant or at ac ground.

Test Your Understanding

10.6 For the Wilson current source in Figure 10.8, the transistor parameters are: $V_{BE}(\text{on}) = 0.7$ V, $\beta = 50$, and $V_A = \infty$. For $I_{REF} = 0.50$ mA, determine all currents shown in the figure. (Ans. $I_O = 0.4996$ mA, $I_{B3} = 9.99\,\mu\text{A}$, $I_{E3} = 0.5096$ mA, $I_{C2} = 0.490$ mA $= I_{C1}$, $I_{B1} = I_{B2} = 9.80\,\mu\text{A}$)

10.1.3 Widlar Current Source

In the current-source circuits considered thus far, the load and reference currents have been nearly equal. For a two-transistor current source, such as that shown in Figure 10.2(a), if we require a load current of $I_O = 10\,\mu\text{A}$, then, for $V^+ = 5$ V and $V^- = -5$ V, the required resistance value is

$$R_1 = \frac{V^+ - V_{BE} - V^-}{I_{REF}} \cong \frac{5 - 0.7 - (-5)}{10 \times 10^{-6}} = 930\,\text{k}\Omega$$

In ICs, resistors on the order of 1 MΩ require large areas and are difficult to fabricate accurately. We therefore need to limit IC resistor values to the low kilohm range.

The transistor circuit in Figure 10.9, called a **Widlar current source**, meets this objective. A voltage difference is produced across resistor R_E, so that the B–E voltage of Q_2 is less than the B–E voltage of Q_1. A smaller B–E voltage

Figure 10.9 Widlar current source

produces a smaller collector current, which in turn means that the load current I_O is less than the reference current I_{REF}.

Current Relationship

If $\beta \gg 1$ for Q_1 and Q_2, and if the two transistors are identical, then

$$I_{\text{REF}} \cong I_{C1} = I_S e^{V_{BE1}/V_T} \qquad (10.26(a))$$

and

$$I_O = I_{C2} = I_S e^{V_{BE2}/V_T} \qquad (10.26(b))$$

Solving for the B–E voltages, we have

$$V_{BE1} = V_T \ln\left(\frac{I_{\text{REF}}}{I_S}\right) \qquad (10.27(a))$$

and

$$V_{BE2} = V_T \ln\left(\frac{I_O}{I_S}\right) \qquad (10.27(b))$$

Combining Equations (10.27(a)) and (10.27(b)) yields

$$V_{BE1} - V_{BE2} = V_T \ln\left(\frac{I_{\text{REF}}}{I_O}\right) \qquad (10.28)$$

From the circuit, we see that

$$V_{BE1} - V_{BE2} = I_{E2} R_E \cong I_O R_E \qquad (10.29)$$

When we combine Equations (10.28) and (10.29), we obtain:

$$I_O R_E = V_T \ln\left(\frac{I_{\text{REF}}}{I_O}\right) \qquad (10.30)$$

This equation gives the relationship between the reference and bias currents.

Design Example 10.4   Objective: Design a Widlar current source to achieve specified reference and load currents.

Design the Widlar current source to produce $I_{REF} = 1\,\text{mA}$ and $I_O = 12\,\mu\text{A}$. Let $V^+ = 5\,\text{V}$ and $V^- = -5\,\text{V}$. Assume $V_{BE1} = 0.7\,\text{V}$ at the reference current of 1 mA.

Solution: Resistance R_1 is

$$R_1 = \frac{V^+ - V_{BE1} - V^-}{I_{REF}} = \frac{5 - 0.7 - (-5)}{1} = 9.3\,\text{k}\Omega$$

Resistance R_E is, from Equation (10.30),

$$R_E = \frac{V_T}{I_O} \ln\left(\frac{I_{REF}}{I_O}\right) = \frac{0.026}{0.012} \ln\left(\frac{1}{0.012}\right) = 9.58\,\text{k}\Omega$$

From Equation (10.29), we can determine the difference between the two B–E voltages, as follows:

$$V_{BE1} - V_{BE2} = I_O R_E = (12 \times 10^{-6})(9.58 \times 10^3) = 0.115\,\text{V}$$

Comment: A difference of 115 mV in the B–E voltages of Q_1 and Q_2 produces approximately two orders of magnitude difference between the reference and load currents. Therefore, we can produce a very low bias current using resistors in the low kilohm range. These resistors can easily be fabricated in an IC. Including the resistor R_E gives the designer additional versatility in adjusting the load to reference current ratio.

In our analysis of constant-current source circuits, we have assumed a piecewise linear approximation for the B–E voltage, $V_{BE}(\text{on})$. However, in the Widlar current source and other current-source circuits, the piecewise linear approximation is not adequate, since the B–E voltages are not all equal. With the exponential relationship between collector current and base–emitter voltage, as shown in Equations (10.26(a)) and (10.26(b)), a small change in B–E voltage produces a large change in collector current. To take this variation into account, either the reverse-biased saturation current I_S or the B–E voltage at a particular collector current must be known.

Also in our analysis, we have assumed that the temperatures of all transistors are equal. Maintaining equal temperatures is important for proper circuit operation.

Test Your Understanding

RD10.7   Consider the Widlar current source in Figure 10.9. The bias voltages are $V^+ = 5\,\text{V}$ and $V^- = 0$. Redesign the circuit such that $I_O = 25\,\mu\text{A}$ and $I_{REF} = 0.75\,\text{mA}$. Assume $V_{BE1} = 0.7\,\text{V}$, and neglect the base currents. What is the difference between the two B–E voltages? (Ans. $R_E = 3.54\,\text{k}\Omega$, $R_1 = 5.73\,\text{k}\Omega$, $V_{BE1} - V_{BE2} = 88.5\,\text{mV}$)

10.8   The Widlar current source in Figure 10.9 is biased at $V^+ = 5\,\text{V}$ and $V^- = -5\,\text{V}$. The resistor values are $R_1 = 12\,\text{k}\Omega$ and $R_E = 6\,\text{k}\Omega$. Neglect the base currents and assume the B–E voltage of Q_1 is 0.7 V. Determine I_{REF} and I_O. (Hint: You may need to use trial and error to find I_O.) (Ans. $I_{REF} = 0.775\,\text{mA}$, $I_O \cong 16.6\,\mu\text{A}$)

Output Resistance

The change in load current with a change in voltage V_{C2} of the Widlar current source in Figure 10.9 can be expressed as

$$\frac{dI_O}{dV_{C2}} = \frac{1}{R_o} \tag{10.31}$$

where R_o is the output resistance looking into the collector of Q_2. This output resistance can be determined by using the small-signal equivalent circuit in Figure 10.10(a). (Again, we use the phasor notation in small-signal analyses.) The base, collector, and emitter terminals of each transistor are indicated on the figure.

First, we calculate the resistance R_{o1} looking into the base of Q_1. Writing a KCL equation at the base of Q_1, we obtain

$$I_{x1} = \frac{V_{x1}}{r_{\pi 1}} + g_{m1} V_{\pi 1} + \frac{V_{x1}}{r_{o1} \| R_1} \tag{10.32}$$

Noting that $V_{\pi 1} = V_{x1}$, we have

$$\frac{1}{R_{o1}} = \frac{I_{x1}}{V_{x1}} = \frac{1}{r_{\pi 1}} + g_{m1} + \frac{1}{r_{o1} \| R_1} \tag{10.33(a)}$$

(a)

(b)   (c)

Figure 10.10 (a) Small-signal equivalent circuit for determining output resistance of Widlar current source, (b) simplified equivalent circuit for determining output resistance, and (c) equivalent circuit after a Norton transformation

or

$$R_{o1} = r_{\pi 1} \left\| \frac{1}{g_{m1}} \right\| r_{o1} \| R_1 \tag{10.33(b)}$$

Next, we calculate the approximate value for R_{o1}. If $I_{REF} = 1 \, mA$, then for $\beta = 100$, $r_{\pi 1} = 2.6 \, k\Omega$ and $g_{m1} = 38.5 \, mA/V$. Assume that $R_1 = 9.3 \, k\Omega$ and $r_{o1} = \infty$. For these conditions, $R_{o1} \cong 0.026 \, k\Omega = 26 \, \Omega$. For a load current of $I_O = 12 \, \mu A$, we find $r_{\pi 2} = 217 \, k\Omega$. Resistance R_{o1} is in series with $r_{\pi 2}$, and since $R_{o1} \ll r_{\pi 2}$, we can neglect the effect of R_{o1}, which means that the base of Q_2 is essentially at signal ground.

Now we determine the output resistance at the collector of Q_2, using the simplified equivalent circuit in Figure 10.10(b). The Norton equivalent of the current source $g_{m2}V_{\pi 2}$ and resistance r_{o2} can be transformed into a Thevenin equivalent circuit, as shown in Figure 10.10(c). Resistances $r_{\pi 2}$ and R_E are in parallel; therefore, we define $R_E' = R_E \| r_{\pi 2}$. Since the current through the parallel combination of R_E and $r_{\pi 2}$ is I_x, we have

$$V_{\pi 2} = -I_x R_E' \tag{10.34}$$

Writing a KVL equation, we obtain

$$V_x = I_x r_{o2} - g_{m2} r_{o2} V_{\pi 2} + I_x R_E' \tag{10.35}$$

Substituting Equation (10.34) into (10.35) yields

$$\frac{V_x}{I_x} = R_o = r_{o2} \left[1 + R_E' \left(g_{m2} + \frac{1}{r_{o2}} \right) \right] \tag{10.36}$$

Normally, $(1/r_{o2}) \ll g_{m2}$; therefore,

$$R_o \cong r_{o2} (1 + g_{m2} R_E') \tag{10.37}$$

The output resistance of the Widlar current source is a factor $(1 + g_{m2}R_E')$ larger than that of the simple two-transistor current source.

Example 10.5   Objective: Determine the change in load current with a change in collector voltage in a Widlar current source.

Consider the circuit in Figure 10.9. The parameters are: $V^+ = 5 \, V$, $V^- = -5 \, V$, $R_1 = 9.3 \, k\Omega$, and $R_E = 9.58 \, k\Omega$. Let $V_A = 80 \, V$ and $\beta = 100$. Determine the change in I_O as V_{C2} changes by $4 \, V$.

Solution: From Example 10.4, we have $I_O = 12 \, \mu A$. The small-signal collector resistance is

$$r_{o2} = \frac{V_A}{I_O} = \frac{80}{0.012} \Rightarrow 6.67 \, M\Omega$$

We can determine that

$$g_{m2} = \frac{I_O}{V_T} = \frac{0.012}{0.026} = 0.462 \, mA/V$$

and

$$r_{\pi 2} = \frac{\beta V_T}{I_O} = \frac{(100)(0.026)}{0.012} = 217 \, k\Omega$$

The output resistance of the circuit is

$$R_o = r_{o2}[1 + g_{m2}(R_E \| r_{\pi2})] = (6.67) \cdot [1 + (0.462)(9.58\|217)] = 34.9\,\text{M}\Omega$$

From Equation (10.31), the change in load current is

$$dI_O = \frac{1}{R_o} dV_{C2} = \frac{1}{34.9 \times 10^6} \times 4 \Rightarrow 0.115\,\mu\text{A}$$

The percentage change in output current is then

$$\frac{dI_O}{I_O} = \frac{0.115}{12} = 0.0096 \Rightarrow 0.96\%$$

Comment: The stability of the load current, as a function of a change in output voltage, is improved in the Widlar current source, compared to the simple two-transistor current source.

Computer Verification: The output resistance and the change in bias current with a change in output voltage were determined by a PSpice analysis for both the two-transistor and Widlar current-source circuits. Figure 10.11(a) shows the PSpice circuit schematic of the two-transistor current source and Figure 10.11(b) shows the PSpice

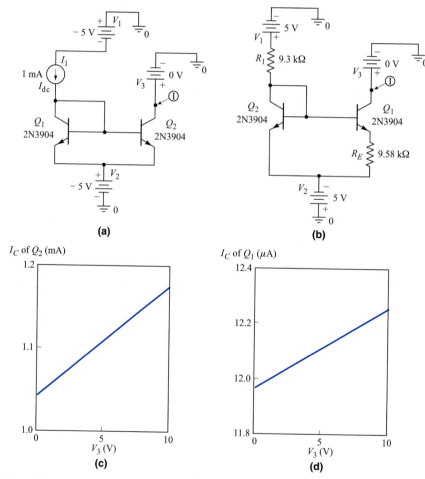

Figure 10.11 (a) The two-transistor current mirror; (b) the Widlar current source; (c) the variation in bias current with output voltage for the two-transistor circuit; (d) the variation in bias current with output voltage for the Widlar circuit

circuit schematic of the Widlar current source used in this analysis. The voltage source V_3 was varied in each circuit from 0 to 10 V.

Figure 10.11(c) shows the change in bias current for the two-transistor current source and Figure 10.11(d) shows the change in bias current for the Widlar current source. The output resistance is 75.8 kΩ for the two-transistor source and is 36.6 MΩ for the Widlar source. The change in bias current for the two-transistor circuit is 13.2 percent and for the Widlar circuit is only 2.28 percent. We, therefore, see the advantage of a large output resistance of a current-source circuit.

Test Your Understanding

*10.9 A Widlar current source is shown in Figure 10.9. The parameters are: $V^+ = 5$ V, $V^- = 0$, $I_{REF} = 0.70$ mA, and $I_O = 25\,\mu$A at $V_{C2} = 1$ V. The transistor parameters are: $\beta = 150$, $V_{BE1}(\text{on}) = 0.7$ V, and $V_A = 100$ V. Determine the change in I_O when V_{C2} changes from 1 V to 4 V. (Ans. $dI_O = 0.176\,\mu$A)

10.1.4 Multitransistor Current Mirrors

In the previous current sources, we established a reference current and one load current. In the two-transistor current source in Figure 10.2(a), the B–E junction of the diode-connected transistor Q_1 is forward biased when the bias voltages V^+ and V^- are applied. Once V_{BE} is established, the voltage is applied to the B–E junction of Q_2, which turns Q_2 on and produces the load current I_O.

The B–E voltage of Q_1 can also be applied to additional transistors, to generate multiple load currents. Consider the circuit in Figure 10.12. Transistor Q_R, which is the reference transistor, is connected as a diode. The resulting B–E voltage of Q_R, established by I_{REF}, is applied to N output transistors, creating N load currents. The relationship between each load current and the reference current, assuming all transistors are matched and $V_A = \infty$, is

$$I_{O1} = I_{O2} = \cdots = I_{ON} = \frac{I_{REF}}{1 + \dfrac{(1+N)}{\beta}} \tag{10.38}$$

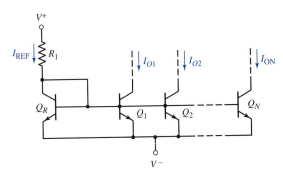

Figure 10.12 Multitransistor current mirror

The collectors of multiple output transistors can be connected together, changing the load current versus reference current relationship. As an example, the circuit in Figure 10.13 has three output transistors with common collectors and a load current I_O. We assume that transistors Q_R, Q_1, Q_2, and Q_3 are all matched. If the current gain β is very large, the base currents can be neglected, $I_1 = I_2 = I_3 = I_{REF}$, and the load current is $I_O = 3I_{REF}$. [Note: This process is not recommended for discrete devices, since a mismatch between devices will generally cause one device to carry more current than the other devices.]

Connecting transistors in parallel increases the effective B–E area of the device. In actual IC fabrication, the B–E area would be doubled or tripled to provide a load current twice or three times the value of I_{REF}.

Figure 10.13   Multioutput transistor current source

Rather than drawing each set of parallel output transistors, we can use the circuit symbols in Figure 10.14. Figure 10.14(a) is the equivalent symbol for two transistors connected in parallel, Figure 10.14(b) is for three transistors in parallel, and Figure 10.14(c) is for N transistors in parallel. Although the transistors appear to be multiemitter devices, we are simply indicating devices with different B–E junction areas.

(a)   (b)   (c)

Figure 10.14   Equivalent circuit symbols (a) two transistors in parallel, (b) three transistors in parallel, and (c) N transistors in parallel

A generalized current mirror is shown in Figure 10.15. We can use pnp transistors to establish the load currents, as shown in the figure. Transistors Q_{R1} and Q_{R2} are connected as diodes. The reference current is established in the branch of the circuit that has the diode-connected transistors, resistor R_1, and bias voltages, and is given by

$$I_{REF} = \frac{V^+ - V_{EB}(Q_{R1}) - V_{BE}(Q_{R2}) - V^-}{R_1}$$

$$(10.39)$$

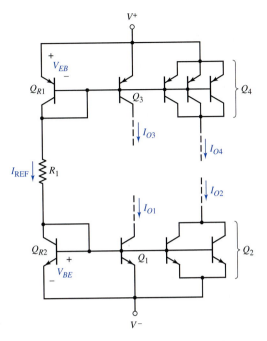

Figure 10.15   Generalized current mirror

If β for each transistor is very large, the base current effects can be neglected. Then the load current I_{O1} generated by output transistor Q_1 is equal to I_{REF}. Likewise, Q_3 generates a load current I_{O3} equal to I_{REF}. Implicitly, all transistors are identical, all load transistors are biased in their forward-active region, and all transistor Early voltages are infinite. Transistor Q_2 is effectively two transistors in parallel; then, since all transistors are identical, $I_{O2} = 2I_{REF}$. Similarly, Q_4 is effectively three transistors connected in parallel, which means that the load current is $I_{O4} = 3I_{REF}$.

In the above discussion, we neglected the effect of base currents. However, a finite β causes the collector currents in each load transistor to be smaller than I_{REF} since the reference current supplies all base currents. This effect becomes more severe as more load transistors are added.

Design Example   10.6   Objective: Design a generalized current mirror.

Consider the current mirror shown in Figure 10.15, with parameters $V^+ = 5$ V and $V^- = -5$ V. Neglect base currents and assume $V_{BE} = V_{EB} = 0.6$ V. Design the circuit such that $I_{O2} = 400\,\mu$A. Determine I_{REF}, I_{O1}, I_{O3}, I_{O4}, and R_1.

Solution: For $I_{O2} = 400\,\mu$A, we have

$$I_{REF} = I_{O1} = I_{O3} = 200\,\mu\text{A} \quad \text{and} \quad I_{O4} = 600\,\mu\text{A}$$

Resistor R_1 is

$$R_1 = \frac{V^+ - V_{EB}(Q_{R1}) - V_{BE}(Q_{R2}) - V^-}{I_{REF}} = \frac{5 - 0.6 - 0.6 - (-5)}{0.2}$$

or

$$R_1 = 44 \, \text{k}\Omega$$

 Comment: If the load and reference currents are to be within a factor of approximately four of each other, it is more efficient, from an IC point of view, to adjust the B–E areas of the transistors to achieve the specified currents rather than use the Widlar current source with its additional resistors.

Design Pointer: This example demonstrates that multiple bias currents can be generated by a single reference current that biases various stages of a complex circuit. We will see specific examples of this technique in Chapter 13 when we consider actual operational amplifier circuits.

Test Your Understanding

*10.10   Figure 10.12 shows the N-output current mirror. Assuming all transistors are matched, with a finite gain and $V_A = \infty$, derive Equation (10.38). If each load current must be within 10 percent of I_{REF}, and if $\beta = 50$, determine the maximum number of load transistors that can be connected. (Ans. $N = 4$)

10.2   FET CURRENT SOURCES

Field-effect transistor integrated circuits are biased with current sources in much the same way as bipolar circuits. We will examine the relationship between the reference and load currents, and will determine the output impedance of the basic two-transistor MOSFET current source. We will then analyze multi-MOSFET current-source circuits to determine reference and load current relationships and output impedance. Finally, we will discuss JFET constant-current source circuits.

10.2.1   Basic Two-Transistor MOSFET Current Source

Current Relationship

Figure 10.16 shows a basic two-transistor NMOS current source. The drain and source terminals of the enhancement-mode transistor M_1 are connected, which means that M_1 is always biased in the saturation region. Assuming $\lambda = 0$, we can write the reference current as

$$I_{\text{REF}} = K_{n1}(V_{GS} - V_{TN1})^2 \qquad \textbf{(10.40)}$$

Solving for V_{GS} yields

$$V_{GS} = V_{TN1} + \sqrt{\frac{I_{\text{REF}}}{K_{n1}}} \qquad \textbf{(10.41)}$$

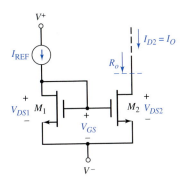

Figure 10.16 Basic two-transistor MOSFET current source

For the drain current to be independent of the drain-to-source voltage (for $\lambda = 0$), transistor M_2 should always be biased in the saturation region. The load current is then

$$I_O = K_{n2}(V_{GS} - V_{TN2})^2 \tag{10.42}$$

Substituting Equation (10.41) into (10.42), we have

$$I_O = K_{n2}\left[\sqrt{\frac{I_{REF}}{K_{n1}}} + V_{TN1} - V_{TN2}\right]^2 \tag{10.43}$$

If M_1 and M_2 are identical transistors, then $V_{TN1} = V_{TN2}$ and $K_{n1} = K_{n2}$, and Equation (10.43) becomes

$$I_O = I_{REF} \tag{10.44}$$

Since there are no gate currents in MOSFETs, the induced load current is identical to the reference current, provided the two transistors are matched. The relationship between the load current and the reference current changes if the width-to-length ratios, or **aspect ratios**, of the two transistors change.

If the transistors are matched except for the aspect ratios, we find

$$I_O = \frac{(W/L)_2}{(W/L)_1} \cdot I_{REF} \tag{10.45}$$

The ratio between the load and reference currents is directly proportional to the aspect ratios and gives designers versatility in their circuit designs.

Output Resistance

The stability of the load current as a function of the drain-to-source voltage is an important consideration in many applications. The drain current versus drain-to-source voltage is similar to the bipolar characteristic shown in Figure 10.4. Taking into account the finite output resistance of the transistors, we can write the load and reference currents as follows:

$$I_O = K_{n2}(V_{GS} - V_{TN2})^2(1 + \lambda_2 V_{DS2}) \tag{10.46(a)}$$

and

$$I_{REF} = K_{n1}(V_{GS} - V_{TN1})^2(1 + \lambda_1 V_{DS1}) \tag{10.46(b)}$$

Since transistors in the current mirror are processed on the same integrated circuit, all physical parameters, such as V_{TN}, μ_n, C_{ox}, and λ, are essentially identical for both devices. Therefore, taking the ratio of I_O to I_{REF}, we have

$$\frac{I_O}{I_{REF}} = \frac{(W/L)_2}{(W/L)_1} \cdot \frac{(1 + \lambda V_{DS2})}{(1 + \lambda V_{DS1})} \tag{10.47}$$

Equation (10.47) again shows that the ratio I_O/I_{REF} is a function of the aspect ratios, which is controlled by the designer, and it is also a function of λ and V_{DS2}.

As before, the stability of the load current can be described in terms of the output resistance. Note from the circuit in Figure 10.16 that $V_{DS1} = V_{GS1} =$ constant for a given reference current. Normally, $\lambda V_{DS1} = \lambda V_{GS1} \ll 1$, and if $(W/L)_2 = (W/L)_1$, then the change in bias current with respect to a change in V_{DS2} is

$$\frac{1}{R_o} \equiv \frac{dI_O}{dV_{DS2}} \cong \lambda I_{REF} = \frac{1}{r_o} \tag{10.48}$$

where r_o is the output resistance of the transistor. As we found with bipolar current-source ciucits, MOSFET current sources require a large output resistance for excellent stability.

Reference Current

The reference current in bipolar current-source circuits is generally established by the bias voltages and a resistor. Since MOSFETs can be configured to act like a resistor, the reference current in MOSFET current mirrors is usually established by using additional transistors.

Consider the current mirror shown in Figure 10.17. Transistors M_1 and M_3 are in series; assuming $\lambda = 0$, we can write,

$$K_{n1}(V_{GS1} - V_{TN1})^2 = K_{n3}(V_{GS3} - V_{TN3})^2 \tag{10.49}$$

If we again assume that V_{TN}, μ_n, and C_{ox} are identical in all transistors, then Equation (10.49) can be rewritten

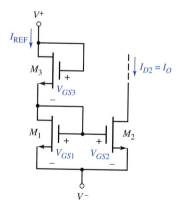

Figure 10.17 MOSFET current source

$$V_{GS1} = \sqrt{\frac{(W/L)_3}{(W/L)_1}} \cdot V_{GS3} + \left(1 - \sqrt{\frac{(W/L)_3}{(W/L)_1}}\right) \cdot V_{TN} \tag{10.50}$$

where V_{TN} is the threshold voltage of both transistors.

From the circuit, we see that

$$V_{GS1} + V_{GS3} = V^+ - V^- \tag{10.51}$$

Therefore,

$$V_{GS1} = \frac{\sqrt{\frac{(W/L)_3}{(W/L)_1}}}{1 + \sqrt{\frac{(W/L)_3}{(W/L)_1}}} \cdot (V^+ - V^-) + \frac{\left(1 - \sqrt{\frac{(W/L)_3}{(W/L)_1}}\right)}{\left(1 + \sqrt{\frac{(W/L)_3}{(W/L)_1}}\right)} \cdot V_{TN} = V_{GS2} \tag{10.52}$$

Finally, the load current, for $\lambda = 0$, is given by

$$I_O = \left(\frac{W}{L}\right)_2 \left(\frac{1}{2}\mu_n C_{ox}\right)(V_{GS2} - V_{TN})^2 \tag{10.53}$$

Since the designer has control over the width-to-length ratios of the transistors, there is considerable flexibility in the design of MOSFET current sources.

Design Example 10.7 **Objective:** Design a MOSFET current source to meet specified current values.

Consider the current source in Figure 10.17, with transistor parameters $\frac{1}{2}\mu_n C_{ox} = 20\,\mu\text{A/V}^2$, $V_{TN} = 1\,\text{V}$, and $\lambda = 0$. Let $V^+ = 5\,\text{V}$ and $V^- = 0$. Design the circuit such that $I_{REF} = 0.25\,\text{mA}$ and $I_O = 0.10\,\text{mA}$.

Solution: If we choose V_{GS2} to be fairly small, yet greater than V_{TN}, then M_2 will remain biased in the saturation region over a fairly large range of V_{DS2} values. Let $V_{GS2} = 1.85\,\text{V}$. Then, from Equation (10.53), we can write

$$\left(\frac{W}{L}\right)_2 = \frac{I_O}{\left(\frac{1}{2}\mu_n C_{ox}\right)(V_{GS2} - V_{TN})^2} = \frac{0.10}{(0.02)(1.85 - 1)^2} = 6.92$$

The reference current is

$$I_{REF} = \left(\frac{W}{L}\right)_1 \left(\frac{1}{2}\mu_n C_{ox}\right)(V_{GS1} - V_{TN})^2$$

Since $V_{GS1} = V_{GS2}$, we have

$$\left(\frac{W}{L}\right)_1 = \frac{I_{REF}}{\left(\frac{1}{2}\mu_n C_{ox}\right)(V_{GS2} - V_{TN})^2} = \frac{0.25}{(0.02)(1.85 - 1)^2} = 17.3$$

The value of V_{GS3} is

$$V_{GS3} = (V^+ - V^-) - V_{GS1} = 5 - 1.85 = 3.15\,\text{V}$$

Then, since $I_{REF} = K_{n3}(V_{GS3} - V_{TN})^2$, we have

$$\left(\frac{W}{L}\right)_3 = \frac{I_{REF}}{\left(\frac{1}{2}\mu_n C_{ox}\right)(V_{GS3} - V_{TN})^2} = \frac{0.25}{(0.02)(3.15 - 1)^2} = 2.70$$

Comment: In this design, the output transistor remains biased in the saturation region for

$$V_{DS} > V_{DS}(\text{sat}) = V_{GS} - V_{TN} = 1.85 - 1 = 0.85\,\text{V}$$

Design Pointer: As with most design problems, there is not a unique solution. The general design criterion was that M_2 was biased in the saturation region over a wide range of V_{DS2} values. Letting $V_{GS2} = 1.85\,\text{V}$ was somewhat arbitrary. If V_{GS2} were smaller, the width-to-length ratios of M_1 and M_2 would need to be larger. Larger values of V_{GS2} would result in smaller width-to-length ratios.

The value of V_{GS3} is the difference between the bias voltage and V_{GS1}. If V_{GS3} becomes too large, the ratio $(W/L)_3$ will become unreasonably small (much less than 1). Two or more transistors in series can be used in place of M_3 to divide the voltage in order to provide reasonable W/L ratios (see Problem 10.49).

Problem-Solving Technique: MOSFET Current-Source Circuit

1. Analyze the reference side of the circuit to determine gate-to-source voltages. Using these gate-to-source voltages, determine the bias current in terms of the reference current.

2. To find the output resistance of the current source circuit, place a test voltage at the output node and analyze the small-signal equivalent circuit. Keep in mind that the reference current is a constant, which may make some of the gate voltages constant or at ac ground.

Test Your Understanding

10.11 Consider the MOSFET current source in Figure 10.17, with $V^+ = 10\,\text{V}$ and $V^- = 0$. The transistor parameters are: $V_{TN} = 1.8\,\text{V}$, $\frac{1}{2}\mu_n C_{ox} = 20\,\mu\text{A/V}^2$, and $\lambda = 0.01\,\text{V}^{-1}$. The transistor width-to-length ratios are: $(W/L)_3 = 3$, $(W/L)_1 = 12$, and $(W/L)_2 = 6$. Determine: (a) I_{REF}, (b) I_O at $V_{DS2} = 2\,\text{V}$, and (c) I_O at $V_{DS2} = 6\,\text{V}$. (Ans. (a) $I_{REF} = 1.13\,\text{mA}$ (b) $I_O = 0.555\,\text{mA}$ (c) $I_O = 0.576\,\text{mA}$)

10.12 Consider the circuit shown in Figure 10.18. The transistor parameters are: $V_{TN} = 2\,\text{V}$, $K_{n1} = K_{n2} = 0.25\,\text{mA/V}^2$, $K_{n3} = 0.10\,\text{mA/V}^2$, and $\lambda = 0$. Determine I_{REF} and I_O. (Note: All transistors labeled M_2 are identical.) (Ans. $I_{REF} = 1.35\,\text{mA}$, $I_O = 4.04\,\text{mA}$)

D10.13 For the circuit shown in Figure 10.17, $V^+ = 10\,\text{V}$ and $V^- = 0$, and the transistor parameters are: $V_{TN} = 2\,\text{V}$, $\frac{1}{2}\mu_n C_{ox} = 20\,\mu\text{A/V}^2$, and $\lambda = 0$. Design the circuit such that $I_{REF} = 0.5\,\text{mA}$ and $I_O = 0.2\,\text{mA}$, and M_2 remains biased in the saturation region for $V_{DS2} \geq 1\,\text{V}$. (Ans. $(W/L)_2 = 10$, $(W/L)_1 = 25$, $(W/L)_3 = 1$)

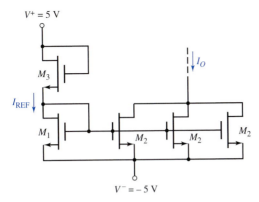

Figure 10.18 Figure for Exercise 10.12

10.2.2 Multi-MOSFET Current-Source Circuits

Cascode Current Mirror

In MOSFET current-source circuits, the output resistance is a measure of the stability with respect to changes in the output voltage. This output resistance can be increased by modifying the circuit, as shown in Figure 10.19, which is a **cascode current mirror**. The reference current is established by including another MOSFET in the reference branch of the circuit as was done in the basic two-transistor current mirror. Assuming all transistors are identical, then $I_O = I_{REF}$.

To determine the output resistance at the drain of M_4, we use the small-signal equivalent circuit. Since I_{REF} is a constant, the gate voltages to M_1 and M_3, and hence to M_2 and M_4, are constant. This is equivalent to an ac short circuit. The ac equivalent circuit for calculating the output resistance is shown

Figure 10.19 MOSFET cascode current mirror

in Figure 10.20(a). The small-signal equivalent circuit is given in Figure 10.20(b). The small-signal resistance looking into the drain of M_2 is r_{O2}.

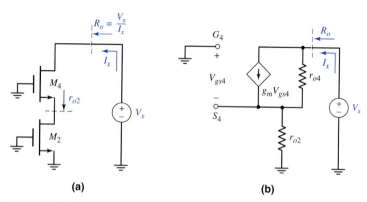

Figure 10.20 Equivalent circuits of the MOSFET cascode current mirror for determining output resistance

Writing a KCL equation, in phasor form, at the output node, we have

$$I_x = g_m V_{gs4} + \frac{V_x - (-V_{gs4})}{r_{o4}} \tag{10.54}$$

Also,

$$V_{gs4} = -I_x r_{o2} \tag{10.55}$$

Substituting Equation (10.55) into (10.54), we obtain

$$I_x + \frac{r_{o2}}{r_{o4}} I_x + g_m r_{o2} I_x = \frac{V_x}{r_{o4}} \tag{10.56}$$

The output resistance is then

$$R_o = \frac{V_x}{I_x} = r_{o4} + r_{o2}(1 + g_m r_{o4}) \tag{10.57}$$

Normally, $g_m r_{o4} \gg 1$, which implies that the output resistance of this cascode configuration is much larger than that of the basic two-transistor current source.

Example 10.8 **Objective:** Compare the output resistance of the cascode MOSFET current source to that of the two-transistor current source.

Consider the two-transistor current source in Figure 10.17 and the cascode current source in Figure 10.19. Assume $I_{REF} = I_O = 100\,\mu A$ in both circuits, $\lambda = 0.01\ \text{V}^{-1}$ for all transistors, and $g_m = 0.5\,\text{mA/V}$.

Solution: The output resistance of the two-transistor current source is, from Equation (10.48)

$$r_o = \frac{1}{\lambda I_{REF}} = \frac{1}{(0.01)(0.10)} \Rightarrow 1\,\text{M}\Omega$$

For the cascode circuit, we have $r_{o2} = r_{o4} = 1\,\text{M}\Omega$. Therefore, the output resistance of the cascode circuit is, from Equation (10.57),

$$R_o = r_{o4} + r_{o2}(1 + g_m r_{o4}) = 1 + (1)[1 + (0.5 \times 10^{-3})(10^6)]$$

or

$$R_o = 502\,\text{M}\Omega$$

Comment: The output resistance of the cascode current source is substantially larger than that of the basic two-transistor circuit. Since $dI_O \propto 1/R_o$, the load current in the cascode circuit is more stable against variations in output voltage.

Design Pointer: Achieving the output resistance of $502\,\text{M}\Omega$ assumes the transistors are ideal. In fact, small leakage currents will begin to be a factor in actual output resistance values, so a value of $502\,\text{M}\Omega$ may not be achieved in reality.

Test Your Understanding

10.14 In the MOSFET cascode current source shown in Figure 10.19, all transistors are identical, with parameters: $V_{TN} = 1\,\text{V}$, $K_n = 80\,\mu\text{A/V}^2$, and $\lambda = 0.02\,\text{V}^{-1}$. Let $I_{REF} = 20\,\mu\text{A}$. The circuit is biased at $V^+ = 5\,\text{V}$ and $V^- = -5\,\text{V}$. Determine: (a) V_{GS} of each transistor, (b) the lowest possible voltage value V_{D4}, and (c) the output resistance R_o. (Ans. (a) $V_{GS} = 1.5\,\text{V}$ (b) $V_{D4}(\text{min}) = -3.0\,\text{V}$ (c) $R_o = 505\,\text{M}\Omega$)

Wilson Current Mirror

Two additional multi-MOSFET current sources are shown in Figures 10.21(a) and 10.21(b). The circuit in Figure 10.21(a) is the **Wilson current source**. Note

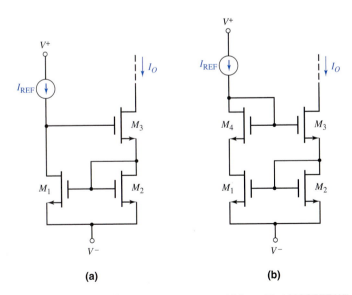

Figure 10.21   (a) MOSFET Wilson current source and (b) modified MOSFET Wilson current source

that the V_{DS} values of M_1 and M_2 are not equal. Since λ is not zero, the ratio I_O/I_{REF} is slightly different from the aspect ratios. This problem is solved in the **modified Wilson current source**, shown in Figure 10.21(b), which includes transistor M_4. For a constant reference current, the drain-to-source voltages of M_1, M_2, and M_4 are held constant. The primary advantage of these circuits is the increase in output resistance, which further stabilizes the load current.

Test Your Understanding

10.15 For the transistors in the MOSFET Wilson current source in Figure 10.21(a), the parameters are: $V_{TN} = 1\,V$, $\lambda = 0$, and $K_{n1} = 2K_{n2} = K_{n3} = 0.15\,mA/V^2$. If $I_{REF} = 200\,\mu A$, determine I_O and V_{GS} for each transistor. (Ans. $V_{GS1} = V_{GS2} = 2.15\,V$, $I_O = 0.10\,mA$, $V_{GS3} = 1.82\,V$)

10.16 All transistors in the MOSFET modified Wilson current source in Figure 10.21(b) are identical. The parameters are: $V_{TN} = 1\,V$, $K_n = 0.2\,mA/V^2$, and $\lambda = 0$. If $I_{REF} = 250\,\mu A$, determine I_O and V_{GS} for each transistor. (Ans. $I_O = I_{REF} = 250\,\mu A$, $V_{GS} = 2.12\,V$)

Wide-Swing Current Mirror

If we consider the cascode current mirror in Figure 10.17, we can determine the minimum value of V_{D4}, which will influence the maximum symmetrical swing of the voltage in the load circuit being biased. The gate voltage of M_4 is

$$V_{G4} = V^- + V_{GS1} + V_{GS2} \tag{10.58}$$

The minimum V_{D4} is then

$$V_{D4}(\text{min}) = V_{G4} - V_{GS4} + V_{DS4}(\text{sat}) \tag{10.59}$$

Assuming matched transistors, $V_{GS1} = V_{GS2} = V_{GS4} \equiv V_{GS}$. We then find

$$V_{D4}(\text{min}) = V^- + \left(V_{GS} + V_{DS4}(\text{sat})\right) \tag{10.60}$$

In considering the simple two-transistor current mirror, the minimum output voltage is

$$V_O(\text{min}) = V^- + V_{DS}(\text{sat}) \tag{10.61}$$

If, for example, $V_{GS} = 0.75\,V$ and $V_{TN} = 0.50\,V$, then from Equation (10.60), $V_{D4}(\text{min}) = 1.0\,V$ above V^-, and from Equation (10.61), $V_O(\text{min})$ is only 0.25 V above V^-. For bias voltages in the range of $\pm 3.5\,V$, this additional required voltage across the output of the cascode current mirror can have a significant effect on the output of the load circuit.

One current mirror circuit that does not limit the output voltage swing as severely as the cascode circuit, but retains the high output resistance, is shown in Figure 10.22. Width-to-length ratios of the transistors are shown. Otherwise, the transistors are assumed to be identical.

The transistor pair M_3 and M_4 acts like a single diode-connected transistor in creating the gate voltage for M_3. By including M_4, the drain-to-source voltage of M_3 is reduced and is matched to the drain-to-source voltage of

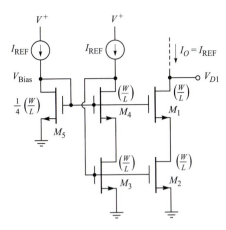

Figure 10.22   A wide-swing MOSFET cascode current mirror

M_2. Since M_5 is one-fourth the size of $M_1 - M_4$ and since all drain currents are equal, we have

$$(V_{GS5} - V_{TN}) = 2(V_{GSi} - V_{TN}) \tag{10.62}$$

where V_{GSi} corresponds to the gate-to-source voltage of $M_1 - M_4$.

The voltage at the gate of M_1 is

$$V_{G1} = V_{GS5} = (V_{GS5} - V_{TN}) + V_{TN} \tag{10.63}$$

The minimum output voltage at the drain of M_1 is

$$
\begin{aligned}
V_{D1}(\text{min}) &= V_{G1} - V_{GS1} + V_{DS1}(\text{sat}) \\
&= [(V_{GS5} - V_{TN}) + V_{TN}] - V_{GS1} + (V_{GS1} - V_{TN})
\end{aligned}
\tag{10.64}
$$

or

$$V_{D1}(\text{min}) = V_{GS5} - V_{TN} = 2(V_{GSi} - V_{TN}) = 2V_{DSi}(\text{sat}) \tag{10.65}$$

If we have $V_{GSi} = 0.75\,\text{V}$ and $V_{TN} = 0.5\,\text{V}$, then $V_{D1}(\text{min}) = 0.50\,\text{V}$, which is one-half the value for the cascode circuit. At the same time, the high output resistance is maintained.

Discussion: In the ideal circuit design in Figure 10.22, the transistors M_3 and M_4 are biased exactly at the transition point between the saturation and non-saturation regions. The analysis has neglected the body effect, so threshold voltages will not be exactly equal. In an actual circuit design, therefore, the size of M_5 will be made slightly smaller to ensure transistors are biased in the saturation region. This design change then means that the minimum output voltage increases by perhaps 0.1 to 0.15 V.

10.2.3   Bias-Independent Current Source

In all of the current mirror circuits considered up to this point (both BJT and MOSFET), the reference current is a function of the applied supply voltages. This implies that the load current is also a function of the supply voltages. In most cases, the supply voltage dependence is undesirable. Circuit designs exist in which the load currents are essentially independent of the bias. One such MOSFET circuit is shown in Figure 10.23. The width-to-length ratios are given.

Since the PMOS devices are matched, the currents I_{D1} and I_{D2} must be equal. Equating the currents in M_1 and M_2, we find

$$I_{D1} = \frac{k_n'}{2}\left(\frac{W}{L}\right)_1 (V_{GS1} - V_{TN})^2 = I_{D2} = \frac{k_n'}{2}\left(\frac{W}{L}\right)_2 (V_{GS2} - V_{TN})^2 \qquad \textbf{(10.66)}$$

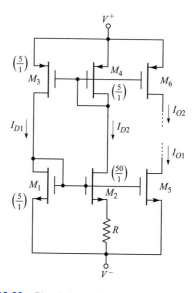

Figure 10.23 Bias-independent MOSFET current mirror

Also

$$V_{GS2} = V_{GS1} - I_{D2}R \qquad \textbf{(10.67)}$$

Substituting Equation (10.67) into Equation (10.66) and solving for R, we obtain

$$R = \frac{1}{\sqrt{K_{n1}I_{D1}}}\left(1 - \sqrt{\frac{(W/L)_1}{(W/L)_2}}\right) \qquad \textbf{(10.68)}$$

This value of resistance R will establish the drain currents $I_{D1} = I_{D2}$. These currents establish the gate-to-source voltage across M_1 and source-to-gate voltage across M_3. These voltages, in turn, can be applied to M_5 and M_6 to establish load currents I_{O1} and I_{O2}.

The currents I_{D1} and I_{D2} are independent of the supply voltages V^+ and V^- as long as M_2 and M_3 are biased in the saturation region. As the difference, $V^+ - V^-$, increases, the values of V_{DS2} and V_{SD3} increase but the currents remain essentially constant.

Similar bipolar bias-independent current mirror designs exist, but will not be covered here.

10.2.4   JFET Current Sources

Current sources are also fundamental elements in JFET integrated circuits. The simplest method of forming a current source is to connect the gate and source terminals of a depletion-mode JFET, as shown in Figure 10.24 for an n-channel device. The device will remain biased in the saturation region as long as

$$v_{DS} \geq v_{DS}(\text{sat}) = v_{GS} - V_P = |V_P| \tag{10.69}$$

In the saturation region, the current is

$$i_D = I_{DSS}\left(1 - \frac{v_{GS}}{V_P}\right)^2 (1 + \lambda v_{DS}) = I_{DSS}(1 + \lambda v_{DS}) \tag{10.70}$$

The output resistance looking into the drain is, from Equation (10.70),

$$\frac{1}{r_o} = \frac{di_D}{dv_{DS}} = \lambda I_{DSS} \tag{10.71}$$

This expression for the output resistance of a JFET current source is the same as that of the MOSFET current source.

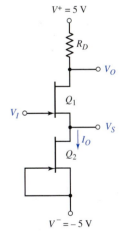

Figure 10.24 Depletion-mode JFET connected as a current source

---

Example 10.9 **Objective:** Determine the currents and voltages in a simple JFET circuit biased with a constant-current source.

Consider the circuit shown in Figure 10.25. The transistor parameters are: $I_{DSS1} = 2\,\text{mA}$, $I_{DSS2} = 1\,\text{mA}$, $V_{P1} = V_{P2} = -1.5\,\text{V}$, and $\lambda_1 = \lambda_2 = 0.05\,\text{V}^{-1}$. Determine the minimum values of V_S and V_I such that Q_2 is biased in the saturation region. What is the value of I_O?

Solution: In order for Q_2 to remain biased in the saturation region, we must have $v_{DS} \geq |V_P| = 1.5\,\text{V}$, from Equation (10.69). The minimum value of V_S is then

$$V_S(\text{min}) - V^- = v_{DS}(\text{min}) = 1.5\,\text{V}$$

or

$$V_S(\text{min}) = 1.5 + V^- = 1.5 + (-5) = -3.5\,\text{V}$$

From Equation (10.70), the output current is

$$i_D = I_O = I_{DSS2}(1 + \lambda v_{DS}) = (1)[1 + (0.05)(1.5)] = 1.08\,\text{mA}$$

As a first approximation in calculating the minimum value of V_I, we neglect the effect of λ in transistor Q_1. Then, assuming Q_1 is biased in the saturation region, we have

$$i_D = I_{DSS1}\left(1 - \frac{v_{GS1}}{V_{P1}}\right)^2$$

or

$$1.08 = 2\left(1 - \frac{v_{GS1}}{(-1.5)}\right)^2$$

which yields

$$v_{GS1} = -0.40\,\text{V}$$

We see that

$$v_{GS1} = -0.40\,\text{V} = V_I - V_S = V_I - (-3.5)$$

Figure 10.25 The dc equivalent circuit of simple JFET amplifier biased with JFET current source

or

$$V_I = -3.90 \, \text{V}$$

Comment: Since Q_1 is an n-channel device, the voltage at the gate is negative with respect to the source.

Figure 10.26 JFET cascode current source

The output resistance of a JFET current source can be increased by using a cascode configuration. A simple JFET cascode current source with two n-channel depletion-mode devices is shown in Figure 10.26. The current–voltage relationship, assuming Q_1 and Q_2 are identical, is given by

$$i_D = I_{DSS}(1 + \lambda v_{DS1}) = I_{DSS}\left(1 - \frac{v_{GS2}}{V_P}\right)^2 (1 + \lambda v_{DS2}) \qquad \textbf{(10.72)}$$

From the circuit, we see that $v_{GS2} = -v_{DS1}$. We define

$$V_{DS} = v_{DS1} + v_{DS2} \qquad \textbf{(10.73(a))}$$

so that

$$v_{DS2} = V_{DS} - v_{DS1} \qquad \textbf{(10.73(b))}$$

From Equation (10.72), we obtain

$$(1 + \lambda v_{DS1}) = \left(1 + \frac{v_{DS1}}{V_P}\right)^2 [1 + \lambda(V_{DS} - v_{DS1})] \qquad \textbf{(10.74)}$$

For a given application, the value of V_{DS} will usually be known, and the value of v_{DS1} can then be determined. The load current i_D can then be calculated by using Equation (10.72).

We can determine the output resistance by using the small-signal equivalent circuit of the composite two-transistor configuration, as shown in Figure 10.27(a), which includes the phasor variables. Since the gate and source of Q_1 are connected together, the small-signal voltage V_{gs1} is zero, which means that

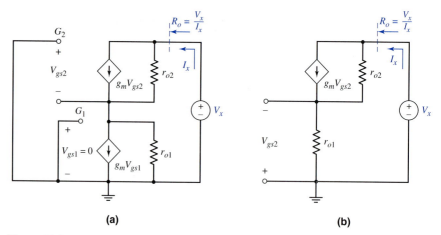

(a)   (b)

Figure 10.27 (a) Equivalent circuit, using phasor notation, of the JFET cascode current source for determining output resistance and (b) final configuration the dependent current source $g_m V_{gs1}$ is zero. This corresponds to an open circuit. Figure 10.27(b) shows the final configuration.

The analysis is the same as for the MOSFET cascode circuit in Figure 10.17. Writing a KCL equation at the output node, we have

$$I_x = g_m V_{gs2} + \frac{V_x - (-V_{gs2})}{r_{o2}} \qquad \text{(10.75)}$$

Noting that

$$V_{gs2} = -I_x r_{o1} \qquad \text{(10.76)}$$

Equation (10.75) becomes

$$I_x = -(g_m r_{o1})I_x + \frac{V_x}{r_{o2}} - \left(\frac{r_{o1}}{r_{o2}}\right)I_x \qquad \text{(10.77)}$$

The output resistance is then

$$R_o = \frac{V_x}{I_x} = r_{o2} + r_{o1} + g_m r_{o1} r_{o2} = r_{o2} + r_{o1}(1 + g_m r_{o2}) \qquad \text{(10.78)}$$

From Equation (10.78), we see that the output resistance relationship for the JFET cascode current source has the same form as that of the MOSFET cascode current source.

Test Your Understanding

*10.17 Consider the JFET circuit in Figure 10.25. The transistor parameters are: $I_{DSS2} = 0.5\,\text{mA}$, $I_{DSS1} = 0.8\,\text{mA}$, $V_{P1} = V_{P2} = -2\,\text{V}$, and $\lambda_1 = \lambda_2 = 0.15\,\text{V}^{-1}$. Determine the minimum values of V_S and V_I such that Q_2 is biased in the saturation region. What is the value of I_O? What is the output impedance looking into the drain of Q_2? (Ans. $V_S(\text{min}) = -3\,\text{V}$, $I_O = 0.65\,\text{mA}$, $V_I(\text{min}) = -3.2\,\text{V}$, $r_o = 1.09\,\text{k}\Omega$)

*10.18 The JFET cascode circuit in Figure 10.26 has identical transistors, with parameters $I_{DSS} = 2\,\text{mA}$, $V_P = -2\,\text{V}$, and $\lambda = 0.1\,\text{V}^{-1}$. If $V_{DS} = v_{DS1} + v_{DS2} = 3\,\text{V}$, determine: (a) v_{DS1}, v_{GS2}, v_{DS2}, and i_D, and (b) the output impedance R_o. (Ans. (a) $v_{DS1} = 0.212\,\text{V}$, $v_{GS2} = -0.212\,\text{V}$, $v_{DS2} = 2.79\,\text{V}$, $i_D = 2.04\,\text{mA}$ (b) $R_o = 54.8\,\text{k}\Omega$)

10.3   CIRCUITS WITH ACTIVE LOADS

In bipolar amplifiers, such as that shown in Figure 10.28, the small-signal voltage gain is directly proportional to the collector resistor R_C. To increase the gain, we need to increase the value of R_C. Typically, circuit parameters are $I_C = 0.5\,\text{mA}$, $V_{CC} = 10\,\text{V}$, and $R_C = 10\,\text{k}\Omega$. However, if we increase R_C, we would also have to increase the supply voltage V_{CC} if we want to maintain the same collector current and collector–emitter voltage. In practice, there is a limited range of values of R_C and V_{CC} that are reasonable.

To get around this limitation, we need a load device that will pass a current of 0.5 mA at a voltage of 5 V, but which will resist a change in current better than a 10 kΩ resistor. This load device can be a transistor, which will also occupy less area in an integrated circuit, another advantage in using transistors

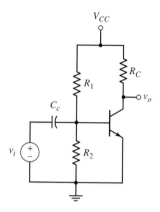

Figure 10.28 Bipolar common-emitter circuit

in place of resistors. In addition, active loads produce a much larger small-signal gain than discrete resistors, as discussed in Chapter 4.

In Chapter 6, we introduced NMOS enhancement load and depletion load devices in MOSFET amplifiers. This was an introduction to active load devices. In this section, we consider the dc analysis of a bipolar active load in a simple BJT circuit and then the dc analysis of a MOSFET active load. Our discussion will include the voltage gains of these active load circuits. The small-signal analysis of active load circuits is covered in the next section.

The discussion of active loads here can be considered an introduction. The use of active loads with differential amplifiers is considered in detail in the next chapter.

10.3.1   DC Analysis: BJT Active Load Circuit

Consider the circuit shown in Figure 10.29. The elements R_1, Q_1, and Q_2 form the active load circuit, and Q_2 is referred to as the **active load device** for driver transistor Q_0. The combination of R_1, Q_1, and Q_2 forms the pnp version of the two-transistor current mirror. For the dc analysis of this circuit, we will use the dc symbols for the currents and voltages. The objective of this analysis is to

Figure 10.29 Simple BJT amplifier with active load, showing currents and voltages

obtain the voltage transfer function V_O versus V_I.

The B–E voltage of Q_0 is the dc input voltage V_I; therefore, the collector current in Q_0 is

$$I_{C0} = I_{S0}\left[e^{V_{i}/V_T}\right]\left(1 + \frac{V_{CE0}}{V_{AN}}\right) \tag{10.79}$$

where I_{S0} is the reverse-saturation current, V_T is the thermal voltage, and V_{AN} is the Early voltage of the npn transistor. Similarly, the collector current in Q_2 is

$$I_{C2} = I_{S2}\left[e^{V_{EB2}/V_T}\right]\left(1 + \frac{V_{EC2}}{V_{AP}}\right) \tag{10.80}$$

where V_{AP} is the Early voltage of the pnp transistors.

If we neglect base currents, then

$$I_{\text{REF}} = I_{C1} = I_{S1}\left[e^{V_{EB1}/V_T}\right]\left(1 + \frac{V_{EC1}}{V_{AP}}\right) \tag{10.81}$$

Assuming Q_1 and Q_2 are identical, then $I_{S1} = I_{S2}$ and the Early voltages of the pnp transistors are equal. Also note that $V_{EC1} = V_{EB1} = V_{EB2}$. We can also assume that $V_{CE} \ll V_{AN}$ and $V_{EC} \ll V_{AP}$. Combining equations, we find the output voltage is given as

$$V_O = \frac{V_{AN}V_{AP}}{V_{AN} + V_{AP}}\left[1 - \frac{I_{S0}e^{V_I/V_T}}{I_{\text{REF}}}\right] + \frac{V_{AN}}{V_{AN} + V_{AP}}(V^+ - V_{EB2}) \tag{10.82}$$

Equation (10.82) is valid as long as Q_0 and Q_2 remain biased in the forward-active region, which means that the output voltage must remain in the range

$$V_{CE0}(\text{sat}) < V_O < \left(V^+ - V_{EC2}(\text{sat})\right) \tag{10.83}$$

A sketch of V_O versus V_I is shown in Figure 10.30. If the circuit is to be used as a small-signal amplifier, a Q-point must be established, as indicated in the figure, for maximum symmetrical swing. Because of the exponential input voltage function, as given in Equation (10.82), the input voltage range over which both Q_0 and Q_2 remain in their active regions is very small. A sinusoidal variation in the input voltage produces a sinusoidal variation in the output voltage as shown in the figure.

Figure 10.30   Voltage transfer characteristics of bipolar circuit with active load

In addition to the voltage transfer function, we can also consider the load curve. Figure 10.31 shows the transistor characteristics of the driver transistor Q_0 for several values of B–E or V_I voltages. Superimposed on these curves is the load curve, which essentially is the I_C versus V_{EC} characteristic of the active load Q_2 at a constant V_{EB} voltage.

The Q-point shown corresponds to a quiescent input voltage V_{IQ}. From the curve, we see that as the input changes between V_{IH} and V_{IL}, the Q-point moves up and down the load curve producing a change in output voltage. Also, as V_I increases to V_{I2}, the driver transistor Q_0 is driven into saturation; as V_I decreases to V_{I1}, the load transistor Q_2 is driven into saturation.

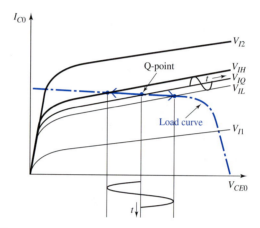

Figure 10.31   Driver transistor characteristics and load curve for BJT circuit with active load

10.3.2   Voltage Gain: BJT Active Load Circuit

The small-signal voltage gain of a circuit is the slope of the voltage transfer function curve at the Q-point. For the bipolar circuit with an active load, the voltage gain can be found by taking the derivative of Equation (10.82) with respect to V_I, as follows:

$$A_v = \frac{dV_O}{dV_I} = -\left(\frac{V_{AN}V_{AP}}{V_{AN}+V_{AP}}\right)\left(\frac{I_{S0}}{I_{REF}}\right)\left(\frac{1}{V_T}\right)e^{V_I/V_T} \tag{10.84}$$

As a good approximation, we can write that

$$I_{REF} \cong I_{S0}e^{V_I/V_T} \tag{10.85}$$

Equation (10.84) then becomes

$$A_v = \frac{dV_O}{dV_I} = -\left(\frac{V_{AN}V_{AP}}{V_{AN}+V_{AP}}\right)\left(\frac{1}{V_T}\right) = \frac{-\left(\dfrac{1}{V_T}\right)}{\dfrac{1}{V_{AN}}+\dfrac{1}{V_{AP}}} \tag{10.86}$$

The small-signal voltage gain is a function of the Early voltages and the thermal voltage. The voltage gain, given by Equation (10.86), relates to the open-circuit condition. When a load is connected to the output, the voltage gain is degraded, as we will see in the next section.

Example 10.10   **Objective:** Calculate the open-circuit voltage gain of a simple BJT amplifier with an active load.

Consider the circuit shown in Figure 10.29. The transistor parameters are $V_{AN} = 120$ V and $V_{AP} = 80$ V. Let $V_T = 0.026$ V.

Solution: From Equation (10.86), the small-signal, open-circuit voltage gain is

$$A_v = \frac{-\left(\dfrac{1}{V_T}\right)}{\dfrac{1}{V_{AN}} + \dfrac{1}{V_{AP}}} = \frac{-\left(\dfrac{1}{0.026}\right)}{\dfrac{1}{120} + \dfrac{1}{80}} = \frac{-38.46}{0.00833 + 0.0125} = -1846$$

Comment: For a circuit with an active load, the magnitude of the small-signal, open-circuit voltage gain is substantially larger than the resulting gain when a discrete resistor load is used.

Computer Verification: The voltage transfer characteristics of the active load circuit in Figure 10.29 were determined for a standard 2N3904 transistor as the npn device and standard 2N3906 transistors as the pnp devices. The circuit was biased at 5 V and the resistor was set at $R = 1$ kΩ. The transfer curve is shown in Figure 10.32.

The input transition region, during which both Q_0 and Q_2 remain biased in the forward-active mode, is indeed very narrow. The slope of the curve, which is the voltage gain, is found to be -572. The reason for the smaller value compared to the hand calculation is that the Early voltages of these standard transistors are smaller than assumed in the previous calculation. The Early voltage of the npn device is 74 V and that of the npn devices is only 18.7 V.

Design Pointer: From the transfer characteristics in Figure 10.32, we can see that, for this circuit, it would be very difficult to apply the required input voltage to bias both Q_0 and Q_2 in the active region. This particular circuit, therefore, is not practical as an amplifier. However, the circuit does demonstrate the basic properties of an active load. In Chapters 11 and 13, we will see how an active load is applied to actual circuits.

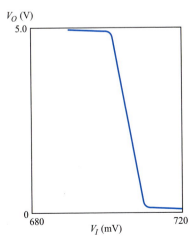

Figure 10.32   Graphical output from a PSpice analysis, showing voltage transfer characteristics of bipolar active load circuit

Test Your Understanding

*10.19 A simple BJT amplifier with active load is shown in Figure 10.29. The transistor parameters are: $I_{S0} = I_{S1} = I_{S2} = 10^{-12}$ A and $V_{AN} = V_{AP} = 100$ V. Let V^+ = 5 V. (a) Determine the value of V_{EB2} such that $I_{REF} = 0.5$ mA. (b) Find the value of R_1. (c) What value of V_I will produce $V_{CE0} = V_{EC2}$? (d) Determine the open-circuit, small-signal voltage gain. (Ans. (a) $V_{EB2} = 0.521$ V (b) $R_1 = 8.96$ kΩ (c) $V_I = 0.521$ V (d) $A_V = -1923$)

*10.20 Repeat Exercise 10.19 if the transistor parameters are $I_{S0} = I_{S1} = I_{S2} = 5 \times 10^{-14}$ A, and if I_{REF} is 0.1 mA. Verify the results with a PSpice analysis. (Ans. (a) $V_{EB2} = 0.557$ V (b) $R_1 = 44.4$ kΩ (c) $V_I = 0.557$ V (d) $A_V = -1923$)

10.3.3   DC Analysis: MOSFET Active Load Circuit

Consider the circuit in Figure 10.33. Transistors M_1 and M_2 form a PMOS active load circuit, and M_2 is the active load device. We will consider the voltage transfer function of V_O versus V_I for this circuit.

Figure 10.33   Simple MOSFET amplifier with active load, showing currents and voltages

The reference current may be written in the form

$$I_{REF} = K_{p1}(V_{SG} + V_{TP1})^2(1 + \lambda_1 V_{SD1}) \tag{10.87}$$

The drain current I_2 is

$$I_2 = K_{p2}(V_{SG} + V_{TP2})^2(1 + \lambda_2 V_{SD2}) \tag{10.88}$$

If we assume that M_1 and M_2 are identical, then $\lambda_1 = \lambda_2 \equiv \lambda_p$, $V_{TP1} = V_{TP2} \equiv V_{TP}$, and $K_{p1} = K_{p2} \equiv K_p$. Combining equations, we find the output voltage as

$$V_O = \frac{[1 + \lambda_p(V^+ - V_{SG})]}{\lambda_n + \lambda_p} - \frac{K_n(V_I - V_{TN})^2}{I_{REF}(\lambda_n + \lambda_p)} \tag{10.89}$$

Equation (10.89) describes the V_O versus V_I characteristic of the circuit, provided that both M_0 and M_2 remain biased in their saturation regions. Figure 10.34 shows a sketch of the voltage transfer characteristics. If the circuit is to be used as a small-signal amplifier, then a Q-point must be established, as indicated on the figure, for maximum symmetrical swing. As before, the input transition region in which both M_0 and M_2 are biased in the saturation region is quite narrow. A sinusoidal variation in the input voltage produces a sinusoidal variation in the output voltage as shown in the figure.

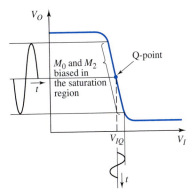

Figure 10.34   Voltage transfer characteristic of MOSFET circuit with active load

We can also consider the load curve for this device. Figure 10.35 shows the transistor characteristics of the driver transistor M_0 for several values of gate-to-source or V_I voltages. Superimposed on these curves is the load curve, which essentially is the I_D versus V_{SD} characteristic of the active load M_2 at a constant V_{SG} voltage.

The Q-point shown corresponds to a quiescent input voltage V_{IQ}. From the curve, we see that as the input changes between V_{IH} and V_{IL}, the Q-point moves up and down the load curve producing a change in output voltage. Also, as V_I increases to V_{I2}, the driver transistor M_0 is driven into the nonsaturation

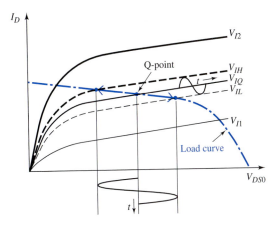

Figure 10.35   Driver transistor characteristics and load curve for MOSFET circuit with active load

region; as V_I decreases to V_{I1}, the load transistor M_2 is driven into the non-saturation region.

10.3.4   Voltage Gain: MOSFET Active Load Circuit

The small-signal voltage gain of a MOSFET circuit with an active load is also the slope of the voltage transfer function curve at the Q-point. Taking the derivative of Equation (10.89) with respect to V_I, we obtain

$$A_v = \frac{dV_O}{dV_I} = \frac{-2K_n(V_I - V_{TN})}{I_{\text{REF}}(\lambda_n + \lambda_p)} \tag{10.90}$$

The transconductance of the driver transistor is $g_m = 2K_n(V_I - V_{TN})$. Since M_1 and M_2 are assumed to be identical, then $I_O = I_{\text{REF}}$, and the small-signal transistor resistances are $r_{on} = 1/\lambda_n I_{\text{REF}}$ and $r_{op} = 1/\lambda_p I_{\text{REF}}$. From Equation (10.90), the small-signal, open-circuit voltage gain can now be written

$$A_v = \frac{-g_m}{\left(\dfrac{1}{r_{on}} + \dfrac{1}{r_{op}}\right)} = -g_m(r_{on}\|r_{op}) \tag{10.91}$$

In general, the transconductance g_m of a MOSFET is less than that of a BJT; therefore, the voltage gain of a MOSFET amplifier with an active load is less than that of a BJT amplifier with an active load. However, the active load still produces a significant increase in the voltage gain.

10.3.5   Discussion

In considering the BJT circuit with active load (Figure 10.29) and MOSFET circuit with active load (Figure 10.33), we could have directly considered the small-signal analysis without the dc analysis. However, it is important to understand how narrow the input transition width is (Figure 10.32) such that the transistors are biased correctly. For this reason, the use of active loads in discrete circuits is almost impossible. The biasing of the circuit with an active load depends to a large extent on the use of matched transistors. Matched transistors can be achieved on an integrated circuit. So in considering the small-signal analysis in the next section, we must keep in mind the very narrow range in which the transistors are biased in the active region.

Test Your Understanding

10.21   Consider the simple MOSFET amplifier with active load in Figure 10.33. The transistor parameters are: $V_{TN} = 1\,\text{V}$, $V_{TP} = -1\,\text{V}$, $K_p = K_n = 0.2\,\text{mA/V}^2$, and $\lambda_p = \lambda_n = 0.015\,\text{V}^{-1}$. Let $V^+ = 10\,\text{V}$ and $I_{\text{REF}} = 0.25\,\text{mA}$. (a) Find V_{SG}. (b) What value of V_I will produce $V_{DS0} = V_{SD2}$? (c) Determine the open-circuit, small-signal voltage gain. (Ans. (a) $V_{SG} = 2.12\,\text{V}$ (b) $V_I = 2.10\,\text{V}$ (c) $A_v = -58.7$)

10.22   Repeat Exercise 10.21 if the transistor parameters are $K_p = K_n = 50\,\mu\text{A/V}^2$, and if I_{REF} is $80\,\mu\text{A}$. Verify the results with a PSpice analysis (Ans. (a) $V_{SG} = 2.26\,\text{V}$ (b) $V_I = 2.24\,\text{V}$ (c) $A_v = -51.7$)

10.4   SMALL-SIGNAL ANALYSIS: ACTIVE LOAD CIRCUITS

The small-signal voltage gain of a circuit with an active load can be determined from the small-signal equivalent circuit. This is probably the easiest and most direct method of obtaining the gain of such circuits. Again, the dc analysis of these circuits, as shown in the previous section, clearly demonstrates the narrow range of input voltages over which the transistors will remain biased in the active region. The load curves in Figure 10.31 for the BJT circuit and in Figure 10.35 for the MOSFET circuit also help in visualizing the operation of these circuits. Even though a small-signal analysis is extremely useful for determining the voltage gain, we must not lose sight of the physical operation of these circuits, which is described through the dc analysis. If the BJTs are not biased in the active region or the MOSFETs are not biased in the saturation region, the small-signal analysis is not valid.

10.4.1   Small-Signal Analysis: BJT Active Load Circuit

To find the small-signal voltage gain of the BJT circuit with an active load, we must determine the resistance looking into the collector of the active load device. Figure 10.36 is the small-signal equivalent circuit of the entire active load circuit in Figure 10.29, which uses pnp transistors. The base, collector, and emitter terminals of the two transistors are indicated on the figure.

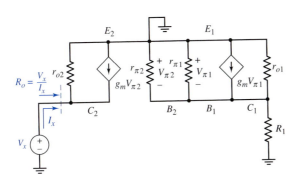

Figure 10.36   Small-signal equivalent circuit of BJT active load circuit

In the Q_1 portion of the equivalent circuit, there are no independent ac sources to excite any currents or voltages. Therefore, $V_{\pi 1} = V_{\pi 2} = 0$, which means that the dependent source $g_m V_{\pi 2}$ is zero and is equivalent to an open circuit. The resistance looking into the collector of Q_2 is just

$$R_o = r_{o2} \tag{10.92}$$

We will use this equivalent resistance to calculate the small-signal voltage gain of the amplifier.

Figure 10.37(a) shows a simple amplifier with an active load and the output voltage capacitively coupled to passive load R_L. The small-signal equivalent circuit, shown in Figure 10.37(b), includes the load resistance R_L, the resistance r_{o2} of the active load, and the output resistance r_o of the amplifying transistor Q_0.

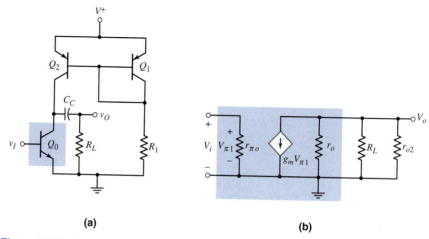

Figure 10.37   (a) Simple BJT amplifier with active load and load resistance and (b) small-signal equivalent circuit

The output voltage is

$$V_o = -(g_m V_{\pi 1})(r_o \| R_L \| r_{o2}) \tag{10.93}$$

Since $V_{\pi 1} = V_i$, where V_i is the ac input voltage, the small-signal voltage gain is

$$A_v = \frac{V_o}{V_i} = -g_m(r_o \| R_L \| r_{o2}) = \frac{-g_m}{\left(\dfrac{1}{r_o} + \dfrac{1}{R_L} + \dfrac{1}{r_{o2}}\right)} \tag{10.94}$$

The small-signal voltage gain can also be written

$$A_v = \frac{-g_m}{g_o + g_L + g_{o2}} \tag{10.95}$$

where g_o and g_{o2} are the output conductances of Q_0 and Q_2, and g_L is the load conductance. The transconductance is $g_m = I_{Co}/V_T$, the small-signal conductances are $g_o = I_{Co}/V_{AN}$ and $g_{o2} = I_{Co}/V_{AP}$, and the load conductance is $g_L = 1/R_L$. Therefore, Equation (10.95) becomes

$$A_v = \frac{-\left(\dfrac{I_{Co}}{V_T}\right)}{\left(\dfrac{I_{Co}}{V_{AN}} + \dfrac{1}{R_L} + \dfrac{I_{Co}}{V_{AP}}\right)} \tag{10.96}$$

If the passive load is an open circuit ($R_L \to \infty$), the small-signal voltage gain is identical to that determined from the dc analysis as given by Equation (10.86). If the load resistance R_L is not an open circuit, then the magnitude of the small-signal voltage gain is reduced.

Example 10.11   **Objective:** Calculate the small-signal voltage gain of an amplifier with an active load and a load resistance R_L.

For the circuit in Figure 10.37(a), the transistor parameters are $V_{AN} = 120\,\text{V}$ and $V_{AP} = 80\,\text{V}$. Let $V_T = 0.026\,\text{V}$ and $I_{Co} = 1\,\text{mA}$. Determine the small-signal voltage gain for load resistances of $R_L = \infty$, $100\,\text{k}\Omega$, and $10\,\text{k}\Omega$.

Solution: For $R_L = \infty$, Equation (10.96) reduces to

$$A_v = \frac{-\left(\dfrac{1}{V_T}\right)}{\left(\dfrac{1}{V_{AN}} + \dfrac{1}{V_{AP}}\right)} = \frac{-\left(\dfrac{1}{0.026}\right)}{\left(\dfrac{1}{120} + \dfrac{1}{80}\right)} = -1846$$

which is the same as that determined for the open-circuit configuration in Example 10.10.

For $R_L = 100\,\text{k}\Omega$, the small-signal voltage gain is

$$A_v = \frac{-\left(\dfrac{1}{0.026}\right)}{\left(\dfrac{1}{120} + \dfrac{1}{100} + \dfrac{1}{80}\right)} = \frac{-38.46}{0.00833 + 0.010 + 0.0125} = -1247$$

and for $R_L = 10\,\text{k}\Omega$, the voltage gain is

$$A_v = \frac{-\left(\dfrac{1}{0.026}\right)}{\left(\dfrac{1}{120} + \dfrac{1}{10} + \dfrac{1}{80}\right)} = \frac{-38.46}{0.00833 + 0.10 + 0.0125} = -318$$

Comment: The small-signal voltage gain is a strong function of the load resistance R_L. As the value of R_L decreases, the loading effect becomes more severe.

Design Pointer: If an amplifier with an active load is to drive another amplifier stage, the loading effect must be taken into account when the small-signal voltage gain is determined. Also, the input resistance of the next stage must be large in order to minimize the loading effect.

Problem-Solving Technique: Active Loads

1. Ensure that the active load devices are biased in the forward-active mode.
2. The small-signal analysis of the circuit with an active load then simply involves considering the output resistance looking into the output of the active load device as well as the equivalent circuit of the amplifying transistor.

Test Your Understanding

10.23 For the circuit in Figure 10.37(a), the transistor parameters are $V_{AN} = V_{AP} = 80\,\text{V}$. Let $I_{Co} = 0.8\,\text{mA}$. (a) Determine the open-circuit small-signal voltage gain. (b) Find the value of R_L that results in a voltage gain of one-half the open-circuit value. (Ans. (a) $A_v = -1540$ (b) $R_L = 50\,\text{k}\Omega$)

10.24 In the circuit shown in Figure 10.37(a), the transistor parameters are $V_{AN} = 120\,\text{V}$ and $V_{AP} = 80\,\text{V}$. Let $I_{Co} = 0.5\,\text{mA}$ and $R_L = 50\,\text{k}\Omega$. (a) Determine the small-signal parameters g_m, r_o, and r_{o2}. (b) Find the small-signal voltage gain. (Ans. (a) $g_m = 19.2\,\text{mA/V}$, $r_o = 240\,\text{k}\Omega$, $r_{o2} = 160\,\text{k}\Omega$ (b) $A_v = -631$)

10.4.2   Small-Signal Analysis: MOSFET Active Load Circuit

The small-signal voltage gain of a MOSFET amplifier with an active load can also be determined from the small-signal equivalent circuit. Figure 10.38 is the small-signal equivalent circuit of the entire MOSFET active load in Figure 10.33. The signal voltages V_{sg1} and V_{sg2} are zero, since there is no ac excitation in this part of the circuit. This means that $g_m V_{sg2} = 0$ and

$$R_o = r_{o2} \tag{10.97}$$

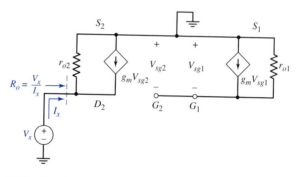

Figure 10.38 Small-signal equivalent circuit of the MOSFET active load circuit

A simple MOSFET amplifier with an active load, and a load resistor R_L capacitively coupled to the output, is shown in Figure 10.39(a). Figure 10.39(b) shows the small-signal equivalent circuit, in which the load R_L, the active load resistance r_{o2}, and the output resistance r_o of transistor M_0 are included.

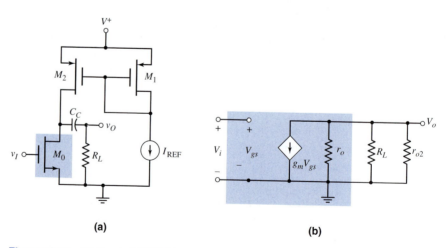

(a)

(b)

Figure 10.39 (a) Simple MOSFET amplifier with active load and load resistance and (b) small-signal equivalent circuit

The output voltage is

$$V_o = -g_m V_{gs}(r_o \| R_L \| r_{o2}) \tag{10.98}$$

and since $V_{gs} = V_i$, where V_i is the ac voltage, the small-signal voltage gain is

$$A_v = \frac{V_o}{V_i} = -g_m(r_o \| R_L \| r_{o2}) = \frac{-g_m}{g_o + g_L + g_{o2}} \tag{10.99}$$

The parameters g_o and g_{o2} are the output conductances of M_0 and M_2, and g_L is the load conductance. This expression for the small-signal voltage gain of a MOSFET amplifier with active load is the same as that of the BJT amplifier.

A load resistance R_L tends to degrade the gain and to cause a loading effect, as it did in the bipolar circuit with an active load. However, in MOSFET amplifiers, the output may be connected to the gate of another MOSFET amplifier in which the effective R_L is very large.

Example 10.12   Objective: Calculate the small-signal voltage gain of an NMOS amplifier with an active load.

For the amplifier shown in Figure 10.39(a) the transistor parameters are: $\lambda_n = \lambda_p = 0.01\,\text{V}^{-1}$, $V_{TN} = 1\,\text{V}$, and $K_n = 1\,\text{mA/V}^2$. Assume M_1 and M_2 are matched and $I_{REF} = 0.5\,\text{mA}$. Calculate the small-signal voltage gain for load resistances of $R_L = \infty$ and $100\,\text{k}\Omega$.

Solution: Since M_1 and M_2 are matched, then $I_O = I_{REF}$, and the transconductance is

$$g_m = 2\sqrt{K_n I_{REF}} = 2\sqrt{(1)(0.5)} = 1.41\,\text{mA/V}$$

The small-signal transistor conductances are

$$g_o = g_{o2} = \lambda I_{REF} = (0.01)(0.5) = 0.005\,\text{mA/V}$$

For $R_L = \infty$, Equation (10.99) reduces to

$$A_v = \frac{-g_m}{g_o + g_{o2}} = \frac{-1.41}{0.005 + 0.005} = -141$$

For $R_L = 100\,\text{k}\Omega$ ($g_L = 0.01\,\text{mA/V}$), the voltage gain is

$$A_v = \frac{-g_m}{g_o + g_L + g_{o2}} = \frac{-1.41}{0.005 + 0.01 + 0.005} = -70.5$$

Comment: The magnitude of the small-signal voltage gain of MOSFET amplifiers with active loads is substantially larger than for those with resistive loads, but it is still smaller than equivalent bipolar circuits, because of the smaller transconductance for the MOSFET.

10.4.3   Small-Signal Analysis: Advanced MOSFET Active Load

The active loads considered in the BJT (Figure 10.37) and MOSFET (Figure 10.39(a)), circuits correspond to the simple two-transistor current mirrors. We may use a more advanced current mirror with a high output resistance as an active load to increase the amplifier gain. Figure 10.40(a) shows a MOSFET cascode amplifying stage with a cascode active load. The small-signal

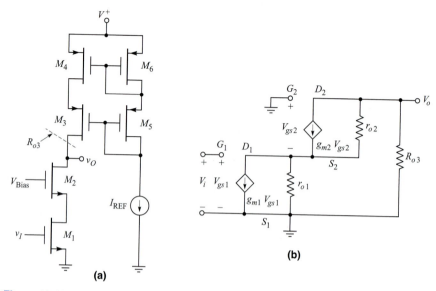

Figure 10.40   (a) MOSFET cascode amplifying stage with cascode active load; (b) small-signal equivalent circuit

equivalent circuit is shown in Figure 10.40(b), where R_{o3} is the effective resistance looking into the drain of M_3. From our discussion of the cascode current mirror, we found $R_{o3} = r_{o3} + r_{o4}(1 + g_m r_{o3})$ (Equation (10.57)).

We can assume all transistors are matched so that the currents in all transistors are equal. Summing currents at D_1, we have

$$g_m V_{gs1} + \frac{(-V_{gs2})}{r_{o1}} = g_m V_{gs2} + \frac{V_o - (-V_{gs2})}{r_{o2}} \tag{10.100}$$

Summing currents at the output node, we find

$$\frac{V_o}{R_{o3}} + \frac{V_o - (-V_{gs2})}{r_{o2}} + g_m V_{gs2} = 0 \tag{10.101}$$

Eliminating V_{gs2} from the two equations, noting that $V_{gs1} = V_i$, and assuming $g_m \gg 1/r_o$, we find the small-signal voltage gain is

$$A_v = \frac{V_o}{V_i} = \frac{-g_m^2}{\dfrac{g_m}{R_{o3}} + \dfrac{1}{r_{o1} r_{o2}}} \tag{10.102}$$

The resistance R_{o3} is approximately $R_{o3} \cong g_m r_{o3} r_{o4}$, so the gain can be written as

$$A_v = \frac{-g_m^2}{\dfrac{1}{r_{o3} r_{o4}} + \dfrac{1}{r_{o1} r_{o2}}} \tag{10.103}$$

For the same transistor parameters given in Example 10.12, the small-signal voltage gain of this circuit would be 39,762! However, a word of warning is in order. As we mentioned previously, output resistances in the hundreds of megohm range are ideal and will, in reality, be limited by leakage currents. For this reason, a voltage gain of 39,000 in a one-stage amplifier will probably not be achieved. However, the voltage gain of this amplifier should be substantially larger than the amplifier using a simple active load.

Test Your Understanding

10.25 For the circuit in Figure 10.39(a), the transistor parameters are: $\lambda_n = \lambda_p = 0.02\,\text{V}^{-1}$, $K_n = K_p = 0.25\,\text{mA/V}^2$, $V_{TN} = 1\,\text{V}$, and $V_{TP} = -1\,\text{V}$. Let $V^+ = 10\,\text{V}$ and $I_{REF} = 0.40\,\text{mA}$. (a) Determine V_{IQ}. (b) Find the open-circuit small-signal voltage gain. (c) Find the value of R_L that results in a voltage gain of one-half the open-circuit value. (Ans. (a) $V_{IQ} = 2.26\,\text{V}$ (b) $A_v = -39.4$ (c) $R_L = 62.5\,\text{k}\Omega$)

10.26 In the circuit in Figure 10.39(a), the transistor parameters are: $K_p = 0.1\,\text{mA/V}^2$, $K_n = 0.2\,\text{mA/V}^2$, $V_{TN} = 1\,\text{V}$, $V_{TP} = -1\,\text{V}$, $\lambda_n = 0.01\,\text{V}^{-1}$, and $\lambda_p = 0.02\,\text{V}^{-1}$. Let $V^+ = 10\,\text{V}$, $I_{REF} = 0.25\,\text{mA}$, and $R_L = 100\,\text{k}\Omega$. (a) Determine the small-signal parameters g_m (for M_0), r_{on}, and r_{op}. (b) Find the small-signal voltage gain. (Ans. (a) $g_m = 0.448\,\text{mA/V}$, $r_{on} = 400\,\text{k}\Omega$, $r_{op} = 200\,\text{k}\Omega$ (b) $A_v = -25.6$)

10.5   SUMMARY

- This chapter addressed the biasing of bipolar and FET circuits with constant-current sources. The basic bipolar current source is the simple two-transistor circuit with a resistor to establish the reference current. The basic FET current source is also a simple two-transistor circuit but includes additional transistors in the reference portion of the circuit.

- One parameter of interest in the current source circuit is the output resistance, which determines the stability of the bias current. More sophisticated current-source circuits, such as the Widlar and Wilson circuits in the BJT configuration and the Wilson and cascode in the FET configuration, have larger output resistance parameters and increased bias-current stability.

- Multitransistor output stages, in both bipolar and FET circuits, are used to bias multiple amplifier stages using a single reference current. These circuits, called current mirrors, reduce the number of elements required to bias amplifier stages throughout an IC.

- Both bipolar and MOSFET active load circuits were analyzed. Active loads are essentially constant current source circuits and replace the discrete collector resistor and drain resistor. The active loads produce a much larger small-signal voltage gain compared to discrete resistor circuits.

CHECKPOINT

After studying this chapter, the reader should have the ability to:

✔ Analyze and design a simple two-transistor BJT current-source circuit to produce a given bias current. (Section 10.1)

✔ Analyze and design more sophisticated BJT current-source circuits, such as the three-transistor circuit, cascode circuit, Wilson circuit, and Widlar circuit. (Section 10.1)

✔ Analyze the output resistance of the various BJT current-source circuits and design a BJT current-source circuit to yield a specified output resistance. (Section 10.1)

✔ Analyze and design a basic two-transistor MOSFET current-source circuit with additional MOSFET devices in the reference portion of the circuit to yield a given bias current. (Section 10.2)

✔ Analyze and design more sophisticated MOSFET current-source circuits, such as the cascode circuit, Wilson circuit, and wide-swing cascode circuit. (Section 10.2)

✔ Analyze the output resistance of the various MOSFET current-source circuits and design a MOSFET current-source circuit to yield a specified output resistance. (Section 10.2)

✔ Describe the operation and characteristics of a BJT and MOSFET active load circuit. (Section 10.3)

✔ Discuss the reason for the increased small-signal voltage gain when an active load is used. (Section 10.3)

REVIEW QUESTIONS

1. Sketch the basic BJT two-transistor current source and explain the operation.
2. Explain the significance of the output resistance of the current-source circuit.
3. Discuss the effect of mismatched transistors on the characteristics of the BJT two-transistor current source.
4. Discuss the advantage of the BJT three-transistor current source.
5. What is the primary advantage of a BJT cascode current source?
6. Sketch a Widlar current source and explain the operation.
7. Can a piecewise linear model of the transistor be used in the analysis of the Widlar current source? Why or why not?
8. Discuss the operation and significance of a multiple-output transistor current mirror.
9. Sketch the basic MOSFET two-transistor current-source circuit and discuss its operation.
10. Discuss the effect of mismatched transistors on the characteristics of the MOSFET two-transistor current source.
11. Discuss how the reference portion of the circuit can be designed with MOSFETs only.
12. Sketch a MOSFET cascode current source circuit and discuss the advantages of this design.
13. Discuss the operation of an active load.
14. What is the primary advantage of using an active load?
15. Sketch the voltage transfer characteristics of a simple amplifier with an active load. Where should the Q-point be placed?
16. What is the impedance seen looking into a simple active load?
17. What is the advantage of using a cascode active load?

PROBLEMS

Section 10.1   Bipolar Transistor Current Sources

10.1 Figure P10.1 shows another form of a bipolar current source. (a) Neglecting base currents, derive the expression for I_C in terms of the circuit, transistor, and diode parameters. (b) If the transistor B–E and diode voltages are equal, show that, for $R_1 = R_2$, the expression for I_C reduces to

$$I_C = \frac{(-V^-)}{2R_3}$$

(c) For $V^- = -10\,\text{V}$ and $V_{BE}(\text{on}) = V_\gamma = 0.7\,\text{V}$, design the circuit such that $I_C = I_1 = I_2 = 2\,\text{mA}$.

10.2 The transistor parameters for the circuit in Figure 10.2(b) are: $V_{BE}(\text{on}) = 0.7\,\text{V}$, $\beta = 50$, and $V_A = \infty$. Let $V^+ = 5\,\text{V}$ and $V^- = 0$. For $I_{\text{REF}} = 1\,\text{mA}$, determine I_{C1}, I_{B1}, I_{B2}, and I_{IC2}.

10.3 (a) For the circuit in Figure 10.2(b), the bias voltages are $V^+ = +15\,\text{V}$ and $V^- = -15\,\text{V}$. Assume Q_1 and Q_2 are matched, and assume $V_{BE}(\text{on}) = 0.7\,\text{V}$. Neglecting base currents, find R_1 such that $I_{\text{REF}} = I_O = 0.5\,\text{mA}$. (b) The upper terminal of R_1 may instead be connected to ground potential. Find R_1 such that $I_{\text{REF}} = I_O = 0.5\,\text{mA}$ for this case. Discuss any advantage of connecting R_1 to ground rather than to the V^+ power supply. (c) Determine the change in I_O if R_1 varies by ± 5 percent from the design values of parts (a) and (b).

D10.4 For the basic two-transistor current source in Figure 10.2(b), the transistor parameters are: $\beta = 100$, $V_{BE}(\text{on}) = 0.7\,\text{V}$, and $V_A = 80\,\text{V}$. Let $V^+ = 15\,\text{V}$ and $V^- = 0$. (a) Design the circuit such that $I_O = 2\,\text{mA}$ when $V_{C2} = 0.7\,\text{V}$. (b) What is the percent change in I_O as V_{C2} varies between $0.7\,\text{V}$ and $10\,\text{V}$?

D10.5 Figure P10.5 shows a basic two-transistor pnp current source. The transistor parameters are: $\beta = 25$, $V_{EB}(\text{on}) = 0.7\,\text{V}$, and $V_A = \infty$. Design the circuit such that $I_O = 0.5\,\text{mA}$. What is the value of I_{REF}?

10.6 In the circuit in Figure P10.5, the transistor parameters are $\beta = 50$, $V_{EB}(\text{on}) = 0.7\,\text{V}$, and $V_A = 50\,\text{V}$. Let $R_1 = 18\,\text{k}\Omega$. Determine I_O for: (a) $V_{EC2} = 0.7\,\text{V}$, (b) $V_{EC2} = 2\,\text{V}$, and (c) $V_{EC2} = 4\,\text{V}$.

Figure P10.1

Figure P10.5

Figure P10.7

Figure P10.9

D10.7 Consider the pnp current source in Figure P10.7, with transistor parameters $\beta = \infty$, $V_A = \infty$, and $V_{EB}(\text{on}) = 0.7\,\text{V}$. (a) Design the circuit such that $I_{\text{REF}} = 1\,\text{mA}$. (b) What is the value of I_O? (c) What is the maximum value of R_{C2} such that Q_2 remains biased in the forward-active mode?

10.8 The transistors in the basic current mirror in Figure 10.2(b) have a finite β and an infinite Early voltage. The B–E area of Q_2 is n times that of Q_1. Derive the expression for I_O in terms of I_{REF}, β, and n.

D10.9 The transistor Q_2 shown in Figure P10.9 is equivalent to two transistors in parallel, each of which is matched to Q_1. Assume the transistor parameters are: $V_{BE}(\text{on}) = 0.7\,\text{V}$, $\beta = 50$, and $V_A = \infty$. For $V^+ = 5\,\text{V}$, design the circuit such that $I_O = 2\,\text{mA}$. What is the value of I_{REF}?

10.10 Using the basic two-transistor topology biased at $V^+ = +5\,\text{V}$ and $V^- = -5\,\text{V}$ with npn transistors that have $v_{BE} = 0.7\,\text{V}$ at $1\,\text{mA}$, design a current source that provides a bias current of $1.5\,\text{mA}$ and a reference current of $0.5\,\text{mA}$. Neglect base currents.

10.11 The values of β for the transistors in Figure P10.11 are very large. (a) If Q_1 is diode-connected with $I_1 = 0.5\,\text{mA}$, determine the collector currents in the other two transistors. (b) Repeat part (a) if Q_2 is diode-connected with $I_2 = 0.5\,\text{mA}$. (c) Repeat part (a) if Q_3 is diode-connected with $I_3 = 0.5\,\text{mA}$.

Figure P10.11   **Figure P10.12**

10.12 Consider the circuit in Figure P10.12. The transistor parameters are: $\beta = 80$, $V_{BE}(\text{on}) = 0.7\,\text{V}$, and $V_A = \infty$. (a) Derive the expression for I_O in terms of I_{REF}, β, and R_2. (b) For $R_2 = 10\,\text{k}\Omega$ and $V^+ = 10\,\text{V}$, design the circuit such that $I_O = 0.70\,\text{mA}$. What is the value of I_{REF}?

10.13 All transistors in the N output current mirror in Figure P10.13 are matched, with a finite β and $V_A = \infty$. (a) Derive the expression for each load current in terms of I_{REF} and β. (b) If the circuit parameters are $V^+ = 5\,\text{V}$ and $V^- = -5\,\text{V}$, and the transistor parameter is $\beta = 50$, determine R_1 such that each load current is $0.5\,\text{mA}$ for $N = 5$. Assume that $V_{EB}(Q_R) = V_{BE}(Q_S) = 0.7\,\text{V}$.

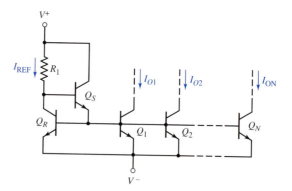

Figure P10.13

D10.14 Design a pnp version of the basic three-transistor current source, using a resistor to establish I_{REF}. The circuit parameters are $V^+ = 5\,\text{V}$ and $V^- = -5\,\text{V}$, and the transistors parameters are: $V_{EB}(\text{on}) = 0.7\,\text{V}$, $\beta = 50$, and $V_A = \infty$. For a load current of $0.5\,\text{mA}$, what is I_{REF}?

D10.15   Design a pnp version of the Wilson current source, using a resistor to establish I_{REF}. The circuit parameters are $V^+ = 9\,V$ and $V^- = -9\,V$, and the transistor parameters are: $V_{EB}(\text{on}) = 0.7\,V$, $\beta = 25$, and $V_A = \infty$. If the load current is 0.8 mA, what is I_{REF}?

*10.16   Consider the Wilson current source in Figure P10.16. The transistors have a finite β and an infinite Early voltage. Derive the expression for I_O in terms of I_{REF} and β.

D10.17   For the transistors in the circuit in Figure P10.17, the parameters are: $V_{BE}(\text{on}) = 0.7\,V$, $\beta = 75$, and $V_A = \infty$. Design the circuit such that $I_O = 2\,mA$. What is the value of I_{REF}?

Figure P10.16

Figure P10.17

10.18   Consider the Wilson current source in Figure 10.8. The reference current is 0.5 mA, and the transistor parameters are: $V_{BE}(\text{on}) = 0.7\,V$, $\beta = 80$, and $V_A = 80\,V$. (a) Determine the output resistance looking into the collector of Q_3. (b) What is the change in I_O as the output voltage changes by 5 V?

D10.19   Design the Widlar current source shown in Figure 10.9 such that $I_{REF} = 2\,mA$ and $I_O = 50\,\mu A$. Let $V^+ = 15\,V$ and $V^- = 0$. The transistors are matched, and $V_{BE} = 0.7\,V$ at 1 mA.

10.20   The circuit parameters of the Widlar current source in Figure 10.9 are $V^+ = 10\,V$, $V^- = 0$, and $R_1 = 20\,k\Omega$. The B–E voltage is $V_{BE} = 0.7\,V$ at 1 mA. (a) Determine I_{REF}. (b) Determine R_E such that $I_O = 100\,\mu A$.

10.21   (a) For the Widlar current source in Figure 10.9, find I_{REF}, I_O, and R_o if $R_1 = 100\,k\Omega$, $R_E = 10\,k\Omega$, $V^+ = +5\,V$, and $V^- = -5\,V$. The transistor parameters are $V_{BE}(\text{on}) = 0.7\,V$ and $V_A = 30\,V$. (b) Determine the voltage difference $V_{BE1} - V_{BE2}$.

*10.22   Consider the Widlar current source in Problem 10.21. For $\beta = 80$ and $V_A = 80\,V$, determine the change in I_O corresponding to a 5 V change in the output voltage.

*10.23   For the Widlar current source in Figure 10.9, the parameters are: $V^+ = 5\,V$, $V^- = 0$, $I_{REF} = 0.75\,mA$, and $I_O = 25\,\mu A$. Assume $V_{BE1} = 0.7\,V$. If $\beta = 80$ and $V_A = 75\,V$, determine the output resistance looking into the collector of Q_2. What is the percent change in I_O if V_{C2} changes by 3 V?

D10.24   Design a Widlar current source to provide a bias current of $I_O = 100\,\mu A$. The power supplies are $V^+ = 12\,V$ and $V^- = -12\,V$. The maximum resistor value is to be limited to 5 kΩ.

10.25 Consider the Widlar current source in Figure 10.9. The circuit parameters are: $V^+ = 10\,\text{V}$, $V^- = -10\,\text{V}$, $R_1 = 40\,\text{k}\Omega$, and $R_E = 12\,\text{k}\Omega$. Neglect base currents and assume $V_{BE1} = 0.7\,\text{V}$ at $1\,\text{mA}$. Determine I_{REF}, I_O, V_{BE1}, and V_{BE2}.

10.26 Consider the circuit in Figure P10.26. The transistors are matched. Assume that base currents are negligible and that $V_A = \infty$. Using the current–voltage relationships given by Equations (10.26(a)) and (10.26(b)), show that

$$I_O R_{E2} - I_{\text{REF}} R_{E1} = V_T \ln\!\left(\frac{I_{\text{REF}}}{I_O}\right)$$

If $R_{E1} = R_{E2} \neq 0$ and $V_A \neq \infty$, explain the advantage of this circuit over the basic two-transistor current source in Figure 10.2(b).

*10.27 Consider the circuit in Figure P10.26, with parameters $V^+ = 5\,\text{V}$, $V^- = -5\,\text{V}$, $R_1 = 13.6\,\text{k}\Omega$, and $R_{E1} = R_{E2} = 5\,\text{k}\Omega$. The transistor parameters are: $\beta = 50$, $V_{BE}(\text{on}) = 0.7\,\text{V}$, and $V_A = 75\,\text{V}$. (a) Determine the output resistance $R_0 = dV_{C2}/dI_O$. (b) Compare this output resistance value to that obtained when $R_{E1} = R_{E2} = 0$ and $R_1 = 18.6\,\text{k}\Omega$.

*10.28 Consider the circuit in Figure P10.28. Neglect base currents and assume $V_A = \infty$. (a) Derive the expression for I_O in terms of I_{REF} and R_E. (b) Determine the value of R_E such that $I_O = I_{\text{REF}} = 100\,\mu\text{A}$. Assume $V_{BE} = 0.7\,\text{V}$ at a collector current of $1\,\text{mA}$.

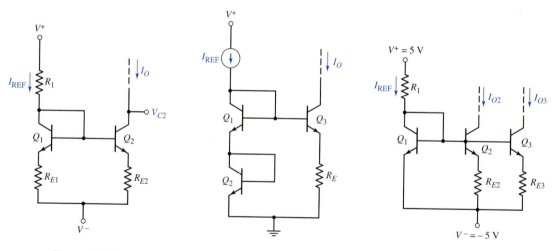

Figure P10.26   **Figure P10.28**   **Figure P10.29**

D10.29 Consider the Widlar current source with multiple output transistors, as shown in Figure P10.29. Assuming $V_{BE1}(\text{on}) = 0.7\,\text{V}$, design the circuit such that $I_{\text{REF}} = 0.5\,\text{mA}$, $I_{O2} = 10\,\mu\text{A}$, and $I_{O3} = 30\,\mu\text{A}$. What are the values of V_{BE2} and V_{BE3}?

10.30 Assume that all transistors in the circuit in Figure P10.30 are matched and that $\beta = \infty$ (neglect base currents). (a) Derive an expression for I_O in terms of bias voltages and resistor values. (b) Show that if $R_1 = R_2$ and $I_O = I_{\text{REF}}$, then $I_O = (V^+ - V^-)/2R_E$, which means that the currents are independent of V_{BE}. (c) For $V^+ = +5\,\text{V}$ and $V^- = -5\,\text{V}$, design the circuit such that $I_O = 0.5\,\text{mA}$.

10.31 In the circuit in Figure P10.31, the transistor parameters are: $\beta = \infty$, $V_A = \infty$, and $V_{BE} = V_{EB} = 0.7\,\text{V}$ at $1\,\text{mA}$. Let $R_{C1} = 2\,\text{k}\Omega$, $R_{C2} = 3\,\text{k}\Omega$, $R_{C3} = 1\,\text{k}\Omega$, and $R_1 = 12\,\text{k}\Omega$. (a) Determine I_{O1}, I_{O2}, and I_{O3}. (b) Calculate V_{CE1}, V_{EC2}, and V_{EC3}.

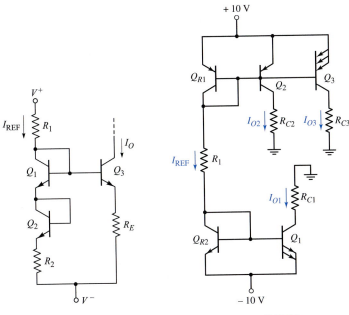

Figure P10.30

Figure P10.31

10.32 Consider the circuit in Figure P10.31, with transistor parameters: $\beta = \infty$, $V_A = \infty$, and $V_{BE} = V_{EB} = 0.7\,\text{V}$ at 1 mA. Let $R_1 = 8\,\text{k}\Omega$. (a) Find I_{O1}, I_{O2}, and I_{O3}. (b) Determine the maximum values of R_{C1}, R_{C2}, and R_{C3} such that Q_1, Q_2, and Q_3 remain biased in the forward-active region.

*D10.33 Consider the circuit in Figure P10.33. The transistor parameters are: $\beta = \infty$, $V_A = \infty$, and $V_{BE} = 0.7\,\text{V}$ at 1 mA. Design the circuit such that the B–E voltages of Q_1, Q_2, and Q_3 are identical to that of Q_R. What are the values of I_{O1}, I_{O2}, and I_{O3}?

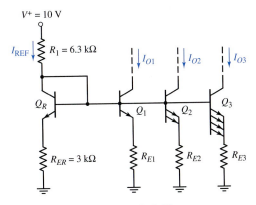

Figure P10.33

Section 10.2 FET Current Sources

10.34 Consider the basic two-transistor NMOS current source in Figure 10.16 with matched transistors. The circuit parameters are: $V^+ = 5\,\text{V}$, $V^- = -5\,\text{V}$, and $I_{REF} = 200\,\mu\text{A}$, and the transistor parameters are: $V_{TN} = 1\,\text{V}$, $K_n = 250\,\mu\text{A/V}^2$, and $\lambda = 0.02\,\text{V}^{-1}$. Find I_O for: (a) $V_{DS2} = 2\,\text{V}$, (b) $V_{DS2} = 4\,\text{V}$, and (c) $V_{DS2} = 6\,\text{V}$.

10.35 In the two-transistor NMOS current source shown in Figure 10.16, the parameters are: $V^+ = 5\,\text{V}$, $V^- = -5\,\text{V}$, and $I_{\text{REF}} = 0.5\,\text{mA}$. The transistor parameters are: $V_{TN1} = 1\,\text{V}$, $K_{n1} = 0.5\,\text{mA/V}^2$, and $\lambda_1 = \lambda_2 = 0$. (a) If $V_{TN2} = 1\,\text{V}$ and $K_{n2} = (0.5 \pm 5\%)\,\text{mA/V}^2$, determine the range in values of I_O. (b) If $K_{n2} = 0.5\,\text{mA/V}^2$ and $V_{TN2} = (1 \pm 5\%)\,\text{V}$, determine the range in values of I_O.

10.36 Consider the two-transistor diode-connected circuit in Figure P10.36. Assume that both transistors are biased in the saturation region, and that $g_{m1} = g_{m2} \equiv g_m$ and $r_{o1} = r_{o2} \equiv r_o$. Neglect the body effect. Derive the expression for the output resistance R_o.

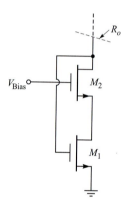

Figure P10.36

10.37 Consider the circuit in Figure 10.22 in the text. Assume $I_{\text{REF}} = 50\,\mu\text{A}$ and assume transistor parameters of $V_{TN} = 0.8\,\text{V}$, $(\frac{1}{2})\mu_n C_{ox} = 48\,\mu\text{A/V}^2$, $\lambda = 0$, and $\gamma = 0$. (a) Find W/L such that $V_{DS3}(\text{sat}) = 0.2\,\text{V}$. (b) What is V_{GS5}? (c) What is the minimum voltage at the drain of M_1 such that all transistors remain biased in the saturation region?

D10.38 For the circuit in Figure 10.17, $V^+ = 5\,\text{V}$ and $V^- = -5\,\text{V}$, and the transistor parameters are: $V_{TN} = 1.5\,\text{V}$, $(\frac{1}{2})\mu_n C_{ox} = 20\,\mu\text{A/V}^2$, and $\lambda = 0$. Design the circuit such that $I_O = 0.25\,\text{mA}$, $I_{\text{REF}} = 0.10\,\text{mA}$, and M_2 remains biased in the saturation region for $V_{DS2} \geq 2\,\text{V}$.

10.39 The parameters for the circuit in Figure 10.17 are $V^+ = 5\,\text{V}$ and $V^- = 0$, and the transistor parameters are: $V_{TN} = 0.5\,\text{V}$, $(\frac{1}{2})\mu_n C_{ox} = 18\,\mu\text{A/V}^2$, and $\lambda = 0.02\,\text{V}^{-1}$. The transistor width-to-length ratios are $(W/L)_3 = 5$, $(W/L)_1 = 25$, and $(W/L)_2 = 15$. Determine: (a) I_{REF}, (b) I_O at $V_{DS2} = 2\,\text{V}$, and (c) I_O at $V_{DS2} = 4\,\text{V}$.

10.40 The circuit in Figure P10.40 is a PMOS version of a two-transistor MOS current mirror. Assume transistor parameters of $V_{TP} = -0.4\,\text{V}$, $(\frac{1}{2})\mu_p C_{ox} = 20\,\mu\text{A/V}^2$, and $\lambda = 0$. The transistor width-to-length ratios are $(W/L)_1 = 25$, $(W/L)_2 = 15$, and $(W/L)_3 = 5$. (a) Determine I_O, I_{REF}, V_{SG1}, and V_{SG3}. (b) What is the largest value of R such that M_2 remains biased in the saturation region?

D10.41 The transistors in Figure P10.40 have the same parameters as in Problem 10.40 except for the (W/L) ratios. Design the circuit such that $I_O = 25\,\mu\text{A}$, $I_{\text{REF}} = 75\,\mu\text{A}$, and $V_{SD2}(\text{sat}) = 0.25\,\text{V}$.

*10.42 For the NMOS cascode current source in Figure 10.19, the parameters are $V^+ = 10\,\text{V}$, $V^- = -10\,\text{V}$, and $I_{\text{REF}} = 100\,\mu\text{A}$. All transistors are matched, with parameters $V_{TN} = 2\,\text{V}$, $K_n = 100\,\mu\text{A/V}^2$, and $\lambda = 0.02\,\text{V}^{-1}$. (a) Determine I_O for $V_{D4} = -3\,\text{V}$. (b) Determine the change in I_O as V_{D4} changes from $-3\,\text{V}$ to $+3\,\text{V}$.

Figure P10.40

Figure P10.43

*10.43   Consider the NMOS current source in Figure P10.43. Let $I_{REF} = 0.2\,\text{mA}$, $K_n = 0.2\,\text{mA/V}^2$, $V_{TN} = 1\,\text{V}$, and $\lambda = 0.02\,\text{V}^{-1}$. (All transistors are matched.) Determine the output resistance looking into the drain of M_6.

10.44   The transistors in the circuit in Figure P10.44 have parameters $V_{TN} = +0.5\,\text{V}$, $V_{TP} = -0.5\,\text{V}$, $(\frac{1}{2})\mu_n C_{ox} = 50\,\mu\text{A/V}^2$, $(\frac{1}{2})\mu_p C_{ox} = 20\,\mu\text{A/V}^2$, and $\lambda_n = \lambda_p = 0$. The transistor width-to-length ratios are $(W/L)_1 = (W/L)_2 = 20$, $(W/L)_3 = 5$, and $(W/L)_4 = 10$. Determine I_O, I_{REF}, and $V_{DS2}(\text{sat})$.

*D10.45   The transistors in the circuit in Figure P10.44 have the same parameters as in Problem 10.44 except for the (W/L) ratios. Design the circuit such that $I_O = 50\,\mu\text{A}$, $I_{REF} = 150\,\mu\text{A}$, $V_{DS2}(\text{sat}) = 0.5\,\text{V}$, and $V_{GS3} = V_{SG4}$.

*10.46   A Wilson current mirror is shown in Figure 10.21(a). The parameters are: $V^+ = 5\,\text{V}$, $V^- = -5\,\text{V}$, and $I_{REF} = 80\,\mu\text{A}$. The transistor parameters are: $V_{TN} = 1\,\text{V}$, $K_n = 80\,\mu\text{A/V}^2$, and $\lambda = 0.02\,\text{V}^{-1}$. Determine I_O at: (a) $V_{D3} = -1\,\text{V}$, and (b) $V_{D3} = +3\,\text{V}$.

*10.47   Repeat Problem 10.46 for the modified Wilson current mirror in Figure 10.21(b).

10.48   Consider the bias-independent current source in Figure 10.23. Assume transistor parameters of $V_{TN} = +0.5\,\text{V}$, $V_{TP} = -0.5\,\text{V}$, $(\frac{1}{2})\mu_n C_{ox} = 50\,\mu\text{A/V}^2$, $(\frac{1}{2})\mu_p C_{ox} = 20\,\mu\text{A/V}^2$, and $\lambda_n = \lambda_p = 0$. The W/L ratios are given for the M_1–M_4 transistors. (a) Determine R such that $I_{D1} = I_{D2} = 50\,\mu\text{A}$. (b) What is the minimum bias voltage difference $(V^+ - V^-)$ that must be applied? (c) Determine $(W/L)_5$ and $(W/L)_6$ such that $I_{O1} = 25\,\mu\text{A}$ and $I_{O2} = 75\,\mu\text{A}$.

D10.49   Consider the multitransistor current source in Figure P10.49. The transistor parameters are: $V_{TN} = 1\,\text{V}$, $(\frac{1}{2})\mu_n C_{ox} = 20\,\mu\text{A/V}^2$, and $\lambda = 0$. Assume M_3, M_4, and M_5 are identical. Design the circuit such that $I_{REF} = 0.1\,\text{mA}$, $I_{O1} = 0.2\,\text{mA}$, and $I_{O2} = 0.3\,\text{mA}$.

10.50   The parameters of the transistors in the circuit in Figure P10.50 are $V_{TN} = 1.2\,\text{V}$, $V_{TP} = -1.2\,\text{V}$, $(\frac{1}{2})\mu_n C_{ox} = 40\,\mu\text{A/V}^2$, $(\frac{1}{2})\mu_p C_{ox} = 18\,\mu\text{A/V}^2$, and $\lambda_n = \lambda_p = 0$. The W/L ratios are given in the figure. For $R = 200\,\text{k}\Omega$, determine I_{REF}, I_1, I_2, I_3, and I_4.

10.51   Repeat Problem 10.50 if the bias voltages are reduced to $V^+ = 5\,\text{V}$ and $V^- = -5\,\text{V}$.

Figure P10.44

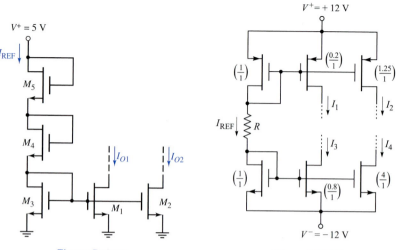

Figure P10.49

Figure P10.50

10.52   For the JFET in Figure P10.52, the parameters are: $I_{DSS} = 2\,\text{mA}$, $V_P = -2\,\text{V}$, and $\lambda = 0.05\,\text{V}^{-1}$. Determine I_O for: (a) $V_D = -5\,\text{V}$, (b) $V_D = 0\,\text{V}$, and (c) $V_D = +5\,\text{V}$.

*10.53   A JFET cascode current source is shown in Figure P10.53. The transistor parameters are: $I_{DSS} = 1\,\text{mA}$, $V_P = -2\,\text{V}$, and $\lambda = 0.05\,\text{V}^{-1}$. Determine I_O, V_{DS1}, and V_{DS2} at: (a) $V_D = -5\,\text{V}$, (b) $V_D = 0\,\text{V}$, and (c) $V_D = +5\,\text{V}$.

D10.54   A JFET circuit is biased with the current source in Figure P10.54. The transistor parameters are: $I_{DSS} = 4\,\text{mA}$, $V_P = -4\,\text{V}$, and $\lambda = 0$. Design the circuit such that $I_O = 2\,\text{mA}$. What is the minimum value of V_D such that the transistor is biased in the saturation region?

Figure P10.52   Figure P10.53   Figure P10.54

Section 10.3   Active Load Circuits

10.55   Consider the simple BJT active load amplifier in Figure 10.29, with transistor parameters: $I_{SO} = 10^{-12}\,\text{A}$, $I_{S1} = I_{S2} = 5 \times 10^{-13}\,\text{A}$, $V_{AN} = 120\,\text{V}$, and $V_{AP} = 80\,\text{V}$. Let $V^+ = 5\,\text{V}$, and neglect base currents. (a) Find the value of V_{EB} that will produce

$I_{REF} = 1\,\text{mA}$. (b) Determine the value of R_1. (c) What value of V_I will produce $V_{CEO} = V_{EC2}$? (d) Determine the open-circuit small-signal voltage gain

10.56 The amplifier shown in Figure P10.56 uses a pnp driver and an npn active load circuit. The transistor parameters are: $I_{S0} = 5 \times 10^{-13}\,\text{A}$, $I_{S1} = I_{S2} = 10^{-12}\,\text{A}$, $V_{AN} = 120\,\text{V}$, and $V_{AP} = 80\,\text{V}$. Let $V^+ = 5\,\text{V}$, and neglect base currents. (a) Find the value of V_{BE} that will produce $I_{REF} = 0.5\,\text{mA}$. (b) Determine the value of R_1. (c) What value of V_I will produce $V_{EC0} = V_{CE2}$? (d) Determine the open-circuit small-signal voltage gain.

Figure P10.56   **Figure P10.57**

D10.57 Consider the basic MOSFET amplifier with active load in Figure P10.57. The transistor parameters are: $V_{TN} = 1\,\text{V}$, $V_{TP} = -1\,\text{V}$, $(\frac{1}{2})\mu_n C_{ox} = 20\,\mu\text{A/V}^2$, $(\frac{1}{2})\mu_p C_{ox} = 10\,\mu\text{A/V}^2$, and $\lambda_n = \lambda_p = 0.02\,\text{V}^{-1}$. (a) Design the circuit such that $I_{REF} = I_O = 0.1\,\text{mA}$. Assume M_1 and M_2 are matched, $(W/L)_3 = 5$, and the quiescent input voltage is $V_{IQ} = 2\,\text{V}$. The quiescent output voltage is to be $V_{OQ} = 2.5\,\text{V}$. (b) Determine the open-circuit small-signal voltage gain.

10.58 For the simple MOSFET amplifier with active load shown in Figure 10.33, the transistor parameters are: $V_{TN} = 1\,\text{V}$, $V_{TP1} = V_{TP2} = -1\,\text{V}$, $K_{p1} = K_{p2} = K_n = 100\,\mu\text{A/V}^2$, and $\lambda_1 = \lambda_2 = \lambda_n = 0.02\,\text{V}^{-1}$. Let $V^+ = 10\,\text{V}$ and $I_{REF} = 100\,\mu\text{A}$. (a) Find V_{SG}. (b) What value of V_I will produce $V_{DS0} = V_{SD2}$? (c) Determine the open-circuit small-signal voltage gain.

Section 10.4   Small-Signal Analysis: Active Load Circuits

10.59 A BJT amplifier with active load is shown in Figure P10.59. The circuit contains emitter resistors R_E and a load resistor R_L. (a) Derive the expression for the output resistance looking into the collector of Q_2. (b) Using the small-signal equivalent circuit, derive the equation for the small-signal voltage gain. Express the relationship in a form similar to Equation (10.94).

10.60 In the circuit in Figure P10.60, the active load circuit is replaced by a Wilson current source. Assume that $\beta = 80$ for all transistors, and that $V_{AN} = 120\,\text{V}$, $V_{AP} = 80\,\text{V}$, and $I_{REF} = 0.2\,\text{mA}$. Determine the open-circuit small-signal voltage gain.

10.61 For the circuit in Figure 10.39(a) the transistor parameters are: $K_p = 0.2\,\text{mA/V}^2$, $K_n = 0.25\,\text{mA/V}^2$, $V_{TN} = 1\,\text{V}$, $V_{TP} = -1\,\text{V}$, $\lambda_n = 0.02\,\text{V}^{-1}$, and $\lambda_p = $

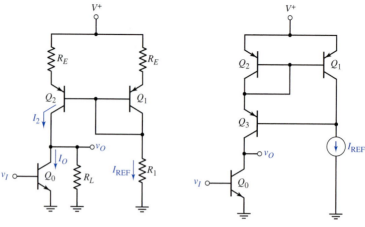

Figure P10.59   **Figure P10.60**

$0.03\,\text{V}^{-1}$. Let $V^+ = 10\,\text{V}$ and $I_{\text{REF}} = 0.2\,\text{mA}$. (a) Determine the small-signal parameters $g_m(M_0)$, r_{on}, and r_{op}. (b) Determine the open-circuit small-signal voltage gain. (c) Determine the value of R_L that results in a voltage gain of one-half the open-circuit value.

10.62 The parameters of the transistors in Figure P10.62 are $V_{TN} = +0.8\,\text{V}$, $V_{TP} = -0.8\,\text{V}$, $(\frac{1}{2})\mu_n C_{ox} = 50\,\mu\text{A/V}^2$, $(\frac{1}{2})\mu_p C_{ox} = 20\,\mu\text{A/V}^2$, and $\lambda_n = \lambda_p = 0.02\,\text{V}^{-1}$. The width-to-length ratios are shown in the figure. The value of V_{GSQ} is such that $I_{D1} = 100\,\mu\text{A}$, and M_1 and M_2 are biased in the saturation region. Find the small-signal voltage gain $A_v = v_o/v_i$.

Figure P10.62   **Figure P10.63**

10.63 The parameters of the transistors in Figure P10.63 are $V_{TN} = +0.8\,\text{V}$, $V_{TP} = -0.8\,\text{V}$, $(\frac{1}{2})\mu_n C_{ox} = 50\,\mu\text{A/V}^2$, $(\frac{1}{2})\mu_p C_{ox} = 20\,\mu\text{A/V}^2$, and $\lambda_n = \lambda_p = 0.02\,\text{V}^{-1}$. The width-to-length ratios of M_1 and M_2 are 20, and those of M_3–M_6 are 40. The value

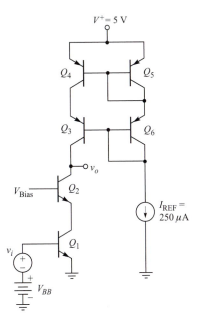

Figure P10.64

of V_{GSQ} is such that $I_{D1} = 80\,\mu A$, and all transistors are biased in the saturation region. Determine the small-signal voltage gain $A_v = v_o/v_i$.

10.64 A BJT cascode amplifier with a cascode active load is shown in Figure P10.64. Assume transistor parameters of $\beta = 120$ and $V_A = 80\,V$. The V_{BB} voltage is such that all transistors are biased in the active region. Determine the small-signal voltage gain $A_v = v_o/v_i$.

D10.65 Design a bipolar cascode amplifier with a cascode active load similar to that in Figure P10.64 except the amplifying transistors are to be pnp and the load transistors are to be npn. Bias the circuit at $V^+ = 10\,V$ and incorporate a reference current of $I_{REF} = 200\,\mu A$. If all transistors are matched with $\beta = 100$ and $V_A = 60\,V$, determine the small-signal voltage gain.

D10.66 Design a MOSFET cascode amplifier with a cascode active load similar to that shown in Figure P10.63 except that the amplifying transistors are to be PMOS and the load transistors are to be NMOS. Assume transistor parameters similar to those in Problem 10.63. Determine the small-signal voltage gain.

COMPUTER SIMULATION PROBLEMS

10.67 Consider the Widlar current source in Figure 10.9, with parameters given in Problem 10.25. Choose appropriate transistor parameters. Connect a $40\,k\Omega$ resistor between V^+ and the collector of Q_2 as a load. Using a PSpice analysis, determine I_{REF}, I_O, V_{BE1}, and V_{BE2}.

10.68 For the circuit in Figure 10.19, the transistor and circuit parameters are given in Problem 10.42. Connect a separate dc voltage source at the drain of M_4. Change the value of the voltage source such that V_{D4} varies between $-3\,V$ and $+3\,V$. From a computer analysis, determine the change in I_O as V_{D4} varies between the two limits.

10.69   In the circuit in Figure P10.59, the parameters are: $V^+ = 10\,\text{V}$, $R_1 = 9\,\text{k}\Omega$, and $R_E = 4\,\text{k}\Omega$. The transistor parameters are: $\beta = 100$ and $V_{AN} = V_{AP} = 100\,\text{V}$. (a) Using a computer simulation, plot the voltage transfer characteristics of v_O versus v_I, similar to those in Figure 10.30, for $R_L = \infty$. What is the voltage gain? (b) Repeat part (a) if $R_L = 100\,\text{k}\Omega$.

10.70   Consider the circuit in Figure P10.60, with parameters $V^+ = 5\,\text{V}$ and $I_{\text{REF}} = 0.5\,\text{mA}$. Assume all transistors are identical, with parameters $\beta = 100$ and $V_A = 100\,\text{V}$. Using a computer analysis, plot the voltage transfer characteristics v_O versus v_I over an appropriate voltage range, and determine the voltage gain.

10.71   A MOSFET active load circuit is shown in Figure P10.57. The circuit and transistor parameters are as described in Problem 10.57. In addition, assume width-to-length ratios of $(W/L)_1 = (W/L)_2 = 5$ and $(W/L)_O = 15$. Using a computer simulation, plot the voltage transfer characteristics of v_O versus v_I, similar to those shown in Figure 10.34. What is the voltage gain?

DESIGN PROBLEMS

[Note: Each design should be verified with a computer analysis.]

*D10.72   Design a generalized Widlar current source (Figure P10.26) to provide a bias current $I_O = 200\,\mu\text{A}$. Assume the output impedance is $R_o = 5\,\text{M}\Omega$, and the circuit is biased at $V^+ = 9\,\text{V}$ and $V^- = -9\,\text{V}$. The transistor parameters are: $I_S = 10^{-14}\,\text{A}$ and $V_A = 120\,\text{V}$.

*D10.73   Consider a MOSFET current source similar to the one shown in Figure 10.17, biased at $V^+ = 10\,\text{V}$ and $V^- = -10\,\text{V}$. The transistor parameters are: $(\frac{1}{2})\mu_n C_{ox} = 20\,\mu\text{A/V}^2$, $V_{TN} = 2\,\text{V}$, and $\lambda = 0$. Design the circuit such that $I_O = 150\,\mu\text{A}$ and $V_{DS}(\text{sat}) = 0.25\,\text{V}$ for M_2.

*D10.74   Using MOSFETs, design a circuit similar to that shown in Figure 10.15, to provide $I_{O1} = 100\,\mu\text{A}$, $I_{O2} = 150\,\mu\text{A}$, $I_{O3} = 200\,\mu\text{A}$, and $I_{O4} = 250\,\mu\text{A}$. Assume the transistor parameters are: $(\frac{1}{2})\mu_n C_{ox} = 20\,\mu\text{A/V}^2$, $(\frac{1}{2})\mu_p C_{ox} = 10\,\mu\text{A/V}^2$, $V_{TN} = |V_{TP}| = 2\,\text{V}$, and $\lambda = 0$. Let $V^+ = 10\,\text{V}$ and $V^- = -10\,\text{V}$.

11

Differential and Multistage Amplifiers

11.0   PREVIEW

In this chapter, we introduce a special multitransistor circuit configuration called the differential amplifier, or diff-amp. We have encountered a diff-amp previously in our discussion of op-amp circuits. However, the diff-amp, in the context of this chapter, is at the basic transistor level.

The diff-amp is a fundamental building block of analog circuits. It is the input stage of virtually every op-amp, and is the basis of a high-speed digital logic circuit family, called emitter-coupled logic, which will be addressed in Chapter 17.

Matched or identical transistor characteristics are critical to the design of the IC diff-amp, as they were in the design of current-source circuits. The design of electronic circuits in this chapter, then, is based on integrated circuit fabrication. The design of IC diff-amps, in general, incorporates current-source biasing and active loads that were analyzed in the last chapter. We consider both BJT and MOSFET differential amplifier designs. At the end of this chapter, the reader should be able to design both BJT and MOSFET diff-amps to meet particular specifications.

Basic BiCMOS analog circuits are considered. BiCMOS circuits combine bipolar and MOS transistors on the same semiconductor chip. The advantages of the MOSFET high input impedance and the bipolar high gain can be utilized in the same circuit.

Up to this point, we have concentrated primarily on the analysis and design of single-stage amplifiers. However, these circuits have limited gain, input resistance, and output resistance. Multistage or cascaded-stage amplifiers can be designed to produce high gain and specified input and output resistance values. In this chapter, we begin to consider these multistage amplifiers.

11.1   THE DIFFERENTIAL AMPLIFIER

In Chapter 4, we discussed the reasons linear amplifiers are necessary in analog electronic systems. In Chapters 4 and 6, we analyzed and designed several configurations of bipolar and MOS transistor amplifiers. In these circuits, there was one input terminal and one output terminal.

In this chapter, we introduce another basic transistor circuit configuration called the differential amplifier. This amplifier, also called a diff-amp, is the input stage to virtually all op-amps and is probably the most widely used amplifier building block in analog integrated circuits. Figure 11.1 is a block diagram of the diff-amp. There are two input terminals and one output terminal. Ideally, the output signal is proportional to only the difference between the two input signals.

11.2   BASIC BJT DIFFERENTIAL PAIR

In this section, we consider the basic bipolar **difference amplifier** or **diff-amp**. We introduce the terminology, qualitatively describe the operation of the circuit, and analyze the dc and small-signal characteristics of the diff-amp.

11.2.1   Terminology and Qualitative Description

Figure 11.1   Difference amplifier block diagram

As mentioned, Figure 11.1 is a block diagram of a difference amplifier. Ideally, the output is proportional only to the difference between the two input signals, or

$$v_o = A_d(v_1 - v_2) \tag{11.1}$$

In the ideal case, if $v_1 = v_2$, the output voltage is zero. We only obtain a nonzero output voltage if v_1 and v_2 are not equal.

We define the **differential-mode input voltage** as

$$v_d = v_1 - v_2 \tag{11.2}$$

and the **common-mode input voltage** as

$$v_{cm} = \frac{v_1 + v_2}{2} \tag{11.3}$$

These equations show that if $v_1 = v_2$, the differential-mode input signal is zero and the common-mode input signal is $v_{cm} = v_1 = v_2$.

If, for example, $v_1 = +10\,\mu\text{V}$ and $v_2 = -10\,\mu\text{V}$, then the differential-mode voltage is $v_d = 20\,\mu\text{V}$ and the common-mode voltage is $v_{cm} = 0$. However, if $v_1 = 110\,\mu\text{V}$ and $v_2 = 90\,\mu\text{V}$, then the differential-mode input signal is still $v_d = 20\,\mu\text{V}$, but the common-mode input signal is $v_{cm} = 100\,\mu\text{V}$. If each pair of input voltages were applied to the ideal difference amplifier, the output voltage in each case would be exactly the same. However, amplifiers are not ideal, and the common-mode input signal does affect the output. One goal of the design of differential amplifiers is to minimize the effect of the common-mode input signal.

Figure 11.2 shows the basic BJT differential-pair configuration. Two identical transistors, Q_1 and Q_2, whose emitters are connected together, are biased by a constant-current source I_Q, which is connected to a negative supply voltage V^-. The collectors of Q_1 and Q_2 are connected through resistors R_C to a positive supply voltage V^+. By design, transistors Q_1 and Q_2 are to remain biased in the forward-active region. We assume that the two collector resistors R_C are equal, and that v_{B1} and v_{B2} are ideal sources, meaning that the output resistances of these sources are negligibly small.

Since both positive and negative bias voltages are used in the circuit, the need for coupling capacitors and voltage divider biasing resistors at the inputs of Q_1 and Q_2 has been eliminated. If the input signal voltages v_{B1} and v_{B2} in the

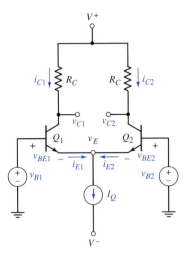

Figure 11.2 Basic BJT differential-pair configuration

circuit shown in Figure 11.2 are both zero, Q_1 and Q_2 are still biased in the active region by the current source I_Q. The common-emitter voltage v_E would be on the order of $-0.7\,\text{V}$. This circuit, then, is referred to as a dc-coupled differential amplifier, so differences in dc input voltages can be amplified. Although the diff-amp contains two transistors, it is considered a single-stage amplifier. The analysis will show that it has characteristics similar to those of the common-emitter amplifier.

First, we consider the circuit in which the two base terminals are connected together and a common-mode voltage v_{cm} is applied as shown in Figure 11.3(a). The transistors are biased "on" by the constant-current source, and

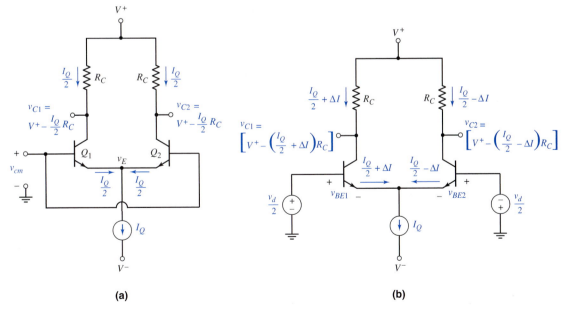

(a)   (b)

Figure 11.3 Basic diff-amp with applied common-mode voltage and (b) basic diff-amp with applied differential-mode voltage the voltage at the common emitters is $v_E = v_{cm} - V_{BE}(\text{on})$. Since Q_1 and Q_2 are matched or identical, current I_Q splits evenly between the two transistors, and

$$i_{E1} = i_{E2} = \frac{I_Q}{2} \tag{11.4}$$

If base currents are negligible, then $i_{C1} \cong i_{E1}$ and $I_{C2} \cong i_{E2}$, and

$$v_{C1} = V^+ - \frac{I_Q}{2} R_C = v_{C2} \tag{11.5}$$

We see from Equation (11.5) that, for an applied common-mode voltage, I_Q splits evenly between Q_1 and Q_2 and the difference between v_{C1} and v_{C2} is zero.

Now, if v_{B1} increases by a few millivolts and v_{B2} decreases by the same amount, or $v_{B1} = v_d/2$ and $v_{B2} = -v_d/2$, the voltages at the bases of Q_1 and Q_2 are no longer equal. Since the emitters are common, this means that the B–E voltages on Q_1 and Q_2 are no longer equal. Since v_{B1} increases and v_{B2} decreases, then $v_{BE1} > v_{BE2}$, which means that i_{C1} increases by ΔI above its quiescent value and i_{C2} decreases by ΔI below its quiescent value. This is shown in Figure 11.3(b). A potential difference now exits between the two collector terminals. We can write

$$v_{C2} - v_{C1} = \left[V^+ - \left(\frac{I_{CQ}}{2} - \Delta I \right) R_C \right] - \left[V^+ - \left(\frac{I_{CQ}}{2} + \Delta I \right) R_C \right] = 2\Delta I R_C \tag{11.6}$$

A voltage difference is created between v_{C2} and v_{C1} when a differential-mode input voltage is applied.

Example 11.1   Objective: Determine the quiescent collector current and collector-emitter voltage in a difference amplifier.

Consider the diff-amp in Figure 11.2, with circuit parameters: $V^+ = 10\,\text{V}$, $V^- = -10\,\text{V}$, $I_Q = 1\,\text{mA}$, and $R_C = 10\,\text{k}\Omega$. The transistor parameters are: $\beta = \infty$ (neglect base currents), $V_A = \infty$, and $V_{BE}(\text{on}) = 0.7\,\text{V}$. Determine i_{C1} and v_{CE1} for common-mode voltages $v_{B1} = v_{B2} = v_{CM} = 0$, $-5\,\text{V}$, and $+5\,\text{V}$.

Solution: We know that

$$i_{C1} = i_{C2} = \frac{I_Q}{2} = 0.5\,\text{mA}$$

therefore,

$$v_{C1} = v_{C2} = V^+ - i_{C1} R_C = 10 - (0.5)(10) = 5\,V$$

From $v_{CM} = 0$, $v_E = -0.7\,\text{V}$ and

$$v_{CE1} = v_{C1} - v_E = 5 - (-0.7) = 5.7\,\text{V}$$

For $v_{CM} = -5\,\text{V}$, $v_E = -5.7\,\text{V}$ and

$$v_{CE1} = v_{C1} - v_E = 5 - (-5.7) = 10.7\,\text{V}$$

For $v_{CM} = +5\,\text{V}$, $v_E = 4.3\,\text{V}$ and

$$v_{CE1} = v_{C1} - v_E = 5 - 4.3 = 0.7\,\text{V}$$

Comment: As the common-mode voltage varies, the ideal constant current I_Q still splits evenly between Q_1 and Q_2, but the collector–emitter voltage varies, which means that the Q-point changes. In this example, if v_{CM} were to increase above $+5\,V$, then Q_1 and Q_2 would be driven into saturation. This demonstrates that there is a limited range of applied common-mode voltage over which Q_1 and Q_2 will remain biased in the forward-active mode.

Test Your Understanding

11.1 Input voltages $v_1 = 2 + 0.005\sin\omega t\,V$ and $v_2 = 0.5 - 0.005\sin\omega t\,V$ are applied to a differential amplifier. Find the differential- and common-mode components of the input signal. (Ans. $V_d = 1.5 + 0.010\sin\omega t\,V$, $V_{cm} = 1.25\,V$)

11.2 For the differential amplifier in Figure 11.2, the parameters are: $V^+ = 10\,V$, $V^- = -10\,V$, $I_Q = 1\,mA$, and $R_C = 10\,k\Omega$. The transistor parameters are: $\beta = 200$, $V_{BE}(\text{on}) = 0.7\,V$, and $V_A = \infty$. Find the voltages v_E, v_{C1}, and v_{C2}, for $v_1 = v_2 = 0$. (Ans. $v_E = -0.7\,V$, $v_{C1} = v_{C2} = 5\,V$)

RD11.3 Consider the diff-amp in Figure 11.2, with parameters: $V^+ = 10\,V$, $V^- = -10\,V$, and $I_Q = 2\,mA$. Redesign the circuit such that the common-mode input voltage is in the range $-4 \leq v_{cm} \leq +4\,V$, while Q_1 and Q_2 remain biased in the forward-active region. (Ans. $R_C = 6\,k\Omega$)

11.2.2   DC Transfer Characteristics

We can perform a general analysis of the differential-pair configuration by using the exponential relationship between collector current and B–E voltage. To begin, we know that

$$i_{C1} = I_S e^{v_{BE1}/V_T} \tag{11.7(a)}$$

and

$$i_{C2} = I_S e^{v_{BE2}/V_T} \tag{11.7(b)}$$

We assume Q_1 and Q_2 are matched and are operating at the same temperature, so the coefficient I_S is the same in each expression.

Neglecting base currents and assuming I_Q is an ideal constant-current source, we have

$$I_Q = i_{C1} + i_{C2} \tag{11.8}$$

where i_{C1} and i_{C2} are the total instantaneous currents, which may include the signal currents. We then have

$$I_Q = I_S\left[e^{v_{BE1}/V_T} + e^{v_{BE2}/V_T}\right] \tag{11.9}$$

Taking the ratios of i_{C1} to I_Q and i_{C2} to I_Q, we obtain

$$\frac{i_{C1}}{I_Q} = \frac{1}{1 + e^{(v_{BE2}-v_{BE1})/V_T}} \tag{11.10(a)}$$

and

$$\frac{i_{C2}}{I_Q} = \frac{1}{1 + e^{-(v_{BE2}-v_{BE1})/V_T}}$$
(11.10(b))

From Figure 11.3(b) we see that

$$v_{BE1} - v_{BE2} \equiv v_d$$
(11.11)

where v_d is the differential-mode input voltage. Equations (11.10(a)) and (11.10(b)) can then be written in terms of v_d, as follows:

$$i_{C1} = \frac{I_Q}{1 + e^{-v_d/V_T}}$$
(11.12(a))

and

$$i_{C2} = \frac{I_Q}{1 + e^{+v_d/V_T}}$$
(11.12(b))

Equations (11.12(a)) and (11.12(b)) describe the basic current–voltage characteristics of the differential amplifier. If the differential-mode input voltage is zero, then the current I_Q splits evenly between i_{C1} and I_{C2}, as we discussed. However, when a differential-mode signal v_d is applied, a difference occurs between i_{C1} and i_{C2} which in turn causes a change in the collector terminal voltage. This is the fundamental operation of the diff-amp. If a common-mode signal $v_{CM} = v_{B1} = v_{B2}$ is applied, the bias current I_Q still splits evenly between the two transistors.

Figure 11.4 is the normalized plot of the **dc transfer characteristics** for the differential amplifier. We can make two basic observations. First, the gain of the differential amplifier is proportional to the slopes of the transfer curves about the point $v_d = 0$. In order to maintain a linear amplifier, the excursion of v_d about zero must be kept small.

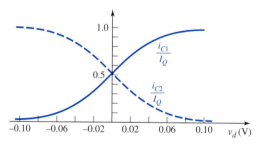

Figure 11.4   Normalized dc transfer characteristics for BJT differential amplifier

Second, as the magnitude of v_d becomes sufficiently large, essentially all of current I_Q goes to one transistor, and the second transistor effectively turns off. This particular characteristic is used in the emitter-coupled logic (ECL) family of digital logic circuits, which is discussed in Chapter 17.

Example   11.2   Objective: Determine the maximum differential-mode input signal that can be applied and still maintain linearity in the differential amplifier.

Figure 11.5 shows an expanded view of the normalized i_{C1} versus v_d characteristic. A linear approximation that corresponds to the slope at $v_d = 0$ is superimposed on the curve. Determine $v_d(\text{max})$ such that the difference between the linear approximation and the actual curve is 1 percent.

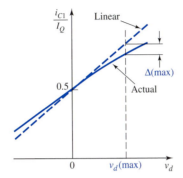

Figure 11.5   Expanded view, normalized i_{C1} versus v_d transfer characteristic

Solution: The actual expression for i_{C1} versus v_d is, from Equation (11.12(a)),

$$i_{C1}(\text{actual}) = \frac{I_Q}{1 + e^{-v_d/V_T}}$$

The slope at $v_d = 0$ is found to be

$$g_f = \left.\frac{di_{C1}}{dv_d}\right|_{v_d=0} = I_Q(-1)\left[1 + e^{-v_d/V_T}\right]^{-2}\left(\frac{-1}{V_T}\right)\left[e^{-v_d/V_T}\right]\Big|_{v_d=0}$$

or

$$g_f = \frac{I_Q}{4V_T} \qquad\qquad (11.13)$$

where g_f is the **forward transconductance**. The linear approximation for i_{C1} versus v_d can be written

$$i_{C1}(\text{linear}) = 0.5I_Q + g_f v_d = 0.5I_Q + \left(\frac{I_Q}{4V_T}\right)v_d \qquad\qquad (11.14)$$

The differential-mode input voltage $v_d(\text{max})$ that results in a 1 percent difference between the ideal linear curve and the actual curve is found from

$$\frac{i_{C1}(\text{linear}) - i_{C1}(\text{actual})}{i_{C1}(\text{linear})} = 0.01$$

or

$$\frac{\left[0.5I_Q + \left(\dfrac{I_Q}{4V_T}\right)v_d(\text{max})\right] - \dfrac{I_Q}{1 + e^{-v_d(\text{max})/V_T}}}{\left[0.5I_Q + \left(\dfrac{I_Q}{4V_T}\right)v_d(\text{max})\right]} = 0.01$$

If we rearrange terms, this expression becomes

$$0.99\left[0.5 + \left(\frac{1}{4V_T}\right)v_d(\text{max})\right] = \frac{1}{1 + e^{-v_d(\text{max})/V_T}}$$

Assuming $V_T = 26\,\text{mV}$, and using trial and error, we find that

$$v_d(\text{max}) \cong 18\,\text{mV}$$

Comment: The differential-mode input voltage must be held to within $\pm 18\,\text{mV}$ in order for the output signal of this diff-amp to be within 1 percent of a linear response.

Test Your Understanding

11.4 Considering the dc transfer characteristics in Figure 11.4, determine the value of the differential-mode input signal such that $i_{C2} = 0.99 I_Q$. (Ans. $v_d = -119.5\,\text{mV} \cong -120\,\text{mV}$)

11.5 Plot the dc transfer characteristics in Figure 11.4 using a computer simulation.

We can now begin to consider the operation of the diff-amp in terms of the small-signal parameters. Figure 11.6 shows the differential-pair configuration with an applied differential-mode input signal. Note that the polarity of the input voltage at Q_1 is opposite to that at Q_2. The forward-transconductance g_f can be written in terms of the individual transistor transconductances g_m. From Equation (11.13), we have

$$g_f = \frac{I_Q}{4V_T} = \frac{1}{2}\frac{I_Q/2}{V_T} = \frac{1}{2}g_m \qquad \textbf{(11.15)}$$

where $(I_Q/2)$ is the quiescent collector current in Q_1 and Q_2. The magnitude of the small-signal collector current in each transistor is then $(g_m v_d)/2$.

Figure 11.6 also shows the linear approximations for the collector currents in terms of the transistor transconductances g_m. The slope of i_{C1} versus v_d is the same magnitude as that of i_{C2} versus v_d, but it has the opposite sign. This is the reason for the negative sign in the expression for i_{C2} versus v_d.

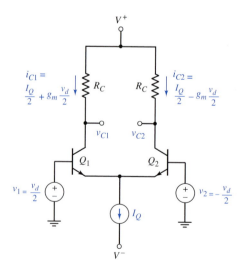

Figure 11.6 BJT differential amplifier with differential-mode input signal

We can define the output signal voltage as

$$v_o = v_{C2} - v_{C1} \tag{11.16}$$

When the output is defined as the difference between the two collector voltages, we have a **two-sided output**. From Figure 11.6, we can write the output voltage as

$$v_o = [V^+ - i_{C2}R_C] - [V^+ - i_{C1}R_C] = (i_{C1} - i_{C2})R_C \tag{11.17(a)}$$

or

$$v_o = \left[\left(\frac{I_Q}{2} + \frac{g_m v_d}{2} \right) - \left(\frac{I_Q}{2} - \frac{g_m v_d}{2} \right) \right] R_C = g_m R_C v_d \tag{11.17(b)}$$

Figure 11.7 shows the ac equivalent circuit of the diff-amp configuration, as well as the signal voltages and currents as functions of the transistor transconductances g_m. Since we are assuming an ideal current source, the output resistance looking into the current source is infinite (represented by the dashed line). Using the equivalent circuit in Figure 11.7(a), we find the signal output voltage to be

$$v_o = v_{c2} - v_{c1} = \left(\frac{g_m v_d}{2} \right) R_C - \left(\frac{-g_m v_d}{2} \right) R_C = g_m R_c v_d \tag{11.18}$$

which is the same as Equation (11.17(b)).

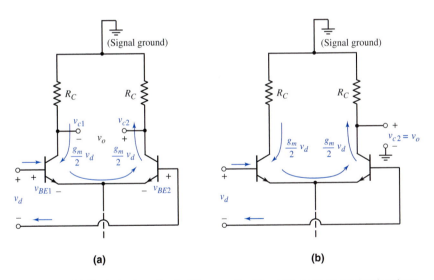

Figure 11.7 (a) Equivalent ac circuit, diff-amp with differential-mode input signal and two-sided output voltage and (b) ac equivalent circuit with one-sided output

The ratio of the output signal voltage to the differential-mode input signal is called the **differential-mode gain**, A_d, which is

$$A_d = \frac{v_o}{v_d} = g_m R_C = \frac{I_Q R_C}{2V_T} \tag{11.19}$$

If the output voltage is the difference between the two collector terminal voltages, then neither side of the output voltage is at ground potential. In many cases, the output voltage is taken at one collector terminal with respect to ground. The resulting voltage output is called a **one-sided output**. If we define the output to be v_{c2}, then from Figure 11.7(b), the signal output voltage is

$$v_o = \left(\frac{g_m v_d}{2}\right) R_C \tag{11.20}$$

The differential gain for the one-sided output is then given by

$$A_d = \frac{v_o}{v_d} = \frac{g_m R_C}{2} = \frac{I_Q R_C}{4 V_T} \tag{11.21}$$

The differential gain for the one-sided output is one-half that of the two-sided output. However, as we will see in our discussion on active loads, only a one-sided output is available.

11.2.3   Small-Signal Equivalent Circuit Analysis

The dc transfer characteristics derived in the last section provide insight into the operation of the differential amplifier. Assuming we are operating in the linear range, we can also derive the gain and other characteristics of the diff-amp, using the small-signal equivalent circuit.

Figure 11.8 shows the small-signal equivalent circuit of the bipolar differential-pair configuration. We assume that the Early voltage is infinite for the two emitter-pair transistors, and that the constant-current source is not ideal but can be represented by a finite output impedance R_o. Resistances R_B are also included. These represent the output resistance of the signal voltage sources. All voltages are represented by their phasor components. Since the two transistors are biased at the same quiescent current, we have

$$r_{\pi 1} = r_{\pi 2} \equiv r_\pi \qquad \text{and} \qquad g_{m1} = g_{m2} \equiv g_m$$

Figure 11.8   Small-signal equivalent circuit, bipolar differential amplifier

Writing a KCL equation at node V_e, using phasor notation, we have

$$\frac{V_{\pi 1}}{r_\pi} + g_m V_{\pi 1} + g_m V_{\pi 2} + \frac{V_{\pi 2}}{r_\pi} = \frac{V_e}{R_o} \tag{11.22(a)}$$

or

$$V_{\pi 1}\left(\frac{1+\beta}{r_\pi}\right) + V_{\pi 2}\left(\frac{1+\beta}{r_\pi}\right) = \frac{V_e}{R_o} \qquad \textbf{(11.22(b))}$$

where $g_m r_\pi = \beta$. From the circuit, we see that

$$\frac{V_{\pi 1}}{r_\pi} = \frac{V_{b1} - V_e}{r_\pi + R_B} \qquad \text{and} \qquad \frac{V_{\pi 2}}{r_\pi} = \frac{V_{b2} - V_e}{r_\pi + R_B}$$

Solving for $V_{\pi 1}$ and $V_{\pi 2}$ and substituting into Equation (11.22(b)), we find

$$(V_{b1} + V_{b2} - 2V_e)\left(\frac{1+\beta}{r_\pi + R_B}\right) = \frac{V_e}{R_o} \qquad \textbf{(11.23)}$$

Solving for V_e, we obtain

$$V_e = \frac{V_{b1} + V_{b2}}{2 + \dfrac{r_\pi + R_B}{(1+\beta)R_o}} \qquad \textbf{(11.24)}$$

If we consider a one-sided output at the collector of Q_2, then

$$V_o = V_{c2} = -(g_m V_{\pi 2})R_C = -\frac{\beta R_C (V_{b2} - V_e)}{r_\pi + R_B} \qquad \textbf{(11.25)}$$

Substituting Equation (11.24) into (11.25) and rearranging terms yields

$$V_o = -g_m R_C \left\{ \frac{V_{b2}\left[1 + \dfrac{r_\pi + R_B}{(1+\beta)R_o}\right] - V_{b1}}{2 + \dfrac{r_\pi + R_B}{(1+\beta)R_o}} \right\} \qquad \textbf{(11.26)}$$

In an ideal constant-current source, the output resistance is $R_o = \infty$, and Equation (11.26) reduces to

$$V_o = -\frac{\beta R_C (V_{b2} - V_{b1})}{2(r_\pi + R_B)} \qquad \textbf{(11.27)}$$

The differential-mode input is

$$V_d = V_{b1} - V_{b2}$$

and the differential-mode gain is

$$A_d = \frac{V_o}{V_d} = \frac{\beta R_C}{2(r_\pi + R_B)} \qquad \textbf{(11.28)}$$

which for $R_B = 0$ is identical to Equation (11.21), which was developed from the voltage transfer characteristics.

Equation (11.26) includes a finite output resistance for the current source. We can see that when a common-mode signal $V_{cm} = V_{b1} = V_{b2}$ is applied, the output voltage is no longer zero.

Differential- and common-mode voltages are defined in Equations (11.2) and (11.3). Using phasor notation, we can solve these equations for V_{b1} and V_{b2} in terms of V_d and V_{cm}. We obtain

$$V_{b1} = V_{cm} + \frac{V_d}{2} \qquad \textbf{(11.29(a))}$$

and

$$V_{b2} = V_{cm} - \frac{V_d}{2} \tag{11.29(b)}$$

Since we are dealing with a linear amplifier, superposition applies. Equations (11.29(a)) and (11.29(b)) then simply state that the two input signals can be written as the sum of a differential-mode input signal component and a common-mode input signal component.

Substituting Equations (11.29(a)) and (11.29(b)) into Equation (11.26) and rearranging terms results in the following:

$$V_o = \frac{\beta R_C}{2(r_\pi + R_B)} \cdot V_d - \frac{g_m R_C}{1 + \frac{2(1+\beta)R_o}{r_\pi + R_B}} \cdot V_{cm} \tag{11.30}$$

We can write the output voltage in the general form

$$V_o = A_d V_d + A_{cm} V_{cm} \tag{11.31}$$

where A_d is the differential-mode gain and A_{cm} is the common-mode gain. Comparing Equations (11.30) and (11.31), we see that the differential-mode gain is

$$A_d = \frac{\beta R_C}{2(r_\pi + R_B)} \tag{11.32(a)}$$

and the common-mode gain is

$$A_{cm} = \frac{-g_m R_C}{1 + \frac{2(1+\beta)R_o}{r_\pi + R_B}} \tag{11.32(b)}$$

We again observe that the common-mode gain goes to zero for an ideal current source in which $R_o = \infty$. For a nonideal current source, R_o is finite and the common-mode gain is not zero for this case of a one-sided output. A nonzero common-mode gain implies that the diff-amp is not ideal. We will see implications of the nonideal effects in later discussions.

Example 11.3 **Objective:** Determine the differential- and common-mode gains of a diff-amp.

Consider the circuit in Figure 11.2, with parameters: $V^+ = 10\,\text{V}$, $V^- = -10\,\text{V}$, $I_Q = 0.8\,\text{mA}$, and $R_C = 12\,\text{k}\Omega$. The transistor parameters are $\beta = 100$ and $V_A = \infty$. Assume the output resistance looking into the constant-current source is $R_o = 25\,\text{k}\Omega$. Assume the source resistors R_B are zero. Use a one-sided output at v_{C2}.

Solution: From Equation (11.32(a)), the differential-mode gain is

$$A_d = \frac{g_m R_C}{2} = \frac{I_{CQ} R_C}{2V_T} = \frac{I_Q R_C}{4V_T} = \frac{(0.8)(12)}{4(0.026)} = 92.3$$

From Equation (11.32(b)), the common-mode gain is

$$A_{cm} = \frac{-\left(\dfrac{I_Q R_C}{2V_T}\right)}{1 + \dfrac{(1+\beta)I_Q R_o}{V_T \beta}} = \frac{-\left[\dfrac{(0.8)(12)}{(2)(0.026)}\right]}{1 + \dfrac{(101)(0.8)(25)}{(0.026)(100)}} = -0.237$$

Comment: The common-mode gain is significantly less than the differential-mode gain, but it is not zero as determined for the ideal diff-amp with an ideal current source.

Design Example 11.4 **Objective:** Design a differential amplifier to meet the specifications of the following experimental system.

Figure 11.9 shows a Hall-effect experiment to measure semiconductor material parameters. A Hall voltage V_H, which is perpendicular to both a current I_X and a magnetic field B_Z, is to be measured by using a diff-amp. The range of V_H is $-8 \le V_H \le +8$ mV and the desired range of the diff-amp output signal is to be $-0.8 \le V_O \le +0.8$ V. The probes that make contact to the semiconductor have an effective resistance of 500 Ω, and each probe has an induced 60 Hz signal with a magnitude of 100 mV. The diff-amp output 60 Hz signal is to be no larger than 10 mV. Typically, $V_X = 5$ V, so that the quiescent or common-mode voltage of the Hall probes is 2.5 V.

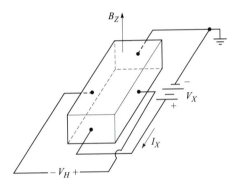

Figure 11.9 Experimental arrangement for measuring Hall voltage

Design Approach: For this example, we will use a bipolar diff-amp with the configuration shown in Figure 11.6. Assume that transistors are available with $\beta = 100$. Assume bias voltages of ± 10 V and choose a bias current of $I_Q = 0.5$ mA.

Solution: Differential-Mode Gain: The differential-mode voltage gain requirement is

$$A_d = \frac{V_o}{V_d} = \frac{0.8}{0.008} = 100$$

The small-signal parameters are then

$$r_\pi = \frac{\beta V_T}{I_{CQ}} = \frac{(100)(0.026)}{0.25} = 10.4 \, \text{k}\Omega$$

and

$$g_m = \frac{I_{CQ}}{V_T} = \frac{0.25}{0.026} = 9.62 \, \text{mA/V}$$

The differential gain is

$$A_d = \frac{\beta R_C}{2(r_\pi + R_B)}$$

or

$$100 = \frac{(100)R_C}{2(10.4 + 0.5)}$$

which means that $R_C = 21.8\,\text{k}\Omega$. We may note that the voltage drop across R_C under quiescent conditions is 5.45 V. With a 2.5 V common-mode input voltage, the quiescent collector-emitter voltages of Q_1 and Q_2 are approximately 3.65 V. The two input transistors will then remain in the active region.

Solution: Common-Mode Gain: The common-mode voltage gain requirement is

$$A_{cm} = \frac{V_o}{V_{cm}} = \frac{10\,\text{mV}}{100\,\text{mV}} = 0.10$$

The common-mode gain is given by

$$|A_{cm}| = \frac{g_m R_C}{1 + \dfrac{2(1+\beta)R_o}{r_\pi + R_B}}$$

or

$$0.10 = \frac{(9.62)(21.8)}{1 + \dfrac{2(101)R_o}{10.4 + 0.5}}$$

which means that $R_o = 113\,\text{k}\Omega$. If we consider a simple two-transistor current source as discussed in the last chapter, the output resistance is $R_o = r_o = V_A/I_Q$, where V_A is the Early voltage. With $I_Q = 0.5\,\text{mA}$, then $V_A = 56.5\,\text{V}$ is the Early voltage requirement. This specification is not difficult to achieve for most bipolar transistors.

Computer Simulation Verification: Figure 11.10 shows the circuit used in the computer simulation for this example. The bias current I_Q supplied by the Q_3 current source transistor is 0.568 mA. A 2.5 V common-mode input voltage is applied, a 500 Ω source (probe) resistance is included, and an 8 mV differential-mode input signal is applied. The differential output signal voltage measured at the collector of Q_2 is 0.84 V, which is just slightly larger than the designed value. The current gains of the standard 2N3904 transistors used in the computer simulation are larger than the values of 100 used in the hand analysis and design. A common-mode signal voltage of 100 mV replaced the differential-mode signals. The common-mode output signal is 7.11 mV, which is within the design specification.

Figure 11.10   Circuit used in the computer simulation of Design Example 11.4

Test Your Understanding

11.6 In the differential amplifier in Figure 11.11, neglect base currents and assume $V_{EB}(\text{on}) = 0.7\,\text{V}$. Determine v_E and v_{EC1} for common-mode input voltages $v_1 = v_2 = v_{cm}$ of: (a) $0\,\text{V}$, (b) $+2.5\,\text{V}$, and (c) $-2.5\,\text{V}$. (Ans. (a) $v_E = +0.7\,\text{V}$, $v_{EC1} = 3.7\,\text{V}$ (b) $v_E = 3.2\,\text{V}$, $v_{EC1} = 6.2\,\text{V}$ (c) $v_E = -1.8\,\text{V}$, $v_{EC1} = 1.2\,\text{V}$)

D11.7 Using the diff-amp configuration in Figure 11.2, design the circuit such that the differential-mode voltage gain at v_{c2} is $+150$ and the differential-mode voltage gain at v_{C1} is -100. (Ans. For example, if $I_Q = 1\,\text{mA}$, then $R_C = 15.6\,\text{k}\Omega$ at collector of Q_2 and $R_C = 10.4\,\text{k}\Omega$ at collector of Q_1)

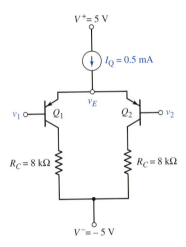

Figure 11.11 Figure for Exercise 11.6

11.2.4 Differential- and Common-Mode Gains

For greater insight into the mechanism that causes differential- and common-mode gains, we reconsider the diff-amp as pure differential- and common-mode signals are applied.

Figure 11.12(a) shows the ac equivalent circuit of the diff-amp with two sinusoidal input signals. The two input voltages are 180 degrees out of phase, so a pure differential-mode signal is being applied to the diff-amp. We see that $v_{b1} + v_{b2} = 0$. From Equation (11.24), the common emitters of Q_1 and Q_2 remain at signal ground. In essence, the circuit behaves like a balanced seesaw. As the base voltage of Q_1 goes into its positive-half cycle, the base voltage of Q_2 is in its negative half-cycle. Then, as the base voltage of Q_1 goes into its negative half-cycle, the base voltage of Q_2 is in its positive half-cycle. The signal current directions shown in the figure are valid for v_{b1} in its positive half-cycle.

Since v_e is always at ground potential, we can treat each half of the diff-amp as a common-emitter circuit. Figure 11.12(b) shows the differential half-circuits, clearly depicting the common-emitter configuration. The differential-

Figure 11.12 (a) Equivalent ac circuit, diff-amp with applied sinusoidal differential-mode input signal, and resulting signal current directions and (b) differential-mode half-circuits

mode characteristics of the diff-amp can be determined by analyzing the half-circuit. In evaluating the small-signal hybrid-π parameters, we must keep in mind that the half-circuit is biased at $I_Q/2$.

Figure 11.13(a) shows the ac equivalent circuit of the diff-amp with a pure common-mode sinusoidal input signal. In this case, the two input voltages are in phase. The current source is represented as an ideal source I_Q in parallel with its output resistance R_O. Current i_q is the time-varying component of the source current. As the two input signals increase, voltage v_e increases and current i_q increases. Since this current splits evenly between Q_1 and Q_2, each collector current also increases. The output voltage v_o then decreases below its quiescent value.

As the two input voltages go through the negative half-cycle, all signal currents shown in the figure reverse direction, and v_o increases above its quiescent value. Consequently, a common-mode sinusoidal input signal produces a sinusoidal output voltage, which means that the diff-amp has a nonzero common-mode voltage gain. If the value of R_o increases, the magnitude of i_q decreases for a given common-mode input signal, producing a smaller output voltage and hence a smaller common-mode gain.

With an applied common-mode voltage, the circuit shown in Figure 11.13(a) is perfectly symmetrical. The circuit can therefore be split into the identical common-mode half-circuits shown in Figure 11.13(b). The common-mode characteristics of the diff-amp can then be determined by analyzing the half-circuit, which is a common-emitter configuration with an emitter resistor. Each half-circuit is biased at $I_Q/2$.

The following examples further illustrate the effect of a nonzero common-mode gain on circuit performance.

(a)

(b)

Figure 11.13 (a) Equivalent ac circuit of diff-amp with common-mode input signal, and resulting signal current directions and (b) common-mode half-circuits

---

Example 11.5 **Objective:** Determine the output voltage of a diff-amp when only a common-mode signal is applied.

Consider the circuit in Figure 11.2. Use the transistor and circuit parameters described in Example 11.3. Assume the common-mode input signal is $v_1 = v_2 = v_{cm} = 200 \sin \omega t \, \mu\text{V}$.

Solution: From Example 11.3, the common-mode gain is $A_{cm} = -0.237$. Since the differential-mode input signal is zero, the output signal voltage is

$$v_o = A_{cm} v_{cm} = -(0.237)(200 \sin \omega t) \, \mu\text{V} = -47.4 \sin \omega t \, \mu\text{V}$$

Comment: When the magnitude of the common-mode gain is less than unity, the common-mode output voltage is less than the common-mode input voltage; yet it is not zero, which would occur in an ideal diff-amp.

Example 11.6 **Objective:** Determine the output of a diff-amp when both differential- and common-mode signals are applied.

Consider the circuit shown in Figure 11.2. Use the transistor and circuit parameters described in Example 11.3. Assume that four sets of inputs are applied, as described in the following table, which also includes the differential- and common-mode voltages.

	Input signal (μV)	Differential- and common-mode input signals (μV)
Case 1	$v_1 = 10 \sin \omega t$ $v_2 = -10 \sin \omega t$	$v_d = 20 \sin \omega t$ $v_{cm} = 0$
Case 2	$v_1 = 20 \sin \omega t$ $v_2 = -20 \sin \omega t$	$v_d = 40 \sin \omega t$ $v_{cm} = 0$
Case 3	$v_1 = 210 \sin \omega t$ $v_2 = 190 \sin \omega t$	$v_d = 20 \sin \omega t$ $v_{cm} = 200 \sin \omega t$
Case 4	$v_1 = 220 \sin \omega t$ $v_2 = 180 \sin \omega t$	$v_d = 40 \sin \omega t$ $v_{cm} = 200 \sin \omega t$

Solution: The output voltage is given by Equation (11.31), as follows:

$$v_o = A_d v_d + A_{cm} v_{cm}$$

From Example 11.3, the differential- and common-mode gains are $A_d = 92.3$ and $A_{cm} = -0.237$. The output voltages for the four sets of inputs are:

	Output signal (mV)
Case 1	$v_o = 1.846 \sin \omega t$
Case 2	$v_o = 3.692 \sin \omega t$
Case 3	$v_o = 1.799 \sin \omega t$
Case 4	$v_o = 3.645 \sin \omega t$

Comment: In cases 1 and 2, the common-mode input is zero, and the output is directly proportional to the differential input signal. Comparing cases 1 and 3 and cases 2 and 4, we see that the output voltages are not equal, even though the differential input signals are the same. This shows that the common-mode signal affects the output. Also, even though the differential signal is doubled, in cases 4 and 3, the ratio of the output signals is not 2.0. If a common-mode signal is present, the output is not exactly linear with respect to the differential input signal.

Problem-Solving Technique: Diff-Amps with Resistive Loads

1. To determine the differential-mode voltage gain, apply a pure differential-mode input voltage and use the differential-mode half-circuit in the analysis.

2. To determine the common-mode voltage gain, apply a pure common-mode input voltage and use the common-mode half-circuit in the analysis.

11.2.5   Common-Mode Rejection Ratio

The ability of a differential amplifier to reject a common-mode signal is described in terms of the common-mode rejection ratio (CMRR). The CMRR is a figure-of-merit for the diff-amp and is defined as

$$\text{CMRR} = \left| \frac{A_d}{A_{cm}} \right| \tag{11.33}$$

For an ideal diff-amp, $A_{cm} = 0$ and $\text{CMRR} = \infty$. Usually, the CMRR is expressed in decibels, as follows:

$$\text{CMMR}_{\text{dB}} = 20 \log_{10} \left| \frac{A_d}{A_{cm}} \right| \tag{11.34}$$

For the diff-amp in Figure 11.2, the one-sided differential- and common-mode gains are given by Equations (11.32(a)) and (11.32(b)). Using these equations, we can express the CMRR as

$$\text{CMRR} = \left| \frac{A_d}{A_{cm}} \right| = \frac{1}{2} \left[1 + \frac{(1 + \beta) I_Q R_o}{V_T \beta} \right] \tag{11.35}$$

The common-mode gain decreases as R_o increases. Therefore, we see that the CMRR increases as R_o increases.

Example 11.7   **Objective:** Determine the CMRR of a differential amplifier.

Consider the circuit shown in Figure 11.2, with circuit and transistor parameters as given in Example 11.3.

Solution: From the results of Example 11.3, we have $A_d = 92.3$ and $A_{cm} = -0.237$. The CMRR is then

$$\text{CMRR} = \left| \frac{A_d}{A_{cm}} \right| = \frac{92.3}{0.237} = 389$$

or, expressed in decibels,

$$\text{CMRR}_{\text{dB}} = 20 \log_{10}(389) = 51.8 \, \text{dB}$$

Comment: For "good" diff-amps, typical values of CMRR are in the range of 80–100 dB. The CMRR of the diff-amp in Figure 11.2 can be improved by increasing the current-source output resistance.

Design Example 11.8   Objective: Design a bipolar current source with the required output resistance parameter to meet a specified CMRR.

Consider the diff-amp in Figure 11.2. Use the circuit and transistor parameters given in Example 11.3. Determine the required value of R_o for $\text{CMRR}_{dB} = 90$ dB.

Solution: If $\text{CMRR}_{dB} = 90$ dB, then $\text{CMRR} = 3.16 \times 10^4$. From Equation (11.35), we have

$$\text{CMRR} = \left| \frac{A_d}{A_{cm}} \right| = \frac{1}{2}\left[1 + \frac{(1+\beta)I_Q R_o}{V_T \beta} \right]$$

or

$$3.16 \times 10^4 = \frac{1}{2}\left[1 + \frac{(101)(0.8)R_o}{(0.026)(100)} \right]$$

which yields

$$R_o = 2.03 \times 10^3 \, \text{k}\Omega = 2.03 \, \text{M}\Omega$$

Comment: This output resistance level can be achieved with a Widlar or Wilson current source.

Computer Simulation Verification: A standard two-transistor current source (Figure 10.2) was designed with 2N3904 bipolar transistors. With a bias current of 0.8 mA, the output resistance of the current source is 93.8 kΩ which is far below the design requirement. Using a modified Widlar current source (Figure P10.26) with 1 kΩ emitter resistors, the output resistance of the current source is 2.22 MΩ, which is within the design specification.

Test Your Understanding

*RD11.8   For the diff-amp shown in Figure 11.2, the parameters are $V^+ = 15$ V and $V^- = -15$ V. Assume $\beta = 200$. The range for the common-mode input voltage is to be $-5 \le v_{cm} \le +5$ V. (a) Redesign the circuit to produce the maximum one-sided differential-mode gain at v_{C2}. (b) If $R_o = 100$ kΩ for the current source, determine the resulting common-mode gain and CMRR_{dB}. (Ans. (a) $I_Q R_C = 20$ V, $A_d(\text{max}) = 192$ (b) For $I_Q = 0.5$ mA and $R_C = 40$ kΩ, $A_{cm} = -0.199$; $\text{CMRR}_{dB} = 59.7$ dB)

*D11.9   Consider a differential amplifier with the configuration in Figure 11.14, biased with a modified Widlar current source. Assume transistor parameters of $\beta = 200$, $V_A = 125$ V for Q_3 and Q_4, and $V_A = \infty$ for Q_1 and Q_2. Design the circuit such that the common-mode input voltage is in the range $-5 \le v_{cm} \le +5$ V, the common-mode rejection ratio is $\text{CMRR}_{dB} = 95$ dB, and the maximum differential-mode voltage gain is achieved. (Ans. For example, let $I_Q = 0.5$ mA and $I_1 = 1$ mA. Then $R_1 = 18.7$ kΩ, $R_2 = 1.31$ kΩ, $R_3 = 0.637$ kΩ, and $R_C = 20$ kΩ.)

Figure 11.14   Figure for Exercise 11.9

11.2.6   Differential- and Common-Mode Input Impedances

The input impedance, or resistance, of an amplifier is as important a property as the voltage gain. The input resistance determines the loading effect of the circuit on the signal source. We will look at two input resistances for the difference amplifier: the **differential-mode input resistance**, which is the resistance seen by a differential-mode signal source; and the **common-mode input resistance**, which is the resistance seen by a common-mode input signal source.

Differential-Mode Input Resistance

The differential-mode input resistance is the effective resistance between the two input base terminals when a differential-mode signal is applied. A diff-amp with a pure differential input signal is shown in Figure 11.15. The applicable differential-mode half-circuits are shown in Figure 11.12(b). For this circuit, we have

$$\frac{v_d/2}{i_b} = r_\pi \tag{11.36}$$

The differential-mode input resistance is therefore

$$R_{id} = \frac{v_d}{i_b} = 2r_\pi \tag{11.37}$$

Another common diff-amp configuration uses emitter resistors, as shown in Figure 11.16. With a pure applied differential-mode voltage, similar differential-mode half-circuits are applicable to this configuration. We can then use the resistance reflection rule to find the differential-mode input resistance. We have

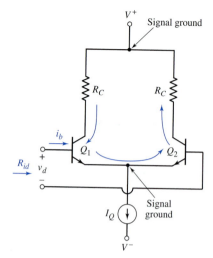

Figure 11.15 BJT differential amplifier with differential-mode input signal, showing differential input resistance

Figure 11.16 BJT differential amplifier with emitter resistors

$$\frac{v_d/2}{i_b} = r_\pi + (1+\beta)R_E \tag{11.38}$$

Therefore,

$$R_{id} = \frac{v_d}{i_b} = 2[r_\pi + (1+\beta)R_E] \tag{11.39}$$

Equation (11.39) implies that differential-mode input resistance increases significantly when emitter resistors are included. Although the differential-mode gain decreases when emitter resistors are included, a larger differential-mode voltage (greater than 18 mV) may be applied and the amplifier remains linear.

Common-Mode Input Resistance

Figure 11.17(a) shows a diff-amp with an applied common-mode voltage. The small-signal output resistance R_o of the constant-current source is also shown. The equivalent common-mode half-circuits are given in Figure 11.13(b). Since the half-circuits are in parallel, we can write

$$2R_{icm} = r_\pi + (1 + \beta)(2R_o) \cong (1 + \beta)(2R_o) \tag{11.40}$$

Equation (11.40) is a first approximation for determining the common-mode input resistance.

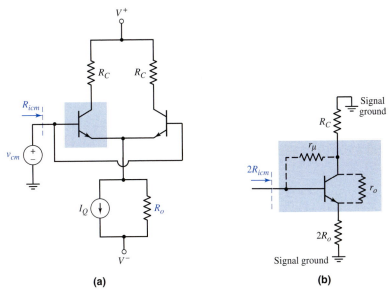

Figure 11.17 (a) BJT differential amplifier with common-mode input signal, including finite current source resistance and (b) equivalent common-mode half-circuit

Normally, R_o is large, and R_{icm} is typically in the megohm range. Therefore, the transistor output resistance r_o and the base–collector resistance r_μ may need to be included in the calculation. Figure 11.17(b) shows the more complete equivalent half-circuit model. For this model, we have

$$2R_{icm} = r_\mu \| [(1 + \beta)(2R_o)] \| [(1 + \beta)r_o] \tag{11.41(a)}$$

Therefore,

$$R_{icm} = \left(\frac{r_\mu}{2}\right) \Big\| [(1 + \beta)(R_o)] \Big\| \left[(1 + \beta)\left(\frac{r_o}{2}\right)\right] \tag{11.41(b)}$$

Example 11.9   **Objective:** Determine the differential- and common-mode input resistances of a differential amplifier.

Consider the circuit in Figure 11.18, with transistor parameters $\beta = 100$, $V_{BE}(\text{on}) = 0.7\,\text{V}$, and $V_A = 100\,\text{V}$. Determine R_{id} and R_{icm}.

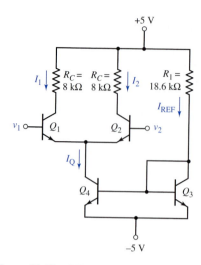

Figure 11.18   BJT differential amplifier for Example 11.9

Solution: From the circuit, we find

$$I_{\mathrm{REF}} = 0.5\,\mathrm{mA} \cong I_Q$$

and

$$I_1 = I_2 \cong I_Q/2 = 0.25\,\mathrm{mA}$$

The small-signal parameters for Q_1 and Q_2 are then

$$r_\pi = \frac{\beta V_T}{I_{CQ}} = \frac{(100)(0.026)}{0.25} = 10.4\,\mathrm{k\Omega}$$

and

$$r_o = \frac{V_A}{I_{CQ}} = \frac{100}{0.25} = 400\,\mathrm{k\Omega}$$

and the output resistance of Q_4 is

$$R_o = \frac{V_A}{I_Q} = \frac{100}{0.5} = 200\,\mathrm{k\Omega}$$

From Equation (11.37), the differential-mode input resistance is

$$R_{id} = 2r_\pi = 2(10.4) = 20.8\,\mathrm{k\Omega}$$

From Equation (11.41(b)), neglecting the effect of r_μ, the common-mode input resistance is

$$R_{icm} = (1 + \beta)\left[(R_o)\left\|\left(\frac{r_o}{2}\right)\right.\right] = (101)\left\{200\left\|\left(\frac{400}{2}\right)\right.\right\}\mathrm{k\Omega} \rightarrow 10.1\,\mathrm{M\Omega}$$

Comment: If a differential-mode input voltage with a peak value of 15 mV is applied, the source must be capable of supplying a current of $15 \times 10^{-3}/20.8 \times 10^{+3} = 0.72\,\mu\mathrm{A}$ without any severe loading effect. However, the input current from a 15 mV common-mode signal would only be approximately 1.5 nA.

Test Your Understanding

11.10 If the differential-mode gain of a diff-amp is $A_d = 60$ and the common-mode gain is $A_{cm} = 0.5$, determine the output voltage for input signals of: (a) $v_1 = 0.505 \sin \omega t$ V, $v_2 = 0.495 \sin \omega t$ V, and (b) $v_1 = 0.5 + 0.005 \sin \omega t$ V, $v_2 = 0.5 - 0.005 \sin \omega t$ V. (Ans. (a) $v_o = 0.85 \sin \omega t$ V (b) $v_o = 0.25 + 0.6 \sin \omega t$ V)

11.11 A differential amplifier is shown in Figure 11.2. The parameters are: $V^+ = 10$ V, $V^- = -10$ V, $I_Q = 2$ mA, and $R_C = 5$ kΩ. The output resistance of the constant-current source is $R_o = 50$ kΩ, and the transistor parameters are: $\beta = 150$, $V_{BE}(\text{on}) = 0.7$ V, and $V_A = \infty$. (a) Determine the dc input base currents. (b) Determine the differential signal input currents if a differential mode input voltage $v_d = 10 \sin \omega t$ mV is applied. (c) If a common-mode input voltage $v_{cm} = 3 \sin \omega t$ V is applied, determine the common-mode signal input currents. (Ans. (a) $I_{B1} = I_{B2} = 6.62$ μA (b) $I_b = 1.28 \sin \omega t$ μA (c) $I_b = 0.199 \sin \omega t$ μA)

11.3 BASIC FET DIFFERENTIAL PAIR

In this section, we will evaluate the basic FET differential amplifier, concentrating on the MOSFET diff-amp. As we did for the bipolar diff-amp, we will develop the dc transfer characteristics, and determine the differential- and common-mode gains. The MOSFET with an active load is then considered.

Differential amplifiers using JFETs are also available. Since the analysis is almost identical to that for the MOSFET diff-amp, we will only briefly consider the JFET differential pair. A few of the problems at the end of this chapter are based on these circuits.

11.3.1 DC Transfer Characteristics

Figure 11.19 shows the basic MOSFET differential pair, with matched transistors M_1 and M_2 biased with a constant current I_Q. We assume that M_1 and

Figure 11.19 Basic MOSFET differential pair configuration

M_2 are always biased in the saturation region. MOSFET current-source circuits were discussed in Chapter 10 in Section 10.2.

Like the basic bipolar configuration, the basic MOSFET diff-amp uses both positive and negative bias voltages, thereby eliminating the need for coupling capacitors and voltage divider biasing resistors at the gate terminals. Even with $v_{G1} = v_{G2} = 0$, the transistors M_1 and M_2 can be biased in the saturation region by the current source I_Q. This circuit, then, is also a **dc-coupled** diff-amp.

Example 11.10   **Objective:** Calculate the dc characteristics of a MOSFET diff-amp.

Consider the differential amplifier shown in Figure 11.20. The transistor parameters are: $K_{n1} = K_{n2} = 0.1 \, \text{mA/V}^2$, $K_{n3} = K_{n4} = 0.3 \, \text{mA/V}^2$, and for all transistors, $\lambda = 0$ and $V_{TN} = 1 \, \text{V}$. Determine the maximum range of common-mode input voltage.

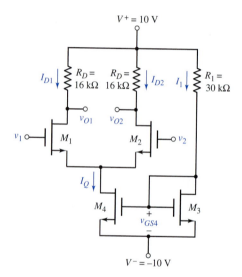

Figure 11.20   MOSFET differential amplifier for Example 11.10

Solution: The reference current can be determined from

$$I_1 = \frac{20 - V_{GS4}}{R_1}$$

and from

$$I_1 = K_{n3}(V_{GS4} - V_{TN})^2$$

Combining these two equations and substituting the parameter values, we obtain

$$9V_{GS4}^2 - 17V_{GS4} - 11 = 0$$

which yields

$$V_{GS4} = 2.40 \, \text{V} \quad \text{and} \quad I_1 = 0.587 \, \text{mA}$$

Since M_3 and M_4 are identical, we also find

$$I_Q = 0.587 \, \text{mA}$$

The quiescent drain currents in M_1 and M_2 are

$$I_{D1} = I_{D2} = I_Q/2 \cong 0.293 \, \text{mA}$$

The gate-to-source voltages are then

$$V_{GS1} = V_{GS2} = \sqrt{\frac{I_{D1}}{K_{n1}}} + V_{TN} = \sqrt{\frac{0.293}{0.1}} + 1 = 2.71 \, \text{V}$$

The quiescent values of v_{O1} and v_{O2} are

$$v_{O1} = v_{O2} = 10 - I_{D1}R_D = 10 - (0.293)(16) = 5.31 \, \text{V}$$

The maximum common-mode input voltage is the value when M_1 and M_2 reach the transition point, or

$$V_{DS1} = V_{DS2} = V_{DS1}(\text{sat}) = V_{GS1} - V_{TN} = 2.71 - 1 = 1.71 \, \text{V}$$

Therefore,

$$v_{CM}(\text{max}) = v_{O1} - V_{DS1}(\text{sat}) + V_{GS1} = 5.31 - 1.71 + 2.71$$

or

$$v_{CM}(\text{max}) = 6.31 \, \text{V}$$

The minimum common-mode input voltage is the value when M_4 reaches the transition point, or

$$V_{DS4} = V_{DS4}(\text{sat}) = V_{GS4} - V_{TN} = 2.4 - 1 = 1.4 \, \text{V}$$

Therefore,

$$v_{CM}(\text{min}) = V_{GS1} + V_{DS4}(\text{sat}) - 10 = 2.71 + 1.4 - 10$$

or

$$v_{CM}(\text{min}) = -5.89 \, \text{V}$$

Comment: For this circuit the maximum range for the common-mode input voltage is $-5.89 \le v_{CM} \le 6.31 \, \text{V}$.

The dc transfer characteristics of the MOSFET differential pair can be determined from the circuit in Figure 11.19. Neglecting the output resistances of M_1 and M_2, and assuming the two transistors are matched, we can write

$$i_{D1} = K_n(v_{GS1} - V_{TN})^2 \qquad \text{(11.42(a))}$$

and

$$i_{D2} = K_n(v_{GS2} - V_{TN})^2 \qquad \text{(11.42(b))}$$

Taking the square roots of Equations (11.42(a)) and (11.42(b)), and subtracting the two equations, we obtain

$$\sqrt{i_{D1}} - \sqrt{i_{D2}} = \sqrt{K_n}(v_{GS1} - v_{GS2}) = \sqrt{K_n} \cdot v_d \qquad \text{(11.43)}$$

where $v_d = v_{G1} - v_{G2} = v_{GS1} - v_{GS2}$ is the differential-mode input voltage. If $v_d > 0$, then $v_{G1} > v_{G2}$ and $v_{GS1} > v_{GS2}$, which implies that $i_{D1} > i_{D2}$. Since

$$i_{D1} + i_{D2} = I_Q \qquad \text{(11.44)}$$

then Equation (11.43) becomes

$$\left(\sqrt{i_{D1}} - \sqrt{I_Q - i_{D1}}\right)^2 = \left(\sqrt{K_n} \cdot v_d\right)^2 = K_n v_d^2 \tag{11.45}$$

when both sides of the equation are squared. After the terms are rearranged, Equation (11.45) becomes

$$\sqrt{i_{D1}(I_Q - i_{D1})} = \frac{1}{2}(I_Q - K_n v_d^2) \tag{11.46}$$

If we square both sides of this equation, we develop the quadratic equation

$$i_{D1}^2 - I_Q i_{D1} + \frac{1}{4}(I_Q - K_n v_d^2)^2 = 0 \tag{11.47}$$

Applying the quadratic formula, rearranging terms, and noting that $i_{D1} > I_Q/2$ and $v_d > 0$, we obtain

$$i_{D1} = \frac{I_Q}{2} + \sqrt{\frac{K_n I_Q}{2}} \cdot v_d \sqrt{1 - \left(\frac{K_n}{2I_Q}\right) v_d^2} \tag{11.48}$$

Using Equation (11.44), we find that

$$i_{D2} = \frac{I_Q}{2} - \sqrt{\frac{K_n I_Q}{2}} \cdot v_d \sqrt{1 - \left(\frac{K_n}{2I_Q}\right) v_d^2} \tag{11.49}$$

The normalized drain currents are

$$\frac{i_{D1}}{I_Q} = \frac{1}{2} + \sqrt{\frac{K_n}{2I_Q}} \cdot v_d \sqrt{1 - \left(\frac{K_n}{2I_Q}\right) v_d^2} \tag{11.50}$$

and

$$\frac{i_{D2}}{I_Q} = \frac{1}{2} - \sqrt{\frac{K_n}{2I_Q}} \cdot v_d \sqrt{1 - \left(\frac{K_n}{2I_Q}\right) v_d^2} \tag{11.51}$$

These equations describe the dc transfer characteristics for this circuit. They are plotted in Figure 11.21 as a function of a normalized differential input voltage $v_d/\sqrt{(2I_Q/K_n)}$.

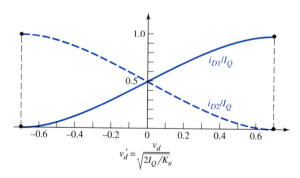

Figure 11.21   Normalized dc transfer characteristics, MOSFET differential amplifier

We can see from Equations (11.50) and (11.51) that, at a specific differential input voltage, bias current I_Q is switched entirely to one transistor or the other. This occurs when

$$|v_d|_{max} = \sqrt{\frac{I_Q}{K_n}} \tag{11.52}$$

The forward transconductance is defined as the slope of the dc transfer characteristic for the i_{D1} curve. From Figure 11.21, we see that the maximum slope, or maximum forward transconductance, occurs at $v_d = 0$, so that

$$g_f(max) = \frac{di_{D1}}{dv_d}\bigg|_{v_d=0} \tag{11.53}$$

Using Equation (11.48), we find that

$$g_f(max) = \sqrt{\frac{K_n I_Q}{2}} = \frac{g_m}{2} \tag{11.54}$$

where g_m is the transconductance of each transistor. The slope of the i_{D2} characteristic curve at $v_d = 0$ is the same, except it is negative.

We can perform an analysis similar to that in Example 11.2 to determine the maximum differential-mode input signal that can be applied and still maintain linearity. If we let $I_Q = 1\,\text{mA}$ and $K_n = 1\,\text{mA/V}^2$, then for differential input voltages less than 0.34 V, the difference between the linear approximation and the actual curve is less than 1 percent. The maximum differential input signal for the MOSFET diff-amp is much larger than for the bipolar diff-amp. The principal reason is that the gain of the MOSFET diff-amp, as we will see, is much smaller than the gain of the bipolar diff-amp.

Figure 11.22 is the ac equivalent circuit of the diff-amp configuration, showing only the differential voltage and signal currents as a function of the transistor transconductance g_m. We assume that the output resistance looking into the current source is infinite. Using this equivalent circuit, we find the one-sided output voltage at v_{o2}, as follows:

$$v_{o2} \equiv v_o = +\left(\frac{g_m v_d}{2}\right) R_D \tag{11.55}$$

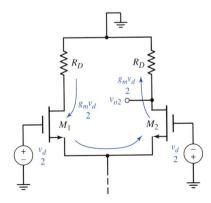

Figure 11.22   AC equivalent circuit, MOSFET differential amplifier

The differential voltage gain is then

$$A_d = \frac{v_o}{v_d} = \frac{g_m R_D}{2} = \sqrt{\frac{K_n I_Q}{2}} \cdot R_D \tag{11.56}$$

Example 11.11   Objective: Compare the forward transconductance of a MOSFET differential pair to that of a bipolar differential pair.

For the MOSFET differential pair, assume $K_n = 0.5\,\text{mA/V}^2$ and $I_Q = 1\,\text{mA}$. For the bipolar differential pair, assume $I_Q = 1\,\text{mA}$.

Solution: From Equation (11.54), the transconductance of the MOSFET in the differential pair is

$$g_m = 2\sqrt{\frac{K_n I_Q}{2}} = 2\sqrt{\frac{(0.5)(1)}{2}} = 1.0\,\text{mA/V}$$

From Equation (11.15), the transconductance of the bipolar transistor in the differential pair is

$$g_m = \frac{I_Q}{2V_T} = \frac{1}{2(0.026)} = 19.2\,\text{mA/V}$$

Comment: The transconductance of the bipolar pair is more than an order of magnitude larger than that of the MOSFET pair. Since the differential-mode voltage gain is directly proportional to the transconductance, the bipolar diff-amp gain is normally larger than the MOSFET diff-amp gain. We observed this same effect in Chapters 4 and 6, when we discussed the single-stage common-emitter and common-source circuits.

Test Your Understanding

11.12 Considering the dc transfer characteristics in Figure 11.21, determine the value of differential-mode input signal such that $i_{D1} = 0.90I_Q$.

*11.13 For the differential amplifier in Figure 11.20, the parameters are: $V^+ = 5\,\text{V}$, $V^- = -5\,\text{V}$, $R_1 = 80\,\text{k}\Omega$, and $R_D = 40\,\text{k}$. The transistor parameters are $\lambda = 0$ and $V_{TN} = 0.8\,\text{V}$ for all transistors, and $K_{n3} = K_{n4} = 100\,\mu\text{A/V}^2$ and $K_{n1} = K_{n2} = 50\,\mu\text{A/V}^2$. Determine the range of the common-mode input voltage. (Ans. $-2.18 \le v_{cm} \le 3.76\,\text{V}$)

11.14 In the diff-amp in Figure 11.19, the transistor parameters are: $K_{n1} = 1\,\text{mA/V}^2$, $V_{TN} = 1\,\text{V}$, and $\lambda = 0$. The circuit is biased at $I_Q = 2\,\text{mA}$, and the drain resistors are $R_D = 5\,\text{k}\Omega$. Determine the maximum forward transconductance $g_f(\text{max})$ and the one-sided differential-mode voltage gain A_d. (Ans. $g_f(\text{max}) = 1\,\text{mA/V}$, $A_d = 5$)

11.3.2   Differential- and Common-Mode Input Impedances

At low frequencies, the input impedance of a MOSFET is essentially infinite, which means that both the differential- and common-mode input resistances of a MOSFET diff-amp are infinite. Also, we know that the differential input resistance of a bipolar pair can be in the low kilohm range. A design trade-off, then, would be to use a MOSFET diff-amp with infinite input resistance, and sacrifice the differential-mode voltage gain.

11.3.3   Small-Signal Equivalent Circuit Analysis

We can determine the basic relationships for the differential-mode gain, common-mode gain, and common-mode rejection ratio from an analysis of the small-signal equivalent circuit.

Figure 11.23 shows the small-signal equivalent circuit of the MOSFET differential pair configuration. We assume the transistors are matched, with $\lambda = 0$ for each transistor, and that the constant-current source is represented by a finite output resistance R_o. All voltages are represented by their phasor components. The two transistors are biased at the same quiescent current, and $g_{m1} = g_{m2} \equiv g_m$.

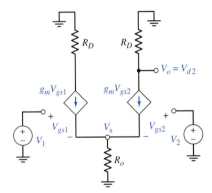

Figure 11.23   Small-signal equivalent circuit, MOSFET differential amplifier

Writing a KCL equation at node V_s, we have

$$g_m V_{gs1} + g_m V_{gs2} = \frac{V_s}{R_o} \qquad (11.57)$$

From the circuit, we see that $V_{gs1} = V_1 - V_s$ and $V_{gs2} = V_2 - V_s$. Equation (11.57) then becomes

$$g_m(V_1 + V_2 - 2V_s) = \frac{V_s}{R_o} \qquad (11.58)$$

Solving for V_s we obtain

$$V_s = \frac{V_1 + V_2}{2 + \dfrac{1}{g_m R_o}} \qquad (11.59)$$

For a one-sided output at the drain of M_2, we have

$$V_o = V_{d2} = -(g_m V_{gs2})R_D = -(g_m R_D)(V_2 - V_s) \qquad (11.60)$$

Substituting Equation (11.59) into (11.60) and rearranging terms yields

$$V_o = -g_m R_D \left[\frac{V_2\left(1 + \dfrac{1}{g_m R_o}\right) - V_1}{2 + \dfrac{1}{g_m R_o}} \right] \tag{11.61}$$

Based on the relationships between the input voltages V_1 and V_2 and the differential- and common-mode voltages, as given by Equation (11.29), Equation (11.61) can be written

$$V_o = \frac{g_m R_D}{2} V_d - \frac{g_m R_D}{1 + 2g_m R_o} V_{cm} \tag{11.62}$$

The output voltage, in general form, is

$$V_o = A_d V_d + A_{cm} V_{cm} \tag{11.63}$$

The transconductance g_m of the MOSFET is

$$g_m = 2\sqrt{K_n I_{DQ}} = \sqrt{2 K_n I_Q}$$

Comparing Equations (11.62) and (11.63), we develop the relationships for the differential-mode gain,

$$A_d = \frac{g_m R_D}{2} = \sqrt{2 K_n I_Q}\left(\frac{R_D}{2}\right) = \sqrt{\frac{K_n I_Q}{2}} \cdot R_D \tag{11.64(a)}$$

and the common-mode gain

$$A_{cm} = \frac{-g_m R_D}{1 + 2g_m R_o} = \frac{-\sqrt{2 K_n I_Q} \cdot R_D}{1 + 2\sqrt{2 K_n I_Q} \cdot R_o} \tag{11.64(b)}$$

We again, see that for an ideal current source, the common-mode gain is zero since $R_o = \infty$.

From Equations (11.64(a)) and (11.64(b)), the common-mode rejection ratio, CMRR $= |A_d/A_{cm}|$, is found to be

$$\text{CMRR} = \frac{1}{2}\left[1 + 2\sqrt{2 K_n I_Q} \cdot R_o\right] \tag{11.65}$$

This demonstrates that the CMRR for the MOSFET diff-amp is also a strong function of the output resistance of the constant-current source.

Example 11.12   **Objective:** Determine the differential-mode voltage gain, common-mode voltage gain, and CMRR for a MOSFET diff-amp.

Consider a MOSFET diff-amp with the configuration in Figure 11.20. Assume the same transistor parameters as given in Example 11.10 except assume $\lambda = 0.01\ \text{V}^{-1}$ for M_4.

Solution: From Example 11.10, we found the bias current to be $I_Q = 0.587\ \text{mA}$. The output resistance of the current source is then

$$R_o = \frac{1}{\lambda I_Q} = \frac{1}{(0.01)(0.587)} = 170\ \text{k}\Omega$$

The differential-mode voltage gain is

$$A_d = \sqrt{\frac{K_n I_Q}{2}} \cdot R_D = \sqrt{\frac{(1)(0.587)}{2}} \cdot (16) = 8.67$$

and the common-mode voltage gain is

$$A_{cm} = -\frac{\sqrt{2K_n I_Q} \cdot R_D}{1 + 2\sqrt{2K_n I_Q} \cdot R_o} = -\frac{\sqrt{2(1)(0.587)} \cdot (16)}{1 + 2\sqrt{2(1)(0.587)} \cdot (170)} = -0.0469$$

The common-mode rejection ratio is then

$$\mathrm{CMRR_{dB}} = 20 \log_{10}\left(\frac{8.67}{0.0469}\right) = 45.3\,\mathrm{dB}$$

Comment: As mentioned earlier, the differential-mode voltage gain of the MOSFET diff-amp is considerably less than that of the bipolar diff-amp, since the value of the MOSFET transconductance is, in general, much smaller than that of the BJT.

The value of the common-mode rejection ratio can be increased by increasing the output resistance of the current source. An increase in the output resistance can be accomplished by using a more sophisticated current source circuit. Figure 11.24 shows a MOSFET cascode current mirror that was discussed in the last chapter. The output resistance, as given by Equation (10.57), is $R_o = r_{o4} + r_{o2}(1 + g_m r_{o4})$. For the parameters of Example 11.12, $r_{o2} = r_{o4} = 170\,\mathrm{k}\Omega$ and $g_m = 2\sqrt{K_n I_Q} = 1.53\,\mathrm{mA/V}$. Then

$$R_o = 170 + 170[1 + (1.53)(170)] \Rightarrow 44.6\,\mathrm{M}\Omega$$

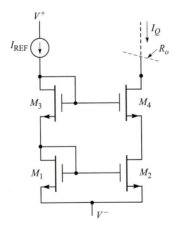

Figure 11.24 MOSFET cascode current source

Again, using the parameters of Example 11.12, the common-mode voltage gain of the diff-amp with a cascode current mirror would be

$$A_{cm} = -\frac{\sqrt{2K_n I_Q} \cdot R_D}{1 + 2\sqrt{2K_n I_Q} \cdot R_o} = -\frac{\sqrt{2(1)(0.587)} \cdot (16)}{1 + 2\sqrt{2(1)(0.587)} \cdot (44600)} = -0.000179$$

so that the CMRR would be

$$\text{CMRR}_{\text{dB}} = 20 \log_{10} \left(\frac{8.67}{0.000179} \right) = 93.7 \, \text{dB}$$

We increased the common-mode rejection ratio dramatically by using the cascode current mirror instead of the single two-transistor current source. Note, however, that the differential-mode voltage gain is unchanged.

To gain an appreciation of the difference in CMRR between 45.3 dB and 93.7 dB, we can reconsider the linear scale. For a $\text{CMRR}_{\text{dB}} = 45.3 \, \text{dB}$, the differential gain is a factor of 185 times larger than the common-mode gain, while for a $\text{CMRR}_{\text{dB}} = 93.7 \, \text{dB}$, the differential gain is a factor of 48,436 times larger than the common-mode gain.

Test Your Understanding

11.15 The diff-amp in Figure 11.19 has a differential gain of $A_d = 8$. The maximum current source available is $I_Q = 4 \, \text{mA}$, and the maximum drain resistance is $R_D = 4 \, \text{k}\Omega$. Determine the required $g_f(\text{max})$ and the transistor conductance K_n. (Ans. $g_f(\text{max}) = 2 \, \text{mA/V}$, $K_n = 2 \, \text{mA/V}^2$)

11.16 Consider the differential amplifier in Figure 11.20. The transistor parameters are given in Example 11.10, except that $\lambda = 0.02 \, \text{V}^{-1}$ for M_3 and M_4. Determine the differential voltage gain $A_d = v_{o2}/v_d$, the common-mode gain $A_{cm} = v_{o2}/v_{cm}$, and the CMRR_{dB}. (Ans. $A_d = 2.74$, $A_{cm} = -0.0925$, $\text{CMRR}_{dB} = 29.4 \, \text{dB}$)

11.17 The diff-amp in Figure 11.19 is biased at $I_Q = 0.2 \, \text{mA}$ and the transistor conduction parameter for all transistors is $K_n = 100 \, \mu\text{A/V}^2$. Determine the minimum output resistance of the current source such that $\text{CMRR}_{dB} = 60 \, \text{dB}$. (Ans. $R_o \cong 5 \, \text{M}\Omega$)

*RD11.18 The differential amplifier in Figure 11.20 is to be redesigned. The current-source biasing is to be replaced with the cascode current source in Figure 11.24. The reference current is $I_{\text{REF}} = 100 \, \mu\text{A}$ and λ for transistors in the current source circuit is $0.01 \, \text{V}^{-1}$. The parameters of the differential pair M_1 and M_2 are the same as described in Example 11.10. The range of the common-mode input voltage is to be $-4 \le v_{cm} \le +4 \, \text{V}$. Redesign the diff-amp to achieve the highest possible differential-mode voltage gain. Determine the values of A_d, A_{cm}, and CMRR_{dB}.

11.3.4   JFET Differential Amplifier

Figure 11.25 shows a basic JFET differential pair biased with a constant-current source. If a pure differential-mode input signal is applied such that $v_{G1} = +v_d/2$ and $v_{G2} = -v_d/2$, then drain currents I_{D1} and I_{D2} increase and decrease, respectively, in exactly the same way as in the MOSFET diff-amp.

We can determine the differential-mode voltage gain by analyzing the small-signal equivalent circuit. Figure 11.26 shows the equivalent circuit, with the output resistance of the constant-current source and the small-signal resistances of Q_1 and Q_2 assumed to be infinite. The small-signal equivalent circuit of the JFET diff-amp is identical to that of the MOSFET diff-amp in Figure 11.23 for the case when the current-source output resistance is infinite. A KCL equation at the common-source node, in phasor notation, is

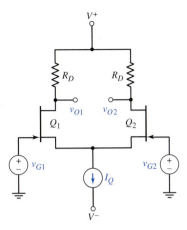

Figure 11.25 Basic JFET differential pair configuration

Figure 11.26 Small-signal equivalent circuit, JFET differential amplifier

$$g_m V_{gs1} + g_m V_{gs2} = 0 \qquad\qquad (11.66(a))$$

or

$$V_{gs1} = -V_{gs2} \qquad\qquad (11.66(b))$$

The differential-mode input voltage is

$$V_d \equiv V_1 - V_2 = V_{gs1} - V_{gs2} = -2V_{gs2} \qquad\qquad (11.67)$$

A one-sided output at V_{o2} is given by

$$V_{o2} = -g_m V_{gs2} R_D = -g_m \left(\frac{-V_d}{2}\right) R_D \qquad\qquad (11.68)$$

and the differential-mode voltage gain is

$$A_d = \frac{V_{o2}}{V_d} = +\frac{g_m R_D}{2} \qquad\qquad (11.69)$$

The expression for the differential-mode voltage gain for the JFET diff-amp (Equation (11.69)) is exactly the same as that of the MOSFET diff-amp

(Equation 11.64(a)). If the constant-current source output resistance is finite, then the JFET diff-amp will also have a nonzero common-mode voltage gain.

11.4   DIFFERENTIAL AMPLIFIER WITH ACTIVE LOAD

In Chapter 10, we considered an active load in conjunction with a simple transistor amplifier. Active loads can also be used in diff-amp circuits to increase the differential-mode gain.

Active loads are essentially transistor current sources used in place of resistive loads. The transistors in the active load circuit are biased at a Q-point in the forward-active mode as shown in Figure 11.27. A change in collector current is induced by the differential-pair, which, in turn, produces a change in the emitter–collector voltage as shown in the figure. The relation between the change in current and change in voltage is proportional to the small-signal output resistance r_o of the transistor. The value of r_o is, in general, much larger than that of a discrete resistive load, so the small-signal voltage gain will be larger with the active load.

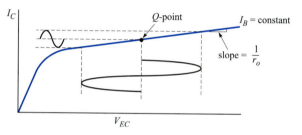

Figure 11.27   Current–voltage characteristic of active load device

11.4.1   BJT Diff-Amp with Active Load

Figure 11.28 shows a differential amplifier with an active load. Transistors Q_1 and Q_2 are the differential pair biased with a constant current I_Q, and transistors Q_3 and Q_4 form the load circuit. From the collectors of Q_2 and Q_4, we obtain a one-sided output.

If we assume all transistors are matched, then a pure applied common-mode voltage means that $v_{B1} = v_{B2} = v_{CM}$, and current I_Q splits evenly between Q_1 and Q_2. Neglecting base currents, $I_4 = I_3$ through the current-source circuit and $I_1 = I_2 = I_3 = I_4 = I_Q/2$ with no load connected at the output.

In actual diff-amp circuits, base currents are not zero. In addition, a second amplifier stage is connected at the diff-amp output. Figure 11.29 shows a diff-amp with an active load circuit, corresponding to a three-transistor current source, as well as a second amplifying stage. In general, the common–emitter current gain β is a function of collector current, as was shown in Figure 4.21(c). However, for simplicity, we assume all transistor current gains are equal, even though the current level in Q_5 is much smaller than in the other transistors. Current I_O is the dc bias current from the gain stage. Assuming all transistors are matched and $v_{B1} = v_{B2} = v_{CM}$, current I_Q splits evenly and $I_1 = I_2$. To

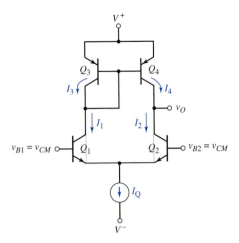

Figure 11.28 BJT differential amplifier with active load

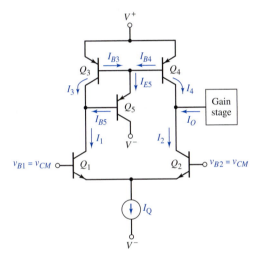

Figure 11.29 BJT differential amplifier with three-transistor active load and second gain stage

ensure that Q_2 and Q_4 are biased in the forward-active mode, the dc currents must be balanced, or $I_3 = I_4$. We see that

$$I_{E5} = I_{B3} + I_{B4} = \frac{I_3}{\beta} + \frac{I_4}{\beta} \tag{11.70}$$

Then

$$I_{B5} = \frac{I_{E5}}{1 + \beta} = \frac{I_3 + I_4}{\beta(1 + \beta)} \tag{11.71}$$

If the base currents and I_O are small, then

$$I_3 + I_4 \cong I_Q \tag{11.72}$$

Therefore,

$$I_{B5} \cong \frac{I_Q}{\beta(1+\beta)} \tag{11.73}$$

For the circuit to be balanced, that is, for $I_1 = I_2$ and $I_3 = I_4$, we must have

$$I_O = I_{B5} = \frac{I_O}{\beta(1+\beta)} \tag{11.74}$$

Equation (11.74) implies that the second amplifying stage must be designed and biased such that the direction of the dc bias current is as shown and is equal to the result of Equation (11.74). To illustrate this condition, we will analyze a second amplifying stage using a Darlington pair, later in this chapter.

11.4.2   Small-Signal Analysis of BJT Active Load

Figure 11.30 shows a diff-amp with a three-transistor active load circuit. The resistance R_L represents the small-signal input resistance of the gain stage. A pure differential-mode input voltage is applied as indicated. The signal voltage at the base of Q_1 produces a signal collector current $i_1 = (g_m v_d)/2$, where g_m is the transistor transconductance for both Q_1 and Q_2. Assuming the base currents are negligible, a signal current $i_3 = i_1$ is induced in Q_3, and the current mirror produces a signal current i_4 equal to i_3. The signal voltage at the base of Q_2 produces a signal collector current $i_2 = (g_m v_d)/2$, with the direction shown. The two signal currents, i_2 and i_4, add to produce a signal current in the load resistance R_L. The discussion is a first-order evaluation of the circuit operation.

Figure 11.30   BJT differential amplifier with three-transistor active load, showing the signal currents

From the above discussion, we know the induced currents in Q_2 and Q_4. To more accurately determine the output voltage, we need to consider the equivalent small-signal collector–emitter output circuit of the two transistors.

Figure 11.31(a) shows the small-signal equivalent circuit at the collector nodes of Q_2 and Q_4. The circuit can be rearranged to combine the signal grounds at a common point, as in Figure 11.31(b). From this figure, we determine that

$$v_o = 2\left(\frac{g_m v_d}{2}\right)(r_{o2} \| r_{o4} \| R_L) \qquad (11.75)$$

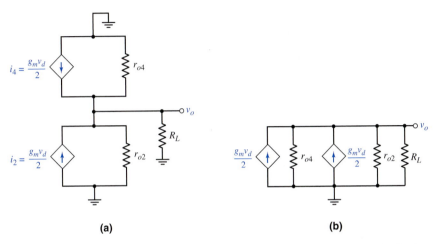

(a)   **(b)**

Figure 11.31   (a) Small-signal equivalent circuit BJT differential amplifier with active load and (b) rearrangement of small-signal equivalent circuit

and the small-signal differential-mode voltage gain is

$$A_d = \frac{v_o}{v_d} = g_m(r_{o2} \| r_{o4} \| R_L) \qquad (11.76)$$

Equation (11.76) can be rewritten in the form

$$A_d = \frac{g_m}{\dfrac{1}{r_{o2}} + \dfrac{1}{r_{o4}} + \dfrac{1}{R_L}} = \frac{g_m}{g_{o2} + g_{o4} + G_L} \qquad (11.77)$$

We recall that $g_m = I_Q/2V_T$, $r_{o2} = V_{A2}/I_2$, and $r_{o4} = V_{A4}/I_4$. The parameters g_{o2}, g_{o4}, and G_L are the corresponding conductances. Assuming $I_2 = I_4 = I_Q/2$, we can write Equation (11.77) in the form

$$A_d = \frac{\dfrac{I_Q}{2V_T}}{\dfrac{I_Q}{2V_{A2}} + \dfrac{I_Q}{2V_{A4}} + \dfrac{1}{R_L}} \qquad (11.78)$$

This expression of the differential-mode voltage gain of the diff-amp with an active load is very similar to that obtained in the last chapter for a simple amplifier with an active load.

Example 11.13   **Objective:** Determine the differential-mode gain of a diff-amp with an active load, taking loading effects into account.

Consider the diff-amp in Figure 11.30, biased with $I_Q = 0.20\,\text{mA}$. Assume an Early voltage of $V_A = 100\,\text{V}$ for all transistors. Determine the open-circuit $(R_L = \infty)$ differential-mode voltage gain, as well as the differential-mode voltage gain when $R_L = 100\,\text{k}\Omega$.

Solution: From Equation (11.78), the open-circuit voltage gain becomes

$$A_d = \frac{\dfrac{1}{V_T}}{\dfrac{1}{V_{A2}} + \dfrac{1}{V_{A4}}} = \frac{\dfrac{1}{0.026}}{\dfrac{1}{100} + \dfrac{1}{100}} = 1923$$

When $R_L = 100\,\text{k}\Omega$, the voltage gain is

$$A_d = \frac{\dfrac{0.02 \times 10^{-3}}{2(0.026)}}{\dfrac{0.20 \times 10^{-3}}{2(100)} + \dfrac{0.20 \times 10^{-3}}{2(100)} + \dfrac{1}{100 \times 10^3}}$$

which can be written

$$A_d = \frac{\dfrac{0.20}{2(0.026)}}{\dfrac{0.20}{2(100)} + \dfrac{0.20}{2(100)} + \dfrac{1}{100}} = \frac{3.85}{0.001 + 0.001 + 0.01} = 321$$

An inspection of this last equation shows that the external load factor, $1/R_L$, dominates the denominator term and thus has a tremendous influence on the gain.

Comment: The open-circuit differential-mode voltage gain, for a diff-amp with an active load, is large. However, a finite load resistance R_L causes severe loading effects, as shown in this example. A $100\,\text{k}\Omega$ load caused almost an order of magnitude decrease in the gain.

The output resistance looking back into the common collector node is $R_o = r_{o2} \| r_{o4}$. To minimize loading effects, we need $R_L > R_o$. However, since R_o is generally large for active loads, we may not be able to satisfy this condition. We can determine the severity of the loading effect by comparing R_L and R_o.

Test Your Understanding

11.19   Consider the diff-amp in Figure 11.29, with parameters: $V^+ = 10\,\text{V}$, $V^- = -10\,\text{V}$, and $I_Q = 0.5\,\text{mA}$. The transistor parameters are: $\beta = 180$, $V_{BE}(\text{on}) = 0.7\,\text{V}$, and $V_A = 100\,\text{V}$. (a) Find I_O such that the circuit is balanced. (b) For the balanced condition, what are the values of V_{EC4} and V_{CE2}, for $v_1 = v_2 = 0$? (Ans. (a) $I_O = 15.3\,\text{nA}$ (b) $V_{EC4} = 0.7\,\text{V}$, $V_{CE2} = 10\,\text{V}$)

11.20   The diff-amp circuit in Figure 11.30 is biased at $I_Q = 0.5\,\text{mA}$. The transistor parameters are: $\beta = 150$, $V_{A1} = V_{A2} = 125\,\text{V}$, and $V_{A3} = V_{A4} = 85\,\text{V}$. (a) Determine the open-circuit $(R_L = \infty)$ differential-mode voltage gain. (b) Find the differential-mode voltage gain when $R_L = 100\,\text{k}\Omega$. (c) Find the differential-mode input resistance. (d) Find the output resistance looking back from the load R_L. (Ans. (a) $A_d = 1947$ (b) $A_d = 644$ (c) $R_{id} = 31.2\,\mathrm{k\Omega}$ (d) $R_o = 202\,\mathrm{k\Omega}$)

11.21   Consider the diff-amp in Figure 11.28, with parameters: $V^+ = 10\,\mathrm{V}$, $V^- = -10\,\mathrm{V}$, and $I_Q = 0.2\,\mathrm{mA}$. The transistor parameters are: $\beta = 120$, $V_{BE}(\mathrm{on}) = V_{EB}(\mathrm{on}) = 0.6\,\mathrm{V}$, $V_{A1} = V_{A2} = 120\,\mathrm{V}$, and $V_{A3} = V_{A4} = 80\,\mathrm{V}$. Assume the output impedance of the current source is $1\,\mathrm{M\Omega}$. Determine the differential-mode gain. (Ans. $A_d = 1843$)

RD11.22   Redesign the circuit in Figure 11.30 using a Widlar current source and bias voltages of $\pm5\,\mathrm{V}$. The bias current I_Q is to be no less than $100\,\mu\mathrm{A}$ and the total power dissipated in the circuit (including the current-source circuit) is to be no more than $10\,\mathrm{mW}$. The diff-amp transistor parameters are the same as in Exercise 11.20. The circuit is to provide a minimum loading effect when a second stage with an input resistance of $R = 90\,\mathrm{k\Omega}$ is connected to the diff-amp. Determine the differential-mode voltage gain for this circuit.

11.23   Consider the diff-amp in Figure 11.28, using the parameters described in Exercise 11.21. (a) For a differential-mode input signal, determine the output resistance R_o at the output terminal. (b) Determine the load resistance R_L that would reduce the differential-mode voltage gain to one-half the open-circuit value. (Ans. (a) $R_o = 0.48\,\mathrm{M\Omega}$ (b) $R_L = 0.48\,\mathrm{M\Omega}$)

11.4.3   MOSFET Differential Amplifier with Active Load

We can use an active load in conjunction with a MOSFET differential pair, as we did for the bipolar differential amplifier. Figure 11.32 shows a MOSFET diff-amp with an active load. Transistors M_1 and M_2 are n-channel devices and form the differential pair biased with I_Q. The load circuit consists of transistors M_3 and M_4, both p-channel devices, connected in a current mirror configuration. A one-sided output is taken from the common drains of M_2 and M_4. When a common-mode voltage of $v_1 = v_2 = v_{cm}$ is applied, the current I_Q splits

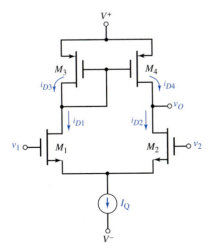

Figure 11.32   MOSFET differential amplifier with active load

evenly between M_1 and M_2, and $i_{D1} = i_{D2} = I_Q/2$. There are no gate currents; therefore, $i_{D3} = i_{D1}$ and $i_{D4} = i_{D2}$.

If a small differential-mode input voltage $v_d = v_1 - v_2$ is applied, then from Equation (11.48) and (11.49), we can write

$$i_{D1} = \frac{I_Q}{2} + i_d \qquad (11.79(a))$$

and

$$i_{D2} = \frac{I_Q}{2} - i_d \qquad (11.79(b))$$

where i_d is the signal current. For small values of v_d, we have $i_d = (g_m v_d)/2$. Since M_1 and M_3 are in series, we see that

$$i_{D3} = i_{D1} = \frac{I_Q}{2} + i_d \qquad (11.80)$$

Finally, the current mirror consisting of M_3 and M_4 produces

$$i_{D4} = i_{D3} = \frac{I_Q}{2} + i_d \qquad (11.81)$$

Figure 11.33 is the ac equivalent circuit of the diff-amp with active load, showing the signal currents. The negative sign for i_{D2} in Equation (11.79(b)) shows up as a change in current direction in M_2, as indicated in the figure.

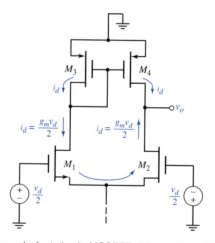

Figure 11.33   The ac equivalent circuit, MOSFET differential amplifier with active load

Figure 11.34(a) shows the small-signal equivalent circuit at the drain node of M_2 and M_4. If the output is connected to the gate of another MOSFET, which is equivalent to an infinite impedance at low frequency, the output terminal is effectively an open circuit. The circuit can be rearranged by combining the signal grounds at a common point, as shown in Figure 11.34(b). Then,

$$v_o = 2\left(\frac{g_m v_d}{2}\right)(r_{o2}\|r_{o4}) \qquad (11.82)$$

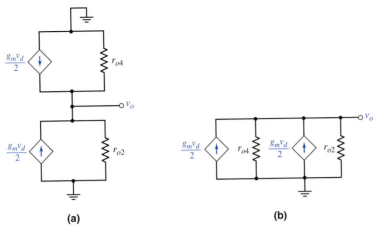

Figure 11.34 (a) Small-signal equivalent circuit, MOSFET differential amplifier with active load and (b) rearranged small-signal equivalent circuit

and the small-signal differential-mode voltage gain is

$$A_d = \frac{v_o}{v_d} = g_m(r_{o2}\|r_{o4}) \tag{11.83}$$

Equation (11.83) can be rewritten in the form

$$A_d = \frac{g_m}{\dfrac{1}{r_{o2}} + \dfrac{1}{r_{o4}}} = \frac{g_m}{g_{o2} + g_{o4}} \tag{11.84}$$

If we recall that $g_m = 2\sqrt{K_n I_D} = \sqrt{2K_n I_Q}$, $g_{o2} = \lambda_2 I_{DQ2} = (\lambda_2 I_Q)/2$, and $g_{o4} = \lambda_4 I_{DQ4} = (\lambda_4 I_Q)/2$, then Equation (11.84) becomes

$$A_d = \frac{2\sqrt{2K_n I_Q}}{I_Q(\lambda_2 + \lambda_4)} = 2\sqrt{\frac{2K_n}{I_Q}} \cdot \frac{1}{\lambda_2 + \lambda_4} \tag{11.85}$$

Design Example 11.14 Objective: Design a MOSFET diff-amp with the configuration in Figure 11.32 to meet the specifications of the experimental system in Example 11.4.

Design Approach: We need not only to try to obtain the necessary differential-mode gain and minimize the common-mode gain in our design, but we must also be cognizant of the swing in the output voltage. In the circuit in Figure 11.32, if the corresponding PMOS and NMOS transistors are matched, then the quiescent value of V_{SD4} is equal to $V_{SG4} = V_{SG3}$. As the signal output voltage increases, the source-to-drain voltage of M_4 decreases. The minimum value of this voltage such that M_4 remains biased in the saturation region is $V_{SD4}(\text{min}) = V_{SD4}(\text{sat}) = V_{SG} + V_{TP}$. This means that the maximum swing in the output voltage is equal to the magnitude of the threshold voltage of M_4. In this example, the maximum swing in the output voltage is 0.8 V, so that the magnitude of the threshold voltages of the PMOS devices must be greater than 0.8 V. Assume that NMOS devices are available with the following parameters: $V_{TN} = 0.5\,\text{V}$, $k_n' = 80\,\mu\text{A/V}^2$, and $\lambda_n = 0.02\,\text{V}^{-1}$. Assume that PMOS devices are available with the following parameters: $V_{TP} = -1.0\,\text{V}$, $k_p' = 40\,\mu\text{A/V}^2$, and $\lambda_p = 0.02\,\text{V}^{-1}$. Choose supply voltages of $\pm 5\,\text{V}$ and choose a bias current of approximately $I_Q = 200\,\mu\text{A}$.

Figure 11.35 is the diff-amp and current-source network used for the design in this example.

Figure 11.35   CMOS differential amplifier and current source network for Example 11.14

Design, Differential Amplifier: Differential-Mode Gain: From Equation (11.85), the differential-mode gain is

$$A_d = 2\sqrt{2\left(\frac{k_n'}{2}\right)\left(\frac{W}{L}\right)_n \frac{1}{I_Q} \cdot \frac{1}{\lambda_n + \lambda_p}}$$

or

$$100 = 2\sqrt{2\left(\frac{80}{2}\right)\left(\frac{W}{L}\right)\frac{1}{200} \cdot \frac{1}{0.02 + 0.02}}$$

which yields a width-to-length ratio of $(W/L)_n = 20$ for the NMOS differential pair. Since the width-to-length ratios of the other transistors do not directly affect the gain of the diff-amp, we may arbitrarily choose width-to-length ratios of 10 for all other transistors except M_5 and M_6. The W/L ratio of 10 means that the other devices are reasonably small and do not lead to a large circuit area.

Design, Current-Source Network: For the transistor M_3 in the current source, we have

$$I_Q = \frac{k'}{2} \cdot \frac{W}{L} \cdot (V_{GS3} - V_{TN})^2$$

or

$$200 = \frac{80}{2}(10)(V_{GS3} - 0.5)^2$$

which means that the required gate-to-source voltage of M_3 is $V_{GS3} = 1.21$ V. We may choose M_4 and M_3 to be identical so that the current in the reference portion of the circuit is also 200 μA. Assuming that M_5 and M_6 are identical, then each transistor must have a gate-to-source voltage of

$$V_{GS5} = V_{GS6} = (10 - 1.21)/2 \cong 4.4\,\text{V}$$

The width-to-length of these transistors is now found from

$$I_{\text{REF}} = I_Q = \frac{k_n'}{2} \cdot \left(\frac{W}{L}\right)_5 (V_{GS5} - V_{TN})^2$$

or

$$200 = \frac{80}{2} \cdot \left(\frac{W}{L}\right)_5 (4.4 - 0.5)^2$$

which yields

$$(W/L)_5 = (W/L)_6 = 0.33$$

Computer Simulation Verification: The circuit in Figure 11.35 was used in the computer simulation verification. In the hand design, the finite output resistance (lambda parameter) was neglected in the dc calculations. These parameters became important in the actual design and in the actual currents developed in the circuit. For $(W/L)_5 = (W/L)_6 = 0.75$, the reference current is $I_{\text{REF}} = 231\,\text{μA}$ and the bias current is $I_Q = 208\,\text{μA}$.

The differential-mode voltage gain is approximately 102 so that the signal output voltage is 0.82 V for a differential-mode input signal voltage of 8 mV. The common-mode output signal is approximately 0.86 mV, which is well within the specified 10 mV maximum value.

Design Pointer: The body effect has been neglected in this design. In actual integrated circuits, the differential pair transistors may actually be fabricated within their own p-type substrate region (for NMOS devices). This p-type substrate region is then directly connected to the source terminals so that the body effect in the NMOS differential pair devices can be neglected.

11.4.4 MOSFET Diff-Amp with Cascode Active Load

The differential-mode voltage gain is proportional to the output resistance looking into the active load transistor. The voltage gain can be increased, therefore, if the output resistance can be increased. An increase in output resistance can be achieved by using, for example, a cascode active load. This configuration is shown in Figure 11.36.

The output resistance R_o was considered in the last section in the discussion of the cascode current source. As applied to Figure 11.36, the output resistance is given by

$$R_o = r_{o4} + r_{o6}(1 + g_m r_{o4}) \cong g_m r_{o4} r_{o6} \qquad \textbf{(11.86)}$$

The small-signal differential-mode voltage gain is then

$$A_d = \frac{v_o}{v_d} = g_m(r_{o2} \| R_o) \tag{11.87}$$

Example 11.15 **Objective:** Calculate the differential-mode voltage gain of a MOSFET diff-amp with a cascode active load.

Consider the diff-amp shown in Figure 11.36. Assume the circuit and transistor parameters are the same as in Example 11.14.

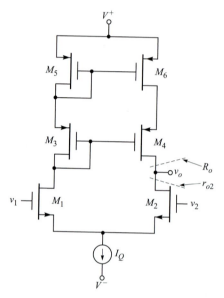

Figure 11.36 MOSFET diff-amp with cascode active load

Solution: The transistor transconductance is

$$g_m = 2\sqrt{K_n I_{DQ}} = 2\sqrt{(0.2)(0.1)} = 0.283 \, \text{mA/V}$$

The output resistance of the individual transistors is

$$r_o = \frac{1}{\lambda I_{DQ}} = \frac{1}{(0.01)(0.1)} = 1000 \, \text{k}\Omega = 1 \, \text{M}\Omega$$

The output resistance of the cascode active load is then

$$R_o = r_{o4} + r_{o6}(1 + g_m r_{o4}) = 1 + (1)[1 + (0.283)(1000)] = 285 \, \text{M}\Omega$$

The differential-mode voltage gain is then found as

$$A_d = g_m(r_{o2} \| R_o) = (0.283)(1000 \| 285000) = 282$$

Comment: Since $R_o \gg r_{o2}$, the voltage gain is now essentially equal to $A_d = g_m r_{o2}$ which is twice as large as the gain calculated in Example 11.14.

The differential-mode voltage gain can be further increased by incorporating a cascode configuration in the differential pair as well as in the active load. One such example is shown in Figure 11.37. Transistors M_3 and M_4 are the cascode transistors for the differential pair M_1 and M_2. The differential-mode voltage gain is now

$$A_d = \frac{v_o}{v_d} = g_m(R_{o4} \| R_{o6})$$

where $R_{o4} \cong g_m r_{o2} r_{o4}$ and $R_{o6} \cong g_m r_{o6} r_{o8}$. The small-signal differential-mode voltage gain of this type of amplifier can be on the order of 10,000.

Other types of MOSFET differential amplifiers will be considered in Chapter 13 when operational amplifier circuits are discussed.

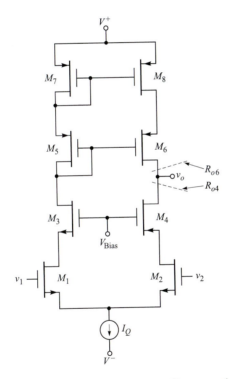

Figure 11.37   A MOSFET cascode diff-amp with a cascode active load

Test Your Understanding

11.24   A differential amplifier is shown in Figure 11.32. The parameters are: $V^+ = 10\,V$, $V^- = -10\,V$, and $I_Q = 0.1\,mA$. The PMOS parameters are: $K_p = 80\,\mu A/V^2$, $\lambda_p = 0.015\,V^{-1}$, and $V_{TP} = -1\,V$. The NMOS parameters are: $K_n = 100\,\mu A/V^2$, $\lambda_n = 0.01\,V^{-1}$, and $V_{TN} = 1\,V$. Determine the differential-mode voltage gain $A_d = v_o/v_d$. (Ans. $A_d = 113$)

11.5   BiCMOS CIRCUITS

Thus far, we have considered two basic amplifier design technologies: the bipolar technology, which uses npn and pnp bipolar junction transistors; and the MOS technology, which uses NMOS and PMOS field-effect transistors. We showed that bipolar transistors have a larger transconductance than MOSFETs biased at the same current levels, and that, in general, bipolar amplifiers have larger voltage gains. We also showed that MOSFET circuits have an essentially infinite imput impedance at low frequencies, which implies a zero input bias current.

These advantages of the two technologies can be exploited by combining bipolar and MOS transistors in the same integrated circuit. The technology is called **BiCMOS**. BiCMOS technology is especially useful in digital circuit design, but also has applications in analog circuits. In this section, we will examine basic BiCMOS analog circuit configurations.

11.5.1   Basic Amplifier Stages

A bipolar multitransistor circuit previously studied is the Darlington pair configuration. Figure 11.38(a) shows a modified Darlington pair configuration, in which the bias current I_{BIAS}, or some equivalent element, is used to control the quiescent current in Q_1. This Darlington pair circuit is used to boost the effective current gain of bipolar transistors. There is no comparable configuration in FET circuits.

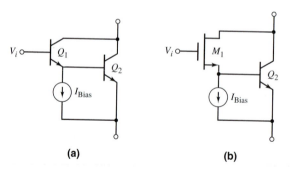

Figure 11.38   (a) Bipolar Darlington pair configuration and (b) BiCMOS Darlington pair configuration

A potentially useful BiCMOS circuit is shown in Figure 11.38(b). Transistor Q_1 in the Darlington pair is replaced with a MOSFET. The advantages of this configuration are an infinite input resistance, and a large transconductance due to the bipolar transistor Q_2.

To analyze the circuit, we consider the small-signal equivalent circuit shown in Figure 11.39. We assume that $r_o = \infty$ in both transistors.

The output signal current is

$$I_o = g_{m1}V_{gs} + g_{m2}V_\pi \tag{11.88}$$

Figure 11.39   Small-signal equivalent circuit, BiCMOS Darlington pair configuration

We see that

$$V_i = V_{gs} + V_\pi \qquad\qquad (11.89)$$

and

$$V_\pi = g_{m1} V_{gs} r_\pi \qquad\qquad (11.90)$$

Combining Equations (11.89) and (11.90) produces

$$V_{gs} = \frac{V_i}{1 + g_{m1} r_\pi} \qquad\qquad (11.91)$$

From Equation (11.88), the output current can now be written

$$I_o = g_{m1} V_{gs} + g_{m2}(g_{m1} r_\pi) V_{gs} = (g_{m1} + g_{m2} g_{m1} r_\pi) V_{gs} \qquad\qquad (11.92)$$

Substituting Equation (11.91) into (11.92), we obtain

$$I_o = \frac{g_{m1}(1 + g_{m2} r_\pi)}{(1 + g_{m1} r_\pi)} \cdot V_i = g_m^c \cdot V_i \qquad\qquad (11.93)$$

where g_m^c is the **composite transconductance**. Since g_{m2} of the bipolar transistor is usually at least an order of magnitude greater than g_{m1} of the MOSFET, the composite transconductance is approximately an order of magnitude larger than that of the MOSFET alone. We now have the advantages of a large transconductance and an infinite input resistance.

A bipolar cascode circuit is shown in Figure 11.40(a); a corresponding BiCMOS configuration is shown in Figure 11.40(b). The output resistance of

(a)   (b)

Figure 11.40   (a) Bipolar cascode configuration and (b) BiCMOS cascode configuration the cascode circuit is very high, as we saw in Chapter 10. Also, the cascode amplifier has a wider frequency bandwidth than the common-emitter circuit, since the input resistance looking into the emitter of Q_2 is very low, thereby minimizing the Miller multiplication effect. This effect was observed in Chapter 7.

Again, the advantage of the BiCMOS circuit is the infinite input resistance of M_1. The equivalent resistance looking into the emitter of a bipolar transistor is much less than the resistance looking into the source of a MOSFET; therefore, the frequency response of a BiCMOS cascode circuit is superior to that of an all-MOSFET cascode circuit.

11.5.2   Current Sources

In our previous discussions of constant-current sources, we mentioned that cascode current sources increase the output resistance, as well as the stability of the bias current. Figure 11.41 shows a bipolar cascode configuration in which the output resistance is $R_o \cong \beta r_{o4}$. The bias current in this circuit is much more stable against variations in output voltage than the basic two-transistor current source.

A BiCMOS double cascode constant-current source is shown in Figure 11.42. The small-signal equivalent circuit for determining output resistance is shown in Figure 11.43(a). The gate voltage to M_6 and the base voltages to Q_2 and Q_4 are constants, equivalent to signal ground. Also, since $V_{\pi 2} = 0$, then $g_{m6}V_{\pi 2} = 0$, and the equivalent circuit can be rearranged as shown in Figure 11.43(b).

The output resistance of this circuit is extremely large. A detailed analysis shows that the output resistance is given approximately by

$$R_o \cong (g_{m6}r_{o6})(\beta r_{o4}) \tag{11.94}$$

The output resistance is increased by a factor $(g_m r_{o6})$ compared to the bipolar cascode circuit in Figure 11.41. If a bipolar transistor were to be used in place of M_6, then a resistance $r_{\pi 6}$ would be connected across the terminals indicated by V_{gs6}. This resistance would effectively eliminate the multiplying constant $(g_{m6}r_{o6})$, and the output resistance would be essentially the same as that of the circuit in Figure 11.41. The BiCMOS circuit, then, increases the output resistance compared to an all-bipolar circuit.

11.5.3   BiCMOS Differential Amplifier

A basic BiCMOS differential amplifier, with a constant-current source bias and a bipolar active load, is shown in Figure 11.44. Again, the primary advantages are the infinite input resistance and the zero input bias current. One disadvantage of a MOSFET input stage is a relatively high offset voltage compared to that of a bipolar input circuit. Offset voltages occur when the differential-pair input transistors are mismatched. In Chapter 14, we will examine the effect of offset voltages, as well as nonzero bias currents, in op-amp circuits.

We will consider additional BiCMOS op-amp circuits in Chapter 13, when we discuss the analysis and design of full op-amp circuits.

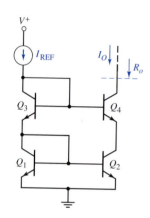

Figure 11.41 Bipolar cascode constant-current source

Figure 11.42 BiCMOS double cascode constant-current source

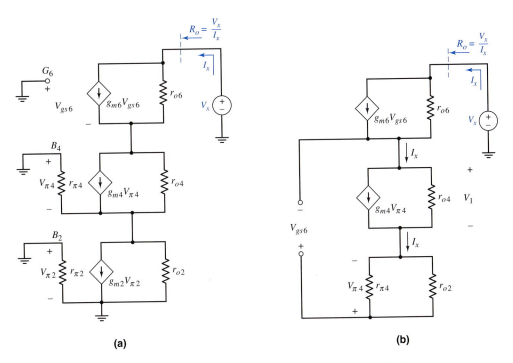

Figure 11.43 (a) Equivalent circuit for determining output impedance of BiCMOS double cascode current source and (b) rearranged equivalent circuit

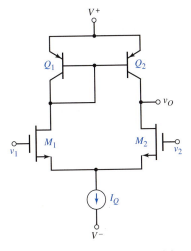

Figure 11.44 Basic BiCMOS differential amplifier

Test Your Understanding

11.25 Consider the BiCMOS Darlington pair in Figure 11.45. The transistor parameters are $K_n = 20 \, \mu A/V^2$, $V_{TN} = 1 \, V$, and $\lambda = 0$ for M_1 and $\beta = 100$, $V_{BE}(\text{on}) = 0.7 \, V$, and $V_A = \infty$ for Q_2. Determine the small-signal parameters for each transistor, as well as the composite transconductance. (Ans. $g_{m1} = 44.8 \, \mu A/V$, $g_{m2} = 2.88 \, mA/V$, $r_{\pi 2} = 34.7 \, k\Omega$, $r_{o1} = r_{o2} = \infty$, $g_m^c = 1.77 \, mA/V$)

$I_{Bias1} = 25\,\mu A$

$I_{Bias2} = 100\,\mu A$

Figure 11.45   Figure for Exercise 11.25

11.26   The reference current in each of the constant-current source circuits shown in Figures 11.41 and 11.42 is $I_{REF} = 0.5\,\text{mA}$. All bipolar transistor parameters are $\beta = 150$ and $V_A = 80\,\text{V}$, and all MOSFET parameters are: $K_n = 500\,\mu\text{A/V}^2$, $V_{TN} = 1\,\text{V}$, and $\lambda = 0.0125\,\text{V}^{-1}$. Neglecting bipolar base currents, determine the output resistance R_o of each constant-current source. (Ans. For Figure 11.41, $R_o \cong 24\,\text{M}\Omega$; for Figure 11.42, $R_o = 3840\,\text{M}\Omega$)

11.6   GAIN STAGE AND SIMPLE OUTPUT STAGE

A diff-amp, including those previously discussed, is the input stage of virtually all op-amps. The second op-amp stage, or gain stage, is often a Darlington pair configuration, and the third, or output, stage is normally an emitter follower.

11.6.1   Darlington Pair and Simple Emitter-Follower Output

Figure 11.46 shows a BJT diff-amp with a three-transistor active load, a Darlington pair connected to the diff-amp output, and a simple emitter-follower output stage.

The differential-pair transistors are biased with a Widlar current source at a bias current I_Q. We noted previously that, for the diff-amp dc currents to be balanced, we must have

$$I_O = I_{B5} = \frac{I_Q}{\beta(1+\beta)} \tag{11.95}$$

From the figure, we see that

$$I_O = \frac{I_{E6}}{(1+\beta)} = \frac{I_{C7}}{\beta(1+\beta)} \tag{11.96}$$

In order for $I_O = I_{B5}$, we must require that $I_{C7} = I_Q$. This means that the emitter resistors of Q_{10} and Q_{11} should have the same value. Transistor Q_{11} also acts as an active load for the Darlington pair gain stage.

Transistor Q_8 and resistor R_4 form the simple emitter-follower output stage. The emitter-follower amplifier minimizes loading effects because its output resistance is small.

Ideally, when the diff-amp input is a pure common-mode signal, the output v_o is zero. The combination of Q_7 and Q_{11} allows the dc level to shift. By slightly changing the bias current I_{C7}, we can vary voltages V_{EC7} and V_{CE11} such that $v_o = 0$. The small variation of I_{C7} required to achieve the necessary dc

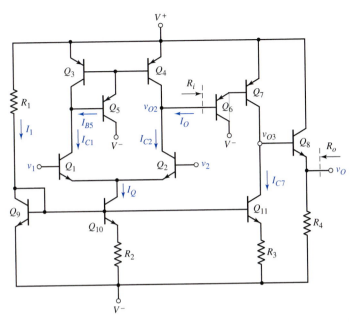

Figure 11.46   BJT diff-amp with three-transistor active load, Darlington pair gain stage, and simple emitter-follower output stage

level shift will not significantly change the balance between I_O and I_{B5}. As we will see in later chapters, other forms of level shifters could also be used.

11.6.2   Input Impedance, Voltage Gain, and Output Impedance

The input resistance of the Darlington pair determines the loading effect on the basic diff-amp. In addition, the gain of the Darlington pair affects the overall gain of the op-amp circuit, and the output resistance of the emitter follower determines any loading effects on the output signal.

Figure 11.47(a) is the ac equivalent circuit of the Darlington pair, where R_{L7} is the effective resistance connected between the collector of Q_7 and signal

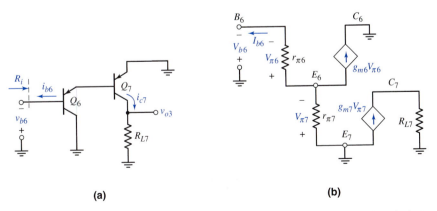

(a)   **(b)**

Figure 11.47   (a) The ac equivalent circuit, Darlington pair, and (b) small-signal equivalent circuit, Darlington pair ground. Figure 11.47(b) shows the simple hybrid-π model of the Darlington pair. We see that the equivalent circuits for Q_6 and Q_7 have been effectively turned upside down compared to the transistors in Figure 11.47(a).

Writing a KVL equation around the B–E loop of Q_6 and Q_7, we have

$$V_{b6} = V_{\pi 6} + V_{\pi 7} \tag{11.97}$$

We can also write that

$$V_{\pi 6} = I_{b6} r_{\pi 6} \tag{11.98}$$

and the KCL equation is

$$\frac{V_{\pi 7}}{r_{\pi 7}} = \frac{V_{\pi 6}}{r_{\pi 6}} + g_{m6} V_{\pi 6} \tag{11.99(a)}$$

or

$$V_{\pi 7} = r_{\pi 7}\left[\frac{(1+\beta)}{r_{\pi 6}}\right] V_{\pi 6} = r_{\pi 7}(1+\beta) I_{b6} \tag{11.99(b)}$$

where $r_{\pi 6} g_{m6} = \beta$. Substituting Equations (11.99(b)) and (11.98) into Equation (11.97), we obtain

$$V_{b6} = I_{b6} r_{\pi 6} + r_{\pi 7}(1+\beta) I_{b6} \tag{11.100}$$

The input resistance is therefore

$$R_i = \frac{V_{b6}}{I_{b6}} = r_{\pi 6} + r_{\pi 7}(1+\beta) \tag{11.101}$$

Assuming $I_{C7} = I_Q$, the hybrid-π parameters are

$$r_{\pi 7} = \frac{\beta V_T}{I_{C7}} = \frac{\beta V_T}{I_Q} \tag{11.102(a)}$$

and

$$r_{\pi 6} = \frac{\beta V_T}{I_{C6}} = \frac{(1+\beta)\beta V_T}{I_Q} \tag{11.102(b)}$$

Combining Equations (11.102(a)), (11.102(b)), and Equation (11.101) yields an expression for the input resistance, as follows:

$$R_i = \frac{(1+\beta)\beta V_T}{I_Q} + \frac{(1+\beta)\beta V_T}{I_Q} = \frac{2(1+\beta)\beta V_T}{I_Q} \tag{11.103}$$

We can determine the small-signal voltage gain of the Darlington pair circuit by using the small-signal equivalent circuit in Figure 11.47(b). We see that

$$v_{o3} = i_{c7} R_{L7} = (\beta i_{b7}) R_{L7} = \beta(1+\beta) i_{b6} R_{L7} \tag{11.104}$$

and

$$i_{b6} = \frac{v_{b6}}{R_i} \tag{11.105}$$

The small-signal voltage gain is therefore

$$A_v = \frac{v_{o3}}{v_{b6}} = \frac{\beta(1+\beta) R_{L7}}{R_i} \tag{11.106}$$

Substituting Equation (11.103) into (11.106), we find that

$$A_v = \frac{\beta(1 + \beta)R_{L7}}{\underbrace{\frac{2(1 + \beta)\beta V_T}{I_Q}}} = \left(\frac{I_Q}{2V_T}\right)R_{L7} \qquad (11.107)$$

In Figure 11.46, we see that resistance R_{L7} is the parallel combination of the resistance looking into the collector of Q_{11} and the resistance looking into the base of Q_8. From Chapter 10, the resistance looking into the collector of Q_{11} is

$$R_{c11} = r_{o11}(1 + g_{m11}R_E') \qquad (11.108)$$

where $R_E' = r_{\pi 11} \| R_3$. The resistance looking into the base of Q_8 is

$$R_{b8} = r_{\pi 8} + (1 + \beta)R_4 \qquad (11.109)$$

Equations (11.108) and (11.109) indicate that resistances R_{c11} and R_{b8} are large, which means that the effective resistance R_{L7} is also large.

Example 11.16   Objective: Calculate the input resistance and the small-signal voltage gain of a Darlington pair.

Consider the circuit shown in Figure 11.46, with parameters $I_{C7} = I_Q = 0.2\,\text{mA}$, $I_{C8} = 1\,\text{mA}$, $R_4 = 10\,\text{k}\Omega$, and $R_3 = 0.2\,\text{k}\Omega$. Assume $\beta = 100$ for all transistors, and the Early voltage for Q_{11} is $100\,\text{V}$.

Solution: The input resistance, given by Equation (11.103), is

$$R_i = \frac{2(1 + \beta)\beta V_T}{I_Q} = \frac{2(101)(100)(0.026)}{0.2} \Rightarrow 2.63\,\text{M}\Omega$$

The small-signal voltage gain is a function of R_{L7}, which in turn is a function of R_{c11} and R_{b8}. We can find that

$$R_{\pi 11} = \beta V_T/I_Q = (100)(0.026)/0.2 = 13\,\text{k}\Omega$$

such that

$$R_E' = 13\|0.2 = 0.197\,\text{k}\Omega$$

Also

$$g_{m11} = I_Q/V_T = 0.2/0.026 = 7.69\,\text{mA/V}$$

and

$$r_{o11} = V_A/I_Q = 100/0.2 = 500\,\text{k}\Omega$$

Therefore,

$$R_{c11} = r_{o11}(1 + g_{m11}R_E') = 500[1 + (7.69)(0.197)] \Rightarrow 1.26\,\text{M}\Omega$$

We can determine that

$$r_{\pi 8} = \beta V_T/I_{C8} = (100)(0.026)/1 = 2.6\,\text{k}\Omega$$

Then

$$R_{b8} = r_{\pi 8} + (1 + \beta)R_4 = 2.6 + (101)(10) \Rightarrow 2.02\,\text{M}\Omega$$

Consequently, resistance R_{L7} is

$$R_{L7} = R_{c11}\|R_{b8} = 1.26\|2.02 = 0.776\,\text{M}\Omega$$

Finally, from Equation (11.107), the small-signal voltage gain is

$$A_v = \left(\frac{I_Q}{2V_T}\right)R_{L7} = \left[\frac{0.2}{2(0.026)}\right](776) = 2985$$

Comment: The input resistance of the Darlington pair is in the megohm range, which should minimize severe loading effects on the diff-amp. In addition, the small-signal gain is large because of the active load (Q_{11}) and the large input resistance of the emitter-follower output stage.

We can use the results of Chapter 4 to determine the output resistance of the emitter follower. The output resistance is

$$R_o = R_4 \left\| \left(\frac{r_{\pi 8} + Z}{(1 + \beta)}\right) \right. \tag{11.110}$$

where Z is the equivalent impedance, or resistance, in the base of Q_8. In this case, $Z = R_{c11} \| R_{c7}$, where R_{c7} is the resistance looking into the collector of Q_7. Because of the factor $(1 + \beta)$ in the denominator, the output resistance of the emitter follower is normally small, as previously determined.

Example 11.17   Objective: Calculate the output resistance of the circuit in Figure 11.46.

Consider the same circuit and transistor parameters described in Example 11.16. Assume the Early voltage of Q_7 is 100 V.

Solution: From Example 11.16, we have that $R_{c11} = 1.26\,\text{M}\Omega$ and $r_{\pi 8} = 2.6\,\text{k}\Omega$. We can then determine that

$$R_{c7} = \frac{V_A}{I_Q} = \frac{100}{0.2} = 500\,\text{k}\Omega$$

Then,

$$Z = R_{c11} \| R_{c7} = 1260 \| 500 = 358\,\text{k}\Omega$$

Therefore,

$$R_o = R_4 \left\| \left[\frac{r_{\pi 8} + Z}{(1 + \beta)}\right] = 10 \left\| \left(\frac{2.6 + 358}{101}\right) = 2.63\,\text{k}\Omega \right.$$

Comment: The output resistance is obviously less than R_4 and is substantially less than the equivalent resistance Z in the base of Q_8. In a later chapter, we will examine a Darlington pair emitter-follower output stage in which the output resistance is on the order of 100 Ω.

A BJT diff-amp with an active load can produce a small-signal differential-mode voltage gain on the order of 10^3, and the Darlington pair can also provide a voltage gain on the order of 10^3. Since the emitter follower has a gain of essentially unity, the overall voltage gain of the op-amp circuit is on the order of 10^6. This value is typical for the low-frequency, open-loop gain of op-amp circuits.

Test Your Understanding

*11.27   Consider the Darlington pair and emitter-follower portions of the circuit in Figure 11.46. The parameters are: $I_{C7} = I_Q = 0.5\,\text{mA}$, $I_{C8} = 2\,\text{mA}$, $R_4 = 5\,\text{k}\Omega$, and $R_3 = 0.1\,\text{k}\Omega$. For all transistors, the current gain is $\beta = 120$, and for Q_{11} and Q_7, the Early voltage is $V_A = 120\,\text{V}$. Calculate the input resistance and small-signal voltage gain of the Darlington pair, and the output resistance of the emitter follower. (Ans. $R_i = 1.51\,\text{M}\Omega$, $A_v = 3115$, $R_o = 1.14\,\text{k}\Omega$)

11.28   In the circuit in Figure 11.46, the Darlington pair and emitter-follower transistor parameters are the same as in Exercise 11.27. Determine the effective resistance R_{L7} (see Figure 11.47(a)) such that the small-signal voltage gain is 10^3. (Ans. $R_{L7} = 104\,\text{k}\Omega$)

11.7   SIMPLIFIED BJT OPERATIONAL AMPLIFIER CIRCUIT

An operational amplifier (op-amp) is a multistage circuit composed of a differential amplifier input stage, a gain stage, and an output stage. In this section, we will consider a simplified BJT op-amp circuit.

Although active load devices increase the gain of an amplifier, in this discussion, we will consider resistive loads, in order to simplify the analysis and design. For the bipolar circuit, all component values are given; we will analyze both the dc and ac circuit characteristics.

Figure 11.48 depicts a simple bipolar operational amplifier. The differential amplifier stage is biased with a Widlar current source, and a one-sided output is connected to the Darlington pair gain stage. An emitter bypass

Figure 11.48   Bipolar operational amplifier circuit

capacitor C_E is included to increase the small-signal voltage gain. The output stage is an emitter follower. In general, we want the dc value of the output voltage to be zero when the input voltage is zero. To accomplish this, we need to insert a dc level shifting circuit between the voltage v_{O3} and the output voltage v_O.

Example 11.18   Objective: Analyze the dc characteristics of the bipolar op-amp circuit.

Consider the circuit in Figure 11.48. Neglect base currents and, as a simplification, assume $V_{BE}(\text{on}) = 0.7\,\text{V}$ for all transistors except Q_8 and Q_9 in the Widlar circuit.

Solution: The reference current I_1 is

$$I_1 = \frac{10 - 0.7 - (-10)}{19.3} = 1\,\text{mA}$$

The bias current I_Q is determined from

$$I_Q R_2 = V_T \ln\left(\frac{I_1}{I_Q}\right)$$

and is

$$I_Q = 0.4\,\text{mA}$$

The collector currents are then

$$I_{C1} = I_{C2} = 0.2\,\text{mA}$$

The dc voltage at the collector of Q_2 is

$$V_{O2} = 10 - I_{c2}R_C = 10 - (0.2)(20) = 6\,\text{V}$$

With these circuit parameters, the common-mode input voltage is limited to the range $-9.3 \le v_{CM} \le 6\,\text{V}$, which will keep all transistors biased in the forward-active mode. The current I_{R4} is determined to be

$$I_{R4} = \frac{V_{O2} - 2V_{BE}(\text{on})}{R_4} = \frac{6 - 1.4}{11.5} = 0.4\,\text{mA}$$

Since base currents are assumed negligible, the current I_{R5} is $I_{R5} \cong I_{R4}$.
The dc voltage at the collectors of Q_3 and Q_4 is then

$$V_{O3} = 10 - I_{R5}R_5 = 10 - (0.4)(5) = 8\,\text{V}$$

This shows us that the dc voltage V_{O3} is midway between the 10 V supply voltage and the dc input voltage $V_{O2} = 6\,\text{V}$ to Q_3. This allows a maximum symmetrical swing in the time-varying voltage at v_{o3}.

Transistor Q_5 and resistor R_6 form the dc voltage level shifting function. Since $R_3 = R_2$, we have

$$I_{R6} = I_Q = 0.4\,\text{mA}$$

The dc voltage at the base of Q_6 is found to be

$$V_{B6} = V_{O3} - V_{BE}(\text{on}) - I_{R6}R_6 = 8 - 0.7 - (0.4)(16.5) = 0.7\,\text{V}$$

This relationship produces a zero dc output voltage when a zero differential-mode voltage is applied at the input.
Finally, current I_{R7} is

$$I_{R7} = \frac{v_o - (-10)}{R_7} = \frac{10}{5} = 2\,\text{mA}$$

Comment: The dc analysis of this simplified op-amp circuit proceeds in much the same way as in previous examples. We observe that all transistors are biased in the forward-active mode.

Example 11.19 **Objective:** Determine the small-signal differential-mode voltage gain of the bipolar op-amp circuit.

Consider the circuit in Figure 11.48, with transistor parameters $\beta = 100$ and $V_A = \infty$.

Solution: The overall differential-mode voltage gain can be written

$$A_d = A_{d1} \cdot A_2 \cdot A_3 = \left(\frac{v_{o2}}{v_1 - v_2}\right) \cdot \left(\frac{v_{o3}}{v_{o2}}\right) \cdot \left(\frac{v_o}{v_{o3}}\right)$$

The overall small-signal voltage gain is the product of the individual stage gains *only if* the load resistance of the following stage is taken into account.

We will rely on previous results to determine the individual voltage gains. The input resistances to the Darlington pair R_{i2} and to the output stage R_{i3} are indicated in Figure 11.48. The one-sided differential-mode voltage gain of the diff-amp is given by

$$A_{d1} = \frac{V_{o2}}{v_d} = \frac{g_m}{2}(R_C \| R_{i2})$$

where R_{i2} is the input resistance of the Darlington pair, as follows:

$$R_{i2} = r_{\pi3} + (1 + \beta)r_{\pi4}$$

where

$$r_{\pi4} = \beta V_T/I_{R4} = (100)(0.026)/0.4 = 6.5\,\text{k}\Omega$$

and

$$r_{\pi3} \cong \beta^2 V_T/I_{R4} = (100)^2(0.026)/0.4 = 650\,\text{k}\Omega$$

Therefore,

$$R_{i2} = 650 + (101)(6.5) = 1307\,\text{k}\Omega$$

The transistor transconductance is

$$g_m = \frac{I_Q}{2V_T} = \frac{0.4}{2(0.026)} = 7.70\,\text{mA/V}$$

The gain of the differential amplifier stage is therefore

$$A_{d1} = \frac{g_m}{2}(R_C \| R_{i2}) = \left(\frac{7.70}{2}\right)[20\|1307] = 75.8$$

Since the load resistance $R_{i2} \gg R_C$, there is no significant loading effect of the second stage on the diff-amp stage.

From previous results, we know the voltage gain of the Darlington pair is given by

$$A_2 = \left(\frac{I_{R4}}{2V_T}\right)(R_5 \| R_{i3})$$

where

$$R_{i3} = r_{\pi5} + (1 + \beta)[R_6 + r_{\pi6} + (1 + \beta)R_7]$$

We find that

$$r_{\pi5} = \beta V_T/I_{R6} = (100)(0.026)/0.4 = 6.5\,\text{k}\Omega$$

and

$$r_{\pi 6} = \beta V_T / I_{R7} = (100)(0.026)/2 = 1.3\,\text{k}\Omega$$

Therefore

$$R_{i3} = 6.5 + (101)[16.5 + 1.3 + (101)(5)] \Rightarrow 52.8\,\text{M}\Omega$$

Since $R_{i3} \gg R_5$, the output stage does not load down the gain stage, and the small-signal voltage gain is approximately

$$A_2 \cong \left(\frac{I_{R4}}{2V_T}\right) R_5 = \left[\frac{0.4}{2(0.026)}\right](5) = 38.5$$

The combination of Q_5 and Q_6 forms an emitter follower, and the gain of the output stage is

$$A_3 = v_o/v_{o3} \cong 1$$

The overall small-signal voltage gain is therefore

$$A_d = A_{d1} \cdot A_2 \cdot A_3 = (75.8)(38.5)(1) = 2918$$

Comment: From our previous discussion, we know that the overall gain can be increased substantially by using active loads. Yet, the analysis of this simplified circuit provides some insight into the design of multistage circuits, as well as the overall small-signal voltage gain of op-amp circuits.

Computer Correlation: A PSpice analysis was performed on the bipolar op-amp circuit in Figure 11.48. The dc output voltage from this analysis was $V_O = -0.333\,\text{V}$, rather than the desired value of zero. This occurred because the B–E voltages were not exactly 0.7 V, as assumed in the hand analysis. A zero output voltage can be obtained by slightly adjusting R_6. The differential voltage gain was $A_d = 2932$, which agrees very well with the hand analysis.

Problem-Solving Technique: Multistage Circuits

1. Perform the dc analysis of the circuit to determine the small-signal parameters of the transistors. In most cases BJT base currents can be neglected. This assumption will normally provide sufficient accuracy for a hand analysis.

2. Perform the ac analysis on each stage of the circuit, *taking into account the loading effect of the following stage.* (In many cases, previous results of small-signal analyses can be used directly.)

3. The overall small-signal voltage gain or current gain is the product of the gains of the individual stages *as long as the loading effect of each stage is taken into account.*

Test Your Understanding

*RD11.29 Consider the bipolar op-amp circuit in Figure 11.48. The transistor parameters are: $\beta = 100$, $V_{BE}(\text{on}) = 0.7\,\text{V}$ (except for Q_8 and Q_9), and $V_A = \infty$. (a) Redesign the circuit such that $I_{C1} = I_{C2} = 0.1\,\text{mA}$, $I_{R7} = 5\,\text{mA}$, $I_1 = I_{R4} = I_{R6} = 0.6\,\text{mA}$,

$V_{CE1} = V_{CE2} = 4\,\text{V}$, $V_{CE4} = 3\,\text{V}$, and $v_o = 0$. (b) Determine the input resistances R_{i2} and R_{i3}. (c) Determine the overall differential-mode voltage gain $A_v = v_o/v_d$. (Ans. (a) $R_1 = 32.2\,\text{k}\Omega$, $R_2 = 143\,\text{k}\Omega$, $R_3 = 0$, $R_C = 67\,\text{k}\Omega$, $R_4 = 3.17\,\text{k}\Omega$, $R_5 = 8.5\,\text{k}\Omega$, $R_6 = 5.83\,\text{k}\Omega$, and $R_7 = 2\,\text{k}\Omega$ (b) $R_{i2} = 870\,\text{k}\Omega$, $R_{i3} = 21.0\,\text{M}\Omega$ (c) $A_d = 11{,}674$)

11.8   DIFF-AMP FREQUENCY RESPONSE

In Chapter 7, we considered the frequency responses of the three basic amplifier configurations. In this section, we will analyze the frequency response of the differential amplifier. Since the diff-amp is a linear circuit, we can determine the frequency response due to: (a) a pure differential-mode input signal, (b) a pure common-mode input signal, and (c) the total or net result, using superposition.

11.8.1   Due to Differential-Mode Input Signal

Consider the basic bipolar diff-amp shown in Figure 11.49(a). The input is a pure differential-mode input signal. We know from Equation (11.24) that the small-signal voltage v_e is at signal ground when a differential-mode input signal is applied. To determine the frequency response, we evaluate the equivalent

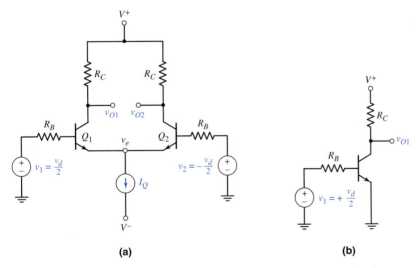

(a)   **(b)**

Figure 11.49   (a) BJT differential amplifier with differential-mode input signal and (b) equivalent common-emitter half-circuit of differential amplifier

common-emitter half-circuit in Figure 11.49(b).

Since the diff-amp is a direct-coupled amplifier, the midband voltage gain extends to zero frequency. This one-sided midband gain is

$$A_{v1} = \frac{V_{o1}}{V_d/2} = -g_m R_C \left(\frac{r_\pi}{r_\pi + R_B} \right) \qquad \textbf{(11.111(a))}$$

or

$$A_{v1} = \frac{-\beta R_C}{r_\pi + R_B} \qquad (11.111(\mathbf{b}))$$

From the high-frequency common-emitter characteristics determined in Chapter 7 we know that the upper 3 dB frequency is

$$f_H = \frac{1}{2\pi[r_\pi \| R_B](C_\pi + C_M)} \qquad (11.112)$$

where C_M is the equivalent Miller capacitance given by

$$C_M = C_\mu(1 + g_m R_C) \qquad (11.113)$$

Equation (11.113) implies that, if the value of R_C is fairly large, the Miller capacitance will significantly affect the bandwidth of the differential amplifier.

11.8.2   Due to Common-Mode Input Signal

Figure 11.50(a) shows the basic diff-amp with a pure common-mode input signal. The circuit is symmetrical, which means that resistors R_B, resistors R_C, and the transistors are effectively in parallel. Figure 11.50(b) is the small-signal equivalent circuit, with the constant-current source replaced by its output resistance R_o and capacitance C_o.

(a)   **(b)**

Figure 11.50   (a) BJT differential amplifier with common-mode input signal and (b) small-signal equivalent circuit, common-mode configuration

We will justify neglecting the transistor parameters C_π and C_μ. The output voltage is

$$V_o = -(2g_m V_\pi)\left(\frac{R_C}{2}\right) \qquad (11.114)$$

A KVL equation around the B–E loop produces

$$V_{cm} = \left(\frac{V_\pi}{r_\pi/2}\right)\left(\frac{R_B}{2}\right) + V_\pi + \left(\frac{V_\pi}{r_\pi/2} + 2g_m V_\pi\right)\left[R_o \middle\| \left(\frac{1}{sC_o}\right)\right] \quad \text{(11.115(a))}$$

or

$$V_{cm} = V_\pi \left\{\frac{R_B}{r_\pi} + 1 + 2\left(\frac{1+\beta}{r_\pi}\right)\left(\frac{R_o}{1+sR_oC_o}\right)\right\} \quad \text{(11.115(b))}$$

Solving for V_π and substituting the result into Equation (11.114) yields the common-mode gain, which is

$$A_{cm} = \frac{V_o}{V_{cm}} = \frac{-g_m R_C}{\dfrac{R_B}{r_\pi} + 1 + \dfrac{2(1+\beta)}{r_\pi}\left(\dfrac{R_o}{1+sR_oC_o}\right)} \quad \text{(11.116(a))}$$

or

$$A_{cm} = \frac{-g_m R_C(1+sR_oC_o)}{\left(1+\dfrac{R_B}{r_\pi}\right)(1+sR_oC_o) + \dfrac{2(1+\beta)R_o}{r_\pi}} \quad \text{(11.116(b))}$$

Equation (11.116(b)) shows that there is a zero in the common-mode gain. To explain, capacitor C_o is in parallel with R_o, and it acts as a bypass capacitor. At very low frequency, C_o is effectively an open circuit and the common-mode signal "sees" R_o. As the frequency increases, the impedance of the capacitor decreases and R_o is effectively bypassed; hence, the zero in Equation (11.116(b)). The frequency analysis of an emitter bypass capacitor also showed the presence of a zero in the voltage gain expression.

The common-mode gain frequency response is shown in Figure 11.51. The frequency of the zero is

$$f_z = \frac{1}{2\pi R_o C_o} \quad \text{(11.117)}$$

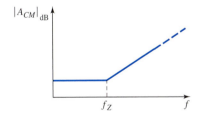

Figure 11.51   Frequency response of common-mode gain

Since the output resistance R_o of a constant-current source is normally large, a small capacitance C_o can result in a small f_z. For frequencies greater than f_z, the common-mode gain increases at the rate of 6 dB/octave.

Equation (11.116(b)) also shows that there is a pole associated with the common-mode gain. Rearranging the terms in that equation, we see that the frequency of the pole is

$$f_p = \frac{1}{2\pi R_{eq} C_o} \qquad \qquad (11.118)$$

where

$$R_{eq} = \frac{R_o\left(1 + \dfrac{R_B}{r_\pi}\right)}{1 + \dfrac{R_B}{r_\pi} + \dfrac{2(1+\beta)R_o}{r_\pi}} \qquad \qquad (11.119)$$

The denominator of Equation (11.119) is very large, because of the term $(1 + \beta)R_o$. This implies that R_{eq} is small, which means that the frequency f_p of the pole is very large.

The differential-mode gain is shown in Figure 11.52. The frequency response of the common-mode rejection ratio is found by combining Figures 11.51 and 11.52, and is shown in Figure 11.53.

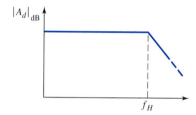

Figure 11.52   Frequency response of differential-mode gain

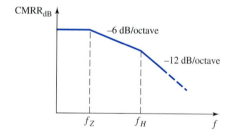

Figure 11.53   Frequency response of common-mode rejection ratio

Example 11.20   **Objective:** Determine the zero and pole frequencies in the common-mode gain.

Consider a diff-amp biased with a constant-current source. The output resistance is $R_o = 10\,\text{M}\Omega$ and the output capacitance is $C_o = 1\,\text{pF}$. Assume the circuit and transistor parameters are $R_B = 0.5\,\text{k}\Omega$, $r_\pi = 10\,\text{k}\Omega$, and $\beta = 100$.

Solution: In the common-mode gain, the frequency of the zero is

$$f_z = \frac{1}{2\pi R_o C_o} = \frac{1}{2\pi(10 \times 10^6)(1 \times 10^{-12})} \Rightarrow 15.9\,\text{kHz}$$

Also in the common-mode gain, the frequency of the pole is

$$f_P = 1/(2\pi R_{eq} C_o)$$

where

$$R_{eq} = \frac{R_o\left(1 + \dfrac{R_B}{r_\pi}\right)}{1 + \dfrac{R_B}{r_\pi} + \dfrac{2(1+\beta)R_o}{r_\pi}} = \frac{(10 \times 10^6)\left(1 + \dfrac{0.5}{10}\right)}{1 + \dfrac{0.5}{10} + \dfrac{2(101)(10 \times 10^6)}{10 \times 10^3}}$$

or

$$R_{eq} = 51.98 \ \Omega$$

The frequency of the pole is therefore

$$f_P = \frac{1}{2\pi(51.98)(1 \times 10^{-12})} \Rightarrow 3.06 \, \text{GHz}$$

Comment: The frequency of the zero in the common-mode gain is fairly low, while the frequency of the pole is extremely large. The relatively low frequency of the zero justifies neglecting the effect of C_π and C_μ. The CMRR frequency response is shown in Figure 11.53, where f_z is the zero frequency of the common-mode gain and f_H is the upper 3 dB frequency of the differential-mode gain.

11.8.3   With Emitter-Degeneration Resistors

Figure 11.54 shows a bipolar diff-amp with two resistances R_E connected in the emitter portion of the circuit. One effect of including an emitter resistor is to reduce the voltage gain, so the presence of these resistors is termed **emitter degeneration**.

In Chapter 7, we found that an emitter-follower circuit, which includes an emitter resistance, is a wide-bandwidth amplifier. Therefore, one effect of

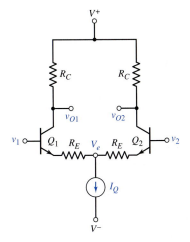

Figure 11.54   BJT differential amplifier with emitter-degeneration resistors

resistors R_E is an increase in the bandwidth of the differential amplifier. We rely on a computer simulation to evaluate emitter degeneration effects.

Figure 11.55 shows the frequency response of a one-sided differential-mode gain, obtained from a PSpice analysis for four R_E resistance values. The diff-amp is biased at $I_Q = 0.5\,\text{mA}$ and the R_C resistors are $R_C = 30\,\text{k}\Omega$. The transistor capacitances are $C_\pi = 34.6\,\text{pF}$ and $C_\mu = 4.3\,\text{pF}$. As the emitter degeneration increases, the differential-mode voltage gain decreases, but the bandwidth increases, as previously indicated. The figure-of-merit for amplifiers, the gain-bandwidth product, is approximately a constant for the results shown in Figure 11.55.

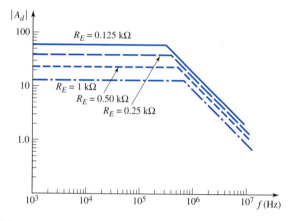

Figure 11.55   PSpice results for frequency response of diff-amp with emitter-degeneration

11.8.4   With Active Load

Figure 11.56 shows a bipolar diff-amp with an active load and a single input at v_1. The base and collector junctions of Q_3 are connected together, and a one-sided output is taken at v_{O2}.

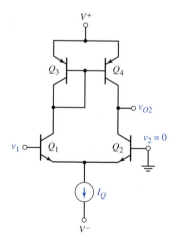

Figure 11.56   BJT diff-amp with active load and single-sided input signal

With the connection of Q_3, the equivalent load resistance in the collector of Q_1 is on the order of $r_\pi/(1 + \beta)$. This small resistance minimizes the Miller multiplication factor in Q_1. Also, with the base of Q_2 at ground potential, one side of $C_{\mu 2}$ is grounded, and the Miller multiplication in Q_2 is zero. Therefore, we expect the bandwidth of the diff-amp with an active load to be relatively wide. At high frequencies, however, the effective impedance in the collector of Q_1 also includes the input capacitances of Q_3 and Q_4. These additional capacitances also affect the frequency response of the diff-amp, potentially narrowing the bandwidth.

Again, we rely on a computer analysis to determine the frequency characteristics of the diff-amp with an active load. Figure 11.57 shows the results of the computer simulation. The diff-amp is biased at $I_Q = 0.5\,\text{mA}$, and the Early voltage of each transistor is assumed to be 80 V. The transistor capacitances are $C_\pi = 34.6\,\text{pF}$ for each transistor, $C_\mu = 3.8\,\text{pF}$ in Q_1 and Q_2, and $C_\mu = 7\,\text{pF}$ and $5.5\,\text{pF}$ in Q_3 and Q_4, respectively.

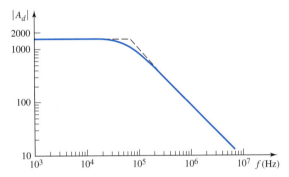

Figure 11.57 PSpice results for frequency response of diff-amp with active load and single-sided input signal

The low-frequency voltage gain is 1560 and the upper 3 dB frequency is 64 kHz. The large gain is as expected for an active load amplifier, but the 3 dB frequency is lower than expected. However, the gain–bandwidth product for the active load diff-amp is approximately four times that of the diff-amp shown in Figure 11.54. The increased gain–bandwidth product implies a reduced Miller multiplication factor in the active load diff-amp, as predicted.

11.9   SUMMARY

- The ideal differential amplifier amplifies only the difference between two input signals.
- The differential-mode input voltage is defined as the difference between the two input signals and the common-mode input voltage is defined as the average of the two input signals.
- When a differential input voltage is applied, one transistor of the differential pair turns on more than the second transistor of the differential pair so that the currents become unbalanced, producing a signal output voltage.

- A common-mode output signal is generated because of a finite output resistance of the current source.
- The common-mode rejection ratio, CMRR, is defined in terms of decibels as $\text{CMRR}_{dB} = 20 \log_{10} |A_d/A_{cm}|$, where A_d and A_{cm} are the differential-mode voltage gain and common-mode voltage gain, respectively.
- Differential amplifiers are usually designed with active loads to increase the differential-mode voltage gain.
- BiCMOS circuits may be designed to incorporate the best parameters and characteristics of BJTs and MOSFETs in the same circuit.
- A BJT Darlington pair is typically used as a second stage to a BJT diff-amp. The input impedance is large, which tends to minimize loading effects on the diff-amp, and the effective current gain of the pair is the product of the individual gains.

CHECKPOINT

After studying this chapter, the reader should have the ability to:

✔ Describe the mechanism by which a differential-mode signal and common-mode signal are produced in a BJT diff-amp. (Section 11.2)

✔ Describe the dc transfer characteristics of a BJT diff-amp. (Section 11.2)

✔ Define common-mode rejection ratio. (Section 11.2)

✔ Describe the mechanism by which a differential-mode signal and common-mode signal are produced in a MOSFET diff-amp. (Section 11.3)

✔ Describe the dc transfer characteristics of a MOSFET diff-amp. (Section 11.3)

✔ Design a MOSFET diff-amp with an active load to yield a specified differential-mode voltage gain. (Section 11.4)

✔ Analyze BiCMOS circuits. (Section 11.5)

✔ Analyze a simplified BJT operational amplifier circuit. (Section 11.7)

REVIEW QUESTIONS

1. Define differential-mode and common-mode input voltages.
2. Sketch the dc transfer characteristics of a BJT differential amplifier.
3. From the dc transfer characteristics, qualitatively define the linear region of operation for a differential amplifier.
4. What is meant by matched transistors?
5. Explain how a differential-mode output signal is generated.
6. Explain how a common-mode output signal is generated.
7. Define the common-mode rejection ratio, CMRR. What is the ideal value?
8. What design criteria will yield a large value of CMRR in an emitter-coupled pair?
9. Sketch the differential-mode and common-mode half-circuit models for an emitter-coupled diff-amp.
10. Define differential-mode and common-mode input resistances.
11. Sketch the dc transfer characteristics of a MOSFET differential amplifier.
12. Sketch and describe the advantages of a MOSFET cascode current source used with a MOSFET differential amplifier.
13. Sketch a simple MOSFET differential amplifier with an active load.
14. Explain the advantages of an active load.
15. Sketch and describe the advantages of a MOSFET cascode active load with a MOSFET differential pair.

16. Discuss one advantage of a BiCMOS circuit.
17. Describe the effect of connecting a second stage to the output of the diff-amp on the differential-mode voltage gain of the first stage.
18. Explain the frequency response of the differential-mode voltage gain.
19. Sketch a BJT Darlington pair circuit and explain the advantages.
20. Describe the three stages of a simple BJT operational amplifier.

PROBLEMS

Section 11.2   Basic BJT Differential Pair

RD11.1 Consider the differential amplifier shown in Figure P11.1, with transistor parameters: $\beta = 200$, $V_{BE}(\text{on}) = 0.7\,\text{V}$, and $V_A = \infty$. (a) Redesign the circuit such that the Q-point values are $I_{C1} = I_{C2} = 2.0\,\text{mA}$ and $V_{O2} = 8\,\text{V}$ when $v_1 = v_2 = 0$. (b) Draw the dc load line and plot the Q-point for transistor Q_2. (c) What are the maximum and minimum values of the common-mode input voltage?

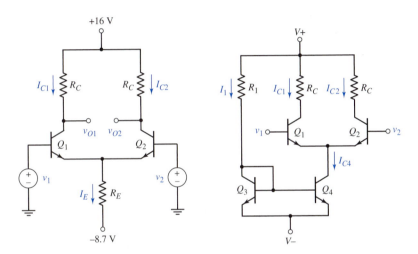

Figure P11.1   **Figure P11.2**

D11.2 The diff-amp configuration shown in Figure P11.2 is biased at $\pm 3\,\text{V}$. The maximum power dissipation in the entire circuit is to be no more than $1.2\,\text{mW}$ when $v_1 = v_2 = 0$. The available transistors have parameters: $\beta = 120$, $V_{BE}(\text{on}) = 0.7\,\text{V}$, and $V_A = \infty$. Design the circuit to produce the maximum possible differential-mode voltage gain, but such that the common-mode input voltage can be within the range $-1 \leq v_{CM} \leq 1\,\text{V}$ and the transistors are still biased in the forward-active region. What is the value of A_d? What are the current and resistor values?

11.3 The differential amplifier in Figure P11.3 is biased with a three-transistor current source. The transistor parameters are: $\beta = 100$, $V_{BE}(\text{on}) = 0.7\,\text{V}$, and $V_A = \infty$. (a) Determine I_1, I_{C2}, I_{C4}, V_{CE2}, and V_{CE4}. (b) Determine a new value of R_1 such that $V_{CE4} = 2.5\,\text{V}$. What are the values of I_{C4}, I_{C2}, I_1, and R_1?

11.4 Consider the circuit in Figure P11.4, with transistor parameters: $\beta = 100$, $V_{BE}(\text{on}) = 0.7\,\text{V}$, and $V_A = \infty$. (a) For $v_1 = v_2 = 0$, find I_{C1}, I_{C2}, I_E, V_{CE1}, and V_{CE2}. (b) Determine the maximum and minimum values of the common-mode input voltage. (c) Calculate A_d for a one-sided output at the collector of Q_2.

Figure P11.3   Figure P11.4

11.5 Consider the differential amplifier in Figure P11.5. Neglect base currents, assume $V_A = \infty$ for all transistors, and let $I_O = 2\,\text{mA}$. The emitter currents can be written as

$$I_{E1} = I_{S1}e^{V_{BE1}/V_T} \quad \text{and} \quad I_{E2} = I_{S2}e^{V_{BE2}/V_T}$$

(a) If $v_1 = v_2 = 0$ and $I_{S1} = I_{S2} = 1 \times 10^{-13}\,\text{A}$, find $(v_{o1} - v_{o2})$ when: (i) $R_{C1} = R_{C2} = 8\,\text{k}\Omega$, and (ii) $R_{C1} = 8\,\text{k}\Omega$, $R_{C2} = 7.9\,\text{k}\Omega$. (b) Repeat part (a) if $I_{S1} = 1 \times 10^{-13}\,\text{A}$ and $I_{S2} = 1.1 \times 10^{-13}\,\text{A}$.

Figure P11.5

11.6 For the diff-amp in Figure 11.2, determine the value of $v_d = v_1 - v_2$ that produces $i_{C2} = 0.90 I_Q$.

*RD11.7 The diff-amp for the experimental system described in Example 11.4 needs to be redesigned. The range of the output voltage has increased to $-2 \leq V_O \leq 2\,\text{V}$ while the differential-mode voltage gain is still $A_d = 100$. The common-mode input voltage has increased to $v_{CM} = 3.5\,\text{V}$. The value of CMRR needs to be increased to 80 dB.

*11.8 The transistor parameters for the circuit in Figure P11.8 are: $\beta = 100$, $V_{BE}(\text{on})$ $= 0.7\,\text{V}$, and $V_A = \infty$. (a) Determine R_E such that $I_E = 150\,\mu\text{A}$. (b) Find A_d, A_{cm}, and CMRR$_{\text{dB}}$ for a one-sided output at v_{O2}. (c) Determine the differential- and common-mode input resistances.

Figure P11.8   **Figure P11.10**

*11.9 The transistor parameters for the circuit in Figure P11.8 are: $\beta = 120$, $V_{BE}(\text{on})$ $= 0.7\,\text{V}$, and $V_A = \infty$. (a) Determine R_E such that $I_E = 0.25\,\text{mA}$. (b) Assume the R_B resistance connected to the base of Q_2 is zero while the R_B resistance connected to the base of Q_1 remains at $0.5\,\text{k}\Omega$. (i) Determine the differential-mode voltage gain for a one-sided output at v_{O2}. (ii) Determine the common-mode voltage gain for a one-sided output at v_{O2}.

*RD11.10 For the transistors in the circuit in Figure P11.10, the parameters are $\beta = 100$ and $V_{BE}(\text{on}) = 0.7\,\text{V}$. The Early voltage is $V_A = \infty$ for Q_1 and Q_2, and is $V_A = 50\,\text{V}$ for Q_3 and Q_4. (a) Redesign resistor values such that $I_3 = 400\,\mu\text{A}$ and $V_{CE1} = V_{CE2} = 10\,\text{V}$. (b) Find A_d, A_{cm}, and CMRR$_{\text{dB}}$ for a one-sided output at v_{O2}. (c) Determine the differential- and common-mode input resistances.

D11.11 Consider the diff-amp in Figure 11.2. Base currents are negligible and $V_A = \infty$ for each transistor. The supply voltages are $V^+ = 10\,\text{V}$ and $V^- = -10\,\text{V}$, and the maximum current source available is $I_Q = 2\,\text{mA}$. Design the circuit such that a differential-mode output voltage of $v_o = v_{C2} - v_{C1} = 2\,\text{V}$ is produced when a differential-mode input voltage of $v_d = v_1 - v_2 = 15\,\text{mV}$ is applied. What is the maximum possible common-mode input voltage for this circuit?

*11.12 Consider the circuit in Figure P11.12. Assume the Early voltage of Q_1 and Q_2 is $V_A = \infty$, and assume the current source I_Q is ideal. Derive the expressions for the one-sided differential-mode gain $A_{v1} = v_{o1}/v_d$ and $A_{v2} = v_{o2}/v_d$, and for the two-sided differential-mode gain $A_d = (v_{o2} - v_{o1})/v_d$.

11.13 The Early voltage of transistors Q_1 and Q_2 in the circuit in Figure P11.13 is $V_A = \infty$. Assming an ideal current source I_Q, derive the expression for the differential-mode gain $A_d = v_o/v_d$.

Figure P11.12   **Figure P11.13**

*11.14   Consider the small-signal equivalent circuit of the differential-pair configuration shown in Figure 11.8. Derive the expressions for the differential- and common-mode voltage gains if the output is a two-sided output defined as $V_o = V_{c2} - V_{c1}$.

*D11.15   Consider a BJT diff-amp with the configuration in Figure P11.15. The signal sources have nonzero source resistances as shown. The transistor parameters are: $\beta = 150$, $V_{BE}(\text{on}) = 0.7\,\text{V}$, and $V_A = \infty$. The range of the common-mode input voltage is to be $-3 \leq v_{CM} \leq 3\,\text{V}$ and the CMRR is to be $75\,\text{dB}$. (a) Design the diff-amp to produce the maximum possible differential-mode voltage gain. (b) Design the current source to produce the desired bias current and CMRR.

Figure P11.15

11.16   A diff-amp has a differential-mode voltage gain of 180 and a CMRR of 85 dB. A differential-mode input signal of $v_d = 2 \sin \omega t\,\text{mV}$ is applied, along with a common-mode voltage of $V_{cm} = 2 \sin \omega t\,\text{V}$. Determine the ideal output voltage and the actual output voltage.

*11.17 The bridge circuit in Figure P11.17 is a temperature transducer in which the resistor R_A is a thermistor (a resistor whose resistance varies with temperature). The value of δ varies over the range of $-0.01 \le \delta \le 0.01$ as temperature varies over a particular range. Assume the value of $R = 10 \, \text{k}\Omega$. The bridge circuit is to be connected to the diff-amp in Figure 11.2. The transistor parameters are: $\beta = 120$, $V_{BE}(\text{on}) = 0.7 \, \text{V}$, and $V_A = \infty$. The circuit parameters are: $I_Q = 0.5 \, \text{mA}$, $R_C = 3 \, \text{k}\Omega$, and dc bias voltages $= \pm 5 \, \text{V}$. Terminal A of the bridge circuit is connected to the base of Q_1 and terminal B is connected to the base of Q_2. Determine the range of output voltage v_{O2} as δ changes. [Hint: Make a Thevenin equivalent circuit at terminals A and B of the bridge circuit.]

Figure P11.17

11.18 A diff-amp is biased with a constant-current source $I_Q = 0.4 \, \text{mA}$, for which the output resistance is $R_o = 1 \, \text{M}\Omega$. The bipolar transistor parameters are $\beta = 180$ and $V_A = 125 \, \text{V}$. Determine: (a) the differential-mode input resistance, and (b) the common-mode input resistance.

D11.19 The transistor parameters for the circuit shown in Figure P11.19 are: $\beta = 180$, $V_{BE}(\text{on}) = 0.7 \, \text{V}$ at $1 \, \text{mA}$, and $V_A = 100 \, \text{V}$. (a) Determine R_1 and R_2 such that $I_1 = 1 \, \text{mA}$ and $I_Q = 100 \, \mu\text{A}$. (b) Determine the common-mode input resistance. (c) For $R_C = 50 \, \text{k}\Omega$, determine the common-mode voltage gain.

Figure P11.19   **Figure P11.20**

D11.20 Figure P11.20 shows a two-stage cascade diff-amp with resistive loads. Power supply voltages of $\pm 10 \, \text{V}$ are available. Assume transistor parameters of: $\beta = 100$, $V_{BE}(\text{on}) = 0.7 \, \text{V}$, and $V_A = \infty$. Design the circuit such that the two-sided differential-mode voltage gain is $A_{d1} = (v_{o2} - v_{o1})/(v_1 - v_2) = 20$ for the first stage, and that the one-sided differential-mode voltage gain is $A_{d2} = v_{o3}/(v_{o2} - v_{o1}) = 30$ for the second stage. The circuit is to be designed such that the maximum differential-mode voltage swing is obtained in each stage.

Section 11.3   Basic FET Differential Pair

11.21 For the differential amplifier in Figure P11.21 the parameters are $R_1 = 50\,\mathrm{k\Omega}$ and $R_D = 24\,\mathrm{k\Omega}$. The transistor parameters are: $K_n = 0.25\,\mathrm{mA/V^2}$, $\lambda = 0$, and $V_{TN} = 2\,\mathrm{V}$. (a) Determine I_1, I_Q, I_{D1}, V_{DS1}, and V_{DS4} when $v_1 = v_2 = 0$. (b) Draw the dc load line and plot the Q-point for transistor M_2. (c) What are the maximum and minimum values of the common-mode input voltage?

RD11.22 The transistor parameters in the differential amplifier in Figure P11.21 are: $K_{n1} = K_{n2} = 100\,\mathrm{\mu A/V^2}$, $K_{n3} = K_{n4} = 200\,\mathrm{\mu A/V^2}$, $\lambda_1 = \lambda_2 = 0$, $\lambda_3 = \lambda_4 = 0.01\,\mathrm{V^{-1}}$, and $V_{TN} = 1.2\,\mathrm{V}$ (all transistors). (a) Design the circuit such that $V_{DS1} = V_{DS2} = 12\,\mathrm{V}$ and $I_{D1} = I_{D2} = 120\,\mathrm{\mu A}$ when $v_1 = v_2 = -5.4\,\mathrm{V}$. What are the values of I_Q and I_1? (b) Calculate the change in I_Q if $v_1 = v_2 = 0$.

11.23 The transistor parameters for the differential amplifier in Figure P11.23 are: $K_n = 0.4\,\mathrm{mA/V^2}$, $\lambda = 0$, and $V_{TN} = 2\,\mathrm{V}$. (a) Find R_D and I_Q such that $I_{D1} = I_{D2} = 0.5\,\mathrm{mA}$ and $v_{O2} = 7\,\mathrm{V}$ when $v_1 = v_2 = 0$. (b) Draw the dc load line, and plot the Q-point for transistor M_2. (c) What are the maximum and minimum common-mode input voltages?

11.24 Consider the differential amplifier in Figure P11.24. Assume $\lambda = 0$ and $V_{TN} = 0.8\,\mathrm{V}$ for all transistors, and let $I_Q = 1\,\mathrm{mA}$. The drain currents can be written as

$$I_{D1} = K_{n1}(V_{GS1} - V_{TN})^2 \quad \text{and} \quad I_{D2} = K_{n2}(V_{GS2} - V_{TN})^2$$

(a) If $v_1 = v_2 = 0$ and $K_{n1} = K_{n2} = 0.4\,\mathrm{mA/V^2}$, find $(v_{o1} - v_{o2})$ when: (i) $R_{D1} = R_{D2} = 6\,\mathrm{k\Omega}$, and (ii) $R_{D1} = 6\,\mathrm{k\Omega}$, $R_{D2} = 5.9\,\mathrm{k\Omega}$. (b) Repeat part (a) if $K_{n1} = 0.4\,\mathrm{mA/V^2}$ and $K_{n2} = 0.44\,\mathrm{mA/V^2}$.

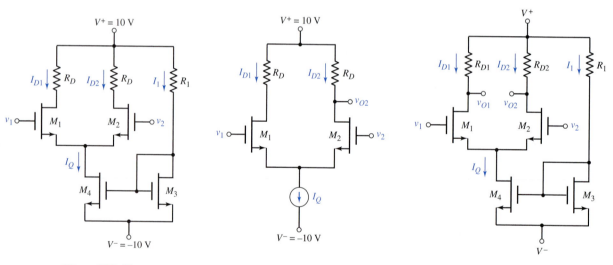

Figure P11.21   Figure P11.23   Figure P11.24

11.25 The transistor parameters for the diff-amp shown in Figure 11.19 are: $K_n = 0.1\,\mathrm{mA/V^2}$, $\lambda = 0$, and $V_{TN} = 1\,\mathrm{V}$. The bias current is $I_Q = 0.25\,\mathrm{mA}$. (a) Determine the value of $v_d = v_{G1} - v_{G2}$ that produces $i_{D2} = 0.90\,I_Q$. (b) At what value of v_d does $i_{D1} = I_Q$?

***D11.26** The Hall effect experimental arrangement was described in Example 11.4. The required diff-amp is to be designed in the circuit configuration in Figure P11.24.

The transistor parameters are $V_{TN} = 0.8\,\text{V}$, $k_n' = 80\,\mu\text{A/V}^2$, $\lambda_1 = \lambda_2 = 0$, and $\lambda_3 = \lambda_4 = 0.01\,\text{V}^{-1}$. If the CMRR requirement cannot be met, a more sophisticated current source may have to be designed.

*11.27 Consider the diff-amp in Figure P11.27. The transistor parameters are: $K_{n1} = K_{n2} = 50\,\mu\text{A/V}^2$, $\lambda_1 = \lambda_2 = 0.02\,\text{V}^{-1}$, and $V_{TN1} = V_{TN2} = 1\,\text{V}$. (a) Determine I_S, I_{D1}, I_{D2}, and v_{O2} for $v_1 = v_2 = 0$. (b) Using the small-signal equivalent circuit, determine the differential-mode voltage gain $A_d = v_{o2}/v_d$, the common-mode voltage gain $A_{cm} = v_{o2}/v_{cm}$, and the CMRR$_{\text{dB}}$.

11.28 Consider the circuit shown in Figure P11.28. Assume that $\lambda = 0$ for M_1 and M_2. Also assume an ideal current source I_Q. Derive the expression for the one-sided differential mode gains $A_{d1} = v_{o1}/v_d$ and $A_{d2} = v_{o2}/v_d$, and the two-sided differential-mode gain $A_d = (v_{o2} - v_{o1})/v_d$.

11.29 Assume $\lambda_1 = \lambda_2 = 0$ for the transistors M_1 and M_2 in the circuit in Figure P11.29. Assuming an ideal current source I_Q, derive the expression for the differential-mode gain $A_d = v_o/v_d$.

*RD11.30 Consider the diff-amp in Figure 11.19. Assume $\lambda = 0$ and $V_{TN} = 1\,\text{V}$ for each transistor. The supply voltages are $V^+ = 10\,\text{V}$ and $V^- = -10\,\text{V}$, and the maximum current source available is $I_Q = 0.5\,\text{mA}$. Redesign the circuit such that a differential-mode output voltage of $v_o = 2\,\text{V}$ is produced when a differential-mode input voltage of $v_d = v_1 - v_2 = 200\,\text{mV}$ is applied. What is the maximum possible common-mode input voltage that can be applied to this circuit?

Figure P11.27   **Figure P11.28**   **Figure P11.29**

11.31 Consider the small-signal equivalent circuit in Figure 11.23. Assume the output is a two-sided output defined as $V_o = V_{d2} - V_{d1}$, where V_{d2} and V_{d1} are the signal voltages at the drains of M_2 and M_1, respectively. Derive expressions for the differential- and common-mode voltage gains.

*D11.32 Consider a MOSFET diff-amp with the configuration in Figure P11.23. The transistor parameters are $V_{TN} = 1\,\text{V}$, $k_n' = 80\,\mu\text{A/V}^2$, $(W/L)_1 = (W/L)_2 = 10$, and $\lambda = 0$. Let $I_Q = 0.2\,\text{mA}$. The range of the common-mode input voltage is $-3 \leq v_{CM} \leq 3\,\text{V}$ and the CMRR is to be 45 dB. (a) Design the diff-amp to produce the maximum possible differential-mode voltage gain. (b) Design an all-MOSFET current source to produce the desired bias current and CMRR. (The minimum W/L ratio of any transistor is to be 0.8.)

11.33   Consider the bridge circuit and diff-amp described in Problem 11.17. The BJT pair is to be replaced with a MOSFET pair whose parameters are $V_{TN} = 0.5\,\text{V}$, $K_n = 0.25\,\text{mA/V}^2$, and $\lambda = 0$. Determine the range of output voltage v_{O2} as δ changes. Explain the advantages and disadvantages of this circuit configuration compared to that in Problem 11.17.

*D11.34   Figure P11.34 shows a two-stage cascade diff-amp with resistive loads. Power supply voltages of $\pm 10\,\text{V}$ are available. Assume transistor parameters of $V_{TN} = 1\,\text{V}$, $k_n' = 60\,\mu\text{A/V}^2$, and $\lambda = 0$. Design the circuit such that the two-sided differential-mode voltage gain is $A_{d1} = (v_{o2} - v_{o1})/(v_1 - v_2) = 20$ for the first stage, and that the one-sided differential-mode voltage gain is $A_{d2} = v_{o3}/(v_{o2} - v_{o1}) = 30$ for the second stage. The circuit is to be designed such that the maximum differential-mode voltage swing is obtained in each stage.

Figure P11.34

*11.35   Figure P11.35 shows a matched JFET differential pair biased with a current source I_Q. (a) Starting with

$$i_D = I_{DSS}\left(1 - \frac{v_{GS}}{V_P}\right)^2$$

show that

$$\frac{i_{D1}}{I_Q} = \frac{1}{2} + \left(\frac{1}{-2V_P}\right)v_d\sqrt{2\left(\frac{I_{DSS}}{I_Q}\right) - \left(\frac{I_{DSS}}{I_Q}\right)^2\left(\frac{v_d}{V_P}\right)^2}$$

and

$$\frac{i_{D2}}{I_Q} = \frac{1}{2} - \left(\frac{1}{-2V_P}\right)v_d\sqrt{2\left(\frac{I_{DSS}}{I_Q}\right) - \left(\frac{I_{DSS}}{I_Q}\right)^2\left(\frac{v_d}{V_P}\right)^2}$$

(b) Show that the I_Q bias current is switched entirely to one transistor or the other when

$$|v_d| = |V_P|\sqrt{\frac{I_Q}{I_{DSS}}}$$

(c) Show that the maximum forward transconductance is given by

$$g_f(\text{max}) = \frac{di_{D1}}{dv_d}\bigg|_{v_d=0} = \left(\frac{1}{-V_P}\right)\sqrt{\frac{I_Q \cdot I_{DSS}}{2}}$$

11.36 A JFET differential amplifier is shown in Figure P11.36. The transistor parameters are: $V_P = -4\,\text{V}$, $I_{DSS} = 2\,\text{mA}$, and $\lambda = 0$. (a) Find R_D and I_Q such that $I_{D1} = I_{D2} = 0.5\,\text{mA}$ and $v_{o2} = 7\,\text{V}$ when $v_1 = v_2 = 0$. (b) Calculate the maximum forward transconductance. (c) Determine the one-sided differential-mode voltage gain $A_d = v_o/v_d$.

*11.37 Consider the JFET diff-amp shown in Figure P11.37. The transistor parameters are: $I_{DSS} = 0.8\,\text{mA}$, $\lambda = 0.02\,\text{V}^{-1}$, and $V_P = -2\,\text{V}$. (a) Determine I_S, I_{D1}, I_{D2}, and v_{o2} for $v_1 = v_2 = 0$. (b) Using the small-signal equivalent circuit, determine the differential-mode voltage gain $A_d = v_{o2}/v_d$, the common-mode voltage gain $A_{cm} = v_o/v_{cm}$, and the CMRR_{dB}.

Figure P11.35   **Figure P11.36**   **Figure P11.37**

*11.38 Consider the circuit in Figure P11.38. Assume that $\lambda = 0$ for the transistors, and assume an ideal current source I_Q. Derive the expressions for the one-sided differential-mode gains $A_{d1} = v_{o1}/v_d$ and $A_{d2} = v_{o2}/v_d$, and for the two-sided differential-mode gain $A_d = (v_{o2} - v_{o1})/v_d$.

Section 11.4   Differential Amplifier with Active Load

*11.39 Consider the diff-amp with active load in Figure P11.39. The Early voltages are $V_{AN} = 120\,\text{V}$ for Q_1 and Q_2 and $V_{AP} = 80\,\text{V}$ for Q_3 and Q_4. (a) Determine the open-circuit differential-mode voltage gain. (b) Compare this value to the gain obtained when $R = 0$. (c) Determine the output resistance R_o for parts (a) and (b).

Figure P11.38

Figure P11.39   **Figure P11.40**

11.40 The diff-amp in Figure P11.40 has a three-transistor active load circuit and a Darlington pair configuration connected to the output. Determine the bias current I_{Q1} in terms of I_Q such that the diff-amp dc currents are balanced.

11.41 For the diff-amp in Figure 11.30, the parameters are: $V^+ = 10\,\text{V}$, $V^- = -10\,\text{V}$, and $I_Q = 0.8\,\text{mA}$. The transistor parameters are: $\beta = 150$, $V_{A1} = V_{A2} = 150\,\text{V}$, and $V_{A3} = V_{A4} = 100\,\text{V}$. (a) Determine the open-circuit differential-mode voltage gain. (b) Find the load resistance R_L that will reduce the differential-mode voltage gain to one-half the open-circuit value.

*11.42 Consider the circuit in Figure P11.42, in which the input transistors to the diff-amp are Darlington pairs. Assume transistor parameters of $\beta(\text{npn}) = 120$, $\beta(\text{pnp}) = 80$, $V_A(\text{npn}) = 100\,\text{V}$, and $V_A(\text{pnp}) = 80\text{V}$. Let the power supply voltages be ± 10 V and let

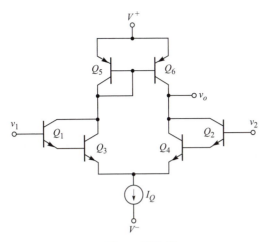

Figure P11.42

$I_Q = 1\,\mathrm{mA}$. (a) Determine the output resistance R_o. (b) Calculate the differential-mode voltage gain. (c) Find the differential-mode input resistance R_{id}.

*D11.43 Design a differential amplifier as shown in Figure 11.28 incorporating a basic two-transistor current source to establish I_Q. The bias voltages are $V^+ = 15\,\mathrm{V}$ and $V^- = -15\,\mathrm{V}$, the transistor parameters are $\beta = 180$ and $V_A = 100\,\mathrm{V}$, and the maximum forward transconductance is to be $8\,\mathrm{mA/V}$. (a) Show the complete circuit, with all component values. (b) What are the values of open-circuit differential-mode voltage gain, differential-mode input resistance, and output resistance? (c) Determine the common-mode input voltage range and the common-mode input resistance.

11.44 The differential amplifier shown in Figure P11.44 has a pair of pnp bipolars as input devices and a pair of npn bipolars connected as an active load. The circuit bias is $I_Q = 0.2\,\mathrm{mA}$, and the transistor parameters are $\beta = 100$ and $V_A = 100\,\mathrm{V}$. (a) Determine I_O such that the dc currents in the diff-amp are balanced. (b) Find the open-circuit differential-mode voltage gain. (c) Determine the differential-mode voltage gain if a load resistance $R_L = 250\,\mathrm{k\Omega}$ is connected to the output.

Figure P11.44

718

Part II Analog Electronics

*11.45   Reconsider the circuit in Figure P11.44 excep that 1 kΩ resistors are inserted at the emitters of the active load transistors Q_3 and Q_4 as in the circuit in Figure P11.39. Assume the same transistor parameters as in Problem 11.44. (a) Determine the output resistance looking into the output of the diff-amp circuit. (b) Find the open-circuit differential-mode voltage gain.

*D11.46   For the transistors in the diff-amp circuit in Figure 11.30 the parameters are: $\beta = 150$, $V_{A1} = V_{A2} = 125\,\text{V}$, and $V_{A3} = V_{A4} = 80\,\text{V}$. The supply voltages are $V^+ = 10\,\text{V}$ and $V^- = -10\,\text{V}$, and the maximum current source available is $I_Q = 2\,\text{mA}$. A load resistance of $R_L = 200\,\text{k}\Omega$ is connected to the output. (a) Design the circuit such that the differential-mode voltage gain is 1000. (b) If $V_{BE}(\text{on}) = 0.6\,\text{V}$, what is the maximum possible common-mode input voltage that can be applied to the circuit?

D11.47   Design a BJT diff-amp with an active load similar to the configuration in Figure P11.42 except that the input devices are to be pnp transistors and the active load will have npn transistors. Using the same parameters as in Problem 11.42, determine the small-signal differential-mode voltage gain.

11.48   The differential amplifier in Figure P11.48 has a pair of PMOS transistors as input devices and a pair of NMOS transistors connected as an active load. The circuit is biased with $I_Q = 0.2\,\text{mA}$, and the transistor parameters are: $K_n = K_p = 0.1\,\text{mA/V}^2$, $\lambda_n = 0.01\,\text{V}^{-1}$, $\lambda_p = 0.015\,\text{V}^{-1}$, $V_{TN} = 1\,\text{V}$, and $V_{TP} = -1\,\text{V}$. (a) Determine the quiescent drain-to-source voltage in each transistor. (b) Find the open-circuit differential-mode voltage gain. (c) What is the output resistance?

11.49   For the differential amplifier in Figure 11.32, the parameters are: $V^+ = 5\,\text{V}$, $V^- = -5\,\text{V}$, and $I_Q = 0.15\,\text{mA}$. The PMOS parameters are: $K_p = 100\,\mu\text{A/V}^2$, $\lambda_p = 0.02\,\text{V}^{-1}$, and $V_{TP} = -0.8\,\text{V}$. The NMOS parameters are: $K_n = 120\,\mu\text{A/V}^2$, $\lambda_n = 0.015\,\text{V}^{-1}$, and $V_{TN} = +0.8\,\text{V}$. Determine the differential-mode voltage gain $A_d = v_o/v_d$.

*11.50   Consider the diff-amp in Figure P11.50. The PMOS parameters are: $K_p = 80\,\mu\text{A/V}^2$, $\lambda_p = 0.02\,\text{V}^{-1}$, $V_{TP} = -2\,\text{V}$. The NMOS parameters are: $K_n = 80\,\mu\text{A/V}^2$, $\lambda_n = 0.015\,\text{V}^{-1}$, $V_{TN} = +2\,\text{V}$. (a) Determine the open-circuit differential-mode voltage

Figure P11.48

Figure P11.50

gain. (b) Compare this value to the gain obtained when $R_1 = 0$. (c) What is the output resistance of the diff-amp for parts (a) and (b)?

D11.51 Reconsider the diff-amp specifications listed in Problem 11.26. (a) Design an all-CMOS diff-amp with the configuration in Figure 11.32 to meet the specifications. Assume NMOS parameters of $V_{TN} = 0.8$ V, $k_n' = 80\,\mu$A/V^2, and $\lambda_n = 0.02$ V^{-1} and PMOS parameters of $V_{TP} = -0.8$ V, $k_p' = 35\,\mu$A/V^2, and $\lambda_p = 0.025$ V^{-1}. (b) Determine the common-mode voltage gain using a computer simulation.

D11.52 Design an all-CMOS diff-amp, including the current source circuit, with the configuration in Figure 11.32 to have a differential-mode voltage gain of $A_d = 80$. The circuit is to be biased at ± 3 V and the total power dissipated in the circuit is to be no more than 0.5 mW. The available transistors have parameters of $V_{TN} = 0.4$ V, $k_n' = 80\,\mu$A/V^2, $\lambda_n = 0.015$ V^{-1}, $V_{TP} = -0.4$ V, $k_p' = 40\,\mu$A/V^2, and $\lambda_p = 0.02$ V^{-1}. Verify the differential-mode voltage gain of the design with a computer simulation. Also, determine the common-mode gain with a computer simulation.

RD11.53 Redesign the cascode active load CMOS diff-amp in Figure 11.36 to achieve a differential-mode voltage gain of $A_d = 400$. Assume $k_n' = 80\,\mu$A/V^2 and $k_p' = 40\,\mu$A/V^2 and use other transistor parameters described in Example 11.15.

*11.54 Consider the fully cascoded diff-amp in Figure 11.37. Assume $I_Q = 80\,\mu$A and transistor parameters of: $V_{TN} = 0.8$ V, $k_n' = 60\,\mu$A/V^2, $\lambda_n = 0.015$ V^{-1}, $V_{TP} = -0.8$ V, $k_p' = 25\,\mu$A/V^2, and $\lambda_p = 0.02$ V^{-1}. The transistor width-to-length ratios are $W/L = 60/4$ for transistors M_1–M_4, $W/L = 40/4$ for transistors M_5–M_6, and $W/L = 4/4$ for transistors M_7–M_8. (a) Determine the output resistance of the diff-amp. (b) Calculate the differential-mode voltage gain of the diff-amp. (c) Find the common-mode voltage gain of the diff-amp using a computer simulation.

Section 11.5 BiCMOS Circuits

11.55 The Darlington pair circuit in Figure 11.45 has new bias current levels of $I_{BIAS1} = 0.25$ mA and $I_{BIAS2} = 1$ mA. The transistor parameters are: $K_n = 0.2$ mA/V^2, $V_{TN} = 1$ V, and $\lambda = 0$ for M_1; and $\beta = 120$, $V_{BE}(\text{on}) = 0.7$ V, and $V_A = \infty$ for Q_2. Determine the small-signal parameters for each transistor, and find the composite transconductance.

11.56 Consider the BiCMOS diff-amp in Figure 11.44, biased at $I_Q = 0.4$ mA. The transistor parameters for M_1 and M_2 are: $K_n = 0.2$ mA/V^2, $V_{TN} = 1$ V, and $\lambda = 0.01$ V^{-1}. The parameters for Q_1 and Q_2 are: $\beta = 120$, $V_{EB}(\text{on}) = 0.7$ V, and $V_A = 80$ V. (a) Determine the differential-mode voltage gain. (b) If the output resistance of the current source is $R_o = 500$ kΩ, determine the common-mode voltage gain using a computer simulation analysis.

*11.57 The BiCMOS circuit in Figure P11.57 is equivalent to a pnp bipolar transistor with an infinite input impedance. The bias current is $I_Q = 900\,\mu$A. The transistor parameters are: $K_p = 1$ mA/V^2, $V_{TP} = -1$ V, and $\lambda = 0$ for M_1; and $\beta = 100$, $V_{BE}(\text{on}) = 0.7$ V, and $V_A = \infty$ for Q_2. (a) Sketch the small-signal equivalent circuit. (b) Determine the small-signal parameters for each transistor. (c) Determine the small-signal voltage gain $A_v = v_o/v_i$.

*11.58 Consider the BiCMOS circuit in Figure P11.57. The bias current is $I_Q = 1.2$ mA, and the transistor parameters are the same as described in Problem 11.57. (a) Determine the small-signal transistor parameters. (b) Find the output impedance R_o.

*11.59 The bias current I_Q is 25$\,\mu$A in each circuit in Figure P11.59. The BJT parameters are $\beta = 100$ and $V_A = 50$ V, and the MOSFET parameters are $V_{TN} = 0.8$ V, $K_n = 0.25$ mA/V^2, and $\lambda = 0.02$ V^{-1}. Assume the two amplifying transistors M_1 and

Figure P11.57

Figure P11.59

Q_1 are biased in the saturation region and forward-active region, respectively. Determine the small-signal voltage gain $A_v = v_o/v_i$ and the output resistance R_o for each circuit.

11.60 For the circuit shown in Figure P11.60, determine the small-signal voltage gain, $A_v = v_o/v_i$. Assume transistor parameters of $V_{TN} = 1\,\text{V}$, $K_n = 0.2\,\text{mA/V}^2$, and $\lambda = 0$ for M_1 and $\beta = 80$ and $V_A = \infty$ for Q_1.

Figure P11.60

Section 11.6   Gain Stage and Simple Output Stage

11.61 Consider the circuit in Figure P11.61. The output stage is a Darlington pair emitter-follower configuration. Assume $\beta = 100$ for all transistors, and let $V_A = 100\,\text{V}$ for Q_7 and Q_{11}. Determine the output resistance R_o.

Figure P11.61

Figure P11.62

*11.62 For the circuit in Figure P11.62, the transistor parameters are $\beta = 100$ and $V_A = \infty$. The bias currents in the transistors are indicated on the figure. Determine the input resistance R_i, the output resistance R_o, and the small-signal voltage gain $A_v = v_o/v_{in}$.

11.63 Consider the circuit in Figure P11.63. The bias currents I_1 and I_2 are such that a zero dc output voltage is established. The transistor parameters are: $K_p = 0.2\,\text{mA/V}^2$, $K_n = 0.5\,\text{mA/V}^2$, $V_{TP} = -0.8\,\text{V}$, $V_{TN} = +0.8\,\text{V}$, and $\lambda_n = \lambda_p = 0.01\,\text{V}^{-1}$. Determine the small-signal voltage gain $A_v = v_o/v_{in}$ and the output resistance R_o.

11.64 The circuit shown in Figure P11.64 has bias currents $I_1 = 0.1\,\text{mA}$ and $I_2 = 0.5\,\text{mA}$. The transistor parameters are: $K_n = 100\,\mu\text{A/V}^2$, $K_p = 250\,\mu\text{A/V}^2$, $V_{TN} = 1\,\text{V}$, $V_{TP} = -1\,\text{V}$, and $\lambda_n = \lambda_p = 0.01\,\text{V}^{-1}$. (a) Determine the resistor values R_1 and R_2 such that the dc value of the output voltage is zero. (b) Find the small-signal voltage gain $A_v = v_o/v_{in}$ and the output resistance R_o.

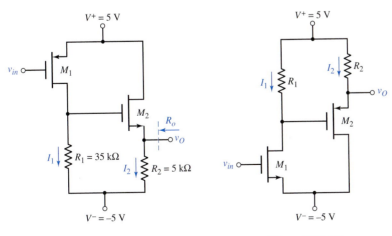

Figure P11.63

Figure P11.64

Section 11.7   Simplified Op-Amp Circuits

*11.65   Consider the multistage bipolar circuit in Figure P11.65, in which base currents are negligible. Assume the transistor parameters are: $\beta = 100$, $V_{BE}(\text{on}) = 0.7\,\text{V}$, and $V_A = \infty$. The output resistance of the constant-current source is $R_o = 100\,\text{k}\Omega$. (a) For $v_1 = v_2 = 0$, design the circuit such that: $v_{o2} = 2\,\text{V}$, $v_{o3} = 3\,\text{V}$, $v_o = 0$, $I_{CQ3} = 0.5\,\text{mA}$, and $I_{CQ4} = 3\,\text{mA}$. (b) Determine the differential-mode voltage gains $A_{d1} = v_{o2}/v_d$ and $A_d = v_o/v_d$. (c) Determine the common-mode voltage gains $A_{cm1} = v_{o2}/v_{cm}$ and $A_{cm} = v_o/v_{cm}$, and the overall CMRR_{dB}.

*D11.66   The circuit in Figure P11.66 has two bipolar differential amplifiers in cascade, biased with ideal current sources I_{Q1} and I_{Q2}. Assume the transistor parameters are $\beta = 180$ and $V_A = \infty$. (a) Design the circuit such that $v_{o1} = v_{o2} = 2\,\text{V}$ and $v_{O4} = 6\,\text{V}$ when $v_1 = v_2 = 0$. (b) Determine the differential-mode voltage gains $A_{d1} = (v_{o1} - v_{o2})/v_d$ and $A_d = v_{o4}/v_d$.

Figure P11.65   Figure P11.66

*11.67   The transistor parameters for the circuit in Figure P11.67 are: $\beta = 200$, $V_{BE}(\text{on}) = 0.7\,\text{V}$, and $V_A = 80\,\text{V}$. (a) Determine the differential-mode voltage gain $A_d = v_{o3}/v_d$ and the common-mode voltage gain $A_{cm} = v_{o3}/v_{cm}$. (b) Determine the output voltage v_{o3} if $v_1 = 2.015 \sin \omega t\,\text{V}$ and $v_2 = 1.985 \sin \omega t\,\text{V}$. Compare this output to the ideal output that would be obtained if $A_{cm} = 0$. (c) Find the differential-mode and common-mode input resistances.

*11.68   For the transistors in the circuit in Figure P11.68, the parameters are: $K_n = 0.2\,\text{mA/V}^2$, $V_{TN} = 2\,\text{V}$, and $\lambda = 0.02\,\text{V}^{-1}$. (a) Determine the differential-mode voltage gain $A_d = v_{o3}/v_d$ and the common-mode voltage gain $A_{cm} = v_{o3}/v_{cm}$. (b) Determine the output voltage v_{o3} if $v_1 = 2.15 \sin \omega t\,\text{V}$ and $v_2 = 1.85 \sin \omega t\,\text{V}$. Compare this output to the ideal output that would be obtained if $A_{cm} = 0$.

Section 11.8   Diff-Amp Frequency Response

11.69   Consider the differential amplifier in Figure 11.49(a), with parameters: $I_Q = 1\,\text{mA}$, $R_C = 10\,\text{k}\Omega$, and $R_B = 0.5\,\text{k}\Omega$. The transistor parameters are: $\beta = 100$, $V_{BE}(\text{on}) = 0.7\,\text{V}$, $V_A = \infty$, $C_\pi = 8\,\text{pF}$, and $C_\mu = 2\,\text{pF}$. Determine the low-frequency

Figure P11.67

Figure P11.68

differential-mode gain and the upper 3 dB frequency. What is the equivalent Miller capacitance of each transistor?

11.70 The differential amplifier in Figure 11.50(a) has the same circuit and transistor parameters as in Problem 11.69. The equivalent impedance parameters of the current source are $R_o = 5\,\text{M}\Omega$ and $C_o = 0.8\,\text{pF}$. (a) Determine the frequency of the zero in the common-mode gain. (b) Plot CMRR_{dB} versus frequency, showing the frequencies f_z and f_H.

11.71 A BJT diff-amp is biased with a current source $I_Q = 2\,\text{mA}$, and the circuit parameters are $R_C = 10\,\text{k}\Omega$ and $R_B = 1\,\text{k}\Omega$. The transistor paramaters are: $\beta = 120$, $f_T = 800\,\text{MHz}$, and $C_\mu = 1\,\text{pF}$. (a) Determine the upper 3 dB frequency of the differential-mode gain. (b) If the current source impedance parameters are $R_o = 10\,\text{M}\Omega$ and $C_o = 1\,\text{pF}$, find the frequency of the zero in the common-mode gain.

11.72 Consider the diff-amp in Figure 11.54. The circuit and transistor parameters are the same as in Problem 11.69. For a one-sided output at v_{o2}, determine the differential-mode gain for: (a) $R_E = 100\,\Omega$, and (b) $R_E = 250\,\Omega$.

COMPUTER SIMULATION PROBLEMS

11.73 For the transistors in the circuit in Figure P11.73, the parameters are: $\beta = 100$, $I_S = 2 \times 10^{-15}\,\text{A}$, and $V_A = 100\,\text{V}$. From a PSpice analysis, determine: (a) the quiescent currents I_1, I_Q, I_{C1}, and I_{C2}, and (b) the differential- and common-mode gains for (i) $R_L = 10\,\text{M}\Omega$, and (ii) $R_L = 200\,\text{k}\Omega$.

11.74 Consider the circuit in Figure P11.74. The transistor parameters are: $V_{TN} = 2\,\text{V}$ (all NMOS devices), $V_{TP} = -2\,\text{V}$ (all PMOS devices), $K_{n5} = K_{n6} = 50\,\mu\text{A/V}^2$, $K_{n7} = K_{n8} = 200\,\mu\text{A/V}^2$, $K_{n1} = K_{n2} = K_{p3} = K_{p4} = 100\,\mu\text{A/V}^2$, and $\lambda = 0.01\,\text{V}^{-1}$ (all devices). From a computer analysis, determine: (a) the quiescent currents I_1 and I_Q; and (b) the differential- and common-mode gains for (i) $R_L = 10\,\text{M}\Omega$ and (ii) $R_L = 400\,\text{k}\Omega$.

Figure P11.73

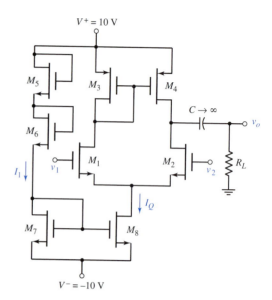

Figure P11.74

11.75 For the circuit in Figure 11.46 the parameters are: $V^+ = 10\,\text{V}$, $V^- = -10\,\text{V}$, $R_1 = 19\,\text{k}\Omega$, $R_2 = R_3 = 0.2\,\text{k}\Omega$, and $R_4 = 10\,\text{k}\Omega$. The transistor parameters are: $\beta = 100$, $I_S = 2 \times 10^{-15}\,\text{A}$, and $V_A = 100\,\text{V}$. (a) From a computer analysis, determine the quiescent currents $I_1, I_Q, I_{C1}, I_{C2}, I_{B5}, I_O$, and I_{C7}. (b) Also from a computer analysis, determine the input resistance R_i, output resistance R_o, and voltage gain $A_v = v_o/v_{o2}$. (c) Compare these results to those obtained in Examples 11.16 and 11.17.

11.76 Consider the circuit in Figure P11.67, with circuit and transistor parameters described in Problem 11.67. Let $I_S = 2 \times 10^{-15}\,\text{A}$. From a computer analysis, determine: (a) the differential-mode voltage gain, (b) the common-mode voltage gain, (c) the input differential-mode resistance, and (d) the input common-mode resistance.

11.77 Consider the diff-amp described in Problems 11.69 and 11.70. Using a computer analysis, determine the CMRR_{dB} versus frequency characteristic.

DESIGN PROBLEMS

[Note: Each design is to be correlated with a computer simulation analysis.]

*D11.78 Design a basic BJT diff-amp with an active load, to provide an open-circuit differential-mode gain of $|A_d| = 2000$ and a common-mode rejection ratio of $\text{CMRR}_{\text{dB}} = 80\,\text{dB}$. Specify the bias currents, minimum Early voltage, and minimum output impedance of the current source. Design the current source to achieve the specified output impedance.

*D11.79 Design a basic MOSFET diff-amp with an active load, to provide an open-circuit differential-mode gain of $|A_d| = 200$ and a common-mode rejection ratio of $\text{CMRR}_{\text{dB}} = 70\,\text{dB}$. Specify the bias currents, conduction parameter values, minimum λ values, and minimum output impedance of the current source. Design the current source to achieve the specified output impedance specification.

Figure P11.82

*D11.80 Consider the BiCMOS diff-amp in Figure 11.44. Design the circuit to provide a differential-mode gain of $|A_d| = 500$ and a common-mode rejection ratio of $CMRR_{dB} = 70$ dB. Specify the bias currents, MOSFET conduction parameter values, minimum bipolar Early voltage, MOSFET λ values, and minimum output resistance of the current source. Design the current source to achieve this specified output resistance.

*D11.81 Consider the bipolar op-amp configuration in Figure 11.48. The bias voltages are ±10 V, as shown, the current I_{R7} is to be $I_{R7} = 3$ mA, and the maximum dc power dissipation in the circuit is to be 120 mW. The output voltage is to be $v_o = 0$ for $v_1 = v_2 = 0$. Design the circuit, using reasonable resistance and current values. What is the overall differential-mode voltage gain?

*D11.82 The transistor parameters for the circuit in Figure P11.82 are: $K_n = 0.2$ mA/V^2, $V_{TN} = 0.8$ V, and $\lambda = 0$. The output resistance of the constant-current source is $R_o = 100$ kΩ. (a) For $v_1 = v_2 = 0$, design the circuit such that: $v_{o2} = 2$ V, $v_{o3} = 3$ V, $v_o = 0$, $I_{DQ3} = 0.25$ mA, and $I_{DQ} = 2$ mA. (b) Determine the differential-mode gains $A_{d1} = v_{o2}/v_d$ and $A_d = v_o/v_d$. (c) Determine the common-mode voltage gains $A_{cm1} = v_{o2}/v_{cm}$ and $A_{cm} = v_o/v_{cm}$, and the overall $CMRR_{dB}$.

12

Feedback and Stability

12.0 PREVIEW

Previously, we found that the small-signal voltage gain and other character-istics of discrete BJT and MOSFET transistor circuit amplifiers are functions of the bipolar current gain and the MOSFET conduction parameter. In general, these transistor parameters vary with temperature and they have a range of values for a given type of transistor group, because of processing and material property tolerances. This means that the Q-point, voltage gain, and other circuit properties can vary from one circuit to another, and can be functions of temperature.

Transistor circuit characteristics can be made essentially independent of the individual transistor parameters by using feedback. The feedback process takes a portion of the output signal and returns it to the input to become part of the input excitation. We previously encountered feedback in our study of ideal op-amps and op-amp circuits. For example, resistors are connected between the output and input terminals of an ideal op-amp to form a feedback network. The voltage gain of these ideal circuits is a function only of the ratio of resistors and not of any individual transistor parameters. In this chapter, we formally study feedback and feedback circuits.

We begin the chapter by presenting general feedback theory and determining general properties of feedback circuits. We then analyze the four basic ideal feedback configurations. For each type of feedback topology, the output-to-input signal transfer function is determined and the expressions for input and output resistances are derived. We analyze various op-amp and discrete transistor circuits representing each of the four basic feedback configurations, and compare the transfer functions, input resistance, and output resistances to the ideal theory. Principal goals of this chapter are to understand the character-istics of the various types of feedback configurations to be able to determine the type of feedback circuit required for a specific design application and to design a stable feedback amplifier.

12.1 INTRODUCTION TO FEEDBACK

Feedback is used in virtually all amplifier systems. Harold Black, an electronics engineer with the Western Electric Company, invented the feedback amplifier in 1928 while searching for methods to stabilize the gain of amplifiers for use in telephone repeaters. In a feedback system, a signal that is proportional to the output is fed back to the input and combined with the input signal to produce a desired system response. As we will see, external feedback is used deliberately to achieve particular system benefits. However, feedback may be unintentional and an undesired system response may be produced.

We have already seen examples of feedback in previous chapters, although the term feedback may not have been used. For example, in Chapters 3 and 5 we introduced resistors at the emitter of BJT common-emitter circuits and at the source of MOSFET common-source circuits to stabilize the Q-point against variations in transistor parameters. This technique introduces *negative feedback* in the circuit. An increase in collector or drain current produces an increase in the voltage across these resistors which produces a decrease in the base-emitter or gate-source voltage. The decrease in these device voltages tends to reduce or oppose the change in collector or drain current. Opposition to change is suggested by use of the term negative feedback.

Feedback can be either negative or positive. In **negative feedback**, a portion of the output signal is subtracted from the input signal; in **positive feedback**, a portion of the output signal is added to the input signal. Negative feedback, for example, tends to maintain a constant value of amplifier voltage gain against variations in transistor parameters, supply voltages, and temperature. Positive feedback is used in the design of oscillators and in a number of other applications. In this chapter, we will concentrate on negative feedback.

12.1.1   Advantages and Disadvantages of Negative Feedback

Before we actually get into the analysis and design of feedback circuits, we will list some of the advantages and disadvantages of negative feedback. Although these characteristics and properties of negative feedback are not obvious at this point, they are listed here so that the reader can anticipate these results during the derivations and analysis.

Advantages

1. *Gain sensitivity.* Variations in the circuit transfer function (gain) as a result of changes in transistor parameters are reduced by feedback. This reduction in sensitivity is one of the most attractive features of negative feedback.
2. *Bandwidth extension.* The bandwidth of a circuit that incorporates negative feedback is larger than that of the basic amplifier.
3. *Noise sensitivity.* Negative feedback may increase the signal-to-noise ratio if noise is generated within the feedback loop.
4. *Reduction of nonlinear distortion.* Since transistors have nonlinear characteristics, distortion may appear in the output signals, especially at large signal levels. Negative feedback reduces this distortion.
5. *Control of impedance levels.* The input and output impedances can be increased or decreased with the proper type of negative feedback circuit.

Disadvantages

1. *Circuit gain.* The overall amplifier gain, with negative feedback, is reduced compared to the basic amplifier used in the circuit.
2. *Stability.* There is a possibility that the feedback circuit may become unstable (oscillate) at high frequencies.

These advantages and disadvantages will be further discussed as we develop the feedback theory.

In the course of our discussion, we will analyze several feedback circuits, in both discrete and op-amp circuit configurations. First, however, we will consider the ideal feedback theory and derive the general characteristics of feedback amplifiers. In this section, we discuss the ideal signal gain, gain sensitivity, bandwidth extension, noise sensitivity, and reduction of nonlinear distortion of a generalized feedback amplifier.

12.1.2 Use of Computer Simulation

Conventional methods of analysis that have been used in the previous chapters apply directly to feedback circuits. That is, the same dc analysis techniques and the same small-signal transistor equivalent circuits apply directly to feedback circuits in this chapter. However, in the analysis of feedback circuits, several simultaneous equations can be obtained, the time involved may be quite long and the probability of introducing errors may become almost certain.

Therefore, computer simulation of feedback circuits may prove to be very useful and is used fairly often throughout this chapter. As always, a word of warning is in order concerning computer simulation. Computer simulation does not replace basic understanding. It is important for the reader to understand the concepts and characteristics of the basic types of feedback circuits. Computer simulation is used only as a tool for obtaining specific results.

12.2 BASIC FEEDBACK CONCEPTS

Figure 12.1 shows the basic configuration of a feedback amplifier. In the diagram, the various signals S can be either currents or voltages. The circuit contains a basic amplifier with an open-loop gain A and a feedback circuit that samples the output signal and produces a feedback signal S_{fb}. The feedback signal is subtracted from the input source signal, which produces an error signal S_ε. The error signal is the input to the basic amplifier and is the signal that is amplified to produce the output signal. The subtraction property produces the negative feedback.

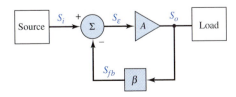

Figure 12.1   Basic configuration of a feedback amplifier

Implicit in the diagram in Figure 12.1 is the assumption that the input signal is transmitted through the amplifier only, none through the feedback network, and that the output signal is transmitted back through the feedback network only, none through the amplifier. Also, there are no loading effects in the ideal feedback system. The feedback network does not load down the output of the basic amplifier, and the basic amplifier and feedback network do not produce a loading effect on the input signal source. In actual feedback circuits, these assumptions and conditions are not entirely accurate. We will see later how nonideal conditions change the characteristics of actual feedback circuits with respect to those of the ideal feedback network.

12.2.1   Ideal Closed-Loop Signal Gain

From Figure 12.1, the output signal is

$$S_o = AS_\varepsilon \tag{12.1}$$

where A is the **amplification factor**, and the feedback signal is

$$S_{fb} = \beta S_o \tag{12.2}$$

where β in this case is the **feedback transfer function**.[1] At the summing node, we have

$$S_\varepsilon = S_i - S_{fb} \tag{12.3}$$

where S_i is the input signal. Equation (12.1) then becomes

$$S_o = A(S_i - \beta S_o) = AS_i - \beta AS_o \tag{12.4}$$

Equation (12.4) can be rearranged to yield the **closed-loop transfer function**, or gain, which is

$$A_f = \frac{S_o}{S_i} = \frac{A}{(1 + \beta A)} \tag{12.5}$$

As mentioned, signals S_i, S_o, S_{fb}, and S_ε can be either currents or voltages; however, they do not need to be all voltages or all currents in a given feedback amplifier. In other words, there may be a combination of current and voltage signals in the same circuit.

Equation (12.5) can be written

$$A_f = \frac{A}{(1 + \beta A)} = \frac{A}{1 + T} \tag{12.6}$$

where $T = \beta A$ is the **loop gain**. For negative feedback, we assume T to be a positive real factor. We will see later that the loop gain can become a complex function of frequency, but for the moment, we will assume that T is positive for negative feedback. We will also see that in some cases the gain will be negative (180 degree phase difference between input and output signals) which means

---

[1]In this chapter, β is the feedback transfer function, rather than the transistor current gain. The parameter h_{FE} will be used as the transistor current gain. Normally, h_{FE} indicates the dc current gain and h_{fe} indicates the ac current gain. However, as usual, we neglect any difference between the two parameters and assume $h_{FE} = h_{fe}$.

that the feedback transfer function β will also be a negative quantity for a negative feedback circuit.

Combining Equations (12.1) and (12.2), we obtain the loop gain relationship

$$T = A\beta = \frac{S_{fb}}{S_{\varepsilon}} \tag{12.7}$$

Normally, the error signal is small, so the expected loop gain is large. If the loop gain is large so that $\beta A \gg 1$, then, from Equation (12.6), we have

$$A_f \cong \frac{A}{\beta A} = \frac{1}{\beta} \tag{12.8}$$

and the gain or transfer function of the feedback amplifier essentially becomes a function of the feedback network only.

The feedback circuit is usually composed of passive elements, which means that the feedback amplifier gain is almost completely independent of the basic amplifier properties, including individual transistor parameters. Since the feedback amplifier gain is a function of the feedback elements only, the closed-loop gain can be designed to be a given value. This property was demonstrated in Chapter 9, where we showed that the closed-loop gain of ideal op-amp circuits is a function of the feedback elements only. The individual transistor parameters may vary widely, and may depend on temperature and frequency, but the feedback amplifier gain is constant. The net results of negative feedback is stability in the amplifier characteristics.

In general, the magnitude and phase of the loop gain are functions of frequency, and they become important when we discuss the stability of feedback circuits.

Example 12.1   **Objective:** Calculate the feedback transfer function β, given A and A_f.

Case A:
Assume that the open-loop gain of a system is $A = 10^5$ and the closed-loop gain is $A_f = 50$.

Solution: From Equation (12.5), the closed-loop gain is

$$A_f = \frac{A}{(1 + \beta A)}$$

Therefore,

$$50 = \frac{10^5}{1 + \beta(10^5)}$$

which yields $\beta = 0.01999$ or $1/\beta = 50.025$.

Case B:
Now assume that the open-loop gain is $A = -10^5$ and the closed-loop gain is $A_f = -50$.

Solution: Again, from Equation (12.5), the closed-loop gain is

$$A_f = \frac{A}{(1 + \beta A)}$$

so that

$$-50 = \frac{-10^5}{1 + \beta(-10^5)}$$

which yields $\beta = -0.01999$ or $1/\beta = -50.025$.

Comment: From these typical parameter values, we see that $A_f \cong 1/\beta$, as Equation (12.8) predicts. We also see that if the open-loop gain A is negative, then the closed-loop gain A_f and feedback transfer function β will also be negative for a negative feedback network.

Assuming a large loop gain, the output signal, from Equation (12.5), becomes

$$S_o = \left(\frac{A}{1 + \beta A}\right) S_i \cong \frac{1}{\beta} \cdot S_i \tag{12.9}$$

Substituting Equation (12.9) into (12.3), we obtain the error signal,

$$S_\varepsilon = S_i - \beta S_o \cong S_i - \beta\left(\frac{S_i}{\beta}\right) = 0 \tag{12.10}$$

With a large loop gain, the error signal decreases to almost zero. We will see this result again as we consider specific feedback circuits throughout the chapter.

Test Your Understanding

12.1 The open-loop gain of an amplifier is $A = 10^4$, and the closed-loop gain is $A_f = 20$. (a) What is the feedback transfer function β? (b) What is the ratio of A_f to $(1/\beta)$? (Ans. (a) $\beta = 0.0499$ (b) 0.998)

12.2 The closed-loop gain of a feedback amplifier is $A_f = 80$, and the feedback transfer function is $\beta = 0.0120$. What is the value of the open-loop gain A? (Ans. $A = 2000$)

12.2.2   Gain Sensitivity

As previously stated, if the loop gain $T = \beta A$ is very large, the overall gain of the feedback amplifier is essentially a function of the feedback network only. We can quantify this characteristic.

If the feedback transfer function β is a constant, then taking the derivative of A_f with respect to A, from Equation (12.5), produces

$$\frac{dA_f}{dA} = \frac{1}{(1 + \beta A)} - \frac{A}{(1 + \beta A)^2} \cdot \beta = \frac{1}{(1 + \beta A)^2} \tag{12.11(a)}$$

or

$$dA_f = \frac{dA}{(1 + \beta A)^2} \tag{12.11(b)}$$

Dividing both sides of Equation (12.11(b)) by the closed-loop gain yields

$$\frac{dA_f}{A_f} = \frac{\dfrac{dA}{(1+\beta A)^2}}{\dfrac{A}{1+\beta A}} = \frac{1}{(1+\beta A)} \cdot \frac{dA}{A} = \left(\frac{A_f}{A}\right)\frac{dA}{A} \tag{12.12}$$

Equation (12.12) shows that the percent change in the closed-loop gain A_f is less than the corresponding percent change in the open-loop gain A by the factor $(1 + \beta A)$. The change in open-loop gain may result from variations in individual transistor parameters in the basic amplifier.

Example 12.2   **Objective:**   Calculate the percent change in the closed-loop gain A_f, given a change in the open-loop gain A.

Using The same parameter values as in Example 12.1, we have $A = 10^5$, $A_f = 50$, and $\beta = 0.01999$. Assume that the change in the open-loop gain is $dA = 10^4$ (a 10 percent change).

Solution: From Equation (12.12), we have

$$dA_f = \frac{A_f}{(1+\beta A)} \cdot \frac{dA}{A} = \frac{50}{[1 + (0.01999)(10^5)]} \cdot \frac{10^4}{10^5} = 2.5 \times 10^{-3}$$

The percent change is then

$$\frac{dA_f}{A_f} = \frac{2.5 \times 10^{-3}}{50} = 5 \times 10^{-5} \Rightarrow 0.005\%$$

compared to the 10 percent change assumed in the open-loop gain.

Comment: From this example, we see that the resulting percent change in the closed-loop gain is substantially less than the percent change in the open-loop gain. This is one of the principal advantages of negative feedback.

From Equation (12.12), the change in A_f is reduced by the factor $(1 + \beta A)$ compared to the change in A. The term $(1 + \beta A)$ is called the **densensitivity factor**.

Test Your Understanding

12.3   Consider a general feedback system with parameters $A = 10^6$ and $A_f = 100$. If the magnitude of A decreases by 20 percent, what is the corresponding percent change in A_f? (Ans. 0.002%)

12.4   The gain factors in a feedback system are $A = 5 \times 10^5$ and $A_f = 100$. Parameter A_f must not change more than ± 0.001 percent because of a change in A. What is the maximum allowable variation in A? (Ans. $\pm 5\%$)

12.2.3   Bandwidth Extension

The amplifier bandwidth is a function of feedback. Assume the frequency response of the basic amplifier can be characterized by a single pole. We can then write

$$A(s) = \frac{A_o}{1 + \dfrac{s}{\omega_H}} \tag{12.13}$$

where A_o is the low-frequency or midband gain, and ω_H is the upper 3 dB or corner frequency.

The closed-loop gain of the feedback amplifier can be expressed as

$$A_f(s) = \frac{A(s)}{(1 + \beta A(s))} \tag{12.14}$$

where we assume that the feedback transfer function β is independent of frequency. Substituting Equation (12.13) into Equation (12.14), we can write the closed-loop gain in the form

$$A_f(s) = \frac{A_o}{(1 + \beta A_o)} \cdot \frac{1}{1 + \dfrac{s}{\omega_H(1 + \beta A_o)}} \tag{12.15}$$

From Equation (12.15), we see that the low-frequency closed-loop gain is smaller than the open-loop gain by a factor of $(1 + \beta A_o)$, but the closed-loop 3 dB frequency is larger than the open-loop value by a factor of $(1 + \beta A_o)$.

If we multiply the low-frequency open-loop gain A_o by the bandwidth (3 dB frequency) ω_H, we obtain $A_o\omega_H$, which is the gain–bandwidth product. The product of the low-frequency closed-loop gain and the closed-loop bandwidth is

$$\frac{A_o}{(1 + \beta A_o)}[\omega_H(1 + \beta A_o)] = A_o\omega_H \tag{12.16}$$

Equation (12.16) states that the gain-bandwidth product of a feedback amplifier is a constant. That is, for a given circuit, we can increase the gain at the expense of a reduced bandwidth, or we can increase the bandwidth at the expense of a reduced gain. This property is illustrated in Figure 12.2.

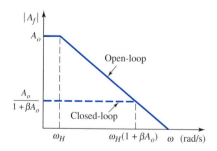

Figure 12.2   Open-loop and closed-loop gain versus frequency, illustrating bandwidth extension

Example 12.3   **Objective:**   Determine the bandwidth of a feedback amplifier.

Consider a feedback amplifier with an open-loop low-frequency gain of $A_o = 10^4$, an open-loop bandwidth of $\omega_H = (2\pi)(100)$ rad/s, and a closed-loop low-frequency gain of $A_f(0) = 50$.

Solution: From Equation (12.15), the low-frequency closed-loop gain is

$$A_f(0) = \frac{A_o}{(1 + \beta A_o)}$$

or

$$50 = \frac{10^4}{(1 + \beta A_o)}$$

which yields

$$(1 + \beta A_o) = \frac{10^4}{50} = 200$$

From Equation (12.15), the closed-loop bandwidth is

$$\omega_{fH} = \omega_H(1 + \beta A_o) = (2\pi)(100)(200) = (2\pi)(20 \times 10^3)$$

Comment: The bandwidth increases from 100 Hz to 20 kHz as the gain decreases from 10^4 to 50.

Test Your Understanding

12.5   A feedback amplifier has an open-loop low-frequency gain of $A_o = 10^5$, an open-loop bandwidth of $\omega_H = (2\pi)(10)$ rad/s, and a closed-loop low-frequency gain of $A_f(0) = 100$. Determine the bandwidth of the closed-loop system. (Ans. $\omega = (2\pi)$ (10^4) rad/s)

12.6   In a feedback amplifier, the open-loop low-frequency gain is $A_o = 10^6$ and the open-loop 3 dB frequency is 8 Hz. If the bandwidth of the closed-loop system is 250 kHz, what is the maximum allowable value of the closed-loop low-frequency gain? (Ans. A_f $(0) = 32$)

12.2.4   Noise Sensitivity

In any electronic system, unwanted random and extraneous signals may be present in addition to the desired signal. These random signals are called **noise**. Electronic noise can be generated within an amplifier, or may enter the amplifier along with the input signal. Negative feedback may reduce the noise level in amplifiers; more accurately, it may increase the **signal-to-noise ratio**. More precisely, feedback can help reduce the effect of noise generated in an amplifier, but it cannot reduce the effect when the noise is part of the input signal.

The input signal-to-noise ratio is defined as

$$(\text{SNR})_i = \frac{S_i}{N_i} = \frac{v_i}{v_n} \tag{12.17}$$

where $S_i = v_i$ is the input source signal and $N_i = v_n$ is the input noise signal. The output signal-to-noise ratio is

$$(\text{SNR})_o = \frac{S_o}{N_o} = \frac{A_{Ti}S_i}{A_{Tn}N_i} \tag{12.18}$$

where the desired output signal is $S_o = A_{Ti}S_i$ and the output noise signal is $N_o = A_{Tn}N_i$. The parameter A_{Ti} is the amplification factor that multiplies the source signal, and the parameter A_{Tn} is the amplification factor that multiplies the noise signal. A large signal-to-noise ratio allows the signal to be detected without any loss of information. This is a desirable characteristic.

The following example compares the signal and noise amplification factors, which may or may not be equal.

Example 12.4 **Objective:** Determine the effect of feedback on the source signal and noise signal levels.

Consider the four possible amplifier configurations shown in Figure 12.3. The amplifiers are designed to provide the same output signal voltage. Determine the effect of the noise signal v_n.

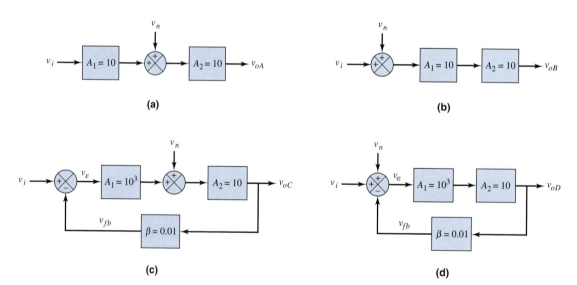

Figure 12.3 Four amplifier configurations with different input noise sources

Solution: Figure 12.3(a): Two open-loop amplifiers are in a cascade configuration, and the noise signal is generated between the two amplifiers. The output voltage is

$$v_{oa} = A_1 A_2 v_i + A_2 v_n = 100 v_i = 10 v_n$$

Therefore, the output signal-to-noise ratio is

$$\frac{S_o}{N_o} = \frac{100 v_i}{10 v_n} = 10 \frac{S_i}{N_i}$$

Solution: Figure 12.3(b): Two open-loop amplifiers are in a cascade configuration, and the noise is part of the input signal. The output voltage is

$$v_{ob} = A_1 A_2 v_i + A_1 A_2 v_n = 100 v_i + 100 v_n$$

Therefore, the output signal-to-noise ratio is

$$\frac{S_o}{N_o} = \frac{100 v_i}{100 v_n} = \frac{S_i}{N_i}$$

Solution: Figure 12.3(c): Two amplifiers are in a feedback configuration, and the noise signal is generated between the two amplifiers. The output voltage is

$$v_{oc} = A_1 A_2 v_\varepsilon + A_2 v_n$$

and the feedback signal is

$$v_{fb} = \beta v_{oc}$$

Then,

$$v_\varepsilon = v_i - v_{fb} = v_i - \beta v_{oc}$$

therefore,

$$v_{oc} = A_1 A_2 (v_i - \beta v_{oc}) + A_2 v_n$$

or

$$v_{oc} = \frac{A_1 A_2}{(1 + \beta A_1 A_2)} \cdot v_i + \frac{A_2}{(1 + \beta A_1 A_2)} \cdot v_n \cong 100 v_i + 0.1 v_n$$

The output signal-to-noise ratio is

$$\frac{S_o}{N_o} = \frac{100 v_i}{0.1 v_n} = 1000 \frac{S_i}{N_i}$$

Solution: Figure 12.3(d): A basic feedback configuration, and the noise is part of the input signal. The output voltage is

$$v_{od} = \frac{A_1 A_2}{(1 + \beta A_1 A_2)} (v_i + v_n) \cong 100 v_i + 100 v_n$$

Therefore, the output signal-to-noise ratio is

$$\frac{S_o}{N_o} = \frac{100 v_i}{100 v_n} = \frac{S_i}{N_i}$$

Comment: Comparing the four configurations, we see that Figure 12.3(c) produces the largest output signal-to-noise ratio. This configuration may occur when amplifier A_2 is an audio power-amplifier stage, in which large currents can produce excessive noise, and when amplifier A_1 corresponds to a low-noise preamplifier, which provides most of the voltage gain.

We must emphasize that the increased signal-to-noise ratio due to feedback occurs only in specific situations. As indicated in Figure 12.3(d), when noise is effectively part of the amplifier input signal, the feedback mechanism does not improve the ratio.

12.2.5   Reduction of Nonlinear Distortion

Distortion in an output signal is caused by a change in the basic amplifier gain or a change in the slope of the basic amplifier transfer function. The change in gain is a function of the nonlinear properties of bipolar and MOS transistors used in the basic amplifier.

Assume the basic amplifier, or open-loop, transfer function is as shown in Figure 12.4(a), which shows changes in gain as the input signal changes. The gain values are shown on the figure. When this amplifier is incorporated in a feedback circuit with a feedback transfer function of $\beta = 0.099$, the resulting closed-loop transfer characteristics are shown in Figure 12.4(b). This transfer function also has changes in gain but, whereas the open-loop gain changes by a factor of 2, the closed-loop gain changes by only 1 percent and 2 percent, respectively. A smaller change in gain means less distortion in the output signal of the negative feedback amplifier.

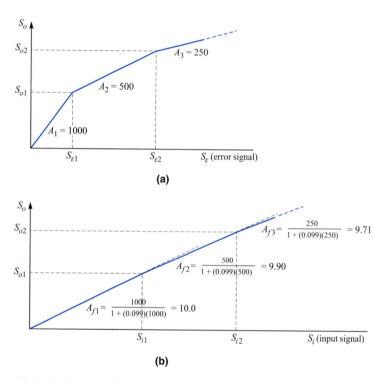

(a)

(b)

Figure 12.4   (a) Basic amplifier (open-loop) transfer characteristics; (b) closed-loop transfer characteristics

12.3   IDEAL FEEDBACK TOPOLOGIES

There are four basic feedback topologies, based on the parameter to be amplified (voltage or current) and the output parameter (voltage or current). The four feedback circuit categories can be described by the types of connections at the input and output of circuit. The four types of connections are shown in Figure 12.5. The four connections are referred to as: series–shunt (voltage amplifier), shunt–series (current amplifier), series–series (transconductance

(a) Series–shunt   (b) Shunt–series

(c) Series–series   (d) Shunt–shunt

Figure 12.5 Basic feedback connections

amplifier), and shunt–shunt (transresistance amplifier). The first term refers to the connection at the amplifier input, and the second term refers to the connection at the output. Also, the type of connection determines which parameter (voltage or current) is sampled at the output and which parameter is amplified. The connections also determine the feedback amplifier characteristics—in particular, the input and output resistances. The resistance parameters become an important circuit property, when, for example, we consider voltage amplifiers versus current amplifiers.

In this section, we will determine the ideal transfer functions and the ideal input and output resistances of each of the four feedback topologies. In later sections, we will compare actual versus ideal feedback circuit characteristics.

As a note, the ideal topologies are small-signal equivalent circuits; therefore, phasor notation is used throughout this analysis.

12.3.1 Series–Shunt Configuration

The configuration of an ideal **series–shunt** feedback amplifier is shown in Figure 12.6. The circuit consists of a basic voltage amplifier with an input resistance R_i and an open-loop voltage gain A_v. The feedback circuit samples the output voltage and produces a feedback voltage V_{fb}, which is in series with the input signal voltage V_i. In this ideal configuration, the input resistance to the feedback circuit is infinite; therefore, there is no loading effect on the output of the basic amplifier due to the feedback circuit.

Voltage V_ε is the difference between the input signal voltage and the feedback voltage and is called an error signal. The error signal is amplified in the basic voltage amplifier. We can recognize the series connection on the input and the shunt connection of the output for this configuration.

The circuit is a voltage-controlled voltage source and is an ideal voltage amplifier. The feedback circuit samples the output voltage and provides a feedback voltage in series with the source voltage. For example, an increase

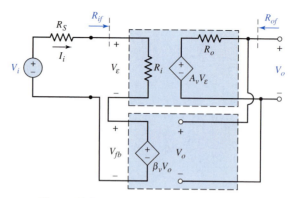

Figure 12.6   Ideal series–shunt feedback topology

in the output voltage produces an increase in the feedback voltage, which in turn decreases the error voltage due to the negative feedback. Then, the smaller error voltage is amplified producing a smaller output voltage, which means that the output signal tends to be stabilized.

If the output of the feedback network is an open circuit, then the output voltage is

$$V_o = A_v V_\varepsilon \tag{12.19}$$

and the feedback voltage is

$$V_{fb} = \beta V_o = \beta_v V_o \tag{12.20}$$

Parameter β_v is the voltage feedback transfer function, which is the ratio of the feedback voltage to the output voltage. The notation is similar to the voltage gain A_v, which is also the ratio of two voltages.

The error voltage, assuming the source resistance R_S is negligible, is

$$V_\varepsilon = V_i - V_{fb} \tag{12.21}$$

Combining Equations (12.19), (12.20), and (12.21), we find the closed-loop voltage transfer function is

$$A_{vf} = \frac{V_o}{V_i} = \frac{A_v}{(1 + \beta_v A_v)} \tag{12.22}$$

Equation (12.22) is the **closed-loop voltage gain** of the feedback amplifier, and it has the same form as the ideal feedback transfer function given by Equation (12.5). Although the magnitude of the closed-loop voltage gain is less than that of the open-loop amplifier, the advantage is that the closed-loop voltage gain becomes essentially independent of the individual transistor parameters. We will examine this characteristic later in this chapter.

The input resistance including feedback is denoted by R_{if}. Starting with Equation (12.21), using Equations (12.19) and (12.20), we find that

$$V_i = V_\varepsilon + V_{fb} = V_\varepsilon + \beta_v V_o = V_\varepsilon + \beta_v (A_v V_\varepsilon) \tag{12.23(a)}$$

or

$$V_\varepsilon = \frac{V_i}{(1 + \beta_v A_v)} \tag{12.23(b)}$$

The input current is

$$I_i = \frac{V_\varepsilon}{R_i} = \frac{V_i}{R_i(1 + \beta_v A_v)}$$ (12.24)

and the input resistance with feedback is then

$$R_{if} = \frac{V_i}{I_i} = R_i(1 + \beta_v A_v)$$ (12.25)

Equation (12.25) shows that a series input connection results in an increased input resistance compared to that of the basic voltage amplifier. A large input resistance is a desirable property of a voltage amplifier. This eliminates loading effects on the input signal source due to the amplifier.

The output resistance of the feedback circuit can be determined from the equivalent circuit in Figure 12.7. The input signal voltage source is set equal to zero (a short circuit), and a test voltage is applied to the output terminals.

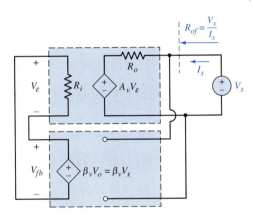

Figure 12.7 Ideal series–shunt feedback configuration for determining output resistance

From the circuit, we see that

$$V_\varepsilon + V_{fb} = V_\varepsilon + \beta_v V_x = 0$$ (12.26(a))

or

$$V_\varepsilon = -\beta_v V_x$$ (12.26(b))

The output current is

$$I_x = \frac{V_x - A_v V_\varepsilon}{R_o} = \frac{V_x - A_v(-\beta_v V_x)}{R_o} = \frac{V_x(1 + \beta_v A_v)}{R_o}$$ (12.27)

and the output resistance, including feedback, is

$$R_{of} = \frac{V_x}{I_x} = \frac{R_o}{(1 + \beta_v A_v)}$$ (12.28)

Equation (12.28) shows that a shunt output connection results in a decreased output resistance compared to that of the basic voltage amplifier. A small output resistance is a desirable property of a voltage amplifier. This eliminates loading effects on the output signal when an output load is connected.

The equivalent circuit of this feedback voltage amplifier is shown in Figure 12.8.

Figure 12.8   Equivalent circuit of the series–shunt feedback circuit or voltage amplifier

Example 12.5   **Objective:** Determine the input resistance of a series input connection and the output resistance of a shunt output connection for an ideal feedback voltage amplifier.

Consider a series–shunt feedback amplifier in which the open-loop gain is $A_v = 10^5$ and the closed-loop gain is $A_{vf} = 50$. Assume the input and output resistances of the basic amplifier are $R_i = 10\,\text{k}\Omega$ and $R_o = 20\,\text{k}\Omega$, respectively.

Solution: The ideal closed-loop voltage transfer function is, from Equation (12.22),

$$A_{vf} = \frac{A_v}{(1 + \beta_v A_v)}$$

or

$$(1 + \beta_v A_v) = \frac{A_v}{A_{vf}} = \frac{10^5}{50} = 2 \times 10^3$$

From Equation (12.25), the input resistance is

$$R_{if} = R_i(1 + \beta_v A_v) = (10)(2 \times 10^3)\,\text{k}\Omega \Rightarrow 20\,\text{M}\Omega$$

and, from Equation (12.28), the output resistance is

$$R_{of} = \frac{R_o}{(1 + \beta_v A_v)} = \frac{20}{2 \times 10^3}\,\text{k}\Omega \Rightarrow 10\,\Omega$$

Comment: With a series input connection, the input resistance increases drastically, and with a shunt output connection, the output resistance decreases substantially, with negative feedback. These are the desired characteristics of a voltage amplifier.

Test Your Understanding

12.7   An ideal series–shunt feedback amplifier is shown in Figure 12.6. Assume R_S is negligibly small. (a) If $V_i = 100\,\text{mV}$, $V_{fb} = 99\,\text{mV}$, and $V_o = 5\,\text{V}$, determine A_v, β_v, and A_{vf}, including units. (b) Using the results of part (a), determine R_{if} and R_{of}, for $R_i = 5\,\text{k}\Omega$ and $R_o = 4\,\text{k}\Omega$. (Ans. (a) $A_v = 5000\,\text{V/V}$, $\beta_v = 0.0198\,\text{V/V}$, $A_{vf} = 50\,\text{V/V}$ (b) $R_{if} = 500\,\text{k}\Omega$, $R_{of} = 40\,\Omega$)

12.3.2   Shunt–Series Configuration

The configuration of an ideal **shunt–series** feedback amplifier is shown in Figure 12.9. The circuit consists of a basic current amplifier with an input resistance R_i and an open-loop current gain A_i. The feedback circuit samples the output current and produces a feedback current I_{fb}, which is in shunt with an input signal current I_i. In this ideal configuration, the feedback circuit does not load down the basic amplifier output; therefore, the load current I_o is not affected.

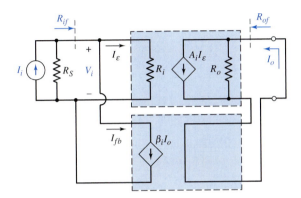

Figure 12.9   Ideal shunt–series feedback topology

Current I_ε is the difference between the input signal current and the feedback current and is the error signal. The error signal is amplified in the basic current amplifier. We can recognize the shunt connection on the input and the series connection on the output for this configuration.

This circuit is a current-controlled current source and is an ideal current amplifier. The feedback circuit samples the output current and provides a feedback signal in shunt with the signal current. An increase in output current produces an increase in feedback current, which in turn decreases the error current. The smaller error current is then amplified, producing a smaller output current and stabilizing the output signal.

The input source shown is a Norton equivalent circuit; it could be converted to a Thevenin equivalent circuit.

If the output is essentially a short circuit, then the output current is

$$I_o = A_i I_\varepsilon \tag{12.29}$$

and the feedback current is

$$I_{fb} = \beta I_o = \beta_i I_o \tag{12.30}$$

The parameter β_i is the feedback current transfer function. The input signal current, assuming R_S is large, is

$$I_i = I_\varepsilon + I_{fb} \tag{12.31}$$

Combining Equations (12.29), (12.30), and (12.31) yields the closed-loop current transfer function

$$A_{if} = \frac{I_o}{I_i} = \frac{A_i}{(1 + \beta_i A_i)} \tag{12.32}$$

Equation (12.32) is the **closed-loop current gain** of the feedback amplifier.

The form of the equation for the current transfer function of the current amplifier (shunt–series connection) is the same as that for the voltage transfer function of the voltage amplifier (series–shunt connection). We will show that this will be the same for the two feedback connections yet to be discussed.

The input resistance of the shunt–series configuration is R_{if}. Starting with Equation (12.31), using Equations (12.29) and (12.30), we find that

$$I_i = I_\varepsilon + I_{fb} = I_\varepsilon + \beta_i I_o = I_\varepsilon + \beta_i (A_i I_\varepsilon) \tag{12.33(a)}$$

or

$$I_\varepsilon = \frac{I_i}{(1 + \beta_i A_i)} \tag{12.33(b)}$$

The input voltage is

$$V_i = I_\varepsilon R_i = \frac{I_i R_i}{(1 + \beta_i A_i)} \tag{12.34}$$

The input resistance with feedback is then

$$R_{if} = \frac{V_i}{I_i} = \frac{R_i}{(1 + \beta_i A_i)} \tag{12.35}$$

Equation (12.35) shows that a shunt input connection decreases the input resistance compared to that of the basic amplifier. A small input resistance is a desirable property of a current amplifier, to avoid loading effects on the input signal current source due to the amplifier.

The output resistance of the feedback circuit can be determined from the equivalent circuit in Figure 12.10. The input signal current is set equal to zero (an open circuit) and a test current is applied to the output terminals. Since the input signal current source is assumed to be ideal we have $R_S = \infty$.

From the circuit, we see that

$$I_\varepsilon + I_{fb} = I_\varepsilon + \beta_i I_x = 0 \tag{12.36(a)}$$

or

$$I_\varepsilon = -\beta_i I_x \tag{12.36(b)}$$

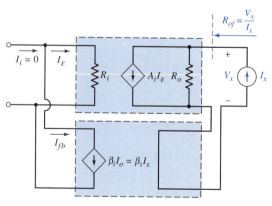

Figure 12.10 Ideal shunt–series feedback configuration for determining output resistance

The output voltage can be written as

$$V_x = (I_x - A_i I_\varepsilon)R_o = [I_x - A_i(-\beta_i I_x)]R_o \qquad (12.37)$$
$$= I_x(1 + \beta_i A_i)R_o$$

Therefore,

$$R_{of} = \frac{V_x}{I_x} = (1 + \beta_i A_i)R_o \qquad (12.38)$$

Equation (12.38) shows that a series output connection increases the output resistance compared to that of the basic amplifier. A large output resistance is a desirable property of a current amplifier, to avoid loading effects on the output signal due to a load connected to the amplifier output.

The equivalent circuit of this feedback current amplifier is shown in Figure 12.11.

Figure 12.11   Equivalent circuit of shunt–series feedback circuit, or current amplifier

Example 12.6   **Objective:** Determine the input resistance of a shunt input connection and the output resistance of a series output connection, for a feedback current amplifier.

Consider a shunt–series feedback amplifier in which the open-loop gain is $A_i = 10^5$ and the closed-loop gain is $A_{if} = 50$. Assume the input and output resistances of the basic amplifier are $R_i = 10\,\text{k}\Omega$ and $R_o = 20\,\text{k}\Omega$, respectively.

Solution: The ideal closed-loop current transfer function, from Equation (12.32), is

$$A_{if} = \frac{A_i}{(1 + \beta_i A_i)}$$

or

$$(1 + \beta_i A_i) = \frac{A_i}{A_{if}} = \frac{10^5}{50} = 2 \times 10^3$$

From Equation (12.35), the input resistance is

$$R_{if} = \frac{R_i}{(1 + \beta_i A_i)} = \frac{10}{2 \times 10^3}\,\text{k}\Omega \Rightarrow 5\,\Omega$$

and from Equation (12.38), the output resistance is

$$R_{of} = (1 + \beta_i A_i)R_o = (2 \times 10^3)(20)\,\text{k}\Omega \Rightarrow 40\,\text{M}\Omega$$

Comment: With a shunt input connection, the input resistance decreases drastically, and with a series output connection, the output resistance increases substantially, assuming negative feedback. These are the desired characteristics of a current amplifier.

Test Your Understanding

12.8   Consider the ideal shunt–series feedback amplifier in Figure 12.9. Assume that the source resistance is $R_S = \infty$. (a) If $I_i = 100\,\mu\text{A}$, $I_{fb} = 99\,\mu\text{A}$, and $I_o = 5\,\text{mA}$, determine A_i, β_i, and A_{if}, including units. (b) Using the results of part (a), determine R_{if} and R_{of}, for $R_i = 5\,\text{k}\Omega$ and $R_o = 4\,\text{k}\Omega$. (Ans. (a) $A_i = 5000\,\text{A/A}$, $\beta_i = 0.0198\,\text{A/A}$, $A_{if} = 50\,\text{A/A}$ (b) $R_{if} = 50\,\Omega$, $R_{of} = 400\,\text{k}\Omega$)

12.3.3   Series–Series Configuration

The configuration of an ideal **series–series** feedback amplifier is shown in Figure 12.12. The feedback samples a portion of the output current and converts it to a voltage. This feedback circuit can therefore be thought of as a voltage-to-current amplifier.

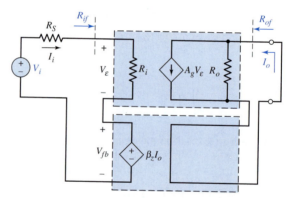

Figure 12.12   Ideal series–series feedback topology

The circuit consists of a basic amplifier that converts the error voltage to an output current with a gain factor A_g and that has an input resistance R_i. The feedback circuit samples the output current and produces a feedback voltage V_{fb}, which is in series with the input signal voltage V_i.

Assuming the output is essentially a short circuit, the output current is

$$I_o = A_g V_\varepsilon$$

and the feedback voltage is

$$V_{fb} = \beta_z I_o$$

where β_z is called a resistance feedback transfer function, with units of resistance. The input signal voltage, neglecting the effect of R_S, is

$$V_i = V_\varepsilon + V_{fb}$$

Combining these equations, as we have in previous analyses, yields the closed-loop current-to-voltage transfer function,

$$A_{gf} = \frac{I_o}{V_i} = \frac{A_g}{(1 + \beta_z A_g)} \qquad (12.39)$$

The units of the transfer function given by Equation (12.39) are amperes/volt, or conductance. We may note that the term $\beta_z A_g$ is dimensionless. This particular feedback circuit is therefore called a **transconductance amplifier**.

The input and output resistances are a function of the specific types of input and output connections, respectively. The input resistance for the series connection is given by Equation (12.25), which shows that with this configuration, the input resistance increases compared to that of the basic amplifier. The output resistance for the series connection is given by Equation (12.38), which shows that with this configuration, the output resistance increases compared to that of the basic amplifier. The equivalent circuit for the series–series feedback amplifier is shown in Figure 12.13.

Figure 12.13   Equivalent circuit of series–series feedback circuit, or transconductance amplifier

Test Your Understanding

12.9   An ideal series–series feedback amplifier is shown in Figure 12.12. Assume R_S is negligibly small. If $V_i = 100\,\text{mV}$, $V_{fb} = 99\,\text{mV}$, and $I_o = 5\,\text{mA}$, determine A_g, β_z, and A_{gf}, including units. (Ans. $A_g = 5\,\text{A/V}$, $\beta_z = 19.8\,\text{V/A}$, $A_{gf} = 50\,\text{mA/V}$)

12.3.4   Shunt–Shunt Configuration

The configuration of the ideal **shunt–shunt** feedback amplifier is shown in Figure 12.14. The feedback samples a portion of the output voltage and converts it to a current. This feedback circuit can therefore be thought of as a current-to-voltage amplifier.

The circuit consists of a basic amplifier that converts the error current to an output voltage with a gain factor A_z and that has an input resistance R_i. The feedback circuit samples the output voltage and produces a feedback current I_{fb}, which is in shunt with the input signal current I_i.

Assuming the output is essentially an open circuit, the output voltage is

$$V_o = A_z I_\varepsilon$$

and the feedback current is

$$I_{fb} = \beta_g V_o$$

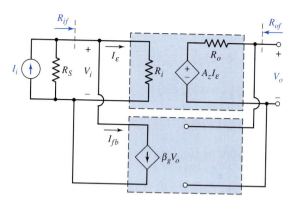

Figure 12.14   Ideal shunt–shunt feedback topology

where β_g is the conductance feedback transfer function, with units of conductance. The input signal current, assuming R_S is very large, is

$$I_i = I_\varepsilon + I_{fb}$$

Combining these equations yields the closed-loop voltage-to-current transfer function,

$$A_{zf} = \frac{V_o}{I_i} = \frac{A_z}{(1 + \beta_g A_z)} \tag{12.40}$$

The units of the transfer function given by Equation (12.40) are volts/ampere, or resistance. We may note that the term $\beta_g A_z$ is dimensionless. This particular feedback circuit is therefore referred to as a **transresistance amplifier**.

The input and output resistances are again a function of only the types of input and output connections, respectively. The input resistance is given by Equation (12.35) and the output resistance is given by Equation (12.28). The equivalent circuit for the shunt–shunt feedback amplifier is shown in Figure 12.15.

Figure 12.15   Equivalent circuit of shunt–shunt feedback circuit or, transresistance amplifier

Test Your Understanding

12.10   Consider the ideal shunt–shunt feedback amplifier in Figure 12.14. Assume that the source resistance is $R_S = \infty$. If $I_i = 100\,\mu\text{A}$, $I_{fb} = 99\,\mu\text{A}$, and $V_o = 5\,\text{V}$, determine A_z, β_g, and A_{zf}, including units. (Ans. $A_z = 5 \times 10^6\,\text{V/A}$, $\beta_g = 1.98 \times 10^{-5}\,\text{A/V}$, $A_{zf} = 50\,\text{V/mA}$)

12.3.5   Summary of Results

Table 12.1 summarizes the ideal relationships, including the transfer functions, input resistances, and output resistances, obtained in the analysis of the four types of feedback amplifiers.

Having analyzed the characteristics of the four ideal feedback topologies, we will next derive the transfer functions and resistance characteristics of op-amp and discrete transistor representations of each type of feedback configuration. We will compare actual results with the ideal results, discussing any deviations from the ideal.

Table 12.1   Summary results of feedback amplifier functions for the ideal feedback circuit

Feedback amplifier	Source signal	Output signal	Transfer function	Input resistance	Output resistance
Series–shunt (voltage amplifier)	Voltage	Voltage	$A_{vf} = \dfrac{V_o}{V_i} = \dfrac{A_v}{(1 + \beta_v A_v)}$	$R_i(1 + \beta_v A_v)$	$\dfrac{R_o}{(1 + \beta_v A_v)}$
Shunt–series (current amplifier)	Current	Current	$A_{if} = \dfrac{I_o}{I_i} = \dfrac{A_i}{(1 + \beta_i A_i)}$	$\dfrac{R_i}{(1 + \beta_i A_i)}$	$R_o(1 + \beta_i A_i)$
Series–series (transconductance amplifier)	Voltage	Current	$A_{gf} = \dfrac{I_o}{V_i} = \dfrac{A_g}{(1 + \beta_z A_g)}$	$R_i(1 + \beta_z A_g)$	$R_o(1 + \beta_z A_g)$
Shunt–shunt (transresistance amplifier)	Current	Voltage	$A_{zf} = \dfrac{V_o}{I_i} = \dfrac{A_z}{(1 + \beta_g A_z)}$	$\dfrac{R_i}{(1 + \beta_g A_z)}$	$\dfrac{R_o}{(1 + \beta_g A_z)}$

12.4   VOLTAGE (SERIES–SHUNT) AMPLIFIERS

In this section, we will analyze an op-amp and a discrete circuit representation of the series–shunt feedback configuration. Since the series–shunt circuit is a voltage amplifier, we will derive the transfer function relating the output signal voltage to the input signal voltage. For the ideal configuration, this function is shown in Equation (12.22) and is

$$A_{vf} = \frac{A_v}{(1 + \beta_v A_v)}$$

where A_v is the basic amplifier voltage gain and β_v is the voltage feedback transfer function. We found that this feedback configuration, the input resistance increases and the output resistance decreases compared to the basic amplifier values.

12.4.1   Op-Amp Circuit Representation

Figure 12.16 shows a noninverting op-amp circuit, which is an example of the series–shunt configuration. The input signal is the input voltage V_i, the feedback voltage is V_{fb}, and the error signal is the voltage V_ε. Since the shunt output samples the output voltage, the feedback voltage is a function of the output voltage.

Figure 12.16   Example of an op-amp series–shunt feedback circuit

In the ideal feedback circuit, the amplification factor A_v is very large; from Equation (12.22), the transfer function is then

$$A_{vf} = \frac{V_o}{V_i} \cong \frac{1}{\beta_v} \qquad (12.41)$$

For the ideal noninverting op-amp amplifier, we found in Chapter 9 that

$$A_{vf} = \frac{V_o}{V_i} = \left(1 + \frac{R_2}{R_1}\right) \qquad (12.42)$$

Therefore, the feedback transfer function β_v is

$$\beta_v = \frac{1}{\left(1 + \dfrac{R_2}{R_1}\right)} \qquad (12.43)$$

We can take a finite amplifier gain into account by considering the equivalent circuit in Figure 12.17. The parameter A_v is the open-loop voltage gain of the basic amplifier. We can write, for $R_o \approx 0$,

$$V_o = A_v V_\varepsilon \qquad (12.44)$$

and

$$V_\varepsilon = V_i - V_{fb} \qquad (12.45)$$

therefore,

$$V_o = A_v(V_i - V_{fb}) \qquad (12.46)$$

Assuming the input resistance R_i is very large, the feedback voltage is given by

$$V_{fb} \cong \left(\frac{R_1}{R_1 + R_2}\right) V_o \qquad (12.47)$$

Substituting Equation (12.47) into (12.46) and rearranging terms, we obtain

$$A_{vf} = \frac{V_o}{V_i} = \frac{A_v}{1 + \dfrac{A_v}{\left(1 + \dfrac{R_2}{R_1}\right)}} \qquad (12.48)$$

The voltage feedback transfer function β_v is given by Equation (12.43), and the closed-loop voltage transfer function can be written

Figure 12.17 Equivalent circuit, op-amp series–shunt feedback configuration

$$A_{vf} = \frac{A_v}{(1 + \beta_v A_v)} \qquad \qquad \textbf{(12.49)}$$

The voltage transfer function for the noninverting op-amp circuit has the same form as that for the ideal series–shunt configuration, assuming the input resistance R_i is very large.

We may note in this case that the voltage gain A_v of the basic amplifier is positive and that the feedback transfer function β_v is also positive, so that the loop gain $T = \beta_v A_v$ is positive for negative feedback.

We can now derive the expression for the input resistance R_{if}. We see from the figure that $V_\varepsilon = I_i R_i$, $V_o = A_v V_\varepsilon$, and $V_i = V_\varepsilon + V_{fb}$. The approximate feedback voltage is given by Equation (12.47). Therefore, the input voltage is

$$V_i = V_\varepsilon + \left(\frac{R_1}{R_1 + R_2}\right) V_o = V_\varepsilon + \frac{A_v V_\varepsilon}{\left(1 + \dfrac{R_2}{R_1}\right)}$$

$$\textbf{(12.50)}$$

$$= V_\varepsilon \left[1 + \frac{A_v}{(1 + R_2/R_1)}\right]$$

The input resistance is then

$$R_{if} = \frac{V_i}{I_i} = \frac{V_i}{(V_\varepsilon/R_i)}$$

$$\textbf{(12.51)}$$

$$= R_i\left[1 + \frac{A_v}{(1 + (R_2/R_1))}\right] = R_i(1 + \beta_v A_v)$$

The expression for the input resistance for the op-amp circuit has the same form as that for the ideal series input connection, as given in Equation (12.25). In the ideal case in which the gain is $A_v = \infty$, the input resistance of the noninverting op-amp is also infinite. However, if the gain is finite, the input resistance will also be finite.

Example 12.7 **Objective:** Determine the expected input resistance of the noninverting op-amp circuit.

Consider the noninverting op-amp in Figure 12.16, with parameters $R_i = 50\,\text{k}\Omega$, $R_1 = 10\,\text{k}\Omega$, $R_2 = 90\,\text{k}\Omega$, and $A_v = 10^4$.

Solution: The feedback transfer function β_v is

$$\beta_v = \frac{1}{\left(1 + \dfrac{R_2}{R_1}\right)} = \frac{1}{\left(1 + \dfrac{90}{10}\right)} = 0.10$$

The input resistance is therefore

$$R_{if} = R_i(1 + \beta_v A_v) = (50)[1 + (0.10)(10^4)]$$

or

$$R_{if} \cong 50 \times 10^3\,\text{k}\Omega = 50\,\text{M}\Omega$$

Comment: Even with a moderate differential input resistance R_i to the op-amp, the closed-loop input resistance R_{if} is very large, because of the series input feedback connection.

The analysis results for the noninverting op-amp circuit are consistent with the ideal series–shunt feedback characteristics.

12.4.2   Discrete Circuit Representation

Figures 12.18(a) and (b) show the basic emitter-follower and source-follower circuits, which we examined in previous chapters. These are examples of discrete-circuit series–shunt feedback topologies. The input signal is the voltage v_i, the error signal is the base-emitter voltage in the emitter follower and the gate-source voltage in the source follower, and the feedback voltage is equal to the output voltage, which means that the feedback transfer function is $\beta_v = 1$.

(a)   **(b)**   **(c)**

Figure 12.18   Discrete transistor series–shunt feedback circuits: (a) emitter-follower, (b) source-follower, and (c) small-signal equivalent circuit of emitter follower

The small-signal equivalent circuit of the emitter follower is shown in Figure 12.18(c). Since we have already analyzed the emitter-follower circuit, we will simply state the results here. The small-signal voltage gain is

$$A_{vf} = \frac{V_o}{V_i} = \frac{\left(\dfrac{1}{r_\pi} + g_m\right) R_E}{1 + \left(\dfrac{1}{r_\pi} + g_m\right) R_E} = \frac{\dfrac{R_E}{r_e}}{1 + \dfrac{R_E}{r_e}} \tag{12.52}$$

where

$$r_e = \frac{r_\pi}{(1 + g_m r_\pi)}$$

The voltage gain of the emitter follower can be written as a voltage divider equation. Since the feedback transfer function is unity, the form of the voltage gain expression is the same as that for the ideal series–shunt configuration, as given in Equation (12.22). The open-loop voltage gain corresponds to

$$A_v = \left(\frac{1}{r_\pi} + g_m\right) R_E = \frac{R_E}{r_e} \tag{12.53}$$

The closed-loop input resistance is[2]

$$R_{if} = r_\pi + (1 + h_{FE})R_E = r_\pi \left[1 + \left(\frac{1}{r_\pi} + g_m \right) R_E \right] \qquad (12.54)$$

The form of the input resistance is also the same as that of the ideal expression, given by Equation (12.25). The input resistance increases with a series input connection.

The output resistance of the emitter-follower circuit is given by

$$R_{of} = R_E \left\| \frac{r_\pi}{1 + h_{FE}} \right. = R_E \| r_e \qquad (12.55)$$

which can be written in the form

$$R_{of} = \frac{R_E}{1 + \left(\dfrac{1}{r_\pi} + g_m \right) R_E} \qquad (12.56)$$

The output resistance decreases with a shunt output connection. For the emitter-follower circuit, the form of the output resistance is also the same as that of the ideal expression, given by Equation (12.28).

Even though the magnitude of the emitter-follower voltage gain is slightly less than unity, this circuit is a classic example of a series–shunt feedback configuration, which represents a voltage amplifier.

Design Example 12.8   **Objective:** Design a feedback amplifier to amplify the output signal of a microphone.

The output signal from the microphone is 10 mV and the output signal from the feedback amplifier must be 0.5 V to drive a power amplifier that in turn drives the speakers. The nominal output resistance of the microphone is $R_S = 5\,\text{k}\Omega$ and the nominal input resistance of the power amplifier is $R_L = 75\,\Omega$. An op-amp with parameters $R_i = 10\,\text{k}\Omega$, $R_o = 100\,\Omega$, and a low-frequency gain of $A_v = 10^4$ is available. [Note: In this simple design, neglect frequency response.]

Solution: Design Approach: Since the source resistance is fairly large, an amplifier with a large input resistance is required to minimize loading at the input. Also, since the load resistance is low, an amplifier with a low output resistance is required to minimize loading at the output. To satisfy these requirements, a series–shunt feedback configuration, or voltage amplifier, should be used.

The closed-loop voltage gain must be $A_{vf} = 0.5/0.01 = 50$. For the ideal case, $A_{vf} = 1/\beta_v$, so the feedback transfer function is $\beta_v = 1/50 = 0.02$. The loop gain is then

$$T = \beta_v A_v = (0.02)(10^4) = 200$$

Referring to Table 12.1, we expect the input resistance to be

$$R_{if} \cong (10)(200)\,\text{k}\Omega \rightarrow 2\,\text{M}\Omega$$

---

[2]Reminder: In this chapter, the parameter h_{FE} is used as the transistor current gain to avoid confusion with β, which is used as the feedback transfer function. Again, we assume that the dc and ac current gains are equal; therefore, $h_{FE} = h_{fe} = g_m r_\pi$.

and the output resistance to be

$$R_{of} \cong (100/200)\Omega = 0.5\,\Omega$$

These input and output resistance values will minimize any loading effects at the amplifier input and output terminals.

If we use the noninverting amplifier configuration in Figure 12.16, then we have

$$\frac{1}{\beta_v} = 1 + \frac{R_2}{R_1} = 50$$

and

$$\frac{R_2}{R_1} = 49$$

The feedback network loads the output of the amplifier; consequently, we need $R_1 + R_2$ to be much larger than R_o. However, the output resistance of the feedback network is in series with the input terminals, so extremely large values of R_1 and R_2 will reduce the actual signal applied to the op-amp because of voltage divider action. Initially, then, we choose $R_1 = 1\,\text{k}\Omega$ and $R_2 = 49\,\text{k}\Omega$.

Computer Simulation Verification: The circuit in Figure 12.19 was used in a PSpice analysis of the voltage amplifier. A standard 741 op-amp was used in the circuit. For a 10 mV input signal, the output signal was 499.6 mV, for a gain of 49.96. This result is within 0.08 percent of the ideal designed value. The input resistance R_{if} was found to be approximately 580 MΩ and the output resistance R_{of} was determined to be approximately 0.042 Ω. The differences between the measured input and output resistances compared to the predicted values are due to the differences between the actual μA-741 op-amp parameters and the assumed parameters. However, the measured input resistance is larger than predicted and the measured output resistance is smaller than predicted, which is desired and more in line with an ideal op-amp circuit.

Comment: An almost ideal feedback voltage amplifier can be realized if an op-amp is used in the circuit.

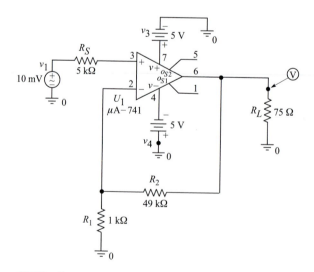

Figure 12.19   Circuit used in the computer simulation analysis in Example 12.8

Test Your Understanding

12.11 Consider the noninverting op-amp circuit in Figure 12.16, with parameters $R_1 = 10\,\text{k}\Omega$, $R_2 = 30\,\text{k}\Omega$, and $A_v = 10^4$. Assume $R_i = \infty$. Determine the closed-loop voltage gain. If the open-loop gain increases by a factor of 10, what is the percent change in the closed-loop gain? (Ans. $A_{vf} = 3.9984$, 0.036%)

12.12 Assume the transistor in the emitter-follower circuit in Figure 12.18(a) is biased such that $I_{CQ} = 0.5\,\text{mA}$. Let $R_E = 2\,\text{k}\Omega$. (a) If the transistor current gain is $h_{FE} = 100$, determine A_{vf}, R_{if}, and R_{of}. (b) Determine the percent change in A_{vf}, R_{if}, and R_{of} if the transistor current gain increases to $h_{FE} = 150$. Assume the quiescent collector current remains unchanged. (Ans. (a) $A_{vf} = 0.97490$, $R_{if} = 207\,\text{k}\Omega$, $R_{of} = 50.2\,\Omega$ (b) A_{vf}, 0.0082%; R_{if}, 1.25%; R_{of}, 0.397%)

12.13 Assume the transistor in the source-follower circuit shown in Figure 12.18(b) is biased such that $I_{DQ} = 250\,\mu\text{A}$. Let $R_S = 5\,\text{k}\Omega$. If the transistor parameters are $K = 200\,\mu\text{A/V}^2$ and $V_{TN} = 1\,\text{V}$, determine A_{vf}, R_{if}, and R_{of}. How do these results agree with the ideal feedback characteristics given in Table 12.1? (Ans. $A_{vf} = 0.691$, $R_{if} = \infty$, $R_{of} = 1.55\,\text{k}\Omega$)

*D12.14 Design a feedback voltage amplifier to provide a voltage gain of 15. The nominal voltage source resistance is $R_S = 2\,\text{k}\Omega$, and the nominal load is $R_L = 100\,\Omega$. An op-amp with parameters $R_i = 5\,\text{k}\Omega$, $R_o = 50\,\Omega$, and a low-frequency open-loop gain of $A_v = 5 \times 10^3$ is available. Correlate the design with a computer simulation analysis to determine the voltage gain, input resistance, and output resistance.

12.5   CURRENT (SHUNT–SERIES) AMPLIFIERS

In this section, we will analzye an op-amp and a discrete circuit representation of the shunt–series feedback amplifier. The shunt–series circuit is a current amplifier; therefore, we must derive the output current to input current transfer function. For the ideal configuration, this function is given in Equation (12.32):

$$A_{if} = \frac{A_i}{(1 + \beta_i A_i)}$$

where A_i is the basic amplifier current gain and β_i is the current feedback transfer function. For this amplifier, the input resistance decreases and the output resistance increases compared to the basic amplifier values.

12.5.1   Op-Amp Circuit Representation

Figure 12.20 shows an op-amp current amplifier, which is a shunt–series configuration. The input signal is the current I_i' from the Norton equivalent source of I_i and R_S. The feedback current is I_{fb}, the error signal is the current I_ε, and the output signal is the current I_o. With the shunt input connection, the input resistance R_{if} is small, as previously stated. Resistance R_S is the output resistance of the current source and is normally large. If $R_S \gg R_{if}$, then $I_i' \cong I_i$.

If we assume initially that I_ε is negligible, then, from Figure 12.20, we have

$$I_i \cong I_i' = I_{fb}$$

Figure 12.20   Example of an op-amp shunt–series feedback circuit

The output voltage V_o, assuming V_1 is at virtual ground, is

$$V_o = -I_{fb}R_F = -I_i R_F$$

and current I_1 is

$$I_1 = -V_o / R_1$$

The output current can be expressed

$$I_o = I_{fb} + I_1 = I_i + \left(-\frac{1}{R_1}\right)(-I_i R_F) = I_i\left(1 + \frac{R_F}{R_1}\right) \tag{12.57}$$

Therefore, the ideal current gain is

$$\frac{I_o}{I_i} = 1 + \frac{R_F}{R_1} \tag{12.58}$$

In the ideal feedback circuit, the amplification factor A_i is very large; consequently, the current transfer function, from Equation (12.32), becomes

$$A_{if} = \frac{I_o}{I_i} \cong \frac{1}{\beta_i} \tag{12.59}$$

Comparing Equation (12.59) with (12.58), we see that the current feedback transfer function for the ideal op-amp current amplifier is

$$\beta_i = \frac{1}{\left(1 + \dfrac{R_F}{R_1}\right)} \tag{12.60}$$

We can take the finite amplifier gain into account by considering the equivalent circuit in Figure 12.21. The parameter A_i is the open-loop current gain. We have

$$I_o = A_i I_\varepsilon \tag{12.61}$$

and

$$I_\varepsilon = I_i' - I_{fb} \cong I_i - I_{fb} \tag{12.62}$$

therefore,

$$I_o = A_i(I_i - I_{fb}) \tag{12.63}$$

If we again assume that V_1 is at virtual ground, voltage V_o is given by

$$V_o = -I_{fb}R_F \tag{12.64}$$

(a)

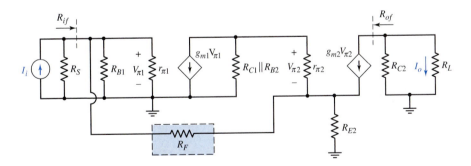

(b)

Figure 12.24 (a) Example of a discrete transistor shunt–series feedback circuit and (b) ac equivalent circuit

Figure 12.25 Small-signal equivalent circuit of circuit in Figure 12.24(a)

derive the expression for the closed-loop current gain by writing and solving a set of simultaneous nodal equations. However, as with most discrete transistor feedback circuits, the transfer function cannot be arranged exactly in the ideal form without several approximations. For this circuit, then, we rely on a computer analysis to provide the required results.

Example 12.9 **Objective:** Determine the closed-loop current gain and input resistance of a discrete shunt–series transistor feedback circuit.

Consider the circuit in Figure 12.24(a), with transistor parameters $h_{FE} = 100$ and $V_A = \infty$. Assume the source resistance is $R_S = 10\,\text{M}\Omega$. The capacitors are large enough to act as short circuits to the signal currents.

Solution: A PSpice analysis shows that the closed-loop current gain is

$$A_{if} = I_o/I_i = 9.58$$

The input resistance R_{if} is defined as the ratio of the signal voltage at the base of Q_1 to the input signal current. The PSpice results show that $R_{if} = 134\,\Omega$. This low input resistance is expected for the shunt input connection.

Comment: The PSpice analysis shows that the closed-loop current gain increases from 9.58 to 10.2 as the transistor current gain h_{FE} increases from 100 to 1000. This result again demonstrates a principal characteristic of feedback circuits, which is that the transfer function is relatively insensitive to changes in the individual transistor parameters.

From the small-signal equivalent circuit in Figure 12.25, we find that the output resistance R_{of} looking into the collector of Q_2 is very large. If r_o of Q_2 is assumed to be infinite, then R_{of} is also infinite. We expect a large output impedance for the series output connection of this feedback circuit.

Design Example 12.10 **Objective:** Design a feedback amplifier to provide a given current gain.

Assume that a signal current source has a nominal output resistance of $R_S = 10\,\text{k}\Omega$ and that the amplifier will drive a nominal load of $R_L = 50\,\Omega$. A current gain of 10 is required. An op-amp with the same characteristics described in Example 12.8 is available.

Solution: Design Approach: An amplifier with a low input resistance and a large output resistance is required, to minimize loading effects at the input and output. For these reasons, a shunt–series feedback configuration, or current amplifier, will be used.

The closed-loop gain is

$$A_{if} = 10 \cong 1/\beta_i$$

and the feedback transfer function is $\beta_i = 0.1$.

The dependent open-loop voltage source of the op-amp, as shown in Figure 12.17, can be transformed to an equivalent dependent open-loop current source, as shown in Figure 12.9. We find that

$$A_i = A_v R_i / R_o$$

Using the parameters specified for the op-amp, we find $A_i = 10^6$. The loop gain for the shunt–series configuration is

$$A_i \beta_i = (10^6)(0.1) = 10^5$$

Referring to Table 12.1, we expect the input resistance to be

$$R_{if} = 10/10^5 \,\text{k}\Omega \rightarrow 0.1\,\Omega$$

and the output resistance to be

$$R_{of} = (100)(10^5)\,\Omega \rightarrow 10\,\text{M}\Omega$$

These resistance values will minimize any loading effects at the amplifier input and output.

For the shunt–series configuration in Figure 12.20, we have

$$\frac{1}{\beta_i} = 1 + \frac{R_F}{R_1} = 10$$

or

$$R_F/R_1 = 9$$

For our purposes, R_1 must be fairly small, to avoid a loading effect at the output. However, R_1 must not be too small, to avoid large currents in the amplifier. Therefore, we choose $R_1 = 1\,\text{k}\Omega$ and $R_F = 9\,\text{k}\Omega$.

Computer Simulation Verification: Figure 12.26 shows the circuit used in the computer simulation. A standard μA-741 op-amp was used in the circuit. The current gain was found to be exactly 10.0. The input resistance R_{if} looking into the op-amp with feedback was found to be 0.056 Ω, which compares favorably to the predicted value of 0.1 Ω. The output resistance seen by the load resistor was found to be approximately 200 MΩ. This value is on the order of 20 times larger than the predicted value, but is closer to the ideal value. The differences between predicted and measured values are due to the differences in assumed op-amp parameters and the μA-741 op-amp parameters.

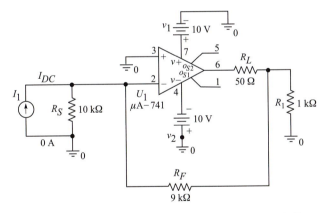

Figure 12.26   Circuit used in the computer simulation analysis in Example 12.10

Comment: This design also produces an almost ideal feedback current amplifier, if reasonable values of feedback resistors are used.

Test Your Understanding

*RD12.15   Consider the common-base circuit in Figure 12.23(a), with transistor parameters $h_{FE} = 80$, $V_{EB}(\text{on}) = 0.7\,\text{V}$, and $V_A = \infty$. Assume the transistor is biased at $I_{CQ} = 0.5\,\text{mA}$. Redesign the circuit such that the closed-loop current gain is greater than 0.95. (Ans. $R_E(\text{min}) = 1.30\,\text{k}\Omega$, and $V^+(\text{min}) = 1.36\,\text{V}$)

*12.16   Consider the shunt–series feedback circuit in Figure 12.24(a). Using a computer simulation analysis, investigate the magnitude of the current gain A_{if} as the emitter resistor R_{E2} is varied between $0.4\,\text{k}\Omega$ and $1.6\,\text{k}\Omega$. What is the relationship between R_F, R_{E2}, and A_{if}?

*12.17   Consider the shunt–series feedback circuit in Figure 12.24(a). Using a computer simulation analysis, investigate the magnitude of the input resistance R_{if} as the feedback resistor R_F is varied between $5\,\text{k}\Omega$ and $50\,\text{k}\Omega$. What is the influence of R_F on the input resistance R_{if}?

*D12.18   Design a feedback current amplifier to provide a current gain of 15. The nominal current source resistance is $R_S = 500\,\Omega$, and the nominal load is $R_L = 200\,\Omega$. An op-amp with parameters $R_i = 5\,\text{k}\Omega$, $R_o = 50\,\Omega$, and a low-frequency open-loop voltage gain of $A_v = 5 \times 10^3$ is available. Correlate the design with a PSpice analysis to determine the current gain, input resistance, and output resistance.

12.6   TRANSCONDUCTANCE (SERIES–SERIES) AMPLIFIERS

In this section, we will analyze an op-amp and a discrete circuit representation of the series–series feedback amplifier. The series–series circuit is a transconductance amplifier; therefore, we must derive the output current to input voltage transfer function. For the ideal configuration, this function is, from Equation (12.39),

$$A_{gf} = \frac{A_g}{(1 + \beta_z A_g)}$$

where A_g is the basic amplifier transconductance gain and β_z is the resistance feedback transfer function. We found that with this feedback configuration, both the input and output resistances increase compared to the basic amplifier values.

12.6.1   Op-Amp Circuit Representation

Figure 12.27 Example of an op-amp series–series feedback circuit

The op-amp circuit in Figure 12.27 is an example of the series–series feedback configuration. The input signal is the input voltage V_i, the feedback voltage is V_{fb}, and the error signal is the voltage V_E. The series output connection samples the output current, which means that the feedback voltage is a function of the output current.

In the ideal feedback circuit, the amplification factor A_g is very large; therefore, from Equation (12.39), the transfer function is

$$A_{gf} = \frac{I_o}{V_i} \cong \frac{1}{\beta_z} \tag{12.70}$$

Assuming an ideal op-amp circuit and neglecting the transistor base current, we have

$$V_i = V_{fb} = I_o R_E$$

and

$$A_{gf} = \frac{I_o}{V_i} = \frac{1}{R_E} \qquad (12.71)$$

Comparing Equations (12.70) and (12.71), we see that the ideal feedback transfer function is

$$\beta_z = R_E \qquad (12.72)$$

We can take a finite amplifier gain into account by considering the equivalent circuit in Figure 12.28. The parameter A_g is the open-loop transconductance gain of the amplifier. Assuming the collector and emitter currents are nearly equal and R_i is very large, we can write that

$$I_o = \frac{V_{fb}}{R_E} = h_{FE}I_b = h_{FE}A_g V_\varepsilon \qquad (12.73)$$

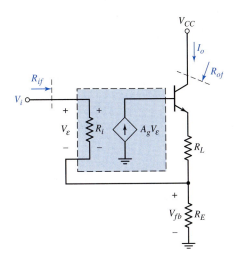

Figure 12.28 Equivalent circuit, op-amp series–series feedback configuration

Also,

$$V_\varepsilon = V_i - V_{fb} = V_i - I_o R_E \qquad (12.74)$$

Substituting Equation (12.74) into Equation (12.73) yields

$$I_o = h_{FE}A_g(V_i - I_o R_E) \qquad (12.75)$$

which can be rearranged to yield the closed-loop transfer function,

$$A_{gf} = \frac{I_o}{V_i} = \frac{(h_{FE}A_g)}{1 + (h_{FE}A_g)R_E} \qquad (12.76)$$

which has the same form as that of the ideal theory. In this example, we see that in this feedback network, the transistor current gain is part of the basic amplifier gain.

12.6.2   Discrete Circuit Representation

Figure 12.29 shows a single bipolar transistor circuit that is an example of a series–series feedback configuration. This circuit is similar to those evaluated in Chapters 3 and 4. The input signal is the input voltage v_i, the feedback voltage is v_{fb}, and the error signal is the base–emitter voltage. The series output connection samples the output current; therefore, the feedback voltage is a function of the output current.

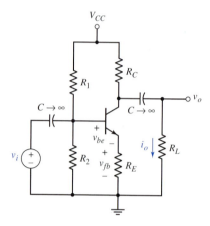

Figure 12.29   Example of a discrete transistor series–series feedback circuit

The small-signal equivalent circuit is shown in Figure 12.30. The Early voltage of the transistor is assumed to be infinite. The output current can be written

$$I_o = -(g_m V_\pi)\left(\frac{R_C}{R_C + R_L}\right) \tag{12.77}$$

and the feedback voltage is

$$V_{fb} = \left(\frac{V_\pi}{r_\pi} + g_m V_\pi\right) R_E \tag{12.78}$$

Figure 12.30   Small-signal equivalent circuit, discrete transistor series–series feedback configuration

A KVL equation around the B–E loop yields

$$V_i = V_\pi + V_{fb} = V_\pi \left[1 + \left(\frac{1}{r_\pi} + g_m \right) R_E \right] \tag{12.79}$$

Solving Equation (12.79) for V_π, substituting that into Equation (12.77), and rearranging terms produces the expression for the transconductance transfer function,

$$A_{gf} = \frac{I_o}{V_i} = \frac{-g_m \left(\dfrac{R_C}{R_C + R_L} \right)}{1 + \left(\dfrac{1}{r_\pi} + g_m \right) R_E} \tag{12.80}$$

Again, the closed-loop transfer function of the discrete transistor feedback circuit cannot be put in exactly the same form as that of the ideal series–series feedback network. Resistor R_C introduces loading on the output, and r_π introduces loading on the input. If both R_C and r_π become large, then Equation (12.80) changes to the ideal form, where the feedback transfer function is $\beta_z = -R_E$ and the basic amplifier transconductance is $A_g = -g_m$.

Example 12.11 **Objective:** Determine the transconductance gain of a transistor feedback circuit.

Consider the circuit in Figure 12.29, with transistor parameters $h_{FE} = 100$, $V_{BE}(\text{on}) = 0.7\,\text{V}$, and $V_A = \infty$. The circuit parameters are: $V_{CC} = 10\,\text{V}$, $R_1 = 55\,\text{k}\Omega$, $R_2 = 12\,\text{k}\Omega$, $R_E = 1\,\text{k}\Omega$, $R_C = 4\,\text{k}\Omega$, and $R_L = 4\,\text{k}\Omega$.

Solution: From a dc analysis of the circuit, the quiescent values are $I_{CQ} = 0.983\,\text{mA}$ and $V_{CEQ} = 5.08\,\text{V}$. The transistor small-signal parameters are

$$r_\pi = \frac{h_{FE} V_T}{I_{CQ}} = \frac{(100)(0.026)}{0.983} = 2.64\,\text{k}\Omega$$

and

$$g_m = \frac{I_{CQ}}{V_T} = \frac{0.983}{0.026} = 37.8\,\text{mA/V}$$

From Equation (12.80), the transconductance transfer function is

$$A_{gf} = \frac{-(37.8)\left(\dfrac{4}{4+4} \right)}{1 + \left(\dfrac{1}{2.64} + 37.8 \right)(1)} = -0.482\,\text{mA/V}$$

As a first approximation, we have

$$A_{gf} = \frac{1}{\beta_z} = \frac{1}{-R_E} = \frac{1}{-1\,\text{k}\Omega} = -1\,\text{mA/V}$$

The term $R_C/(R_C + R_L)$ introduces the largest discrepancy between the actual and ideal transconductance values.

This circuit is often used as a voltage amplifier. The output voltage is directly proportional to the output current. Therefore,

$$A_{vf} = \frac{v_o}{v_i} = \frac{i_o R_L}{v_i} = A_{gf} R_L$$

which yields

$$A_{vf} = (-0.482)(4) = -1.93$$

Comment: The circuit in Figure 12.29 is an example of a series–series feedback topology, even though in many cases we treat this circuit as a voltage amplifier. When an emitter resistor is included, the small-signal voltage gain decreases, because of the feedback effect of R_E. However, the transconductance and voltage gain become insensitive to the transistor parameters, also a result of the feedback effect of R_E. A 100 percent increase in the transistor current gain h_{FE} produces a 0.5 percent change in the closed-loop voltage gain.

The input resistance R_{if} of the series input feedback connection includes R_E multiplied by $(1 + h_{FE})$, where h_{FE} is the transistor current gain. The input resistance increases significantly because of the series connection.

The output resistance of a series output feedback connection is usually very large. However, resistance R_C reduces the output resistance and introduces a loading effect. The reduced output resistance demonstrates that discrete transistor feedback circuits do not conform exactly to ideal feedback circuits. Nevertheless, overall circuit characteristics improve when feedback is used.

Design Example   12.12   Objective: Design a driver amplifier to supply current to an LED.

The available voltage source is variable from 0 to 5 V and has an output resistance of 200 Ω. The required diode current is 10 mA when the maximum input voltage is applied. The required closed-loop transconductance gain is then $A_{gf} = I_o/V_i = (10 \times 10^{-3})/5 \rightarrow 2$ mS. An op-amp with the characteristics described in Example 12.8 and a BJT with $h_{FE} = 100$ are available.

Solution: Design Approach: To minimize loading effects on the input, an amplifier with a large input resistance is required; to minimize loading effects on the output, a large output resistance is required. For these reasons, a series–series feedback configuration, or transconductance amplifier, is selected.

The closed-loop gain is

$$A_{gf} = 2 \times 10^{-3} \cong 1/\beta_z$$

and the resistance feedback transfer function is

$$\beta_z = 500 \, \Omega$$

The dependent open-loop voltage source of the op-amp, as shown in Figure 12.17, can be transformed to an equivalent dependent op-loop transconductance source for the transconductance amplifier, as shown in Figure 12.12. We find that

$$A_g = A_v/R_o$$

The parameters specified for the op-amp yield

$$A_g = 100 \, \text{A/V}$$

The loop gain for the series–series configuration is

$$A_g\beta_z = (100)(500) = 5 \times 10^4$$

Referring to Table 12.1, the expected input resistance is

$$R_{if} = (10)(5 \times 10^4)\,k\Omega \rightarrow 500\,M\Omega$$

and the expected output resistance is

$$R_{of} = (100)(5 \times 10^4)\,\Omega \rightarrow 5\,M\Omega$$

These input and output resistances should minimize any loading effects at the amplifier input and output.

For this example, we may use the amplifier configuration shown in Figure 12.27, in which the load resistor R_L is replaced by an LED. In the ideal case,

$$\beta_z = R_E = 500\,\Omega$$

Computer Simulation Verification: Figure 12.31 shows the circuit used in the computer simulation. Again, a standard µA-741 op-amp was used in the circuit and a standard diode was used in place of an LED. When the input voltage reached 5 V, the current through the diode was 10.0 mA, which was the design value. The input resistance R_{if} was found to be approximately 2400 MΩ and the output resistance R_{of} was found to be approximately 60 MΩ. Both of these values are larger than predicted because of the differences in the assumed op-amp parameters and those of the µA-741 op-amp.

Figure 12.31   Circuit used in the computer simulation analysis for Example 12.12

Comment: Again, an almost ideal feedback circuit can be designed by using an op-amp.

Test Your Understanding

12.19   Consider the op-amp circuit in Figure 12.27, with parameters $R_E = 1\,k\Omega$ and $A_g = 10^3\,A/V$. Assume the transistor current gain is $h_{FE} = 200$. (a) Determine the transfer function $A_{gf} = I_o/V_i$. (b) If the amplifier gain increases by a factor of 10, determine the percent change in the transconductance transfer function. (Ans. (a) $A_{gf} = 1\,mA/V$ (b) $4.5 \times 10^{-7}\,\% \cong 0\%$)

12.20 For the circuit in Figure 12.32, the transistor parameters are: $K_n = 1.5\,\text{mA/V}^2$, $V_{TN} = 2\,\text{V}$, and $\lambda = 0$. (a) Determine the transconductance transfer function $A_{gf} = i_o/v_i$. (b) Determine the percent change in A_{gf} if the transistor conduction parameter decreases to $K_n = 1\,\text{mA/V}^2$. (Ans. (a) $A_{gf} = -0.732\,\text{mA/V}$ (b) A_{gf}, 12.7% decrease)

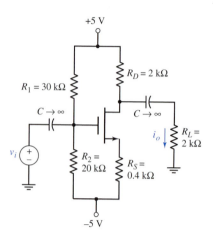

Figure 12.32   Figure for Exercise 12.20

*D12.21 Design a transconductance feedback amplifier with a gain of $A_{gf} = 10\,\text{mS}$. The source resistance is $R_S = 500\,\Omega$, and the load is an LED. State any necessary assumptions. Use an op-amp with the characteristics described in Example 12.8. From a computer simulation analysis, determine the closed-loop transconductance, input resistance, and output resistance of your design.

12.7   TRANSRESISTANCE (SHUNT–SHUNT) AMPLIFIERS

In this section, we will analyze an op-amp and a discrete circuit representation of the shunt–shunt feedback amplifier. The shunt–shunt circuit is a transresistance amplifier; therefore, we must derive the output voltage to input current transfer function. For the ideal configuration, this function is given by Equation (12.40) as

$$A_{zf} = \frac{A_z}{(1 + \beta_g A_z)}$$

where A_z is the basic amplifier transresistance gain, and β_g is the feedback transfer function. With this feedback connection, both the input and output resistance decrease compared to the basic amplifier values.

12.7.1   Op-Amp Circuit Representation

Figure 12.33 shows an inverting op-amp circuit, which is an example of the shunt–shunt configuration. The input signal is the input current I_i, the feedback current is I_{fb}, and the error signal is the current I_ε. The shunt output

Figure 12.33 Example of an op-amp shunt–shunt feedback circuit

samples the output voltage; therefore, the feedback current is a function of the output voltage.

In the ideal feedback circuit, the amplification factor A_z is very large, and the transresistance transfer function is, from Equation (12.40),

$$A_{zf} = \frac{V_o}{I_i} \cong \frac{1}{\beta_g} \tag{12.81}$$

For the ideal inverting op-amp circuit, V_1 is at virtual ground, and

$$V_o = -I_{fb}R_2$$

Also for the ideal op-amp, $I_{fb} = I_i$, and the ideal transresistance transfer function is

$$A_{zf} = \frac{V_o}{I_i} = -R_2 \tag{12.82}$$

Comparing Equation (12.82) to Equation (12.81), we see that the feedback transfer function for the ideal inverting op-amp circuit is

$$\beta_g = -\frac{1}{R_2} \tag{12.83}$$

We can take a finite amplifier gain into account by considering the equivalent circuit in Figure 12.34. The parameter A_z is the open-loop transresistance gain factor, and the minus sign indicates that the error signal current is entering the inverting terminal. Therefore, we can write $V_o = -A_z I_\varepsilon$, $I_\varepsilon = I_i - I_{fb}$, and $V_o = -A_z(I_i - I_{fb})$. If we assume that voltage V_1 is at virtual ground, then

$$I_{fb} = -V_o/R_2$$

Figure 12.34 Equivalent circuit, op-amp shunt–shunt feedback configuration

Combining equations, we see that the closed-loop transresistance transfer function is

$$A_{zf} = \frac{V_o}{I_i} = \frac{-A_z}{\left(1 + \dfrac{A_z}{R_2}\right)}$$

(12.84)

From Equation (12.83), the feedback transfer function is $\beta_g = -1/R_2$, and Equation (12.84) becomes

$$A_{zf} = \frac{V_o}{I_i} = \frac{(-A_z)}{1 + (-A_z)\beta_g}$$

(12.85)

This feedback circuit is one example in which the gain of the basic amplifier, $A_z = V_o/I_\varepsilon$, is negative and the feedback transfer function, $\beta_g = -1/R_2$, is also negative, but the loop gain $T = \beta_g A_z$ is positive for a negative feedback circuit.

The transresistance transfer function for the inverting op-amp circuit has the same form as that for the ideal shunt–shunt configuration. In addition, since V_1 is at virtual ground, the input resistance including feedback, R_{if}, is essentially zero, and we have shown that the output resistance with feedback, R_{of}, is very small. These small resistance values are a result of the shunt–shunt configuration. Therefore, our analysis of the inverting op-amp circuit produces results consistent with ideal shunt–shunt feedback characteristics.

The inverting amplifier circuit in Figure 12.33 is most often thought of as a voltage amplifier. The input current I_i is directly proportional to the input voltage V_i, which means that the voltage transfer function (gain) and transresistance transfer function have the same characteristics. Even though we are usually concerned with the voltage gain, the inverting amplifier is an example of a shunt–shunt feedback topology which is a transresistance amplifier.

12.7.2 Discrete Circuit Representation

Figure 12.35 shows a single bipolar transistor circuit, which is an example of a shunt–shunt feedback configuration. The input signal current is I_i, the feedback current is i_{fb}, and the error signal current is i_ε and is the signal base

Figure 12.35 Example of a discrete transistor shunt–shunt feedback circuit

current. The shunt output samples the output voltage; therefore, the feedback current is a function of v_o.

The small-signal equivalent circuit is shown in Figure 12.36. The input signal is assumed to be an ideal signal current source. Also the Early voltage of the transistor is assumed to be infinite.

Figure 12.36   Small-signal equivalent circuit, discrete transistor shunt–shunt feedback configuration

Writing a KCL equation at the output node, we find

$$\frac{V_o}{R_C} + g_m V_\pi + \frac{V_o - V_\pi}{R_F} = 0 \tag{12.86}$$

A KCL equation at the input node yields

$$I_i = \frac{V_\pi}{r_\pi} + \frac{V_\pi - V_o}{R_F} \tag{12.87}$$

Solving Equation (12.87) for V_π and substituting that result into Equation (12.86), we obtain

$$V_o\left(\frac{1}{R_C} + \frac{1}{R_F}\right)\left(\frac{1}{r_\pi} + \frac{1}{R_F}\right) + \left(g_m - \frac{1}{R_F}\right)\left(I_i + \frac{V_o}{R_F}\right) = 0 \tag{12.88}$$

The transresistance transfer function is then

$$A_{zf} = \frac{V_o}{I_i} = \frac{-\left(g_m - \dfrac{1}{R_F}\right)}{\left(\dfrac{1}{R_C} + \dfrac{1}{R_F}\right)\left(\dfrac{1}{r_\pi} + \dfrac{1}{R_F}\right) + \dfrac{1}{R_F}\left(g_m - \dfrac{1}{R_F}\right)} \tag{12.89}$$

The open-loop transresistance gain factor A_z is found by setting $R_F = \infty$. We find

$$A_z = \frac{-g_m}{\left(\dfrac{1}{R_C}\right)\left(\dfrac{1}{r_\pi}\right)} = -g_m r_\pi R_C = -h_{FE} R_C \tag{12.90}$$

where h_{FE} is the common-emitter transistor current gain. Multiplying both numerator and denominator of Equation (12.89) by $(r_\pi R_C)$, we obtain the closed-loop transresistance gain,

$$A_{zf} = \frac{V_o}{I_i} = \frac{+\left(A_z + \dfrac{r_\pi R_C}{R_F}\right)}{\left(1 + \dfrac{R_C}{R_F}\right)\left(1 + \dfrac{r_\pi}{R_F}\right) - \dfrac{1}{R_F}\left(A_z + \dfrac{r_\pi R_C}{R_F}\right)}$$

(12.91)

The closed-loop transresistance gain for the single-transistor feedback circuit cannot be put into the ideal form, as given in Equation (12.40), without further approximations. To explain, in an ideal feedback circuit, the feedback network does not load the basic amplifier. Also, the forward transmission occurs entirely through the basic amplifier. However, in a discrete transistor feedback circuit, these ideal assumptions are not entirely valid; therefore, the form of the transfer function is usually not exactly the same as that of the ideal configuration.

We may assume that the feedback resistor is fairly large, which means that the feedback does not drastically perturb the circuit. We may then assume

$$h_{FE} = g_m r_\pi \gg (r_\pi / R_F)$$

If we also assume that $R_C \ll R_F$ and $r_\pi \ll R_F$, then Equation (12.91) reduces to

$$A_{zf} = \frac{V_o}{I_i} \cong \frac{A_z}{1 + (A_z)\left(\dfrac{-1}{R_F}\right)}$$

(12.92)

Consequently, the feedback transfer function is approximately

$$\beta_g \cong \frac{-1}{R_F}$$

(12.93)

Equation (12.93) demonstrates that the approximate value of the feedback transfer function depends only on a resistance value.

Although the actual closed-loop transfer function does not fit the ideal form, the magnitude of that function depends less on the individual transistor parameters than does the open-loop gain. This characteristic is one of the general properties of feedback circuits.

Also, since the input current is proportional to the input voltage, we can use this circuit as a voltage amplifier.

Example 12.13 **Objective:** Determine the transresistance and voltage gain of a single-transistor shunt–shunt feedback circuit.

Consider the circuit in Figure 12.37(a). The transistor parametes are: $h_{FE} = 100$, $V_{BE}(\text{on}) = 0.7\,\text{V}$, and $V_A = \infty$. Since the input signal current is directly proportional to the input voltage, the voltage gain of this shunt–shunt configuration has the same general properties as the transresistance transfer function.

As with many circuits considered in this chapter, several capacitors are included. In the circuit in Figure 12.37(a), R_1 and C_{C2} may be removed. Resistor R_F can be used for biasing, and the circuit can be redesigned to provide the same feedback properties.

Solution: By including C_{C2} in the circuit, the feedback is a function of the ac signal only, which means that the transistor quiescent values are not affected by feedback. The quiescent parameters are found to be

(a)

(b)

Figure 12.37 (a) Circuit for Example 12.13 and (b) small-signal equivalent circuit

$I_{CQ} = 0.492\,\text{mA}$ and $V_{CEQ} = 5.08\,\text{V}$

The small-signal transistor parameters are

$$r_\pi = \frac{h_{FE}V_T}{I_{CQ}} = \frac{(100)(0.026)}{0.492} = 5.28\,\text{k}\Omega$$

and

$$g_m = \frac{I_{CQ}}{V_T} = \frac{0.492}{0.026} = 18.9\,\text{mA/V}$$

In the small-signal equivalent circuit, which is shown in Figure 12.37(b), the Thevenin equivalent input source is converted to a Norton equivalent circuit. Writing a KCL equation at the output, we obtain

$$\frac{V_o}{10} + (18.9)V_\pi + \frac{V_o - V_\pi}{82} = 0$$

A KCL equation at the input yields

$$I_i = \frac{V_\pi}{10} + \frac{V_\pi}{4.96} + \frac{V_\pi}{5.28} + \frac{V_\pi - V_o}{82}$$

Combining these two equations and eliminating V_π, we find the small-signal transresistance gain, which is

$$A_{zf} = \frac{V_o}{I_i} = -65.8\,\text{k}\Omega$$

Since this unit of gain is not as familiar as voltage gain, we determine the voltage gain from

$$I_i = V_i/R_S = V_i/10$$

Therefore,

$$\frac{V_o}{V_i} = -(65.8)(0.10) = -6.58$$

If the current gain h_{FE} of the transistor decreases from 100 to 75, the transistor quiescent values change slightly to

$$I_{CQ} = 0.478 \, \text{mA} \qquad \text{and} \qquad V_{CEQ} = 5.22 \, \text{V}$$

The small-signal parameters also change, to

$$r_\pi = 4.08 \, \text{k}\Omega \qquad \text{and} \qquad g_m = 18.4 \, \text{mA/V}$$

and the closed-loop small-signal voltage gain becomes

$$V_o/V_i = -6.41$$

Comment: With a 25 percent decrease in the transistor current gain h_{FE}, the closed-loop voltage gain decreases by only 2.6 percent. If no feedback were present, the voltage gain would be directly proportional to h_{FE}. The ideal closed-loop voltage gain of the feedback circuit, which is determined as h_{FE} approaches infinity, is

$$A_v(h_{FE} \to \infty) = -R_F/R_S = -7.20$$

Computer Simulation Verification: Additional results of the PSpice analysis are shown in Figure 12.38. The magnitude of the voltage gain is plotted as a function of the transistor current gain h_{FE}, for three values of feedback resistance. The results for $R_F = 82 \, \text{k}\Omega$ agree very well with the results from the hand analysis. As R_F increases to $160 \, \text{k}\Omega$, there is less feedback, and the magnitude of the voltage gain increases. However, the variation in the closed-loop gain is substantially greater as the transistor gain changes. In contrast, when R_F decreases to $47 \, \text{k}\Omega$, there is increased feedback, and

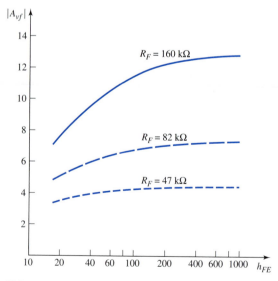

Figure 12.38 Voltage gain magnitude versus transistor current gain, for three values of feedback resistance, from a PSpice analysis of the circuit in Figure 12.37(a)

the magnitude of the voltage gain decreases. However, there is very little variation in closed-loop gain as the transistor gain changes. In all cases, as the gain of the transistor increases, there is less change in closed-loop gain. This result demonstrates the need for a large gain in the basic amplifier in the feedback network.

Expressions for the input and output resistances of the ideal shunt–shunt configuration are given in Equations (12.35) and (12.28), respectively. As with the loop gain function, the input and output resistance expressions for the single-transistor feedback circuit cannot be put in exactly the same form as that for the ideal configuration. However, the same general characteristics are obtained; that is, both input and output resistances decrease, predicted by the ideal case.

Example 12.14 **Objective:** Determine the input and output resistances of a single-transistor shunt–shunt feedback circuit.

Consider the circuit in Figure 12.37(a), with transistor parameters: $h_{FE} = 100$, $V_{BE}(\text{on}) = 0.7\,\text{V}$, and $V_A = \infty$.

Solution: The small-signal equivalent circuit for calculating the input resistance R_{if} is shown in Figure 12.39(a). The small-signal transistor parameters were determined in Example 12.13.

Figure 12.39 Small-signal equivalent circuits of the circuit in Figure 12.37(a) for calculating (a) input resistance and (b) output resistance

Writing a KCL equation at the input, we have

$$I_x = \frac{V_\pi}{r_\pi} + \frac{V_\pi - V_o}{R_F} = \frac{V_\pi}{5.28} + \frac{V_\pi - V_o}{82}$$

From a KCL equation at the output node, we have

$$\frac{V_o}{R_C} + g_m V_\pi + \frac{V_o - V_\pi}{R_F} = \frac{V_o}{10} + (18.9)V_\pi + \frac{V_o - V_\pi}{82} = 0$$

Combining these two equations, eliminating V_o, and noting that $V_\pi = V_x$, we find that

$$R_{if} = \frac{V_x}{I_x} = 0.443\,\text{k}\Omega$$

The small-signal equivalent circuit for calculating the output resistance R_{of} is shown in Figure 12.39(b). If we define

$$R_{eq} = r_\pi \| R_1 \| R_2 \| R_S$$

then a KCL equation at node V_x yields

$$I_x = \frac{V_x}{R_C} + g_m V_\pi + \frac{V_x}{R_F + R_{eq}}$$

From a voltage divider equation, we find that

$$V_\pi = \left(\frac{R_{eq}}{R_{eq} + R_F}\right) V_x$$

Combining these two equations, we find the output resistance to be

$$R_{of} = \frac{V_x}{I_x} = 1.75\,\text{k}\Omega$$

Comment: The input resistance with no feedback would be $r_\pi = 5.28\,\text{k}\Omega$. The shunt input feedback connection has lowered the input resistance to $R_{if} = 0.443\,\text{k}\Omega$. Similarly, the output resistance with no feedback would be $R_C = 10\,\text{k}\Omega$. The shunt output feedback connection has lowered the output resistance to $R_{of} = 1.75\,\text{k}\Omega$. The decrease in both the input and output resistances agrees with the ideal feedback theory.

The magnitude of the transfer function, input resistance, and output resistance of the discrete transistor feedback circuit all tend to approach the ideal values if additional transistor stages are included to increase the basic amplifier gain. As an example, a multistage shunt–shunt connection is shown in Figure 12.40. Once again, several capacitors are included, which simplifies the dc analysis. However, the capacitors may adversely affect the circuit frequency response.

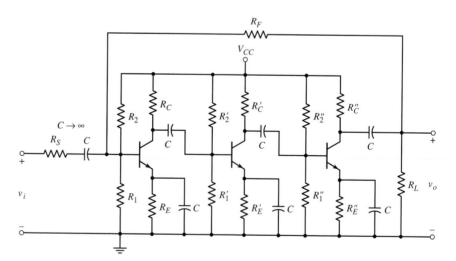

Figure 12.40 Example of multistage shunt–shunt feedback circuit

Since negative feedback is desired, there must be an odd number of negative gain stages. As the number of stages increases, the open-loop gain increases, and the circuit characteristics approach those of the ideal shunt–shunt configuration. The analysis of this circuit is left as a computer simulation problem at the end of the chapter.

Design Example 12.15   Objective: Design an amplifier that converts a photo-diode current to an output voltage.

Assume the photodiode signal is variable from 0 to 1 mA, the source resistance is $R_S = 100\,\Omega$, and the amplifier is to drive a nominal load of $R_L = 1\,k\Omega$. The required output voltage is $V_o = \pm 5 \times 10^3 I_i$ (the phase of the output is not important), which means that the amplifier transresistance is to be $A_{zf} = 5 \times 10^3\,\Omega$. An op-amp with the characteristics described in Example 12.8 is available.

Solution: Design Approach: To minimize loading effects on the amplifier input, a small input resistance is required; to minimize loading effects on the amplifier output, a small output resistance is also required. For these reasons, a shunt–shunt feedback, or transresistance, amplifier should be used.

The closed-loop gain is

$$A_{zf} = 5 \times 10^3 \cong 1/\beta_g$$

therefore, the conductance feedback transfer function is

$$\beta_g = 2 \times 10^{-4}\,S$$

The dependent open-loop voltage source of the op-amp, as shown in Figure 12.17, can be transformed to an equivalent dependent open-loop transresistance source for the transresistance amplifier, as shown in Figure 12.14. We find that

$$A_z = A_v R_i$$

Using the parameters specified for the op-amp, we find

$$A_z = (10^4)(10^4) = 10^8\,\Omega$$

The loop gain for the shunt–shunt configuration is

$$A_z \beta_g = (10^8)(2 \times 10^{-4}) = 2 \times 10^4$$

Referring to Table 12.1, we expect the input resistance to be

$$R_{if} = 10/2 \times 10^4 = 5 \times 10^{-4}\,k\Omega \rightarrow 0.5\,\Omega$$

and the output resistance to be

$$R_{of} = 100/2 \times 10^4 = 5 \times 10^{-3}\,\Omega$$

These input and output resistances should minimize any loading effects at the amplifier input and output.

For our design we may use the amplifier configuration in Figure 12.41. In the ideal case, we have,

$$\frac{V_o}{I_i} = -R_F = \frac{1}{\beta_g}$$

Figure 12.41 Transresistance amplifier for Example 12.15

or

$$R_F = \frac{1}{2 \times 10^{-4}} \rightarrow 5\,\text{k}\Omega$$

 Comment: The design produces a transresistance amplifier that is extremely close to the ideal.

Test Your Understanding

12.22 Consider the circuit in Figure 12.42, with transistor parameters $V_{TN} = 1.5\,\text{V}$, $K_n = 1\,\text{mA/V}^2$, and $\lambda = 0$. (a) Determine the closed-loop small-signal voltage gain $A_{vf} = V_o/V_i$. (b) If the transistor conduction parameter K_n increases to $1.5\,\text{mA/V}^2$, determine the new value of voltage gain. By what percentage does the voltage gain change? (Ans. (a) $A_{vf} = -1.48$ (b) $A_{vf} = -1.62$, 9.46% change)

Figure 12.42   Figure for Exercises 12.22 and 12.23

12.23 Consider the feedback circuit in Figure 12.42, with transistor parameters $V_{TN} = 1.5\,\text{V}$, $K_n = 1\,\text{mA/V}^2$, and $\lambda = 0$. (a) Determine the input and output resistances R_{if} and R_{of}. (b) Repeat part (a) if the transistor conduction parameter increases to $K_n = 1.5\,\text{mA/V}^2$. (Ans. (a) $R_{if} = 7.0\,\text{k}\Omega$, $R_{of} = 1.58\,\text{k}\Omega$ (b) $R_{if} = 5.56\,\text{k}\Omega$, $R_{of} = 1.32\,\text{k}\Omega$)

*D12.24 Design a feedback transresistance amplifier to provide a gain of $-10\,\text{k}\Omega$. The nominal current signal source resistance is $50\,\Omega$, and the nominal load is $500\,\Omega$. An op-amp with parameters $R_i = 5\,\text{k}\Omega$, $R_o = 50\,\Omega$, and a low-frequency open-loop gain of $A_v = 5 \times 10^3$ is available. Correlate the design with a computer simulation analysis to determine the gain, input resistance, and output resistance.

12.8   LOOP GAIN

In previous sections, the loop gain T was easily determined for circuits involving ideal op-amps. For discrete transistor circuits, however, the loop gain usually cannot be obtained directly from the closed-loop transfer function. As we will see later in this chapter, loop gain is an important parameter in the stability of a feedback circuit; we will describe a number of techniques for determining the loop gain.

12.8.1   Basic Approach

The general feedback network was shown in Figure 12.1 and is repeated in Figure 12.43(a). To find the loop gain, set the source S_i equal to zero, and break the feedback loop at some point. Figure 12.43(b) shows a feedback network in which the loop is broken at the amplifier input and a test signal S_t is applied at this point. The amplifier output signal is $S_o = AS_t$, and the feedback signal is

$$S_{fb} = \beta S_o = A\beta S_t$$

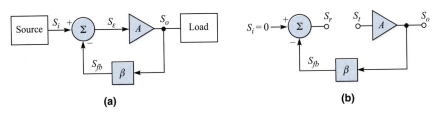

Figure 12.43   (a) Ideal configuration of a feedback amplifier; (b) basic feedback network with loop broken at amplifier input

The return signal S_r, which was previously the error signal, is now $-S_{fb}$ (the minus sign indicates negative feedback). Therefore,

$$\frac{S_r}{S_t} = -A\beta \tag{12.94}$$

The ratio of the return signal S_r to the test signal S_t is the negative of the loop gain factor.

As the feedback loop is broken, the conditions that existed prior to the loop being broken must remain unchanged. These conditions include: maintaining the same transistor biasing and maintaining the same impedance at the return point. An equivalent impedance must therefore be inserted at the point where the loop is broken. This is shown in Figure 12.44. Figure 12.44(a) shows the amplifier input impedance R_{in} prior to the loop being broken. Figure 12.44(b) shows the configuration after the loop is broken. A test voltage V_t is applied, and a load impedance R_{in} is inserted at the output of the broken

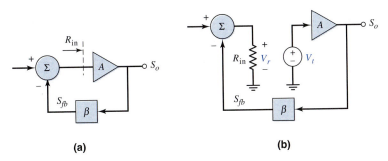

Figure 12.44   (a) Basic feedback network, showing amplifier input resistance and (b) feedback network after the loop is broken, showing test voltage and load resistance

loop. The return voltage is then measured at this output terminal. The loop
gain is found to be

$$T = A\beta = -\frac{V_r}{V_t} \tag{12.95}$$

Also, a test current I_t may be applied and a return current signal I_r measured, to find the loop gain as

$$T = -\frac{I_r}{I_t} \tag{12.96}$$

As an example, consider the circuit shown in Figure 12.45(a). The circuit is
similar to the one considered in Examples 12.13 and 12.14. The feedback loop
is broken at the input to the transistor, at the point marked X. The small-signal
equivalent circuit is shown in Figure 12.45(b). A test voltage is applied to the
base of the transistor and the equivalent load resistance r_π is connected at the
return point. The input signal current is set equal to zero.

Since $V_\pi = V_t$, if we define $R_{eq} = R_S \| R_1 \| R_2 \| r_\pi$, then the output voltage
can be written

$$V_o = -g_m V_t [R_C \| (R_F + R_{eq})] \tag{12.97}$$

(a)

(b)

Figure 12.45   (a) Feedback circuit prior to breaking the loop and (b) small-signal equivalent
circuit after breaking the loop

From a voltage divider, the return voltage V_r expression is

$$V_r = \left(\frac{R_{eq}}{R_F + R_{eq}}\right)V_o \tag{12.98}$$

Substituting Equation (12.97) into Equation (12.98) yields the loop gain

$$T = -\frac{V_r}{V_t} = +g_m\left(\frac{R_{eq}}{R_F + R_{eq}}\right)[R_C\|(R_F + R_{eq})] \tag{12.99(a)}$$

which can be written as

$$T = (g_m R_c)\left(\frac{R_{eq}}{R_C + R_F + R_{eq}}\right) \tag{12.99(b)}$$

Example 12.16 **Objective:** Determine the loop gain for a feedback circuit.

Consider the circuit shown in Figure 12.45(a), with transistor parameters: $h_{FE} = 100$, $V_{BE}(\text{on}) = 0.7\,\text{V}$, and $V_A = \infty$. From Example 12.13, the quiescent collector current is $I_{CQ} = 0.492\,\text{mA}$, and the resulting small-signal parameters are $r_\pi = 5.28\,\text{k}\Omega$ and $g_m = 18.9\,\text{mA/V}$.

Solution: The equivalent resistance is

$$R_{eq} = R_S\|R_1\|R_2\|r_\pi = (10)\|(51)\|(5.5)\|(5.28) = 2.04\,\text{k}\Omega$$

From Equation (12.99(b)), the loop gain is

$$T = (g_m R_C)\left(\frac{R_{eq}}{R_C + R_F + R_{eq}}\right)$$
$$= [(18.9)(10)]\left(\frac{2.04}{10 + 82 + 2.04}\right) = 4.10$$

If the transistor current gain h_{FE} increases to 1000, then $I_{CQ} = 0.547\,\text{mA}$, $r_\pi = 47.5\,\text{k}\Omega$, and $g_m = 21.0\,\text{mA/V}$. The new value of R_{eq} becomes $3.10\,\text{k}\Omega$ and the loop gain is $T = 6.85$.

Comment: Since the loop gain is a function of the basic amplifier gain, we expect this parameter to change as the transistor current gain changes. Also, since no capacitance effects were considered, the loop gain is a positive, real number that corresponds to negative feedback.

12.8.2 Computer Analysis

The loop gain can also be determined from a computer analysis of the feedback circuit. In Example 12.17, we demonstrate a direct approach to determining the loop gain. First, we consider the circuit analyzed in the last example, to correlate the results of a computer analysis to those of a hand analysis. Then, we determine the loop gain of a feedback circuit when taking capacitance effects into account.

Example 12.17 **Objective:** Determine the loop gain factor for a feedback circuit, using a computer simulation analysis.

Consider the circuit in Figure 12.45(a).

Solution: We determine the loop gain factor by using the circuit in Figure 12.46, in which the loop is effectively broken at the base of the transistor. The circuit conditions, however, must remain unchanged from those prior to breaking the loop. This includes maintaining the same bias currents in the transistor and terminating the broken loop with the proper impedance.

A large inductance is inserted in the transistor base connection, to act as a short circuit for dc signals, so that the proper dc bias can be maintained on the transistor, and to act as an open circuit for ac signals, so that the loop appears to be broken for the ac signal. A test voltage V_t is applied to the base of the transistor through a coupling capacitor, and a load resistance R_p is connected through a coupling capacitor at the return point. These coupling capacitors act as short circuits to the ac signals, but as open circuits to dc signals, so that the dc bias is not disturbed by these elements.

Figure 12.46 Feedback circuit with the loop effectively broken, for determining the loop gain from a computer analysis

From the computer simulation, the loop gain for a transistor current gain of $h_{FE} = 100$ is

$$T = -V_r/V_t = 5.04$$

For a current gain of 1000, the loop gain is $T = 9.37$. These values differ slightly from the hand analysis results in Example 12.16. The slight difference arises because the quiescent collector currents determined in the hand analysis and the computer analysis are not quite the same, leading to different values of g_m and r_π.

Comment: The analysis of this circuit is straightforward. In the next example, we demonstrate another advantage of a computer analysis.

When capacitances are part of the feedback circuit, the phase of the loop gain becomes a factor in determining whether the feedback is negative or positive. Figure 12.47 shows a three-stage amplifier with feedback. Each stage is the same as the circuit given in Figure 12.45(a). For an odd number of stages at low frequency, the loop gain is a positive, real quantity, and negative feedback is applied. The coupling and emitter bypass capacitors are assumed to be very large, and capacitors C_1, C_2, and C_3 between the stages can represent either load capacitances or transistor input capacitances. As the frequency increases, the magnitude of the loop gain decreases, because of decreasing capacitor impedances, and the phase of the loop gain also changes.

Figure 12.47   The ac equivalent circuit of three-stage feedback amplifier, including load capacitors

---

Example   12.18   Objective: Determine the magnitude and phase of the loop gain of a multistage feedback circuit.

Consider the circuit in Figure 12.47, with parameters: $R_S = 10\,\text{M}\Omega$, $R_A = 51\,\text{k}\Omega$, $R_B = 5.5\,\text{k}\Omega$, $R_F = 82\,\text{k}\Omega$, $R_C = 10\,\text{k}\Omega$, and $C = 100\,\text{pF}$. The transistor current gains are assumed to be $h_{FE} = 15$, which keeps the overall gain fairly small.

Solution: The loop is broken at the base of Q_1, and the ratio of the return signal to the test signal is measured by the same technique shown in Figure 12.46.

The magnitude of V_r/V_t versus frequency is shown in Figure 12.48(a). The magnitude of loop gain drops off with frequency, as expected, and is equal to unity at approximately 5.5 MHz.

The phase of the return signal is shown in Figure 12.48(b). Since the loop gain is given by $T = -V_r/V_t$, then the phase of the loop gain is $\angle T = -180° + \angle V_r - \angle V_t$ where the $-180°$ corresponds to the minus sign. Since the phase of the input signal was set to zero, then the phase of the loop gain is $\angle T = -180° + \angle V_r$. At low frequencies, where the phase of the return signal is approximately $+180°$, the phase of the loop gain is essentially zero, corresponding to negative feedback. At approximately $f = 2.5\,\text{MHz}$, the phase of the return signal is zero so that the phase of the loop gain is $-180°$, which corresponds to positive feedback.

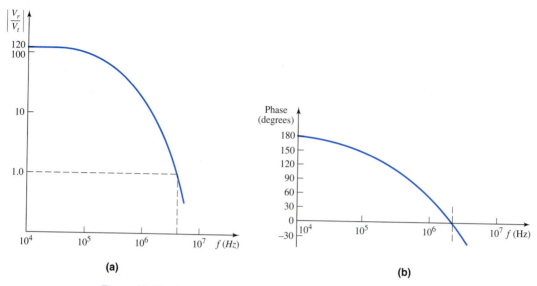

Figure 12.48 (a) Bode plot of loop gain magnitude for three-stage feedback amplifier, from Example 12.17; (b) phase of the return signal for the three-stage amplifier

Comment: For this circuit, the loop gain magnitude is greater than unity at the frequency at which the phase of T is $-180°$. As discussed in the next section, this condition means that the circuit is unstable and will oscillate.

A hand analysis of the three-stage amplifier just considered would be tedious, especially taking the frequency response into account. In this case, a computer analysis is more suitable.

Test Your Understanding

12.25 Consider the circuit in Figure 12.45(a) with a new value of $R_E = 1\,\text{k}\Omega$. The transistor parameters are: $h_{FE} = 120$, $V_{BE}(\text{on}) = 0.7\,\text{V}$, and $V_A = \infty$. Determine the loop gain T. (Ans. $T = 3.68$)

12.26 Consider the feedback circuit in Figure 12.16, with the equivalent circuit given in Figure 12.17. Break the feedback loop at an appropriate point, and derive the expression for the loop gain. (Ans. $T = A_v/[1 + R_2/(R_1 \| R_i)]$)

12.9   STABILITY OF THE FEEDBACK CIRCUIT

In negative feedback, a portion of the output signal is subtracted from the input signal to produce the error signal. However, as we found in the last section, this subtraction property, or the loop gain, may change as a function of frequency. At some frequencies, the subtraction may actually be addition; that is, the negative feedback may become positive, producing an unstable system. In this section, we will examine the stability of feedback circuits.

12.9.1   The Stability Problem

The basic feedback configuration is shown in Figure 12.1, and the ideal closed-loop transfer function is given by Equation (12.5), which is repeated here:

$$A_f = \frac{S_o}{S_i} = \frac{A}{(1 + \beta A)} \tag{12.5}$$

The open-loop gain is a function of the individual transistor parameters and capacitances, and is therefore a function of frequency. The closed-loop gain can then be written as

$$A_f(s) = \frac{A(s)}{(1 + \beta A(s))} = \frac{A(s)}{1 + T(s)} \tag{12.100}$$

where $T(s)$ is the loop gain. For physical frequencies, $s = j\omega$, and the loop gain is $T(j\omega)$, which is a complex function. The loop gain can be represented by its magnitude and phase, as follows:

$$T(j\omega) = |T(j\omega)| \angle \phi \tag{12.101}$$

The closed-loop gain can be written

$$A_f(j\omega) = \frac{A(j\omega)}{1 + T(j\omega)} \tag{12.102}$$

The stability of the feedback circuit is a function of the loop gain $T(j\omega)$. If the loop gain magnitude is unity when the phase is 180 degrees, then $T(j\omega) = -1$ and the closed-loop gain goes to infinity. This implies that an output will exist for a zero input, which means that the circuit will oscillate. If we are trying to build a linear amplifier, an oscillator is considered an unstable circuit. We will show that if $|T(j\omega)| < 1$ when the phase is 180 degrees, the system is stable, whereas if $|T(j\omega)| \geq 1$ when the phase is 180 degrees, the system is unstable. To study the stability of feedback circuits, we must therefore analyze the frequency response of the loop gain factor.

12.9.2   Bode Plots: One-, Two-, and Three-Pole Amplifiers

Figure 12.49(a) shows a simple single-stage common-emitter current amplifier. The high-frequency small-signal equivalent circuit is shown in Figure 12.49(b). The capacitance C_1 includes the forward-biased base-emitter junction capacitance as well as the effective Miller capacitance. The Miller capacitance and Miller effect were discussed in Chapter 7. The equivalent circuit shown in Figure 12.49(b) is identical to that developed in Figure 7.39. The output current in Figure 12.49(b) is given by

$$I_o = \left(\frac{R_C}{R_C + R_L} \right) g_m V_\pi \tag{12.103}$$

and the voltage V_π is

$$V_\pi = I_i \left[R_\pi \middle\| \left(\frac{1}{sC_1} \right) \right] \tag{12.104}$$

Figure 12.49   (a) Single-stage common-emitter amplifier and (b) small-signal equivalent circuit, including input capacitance

where $R_\pi = r_\pi \| R_B = r_\pi \| R_1 \| R_2$. Equation (12.104) can be expanded to

$$V_\pi = I_i \left[\frac{R_\pi}{1 + sR_\pi C_1} \right] \qquad (12.105)$$

Substituting Equation (12.105) into (12.103), we get an expression for the small-signal current gain,

$$A_i = g_m R_\pi \left(\frac{R_C}{R_C + R_L} \right) \left[\frac{1}{1 + sR_\pi C_1} \right] \qquad (12.106)$$

When we set $s = j\omega = j(2\pi f)$, Equation (12.106) can be written as

$$A_i = \frac{A_{io}}{1 + j\left(\dfrac{f}{f_1} \right)} \qquad (12.107)$$

where A_{io} is the low-frequency or midband gain and f_1 is the upper 3 dB frequency. The gain is a complex function that can be written

$$A_i = \frac{A_{io}}{\sqrt{1 + \left(\dfrac{f}{f_1} \right)^2}} \angle -\tan^{-1} \left(\frac{f}{f_1} \right) \qquad (12.108)$$

Figure 12.50(a) is a Bode plot of the current gain magnitude, and Figure 12.50(b) is a Bode plot of the current gain phase. Note that, from the definition of the directions of input and output currents, the output current is in phase with the input current at low frequencies. At high frequencies, the output current becomes 90 degrees out of phase with respect to the input current. This single-stage circuit is an example of a one-pole amplifier. As we have previously shown, similar expressions can be obtained for voltage gain, the transresistance transfer function, and the transconductance transfer function.

   Figure 12.51 shows the small-signal equivalent circuit of a two-stage amplifier, using the same hybrid-π configuration for the transistors. The capacitance

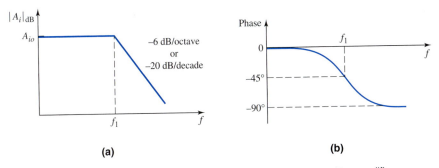

Figure 12.50 Bode plots of current gain for single-stage common-emitter amplifier: (a) magnitude and (b) phase

Figure 12.51 Small-signal equivalent circuit, two-stage amplifier including input capacitances

C_2 is the input capacitance of the second transistor, including the effective Miller capacitance. The output current is

$$I_o = -g_{m2}V_{\pi 2} \tag{12.109}$$

and $V_{\pi 2}$ is

$$V_{\pi 2} = -g_{m1}V_{\pi 1}\left[R_{L1} \left\| R_{\pi 2} \right\| \left(\frac{1}{sC_2} \right) \right] \tag{12.110}$$

The voltage $V_{\pi 1}$ is

$$V_{\pi 1} = I_i\left[R_{\pi 1} \left\| \left(\frac{1}{sC_1} \right) \right. \right] \tag{12.111}$$

Combining Equations (12.109), (12.110), and (12.111) yields an expression for the small-signal current gain, as follows:

$$A_i = \frac{I_o}{I_i} = (g_{m1}g_{m2})(R_{\pi 1})(R_{L1}\|R_{\pi 2})\left[\frac{1}{1 + sR_{\pi 1}C_1} \right]\left[\frac{1}{1 + s(R_{L1}\|R_{\pi 2})C_2} \right] \tag{12.112}$$

Setting $s = j\omega = j(2\pi f)$, we can write Equation (12.112)

$$A_i = \frac{A_{io}}{\left(1 + j\dfrac{f}{f_1} \right)\left(1 + j\dfrac{f}{f_2} \right)} \tag{12.113}$$

where $f_1 = 1/2\pi R_{\pi 1} C_1$ and $f_2 = 1/2\pi (R_{L1} \| R_{\pi 2}) C_2$. Frequency f_1 is the upper 3 dB frequency of the first stage, and f_2 is the upper 3 dB frequency of the second stage. This two-stage circuit is an example of a two-pole amplifier.

Equation (12.113) can be written

$$A_i = \frac{A_{io}}{\sqrt{1 + \left(\frac{f}{f_1}\right)^2}\sqrt{1 + \left(\frac{f}{f_2}\right)^2}} \angle - \left[\tan^{-1}\left(\frac{f}{f_1}\right) + \tan^{-1}\left(\frac{f}{f_2}\right)\right] \qquad \textbf{(12.114)}$$

Figure 12.52(a) is a Bode plot of the current gain magnitude, assuming $f_1 \ll f_2$. This assumption implies that the two poles are far apart. The Bode plot of the current gain phase is shown in Figure 12.52(b). Again the phase of the output current is in phase with the input current at low frequency. This phase relation is a direct result of the way the directions of current were defined. At high frequencies, the output current becomes 180 degrees out of phase with respect to the input current.

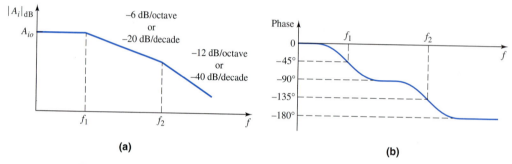

(a)   **(b)**

Figure 12.52 Bode plots of current gain for two-stage amplifier: (a) magnitude and (b) phase

An op-amp is a three-stage amplifier, as shown in Figure 12.53. Since each stage has an equivalent input resistance and capacitance, this circuit is an example of a three-pole amplifier. The overall gain can be expressed as

$$A = \frac{A_o}{\left(1 + j\frac{f}{f_1}\right)\left(1 + j\frac{f}{f_2}\right)\left(1 + j\frac{f}{f_3}\right)} \qquad \textbf{(12.115)}$$

where A_o is the low-frequency gain factor. Assuming the poles are far apart (let $f_1 \ll f_2 \ll f_3$), the Bode plots of the gain magnitude and phase are shown in Figure 12.54. At very high frequencies, the phase difference between the output and input signals is -270 degrees.

If we assume an ideal feedback amplifier, the loop gain is

$$T(j\omega) = \beta A(j\omega) \qquad \textbf{(12.116)}$$

Figure 12.53 Three-stage amplifier

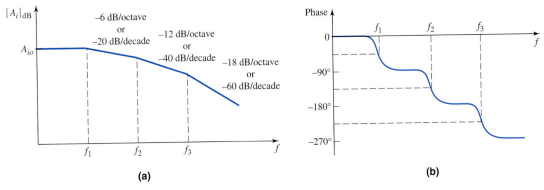

Figure 12.54 Bode plots of three-stage amplifier gain: (a) magnitude and (b) phase

where the feedback transfer function β is assumed to be independent of frequency. For op-amp feedback circuits, we can determine the feedback transfer function β, as previously shown, and the basic amplifier characteristics are assumed to be known. For a three-stage amplifier, the loop gain is therefore

$$T(f) = \frac{\beta A_o}{\left(1 + j\dfrac{f}{f_1}\right)\left(1 + j\dfrac{f}{f_2}\right)\left(1 + j\dfrac{f}{f_3}\right)} \qquad (12.117)$$

Both the magnitude and phase of the loop gain are functions of frequency. For the three-stage amplifier, the phase will be -180 degrees at some particular frequency, which means that the amplifier may become unstable.

12.9.3   Nyquist Stability Criterion

In the last section, we saw that a feedback system can become unstable. Several methods can be used to determine whether a system is stable or unstable. The method we will consider is called the **Nyquist stability criterion**. This method not only determines if a system is stable, it also indicates the degree of system stability.

To apply this method, we must plot a **Nyquist diagram**, which is a polar plot of the loop gain factor $T(j\omega)$. The loop gain, which is a complex function, can be written in terms of its magnitude and phase, $T(j\omega) = |T(j\omega)|\angle\phi$, as shown in Equation (12.101). The Nyquist diagram is a plot of the real and imaginary components of $T(j\omega)$ as the frequency ω varies from minus infinity to plus infinity. Although negative frequencies have no physical meaning, they are not mathematically excluded in the loop gain function. The polar plot for negative frequencies, as we will see, is the complex conjugate of the polar plot for positive frequencies.

The loop gain for a two-pole amplifier is, from Equation (12.113),

$$T(j\omega) = \frac{\beta A_{io}}{\left(1 + j\dfrac{\omega}{\omega_1}\right)\left(1 + j\dfrac{\omega}{\omega_2}\right)} \qquad (12.118)$$

where ω_1 and ω_2 are the upper 3 dB radian frequencies of the first and second stages, respectively. We can also write Equation (12.118) in the form

$$T(j\omega) = \frac{\beta A_{io}}{\sqrt{1 + \left(\frac{\omega}{\omega_1}\right)^2}\sqrt{1 + \left(\frac{\omega}{\omega_2}\right)^2}} \angle -\left[\tan^{-1}\left(\frac{\omega}{\omega_1}\right) + \tan^{-1}\left(\frac{\omega}{\omega_2}\right)\right]$$

(12.119)

The Nyquist plot of Equation (12.119) is shown in Figure 12.55. At $\omega = 0$, the magnitude of $T(j\omega)$ is βA_{io} and the phase is zero. As ω increases, the magnitude decreases and the phase is negative. From Equation (12.119), we see that for negative values of ω, the magnitude also decreases, but the phase becomes positive. This means that the loop gain function for negative frequencies is the complex conjugate of the loop gain function for positive frequencies, and the real axis is the axis of symmetry. As ω approaches $+\infty$, the magnitude approaches zero and the phase approaches -180 degrees.

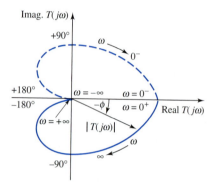

Figure 12.55 Nyquist plot, loop gain for two-stage amplifier

The loop gain for a three-pole amplifier is, from Equation (12.117),

$$T(j\omega) = \frac{\beta A_o}{\left(1 + j\frac{\omega}{\omega_1}\right)\left(1 + j\frac{\omega}{\omega_2}\right)\left(1 + j\frac{\omega}{\omega_3}\right)}$$

(12.120)

This loop gain function can also be written in the form

$$T(j\omega) = \frac{\beta A_o}{\sqrt{1 + \left(\frac{\omega}{\omega_1}\right)^2}\sqrt{1 + \left(\frac{\omega}{\omega_2}\right)^2}\sqrt{1 + \left(\frac{\omega}{\omega_3}\right)^2}} \angle \phi$$

(12.121(a))

where ϕ is the phase, given by

$$\phi = -\left[\tan^{-1}\left(\frac{\omega}{\omega_1}\right) + \tan^{-1}\left(\frac{\omega}{\omega_2}\right) + \tan^{-1}\left(\frac{\omega}{\omega_3}\right)\right]$$

(12.121(b))

Figure 12.56(a) shows one possible Nyquist plot. For $\omega = 0$, the magnitude is βA_o and the phase is zero. As ω increases in the positive direction, the magnitude decreases and the phase becomes negative. As the Bode plot in Figure 12.54 shows, the phase goes through -90 degrees, then through -180 degrees, and finally approaches -270 degrees as the magnitude approaches zero. This same effect is shown in the Nyquist diagram. The plot approaches

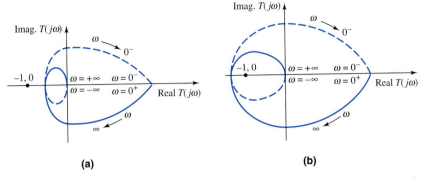

Figure 12.56 Nyquist plot, loop gain for three-stage amplifier, for: (a) stable system and (b) unstable system

the origin and is tangent to the imaginary axis as $\omega \to \infty$. Again, the plot for negative frequencies is the mirror image of the positive frequency plot about the real axis.

Another possible Nyquist plot for the three-pole loop gain function is shown in Figure 12.56(b). The basic plot is the same as that in Figure 12.56(a), except that the position of the point $(-1, 0)$ is different. At the frequency at which the phase is -180 degrees, the curve crosses the negative real axis. In Figure 12.56(a), $|T(j\omega)| < 1$ when the phase is -180 degrees, whereas in Figure 12.56(b), $|T(j\omega)| > 1$ when the phase is -180 degrees. The Nyquist diagram encircles the point $(-1, 0)$ in Figure 12.56(b), and this has particular significance for stability. For this treatment of a three-pole amplifier, the Nyquist criterion for stability of the amplifier can be stated as follows: "If the Nyquist plot encircles or goes through the point $(-1, 0)$, the amplifier is unstable."

Using the criterion, a simpler test for stability can be used in most cases. If $|T(j\omega)| \geq 1$ at the frequency at which the phase is -180 degrees, then the amplifier is unstable. This simpler test allows us to use the Bode plots considered previously, instead of explicitly constructing the Nyquist diagram.

Example 12.19 Objective: Determine the stability of an amplifier, given the loop gain function.

Consider a three-pole feedback amplifier with a loop gain given by

$$T(f) = \frac{\beta(100)}{\left(1 + j\,\dfrac{f}{10^5}\right)^3}$$

In this case, the three poles all occur at the same frequency. Determine the stability of the amplifier for $\beta = 0.20$ and $\beta = 0.02$.

Solution: The loop gain can be written in terms of its magnitude and phase,

$$T(f) = \frac{\beta(100)}{\left[\sqrt{1 + \left(\dfrac{f}{10^5}\right)^2}\right]^3} \angle - 3\tan^{-1}\left(\frac{f}{10^5}\right)$$

The frequency f_{180} at which the phase becomes -180 degrees is

$$-3\tan^{-1}\left(\frac{f_{180}}{10^5}\right) = -180°$$

which yields

$$f_{180} = 1.73 \times 10^5 \text{ Hz}$$

The magnitude of the loop gain at this frequency for, $\beta = 0.20$, is then

$$|T(f_{180})| = \frac{(0.20)(100)}{8} = 2.5$$

For $\beta = 0.02$, the magnitude is

$$|T(f_{180})| = \frac{(0.020)(100)}{8} = 0.25$$

Comment: The loop gain magnitude at the frequency at which the phase is -180 degrees is 2.5 when $\beta = 0.20$ and 0.25 when $\beta = 0.02$. The system is therefore unstable for $\beta = 0.20$ and stable for $\beta = 0.02$.

We can also consider the stability of the feedback system in terms of Bode plots. The Bode plot of the loop gain magnitude from the previous example is shown in Figure 12.57(a), for $\beta = 0.20$ and $\beta = 0.02$. The low-frequency loop gain magnitude is dependent on β, but the 3 dB frequency is the same in both cases. Since the three poles all occur at the same frequency, the magnitude of $T(f)$ decreases at the rate of -18 dB/octave at the higher frequencies. The frequencies at which $|T(f)| = 1$ are indicated on the figure.

The phase of the loop gain function is shown in Figure 12.57(b). The two frequencies at which $|T(f)| = 1$, for the two values of β, are also indicated. We see that $|\phi| > 180°$ at $|T(f)| = 1$, when $\beta = 0.20$. This is equivalent to $|T(f)| > 1$ when $\phi = -180°$, which makes the system unstable. However, $|\phi| < 180°$ at

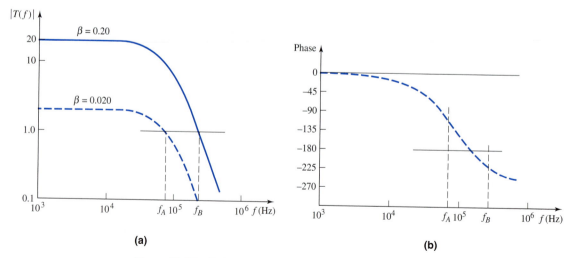

Figure 12.57 Bode plots of loop gain of function described in Example 12.19, for two values of feedback transfer function: (a) magnitude and (b) phase

$|T(f)| = 1$, when $\beta = 0.02$, so the feedback circuit is stable for this feedback transfer factor.

12.9.4 Phase and Gain Margins

From the discussion in the previous section, we can determine whether a feedback amplifier is stable or unstable by examining the loop gain as a function of frequency. This can be done from a Nyquist diagram or from the Bode plots. We can also use this technique to determine the degree of stability of a feedback amplifier.

At the frequency at which the loop gain magnitude is unity, if the magnitude of the phase is less than 180 degrees, the system is stable. This is illustrated in Figure 12.58. The difference (magnitude) between the phase angle at this frequency and 180 degrees is called the **phase margin**. The loop gain can change due, for example, to temperature variations, and the phase margin indicates how much the loop gain can increase and still maintain stability. A typical desired phase margin is in the range of 45 to 60 degrees.

A second term that describes the degree of stability is the **gain margin**, which is also illustrated in Figure 12.58. This function is defined to be $|T(j\omega)|$ in decibels at the frequency where the phase is -180 degrees. This value is usually expressed in dB and also gives an indication of how much the loop gain can increase and still maintain stability.

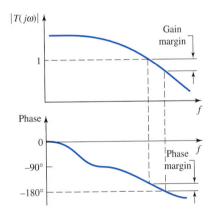

Figure 12.58 Bode plots of loop gain magnitude and phase, indicating phase margin and gain margin

---

Example 12.20 **Objective:** Determine the required feedback transfer function β to yield a specific phase margin.

Consider a three-pole feedback amplifier with a loop gain function given by

$$T(f) = \frac{\beta(100)}{\left(1 + j\dfrac{f}{10^3}\right)\left(1 + j\dfrac{f}{5 \times 10^4}\right)\left(1 + j\dfrac{f}{10^6}\right)}$$

Determine the value of β that yields a phase margin of 45 degrees.

Solution: A phase margin of 45 degrees implies that the phase of the loop gain is -135 degrees at the frequency at which the magnitude of the loop gain is unity. The phase of the loop gain is

$$\phi = -\left[\tan^{-1}\left(\frac{f}{10^3}\right) + \tan^{-1}\left(\frac{f}{5\times10^4}\right) + \tan^{-1}\left(\frac{f}{10^6}\right)\right]$$

Since the three poles are far apart, the frequency at which the phase is -135 degrees is approximately equal to the frequency of the second pole, as shown in Figure 12.54. In this example, $f_{135} \cong 5 \times 10^4$ Hz, so we have that

$$\phi = -\left[\tan^{-1}\left(\frac{5\times10^4}{10^3}\right) + \tan^{-1}\left(\frac{5\times10^4}{5\times10^4}\right) + \tan^{-1}\left(\frac{5\times10^4}{10^6}\right)\right]$$

or

$$\phi = -[88.9° + 45° + 2.86°] \cong -135°$$

Since we want the loop gain magnitude to be unity at this frequency, we have

$$|T(f)| = 1 = \frac{\beta(100)}{\sqrt{1+\left(\frac{5\times10^4}{10^3}\right)^2}\sqrt{1+\left(\frac{5\times10^4}{5\times10^4}\right)^2}\sqrt{1+\left(\frac{5\times10^4}{10^6}\right)^2}}$$

or

$$1 \cong \frac{\beta(100)}{(50)(1.41)(1)}$$

which yields $\beta = 0.705$.

Comment: If the frequency is greater than 5×10^4 Hz, the loop gain magnitude is less than unity, and the system remains stable.

Test Your Understanding

12.27   Consider a feedback amplifier with a single pole and an open-loop gain given by Equation (12.107). Assume the parameters are $A_{io} = 10^5$ A/A and $f_1 = 10$ Hz. The basic amplifier is connected to a feedback circuit for which the feedback transfer function is $\beta = 0.01$ A/A. Find the frequency at which $|T(f)| = 1$, and determine the phase margin. (Ans. $f = 10^4$ Hz, 90 degrees)

*12.28   A two-pole feedback amplifier has an open-loop gain given by Equation (12.113), with parameters: $A_{io} = 10^5$ A/A, $f_1 = 10^4$ Hz, and $f_2 = 10^5$ Hz. The basic amplifier is connected to a feedback circuit, for which the feedback transfer ratio is β. Determine the value of β that results in a phase margin of 60 degrees. (Ans. $\beta = 9.73 \times 10^{-5}$ A/A)

12.29   Consider the loop gain function described in Example 12.19. Determine the value of β at which the amplifier becomes unstable. (Ans. $\beta = 0.08$)

12.30   For the loop gain function given in Example 12.19, determine the value of β that produces a phase margin of 60 degrees. (Ans. $\beta = 0.0222$)

12.10   FREQUENCY COMPENSATION

In the previous section, we presented a method for determining whether a feedback system is stable or unstable. In this section, we will discuss a method for modifying the loop gain of a feedback amplifier, to make the system stable. The general technique of making a feedback system stable is called **frequency compensation**.

12.10.1   Basic Theory

One basic method of frequency compensation involves introducing a new pole in the loop gain function, at a sufficiently low frequency that $|T(f)| = 1$ occurs when $|\phi| < 180°$. As an example, consider the Bode plots of a three-pole loop gain magnitude and phase given in Figure 12.59 and shown by the solid lines. In this case, when the magnitude of the loop gain is unity, the phase is nearly -270 degrees and the system is unstable.

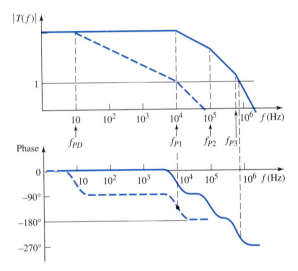

Figure 12.59   Bode plots of loop gain magnitude and phase for three-stage amplifier, before frequency compensation (solid curves), and after frequency compensation (dotted curves)

If we introduce a new pole f_{PD} at a very low frequency, and if we assume that the original three poles do not change, the new Bode plots of the magnitude and phase will be as shown by the dotted lines in Figure 12.59. In this situation, the magnitude of the loop gain becomes unity when the phase is $|\phi| < 180°$, and the system is stable. Since the pole is introduced at a low frequency and since it dominates the frequency response, it is called a **dominant pole**. This fourth pole can be introduced by adding a fourth stage with an extremely large input capacitance. Though not practical, this method demonstrates the basic idea of stabilizing a circuit.

Example 12.21 **Objective:** Determine the dominant pole required to stabilize a feedback system.

Consider a three-pole feedback amplifier with a loop gain given by

$$T(f) = \frac{5 \times 10^5}{\left(1 + j\dfrac{f}{10^6}\right)\left(1 + j\dfrac{f}{10^7}\right)\left(1 + j\dfrac{f}{10^8}\right)}$$

Insert a dominant pole, assuming the original poles do not change, such that the phase margin is at least 45 degrees.

Solution: By inserting a dominant pole, we change the loop gain function to

$$T_{PD}(f) = \frac{5 \times 10^5}{\left(1 + j\dfrac{f}{f_{PD}}\right)\left(1 + j\dfrac{f}{10^6}\right)\left(1 + j\dfrac{f}{10^7}\right)\left(1 + j\dfrac{f}{10^8}\right)}$$

We assume that $f_{PD} \ll 10^6$ Hz. A phase of -135 degrees, giving a phase margin of 45 degrees, occurs approximately at $f_{135} = 10^6$ Hz.

Since we want the loop gain magnitude to be unity at this frequency, we have

$$|T_{PD}(f_{135})| = 1 = \frac{5 \times 10^5}{\sqrt{1 + \left(\dfrac{10^6}{f_{PD}}\right)^2}\sqrt{1 + \left(\dfrac{10^6}{10^6}\right)^2}\sqrt{1 + \left(\dfrac{10^6}{10^7}\right)^2}\sqrt{1 + \left(\dfrac{10^6}{10^8}\right)^2}}$$

or

$$1 = \frac{5 \times 10^5}{\sqrt{1 + \left(\dfrac{10^6}{f_{PD}}\right)^2}(1.414)(1.0)(1.0)}$$

Solving for f_{PD} yields

$$f_{PD} = 2.83\,\text{Hz}$$

Comment: With high-gain amplifiers, the dominant pole must be at a very low frequency to ensure stability of the feedback circuit.

Problem-Solving Technique: Frequency Compensation

1. To stabilize a circuit, insert a dominant pole or move an existing pole to a dominant pole position (see next section). Assume that the dominant pole frequency is small. Determine the frequency of the resulting loop gain function to achieve the required phase margin.

2. Set the magnitude of the loop gain function equal to unity at the frequency determined in step 1 to find the required dominant pole frequency.

3. To actually achieve the required dominant pole frequency in the circuit, a number of techniques are available (for example, see Miller compensation).

One disadvantage of this frequency compensation method is that the loop gain magnitude, and in turn the open-loop gain magnitude, is drastically reduced over a very wide frequency range. This affects the closed-loop response of the feedback amplifier. However, the advantage of maintaining a stable amplifier greatly outweighs the disadvantage of a reduced gain, demonstrating another trade-off in design criteria.

12.10.2   Closed-Loop Frequency Response

Inserting a dominant pole to obtain the open-loop characteristics (dotted lines, Figure 12.59) is not as extreme or devastating to the circuit as it might first appear. Amplifiers are normally used in a closed-loop configuration, for which we briefly considered the bandwidth extension, in Section 12.2.3.

For the region in which the frequency response is characterized by the dominant pole, the open-loop amplifier gain is

$$A(f) = \frac{A_o}{1 + j\dfrac{f}{f_{PD}}} \qquad\qquad (12.122)$$

where A_o is the low-frequency gain and f_{PD} is the dominant-pole frequency. The feedback amplifier closed-loop gain can be expressed as

$$A_f(f) = \frac{A(f)}{(1 + \beta A(f))} \qquad\qquad (12.123)$$

where β is the feedback transfer ratio, which is assumed to be independent of frequency. Substituting Equation (12.122) into (12.123), we can write the closed-loop gain as

$$A_f(f) = \frac{A_o}{(1 + \beta A_o)} \times \frac{1}{1 + j\dfrac{f}{f_{PD}(1 + \beta A_o)}} \qquad\qquad (12.124)$$

The term $A_o/(1 + \beta A_o)$ is the closed-loop low-frequency gain, and $f_{PD}(1 + \beta A_o) = f_C$ is the 3 dB frequency of the closed-loop system.

Figure 12.60 shows the Bode plot of the gain magnitude for the open-loop parameters $A_o = 10^6$ and $f_{PD} = 10$ Hz, at several feedback transfer ratios. As the closed-loop gain decreases, the bandwidth increases. As previously determined, the gain–bandwidth product is essentially a constant.

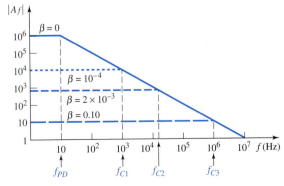

Figure 12.60   Bode plot, gain magnitude for open-loop and three closed-loop conditions

Example   12.22   **Objective:**   Determine the shift in the 3 dB frequency when an amplifier is operated in a closed-loop system.

Consider an amplifier with a low-frequency open-loop gain of $A_o = 10^6$ and an open-loop 3 dB frequency of $f_{PD} = 10$ Hz. The feedback transfer ratio is $\beta = 0.01$.

Solution: The low-frequency closed-loop gain is

$$A_f(0) = \frac{A_o}{(1 + \beta A_o)} = \frac{10^6}{1 + (0.01)(10^6)} \cong 100$$

From Equation (12.124), the closed-loop 3 dB frequency is

$$f_C = f_{PD}(1 + \beta A_o) = (10)[1 + (0.01)(10^6)]$$

or

$$f_C \cong 10^5 \, \text{Hz} = 100 \, \text{kHz}$$

Comment: Even though the open-loop 3 dB frequency is only 10 Hz, the closed-loop bandwidth is extended to 100 kHz. This effect is due to the fact that the gain–bandwidth product is a constant.

12.10.3   Miller Compensation

As previously discussed, an op-amp consists of three stages, with each stage normally responsible for one of the loop gain poles. Assume, for purposes of discussion, that the first pole f_{P1} is created by the capacitance effects in the second gain stage. Instead of adding a fourth dominant pole to achieve a stable system, we can move pole f_{P1} to a low frequency. This can be done by increasing the effective input capacitance to the gain stage.

Previously in Chapter 7, we determined that the effective Miller input capacitance to a transistor amplifier is a feedback capacitance multiplied by the magnitude of the gain of the amplifier stage. We can use this Miller multiplication factor to stabilize a feedback system. The three-stage op-amp circuit is shown in Figure 12.61. The second stage, an inverting amplifier, has a feedback capacitor connected between the output and input. This capacitor C_F is called a **compensation capacitor**.

Figure 12.61   Three-stage amplifier, including Miller compensation capacitor

The effective input Miller capacitance is

$$C_M = C_F(1 + A) \tag{12.125}$$

Since the gain of the second stage is large, the equivalent Miller capacitance will normally be very large. The pole introduced by the second stage is approximately

$$f_{P1} = \frac{1}{2\pi R_2 C_M}$$

(12.126)

where R_2 is the effective resistance between the amplifier input node and ground. Resistance R_2, then, is the parallel combination of the input resistance to the amplifier and the output resistance of the diff-amp stage.

Example · 12.23   **Objective:** Determine the pole of the gain stage that includes a feedback capacitor.

Consider a gain stage with an amplification $A = 10^3$, a feedback capacitor $C_F = 30\,\text{pF}$, and a resistance $R_2 = 5 \times 10^5\,\Omega$.

Solution: The effective input Miller capacitance is

$$C_M = C_F(1 + A) \cong (30)(1000)\,\text{pF} = 3 \times 10^{-8}\,\text{F}$$

The dominant-pole frequency is therefore

$$f_{P1} = \frac{1}{2\pi R_2 C_M} = \frac{1}{2\pi(5 \times 10^5)(3 \times 10^{-8})} = 10.6\,\text{Hz}$$

Comment: The pole of the second stage can be moved to a significantly lower frequency by using the Miller effect.

The effect of moving pole f_{P1}, using the Miller compensation technique, is shown in Figure 12.62. We assume at this point that the other two poles f_{P2} and f_{P3} are not affected. Moving the pole f_{P1} to f'_{P1} means that the frequency at which $|T(f)| = 1$ is lower, and that the phase is $|\phi| < 180°$, which means that the amplifier is stabilized.

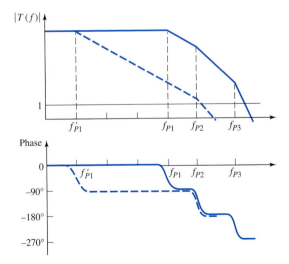

Figure 12.62 Bode plots of loop gain for three-stage amplifier, before (solid curves) and after (dotted curves) incorporating Miller compensation capacitor: (a) magnitude and (b) phase

A detailed analysis of the system using Miller compensation shows that pole f_{P2} does not remain constant; it increases. This phenomenon is called **pole-splitting**. The increase in f_{P2} is actually beneficial, because it increases the phase margin, or the frequency at which a particular phase margin is achieved.

Test Your Understanding

12.31 Consider a three-pole amplifier with a loop gain function given by

$$T(f) = \frac{10^5}{\left(1 + j\dfrac{f}{5 \times 10^5}\right)\left(1 + j\dfrac{f}{10^7}\right)\left(1 + j\dfrac{f}{5 \times 10^8}\right)}$$

Stabilize the circuit by inserting a new dominant pole. Assume the original poles are not altered. At what frequency must the new pole be placed to achieve a phase margin of 45 degrees? (Ans. $f_{PD} = 7.07\,\text{Hz}$)

12.32 The loop gain function for an amplifier is described in Exercise 12.31. To stabilize the circuit, move the first pole $f_{P1} = 5 \times 10^5$ Hz by introducing a compensation capacitor. Assume the second pole remains fixed. Determine the frequency to which the first pole must be moved to achieve a phase margin of 45 degrees. (Ans. $f_{PD} = 141\,\text{Hz}$)

12.33 A dc amplifier has a single-pole response with a pole frequency of $f_{PD} = 100$ Hz and a low-frequency gain of $A_o = 2 \times 10^5$. The amplifier is operated in a closed-loop system with $\beta = 0.05$. Find the closed-loop low-frequency gain and bandwidth. (Ans. $A_f(0) \cong 20$, $f_C \cong 1\,\text{MHz}$)

12.11   SUMMARY

- In a feedback circuit, a portion of the output signal is fed back to the input and combined with the input signal. In negative feedback, a portion of the output signal is subtracted from the input signal. In positive feedback, a portion of the output signal is added to the input signal.

- An important advantage of negative feedback is that the closed-loop amplifier gain is essentially independent of individual transistor parameters and is a function only of the feedback elements.

- Negative feedback increases bandwidth, may increase the signal-to-noise ratio, reduces nonlinear distortion, and controls input and output impedance values at the expense of reduced gain magnitude.

- A series input connection is used when the input signal is a voltage, and a shunt input connection is used when the input signal is a current. A series output connection is used when the output signal is a current, and a shunt output connection is used when the output signal is a voltage.

- The loop gain factor of a feedback amplifier is defined as $T = A\beta$, which is dimensionless and where A is the gain of the basic amplifier and β is the feedback factor. The loop gain is a function of frequency and is complex when the input capacitance of each transistor stage is taken into account.

- A three-stage negative feedback amplifier is guaranteed to be stable if, at the frequency for which the phase of the loop gain is -180 degrees, the magnitude is less than unity.

- A common technique of frequency compensation utilizes the Miller multiplication effect by incorporating a feedback capacitor across, usually, the second stage of the basic amplifier.

CHECKPOINT

After studying this chapter, the reader should have the ability to:

✔ Describe some of the advantages and disadvantages of negative feedback. (Section 12.2)

✔ Discuss the general characteristics of the four basic feedback configurations in terms of input and output signals and input and output resistances. (Section 12.3)

✔ Analyze feedback circuits. (Sections 12.4–12.7)

✔ Design a feedback circuit given the input signal and desired output signal. (Sections 12.4–12.7)

✔ Determine the loop gain of a feedback circuit. (Section 12.8)

✔ Determine whether or not a three-stage feedback amplifier is stable. (Section 12.9)

✔ Stabilize a three-stage amplifier using frequency compensation techniques. (Section 12.10)

REVIEW QUESTIONS

1. What are the two general types of feedback and what are the advantages and disadvantages of each type?

2. Write the ideal form of the general feedback transfer function.

3. Define the loop gain factor.

4. What is the difference between open-loop gain and closed-loop gain?

5. Describe what is meant by the terms (a) gain sensitivity and (b) bandwidth extension.

6. Sketch an ideal series input connection. What is the input signal?

7. Sketch an ideal shunt input connection. What is the input signal?

8. Sketch an ideal series output connection. What is the output signal?

9. Sketch an ideal shunt output connection. What is the output signal?

10. Is the input resistance of a series input connection smaller or larger than that of the basic amplifier? Explain why from the input connection.

11. Is the input resistance of a shunt input connection smaller or larger than that of the basic amplifier? Explain why from the input connection.

12. Is the output resistance of a series output connection smaller or larger than that of the basic amplifier? Explain why from the output connection.

13. Is the output resistance of a shunt output connection smaller or larger than that of the basic amplifier. Explain why from the output connection.

14. Describe the characteristics of a voltage amplifier.

15. Describe the characteristics of a current amplifier.

16. Describe the characteristics of a transconductance amplifier.

17. Describe the characteristics of a transresistance amplifier.

18. Consider a noninverting op-amp circuit. Describe the type of input and output feedback connections.

19. Consider an inverting op-amp circuit. Describe the type of input and output feedback connections.

20. What is the Nyquist stability criterion for a feedback amplifier?

21. Using Bode plots, describe the conditions of stability and instability in a feedback amplifier.
22. What is phase margin?
23. What is meant by frequency compensation?
24. What is a dominant pole?
25. What is a common technique of frequency compensation in a feedback amplifier?

PROBLEMS

Section 12.2   Basic Feedback Concepts

12.1 A negative feedback amplifier has a closed-loop gain of $A_f = 80$ and an open-loop gain of $A = 10^5$. (a) What is the feedback transfer function β? (b) If the open-loop gain decreases by 20 percent, determine the percent change in the closed-loop gain. What is the new value of A_f? (c) Repeat parts (a) and (b) for $A = 10^3$.

12.2 (a) A feedback amplifier is connected as shown in Figure P12.2. Each basic amplifier stage has an open-loop gain of $A = 10$. The closed-loop gain is $A_f = 100$. Determine the required feedback transfer function β. (b) If the gain of each stage increases by 10 percent, determine the percent change in the closed-loop voltage gain.

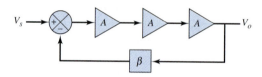

Figure P12.2

12.3 Three voltage amplifiers are in cascade as shown in Figure P12.3 with various amplification factors. The 180 degree phase shift for negative feedback actually occurs in the basic amplifier itself. (a) Determine the value of β such that the closed-loop voltage gain is $A_{vf} = V_o/V_s = -120$. (b) Using the results of part (a), determine the percent change in A_{vf} if each individual amplifier gain decreases by 10 percent.

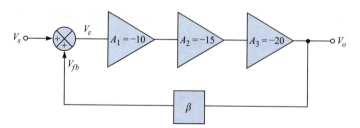

Figure P12.3

12.4 In a voltage-follower application, the feedback transfer function is $\beta = 1$ and the ideal closed-loop voltage gain is $A_f = 1$. Determine the magnitude of the open-loop voltage gain A such that the closed-loop gain in an actual feedback circuit is within 0.02 percent of the ideal value (see Equation 12.5).

12.5 An op-amp has an open-loop low-frequency gain of $A = 10^5$ and an open-loop 3 dB frequency $f_H = 4\,\text{Hz}$. If an inverting amplifier with a closed-loop low-frequency gain of $|A_{vf}| = 50$ uses this op-amp, determine the closed-loop bandwidth.

12.6 (a) Determine the closed-loop bandwidth of a noninverting amplifier with a gain of 50. The op-amp has the characterisics described in Problem 12.5. (b) If the noninverting amplifier gain is reduced to 10, determine the bandwidth.

12.7 An inverting amplifier uses an op-amp with an open-loop 3 dB frequency of 5 Hz, and has a gain of $|A_{vf}| = 50$ and a bandwidth of 20 kHz. Determine the required open-loop low-frequency op-amp gain.

12.8 Consider two open-loop amplifiers in cascade, with a noise signal generated between the two amplifiers as in Figure 12.3(a). Assume the amplification of the first stage is $A_2 = 100$ and that of the second stage is $A_1 = 1$. If $V_{\text{in}} = 10\,\text{mV}$ and $V_n = 1\,\text{mV}$, determine the signal-to-noise ratio at the output.

12.9 Two feedback configurations are shown in Figures P12.9(a) and (b). At low input voltages, the two gains are $A_1 = A_2 = 90$ and at higher input voltages, the gains change to $A_1 = A_2 = 60$. Determine the change in closed-loop gain, $A_f = V_o/V_i$, for the two feedback circuits. (See Figure 12.4.) Which feedback configuration will result in less distortion in the output signal?

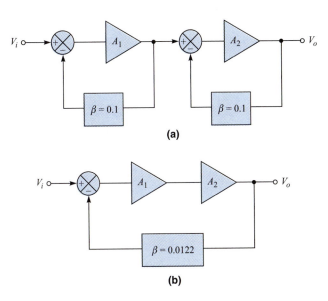

(a)

(b)

Figure P12.9

D12.10 Determine the type of feedback configuration that should be used in a design to achieve the following objectives: (a) low input resistance and low output resistance, (b) high input resistance and high output resistance, (c) low input resistance and high output resistance, and (d) high input resistance and low output resistance.

12.11 Consider a series of amplifiers and feedback circuits connected in the ideal feedback configurations. In each case the input resistance to the basic amplifier is $R_i = 10\,\text{k}\Omega$, the output resistance of the basic amplifier is $R_o = 1\,\text{k}\Omega$, and the loop gain is $T = 10^4$. (a) Determine the maximum possible input resistance and minimum possible input resistance to the feedback circuit. (b) Determine the maximum possible output resistance and minimum possible output resistance to the feedback circuit.

D12.12 A compound transconductance amplifier is to be designed by connecting two basic feedback amplifiers in cascade. What two amplifiers should be connected in cascade to form the compound circuit? Is there more than one possible design?

Section 12.3   Ideal Feedback Topologies

12.13 Consider the noninverting op-amp circuit in Figure P12.13. The input resistance of the op-amp is $R_i = \infty$ and the output resistance is $R_o = 0$, but the op-amp has a finite gain A. (a) Write the closed-loop transfer function in the form

$$A_{vf} = \frac{v_o}{v_s} = \frac{A}{(1 + \beta A)}$$

(b) What is the expression for β? (c) If $A = 10^5$ and $A_{vf} = 20$, what is the required β and R_2/R_1? (d) If A decreases by 10 percent, what is the percent change in A_{vf}?

Figure P12.13

12.14 For the noninverting op-amp circuit in Figure P12.13, the parameters are: $A = 10^5$, $A_{vf} = 20$, $R_i = 100\,\text{k}\Omega$, and $R_o = 100\,\Omega$. Determine the closed-loop input and output resistances, R_{if} and R_{of}, respectively.

12.15 Consider the op-amp circuit in Figure P12.15. The op-amp has a finite gain, so that $i_o = Ai_\varepsilon$, and a zero output impedance. (a) Write the closed-loop transfer function in the form

$$A_{if} = \frac{i_o}{i_s} = \frac{A_i}{(1 + \beta_i A_i)}$$

(b) What is the expression for β_i? (c) If $A_i = 10^5$ and $A_{if} = 25$, what is the required β_i and R_F/R_3? (d) If A_i decreases by 15 percent, what is the percent change in A_{if}?

12.16 An op-amp circuit is shown in Figure P12.15. Its parameters are as described in Problem 12.15, except that $R_i = 2\,\text{k}\Omega$ and $R_o = 20\,\text{k}\Omega$. Determine the closed-loop input and output resistances, R_{if} and R_{of}, respectively.

Figure P12.15

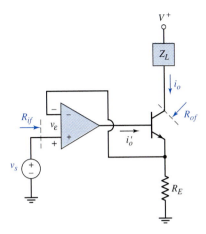

Figure P12.17

12.17 Consider the circuit in Figure P12.17. The input resistance of the op-amp is $R_i = \infty$ and the output resistance is $R_o = 0$. The op-amp has a finite gain, so that $i_o' = A_g v_\varepsilon$. The current gain of the transistor is h_{FE}. (a) Write the closed-loop transfer function in the form

$$A_{gf} = \frac{i_o}{v_s} = \frac{A_g}{(1 + \beta_z A_g)}$$

where A_g is the open-loop gain of the system. (b) What is the expression for β_z? (c) If $A_g = 5 \times 10^5$ mS and $A_{gf} = 10$ mS, what is the required β_z and R_E? (d) If A_g increases by 10 percent, what is the corresponding percent change in A_{gf}?

12.18 The circuit shown in Figure P12.17 has the same parameters as described in Problem 12.17, except that $R_i = 20\,\text{k}\Omega$ and $R_o = 50\,\text{k}\Omega$. Determine the closed-loop input and output resistances, R_{if} and R_{of}, respectively.

12.19 Consider the current-to-voltage converter circuit shown in Figure P12.19. The input resistance R_{if} is assumed to be small, the output resistance is $R_o = 0$, and the op-amp gain A_z is large. (a) Write the closed-loop transfer function in the form

$$A_{zf} = \frac{v_o}{i_s} = \frac{A_z}{(1 + \beta_g A_z)}$$

(b) What is the expression for β_g? (c) If $A_z = 5 \times 10^6\,\Omega$ and $A_{zf} = 5 \times 10^4\,\Omega$, what is the required β_g and R_F? (d) If A_z decreases by 10 percent, what is the percent change in A_{zf}?

12.20 For the current-to-voltage converter circuit in Figure P12.19, the parameters are as described in Problem 12.19. If $R_i = 10\,\text{k}\Omega$, determine the closed-loop input resistance R_{if}.

Figure P12.19

Section 12.4   Voltage (Series–Shunt) Amplifiers

*12.21 Consider the voltage amplifier in Figure P12.21. The op-amp parameters are $A_v = 5 \times 10^3$, $R_i = 10\,\text{k}\Omega$, and $R_o = 1\,\text{k}\Omega$, and the transistor parameters are $h_{FE} = 100$ and $V_A = 80\,\text{V}$. Determine A_{vf}, R_{if}, and R_{of}.

Figure P12.21

Figure P12.22

12.22 The circuit in Figure P12.22 is an example of a series–shunt feedback circuit. Assume the transistor parameters are: $h_{FE} = 100$, $V_{BE}(\text{on}) = 0.7\,\text{V}$, and $V_A = \infty$. (a) Determine the quiescent collector currents and the dc voltage at the output. (b) Determine the small-signal voltage gain $A_{vf} = v_o/v_i$.

12.23 Consider the series–shunt feedback circuit in Figure P12.23, with transistor parameters: $h_{FE} = 120$, $V_{BE}(\text{on}) = 0.7\,\text{V}$, and $V_A = \infty$. (a) Determine the small-signal parameters for Q_1, Q_2, and Q_3. Using nodal analysis, determine: (b) the small-signal voltage gain $A_{vf} = v_o/v_i$, (c) the input resistance R_{if}, and (d) the output resistance R_{of}.

Figure P12.23

12.24 Consider the BiCMOS circuit in Figure P12.24. The transistor parameters are: $K_n = 0.2\,\mathrm{mA/V^2}$, $V_{TN} = 1\,\mathrm{V}$, $\lambda = 0$ for M_1; and $h_{FE} = 100$, $V_{EB}(\mathrm{on}) = 0.7\,\mathrm{V}$, $V_A = \infty$ for Q_2. (a) Determine the small-signal parameters for M_1 and Q_2. (b) Find the small-signal voltage gain $A_{vf} = v_o/v_i$. (c) Determine the output resistance R_{of}.

Figure P12.24

12.25 Figure P12.25 shows a basic source-follower circuit. Assume the transistor is biased such that $I_{DQ} = 0.5\,\mathrm{mA}$. Assume the transistor parameters are $V_{TN} = 1\,\mathrm{V}$ and $\lambda = 0$, and let $R_S = 2\,\mathrm{k\Omega}$. (a) If the transistor conduction parameter is $K_n = 0.5\,\mathrm{mA/V^2}$, determine $A_{vf} = v_o/v_i$ and R_{of}. (b) Determine the percent change in A_{vf} and R_{of} if the conduction parameter increases to $K_n = 0.8\,\mathrm{mA/V^2}$.

12.26 The transistor parameters for the circuit in Figure P12.26 are: $h_{FE} = 50$, $V_{BE}(\mathrm{on}) = 0.7\,\mathrm{V}$, and $V_A = \infty$. Using nodal analysis, determine the closed-loop small-signal voltage gain $A_{vf} = v_o/v_s$ at the midband frequency.

Figure P12.25   **Figure P12.26**

*D12.27 Design a discrete transistor feedback voltage amplifier to provide a voltage gain of 50. Assume the available transistors have parameters: $h_{FE} = 120$ and $V_A = \infty$. The signal voltage source has a source resistance of $R_S = 2\,k\Omega$ and the load is $R_L = 3\,k\Omega$. Verify the design with a computer simulation. Determine R_{if} and R_{of}.

*RD12.28 Redesign the feedback circuit in Figure P12.22 using MOSFETs to provide a voltage gain of $A_{vf} = 10$. Assume transistor parameters of $V_{TN} = 2\,V$, $k_n' = 80\,\mu A/V^2$, and $\lambda = 0$.

Section 12.5   Current (Shunt–Series) Amplifiers

D12.29 An op-amp current gain amplifier (shunt–series configuration) is shown in Figure P12.29. Design the circuit such that the load current is $I_o = 20\,mA$ when the input current is $I_s = 200\,\mu A$.

Figure P12.29

Figure P12.30

12.30 The circuit in Figure P12.30 has transistor parameters: $h_{FE} = 100$, $V_{BE}(\text{on}) = 0.7\,V$, and $V_A = \infty$. (a) From the quiescent values, determine the small-signal parameters for Q_1 and Q_2. (b) Using nodal analysis, determine the small-signal closed-loop current gain $A_{if} = i_o/i_s$. (c) Using nodal analysis, find the input resistance R_{if}.

12.31 (a) Using the small-signal equivalent circuit in Figure 12.25 for the circuit in Figure 12.24(a), derive the expression for the small-signal current gain $A_{if} = I_o/I_s$. (b) Using the circuit parameters given in Figure 12.24(a) and assuming transistor parameters $h_{FE} = 100$ and $V_A = \infty$, calculate the value of A_{if}. Compare this answer with the results of Example 12.9.

*12.32 The circuit in Figure P12.32 is an example of a shunt–series feedback circuit. A signal proportional to the output current is fed back to the shunt connection at the base of Q_1. However, the circuit may be used as a voltage amplifier. Assume transistor parameters of $h_{FE} = 120$, $V_{BE}(\text{on}) = 0.7\,V$, and $V_A = \infty$. (a) Determine the small-signal parameters for Q_1 and Q_2. (b) Using nodal analysis, determine the small-signal voltage gain $A_v = v_o/v_s$.

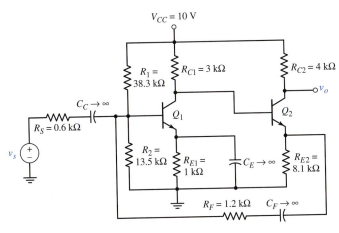

Figure P12.32

12.33 Consider the circuit in Figure P12.32 with transistor parameters, $h_{FE} = 120$, $V_{BE}(\text{on}) = 0.7\,\text{V}$, and $V_A = \infty$. Using nodal analysis, determine the input resistance R_{if}.

12.34 For the transistors in the circuit in Figure P12.34, the parameters are: $h_{FE} = 50$, $V_{BE}(\text{on}) = 0.7\,\text{V}$, and $V_A = \infty$. Using nodal analysis, determine the closed-loop current gain $A_{if} = i_o/i_s$.

Figure P12.34

*D12.35 Design a discrete transistor feedback current amplifier to provide a current gain of 30. Assume the available transistors have parameters $h_{FE} = 120$ and $V_A = \infty$. The signal current source has a source resistance of $R_S = 25\,\text{k}\Omega$ and the load is $R_L = 500\,\Omega$. Verify the design with a computer simulation. Determine R_{if} and R_{of}.

Section 12.6   Transconductance (Series–Series) Amplifiers

12.36 The circuit in Figure P12.36 is the ac equivalent circuit of a series–series feedback amplifier. Assume that the bias circuit, which is not shown, results in quiescent collector currents of $I_{C1} = 0.5\,\text{mA}$, $I_{C2} = 1\,\text{mA}$, and $I_{C3} = 2\,\text{mA}$. Assume transistor

Figure P12.36

parameters of $h_{FE} = 120$ and $r_o = \infty$. Determine the transconductance transfer function $A_{gf} = I_o/V_s$.

RD12.37 Using a computer simulation analysis, redesign the circuit in Figure P12.36 by changing the value of R_F to achieve a transconductance gain of $A_{gf} = I_o/V_s = 120$ mA/V.

12.38 In the circuit in Figure P12.38, the transistor parameters are: $h_{FE} = 100$, $V_{BE}(\text{on}) = 0.7\,\text{V}$, and $V_A = \infty$. Determine the transconductance transfer function $A_{gf} = i_o/v_s$.

Figure P12.38

D12.39 Design a feedback amplifier to supply a current to an LED. The diode current should be $I_o = 10^{-3}V_i$, where V_i is the amplifier input voltage, which has a range of 0 to 10 V. The voltage source has an output resistance of $R_S = 1\,\text{k}\Omega$. The op-amp parameters are $R_i = 5\,\text{k}\Omega$, $R_o = 50\,\Omega$, and the low-frequency open-loop voltage gain is 5×10^3. Determine the gain, input resistance, and output resistance, from a computer simulation.

Section 12.7   Transresistance (Shunt–Shunt) Amplifiers

12.40 Consider the common-emitter circuit in Figure P12.40, driven by an ideal signal current source. The transistor parameters are: $h_{FE} = 50$, $V_{EB}(\text{on}) = 0.7\,\text{V}$, and $V_A = 100\,\text{V}$. (a) Determine the input and output resistances, R_{if} and R_{of}, respectively. (b) Find the transresistance transfer function $A_{zf} = v_o/i_s$. (c) What happens in the feedback network if the capacitance is finite?

Figure P12.40

12.41 For the circuit shown in Figure P12.41, the transistor parameters are: $V_{TN} = 2\,\text{V}$, $K_n = 0.20\,\text{mA/V}^2$, and $\lambda = 0$. Determine: (a) the voltage gain $A_v = V_o/V_s$, (b) the transresistance transfer function $A_{zf} = V_o/I_s$, (c) the input impedance R_{if}, and (d) the output impedance R_{of}.

Figure P12.41

12.42 Consider the circuit in Figure P12.41. The transistor parameters are $V_{TN} = 1.5\,\text{V}$ and $\lambda = 0$. Determine the required value of transconductance g_m such that the magnitude of the closed-loop voltage gain is within 10 percent of the ideal value when $g_m \to \infty$.

12.43 For the circuit in Figure P12.43, the transistor parameters are: $h_{FE} = 150$, $V_{BE}(\text{on}) = 0.7\,\text{V}$, and $V_A = \infty$. Determine the value of R_F that will result in a closed-loop voltage gain of $A_v = V_o/V_s = -5.0$.

Figure P12.43

12.44 Consider the three-stage cascade feedback circuit in Figure 12.40. Each stage corresponds to the circuit in Figure P12.44, with transistor parameters: $h_{FE} = 180$, $V_{BE}(\text{on}) = 0.7\,\text{V}$, and $V_A = \infty$. The source resistor is $R_S = 10\,\text{k}\Omega$, and the load resistor is $R_L = 4\,\text{k}\Omega$. Determine the value of R_F such that the closed-loop gain is $A_v = v_o/v_i = -80$.

Figure P12.44   **Figure P12.45**

12.45 The op-amp in the circuit in Figure P12.45 has an open-loop differential voltage gain of $A_d = 10^4$. Neglect the current into the op-amp, and assume the output resistance looking back into the op-amp is zero. Determine: (a) the closed-loop voltage gain $A_v = V_o/V_s$, (b) the input resistance R_{if}, and (c) the output resistance R_{of}.

D12.46 Design a feedback transresistance amplifier using an op-amp with parameters $R_i = 10\,\text{k}\Omega$, $R_o = 100\,\Omega$, and a low-frequency open-loop gain of $A_v = 10^4$ to produce a gain of $5\,\text{k}\Omega$. The source resistance is $R_S = 500\,\Omega$ and the load resistance is $R_L = 2\,\text{k}\Omega$. Determine the actual gain, input resistance, and output resistance using a computer simulation.

Section 12.8 Loop Gain

12.47 The op-amp in Figure 12.20 has an open-loop differential input resistance R_i, an open-loop current gain A_i, and a zero output resistance. Break the feedback loop at an appropriate point, and derive the expression for the loop gain.

12.48 The small-signal parameters of the transistors in the circuit in Figure P12.23 are h_{FE} and $V_A = \infty$. Derive the expression for the loop gain.

12.49 Determine the loop gain T for the circuit in Figure P12.30. The transistor parameters are: $h_{FE} = 100$, $V_{BE}(\text{on}) = 0.7\,\text{V}$, and $V_A = \infty$.

12.50 The transistor parameters for the circuit shown in Figure P12.40 are: $h_{FE} = 50$, $V_{BE}(\text{on}) = 0.7\,\text{V}$, and $V_A = 100\,\text{V}$. Find the loop gain T.

Section 12.9 Stability of the Feedback Circuit

12.51 A three-pole feedback amplifier has a loop gain given by

$$T(f) = \frac{\beta(10^5)}{\left(1 + j\dfrac{f}{5 \times 10^2}\right)\left(1 + j\dfrac{f}{10^4}\right)^2}$$

(a) Determine the frequency f_{180} at which the phase is -180 degrees. (b) At the frequency f_{180}, determine the value of β such that $|T(f_{180})| = 1$.

12.52 The open-loop gain of an amplifier is given by

$$A = \frac{5 \times 10^3}{\left(1 + j\dfrac{f}{10^4}\right)\left(1 + j\dfrac{f}{10^5}\right)^2}$$

Assuming the feedback function is not a function of frequency, determine the frequency at which the phase is 180 degrees. Determine the value of the feedback transfer function at which the amplifier can break into oscillation.

12.53 A loop gain function is given by

$$T(f) = \frac{\beta(10^3)}{\left(1 + j\dfrac{f}{10^4}\right)\left(1 + j\dfrac{f}{5 \times 10^4}\right)\left(1 + j\dfrac{f}{10^5}\right)}$$

Sketch the Nyquist plot for: (a) $\beta = 0.005$, and (b) $\beta = 0.05$. (c) Is the system stable or unstable in each case?

12.54 A three-pole feedback amplifier has a loop gain function given by

$$T(f) = \frac{\beta(5 \times 10^3)}{\left(1 + j\dfrac{f}{10^3}\right)^2\left(1 + j\dfrac{f}{5 \times 10^4}\right)}$$

(a) Sketch the Nyquist diagram for $\beta = 0.20$. (b) Determine the value of β that produces a phase margin of 80 degrees.

12.55   A three-pole feedback amplifier has a loop gain given by

$$T(f) = \frac{\beta(10^4)}{\left(1+j\dfrac{f}{10^3}\right)\left(1+j\dfrac{f}{10^4}\right)\left(1+j\dfrac{f}{10^5}\right)}$$

Sketch Bode plots of the loop gain magnitude and phase for: (a) $\beta = 0.005$, and (b) $\beta = 0.05$. (c) Is the system stable or unstable in each case? If the system is stable, what is the phase margin?

12.56   An amplifier with a low-frequency open-loop gain of 10^5 has poles at 5×10^4 Hz, 10^5 Hz, and 5×10^5 Hz. Determine the feedback transfer function β and the low-frequency closed-loop gain for which the phase margin is 60 degrees.

12.57   A two-pole loop gain function is given by

$$T(f) = \frac{\beta(10^3)}{\left(1+j\dfrac{f}{10^3}\right)^2}$$

(a) Determine the value of β that produces a phase margin of 60 degrees. (b) Using the results of part (a), sketch the Bode plots of the loop gain magnitude and phase.

12.58   The open-loop gain of an amplifier has pole frequencies at 10 kHz, 100 kHz, and 1 MHz. The low-frequency open-loop gain is 500 and the feedback transfer function is $\beta = 0.6$. Find the phase margin.

12.59   Sketch the Bode plots of the magnitude and phase of the function

$$T(f) = \frac{K}{\left(1+j\dfrac{f}{10^2}\right)^2}$$

for: (a) $K = 1$, and (b) $K = 10^3$.

12.60   Consider a four-pole feedback system with a loop gain given by

$$T(f) = \frac{\beta(10^3)}{\left(1+j\dfrac{f}{10^3}\right)\left(1+j\dfrac{f}{10^4}\right)\left(1+j\dfrac{f}{10^5}\right)\left(1+j\dfrac{f}{10^6}\right)}$$

Determine the value of β that produces a phase margin of 45 degrees.

Section 12.10   Frequency Compensation

12.61   A feedback amplifier has a low-frequency loop gain of 5000 and three poles at $f_{P1} = 300$ kHz, $f_{P2} = 2$ MHz, and $f_{P3} = 25$ MHz. A dominant pole is to be added such that the phase margin if 45 degrees. Assuming the original poles remain fixed, determine the dominant pole frequency.

12.62   A feedback amplifier with a compensation capacitor has a low-frequency loop gain of $T(0) = 100$ dB and poles at $f'_{P1} = 10$ Hz, $f_{P2} = 5$ MHz, and $f_{P3} = 10$ MHz. (a) Find the frequency at which $|T(f)| = 1$, and determine the phase margin. (b) If the frequency f'_{P1} is due to a compensation capacitor $C_F = 20$ pF, determine the new dominant pole frequency f'_{P1} and phase margin if the compensation capacitor is increased to $C_F = 75$ pF.

12.63   The equivalent circuit at the interface between the first and second stages of an op-amp is shown in Figure P12.63. The parameters are $R_{o1} = 500$ kΩ, $R_{i2} = 1$ MΩ, and $C_i = 2$ pF. (a) Determine the pole frequency for this part of the circuit. (b) Determine the additional Miller capacitance C_M that would need to be added so that the pole frequency is moved to $f_{PD} = 10$ Hz.

Figure P12.63

12.64 The loop gain function of a feedback amplifier has its first two poles at $f_{p1} = 2\,\text{MHz}$ and $f_{p2} = 12\,\text{MHz}$, and has a low-frequency gain of $|T(0)| = 80\,\text{dB}$. The amplifier is to be stabilized by moving the first pole by using Miller compensation. Assuming that the second pole f_{p2} remains fixed, find the frequency to which f_{p1} must be changed to produce a 45 degree phase margin.

12.65 A three-pole amplifier has its first two poles at $f_{P1} = 1\,\text{MHz}$ and $f_{P2} = 10\,\text{MHz}$, and has a low-frequency open-loop gain of $|A_o| = 80\,\text{dB}$. A dominant pole is to be inserted such that the closed-loop system remains stable when the closed-loop, low-frequency gain is $|A_f(0)| = 5$. Determine the dominant pole frequency assuming the initial poles remain constant.

12.66 The amplifier described in Problem 12.61 is to be stabilized by moving the first pole by using Miller compensation. Assuming f_{P2} remains constant, determine the frequency to which f_{P1} must be moved.

COMPUTER SIMULATION PROBLEMS

12.67 Using a computer analysis, investigate the loop gain factor for the circuit in Figure 12.24(a). Investigate the loop gain as a function of R_F and of h_{FE}.

12.68 Consider the multistage feedback circuit in Figure 12.40. Assume each stage corresponds to the circuit in Figure P12.44. Let $R_F = 200\,\text{k}\Omega$, $R_S = 10\,\text{k}\Omega$, and $R_L = 4\,\text{k}\Omega$. (a) Investigate the open-loop voltage gain $A_v = v_o/v_\varepsilon$ as a function of the individual transistor current gains h_{FE}. (b) Determine the required value of open-loop gain and transistor current gain needed to achieve a closed-loop gain that is within 2 percent of the ideal value.

12.69 Consider the circuit in Figure P12.32. From a computer analysis, determine the loop gain and the closed-loop transfer gain.

12.70 Consider the circuit in Figure 12.47 with parameters given in Example 12.18. The circuit is biased with $V_{CC} = 10\,\text{V}$, and it includes $0.5\,\text{k}\Omega$ emitter resistors. Insert coupling and emitter bypass capacitors where appropriate. (a) Determine the loop gain versus frequency characteristic. (b) Insert a compensation capacitor, $C_1 = 30\,\text{pF}$, between the collector and base of Q_2. Replot the loop gain versus frequency characteristic and determine whether the system is stable or unstable.

12.71 Consider the circuit in Figure 12.16 with parameters: $A_v = 10^4$, $R_i = 100\,\text{k}\Omega$, $R_o = 50\,\Omega$, $R_2 = 20\,\text{k}\Omega$, and $R_1 = 1\,\text{k}\Omega$. Determine the exact values of voltage gain A_{vf}, input resistance R_{if}, and output resistance R_{of}, from a computer analysis.

DESIGN PROBLEMS

[Note: Each design should be correlated with a computer simulation analysis.]

*D12.72   Redesign the circuit shown in Figure 12.45(a) to provide a loop gain of at least 100. What are the values of I_{CQ} and V_{CEQ}?

*D12.73   An op-amp has a low-frequency open-loop gain of 10^5 and a dominant-pole frequency of 5 Hz. Design a cascade of noninverting amplifiers with an overall minimum gain of 800 and a minimum bandwidth of 12 kHz.

*D12.74   An op-amp has a low-frequency open-loop gain of 5×10^4 and a dominant-pole frequency of 10 Hz. Using this op-amp, design a preamplifier system that can amplify the output of a microphone and produce a 1 V peak signal over a frequency range from 10 Hz to 15 kHz. The equivalent circuit of the microphone is a voltage source in series with an output resistance. The voltage source produces a 5 mV peak signal and the output resistance is 10 kΩ.

*D12.75   The equivalent circuit of a transducer that measures the speed of a motor is a current source in parallel with an output resistance. The current source produces an output of 1 μA per revolution of the motor and the output resistance is 50 kΩ. Design a discrete transistor circuit that produces a full-scale output of 5 V for a maximum motor speed of 60 revolutions per second. The nominal transistor current gain is $h_{FE} = 100$ with tolerances of ±20 percent. The accuracy of the output signal is to remain within ±1 percent.

13

Operational Amplifier Circuits

13.0 PREVIEW

Thus far, we have considered basic circuit configurations, such as the common emitter, emitter follower, and diff-amp, among others. We have discussed the basic concepts in design and analysis, including biasing techniques, frequency response, and feedback effects. In this chapter, we combine basic circuit configurations to form larger analog circuits that are fabricated as integrated circuits. Operational amplifiers are used extensively in electronic systems, so we concentrate on several forms of the operational amplifier circuit in this chapter.

We introduced the ideal op-amp in Chapter 9. Now, we analyze and design the circuitry of the op-amp, to determine how the various circuit configurations can be combined to form a nearly ideal op-amp.

The LM741 is an example of an all-bipolar general-purpose op-amp. Even though this op-amp is considered a classic, it still provides a good case study in which we perform a detailed analysis to determine both the dc and the small-signal characteristics of the circuit. Since the 741 is an internally compensated op-amp, we determine the dominant-pole frequency and the unity-gain bandwidth.

All-CMOS op-amps can be designed for special on-chip applications. In general, these op-amps are designed to drive other CMOS circuits, which form high capacitive loads. A goal of this chapter is to enable the reader to design CMOS op-amp circuits, including choosing transistor width-to-width ratios, to meet particular specifications, including power and gain parameters.

13.1 GENERAL OP-AMP CIRCUIT DESIGN

An operational amplifier, in general, is a three-stage circuit, as shown in Figure 13.1, and is fabricated as an integrated circuit. The first stage is a differential amplifier, the second stage provides additional voltage gain, and the third stage provides current gain and low output impedance. A feedback capacitor is often included in the second stage to provide frequency compensation as discussed in the last chapter. In some cases, in particular with MOSFET op-amp circuits, only the first two stages are used.

We have on numerous occasions made reference to the op-amp. In Chapter 9, we analyzed and designed op-amp circuits using the ideal op-amp

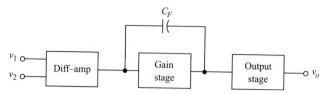

Figure 13.1   General block diagram of an operational amplifier

model. In Chapter 10, we introduced current-source biasing and introduced the active load. The differential amplifier, using current source biasing and active loads was considered in Chapter 11. We also introduced the bipolar Darlington pair in Chapter 11, which is often used as a second gain stage. Previously, in Chapter 8, we considered the class-AB output stage that is often used in operational amplifier circuits. These individual building blocks will now be combined to form the operational amplifier.

In Chapter 9, as mentioned, we analyzed and designed ideal op-amp circuits. Practical operational amplifiers, as we will see in this chapter, exhibit characteristics that deviate from the ideal characteristics. Once we have analyzed these practical op-amp circuits and determined some of their nonideal properties, we will then consider, in the next chapter, the effect of these nonideal characteristics on the op-amp circuits.

13.1.1   General Design Philosophy

All stages of the operational amplifier circuit are direct coupled. There are no coupling capacitors and there are also no bypass capacitors. These types of capacitors would require extremely large areas on the IC chip and hence are impractical. In addition, resistors whose values are over approximately $50\,k\Omega$ are avoided in ICs, since they also require large areas and introduce parasitic effects. Op-amp circuits are designed with transistors having matching characteristics.

We may begin to design a simple bipolar operational amplifier by using the knowledge gained in the previous chapters. Figure 13.2 shows the general configuration of the circuit. The first stage will be a differential pair, Q_1 and Q_2, biased with a Widlar current source, Q_3, Q_4, and R_2, and using a three-transistor active load. Assuming matched transistors, we expect the dc voltage at the collector of Q_6 to be two base-emitter voltage drops below the positive bias voltage. Therefore, the Darlington pair, Q_8 and Q_9, that forms the second stage should be properly biased. The bias current for Q_8 is supplied by the Widlar current source, Q_4, Q_{10}, and R_3. The output stage is the complementary push–pull, emitter-follower configuration of Q_{11} and Q_{12}. The crossover distortion is eliminated by including the diodes D_1 and D_2. The emitter-follower configuration provides low output resistance so that the op-amp can drive a load with minimal loading effect. By changing the value of R_3 slightly, the current through Q_{10} and Q_8 can be changed, which will change the collector–emitter voltages across these transistors. This part of the circuit then acts as a dc voltage shifter such that the output voltage, v_O, can be set equal to zero for zero input voltages.

From results that we have derived previously, we expect the differential-mode voltage gain of the first stage to be in the range of 10^2–10^3, depending on

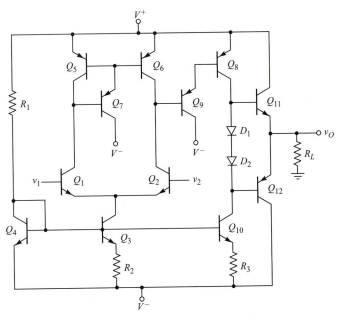

Figure 13.2 A simple bipolar operational amplifier

the specific transistor parameters and the voltage gain of the second stage to also be the range of 10^2–10^3. The voltage gain of the output stage, an emitter follower, is essentially unity. The overall voltage gain of the op-amp circuit is then expected to be in the range of 10^4–10^6. From our study in Chapter 9, this magnitude of voltage gain is required for the circuit to act essentially as an ideal op-amp.

The same op-amp configuration can be designed with MOS transistors. In general, as we have seen, BJT circuits have higher voltage gains, whereas MOSFET circuits have higher input resistances. So, whether a bipolar or MOSFET design is used depends to a large extent on the specific application of the op-amp.

13.1.2 Circuit Element Matching

Integrated circuit design is based directly on the ability to fabricate transistors on a chip that have nearly identical characteristics. In the analysis of current mirrors in Chapter 10 and differential amplifiers in Chapter 11, we assumed that transistors in a given circuit were matched. Transistors are **matched** when they have identical parameters. For bipolar transistors, the parameters are I_S, β, and V_A. Recall that I_S includes the electrical parameters of the semiconductor material as well as the cross-sectional area (geometry) of the base-emitter junction. For NMOS transistors, the parameters are V_{TN}, K_n, and λ_n, and for PMOS transistors, the same corresponding parameters must be identical. Again, recall that the parameter K_n contains semiconductor parameters as well as the width-to-length (geometry) of the transistor.

The absolute parameter values of transistors on an IC chip may vary substantially (on the order of ± 25 percent) from one IC chip to the next because of processing variations. However, the variation in parameter values of adjacent or nearby transistors on a given IC chip are usually within a fraction of a percent. In general, much of an amplifier design is based on the ratio of transistor parameters and on the ratio of resistor values rather than on the absolute values. For this reason, the operational amplifiers described in this chapter can be fabricated as ICs, but are almost impossible to fabricate with discrete circuit elements.

Test Your Understanding

*13.1 Using a computer simulation, determine the dc voltages and currents in the bipolar op-amp circuit in Figure 13.2. Use reasonable resistor values. Adjust the value of R_3 such that the output voltage is nearly zero for zero input voltages.

13.2 Consider the basic diff-amp with active load and current biasing in Figure 13.2. Using a computer simulation, investigate the change in the voltage at the collector of Q_2 as Q_1 and Q_2, and also Q_5 and Q_6, become slightly mismatched.

13.2 A BIPOLAR OPERATIONAL AMPLIFIER CIRCUIT

The **741 op-amp** has been produced since 1966 by many semiconductor device manufacturers. Since then, there have been many advances in op-amp design, but the 741 is still a widely used general-purpose op-amp. Even though the 741 is a fairly old design, it still provides a useful case study to describe the general circuit configuration and to perform a detailed dc and small-signal analysis. From the ac analysis, we determine the voltage gain and the frequency response of this circuit.

13.2.1 Circuit Description

Figure 13.3 shows the equivalent circuit of the 741 op-amp. For easier analysis, we break the overall circuit down into its basic circuits and consider each one individually.

As with most op-amps, this circuit consists of three stages: the input differential amplifier, the gain stage, and the output stage. Figure 13.3 also shows a separate bias circuit, which establishes the bias currents throughout the op-amp. Like most op-amps, the 741 is biased with both positive and negative supply voltages. This eliminates the need for input coupling capacitors, which in turn means that the circuit is also a dc amplifier. The dc output voltage is zero when the applied differential input signal is zero. Typical supply voltages are $V^+ = 15\,\text{V}$ and $V^- = -15\,\text{V}$, although input voltages as low as $\pm 5\,\text{V}$ can be used.

Input Diff-Amp

The input diff-amp stage is more complex than those previously covered. The input stage consists of transistors Q_1 through Q_7, with biasing established by transistors Q_8 through Q_{12}. The two input transistors Q_1 and Q_2 act as emitter

Figure 13.3 Equivalent circuit, 741 op-amp

followers, which results in a high differential input resistance. The differential output currents from Q_1 and Q_2 are the inputs to the common-base amplifier formed by Q_3 and Q_4, which provides a relatively large voltage gain.

Transistors Q_5, Q_6, and Q_7, with associated resistors R_1, R_2, and R_3, form the active load for the diff-amp. A single-sided output at the common collectors of Q_4 and Q_6 is the input signal to the following gain stage.

The dc output voltage at the collector of Q_6 is at a lower potential than the inputs at the bases of Q_1 and Q_2. As the signal passes through the op-amp, the dc voltage level shifts several times. By design, when the signal reaches the output terminal, the dc voltage should be zero if a zero differential input signal is applied. The two null terminals on the input stage are used to make appropriate adjustments to accomplish this design goal. The "null technique" and the corresponding portion of the circuit will be discussed in detail in the next chapter.

The dc current biasing is initiated through the diode-connected transistors Q_{12} and Q_{11} and resistor R_5. Transistors Q_{11} and Q_{10}, with resistor R_4, form a Widlar current source that establishes the bias currents in the common-base transistors Q_3 and Q_4, as well as the current mirror formed by Q_9 and Q_8.

Transistors Q_3 and Q_4 are lateral pnp devices, which refers to the fabrication process and the geometry of the transistors. Lateral pnp transistors provide added protection against voltage breakdown, although the current gain is smaller than in npn devices.

Figure 13.4(a) shows a basic common-emitter differential pair used as the input to a diff-amp. If the input voltage V_1 were to be connected to a supply voltage of 15 V, with V_2 at ground potential, then the B–E junction of Q_2 would be reverse biased by approximately 14.3 V. Since the breakdown voltage of an npn B–E junction is typically in the range of 3–6 V, transistor Q_2 in Figure 13.4(a) would probably enter breakdown and suffer permanent damage.

(a)   (b)

Figure 13.4   (a) Basic common-emitter differential pair, with a large differential voltage and (b) the 741 input stage, with a large differential voltage

By comparison, Figure 13.4(b) shows the input stage of the 741 op-amp with the same input voltages. The B–E junctions of Q_1 and Q_3 are forward biased, which means that the series combination of B–E junctions of Q_2 and Q_4 is reverse biased by approximately 13.6 V. The breakdown voltage of a lateral pnp B–E junction is typically on the order of 50 V, which means that for this input voltage polarity, the B–E junction of Q_4 provides the necessary breakdown protection for the input diff-amp stage.

Gain Stage

The second, or gain, stage consists of transistors Q_{16} and Q_{17}. Transistor Q_{16} operates as an emitter follower; therefore, the input resistance of the gain stage is large. As previously discussed, a large input resistance to the gain stage minimizes loading effects on the diff-amp stage.

Transistor Q_{13} is effectively two transistors connected in parallel, with common base and emitter terminals. The area of Q_{13A} is effectively one-fourth the area of Q_{12}, and the area of Q_{13B} is effectively three-fourths that of Q_{12}. Transistor Q_{13B} provides the bias current for Q_{17} and also acts as an active load to produce a high-voltage gain. Transistor Q_{17} operates in a common-emitter configuration; therefore, the voltage at the collector of Q_{17} is the input signal to the output stage. The signal undergoes another dc level shift as it goes through this gain stage.

The 741 is internally compensated by the feedback capacitor C_1 connected between the output and input terminals of the gain stage. This Miller compensation technique assures that the 741 op-amp forms stable feedback circuits.

Output Stage

The output stage of an op-amp should provide a low output resistance, as well as a current gain, if it is to drive relatively large load currents. The output stage is therefore a class-AB circuit consisting of the complementary emitter-follower pair Q_{14} and Q_{20}.

The output of the gain stage is connected to the base of Q_{22}, which operates as an emitter follower and provides a very high input resistance; the gain stage therefore suffers no significant loading effects due to the output stage. Transistor Q_{13A} provides a bias current for Q_{22}, as well as for Q_{18} and Q_{19}, which are used to establish a quiescent bias current in the output transistors Q_{14} and Q_{20}. Transistors Q_{15} and Q_{21} are referred to as short-circuit protection devices. These transistors are normally off; they conduct only if the output is inadvertently connected to ground, resulting in a very large output current. We will consider the characteristics of the output stage in Section 13.2.2.

An abbreviated data sheet for the 741 is shown in Table 13.1. During our discussions in this chapter, we will compare our analysis results to the values in the table. A more complete data sheet for the 741 op-amp is given in Appendix C.

Table 13.1   Data for 741 at $T = 300\,^\circ$K and supply voltage of ± 15 V

Parameter	Minimum	Typical	Maximum	Units
Input bias current		80	500	nA
Differential-mode input resistance	0.3	2.0		MΩ
Input capacitance		1.4		pF
Output short-circuit current		25		mA
Open-loop gain ($R_L \ge 2\,\text{k}\Omega$)	50,000	200,000		V/V
Output resistance		75		Ω
Unity-gain frequency		1		MHz

Test Your Understanding

13.3   The 741 op-amp in Figure 13.3 is biased at $V^+ = 15$ V and $V^- = -15$ V. Assume $V_{BE}(\text{npn}) = V_{EB}(\text{pnp}) = 0.6$ V. Determine the input common-mode voltage range, neglecting voltage drops across R_1 and R_2. (Ans. $-12.6 < v_{\text{in}}(\text{cm}) \le 14.4$ V)

13.4   (a) If the 741 op-amp in Figure 13.3 is biased at $V^+ = 15$ and $V^- = -15$ V, estimate the maximum and minimum output voltages such that the op-amp remains biased in its linear region. (b) Repeat part (a) if $V^+ = 5$ V and $V^- = -5$ V. (Ans. (a) $-13.2 \le v_O \le 13.8$ V (b) $-3.2 \le v_O \le 3.8$ V)

13.2.2   DC Analysis

In this section, we will analyze the dc characteristics of the 741 op-amp to determine the dc bias currents. We assume that both the noninverting and inverting input terminals are at ground potential, and that the dc supply voltages are $V^+ = 15\,\text{V}$ and $V^- = -15\,\text{V}$. As an approximation, we assume $V_{BE} = 0.6\,\text{V}$ for npn transistors and $V_{EB} = 0.6\,\text{V}$ for pnp transistors. In most dc calculations, we neglect dc base currents, although we include base current effects in a few specific cases.

Bias Circuit and Input Stage

Figure 13.5 shows the bias circuit and input stage portion of the 741 circuit. The reference current, which is established in the bias circuit branch composed of Q_{12}, Q_{11}, and R_5, is

$$I_{\text{REF}} = \frac{V^+ - V_{EB12} - V_{BE11} - V^-}{R_5} \tag{13.1}$$

Transistors Q_{11} and Q_{10} and resistor R_4 form a Widlar current source. Therefore, I_{C10} is determined from the relationship

$$I_{C10}R_4 = V_T \ln\left(\frac{I_{\text{REF}}}{I_{C10}}\right) \tag{13.2}$$

where V_T is the thermal voltage and Q_{10} and Q_{11} are assumed to be matched transistors.

Neglecting base currents, $I_{C8} = I_{C9} = I_{C10}$. The quiescent collector currents in Q_1 through Q_4 are then

$$I_{C1} = I_{C2} = I_{C3} = I_{C4} = \frac{I_{C10}}{2} \tag{13.3}$$

Assuming the dc currents in the input stage are exactly balanced, the dc voltage at the collector of Q_6, which is the input to the second stage, is the same as the dc voltage at the collector of Q_5. We can write

$$V_{C6} = V_{BE7} + V_{BE6} + I_{C6}R_2 + V^- \tag{13.4}$$

As previously discussed, the dc level shifts through the op-amp.

Example 13.1   **Objective:**   Calculate the dc currents in the bias circuit and input stage of the 741 op-amp.

The bias circuit and input stage are shown in Figure 13.5.

Solution: From Equation (13.1), the reference current is

$$I_{\text{REF}} = \frac{V^+ - V_{EB12} - V_{BE11} - V^-}{R_5} = \frac{15 - 0.6 - 0.6 - (-15)}{40} = 0.72\,\text{mA}$$

Current I_{C10} is found from Equation (13.2), as follows:

$$I_{C10}(5) = (0.026)\ln\left(\frac{0.72}{I_{C10}}\right)$$

By trial and error, we find that $I_{C10} = 19\,\mu\text{A}$. The bias currents in the input stage are then

$$I_{C1} = I_{C2} = I_{C3} = I_{C4} = 9.5\,\mu\text{A}$$

From Equation (13.4), the voltage at the collector of Q_6 is

$$V_{C6} = V_{BE7} + V_{BE6} + I_{C6}R_2 + V^- = 0.6 + 0.6 + (0.0095)(1) + (-15)$$

Figure 13.5 Bias circuit and input stage portion of 741 op-amp circuit

or

$$V_{C6} \cong -13.8 \, \text{V}$$

Comment: The bias currents in the input stage are quite small; the input base currents at the noninverting and inverting terminals are generally in the nanoampere range. Small bias currents mean that the differential input resistance is large.

The transistor current gain of the lateral pnp transistors Q_3, Q_4, Q_8, and Q_9 may be relatively small, which means that the base currents in these transistors may not be negligible. To determine the effect of the base currents, consider the expanded input stage shown in Figure 13.6. The base currents in the npn transistors are still assumed to be negligible. Current I_{C10} establishes the base currents in Q_3 and Q_4, which then establish the emitter currents designated as I. At the Q_8 collector, we have

$$2I = I_{C8} + \frac{2I_{C9}}{\beta_p} = I_{C9}\left(1 + \frac{2}{\beta_p}\right) \tag{13.5}$$

Since Q_8 and Q_9 are matched, $I_{C8} = I_{C9}$. Then,

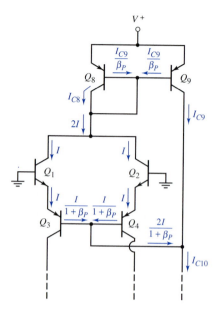

Figure 13.6   Expanded input stage, 741 op-amp, showing base currents

$$I_{C10} = \frac{2I}{1 + \beta_p} + I_{C9} = \frac{2I}{1 + \beta_p} + \frac{2I}{\left(1 + \dfrac{2}{\beta_p}\right)} = 2I\left[\frac{\beta_p^2 + 2\beta_p + 2}{\beta_p^2 + 3\beta_p^2 + 2}\right] \qquad (13.6)$$

Even if the pnp transistor base currents are not negligible, the bias currents in Q_1 and Q_2 are, from Equation (13.6), very nearly

$$I = \frac{I_{C10}}{2} \qquad (13.7)$$

This bias current is essentially the same as originally assumed in Equation (13.3).

Test Your Understanding

13.5   The current gain β_n of the npn transistors in the 741 op-amp input stage in Figure 13.5 is $\beta_n = 200$. Determine the input base currents to Q_1 and Q_2. (Ans. 47.5 nA)

*13.6   Consider the input stage and bias circuit in Figure 13.5, with $V^+ = 15$ V and $V^- = -15$ V. If $I_S = 10^{-14}$ A for each transistor, determine I_{REF}, V_{BE11}, V_{BE10}, and V_{BE6}. (Ans. $I_{REF} = 0.718$ mA, $V_{BE11} = 0.650$ V, $V_{BE10} = 0.556$ V, $V_{BE6} = 0.537$ V)

Gain Stage

Figure 13.7 shows the reference portion of the bias circuit and the gain stage. The reference current is given by Equation (13.1). Transistors Q_{12} and Q_{13}

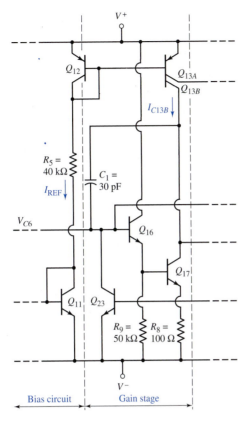

Figure 13.7   Reference circuit and gain stage, 741 op-amp

form a current mirror, and Q_{13B} has a scale factor 0.75 times that of Q_{12}. Neglecting base currents, current I_{C13B} is then

$$I_{C13B} = 0.75I_{REF} \tag{13.8}$$

The emitter current in Q_{16} is the sum of the base current in Q_{17} and the current in R_9, as follows:

$$I_{C16} \cong I_{E16} = I_{B17} + \frac{I_{E17}R_8 + V_{BE17}}{R_9} \tag{13.9}$$

Example 13.2   **Objective:** Calculate the bias currents in the gain stage of the 741 op-amp in Figure 13.7.

Solution: In Example 13.1, we determined the reference current to be $I_{REF} = 0.72\,\text{mA}$. From Equation (13.8), the collector current in Q_{17} is

$$I_{C17} = I_{C13B} = 0.75I_{REF} = (0.75)(0.72) = 0.54\,\text{mA}$$

Assuming $\beta = 200$ for the npn transistor, the collector current in Q_{16} is, from Equation (13.9),

$$I_{C16} \cong I_{B17} + \frac{I_{E17} R_8 + V_{BE17}}{R_9} = \frac{0.54}{200} + \frac{(0.54)(0.1) + 0.6}{50}$$

or

$$I_{C16} = 15.8 \, \mu A$$

Comment: The small bias current in Q_{16}, in conjunction with the resistor R_9, ensures that the input resistance to the gain stage is large, which minimizes loading effects on the diff-amp stage. The small bias current in Q_{16} also means that the base current in Q_{16} is negligible, as assumed in the dc analysis of the input stage.

Output Stage

Figure 13.8 shows the basic output stage of the 741 op-amp. This is a class-AB configuration, discussed in Chapter 8. The I_{Bias} is supplied by Q_{13A}, and the input signal is applied to the base of Q_{22}, which operates as an emitter follower. The combination of Q_{18} and Q_{19} establishes two B–E voltage drops between the base terminals of Q_{14} and Q_{20}, causing the output transistors to be biased slightly in the conducting state. This V_{BB} voltage produces quiescent collector currents in Q_{14} and Q_{20}. Biasing both Q_{14} and Q_{20} "on" with no signal present at the input ensures that the output stage will respond linearly when a signal is applied to the op-amp input.

Figure 13.8   Basic output stage, 741 op-amp, showing currents and voltages

The collector of Q_{13A} has a scale factor of 0.25 times that of Q_{12}. Neglecting base currents, current I_{C13A} is

$$I_{C13A} = 0.25 I_{\text{REF}} = I_{\text{Bias}} \tag{13.10}$$

where I_{REF} is given by Equation (13.1). Neglecting base currents, the collector current in Q_{22} is also equal to I_{Bias}. The collector current in Q_{18} is

$$I_{C18} \cong \frac{V_{BE19}}{R_{10}} \qquad (13.11)$$

Therefore,

$$I_{C19} = I_{Bias} - I_{C18} \qquad (13.12)$$

Example 13.3 **Objective:** Calculate the bias currents in the output stage of the 741 op-amp.

Consider the output stage shown in Figure 13.8. Assume the reverse saturation currents of Q_{18} and Q_{19} are $I_S = 10^{-14}$ A, and the reverse saturation currents of Q_{14} and Q_{20} are $I_S = 3 \times 10^{-14}$ A. Neglect base currents.

Solution: The reference current, from Example 13.1, is $I_{REF} = 0.72$ mA. Current I_{C13A} is then

$$I_{C13A} = (0.25)I_{REF} = (0.25)(0.72) = 0.18 \text{ mA} \cong I_{Bias}$$

If we assume $V_{BE19} = 0.6$ V, then the current in R_{10} is

$$I_{R10} = \frac{V_{BE19}}{R_{10}} = \frac{0.6}{50} = 0.012 \text{ mA}$$

The current in Q_{19} is

$$I_{C19} \cong I_{E19} = I_{C13A} - I_{R10} = 0.18 - 0.012 = 0.168 \text{ mA}$$

For this value of collector current, the B–E voltage of Q_{19} is

$$V_{BE19} = V_T \ln\left(\frac{I_{C19}}{I_S}\right) = (0.026) \ln\left(\frac{0.168 \times 10^{-3}}{10^{-14}}\right) = 0.612 \text{ V}$$

which is close to the assumed value of 0.6 V. Assuming $\beta_n = 200$ for the npn devices, the base current in Q_{19} is

$$I_{B19} = \frac{I_{C19}}{\beta_n} = \frac{169 \,\mu\text{A}}{200} \cong 0.8 \,\mu\text{A}$$

The current in Q_{18} is now

$$I_{C18} \cong I_{E18} = I_{R10} + I_{B19} = 12 + 0.8 = 12.8 \,\mu\text{A}$$

The B–E voltage of Q_{18} is therefore

$$V_{BE18} = V_T \ln\left(\frac{I_{C18}}{I_S}\right) = (0.026) \ln\left(\frac{12.8 \times 10^{-6}}{10^{-14}}\right) = 0.545 \text{ V}$$

The voltage difference V_{BB} is thus

$$V_{BB} = V_{BE18} + V_{BE19} = 0.545 + 0.612 = 1.157 \text{ V}$$

Since the output transistors Q_{14} and Q_{20} are identical, one-half of V_{BB} is across each B–E junction. The quiescent currents in Q_{14} and Q_{20} are

$$I_{C14} = I_{C20} = I_S e^{(V_{BB}/2)/V_T} = 3 \times 10^{-14} e^{(1.157/2)/0.026}$$

or

$$I_{C14} = I_{C20} = 138 \,\mu\text{A}$$

Comment: Using the piecewise linear approximation of 0.6 V for the B–E junction voltage does not allow us to determine the quiescent currents in Q_{14} and Q_{20}. For a more accurate analysis, the exponential relationship must be used, since the base–emitter areas of the output transistors are larger than those of the other transistors, and because the output transistors are biased at a low quiescent current.

Test Your Understanding

13.7   In Figure 13.8, replace the Q_{18}, Q_{19}, and R_{10} combination by two series diodes with $I_S = 10^{-14}$ A. Assume that I_{C13A} is the same as previously determined, and let $I_S = 3 \times 10^{-14}$ A for Q_{14} and Q_{20}. Calculate V_{BB}, I_{C14}, and I_{C20}. (Ans. $V_{BB} = 1.228$ V, $I_{C14} = I_{C20} = 0.541$ mA)

As the input signal v_I increases, the base voltage of Q_{14} increases since the V_{BB} voltage remains almost constant. The output voltage increases at approximately the same rate as the input signal. As v_I decreases, the base voltage of Q_{20} decreases, and the output voltage also decreases, again at approximately the same rate as the input signal. The small-signal voltage gain of the output stage is essentially unity.

Short-Circuit Protection Circuitry

The output stage includes a number of transistors that are off during the normal operation of the amplifier. If the output terminal is at a positive voltage because of an applied input signal, and if the terminal is inadvertently shorted to ground potential, a large current will be induced in output transistor Q_{14}. A large current can produce sufficient heating to cause transistor burnout.

The complete output stage of the 741, including the **short-circuit protection devices**, is shown in Figure 13.9. Resistor R_6 and transistor Q_{15} limit the current in Q_{14} in the event of a short circuit. If the current in Q_{14} reaches 20 mA, the voltage drop across R_6 is 540 mV, which is sufficient to bias Q_{15} in the conducting stage. As Q_{15} turns on, excess base current into Q_{14} is shunted through the collector of Q_{15}. The base current into Q_{14} is then limited to a maximum value, which limits the collector current.

The maximum current in Q_{20} is limited by components R_7, Q_{21}, and Q_{24}, in much the same way as just discussed. A large output current will result in a voltage drop across R_7, which will be sufficient to bias Q_{21} in its conducting state. Transistors Q_{21} and Q_{24} will shunt excessive output current away from Q_{20}, to protect this output transistor.

13.2.3   Small-Signal Analysis

We can analyze the small-signal voltage gain of the 741 op-amp by dividing it into its basic circuits and using results previously obtained.

Figure 13.9   Output stage, 741 op-amp with short-circuit protection devices

Input Stage

Figure 13.10 shows the ac equivalent circuit of the input stage with a differential voltage v_d applied between the input terminals. The constant-current biasing at the base of Q_3 and Q_4 means that the effective impedance connected to the base terminal of Q_3 and Q_4 is ideally infinite, or an open circuit. Resistance R_{act1} is the effective resistance of the active load and R_{i2} is the input resistance of the gain stage.

Figure 13.10   The ac equivalent circuit, input stage of 741 op-amp

From the results in Chapter 11, the small-signal differential voltage gain can be written as

$$A_d = \frac{v_{o1}}{v_d} = -g_m(r_{o4}\|R_{act1}\|R_{i2}) = -\left(\frac{I_{CQ}}{V_T}\right)(r_{o4}\|R_{act1}\|R_{i2}) \tag{13.13}$$

where I_{CQ} is the quiescent collector current in each of the transistors Q_1 through Q_4, and r_{o4} is the small-signal output resistance looking into the collector of Q_4. Using r_{o4} as the resistance looking into the collector of Q_4 neglects the effective resistance in the emitter of Q_4. This effective resistance is simply the resistance looking into the emitter of Q_2, which is normally very small. The minus sign in the voltage gain expression results from the applied signal voltage polarity and resulting current directions.

The effective resistance of the active load is given by

$$R_{act1} = r_{o6}[1 + g_{m6}(R_2\|r_{\pi6})] \tag{13.14}$$

as determined in Chapter 10 for the output resistance of a Widlar current source. From Figure 13.7, the input resistance of the gain stage is

$$R_{i2} = r_{\pi16} + (1 + \beta_n)R_E' \tag{13.15}$$

where R_E' is the effective resistance in the emitter of Q_{16}, as given by

$$R_E' = R_9\|[r_{\pi17} + (1 + \beta_n)R_8] \tag{13.16}$$

Example 13.4   **Objective:** Determine the small-signal differential voltage gain of the 741 op-amp input stage.

Assume npn transistor gains of $\beta_n = 200$ and Early voltages of $V_A = 50\,\text{V}$.

Solution: The quiescent collector currents were determined previously in this chapter. The input resistance to the gain stage is found from Equations (13.15) and (13.16), as follows:

$$r_{\pi17} = \frac{\beta_n V_T}{I_{C17}} = \frac{(200)(0.026)}{0.54} = 9.63\,\text{k}\Omega$$

Therefore,

$$R_E' = R_9\|[r_{\pi17} + (1 + \beta_n)R_8] = 50\|[9.63 + (201)(0.1)] = 18.6\,\text{k}\Omega$$

Also,

$$r_{\pi16} = \frac{\beta_n V_T}{I_{C16}} = \frac{(200)(0.026)}{0.0158} = 329\,\text{k}\Omega$$

Consequently,

$$R_{i2} = r_{\pi16} + (1 + \beta_n)R_E' = 329 + (201)(18.6) \Rightarrow 4.07\,\text{M}\Omega$$

The resistance of the active load is determined from Equation (13.14). We find

$$r_{\pi6} = \frac{\beta_n V_T}{I_{C6}} = \frac{(200)(0.026)}{0.0095} = 547\,\text{k}\Omega$$

$$g_{m6} = \frac{I_{C6}}{V_T} = \frac{0.0095}{0.026} = 0.365\,\text{mA/V}$$

and

$$r_{o6} = \frac{V_A}{I_{C6}} = \frac{50}{0.0095} \Rightarrow 5.26 \, \text{M}\Omega$$

Then,

$$R_{act1} = r_{o6}[1 + g_{m6}(R_2 \| r_{\pi6})] = 5.26[1 + (0.365)(1 \| 547)] = 7.18 \, \text{M}\Omega$$

Resistance r_{o4} is

$$r_{o4} = \frac{V_A}{I_{C4}} = \frac{(50)}{(0.0095)} \Rightarrow 5.26 \, \text{M}\Omega$$

Finally, from Equation (13.13), the small-signal differential voltage gain is

$$A_d = -\left(\frac{I_{CQ}}{V_T}\right)(r_{o4} \| R_{act1} \| R_{i2}) = -\left(\frac{9.5}{0.026}\right)(5.26 \| 7.18 \| 4.07)$$

or

$$A_d = -636$$

Comment: The relatively large gain results from the use of an active load and the fact that the gain stage does not drastically load the input stage.

Gain Stage

Figure 13.11 shows the ac equivalent circuit of the gain stage. Resistance R_{act2} is the effective resistance of the active load and R_{i3} is the input resistance of the output stage.

Figure 13.11   The ac equivalent circuit, gain stage of 741 op-amp

We develop the small-signal voltage gain using Figure 13.11 directly. The input base current to Q_{16} is

$$i_{b16} = \frac{v_{o1}}{R_{i2}} \tag{13.17}$$

where R_{i2} is the input resistance to the gain stage. The base current into Q_{17} is

$$i_{b17} = \frac{R_9}{R_9 + [r_{\pi17} + (1 + \beta_n)R_8]} \times i_{e16} \tag{13.18}$$

where i_{e16} is the emitter current from Q_{16}. The output voltage is

$$v_{o2} = -i_{c17}(R_{act2} \| R_{i3} \| R_{o17}) \tag{13.19}$$

where i_{c17} is the ac collector current in Q_{17} and R_{o17} is the output impedance looking into the collector of Q_{17}. Combining Equations (13.17), (13.18), and (13.19), we get the following expression for the small-signal voltage gain:

$$A_{v2} = \frac{v_{o2}}{v_{o1}} = \frac{-\beta_n(1+\beta_n)R_9(R_{act2} \| R_{i3} \| R_{o17})}{R_{i2}\{R_9 + [r_{\pi17} + (1+\beta_n)R_8]\}} \tag{13.20}$$

The effective resistance of the active load is the resistance looking into the collector of Q_{13B}, or

$$R_{act2} = r_{o13B} = \frac{V_A}{I_{C13B}} \tag{13.21}$$

The input resistance of the output stage can be determined from the ac equivalent circuit in Figure 13.12. In this figure, we assume that the pnp output transistor Q_{20} is active and the npn output transistor Q_{14} is cut off. A load resistor R_L is also included. Transistor Q_{22} operates as an emitter follower, which means that the input resistance is

Figure 13.12   The ac equivalent circuit, 741 op-amp output stage, for calculating input resistance

$$R_{i3} = r_{\pi22} + (1+\beta_p)[R_{19} \| R_{20}] \tag{13.22}$$

Resistance R_{19} is the series combination of the resistance looking into the emitters of Q_{19} and Q_{18}, and the resistance looking into the collector of Q_{13A}. The effective resistance of the combination of Q_{18} and Q_{19} is small compared to R_{13A}; therefore,

$$R_{19} \cong R_{13A} = r_{o13A} = \frac{V_A}{I_{C13A}} \tag{13.23}$$

The output transistor Q_{20} is also an emitter follower; therefore,

$$R_{20} = r_{\pi 20} + (1 + \beta_p)R_L \qquad\qquad (13.24)$$

where the load resistance R_L is assumed to be much larger than R_7.

Example 13.5 **Objective:** Determine the small-signal voltage gain of the second stage of the 741 op-amp.

Assume the current gains of the pnp transistors are $\beta_p = 50$ and the gains of the npn transistors are $\beta_n = 200$. Also assume the Early voltage is 50 V for all transistors and the load resistance connected to the output is $R_L = 2\,k\Omega$. The dc quiescent currents were determined previously.

Solution: First, we calculate the various resistances. To begin,

$$r_{\pi 20} = \frac{\beta_p V_T}{I_{C20}} = \frac{(50)(0.026)}{0.138} = 9.42\,k\Omega$$

which means that

$$R_{20} = r_{\pi 20} + (1 + \beta_p)R_L = 9.42 + (51)(2) \cong 111\,k\Omega$$

Also,

$$R_{19} = r_{o13A} = \frac{V_A}{I_{C13A}} = \frac{50}{0.18} = 278\,k\Omega$$

and

$$r_{\pi 22} = \frac{\beta_p V_T}{I_{C13A}} = \frac{(50)(0.026)}{0.18} = 7.22\,k\Omega$$

The input resistance to the output stage is therefore

$$R_{i3} = r_{\pi 22} + (1 + \beta_p)[R_{19} \| R_{20}] = 7.22 + (51)[278\|111] \Rightarrow 4.05\,M\Omega$$

The effective resistance of the active load is

$$R_{act2} = \frac{V_A}{I_{C13B}} = \frac{50}{0.54} = 92.6\,k\Omega$$

and the output resistance R_{o17} is

$$R_{o17} \cong \frac{V_A}{I_{C17}} = \frac{50}{0.54} = 92.6\,k\Omega$$

This calculation neglects the very small value of R_8 in the emitter.

From Equation (13.20), the small-signal voltage gain is as follows (all resistances are given in kilohms):

$$A_{v2} = \frac{-\beta_n(1 + \beta_n)R_9(R_{act2}\|R_{i3}\|R_{o17})}{R_{i2}\{R_9 + [r_{\pi 17} + (1 + \beta_n)R_8]\}} = \frac{-(200)(201)(50)(92.6\|4050\|92.6)}{4070\{50 + [9.63 + (201)(0.1)]\}}$$

or

$$A_{v2} = -285$$

Comment: The voltage gain of the second stage is fairly large, again because an active load is used and because there is no severe loading effect from the output stage.

Overall Gain

In calculating the voltage gain of each stage, we took the loading effect of the following stage into account. Therefore, the overall voltage gain is the product of the individual gain factors, or

$$A_v = A_d A_{v2} A_{v3} \tag{13.25}$$

where A_{v3} is the voltage gain of the output stage. If we assume that $A_{v3} \approx 1$, as previously discussed, then the overall gain of the 741 op-amp is

$$A_v = A_d A_{v2} A_{v3} = (-636)(-285)(1) = 181{,}260 \tag{13.26}$$

Typical voltage gain values for the 741 op-amp are in the range of 200,000. The value determined in our calculations illustrates the magnitude of voltage gains that can be obtained in op-amp circuits.

Output Resistance

The output resistance can be determined by using the ac equivalent circuit in Figure 13.13. In this case, we assume the output transistor Q_{20} is conducting and Q_{14} is cut off. The same basic result is obtained if Q_{14} is conducting and Q_{20} is cut off. We again rely on results obtained previously for output resistances of basic amplifier stages.

The output resistance is

$$R_o = R_7 + R_{e20} \tag{13.27}$$

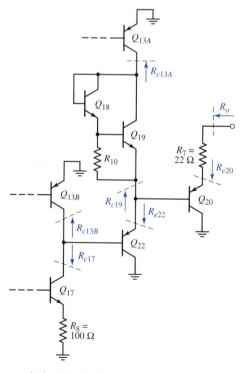

Figure 13.13   The ac equivalent circuit, 741 op-amp output stage, for calculating output resistance

where

$$R_{e20} = \frac{r_{\pi20} + R_{e22}\|R_{c19}}{(1+\beta_p)} \tag{13.28}$$

Previously we argued that the series resistance due to Q_{18} and Q_{19} is small compared to R_{c13A}, so that $R_{c19} \cong R_{c13A}$. We also have

$$R_{e22} = \frac{r_{\pi22} + R_{c17}\|R_{c13B}}{(1+\beta_p)} \tag{13.29}$$

where

$$R_{c13B} = r_{o13B}$$

and

$$R_{c17} = r_{o17}[1 + g_{m17}(R_8\|r_{\pi17})]$$

The output resistance of the op-amp is then found by combining all the resistance terms.

Example 13.6 **Objective:** Calculate the output resistance of the 741 op-amp.

Consider the output stage configuration in Figure 13.13. Assume the output current is $I_{c20} = 2\,\text{mA}$ and all other bias currents are as previously determined.

Solution: Using $\beta_n = 200$, $\beta_p = 50$, and $V_A = 50\,\text{V}$, we find the following:

$$r_{\pi17} = 9.63\,\text{k}\Omega \qquad r_{\pi22} = 7.22\,\text{k}\Omega \qquad r_{\pi20} = 0.26\,\text{k}\Omega$$
$$g_{m17} = 20.8\,\text{mA/V} \qquad r_{o17} = 92.6\,\text{k}\Omega \qquad r_{o13B} = 92.6\,\text{k}\Omega$$

Then,

$$R_{c17} = r_{o17}[1 + g_{m17}(R_8\|r_{\pi17})] = 92.6[1 + (20.8)(0.1\|9.63)] = 283\,\text{k}\Omega$$

and

$$R_{e22} = \frac{r_{\pi22} + R_{c17}\|R_{c13B}}{(1+\beta_p)} = \frac{7.22 + 283\|92.6}{51} = 1.51\,\text{k}\Omega$$

Also,

$$R_{c19} \cong R_{c13A} = r_{o13A} = \frac{V_A}{I_{C13A}} = \frac{50}{0.18} = 278\,\text{k}\Omega$$

Therefore

$$R_{e20} = \frac{r_{\pi20} + R_{e22}\|R_{c19}}{(1+\beta_p)} = \frac{0.26 + 1.51\|278}{51} = 0.0345\,\text{k}\Omega \Rightarrow 34.5\,\Omega$$

Consequently, the output resistance is

$$R_o = R_7 + R_{e20} = 22 + 34.5 = 56.5\,\Omega$$

Comment: We showed previously that the output resistance of an emitter-follower circuit is low. For comparison, typical output resistance values for the 741 op-amp are $75\,\Omega$. This correlates well with our analysis.

13.2.4   Frequency Response

The 741 op-amp is internally compensated by the Miller compensation technique to introduce a dominant low-frequency pole. From Miller's theorem, the effective input capacitance of the second gain stage is

$$C_i = C_1(1 + |A_{v2}|) \qquad (13.30)$$

The dominant low-frequency pole is

$$f_{PD} = \frac{1}{2\pi R_{eq} C_i} \qquad (13.31)$$

where R_{eq} is the equivalent resistance between the second-stage input node and ground, and is

$$R_{eq} = R_{o1} \| R_{i2} \qquad (13.32)$$

Here R_{i2} is the input resistance of the gain stage and R_{o1} is the output resistance of the diff-amp stage. From Figure 13.10, we see that

$$R_{o1} = R_{act1} \| r_{o4} \qquad (13.33)$$

Example   13.7   Objective: Determine the dominant-pole frequency of the 741 op-amp.

Use appropriate results from previous calculations.

Solution: Previously, we determined that $|A_{v2}| = 285$, which means that the effective input capacitance is

$$C_i = C_1(1 + |A_{v2}|) = (30)(1 + 285) = 8580\,\text{pF}$$

The gain stage input resistance was found to be $R_{i2} = 4.07\,\text{M}\Omega$. We find

$$R_{o1} = R_{act1} \| r_{o4} = 7.18 \| 5.26 = 3.04\,\text{M}\Omega$$

The equivalent resistance is then

$$R_{eq} = R_{o1} \| R_{i2} = 3.04 \| 4.07 = 1.74\,\text{M}\Omega$$

Finally, the dominant-pole frequency is

$$f_{PD} = \frac{1}{2\pi R_{eq} C_i} = \frac{1}{2\pi(1.74 \times 10^6)(8580 \times 10^{-12})} = 10.7\,\text{Hz}$$

Comment: The very large equivalent input capacitance C_i justifies neglecting any other capacitance effects at the gain stage input.

If all other poles of the op-amp circuit are at very high frequencies, then the unity-gain bandwidth is

$$f_T = A_o f_{PD} \qquad (13.34)$$

Using our results, we find that

$$f_T = (181,260)(10.7) \cong 1.9\,\text{MHz} \qquad (13.35)$$

A typical unity-gain bandwidth value for the 741 op-amp is 1 MHz. With all the approximations and assumptions, such as the value of reverse saturation current and Early voltage, used in the calculations, a factor of two between the actual and predicted cutoff frequency is not significant.

If the frequencies of the other poles of the 741 op-amp are greater than 1.9 MHz, the phase margin is 90 degrees. This phase margin ensures that any closed-loop amplifier circuit using the 741 op-amp will be stable for any feedback transfer function.

Problem-Solving Technique: Operational Amplifier Circuits

1. *DC analysis.* The bias portion of the op-amp circuit must be identified. A reference current must be determined and then the bias currents in the individual building blocks of the overall circuit can be determined.

2. *AC analysis.* The small-signal properties of the building blocks of the overall circuit can be analyzed individually, provided that the loading effects of follow-on stages are taken into account.

Test Your Understanding

13.8 The power supply voltages for the 741 op-amp in Figure 13.3 are $V^+ = 10\,\text{V}$ and $V^- = -10\,\text{V}$. Neglect base currents and assume $V_{BE}(\text{npn}) = V_{EB}(\text{pnp}) = 0.6\,\text{V}$. Calculate the bias currents I_{REF}, I_{C10}, I_{C6}, I_{C13B}, and I_{C13A}. (Ans. $I_{\text{REF}} = 0.47\,\text{mA}$, $I_{C10} = 17.2\,\mu\text{A}$, $I_{C6} = 8.6\,\mu\text{A}$, $I_{C13B} = 0.353\,\text{mA}$, $I_{C13A} = 0.118\,\text{mA}$)

*13.9 In the 741 op-amp output stage in Figure 13.3, the combination of Q_{18}, Q_{19}, and R_{10} is replaced by two series diodes with $I_S = 10^{-14}\,\text{A}$. The transistor parameters are: $\beta_n = 200$, $\beta_p = 50$, and $V_A = 50\,\text{V}$. Assume the same dc bias currents calculated previously. Calculate the output resistance, assuming Q_{14} is conducting, producing a load current of 5 mA. (Ans. $41\,\Omega$)

13.3 CMOS OPERATIONAL AMPLIFIER CIRCUITS

The 741 bipolar op-amp is a general-purpose op-amp capable of sourcing and sinking reasonably large load currents. The output stage is an emitter follower capable of supplying the necessary load current, with a low output resistance to minimize loading effects.

In contrast, most CMOS op-amps are designed for specific on-chip applications and are only required to drive capacitive loads of a few picofarads. Most CMOS op-amps therefore do not need a low-resistance output stage, and, if the op-amp inputs are not connected directly to the IC external terminals, they also do not need electrostatic input protection devices.

In this section, we will initially consider a simple CMOS design to begin to understand the basic concepts of a CMOS op-amp. We will then consider a more sophisticated CMOS op-amp design, called a folded cascode op-amp, and then analyze a current-mirror CMOS op-amp circuit. In each case we will do a dc analysis/design and then a small-signal analysis/design.

13.3.1   MC14573 CMOS Operational Amplifier Circuit

Circuit Description

An example of an all-CMOS op-amp is the MC14573, for which a simplified circuit diagram is shown in Figure 13.14. The p-channel transistors M_1 and M_2 form the input differential pair, and the n-channel transistors M_3 and M_4 form the active load. The diff-amp input stage is biased by the current mirror M_5 and M_6, in which the reference current is determined by an external resistor R_{set}.

The second stage, which is also the output stage, consists of the common-source-connected transistor M_7. Transistor M_8 provides the bias current for M_7 and acts as the active load. An internal compensation capacitor C_1 is included to provide stability.

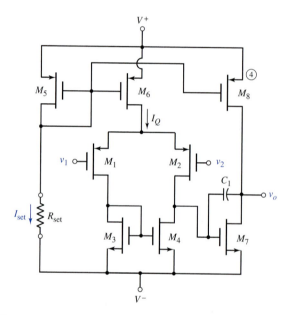

Figure 13.14   MC14573 CMOS op-amp equivalent circuit

DC Analysis

Assuming transistors M_5 and M_6 are matched, the reference and input-stage bias currents are given by

$$I_{set} = I_Q = \frac{V^+ - V^- - V_{SG5}}{R_{set}} \tag{13.36}$$

The reference current and source-to-gate voltage are also related by

$$I_{set} = K_{p5}(V_{SG5} + V_{TP})^2 \tag{13.37}$$

where V_{TP} and K_{p5} are the threshold voltage and conduction parameter of the p-channel transistor M_5.

Example 13.8   Objective: Determine the dc bias currents in the MC14573 op-amp.

Assume transistor parameters of $|V_T| = 0.5\,\text{V}$ (all transistors), $(\frac{1}{2})\mu_n C_{ox} = 20\,\mu\text{A/V}^2$, $(\frac{1}{2})\mu_p C_{ox} = 10\,\mu\text{A/V}^2$, and circuit parameters of $V^+ = 5\,\text{V}$, $V^- = -5\,\text{V}$, and $R_{\text{set}} = 225\,\text{k}\Omega$. Assume transistor width-to-length ratios of 6.25 for M_3 and M_4, and 12.5 for all other transistors.

Solution: For transistors M_5 and M_6, the conduction parameters are:

$$K_p = \left(\frac{W}{L}\right)\left(\frac{1}{2}\mu_p C_{ox}\right) = (12.5)(10) = 125\,\mu\text{A/V}^2$$

Combining Equations (13.36) and (13.37) yields the source-to-gate voltage of M_5:

$$K_p(V_{SG5} + V_{TP})^2 = \frac{V^+ - V^- - V_{SG5}}{R_{\text{set}}}$$

or

$$0.125(V_{SG5} - 0.5)^2 = \frac{5 + 5 - V_{SG5}}{225}$$

which yields

$$V_{SG5} = 1.06\,\text{V}$$

From Equation (13.36), we have

$$I_{\text{REF}} = I_Q = \frac{10 - 1.06}{225} \Rightarrow 39.7\,\mu\text{A}$$

The quiescent drain currents in M_7 and M_8 are then also $39.7\,\mu\text{A}$, and the currents in M_1 through M_4 are $19.9\,\mu\text{A}$.

Comment: The quiescent bias currents can be changed easily by changing the external resistor R_{set}. Transistors M_5, M_6, and M_8 are identical, so the currents in these three devices are equal since the source-to-gate voltages are the same. The width-to-length ratio of M_7 is twice that of M_3 and M_4, which means the current in M_7 is twice that in M_3 and M_4. However, this is consistent with the current-source transistor currents.

Test Your Understanding

*13.10 Using the parameters given in Example 13.8, determine the input common-mode voltage range for the MC14573 op-amp. (Ans. $-4.60 \le v_{\text{in}}(\text{cm}) \le 3.54\,\text{V}$)

13.11 Using the parameters given in Example 13.8, determine the maximum and minimum output voltage in the MC14573 circuit such that the op-amp remains biased in its linear region. (Ans. $-4.44 \le v_O \le +4.44\,\text{V}$)

Small-Signal Analysis

The small-signal differential voltage gain of the input stage can be written as

$$A_d = \sqrt{2K_{p1}I_Q}\,(r_{o2}\|r_{o4}) \tag{13.38}$$

where r_{o2} and r_{o4} are the output resistances of M_2 and M_4, respectively. The input impedance to the second stage is essentially infinite; therefore, there is no loading effect due to the second stage. If we assume that the parameter λ is the same for all transistors, then

$$r_{o2} = r_{o4} = \frac{1}{\lambda I_D} \tag{13.39}$$

where I_D, which is the quiescent drain current in M_2 and M_4, is $I_D = I_Q/2$.

The gain of the second stage is

$$A_{v2} = g_{m7}(r_{o7}\|r_{o8}) \tag{13.40}$$

where

$$g_{m7} = 2\sqrt{K_{n7}I_{D7}}$$

and

$$r_{o7} = r_{o8} = 1/\lambda I_{D7}$$

Equation (13.40) implies that there is no loading effect due to an external load connected at the output.

Example   13.9   **Objective:** Determine the small-signal voltage gains of the input and second stages, and the overall voltage gain, of the MC14573 op-amp.

Assume the same transistor and circuit parameters as in Example 13.8. Let $\lambda = 0.02\,\text{V}^{-1}$ for all transistors.

Solution: The conduction parameters of M_1 and M_2 are

$$K_{p1} = K_{p2} = \left(\frac{W}{L}\right)\left(\frac{1}{2}\mu_p C_{ox}\right) = (12.5)(10) = 125\,\mu\text{A/V}^2$$

and the output resistances are

$$r_{o2} = r_{o4} = \frac{1}{\lambda I_D} = \frac{1}{(0.02)(0.0199)} \Rightarrow 2.51\,\text{M}\Omega$$

From Equation (13.38), the gain of the input stage is then

$$A_d = \sqrt{2K_{p1}I_Q}\,(r_{o2}\|r_{o4}) = \sqrt{2(0.125)(0.0397)}\,(2510\|2510)$$

or

$$A_d = 125$$

The transconductance of M_7 is

$$g_{m7} = 2\sqrt{K_{n7}I_{D7}} = 2\sqrt{(0.250)(0.0397)} = 0.199\,\text{mA/V}$$

and the output resistances of M_7 and M_8 are

$$r_{o7} = r_{o8} = \frac{1}{\lambda I_{D7}} = \frac{1}{(0.02)(0.0397)} \Rightarrow 1.26\,\text{M}\Omega$$

From Equation (13.40), the gain of the second stage is then

$$A_{v2} = g_{m7}(r_{o7}\|r_{o8}) = (0.199)(1260\|1260) = 125$$

Finally, the overall voltage gain of the op-amp is

$$A_v = A_d A_{v2} = (125)(125) = 15,625$$

Comment: The calculated overall voltage gain is 84 dB, which correlates fairly well with typical values of 90 dB, as listed in the data sheet for the MC14573 op-amp. The open-loop gain of a CMOS op-amp is generally less than that of a bipolar op-amp, but the use of active loads provides acceptable results.

Test Your Understanding

*13.12 Consider the MC14573 op-amp in Figure 13.14. Assume the same circuit and transistor parameters as given in Examples 13.8 and 13.9, except change R_{set} to $100\,k\Omega$. (a) Calculate all dc bias currents. (b) Determine the overall voltage gain of the op-amp. (Ans. (a) $I_{set} = I_Q = I_{D8} = I_{D7} = 86.7\,\mu A$, $I_{D1} = I_{D2} = I_{D3} = I_{D4} = I_Q/2 = 43.35\,\mu A$ (b) 7189)

13.3.2 Folded Cascode CMOS Operational Amplifier Circuit

As we have mentioned previously, the voltage gain of an amplifier can be increased by using a cascode configuration. In its simplest form, the conventional cascode configuration consists of two transistors in series, as shown in Figure 13.15(a). The transistor M_1 is the common-source amplifying device whose current is determined by the input voltage. This current is the input signal to M_2, which is connected in a common-gate configuration. The output is taken off the drain of the cascode transistor. The circuit in Figure 13.15(b) has a slightly different configuration. The dc current I_1 in M_1 is determined by the input voltage. The dc current in M_2 is the difference between the bias current I_Q and I_1.

Figure 13.15 (a) Classical cascode stage; (b) folded-cascode stage

The ac current in the conventional cascode circuit of Figure 13.15(a) is through both transistors and the dc power supply. The ac current in the cascode circuit in Figure 13.15(b) is through both transistors and ground as indicated in the figure. The ac current in M_2 of this circuit is equal in magnitude but in the opposite direction to M_1. Thus the current is said to be folded back and the circuit in Figure 13.15(b) is called a folded cascode circuit.

The folded cascode configuration can be applied to the diff-amp as shown in Figure 13.16. The transistors M_1 and M_2 are the differential pair, as usual, and transistors M_5 and M_6 are the cascode transistors. Transistors M_7–M_{10} form a modified Wilson current mirror acting as an active load. This configuration was discussed in Chapter 10. The biasing V_{B1} and V_{B2} must be provided by a separate network.

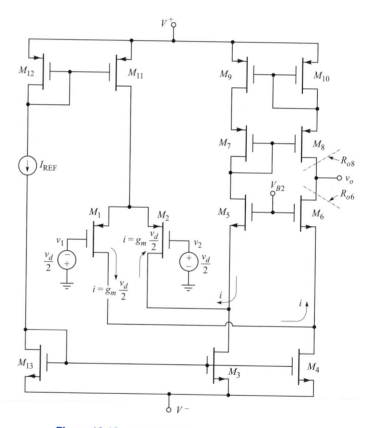

Figure 13.16　CMOS folded cascode amplifier

Assuming that transistors M_3, M_4, and M_{11}–M_{13} are all matched, then the dc currents in M_1 and M_2 are $I_{REF}/2$ and those in M_3 and M_4 are I_{REF}. This means that the dc currents in the cascode transistors M_5 and M_6 are $I_{REF}/2$.

If a differential-mode input voltage is applied, then ac currents are induced in the differential pair as shown in the figure. The ac current in M_1 flows through M_6 to the output. The ac current in M_2 flows through M_5 and is induced in M_8 by the current-mirror action of the active load. From previous work on diff-amps, the differential-mode voltage gain is

$$A_d = g_{m1}(R_{o6}\|R_{o8}) \tag{13.41}$$

where

$$R_{o8} = g_{m8}(r_{o8}r_{o10})$$ (13.42(a))

and

$$R_{o6} = g_{m6}(r_{o6})(r_{o4}\|r_{o1})$$ (13.42(b))

We may note that we are neglecting the body effect. Normally the substrates of all NMOS devices are tied to V^- and the substrates of all PMOS devices are tied to V^+.

Example 13.10 **Objective:** Determine the differential-mode voltage gain of the folded cascode diff-amp in Figure 13.16.

Assume circuit and transistor parameters: $I_{REF} = 100\,\mu A$, $k_n' = 80\,\mu A/V^2$, $k_p' = 40\,\mu A/V^2$, $(W/L) = 25$, and $\lambda_n = \lambda_p = 0.02\,V^{-1}$.

Solution: The transconductances are determined to be

$$g_{m1} = g_{m8} = 2\sqrt{\frac{k_p'}{2}\cdot\frac{W}{L}\cdot I_D} = 2\sqrt{\frac{40}{2}\cdot(25)(50)} = 316\,\mu A/V$$

and

$$g_{m6} = 2\sqrt{\frac{k_n'}{2}\cdot\frac{W}{L}\cdot I_D} = 2\sqrt{\frac{80}{2}\cdot(25)(50)} = 447\,\mu A/V$$

The transistor output resistances are found to be

$$r_{o1} = r_{o6} = r_{o8} = r_{o10} = \frac{1}{\lambda I_D} = \frac{1}{(0.02)(50)} = 1\,M\Omega$$

and

$$r_{o4} = \frac{1}{\lambda I_{D4}} = \frac{1}{(0.02)(100)} = 0.5\,M\Omega$$

The composite output resistances can be determined as

$$R_{o8} = g_{m8}(r_{o8}r_{o10}) = (316)(1)(1) = 316\,M\Omega$$

and

$$R_{o6} = g_{m6}(r_{o6})(r_{o4}\|r_{o1}) = (447)(1)(0.5\|1) = 149\,M\Omega$$

The differential-mode voltage gain is then

$$A_d = g_{m1}(R_{o6}\|R_{o8}) = (316)(149\|316) \cong 32{,}000$$

Comment: This example shows that very high differential-mode voltage gains can be achieved in a folded cascode CMOS circuit. In actual circuits, the output resistances may be limited by leakage currents so the very ideal values may not be realizable. However, substantially higher differential-mode voltage gains can be achieved in the folded cascode configuration than in the simpler diff-amp circuits.

Test Your Understanding

13.13 Assume the reference current in the folded cascode circuit shown in Figure 13.16 is $I_{REF} = 50\,\mu A$. Assume the transistor parameters are the same as given in Example 13.10. Determine the differential-mode voltage gain. (Ans. $\cong 64{,}000$)

13.3.3 CMOS Current-Mirror Operational Amplifier Circuit

Another CMOS op-amp circuit is shown in Figure 13.17. The differential pair is formed by M_1 and M_2. The induced ac currents from these transistors drive transistors M_3 and M_4, which are the inputs of two current mirrors with a current multiplication factor B. The current output of M_5 is then induced in M_8 by the current-mirror action of M_7 and M_8. The output signal currents then have a multiplication factor B. The differential-mode voltage gain is then given by

$$A_d = \frac{v_o}{v_d} = Bg_{m1}(r_{o6}\|r_{o8})$$

(13.43)

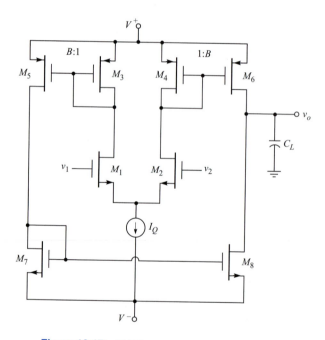

Figure 13.17 CMOS current-mirror op-amp

The factor of B in the gain expression of Equation (13.43) may be slightly misleading. Recall that the individual transistor output resistance is inversely proportional to the drain current. If the current in the output transistors increases by the factor B, then $R_o = r_{o6}\|r_{o8}$ decreases by the factor B so the differential-mode voltage gain remains unchanged.

The advantage of the current-mirror op-amp is an increase in the gain–bandwidth product. The dominant-pole frequency will be determined by the parameters at the output node. The dominant-pole frequency is given by

$$f_{pd} = \frac{1}{2\pi R_o(C_L + C_p)} \tag{13.44}$$

where R_o is the output resistance, C_L is the load capacitance, and C_p is the sum of all other capacitances at the output node. If R_o decreases by the factor B, then the dominant-pole frequency increases by the same factor B. The gain–bandwidth product is

$$\text{GBW} = A_d \cdot f_{pd} \tag{13.45}$$

Since A_d is now independent of B and f_{pd} increases by B, then the gain–bandwidth product increases by B.

Further analysis of this circuit shows that the phase margin decreases with increasing B. As a practical limit, the maximum value of B is limited to approximately 3.

Test Your Understanding

13.14 Consider the CMOS current-gain op-amp in Figure 13.17. Assume the bias current is $I_Q = 100\,\mu\text{A}$ and assume transistor parameters $k_n' = 80\,\mu\text{A/V}^2$, $k_p' = 40\,\mu\text{A/V}^2$, and $\lambda_n = \lambda_p = 0.02\,\text{V}^{-1}$. Assume the basic W/L ratio of the transistors is 20 and let $B = 3$. (a) Determine the small-signal voltage gain. (b) If the effective capacitance at the output node is $C_L + C_p = 2\,\text{pF}$, determine the dominant-pole frequency and the gain–bandwidth product. (Ans. (a) 200, (b) 477 kHz, 95 MHz)

13.3.4   CMOS Cascode Current-Mirror OP-Amp Circuit

As we have already seen, the differential-mode gain can be increased by adding cascode transistors in the output portion of the circuit. Figure 13.18 shows the same current-mirror configuration considered previously but with cascode transistors added to the output. Transistors M_9–M_{12} are the cascode transistors. The differential-mode voltage gain is given by

$$A_d = \frac{v_o}{v_d} = Bg_{m1}(R_{o10} \| R_{o12}) \tag{13.46}$$

where

$$R_{o10} = g_{m10}(r_{o10}r_{o6}) \tag{13.47}$$

and

$$R_{o12} = g_{m12}(r_{o12}r_{o8}) \tag{13.48}$$

The advantage of this circuit is the increased gain at low frequency. The gain–bandwidth product of this circuit is not changed from that of the simple current-mirror op-amp considered previously.

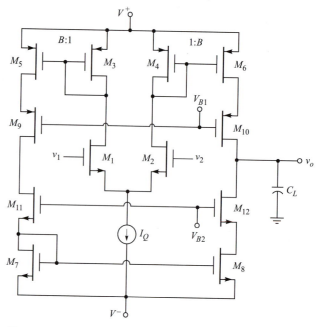

Figure 13.18 CMOS cascode current-mirror op-amp

Test Your Understanding

13.15 Consider the CMOS cascode current-mirror op-amp in Figure 13.18. Assume the bias current and transistor parameters are the same as in Exercise 13.14. Repeat parts (a) and (b) of Exercise 13.14 for this circuit. (Ans. (a) 38,171, (b) 2.50 kHz, 95.4 MHz)

13.4   BiCMOS OPERATIONAL AMPLIFIER CIRCUITS

As discussed in Chapter 11, BiCMOS circuits combine the advantages of bipolar and MOSFET devices in the same circuit. One advantage of MOSFETs is the very high input impedance. Therefore, when MOSFETs form the input differential pair of an op-amp, the input bias currents are extremely small. However, the equivalent noise of the input stage may be greater than for an all-BJT op-amp.

In this section, we will examine two BiCMOS op-amp circuits. The first is a variation of the folded cascode configuration analyzed in the last section and the second is the CA3140 BiCMOS op-amp. Since we previously fully analyzed the folded cascode circuit, we will discuss, here, the advantages of using the BiCMOS technology. Many features of the CA3140 BiCMOS op-amp are similar to those of the 741. Therefore, we will not analyze this op-amp in as great a detail as we did the 741. Instead, we will concentrate on some of its unique features.

13.4.1   BiCMOS Folded Cascode Op-Amp

Figure 13.19 shows an example of a BiCMOS folded cascode op-amp. The cascode transistors, Q_5 and Q_6, are now bipolar devices, replacing n-channel MOSFETs. The small-signal voltage gain expression for this circuit is identical to that of the all-CMOS design. We have mentioned that the dominant-pole frequency is determined by the circuit parameters at the output node because of the very large output resistance. Nondominant-pole frequencies are then a function of the parameters at the other circuit nodes. In particular, one node of interest is at the drain of an input transistor and emitter of a cascode transistor. The nondominant-pole frequency can be written as

$$f_{3-\text{dB}} = \frac{g_{m6}}{2\pi C_{p6}} \tag{13.49}$$

Figure 13.19   BiCMOS folded cascode amplifier

where g_{m6} is the transconductance of the cascode transistor Q_6 and C_{p6} is the effective capacitance at this node. Since the transconductance of a bipolar is usually greater than that of a MOSFET, this 3 dB frequency is larger for the BiCMOS circuit than for the all-CMOS design. This result means that the phase margin of the BiCMOS op-amp circuit is larger than that of the all-CMOS op-amp.

Test Your Understanding

13.16   Consider the BiCMOS folded cascode amplifier in Figure 13.19. Assume the circuit and MOS transistor parameters are the same as in Example 13.10. Assume BJT parameters of $\beta = 120$ and $V_A = 80\,\text{V}$. (a) Determine the small-signal voltage gain. (b) If the effective capacitance at the output node is $2\,\text{pF}$, determine the dominant-pole frequency and the gain–bandwidth product. (Ans. (a) 76,343, (b) 329 Hz, 25.1 MHz)

13.4.2   CA3140 BiCMOS Circuit Description

Figure 13.20 shows the basic equivalent circuit of the CA3140 op-amp. Like the 741, this op-amp consists of three basic stages: the input differential stage, the gain stage, and the output stage. Also shown in the figure are: the bias circuit, which establishes the dc bias currents in the op-amp; and a section referred to as a dynamic current sink, which will be explained later. Typical supply voltages are $V^+ = 15\,\text{V}$ and $V^- = -15\,\text{V}$.

Input Diff-Amp

The input differential pair consists of p-channel transistors M_9 and M_{10}, and transistors Q_{11} and Q_{12} form the active load for the diff-amp. A single-sided output at the collector of Q_{12} is the input signal to the following gain stage. Two offset null terminals are also shown, and will be discussed in the next chapter.

MOS transistors are very susceptible to damage from electrostatic charge. For example, electrostatic voltage can be inadvertently induced on the gate of a MOSFET during routine handling. These voltages may be great enough to induce breakdown in the gate oxide, destroying the device. Therefore, input protection against electrostatic damage is provided by the Zener diodes D_3, D_4, and D_5. If the gate voltage becomes large enough, these diodes will provide a discharge path for the electrostatic charge, thus protecting the gate oxide from breakdown.

The dc current biasing is initiated in the bias circuit. The elements labeled D_1 and D_2 are diode-connected transistors. Transistor Q_1 and diode D_1 are matched, which forces the currents in the two branches of the bias circuit to be equal. The current is determined from Q_7, R_1, and M_8. The combination of Q_6 and Q_7 makes the bias current essentially independent of the power supply voltages.

Gain Stage

The second stage consists of Q_{13} connected in a common-emitter configuration. The cascode configuration of transistors Q_3 and Q_4 provides the bias current for Q_{13}, in addition to acting as the active load. Since Q_3 and Q_4 are connected in a cascode configuration, the resistance looking into the collector of Q_4 is very high.

Figure 13.20   CA3140 BiCMOS op-amp equivalent circuit

Output Stage

The basic output stage consists of the npn transistors Q_{17} and Q_{18}. During the positive portion of the output voltage cycle, Q_{18} acts as an emitter follower, supplying a load current. During the negative portion of the output voltage cycle, Q_{16} sinks current from the load. As the output voltage decreases, the source-to-gate voltage on the p-channel M_{21} MOSFET increases, producing a larger current in D_6 and R_7 so that the base voltage on Q_{16} increases. The increase B–E voltage of Q_{16} allows increased load current sinking. Short-circuit protection is provided by the combination of R_{11} and Q_{19}. If a sufficiently large voltage is developed across R_{11}, Q_{19} turns on and shunts excess base current away from Q_{17}.

An abbreviated data sheet for the CA3140 op-amp is in Table 13.2. As before, we will compare the results of our analysis to the values listed in the table.

Table 13.2   CA3140 BiCMOS data

Parameter	Minimum	Typical	Maximum	Units
Input bias current		10	50	pA
Open-loop gain	20,000	100,000		V/V
Unity-gain frequency		4.5		MHz

13.4.3   CA3140 DC Analysis

In this section, we will determine the dc bias currents in the CA3140 op-amp. As previously stated, we will concentrate on the features that are unique to the CA3140 compared to the 741.

The basic bias circuit is shown in Figure 13.21. The current mirror consisting of Q_1 and D_1 ensures that the two branch currents I_1 and I_2 are equal, since Q_1 and D_1 are matched. The p-channel MOSFET M_8 is to operate in the saturation region, so that we must have

$$V_{SD} > V_{SG} - |V_{TP}| \tag{13.50}$$

Figure 13.21   Bias circuit, CA3140 BiCMOS op-amp

From the figure, we see that

$$V_{SG} = V_{SD} + V_D \tag{13.51}$$

or

$$V_{SD} = V_{SG} - V_D \qquad (13.52)$$

Combining Equations (13.52) and (13.50) yields

$$V_{SG} - V_D > V_{SG} - |V_{TP}| \qquad (13.53)$$

which implies that $|V_{TP}| > V_D$. In other words, for M_8 to remain biased in the saturation region, the magnitude of the threshold voltage must be greater than the diode voltage.

From the left branch of the bias circuit, we see that the current can be written

$$I_2 \cong I_{R1} = \frac{V_{SG} - V_{BE7}}{R_1} \qquad (13.54)$$

and from the right branch, we have

$$I_1 = K_p(V_{SG} - |V_{TP}|)^2 \qquad (13.55)$$

Since $I_1 = I_2$, a simultaneous solution of Equations (13.54) and (13.55) determines the currents and voltages in this bias circuit.

Example 13.11 **Objective:** Determine the currents and voltages in the bias circuit of the CA3140 op-amp.

Consider the bias circuit in Figure 13.21, with parameters: $V^+ = 15\,\text{V}$, $V^- = -15\,\text{V}$, and $R_1 = 8\,\text{k}\Omega$. Assume transistor parameters of $V_{BE}(\text{npn}) = V_{EB}(\text{pnp}) = 0.6\,\text{V}$ for the bipolars, and $K_p = 0.2\,\text{mA/V}^2$ and $|V_{TP}| = 1.4\,\text{V}$ for the MOSFET M_8.

Solution: Set $I_1 = I_2$. Then, from Equations (13.54) and (13.55), we find

$$V_{SG} = 2.49\,\text{V} \qquad \text{and} \qquad I_1 = I_2 = 0.236\,\text{mA}$$

The voltage at the collector of Q_6 is

$$V_{C6} = V_{SG8} + V^- = 2.49 - 15 = -12.5\,\text{V}$$

and the voltage at the collector of Q_7 is

$$V_{C7} = V^+ - V_{EB1} - V_{EB6} = 15 - 0.6 - 0.6 = 13.8\,\text{V}$$

Therefore, the collector–base junctions of both Q_6 and Q_7 are reverse biased by $13.8 - (-12.5) = 26.3\,\text{V}$, and both Q_6 and Q_7 are biased in the active region.

Comment: The nominal bias current listed in Table 13.2 is 200 µA, which correlates well with our calculated value of 236 µA. As long as the B–C junctions of Q_6 and Q_7 remain reverse biased, the bias currents remain constant. This means that the bias current is independent of V^+ and V^- over a wide range of voltages.

The PSpice analysis, using $I_S = 2 \times 10^{-15}\,\text{A}$ for the BJTs shows that the currents in the two branches of the current source are essentially 220 µA. This compares very favorably with the 236 µA obtained by the hand analysis.

Transistors Q_1 through Q_6 and diode D_1 in Figure 13.20 are all matched, which means that $I_{C5} = I_{C4} \cong 200\,\text{µA}$. The current in D_2 establishes the diode voltage that also biases Q_{14} and Q_{15}. The nominal value of I_{C18} is 2 mA.

13.4.4   CA3140 Small-Signal Analysis

We analyze the small-signal voltage gain of the CA3140 op-amp by dividing the configuration into its basic circuits and using results previously obtained.

Input Stage

From the results in Chapter 11, the small-signal differential voltage gain can be written

$$A_d = \sqrt{2K_pI_{Q5}}\,(r_{o10}\|R_{act1}\|R_{i2}) \tag{13.56}$$

where I_{Q5} is the bias current supplied by Q_2 and Q_5. Resistance r_{o10} is the output resistance looking into the drain of M_{10}, R_{act1} is the effective resistance of the active load, and R_{i2} is the input resistance of the gain stage.

Example 13.12   Objective: Calculate the small-signal differential voltage gain of the CA3140 op-amp input stage.

Assume a conduction parameter value of $K_p = 0.6\,\text{mA/V}^2$ for M_{10}, an npn bipolar current gain of $\beta_n = 200$, and a bipolar Early voltage of $V_A = 50\,\text{V}$.

Solution: The input resistance to the gain stage is $R_{i2} = r_{\pi13}$; therefore,

$$R_{i2} = r_{\pi13} = \frac{\beta_n V_T}{I_{C13}} = \frac{(200)(0.026)}{0.20} = 26\,\text{k}\Omega$$

Resistances r_{o10} and R_{act1} are normally in the hundreds of kilohms or megohm range, so the small value of R_{i2} dominates the parallel resistance value in the gain expression. We then have

$$A_d \cong \sqrt{2K_pI_{Q5}}\,(R_{i2}) = \sqrt{2(0.6)(0.2)}\,(26) = 12.7$$

Comment: The low input resistance of the gain stage severely loads the input stage, which in turn results in a relatively low voltage gain for the input stage.

Gain Stage

The magnitude of the small-signal voltage gain for the second stage is

$$|A_{v2}| = g_{m13}(r_{o13}\|R_{o4}\|R_{i3}) \tag{13.57}$$

where R_{i3} is the input resistance of the output stage and R_{o4} is the output resistance of the cascode configuration of Q_3 and Q_4. Transistor Q_{17}, which is the input transistor of the output stage, is connected as an emitter follower, which means that R_{i3} is typically in the megohm range. Similarly, the output resistance R_{o4} of the cascode configuration is typically in the megohm range.

The voltage gain of the second stage is then approximately

$$|A_{v2}| \cong g_{m13}r_{o13} \tag{13.58}$$

Example 13.13 **Objective:** Calculate the small-signal voltage gain of the second stage of the CA3140 op-amp.

Assume an Early voltage of $V_A = 150\,\text{V}$ for Q_{13}.

Solution: The transconductance is

$$g_{m13} = \frac{I_{C13}}{V_T} = \frac{0.20}{0.026} = 7.69\,\text{mA/V}$$

and the output resistance is

$$r_{o13} = \frac{V_A}{I_{C13}} = \frac{150}{0.20} = 750\,\text{k}\Omega$$

The voltage gain is therefore

$$|A_{v2}| = g_{m13} r_{o13} = (7.69)(750) = 5768$$

Comment: The second stage of the CA3140 operational amplifier provides the majority of the voltage gain.

Overall Gain

Since we have taken the loading effects of each following stage into account, the overall voltage gain is the product of the individual gain factors, or

$$A_v = A_d A_{v2} A_{v3} \tag{13.59}$$

where A_{v3} is the voltage gain of the output stage. If we assume that $A_{v3} \cong 1$ for the emitter-follower output stage, then the overall gain of the CA3140 op-amp is

$$A_v = A_d A_{v2} A_{v3} = (12.7)(5768)(1) = 73,254 \tag{13.60}$$

Typical values of the gain of the CA3140 op-amp are in the area of 100,000; thus, our calculations are in reasonable agreement with this value.

Frequency Response

The CA3140 op-amp is internally compensated by the Miller compensation technique to introduce a dominant pole, as was done in the 741 op-amp. The feedback capacitor C_1 is 12 pF and is connected between the collector and the base of Q_{13}, as shown in Figure 13.20. From Miller's theorem, the effective input capacitance of the second stage is

$$C_i = C_1(1 + |A_{v2}|) \tag{13.61}$$

The low-frequency dominant pole is

$$f_{PD} = \frac{1}{2\pi R_{eq} C_i} \tag{13.62}$$

where R_{eq} is the equivalent resistance between the second-stage input node and ground. Since this resistance is dominated by the input resistance to Q_{13}, we have

$$R_{eq} \cong R_{i2} = r_{\pi13} \tag{13.63}$$

Example 13.14   **Objective:** Determine the dominant-pole frequency and unity-gain bandwidth of the CA3140 op-amp.

Again, we will use results from previous calculations.

Solution: Previously, we determined that $|A_{v2}| = 5768$; therefore, the effective input capacitance is

$$C_i = C_1(1 + |A_{v2}|) = 12(1 + 5768) = 69{,}228 \text{ pF}$$

The gain stage input resistance is

$$R_{i2} = r_{\pi13} = 26 \text{ k}\Omega$$

which means that

$$f_{PD} \cong \frac{1}{2\pi R_{i2}C_i} = \frac{1}{2\pi(26 \times 10^3)(69{,}228 \times 10^{-12})} = 88 \text{ Hz}$$

Finally, the unity-gain bandwidth is

$$f_T = f_{PD}A_v = (88)(73{,}254) \Rightarrow 6.4 \text{ MHz}$$

Comment: This unity-gain bandwidth value compares favorably with typical values of 4.5 MHz listed in the data sheet.

Figure 13.22   Figure for Exercise 13.19

Test Your Understanding

13.17   Using the CA3140 op-amp circuit and the transistor parameters given in Example 13.11, determine the minimum supply voltages that will still mainain Q_6 and Q_7 in the active region. Assume $V^+ = -V^-$. (Ans. $V^+ = -V^- = 1.86$ V)

13.18   Consider the CA3140 op-amp bias circuit in Figure 13.21. Assume that $V_{BE7} = 0.6$ V and $R_1 = 5$ kΩ. If the p-channel MOSFET parameters are $K_p = 0.3$ mA/V^2 and $|V_{TP}| = 1.4$ V, determine I_1, I_2, and V_{SG}. (Ans. $V_{SG} = 2.54$ V, $I_1 = I_2 = 0.388$ mA)

*13.19   Assume the gain stage of the CA3140 op-amp is modified to include an emitter resistor, as shown in Figure 13.22. Let $\lambda = 0.02$ V^{-1} for M_{10}. Assume all other transistor parameters are the same as those in Example 13.12. If the transistor bias currents in M_{10} and Q_{12} are 100 μA and the current in Q_{13} is 200 μA, determine the new value of the small-signal differential voltage gain of the input stage. (Ans. 69.1)

13.5   JFET OPERATIONAL AMPLIFIER CIRCUITS

The advantage of using MOSFETs as input devices in a BiCMOS op-amp is that extremely small input bias currents can be achieved. However, MOSFET gates connected to outside terminals of an IC must be protected against electrostatic damage. Typically, this is accomplished by using back-biased diodes on the input, as was shown in Figure 13.20. Unfortunately, the input op-amp bias currents are then dominated by the leakage currents in the protection diodes, which means that the small input bias currents cannot be fully realized. JFETs as input devices also offer the advantage of low input currents, and they do not need electrostatic protection devices. Input gate currents in a JFET are usually well below 1 nA, and are often on the order of 10 pA. In addition, JFETs offer greatly reduced noise properties.

In this section, we will examine two op-amp configurations using JFETs as input devices. Since the analysis is essentially identical to that given in the last two sections, we will limit ourselves to a general discussion of the circuit characteristics.

13.5.1   Hybrid FET Op-Amp, LH002/42/52 Series

Figure 13.23 is a simplified circuit diagram of an LH002/42/52 series op-amp, which uses a pair of JFETs for the input differential pair. Note that the general layout of the circuit is essentially the same as that of the 741 op-amp.

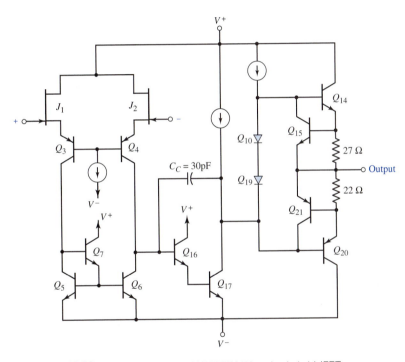

Figure 13.23   Equivalent circuit, LH0022/42/52 series hybrid JFET op-amp

The input diff-amp stage consists of transistors J_1, J_2, Q_3, and Q_4; J_1 and J_2 are n-channel JFETs operating in a source-follower configuration. The differential output signal from J_1 and J_2 is the input to the common-base amplifier formed by Q_3 and Q_4, which provides a large voltage gain. Transistors Q_5, Q_6, and Q_7 form the active load for the input stage.

The gain stage is composed of Q_{16} and Q_{17} connected in a Darlington pair configuration. This stage also includes a 30 pF compensation capacitor. The output stage consists of the complementary push–pull emitter-follower configuration of Q_{14} and Q_{20}. Transistors Q_{14} and Q_{20} are biased slightly "on" by diodes Q_{10} and Q_{19}, to minimize crossover distortion. Transistors Q_{15} and Q_{21} and the associated 27 Ω and 22 Ω resistors provide the short-circuit protection.

An abbreviated data sheet for an LH0042C op-amp is shown in Table 13.3. Note the very large differential-mode input resistance and the low input bias current.

Table 13.3 LH0042C data

Parameter	Minimum	Typical	Maximum	Units
Input bias current		15	50	pA
Differential-mode input resistance		10^{12}		Ω
Input capacitance		4		pF
Open-loop gain ($R_L = 1\,\text{k}\Omega$)	25,000	100,000		V/V
Unity-gain frequency		1		MHz

13.5.2   Hybrid FET Op-Amp, LF155 Series

Another example of a JFET op-amp is the LF155 BiFET op-amp. A simplified circuit diagram showing the input stage is in Figure 13.24. The input BiFET op-amp stage consists of p-channel JFETs J_1 and J_2 biased by the bipolar transistor Q_1. The active load for the input diff-amp consists of the p-channel JFETs J_3 and J_4, for which $V_{GS} = 0$.

Figure 13.24   Equivalent circuit, LF155 BiFET op-amp input stages

A two-sided output from the input diff-amp stage is connected to a second diff-amp stage consisting of Darlington pairs Q_7 through Q_{10}. The second, or gain, stage is biased by bipolar transistor Q_5. The cascode configuration of J_5 and Q_2 form the active load for the gain stage.

The circuit has a common-mode feedback loop in the bias circuit. The base of Q_6 is connected to the collector of Q_5. If the drain voltages of J_1 and J_2 increase, the Darlington second stage drives the base voltage of Q_6 higher. The current in Q_6 then increases, reducing the drain currents in J_1 and J_2, since I_{C1} is a constant current. Smaller drain currents cause the voltages at the J_1 and J_2 drains to decrease, which then stabilizes the drain voltages.

JFET J_6 is connected as a current source, which establishes a reference current in Q_3, Q_4, and J_6. This reference current then produces the bias currents in the current mirrors Q_4–Q_5 and Q_1–Q_2–Q_3.

In this BiFET op-amp, we see the advantages of incorporating both JFET and bipolars in the same circuit. The JFET input devices provide a very high input impedance, normally in the range of $10^{12}\,\Omega$. The current-connected transistor J_6 allows the reference bias current to be controlled without the use of a resistor. Incorporating bipolar transistors in the second stage takes advantage of their higher transconductance values compared to JFETs, to produce a high second-stage gain.

Test Your Understanding

13.20   Consider the LF155 BiFET input stage in Figure 13.24. The p-channel JFET parameters are $I_{DSS} = 300\,\mu\text{A}$, $V_P = 1\,\text{V}$, and $\lambda = 0.01\,\text{V}^{-1}$. The supply voltages are $V^+ = 5\,\text{V}$ and $V^- = -5\,\text{V}$. Let $V_{BE}(\text{npn}) = 0.6\,\text{V}$ and $V_{EB}(\text{pnp}) = 0.6\,\text{V}$. Determine the bias currents I_{C3}, I_{C2}, and I_{C1}. (Ans. $I_{C1} = I_{C2} = I_{C3} = 300\,\mu\text{A}$)

13.6   SUMMARY

- In this chapter, we have combined various basic circuit configurations to form larger operational amplifier circuits. In general, an op-amp circuit consists of a diff-amp input stage, a second or gain stage, and an output stage. The design of integrated circuit operational amplifier circuits depends on the use of matched devices.

- The LM741 op-amp is a widely used, general-purpose, bipolar op-amp. This circuit serves as a good case study for a detailed discussion of the circuit design, including a discussion of the input stage design, the Darlington pair gain stage, and the class-AB complementary output stage with the protection circuitry.

- A detailed dc analysis of each stage of the 741 was performed to determine the dc currents and voltages. A detailed small-signal analysis determined the gain of each stage and the overall small-signal voltage gain. The calculated voltage gain of approximately 200,000 agrees well with the typical value given in data sheets. The output resistance is approximately $56\,\Omega$. The 741 is internally compensated, and the dominant pole frequency is on the order of $10\,\text{Hz}$. The unity-gain bandwidth is approximately $1.9\,\text{MHz}$.

- In many cases, all-CMOS operational amplifier circuits require only two stages. These circuits typically drive only low capacitive loads on an IC chip, so the low output impedance of a third output stage is not required. The MC14573 all-CMOS op-amp circuit was considered. The calculated small-signal voltage gain of this two-stage circuit was approximately 84 dB, which agrees well with data sheets. Even though the gain is smaller than that of typical bipolar op-amps, this circuit is useful in specialized on-chip applications.

- An all-CMOS folded cascode operational amplifier circuit was analyzed. The advantage of this circuit is a very high output resistance that produces a very large differential-mode voltage gain.

- An all-CMOS current-mirror operational amplifier circuit was considered. The advantage of this circuit is an increased gain–bandwidth product. A cascode version of the current mirror op-amp was briefly considered.

- Two BiCMOS operational amplifier circuits were discussed. The first was a modified version of the folded cascode design. The use of bipolar cascode transistors in this circuit produces an increased phase margin. A CA3140 BiCMOS op-amp was analyzed. A unique aspect of this circuit is that the bias current generated from the bias circuitry is independent of bias voltage over a wide range of applied bias voltages.

- Two examples of a hybrid JFET or BiFET op-amp circuit were considered. The input stage is composed of a JFET differential pair, while the remainder of the circuits are designed primarily with bipolar transistors. Using JFETs as the input devices keeps the input bias currents extremely small, usually in the picoampere range.

CHECKPOINT

After studying this chapter, the reader should have the ability to:

✔ Design a basic bipolar or MOSFET operational amplifier circuit. (Section 13.1)

✔ Analyze and understand the operation and characteristics of the LM741 op-amp circuit. (Section 13.2)

✔ Analyze and understand the operation and characteristics of CMOS op-amp circuits, including the folded cascode and the CMOS current-mirror circuits. (Section 13.3)

✔ Analyze and understand the operation and characteristics of BiCMOS operational amplifier circuits. (Section 13.4)

REVIEW QUESTIONS

1. Describe the principal stages of a general-purpose operational amplifier.
2. What is meant by the term matched transistors? What parameters in BJTs and MOSFETs are identical in matched devices?
3. Describe the operation and characteristics of a BJT complementary push–pull output stage. What are the advantages of this circuit?
4. Describe the operation and characteristics of a MOSFET complementary push–pull output stage. What are the advantages of this circuit?
5. Describe the advantages and disadvantages of an all-BJT op-amp circuit.
6. Describe the advantages and disadvantages of an all-CMOS op-amp circuit.
7. Describe the advantages and disadvantages of a BiCMOS op-amp circuit.
8. Describe the advantages and disadvantages of a JFET op-amp circuit.
9. Sketch and describe the characteristics of the 741 input stage.
10. Describe what is meant by output short-circuit protection.
11. Describe the frequency compensation technique in the 741 op-amp circuit.
12. Sketch and describe the general characteristics of a folded cascode circuit.

13. Sketch and describe the general characteristics of a current–mirror op-amp circuit. Why is the gain not increased? What is the principal advantage of this circuit?

14. Sketch and describe the principal advantage of a BiCMOS folded cascode op-amp circuit.

15. Explain why an output resistance on the order of five hundred megohms may not be achieved in practice.

16. What are the principal factors limiting the unity-gain bandwidth of an op-amp circuit?

PROBLEMS

Section 13.1   General Op-Amp Circuit Design

D13.1 Design the circuit in Figure 13.2 such that the maximum power dissipated in the circuit is 15 mW and such that the common-mode input voltage is in the range $-3 \leq v_{CM} \leq 3$ V. Using a computer simulation, adjust the value of R_3 such that the output voltage is zero for zero input signal voltages.

13.2 Using the results of Problem 13.1, determine, from a computer simulation, the differential-mode voltage gain of the diff-amp and the voltage gain of the second stage of the op-amp circuit in Figure 13.2. Use standard transistor models in the circuit.

Figure P13.3

*13.3 Consider the BJT op-amp circuit in Figure P13.3. The transistor parameters are: $\beta(\text{npn}) = 120$, $\beta(\text{pnp}) = 80$, $V_A = 80$ V (all transistors), and base–emitter turn-on voltage = 0.6 V (all transistors). (a) Determine the small-signal differential-mode voltage gain. (b) Find the differential-mode input resistance. (c) Determine the unity-gain bandwidth.

Section 13.2   A Bipolar Operational Amplifier Circuit

13.4 Consider the input stage of the 741 op-amp in Figure 13.4(b). (a) Assume the input voltages are $V_1 = 0$ and $V_2 = +15$ V. Consider the B–E voltage of each transistor and determine which transistor acts as the protection device. (b) Repeat part (a) for $V_1 = -15$ V and $V_2 = 0$.

13.5 For the input stage of the 741 op-amp, assume B–E breakdown voltages of 5 V for the npn devices and 50 V for the pnp devices. Estimate the differential input voltage at which breakdown will occur.

RD13.6 Consider the bias circuit portion of the 741 op-amp in Figure 13.5. (a) Redesign the resistor values of R_5 and R_4 such that $I_{REF} = 0.50$ mA and $I_{C10} = 30\,\mu$A for bias voltages of ± 15 V. Assume base–emitter turn-on voltages of 0.6 V. (b) Using the results of part (a), determine I_{REF} and I_{C10} if the bias voltages change to ± 5 V.

13.7 Repeat Problem 13.6 using the exponential relationship between collector current and base–emitter voltage in which $I_S = 10^{-14}$ A. What are the actual values of base–emitter voltage in each case?

13.8 The minimum recommended supply voltages for the 741 op-amp are $V^+ = 5$ V and $V^- = -5$ V. Using these lower supply voltages, calculate: (a) I_{REF}, I_{C10}, I_{C6}, I_{C17}, and I_{C13A}; and (b) the voltage gains of the input and gain stages for the circuit values in Figure 13.3.

13.9 An expanded circuit diagram of the 741 input stage is shown in Figure 13.6. Assume $I_{C10} = 19\,\mu$A. If the current gain of the npn transistors is $\beta_n = 200$ and the current gain of the pnp transistors is $\beta_p = 10$, determine I_{C9}, I_{C2}, I_{C4}, I_{B9}, and I_{B4}.

13.10 Consider the 741 op-amp in Figure 13.3, biased with $V^+ = 15$ V and $V^- = -15$ V. Assume that no load is connected at the output, and let the input voltages be zero. Calculate the total power dissipated in the op-amp circuit. What are the currents supplied by V^+ and V^-?

13.11 Consider the 741 circuit in Figure 13.3. (a) Determine the maximum range of common-mode input voltage if the bias voltages are ± 15 V. (b) Repeat part (a) if the bias voltages are ± 5 V.

13.12 For Q_{15} in the output stage of the 741 op-amp, assume $I_S = 10^{-14}$ A. If the output is inadvertently connected to $V^- = -15$ V and the inputs are at zero, estimate the currents I_{C14} and I_{C15}.

13.13 Consider the output stage in Figure P13.13, with parameters $V^+ = 10$ V, $V^- = -10$ V, $R_L = 4$ kΩ, and $I_{Bias} = 0.25$ mA. Assume the diode parameters are $I_S = 2 \times 10^{-14}$ A and the transistor parameters are $I_S = 5 \times 10^{-14}$ A. (a) For $v_I = 0$, determine V_{BB}, I_{CN}, and I_{CP}. (b) For $v_I = 5$ V, determine v_o, i_L, V_{BB}, I_{CN}, and I_{CP}.

D13.14 Figure P13.14 shows a circuit often used to provide the V_{BB} voltage in the op-amp output stage. Assume $I_S = 10^{-14}$ A for the transistor, $I_{Bias} = 180\,\mu$A, and $I_C = 0.9 I_{Bias}$. Neglect the base current. Design the circuit such that $V_{BB} = 1.157$ V.

13.15 Assume bias voltages on the 741 op-amp of ± 15 V. (a) Determine the differential-mode voltage gain of the first stage if $R_1 = R_2 = 0$. (b) Determine the voltage gain of the second stage if $R_8 = 0$.

13.16 Recalculate the voltage gain of the 741 op-amp input stage if $I_{C10} = 40\,\mu$A.

13.17 Calculate the output resistance of the 741 op-amp if Q_{14} is conducting and Q_{20} is cut off. Assume an output current of 2 mA.

13.18 Determine the differential input resistance of the 741 op-amp.

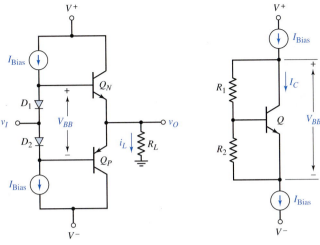

Figure P13.13   **Figure P13.14**

13.19 The frequency response of a particular 741 op-amp shows that the op-amp has a phase margin of 70 degrees. If a second single pole exists, in addition to the dominant pole, determine the frequency of the second pole. Use the overall gain and dominant-pole parameters calculated in Section 13.2.

Section 13.3   CMOS Operational Amplifier Circuits

RD13.20 Consider the MC14573 op-amp in Figure 13.14. The dc bias currents and small-signal voltage gains were determined in Examples 13.8 and 13.9. Redesign the circuit such that the width-to-length ratio of M_1 and M_2 is increased from 12.5 to 50. All other circuit and transistor parameters remain the same. (a) Determine the original transconductance of M_1 and M_2, and the new transconductance value. (b) Determine the new values of voltage gain for the input and second stages, and the overall voltage gain.

13.21 Consider the basic diff-amp with active load and current biasing in Figure 13.14. Using a computer simulation, investigate the change in the voltage at the drain of M_2 as M_1 and M_2 and also as M_3 and M_4 become slightly mismatched.

13.22 The CMOS op-amp in Figure 13.14 is biased at $V^+ = 10\,\text{V}$ and $V^- = -10\,\text{V}$. Assume transistor parameters of $|V_T| = 1.5\,\text{V}$ (all transistors), $(\frac{1}{2})\mu_n C_{ox} = 20\,\mu\text{A/V}^2$, $(\frac{1}{2})\mu_p C_{ox} = 10\,\mu\text{A/V}^2$, $\lambda_p = 0.02\,\text{V}^{-1}$, and $\lambda_n = 0.01\,\text{V}^{-1}$. Let $R_{\text{set}} = 200\,\text{k}\Omega$. Assume transistor width-to-length ratios of 10 for M_3 and M_4, and 20 for all other transistors. (a) Determine I_{REF}, I_Q, and I_{D7}. (b) Find the small-signal voltage gain of the input and second stages, and the overall voltage gain.

13.23 For the CMOS op-amp in Figure 13.14, the dc biasing is designed such that $I_{\text{REF}} = I_Q = I_{D8} = 200\,\mu\text{A}$. The transistor parameters are $|V_T| = 1\,\text{V}$ (all transistors), $\lambda_n = 0.005\,\text{V}^{-1}$, $\lambda_p = 0.01\,\text{V}^{-1}$, $(\frac{1}{2})\mu_n C_{ox} = 20\,\mu\text{A/V}^2$, and $(\frac{1}{2})\mu_p C_{ox} = 10\,\mu\text{A/V}^2$. The transistor width-to-length ratios are 5 for M_5, M_6, and M_8; 10 for M_1 and M_2; and 20 for M_3, M_4, and M_7. Determine the small-signal voltage gains of the input and second stages, and the overall voltage gain.

13.24 Consider the MC14573 op-amp in Figure 13.14, with circuit and transistor parameters as given in Examples 13.8 and 13.9. If the compensation capacitor is $C_1 = 12\,\text{pF}$, determine the dominant-pole frequency.

13.25 The CMOS op-amp in Figure 13.14 has circuit and transistor parameters as given in Problem 13.22. Determine the compensation capacitor required such that the dominant-pole frequency is $f_{PD} = 8\,\text{Hz}$.

13.26 Consider the CMOS op-amp in Figure 13.14, with transistor and circuit parameters as given in Examples 13.8 and 13.9. Determine the output resistance R_o of the open-loop circuit.

13.27 A simple output stage for an NMOS op-amp is shown in Figure P13.27. Device M_1 operates as a source follower. Assume that M_1 and M_2 are biased at $I_D = 0.5\,\text{mA}$. (a) Calculate the small-signal open-circuit voltage gain $A_v = v_O/v_I$. (b) If the output resistance of source v_I is $10\,\text{k}\Omega$, determine the output resistance of this output stage.

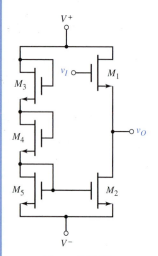

Figure P13.27

*13.28 The CMOS folded cascode circuit in Figure 13.16 is biased at $\pm 5\,\text{V}$ and the reference current is $I_{REF} = 50\,\mu\text{A}$. The transistor parameters are $V_{TN} = 0.5\,\text{V}$, $V_{TP} = -0.5\,\text{V}$, $K_n = K_p = 0.5\,\text{mA/V}^2$, and $\lambda_n = \lambda_p = 0.015\,\text{V}^{-1}$. (a) Determine the small-signal differential voltage gain. (b) Find the output resistance of the circuit. (c) If the capacitance at the output node is $C_L = 5\,\text{pF}$, determine the unity-gain bandwidth of the amplifier.

*RD13.29 The CMOS folded cascode amplifier in Figure 13.16 is to be redesigned to provide a differential voltage gain of 10,000. The biasing is the same as described in Problem 13.28. The transistor parameters are $V_{TN} = 0.5\,\text{V}$, $V_{TP} = -0.5\,\text{V}$, $k_n' = 80\,\mu\text{A/V}^2$, $k_p' = 35\,\mu\text{A/V}^2$, $\lambda_n = 0.015\,\text{V}^{-1}$, and $\lambda_p = 0.02\,\text{V}^{-1}$. Assume $(W/L)_p = 2.2(W/L)_n$ where appropriate so that the electrical parameters of PMOS and NMOS devices are nearly identical.

*D13.30 The CMOS folded cascode amplifier of Figure 13.16 is to be designed to provide a differential voltage gain of 25,000. The maximum power dissipated in the circuit is to be limited to $3\,\text{mW}$. Assume transistor parameters as described in Problem 13.29, except the relation between NMOS and PMOS width-to-length ratios need not be maintained.

13.31 The bias current in the CMOS current-gain op-amp in Figure 13.17 is $I_Q = 60\,\mu\text{A}$. The transistor parameters are $V_{TN} = 0.5\,\text{V}$, $V_{TP} = -0.5\,\text{V}$, $K_n = K_p = 0.5\,\text{mA/V}^2$ (all transistors except M_5 and M_6), and $\lambda_n = \lambda_p = 0.015\,\text{V}^{-1}$. Let $B = 3$. (a) Determine the small-signal differential voltage gain. (b) Find the output resistance of the circuit. (c) If the total capacitance at the output terminal is $5\,\text{pF}$, determine the dominant-pole frequency and the unity-gain bandwidth.

RD13.32 The CMOS current gain op-amp in Figure 13.17 is to be redesigned to provide a differential voltage gain of 400. The transistor parameters are $V_{TN} = 0.5\,\text{V}$, $V_{TP} = -0.5\,\text{V}$, $k_n' = 80\,\mu\text{A/V}^2$, $k_p' = 35\,\mu\text{A/V}^2$, $\lambda_n = 0.015\,\text{V}^{-1}$, and $\lambda_p = 0.02\,\text{V}^{-1}$. The bias current is to be $I_Q = 80\,\mu\text{A}$. Let $B = 2.5$. (a) Design the basic amplifier to provide the specified voltage gain. (b) Design a current source to provide the necessary bias current. (c) Determine the unity-gain bandwidth if the capacitance at the output terminal is $3\,\text{pF}$.

RD13.33 Redesign the CMOS cascode current mirror in Figure 13.18 to provide a differential voltage gain of 20,000. The bias current and transistor parameters are the same as in Problem 13.32. (a) Design the basic amplifier to provide the specified voltage gain. (b) Design a current source to provide the necessary bias current. (c) Determine the unity gain bandwidth if the capacitance at the output terminal is $3\,\text{pF}$.

Section 13.4   BiCMOS Operational Amplifier Circuits

13.34 A BiCMOS amplifier is shown in Figure P13.34. The transistor parameters are $V_{TP} = -0.7\,\text{V}$, $k_p' = 40\,\mu\text{A/V}^2$, $(W/L) = 25$, $\lambda = 0.02\,\text{V}^{-1}$, $\beta = 120$, and $V_A = 120\,\text{V}$. The bias current is $I_Q = 200\,\mu\text{A}$. Determine the small-signal differential voltage gain.

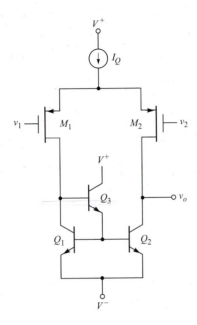

Figure P13.34

D13.35 Design a BiCMOS amplifier that is complementary to the one in Figure P13.34 in that the input devices are NMOS and the load transistors are pnp. Assume transistor parameters of $V_{TN} = 0.5\,\text{V}$, $k_n' = 80\,\mu\text{A/V}^2$, $(W/L) = 25$, $\lambda = 0.015\,\text{V}^{-1}$, $\beta = 80$, and $V_A = 80\,\text{V}$. Assume the bias current is $I_Q = 200\,\mu\text{A}$. Determine the small-signal differential voltage gain.

*13.36 The reference current in the BiCMOS folded cascode amplifier in Figure 13.19 is $I_{\text{REF}} = 200\,\mu\text{A}$ and the circuit bias voltages are $\pm10\,\text{V}$. The MOS transistor parameters are the same as in Problem 13.28. The BJT parameters are $\beta = 120$ and $V_A = 80\,\text{V}$. (a) Determine the small-signal differential voltage gain. (b) Find the output resistance of the circuit. (c) If the capacitance at the output node is $5\,\text{pF}$, determine the unity-gain bandwidth of the amplifier.

*D13.37 The BiCMOS folded cascode amplifier in Figure 13.19 is to be designed to provide a differential voltage gain of 25,000. The maximum power dissipated in the circuit is to be limited to $10\,\text{mW}$. Assume MOS transistor parameters as described in Problem 13.29. The BJT parameters are $\beta = 120$ and $V_A = 80\,\text{V}$.

13.38 If the CA3140 op-amp is biased at $V^+ = 15\,\text{V}$ and $V^- = -15\,\text{V}$, determine the input common-mode voltage range. Assume B–E voltages of $0.6\,\text{V}$ for the bipolar transistors and $|V_{TP}| = 1.4\,\text{V}$ for the MOSFETs.

13.39 Consider the bias circuit portion of the CA3140 op-amp in Figure 13.21. If $V_{BE7} = 0.6\,\text{V}$ for Q_7 and $V_{TP} = -1.4\,\text{V}$ for M_8, determine the necessary conduction parameter for M_8 such that $I_1 = I_2 = 300\,\mu\text{A}$.

13.40 In the bias circuit portion of the CA3140 op-amp in Figure 13.21, the bipolar transistor parameters are $V_{BE}(\text{npn}) = 0.6\,\text{V}$ and $V_{EB}(\text{pnp}) = 0.6\,\text{V}$, and the MOSFET parameters are $|V_{TP}| = 1.4\,\text{V}$ and $K_p = 0.25\,\text{mA/V}^2$. If the power supply voltages are $V^+ = -V^- \equiv V_S$, determine the minimum value of V_S such that the bias currents are independent of the supply voltage.

13.41 Consider the CA3140 op-amp in Figure 13.20. If the bias currents change such that $I_{C5} = I_{C4} = 300\,\mu\text{A}$, determine the voltage gains of the input and second stages, and find the overall voltage gain.

13.42 Assume the gain stage of the CA3140 op-amp is modified to include an emitter resistor, as shown in Figure 13.22. Let $\lambda = 0.02\,\text{V}^{-1}$ for M_{10}. If the transistor bias currents in M_{10} and Q_{12} are $150\,\mu\text{A}$ and the current in Q_{13} is $300\,\mu\text{A}$, determine the dominant-pole frequency and unity-gain bandwidth.

Section 13.5   JFET Operational Amplifier Circuits

13.43 In the LF155 BiFET op-amp in Figure 13.24, the combination of Q_3, J_6, and Q_4 establishes the reference bias current. Assume the power supply voltages are $V^+ = 10\,\text{V}$ and $V^- = -10\,\text{V}$. The transistor parameters are $V_{EB}(\text{on}) = 0.6\,\text{V}$, $V_{BE}(\text{on}) = 0.6\,\text{V}$, and $V_P = 4\,\text{V}$ for Q_3, Q_4, and J_6, respectively. Determine the required I_{DSS} value for J_6 to establish a reference current of $I_{\text{REF}} = 0.8\,\text{mA}$.

13.44 Consider the circuit in Figure P13.44. A JFET diff-amp input stage drives a bipolar Darlington second stage. The p-channel differential pair J_1 and J_2 are connected to the bipolar active load transistors Q_3 and Q_4. Assume JFET parameters of $V_P = 3\,\text{V}$, $I_{DSS} = 200\,\mu\text{A}$, and $\lambda = 0.02\,\text{V}^{-1}$. The bipolar transistor parameters are $\beta = 100$ and $V_A = 50\,\text{V}$. (a) Determine the input resistance R_{i2} to the second stage. (b) Calculate the small-signal differential-mode voltage gain of the input stage. Compare this value to the 741 and CA3140 input stage results.

Figure P13.44

D13.45 Consider the BiFET differential input stage in Figure P13.45, biased with power supply voltages V^+ and V^-. Let $V^+ = -V^- \equiv V_S$. (a) Design the bias circuit such that $I_{\text{REF2}} = 100\,\mu\text{A}$ for supply voltages in the range $3 \leq V_S \leq 12\,\text{V}$. Determine

Figure P13.45

V_{ZK}, R_3, and the JFET parameters. (b) Determine the value of R_4 such that $I_{O1} = 500\,\mu A$ when $V^+ = 12\,V$.

13.46 The BiFET diff-amp input stage in Figure P13.45 is biased at $I_{O1} = 1\,mA$. The JFET parameters are $V_P = 4\,V$, $I_{DSS} = 1\,mA$, and $\lambda = 0.02\,V^{-1}$. The bipolar transistor parameters are $\beta = 200$ and $V_A = 100\,V$. (a) For $R_1 = R_2 = 500\,\Omega$, determine the minimum load resistance R_L such that a differential-mode voltage gain of $A_d = 500$ is obtained in the input stage. (b) If $R_L = 500\,k\Omega$, determine the range of resistance values $R_1 = R_2$ such that a differential-mode voltage gain of $A_d = 700$ is obtained in this input stage.

COMPUTER SIMULATION PROBLEMS

13.47 Consider the input stage and bias circuit of the 741 op-amp in Figure 13.5. Transistor Q_{10} may be replaced by a constant-current source equal to $19\,\mu A$. Assume: the npn devices have parameters $\beta = 200$ and $V_A = 150\,V$; the pnp devices have parameters $\beta = 50$ and $V_A = 50\,V$; and all transistors have $I_S = 10^{-14}\,A$. (a) Using an appropriate ac load at the collector of Q_6, determine the differential gain of the input stage. (b) Determine the differential-mode input resistance. (c) Determine the common-mode input resistance.

13.48 The output stage of the 741 op-amp is shown in Figure 13.9. Transistor Q_{13} may be replaced with a constant-current source equal to $0.18\,mA$. The transistor parameters are as given in Problem 13.47. (a) Plot the voltage transfer function v_O verus v_{I3}. What is the voltage gain? Has the crossover distortion been eliminated? (b) Apply an input voltage v_{I3} that establishes an output voltage of $v_O = 10\,V$, for example, and set $R_L = 0$. Find the output short-circuit current and the transistor currents.

13.49 The bias circuit and gain stage, including the compensation capacitor, of the 741 op-amp is shown in Figure 13.7. Transistor Q_{13} can be simulated by connecting two pnp transistors in parallel, with relative B–E junction areas of 0.25 and 0.75 compared to all other pnp transistors. (a) Determine the low-frequency voltage gain. (b) Plot the magnitude of the voltage gain versus frequency. Compare the 3 dB frequency to the dominant-pole frequency found in Example 13.7.

13.50 Consider the BiCMOS input stage of the CA3140 op-amp in Figure 13.20. Transistor Q_5 can be replaced with a constant-current source of 200 μA. Assume: bipolar transistor parameters of $\beta = 200$, $I_{EO} = 10^{-14}$ A, and $V_A = 50$ V; and MOSFET parameters of $K_p = 0.6\,\text{mA/V}^2$, $|V_{TP}| = 1$ V, and $\lambda = 0.01\,\text{V}^{-1}$. Using an appropriate ac load at the collector of Q_{12}, determine the differential gain of the input stage. Compare the computer analysis results with those in Example 13.12.

13.51 Consider the CMOS op-amp in Figure 13.14. Assume the circuit and transistor parameters are as given in Example 13.8. In addition, let $\lambda = 0.01\,\text{V}^{-1}$ for all transistors. (a) Determine the overall low-frequency differential voltage gain. Compare these results with those in Example 13.9. (b) If the compensation capacitor is $C_1 = 12\,\text{pF}$, plot the magnitude of the voltage gain versus frequency. What is the 3 dB frequency?

DESIGN PROBLEMS

[Note: Each design should be correlated with a computer analysis.]

*D13.52 Redesign the bias circuit of the 741 op-amp such that a current $I_{C10} = 25\,\mu\text{A}$ is established when $V^+ = -V^- = 5$ V. Limit the power dissipated in the input stage and the bias circuit to 2.5 mW.

*D13.53 Consider the bipolar op-amp circuit in Figure P13.53. Design the circuit such that the differential gain is at least 800, and the output voltage is zero when the input

Figure P13.53

voltages are zero. The transistor current gains are 120 for all transistors, and the base–emitter voltages are 0.6 V, where appropriate.

*D13.54 Redesign the CMOS op-amp in Figure 13.14 to provide a minimum overall voltage gain of at least 50,000. The bias voltages are $V^+ = 10$ V and $V^- = -10$ V. The threshold voltage is $|V_T| = 1$ V for all transistors, and $\lambda = 0.01$ V^{-1} for all transistors. Design reasonable width-to-length ratios and bias currents.

*D13.55 Consider the CMOS op-amp in Figure 13.14. Design a complementary CMOS circuit in which each element is replaced by its complement. The bias voltages are ± 5 V. The threshold voltage is $|V_T| = 0.7$ V for all transistors, and $\lambda = 0.01$ V^{-1} for all transistors. Design reasonable width-to-length ratios and bias currents to provide a minimum overall voltage gain of at least 20,000.

14

Nonideal Effects in Operational Amplifier Circuits

14.0 PREVIEW

Chapter 9 introduced the ideal operational amplifier and covered a few of its many applications. In the previous chapter, we analyzed actual operational amplifier circuits, including the classic 741 op-amp. From those discussions, we can identify sources of nonideal properties in actual op-amps. Although nonideal effects could have been introduced in Chapter 9, that discussion would have been less meaningful since the source of any nonideal effect would not have been completely understood at that time. In particular, the reason for a very low dominant-pole frequency in the basic amplifier would have been a mystery. Therefore, the discussion of nonideal effects in op-amp circuits has been postponed until now.

This chapter opens by discussing and defining several practical op-amp parameters that will be further analyzed as to the effect they have on the nonideal characteristics of op-amp circuits.

We have seen how matched transistor characteristics are utilized in the design of diff-amp circuits. However, slight mismatches may occur. One part of this chapter is devoted to determining the effect of these slight transistor mismatched characteristics on the op-amp properties.

A general goal of this chapter is for the reader to understand the source of nonideal effects in op-amps and to be able to minimize their effects in the design of op-amp circuits.

14.1 PRACTICAL OP-AMP PARAMETERS

In ideal op-amps, we assume, for example, that the differential voltage gain is infinite, the input resistance is infinite, and the output resistance is zero. In practical op-amp circuits, these ideal parameter values are not realized. In this section, we define some of the practical op-amp parameters that will be considered in detail throughout the chapter. We will discuss and analzye the effect of these nonideal parameters in op-amp circuits.

14.1.1   Practical Op-Amp Parameter Definitions

Input voltage limits. Two input voltage limitations must be considered—a dc input voltage limit and a differential signal input voltage. All transistors in the input diff-amp stage must be properly biased, so there is a limit in the range of common-mode input voltage that can be applied and still maintain the proper transistor biasing. The maximum differential input signal voltage that can be applied and still maintain linear circuit operation is limited primarily by the maximum allowed output signal voltage.

Output voltage limits. The output voltage of the op-amp can never exceed the limits of the dc supply voltages. In practice, the difference between the bias voltage and output voltage must be greater than 1 to 4 V, depending on the design of the output stage. Otherwise, the output voltage saturates and is no longer a function of input voltage.

Output current limitation. The maximum current out of or into the op-amp is determined by the current ratings of the output transistors. Practical op-amp circuits cannot source or sink an infinite amount of current.

Finite open-loop voltage gain. The open-loop gain of the ideal op-amp is assumed to be infinite. In practice, the open-loop gain of any op-amp circuit is always finite. This nonideal parameter value will affect circuit performance.

Input resistance. The input resistance R_i is the small-signal resistance between the inverting and noninverting terminals when a differential voltage is applied. Ideally, this parameter is infinite, but, especially for BJT circuits, this parameter is finite.

Output resistance. The output resistance is the Thevenin equivalent small-signal resistance looking back into the output terminal of the op-amp measured with respect to ground. The ideal output resistance is zero, which means there is no loading effect at the output. In practice, this value is not zero.

Finite bandwidth. In the ideal op-amp, the bandwidth is infinite. In practical op-amps, the bandwidth is finite because of capacitances within the op-amp circuit.

Slew rate. The slew rate is defined as the maximum rate of change in output voltage per unit of time. The maximum rate at which the output voltage can change is also a function of capacitances within the op-amp circuit.

Input offset voltage. In an ideal op-amp, the output voltage is zero for zero differential input signal voltage. However, mismatches between input devices, for example, may create an output voltage with zero input. The input offset voltage is the applied differential input voltage required to induce a zero output voltage.

Input bias currents. In an ideal op-amp, the input current to the op-amp circuit is assumed to be zero. However, in practical op-amps, especially with BJT input devices, the input bias currents are not zero.

The cause of these nonideal op-amp parameters will be discussed in the following sections, as well as the effect these nonideal parameters have on op-amp circuit performance. A few other nonideal parameters will be considered in the last section of the chapter.

Table 14.1 Nonideal parameter values for three op-amp circuits

	741E			CA3140			LH0042C		
	Typ.	Max.	Unit	Typ.	Max.	Unit	Typ.	Max.	Unit
Input offset voltage	0.8	3	mV	5	15	mV	6	20	mV
Average input offset voltage drift		15	μV/°C				10		μV/°C
Input offset current	3.0	30	nA	0.5	30	pA	2		pA
Average input offset current drift		0.5	nA/°C						
Input bias current	30	80	nA	10	50	pA	2	10	pA
Slew rate	0.7		V/μs	9		V/μs	3		V/μs
CMRR	95		dB	90		dB	80		dB

Table 14.1 lists a few of the nonideal parameter values for three of the op-amps considered in the previous chapter. We will refer to this table as we discuss each of the nonideal parameters.

14.1.2   Input and Output Voltage Limitations

For linear circuit operation, all BJTs in an op-amp circuit must be biased in the forward-active region and all MOSFETs must be biased in the saturation region. For these reasons, there are limitations to the range of input and output voltages in op-amp circuits.

Figure 14.1(a) shows the simple all-BJT op-amp circuit discussed at the beginning of Chapter 13 and Figure 14.1(b) shows the all-CMOS folded cascode op-amp circuit discussed in the last chapter. We will use these two circuits to discuss the input and output voltage limitations.

Input Voltage Limitations

Assume that in the BJT circuit of Figure 14.1(a) we apply a common-mode input voltage such that $v_{cm} = v_1 = v_2$. As v_{cm} increases, the base–collector voltages of Q_1 and Q_2 decrease, since the collector voltages are fixed at two base–emitter voltage drops below V^+. If we assume the minimum base–collector voltage is zero so that the transistor is still biased in the active mode, then the maximum value of v_{cm} is $v_{cm}(\text{max}) = V^+ - 2V_{EB}(\text{on})$.

As v_{cm} decreases, the collector–emitter voltage of Q_3 decreases. If we again assume the minimum base–collector voltage is zero, or the minimum collector–emitter voltage is $V_{BE}(\text{on})$, then, taking into account the base–emitter voltage of the input transistors, the minimum value of v_{cm} is $v_{cm}(\text{min}) = V^- + 2V_{BE}(\text{on})$. So the maximum range of v_{cm} is within approximately 1.4 V of each bias voltage.

The same range of common-mode input voltage can be found for the all-MOSFET diff-amp in Figure 14.1(b). In this case, all MOSFETs must be biased in the saturation region. We can again define the common-mode input voltage as $v_{cm} = v_1 = v_2$. Now, as v_{cm} increases, V_{SD} of M_{11} decreases. The minimum value of V_{SD} is $V_{SD11}(\text{sat}) = V_{SG11} + V_{TP11}$. The maximum value of v_{cm} is then $v_{cm}(\text{max}) = V^+ - [V_{SG1} + (V_{SG11} + V_{TP11})]$. The gate-to-source voltages can be determined from the transistor parameters and currents.

As v_{cm} decreases, the source-to-drain voltage of the input transistors decreases. Assuming that M_3 and M_4 are matched to M_{13}, then the drain-to-source voltage of these transistors is equal to V_{GS13}. The minimum common-

(a)

(b)

Figure 14.1 (a) Simple all-bipolar op-amp circuit; (b) all-CMOS folded cascode op-amp circuit mode input voltage is then $v_{cm}(\text{min}) = V^- + [V_{GS11} + (V_{SG1} + V_{TP1}) - V_{SG1}]$. The V_{SG1} terms cancel, so $v_{cm}(\text{min}) = V^- + [V_{GS11} + V_{TP1}]$.

Output Voltage Limitations

As the output voltage of the BJT circuit in Figure 14.1(a) increases or decreases, the collector–emitter voltages of the output transistors change. Again, assuming the minimum base–collector voltage is zero for a BJT biased in the forward active region, then the maximum output voltage is $v_O(\text{max}) = V^+ - [V_{EB8}(\text{on}) + V_{BE11}(\text{on})]$. The minimum output voltage is similarly found to be $v_O(\text{min}) = V^- + [V_{BE4}(\text{on}) + V_{EB12}(\text{on})]$.

For the all-CMOS circuit in Figure 14.1(b), the maximum output voltage is $v_O(\text{max}) = V^+ - [(V_{SG8} + V_{TP8}) + V_{SG10}]$. The minimum output voltage is $v_O(\text{min}) = V^- + [(V_{GS6} - V_{TN6}) + V_{GS13}]$.

Test Your Understanding

14.1 Using the circuit and transistor parameters of Example 13.10, and assuming threshold voltages of $V_{TN} = 0.5\,\text{V}$ and $V_{TP} = -0.5\,\text{V}$, determine the maximum range of common-mode input voltage for the all-CMOS folded cascode circuit of Figure 14.1(b).

14.2 Using the same circuit and transistor parameters as in Exercise 14.1, calculate the maximum range of output voltage for the all-CMOS folded cascode circuit of Figure 14.1(b).

14.2 FINITE OPEN-LOOP GAIN

In the ideal op-amp, the open-loop gain is infinite, the input differential resistance is infinite, and the output resistance is zero. None of these conditions exists in actual operational amplifiers. In the last chapter, we determined that the open-loop gain and input differential resistance may be large but finite, and the output resistance may be small but nonzero. In this section, we will determine the effect of a finite open-loop gain and input resistance on both the inverting and noninverting amplifier characteristics. We will then calculate the output resistance.

In this section, we limit our discussion of the finite open-loop gain to low frequency. In the next section, we consider the effect of finite gain as well as the frequency response of the amplifier.

14.2.1 Inverting Amplifier Closed-Loop Gain

The equivalent circuit of the inverting amplifier with a finite open-loop gain is shown in Figure 14.2. If the open-loop input resistance is assumed to be infinite, then $i_1 = i_2$, or

$$\frac{v_I - v_1}{R_1} = \frac{v_1 - v_O}{R_2} \qquad\qquad \textbf{(14.1(a))}$$

Figure 14.2   Equivalent circuit, inverting amplifier with finite open-loop gain

or

$$\frac{v_I}{R_1} = v_1\left(\frac{1}{R_1} + \frac{1}{R_2}\right) - \frac{v_O}{R_2}$$ (14.1(b))

Since $v_2 = 0$, the output voltage is

$$v_O = -A_{OL}v_1$$ (14.2)

where A_{OL} is the low-frequency open-loop gain. Solving for v_1 from Equation (14.2) and substituting the result into Equation (14.1(b)), we find

$$\frac{v_I}{R_1} = -\left(\frac{v_O}{A_{OL}}\right)\left(\frac{1}{R_1} + \frac{1}{R_2}\right) - \frac{v_O}{R_2}$$ (14.3)

The closed-loop voltage gain is then

$$A_{CL} = \frac{v_O}{v_I} = \frac{-\dfrac{R_2}{R_1}}{1 + \dfrac{1}{A_{OL}}\left(1 + \dfrac{R_2}{R_1}\right)}$$ (14.4)

Example 14.1   Objective: Determine the minimum open-loop voltage gain to achieve a particular accuracy.

A pressure transducer produces a maximum dc voltage signal of 2 mV and has an output resistance of $R_S = 2\,\text{k}\Omega$. The maximum dc current from the transducer is to be limited to $0.2\,\mu\text{A}$. An inverting amplifier is to be used in conjunction with the transducer to produce an output voltage of -0.10 V for a 2 mV transducer signal. The error in the output voltage cannot be greater than 0.1 percent. Determine the minimum open-loop gain of the amplifier to meet this specification.

Solution: We must first determine the resistor values to be used in the inverting amplifier. The source resistor is in series with R_1, so let

$$R_1' = R_1 + R_S$$

The minimum input resistance is found from the maximum input current as

$$R_1'(\text{min}) = \frac{v_i}{i_i(\text{max})} = \frac{2 \times 10^{-3}}{0.2 \times 10^6} = 10 \times 10^3\,\Omega = 10\,\text{k}\Omega$$

The resistor R_1 then needs to be $8\,\text{k}\Omega$. The closed-loop voltage gain required is

$$A_{CL} = \frac{v_O}{v_i} = \frac{-0.10}{2 \times 10^{-3}} = -50 = \frac{-R_F}{R_1'}$$

The required value of the feedback resistor is then $R_F = 500\,\text{k}\Omega$.

For the voltage gain to be within 0.1 percent, the minimum gain (magnitude) is 49.95. Using Equation (14.4), we can determine the minimum value of the open-loop gain. We have

$$A_{CL} = \frac{\dfrac{-R_2}{R_1'}}{1 + \dfrac{1}{A_{OL}}\left(1 + \dfrac{R_2}{R_1'}\right)} = -49.95 = \frac{-50}{1 + \dfrac{1}{A_{OL}}(51)}$$

which yields $A_{OL}(\text{min}) = 50{,}949$.

Comment: If the open-loop gain is greater than the value of $A_{OL}(\text{min}) = 50{,}949$, then the error in the voltage gain will be less than 0.1 percent.

In the limit as $A_{OL} \to \infty$, the closed-loop gain is equal to the ideal value, designated $A_{CL}(\infty)$, which for the inverting amplifier is

$$A_{CL}(\infty) = -\frac{R_2}{R_1} \qquad (14.5)$$

as previously determined. Equation (14.4) is then

$$A_{CL} = \frac{A_{CL}(\infty)}{1 + \dfrac{1 - A_{CL}(\infty)}{A_{OL}}} \qquad (14.6)$$

To determine the variation in closed-loop gain with changes in open-loop gain, we take the derivative of A_{CL} with respect to A_{OL}. We find

$$\frac{dA_{CL}}{dA_{OL}} = \frac{A_{CL}(\infty)(1 - A_{CL}(\infty))}{[A_{OL} + (1 - A_{CL}(\infty))]^2} \qquad (14.7)$$

which can be rearranged in the form

$$\frac{dA_{CL}}{A_{CL}} = \frac{dA_{OL}}{A_{OL}} \frac{\dfrac{1 - A_{CL}(\infty)}{A_{OL}}}{1 + \left(\dfrac{1 - A_{CL}(\infty)}{A_{OL}}\right)} \qquad (14.8)$$

Normally, $A_{CL}(\infty)| \ll |A_{OL}|$ and Equation (14.8) is approximately

$$\frac{dA_{CL}}{A_{CL}} \cong \frac{dA_{OL}}{A_{OL}} \frac{1 - A_{CL}(\infty)}{A_{OL}} \qquad (14.9)$$

Equation (14.9) relates the percent change in the closed-loop gain of the inverting amplifier as the result of a change in open-loop gain. Open-loop gain variations occur when individual transistor parameters change from one circuit to another or with temperature.

From Equation (14.9), we see that changes in closed-loop gain become smaller as the open-loop gain becomes larger.

Test Your Understanding

14.3 Consider an inverting amplifier in which the op-amp open-loop gain is $A_{OL} = 5 \times 10^4$ and the ideal closed-loop amplifier gain is $A_{CL}(\infty) = -50$. (a) Determine the actual closed-loop gain. (b) If the open-loop gain decreases by 10 percent, find the percent change in closed-loop gain and determine the actual closed-loop gain. (Ans. (a) $A_{CL} = -49.949$ (b) 0.0102%, $A_{CL} = -49.943$)

14.4 In an inverting amplifier, the resistors are $R_2 = 500\,\text{k}\Omega$ and $R_1 = 20\,\text{k}\Omega$. If the closed-loop gain must be within 0.1 percent of the ideal value, determine the minimum required open-loop op-amp gain. (Ans. $A_{OL} = 25{,}974$)

14.2.2   Noninverting Amplifier Closed-Loop Gain

Figure 14.3 shows the equivalent circuit of the noninverting amplifier with a finite open-loop gain. Again, the open-loop input differential resistance is assumed to be infinite. The analysis proceeds in much the same way as in the previous section. We have $i_1 = i_2$, and

$$-\frac{v_1}{R_1} = \frac{v_1 - v_O}{R_2}$$

(**14.10(a)**)

or

$$\frac{v_O}{R_2} = v_1\left(\frac{1}{R_1} + \frac{1}{R_2}\right)$$

(**14.10(b)**)

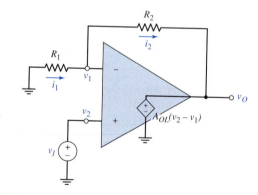

Figure 14.3   Equivalent circuit, noninverting amplifier with finite open-loop gain

The output voltage is

$$v_O = A_{OL}(v_2 - v_1)$$

(**14.11**)

Since $v_2 = v_I$, voltage v_1 can be written

$$v_1 = v_I - \frac{v_O}{A_{OL}}$$

(**14.12**)

Combining Equations (14.12) and (14.10(b)) and rearranging terms, we have an expression for the closed-loop voltage gain:

$$A_{CL} = \frac{v_O}{v_I} = \frac{1 + \dfrac{R_2}{R_1}}{1 + \dfrac{1}{A_{OL}}\left(1 + \dfrac{R_2}{R_1}\right)} \tag{14.13}$$

In the limit as $A_{OL} \to \infty$, the ideal closed-loop gain is

$$A_{CL}(\infty) = 1 + \frac{R_2}{R_1} \tag{14.14}$$

and Equation (14.13) becomes

$$A_{CL} = \frac{A_{CL}(\infty)}{1 + \dfrac{A_{CL}(\infty)}{A_{OL}}} \tag{14.15}$$

Taking the derivative of the closed-loop gain with respect to the open-loop gain and rearranging terms, we obtain

$$\frac{dA_{CL}}{A_{CL}} = \frac{dA_{OL}}{A_{OL}}\left(\frac{A_{CL}}{A_{OL}}\right) \tag{14.16}$$

Equation (14.16) yields the fractional change in the closed-loop gain of the noninverting amplifier as a result of a change in the open-loop gain. The result for the noninverting amplifier is very similar to that for the inverting amplifier.

Test Your Understanding

14.5 An operational amplifier connected in a noninverting configuration has an open-loop gain of $A_{OL} = 10^5$. The resistors are $R_2 = 495\,\text{k}\Omega$ and $R_1 = 5\,\text{k}\Omega$. (a) Determine the actual and ideal closed-loop gains. (b) If the open-loop gain decreases by 10 percent, determine the percent change in closed-loop gain and the actual closed-loop gain. (Ans. (a) $A_{CL} = 99.90$, $A_{CL}(\infty) = 100$ (b) 0.01%, $A_{CL} = 99.89$)

14.6 A noninverting amplifier has an op-amp with an open-loop gain of $A_{OL} = 10^4$. The closed-loop gain must be within 0.1 percent of the ideal value. Determine the maximum closed-loop gain that will still meet the specification. (Ans. $A_{CL} = 10.0$)

14.2.3 Inverting Amplifier Closed-loop Input Resistance

The closed-loop input resistance R_{if} of the inverting amplifier is defined in Figure 14.4(a), and it includes the effect of feedback. The equivalent circuit, including a finite open-loop gain A_{OL}, finite open-loop input differential resistance R_i, and nonzero output resistance R_o, is shown in Figure 14.4(b).

A KCL equation at the output node yields

$$\frac{v_O}{R_L} + \frac{v_O - (-A_{OL}v_1)}{R_o} + \frac{v_o - v_1}{R_2} = 0 \tag{14.17}$$

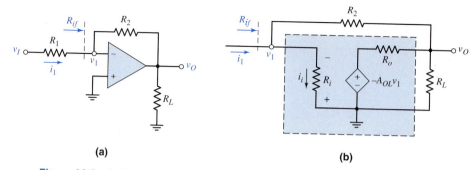

(a)   (b)

Figure 14.4 (a) Inverting amplifier and (b) inverting amplifier equivalent circuit, for calculating closed-loop input resistance

Solving for the output voltage, we have

$$v_O = \frac{-v_1 \left(\dfrac{A_{OL}}{R_o} - \dfrac{1}{R_2} \right)}{\dfrac{1}{R_L} + \dfrac{1}{R_o} + \dfrac{1}{R_2}}$$ (14.18)

A KCL equation at the input node yields

$$i_1 = \frac{v_1}{R_i} + \frac{v_1 - v_O}{R_2}$$ (14.19)

Combining Equations (14.18) and (14.19) and rearranging terms produces

$$\frac{i_1}{v_1} = \frac{1}{R_{if}} = \frac{1}{R_i} + \frac{1}{R_2} \frac{1 + A_{OL} + \dfrac{R_o}{R_L}}{1 + \dfrac{R_o}{R_L} + \dfrac{R_o}{R_2}}$$ (14.20)

Equation (14.20) describes the closed-loop input resistance of the inverting amplifier, with a finite open-loop gain, finite open-loop input resistance, and nonzero output resistance. In the limit as $A_{OL} \to \infty$, we see that $1/R_{if} \to \infty$, or $R_{if} \to 0$, which means that $v_1 \to 0$, or v_1 is at virtual ground. This is a characteristic of an ideal inverting op-amp.

Example 14.2   **Objective:** Determine the closed-loop input resistance at the inverting terminal of an inverting amplifier.

Consider an inverting amplifier with a feedback resistor $R_2 = 10\,k\Omega$, and an op-amp with parameters $A_{OL} = 10^5$ and $R_i = 10\,k\Omega$. Assume the output resistance R_o of the op-amp is negligible.

Solution: If $R_o = 0$, then Equation (14.20) becomes

$$\frac{1}{R_{if}} = \frac{1}{R_i} + \frac{1 + A_{OL}}{R_2} = \frac{1}{10^4} + \frac{1 + 10^5}{10^4} \cong 10^{-4} + 10$$ (14.21)

The closed-loop input resistance is then $R_{if} \cong 0.1\,\Omega$.

Comment: The closed-loop input resistance of the inverting amplifier is a very strong function of the finite open-loop gain. Equation (14.21) shows that the open-loop input resistance R_i essentially does not affect the closed-loop input resistance.

A nonzero closed-loop input resistance R_{if} in conjunction with a finite open-loop input resistance R_i implies that the signal current into the op-amp is not zero, as assumed in the ideal case. From Figure 14.4(b), we see that

$$v_1 = i_1 R_{if} \tag{14.22}$$

Therefore,

$$i_i = \frac{v_1}{R_i} = i_1 \left(\frac{R_{if}}{R_i} \right) \tag{14.23}$$

The fraction of input signal current shunted away from R_2 and into the op-amp is (R_{if}/R_i).

Test Your Understanding

14.7   Determine the closed-loop input resistance at the inverting terminal of an inverting amplifier if $A_{OL} = 10^4$, $R_2 = R_i = R_L = 10\,\text{k}\Omega$, and if: (a) $R_o = 0$, and (b) $R_o = 10\,\text{k}\Omega$. (Ans. (a) $R_{if} = 1\,\Omega$ (b) $R_{if} = 3\,\Omega$)

14.8   Consider the equivalent circuit in Figure 14.4(b). If $R_i = 10\,\text{k}\Omega$, determine the percentage of input signal current i_1 shunted from R_2 for: (a) $R_{if} = 0.1\,\Omega$, and (b) $R_{if} = 10\,\Omega$. (Ans. (a) $10^{-3}\%$ (b) 0.1%)

14.2.4   Noninverting Amplifier Closed-Loop Input Resistance

A noninverting amplifier is shown in Figure 14.5(a). The input resistance seen by the signal source is designated R_{if}. The equivalent circuit, including a finite

(a)   (b)

Figure 14.5   (a) Noninverting amplifier and (b) noninverting amplifier equivalent circuit, for calculating closed-loop input resistance open-loop gain A_{OL}, finite open-loop input differential resistance R_i and non-zero output resistance R_o, is shown in Figure 14.5(b).

Writing a KCL equation at the output node yields

$$\frac{v_O}{R_L} + \frac{v_O - A_{OL}v_d}{R_o} + \frac{v_O - v_1}{R_2} = 0 \tag{14.24}$$

Solving for the output voltage, we have

$$v_O = \frac{\dfrac{v_1}{R_2} + \dfrac{A_{OL}v_d}{R_o}}{\dfrac{1}{R_L} + \dfrac{1}{R_o} + \dfrac{1}{R_2}} \tag{14.25}$$

A KCL equation at the v_1 node yields

$$i_I = \frac{v_1}{R_1} + \frac{v_1 - v_O}{R_2} \tag{14.26}$$

Combining Equations (14.25) and (14.26) and rearranging terms, we obtain

$$i_I\left(1 + \frac{R_o}{R_L} + \frac{R_o}{R_2}\right) = v_1\left\{\left(\frac{1}{R_1} + \frac{1}{R_2}\right)\left(1 + \frac{R_o}{R_L} + \frac{R_o}{R_2}\right) - \frac{R_o}{R_2^2}\right\} - \frac{A_{OL}v_d}{R_2} \tag{14.27}$$

From Figure 14.5(b), we see that

$$v_d = i_1 R_i \tag{14.28}$$

and

$$v_1 = v_I - i_I R_i \tag{14.29}$$

Substituting Equations (14.28) and (14.29) into (14.27) we obtain an equation in i_I and v_I so that the input resistance R_{if} can be found as

$$R_{if} = v_I/i_I$$

In order to simplify the algebra, we neglect the effect of R_o, which is normally small. Setting $R_o = 0$ reduces Equation (14.27) to

$$i_I = v_1\left(\frac{1}{R_1} + \frac{1}{R_2}\right) - \frac{A_{OL}v_d}{R_2} \tag{14.30}$$

Substituting Equations (14.28) and (14.29) into (14.30), we find that the input resistance can be written in the form

$$R_{if} = \frac{v_I}{i_I} = \frac{R_i(1 + A_{OL}) + R_2\left(1 + \dfrac{R_i}{R_1}\right)}{1 + \dfrac{R_2}{R_1}} \tag{14.31}$$

Equation (14.31) describes the closed-loop input resistance of the noninverting amplifier with a finite open-loop gain and a finite open-loop input resistance. In the limit as $A_{OL} \to \infty$, or as the open-loop input resistance approaches infinity, we see that $R_{if} \to \infty$, which is a property of the ideal noninverting amplifier.

Example 14.3 Objective: Determine the closed-loop input resistance at the non-inverting terminal of a noninverting amplifier.

Consider an op-amp with an open-loop gain of $A_{OL} = 10^5$ and an input resistance of $R_i = 10\,\text{k}\Omega$ in a noninverting amplifier configuration with resistor values of $R_1 = R_2 = 10\,\text{k}\Omega$.

Solution: From Equation (14.31), the input resistance is

$$R_{if} = \frac{R_i(1 + A_{OL}) + R_2\left(1 + \dfrac{R_i}{R_1}\right)}{1 + \dfrac{R_2}{R_1}} = \frac{10(1 + 10^5) + 10\left(1 + \dfrac{10}{10}\right)}{1 + \dfrac{10}{10}} \qquad (14.32)$$

or

$$R_{if} \cong 5 \times 10^5\,\text{k}\Omega \Rightarrow 500\,\text{M}\Omega$$

Comment: As expected, the closed-loop input resistance of the noninverting amplifier is very large. Equation (14.32) shows that the input resistance is dominated by the term $R_i(1 + A_{OL})$. The combination of a large R_i and large A_{OL} produces an extremely large input resistance, as predicted by ideal feedback theory.

Test Your Understanding

14.9 For a noninverting amplifier, the resistances are $R_2 = 99\,\text{k}\Omega$ and $R_1 = 1\,\text{k}\Omega$. The op-amp properties are: $A_{OL} = 10^4$, $R_i = 40\,\text{k}\Omega$, and $R_o = 0$. Determine the closed-loop input resistance. (Ans. $R_{if} = 4.04\,\text{M}\Omega$)

14.10 Find the closed-loop input resistance of a voltage follower with op-amp characteristics $A_{OL} = 5 \times 10^5$, $R_i = 10\,\text{k}\Omega$, and $R_o = 0$. (Ans. $R_{if} = 5000\,\text{M}\Omega$)

14.2.5 Nonzero Output Resistance

Since the ideal op-amp has a zero output resistance, the output voltage is independent of the load impedance. The op-amp acts as an ideal voltage source and there is no loading effect. An actual op-amp circuit has a nonzero output resistance, which means that the output voltage, and therefore the closed-loop gain, is a function of the load impedance.

Figure 14.6 is the equivalent circuit of both an inverting and noninverting amplifier and is used to find the output resistance. The op-amp has a finite open-loop gain A_{OL}, a nonzero output resistance R_o, and an infinite input resistance R_i. To determine the output resistance, we set the independent input voltages equal to zero. A KCL equation at the output node yields

$$i_o = \frac{v_o - A_{OL}v_d}{R_o} + \frac{v_o}{R_1 + R_2} \qquad (14.33)$$

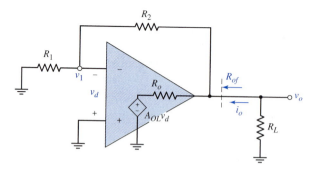

Figure 14.6　Equivalent circuit for calculating closed-loop output resistance

The differential input voltage is $v_d = -v_1$, where

$$v_1 = \left(\frac{R_1}{R_1 + R_2}\right)v_o \tag{14.34}$$

Combining Equations (14.34) and (14.33), we have

$$i_o = \frac{v_o}{R_o} - \frac{A_{OL}}{R_o}\left[-\left(\frac{R_1}{R_1 + R_2}\right)v_o\right] + \frac{v_o}{R_1 + R_2} \tag{14.35(a)}$$

or

$$\frac{i_o}{v_o} = \frac{1}{R_{of}} = \frac{1}{R_o}\left[1 + \frac{A_{OL}}{(1 + R_2/R_1)}\right] + \frac{1}{R_1 + R_2} \tag{14.35(b)}$$

Since R_o is normally small and A_{OL} is normally large, Equation (14.35b), to a good approximation, is as follows:

$$\frac{1}{R_{of}} \cong \frac{1}{R_o}\left[\frac{A_{OL}}{1 + R_2/R_1}\right] \tag{14.36}$$

In most op-amp circuits, the open-loop output resistance R_o is on the order of 100 Ω. Since A_{OL} is normally much larger than $(1 + R_2/R_1)$, the closed-loop output resistance can be very small. Output resistance values in the milliohm range are easily attained.

Example　14.4　**Objective:**　Determine the output resistance of an op-amp circuit.

Computer Simulation Solution:　Figure 14.7 shows an inverting amplifier circuit with a standard 741 op-amp. One method of determining the output resistance is to measure the output voltage for two different values of load resistance connected to the output. Then, treating the amplifier as a Thevenin equivalent circuit with a fixed source in series with an output resistance, the output resistance can be determined. A 1 mV signal was applied. For a 10 Ω load, the output voltage is 0.999837 mV, and for a 20 Ω load, the output voltage is 0.9999132 mV. This gives an output resistance of 1.53 mΩ.

Comment:　As mentioned, the output resistance of a voltage amplifier with negative feedback can be very small. The ideal output resistance is zero, but a practical op-amp circuit can have an output resistance in the milliohm range.

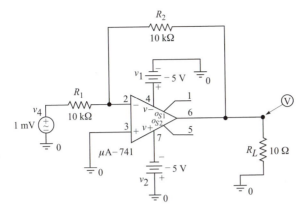

Figure 14.7   Circuit using 741 op-amp to measure output resistance

---

Test Your Understanding

14.11   An op-amp with an open-loop gain of $A_{OL} = 10^5$ is used in a noninverting amplifier configuration with a closed-loop gain of $A_{CL} = 100$. Determine the closed-loop output resistance R_{of} for: (a) $R_o = 100\,\Omega$, and (b) $R_o = 10\,k\Omega$. (Ans. (a) $R_{of} = 01.\,\Omega$ (b) $R_{of} = 10\,\Omega$)

14.3   FREQUENCY RESPONSE

In the last chapter, we considered the basic op-amp frequency response. Frequency compensation was included as a means of stabilizing the circuit. In this section, we will consider the bandwidth and the transient response of the closed-loop amplifier.

When a step function is applied at the op-amp input, the output voltage cannot change instantaneously with time because of capacitance effects within the op-amp circuit. The maximum rate at which the output changes with time is called the **slew rate**. We will determine the factors that limit the slew rate.

14.3.1   Open-Loop and Closed-Loop Frequency Response

The frequency response of the open-loop gain can be written as

$$A_{OL}(f) = \frac{A_O}{1 + j\dfrac{f}{f_{PD}}} \tag{14.37}$$

where A_O is the low-frequency open-loop gain and f_{PD} is the dominant-pole frequency. Figure 14.8 shows the Bode plot of the open-loop gain magnitude. The dominant-pole frequency f_{PD} is shown as well as the unity-gain bandwidth f_T. We showed previously that the unity-gain bandwidth is

$$f_T = f_{PD} A_O \tag{14.38}$$

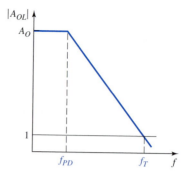

Figure 14.8 Bode plot, open-loop gain magnitude

and is also called the gain–bandwidth product. Equation (14.38) assumes that additional poles of the open-loop frequency response occur at higher frequencies than f_T.

Figure 14.9 shows a noninverting amplifier. In our discussion on feedback theory in Chapter 12, we found that, assuming ideal feedback, the closed-loop gain A_{CL} can be written

$$A_{CL} = \frac{A_{OL}}{(1 + \beta A_{OL})} \tag{14.39}$$

where β is the feedback transfer function. For the noninverting amplifier, this feedback transfer function is

$$\beta = \frac{1}{1 + \dfrac{R_2}{R_1}} \tag{14.40}$$

Combining Equations (14.37), (14.40) and (14.39), we find the expression

Figure 14.9 Noninverting amplifier

for the closed-loop gain as a function of frequency, as follows:

$$A_{CL}(f) = \frac{A_O}{1 + \dfrac{A_O}{1 + (R_2/R_1)}} \times \frac{1}{1 + j \dfrac{f}{f_{PD}\left[1 + \dfrac{A_O}{1 + (R_2/R_1)}\right]}} \tag{14.41}$$

Normally, $A_O \gg [1 + (R_2/R_1)]$; therefore, the low-frequency closed-loop gain is

$$A_{CLO} = 1 + \frac{R_2}{R_1} \tag{14.42}$$

as previously determined. For $A_O \gg A_{CLO}$, Equation (14.41) is approximately

$$A_{CL}(f) = \frac{A_{CLO}}{1 + j\dfrac{f}{f_{PD}\left(\dfrac{A_O}{A_{CLO}}\right)}} \tag{14.43}$$

The 3 dB frequency, or small-signal bandwidth, is then

$$f_{3\,\mathrm{dB}} = f_{PD}\left(\frac{A_O}{A_{CLO}}\right) \tag{14.44}$$

Since in most cases $A_O \gg A_{CLO}$, the bandwidth of the closed-loop system is substantially larger than the open-loop dominant-pole frequency f_{PD}. Note also that Equation (14.44) applies to the inverting, as well as the noninverting, amplifier in which A_{CLO} is the magnitude of the closed-loop gain. We have seen this same bandwidth extension for negative feedback several times previously.

14.3.2 Gain–Bandwidth Product

We can also determine the unity-gain bandwidth of the closed-loop system. From Equation (14.43), we can write

$$|A_{CL}(f = f_{\text{unity}})| = 1 = \frac{A_{CLO}}{\sqrt{1 + \left[\dfrac{f_{\text{unity}}}{f_{PD}(A_O/A_{CLO})}\right]^2}} \tag{14.45}$$

where f_{unity} is the unity-gain frequency of the closed-loop system.

If $A_{CLO} \gg 1$, then Equation (14.45) yields

$$\frac{f_{\text{unity}}}{f_{PD}\left(\dfrac{A_O}{A_{CLO}}\right)} \cong A_{CLO} \tag{14.46(a)}$$

which reduces to

$$f_{\text{unity}} = A_{CLO} f_{PD}\left(\frac{A_O}{A_{CLO}}\right) = f_{PD} A_O = f_T \tag{14.46(b)}$$

The unity-gain frequency or bandwidth of the closed-loop system is essentially the same as that of the open-loop amplifier.

The open-loop and closed-loop frequency response curves are shown in Figure 14.10. We observed these same results in Chapter 12 in the discussion on ideal feedback theory.

Figure 14.10   Bode plot, open-loop and closed-loop gain magnitude

Example 14.5 **Objective:** Determine the unity-gain bandwidth and the maximum closed-loop gain for a specified closed-loop bandwidth.

An audio amplifier system is to use an op-amp with an open-loop gain of $A_O = 2 \times 10^5$ and a dominant-pole frequency of 5 Hz. The bandwidth of the audio system is to be 20 kHz. Determine the maximum closed-loop gain for the audio amplifier.

Solution: The unity-gain bandwidth is found as

$$f_T = f_{PD} A_O = (5)(2 \times 10^5) = 10^6 \, \text{Hz} \Rightarrow 1 \, \text{MHz}$$

Since the gain–bandwidth product is a constant, we have

$$f_{3\text{-dB}} \cdot A_{CL} = f_T$$

where $f_{3\text{-dB}}$ is the closed-loop bandwidth and A_{CL} is the closed-loop gain. The maximum closed-loop gain is then

$$A_{CL} = \frac{f_T}{f_{3\text{-dB}}} = \frac{10^6}{20 \times 10^3} = 50$$

Comment: If the closed-loop gain is less than or equal to 50, then the required bandwidth of 20 kHz for the audio amplifier will be realized.

Test Your Understanding

14.12 An op-amp with open-loop parameters of $A_{OL} = 10^4$ and $f_{PD} = 50 \, \text{Hz}$ is connected in a noninverting amplifier configuration with a low-frequency closed-loop gain of $A_{CLO} = 25$. If an input voltage of $v_I = 50 \sin(2\pi f t) \, \mu\text{V}$ is applied, determine the output voltage peak amplitude for: (a) $f = 2 \, \text{kHz}$, (b) $f = 20 \, \text{kHz}$, and (c) $f = 100 \, \text{kHz}$. (Ans. (a) 1.25 mV (b) 0.884 mV (c) 0.245 mV)

14.3.3 Slew Rate

Implicit in the frequency response analysis for the closed-loop amplifier is the assumption that the sinusoidal input signals are small. If a large sinusoidal signal or step function is applied to an op-amp circuit, the input stage can be overdriven and the small-signal model will no longer apply.

Figure 14.11 shows a simplified op-amp circuit. If a large step voltage (greater than 120 mV) is applied at v_2 with v_1 held at ground potential, then Q_2 is effectively cut off, which means $i_{C2} \cong 0$ and $i_{C1} \cong I_Q$. The entire bias current is switched to Q_1. Since $i_{C3} \cong i_{C1}$, then $i_{C3} \cong I_Q$; since Q_3–Q_4 form a current mirror, then we also have $i_{C4} \cong I_Q$.

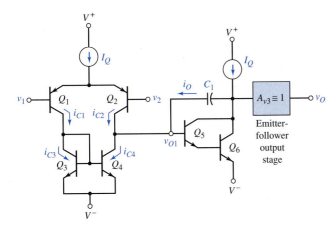

Figure 14.11   Simplified op-amp for calculating slew rate

The base current into Q_5 is very small; therefore, the current through the compensation capacitor C_1 is $i_O = i_{C4} = I_Q$. Since the voltage gain of the emitter-follower output stage is essentially unity, the capacitor current can be written as

$$i_O = C_1 \frac{d(v_O - v_{O1})}{dt} \tag{14.47}$$

The gain of the second stage is large, which means that $v_{O1} \ll v_O$. Equation (14.47) then becomes

$$i_O \cong C_1 \frac{dv_O}{dt} = I_Q \tag{14.48}$$

or

$$\frac{dv_O}{dt} = \frac{I_Q}{C_1} \tag{14.49}$$

The maximum current through the compensation capacitor is limited to the bias current I_Q; consequently, the maximum rate at which the output voltage can change is also limited by the bias current I_Q.

The maximum rate of change of the output voltage is the slew rate of the op-amp, the units of which are usually given as volts per microsecond. From Equation (14.49), we have

$$\text{Slew rate (SR)} = \left(\frac{dv_O}{dt} \right)_{max} = \frac{I_Q}{C_1} \tag{14.50}$$

Although the rate of change in output voltage can be either positive or negative, the slew rate is *defined* as a positive quantity.

Figure 14.12 shows the slew-rate limited response of an op-amp voltage follower to a rectangular input voltage pulse. Note the trapezoidal shaped output response. The time needed to reach the full-scale response is approximately $V_O(\text{max})/\text{SR}$.

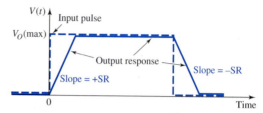

Figure 14.12   Slew-rate-limited response of voltage follower to rectangular input voltage pulse

Example   14.6   **Objective:**   Calculate the slew rate of the 741 op-amp.

From the previous chapter, the bias current in the 741 op-amp is $I_Q = 19\,\mu\text{A}$ and the internal frequency compensation capacitor is $C_1 = 30\,\text{pF}$.

Solution: From Equation (14.50), the slew rate is

$$\text{SR} = \frac{I_Q}{C_1} = \frac{19 \times 10^{-6}}{30 \times 10^{-12}} = 0.63 \times 10^6\,\text{V/s} \Rightarrow 0.63\,\text{V/}\mu\text{s}$$

Comment: The partial data sheet in Table 14.1 for the 741 op-amp lists the typical slew rate as $0.7\,\text{V/}\mu\text{s}$, which is in close agreement with our calculated value.

Typical slew-rate values for the CA3140 BiCMOS and LH0042C BiFET op-amps are also given in Table 14.1. The BiCMOS circuit has a typical slew rate of $9\,\text{V/}\mu\text{s}$, and the BiFET op-amp has a typical value of $3\,\text{V/}\mu\text{s}$. The slew rates are larger in the FET op-amps because the bias currents are larger than in the 741 circuit and the gain of the FET input stage is smaller than that of the 741 input stage.

The slew rate is directly related to the unity-gain bandwidth. To explain, the unity-gain bandwidth is directly proportional to the dominant-pole frequency, or $f_T \propto f_{PD}$. In turn, the dominant-pole frequency is inversely proportional to $R_{eq}C_1$, where R_{eq} is the equivalent resistance at the node of the second stage input and C_1 is the compensation capacitance. The equivalent resistance R_{eq} is a function of the second stage input resistance and the diff-amp stage output resistance, both of which are inversely proportional to I_Q. Then,

$$f_T \propto f_{PD} \propto \frac{1}{R_{eq}C_1} \propto \frac{1}{\left(\dfrac{1}{I_Q}\right)C_1} \propto \frac{I_Q}{C_1} \tag{14.51}$$

where I_Q/C_1 is the slew rate. Equation (14.51) shows that the slew rate is directly proportional to the unity-gain bandwidth.

Now consider what happens when a sinosoidal input signal is applied, for example, to the noninverting amplifier shown in Figure 14.9. If $v_I = V_p \sin \omega t$, then

$$v_O(t) = V_P\left(1 + \frac{R_2}{R_1}\right) \sin \omega t = V_{po} \sin \omega t \qquad \textbf{(14.52)}$$

where V_{po} is the ideal peak value of the sinusoidal output voltage.

The rate at which the output voltage changes is

$$\frac{dv_O(t)}{dt} = \omega V_{po} \cos \omega t \qquad \textbf{(14.53)}$$

Therefore, the maximum rate of change is ωV_{po}. Figure 14.13 shows two sinusoidal waveforms of the same frequency but different peak amplitudes. The maximum rate of change, or slope, occurs as the curves cross the zero axis. The waveform with the larger peak value has a larger maximum slope. Curve a in Figure 14.13 has a maximum slope corresponding to the slew rate; curve b, with a smaller peak value, has a maximum slope less than the slew rate. If the maximum slope, ωV_{po}, is greater than the slew rate SR, then the op-amp is slew-rate-limited and the output signal is distorted.

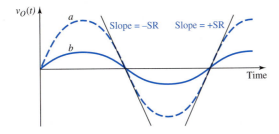

Figure 14.13   Two sinusoidal waveforms of the same frequency with different peak voltages, showing different maximum slopes

Thus, the maximum frequency at which the op-amp can operate without being slew-rate-limited is a function of both the frequency and peak amplitude of the signal. We have that

$$\omega_{\max} V_{po} = 2\pi f_{\max} V_{po} = \text{SR} \qquad \textbf{(14.54(a))}$$

or

$$f_{\max} = \frac{\text{SR}}{2\pi V_{po}} \qquad \textbf{(14.54(b))}$$

As the output voltage peak amplitude increases, the maximum frequency at which slew-rate-limiting occurs decreases. The **full-power bandwidth (FPBW)** is the frequency at which the op-amp output becomes slew-rate-limited. The FPBW is the $f_{\max}$ frequency from Equation (14.54(b)), or

$$\text{FPBW} = \frac{\text{SR}}{2\pi V_{po}} \qquad \textbf{(14.55)}$$

The full-power bandwidth can be considerably less than the small-signal bandwidth.

Example 14.7 **Objective:** Determine the small-signal bandwidth of an amplifier and the full-power bandwidth that will produce an undistorted output voltage.

Consider an amplifier with a unity-gain bandwidth of $f_T = 1\,\text{MHz}$ and a low-frequency closed-loop gain of $A_{CLO} = 10$. Assume the op-amp slew rate is $\text{SR} = 1\,\text{V/}\mu\text{s}$ and the desired peak output voltage is $V_{po} = 10\,\text{V}$.

Solution: The small-signal closed-loop bandwidth is, from Equation (14.44),

$$f_{3\text{-dB}} = \frac{f_T}{A_{CLO}} = \frac{10^6}{10} \Rightarrow 100\,\text{kHz}$$

The full-power bandwidth, based on slew-rate limitation, from Equations (14.54(b)) and (14.55), is

$$f_{\text{max}} = \text{FPBW} = \frac{\text{SR}}{2\pi V_{po}} = \frac{(1\,\text{V/}\mu\text{s})(10^6\,\mu\text{s/s})}{2\pi(10)} \Rightarrow 15.9\,\text{kHz}$$

Comment: The full-power bandwidth, or the actual maximum frequency at which the system can be operated and still produce a large, undistorted output signal, is considerably smaller than the bandwidth under small-signal nonslew-rate-limiting conditions.

Test Your Understanding

14.13 A 1 V input step function is applied to a noninverting amplifier with a closed-loop gain of 5. The slew rate of the op-amp is $2\,\text{V/}\mu\text{s}$. Determine the time needed for the output voltage to reach its full-scale response. (Ans. $2.5\,\mu\text{s}$)

14.14 For a 741 op-amp with a slew rate of $0.63\,\text{V/}\mu\text{s}$, find the full-power bandwidth for a peak undistorted output voltage of: (a) 1 V, and (b) 10 V. (Ans. (a) 100 kHz (b) 10 kHz)

14.15 An op-amp with a low-frequency open-loop gain of $A_{OL} = 10^5$ and a dominant-pole frequency of $f_{PD} = 10\,\text{Hz}$ is used in a noninverting amplifier configuration with a low-frequency closed-loop gain of $A_{CLO} = 50$. The slew rate of the op-amp is $0.8\,\text{V/}\mu\text{s}$. Determine the maximum undistorted output voltage amplitude such that $f_{\text{max}} = f_{3\text{-dB}}$. (Ans. 6.37 V)

14.4 OFFSET VOLTAGE

In Chapter 11, we analyzed the basic difference amplifier, which is the input stage of the op-amp. In that analysis, we assumed the input differential-pair transistors to be identical, or matched. If the two input devices are mismatched, the currents in the two branches of the diff-amp are unequal and this affects the diff-amp dc output voltage. In fact, the internal circuitry of the entire op-amp usually contains imbalances and asymmetries, all of which can cause a nonzero output voltage for a zero input differential voltage.

The **output dc offset voltage** is the measured open-loop output voltage when the input voltage is zero. This configuration is shown in Figure 14.14. The **input dc offset voltage** is defined as the input differential voltage that must be applied to the open-loop op-amp to produce a zero output voltage. This

Figure 14.14 Circuit for measuring output offset voltage

Figure 14.15 Circuit for measuring input offset voltage

configuration is shown in Figure 14.15. The input offset voltage is the parameter most often specified and is usually referred to simply as the offset voltage.

Offset voltage values have a statistical distribution among op-amps of the same type, and the offset voltage polarity may vary from one op-amp to another. The offset voltage specification for an op-amp is the magnitude of the maximum offset voltage for a particular type of op-amp. The offset voltage is a dc value, generally in the range of 1–2 mV for bipolar op-amps, although some op-amps may have offset voltages in the range of 5–10mV. Further, the maximum offset voltage specification for a precision op-amp may be as low as 10 μV.

In this section we will analyze offset voltage effects in the input diff-amp stage and will then consider various techniques used to compensate for offset voltage.

14.4.1   Input Stage Offset Voltage Effects

Several possible mismatches in the input diff-amp stage can produce offset voltages. We will analyze offset voltage effects in two bipolar input stages and in a MOSFET input diff-amp circuit.

Basic Bipolar Diff-Amp Stage

A basic bipolar diff-amp is shown in Figure 14.16. The differential pair is biased with a constant-current source. If Q_1 and Q_2 are matched, then for $v_1 = v_2 = 0$, I_Q splits evenly between the two transistors and $i_{C1} = i_{C2}$. If a two-sided output is defined as the difference in voltage between the two collector terminals, then $v_O = 0$ when the transistors are matched and the collector resistors are matched, which means that the offset voltage is zero.

The collector currents can be written as

$$i_{C1} = I_{S1} e^{v_{BE1}/V_T} \tag{14.56(a)}$$

and

$$i_{C2} = I_{S2} e^{v_{BE2}/V_T} \tag{14.56(b)}$$

where I_{S1} and I_{S2} are related to the reverse-saturation currents in the B–E junctions and are functions of the electrical and geometric transistor properties. If the two transistors are exactly matched, then $I_{S1} = I_{S2}$; if there is any mismatch in the electrical or geometric parameters, then $I_{S1} \neq I_{S2}$.

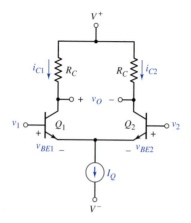

Figure 14.16 Basic bipolar difference amplifier

The input offset voltage is defined as the input differential voltage required to produce a zero output voltage, or in this case to produce $i_{C1} = i_{C2}$. Figure 14.17 shows the i_C versus v_{BE} characteristics of two unmatched transistors. Slightly different B–E voltages must be applied to produce equal collector currents that will result in a zero output voltage in the diff-amp.

Figure 14.17 The i_C versus v_{BE} characteristics for two unmatched bipolar transistors

For $i_{C1} = i_{C2}$, we have

$$I_{S1}e^{v_{BE1}/V_T} = I_{S2}e^{v_{BE2}/V_T} \tag{14.57}$$

or

$$e^{(v_{BE1}-v_{BE2})/V_T} = \frac{I_{S2}}{I_{S1}} \tag{14.58}$$

We define the offset voltage as

$$v_{BE1} - v_{BE2} \equiv V_{OS}$$

Since $v_1 - v_2 = v_{BE1} - v_{BE2}$, then the offset voltage V_{OS} is the differential input voltage that must be applied to produce $i_{C1} = i_{C2}$.

Equation (14.58) can then be written as

$$e^{V_{OS}/V_T} = \frac{I_{S2}}{I_{S1}} \tag{14.59(a)}$$

or

$$V_{OS} = V_T \ln\left(\frac{I_{S2}}{I_{S1}}\right) \tag{14.59(b)}$$

Example 14.8   Objective: Calculate the offset voltage in a bipolar diff-amp for a given mismatch between the input transistors.

Consider the diff-amp in Figure 14.16 with transistor parameters $I_{S1} = 10^{-14}$ A and $I_{S2} = 1.05 \times 10^{-14}$ A.

Solution: From Equation (14.59(b)), the offset voltage is

$$V_{OS} = V_T \ln\left(\frac{I_{S2}}{I_{S1}}\right) = (0.026)\ln\left(\frac{1.05 \times 10^{-14}}{1 \times 10^{-14}}\right) = 0.00127\,\text{V} \Rightarrow 1.27\,\text{mV}$$

Comment: A 5 percent difference in I_S for Q_1 and for Q_2 produces an offset voltage of 1.27 mV. Since the offset voltage is defined as a positive quantity, if in the previous example I_{S1} were 5 percent larger than I_{S2}, the offset voltage would also be 1.27 mV.

It should be cautioned that the offset voltage in this example is one component of the offset voltage for the entire op-amp. For example, if the two collector resistors are not equal, then the two-sided output voltage v_O will not be zero even if the two transistors are identical. Nevertheless, the calculation provides information on one source of offset voltage, as well as the resulting magnitude of V_{OS}.

Test Your Understanding

14.16   Consider the bipolar diff-amp in Figure 14.16 with transistor parameters $I_{S1} = 2 \times 10^{-14}$ A and $I_{S2} = 1.85 \times 10^{-14}$ A. Calculate the offset voltage. (Ans. 2.03 mV)

Bipolar Active Load Diff-Amp Stage

Figure 14.18 shows a bipolar diff-amp with a simple two-transistor active load. As before, this input stage is biased with a constant-current source. If Q_1 and Q_2 are matched and if Q_3 and Q_4 are matched, then I_Q splits evenly between Q_1 and Q_2 for $v_1 = v_2$, and the E–C voltages of Q_3 and Q_4 are equal. The one-sided dc output voltage v_O will therefore be one E–B voltage below V^+.

If, however, Q_3 and Q_4 are not exactly matched, then i_{C1} and i_{C2} may not be equal since the active load influences the split in the bias current, even if Q_1 and Q_2 are matched. This effect is caused by a finite Early voltage. Taking the

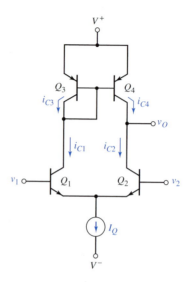

Figure 14.18   Basic bipolar diff-amp with active load

Early voltages into account, but neglecting base currents, we can write the collector currents as

$$i_{C1} = i_{C3} = I_{S1}\left(e^{v_{BE1}/V_T}\right)\left(1 + \frac{v_{CE1}}{V_{A1}}\right)$$

$$= I_{S3}\left(e^{v_{EB3}/V_T}\right)\left(1 + \frac{v_{EC3}}{V_{A3}}\right)$$

(14.60(a))

and

$$i_{C2} = i_{C4} = I_{S2}\left(e^{v_{BE2}/V_T}\right)\left(1 + \frac{v_{CE2}}{V_{A2}}\right)$$

$$= I_{S4}\left(e^{v_{EB4}/V_T}\right)\left(1 + \frac{v_{EC4}}{V_{A4}}\right)$$

(14.60(b))

If we assume that Q_1 and Q_2 are matched, then $I_{S1} = I_{S2} \equiv I_S$ and $V_{A1} = V_{A2} \equiv V_{AN}$. Assume that Q_3 and Q_4 are slightly mismatched, so that $I_{S3} \neq I_{S4}$ but still assume that $V_{A3} = V_{A4} \equiv V_{AP}$. For $v_1 = v_2$, we have $v_{BE1} = v_{BE2}$; also, $v_{EB3} = v_{EB4} = v_{EC3} \equiv v_{EB}$. Taking the ratio of Equations (14.60(a)) and (14.60(b)) produces

$$\frac{i_{C1}}{i_{C2}} = \frac{1 + \dfrac{v_{CE1}}{V_{AN}}}{1 + \dfrac{v_{CE2}}{V_{AN}}} = \frac{I_{S3}}{I_{S4}} \frac{1 + \dfrac{v_{EB}}{V_{AP}}}{1 + \dfrac{v_{EC4}}{V_{AP}}}$$

(14.61)

Equation (14.61) can be rearranged in the form

$$\frac{1 + \dfrac{v_{CE1}}{V_{AN}}}{1 + \dfrac{v_{EB}}{V_{AP}}} = \frac{I_{S3}}{I_{S4}} \frac{1 + \dfrac{v_{CE2}}{V_{AN}}}{1 + \dfrac{v_{EC4}}{V_{AP}}}$$

(14.62)

Since Q_3 is connected as a diode, v_{CE1} is a constant for a given bias current and supply voltage, which means that the left side of Equation (14.62) is a constant. If $I_{S3} = I_{S4}$, then $v_{CE2} = v_{CE1}$ and $v_{EC4} = v_{EB} = v_{EC3}$. However, if $I_{S3} \neq I_{S4}$, then the collector–emitter voltages on Q_2 and Q_4 must change. If, for example, $I_{S3} > I_{S4}$, then v_{EC4} is larger than v_{CE2}. If, on the other hand, $I_{S4} > I_{S3}$, then v_{EC4} is smaller than v_{CE2}, and Q_4 may be driven into saturation by the mismatch.

Example 14.9   Objective: Calculate the change in output voltage for a given mismatch in the active load transistors.

Consider the diff-amp in Figure 14.18 with $V^+ = 10$ V. Assume that Q_1 and Q_2 are matched with $v_{BE1} = v_{BE2} = 0.6$ V, and assume that $v_{EB3} = v_{EB4} = v_{EC3} = 0.6$ V. Let $I_{S3} = 1.05 I_{S4}$. Also assume that $V_{AN} = V_{AP} = 50$ V.

Solution: Since $v_{EB3} = 0.6$ V $= v_{BE1}$, then for $v_1 = v_2 = 0$,

$$v_{CE1} = V^+ = 10 \text{ V}$$

The left side of Equation (14.62) is therefore

$$\frac{1 + \dfrac{v_{CE1}}{V_{AN}}}{1 + \dfrac{v_{EB}}{V_{AP}}} = \frac{1 + \dfrac{10}{50}}{1 + \dfrac{0.6}{50}} = 1.186$$

We have that

$$v_{EC4} + v_{CE2} = V^+ + v_{BE2} = 10.6 \text{ V}$$

or

$$v_{CE2} = 10.6 - v_{EC4}$$

Equation (14.62) then becomes

$$1.186 = 1.05 \frac{1 + \dfrac{10.6 - v_{EC4}}{50}}{1 + \dfrac{v_{EC4}}{50}}$$

which yields

$$v_{EC4} = 1.94 \text{ V}$$

Comment: A 5 percent difference between the properties of Q_3 and Q_4 produces a change from 0.6 to 1.94 V in the E–C voltage of Q_4.

Computer Simulation Verification: A PSpice analysis of the offset voltage effects in the active load diff-amp was performed. The two input terminals are at ground potential. Using $I_S = 5 \times 10^{-15}$ A for all transistors, the PSpice analysis shows that $v_{EB3} = 0.654$ V rather than the assumed value of 0.6 V. Also, v_{EC4} is 1.19 V rather than equal to v_{EB3}. This occurs because the circuit is slightly unbalanced; that is, i_{C1} includes the base currents of Q_3 and Q_4, and i_{C4} does not. When Q_3 and Q_4 are not matched and $I_{S3} = 1.05 I_{S4} = 5.25 \times 10^{-15}$ A, then v_{EC4} increases to 2.51 V, compared to 1.94 V from the hand analysis. If, however, $I_{S3} = 0.95 I_{S4} = 4.75 \times 10^{-15}$ A, then Q_4 goes into saturation.

An offset voltage that will slightly change i_{C1} and i_{C2} will allow the E–C voltage of Q_4 to be adjusted back to its original value.

As shown in actual op-amp circuits, resistors are usually included in the emitters of the active load transistors. By producing a slight imbalance in the two resistor values, we can change the ratio of i_{C1} to i_{C2}, causing a change in the output voltage. This is discussed in the next section when offset voltage null adjustment is discussed.

Test Your Understanding

⁎14.17 Consider the active load bipolar diff-amp stage in Figure 14.18. Assume the circuit and transistor parameters are as given in Example 14.9. Using Equations (14.60(a)) and (14.60(b)), determine the offset voltage $V_{OS} = |v_{BE2} - v_{BE1}|$ such that $v_{EC3} = v_{EC4}$ and $v_{CE1} = v_{CE2}$. (Ans. 1.27 mV)

MOSFET Diff-Amp Stage

Figure 14.19 shows a basic MOSFET diff-amp in which the differential pair is biased with a constant-current source. If M_1 and M_2 are matched, then for $v_1 = v_2 = 0$, I_Q splits evenly between the two transistors and $i_{D1} = i_{D2}$. Since a two-sided output is the voltage difference between the two drain terminals, then for this symmetrical situation, $v_O = 0$ and the offset voltage is zero.

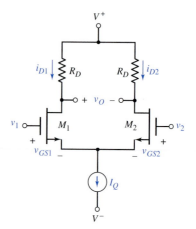

Figure 14.19   Basic MOSFET diff-amp

The drain currents can be written as

$$i_{D1} = K_{n1}(v_{GS1} - V_{TN1})^2 \tag{14.63(a)}$$

and

$$i_{D2} = K_{n2}(v_{GS2} - V_{TN2})^2 \tag{14.63(b)}$$

As previously stated, the conduction parameters K_{n1} and K_{n2} are functions of the electrical and geometric properties of the two transistors, and the threshold voltages V_{TN1} and V_{TN2} are also functions of the transistor electrical properties. If there is a mismatch in electrical or geometric parameters, then we may have $K_{n1} \neq K_{n2}$ and $V_{TN1} \neq V_{TN2}$.

As with the bipolar diff-amp, the input offset voltage is defined as the input differential voltage that must be applied to produce a zero output voltage, or

$$V_{OS} = v_{GS1} - v_{GS2} \tag{14.64}$$

When the offset voltage is applied, $i_{D1} = i_{D2} = I_Q/2$; when the two drain resistors are equal, then $v_O = 0$. Solving Equations (14.63(a)) and (14.63(b)) for v_{GS1} and v_{GS2} and substituting the results into Equation (14.64), we find

$$V_{OS} = \sqrt{\frac{i_{D1}}{K_{n1}}} + V_{TN1} - \left(\sqrt{\frac{i_{D2}}{K_{n2}}} + V_{TN2} \right) \tag{14.65}$$

The various difference and average quantities are defined as follows:

$$\Delta K_n = K_{n1} - K_{n2} \tag{14.66(a)}$$

$$K_n = \frac{K_{n1} + K_{n2}}{2} \tag{14.66(b)}$$

$$\Delta V_{TN} = V_{TN1} - V_{TN2} \tag{14.67(a)}$$

and

$$V_{TN} = \frac{V_{TN1} + V_{TN2}}{2} \tag{14.67(b)}$$

Combining Equations (14.66(a)) and (14.66(b)), we have

$$K_{n1} = K_n + \frac{\Delta K_n}{2} \tag{14.68(a)}$$

and

$$K_{n2} = K_n - \frac{\Delta K_n}{2} \tag{14.68(b)}$$

Similarly,

$$V_{TN1} = V_{TN} + \frac{\Delta V_{TN}}{2} \tag{14.69(a)}$$

and

$$V_{TN2} = V_{TN} - \frac{\Delta V_{TN}}{2} \tag{14.69(b)}$$

Noting that $i_{D1} = i_{D2} = I_Q/2$ and substituting Equations (14.68(a)) through (14.69(b)) into Equation (14.65), we obtain

$$V_{OS} = \sqrt{\frac{I_Q}{2}} \left[\frac{1}{\sqrt{K_n + (\Delta K_n/2)}} - \frac{1}{\sqrt{K_n - (\Delta K_n/2)}} \right] + \Delta V_{TN} \tag{14.70}$$

If we assume that $\Delta K_n \ll K_n$ then Equation (14.70) reduces to

$$V_{OS} = -\frac{1}{2}\sqrt{\frac{I_Q}{2K_n}} \cdot \left(\frac{\Delta K_n}{K_n}\right) + \Delta V_{TN} \qquad \textbf{(14.71)}$$

Equation (14.71) is the offset voltage in a MOSFET diff-amp as a function of the differences in conduction parameters and threshold voltages.

Example 14.10   Objective: Calculate the offset voltage in a MOSFET diff-amp stage for a given mismatch between input transistors.

Consider the diff-amp in Figure 14.19 with transistor parameters $K_{n1} = 105\,\mu\text{A/V}^2$, $K_{n2} = 100\,\mu\text{A/V}^2$, and $V_{TN1} = V_{TN2}$. Assume $I_Q = 200\,\mu\text{A}$.

Solution: From Equation (14.66(a)), the difference in conduction parameters is

$$\Delta K_n = K_{n1} - K_{n2} = 105 - 100 = 5\,\mu\text{A/V}^2$$

From Equation (14.66(b)), the average of the conduction parameters is

$$K_n = \frac{K_{n1} + K_{n2}}{2} = \frac{105 + 100}{2} = 102.5\,\mu\text{A/V}^2$$

The magnitude of the offset voltage is, from Equation (14.71),

$$|V_{OS}| = \frac{1}{2}\sqrt{\frac{I_Q}{2K_n}} \cdot \left(\frac{\Delta K_n}{K_n}\right) = \frac{1}{2}\sqrt{\frac{200}{2(102.5)}}\left(\frac{5}{102.5}\right) = 0.0241\,\text{V} \Rightarrow 24.1\,\text{mV}$$

Comment: A 5 percent difference in conduction parameter values between the input MOS transistors produces an offset voltage of 24.1 mV.

Test Your Understanding

14.18 Assume the MOSFET diff-amp shown in Figure 14.19 is biased with a current $I_Q = 150\,\mu\text{A}$. Let $V_{TN1} = V_{TN2}$. Assume the nominal conduction parameter value is $K_n = 50\,\mu\text{A/V}^2$. Determine the maximum variation ΔK_n such that the offset voltage is limited to $V_{OS} = 20\,\text{mV}$. (Ans. $\Delta K_n = 1.63\,\mu\text{A/V}^2$)

Comparing the results of Examples 14.8 and 14.10 shows that typically the offset voltage for a MOSFET diff-amp is substantially larger than that of a bipolar diff-amp. The difference can be explained by comparing Equation (14.71) for the MOSFET diff-amp and Equation (14.59(b)) for the bipolar diff-amp. The offset voltage for the MOSFET diff-amp is directly proportional to the percent change in conduction parameter values, whereas the offset voltage for the bipolar diff-amp is proportional to the logarithm of the percent change in the I_S current parameters. In addition, the offset voltage for the MOSFET pair is proportional to

$$\sqrt{I_Q/K_n} = V_{GS} - V_{TN}$$

which is typically in the range of 1–2 V. In contrast, the offset voltage for the bipolar pair is proportional to

$$V_T \approx 26\,\text{mV}$$

which is substantially smaller than $(V_{GS} - V_{TN})$. Thus, a MOSFET diff-amp inherently displays a higher input offset voltage than a bipolar pair for the same level of mismatch.

Partial data sheets showing some of the nonideal characteristics for the op-amps considered in the last chapter are in Table 14.1. The 741 op-amp, an all-bipolar circuit, has a maximum input offset voltage of 3 mV. The CA3140, which has a MOSFET input differential pair, has a maximum input offset voltage of 15 mV; and the LH0042C, which has a JFET input differential pair, has a maximum input offset voltage of 20 mV. This supports our conclusion that op-amps with FET input transistors have substantially larger input offset voltages than the all-bipolar circuit discussed.

14.4.2   Offset Voltage Compensation

In many applications, especially those for which the input signal is large compared to the offset voltage V_{OS}, the effect of the offset voltage is negligible. However, there are situations in which it is necessary to compensate for, or "null out," the offset voltage. Two such methods are: (a) an externally connected **offset compensation network**, and (2) an operational amplifier with **offset-null terminals**.

External Offset Compensation Network

Figure 14.20 shows a simple network for offset voltage compensation in an inverting amplifier. The resistive voltage divider of R_4 and R_5, in conjunction with potentiometer R_3, is used to make voltage adjustments of either polarity at the noninverting terminal to cancel the effects of V_{OS}. If $R_5 \ll R_4$, then the compensating voltage applied to the noninverting terminal can be in the milli-volt range, which is typical of offset voltage values.

Figure 14.20   Offset voltage compensation circuit for inverting amplifier

Example   14.11   **Objective:**   Determine the range of voltage produced by an offset voltage compensation network.

Consider the compensation network in Figure 14.20 with $R_5 = 100\,\Omega$, $R_4 = 100\,\text{k}\Omega$, and a $100\,\text{k}\Omega$ potentiometer R_3. Let $V^+ = 15\,\text{V}$ and $V^- = -15\,\text{V}$. Determine the voltage range at V_Y.

Solution: Assume the potentiometer wiper arm is connected to the V^+ supply voltage. The voltage V_Y is then

$$V_Y = \left(\frac{R_5}{R_5 + R_4}\right)V^+ = \left(\frac{0.1}{0.1 + 100}\right)(15) \Rightarrow 15\,\text{mV}$$

Comment: For this particular circuit, the compensation voltage range is $-15\,\text{mV}$ to $+15\,\text{mV}$. A larger resistance R_5 will increase the offset voltage compensation range, and a smaller resistance R_5 will increase the sensitivity of offset voltage compensation.

Test Your Understanding

D14.19 Consider the compensation network in Figure 14.20. Assume $V^+ = 10\,\text{V}$, $V^- = -10\,\text{V}$, $R_3 = 100\,\text{k}\Omega$, and $R_4 = 100\,\text{k}\Omega$. Design R_5 such that the circuit can compensate for an offset voltage of $V_{OS} = 5\,\text{mV}$. (Ans. $50\,\Omega$)

Figure 14.21 shows a compensation network that can be used with a non-inverting op-amp circuit. The same R_4–R_5 voltage divider is used with the potentiometer R_3. Typically, R_5 is on the order of $100\,\Omega$ and R_4 on the order of $100\,\text{k}\Omega$. If $V^+ = 15\,\text{V}$ and $V^- = -15\,\text{V}$, then the compensation voltage is again in the range of $-15\,\text{mV}$ to $+15\,\text{mV}$.

Figure 14.21 Offset voltage compensation circuit for noninverting amplifier

The voltage gain of the noninverting amplifier becomes a function of the compensation network. Since $R_5 \ll R_4$, then the gain of the amplifier, to a good approximation, is

$$A_v = \frac{v_O}{v_I} = \left(1 + \frac{R_2}{R_1 + R_5}\right) \tag{14.72}$$

Since R_5 is small, Equation (14.72) shows that the gain is not a strong function of the compensation network; however, it may still need to be taken into account.

Offset-Null Terminals

Many op-amps, including the 741 bipolar and the CA3140 BiCMOS circuits studied in Chapter 13, include a pair of external offset-null terminals, which are used to compensate for the offset voltage. Figure 14.22 shows a basic bipolar

Figure 14.22 Basic bipolar input diff-amp stage, including a pair of offset-null terminals connected to a potentiometer

input diff-amp stage, including a pair of offset-null terminals. An external potentiometer R_x is connected between these terminals, and the wiper arm is connected to supply voltage V^-.

If the wiper arm of R_x is centered, then R_1 and R_2 will each have a resistance $R_x/2$ connected in parallel. When the wiper arm is moved off center, then R_1 and R_2 will each have a different resistance connected in parallel, and an asymmetry will be introduced into the circuit. This asymmetry in turn introduces an offset voltage, which cancels the input offset voltage effects. In practice, to adjust for offset voltage effects, the op-amp is connected in a feedback configuration with the input differential voltage set equal to zero. The wiper arm of potentiometer R_x is then adjusted until the output voltage becomes zero.

To demonstrate the offset-null technique, we first write a KVL equation between the base terminals of Q_3 and Q_4 and voltage V^- in Figure 14.22, as follows:

$$v_{BE3} + i_{C1}R_1' = v_{BE4} + i_{C2}R_2' \tag{14.73}$$

where R_1' and R_2' are the effective resistances in the emitters of Q_3 and Q_4, including the parallel effects of potentiometer R_x. We have that

$$R_1' = R_1 \| x R_x \qquad \text{and} \qquad R_2' = R_2 \| (1 - x) R_x$$

The base–emitter voltages are

$$v_{BE3} = V_T \ln\left(\frac{i_{C1}}{I_{S3}}\right) \tag{14.74(a)}$$

and

$$v_{BE4} = V_T \ln\left(\frac{i_{C2}}{I_{S4}}\right)$$ (14.74(b))

Substituting Equations (14.74(a)) and (14.74(b)) into Equation (14.73) yields

$$V_T \ln\left(\frac{i_{C1}}{I_{S3}}\right) + i_{C1}R_1' = V_T \ln\left(\frac{i_{C2}}{I_{S4}}\right) + i_{C2}R_2'$$ (14.75)

If a mismatch occurs between Q_3 and Q_4, meaning $I_{S3} \neq I_{S4}$, then a deliberate mismatch between R_1' and R_2' can be introduced to compensate for the transistor mismatch and the adjustment can make $i_{C1} = i_{C2}$. Similarly, a deliberate mismatch between R_1' and R_2' can be used to compensate for a mismatch between Q_1 and Q_2.

Example 14.12 **Objective:** Determine the required difference between R_1' and R_2', and the value of x in the potentiometer to compensate for a mismatch between active load transistors Q_3 and Q_4 in the diff-amp in Figure 14.22.

Assume that $I_Q = 200\,\mu A$, which means that we want $i_{C1} = i_{C2} = 100\,\mu A$. Let $I_{S3} = 10^{-14}$ A and $I_{S4} = 1.05 \times 10^{-14}$ A. Also assume $R_1 = R_2 = 1\,k\Omega$ and $R_x = 100\,k\Omega$.

Solution: The difference between R_2' and R_1' is determined from Equation (14.75), as follows:

$$V_T \ln\left(\frac{i_{C1}}{I_{S3}}\right) + i_{C1}R_1' = V_T \ln\left(\frac{i_{C2}}{I_{S4}}\right) + i_{C2}R_2'$$

or

$$(0.026)\ln\left(\frac{100 \times 10^{-6}}{10^{-14}}\right) + (0.10)R_1' = (0.026)\ln\left(\frac{100 \times 10^{-6}}{1.05 \times 10^{-14}}\right) + (0.10)R_2'$$

which yields

$$R_2' - R_1' = 0.0127\,k\Omega \Rightarrow 12.7\,\Omega$$

We can also write the difference between R_2' and R_1' as

$$\frac{R_2(1-x)R_x}{R_2 + (1-x)R_x} - \frac{R_1 x R_x}{R_1 + x R_x} = 0.0127\,k\Omega$$

Substituting the values for R_1, R_2, and R_x, we find that

$$x = 0.349$$

Comment: On the basis of this analysis, the value of R_1' is $1\|34.9 = 0.9721\,k\Omega$, and the value of R_2' is $1\|(100 - 34.9) = 0.9849\,k\Omega$.

Computer Simulation Verification: Figure 14.23 is the circuit used in PSpice simulation. The values of R_X and R_Y were varied to simulate a change in the variable x in the potentiometer in the circuit in Figure 14.22. The output voltage v_O is taken off the common collectors of Q_1 and Q_3. This voltage would correspond to the input voltage of a second stage.

A change in the values of R_X and R_Y causes a slight change in the currents in the two sides of the circuit. A change in current causes a change in the collector–emitter

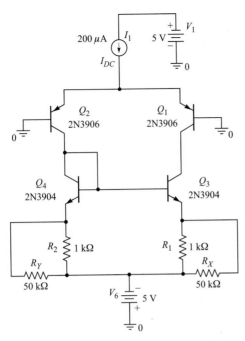

Figure 14.23 Circuit used in the computer simulation analysis for Example 14.12

voltages of Q_1 and Q_3, or a change in the output voltage. Figure 14.24 shows the output voltage as a function of x, or as a function of the position of the potentiometer. The results show that a change of approximately 0.7 V is possible for this range in potentiometer setting. This change in voltage would represent a large change in input voltage for the second stage, which in turn would cause a large change in the dc value of the output voltage. The dc output voltage could therefore be set to zero by adjusting the potentiometer setting.

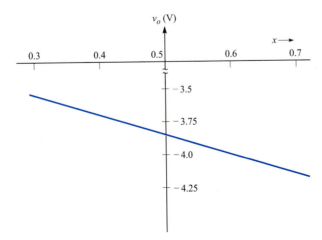

Figure 14.24 Output voltage versus potentiometer setting

Test Your Understanding

*14.20   Consider the diff-amp in Figure 14.22 with a pair of offset-null terminals. Let $R_1 = R_2 = 1\,\text{k}\Omega$. Let R_x be a $100\,\text{k}\Omega$ potentiometer. Assume $I_Q = 100\,\mu\text{A}$ and $I_{S3} = 10^{-14}\,\text{A}$. If the wiper arm on the potentiometer is adjusted such that $25\,\text{k}\Omega$ is in parallel with R_1 and $75\,\text{k}\Omega$ is in parallel with R_2, determine the value of I_{S4} for $i_{C1} = i_{C2}$. (Ans. $1.05 \times 10^{-14}\,\text{A}$)

14.5   INPUT BIAS CURRENT

The input currents to an ideal op-amp are zero. In actual operational amplifiers, however, the input bias currents are not zero. If the input stage consists of a pair of npn transistors, as shown in Figure 14.25(a), the bias currents enter the input terminals. However, if the input state consists of a pair of pnp transistors, as shown in Figure 14.25(b), the bias currents leave the input terminals.

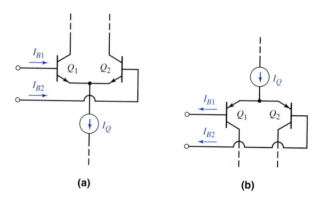

(a)   (b)

Figure 14.25   (a) Pair of npn transistors, showing input bias currents, and (b) pair of pnp transistors, showing input bias currents

If the input diff-amp consists of a pair of JFETs, the input bias currents are normally much smaller than those in a bipolar differential pair. A MOSFET input differential pair, generally, must include protection devices as discussed in Chapter 13, so the input bias currents are also not zero even in this case.

For op-amps with a bipolar input stage, the input bias currents may be as high as $10\,\mu\text{A}$ and as low as a few nanoamperes. For op-amps with an FET input stage, the bias currents may be as low as a few picoamperes. Table 14.1 lists the typical input bias current. For the 741 op-amp it is $30\,\text{nA}$, and for the FET input op-amps it is in the low picoampere range.

14.5.1   Bias Current Effects

Figure 14.26 schematically shows an op-amp with input bias currents. If the input stage is symmetrical, with all corresponding elements matched, then

$I_{B1} = I_{B2}$. However, if the input transistors are not exactly identical, then $I_{B1} \neq I_{B2}$. The **input bias current** is then defined as the average of the two input currents, or

$$I_B = \frac{I_{B1} + I_{B2}}{2} \qquad\qquad (14.76)$$

Figure 14.26 Op-amp with input bias currents

The difference between the two input currents is called the **input offset current** I_{OS} and is given by

$$I_{OS} = |I_{B1} - I_{B2}| \qquad\qquad (14.77)$$

The algebraic sign of the offset current is usually not important, just as the offset voltage polarity is not critical. The typical input offset current is on the order of 10 percent of the input bias current, although data sheets may list larger values. The typical and maximum input offset currents for the three op-amps analyzed in the last chapter are given in Table 14.1.

Figure 14.27 shows an op-amp and associated resistors for a zero input voltage. Even if $I_{B2} \neq 0$, the noninverting terminal is still at zero volts, or $V_Y = 0$. From the virtual ground concept, we have $V_X = 0$, which means that the current in R_1 must be zero. Bias current I_{B1} is therefore supplied by the output of the op-amp and flows through R_2, producing an output voltage. If, for example, $I_{B1} = 5\,\mu\text{A}$ and $R_2 = 100\,\text{k}\Omega$, then $v_O = 0.5\,\text{V}$, which is unacceptable in most applications. Smaller input bias currents and a smaller feedback resistor will reduce the bias current effects.

Figure 14.27 Op-amp with grounded noninverting terminal

14.5.2 Bias Current Compensation

The effect of bias currents in op-amp circuits can be minimized with a simple compensation technique. Consider the circuit in Figure 14.28. We determine v_O as a function of I_{B1} and I_{B2} using superposition. For $I_{B2} = 0$, then $V_Y = V_X = 0$, and the output voltage due to I_{B1} is

$$v_O(I_{B1}) = I_{B1} R_2 \qquad\qquad (14.78(\text{a}))$$

For $I_{B1} = 0$, we find

$$V_Y = -I_{B2} R_3 = V_X$$

Since

$$v_O = (1 + R_2/R_1) V_X$$

Figure 14.28 Op-amp circuit with resistor connected to noninverting terminal, for input bias current compensation

the output voltage due to I_{B2} is

$$v_O(I_{B2}) = -I_{B2}R_3\left(1 + \frac{R_2}{R_1}\right) \tag{14.78(b)}$$

The net output voltage due to both I_{B1} and I_{B2} is the sum of Equations (14.78(a)) and (14.78(b)), or

$$v_O = I_{B1}R_2 - I_{B2}R_3\left(1 + \frac{R_2}{R_1}\right) \tag{14.79}$$

If $I_{B1} = I_{B2} \equiv I_B$ and if the combination of the three resistances can be adjusted to produce $v_O = 0$, then Equation (14.79) becomes

$$0 = I_B\left[R_2 - R_3\left(1 + \frac{R_2}{R_1}\right)\right] \tag{14.80}$$

which means that

$$R_2 = R_3\left(1 + \frac{R_2}{R_1}\right) \tag{14.81}$$

Equation (14.81) can be rearranged as follows:

$$R_3 = \frac{R_1 R_2}{R_1 + R_2} = R_1 \| R_2 \tag{14.82}$$

Equation (14.82) shows that R_3 should be made equal to the parallel combination of R_1 and R_2, to eliminate the effect of equal input bias currents.

If $R_3 = R_1 \| R_2$ and if the bias currents are not equal, then from Equation (14.79), we have

$$v_O = R_2(I_{B1} - I_{B2}) = R_2 I_{OS} \tag{14.83}$$

Since the input offset current is normally a fraction of the input bias current, Equation (14.83) shows that the bias current effect can be reduced by making $R_3 = R_1 \| R_2$.

Example 14.13   **Objective:** Determine the bias current effect in an op-amp circuit, with and without bias current compensation.

Consider the op-amp circuits in Figures 14.27 and 14.28. Let $R_1 = 10\,\text{k}\Omega$ and $R_2 = 100\,\text{k}\Omega$. Assume $I_{B1} = 1.1\,\mu\text{A}$ and $I_{B2} = 1.0\,\mu\text{A}$.

Solution: For the op-amp circuit in Figure 14.27, the output voltage due to the bias currents is

$$v_O = I_{B1}R_2 = (1.1 \times 10^{-6})(100 \times 10^3) = 0.11\,\text{V}$$

For the circuit in Figure 14.28, we design R_3 such that

$$R_3 = R_1 \| R_2 = 10 \| 100 = 9.09\,\text{k}\Omega$$

Then, from Equation (14.83), we find

$$v_O = R_2(I_{B1} - I_{B2}) = (100 \times 10^3)(1.1 - 1.0) \times 10^{-6} = 0.010\,\text{V}$$

Comment: Even if the input offset current is not zero, the effect of the input bias currents can be reduced substantially by incorporating resistor R_3.

Usually the effect of bias currents in op-amp circuits is significant only for circuits with large resistor values. For these situations, an op-amp with an FET input stage may be necessary.

Test Your Understanding

14.21 For the op-amp in Figure 14.28, the parameters are: $R_1 = 10\,\text{k}\Omega$ and $R_2 = 100\,\text{k}\Omega$. If $I_{B1} = 1.1\,\mu\text{A}$ and $I_{B2} = 1.0\,\mu\text{A}$, can R_3 be adjusted such that $v_O = 0$? If so, what is the value of R_3? (Ans. $R_3 = 10\,\text{k}\Omega$)

14.22 Consider the inverting summing amplifier in Figure 14.29. Assume input bias currents of $I_{B1} = I_{B2} = 1\,\mu\text{A}$. (a) For $v_{i1} = v_{i2} = 0$ and $R_4 = 0$, determine v_O due to the bias currents. (b) Find the value of R_4 that compensates for the effects of the bias currents. (Ans. (a) $v_O = 0.20\,\text{V}$ (b) $R_4 = 28.6\,\text{k}\Omega$)

Figure 14.29 Figure for Exercise 14.22

14.6 ADDITIONAL NONIDEAL EFFECTS

Two additional nonideal effects in op-amps are: temperature effects and common-mode rejection ratio. We will look at each of these in this section.

14.6.1 Temperature Effects

Individual transistor parameters are functions of temperature. For bipolar transistors, the collector current is

$$i_C = I_S e^{v_{BE}/V_T} \tag{14.84}$$

where both I_S and V_T are functions of temperature. We expect the open-loop gain to vary with temperature, but as we saw in Section 14.2, the fractional change in the closed-loop gain is orders of magnitude less than the fractional change in the open-loop gain. This then makes the closed-loop gain very insensitive to temperature variations.

Offset Voltage Temperature Coefficient

The electrical properties of transistors are functions of temperature, which means that the input offset voltage is a function of temperature. The rate of change of offset voltage with temperature is defined as the **temperature coefficient of offset voltage**, or **input offset voltage drift**, and is given by

$$\text{TC}v_{OS} = \frac{dV_{OS}}{dT} \tag{14.85}$$

For a bipolar diff-amp input stage, the offset voltage, from Equation (14.59(b)), is

$$V_{OS} = V_T \ln(I_{S2}/I_{S1})$$

The temperature variations of the I_S parameters cancel; therefore, the offset voltage is directly proportional to the thermal voltage V_T, which in turn is directly proportional to temperature. From Equation (14.59(b)), the temperature coefficient is then

$$\text{TC}v_{OS} = \frac{V_{OS}}{T} \tag{14.86}$$

where T is the absolute temperature. Thus, for $V_{OS} = 1\,\text{mV}$, the temperature coefficient is $\text{TC}v_{OS} = 1\,\text{mV}/300\,^\circ\text{K} \Rightarrow 3.3\,\mu\text{V}/^\circ\text{C}$. A change of $10\,^\circ\text{C}$ will therefore result in an offset voltage change of approximately $33\,\mu\text{V}$. The temperature coefficients of offset voltage listed in Table 14.1 are in the range of 10–15 $\mu\text{V}/^\circ\text{C}$.

Consequently, the offset voltage compensation techniques discussed previously are completely effective at only one temperature. As the device temperature drifts in either direction from the temperature at which the compensation network was designed, the offset voltage effect is not completely compensated. However, the offset voltage drift is substantially less than the initial offset voltage, so offset voltage compensation is still desirable.

Input Offset Current Temperature Coefficient

The input bias currents are functions of temperature. For example, the input bias current of a bipolar input stage has the same functional dependence as the collector current, as given by Equation (14.84). If the input devices are not matched, then an input offset current I_{OS} exists, which is also a function of temperature. The input offset current temperature coefficient is dI_{OS}/dT. For the 741 op-amp, the maximum value given in Table 14.1 is $0.5\,\text{nA}/^\circ\text{C}$. If the input offset current becomes a problem in a particular design, then a JFET of MOSFET input stage of-amp may be required.

14.6.2   Common-Mode Rejection Ratio

We considered the common-mode gain (A_{cm}) and common-mode rejection ratio (CMRR) of the difference amplifier in Chapter 11. Since a diff-amp is the op-amp input stage, any common-mode signal produced at the input stage will propagate through the op-amp to the output. Therefore, the CMRR of the op-amp is essentially the same as the CMRR of the input diff-amp.

Figure 14.30(a) shows the open-loop op-amp with a pure differential-mode input signal. The differential-mode gain A_d is the same as the open-loop gain A_{OL}. Figure 14.30(b) shows the open-loop op-amp with a pure common-mode input signal. The common-mode rejection ratio, in dB, is

$$\text{CMRR}_{\text{dB}} = 20 \log_{10} \left| \frac{A_d}{A_{cm}} \right| \qquad\qquad (14.87)$$

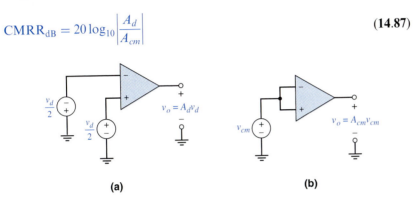

(a)   **(b)**

Figure 14.30   Open-loop op-amp (a) with pure differential-mode input signal and (b) with pure common-mode input signal

Typical values of CMRR_{dB} range from 80 to 100 dB. Table 14.1 lists typical CMRR_{dB} values for three op-amps.

14.7   SUMMARY

- A finite open-loop amplifier gain results in the magnitudes of the inverting amplifier and noninverting amplifier gains being smaller than the ideal values.

- A finite open-loop amplifier gain plus finite input amplifier resistance and nonzero output resistance results in nonideal op-amp input and output resistance values. In the case of a shunt input feedback connection (e.g., inverting op-amp), the input resistance is small but not ideally zero. In the case of a series input feedback connection (e.g., noninverting op-amp), the input resistance is large but not ideally infinite. For a shunt output feedback connection, the output resistance is small (may be in the milliohm range) but not zero.

- The practical op-amp circuit has a finite bandwidth. With negative feedback, the gain–bandwidth product is essentially constant, so an op-amp circuit with negative feedback has a reduced gain magnitude but an increased small-signal bandwidth.

- Slew rate is defined as the maximum rate at which the op-amp output signal can change per unit time. In general, the slew rate is limited by the internal frequency compensation capacitor. The slew rate is also a function of the bias current in the input diff-amp stage. Slew rates are typically in the 0.5–3 V/μs range. Full-power bandwidth is the maximum frequency at which an op-amp circuit can operate without being slew-rate limited. This frequency is a function of both the slew rate and the peak value of output voltage.

- An input offset voltage means that the output voltage is not zero when the input signal voltages are zero. One source of an offset voltage is a mismatch in the differential pair transistor parameters and/or mismatches in active load transistor parameters. Typically, an offset voltage of a few millivolts may occur in a bipolar circuit, whereas an offset voltage of tens of millivolts may occur in a MOSFET circuit.

- Two techniques of offset voltage compensation were analyzed. The first technique involves using an external potentiometer network at the input to the amplifier to null out the offset voltage. The second technique uses a potentiometer connected to a pair of offset-null terminals connected to the input diff-amp circuit.

- Input bias currents of an ideal op-amp are zero. However, actual bias currents may range from a few picoamperes for FET input stages to as high as a few microamperes for some bipolar input stages. The input bias currents can produce an unwanted component of output voltage. We analyzed the design of input bias current compensation circuits that eliminate or at least minimize these bias current effects.

- Variations in temperature produce variations in offset voltage and input bias currents. Therefore, the offset voltage and input bias current compensation circuits are completely effective only at one temperature. Typical offset voltage temperature coefficients are in the range of a few $\mu V/°C$ and input bias current temperature coefficients may be in the range of a few $nA/°C$.

CHECKPOINT

After studying this chapter the reader should have the ability to:

✔ Understand differences between ideal and practical values of various parameters of the operational amplifier circuit. (Section 14.1)

✔ Understand the effect of a finite open-loop amplifier gain on the characteristics of the op-amp. (Section 14.2)

✔ Understand the small-signal frequency response and the large-signal slew-rate response of op-amps. (Section 14.3)

✔ Understand offset voltage characteristics and design offset voltage compensation circuits for an op-amp. (Section 14.4)

✔ Understand input bias current effects and design input bias current compensation circuits for an op-amp. (Section 14.5)

REVIEW QUESTIONS

1. List and describe five practical op-amp parameters and discuss the effect they have on op-amp circuit characteristics.

2. What is a typical value of open-loop, low-frequency gain of an op-amp circuit? How does this compare to the ideal value?

3. How does a finite open-loop gain affect the closed-loop gains of the inverting and noninverting amplifiers?

4. How does a finite open-loop gain affect the (a) input resistance of an op-amp circuit and (b) the output resistance of an op-amp circuit? Consider the inverting and noninverting amplifiers.

5. Describe the open-loop amplifier frequency response and define the unity-gain bandwidth.

6. What is a typical corner frequency value, or dominant-pole frequency, in an open-loop frequency characteristic?

7. Describe the gain–bandwidth product property on a closed-loop amplifier response.

8. Define slew rate.

9. What is meant by full-power bandwidth?

10. What is the primary source of slew-rate limitation in an op-amp circuit?

11. What is one cause of an offset voltage in the input stage of a BJT op-amp?

12. What is one cause of an offset voltage in the input stage of a CMOS op-amp?

13. Describe an offset voltage compensation technique.

14. What is the source of input bias current in the 741 op-amp?

15. What can be the effect of an input bias current?

16. Describe any difference in input bias current effects between a pnp BJT input differential pair and an npn BJT input differential pair.

17. Describe the effect of input bias currents on an integrator.

18. Describe an input bias current compensation technique.

19. Define and explain common-mode rejection ratio.

PROBLEMS

Section 14.2   Finite Open-Loop Gain

14.1   For the op-amp used in the inverting amplifier configuration in Figure P14.1, the open-loop parameters are $A_{OL} = 10^3$ and $R_o = 0$. Determine the closed-loop gain $A_{CL} = v_O/v_I$ and input resistance R_{if} for an open-loop input differential-mode resistance of: (a) $R_i = 1\,k\Omega$, (b) $R_i = 10\,k\Omega$, and (c) $R_i = 100\,k\Omega$.

Figure P14.1

14.2   A pressure transducer, as described in Example 14.1, is to be used in conjunction with a noninverting op-amp circuit. The ideal output voltage is to be +0.10 V for a transducer voltage of 2 mV. Determine the minimum open-loop gain required so that the actual output voltage is within 0.1 percent of the ideal.

14.3   Consider the two inverting amplifiers in cascade in Figure P14.3. The op-amp parameters are $A_{OL} = 5 \times 10^3$, $R_i = 10\,k\Omega$, and $R_o = 1\,k\Omega$. Determine the actual

Figure P14.3

closed-loop gains $A_{vf1} = v_{o1}/v_i$ and $A_{vf} = v_{o2}/v_i$. What is the percent error from the ideal values?

14.4 The noninverting amplifier in Figure P14.4 has an op-amp with open-loop properties: $A_{OL} = 10^3$, $R_i = 20\,\text{k}\Omega$, and $R_o = 0.5\,\text{k}\Omega$. (a) Determine the closed-loop values of $A_{CL} = v_O/v_I$, R_{if}, and R_{of}. (b) If A_{OL} decreases by 10 percent, determine the percentage change in A_{CL}.

14.5 For the op-amp in the voltage follower circuit in Figure P14.5, the open-loop parameters are: $A_{OL} = 10^4$, $R_i = 100\,\text{k}\Omega$, and $R_o = 200\,\Omega$. Determine: (a) the closed-loop voltage gain $A_v = v_O/v_I$, and (b) the output resistance R_{of}.

14.6 The summing amplifier in Figure P14.6 has an op-amp with open-loop parameters: $A_{OL} = 2 \times 10^3$, $R_i = \infty$, and $R_o = 0$. Determine the actual output voltage as a function of v_{I1} and v_{I2}. What is the percent error from the ideal value?

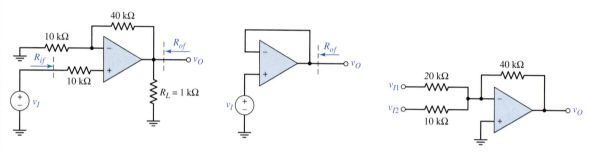

Figure P14.4   Figure P14.5   Figure P14.6

14.7 For the op-amp in the differential amplifier in Figure P14.7, the open-loop parameters are: $A_{OL} = 10^3$, $R_i = \infty$, and $R_o = 0$. Determine the actual differential voltage gain $A_d = v_O/(v_{I2} - v_{I1})$. What is the percentage error from the ideal value?

14.8 Because of a manufacturing error, the open-loop gain of each op-amp in the circuit in Figure P14.8 is only $A_{OL} = 100$. The open-loop input and output resistances are $R_i = 10\,\text{k}\Omega$ and $R_o = 1\,\text{k}\Omega$, respectively. Determine the closed-loop parameters: (a) R_{if}, (b) R_{of}, and (c) $A_{CL} = v_{O2}/v_I$. (d) What is the ratio of the actual closed-loop gain to the ideal value?

Figure P14.7   Figure P14.8

Section 14.3   Frequency Response

14.9   The low-frequency open-loop gain of an op-amp is 2×10^5 and the second pole occurs at a frequency of 5 MHz. An amplifier using this op-amp has a low-frequency closed-loop gain of 100 and a phase margin of 80 degrees. Determine the dominant-pole frequency.

14.10   Two inverting amplifiers are connected in cascade to provide an overall voltage gain of 500. The gain of the first amplifier is -10 and the gain of the second amplifier is -50. The unity-gain bandwidth of each op-amp is 1 MHz. (a) What is the bandwidth of the overall amplifier system? (b) Redesign the system to achieve the maximum bandwidth. What is the maximum bandwidth?

14.11   The open-loop low-frequency gain of an op-amp is found to be $A_o = 5 \times 10^4$. At a frequency of $f = 10^4$ Hz, the open-loop gain is 200. Determine the dominant-pole frequency and the unity-gain bandwidth.

14.12   An inverting amplifier circuit has a voltage gain of -25. The op-amp used in the circuit has a low-frequency voltage gain of 5×10^4 and a unity-gain bandwidth of 1 MHz. Determine the dominant pole frequency of the op-amp and the small-signal bandwidth, $f_{3\text{-dB}}$, of the inverting amplifier. What is the magnitude of the closed-loop voltage gain at $0.5 f_{3\text{-dB}}$ and at $2 f_{3\text{-dB}}$?

14.13   An audio amplifier system, using a noninverting op-amp circuit, needs to have a small-signal bandwidth of 20 kHz. The open-loop low-frequency voltage gain of the op-amp is 10^5 and the unity-gain bandwidth is 1 MHz. What is the maximum closed-loop voltage gain that can be obtained for these specifications?

14.14   If an op-amp has a slew rate of 10 V/μs, find the full-power bandwidth for a peak output voltage of 10 V.

14.15   (a) An op-amp with a slew rate of 8 V/μs is driven by a 250 kHz sine wave. What is the maximum output amplitude at which slew-rate limiting is reached? (b) Repeat part (a) for a 250 kHz zero time-average triangular wave.

14.16   The op-amp in the noninverting amplifier configuration in Figure P14.16 has a slew rate of 1 V/μs. Sketch the output voltage versus time for each of the three inputs shown. The op-amp is biased at ± 10 V.

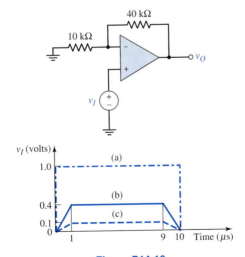

Figure P14.16

14.17 For each op-amp in the circuit shown in Figure P14.17, the bias is ± 15 V and the slew rate is 3 V/µs. Sketch the output voltages v_{O1} and v_{O2} versus time for each input shown.

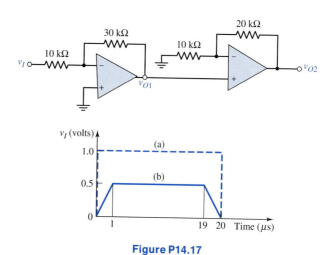

Figure P14.17

14.18 The op-amp to be used in the audio amplifier system in Problem 14.13 has a slew rate of 0.8 V/µs. Determine the maximum peak value of output voltage that can be obtained for these specifications.

Section 14.4   Offset Voltage

14.19 For the transistors in the diff-amp in Figure 14.16 in the text, the current parameters I_{S1} and I_{S2} can be written as $5 \times 10^{-14}(1 + x)$ A, where x represents the deviation from the ideal due to variations in electrical and geometric characteristics. (The value of x is positive for one transistor and negative for the other transistor.) Determine the maximum value of x such that the maximum offset voltage is limited to $V_{OS} = 2.5$ mV.

14.20 The bipolar active load diff-amp in Figure 14.18 in the text is biased at $V^+ = 5$ V. The transistor parameters are: $v_{BE}(\text{npn}) = v_{EB}(\text{pnp}) = 0.6$ V, $V_{AN} = V_{AP} = 80$ V, $I_{S1} = I_{S2}$, and $I_{S4} = 10^{-14}$ A. Determine the value of I_{S3} for which Q_2 has a C–E voltage of $v_{CE2} = 0.6$ V.

14.21 An inverting op-amp circuit has a gain of -50. The op-amp used in the circuit has an offset voltage of ± 2.5 mV. If the input signal voltage to the circuit is 20 mV, determine the possible range in the output voltage.

14.22 Consider the integrator circuit in Figure P14.22. The circuit parameters are $R = 10$ kΩ and $C = 10$ µF. The op-amp offset voltage is ± 5 mV. For $v_i = 0$, determine the output voltage versus time. For the worst-case offset voltage, determine the time that it would take for the output voltage to reach ± 5 V.

D14.23 In the circuit in Figure P14.23, the offset voltage of each op-amp is $V_{OS} = 10$ mV. (a) Find the worst-case output voltages v_{O1} and v_{O2} for $v_I = 0$. (b) Design offset voltage compensation circuit(s) to adjust both v_{O1} and v_{O2} to zero when $v_I = 0$.

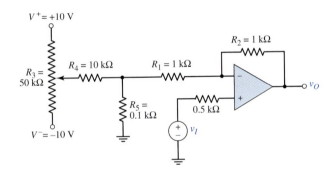

Figure P14.22

Figure P14.23

14.24 In the circuit shown in Figure P14.24, the op-amp is ideal. For $v_I = 0.5\,\text{V}$, determine v_O when the wiper arm of the potentiometer is at the V^+ node, in the center, and at the V^- node.

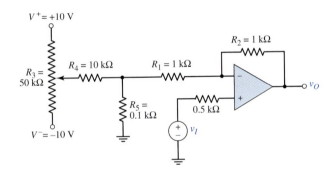

Figure P14.24

14.25 Consider the bipolar diff-amp with an active load and a pair of offset-null terminals as shown in Figure 14.22 in the text. Let $R_1 = R_2 = 500\,\Omega$ and let R_x be a $50\,\text{k}\Omega$ potentiometer. (a) If the wiper arm of the potentiometer is exactly in the center, determine the effective resistances R_1' and R_2'. (b) Assume $I_Q = 250\,\mu\text{A}$ meaning that $i_{C1} = i_{C2} = 125\,\mu\text{A}$. Let $I_{S3} = 2 \times 10^{-14}\,\text{A}$ and $I_{S4} = 2.2 \times 10^{-14}\,\text{A}$. Determine the required values of x and $(1 - x)$ of the potentiometer to compensate for the transistor mismatches.

14.26 The bipolar diff-amp in Figure 14.22 in the text is biased at $I_Q = 500\,\mu\text{A}$. Assume all transistors are matched, with $I_S = 10^{-14}\,\text{A}$. Let $R_1 = R_2 = 500\,\Omega$, and assume R_x is a $50\,\text{k}\Omega$ potentiometer. If the wiper arm of the potentiometer is off center such that $x = 15\,\text{k}\Omega$ and $(1 - x) = 35\,\text{k}\Omega$, determine the ratio of i_{C1}/i_{C2}. What is the corresponding offset voltage?

Section 14.5   Input Bias Current

D14.27 An op-amp used in a voltage follower configuration is ideal except that the input bias currents are $I_{B1} = I_{B2} = 1\,\mu\text{A}$. The source driving the voltage follower has an output resistance of $10\,\text{k}\Omega$. (a) Find the output voltage due to the bias current effects when $v_I = 0$. (b) Can the circuit be designed to compensate for the input bias currents? If so, how?

14.28 In the differential amplifier in Figure P14.7, the op-amp is ideal except that the average input bias current is $I_B = 10\,\mu A$ and the input offset current is $I_{OS} = 3\,\mu A$. If $v_{i1} = v_{i2} = 0$, determine the worst-case output voltage v_O due to the input bias current effects.

D14.29 The op-amp bias currents for the circuit in Figure P14.23 are equal at $I_{B1} = I_{B2} = 1\,\mu A$. (a) Find the worst-case output voltages v_{O1} and v_{O2} for $v_I = 0$. (b) Design input bias current compensation circuit(s) to adjust both v_{O1} and v_{O2} to zero when $v_I = 0$.

14.30 (a) For the integrator circuit in Figure P14.30, let the input bias currents be $I_{B1} = I_{B2} = 0.1\,\mu A$. Assume that switch S opens at $t = 0$. Derive an expression for the output voltage versus time for $v_I = 0$. (b) Plot v_O versus time for $0 \le t \le 10\,s$. (c) Repeat part (b) for $I_{B1} = I_{B2} = 100\,pA$.

Figure P14.30

14.31 For the circuit in Figure P14.31, the op-amps are ideal except that each op-amp has input bias currents $I_{B1} = I_{B2} = 10\,\mu A$. (a) For $v_I = 0$ and $R_A = R_B = 0$, determine the worst-case values of v_{O1}, v_{O2}, and v_{O3} due to bias currents. (b) Determine the values of R_A and R_B for input bias current compensation. (c) If the average input bias current is $I_B = 10\,\mu A$ and the input offset current is $I_{OS} = 2\,\mu A$, determine the worst-case output values of v_{O1}, v_{O2}, and v_{O3} using the results of part (b).

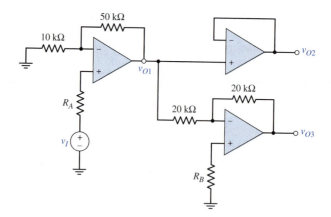

Figure P14.31

14.32 For each circuit in Figure P14.32, the input bias current is $I_B = 0.8\,\mu A$ the input offset current is $I_{OS} = 0.2\,\mu A$. (a) Determine the output voltage due to the average bias current I_B. (b) Determine the worst-case output voltage, including the effect of the input offset current.

Figure P14.32

Sections 14.4 and 14.5   Offset Voltage and Input Bias Current: Total Effects

D14.33   For the op-amp in Figure P14.33, the input offset voltage is $V_{OS} = 10\,\text{mV}$, the input bias current is $I_B = 2\,\mu\text{A}$, and the input offset current is $I_{OS} = 0.2\,\mu\text{A}$. (a) Determine the worst-case, or maximum, output voltage when $v_I = 0$. (b) Design compensation circuit(s) to minimize v_O when $v_I = 0$.

D14.34   Consider the op-amp circuit in Figure P14.34. (a) Find the value of R_2 needed for a $\pm 10\,\text{mV}$ offset voltage adjustment. (b) Determine R_1 to minimize bias current effects. (Assume $R_2 \gg R_i$.)

Figure P14.33   **Figure P14.34**

D14.35   For each op-amp in the circuit in Figure P14.23, the offset voltage is $V_{OS} = 10\,\text{mV}$ and the input bias currents are $I_{B1} = I_{B2} = 2\,\mu\text{A}$. (a) Find the worst-case output voltages v_{O1} and v_{O2} for $v_I = 0$. (b) Design compensation circuits to adjust both v_{O1} and v_{O2} to zero when $v_I = 0$.

D14.36   The op-amps in the circuit in Figure P14.31 have an offset voltage of $V_{OS} = 5\,\text{mV}$, and average input bias current of $I_B = 5\,\mu\text{A}$, and an input offset current of $I_{OS} = 1\,\mu\text{A}$. (a) For $v_I = 0$ and $R_A = R_B = 0$, determine the worst-case output voltages v_{O1}, v_{O2}, and v_{O3}. (b) Design compensation circuits to minimize the effects of the offset voltage and input bias current.

14.37   Each op-amp in Figure P14.32 has an offset voltage of $V_{OS} = 2\,\text{mV}$, an average input bias current of $I_B = 500\,\text{nA}$, and an input offset current of $I_{OS} = 100\,\text{nA}$. Determine the worst-case output voltage for each circuit.

Section 14.6   Additional Nonideal Effects

14.38 For each op-amp in Figure P14.32, the input offset voltage is $V_{OS} = 2\,\text{mV}$ at $T = 25\,^\circ\text{C}$ and the input offset voltage temperature coefficient is $\text{TC}v_{OS} = 6.7\,\mu\text{V}/^\circ\text{C}$. Find the output voltage v_O due to the input offset voltage effects at: (a) $T = 25\,^\circ\text{C}$ and (b) $T = 50\,^\circ\text{C}$.

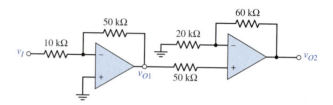

Figure P14.39

14.39 The input offset voltage in each op-amp in Figure P14.39 is $V_{OS} = 1\,\text{mV}$ at $T = 25\,^\circ\text{C}$ and the input offset voltage coefficient is $\text{TC}v_{OS} = 3.3\,\mu\text{V}/^\circ\text{C}$. Find the worst-case output voltages v_{O1} and v_{O2} at: (a) $T = 25\,^\circ\text{C}$ and (b) $T = 50\,^\circ\text{C}$.

14.40 For each op-amp in Figure P14.32, the input bias current is $I_B = 500\,\text{nA}$ at $T = 25\,^\circ\text{C}$, the input offset current is $I_{OS} = 200\,\text{nA}$ at $T = 25\,^\circ\text{C}$, the input bias current temperature coefficent is $8\,\text{nA}/^\circ\text{C}$, and the input offset current temperature coefficient is $2\,\text{nA}/^\circ\text{C}$. (a) Find the output voltage due to the average input bias currents at $T = 25\,^\circ\text{C}$. (b) Find the worst-case output voltage due to the input bias current and input offset current at $T = 25\,^\circ\text{C}$. (c) Repeat parts (a) and (b) for $T = 50\,^\circ\text{C}$.

14.41 For each op-amp in Figure P14.39, the input bias current is $I_B = 2\,\mu\text{A}$ at $T = 25\,^\circ\text{C}$, the input offset current is $I_{OS} = 0.2\,\mu\text{A}$ at $T = 25\,^\circ\text{C}$, the input bias current temperature coefficient is $20\,\text{nA}/^\circ\text{C}$, and the input offset current temperature coefficient is $5\,\text{nA}/^\circ\text{C}$. (a) Find the worst-case output voltages v_{O1} and v_{O2} due to the average input bias currents at $T = 25\,^\circ\text{C}$. (b) Find the worst-case output voltages v_{O1} and v_{O2} due to the input bias currents and input offset current at $T = 25\,^\circ\text{C}$. (c) Repeat parts (a) and (b) for $T = 50\,^\circ\text{C}$.

Figure P14.42

14.42 The op-amp in the diff-amp in Figure P14.42 is ideal. If the tolerance of each resistor is $\pm 2\%$, determine the minimum value of CMRR_{dB}.

14.43 If the tolerances of each resistor in the diff-amp in Figure P14.42 are $\pm x\%$, what is the maximum value of x if the minimum CMRR_{dB} is: (a) $90\,\text{dB}$ and (b) $60\,\text{dB}$.

COMPUTER SIMULATION PROBLEMS

14.44 Consider the reference circuit and gain stage of the 741 op-amp in Figure 13.7. Using a computer analysis, determine the slew rate of the gain stage.

14.45 The equivalent circuit of the all-CMOS MC14573 op-amp was given in Figure 13.14. Using a computer analysis, determine the slew rate of the op-amp, assuming $C_1 = 12\,\text{pF}$. Use the transistor and circuit parameters given in Example 13.8 and 13.9.

14.46 A basic bipolar input diff-amp stage is shown in Figure 14.22. Assume the circuit parameters are: $I_Q = 0.2\,\text{mA}$, $V^+ = 10\,\text{V}$, $V^- = -10\,\text{V}$, and $R_1 = R_2 = 500\,\Omega$. Let R_x be a $100\,\text{k}\Omega$ potentiometer and assume transistor Early voltages of 80 V. (a) Plot the collector voltage at Q_4 as a function of the wiper arm position. (b) Assume the I_S parameters of Q_1 and Q_2 vary such that Q_1 and Q_2 are mismatched by $\pm 5\%$. Repeat part (a). (c) Repeat part (b) for a $\pm 5\%$ mismatch in Q_3 and Q_4.

14.47 Consider the input stage and bias circuit of the 741 op-amp in Figure 13.5. Assume the transistor Early voltages are 50 V. Using a computer analysis, determine the diff-amp common-mode rejection ratio.

15

Applications and Design of Integrated Circuits

15.0 PREVIEW

In Chapter 9, we introduced the ideal operational amplifier and analyzed and designed basic op-amp circuits. In this chapter, we consider additional applications and designs of op-amp and comparator circuits that may be fabricated as integrated circuits. A comparator is essentially an op-amp operated in an open-loop configuration with either a high or low saturated output signal.

A general goal of this chapter is to increase our skill at designing electronic circuits to meet particular specifications and to perform particular functions.

Five types of circuits are considered. These circuits and their purpose are as follows:

1. *Active filters.* The primary functions of a filter are transmission of the desired frequency components of an input signal and attenuation of any undesired frequency components.

2. *Oscillators.* These circuits provide sinusoidal signals at a specified frequency for communication systems, for example.

3. *Multivibrator circuits.* Some electronic systems, for example, digital systems, require signals with particular waveforms, such as square-wave, triangular-wave, or single-pulse signals. Multivibrator circuits generate these types of signals.

4. *Power amplifiers.* An IC power amplifier usually consists of a high-gain small-signal amplifier in cascade with a class-AB output stage.

5. *Voltage regulators.* This circuit establishes a relatively constant dc voltage from an ac signal source.

We present three examples of integrated circuit power amplifiers. The IC power amplifier usually consists of a high-gain small-signal amplifier in cascade with a class-AB output stage. The output stage and feedback may be part of the integrated circuit, or may be external to the chip.

In the last section, we discuss the basis of voltage regulator circuits. The voltage regulator generally consists of an amplifier and other basic circuits and it uses feedback techniques. We will discuss one example of a basic voltage regulator IC.

15.1 ACTIVE FILTERS

An important application of op-amp is the **active filter**. The word filter refers to the process of removing undesired portions of the frequency spectrum. The word *active* implies the use of one or more active devices, usually an operational amplifier, in the filter circuit. As an example of the application of op-amps in the area of active filters, we will discuss the Butterworth filter. The discussion is only an introduction to the subject of filter theory design.

Two advantages of active filters over passive filters are:

1. The maximum gain or the maximum value of the transfer function may be greater than unity.
2. The loading effect is minimal, which means that the output response of the filter is essentially independent of the load driven by the filter.

15.1.1 Active Network Design

From our discussions of frequency response in Chapter 7, we know that *RC* networks form filters. Figure 15.1(a) is a simple example of a coupling-capacitor circuit. The voltage transfer function for this circuit is

$$T(s) = \frac{V_o(s)}{V_i(s)} = \frac{R}{R + \dfrac{1}{sC}} = \frac{sRC}{1 + sRC} \tag{15.1}$$

The Bode plot of the voltage gain magnitude $|T(j\omega)|$ is shown in Figure 15.1(b). The circuit is called a **high-pass filter**.

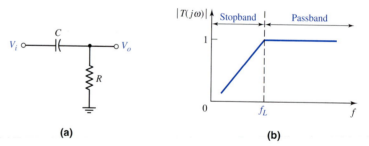

(a)   **(b)**

Figure 15.1   (a) Simple high-pass filter and (b) Bode plot of transfer function magnitude

Figure 15.2(a) is another example of a simple RC network. Here, the voltage transfer function is

$$T(s) = \frac{V_o(s)}{V_i(s)} = \frac{\dfrac{1}{sC}}{\dfrac{1}{sC} + R} = \frac{1}{1 + sRC} \tag{15.2}$$

The Bode plot of the voltage gain magnitude $|T(j\omega)|$ for this circuit is shown in Figure 15.2(b). This circuit is called a **low-pass filter**.

Although these circuits both perform a basic filtering function, they may suffer from loading effects, substantially reducing the maximum gain from the

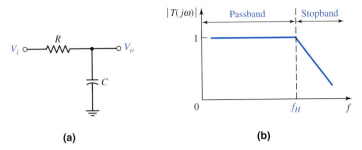

(a)   **(b)**

Figure 15.2   (a) Simple low-pass filter and (b) Bode plot of transfer function magnitude

unity value shown in Figures 15.1(b) and 15.2(b). Also, the cutoff frequencies f_L and f_H may change when a load is connected to the output. The loading effect can essentially be eliminated by using a voltage follower as shown in Figure 15.3. In addition, a noninverting amplifier configuration can be incorporated to increase the gain, as well as eliminate the loading effects.

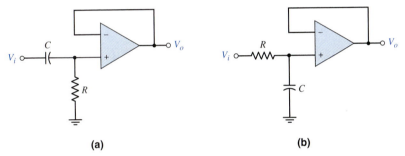

(a)   **(b)**

Figure 15.3   (a) High-pass filter with voltage follower and (b) low-pass filter with voltage follower

These two filter circuits are called one-pole filters; the slope of the voltage gain magnitude curve outside the passband is 6 dB/octave or 20 dB/decade. This characteristic is called the rolloff. The rolloff becomes sharper or steeper with higher-order filters and is usually one of the specifications given for active filters.

Two other categories of filters are **bandpass** and **band-reject**. The desired ideal frequency characteristics are shown in Figure 15.4.

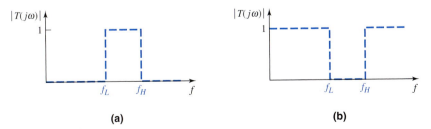

(a)   **(b)**

Figure 15.4   Ideal frequency characteristics: (a) bandpass filter and (b) band-reject filter

15.1.2   General Two-Pole Active Filter

Consider Figure 15.5 with admittances Y_1 through Y_4 and an ideal voltage follower. We will derive the transfer function for the general network and will then apply specific admittances to obtain particular filter characteristics.

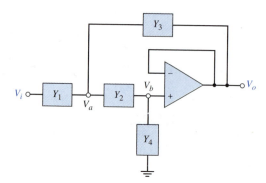

Figure 15.5   General two-pole active filter

A KCL equation at node V_a yields

$$(V_i - V_a)Y_1 = (V_a - V_b)Y_2 + (V_a - V_o)Y_3 \tag{15.3}$$

A KCL equation at node V_b produces

$$(V_a - V_b)Y_2 = V_b Y_4 \tag{15.4}$$

From the voltage follower characteristics, we have $V_b = V_o$. Therefore, Equation (15.4) becomes

$$V_a = V_b\left(\frac{Y_2 + Y_4}{Y_2}\right) = V_o\left(\frac{Y_2 + Y_4}{Y_2}\right) \tag{15.5}$$

Substituting Equation (15.5) into (15.3) and again noting that $V_b \triangleq V_o$, we have

$$V_i Y_1 + V_o(Y_2 + Y_3) = V_a(Y_1 + Y_2 + Y_3)$$
$$= V_o\left(\frac{Y_2 + Y_4}{Y_2}\right)(Y_1 + Y_2 + Y_3) \tag{15.6}$$

Multiplying Equation (15.6) by Y_2 and rearranging terms, we get the following expression for the transfer function:

$$T(s) = \frac{V_o(s)}{V_i(s)} = \frac{Y_1 Y_2}{Y_1 Y_2 + Y_4(Y_1 + Y_2 + Y_3)} \tag{15.7}$$

To obtain a low-pass filter, both Y_1 and Y_2 must be conductances, allowing the signal to pass into the voltage follower at low frequencies. If element Y_4 is a capacitor, then the output rolls off at high frequencies.

To produce a two-pole function, element Y_3 must also be a capacitor. On the other hand, if elements Y_1 and Y_2 are capacitors, then the signal will be blocked at low frequencies but will be passed into the voltage follower at high frequencies, resulting in a high-pass filter. Therefore, admittances Y_3 and Y_4 must both be conductances to produce a two-pole high-pass transfer function.

15.1.3 Two-Pole Low-Pass Butterworth Filter

To form a low-pass filter, we set $Y_1 = G_1 = 1/R_1$, $Y_2 = G_2 = 1/R_2$, $Y_3 = sC_3$, and $Y_4 = sC_4$, as shown in Figure 15.6. The transfer function, from Equation (15.7), becomes

$$T(s) = \frac{V_o(s)}{V_i(s)} = \frac{G_1 G_2}{G_1 G_2 + sC_4(G_1 + G_2 + sC_3)} \qquad (15.8)$$

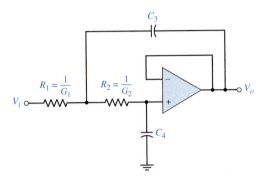

Figure 15.6 General two-pole low-pass filter

At zero frequency, $s = j\omega = 0$ and the transfer function is

$$T(s = 0) = \frac{G_1 G_2}{G_1 G_2} = 1 \qquad (15.9)$$

In the high-frequency limit, $s = j\omega \rightarrow \infty$ and the transfer function approaches zero. This circuit therefore acts as a low-pass filter.

A **Butterworth filter** is a **maximally flat magnitude filter**. The transfer function is designed such that the magnitude of the transfer function is as flat as possible within the passband of the filter. This objective is achieved by taking the derivatives of the transfer function with respect to frequency and setting as many as possible equal to zero at the center of the passband, which is at zero frequency for the low-pass filter.

Let $G_1 = G_2 \equiv G = 1/R$. The transfer function is then

$$T(s) = \frac{\dfrac{1}{R^2}}{\dfrac{1}{R^2} + sC_4\left(\dfrac{2}{R} + sC_3\right)} = \frac{1}{1 + sRC_4(2 + sRC_3)} \qquad (15.10)$$

We define time constants at $\tau_3 = RC_3$ and $\tau_4 = RC_4$. If we then set $s = j\omega$, we obtain

$$T(j\omega) = \frac{1}{1 + j\omega\tau_4(2 + j\omega\tau_3)} = \frac{1}{(1 - \omega^2\tau_3\tau_4) + j(2\omega\tau_4)} \qquad (15.11)$$

The magnitude of the transfer function is therefore

$$|T(j\omega)| = \left[(1 - \omega^2\tau_3\tau_4)^2 + (2\omega\tau_4)^2\right]^{-1/2} \qquad (15.12)$$

For a maximally flat filter (that is, a filter with a minimum rate of change), which defines a Butterworth filter, we set

$$\left. \frac{d|T|}{d\omega} \right|_{\omega=0} = 0 \qquad\qquad (15.13)$$

Taking the derivative, we find

$$\frac{d|T|}{d\omega} = -\frac{1}{2}[(1 - \omega^2\tau_3\tau_4)^2 + (2\omega\tau_4)^2]^{-3/2}[-4\omega\tau_3\tau_4(1 - \omega^2\tau_3\tau_4) + 8\omega\tau_4^2]$$

$$(15.14)$$

Setting the derivative equal to zero at $\omega = 0$ yields

$$\left. \frac{d|T|}{d\omega} \right|_{\omega=0} = \left[-4\omega\tau_3\tau_4(1 - \omega^2\tau_3\tau_4) + 8\omega\tau_4^2 \right]$$

$$= 4\omega\tau_4\left[-\tau_3(1 - \omega^2\tau_3\tau_4) + 2\tau_4 \right] \qquad (15.15)$$

Equation (15.15) is satisfied when $2\tau_4 = \tau_3$, or

$$C_3 = 2C_4 \qquad\qquad (15.16)$$

For this condition, the transfer magnitude is, from Equation (15.12),

$$|T| = \frac{1}{\left[1 + 4(\omega\tau_4)^4\right]^{1/2}} \qquad\qquad (15.17)$$

The 3 dB, or cutoff, frequency occurs when $|T| = 1/\sqrt{2}$, or when $4(\omega_{3\,dB}\tau_4)^4 = 1$. We then find that

$$\omega_{3\,dB} = 2\pi f_{3\,dB} = \frac{1}{\tau_4\sqrt{2}} = \frac{1}{\sqrt{2}RC_4} \qquad\qquad (15.18)$$

In general, we can write the cutoff frequency in the form

$$\omega_{3\,dB} = \frac{1}{RC} \qquad\qquad (15.19)$$

Finally, comparing Equations (15.19), (15.18), and (15.16) yields

$$C_4 = 0.707C \qquad\qquad (15.20(a))$$

and

$$C_3 = 1.414C \qquad\qquad (15.20(b))$$

The two-pole low-pass Butterworth filter is shown in Figure 15.7(a). The Bode plot of the transfer function magnitude is shown in Figure 15.7(b). From Equation (15.17), the magnitude of the voltage transfer function for the two-pole low-pass Butterworth filter can be written as

$$|T| = \frac{1}{\sqrt{1 + \left(\dfrac{f}{f_{3\,dB}}\right)^4}} \qquad\qquad (15.21)$$

Equation (15.15) shows that the derivative of the voltage transfer function magnitude at $\omega = 0$ is zero even without setting $2\tau_4 = \tau_3$. However, the added condition of $2\tau_4 = \tau_3$ produces the maximally flat transfer characteristics of the Butterworth filter.

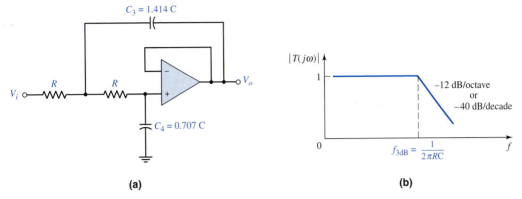

(a)   **(b)**

Figure 15.7   (a) Two-pole low-pass Butterworth filter and (b) Bode plot, transfer function magnitude

Design Example 15.1   **Objective:**   Design a two-pole low-pass Butterworth filter for an audio amplifier application.

Consider the circuit shown in Figure 15.7(a). Design the circuit such that the bandwidth is 20 kHz.

Solution: From Equation (15.19), we have

$$f_{3\,dB} = \frac{1}{2\pi RC}$$

or

$$RC = \frac{1}{2\pi f_{3\,dB}} = \frac{1}{2\pi(20 \times 10^3)} = 7.96 \times 10^{-6}$$

If we let $R = 100\,k\Omega$, then $C = 79.6\,pF$, which means tht $C_3 = 1.414C = 113\,pF$ and $C_4 = 0.707C = 56.3\,pF$.

Comment: These resistance and capacitance values are generally too large to be fabricated conveniently on an IC. Instead, discrete resistors and capacitors, in conjunction with the IC op-amp, would need to be used.

Computer Simulation Verification: Figure 15.8(a) shows the circuit used in the computer simulation. A standard LM324 op-amp is used. A 1 V sinusoidal input signal is applied. Figure 15.8(b) shows the output signal as a function of frequency. The 3 dB frequency, the frequency at which the output signal is 0.707 V, is 20 kHz, as designed. The slope of the rolloff at high frequency is also $-12\,dB/octave$, as predicted from theory.

15.1.4   Two-Pole High-Pass Butterworth Filter

To form a high-pass filter, the resistors and capacitors are interchanged from those in the low-pass filter. A two-pole high-pass Butterworth filter is shown in Figure 15.9(a). The analysis proceeds exactly the same as in the last section,

(a)

Figure 15.8 (a) Circuit used in the computer simulation of the design in Example 15.1; (b) output versus frequency

Figure 15.9 (a) Two-pole high-pass Butterworth filter and (b) Bode plot, transfer function magnitude except that the derivative is set equal to zero at $s = j\omega = \infty$. Also, the two capacitors are set equal to each other. The 3 dB or cutoff frequency can be written in the general form

$$\omega_{3\,dB} = 2\pi f_{3\,dB} = \frac{1}{RC} \qquad (15.22)$$

We find that $R_3 = 0.707R$ and $R_4 = 1.414R$. The magnitude of the voltage transfer function for the two-pole high-pass Butterworth is

$$|T| = \frac{1}{\sqrt{1 + \left(\dfrac{f_{3\,dB}}{f}\right)^4}} \qquad (15.23)$$

The Bode plot of the transfer function magnitude for the two-pole high-pass Butterworth filter is shown in Figure 15.9(b).

15.1.5 Higher-Order Butterworth Filters

The filter order is the number of poles and is usually dictated by the application requirements. An N-pole active low-pass filter has a high-frequency rolloff rate of $N \times 6$ dB/octave. Similarly, the response of an N-pole high-pass filter increases at a rate of $N \times 6$ dB/octave, up to the cutoff frequency. In each case, the 3 dB frequency is defined as

$$f_{3\,dB} = \frac{1}{2\pi RC} \qquad (15.24)$$

The magnitude of the voltage transfer function for a Butterworth Nth-order low-pass filter is

$$|T| = \frac{1}{\sqrt{1 + \left(\dfrac{f}{f_{3\,dB}}\right)^{2N}}} \qquad (15.25)$$

For a Butterworth Nth-order high-pass filter, the voltage transfer function magnitude is

$$|T| = \frac{1}{\sqrt{1 + \left(\dfrac{f_{3\,dB}}{f}\right)^{2N}}} \qquad (15.26)$$

Figure 15.10(a) shows a three-pole low-pass Butterworth filter. The three resistors are equal, and the relationship between the capacitors is found by taking the first and second derivatives of the voltage gain magnitude with respect to frequency and setting those derivatives equal to zero at $s = j\omega = 0$. Figure 15.10(b) shows a three-pole high-pass Butterworth filter. In this case, the three capacitors are equal and the relationship between the resistors is also found through the derivatives.

Higher-order filters can be created by adding additional RC networks. However, the loading effect on each additional RC circuit becomes more severe. The usefulness of active filters is realized when two or more op-amp filter circuits are cascaded to produce one large higher-order active filter.

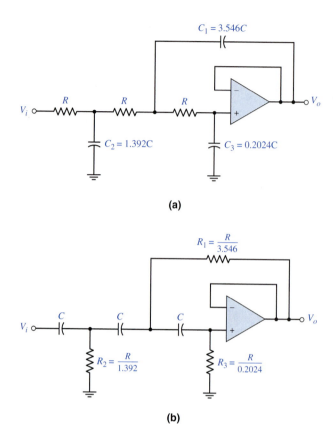

Figure 15.10   (a) Three-pole low-pass Butterworth filter and (b) three-pole high-pass Butterworth filter

Because of the low output impedance of the op-amp, there is virtually no loading effect between cascaded stages.

Figure 15.11(a) shows a four-pole low-pass Butterworth filter. The maximally flat response of this filter is *not* obtained by simply cascading two two-pole filters. The relationship between the capacitors is found through the first three derivatives of the transfer function. The four-pole high-pass Butterworth filter is shown in Figure 15.11(b).

Higher-order filters can be designed but are not considered here. Bandpass and band-reject filters use similar circuit configurations.

Test Your Understanding

D15.1   Design a three-pole low-pass Butterworth active filter with a cutoff frequency of 10 kHz and unity gain at low frequency. What is the magnitude of the voltage transfer function at 20 kHz? (Ans. For example, $R = 1.59\,k\Omega$, $C_1 = 0.03546\,\mu F$, $C_2 = 0.01392\,\mu F$, $C_3 = 0.002024\,\mu F$, $|T| = -18.1\,dB$)

D15.2   Design a four-pole high-pass Butterworth active filter with a 3 dB frequency of 50 kHz. Determine the frequency at which the voltage transfer function magnitude is 1 percent of its maximum value. (Ans. For example, $C = 0.001\,\mu F$, $R_1 = 2.94\,k\Omega$, $R_2 = 3.44\,k\Omega$, $R_3 = 1.22\,k\Omega$, $R_4 = 8.31\,k\Omega$, $f \cong 15.8\,kHz$)

(a)

(b)

Figure 15.11   (a) Four-pole low-pass Butterworth filter and (b) four-pole high-pass Butterworth filter

15.3   One-, two-, three-, and four-pole low-pass Butterworth active filters are all designed with a cutoff frequency of 10 kHz and unity gain at low frequency. Determine the voltage transfer function magnitude, in dB, at 12 kHz for each filter. (Ans. -3.87 dB, -4.88 dB, -6.0 dB, and -7.24 dB)

15.1.6   Switched-Capacitor Filter

The results of Example 15.1 demonstrated that discrete resistors and capacitors may be needed in active filters, since the required resistance and capacitance values are too large to be conveniently fabricated on a monolithic IC chip. Large-value resistors ($R > 10$ kΩ) require a large chip area, and the absolute-value tolerance is difficult to maintain. In addition, the maximum capacitance for a monolithic IC capacitor is approximately 100 pF, which is also limited by the large chip area required and the absolute-value tolerance. In these cases, accurate RC time constants may be difficult to maintain.

Conventional active filters usually combine an IC op-amp and discrete resistors and capacitors. However, even with discrete resistors and capacitors, standard components may not be available for the design of a specific cutoff frequency. Design accuracy for a specific cutoff frequency may therefore have to be sacrificed.

Switched-capacitor filters have the advantage of an all-IC circuit. The filter uses small capacitance values and realizes large effective resistance values by using a combination of capacitors and MOS switching transistors.

The Basic Principle of the Switched Capacitor

Figure 15.12 Voltages applied to resistor terminals, and the current

Figure 15.12 shows a simple circuit in which voltages V_1 and V_2 are applied at the terminals of a resistance R. The current in the resistor is

$$I = \frac{V_1 - V_2}{R} \qquad (15.27(a))$$

The resistance is therefore

$$R = \frac{V_1 - V_2}{I} \qquad (15.27(b))$$

Since the current is the rate of charge flow, Equation (15.27(b)) states that the resistance is a voltage difference divided by the rate of charge flow. We use this basic definition in switched-capacitor circuits.

The circuit in Figure 15.13(a) consists of two MOSFETs and a capacitor. A two-phase clock provides complementary but nonoverlapping ϕ_1 and ϕ_2 gate pulses, as shown in Figure 15.13(b). When a clock pulse is high, the corresponding transistor turns on; when the gate pulse is low, the transistor is off.

When ϕ_1 goes high, M_1 turns on and capacitor C charges up to V_1. When ϕ_2 goes high, M_2 turns on and capacitor C discharges to V_2 (assuming $V_1 > V_2$). The amount of charge transferred during this process is $Q = C(V_1 - V_2)$ and the transfer occurs during one clock period T_C. The equivalent current is then

$$I_{eq} = \frac{Q}{T_C} = \frac{C(V_1 - V_2)}{T_C} = f_C C(V_1 - V_2) = \frac{V_1 - V_2}{R_{eq}} \qquad (15.28)$$

where f_C is the clock frequency and R_{eq} is the equivalent resistance given by

$$R_{eq} = \frac{1}{f_C C} \qquad (15.29)$$

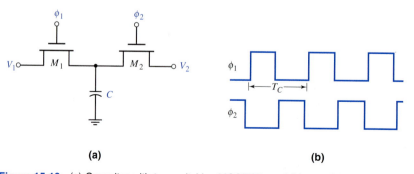

(a)   **(b)**

Figure 15.13   (a) Capacitor with two switching MOSFETs and (b) two-phase clock pulses

Using this technique, we can simulate an equivalent resistance by alternately charging and discharging a capacitor between two voltage levels. A large equivalent resistance can be simulated by using a small capacitance and an appropriate clock frequency. The circuit in Figure 15.13(a) is therefore called a switched-capacitor circuit.

Example 15.2   **Objective:** Determine the clock frequency required to simulate a specific resistance.

Consider the switched-capacitor circuit in Figure 15.13(a). Assume a capacitance of $C = 20\,\text{pF}$. Determine the clock frequency required to simulate a $1\,\text{M}\Omega$ resistance.

Solution: From Equation (15.29), we find that

$$f_C = \frac{1}{CR_{eq}} = \frac{1}{(20 \times 10^{-12})(10^6)} \Rightarrow 50\,\text{kHz}$$

Comment: A very large resistance can be readily simulated by a small capacitance and a reasonable clock frequency.

Various classes of active filters, such as low-pass, high-pass, bandpass, and band-reject circuits, can be implemented by the switched-capacitor technique, which then results in an all-capacitor filter circuit.

Example of Switched-Capacitor Filter

Consider the one-pole low-pass filter in Figure 15.14(a). The transfer function is

$$T(s) = \frac{V_o(s)}{V_{in}(s)} = -\frac{R_F}{R_1}\frac{1}{1 + sR_FC_F} \tag{15.30}$$

and the cutoff frequency is

$$f_{3\,\text{dB}} = \frac{1}{2\pi R_F C_F} \tag{15.31}$$

(a)   (b)

Figure 15.14   (a) One-pole low-pass filter and (b) equivalent switched-capacitor circuit

If a 10 kHz cutoff frequency is required and if $C_F = 10\,\text{pF}$, then the R_F resistance required is approximately 1.6 MΩ. In addition, if a gain of -10 is desired, then resistance R_1 must be 160 kΩ.

The equivalent switched-capacitor filter is shown in Figure 15.14(b). The transfer function is still given by Equation (15.30), where $R_{Feq} = 1/(f_C C_2)$ and $R_{1eq} = 1/(f_C C_1)$. The transfer function is then

$$T(j\omega) = -\frac{(1/f_C C_2)}{(1/f_C C_1)} \cdot \frac{1}{1 + j\dfrac{(2\pi f)C_F}{f_C C_2}} = -\frac{C_1}{C_2} \cdot \frac{1}{1 + j\dfrac{f}{f_{3\,\text{dB}}}} \qquad (15.32)$$

The low-frequency gain is $-C_1/C_2$, which is just the ratio of two capacitances, and the 3 dB frequency is

$$f_{3\,\text{dB}} = (f_C C_2)/(2\pi C_F)$$

which is also proportional to the ratio of two other capacitances. For MOS IC capacitance values of approximately 10 pF, the ratio tolerance is on the order of 0.1 percent. This means that switched-capacitor filter characteristics can be precisely controlled.

Design Example 15.3 **Objective:** Design a one-pole low-pass switched-capacitor filter.

Design the circuit in Figure 15.14(b) such that the low-frequency gain is -1 and the cutoff frequency is 1 kHz.

Solution: From Equation (15.32), the low-frequency gain is $-(C_1/C_2)$, and the capacitance ratio must be $(C_1/C_2) = 1$. From Equation (15.32), the cutoff frequency is

$$f_{3\,\text{dB}} = \frac{f_C C_2}{2\pi C_F}$$

If we set the clock frequency to $f_C = 10\,\text{kHz}$, then

$$\frac{C_2}{C_F} = \frac{2\pi f_{3\,\text{dB}}}{f_C} = \frac{2\pi(10^3)}{10 \times 10^3} = 0.628$$

Comment: Since the low-frequency gain and cutoff frequency are both functions of capacitor ratios, the absolute capacitor values can be designed for compatibility with IC fabrication.

This discussion of switched-capacitor filters is a short introduction to the topic and is intended only to show another application of operational amplifiers. Switched-capacitor filters are "sampled-data systems"; that is, the analog input signal is not transmitted through the circuit as a continuous signal but passes through the system as a series of pulses. The equivalent resistance given by Equation (15.29) is valid only for clock frequencies much greater than the analog input signal frequency. Switched-capacitor systems can be analyzed and designed by z-transform techniques.

Test Your Understanding

15.4 Simulate a $5\,M\Omega$ resistance using the circuit in Figure 15.13(a). What capacitor value and clock frequency are required? (Ans. For example, $C = 10\,pF$, $f_C = 20\,kHz$)

15.5 For the switched-capacitor circuit in Figure 15.14(b), the parameters are: $C_1 = 30\,pF$, $C_2 = 5\,pF$, and $C_F = 12\,pF$. The clock frequency is $100\,kHz$. Determine the low-frequency gain and the cutoff frequency. (Ans. $-C_1/C_2 = -6$, $f_{3\,dB} = 6.63\,kHz$)

15.2　OSCILLATORS

In this section, we will look at the basic principles of sine-wave oscillators. In our study of feedback in Chapter 12, we emphasized the need for negative feedback to provide a stable circuit. Oscillators, however, use positive feedback and, therefore, are actually nonlinear circuits in some cases. The analysis and design of oscillator circuits are divided into two parts. In the first part, the condition and frequency for oscillation are determined; in the second part, means for amplitude control is addressed. We consider only the first step in this section to gain insight into the basic operation of oscillators.

15.2.1　Basic Principles for Oscillation

The basic **oscillator** consists of an amplifier and a **frequency-selective network** connected in a feedback loop. Figure 15.15 shows a block diagram of the fundamental feedback circuit, in which we are implicitly assuming that negative feedback is employed. Although actual oscillator circuits do not have an input signal, we initially include one here to help in the analysis. In previous feedback circuits, we assumed the feedback transfer function β was independent of frequency. In oscillator circuits, however, β is the principal portion of the loop gain that is dependent on frequency.

For the circuit shown, the ideal closed-loop transfer function is given by

$$A_f(s) = \frac{A(s)}{1 + A(s)\beta(s)} \tag{15.33}$$

and the loop gain of the feedback circuit is

$$T(s) = A(s)\beta(s) \tag{15.34}$$

From our discussion of feedback in Chapter 12, we know that the loop gain $T(s)$ is positive for negative feedback, which means that the feedback

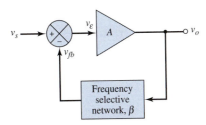

Figure 15.15　Block diagram of the fundamental feedback circuit

signal v_{fb} subtracts from the input signal v_s. If the loop gain $T(s)$ becomes negative, then the feedback signal phase causes v_{fb} to add to the input signal, increasing the error signal v_ε. If $T(s) = -1$, the closed-loop transfer function goes to infinity, which means that the circuit can have a finite output for a zero input signal.

As $T(s)$ approaches -1, an actual circuit becomes nonlinear, which means that the gain does not go to infinity. Assume that $T(s) \approx -1$ so that positive feedback exists over a particular frequency range. If a spontaneous signal (due to noise) is created at v_s in this frequency range, the resulting feedback signal v_{fb} is in phase with v_s, and the error signal v_ε is reinforced and increased. This reinforcement process continues at only those frequencies for which the total phase shift around the feedback loop is zero. Therefore, the condition for oscillation is that, at a specific frequency, we have

$$T(j\omega_o) = A(j\omega_o)\beta(j\omega_o) = -1 \tag{15.35}$$

The condition that $T(j\omega_o) = -1$ is called the **Barkhausen criterion**.

Equation (15.35) shows that two conditions must be satisfied to sustain oscillation:

1. The total phase shift through the amplifier and feedback network must be $N \times 360°$, where $N = 0, 1, 2, \ldots$.
2. The magnitude of the loop gain must be unity.

In the feedback circuit block diagram in Figure 15.15, we implicitly assume negative feedback. For an oscillator, the feedback transfer function, or the frequency-selective network, must introduce an additional 180 degree phase shift such that the net phase around the entire loop is zero. For the circuit to oscillate at a single frequency ω_o, the condition for oscillation, from Equation (15.35), should be satisfied at only that one frequency.

15.2.2 Phase-Shift Oscillator

An example of an op-amp oscillator is the **phase-shift oscillator**. One configuration of this oscillator circuit is shown in Figure 15.16. The basic amplifier of the circuit is the op-amp A_3, which is connected as an inverting amplifier with its output connected to a three-stage RC filter. The voltage followers in the circuit eliminate loading effects between each RC filter stage.

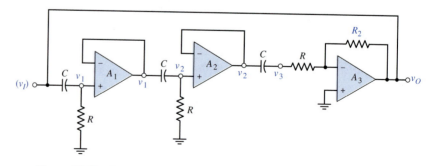

Figure 15.16 Phase-shift oscillator circuit with voltage-follower buffer stages

The inverting amplifier introduces a -180 degree phase shift, which means that each RC network must provide 60 degrees of phase shift to produce the 180 degrees required of the frequency-sensitive feedback network in order to produce positive feedback. Note that the inverting terminal of op-amp A_3 is at virtual ground; therefore, the RC network between op-amps A_2 and A_3 functions exactly as the other two RC networks. We assume that the frequency effects of the op-amps themselves occur at much higher frequencies than the response due to the RC networks. Also, to aid in the analysis, we assume an input signal (v_I) exists at one node as shown in the figure.

The transfer function of the first RC network is

$$v_1 = \left(\frac{sRC}{1+sRC}\right)(v_I) \tag{15.36}$$

Since the RC networks are assumed to be identical, and since there is no loading effect of one RC stage on another, we have

$$\frac{v_3}{(v_I)} = \left(\frac{sRC}{1+sRC}\right)^3 = \beta(s) \tag{15.37}$$

where $\beta(s)$ is the feedback transfer function. The amplifier gain $A(s)$ in Equation (15.33) and (15.34) is actually the magnitude of the gain, or

$$A(s) = \left|\frac{v_O}{v_3}\right| = \frac{R_2}{R} \tag{15.38}$$

The loop gain is then

$$T(s) = A(s)\beta(s) = \left(\frac{R_2}{R}\right)\left(\frac{sRC}{1+sRC}\right)^3 \tag{15.39}$$

From Equation (15.35), the condition for oscillation is that $|T(j\omega_o)| = 1$ and the phase of $T(j\omega_o)$ must be 180 degrees. When these requirements are satisfied, then v_O will equal (v_I) and a separate input signal will not be required.

If we set $s = j\omega$, Equation (15.39) becomes

$$\begin{aligned} T(j\omega) &= \left(\frac{R_2}{R}\right)\frac{(j\omega RC)^3}{(1+j\omega RC)^3} \\ &= -\left(\frac{R_2}{R}\right)\frac{(j\omega RC)(\omega RC)^2}{\left[1-3\omega^2R^2C^2\right]+j\omega RC\left[3-\omega^2R^2C^2\right]} \end{aligned} \tag{15.40}$$

To satisfy the condition $T(j\omega_o) = -1$, the imaginary component of Equation (15.40) must equal zero. Since the numerator is purely imaginary, the denominator must become purely imaginary, or

$$\left[1-3\omega_o^2R^2C^2\right] = 0$$

which yields

$$\omega_o = \frac{1}{\sqrt{3}\,RC} \tag{15.41}$$

where ω_o is the oscillation frequency. At this frequency, Equation (15.40) becomes

$$T(j\omega_o) = -\left(\frac{R_2}{R}\right)\frac{(j/\sqrt{3})(1/3)}{0 + (j/\sqrt{3})[3 - (1/3)]} = -\left(\frac{R_2}{R}\right)\left(\frac{1}{8}\right) \tag{15.42}$$

Consequently, the condition $T(j\omega_o) = -1$ is satisfied when

$$\frac{R_2}{R} = 8 \tag{15.43}$$

Equation (15.43) implies that if the magnitude of the inverting amplifier gain is greater than 8, the circuit will spontaneously begin oscillating and will sustain oscillation.

Example 15.4   **Objective:** Determine the oscillation frequency and required amplifier gain for a phase-shift oscillator.

Consider the phase-shift oscillator in Figure 15.16 with parameters $C = 0.1\,\mu\text{F}$ and $R = 1\,\text{k}\Omega$.

Solution: From Equation (15.41), the oscillation frequency is

$$f_o = \frac{1}{2\pi\sqrt{3}RC} = \frac{1}{2\pi\sqrt{3}(10^3)(0.1 \times 10^{-6})} = 919\,\text{Hz}$$

The minimum amplifier gain magnitude is 8 from Equation 15.43; therefore, the minimum value of R_2 is $8\,\text{k}\Omega$.

Comment: Higher oscillation frequencies can easily be obtained by using smaller capacitor values.

Using Equation (15.36), we can determine the effect of each RC network in the phase-shift oscillator. At the oscillation frequency ω_o, the transfer function of each RC network stage is

$$\frac{j\omega_o RC}{1 + j\omega_o RC} = \frac{(j/\sqrt{3})}{1 + (j/\sqrt{3})} = \frac{j}{\sqrt{3} + j} \tag{15.44}$$

which can be written in terms of the magnitude and phase, as follows:

$$\frac{1}{\sqrt{3} + 1} \times \frac{\angle 90°}{\angle \tan^{-1}(1/\sqrt{3})} = \frac{1}{2} \times [\angle 90° - \angle \tan^{-1}(0.577)] \tag{15.45(a)}$$

or

$$\frac{1}{2} \times (\angle 90° - \angle 30°) = \frac{1}{2} \times \angle 60° \tag{15.45(b)}$$

As required, each RC network introduces a 60 degree phase shift, but they each also introduce an attenuation factor of $(\frac{1}{2})$ for which the amplifier must compensate.

Test Your Understanding

D15.6 Design the phase-shift oscillator shown in Figure 15.16 to oscillate at $f_o = $ 15 kHz. Choose appropriate component values. (Ans. For example, $C = 0.001 \mu F$, $R = 6.13 k\Omega$, $R_2 = 49 k\Omega$)

The two voltage followers in the circuit in Figure 15.16 need not be included in a practical phase-shift oscillator. Figure 15.17 shows a phase-shift oscillator without the voltage-follower buffer stages. The three RC network stages and the inverting amplifier are still included. The loading effect of each successive RC network complicates the analysis, but the same principle of operation applies. The analysis shows that the oscillation frequency is

$$\omega_o = \frac{1}{\sqrt{6}\, RC} \tag{15.46}$$

and the amplifier resistor ratio must be

$$\frac{R_2}{R} = 29 \tag{15.47}$$

in order to sustain oscillation.

Figure 15.17   Phase-shift oscillator circuit

Test Your Understanding

15.7   For the phase-shift oscillator in Figure 15.17, the parameters are $R = 10 k\Omega$ and $C = 100 pF$. Determine the frequency of oscillation and the required value of R_2. (Ans. $f_o \cong 65 kHz$, $R_2 = 290 k\Omega$)

15.2.3   Wien-Bridge Oscillator

Another basic oscillator is the **Wien-bridge circuit**, shown in Figure 15.18. The circuit consists of an op-amp connected in a noninverting configuration and two RC networks connected as the frequency-selecting feedback circuit.

Figure 15.18   Wien-bridge oscillator

Again, we initially assume that an input signal exists at the noninverting terminals of the op-amp. Since the noninverting amplifier introduces zero phase shift, the frequency-selective feedback circuit must also introduce zero phase shift to create the positive feedback condition.

The loop gain is the product of the amplifier gain and the feedback transfer function, or

$$T(s) = \left(1 + \frac{R_2}{R_1}\right)\left(\frac{Z_p}{Z_p + Z_s}\right) \tag{15.48}$$

where Z_p and Z_s are the parallel and series RC network impedances, respectively. These impedances are

$$Z_p = \frac{R}{1 + sRC} \tag{15.49(a)}$$

and

$$Z_s = \frac{1 + sRC}{sC} \tag{15.49(b)}$$

Combining Equations (15.49(a)), (15.49(b)), and (15.48), we get an expression for the loop gain function,

$$T(s) = \left(1 + \frac{R_2}{R_1}\right)\left[\frac{1}{3 + sRC + (1/sRC)}\right] \tag{15.50}$$

Since this circuit has no explicit negative feedback, as was assumed in the general network shown in Figure 15.15, the condition for oscillation is given by

$$T(j\omega_o) = 1 = \left(1 + \frac{R_2}{R_1}\right)\left[\frac{1}{3 + j\omega_o RC + (1/j\omega_o RC)}\right] \tag{15.51}$$

Since $T(j\omega_o)$ must be real, the imaginary component of Equation (15.51) must be zero; therefore,

$$j\omega_o RC + \frac{1}{j\omega_o RC} = 0 \tag{15.52(a)}$$

which gives the frequency of oscillation as

$$\omega_o = \frac{1}{RC}$$ **(15.52(b))**

The magnitude condition is then

$$1 = \left(1 + \frac{R_2}{R_1}\right)\left(\frac{1}{3}\right)$$ **(15.53(a))**

or

$$\frac{R_2}{R_1} = 2$$ **(15.53(b))**

Equation (15.53(b)) states that to ensure the startup of oscillation, we must have $(R_2/R_1) > 2$.

Design Example 15.5   Objective: Design a Wien-bridge circuit to oscillate at a specified frequency.

Consider the Wien-bridge oscillator in Figure 15.18. Design the circuit to oscillate at $f_o = 20\,\text{kHz}$.

Solution: The oscillation frequency given by Equation (15.52(b)) yields

$$RC = \frac{1}{2\pi f_o} = \frac{1}{2\pi(20 \times 10^3)} = 7.96 \times 10^{-6}$$

a $10\,\text{k}\Omega$ resistor and $796\,\text{pF}$ capacitor satisfy this requirement. Since the amplifier resistor ratio must be $R_2/R_1 = 2$, we could, for example, have $R_2 = 20\,\text{k}\Omega$ and $R_1 = 10\,\text{k}\Omega$, which would satisfy the requirement.

Comment: As usual in any electronic circuit design, there is no unique solution. Reasonably sized component values should be chosen whenever possible.

Computer Simulation Verification: A computer simulation was performed using the circuit in Figure 15.19(a). Figure 15.19(b) shows the output voltage versus time. Since the ratio of resistances is $R_2/R_1 = 22/10 = 2.2$, the overall gain is greater than unity so the output increases as a function of time. This increase shows the oscillation nature of the circuit. Another characteristic of the circuit is shown in Figure 15.19(c). A $1\,\text{mV}$ sinusoidal signal was applied to the input of R_1 and the output voltage measured as the frequency was swept from $10\,\text{kHz}$ to $30\,\text{kHz}$. The resonant nature of the circuit is observed. The oscillation frequency and the resonant frequency are both at approximately $18.2\,\text{kHz}$, which is below the design value of $20\,\text{kHz}$.

If the capacitor in the circuit is reduced from $796\,\text{pF}$ to $720\,\text{pF}$, the resonant frequency is exactly $20\,\text{kHz}$. This example is one case, then, when the design parameters need to be changed slightly in order to meet the design specifications.

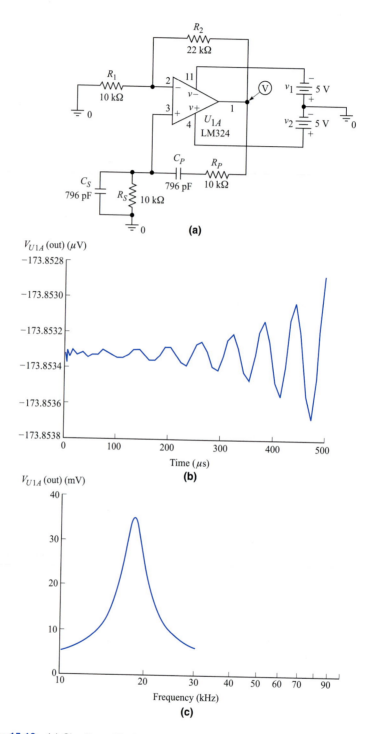

Figure 15.19 (a) Circuit used in the computer simulation for Example 15.5, (b) output voltage versus time, and (c) output voltage versus input frequency

Test Your Understanding

D15.8 Design the Wien-bridge circuit in Figure 15.18 to oscillate at $f_o = 800\,\text{Hz}$. Assume $R = R_1 = 10\,\text{k}\Omega$. (Ans. $C \cong 0.02\,\mu\text{F}$, $R_2 = 20\,\text{k}\Omega$)

15.2.4   Additional Oscillator Configurations

Oscillators that use transistors and LC tuned circuits or crystals in their feedback networks can be used in the hundreds of kHz to hundreds of MHz frequency range. Although these oscillators do not typically contain an opamp, we include a brief discussion of such circuits for completeness. We will examine the Colpitts, Hartley, and crystal oscillators.

Colpitts Oscillator

The ac equivalent circuit of the **Colpitts oscillator** with an FET is shown in Figure 15.20. A circuit with a BJT can also be designed. A parallel LC resonant circuit is used to establish the oscillator frequency, and feedback is provided by a voltage divider between capacitors C_1 and C_2. Resistor R in conjunction with the transistor provides the necessary gain at resonance. We assume that the transistor frequency response occurs at a high enough frequency that the oscillation frequency is determined by the external elements only.

Figure 15.20   The ac equivalent circuit, MOSFET Colpitts oscillator

Figure 15.21 shows the small-signal equivalent circuit of the Colpitts oscillator. The transistor output resistance r_o can be included in R. A KCL equation at the output node yields

$$\frac{V_o}{\dfrac{1}{sC_1}} + \frac{V_o}{R} + g_m V_{gs} + \frac{V_o}{sL + \dfrac{1}{sC_2}} = 0 \tag{15.54}$$

and a voltage divider produces

$$V_{gs} = \left(\frac{\dfrac{1}{sC_2}}{\dfrac{1}{sC_2} + sL}\right) \cdot V_o \tag{15.55}$$

Figure 15.21   Small-signal equivalent circuit, MOSFET Colpitts oscillator

Substituting Equation (15.55) into Equation (15.54), we find that

$$V_o\left[g_m + sC_2 + (1 + s^2LC_2)\left(\frac{1}{R} + sC_1\right)\right] = 0 \tag{15.56}$$

If we assume that oscillation has started, then $V_o \neq 0$ and can be eliminated from Equation (15.56). We then have

$$s^3LC_1C_2 + \frac{s^2LC_2}{R} + s(C_1 + C_2) + \left(g_m + \frac{1}{R}\right) = 0 \tag{15.57}$$

Letting $s = j\omega$, we obtain

$$\left(g_m + \frac{1}{R} - \frac{\omega^2LC_2}{R}\right) + j\omega[(C_1 + C_2) - \omega^2LC_1C_2] = 0 \tag{15.58}$$

The condition for oscillation implies that both the real and imaginary components of Equation (15.58) must be zero. From the imaginary component, the oscillation frequency is

$$\omega_o = \frac{1}{\sqrt{L\left(\dfrac{C_2C_2}{C_1 + C_2}\right)}} \tag{15.59}$$

which is the resonant frequency of the LC circuit. From the real part of Equation (15.58), the condition for oscillation is

$$\frac{\omega_o^2LC_2}{R} = g_m + \frac{1}{R} \tag{15.60}$$

Combining Equations (15.59) and (15.60) yields

$$\frac{C_2}{C_1} = g_mR \tag{15.61}$$

where g_mR is the magnitude of the gain. Equation (15.61) states that to initiate oscillations spontaneously, we must have $g_mR > (C_2/C_1)$.

Hartley Oscillator

Figure 15.22 shows the ac equivalent circuit of the Hartley oscillator with a BJT. An FET can also be used. Again, a parallel LC resonant circuit establishes the oscillator frequency, and feedback is provided by a voltage divider between inductors L_1 and L_2.

The analysis of the Hartley oscillator is essentially identical to that of the Colpitts oscillator. The frequency of oscillation, neglecting transistor frequency effects, is

Figure 15.22 The ac equivalent, BJT Hartley oscillator

$$\omega_o = \frac{1}{\sqrt{(L_1 + L_2)C}} \tag{15.62}$$

Equation (15.62) also assumes that $r_\pi \gg 1/(\omega C_2)$.

Crystal Oscillator

A piezoelectric crystal, such as quartz, exhibits electromechanical resonance characteristics in response to a voltage applied across the crystal. The oscillations are very stable over time and temperature, with temperature coefficients on the order of 1 ppm per °C. The oscillation frequency is determined by the crystal dimensions. This means that crystal oscillators are fixed-frequency devices.

The circuit symbol for the piezoelectric crystal is shown in Figure 15.23(a), and the equivalent circuit is shown in Figure 15.23(b). The inductance L can be as high as a few hundred henrys, the capacitance C_s can be on the order of 0.001 pF, and the capacitance C_p can be on the order of a few pF. Also, the Q-factor can be on the order of 10^4, which means that the series resistance r can be neglected.

The impedance of the equivalent circuit in Figure 15.23(b) is

(a)   **(b)**

Figure 15.23   (a) Piezoelectric crystal circuit symbol and (b) piezoelectric crystal equivalent circuit

$$Z(s) = \frac{1}{sC_p} \cdot \frac{s^2 + (1/LC_s)}{s^2 + [(C_p + C_s)/(LC_sC_p)]} \qquad (15.63)$$

Equation (15.63) indicates that the crystal has two resonant frequencies, which are very close together. At the series-resonant frequency f_s, the reactance of the series branch is zero; at the parallel-resonant frequency f_p, the reactance of the crystal approaches infinity.

Between the resonant frequencies f_s and f_p, the crystal reactance is inductive, so the crystal can be substituted for an inductance, such as that in a Colpitts oscillator. Figure 15.24 shows the ac equivalent circuit of a Pierce oscillator, which is similar to the Colpitts oscillator in Figure 15.20 but with the inductor replaced by the crystal. Since the crystal reactance is inductive over a very narrow frequency range, the frequency of oscillation is also confined to this narrow range and is quite constant relative to changes in bias current or temperature. Crystal oscillator frequencies are usually in the range of tens of kHz to tens of MHz.

Figure 15.24   Pierce oscillator in which the inductor in a Colpitts oscillator is replaced by a crystal

Test Your Understanding

*15.9   For the Colpitts oscillator in Figure 15.20, assume parameters of $L = 1\,\mu H$, C_1 and $C_2 = 1\,nF$, and $R = 4\,k\Omega$. Determine the oscillator frequency and the required value of g_m. Is this value of g_m reasonable for a MOSFET? Why? (Ans. $f_o = 7.12\,MHz$, $g_m = 0.25\,mA/V$)

15.3   SCHMITT TRIGGER CIRCUITS

In this section, we will analyze another class of circuits that utilize positive feedback. The basic circuit is commonly called a **Schmitt trigger**, which can be used in the class of waveform generators called multivibrators. The three general types of multivibrators are: bistable, monostable, and astable. In this section, we will examine the bistable multivibrator, which has a comparator with positive feedback and has two stable states. We will discuss the comparator first, and will then describe various applications of the Schmitt trigger.

15.3.1 Comparator

The comparator is essentially an op-amp operated in an open-loop configuration, as shown in Figure 15.25(a). As the name implies, a comparator compares two voltages to determine which is larger. The comparator is usually biased at voltages $+V_S$ and $-V_S$, although other biases are possible.

Figure 15.25   (a) Open-loop comparator and (b) voltage transfer characteristics, open-loop comparator

The voltage transfer characteristics, neglecting any offset voltage effects, are shown in Figure 15.25(b). When v_2 is slightly greater than v_1, the output is driven to a high saturated state V_H; when v_2 is slightly less than v_1, the output is driven to a low saturated state V_L. The saturated output voltages V_H and V_L may be close to the supply voltages $+V_S$ and $-V_S$, respectively, which means that V_L may be negative. The transition region is the region in which the output voltage is in neither of its saturation states. This region occurs when the input differential voltage is in the range $-\delta < (v_2 - v_1) < +\delta$. If, for example, the open-loop gain is 10^5 and the difference between the two output states is $(V_H - V_L) = 10\,\text{V}$, then

$$2\delta = 10/10^5 = 10^{-4}\,\text{V} = 0.1\,\text{mV}$$

The range of input differential voltage in the transition region is normally very small.

One major difference between a comparator and op-amp is that a comparator need not be frequency compensated. Frequency stability is not a consideration since the comparator is being driven into one of two states. Since a comparator does not contain a frequency compensation capacitor, it is not slew-rate-limited by the compensation capacitor as is the op-amp. Typical response times for the comparator output to change states are in the range of 30 to 200 ns. An expected response time for a 741 op-amp with a slew rate of $0.7\,\text{V}/\mu\text{s}$ would be on the order of $30\,\mu\text{s}$, which is a factor of 1000 times greater.

Figure 15.26 shows two comparator configurations along with their voltage transfer characteristics. In both, the input transition region width is assumed to be negligibly small. The reference voltage may be either positive or negative, and the output saturation voltages are assumed to be symmetrical about zero. The crossover voltage is defined as the input voltage at which the output changes states.

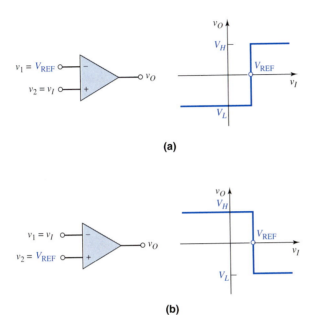

(a)

(b)

Figure 15.26   (a) Noninverting comparator circuit and (b) inverting comparator circuit

Two other comparator configurations, in which the crossover voltage is a function of resistor ratios, are shown in Figure 15.27. Input bias current compensation is also included in this figure. From Figure 15.27(a), we use superposition to obtain

$$v_+ = \left(\frac{R_2}{R_1 + R_2}\right)V_{\text{REF}} + \left(\frac{R_1}{R_1 + R_2}\right)v_I \tag{15.64}$$

The ideal crossover voltage occurs when $v_+ = 0$, or

$$R_2 V_{\text{REF}} + R_1 v_I = 0 \tag{15.65(a)}$$

which can be written as

$$v_I = -\frac{R_2}{R_1}V_{\text{REF}} \tag{15.65(b)}$$

The output goes high when $v_+ > 0$. From Equation (15.64), we see that $v_o =$ High when v_I is greater than the crossover voltage. A similar analysis produces the characteristics shown in Figure 15.27(b).

Figure 15.28 shows one application of a comparator, to control street lights. The input signal is the output of a photodetector circuit. Voltage v_I is directly proportional to the amount of light incident on the photodetector.

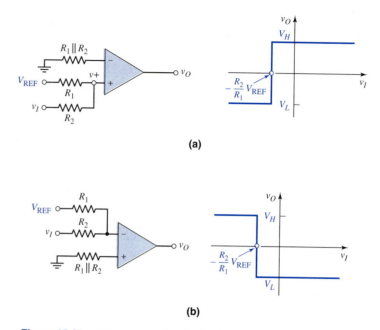

(a)

(b)

Figure 15.27   Other comparator circuits: (a) noninverting and (b) inverting

During the night, $v_I < V_{REF}$, and v_O is on the order of $V_S = +15\,\text{V}$; the transistor turns on. The current in the relay switch then turns the street lights on. During the day, the light incident on the photodetector produces an output signal such that $v_I > V_{REF}$. In this case, v_O is on the order of $-V_S = -15\,\text{V}$, and the transistor turns off.

Diode D_1 is used as a protection device, preventing reverse-bias breakdown in the B–E junction. With zero output current, the relay switch is open and the street lights are off. At dusk and dawn, $v_I = V_{REF}$.

The open-loop comparator circuit in Figure 15.28 may exhibit unacceptable behavior in response to noise in the system. Figure 15.29(a) shows the same comparator circuit, but with a variable light source, such as clouds causing the

Figure 15.28   Comparator application

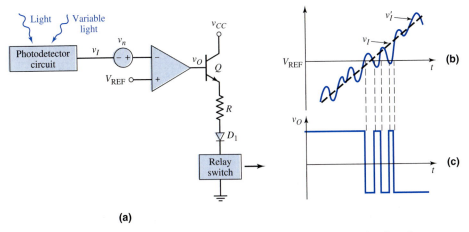

Figure 15.29 (a) Comparator circuit including input noise source, (b) input signal, and (c) output signal, showing chatter effect

light intensity to fluctuate over a short period of time. A variable light intensity would be equivalent to a noise source v_n in series with the signal source v_I. If we assume that v_I is increasing linearly with time (corresponding to dawn), then the total input signal v_I' versus time is shown in Figure 15.29(b). When $v_I' > V_{REF}$, the output switches low; when $v_I' < V_{REF}$, the output switches high, producing a chatter effect in the output signal as shown in Figure 14.29(c). This effect would turn the street lights off and on over a relatively short time period. If the amplitude of the noise signal increases, the chatter effect becomes more severe. This chatter can be eliminated by using a Schmitt trigger.

15.3.2 Basic Inverting Schmitt Trigger

The Schmitt trigger or **bistable multivibrator** uses positive feedback with a loop-gain greater than unity to produce a bistable characteristic. Figure 15.30(a) shows one configuration of a Schmitt trigger. Positive feedback occurs because the feedback resistor is connected between the output and noninverting input terminals. Voltage v_+, in terms of the output voltage, can be found by using a voltage divider equation to yield

$$v_+ = \left(\frac{R_1}{R_1 + R_2} \right) v_O \qquad (15.66)$$

Voltage v_+ does not remain constant; rather, it is a function of the output voltage. Input signal v_I is applied to the inverting terminal.

Voltage Transfer Characteristics

To determine the voltage transfer characteristics, we assume that the output of the comparator is in one state, namely $v_O = V_H$, which is the high state. Then

$$v_+ = \left(\frac{R_1}{R_1 + R_2} \right) V_H \qquad (15.67)$$

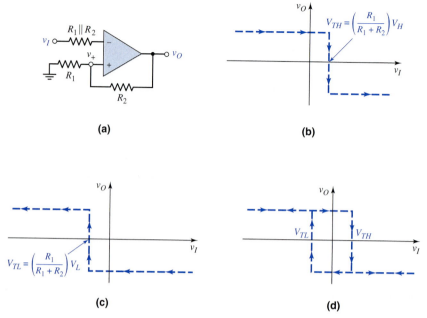

Figure 15.30   (a) Schmitt trigger circuit, (b) voltage transfer characteristic as input voltage increases, (c) voltage transfer characteristic as input voltage decreases, and (d) net voltage transfer characteristics, showing hysteresis effect

As long as the input signal is less than v_+, the output remains in its high state. The crossover voltage occurs when $v_I = v_+$ and is defined as V_{TH}. We have

$$V_{TH} = \left(\frac{R_1}{R_1 + R_2}\right) V_H \tag{15.68}$$

When v_I is greater than V_{TH}, the voltage at the inverting terminal is greater than that at the noninverting terminal. The differential input voltage $(v_I - V_{TH})$ is amplified by the open-loop gain of the comparator, and the output switches to its low state, or $v_O = V_L$. Voltage v_+ then becomes

$$v_+ = \left(\frac{R_1}{R_1 + R_2}\right) V_L \tag{15.69}$$

Since $V_L < V_H$, the input voltage v_I is still greater than v_+, and the output remains in its low state as v_I continues to increase. This voltage transfer characteristic is shown in Figure 15.30(b). Implicit in these transfer characteristics is the assumption that V_H is positive and V_L is negative.

Now consider the transfer characteristic as v_I decreases. As long as v_I is larger than $v_+ = [R_1/(R_1 + R_2)]V_L$, the output remains in its low saturation state. The crossover voltage now occurs when $v_I = v_+$ and is defined as V_{TL}. We have

$$V_{TL} = \left(\frac{R_1}{R_1 + R_2}\right) V_L \tag{15.70}$$

As v_I drops below this value, the voltage at the noninverting terminal is greater than that at the inverting terminal. The differential voltage at the comparator terminals is amplified by the open-loop gain, and the output switches to its high state, or $v_O = V_H$. As v_I continues to decrease, it remains less than v_+; therefore, v_O remains in its high state. This voltage transfer characteristic is shown in Figure 15.30(c).

Complete Voltage Transfer and Bistable Characteristics

The complete voltage transfer characteristics of the Schmitt trigger in Figure 15.30(a) combine the characteristics in Figures 15.30(b) and 15.30(c). These complete characteristics are shown in Figure 15.30(d). As shown, the crossover voltages depend on whether the input voltage is increasing or decreasing. The complete transfer characteristics therefore show a **hysteresis effect**. The width of the hysteresis is the difference between the two crossover voltages V_{TH} and V_{TL}.

The bistable characteristic of the circuit occurs around the point $v_I = 0$, at which the output may be in either its high or low state. The output remains in either state as long as v_I remains in the range $V_{TL} < v_I < V_{TH}$. The output switches states only if the input increases above V_{TH} or decreases below V_{TL}.

Example 15.6 **Objective:** Determine the hysteresis width of a particular Schmitt trigger.

Consider the Schmitt trigger in Figure 15.30(a), with parameters $R_1 = 10 \, k\Omega$ and $R_2 = 90 \, k\Omega$. Let $V_H = 10 \, V$ and $V_L = -10 \, V$.

Solution: From Equation (15.68), the upper crossover voltage is

$$V_{TH} = \left(\frac{R_1}{R_1 + R_2}\right) V_H = \left(\frac{10}{10 + 90}\right)(10) = 1 \, V$$

and from Equation (15.70), the lower crossover voltage is

$$V_{TL} = \left(\frac{R_1}{R_1 + R_2}\right) V_L = \left(\frac{10}{10 + 90}\right)(-10) = -1 \, V$$

The hysteresis width is therefore $(V_{TH} - V_{TL}) = 2 \, V$.

Comment: The hysteresis width can be designed to be larger or smaller for specific applications by adjusting the voltage divider ratio of R_1 and R_2.

The complete voltage transfer characteristics in Figure 15.30(d) show the inverting characteristics of this particular Schmitt trigger. When the input signal becomes sufficiently positive, the output is in its low state; when the input signal is sufficiently negative, the output is in its high state. Since the input signal is applied to the inverting terminal of the comparator, this characteristic is as expected.

Test Your Understanding

15.10 For the comparator in Figure 15.30(a), the high and low saturated output states are $+12 \, V$ and $-12 \, V$, respectively. If $R_2 = 20 \, k\Omega$, find R_1 such that the crossover voltages are $\pm 2 \, V$. (Ans. $R_1 = 4 \, k\Omega$)

15.3.3   Additional Schmitt Trigger Configurations

A noninverting Schmitt trigger can be designed by applying the input signal to the network connected to the comparator noninverting terminal. Also, both crossover voltages of a Schmitt trigger circuit can be shifted in either a positive or negative direction by applying a reference voltage. We will study these general circuit configurations, the resulting voltage transfer characteristics, and an application of a Schmitt trigger circuit in this section.

Noninverting Schmitt Trigger Circuit

Consider the circuit in Figure 15.31(a). The inverting terminal is held essentially at ground potential, and the input signal is applied to resistor R_1, which is connected to the comparator noninverting terminal. Voltage v_+ at the noninverting terminal then becomes a function of both the input signal v_I and the output voltage v_O. Using superposition, we find that

$$v_+ = \left(\frac{R_2}{R_1 + R_2}\right)v_I + \left(\frac{R_1}{R_1 + R_2}\right)v_O \qquad \textbf{(15.71)}$$

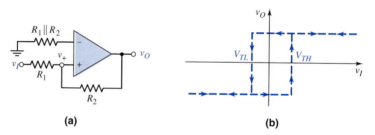

(a)   **(b)**

Figure 15.31   (a) Noninverting Schmitt trigger circuit and (b) voltage transfer characteristics

If v_I is negative, and the output is in its low state, then $v_O = V_L$ (assumed to be negative), v_+ is negative, and the output remains in its low saturation state. Crossover voltage $v_I = V_{TH}$ occurs when $v_+ = 0$ and $v_O = V_L$, or, from Equation (15.71),

$$0 = R_2 V_{TH} + R_1 V_L \qquad \textbf{(15.72(a))}$$

which can be written

$$V_{TH} = -\left(\frac{R_1}{R_2}\right)V_L \qquad \textbf{(15.72(b))}$$

Since V_L is negative, V_{TH} is positive.

If we let $v_I = V_{TH} + \delta$, where δ is a small positive voltage, the input voltage is just greater than the crossover voltage and Equation (15.71) becomes

$$v_+ = \left(\frac{R_2}{R_1 + R_2}\right)(V_{TH} + \delta) + \left(\frac{R_1}{R_1 + R_2}\right)V_L \qquad \textbf{(15.73)}$$

Equation (15.73) then becomes

$$v_+ = \left(\frac{R_2}{R_1 + R_2}\right)\left(\frac{-R_1}{R_2}\right)V_L + \left(\frac{R_2}{R_1 + R_2}\right)\delta + \left(\frac{R_1}{R_1 + R_2}\right)V_L \qquad (15.74(\mathbf{a}))$$

or

$$v_+ = \left(\frac{R_2}{R_1 + R_2}\right)\delta > 0 \qquad (15.74(\mathbf{b}))$$

When $v_+ > 0$, the output switches to its high saturation state.

The lower crossover voltage $v_I = V_{TL}$ occurs when $v_+ = 0$ and $v_O = V_H$. From Equation (15.71), we have

$$0 = R_2 V_{TL} + R_1 V_H \qquad (15.75(\mathbf{a}))$$

which can be written

$$V_{TL} = -\left(\frac{R_1}{R_2}\right)V_H \qquad (15.75(\mathbf{b}))$$

Since $V_H > 0$, then $V_{TL} < 0$.

The complete voltage transfer characteristics are shown in Figure 15.31(b). We again note the hysteresis effect and the bistable characteristic around $v_I = 0$. With v_I sufficiently positive, the output is in its high state; with v_I sufficiently negative, the output is in its low state. The circuit thus exhibits the noninverting transfer characteristic.

Test Your Understanding

D15.11 A noninverting Schmitt trigger is shown in Figure 15.31(a). Its saturated output voltages are $\pm 10\,$V. Design the circuit to obtain $\pm 100\,$mV crossover voltages. Choose suitable component values. (Ans. $R_1/R_2 = 0.010$)

Schmitt Trigger Circuits with Applied Reference Voltages

The switching voltage of a Schmitt trigger is defined as the average value of V_{TH} and V_{TL}. For the two circuits in Figure 15.30(a) and 15.31(a), the switching voltages are zero, assuming $V_{TL} = -V_{TH}$. In some applications, the switching voltage must be either positive or negative. Both crossover voltages can be shifted in either a positive or negative direction by applying a reference voltage.

Figure 15.32(a) shows an inverting Schmitt trigger with a reference voltage V_{REF}. The complete voltage transfer characteristics are shown in Figure 15.32(b). The switching voltage V_S, assuming V_H and V_L are symmetrical about zero, is given by

$$V_S = \left(\frac{R_2}{R_1 + R_2}\right)V_{\text{REF}} \qquad (15.76)$$

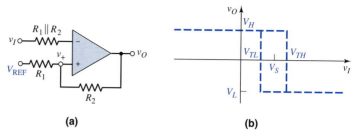

(a)   **(b)**

Figure 15.32 (a) Inverting Schmitt trigger circuit with applied reference voltage and
(b) voltage transfer characteristics

Note that the switching voltage is not the same as the reference voltage. The
upper and lower crossover voltages are

$$V_{TH} = V_S + \left(\frac{R_1}{R_1 + R_2}\right)V_H \qquad\qquad (15.77(\text{a}))$$

and

$$V_{TL} = V_S + \left(\frac{R_1}{R_1 + R_2}\right)V_L \qquad\qquad (15.77(\text{b}))$$

Test Your Understanding

15.12   For the Schmitt trigger in Figure 15.32(a), the parameters are: $V_{\text{REF}} = 2\,\text{V}$,
$V_H = 10\,\text{V}$, $V_L = -10\,\text{V}$, $R_1 = 1\,\text{k}\Omega$, and $R_2 = 10\,\text{k}\Omega$: (a) Determine V_S, V_{TH}, and V_{TL}.
(b) Let v_I be a triangular wave with a zero average voltage, a 10 V peak amplitude, and a
10 ms period. Sketch v_O versus time over two periods. Label the appropriate voltages
and times. (Ans. (a) $V_S = 1.82\,\text{V}$, $V_{TH} = 2.73\,\text{V}$, $V_{TL} = 0.91\,\text{V}$)

A noninverting Schmitt trigger with a reference voltage is shown in Figure
15.33(a), and the complete voltage transfer characteristics are shown in Figure
15.33(b). The switching voltage V_s, again assuming V_H and V_L are symmetrical
about zero, is given by

$$V_S = \left(1 + \frac{R_1}{R_2}\right)V_{\text{REF}} \qquad\qquad (15.78)$$

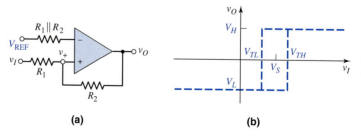

(a)   **(b)**

Figure 15.33 (a) Noninverting Schmitt trigger circuit with applied reference voltage and
(b) voltage transfer characteristics

and the upper and lower crossover voltages are

$$V_{TH} = V_S - \left(\frac{R_1}{R_2}\right)V_L \qquad (15.79(a))$$

and

$$V_{TL} = V_S - \left(\frac{R_1}{R_2}\right)V_H \qquad (15.79(b))$$

If the output saturation voltages are symmetrical such that $V_L = -V_H$, then the crossover voltages are symmetrical about the switching voltage V_S.

Test Your Understanding

D15.13   Consider the Schmitt trigger in Figure 15.33(a). Let $V_H = 5\,\text{V}$ and $V_L = -5\,\text{V}$. Design the circuit such that $V_S = -1\,\text{V}$ and the hysteresis width is 2.5 V. What are the values of V_{TL} and V_{TH}? (Ans. $R_1/R_2 = 0.25$, $V_{\text{REF}} = -0.8\,\text{V}$, $V_{TH} = 0.25\,\text{V}$, $V_{TL} = -2.25\,\text{V}$)

Schmitt Trigger Application

Let us reconsider the street light control in Figure 15.29(a), which included a noise source. Figure 15.34(a) shows the same basic circuit, except that a Schmitt trigger is used instead of a simple comparator.

The input signal v_I is again assumed to increase linearly with time. The total input signal v_I' is v_I with the noise signal superimposed, as shown in Figure 15.34(b). At time t_1, the input signal becomes greater than the switching voltage V_S. The output, however, does not switch, since $v_I' < V_{TH}$. This means that the input signal is less than the upper crossover voltage. At time t_2, the input signal becomes larger than the crossover voltage, or $v_I' > V_{TH}$, and the

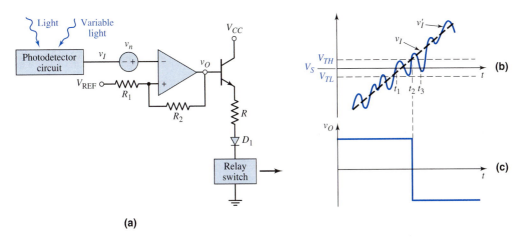

(a)

Figure 15.34   (a) Application of Schmitt trigger circuit including input noise source, (b) input signal, and (c) output signal, showing elimination of chatter effect output signal switches from its high to its low state. At time t_3, the input signal drops below V_S, but the output does not switch states since $v_I' > V_{TL}$. This means that the input signal remains greater than the lower crossover voltage. The Schmitt trigger circuit thus eliminates the chatter effect that occurs in the output voltage in Figure 15.29(c). Elimination of the chatter in the output voltage response results directly from the hysteresis effect in the Schmitt trigger characteristics.

Design Example 15.7 **Objective:** Design a Schmitt trigger circuit for the photo-detector switch circuit.

Consider the Schmitt trigger circuit in Figure 15.34(a). Design the circuit such that the switching voltage is $V_S = 2$ V and the hysteresis width is 60 mV. Assume $V_H = 15$ V and $V_L = -15$ V.

Solution: The Schmitt trigger circuit is the inverting type, for which the voltage transfer characteristics are shown in Figure 15.32(b). From Equations (15.77(a)) and (15.77(b)), the hysteresis width is

$$V_{TH} - V_{TL} = \left(\frac{R_1}{R_1 + R_2}\right)(V_H - V_L)$$

so

$$0.060 = \left(\frac{R_1}{R_1 + R_2}\right)[15 - (-15)] = 30\left(\frac{R_1}{R_1 + R_2}\right)$$

which yields $R_2/R_1 = 499$. We can find the reference voltage from Equation (15.76), which can be rewritten to obtain

$$V_{REF} = \left(1 + \frac{R_1}{R_2}\right)V_S = \left(1 + \frac{1}{499}\right)(2) = 2.004 \text{ V}$$

Resistor values of $R_1 = 100 \, \Omega$ and $R_2 = 49.9 \, \text{k}\Omega$ will satisfy the requirements. The crossover voltages are thus $V_{TH} = 2.03$ V and $V_{TL} = 1.97$ V.

 Comment: In this case, the output chatter effect is eliminated for noise signals with amplitudes lower than 30 mV. The hysteresis width can be adjusted up or down to fit specific application requirements in which the noise signal is larger or smaller than that given in this example.

Test Your Understanding

D15.14 Redesign the street light control circuit shown in Figure 15.34(a) such that the switching voltage is $V_S = 1$ V and the hysteresis width is 100 mV. Assume $V_H = +10$ V and $V_L = -10$ V. Also, find R such that $I = 200$ mA when $v_O = V_H$. Assume $V_{BE}(\text{on}) = 0.7$ V and $V_\gamma = 0.7$ V, and assume the relay switch resistance is $100 \, \Omega$. (Ans. $R_2/R_1 = 199$, $V_{REF} = 1.005$ V, $R = 42.9 \, \text{k}\Omega$)

15.3.4 Schmitt Triggers with Limiters

In the Schmitt trigger circuits we have thus far considered, the open-loop saturation voltages of the comparator may not be very precise and may also vary from one comparator to another. The output saturation voltages can be controlled and made more precise by adding limiter networks.

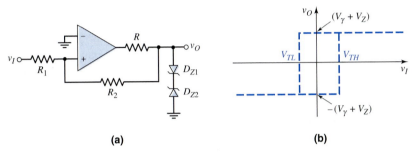

Figure 15.35 (a) Schmitt trigger with Zener diode limiters and (b) voltage transfer characteristics

A direct approach at limiting the output is shown in Figure 15.35. Two back-to-back Zener diodes are connected between the output and ground. Assuming the two diodes are matched, the output is limited to either the positive or negative value of $(V_\gamma + V_Z)$, where V_γ is the forward diode voltage and V_Z is the reverse Zener voltage. Resistor R is chosen to produce a specified current in the diodes.

Another Schmitt trigger with a limiter is shown in Figure 15.36(a). If we assume that $v_I = 0$ and v_O is in its high state, then D_2 is on and D_1 is off.

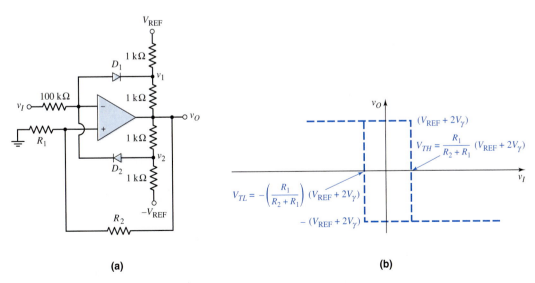

Figure 15.36 (a) Inverting Schmitt trigger with diode limiters and (b) voltage transfer characteristics

Neglecting currents in the $100\,k\Omega$ resistor, we have $v_2 = +V_\gamma$, where V_γ is the forward diode voltage. We can write

$$\frac{v_O - v_2}{1} = \frac{v_2 - (-V_{REF})}{1} \tag{15.80}$$

Solving for v_O yields

$$v_O = V_{REF} + 2V_\gamma \tag{15.81}$$

which means that the output voltage can be controlled and can be designed more accurately. The ideal hysteresis characteristics for this Schmitt trigger are shown in Figure 15.36(b). As v_I increases or decreases, a small current flows in the $100\,k\Omega$ resistor, producing a nonzero slope in the voltage transfer characteristics. The slope is on the order of $1/100$, which is quite small.

A noninverting Schmitt trigger with a limiting network is shown in Figure 15.37(a), and the resulting voltage transfer characteristics are given in Figure 15.37(b).

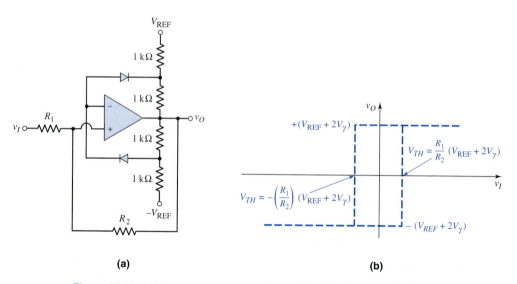

(a)   **(b)**

Figure 15.37   (a) Noninverting Schmitt trigger with diode limiters and (b) voltage transfer characteristics

15.4   NONSINUSOIDAL OSCILLATORS AND TIMING CIRCUITS

Many applications, especially digital electronic systems, use a nonsinusoidal square-wave oscillator to provide a clock signal for the system. This type of oscillator is called an astable multivibrator. In other applications, a single pulse of known height and width is used to initiate a particular set of functions. This type of oscillator is called a monostable multivibrator. First, we will examine the Schmitt trigger connected as an oscillator. Then we will analyze the 555 timer circuit. Although used extensively in digital electronic systems, these circuits are included here as comparator circuit applications.

15.4.1 Schmitt Trigger Oscillator

The Schmitt trigger can be used in an oscillator circuit to generate a square-wave output signal. This is accomplished by adding an RC network to the negative feedback loop of the Schmitt trigger as shown in Figure 15.38. As we will see, this circuit has no stable states. It is therefore called an **astable multivibrator**.

Initially, we set R_1 and R_2 equal to the same value, or $R_1 = R_2 \equiv R$. We assume that the output switches symmetrically about zero volts, with the high saturated output denoted by $V_H = V_P$ and the low saturated output denoted by $V_L = -V_P$. If v_O is low, or $v_O = -V_P$, then $v_+ = -(\frac{1}{2})V_P$. When v_X drops just slightly below v_+, the output switches high so that $v_O = +V_P$ and $v_+ = +(\frac{1}{2})V_P$. The $R_X C_X$ network sees a positive step-increase in voltage, so capacitor C_X begins to charge and voltage v_X starts to increase toward a final value of V_P.

The general equation for the voltage across a capacitor in an RC network is

Figure 15.38 Schmitt trigger oscillator

$$v_X = v_{\text{Final}} + (v_{\text{Initial}} - v_{\text{Final}})e^{-t/\tau} \tag{15.82}$$

where v_{Initial} is the initial capacitor voltage at $t = 0$, v_{Final} is the final capacitor voltage at $t = \infty$, and τ is the time constant. We can now write

$$v_X = V_P + \left(-\frac{V_P}{2} - V_P\right)e^{-t/\tau_x} \tag{15.83(a)}$$

or

$$v_X = V_P - \frac{3V_P}{2}e^{-t/\tau_x} \tag{15.83(b)}$$

where $\tau_x = R_X C_X$. Voltage v_X increases exponentially with time toward a final voltage V_P. However, when v_X becomes just slightly greater than $v_+ = +(1/2)V_P$, the output switches to its low state of $v_O = -V_P$ and $v_+ = -(\frac{1}{2})V_P$. The $R_X C_X$ network sees a negative step change in voltage, so capacitor C_X now begins to discharge and voltage v_X starts to decrease toward a final value of $-V_P$. We can now write

$$v_X = -V_P + \left[+\frac{V_P}{2} - (-V_P)\right]e^{-(t-t_1)/\tau_x} \tag{15.84(a)}$$

or

$$v_X = -V_P + \frac{3V_P}{2}e^{-(t-t_1)/\tau_x} \tag{15.84(b)}$$

where t_1 is the time at which the output switches to its low state. The capacitor voltage then decreases exponentially with time. When v_X decreases to $v_+ = -(\frac{1}{2})V_P$, the output again switches to its high state. The process continues to repeat itself, which means that this positive-feedback circuit oscillates producing a square-wave output signal. Figure 15.39 shows the output voltage v_O and the capacitor voltage v_X versus time.

Time t_1 can be found from Equation (15.83(b)) by setting $t = t_1$ when $v_X = V_P/2$, or

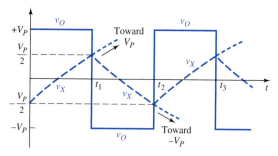

Figure 15.39 Output voltage and capacitor voltage versus time for Schmitt trigger oscillator

$$\frac{V_P}{2} = V_P - \frac{3V_P}{2}e^{-t_1/\tau_x} \tag{15.85}$$

Solving for t_1, we find that

$$t_1 = \tau_x \ln 3 = 1.1 R_X C_X \tag{15.86}$$

From a similar analysis using Equation (15.84(b)), we find that the difference between t_2 and t_1 is also $1.1 R_X C_X$; therefore, the period of oscillation T is

$$T = 2.2 R_X C_X \tag{15.87}$$

and the frequency of oscillation is

$$f = \frac{1}{T} = \frac{1}{2.2 R_X C_X} \tag{15.88}$$

As an example of an application of this circuit, a variable frequency oscillator is created by letting R_X be a variable resistor.

Design Example 15.8   Objective: Design a Schmitt trigger oscillator for a particular frequency.

Consider the oscillator in Figure 15.38. Design the circuit to oscillate at $f_o = 1$ kHz.

Solution: Using Equation (15.88), we can write

$$R_X C_X = \frac{1}{2.2 f_o} = \frac{1}{2.2(10^3)} = 4.55 \times 10^{-4}$$

If $C_X = 0.1\,\mu\text{F}$, then $R_X = 4.55\,\text{k}\Omega$.

Comment: A larger frequency of oscillation can easily be obtained by using a smaller capacitor value.

The **duty cycle** of the oscillator is defined as the percentage of time that the output voltage v_O is in its high state. For the circuit just considered, the duty cycle is 50 percent, as seen in Figure 15.39. This is a result of the symmetrical output voltages $+V_P$ and $-V_P$. If asymmetrical output voltages are used, then the duty cycle changes from the 50 percent value.

Test Your Understanding

*15.15 For the Schmitt trigger oscillator in Figure 15.38, the saturation output voltages are $+10\,\text{V}$ and $-5\,\text{V}$. $R_1 = R_2 = 20\,\text{k}\Omega$, $R_X = 50\,\text{k}\Omega$, and $C_X = 0.01\,\mu\text{F}$. Determine the frequency of oscillation and the duty cycle. Sketch v_O and v_X versus time over two periods of the oscillation. (Ans. $f = 866\,\text{Hz}$, duty cycle $= 39.7\%$)

15.16 The Schmitt trigger oscillator is shown in Figure 15.38. The saturation output voltages are $\pm 10\,\text{V}$, and $R_1 = 10\,\text{k}\Omega$, $R_2 = 20\,\text{k}\Omega$, $R_X = 10\,\text{k}\Omega$, and $C_X = 0.1\,\mu\text{F}$. Determine the frequency of oscillation and the duty cycle. Sketch v_O and v_X versus time over two periods of oscillation. (Ans. $f = 722\,\text{Hz}$, duty cycle $= 50\%$)

15.4.2   Monostable Multivibrator

A **monostable multivibrator** has one stable state, in which it can remain indefinitely if not disturbed. However, a trigger pulse can force the circuit into a quasi-stable state for a definite time, producing an output pulse with a particular height and width. The circuit then returns to its stable state until another trigger pulse is applied. The monostable multivibrator is also called a **one-shot**.

A monostable multivibrator is created by modifying the Schmitt trigger oscillator as shown in Figure 15.40. A clamping diode D_1 is connected in parallel with C_X. In the stable state, the output is high and voltage v_X is held low by the conducting diode D_1.

Figure 15.40   Schmitt trigger monostable multivibrator

The trigger circuit is composed of the capacitor C, resistor R_3, and diode D_2, and is connected to the noninverting terminal of the comparator. The value of R_3 is chosen to be much larger than R_1, so that voltage v_Y is determined primarily by a voltage divider of R_1 and R_2. We then have

$$v_Y \cong \left(\frac{R_1}{R_1 + R_2}\right) V_P \equiv \beta V_P \tag{15.89}$$

where V_P is the sum of the forward and breakdown voltages of D_{Z1} and D_{Z2}, or $V_P = (V_{\gamma 1} + V_{Z2})$. This voltage is the positive saturated output voltage.

The circuit is triggered by a negative-going step voltage applied to capacitor C. This action forward-biases diode D_2 and pulls the voltage v_Y below v_X. Since the comparator then sees a larger voltage at the inverting terminal, the output switches to its low state of

$$v_O = -V_P = -(V_{\gamma2} + V_{Z1})$$

Voltage v_Y then becomes

$$v_Y \cong -\left(\frac{R_1}{R_1 + R_2}\right)V_P \equiv -\beta V_P \tag{15.90}$$

causing D_2 to become reverse biased, thus isolating the oscillator circuit from the input triggering network. The negative-step change in v_O causes voltage v_X to decrease exponentially with a time constant of $\tau_x = R_X C_X$ toward a final value of $-V_P$. Diode D_1 is reverse biased during this time. When v_X drops just below the value of v_Y given by Equation (15.90), the output switches back to its positive saturated value of $+V_P$. The capacitor voltage v_X then starts to increase exponentially toward a final value of $+V_P$. When v_X reaches V_γ, diode D_1 again becomes forward biased, v_X is clamped at V_γ, and the output remains in its high state.

The waveforms of v_O and v_X versus time are shown in Figure 15.41. After the output has switched back to its high state, the capacitor voltage v_X must return to its quiescent value of $v_X = V_\gamma$. This implies that there is a **recovery time** of $(T' - T)$ during which the circuit should not be retriggered.

For $t > 0$, voltage v_X can be written in the same general form as Equation (15.82), as follows:

$$v_X = -V_P + (V_\gamma - (-V_P))e^{-t/\tau_x} \tag{15.91}$$

where $\tau_x = R_X C_X$. At $t = T$, $v_X = -\beta V_P$ and the output switches high. The pulse width is then

$$T = \tau_x \ln\left[\frac{1 + (V_\gamma/V_P)}{(1 - \beta)}\right] \tag{15.92}$$

If we assume $V_\gamma \ll V_P$ and if we let $R_1 = R_2$ such that $\beta = 1/2$, then the pulse width is $T = 0.69\tau_x$. We can show that for $V_\gamma \ll V_P$ and $\beta = 1/2$, the recovery time is $(T' - T) = 0.4\tau_x$. There are alternative circuits with shorter recovery times, but we will not consider them here.

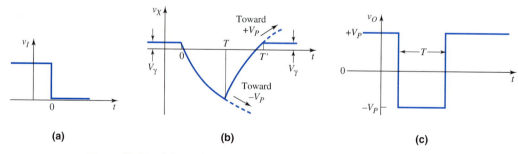

(a)   **(b)**   **(c)**

Figure 15.41   Schmitt trigger monostable multivibrator voltages versus time (a) input trigger pulse, (b) capacitor voltage, and (c) output pulse

Design Example 15.9 **Objective:** Design a monostable multivibrator to produce a given pulse width.

Consider the circuit in Figure 15.40 with parameters $V_P = 10\,\text{V}$, $V_\gamma = 0.7\,\text{V}$, and $R_1 = R_2 = 20\,\text{k}\Omega$. Design the circuit to produce a pulse that is $1\,\mu\text{s}$ wide.

Solution: Since $V_\gamma \ll V_P$ and $R_1 = R_2$, then from Equation (15.92), we have

$$T = 0.69\tau_x$$

or

$$\tau_x = R_X C_X = \frac{T}{0.69} = \frac{1}{0.69} = 1.45\,\mu\text{s}$$

If $R_X = 10\,\text{k}\Omega$, then $C_X = 145\,\text{pF}$.

Comment: In actual monostable multivibrator ICs, R_X and C_X are external elements to allow for variable times.

Test Your Understanding

*15.17 For the monostable circuit shown in Figure 15.40, the parameters are: $V_P = 12\,\text{V}$, $V_\gamma = 0.7\,\text{V}$, $C_X = 0.1\,\mu\text{F}$, $R_1 = 10\,\text{k}\Omega$, and $R_2 = 90\,\text{k}\Omega$. (a) Find the value of R_X that will result in a $50\,\mu\text{s}$ output pulse. (b) Using the results of part (a), find the recovery time. (Ans. (a) $R_X = 3.09\,\text{k}\Omega$ (b) $47.9\,\mu\text{s}$)

15.18 Consider the monostable multivibrator in Figure 15.40 with parameters: $V_P = 8\,\text{V}$, $V_\gamma = 0.7\,\text{V}$, $C_X = 0.01\,\mu\text{F}$, $R_X = 10\,\text{k}\Omega$, $R_1 = 20\,\text{k}\Omega$, and $R_2 = 40\,\text{k}\Omega$. Determine the output pulse width and recovery time. (Ans. $T = 48.9\,\mu\text{s}$, $t_2 = 37.8\,\mu\text{s}$)

15.4.3 The 555 Circuit

The **555 monolithic integrated circuit timer** was first introduced by Signetics Corporation in 1972 in bipolar technology. It quickly became an industry standard for timing and oscillation functions. Many manufacturers produce a version of a 555 IC, some in CMOS technology. The 555 is a general-purpose IC that can be used for precision timing, pulse generation, sequential timing, time delay generation, pulse width modulation, pulse position modulation, and linear ramp generation. The 555 can operate in both astable and monostable modes, with timing pulses ranging from microseconds to hours. It also has an adjustable duty cycle and can generally source or sink output currents up to 200 mA.

Basic Operation

The basic block diagram of the 555 IC is shown in Figure 15.42(a). The circuit consists of two comparators, which drive an RS flip-flop, an output buffer, and a transistor that discharges an external timing capacitor. The actual circuit of an LM555 timer is shown in Figure 15.42(b).

The **RS flip-flop** is a digital circuit that will be considered in detail in a later chapter. Here, we will describe only the basic digital function of the flip-flop, so

(a)

Figure 15.42   (a) Basic block diagram, 555 IC timer circuit and (b) circuit diagram, LM555 timer circuit

that the operation of the 555 timer can be explained. When the input R is high and input S is low, output $\overline{Q}$ is high. The complementary state occurs when R is low and S is high, producing a low $\overline{Q}$ output. If both R and S are low, then output $\overline{Q}$ remains in its previous state.

Comparator 1 is called the **threshold comparator**, which compares its input with an internal voltage reference set at $(\frac{2}{3})V^+$ by the voltage divider R_3, R_4, and R_5. When the input level exceeds this reference level, the threshold comparator output goes high, producing a high output at flip-flop terminal $\overline{Q}$. This turns the discharge transistor on and an external timing capacitor (not shown in this figure) starts to discharge.

The internal control voltage node is connected to an external terminal. This provides external control of the reference level, should the timing period need to be modified. When not in active use, this terminal should be bypassed to ground with a 0.01 µF capacitor, to improve the circuit's noise immunity.

Comparator 2, called the **trigger comparator**, compares its input trigger voltage to an internal voltage reference set to $(\frac{1}{3})V^+$ by the same voltage divider as before. When the output trigger level is reduced below this reference level, the trigger comparator output goes high, causing the RS flip-flop to reset. Output $\overline{Q}$ goes low and the discharge transistor turns off. This comparator triggers on the leading edge of a negative-going input pulse.

The output stage of the 555 IC is driven by output $\overline{Q}$ of the RS flip-flop. This output is usually a totem-pole push-pull circuit, or a simple buffer, and is generally capable of sourcing or sinking 200 mA.

Figure 15.42 (continued)

(b)

An external reset input to the RS flip-flop overrides all other inputs and is used to initiate a new timing cycle by turning the discharge transistor on. The reset input must be less than 0.4 V to initiate a reset. When not actively in use, the reset terminal should be connected to V^+ to prevent a false reset.

Monostable Multivibrator

A monostable multivibrator, also called a one-shot, operates by charging a timing capacitor with a current set by an external resistance. When the one-shot is triggered, the charging network cycles only once during the timing interval. The total timing interval includes the recovery time needed for the capacitor to charge up to the threshold level.

The external circuitry and connections for the 555 to be used as a one-shot multivibrator are shown in Figure 15.43. With a high voltage V^+ applied to the trigger input, the trigger comparator output is low, the flip-flop output $\overline{Q}$ is high, the discharge transistor is turned on, and the timing capacitor C is discharged to nearly ground potential. The output of the 555 circuit is then low, which is the quiescent state of the one-shot.

Figure 15.43 The 555 circuit connected as a monostable multivibrator

When a negative-going pulse is applied to the trigger input, the output of the trigger comparator goes high when the trigger pulse drops below $(\frac{1}{3})V^+$. Output $\overline{Q}$ goes low, which means that the output of the 555 goes high, and the discharge transistor turns off. The output of the 555 remains high even if the trigger pulse returns to its initial high value, because the reset input to the flip-flop is still low. The timing capacitor charges up exponentially toward a final value of V^+ through resistor R. The capacitor voltage is given by

$$v(t) = V^+\left(1 - e^{-t/RC}\right) \tag{15.93}$$

When $v(t) = (\frac{2}{3})V^+$, the threshold comparator output goes high, resetting the flip-flop. Output $\overline{Q}$ then goes high and the output of the 555 goes low. The high output at $\overline{Q}$ also turns on the discharge transistor, allowing the timing capacitor to discharge to near zero volts. The circuit thus returns to its quiescent state.

The width of the output pulse is determined from Equation (15.93). If we set $v(t) = \left(\frac{2}{3}\right)V^+$ and $t = T$, then

$$\left(\frac{2}{3}\right)V^+ = V^+\left(1 - e^{-T/RC}\right) \tag{15.94}$$

Solving for T, we have

$$T = RC\ln(3) = 1.1RC \tag{15.95}$$

The width of the output pulse is a function of only the external time constant RC; it is independent of the supply voltage V^+ and any internal circuit parameters. The triggering input pulse must be of a shorter duration than T. The output pulse height is a function of V^+ as well as of the internal circuitry. For a bipolar 555, the output pulse amplitude is approximately 1.7 V below supply voltage V^+.

When the output is high and the timing capacitor is charging, another trigger input pulse will have no effect on the circuit. If desired, the circuit can be reset during this period by applying a low input to the reset terminal. The output will return to zero and will remain in this quiescent state until another trigger pulse is applied.

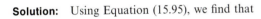

Design Example 15.10 **Objective:** Design the 555 IC as a monostable multi-vibrator with a 100 μs output pulse.

Consider the circuit in Figure 15.43. Let $C = 15\,\text{nF}$.

Solution: Using Equation (15.95), we find that

$$R = \frac{T}{1.1C} = \frac{100 \times 10^{-6}}{(1.1)(15 \times 10^{-9})} \Rightarrow 6.06\,\text{k}\Omega$$

Comment: To a very good approximation, the pulse width is a function of only the external resistor and capacitance values. A wide range of pulse widths can be obtained by changing these component values.

Astable Multivibrator

Figure 15.44 shows a typical external circuit connection for the 555 operating as an astable multivibrator, also called a timer circuit or clock. The threshold input and trigger input terminals are connected together. In the astable mode, the timing capacitor C charges through $R_A = R_B$ until $v(t)$ reaches $\left(\frac{2}{3}\right)V^+$. The threshold comparator output then goes high, forcing the flip-flop output $\overline{Q}$ to go high. The discharge transistor turns on, and the timing capacitor C discharges through R_B and the discharge transistor. The capacitor voltage decreases until it reaches $\left(\frac{1}{3}\right)V^+$, at which point the trigger comparator switches states and sends $\overline{Q}$ low. The discharge transistor turns off, and the timing capacitor begins to recharge. When $v(t)$ reaches the threshold level of $\left(\frac{2}{3}\right)V^+$, the cycle repeats itself.

Figure 15.44 Astable multivibrator 555 circuit

When the timing capacitor is charging, during the time $0 < t < T_C$, the capacitor voltage is

$$v(t) = \frac{1}{3}V^+ + \frac{2}{3}V^+\left(1 - e^{-t/\tau_A}\right) \tag{15.96}$$

where $\tau_A = (R_A + R_B)C$. At time $t = T_C$, the capacitor voltage reaches the threshold level, or

$$v(T_C) = \frac{2}{3}V^+ = \frac{1}{3}V^+ + \frac{2}{3}V^+\left(1 - e^{-T_C/\tau_A}\right) \tag{15.97}$$

Solving Equation (15.97) for the timing capacitor charging time T_C yields

$$T_C = \tau_A \ln(2) = 0.693(R_A + R_B)C \tag{15.98}$$

When the timing capacitor is discharging, during the time $0 < t' < T_D$, the capacitor voltage is

$$v(t') = \frac{2}{3}V^+ e^{-t'/\tau_B} \tag{15.99}$$

where $\tau_B = R_B C$. At time $t' = T_D$, the capacitor voltage reaches the trigger level and

$$v(T_D) = \frac{1}{3}V^+ = \frac{2}{3}V^+ e^{-T_D/\tau_B} \tag{15.100}$$

Solving Equation (15.100) for the timing capacitor discharge time T_D yields

$$T_D = \tau_B \ln(2) = 0.693 R_B C \tag{15.101}$$

The period T of the astable multivibrator cycle is the sum of the charging period T_C and the discharging period T_D. The frequency of oscillation is therefore

$$f = \frac{1}{T} = \frac{1}{T_C + T_D} = \frac{1}{0.693(R_A + 2R_B)C} \tag{15.102}$$

The duty cycle is defined as the percentage of time the output is high during one period of oscillation. During the charging time T_C, the output is high; during the discharging time, the output is low. The duty cycle is therefore

$$\text{duty cycle} = \frac{T_C}{T} \times 100\% = \frac{R_A + R_B}{R_A + 2R_B} \times 100\% \qquad \textbf{(15.103)}$$

Equation (15.103) shows that the duty cycle for this circuit is always greater than 50 percent. The duty cycle approaches 50 percent for $R_A \ll R_B$ and 100 percent for $R_B \ll R_A$. Alternative circuits can provide duty cycles of less than 50 percent.

Design Example 15.11   Objective: Design the 555 IC as an astable multivibrator for a specific frequency and duty cycle.

Consider the circuit in Figure 15.44. Design the circuit such that the frequency is 50 kHz and the duty cycle is 75 percent. Let $C = 1\,\text{nF}$.

Solution: The frequency of oscillation, as given by Equation (15.102), is

$$f = \frac{1}{0.693(R_A + 2R_B)C}$$

Therefore,

$$R_A + 2R_B = \frac{1}{(0.693)fC} = \frac{1}{(0.693)(50 \times 10^3)(1 \times 10^{-9})} \Rightarrow 28.9\,\text{k}\Omega \qquad \textbf{(15.104)}$$

The duty cycle, given by Equation (15.103), is

$$\text{Duty cycle} = 0.75 = \frac{R_A + R_B}{R_A + 2R_B}$$

which yields

$$R_A = 2R_B \qquad \textbf{(15.105)}$$

Combining Equations (15.104) and (15.105), we find that

$$R_A = 14.5\,\text{k}\Omega \quad \text{and} \quad R_B = 7.23\,\text{k}\Omega$$

Comment: A wide range of oscillation frequencies can be obtained by changing the resistance and capacitance values.

Other Applications

When the 555 is connected in the monstable mode, an external signal applied to the control voltage terminal will change the charging time of the timing capacitor and the pulse width. If the one-shot is triggered with a continuous pulse train, the output pulse width will be modulated by the external signal. This circuit is known as a **pulse width modulator (PWM)**.

A **pulse position modulator** can also be developed using the astable mode. A modulating signal applied to the control voltage terminal will vary the pulse position, which will be controlled by the modulating signal in a manner similar to the PWM.

Finally, a **linear ramp generator** can be constructed, again using the 555 monostable mode. The normal charging pattern of the timing capacitor is exponential because of the RC circuit. If resistor R is replaced by a constant current source, a linear ramp will be generated.

Test Your Understanding

15.19   The 555 IC is connected as an astable multivibrator. Let $R_A = 20\,\text{k}\Omega$, $R_B = 80\,\text{k}\Omega$, and $C = 0.01\,\mu\text{F}$. Determine the frequency of oscillation and the duty cycle. (Ans. $f = 802\,\text{Hz}$, duty cycle $= 55.6\%$)

D15.20   Design the 555 IC as an astable multivibrator to deliver a 1 kHz signal with a 55 percent duty cycle. (Ans. For example, $C = 0.01\,\mu\text{F}$, $R_A = 26\,\text{k}\Omega$, $R_B = 118\,\text{k}\Omega$)

15.5   INTEGRATED CIRCUIT POWER AMPLIFIERS

Most IC power amplifiers consist of a high-gain small-signal amplifier cascaded with a class-AB output stage. Some IC power amplifiers are a fixed-gain circuit with negative feedback incorporated on the chip, while others use a current gain output stage and negative feedback external to the chip. We consider three examples of IC power amplifiers in this section.

15.5.1   LM380 Power Amplifier

The LM380 is a popular fixed-gain power amplifier capable of an ac power output up to 5 W. Figure 15.45 is a simplified circuit diagram of the amplifier. The input stage is a Darlington pair configuration composed of Q_1 through Q_4 and an active load formed by Q_5 and Q_6.

The input stage is biased by currents through resistors R_{1A}, R_{1B}, and R_2. Transistor Q_3 is biased by a current from power supply V^+, through the diode-connected transistor Q_{10} and resistors R_{1A} and R_{1B}. Transistor Q_4 is biased by a current from the output terminal through R_2. For zero input voltages, the currents in Q_3 and Q_4 are nearly equal. Assuming matched input transistors and neglecting base currents, we find that

$$I_{C3} = \frac{V^+ - 3V_{EB}}{R_{1A} + R_{2A}}$$
(15.106)

and

$$I_{C4} = \frac{V_O - 2V_{EB}}{R_2}$$
(15.107)

Since $I_{C3} = I_{C4}$, we can find the quiescent output voltage by combining Equations (15.106) and (15.107), or

$$V_O = 2V_{EB} + \frac{R_2}{R_{1A} + R_{2B}}(V^+ - 3V_{EB}) = \frac{1}{2}V^+ + \frac{1}{2}V_{EB}$$
(15.108)

Figure 15.45 The LM380 power amplifier

The quiescent output voltage is approximately half the power supply voltage, which allows for a maximum output voltage swing and for maximum power to be delivered to a load. The feedback from the output to the emitter of Q_4, through R_2, stabilizes the quiescent output voltage at this value.

The output signal of the diff-amp is the input signal to the base of Q_{12}, which is connected in a common-emitter configuration in which Q_{11} acts as an active load. The output signal from the collector of Q_{12} is the input to the class-AB output stage, and capacitor C_F provides frequency compensation.

The class-AB complementary push-pull emitter-follower output stage comprises transistors Q_7, Q_8, and Q_9 and diodes D_1 and D_2. Transistor Q_7, which is the npn half of the push-pull output stage, sources current to the load. Transistors Q_8 and Q_9 operate as a composite pnp transistor, with the overall current gain equal to the product of the current gains of each transistor. This composite transistor is the pnp half of the push-pull output stage sinking current from the load. Diodes D_1 and D_2 provide the quiescent bias for class-AB operation.

The closed-loop gain is determined from the ac equivalent circuit in Figure 15.46. A differential-input voltage is applied at the input, with $V_{id}/2$ applied at the noninverting terminal and $-V_{id}/2$ applied at the inverting terminal. An external bypass capacitor is connected at the node between R_{1A} and R_{1B}, putting this node at signal ground. The second stage and output stage are

Figure 15.46   The ac equivalent circuit, LM380 power amplifier

represented by amplifier A. The input impedance is assumed to be large, which means that the input current is assumed to be negligible.

Since the input stage is an emitter-follower configuration, the signal voltage is approximately $+V_{id}/2$ at the emitter of Q_4 and is approximately $-V_{id}/2$ at the emitter of Q_3. Comparing the resistor values of R_3 and R_{1B}, we see the signal current in R_{1B} is negligible. The signal current in Q_3 is equal to that in R_3, and the current-mirror configuration of Q_5 and Q_6 implies that the current in Q_6 is also V_{id}/R_3. Summing the currents at the emitter of Q_4, we obtain

$$\frac{V_o - V_{id}/2}{R_2} = \frac{V_{id}}{R_3} + \frac{V_{id}}{R_3} \tag{15.109}$$

which yields the closed-loop voltage gain

$$\frac{V_o}{V_{id}} = \frac{1}{2} + \frac{2R_2}{R_3} \cong 50 \tag{15.110}$$

Equation (15.110) shows that the LM380 has a fixed gain of approximately 50.

The LM380 is designed to operate in the range of 12–22 V from a single supply V^+. The value of V^+ depends on the power requirements. Figure 15.47 shows the relationship between device dissipation, output power, and supply voltage for an 8 Ω load. As the output signal increases, harmonic distortion in the sinusoidal signal increases because the output transistor is approaching the saturation region. The lines marked 3% and 10% are the points at which harmonic distortion reaches 3% and 10%, respectively.

Example   15.12   **Objective:** Determine the output voltage and conversion efficiency for an LM380 power amplifier.

The required power for an 8 Ω is to be 4 W, with minimum distortion in the output signal.

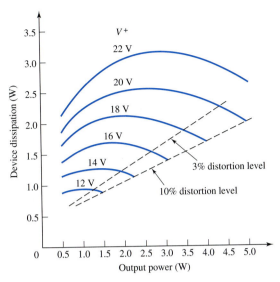

Figure 15.47 LM380 power amplifier characteristics

Solution: From the curves in Figure 15.47, for an output of 4 W, minimum distortion occurs when the supply voltage is a maximum, or $V^+ = 22$ V. For 4 W to be delivered to the 8 Ω load, the peak output signal voltage is determined by

$$\overline{P}_L = 4 = \frac{V_P^2}{2R_L} = \frac{V_P^2}{2(8)}$$

which yields $V_P = 8$ V.

The power dissipated in the device is 3 W, which means that the conversion efficiency is $4/(3 + 4) \rightarrow 57$ percent.

Comment: A reduction in the harmonic distortion means that the conversion efficiency is less than the theoretical value of 78.5 percent for the class-B output stages. However, a conversion efficiency of 57 percent is still substantially larger than would be obtained in any class-A amplifier.

Test Your Understanding

15.21 The supply voltage to an LM380 power amplifier, as shown in Figure 15.45, is 12 V. With a sinuosidal input signal, an average output power of 1 W must be delivered to an 8 Ω load. (a) Determine the peak output voltage and peak output current. (b) When the output voltage is at its peak value, calculate the instantaneous power being dissipated in Q_7. (Ans. (a) $V_P = 4$ V, $I_p = 0.5$ A (b) $P_Q = 4$ W)

15.5.2 PA12 Power Amplifier

The basic circuit diagram of the PA12 amplifier is shown in Figure 15.48. The input signal to the class-AB output stage is from a small-signal high-gain

Figure 15.48 PA12 power amplifier

op-amp. The power supply voltages are in the range of $10 \leq V_S \leq 50$ V, the peak output current is in the range $-15 \leq I_L \leq +15$ A, and the maximum internal power dissipation is 125 W. The output stage is a class-AB configuration using npn and pnp Darlington pair transistors. The bias for the output transistors is established by the V_{BE} multiplier circuit composed of R_1, R_2, and Q_4. Also, external feedback is required.

Design Example 15.13   Objective: Design the supply voltage required to meet a specific conversion efficiency in the PA12 power amplifier.

Consider the power amplifier in Figure 15.48 with a load resistance of 10 Ω. The required average power delivered to the load is 20 W. Determine the power supply voltage such that the conversion efficiency is 50 percent.

Solution: For an average of 20 W delivered to the load, the peak output voltage is

$$V_p = \sqrt{2 R_L \overline{P_L}} = \sqrt{2(10)(20)} = 20 \text{ V}$$

and the peak load current is

$$I_p = \frac{V_p}{R_L} = \frac{20}{10} = 2 \text{ A}$$

Assuming an ideal class-B condition, for a 50 percent conversion efficiency, the average power supplied by each V_S source must be 20 W. If we neglect power dissipation in the bias circuit, the average power supplied by each source is

$$P_S = V_S \left(\frac{V_p}{\pi R_L} \right)$$

and the required supply voltage is then

$$V_S = \frac{\pi R_L P_S}{V_p} = \frac{\pi (10)(20)}{20} = 31.4 \, \text{V}$$

Comment: The actual conversion efficiency for class-AB operation is less than 50 percent. This reduced conversion efficiency ensures that harmonic distortion in the output signal is not severe.

Computer Simulation Verification: A computer simulation analysis of the circuit in Figure 15.48 was performed. The supply voltages were set at ± 31.4 V and the input sinusoidal signal was adjusted so that the peak sinusoidal output voltage was 19.7 V across a 10 Ω load resistor. For these settings, the bias supply currents were 1.971 A. The average power delivered by the supply voltage sources is 39.4 W, so that the conversion efficiency is 49.25 percent, which is just slightly below the design value of 50 percent.

15.5.3 Bridge Power Amplifier

Figure 15.49 shows a bridge power amplifier that uses two op-amps. Amplifier A_1 is connected in a noninverting configuration; A_2 is connected in an inverting configuration. The magnitudes of the two gains are equal to each other. The load, such as an audio speaker, is connected between the two output terminals and is floating. A sinusoidal input signal produces output voltages v_{o1} and v_{o2}, which are equal in magnitude but 180 degrees out of phase. The voltage across the load is therefore twice as large as it would be if produced from a single op-amp.

Figure 15.49 Bridge power amplifier

Test Your Understanding

15.22 Consider the bridge amplifier in Figure 15.49 with parameters $R_1 = R_3 = 20 \, \text{k}\Omega$, $R_2 = 30 \, \text{k}\Omega$, $R_4 = 50 \, \text{k}\Omega$, and $R_L = 1.2 \, \text{k}\Omega$. Assume the op-amps are biased at ± 15 V, and the peak output voltage of each op-amp is limited to ± 12 V. Determine: (a) the voltage gain of each op-amp circuit, (b) the average power that can be delivered to the load, and (c) the peak amplitude of the input voltage. (Ans. (a) $A_{v1} = 2.5$, $A_{v2} = -2.5$ (b) $\overline{P}_L = 0.24 \, \text{W}$ (c) $V_{pi} = 4.8$ V)

15.6   VOLTAGE REGULATORS

Another class of analog circuits that is used extensively in electronic systems is the voltage regulator. We briefly considered constant-voltage circuits, or voltage regulators, when we studied diode circuits and when we considered ideal op-amp circuits in Chapter 9. In this section, we will discuss examples of IC voltage regulators.

15.6.1   Basic Regulator Description

A **voltage regulator** is a circuit or device that provides a constant voltage to a load. The output voltage is controlled by the internal circuitry and is relatively independent of the load current supplied by the regulator.

A basic diagram of a voltage regulator is shown in Figure 15.50. It consists of three basic parts: a reference voltage circuit; an error amplifier, which is part of a feedback circuit; and a current amplifier, which supplies the required load current. The reference voltage circuit produces a voltage that is essentially independent of both supply voltage V^+ and temperature. As shown in the basic circuit of Figure 15.50, a fraction of the output voltage is fed back to the error amplifier which, through negative feedback, maintains the feedback voltage at a value equal to the reference voltage.

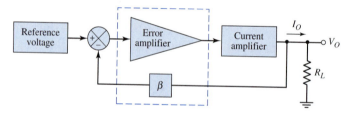

Figure 15.50   Basic circuit diagram of a voltage regulator

Since the regulator output voltage is derived from the reference voltage, any variation in that reference voltage, as the power supply voltage V^+ changes, also affects the output voltage. **Line regulation** is defined as the ratio of the change in output voltage to a given change in the input supply voltage, or

$$\text{Line regulation} = \frac{\Delta V_O}{\Delta V^+} \tag{15.111}$$

Line regulation is one figure of merit of voltage regulators. In many cases, the reference voltage circuit contains one or more Zener diodes. Line regulation is then a function of the Zener diode resistance and the effective resistance of the circuit biasing the diode.

15.6.2   Output Resistance and Load Regulation

The ideal voltage regulator is equivalent to an ideal voltage source in that the output voltage is independent of the output current and any output load impedance. In actual voltage regulators, however, the output voltage is a slight function of output current. This dependence is related to the output resistance of the regulator.

The output resistance is defined as the rate of change of output voltage with output current, or

$$R_{of} = -\frac{\Delta V_O}{\Delta I_O} \tag{15.112}$$

The change in V_O and I_O is caused by a change in the load resistance R_L. Everything else in the circuit remains constant. The negative sign in Equation (15.112) results from the voltage polarity and current direction, as shown in Figure 15.50. An increase in I_O produces a decrease in V_O; therefore, the output resistance R_{of} is positive. The output resistance of a voltage regulator should be small, so that a change in output current ΔI_O will result in only a small change in output voltage ΔV_O.

The notation R_{of} for the output resistance of the voltage regulator is the same as the term for the output resistance of a feedback circuit. This is appropriate since voltage regulators use feedback.

A second figure-of-merit for voltage regulators is load regulation. **Load regulation** is defined as the change in output voltage between a no-load current condition and a full-load current condition. Load regulation can be expressed as a percentage, or

$$\text{Load regulation} = \frac{V_O(\text{NL}) - V_O(\text{FL})}{V_O(\text{NL})} \times 100\% \tag{15.113}$$

where $V_O(\text{NL})$ is the output voltage for a zero-load current condition and $V_O(\text{FL})$ is the output voltage for a full-load or maximum load current condition.

In some applications, a zero-load current is impractical, and a load current that is approximately 1 percent of the full-load current is used as the no-load condition. In most cases, this condition provides an adequate definition for load regulation.

Example 15.14 **Objective:** Determine the output resistance and load regulation of a voltage regulator.

Assume the output voltage of a regulator is 5.0 V for a load current of 5 mA, and is 4.96 V for a load current of 1.5 A.

Solution: If we assume that the output voltage decreases linearly with load current, then the output resistance is

$$R_{of} = -\frac{\Delta V_O}{\Delta I_O} = -\left(\frac{5.0 - 4.96}{0.005 - 1.5}\right) \cong 0.0267\,\Omega$$

or

$$R_{of} \cong 27\,\text{m}\Omega$$

The load regulation is then

$$\text{Load regulation} = \frac{V_O(\text{NL}) - V_O(\text{FL})}{V_O(\text{NL})} \times 100\% = \frac{5.0 - 4.96}{5.0} \times 100\% = 0.80\%$$

Comment: The output resistance of a voltage regulator is usually not constant at all load currents, but the values are typically in the milliohm range. Also, a load regulation of 0.8% is typical of many voltage regulators.

15.6.3   Simple Series-Pass Regulator

Figure 15.51 shows a simple voltage regulator that includes an error amplifier (comparator) and series-pass transistors. The series-pass transistors, which are connected in a Darlington emitter-follower configuration, form the current amplifier. A resistive voltage divider allows a portion of the output voltage to be fed back to the error amplifier. The closed-loop feedback system acts to maintain this fraction of the output voltage at a value equal to the reference voltage.

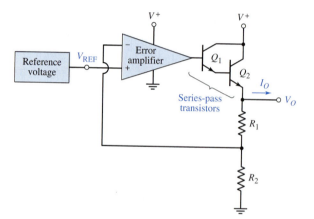

Figure 15.51   Basic series-pass voltage regulator

For an ideal system, we can write

$$\left(\frac{R_2}{R_1 + R_2}\right) V_O = V_{\text{REF}} \tag{15.114(a)}$$

or

$$V_O = V_{\text{REF}}\left(1 + \frac{R_1}{R_2}\right) \tag{15.114(b)}$$

Since the output of the feedback circuit is a shunt connection, the output resistance can be written, according to the results from Chapter 12, as

$$R_{of} = \frac{R_o}{1 + T} \tag{15.115}$$

where R_o is the output resistance of the open-loop system and T is the loop gain.

From feedback theory, the closed-loop and open-loop gains are related by

$$A_{CL} = \frac{A_{OL}}{1+T} \tag{15.116}$$

Combining Equation (15.115) and (15.116), we can write the closed-loop output resistance of the voltage regulator in the form

$$R_{of} = R_o \left(\frac{A_{CL}}{A_{OL}} \right) \tag{15.117}$$

From the circuit in Figure 15.51, the closed-loop gain is

$$A_{CL} = \frac{V_O}{V_{REF}} \tag{15.118}$$

The open-loop output resistance is the output resistance of the series-pass transistors, which are operating in an emitter-follower configuration. From previous results, we can write

$$R_o = \frac{r_{\pi 2} + R_{o1}}{(1+\beta_2)} \tag{15.119}$$

where

$$R_{o1} = \frac{r_{\pi 1} + R_{oa}}{(1+\beta_1)} \tag{15.120}$$

in which R_{oa} is the output resistance of the error amplifier. If the current in Q_2 is essentially equal to I_O and if β_1 and β_2 are large, then combining Equations (15.119) and (15.120) yields

$$R_o \cong \frac{2V_T}{I_O} + \frac{R_{oa}}{\beta_1 \beta_2} \tag{15.121}$$

Since the product $\beta_1\beta_2$ is large, the second term in Equation (15.121) is generally negligible.

The closed-loop output resistance, given by Equation (15.117), is then

$$R_{of} \cong \left(\frac{2V_T}{I_O} \right)\left(\frac{A_{CL}}{A_{OL}} \right) = \left(\frac{2V_T}{I_O} \right)\left(\frac{V_O}{V_{REF}} \right)\left(\frac{1}{A_{OL}} \right) \tag{15.122}$$

Equation (15.122) shows that the output resistance of the voltage regulator is not constant, but varies inversely with load current. Also, for very small values of load current, the output resistance may be unacceptably high.

The basic definition of output resistance is given in Equation (15.112). Using this definition and Equation (15.122), and rearranging terms, we obtain

$$\frac{\Delta V_O}{V_O} = -\left(\frac{\Delta I_O}{I_O} \right)\left(\frac{2V_T}{V_{REF}} \right)\left(\frac{1}{A_{OL}} \right) \tag{15.123}$$

Equation (15.123) relates the fractional change in output voltage to a fractional change in output current. Although valid for only small variations in voltage and current, this equation provides insight into the concept of load regulation.

Example 15.15 **Objective:** Determine the output resistance and the variation in output voltage of a series-pass regulator.

Assume an open-loop gain of $A_{OL} = 1000$, a reference voltage of $V_{REF} = 5\,\text{V}$, a nominal output voltage of $V_O = 10\,\text{V}$, and a nominal output current of $I_O = 100\,\text{mA}$.

Solution: From Equation (15.122), the output resistance is

$$R_{of} = \left(\frac{2V_T}{I_O}\right)\left(\frac{V_O}{V_{REF}}\right)\left(\frac{1}{A_{OL}}\right) = \left[\frac{2(0.026)(10)}{(0.10)(5)(1000)}\right] \Rightarrow 1.04\,\text{m}\Omega$$

From Equation (15.123), the relative change in output voltage is

$$\frac{\Delta V_O}{V_O} = -\left(\frac{\Delta I_O}{I_O}\right)\left(\frac{2V_T}{V_{REF}}\right)\left(\frac{1}{A_{OL}}\right) = -\left(\frac{\Delta I_O}{I_O}\right)\left[\frac{2(0.026)}{(5)(1000)}\right]$$

or

$$\frac{\Delta V_O}{V_O} = -\left(\frac{\Delta I_O}{I_O}\right)(1.04 \times 10^{-5})$$

A 10 percent change in output current results in only a 1.04×10^{-4} percent change in output voltage.

Comment: An output resistance in the $\text{m}\Omega$ range is typical of voltage regulators, and a change of only 10^{-4} percent in output for a 10 percent change in current is a good load regulation value.

15.6.4 Positive Voltage Regulator

In this section, we will analyze an example of a three-terminal positive voltage regulator fabricated as an IC. The equivalent circuit, shown in Figure 15.52, is part of the LM78LXX series, in which the XX designation indicates the output voltage of the regulator. For example, an LM78L08 is an 8 V regulator.

Basic Circuit Description

Once the bias current is established, Zener diode D_2 provides the basic reference voltage. Transistors Q_{15} and Q_{16} and diode D_1 form a start-up circuit that applies the initial bias to the reference voltage circuit. As the voltage across D_2 reaches the Zener voltage, transistor Q_{15} turns off, since the B–E voltage goes to zero (D_1 and D_2 are identical) and, the start-up circuit is then effectively disconnected from the reference voltage circuit.

The reference portion of the circuit is composed of Zener diode D_2 and transistors Q_3, Q_2, and Q_1, which are used for temperature compensation. The temperature compensation aspects of the circuit are discussed later in this section. Zener diode D_2 is biased by the current-source transistor Q_4. The temperature-compensated portion of the reference voltage at the node between R_1 and R_2 is applied to the base of Q_7, which is part of the error amplifier.

The bias current in Q_4 is established by the current in Q_5, which is a multiple-collector, multiple-emitter transistor. Transistor Q_5 is biased by the current in Q_3, which is controlled by the Zener voltage across D_2 and the B–E junction voltages of Q_3, Q_2, and Q_1. Consequently, the bias currents in the reference portion of the circuit become almost independent of the input supply

Figure 15.52 Equivalent circuit, LM78LXX series three-terminal positive voltage regulator

voltage. This in turn means that the reference voltage, and thus the output voltage are essentially independent of the power supply voltage. The overall result is very good line regulation.

The error amplifier is the differential pair Q_7 and Q_8, biased by Q_6 and R_6. The error amplifier output is the input to the base of Q_9, which is connected as an emitter follower and forms part of the drive for the series-pass transistors. The series-pass output transistors Q_{10} and Q_{11} are connected in a Darlington emitter-follower configuration.

A fraction of the output voltage, determined by the voltage divider R_{12} and R_{13}, is fed back to the base of Q_8, which is the error-amplifier inverting terminal. If the output voltage is slightly *below* its nominal value, then the base voltage at Q_8 is smaller than that at Q_7, and the current in Q_7 becomes a larger fraction of the total diff-amp bias current. The increased current in Q_7 induces a larger current in Q_{10}, which in turn produces a larger current in Q_{11} and increases the output voltage to the proper value. The opposite process occurs if the output voltage is *above* its nominal value.

Example 15.16   **Objective:**   Determine the bias current, temperature-compensated reference voltage, and required resistor R_{12} in a particular LM78LXX voltage regulator.

Consider the voltage regulator circuit in Figure 15.52. Assume Zener diode voltages of $V_Z = 6.3\,\text{V}$ and transistor parameters of $V_{BE}(\text{npn}) = V_{EB}(\text{pnp}) = 0.6\,\text{V}$. Design R_{12} such that $V_O = 8\,\text{V}$.

Solution: The bias current, neglecting base currents, is found as

$$I_{C3} = I_{C5} = \frac{V_Z - 3V_{BE}(\text{npn})}{R_3 + R_2 + R_1} = \frac{6.3 - 3(0.6)}{0.576 + 3.4 + 3.9} = 0.571\,\text{mA}$$

The temperature-compensated portion of the reference voltage, which is the input to the base of Q_7, is

$$V_{B7} = I_{C3}R_1 + 2V_{BE}(\text{npn}) = (0.571)(3.9) + 2(0.6) = 3.43\,\text{V}$$

From the voltage divider network, we have

$$\left(\frac{R_{13}}{R_{12} + R_{13}}\right)V_O = V_{B8} = V_{B7}$$

or

$$\left(\frac{2.23}{R_{12} + 2.23}\right)(8) = 3.43$$

which yields

$$R_{12} = 2.97\,\text{k}\Omega$$

Comment: The voltage divider of R_{12} and R_{13} is internal to the IC. This means the output voltage of a voltage regulator is fixed.

Temperature Compensation

Zener diodes with breakdown voltages greater than approximately 5 V have positive temperature coefficients, and forward-biased pn junctions have negative temperature coefficients. The magnitude of the temperature coefficients in the two devices is nearly the same.

For a given increase in temperature, V_{Z2} increases by ΔV and each B–E voltage decreases by ΔV, which means that I_{C3} in Figure 15.52 increases by approximately

$$\Delta I_{C3} \cong \frac{4\Delta V}{R_1 + R_2 + R_3} \tag{15.124}$$

The total voltage across the B–E junctions of Q_1 and Q_2 decreases by approximately $2\Delta V$, and the change in voltage at the base of Q_7 is

$$\Delta V_{B7} \cong \Delta I_{C3}R_1 - 2\Delta V = 4\Delta V\left(\frac{R_1}{R_1 + R_2 + R_3}\right) - 2\Delta V \approx 0 \tag{15.125}$$

This indicates that the voltage divider across R_1 effectively cancels any temperature variation. The input signal to the error amplifier is thus temperature compensated.

Protection Devices

Transistors Q_{13} and Q_{14} and resistor R_3 in the regulator in Figure 15.52 provide thermal protection. From the results of Example 15.16, the B–E voltage of Q_{14} is approximately 330 mV, which means that both Q_{14} and Q_{13} are effectively cut off. As the temperature increases, the combination of a negative B–E temperature coefficient and an increase in I_{C3} causes Q_{14} to begin conducting, which in turn causes Q_{13} to conduct. The current in Q_{13} shunts current away from the output series-pass transistors and produces thermal shutdown.

Output current limiting is provided by transistor Q_{12} and resistor R_{11}, as we saw previously in op-amp output stages. The combination of resistors R_{14} and R_{15} and diodes D_3 and D_4 produces what is called a **foldback characteristic**. The vast majority of the power dissipated in the regulator is usually due to the output current, or

$$P_D \cong (V^+ - V_O)I_O \tag{15.126}$$

The output current limit, to prevent power dissipation from reaching its maximum value $P_D(\text{max})$, is given by

$$I_O(\text{max}) = \frac{P_D(\text{max})}{V^+ - V_O} \tag{15.127}$$

A current-limiting characteristic of the type described by Equation (15.127) will protect the regulator and allow the maximum output current possible. This type of current limiting is called foldback current limiting.

Three-Terminal Regulator

The three-terminal voltage regulator is designed with an output voltage set at a predetermined value; external feedback elements and connections are not required. Figure 15.53 shows the basic circuit configuration of a three-terminal regulator. In some applications, capacitors may be inserted across the input and output terminals. The lead inductance between the voltage supply and regulator may cause stability problems. The capacitor across the input terminals is used only if the power supply and regulator are separated by a few centimeters. The load capacitor may improve the response of the regulator to transient changes in load current.

Figure 15.53   Basic circuit configuration of a three-terminal voltage regulator

Test Your Understanding

15.23   The reference voltage for a constant-voltage source is established by the simple combination of V^+, R_1, and D_1, as shown in the regulator circuit in Figure

15.54. If the Zener diode resistance is $R_Z = 10\,\Omega$ and the zero-current diode voltage is $V_{Zo} = 5.6\,V$, determine the line regulation of the voltage regulator. Assume an ideal op-amp. (Ans. 0.454%)

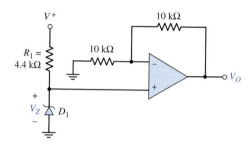

Figure 15.54   Figure for Exercise 15.23

*15.24   Consider the voltage regulator in Figure 15.55. The Zener diode is ideal, with $V_Z = 6.3\,V$, and the op-amp has a finite open-loop gain of $A_{OL} = 1000$. The no-load current is $I_O = 1\,mA$, and the full-load current is $I_O = 100\,mA$. Determine the load regulation. (Ans. 0.786%)

Figure 15.55   Figure for Exercise 15.24

15.25   Consider the voltage regulator circuit shown in Figure 15.52 with Zener diode voltages of $V_Z = 5.6\,V$. Assume transistor parameters of $V_{BE}(\text{npn}) = V_{EB}(\text{pnp}) = 0.6\,V$, neglect base currents, and let the resistor in the emitter of Q_4 be $R_4 = 100\,\Omega$. (a) Determine the bias currents I_{C3} and I_{C4}, and the temperature-compensated portion of the reference voltage V_{B7}. (b) Determine R_{12} such that $V_O = 5\,V$. (Ans. (a) $I_{C3} = 0.482\,mA$, $I_{C4} = 0.213\,mA$, $V_{B7} = 3.08\,V$ (b) $R_{12} = 1.39\,k\Omega$)

15.7   SUMMARY

- This chapter has presented several applications of op-amps and comparators that may be fabricated as integrated circuits.
- An active filter uses an active device, such as an op-amp, so as to minimize the effect of loading on the frequency characteristics of the filter.
- A Butterworth filter has a maximally flat response in the passband. The maximally flat response is obtained by setting the derivative of the transfer function with respect

to frequency equal to zero in the center of the passband. This procedure establishes the relationships between the various resistor and capacitor values.

- A switched-capacitor filter offers the advantage of an all-IC configuration, since this uses small capacitance values in conjunction with MOS switching transistors that simulate large resistance values.

- The basic principles of oscillation are: (1) the net phase through the amplifier and feedback network must be zero and (2) the magnitude of the loop gain must be unity. For an oscillator to function, the loop gain of a feedback network must provide sufficient phase shift to produce positive feedback.

- A phase shift oscillator consists of three RC networks, each providing a phase shift of 60 degrees, and an inverting op-amp, providing a phase shift of 180 degrees, for a total phase shift of 360 degrees.

- A Wien-bridge oscillator uses two RC networks as positive feedback in an op-amp circuit.

- The Colpitts, Harley, and crystal oscillator circuits use discrete transistors rather than op-amps, but have the potential of being very high frequency oscillators.

- A comparator is essentially an op-amp operated in an open-loop configuration. The output signal is either a high or low saturated voltage.

- A Schmitt trigger uses a comparator with positive feedback, which produces a hysteresis in the voltage transfer characteristics. This circuit, with its hysteresis characteristic, can eliminate the chatter effect in an output signal during switching applications in which noise is superimposed on the input signal.

- A square-wave generator or oscillator can be produced by incorporating an RC network in the negative feedback loop of a Schmitt trigger. This type of oscillator is called an astable multivibrator.

- The 555 IC timer uses two comparators and can operate in either astable or monostable modes. The frequency and duty cycle of the astable output signal, and the output pulse width of the monostable output signal, can be adjusted over a wide range by varying external resistor and capacitor values.

- Three examples of IC power amplifiers were discussed. The LM380 power amplifier is an all-IC device capable of delivering 5 W of ac power to a load. The PA12 power amplifier consists of a high-gain amplifier in conjunction with an external class-AB output stage and is capable of supplying peak output currents in the range of ±15 A. The bridge power amplifier uses two op-amps connected to an external load.

- A simple series-pass voltage regulator was analyzed to determine the basic characteristics of a regulator. The line regulation and load regulation were defined for regulators. Finally, an all-IC LM78L08 voltage regulator was discussed.

CHECKPOINT

After studying this chapter, the reader should have the ability to:

✔ Design a basic active filter. (Section 15.1)

✔ Design a basic oscillator. (Section 15.2)

✔ Design a basic Schmitt trigger circuit. (Section 15.3)

✔ Design a Schmitt trigger square-wave oscillator and use a 555 timer circuit. (Section 15.4)

✔ Understand the operation and characteristics of examples of integrated circuit power amplifiers. (Section 15.5)

REVIEW QUESTIONS

1. Describe the difference between an active filter and a passive filter. What is the primary advantage of an active filter?
2. Sketch the general characteristics of a low-pass filter, a high-pass filter, and a band-pass filter.
3. Consider a low-pass filter. What is the slope of the roll-off with frequency for a (a) one-pole filter, (b) two-pole filter, (c) three-pole filter, and (d) four-pole filter?
4. What characteristic defines a Butterworth filter?
5. Describe how a capacitor in conjunction with two switching transistors can behave as a resistor.
6. Sketch a one-pole low-pass switched-capacitor filter circuit.
7. Describe the characterisics of an oscillator.
8. Describe and explain the operation of a phase-shift oscillator.
9. Describe and explain the operation of a Wien-bridge oscillator.
10. What is the advantage of a Colpitts oscillator or Hartley oscillator compared to a phase-shift oscillator.
11. Sketch the circuits and characteristics of inverting and noninverting comparators.
12. Sketch the circuit and characteristics of a basic inverting Schmitt trigger.
13. What is meant by a bistable circuit?
14. What is the primary advantage of a Schmitt trigger circuit.
15. Sketch the circuit and explain the operation of a Schmitt trigger oscillator.
16. Describe the characteristics of a monostable multivibrator.
17. Describe how an op-amp in conjunction with a class-AB output stage can be used as a power amplifier.
18. Sketch a bridge power amplifier and describe its operation.
19. Sketch the basic circuit block diagram of a voltage regulator and explain the principle of operation.
20. Define load regulation of a voltage regulator.
21. Sketch the basic circuit of a series-pass voltage regulator.

PROBLEMS

Section 15.1   Active Filters

D15.1 (a) Design a single-pole low-pass filter with a gain of 10 in the passband and a 3 dB frequency of 5 kHz. (b) Repeat part (a) for a gain of -15 in the passband and a 3 dB frequency of 10 kHz. The minimum input resistance in the passband for this filter is to be 10 kΩ.

15.2 Determine the reduction in gain at $f = 2f_{3\,dB}$ for a (a) one-pole, (b) two-pole, and (c) three-pole low-pass filter.

D15.3 Design a two-pole high-pass Butterworth active filter with a cutoff frequency of 10 kHz and a unity-gain magnitude at high frequency.

D15.4 Design a three-pole high-pass Butterworth active filter with a cutoff frequency of 50 kHz. What is the magnitude of the transfer function at frequencies of 30, 35, 40, and 45 kHz?

15.5 Starting with the general transfer function given by Equation (15.7), derive the relationship between R_1 and R_2 in the two-pole high-pass Butterworth active filter.

15.6 A low-pass filter is to have a cutoff frequency of 10 kHz and is to have a gain at 20 kHz, which is reduced by at least 25 dB from its maximum value. Find the minimum number of poles required for a Butterworth filter.

*D15.7 Design a special type of first-order filter (one capacitor) in which the gain magnitude is 25 for frequencies less than approximately 25 kHz and is 1 for frequencies greater than approximately 25 kHz.

*D15.8 An amplitude-modulated radio signal consists of an 80 Hz to 12 kHz audio signal superimposed on a 770 kHz carrier signal. A low-pass filter is to be designed in which the gain in the passband is unity and the carrier signal is attenuated by at least −100 dB. What order of filter is required?

D15.9 A band-reject filter may be designed by combining a low-pass filter and a high-pass filter with a summing amplifier. A 60 Hz signal is to be at least −50 dB below the maximum gain value of 0 dB with a two-pole low-pass Butterworth filter and a two-pole high-pass Butterworth filter. What is the bandwidth of the reject filter?

15.10 Consider the bandpass filter in Figure P15.10. (a) Show that the voltage transfer function is

$$A_v(s) = \frac{v_O}{v_I} = \frac{-1/R_4}{(1/R_1) + sC + 1/(sCR_2R_3)}$$

(b) For $C = 0.1\,\mu\text{F}$, $R_1 = 85\,\text{k}\Omega$, $R_2 = R_3 = 300\,\Omega$, $R_4 = 3\,\text{k}\Omega$, and $R_5 = 30\,\text{k}\Omega$, determine: (i) $|A_v(\text{max})|$; (ii) the frequency f_o at which $|A_v(\text{max})|$ occurs; and (iii) the two 3 dB frequencies.

Figure P15.10

15.11 Consider the circuit in Figure P15.11. (a) Derive the expressions for the magnitude and phase of the voltage transfer function. (b) Plot the phase versus frequency for $R = 10\,\text{k}\Omega$ and $C = 15.9\,\text{nF}$. [Note: this filter is referred to as an all-pass filter in that the magnitude of the voltage gain is constant, but the phase of the output voltage changes with frequency.]

Figure P15.11

15.12 For each of the circuits in Figures P15.12, derive the expressions for the voltage transfer function $T(s) = V_o(s)/V_i(s)$ and the cutoff frequency $f_{3\,dB}$.

(a) **(b)**

Figure P15.12 **Figure P15.13**

15.13 The circuit in Figure P15.13 is a bandpass filter. (a) Derive the expression for the voltage transfer function $T(s)$. (b) If $R_1 = 10\,k\Omega$, determine R_2, C_1, and C_2 such that the magnitude of the midband gain is 50 and the cutoff frequencies are 200 Hz and 5 kHz.

D15.14 A simple bandpass filter can be designed by cascading one-pole high-pass and one-pole low-pass filters. Using op-amp circuits similar to those in Figure 15.3, design a bandpass filter with cutoff frequencies of 200 Hz and 50 kHz and with a midband gain of 10 dB. Resistor values must be no larger than 200 kΩ, but the input resistance must be as large as possible.

15.15 The clock frequency in the switched-capacitor circuit in Figure 15.13(a) is 100 kHz. Find the equivalent resistance when: (a) $C = 1\,pF$, (b) $C = 10\,pF$, and (c) $C = 30\,pF$.

15.16 In the switched-capacitor circuit in Figure 15.13(a), the voltages are $V_1 = 2\,V$ and $V_2 = 1\,V$, the capacitor value is $C = 10\,pF$, and the clock frequency is $f_C = 100\,kHz$. (a) Determine the charge transferred from V_1 to V_2 during each clock pulse. (b) What is the average current that source V_1 supplies? (c) If the "on" resistance of each MOSFET is 1000 Ω, determine the time required to transfer 99 percent of the charge during each half-clock period.

D15.17 Consider the switched-capacitor filter in Figure 15.14(b). Design the circuit for a low-frequency gain of -10 and a cutoff frequency of 10 kHz. The clock frequency must be 10 times the cutoff frequency and the largest capacitance is to be 30 pF. Find the required values of C_1, C_2, and C_F.

15.18 The circuit in Figure P15.18 is a switched-capacitor integrator. Let $C_F = 30\,pF$ and $C_1 = 5\,pF$, and assume the clock frequency is 100 kHz. Also, let $v_I = 1\,V$. (a) Determine the integrating RC time constant. (b) Find the change in output voltage during each clock period. (c) If C_F is initially uncharged, how many clock pulses are required for v_O to change by 13 V?

Figure P15.18

Section 15.2   Oscillators

15.19   Consider the phase-shift oscillator in Figure 15.16 with parameters $R = 4\,\text{k}\Omega$ and $C = 10\,\text{nF}$. Determine the frequency of oscillation and the required value of R_2.

15.20   In the phase-shift oscillator in Figure 15.16, the capacitor at the noninverting terminal of op-amp A_1 is replaced by a variable capacitor C_V. (a) Derive the expression for the frequency of oscillation. (b) If $C = 10\,\text{pF}$, $R = 10\,\text{k}\Omega$, and C_V is variable between 10 and 50 pF, determine the range of oscillation frequency.

D15.21   Design the phase-shift oscillator in Figure 15.17 to operate at $f_o = 80\,\text{kHz}$. Let $C = 100\,\text{pF}$.

15.22   Analyze the phase-shift oscillator in Figure 15.17. Show that the frequency of oscillation is given by Equation (15.46) and that the condition for oscillation is given by Equation (15.47).

15.23   The circuit in Figure P15.23 is an alternative configuration of a phase-shift oscillator. (a) Assume that $R_1 = R_2 = R_3 = R_{A_1} = R_{A_2} = R_{A_3} \equiv R$ and $C_1 = C_2 = C_3 \equiv C$. Show that the frequency of oscillation is $\omega_o = \sqrt{3}/RC$. (b) Assume equal magnitudes of gain in each amplifier stage. What is the minimum magnitude of gain required in each stage to sustain oscillation?

Figure P15.23

Figure P15.24

15.24   Consider the phase-shift oscillator in Figure P15.24. (a) Derive the expression for the frequency of oscillation. (b) If $R = 5\,\text{k}\Omega$, find the values of C and R_F that will produce sustained oscillations at 5 kHz.

15.25   A Wien-bridge oscillator is shown in Figure P15.25. (a) Derive the expression for the frequency of oscillation. (b) What is the condition for sustained oscillations?

15.26   Consider the oscillator circuit in Figure P15.26. (a) Derive the expression for the loop gain $T(s)$. (b) Determine the expression for the frequency of oscillation. (c) Find the condition for oscillation.

Figure P15.25   Figure P15.26

D15.27 Design the Wien-bridge oscillator in Figure 15.18 to oscillate at $f_o = 80$ kHz. Choose appropriate component values.

D15.28 The Colpitts oscillator in Figure 15.20 is biased at $I_D = 1$ mA. The transistor parameters are $V_{TN} = 1$ V and $K_n = 0.5$ mA/V^2. Let $C_1 = 0.01$ μF and $R_L = 4$ kΩ. Design the circuit to oscillate at $f_o = 400$ kHz.

15.29 Figure P15.29 shows a Colpitts oscillator with a BJT. Assume r_π and r_o are both very large. Derive the expressions for the frequency of oscillation and the condition of oscillation.

15.30 Consider the ac equivalent circuit of the Hartley oscillator in Figure 15.22. (a) Derive the expression for the frequency of oscillation. (b) Determine the condition for sustained oscillations.

D15.31 For the Hartley oscillator in Figure 15.22, assume $r_\pi \to \infty$ and let $g_m = 20$ mA/V. Design the circuit to oscillate at $f_o = 800$ kHz and verify that the circuit will sustain oscillations.

15.32 Find the loop gain functions $T(s)$ and $T(j\omega)$, the frequency of oscillation, and the R_2/R_1 required for oscillation for the circuit in Figure P15.32.

15.33 Repeat Problem 15.32 for the circuit in Figure P15.33.

Figure P15.29   Figure P15.32   Figure P15.33

Section 15.3 Schmitt Trigger Circuits

D15.34 For the comparator in the circuit in Figure 15.27(a), the output saturation voltages are ± 10 V. Let $R_1 = 50$ kΩ. Design R_2 as a potentiometer in series with a fixed resistor, and find a reference voltage such that the crossover voltage can easily be varied over the range of 1 to 5 V.

D15.35 Consider the Schmitt trigger in Figure 15.30(a). Assume the saturated output voltages are $V_H = +10$ V and $V_L = -10$ V. Neglecting input bias current effects, design the circuit such that the maximum current in R_1 and R_2 is 100 µA and the hysteresis width is 0.1 V.

15.36 A Schmitt trigger is shown in Figure 15.30(a). The parametes are: $V_H = +10$ V, $V_L = -10$ V, $R_1 = 10$ kΩ, and $R_2 = 40$ kΩ. (a) Determine the crossover voltages V_{TH} and V_{TL}. (b) Assume a sinusoidal voltage $v_I = 5 \sin[2\pi(60)t]$ V is applied at the input. Sketch the steady-state output voltage versus time over two periods of the waveform.

15.37 Consider the Schmitt trigger in Figure P15.37. Assume the saturated output voltages are $\pm V_P$. (a) Derive the expression for the crossover voltages V_{TH} and V_{TL}. (b) Let $R_A = 10$ kΩ, $R_B = 20$ kΩ, $R_1 = 5$ kΩ, $R_2 = 20$ kΩ, $V_P = 10$ V, and $V_{REF} = 2$ V. (a) Find V_{TH} and V_{TL}. (b) Sketch the voltage transfer characteristics.

15.38 The saturated output voltages are $\pm V_P$ for the Schmitt trigger in Figure P15.38. (a) Derive the expressions for the crossover voltages V_{TH} and V_{TL} (b) If $V_P = 12$ V, $V_{REF} = -10$ V, and $R_3 = 10$ kΩ, find R_1 and R_2 such that the switching point is $V_S = -5$ V and the hysteresis width is 0.2 V. (c) Sketch the voltage transfer characteristics.

Figure P15.37	**Figure P15.38**	**Figure P15.39**

15.39 (a) Plot the voltage transfer characteristics of the comparator circuit in Figure P15.39 assuming the open-loop gain is infinite. Let the reverse Zener voltage be $V_Z = 5.6$ V and the forward diode voltage be $V_\gamma = 0.6$ V. (b) Repeat part (a) for an open-loop gain of 10^3. (c) Repeat part (a) for 2.5 V applied to the inverting terminal of the comparator.

15.40 Consider the Schmitt trigger in Figure 15.32(a). (a) Derive the expressions for the switching point and crossover voltages, as given in Equations (15.76) and (15.77). (b) Let $V_H = +10$ V, $V_L = -10$ V, and $R_1 = 10$ kΩ. Determine R_2 and V_{REF} such that $V_{TH} = 2$ V and $V_{TL} = 1$ V.

15.41 Consider the Schmitt trigger in Figure 15.33(a). (a) Derive the expressions for the switching point and crossover voltages, as given in Equations (15.78) and (15.79). (b) Let $V_H = 12$ V, $V_L = -12$ V, and $R_2 = 20$ kΩ. Determine R_1 and V_{REF} such that $V_{TH} = -1$ V and $V_{TL} = -2$ V.

15.42 For the comparator in the circuit in Figure 15.35, the nominal output saturation voltages are ± 12 V. Assume forward diode voltage drops of 0.7 V and reverse Zener voltages of 5.6 V. (a) If $R_1 = 1$ kΩ, find R_2 such that the hysteresis width is 1 V. (b) Find R such that the average diode current is 1 mA.

15.43 Consider the Schmitt trigger with limiter, as shown in Figure 15.36. Assume the forward diode turn-on voltage V_γ is 0.7 V. (a) Determine V_{REF} such that the bistable output voltages at $v_I = 0$ are ± 5 V. (b) Find values of R_1 and R_2 such that the crossover voltages are ± 0.5 V. (c) Taking R_1, R_2, and the 100 kΩ resistors into account, find v_O when $v_I = 10$ V.

15.44 Consider the inverting Schmitt trigger with limiting network, as shown in Figure 15.36(a). Show that the crossover voltages are those given in Figure 15.36(b).

15.45 (a) For the Schmitt trigger with limiter in Figure 15.37(a), find the two output voltage values at $v_I = 0$ and the two crossover voltages. (b) Derive the expression for the slope of v_O versus v_I for $v_I > V_{TH}$.

Section 15.4   Nonsinusoidal Oscillators and Timing Circuits

D15.46 Using the Schmitt trigger circuit in Figure 15.38, design a square-wave oscillator with a frequency of $f_o = 5$ kHz and a 50 percent duty cycle. Choose reasonable component values.

15.47 For the Schmitt trigger oscillator in Figure 15.38, the parameters are: $C_x = 0.1$ μF, $R_x = 10$ kΩ, $R_2 = 10$ kΩ, and $R_1 = 30$ kΩ. The saturated output voltages are ± 10 V. (a) Plot v_o and v_x versus time over two periods of oscillation. (b) Find the frequency of oscillation and the duty cycle.

15.48 Repeat Problem 15.47 for saturated output voltages of $V_H = +15$ V and $V_L = -10$ V.

15.49 Consider the circuit in Figure P15.49. The saturated output voltages of the Schmitt trigger comparator are ± 10 V. Assume that at $t = 0$, output v_{o1} switches from its low state to its high state and C_y is uncharged. Plot v_{o1} and v_o versus time over two periods of oscillation.

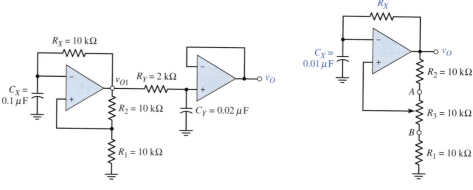

Figure P15.49   **Figure P15.50**

15.50 The saturated output voltages of the comparator in Figure P15.50 are ± 10 V. (a) Find R_x such that the frequency of oscillation is 500 Hz when the potentiometer is connected to point A. (b) Using the results of part (a), determine the oscillator frequency when the potentiometer is connected to point B.

15.51 The monostable multivibrator in Figure 15.40 is to be designed to produce a 100 μs pulse. Assume the saturated output voltages are ±5 V, and let $V_\gamma = 0.7$ V, $R_1 = 10$ kΩ, and $R_2 = 25$ kΩ. What is the minimum input triggering voltage required? What is the recovery time?

15.52 A monostable multivibrator is shown in Figure 15.40. The parameters are: $R_x = 50$ kΩ, $C_x = 01.$ μF, and $R_1 = R_2 = 20$ kΩ. The saturated output voltages are ±10 V. Let $V_\gamma = 0.7$ V for D_1 and D_2. What is the width of the output pulse? What is the recovery time?

D15.53 Figure 15.43 shows the 555 timer connected in the monostable multivibrator mode. (a) Design the circuit to provide an output pulse 60 seconds wide. (b) Determine the recovery time.

D15.54 Design a 555 monostable multivibrator to provide a 5 μs pulse. What is the recovery time?

15.55 A 555 timer is connected in the astable mode as shown in Figure 15.44. The parameters are $R_A = R_B = 20$ kΩ and $C = 0.1$ μF. Determine the frequency of oscillation and the duty cycle.

15.56 A 555 ICC is connected as shown in Figure P15.56. Determine the range of oscillation frequency and the duty cycle.

15.57 Repeat Problem 15.56 for the circuit in Figure P15.57.

Figure P15.56   **Figure P15.57**

Section 15.5   Integrated Circuit Power Amplifiers

15.58 The LM380 power amplifier in Figure 15.45 is biased at $V^+ = 22$ V. Let $\beta_n = 100$ and $\beta_p = 20$ for the npn and pnp transistors, respectively. (a) Determine the quiescent collector currents in transistors Q_1 through Q_6. (b) Assume that diodes D_1 and D_2 and transistors Q_7, Q_8, and Q_9 are all matched, with parameters $I_S = 10^{-13}$ A. For zero input voltages, determine the quiescent currents in D_1, D_2, Q_7, Q_8, and Q_9. (c) For no load, calculate the quiescent power dissipated in the amplifier.

15.59 An LM380 must deliver ac power to a 10 Ω load. The maximum power dissipated in the amplifier must be limited to 2 W and the maximum allowed distortion must be limited to 3 percent. Determine: (a) the maximum power that can be delivered to the load, (b) the maximum supply voltage, and (c) the peak amplitude of the sinusoidal output voltage.

D15.60 Design the bridge circuit in Figure 15.49 such that it can deliver an average ac power of 20 W to a 10 Ω speaker. Design each op-amp to have a gain magnitude of 15. Each supply voltage must be approximately 20 percent larger than the peak amplitude of the output voltage. What is the peak amplitude of the output voltage and current for each op-amp?

D15.61 Another form of the bridge power amplifier is shown in Figure P15.61. This amplifier has a very high input resistance since the input is to the noninverting terminal of an op-amp. (a) Derive the expression for the voltage gain $A_v = v_L/v_I$. (b) Design the circuit to provide a gain of $A_v = 10$ so that the magnitudes of v_{o1} and v_{o2} are equal. Let $R_1 = 50\,k\Omega$. (c) If $R_L = 20\,\Omega$ and if the average power delivered to the load is 10 W, determine the peak amplitude of v_{o1} and v_{o2} and the peak load current.

D15.62 Figure P15.62 shows an audio power amplifier using two identical op-amps connected in a bridge configuration. (a) Derive the expression for the voltage gain $A_v = v_L/v_I$. (b) Design the circuit to provide a gain of $A_v = 15$ and so that the magnitudes of v_{o1} and v_{o2} are equal. (c) If $R_L = 8\,\Omega$ and if the average power delivered to the load is 50 W, determine the peak amplitudes of v_{o1} and v_{o2} and the peak load current.

Figure P15.61   **Figure P15.62**

Section 15.6   Voltage Regulators

15.63 Transistors Q_1 and Q_2 in the voltage regulator circuit in Figure P15.63 have parameters $\beta = 200$, $V_{EB}(\text{on}) = 0.7\,V$, and $V_A = 100\,V$. The zero-current Zener voltage is $V_{ZO} = 6.3\,V$ and the Zener resistance is $r_z = 15\,\Omega$. Assuming an ideal op-amp, calculate the line regulation.

15.64 The output voltage of a voltage regulator decreases by 10 mV as the load current changes from a no-load current of zero to a full-load current of 1 A. If the output voltage changes linearly with load current, determine the output resistance of the regulator.

15.65 Consider the three-terminal voltage regulator in Figure 15.52, with parameters as given in Example 15.16. If the maximum load current is $I_O(\text{max}) = 100\,mA$, determine the minimum applied power supply voltage V^+ that will still maintain all transistors biased in the active region.

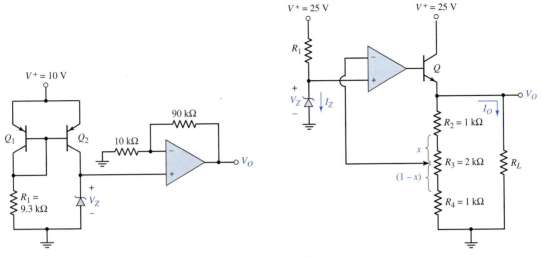

Figure P15.63   **Figure P15.68**

D15.66 Consider the three-terminal voltage regulator in Figure 15.52, with Zener diode voltages of $V_Z = 6.3\,\text{V}$. Assume transistor parameters of $V_{BE}(\text{npn}) = V_{EB}(\text{pnp}) = 0.6\,\text{V}$, and neglect base currents. (a) Determine resistance R_4 such that $I_{Z2} = 0.25\,\text{mA}$. (b) Determine R_{12} such that $V_O = 12\,\text{V}$.

15.67 The three-terminal voltage regulator in Figure 15.52 has parameters as described in Example 15.16. Assume $R_4 = 0$, $V_A = 50\,\text{V}$ for Q_4, and $r_z = 15\,\Omega$ for D_2. Determine the line regulation.

15.68 The voltage regulator in Figure P15.68. is a variable voltage, 0-to-1 A power supply. The transistor parameters are $\beta = 100$ and $V_{BE}(\text{on}) = 0.7\,\text{V}$. The op-amp has a finite open-loop gain of $A_{OL} = 10^4$. The zero-current Zener voltage is $V_{ZO} = 5\,\text{V}$ and the Zener resistance is $r_z = 10\,\Omega$. (a) For $I_Z = 10\,\text{mA}$, find R_1. (b) Determine the range of output voltage as the potentiometer R_3 is varied. (c) If the potentiometer is varied such that $x = 0$, determine the load regulation. Assume R_o of the op-amp is zero.

15.69 For the transistor in the circuit in Figure P15.69, the parameters are $\beta = 100$ and $V_{EB}(\text{on}) = 0.6\,\text{V}$. The diode is an idea Zener with $V_Z = 5.6\,\text{V}$, and the op-amp is ideal. Determine the range of load resistance R_L such that the load current is a constant. What is the value of that constant load current?

Figure P15.69

COMPUTER SIMULATION PROBLEMS

15.70 Simulate the three-pole low-pass Butterworth filter in Figure 15.10(a) using parameters $R = 1.59 \, \text{k}\Omega$, $C_1 = 0.03546 \, \mu\text{F}$, $C_2 = 0.01392 \, \mu\text{F}$, and $C_3 = 0.002024 \, \mu\text{F}$. Plot the magnitude of the voltage transfer function versus frequency and compare the computer results to the results obtained in Exercise 15.1.

15.71 Simulate the switched-capacitor filter in Figure 15.14(b) using parameters $C_1 = 30 \, \text{pF}$, $C_2 = 5 \, \text{pF}$, and $C_F = 12 \, \text{pF}$. Assume a clock frequency of 100 kHz. Plot the magnitude of the voltage transfer function versus frequency. Determine the 3 dB frequency and low-frequency gain. Compare these results with those obtained in Exercise 15.3.

15.72 Simulate the phase-shift oscillator in Figure 15.17 using parameters $R = 10 \, \text{k}\Omega$, $C = 100 \, \text{pF}$, and $R_2 = 300 \, \text{k}\Omega$. Plot the output voltage versus time. What is the frequency of oscillation?

15.73 Simulate the Schmitt trigger with limiters in Figure 15.36(a). Let $V_{\text{REF}} = 5 \, \text{V}$. Plot v_O versus v_I as v_I increases from -5 to $+5 \, \text{V}$, and then as v_I decreases from $+5$ to $-5 \, \text{V}$.

15.74 Simulate the ac equivalent circuit of the LM380 power amplifier in Figure 15.46. Determine the small-signal differential voltage gain.

15.75 Consider the reference voltage and error amp sections of the LM78LXX voltage regulator in Figure 15.52. Use the parameters described in Example 15.16. From a PSpice analysis, determine the temperature sensitivity and load regulation.

DESIGN PROBLEMS

[Note: Each design should be correlated with a computer analysis.]

*D15.76 Design a low-pass Butterworth filter to have a cutoff frequency at 15 kHz and a gain at 20 kHz, which is reduced by at least 20 dB from its maximum value. Determine the minimum number of poles and specify all component values.

*D15.77 Consider the Colpitts oscillator in Figure P15.77. The capacitors C_E and C_C are very large bypass and coupling capacitors. Let $V_{CC} = 10 \, \text{V}$. (a) Design the circuit such that the quiescent collector current is $I_{CQ} = 1 \, \text{mA}$. (b) Design the circuit to oscillate at $f_o = 800 \, \text{kHz}$.

*D15.78 Design a Schmitt trigger oscillator to produce a square-wave output at a frequency of $f_o = 5 \, \text{kHz}$ with peak output voltages of $\pm 5 \, \text{V}$.

*D15.79 Design a 555 timer as an astable multivibrator with an output signal frequency of 800 Hz and a 60 percent duty cycle.

*D15.80 Consider the power amplifier in Figure P15.80 with parameters $V^+ = 15 \, \text{V}$, $V^- = -15 \, \text{V}$, and $R_L = 20 \, \Omega$. The closed-loop gain must be 10. Design the circuit such that the power delivered to the load is 5 W when $v_I = -1 \, \text{V}$. If the four transistors are matched, determine the minimum β required such that the op-amp output current is limited to 2 mA when 5 W is delivered to the load.

*D15.81 Consider the simple series-pass regulator circuit in Figure P15.81. Assume an ideal Zener diode with $V_Z = V_{\text{REF}} = 4.7 \, \text{V}$. Let $\beta = 100$ and $V_{BE}(\text{on}) = 0.7 \, \text{V}$ for all transistors. (a) Design the circuit such that $V_O = 10 \, \text{V}$ and $I_Z = 10 \, \text{mA}$ for a nominal supply voltage of $V^+ = 20 \, \text{V}$. (b) Determine the regulator output resistance R_{of}.

Figure P15.77 Figure P15.80

Figure P15.81

Industry Insight

PETER J. STARVASKI
Program Manager
Digital Equipment Corporation

''The equations presented in the following chapters have found (and will continue to find) use throughout my career. Two examples come quickly to mind. The first stems from my involvement with a team whose goal was to improve the yield of one of our NMOS FET chips. This was an important project, needed to meet our customer's requirements. Engineers from a variety of disciplines were on the team. One hurdle everyone had to overcome was the fact that manufacturing engineers speak in terms of statistical process controls, product engineers speak in terms of specific failures for that product, design engineers speak in terms of their circuit models, etc.

The common language across all of these functions was the performance of the individual transistor. With only slight modifications, resulting from the idiosyncrasises of our manufacturing process, we used the FET equations in the following chapters. Everyone on the team needed to know these equations, as they dictate the parameters on the manufacturing floor, for the tester, in the circuit models, etc.

The second example occurred while I was designing the interrupt logic for a microprocessor. This was one of our early designs using CMOS FET technology. The input/output section of the chip was especially susceptible to a failure mechanism known as ''latchup.'' Latchup is a bipolar phenomenon that occurs in a CMOS field effect transistor circuit (quite inadvertently—it's a result of the physical layout of the FET devices). None of our models correctly handled this effect, and we found ourselves using the BJT equations presented in the following chapters to understand and solve the problem. FET designers need to know bipolar theory!

Everything comes back to the basics. Whether you're working as part of a team with a variety of engineering disciplines, or tackling infant technologies, the equations and concepts you're about to study will be tools you'll need to call upon often. Treat each section with equal importance, for, as I've shown, being one type of designer does not preclude you from having to use another's equations.

Good luck with your studies.''

DIGITAL ELECTRONICS

Part II of the text dealt with analog electronic circuits. Part III deals with digital electronics, another important category of electronics.

Chapter 16 examines field-effect transistor digital circuits. MOSFET digital circuits have revolutionized digital electronics, with the CMOS technology producing high-density, low-power digital circuits. Initially, we analyze basic NMOS and CMOS inverters, and then we develop NMOS and CMOS logic gates. Finally in this chapter, we analyze FET shift registers and flip-flops.

Bipolar digital circuits are considered in Chapter 17. We initially examine emitter-coupled logic, which is primarily used in specialized high-speed applications. Then, because transistor–transistor logic (TTL) circuits were the mainstay of logic design for many years, we analyze basic TTL and low-power Schottky TTL circuits, in order to obtain a good comparison between the FET and bipolar digital technologies.

16

MOSFET Digital Circuits

16.0 PREVIEW

This chapter presents the basic concepts of MOSFET digital integrated circuits, which is the most widely used technology for the fabrication of digital systems. The small transistor size and low power dissipation of CMOS circuits allows for a high level of integration for logic and memory circuits. We initially examine NMOS logic circuits, which contain only n-channel transistors, and then complementary MOS, or CMOS, logic circuits, which contain both n-channel and p-channel transistors. JFET logic circuits are very specialized and are therefore not considered here.

The discussion of NMOS logic circuits will serve as an introduction to the analysis and design of digital circuits. This technology deals with only one type of transistor (n-channel) and therefore makes the analysis more straightforward than dealing with two types of transistors in the same circuit. This discussion will also serve as a baseline to point out the advantages of CMOS technology.

The CMOS inverter is the basis of CMOS logic gates. We will analyze the inverter dc voltage transfer characteristics and will determine the power dissipation in the CMOS inverter, demonstrating the principal advantage of CMOS circuits over NMOS circuits. The CMOS inverter leads to the basic CMOS NOR and NAND logic gates. We also look at more advanced clocked CMOS logic circuits, which eliminate almost half of the transistors in a conventional CMOS logic design while maintaining the lower power advantage of the CMOS technology.

In addition to the basic logic gates that are discussed, we consider additional logic circuits such as flip-flops, shift registers, and adders. Finally, a whole class of digital systems, called memories, is considered. Static memory cells, dynamic memory cells, and read-only memory cells are analyzed. Sense amplifiers and read/write circuitry are briefly discussed.

16.1 NMOS INVERTERS

The inverter is the basic circuit of most MOS logic circuits. The design techniques used in NMOS logic circuits are developed from the dc analysis results for the MOS inverter. Extending the concepts developed from the inverter to NOR and NAND gates is then direct. Alternative inverter load elements are compared in terms of power consumption, packing density, and transfer characteristics. The transient analysis and switching characteristics of the inverters give an indication of the propagation delay times of NMOS logic circuits.

16.1.1   n-Channel MOSFET Revisited

We studied the structure, operation, and characteristics of MOS transistors in Chapter 5. In this section, we will quickly review the n-channel MOSFET characteristics, emphasizing specific properties important in digital circuit design.

A simplified n-channel MOSFET is shown in Figure 16.1(a). The body, or substrate, is a single-crystal silicon wafer, which is the starting material for circuit fabrication and provides physical support for the integrated circuit. The active transistor region is the surface of the semiconductor and comprises the heavily doped n^+ source and drain regions and p-type channel region. The channel length is L and the channel width is W. Normally, in any given fabrication process, the channel length is the same for all transistors, while the channel width is variable.

Figure 16.1(b) shows a more detailed view of the n-channel MOSFET. This figure demonstrates that the actual device geometry is more complicated than that indicated by the simplified cross section.

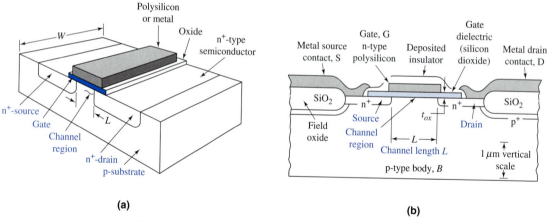

(a)

(b)

Figure 16.1   (a) n-channel MOSFET simplified view and (b) n-channel MOSFET detailed cross section

Figure 16.2(a) shows the simplified circuit symbols for the n-channel enhancement- and depletion-mode devices. When we explicitly consider the body or substrate connection, we will use the symbols shown in Figure 16.2(b).

In an integrated circuit, all n-channel transistors are fabricated in the same p-type substrate material. The substrate is connected to the most negative potential in the circuit, which for digital circuits, is normally at ground potential or zero volts. However, the source terminal of many of the transistors will not be at zero volts, which means that a reverse-biased pn junction will exist between the source and substrate.

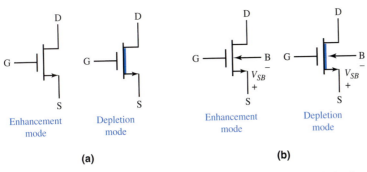

Figure 16.2 (a) Simplified circuit symbols for n-channel MOSFETs and (b) circuit symbols, showing substrate or body terminal

When the source and body terminals are connected together, the threshold voltage, to a first approximation, is independent of the applied voltages. However, when the source and body voltages are not equal, as when transistors are used for active loads, for instance, the threshold voltage is a function of difference between these voltages. We can write

$$V_{TN} = V_{TNO} + \frac{\sqrt{2e\varepsilon_s N_a}}{C_{ox}}\left[\sqrt{2\phi_{fp} + V_{SB}} - \sqrt{2\phi_{fp}}\right]$$

$$= V_{TNO} + \gamma\left[\sqrt{2\phi_{fp} + V_{SB}} - \sqrt{2\phi_{fp}}\right]$$

(16.1)

where V_{SB} is the source-to-body voltage, and V_{TNO} is the threshold voltage for zero source-to-body voltage or $V_{SB} = 0$. The parameter N_a is the p-type substrate doping concentration, ε_s is the semiconductor permittivity, C_{ox} is the oxide capacitance per unit area, ϕ_{fp} is a potential related to the substrate doping concentration, and γ is the body-effect coefficient.

Example 16.1 **Objective:** Determine the threshold voltage change due to a source-to-body voltage.

Consider a silicon n-channel MOSFET with the following parameters: $N_a = 1 \times 10^{16}$ cm^{-3}, t_{ox} (oxide thickness) = 500 Å, and $\phi_{fp} = 0.347$ V.

Solution: The oxide capacitance is

$$C_{ox} = \frac{\varepsilon_{ox}}{t_{ox}} = \frac{(3.9)(8.85 \times 10^{-14})}{500 \times 10^{-8}} = 6.9 \times 10^{-8} \text{ F/cm}^2$$

The change in threshold voltage is therefore

$$\Delta V_{TN} = V_{TN} - V_{TNO} = \frac{\sqrt{2e\varepsilon_s N_a}}{C_{ox}}\left[\sqrt{2\phi_{fp} + V_{SB}} - \sqrt{2\phi_{fp}}\right]$$

$$= \frac{\sqrt{2(1.6 \times 10^{-19})(11.7)(8.85 \times 10^{-14})(1 \times 10^{16})}}{6.9 \times 10^{-8}}\left[\sqrt{0.694 + V_{SB}} - \sqrt{0.694}\right]$$

$$= 0.834\left[\sqrt{0.694 + V_{SB}} - \sqrt{0.694}\right]$$

For this case, the body-effect coefficient is $\gamma = 0.834$ V$^{1/2}$. The threshold voltage change resulting from a source-to-body voltage V_{SB} is shown in Figure 16.3.

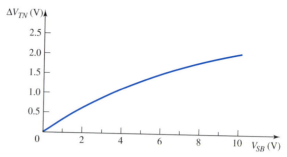

Figure 16.3   Change in threshold voltage versus source-to-body voltage for n-channel MOSFET in Example 16.1

Comment: The threshold voltage change with a change in V_{SB} will alter the current–voltage characteristics of the device and can alter the output voltage of an inverter.

The current–voltage characterisics of the n-channel MOSFET are functions of both the electrical and geometric properties of the device. When the transistor is biased in the nonsaturation region, for $v_{GS} \geq V_{TN}$ and $v_{DS} \leq (v_{GS} - V_{TN})$, we can write

$$i_D = K_n \left[2(v_{GS} - V_{TN})v_{DS} - v_{DS}^2 \right] \tag{16.2(a)}$$

In the saturation region, for $v_{GS} \geq V_{TN}$ and $v_{DS} \geq (v_{GS} - V_{TN})$, we have

$$i_D = K_n (v_{GS} - V_{TN})^2 \tag{16.2(b)}$$

The transition point separates the nonsaturation and saturation regions and is the drain-to-source saturation voltage, which is given by

$$v_{DS} = v_{DS}(\text{sat}) = v_{GS} - V_{TN} \tag{16.3}$$

The term $(1 + \lambda v_{DS})$ is sometimes included in Equation (16.2(b)) to account for channel length modulation and the finite output resistance. In most cases, it has little effect on the operating characteristics of MOS digital circuits. In our analysis, the term λ is assumed to be zero unless otherwise stated.

The parameter K_n is the NMOS transistor conduction parameter and is given by

$$K_n = \left(\frac{1}{2} \mu_n C_{ox} \right) \left(\frac{W}{L} \right) = \frac{k_n'}{2} \frac{W}{L} \tag{16.4}$$

The electron mobility μ_n and oxide capacitance C_{ox} are assumed to be constant for all devices in a particular IC.

The current–voltage characteristics are directly related to the channel width-to-length ratio, or the size of the transistor. In general, in a given IC, the length L is fixed, but the designer can control the channel width W.

Since the MOS transistor is a majority carrier device, the switching speed of MOS digital circuits is limited by the time required to charge and discharge the capacitances between device electrodes and between interconnect lines and ground. Figure 16.4 shows the significant capacitances in a MOSFET. The capacitances C_{sb} and C_{db} are the source-to-body and drain-to-body n^+p junction capacitances. The total input gate capacitance, to a first approximation, is a constant equal to

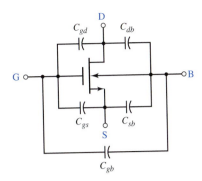

Figure 16.4   n-channel MOSFET and device capacitances

$$C_g = WLC_{ox} = WL\left(\frac{\varepsilon_{ox}}{t_{ox}}\right) \tag{16.5}$$

where C_{ox} is the oxide capacitance per unit area, and is a function of the oxide thickness. The parameter C_{ox} also appears in the expression for the conduction parameter.

Small Geometry Effects

The current–voltage relationships given by Equations (16.2(a)), (16.2(b)), and (16.3) are first-order approximations that apply to "long" channel devices. The tendency in device design is to make the devices as small as possible, which means the channel length is being reduced to values on the order of 0.25 µm or less. The corresponding channel widths are also being reduced. As the channel length is reduced, several effects alter the current–voltage characteristics. First, the threshold voltage becomes a function of the geometry of the device and is dependent on the channel length. This effect must be taken into account in the design of the transistor. Second, carrier velocity saturation reduces the saturation-mode current below the current value predicted by Equation (16.2(b)). The current is no longer a quadratic function of gate-to-source voltage, and tends to become a linear function of voltage. Channel length modulation means that the current tends to be larger than that predicted by the ideal equation. Third, the electron mobility is a function of the gate voltage so that the current tends to be smaller than the predicted value as the gate-to-source voltage increases. All of these effects complicate the analysis considerably.

We can, however, determine the basic operation and behavior of MOSFET logic circuits by using the first-order equations. We will use these first-order equations in our design of logic circuits. To determine the effect of small device size, a computer simulation may be performed in which the appropriate device models are incorporated in the simulation.

16.1.2   NMOS Inverter Transfer Characteristics

Since the inverter is the basis for most logic circuits, we will describe the NMOS inverter and will develop the dc transfer characteristics for three types of inverters with different load devices. This discussion will introduce voltage transfer functions, noise margins, and the transient characteristics of FET digital circuits.

NMOS Inverter with Resistor Load

Figure 16.5(a) shows a single NMOS transistor connected to a resistor to form an inverter. The transistor characteristics and load line are shown in Figure 16.5(b), along with the parametric curve separating the saturation and non-saturation regions. We determine the voltage transfer characteristics of the inverter by examining the various regions in which the transistor can be biased.

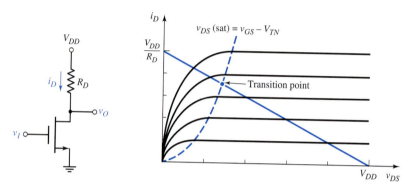

Figure 16.5   (a) NMOS inverter with resistor load and (b) transistor characteristics and load line

When the input voltage is less than or equal to the threshold voltage, or $v_I \leq V_{TN}$, the transistor is cut off, $i_D = 0$, and the output voltage is $v_O = V_{DD}$. The maximum output voltage is defined as the logic 1 level. As the input voltage becomes just greater than V_{TN}, the transistor turns on and is biased in the saturation region. The output voltage is then

$$v_O = V_{DD} - i_D R_D \tag{16.6}$$

where the drain current is given by

$$i_D = K_n(v_{GS} - V_{TN})^2 = K_n(v_I - V_{TN})^2 \tag{16.7}$$

Combining Equations (16.6) and (16.7) yields

$$v_O = V_{DD} - K_n R_D(v_I - V_{TN})^2 \tag{16.8}$$

which relates the output and input voltages as long as the transistor is biased in the saturation region.

As the input voltage increases, the Q-point of the transistor moves up the load line. At the transition point, we have

$$V_{Ot} = V_{It} - V_{TN} \tag{16.9}$$

where V_{Ot} and V_{It} are the drain-to-source and gate-to-source voltages, respectively, at the transition point. Substituting Equation (16.9) into (16.8), we determine the input voltage at the transition point from

$$K_n R_D(V_{It} - V_{TN})^2 + (V_{It} - V_{TN}) - V_{DD} = 0 \tag{16.10}$$

As the input voltage becomes greater than V_{It}, the Q-point continues to move up the load line, and the transistor becomes biased in the nonsaturation region. The drain current is then

$$i_D = K_n[2(v_{GS} - V_{TN})v_{DS} - v_{DS}^2] = K_n[2(v_I - V_{TN})v_O - v_O^2] \tag{16.11}$$

Combining Equations (16.6) and (16.11) yields

$$v_O = V_{DD} - K_n R_D[2(v_I - V_{TN})v_O - v_O^2] \tag{16.12}$$

which relates the input and output voltages as long as the transistor is biased in the nonsaturation region.

Figure 16.6 shows the voltage transfer characteristics of this inverter for three resistor values. Also shown is the line, given by Equation (16.9), which separates the saturation and nonsaturation bias regions of the transistor. The figure shows that the minimum output voltage, or the logic 0 level, for a high input decreases with increasing load resistance, and the sharpness of the transition region between a low input and a high input increases with increasing load resistance.

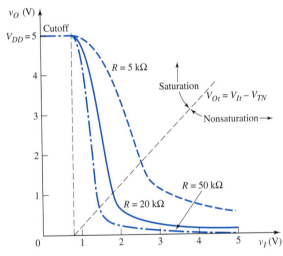

Figure 16.6 Voltage transfer characteristics, NMOS inverter with resistor load, for three resistor values

It should be noted that a large resistance is difficult to fabricate in an IC. A large resistor value in the inverter will limit current and power consumption as well as provide a small V_{OL} value. But it would also require a large chip area if fabricated in a standard MOS process. To avoid this problem MOS transistors can be used as load devices, replacing the resistor, as discussed in subsequent paragraphs.

Example 16.2   **Objective:** Determine the transition point and minimum output voltage of an NMOS inverter with resistor load.

Consider the circuit in Figure 16.5(a) with parameters $V_{DD} = 5\,\text{V}$ and $R_D = 20\,\text{k}\Omega$. The transistor parameters are $V_{TN} = 0.8\,\text{V}$ and $K_n = 0.2\,\text{mA/V}^2$.

Solution: The input voltage at the transition point is found from Equation (16.10). We have

$$(0.2)(20)(V_{It} - 0.8)^2 + (V_{it} - 0.8) - 5 = 0$$

which yields

$$V_{It} - 0.8 = 1 \quad \text{or} \quad V_{It} = 1.8 \text{ V}$$

The output voltage at the transition point is

$$V_{Ot} = V_{It} - V_{TN} = 1.8 - 0.8 = 1 \text{ V}$$

When v_I is high at $v_I = 5$ V, the output voltage is found from Equation (16.12). We find

$$v_O = 5 - (0.2)(20)\left[2(5 - 0.8)v_O - v_O^2\right]$$

which yields the output low level as

$$v_O = V_{OL} = 0.147 \text{ V}$$

Only the negative root of the quadratic has physical significance because the positive root yields an output voltage greater than the supply voltage V_{DD}.

Comment: The level of V_{OL} is less than the threshold voltage V_{TN}; therefore, if the output of this inverter is used to drive a similar inverter, the driver transistor of the load inverter would be cut off and its output would be high, which is the desired condition.

Test Your Understanding

[Note: In the following exercise, assume $k_n' = 35\ \mu\text{A/V}^2$ for all NMOS transistors.]

16.1   Consider the NMOS inverter with resistor load in Figure 16.5(a) biased at $V_{DD} = 5$ V. Assume transistor parameters of $W/L = 5$ and $V_{TN} = 0.8$ V. (a) Find the value of R_D such that $v_O = 0.15$ V when $v_I = 5$ V. (b) Using the results of part (a), determine the transition point for the driver transistor. (Ans. (a) $R_D = 44.8\ \text{k}\Omega$ (b) $V_{It} = 1.8$ V, $V_{Ot} = 1.0$ V)

NMOS Inverter with Enhancement Load

An n-channel enhancement-mode MOSFET with the gate connected to the drain can be used as a load device in an NMOS inverter. Figure 16.7(a) shows such a device. For $v_{GS} = v_{DS} \leq V_{TN}$, the drain current is zero. For

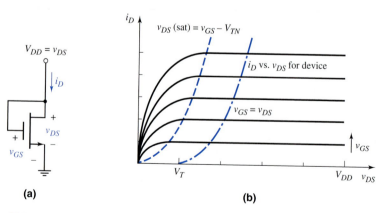

(a)   **(b)**

Figure 16.7   (a) n-channel MOSFET connected as saturated load device and (b) current–voltage characteristics of saturated load device

$v_{GS} = v_{DS} > V_{TN}$, a nonzero drain current is induced in the device. We can see that the following condition is satisfied:

$$v_{DS} > (v_{GS} - V_{TN}) = (v_{DS} - V_{TN}) = v_{DS}(\text{sat}) \tag{16.13}$$

A transistor with this connection always operates in the saturation region when not in cutoff.

The drain current is

$$i_D = K_n(v_{GS} - V_{TN})^2 = K_n(v_{DS} - V_{TN})^2 \tag{16.14}$$

We continue to neglect the effect of the output resistance and the λ parameter. The i_D versus v_{DS} characteristic is shown in Figure 16.7(b), which indicates that this device acts as a nonlinear resistor.

Figure 16.8(a) shows an NMOS inverter with the enhancement load device. The driver transistor parameters are denoted by V_{TND} and K_D, and the load transistor parameters are denoted by V_{TNL} and K_L. The substrate connections are not shown. In the following analysis, we neglect the body effect and we assume all threshold voltages are constant. These assumptions do not seriously affect the basic analysis, nor the inverter characteristics.

The driver transistor characteristics and the load curve are shown in Figure 16.8(b). When the inverter input voltage is less than the driver threshold voltage, the driver is cut off and the drain currents are zero. From Equation (16.14), we have

$$i_{DL} = 0 = K_L(v_{DSL} - V_{TNL})^2 \tag{16.15}$$

From Figure 16.8(a), we see that $v_{DSL} = V_{DD} - v_O$, which means that

$$v_{DSL} - V_{TNL} = V_{DD} - v_O - V_{TNL} = 0 \tag{16.16(a)}$$

The maximum output voltage is then

$$v_{O,\max} \equiv V_{OH} = V_{DD} - V_{TNL} \tag{16.16(b)}$$

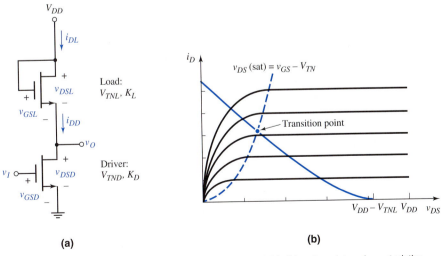

(a)   **(b)**

Figure 16.8 (a) NMOS inverter with saturated load and (b) driver transistor characteristics and load curve

For the enhancement-load NMOS inverter, the maximum output voltage, which is the logic 1 level, does not reach the full V_{DD} value. This cutoff point is shown in the load curve in Figure 16.8(b).

As the input voltage becomes just greater than the driver threshold voltage V_{TND}, the driver transistor turns on and is biased in the saturation region. In steady-state, the two drain currents are equal since the output will be connected to the gates of other MOS transistors. We have $i_{DD} = i_{DL}$, which can be written as

$$K_D(v_{GSD} - V_{TND})^2 = K_L(v_{GSL} - V_{TNL})^2 \tag{16.17}$$

Equation (16.17) is expressed in terms of the individual transistor parameters. In terms of the input and output voltages, the expression becomes

$$K_D(v_I - V_{TND})^2 = K_L(V_{DD} - v_O - V_{TNL})^2 \tag{16.18}$$

Solving for the output voltage yields

$$v_O = V_{DD} - V_{TNL} - \sqrt{\frac{K_D}{K_L}}(v_I - V_{TND}) \tag{16.19}$$

As the input voltage increases, the driver Q-point moves up the load curve and the output voltage decreases linearly with v_I.

At the driver transition point, we have

$$v_{DSD}(\text{sat}) = v_{GSD} - V_{TND}$$

or

$$V_{Ot} = V_{It} - V_{TND} \tag{16.20}$$

Substituting Equation (16.20) into (16.19), we find the input voltage at the transition point, which is

$$V_{It} = \frac{V_{DD} - V_{TNL} + V_{TND}\left(1 + \sqrt{\frac{K_D}{K_L}}\right)}{1 + \sqrt{\frac{K_D}{K_L}}} \tag{16.21}$$

As the input voltage becomes greater than V_{It}, the driver transistor Q-point continues to move up the load curve and the driver becomes biased in the nonsaturation region. Since the driver and load drain currents are still equal, or $i_{DD} = i_{DL}$, we now have

$$K_D\left[2(v_{GSD} - V_{TND})v_{DSD} - v_{DSD}^2\right] = K_L(v_{DSL} - V_{TNL})^2 \tag{16.22}$$

Writing Equation (16.22) in terms of the input and output voltages produces

$$K_D\left[2(v_I - V_{TND})v_O - v_O^2\right] = K_L(V_{DD} - v_O - V_{TNL})^2 \tag{16.23}$$

Obviously, the relationship between v_I and v_O in this region is not linear.

Figure 16.9 shows the voltage transfer characteristics of this inverter for three K_D-to-K_L ratios. The ratio K_D/K_L is the aspect ratio and is related to the width-to-length parameters of the driver and load transistors.

The line, given by Equation (16.20), separating the driver saturation and nonsaturation regions is also shown in the figure. We see that the minimum

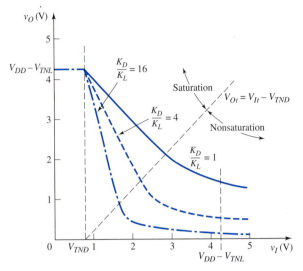

Figure 16.9 Voltage transfer characteristics, NMOS inverter with saturated load, for three aspect ratios

output voltage, or the logic 0 level, for a high input decreases with an increasing K_D/K_L ratio. As the width-to-length ratio of the load transistor decreases, the effective resistance increases, which means that the general behavior of the transfer characteristics is the same as for the resistor load. However, the high output voltage is

$$V_{OH} = V_{DD} - V_{TNL}$$

When the driver is biased in the saturation region, we find the slope of the transfer curve, which is the **inverter gain**, by taking the derivative of Equation (16.19) with respect to v_I. We see that

$$dv_O/dv_I = -\sqrt{K_D/K_L}$$

When the aspect ratio is greater than unity, the inverter gain magnitude is greater than unity. A logic circuit family with an inverter transfer curve that exhibits a gain greater than unity for some region is called a **restoring logic family**. Restoring logic is so named because logic signals that are degraded for some reason in one circuit can be restored by the gain of subsequent logic circuits.

Design Example 16.3 **Objective:** Design the aspect ratio K_D/K_L to produce a specified low output voltage, and determine the power dissipation in the inverter with enhancement load for a minimum W/L ratio for the load transistor. (Neglect the body effect.)

Consider the inverter shown in Figure 16.8(a) biased at $V_{DD} = 5$ V. The transistor parameters are: $V_{TND} = V_{TNL} = 0.8$ V and $k'_n = 35\,\mu A/V^2$. Determine K_D/K_L such that $v_O = 0.10$ V when $v_I = $ Logic $1 = 4.2$ V, and determine $(W/L)_D$ and the power dissipation in the inverter for $(W/L)_L = 0.5$ and $v_I = 4.2$ V.

Solution: For $v_I = 4.2 \, \text{V}$, the driver transistor is biased in the nonsaturation region. Using Equation (16.23), we have

$$\frac{K_D}{K_L}\left[2(4.2 - 0.8)(0.1) - (0.1)^2\right] = (5 - 0.1 - 0.8)^2$$

which yields

$$\frac{K_D}{K_L} = 25.1$$

Since

$$\frac{K_D}{K_L} = \frac{(W/L)_D}{(W/L)_L}$$

then

$$(W/L)_D = 12.6$$

when

$$(W/L)_L = 0.5$$

The power dissipated in the inverter is $P = i_D V_{DD}$, and the drain current can be found from the load transistor, as follows:

$$i_D = K_L(V_{DD} - v_O - V_{TNL})^2 = \frac{k_n'}{2}\left(\frac{W}{L}\right)_L (V_{DD} - v_O - V_{TNL})^2$$

Therefore,

$$i_D = \left(\frac{35}{2}\right)(0.5)(5 - 0.1 - 0.8)^2 = 147 \, \mu\text{A}$$

The power dissipation is

$$P = i_D V_{DD} = (147)(5) = 735 \, \mu\text{W}$$

 Comment: In the NMOS inverter with enhancement load, producing a relatively low output voltage V_{OL} requires a large difference in the sizes of the driver and load transistors. The load transistor size cannot be substantially reduced, so the power consumption also cannot be substantially reduced from the $735 \, \mu\text{W}$ value. If an IC contained a modest 100,000 inverters and all inverters were conducting, the total required current to the IC would be $14.7 \, \text{A}$ and the total power dissipated would be $73.5 \, \text{W}$! We thus see the need to drastically reduce the power dissipation in each inverter.

Test Your Understanding

[Assume $k_n' = 35 \, \mu\text{A/V}^2$.]

16.2   The enhancement-load NMOS inverter shown in Figure 16.8(a) is biased at $V_{DD} = 5 \, \text{V}$. The threshold voltages are $V_{TND} = V_{TNL} = 1 \, \text{V}$, and the width-to-length ratios are $(W/L)_D = 16$ and $(W/L)_L = 2$. (a) Find v_O when: (i) $v_I = 0$, and (ii) $v_I = 4 \, \text{V}$. (b) Calculate the power dissipated in the inverter when $v_I = 4 \, \text{V}$. (Ans. (a) $v_O = 4 \, \text{V}$, $v_O = 0.30 \, \text{V}$ (b) $P = 2.4 \, \text{mW}$)

D16.3 Consider the NMOS inverter with enhancement load, as shown in Figure 16.8(a), biased at $V_{DD} = 5\,\text{V}$. The transistor threshold voltages are $V_{TND} = V_{TNL} = 0.8\,\text{V}$. Design the width-to-length ratios such that the output voltage is 0.2 V and the inverter power dissipation is $750\,\mu\text{W}$ when $v_I = 4.2\,\text{V}$. (Ans. $(W/L)_L = 0.536$, $(W/L)_D = 6.49$)

NMOS Inverter with Depletion Load

Depletion-mode MOSFETs can also be used as load elements in NMOS inverters. Figure 16.10(a) shows the NMOS inverter with depletion load. The gate and source of the depletion-mode transistor are connected together. The driver transistor is still an enhancement-mode device. As before, the driver transistor parameters are $V_{TND}(V_{TND} > 0)$ and K_D, and the load transistor parameters are $V_{TNL}(V_{TNL} < 0)$ and K_L. Again, the substrate connections are not shown. The fabrication process for this inverter is slightly more complicated than for

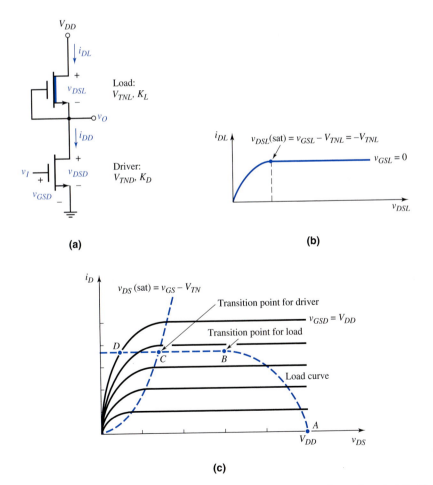

Figure 16.10 (a) NMOS inverter with depletion load, (b) current–voltage characteristic of depletion load, and (c) driver transistor characteristics and load curve

the enhancement-load inverter, since the threshold voltages of the two devices are not equal. However, as we will see, the advantages of this inverter make the extra processing steps worthwhile. This inverter has been the basis of many microprocessor and static memory designs.

The current–voltage characteristic curve for the depletion load, neglecting the body effect, is shown in Figure 16.10(b). Since the gate is connected to the source, $v_{GSL} = 0$, and the Q-point of the load is on this particular curve.

The driver transistor characteristics and the ideal load curve are shown in Figure 16.10(c). When the inverter input is less than the driver threshold voltage, the driver is cut off and the drain currents are zero. From Figure 16.10(b), we see that for $i_D = 0$, the drain-to-source voltage of the load transistor must be zero; therefore, $v_O = V_{DD}$ for $v_I \leq V_{TND}$. An advantage of the depletion-load inverter over the enhancement-load inverter is that the high output voltage, or the logic 1 level, is at the full V_{DD} value.

As the input voltage becomes just greater than the driver threshold voltage V_{TND}, the driver turns on and is biased in the saturation region; however, the load is biased in the nonsaturation region. The Q-point lies between points A and B on the load curve shown in Figure 16.10(c). We again set the two drain currents equal, or $i_{DD} = i_{DL}$, which means that

$$K_D[v_{GSD} - V_{TND}]^2 = K_L[2(v_{GSL} - V_{TNL})v_{DSL} - v_{DSL}^2] \tag{16.24}$$

Writing Equation (16.24) in terms of the input and output voltages yields

$$K_D[v_I - V_{TND}]^2 = K_L[2(-V_{TNL})(V_{DD} - v_O) - (V_{DD} - v_O)^2] \tag{16.25}$$

This equation relates the input and output voltages as long as the driver is biased in the saturation region and the load is biased in the nonsaturation region.

There are two transition points for the NMOS inverter with a depletion load: one for the load and one for the driver. These are points B and C, respectively, in Figure 16.10(c). The transition point for the load is given by

$$v_{DSL} = V_{DD} - V_{Ot} = v_{GSL} - V_{TNL} = -V_{TNL} \tag{16.26(a)}$$

or

$$V_{Ot} = V_{DD} + V_{TNL} \tag{16.26(b)}$$

Since V_{TNL} is negative, the output voltage at the transition point is less than V_{DD}. The transition point for the driver is given by

$$v_{DSD} = v_{GSD} - V_{TND}$$

or

$$V_{Ot} = V_{It} - V_{TND} \tag{16.27}$$

When the Q-point lies between points B and C on the load curve, both devices are biased in the saturation region, and

$$K_D(v_{GSD} - V_{TND})^2 = K_L(v_{GSL} - V_{TNL})^2 \tag{16.28(a)}$$

or

$$\sqrt{\frac{K_D}{K_L}}(v_I - V_{TND}) = -V_{TNL} \tag{16.28(b)}$$

Equation (16.28(b)) demonstrates that the input voltage is a constant as the Q-point passes through this region. This effect is also shown in Figure 16.10(c); the load curve between points B and C lies on a constant v_{GSD} curve. (This characteristic will change when the body effect is taken into account.)

For an input voltage greater than the value given by Equation (16.28(b)), the driver is biased in the nonsaturation region while the load is biased in the saturation region. The Q-point is now between points C and D on the load curve in Figure 16.10(c). Equating the two drain currents, we obtain

$$K_D\left[2(v_{GSD} - V_{TND})v_{DSD} - v_{DSD}^2\right] = K_L(v_{GSL} - V_{TNL})^2 \qquad \textbf{(16.29(a))}$$

which becomes

$$\frac{K_D}{K_L}\left[2(v_I - V_{TND})v_O - v_O^2\right] = (-V_{TNL})^2 \qquad \textbf{(16.29(b))}$$

This equation implies that the relationship between the input and output voltages are not linear in this region.

Figure 16.11 shows the voltage transfer characteristics of this inverter for three values of K_D/K_L. Also shown are the locus of transition points for the load and driver transistors as given by Equations (16.26(b)) and (16.27), respectively.

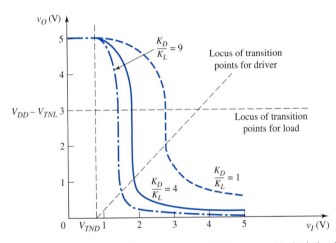

Figure 16.11 Voltage transfer characteristics, NMOS inverter with depletion load, for three aspect ratios

---

Design Example   16.4   Objective: Design the aspect ratio K_D/K_L to produce a specified low output voltage, and determine the power dissipation in the inverter with depletion load for a minimum W/L ratio for the load transistor.

Consider the inverter in Figure 16.10(a) biased at $V_{DD} = 5\,\text{V}$. The transistor parameters are: $V_{TND} = 0.8\,\text{V}$, $V_{TNL} = -2\,\text{V}$, and $k_n' = 35\,\mu\text{A/V}^2$. Determine K_D/K_L such that $v_O = 0.10\,\text{V}$ when $v_I = 5\,\text{V}$. Determine $(W/L)_D$ and the power dissipation in the inverter for $(W/L)_L = 0.5$.

Solution: For $v_I = 5\,\text{V}$, we assume the driver transistor is biased in the nonsaturation region and the load is in the saturation region. Using Equation (16.29(b)), we have

$$\frac{K_D}{K_L}[2(5 - 0.8)(0.1) - (0.1)^2] = [-(-2)]^2$$

which yields

$$\frac{K_D}{K_L} = 4.82$$

Since

$$\frac{K_D}{K_L} = \frac{(W/L)_D}{(W/L)_L}$$

then

$$(W/L)_D = 2.41$$

when

$$(W/L)_L = 0.5$$

The power dissipated in the inverter is $P = i_D V_{DD}$, and the drain current can be found from the load transistor, as follows:

$$i_D = K_L(-V_{TNL})^2 = \frac{k_n'}{2}\left(\frac{W}{L}\right)_L (-V_{TNL})^2 = \left(\frac{35}{2}\right)(0.5)[-(-2)]^2 = 35\,\mu\text{A}$$

The power dissipation is therefore

$$P \doteq i_D V_{DD} = (35)(5) = 175\,\mu\text{W}$$

 Comment: A relatively low output voltage V_{OL} can be produced in the NMOS inverter with depletion load, even when the load and driver transistors are not vastly different in size. The power dissipation in this inverter is also substantially less than in the enhancement-load inverter since the aspect ratio is smaller.

Design Consideration: The static analysis of the three types of NMOS inverters clearly demonstrates the advantage of the depletion load inverter. The size of the driver transistor is smaller for a given load device size to produce a given low output state. This allows a greater number of inverters to be fabricated in a given chip area. In addition, since the power dissipation is less, more inverters can be fabricated on a chip for a given total power dissipation.

Test Your Understanding

[Assume $k_n' = 35\,\mu\text{A/V}^2$.]

*16.4 The depletion load NMOS inverter shown in Figure 16.10(a) is biased at $V_{DD} = 5\,\text{V}$. the transistor parameters are: $V_{TND} = 0.7\,\text{V}$, $V_{TNL} = -1.5\,\text{V}$, $(W/L)_D = 6$, and $(W/L)_L = 2$. (a) Determine v_O for $v_I = 5\,\text{V}$. (b) Find the transition points for the driver and the load. (c) Calculate the power dissipation in the inverter when $v_I = 5\,\text{V}$. (Ans. (a) $v_O = 0.0881\,\text{V}$ (b) Load: $v_{It} = 1.57\,\text{V}$, $v_{Ot} = 3.5\,\text{V}$. Driver: $v_{It} = 1.57\,\text{V}$, $v_{Ot} = 0.87\,\text{V}$ (c) $P = 394\,\mu\text{W}$)

D16.5 Consider the depletion load inverter in Figure 16.10(a) biased at $V_{DD} = 5\,\text{V}$. The threshold voltages are $V_{TND} = 0.8\,\text{V}$ and $V_{TNL} = -2\,\text{V}$. Design the inverter such that the maximum power dissipation is $350\,\mu\text{W}$ and the output voltage is $0.05\,\text{V}$ when $v_I = 5\,\text{V}$. (Ans. $(W/L)_L = 1$, $(W/L)_D = 9.58$)

16.1.3 Noise Margin

The word "noise" means transient, unwanted variations in voltages or currents. In digital circuits, if the magnitude of the noise at a logic node is too large, logic errors can be introduced into the system. However, if the noise amplitude is less than a specified value, called the **noise margin**, the noise signal will be attenuated as it passes through a logic gate or circuit, while the logic signals will be transmitted without error.

Noise signals are usually generated outside the digital circuit and transferred to logic nodes or interconnect lines through parasitic capacitances or inductances. The coupling process is usually time dependent, leading to dynamic conditions in the circuit. In digital systems, however, the noise margins are usually defined in terms of static voltages.

Noise Margin Definition

For static noise margins, the type of noise usually considered is called series-voltage noise. Figure 16.12 shows two inverters in series in which the output of the second is connected back to the input of the first. Also included are series-voltage noise sources δV_L and δV_H. This type of noise can be developed by inductive coupling. The input voltage levels are indicated by H (high) and L (low). The noise amplitudes δV_L and δV_H can be different, and the polarities may be such as to increase the low output and reduce the high output. The noise margins are defined as the maximum values of δV_L and δV_H at which the inverters will remain in the correct state.

Figure 16.12 Two-inverter flip-flop, including series-voltage noise sources

The actual definitions of the noise margins NM_L and NM_H are not unique. In addition other types of noise, other than series-voltage source noise, may be present in the system. Dynamic noise sources also complicate the issue. However, in this text, in order to provide some measure of noise margin in a logic circuit, we will use the unity-gain approach to determine the logic threshold levels V_{IL} and V_{IH} and the corresponding noise margins.

Figure 16.13 shows a general voltage transfer function for an inverter. The expected logic 1 and logic 0 output voltages of the inverter are V_{OH} and V_{OL}, respectively. The parameters V_{IH} and V_{IL}, which determine the noise margins, are defined as the points at which

$$\frac{dv_O}{dv_I} = -1 \tag{16.30}$$

For $v_I \le V_{IL}$, the inverter gain magnitude is less than unity, and the output changes slowly with a change in the input voltage. Similarly, for $v_I \ge V_{IH}$, the output again changes slowly with input voltage since the gain magnitude is less than unity. However, when the input voltage is in the range $V_{IL} < v_I < V_{IH}$, the gain magnitude is greater than one, and the output signal changes rapidly.

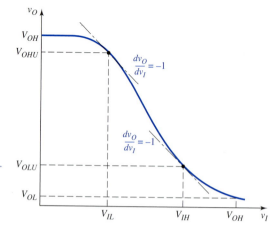

Figure 16.13 Generalized inverter voltage curve and defined voltage limits V_{IL} and V_{IH}

This region is called the **undefined range**. If the input voltage is inadvertently pushed into this range by a noise signal, the output may change logic states, and a logic error could be introduced into the system. The corresponding output voltages at the unity-gain points are denoted V_{OHU} and V_{OLU}, where the last subscript U signifies the unity-gain values.

The noise margins are defined as

$$NM_L = V_{IL} - V_{OLU} \tag{16.31(a)}$$

and

$$NM_H = V_{OHU} - V_{IH} \tag{16.31(b)}$$

We will see how these noise margin definitions correspond to the flip-flop conditions just discussed.

Since an inverter with a resistor load is rarely used in practice, we will determine the noise margins for NMOS inverters with enhancement and depletion loads.

Enhancement-Load Inverter

The general voltage transfer characteristic for the NMOS inverter with enhancement load is shown in Figure 16.14. Normally, $K_D > K_L$ and the gain magnitude for $v_I > V_{TND}$ is greater than one. Since the slope for this ideal curve is discontinuous at $v_I = V_{TND}$, we define the parameter $V_{IL} \equiv V_{TND}$. This is the threshold voltage of the driver. In this case, the output voltage V_{OHU} corresponds to V_{OH}.

At voltage V_{IH}, the driver is biased in the nonsaturation region. The relationship between the input and output voltages is given by Equation (16.23). Taking the derivative with respect to v_I, we obtain

$$2K_D v_O + 2K_D v_I \frac{dv_O}{dv_I} - 2K_D V_{TND} \frac{dv_O}{dv_I} - 2K_D v_O \frac{dv_O}{dv_I}$$

$$= -2K_L(V_{DD} - v_O - V_{TNL})\frac{dv_O}{dv_I} \tag{16.32}$$

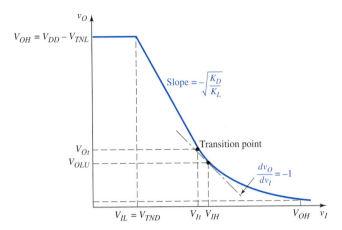

Figure 16.14 Voltage transfer characteristic, NMOS inverter with saturated load, and defined voltage limits V_{IL} and V_{IH}

Setting the derivative equal to -1 yields

$$K_D v_O - K_D v_I + K_D V_{TND} + K_D v_O = +K_L(V_{DD} - v_O - V_{TNL}) \tag{16.33}$$

Solving for v_O, we obtain

$$v_O = V_{OLU} = \frac{(V_{DD} - V_{TNL}) + \dfrac{K_D}{K_L}(v_I - V_{TND})}{1 + 2\dfrac{K_D}{K_L}} \tag{16.34}$$

Finally, combining Equations (16.34) and (16.23) and solving for v_I, we have

$$v_I = V_{IH} = V_{TND} + \left(\frac{V_{DD} - V_{TNL}}{\dfrac{K_D}{K_L}}\right)\left(\frac{1 + 2\dfrac{K_D}{K_L}}{\sqrt{1 + 3\dfrac{K_D}{K_L}}} - 1\right) \tag{16.35}$$

Example 16.5 **Objective:** Determine the noise margins of an inverter with enhancement load.

Consider the inverter shown in Figure 16.8(a) with the parameters given in Example 16.3. We have that $V_{OH} = 4.2\,\text{V}$, $V_{OL} = 0.10\,\text{V}$, and $K_D/K_L = 25.1$.

Solution: We know that $V_{IL} = V_{TND} = 0.8\,\text{V}$ and $V_{OHU} = V_{OH} = 4.2\,\text{V}$. The value of V_{IH} is, from Equation (16.35),

$$V_{IH} = 0.8 + \left(\frac{5 - 0.8}{25.1}\right)\left[\frac{1 + 2(25.1)}{\sqrt{1 + 3(25.1)}} - 1\right] = 1.61\,\text{V}$$

The output voltage corresponding to V_{IH} is, from Equation (16.34),

$$v_O = V_{OLU} = \frac{(5 - 0.8) + (25.1)(1.61 - 0.8)}{1 + 2(25.1)} = 0.479\,\text{V}$$

The noise margins are

$$\text{NM}_L = V_{IL} - V_{OLU} = 0.8 - 0.479 = 0.321\,\text{V}$$

and

$$\text{NM}_H = V_{OHU} - V_{IH} = 4.2 - 1.61 = 2.59 \text{ V}$$

Comment: This example shows that the two noise margins are not necessarily equal. In addition, the output voltage corresponding to V_{IH} is less than the threshold voltage of a driver transistor driven by the inverter. This means that as long as v_I remains in the range $V_{IH} \le v_I \le V_{OH}$, no logic error will be transmitted through a digital system.

The results of this example can be compared to the flip-flop conditions in Figure 16.12. Figure 16.15 shows two inverters in series, including series-voltage noise sources. If the input is high at $V_{OHU} = 4.2$ V and if a noise source of $\delta V_H = \text{NM}_H = 2.59$ V is included, the input to inverter G_1 is 1.61 V corresponding to V_{IH}, which is the minimum value corresponding to a logic 1 level. With an input of $V_{IH} = 1.61$ V to G_1, the output is 0.479 V. If a noise source of $\delta V_L = \text{NM}_L = 0.321$ V is present, then input to G_2 is 0.8 V corresponding to V_{IL}, which is the minimum value corresponding to a logic 0 level. With an input of $V_{IL} = 0.8$ V to G_2, the output is 4.2 V. If the output is connected back to the input, the resulting flip-flop configuration is in a stable state, although at the edge of switching. If the noise sources δV_H and δV_L increase slightly above the noise margins, the flip-flop will switch states.

Figure 16.15 Two inverters and two series-voltage noise sources with the noise margins from Example 16.5

Test Your Understanding

16.6 For the transistors in the NMOS inverter with enhancement load, the parameters are $V_{TN} = 0.85$ V and $K_D/K_L = 16$. Let $V_{DD} = 5$ V. Determine V_{IL}, V_{IH}, NM_L, and NM_H. (Ans. $V_{IH} = 1.81$ V, $V_{IL} = 0.85$ V, $\text{NM}_L = 0.259$ V, $\text{NM}_H = 2.34$ V)

Depletion-Load Inverter

The general transfer characteristic of the NMOS inverter with depletion load is shown in Figure 16.16. The point V_{IL} occurs when the load is biased in the nonsaturation region and the driver is biased in the saturation region. The relationship between the input and output voltages is given by Equation (16.25). Taking the derivative with respect to v_I, we obtain

$$2K_D[v_I - V_{TND}] = K_L\left[2(-V_{TNL})\left(-\frac{dv_O}{dv_I}\right) - 2(V_{DD} - v_O)\left(-\frac{dv_O}{dv_I}\right)\right]$$

(16.36)

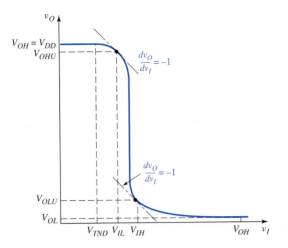

Figure 16.16 Voltage transfer characteristic, inverter with depletion load, and the defined voltage limits V_{IL} and V_{IH}

Setting the derivative equal to -1 and solving for v_O yields

$$v_O = V_{OHU} = (V_{DD} + V_{TNL}) + \left(\frac{K_D}{K_L}\right)(v_I - V_{TND}) \qquad (16.37)$$

Combining Equations (16.37) and (16.25), we then have

$$v_I = V_{IL} = V_{TND} + \frac{(-V_{TNL})}{\sqrt{\left(\frac{K_D}{K_L}\right)\left(1 + \frac{K_D}{K_L}\right)}} \qquad (16.38)$$

The point V_{IH} occurs when the load is biased in the saturation region and the driver is biased in the nonsaturation region. The relationship between the input and output voltages is given by Equation (16.29(b)). Taking the derivative with respect to v_I, we find

$$\frac{K_D}{K_L}\left[2(v_I - V_{TND})\frac{dv_O}{dv_I} + 2v_O - 2v_O\frac{dv_O}{dv_I}\right] = 0 \qquad (16.39)$$

Setting the derivative equal to -1 and solving for v_O yields

$$v_O = V_{OLU} = \frac{(v_I - V_{TND})}{2} \qquad (16.40)$$

Combining Equations (16.40) and (16.29(b)), we then have

$$v_I = V_{IH} = V_{TND} + \frac{2(-V_{TNL})}{\sqrt{3\left(\frac{K_D}{K_L}\right)}} \qquad (16.41)$$

The noise margins are then determined from Equations (16.31(a)) and (16.31(b)). The calculation of noise margins in an inverter with depletion load is considered in Exercise 16.7.

Test Your Understanding

16.7 For the transistors in the NMOS inverter with depletion load, the parameters are: $V_{TND} = 1\,\text{V}$, $V_{TNL} = -1.7\,\text{V}$, and $K_D/K_L = 5$. Let $V_{DD} = 5\,\text{V}$. Determine V_{IL}, V_{IH}, NM_L and NM_H. (Ans. $V_{IL} = 1.31\,\text{V}$, $V_{IH} = 1.88\,\text{V}$, $\text{NM}_L = 0.87\,\text{V}$, $\text{NM}_H = 2.97\,\text{V}$)

16.1.4 Body Effect

Up to this point, we have neglected the body effect and assumed that all threshold voltages are constant. Figure 16.17 shows enhancement-load and depletion-load NMOS inverters with the substrates of all transistors tied to ground. A nonzero source-to-body voltage will then exist in the load devices. In fact, the source terminal of the depletion load can increase to V_{DD}. The threshold voltage given by Equation (16.1) must be used in the circuit calculations for the load transistor. This significantly complicates the equations for the voltage transfer calculations, making them very cumbersome for hand analyses.

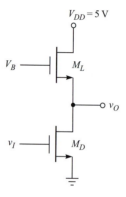

Figure 16.17 NMOS inverters, showing substrate connections to ground potential: (a) enhancement-load inverter and (b) depletion-load inverter

Example 16.6 Objective: Determine the change in the high output voltage of an NMOS inverter with enhancement load, taking the body effect into account.

Consider the NMOS inverter with enhancement load in Figure 16.17(a). The transistor parameters are $V_{TNDO} = V_{TNLO} = 0.8\,\text{V}$ and $K_D/K_L = 16$. Assume the inverter is biased at $V_{DD} = 5\,\text{V}$, assume the body effect coefficient is $\gamma = 0.90\,\text{V}^{1/2}$, and let $\phi_{fp} = 0.365\,\text{V}$.

Solution: When $v_I < V_{TNDO}$, the driver is cut off and the output goes high. From Equation (16.16(b)), the maximum output voltage is

$$v_{O,\text{max}} = V_{OH} = V_{DD} - V_{TNL}$$

where V_{TNL} is, from Equation (16.1),

$$V_{TNL} = V_{TNLO} + \gamma\left[\sqrt{2\phi_{fp} + V_{SB}} - \sqrt{2\phi_{fp}}\right]$$

From Figure 16.17(a), we see that $V_{SB} = v_O$; therefore, Equation (16.16(b)) can be written

$$v_{O,max} = V_{DD} - \left\{ V_{TNLO} + \gamma \left[\sqrt{2\phi_{fp} + v_{O,max}} - \sqrt{2\phi_{fp}} \right] \right\}$$

Defining $v_{O,max} \equiv V_{OH}$, we have

$$V_{OH} - 4.97 = -0.90\sqrt{0.73 + V_{OH}}$$

Squaring both sides and rearranging terms yields

$$V_{OH}^2 - 9.29V_{OH} + 24.9 = 0$$

Consequently, the maximum output voltage, or the logic 1 level, is

$$V_{OH} = 3.19 \, \text{V}$$

Comment: Neglecting the body effect, the logic 1 output level is

$$V_{OH} = V_{DD} - V_{TNL} = 5 - 0.8 = 4.2 \, \text{V}$$

The body effect, then, can significantly influence the logic high state of the NMOS inverter with enhancement load. These results also impact the inverter noise margins.

The source and body terminals of the depletion load device in the NMOS inverter shown in Figure 16.17(b) are not at the same potential when the output goes high. However, when the driver is cut off, the drain-to-source voltage of the depletion device must be zero in order that $v_{O,max} = V_{OH} = V_{DD}$.

Computer Simulation: A computer analysis of the inverters in Figure 16.17 was performed, neglecting the body effect and taking the body effect into account. The threshold voltage of the depletion load device is $V_{TNLO} = -2 \, \text{V}$ and the ratio K_D/K_L of the depletion load inverter is 4.82.

The body effect changes the voltage transfer characteristics of both the enhancement load and depletion load inverters. Figure 16.18(a) shows the voltage transfer characteristics for the enhancement-load inverter. The circuit and transistor parameters are the same as given in this example. For $v_I = 0$, the output voltage is 3.15 V when the

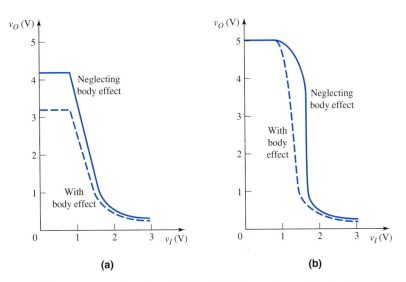

(a)   **(b)**

Figure 16.18   Voltage transfer characteristics of NMOS inverters with and without the body effect (a) enhancement load and (b) depletion load body effect is taken into account. This compares favorably with the 3.19 V from the hand analysis.

Figure 16.18(b) shows the voltage transfer characteristics for the depletion-load inverter. As discussed, the output voltage is 5 V in the high state, which is independent of the body effect. However, the characteristics during the transition region are a function of the body effect.

16.1.5　Transient Analysis of NMOS Inverters

Figure 16.19 shows three NMOS inverters in cascade with generalized load devices. The output of the first inverter sees an effective capacitance looking into the gate of M_{D2}, and the output of the second inverter sees an effective capacitance looking into the gate of M_{D3}. The capacitances are C_{T2} and C_{T3}, respectively, and they include the transistor input capacitances, as well as any parasitic capacitances due to the interconnect lines between the inverter stages.

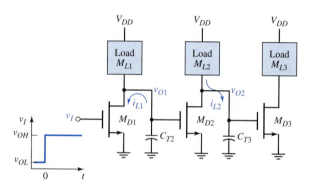

Figure 16.19　Three generalized NMOS inverters in cascade, for calculating transient effects

If input v_I is switched from high to low at time $t = 0$, driver M_{D1} cuts off, and output v_{O1} begins to go high. The effective load capacitance C_{T2} must then be charged by a current through load device M_{L1}. As v_{O1} increases, M_{D2} turns on and v_{O2} begins to go low. For v_{O2} to go low, the effective load capacitance C_{T3} must discharge. The discharge current i_{L2} is the difference between the driver current in M_{D2} and the load current in M_{L2}.

The rate at which capacitance C_{T2} charges is a function of the current–voltage characteristics of the load device. Figure 16.20 shows the characteristics of the driver transistor M_{D1}, with superimposed load curves for the three basic load components. These load curves neglect the body effect.

The load devices are assumed to be scaled such that $I_{D,\max}$ is the same for each device. The constant current over a wide range of v_{DS} provided by the depletion load implies that this type of inverter will switch a capacitive load more rapidly than the other two types of inverter configurations. The rate at which the voltage across a load capacitance changes is a direct function of the current through the capacitance.

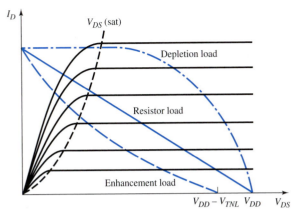

Figure 16.20   Driver transistor characteristics and load curves for the three types of NMOS inverters

Figure 16.21 shows the enhancement- and depletion-load inverter load curves that take the body effect into account. For the enhancement-load inverter, the resulting high output voltage is less than that when the body effect is neglected, as previously determined. Also, the current in the inverter is less over the entire voltage range, which implies that the switching times will be longer. For the depletion-load inverter, the resulting current is not a constant over a wide voltage range, and it is less than that when the body effect is neglected. Therefore, the time required to charge the load capacitance is longer.

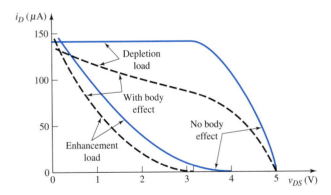

Figure 16.21   Load curves, enhancement-load and depletion-load NMOS inverters, with and without body effect

Figure 16.22 shows the switching characteristics of NMOS inverters with depletion-loads as determined from a PSpice analysis. A series of inverters, such as shown in Figure 16.19, was used. The width-to-length ratio of the load devices was $(W/L)_L = 1$ and that of the driver transistors was $(W/L)_D = 4$. The effective load capacitances were assumed to be 0.5 pF, which is larger than would normally be encountered in an IC.

Figure 16.22(a) shows the input (v_{O1}) and output (v_{O2}) voltage characteristics when the body effect is neglected. The fall time is relatively short, since the load capacitance discharges through the larger driver transistor. The rise

(a)

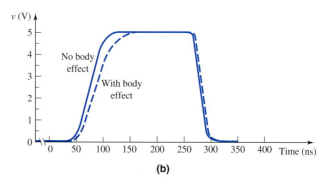

(b)

Figure 16.22 (a) Switching characteristics of an NMOS inverter with depletion load and (b) switching characteristics with and without body effect

time is longer, since the load capacitance is charged by the current through the smaller load transistor. The propagation delay times are shown in the figure.

The rise and fall times of the inverter with and without the body effect are shown in Figure 16.22(b). The rise time is longer when the body effect is taken into account, since the current in the load device is smaller. This was shown by the load curve in Figure 16.21. The fall time is not affected by the body effect, since the substrate of the driver is connected to the source terminal, which is at ground potential.

16.2   NMOS LOGIC CIRCUITS

NMOS logic circuits are formed by combining driver transistors in parallel, series, or series–parallel combinations to produce a desired output logic function.

16.2.1   NMOS NOR and NAND Gates

The NMOS NOR logic gate contains additional driver transistors connected in parallel. Figure 16.23 shows a two-input NMOS NOR logic gate with a depletion load. If $A = B = $ logic 0, then both M_{DA} and M_{DB} are cut off and $v_O = V_{DD}$. If $A = $ logic 1 and $B = $ logic 0, then M_{DB} is cut off and the NMOS inverter configuration with M_L and M_{DA} is the same as previously considered,

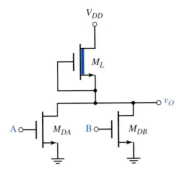

Figure 16.23 Two-input NMOS NOR logic gate with depletion load

and the output voltage goes low. Similarly, if $A = $ logic 0 and $B = $ logic 1, we again have the same inverter configuration.

If $A = B = $ logic 1, then both M_{DA} and M_{DB} turn on and the two driver transistors are effectively in parallel. The value of the output voltage now changes slightly. Figure 16.24 shows the NOR gate when both input voltages are a logic 1. From our previous analysis, we can assume that the two driver transistors are biased in the nonsaturation region and the load device is biased in the saturation region. We then have

$$i_{DL} = i_{DA} + i_{DB}$$

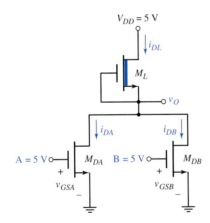

Figure 16.24 Two-input NMOS NOR logic gate for Example 16.7

which in general terms can be written

$$K_L[v_{GSL} - V_{TNL}]^2 = K_{DA}\left[2(v_{GSA} - V_{TNA})v_{DSA} - v_{DSA}^2\right]$$
$$+ K_{DB}\left[2(v_{GSB} - V_{TNB})v_{DSB} - v_{DSB}^2\right] \qquad \textbf{(16.42)}$$

If we assume the two driver transistors are identical, then the driver conduction parameters and threshold voltages are also identical, or $K_{DA} = K_{DB} \equiv K_D$ and $V_{TNA} = V_{TNB} \equiv V_{TND}$. Noting that $v_{GSL} = 0$, $v_{GSA} = v_{GSB} = V_{DD}$, and $v_{DSA} = v_{DSB} = v_O$, we can write Equation (16.42) as

$$[-V_{TNL}]^2 = 2\left(\frac{K_D}{K_L}\right)\left[2(V_{DD} - V_{TND})v_O - v_O^2\right] \tag{16.43}$$

Equation (16.43) shows that when both drivers are conducting, the effective width-to-length ratio of the composite driver transistor doubles. This means that the output voltage becomes slightly smaller when both inputs are high.

Example 16.7 Objective: Determine the low output voltage of an NMOS NOR circuit.

Consider the NOR circuit in Figure 16.24 biased at $V_{DD} = 5\,\text{V}$. Assume that $k_n' = 35\,\mu\text{A/V}^2$. Also assume the width-to-length ratios of the load and driver transistors are $(W/L)_L = 1$ and $(W/L)_D = 4$, respectively. Let $V_{TND} = 0.8\,\text{V}$ and $V_{TNL} = -2\,\text{V}$. Neglect the body effect.

Solution: If, for example, $A = \text{logic } 1 = 5\,\text{V}$ and $B = \text{logic } 0$, then M_{DB} is cut off. The output voltage is determined from Equation (16.29(b)), which is

$$\frac{K_D}{K_L}[2(v_I - V_{TND})v_O - v_O^2] = (-V_{TNL})^2$$

or

$$\left(\frac{4}{1}\right)[2(5 - 0.8)v_O - v_O^2] = (2)^2$$

The output voltage is found to be

$$v_O = 0.121\,\text{V}$$

If both inputs go high, then

$$A = B = V_{DD} = 5\,\text{V}$$

and the output voltage can be found using Equation (16.43), which is

$$[-V_{TNL}]^2 = 2\left(\frac{K_D}{K_L}\right)\left[2(V_{DD} - V_{TND})v_O - v_O^2\right]$$

or

$$(2)^2 = 2\left(\frac{4}{1}\right)[2(5 - 0.8)v_O - v_O^2]$$

The output voltage is found to be

$$v_O = 0.060\,\text{V}$$

Comment: An NMOS NOR gate must be designed to achieve a specified V_{OL} output voltage when only one input is high. This will give the largest logic 0 value. When more than one input is high, the output voltage is smaller than the specified V_{OL} value, since the effective width-to-length ratio of the composite driver transistor increases.

The NMOS NAND logic gate contains additional driver transistors connected in series. Figure 16.25 shows a two-input NMOS NAND logic gate with a depletion load. If both $A = B = \text{logic } 0$, or if either A or B is a logic 0, at least

Figure 16.25 Two-input NMOS NAND logic gate with depletion load

one driver is cut off, and the output is high. If both $A = B = $ logic 1, then the composite driver of the NMOS inverter conducts and the output goes low.

Since the gate-to-source voltages of M_{DA} and M_{DB} are not equal, determining the actual voltage V_{OL} of a NAND gate is difficult. The drain-to-source voltages of M_{DA} and M_{DB} must adjust themselves to produce the same current. In addition, if the body effect is also included, the analysis becomes even more difficult. Since the two driver transistors are in series, a good approximation assumes that the width-to-length ratio of the drivers must be twice that of a single driver in an NMOS inverter to achieve a given V_{OL} value.

The composite width-to-length ratios of the driver transistors in the two-input NMOS NOR and NAND gates are shown schematically in Figure 16.26. For the NOR gate, the effective *width* doubles; for the NAND gates, the effective *length* doubles.

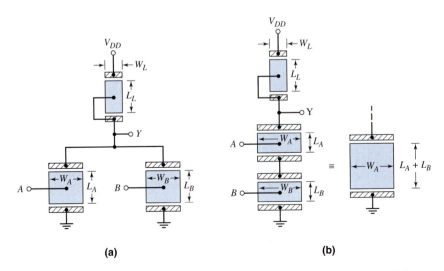

(a) (b)

Figure 16.26 Composite width-to-length ratios of driver transistors in two-input NMOS logic configurations (a) NOR and (b) NAND

Example   16.8   **Objective:**   Determine the low output voltage of an NMOS NAND circuit.

Consider the NAND circuit in Figure 16.25 biased at $V_{DD} = 5\,\text{V}$. Assume $k_n' = 35\,\mu\text{A/V}^2$. Also assume the width-to-length ratio of the load transistor is $(W/L)_L = 1$. Let $V_{TND} = 0.8\,\text{V}$ and $V_{TNL} = -2\,\text{V}$. Neglect the body effect.

Solution: From a PSpice analysis for $A = B = $ logic $1 = 5\,\text{V}$, the output voltage is $0.060\,\text{V}$ when the width-to-length ratio of each driver transistor is $(W/L)_D = 16$.

This result correlates very well with the results of Example 16.7. For the two-input NOR gate, the effective width of the composite driver doubles, or $(W/L)_C = 2 \times 4 = 8$, which results in an output voltage of $0.060\,\text{V}$. For the two input NAND gate, the effective length of the composite driver doubles, or $(W/L)_C = (1/2) \times 16 = 8$, which also results in an output voltage of $0.060\,\text{V}$.

Comment: If an N-input NMOS NAND logic gate were to be fabricated, then the width-to-length ratio of the drivers would need to be N times that of a single driver in an NMOS inverter to achieve a given value of V_{OL}. The increase in the required area of the driver transistors in a NAND logic gate means that logic gates with more than three or four inputs are not attractive.

Test Your Understanding

[Note: In the following exercises, assume $k_n' = 35\,\mu\text{A/V}^2$ for all NMOS transistors.]

*16.8   Consider the two-input NMOS NOR logic gate shown in Figure 16.23 biased at $V_{DD} = 10\,\text{V}$. Let $(W/L)_L = 2$, $(W/L)_D = 10$, $V_{TND} = 1.5\,\text{V}$, and $V_{TNL} = -3\,\text{V}$. Neglect the body effect. (a) Determine V_{OL} when: (i) $A = $ logic 1, $B = $ logic 0, and (ii) $A = B = $ logic 1. (b) Calculate the power dissipation in the circuit when: (i) $A = $ logic 1, $B = $ logic 0, and (ii) $A = B = $ logic 1. (Ans. (a) $V_{OL} = 0.107\,\text{V}$, $V_{OL} = 0.0531\,\text{V}$ (b) $P = 3.15\,\text{mW}$)

D16.9   Design a three-input NMOS NOR logic gate with depletion load such that $V_{OL}(\text{max}) = 0.12\,\text{V}$. Let $V_{DD} = 5\,\text{V}$, $V_{TND} = 0.8\,\text{V}$, and $V_{TNL} = -1.4\,\text{V}$. The maximum power dissipation in the circuit must be $0.8\,\text{mW}$. Determine (W/L) for the load and (W/L) for each driver. (Ans. $(W/L)_L = 4.66$, $(W/L)_D = 9.20$)

D16.10   Repeat Exercise 16.9 for a three-input NMOS NAND logic gate with depletion load. (Ans. $(W/L)_L = 4.66$, $(W/L)_D = 27.6$)

16.2.2   NMOS Logic Circuits

The series–parallel combination of drivers can be expanded to synthesize more complex logic functions. Consider the circuit in Figure 16.27. We can show that the Boolean output function is

$$f = \overline{(A \cdot B + C)}$$

Also, the individual transistor width-to-length ratios shown produce an effective K_D/K_L ratio of 4 for an effective single inverter when only M_{DA} and M_{DB} are conducting, or only M_{DC} is conducting. The actual complexity of the Boolean function is limited since the required width-to-length ratios of individual transistors may become unreasonably large.

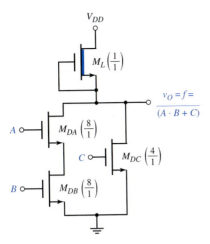

Figure 16.27   NMOS logic circuit example

Two additional logic functions are the exclusive-OR and exclusive-NOR. Figure 16.28 shows a circuit configuration that produces the exclusive-OR function. If $A = B =$ logic 1, a path exists from the output to ground through drivers M_{DA} and M_{DB}, and the output goes low. Similarly, if $A = B =$ logic 0, which means that $\bar{A} = \bar{B} =$ logic 1, a path exists from the output to ground through the drivers $M_{D\bar{B}}$ and $M_{D\bar{A}}$, and the output gain goes low. For all other input logic signal combinations, the output is isolated from ground so the output goes high.

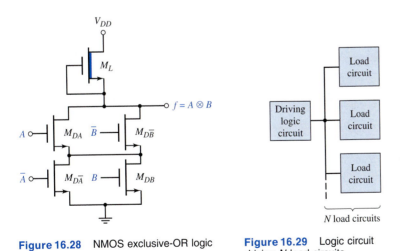

Figure 16.28   NMOS exclusive-OR logic gate

Figure 16.29   Logic circuit driving N load circuits

16.2.3   Fanout

An NMOS inverter or NMOS logic gate must be capable of driving more than one load, as shown in Figure 16.29. It is assumed that each load is identical to the driver logic circuit. The number of identical-load circuits connected to the output of a driver logic circuit is defined as the **fanout**. For MOS logic circuits, the inputs to the load circuits are the oxide-insulated gates of the MOS transistors; therefore, the static loading caused by multiple driver loads is so small that the dc transfer curve is essentially identical to a no-load condition. The dc characteristics of MOS logic circuits are unaffected by the fanout to other MOS logic inputs. However, the load capacitance due to a large fanout seriously degrades the switching speed and propagation delay times. Consequently, maintaining the propagation delay time below a specified maximum value determines the fanout of MOS digital circuits.

Test Your Understanding

[Assume $k_n' = 35\,\mu\text{A/V}^2$.]

*16.11   Consider the NMOS logic circuit in Figure 16.30. Let $V_{TN} = 0.7\,\text{V}$ for each transistor and assume all driver transistors are identical. (a) If $(W/L)_L = 0.5$, determine (W/L) for the drivers such that $V_{OL}(\text{max}) = 0.15\,\text{V}$. (b) Determine the maximum power dissipation in the logic circuit. (Ans. (a) $(W/L)_D = 13.6$ (b) $P = 753\,\mu\text{W}$)

16.12   Repeat Exercise 16.11 for the NMOS logic circuit in Figure 16.31, except assume that the threshold voltage of the load device is $V_{TN} = -1.2\,\text{V}$. (Ans. (a) $(W/L)_D = 1.14$ (b) $P = 63\,\mu\text{W}$)

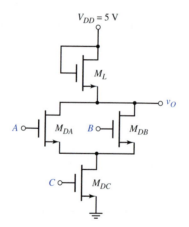

Figure 16.30   Figure for Exercise 16.11

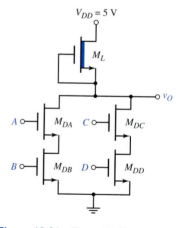

Figure 16.31   Figure for Exercise 16.12

16.3   CMOS INVERTER

Complementary MOS, or CMOS, circuits contain both n-channel and p-channel MOSFETs. As we will see, the power dissipation in CMOS logic circuits is much smaller than in NMOS circuits, which makes CMOS very attractive. We briefly review the characteristics of p-channel transistors, and will then analyze the CMOS inverter, which is the basis of most CMOS logic circuits. We will examine the CMOS NOR and NAND gates and other basic CMOS logic circuits, covering power dissipation, noise margin, fanout, and switching characteristics.

16.3.1 p-Channel MOSFET Revisited

Figure 16.32 shows a simplified view of a p-channel MOSFET. The p- and n-regions are reversed from those in an n-channel device. Again, the channel length is L and the channel width is W. Usually in any given fabrication process, the channel length is the same for all devices, so the channel width W is the variable in logic circuit design.

Figure 16.32 Simplified cross section of p-channel MOSFET

Figure 16.33(a) shows the simplified circuit symbol for the p-channel enhancement-mode device. When the body or substrate connection is needed, we will use the symbol shown in Figure 16.33(b). Usually, the p-channel depletion-mode device is not used in CMOS digital circuits; therefore, it is not addressed here.

(a)   **(b)**

Figure 16.33 (a) Simplified circuit symbol, p-channel enhancement-mode MOSFET and (b) circuit symbol showing substrate connection

Normally, in an integrated circuit, more than one p-channel device will be fabricated in the same n-substrate so the p-channel transistors will exhibit a body effect. The n-substrate is connected to the most positive potential. The source terminal may be negative with respect to the substrate; therefore, voltage V_{BS} may exist between the body and the source. The threshold voltage is

$$V_{TP} = V_{TPO} - \frac{\sqrt{2e\varepsilon_s N_d}}{C_{ox}}\left[\sqrt{2\phi_{fn} + V_{BS}} - \sqrt{2\phi_{fn}}\right]$$

$$= V_{TPO} - \gamma\left[\sqrt{2\phi_{fn} + V_{BS}} - \sqrt{2\phi_{fn}}\right]$$

(16.44)

where V_{TPO} is the threshold voltage for zero body-to-source voltage, or $V_{BS} = 0$. The parameter N_d is the n-substrate doping concentration and ϕ_{fn} is a potential related to the substrate doping. The parameter γ is the body effect coefficient.

The current–voltage characteristic of the p-channel MOSFET are functions of both the electrical and geometric properties of the device. When the transistor is biased in the nonsaturation region, we have $v_{SD} \leq v_{SG} + V_{TP}$. Therefore,

$$i_D = K_P\left[2(v_{SG} + V_{TP})v_{SD} - v_{SD}^2\right] \tag{16.45(a)}$$

In the saturation region, we have $v_{SD} \geq v_{SG} + V_{TP}$, which means that

$$i_D = K_P(v_{SG} + V_{TP})^2 \tag{16.45(b)}$$

The gate potential is negative with respect to the source. For the p-channel transistor to conduct, we must have $v_{GS} < V_{TP}$, where V_{TP} is negative for an enhancement-mode device. We also see that $v_{SG} > |V_{TP}|$ when the p-channel device is conducting.

In most cases, the channel length modulation factor λ has very little effect on the operating characteristics of MOS digital circuits. Therefore, the term λ is assumed to be zero unless otherwise stated.

The transition point, which separates the nonsaturation and saturation bias regions, is given by

$$v_{SD} = v_{SD}(\text{sat}) = v_{SG} + V_{TP} \tag{16.46}$$

The parameter K_P is the conduction parameter and is given by

$$K_P = \left(\frac{1}{2}\mu_p C_{ox}\right)\left(\frac{W}{L}\right) = \frac{k_p'}{2}\frac{W}{L} \tag{16.47}$$

As before, the hole mobility μ_p and oxide capacitance C_{ox} are assumed to be constant for all devices. The hole mobility in p-channel silicon MOSFETs is approximately one-half the electron mobility μ_n in n-channel silicon MOSFETs. This means that a p-channel device width must be approximately twice as large as that of an n-channel device in order that the two devices be electrically equivalent (that is, that they have the same conduction parameter values).

Small Geometry Effects

The same small geometry effects apply to the p-channel devices as we discussed for the n-channel devices in Section 16.1.1. As with the NMOS inverters and logic circuits, we can use Equations (16.45(a)), (16.45(b)), and (16.46) as first-order equations in the design of NMOS logic circuits. The basic operation and behavior of CMOS logic circuits can be predicted using these first-order equations.

16.3.2   DC Analysis of the CMOS Inverter

The **CMOS inverter**, shown in Figure 16.34, is a series combination of a p-channel and an n-channel MOSFET. The gates of the two MOSFETs are connected together to form the input and the two drains are connected together

Figure 16.34 CMOS inverter

to form the output. Both transistors are enhancement-mode devices. The parameters of the NMOS are denoted by K_N and V_{TN}, where $V_{TN} > 0$, and the parameters of the PMOS are denoted by K_P and V_{TP}, where $V_{TP} < 0$.

Figure 16.35 shows a simplified cross section of a CMOS inverter. In this process, a separate p-well region is formed within the starting n-substrate. The n-channel device is fabricated in the p-well region and the p-channel device is fabricated in the n-substrate. Although other approaches, such as an n-well in a p-substrate, are also used to fabricate CMOS circuits, the important point is that the processing is more complicated for CMOS circuits than for NMOS circuits. However, the advantages of CMOS digital logic circuits over NMOS circuits justify their use.

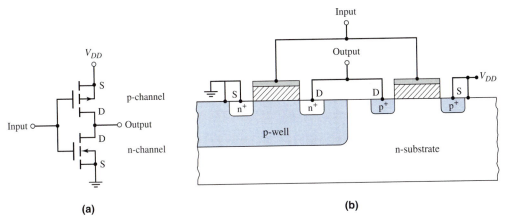

(a) (b)

Figure 16.35 Simplified cross section, CMOS inverter

Voltage Transfer Curve

Figure 16.36 shows the transistor characteristics for both the n- and p-channel devices. We can determine the voltage transfer characteristics of the inverter by evaluating the various transistor bias regions. For $v_I = 0$, the NMOS device is cut off, $i_{DN} = 0$, and $i_{DP} = 0$. The PMOS source-to-gate voltage is V_{DD}, which

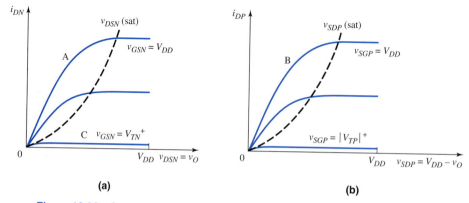

Figure 16.36   Current–voltage characteristics, (a) NMOS transistor and (b) PMOS transistor

means that the PMOS is biased on the curve marked B in Figure 16.36(b). Since the only point on the curve corresponding to $i_{DP} = 0$ occurs at $v_{SDP} = 0 = V_{DD} - v_O$, the output voltage is $v_O = V_{DD}$. This condition exists as long as the NMOS transistor is cut off, or $v_I \leq V_{TN}$.

For $v_I = V_{DD}$, the PMOS device is cut off, $i_{DP} = 0$, and $i_{DN} = 0$. The NMOS gate-to-source voltage is V_{DD} and the NMOS is biased on the curve marked A in Figure 16.36(a). The only point on the curve corresponding to $i_{DN} = 0$ occurs at $v_{DSN} = v_O = 0$. The output voltage is zero as long as the PMOS transistor is cut off, or $v_{SGP} = V_{DD} - v_I \leq |V_{TP}|$. This means that the input voltage is in the range $V_{DD} - |V_{TP}| \leq v_I \leq V_{DD}$.

Figure 16.37 shows the voltage transfer characteristics generated thus far for the CMOS inverter. The more positive output voltage corresponds to a logic 1, or $V_{OH} = V_{DD}$, and the more negative output voltage corresponds to a logic 0, or $V_{OL} = 0$. When the output is in the logic 1 state, the NMOS transistor is cut off; when the output is in the logic 0 state, the PMOS transistor is cut off.

Ideally, the current in the CMOS inverter in either steady-state condition is zero, which means that, ideally, the quiescent power dissipation is zero. This result is the attractive feature of CMOS digital circuits. In actuality, CMOS inverter circuits exhibit a small leakage current in both steady-state conditions,

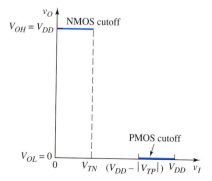

Figure 16.37   CMOS inverter output voltage for input voltage in either high state or low state

due to the reverse-biased pn junctions. However, the power dissipation may be in the nanowatt range rather than in the milliwatt range of NMOS inverters. Without this feature, VLSI would not be possible.

When the input voltage is just greater than V_{TN}, or

$$v_I = v_{GSN} = V_{TN}^+$$

the NMOS begins to conduct and the Q-point falls on the curve marked C in Figure 16.36(a). The current is small and $v_{DSN} \cong V_{DD}$, which means that the NMOS is biased in the saturation region. The PMOS source-to-drain voltage is small, so the PMOS is biased in the nonsaturation region. Setting $i_{DN} = i_{DP}$, we can write

$$K_N[v_{GSN} - V_{TN}]^2 = K_P[2(v_{SGP} + V_{TP})v_{SDP} - v_{SDP}^2] \qquad \textbf{(16.48)}$$

Relating the gate-to-source and drain-to-source voltages in each transistor to the inverter input and output voltages, respectively, we can rewrite Equation (16.48) as follows:

$$K_N[v_I - V_{TN}]^2 = K_P[2(V_{DD} - v_I + V_{TP})(V_{DD} - v_O) - (V_{DD} - v_O)^2]$$
$$\textbf{(16.49)}$$

Equation (16.49) relates the input and output voltages as long as the NMOS is biased in the saturation region and the PMOS is biased in the nonsaturation region.

The transition point for the PMOS is defined from

$$v_{SDP}(\text{sat}) = v_{SGP} + V_{TP} \qquad \textbf{(16.50)}$$

Using Figure 16.38, Equation (16.50) can be written

$$V_{DD} - V_{OPt} = V_{DD} - V_{IPt} + V_{TP} \qquad \textbf{(16.51(a))}$$

or

$$V_{OPt} = V_{IPt} - V_{TP} \qquad \textbf{(16.51(b))}$$

where V_{OPt} and V_{IPt} are the PMOS output and input voltages, respectively, at the transition point.

The transition point for the NMOS is defined from

$$v_{DSN}(\text{sat}) = v_{GSN} - V_{TN} \qquad \textbf{(16.52(a))}$$

or

$$V_{ONt} = V_{INt} - V_{TN} \qquad \textbf{(16.52(b))}$$

where V_{ONt} and V_{INt} are the NMOS output and input voltages, respectively, at the transition point.

On the basis that V_{TP} is negative for an enhancement-mode PMOS, Equations (16.51(b)) and (16.52(b)) are plotted in Figure 16.38. We determine the input voltage at the transition points by setting the two drain currents equal to each other when both transistors are biased in the saturation region. The result is

$$K_N(v_{GSN} - V_{TN})^2 = K_P(v_{SGP} + V_{TP})^2 \qquad \textbf{(16.53)}$$

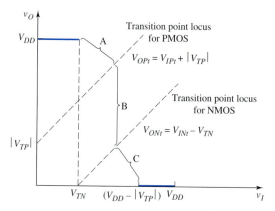

Figure 16.38 Regions of the CMOS transfer characteristics indicating NMOS and PMOS transistor bias conditions. The NMOS device is biased in the saturation region in areas A and B and in the nonsaturation region in area C. The PMOS device is biased in the saturation region in areas B and C and in the nonsaturation region in area A.

With the gate-to-source voltages related to the input voltage, Equation (16.53) becomes

$$K_N(v_I - V_{TN})^2 = K_P(V_{DD} - v_I + V_{TP})^2 \qquad (16.54)$$

For this ideal case, the output voltage does not appear in Equation (16.54), and the input voltage is a constant, as long as the two transistors are biased in the saturation region.

Voltage v_I from Equation (16.54) is the input voltage at the PMOS and NMOS transition points. Solving for v_I, we find that

$$v_I = v_{It} = \frac{V_{DD} + V_{TP} + \sqrt{\dfrac{K_N}{K_P}}V_{TN}}{1 + \sqrt{\dfrac{K_N}{K_P}}} \qquad (16.55)$$

For $v_I > V_{It}$, the NMOS is biased in the nonsaturation region and the PMOS is biased in the saturation region. Again equating the two drain currents, we have

$$K_N\big[2(v_{GSN} - V_{TN})v_{DSN} - v_{DSN}^2\big] = K_P(v_{SGP} + V_{TP})^2 \qquad (16.56)$$

Also, relating the gate-to-source and drain-to-source voltages to the input and output voltages, respectively, modifies Equation (16.56) as follows:

$$K_N\big[2(v_I - V_{TN})v_O - v_O^2\big] = K_P(V_{DD} - v_I + V_{TP})^2 \qquad (16.57)$$

Equation (16.57) relates the input and output voltages as long as the NMOS is biased in the nonsaturation region and the PMOS in the saturation region. Figure 16.39 shows the complete voltage transfer curve.

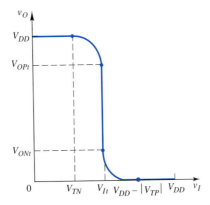

Figure 16.39   Complete voltage transfer characteristics, CMOS inverter

Example 16.9 **Objective:** Determine the critical voltages on the voltage transfer curve of a CMOS inverter.

Consider a CMOS inverter biased at $V_{DD} = 5$ V with transistor parameters of $K_N = K_P$ and $V_{TN} = -V_{TP} = 1$ V. Then consider another CMOS inverter biased at $V_{DD} = 10$ V with the same transistor parameters.

Solution: For $V_{DD} = 5$ V, the input voltage at the transition points is, from Equation (16.55),

$$V_{It} = \frac{5 + (-1) + \sqrt{1 \cdot 1}}{1 + \sqrt{1}} = 2.5 \text{ V}$$

The output voltage at the transition point for the PMOS is, from Equation (16.51(b)),

$$V_{OPt} = V_{IPt} - V_{TP} = 2.5 - (-1) = 3.5 \text{ V}$$

and the output voltage at the transition point for the NMOS is, from Equation (16.52(b)),

$$V_{ONt} = V_{INt} - V_{TN} = 2.5 - 1 = 1.5 \text{ V}$$

For $V_{DD} = 10$ V and the same transistor parameters, the critical voltages are:

$$V_{It} = 5 \text{ V} \qquad V_{OPt} = 6 \text{ V} \qquad V_{ONt} = 4 \text{ V}$$

Comment: The two voltage transfer curves are shown in Figure 16.40. The figure depicts another advantage of CMOS technology, that is, CMOS circuits can be biased over a wide range of voltages.

Test Your Understanding

16.13 The CMOS inverter in Figure 16.34 is biased at $V_{DD} = 10$ V, and the transistor threshold voltages are $V_{TN} = 2$ V and $V_{TP} = -2$ V. Sketch the voltage transfer curve and show the critical points, as in Figure 16.39, for: (a) $K_N/K_P = 1$, (b) $K_N/K_P = 0.5$, and (c) $K_N/K_P = 2$. (Ans. (a) $V_{It} = 5$ V, $V_{OPt} = 7$ V, $V_{ONt} = 3$ V (b) $V_{It} = 5.51$ V, $V_{OPt} = 7.51$ V, $V_{ONt} = 3.51$ V (c) $V_{It} = 4.49$ V, $V_{OPt} = 6.49$ V, $V_{ONt} = 2.49$ V)

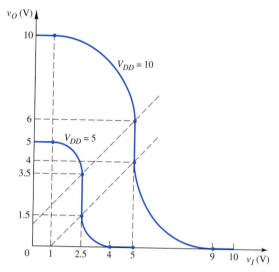

Figure 16.40 Voltage transfer characteristics, CMOS inverter biased at either $V_{DD} = 5\,\text{V}$ or $V_{DD} = 10\,\text{V}$

CMOS Inverter Currents

When the CMOS inverter input voltage is either a logic 0 or a logic 1, the current in the circuit is zero, since one of the transistors is cut off. When the input voltage is in the range $V_{TN} < v_I < V_{DD} - |V_{TP}|$, both transistors are conducting and a current exists in the inverter.

When the NMOS transistor is biased in the saturation region, the current in the inverter is controlled by v_{GSN} and the PMOS source-to-drain voltage adjusts such that $i_{DP} = i_{DN}$. This condition is demonstrated in Equation (16.48). We can write

$$i_{DN} = i_{DP} = K_N(v_{GSN} - V_{TN})^2 = K_N(v_I - V_{TN})^2 \qquad (16.58(a))$$

Taking the square root yields

$$\sqrt{i_{DN}} = \sqrt{i_{DP}} = \sqrt{K_N}\,(v_I - V_{TN}) \qquad (16.58(b))$$

As long as the NMOS transistor is biased in the saturation region, the square root of the CMOS inverter current is a linear function of the input voltage.

When the PMOS transistor is biased in the saturation region, the current in the inverter is controlled by v_{SGP} and the NMOS drain-to-source voltage adjusts such that $i_{DP} = i_{DN}$. This condition is demonstrated in Equation (16.56). Using Equation (16.57), we can write that

$$i_{DN} = i_{DP} = K_P(V_{DD} - v_I + V_{TP})^2 \qquad (16.59(a))$$

Taking the square root yields

$$\sqrt{i_{DN}} = \sqrt{i_{DP}} = \sqrt{K_P}\,(V_{DD} - v_I + V_{TP}) \qquad (16.59(b))$$

As long as the PMOS transistor is biased in the saturation region, the square root of the CMOS inverter current is also a linear function of the input voltage.

Figure 16.41 shows plots of the square root of the inverter current for two values of V_{DD} bias. These curves are quasi-static characteristics in that no

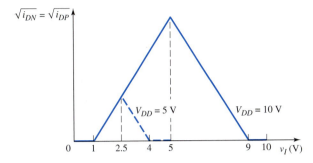

Figure 16.41 Square root of inverter current versus input voltage, CMOS inverter biased at either $V_{DD} = 5$ V or $V_{DD} = 10$ V

current is diverted into a capacitive load. At the inverter switching point, both transistors are biased in the saturation region and both transistors influence the current. At the switching point, the actual current characteristic does not have a sharp discontinuity in the slope. The channel length modulation parameter λ also influences the current characteristics at the peak value. However, the curves in Figure 16.41 are excellent approximations.

Test Your Understanding

16.14 Consider a CMOS inverter biased at $V_{DD} = 5$ V, with transistor threshold voltages of $V_{TN} = +0.8$ V and $V_{TP} = -0.8$ V. Calculate the peak current in the inverter for: (a) $K_N = K_P = 50\,\mu\text{A/V}^2$, and (b) $K_N = K_P = 200\,\mu\text{A/V}^2$. (Ans. (a) $i_D(\text{max}) = 145\,\mu\text{A}$ (b) $i_D(\text{max}) = 578\,\mu\text{A}$)

16.3.3 Power Dissipation

In the quiescent or static state, in which the input is either a logic 0 or a logic 1, power dissipation in the CMOS inverter is virtually zero. However, during the switching cycle from one state to another, current flows and power is dissipated. The CMOS inverter and logic circuits are used to drive other MOS devices for which the input impedance is a capacitance. During the switching cycle, then, this load capacitance must be charged and discharged.

In Figure 16.42(a), the output switches from its low to its high state. The input is switched low, the PMOS gate is at zero volts, and the NMOS is cut off. The load capacitance C_L must be charged through the PMOS device. Power dissipation in the PMOS transistor is given by

$$P_P = i_L v_{SD} = i_L(V_{DD} - v_O) \tag{16.60}$$

The current and the output voltage are related by

$$i_L = C_L \frac{dv_O}{dt} \tag{16.61}$$

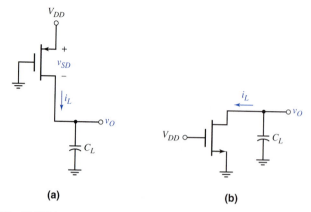

Figure 16.42   CMOS inverter when the output switches (a) low to high and (b) high to low

The energy dissipated in the PMOS device as the output switches from low to high is

$$E_P = \int_0^\infty P_P dt = \int_0^\infty C_L(V_{DD} - v_O)\frac{dv_O}{dt}dt$$
$$= C_L V_{DD}\int_0^{V_{DD}} dv_O - C_L\int_0^{V_{DD}} v_O dv_O \tag{16.62}$$

which yields

$$E_P = C_L V_{DD} v_O\Big|_0^{V_{DD}} - C_L\frac{v_O^2}{2}\Big|_0^{V_{DD}} = \frac{1}{2}C_L V_{DD}^2 \tag{16.63}$$

After the output has switched high, the energy stored in the load capacitance is $(\frac{1}{2})C_L V_{DD}^2$. When the inverter input goes high, the output switches low, as shown in Figure 16.42(b). The PMOS device is cut off, the NMOS transistor conducts, and the load capacitance discharges through the NMOS device. All the energy stored in the load capacitance is dissipated in the NMOS device. As the output switches from high to low, the energy dissipated in the NMOS transistor is

$$E_N = \tfrac{1}{2}C_L V_{DD}^2 \tag{16.64}$$

The total energy dissipated in the inverter during one switching cycle is therefore

$$E_T = E_P + E_N = \tfrac{1}{2}C_L V_{DD}^2 + \tfrac{1}{2}C_L V_{DD}^2 = C_L V_{DD}^2 \tag{16.65}$$

If the inverter is switched at frequency f, the power dissipated in the inverter is

$$P = fE_T = fC_L V_{DD}^2 \tag{16.66}$$

Equation (16.66) shows that the power dissipated in a CMOS inverter is directly proportional to the switching frequency and to V_{DD}^2. The drive in digital IC design is toward lower supply voltages, such as 3 V or less.

Example 16.10   Objective: Calculate the power dissipation in a CMOS inverter.

Consider a CMOS inverter with a load capacitance of $C_L = 2\,\text{pF}$ biased at $V_{DD} = 5\,\text{V}$. The inverter switches at a frequency of $f = 100\,\text{kHz}$.

Solution: From Equation (16.66), power dissipation in the CMOS inverter is

$$P = f C_L V_{DD}^2 = (10^5)(2 \times 10^{-12})(5)^2 \Rightarrow 5\,\mu\text{W}$$

Comment: Previously determined values of static power dissipation in NMOS inverters were on the order of $500\,\mu\text{W}$; therefore, power dissipation in a CMOS inverter is substantially smaller. In addition, in most digital systems, only a small fraction of the logic gates change state during each clock cycle; consequently, the power dissipation in a CMOS digital system is substantially less than in an NMOS digital system of similar complexity.

The power dissipation is proportional to V_{DD}^2. In some digital circuits, such as digital watches, the CMOS logic circuits are biased at $V_{DD} = 1.5\,\text{V}$, so the power dissipation is substantially reduced.

Test Your Understanding

16.15 A CMOS inverter is biased at $V_{DD} = 3\,\text{V}$. The inverter drives an effective load capacitance of $C_L = 0.5\,\text{pF}$. Determine the maximum switching frequency such that the power dissipation is limited to $P = 0.10\,\mu\text{W}$. (Ans. $f = 22.2\,\text{kHz}$)

16.3.4   Noise Margin

Figure 16.43 shows the general voltage transfer function of a CMOS inverter. The parameters V_{IH} and V_{IL} determine the noise margins and are defined as the points at which

$$\frac{dv_O}{dv_I} = -1 \tag{16.67}$$

which is the same as for the NMOS inverters. For $v_I \le V_{IL}$ and $v_I \ge V_{IH}$, the gain is less than unity and the output changes slowly with input voltage. However, when the input voltage is in the range $V_{IL} < v_I < V_{IH}$, the inverter gain is greater than unity, and the output signal changes rapidly with a change in the input voltage. This is the undefined range.

Point V_{IL} occurs when the NMOS is biased in the saturation region and the PMOS is biased in the nonsaturation region. The relationship between the input and output voltages is given by Equation (16.49). Taking the derivative with respect to v_I yields

$$2K_N[v_I - V_{TN}] = K_P\left[-2(V_{DD} - v_O) - 2(V_{DD} - v_I + V_{TP})\frac{dv_O}{dv_I}\right.$$
$$\left. - 2(V_{DD} - v_O)\left(-\frac{dv_O}{dv_I}\right)\right] \tag{16.68}$$

Figure 16.43 CMOS inverter voltage transfer characteristics with defined noise margins

Setting the derivative equal to −1, we have

$$K_N[v_I - V_{TN}] = -K_P[(V_{DD} - v_O) - (V_{DD} - v_I + V_{TP}) + (V_{DD} - v_O)]$$

(16.69)

Solving for v_O produces

$$v_O = V_{OHU} = \frac{1}{2}\left\{\left(1 + \frac{K_N}{K_P}\right)v_I + V_{DD} - \left(\frac{K_N}{K_P}\right)V_{TN} - V_{TP}\right\}$$

(16.70)

Combining Equations (16.70) and (16.49), we see that voltage V_{IL} is

$$v_I = V_{IL} = V_{TN} + \frac{(V_{DD} + V_{TP} - V_{TN})}{\left(\frac{K_N}{K_P} - 1\right)}\left[2\sqrt{\frac{\frac{K_N}{K_P}}{\frac{K_N}{K_P} + 3}} - 1\right]$$

(16.71)

If $K_N = K_P$, Equation (16.71) becomes indefinite, since a zero would exist in both the numerator and the denominator. However, when $K_N = K_P$, Equation (16.70) becomes

$$v_O = V_{OHU(K_N=K_P)} = \frac{1}{2}\{2v_I + V_{DD} - V_{TN} - V_{TP}\}$$

(16.72)

Substituting Equation (16.72) into Equation (16.49) yields a voltage V_{IL} of

$$v_I = V_{IL(K_N=K_P)} = V_{TN} + \frac{3}{8}(V_{DD} + V_{TP} - V_{TN})$$

(16.73)

for $K_N = K_P$.

Point V_{IH} occurs when the NMOS is biased in the nonsaturation region and the PMOS is biased in the saturation region. The relationship between the input and output voltages is given by Equation (16.57). Taking the derivative with respect to v_I yields

$$K_N\left[2(v_I - V_{TN})\frac{dv_O}{dv_I} + 2v_O - 2v_O\frac{dv_O}{dv_I}\right] = 2K_P(V_{DD} - v_I + V_{TP})(-1)$$

(16.74)

Setting the derivative equal to -1, we find that

$$K_N[-(v_I - V_{TN}) + v_O + v_O] = -K_P[V_{DD} - v_I + V_{TP}] \qquad \textbf{(16.75)}$$

The output voltage v_O is then

$$v_O = V_{OLU} = \frac{v_I\left(1 + \dfrac{K_N}{K_P}\right) - V_{DD} - \left(\dfrac{K_N}{K_P}\right)V_{TN} - V_{TP}}{2\left(\dfrac{K_N}{K_P}\right)} \qquad \textbf{(16.76)}$$

Combining Equations (16.76) and (16.57), yields voltage V_{IH} as

$$v_I = V_{IH} = V_{TN} + \frac{(V_{DD} + V_{TP} - V_{TN})}{\left(\dfrac{K_N}{K_P} - 1\right)}\left[\frac{2\dfrac{K_N}{K_P}}{\sqrt{3\dfrac{K_N}{K_P} + 1}} - 1\right] \qquad \textbf{(16.77)}$$

Again, if $K_N = K_P$, Equation (16.77) becomes indefinite, since a zero would exist in both the numerator and the denominator. However, when $K_N = K_P$, Equation (16.76) becomes

$$v_O = V_{OLU(K_N=K_P)} = \tfrac{1}{2}\{2v_I - V_{DD} - V_{TN} - V_{TP}\} \qquad \textbf{(16.78)}$$

Substituting Equation (16.78) into Equation (16.57) yields a voltage V_{IH} of

$$v_I = V_{IH(K_N=K_P)} = V_{TN} + \tfrac{5}{8}(V_{DD} + V_{TP} - V_{TN}) \qquad \textbf{(16.79)}$$

Example 16.11 **Objective:** Determine the noise margins of a CMOS inverter.

Consider a CMOS inverter biased at $V_{DD} = 5\,\text{V}$. Assume the transistors are matched with $K_N = K_P$ and $V_{TN} = -V_{TP} = 1\,\text{V}$.

Solution: From Equation (16.57), the input voltage at the transition points, or the inverter switching point, is 2.5 V. Since $K_N = K_P$, V_{IL} is, from Equation (16.73)

$$V_{IL} = V_{TN} + \tfrac{3}{8}(V_{DD} + V_{TP} - V_{TN}) = 1 + \tfrac{3}{8}(5 - 1 - 1) = 2.125\,\text{V}$$

Point V_{IH} is, from Equation (16.79)

$$V_{IH} = V_{TN} + \tfrac{5}{8}(V_{DD} + V_{TP} - V_{TN}) = 1 + \tfrac{5}{8}(5 - 1 - 1) = 2.875\,\text{V}$$

The output voltages at points V_{IL} and V_{IH} are determined from Equations (16.72) and (16.78), respectively. They are

$$V_{OHU} = \tfrac{1}{2}[2V_{IL} + V_{DD} - V_{TN} - V_{TP}]$$
$$= \tfrac{1}{2}[2(2.125) + 5 - 1 + 1] = 4.625\,\text{V}$$

and

$$V_{OLU} = \tfrac{1}{2}[2V_{IH} - V_{DD} - V_{TN} - V_{TP}]$$
$$= \tfrac{1}{2}[2(2.875) - 5 - 1 + 1] = 0.375 \text{ V}$$

The noise margins are therefore

$$\text{NM}_L = V_{IL} - V_{OLU} = 2.125 - 0.375 = 1.75 \text{ V}$$

and

$$\text{NM}_H = V_{OHU} - V_{IH} = 4.625 - 2.875 = 1.75 \text{ V}$$

Comment: The results of this example are shown in Figure 16.43. Since the two transistors are electrically identical, the voltage transfer curve and the resulting critical voltages are symmetrical. Also, $(V_{OH} - V_{OHU}) = 0.375$ V, which is less than $|V_{TP}|$, and $(V_{OLU} - V_{OL}) = 0.375$ V, which is less than V_{TN}. As long as the input voltage remains within the limits of the noise margins, no logic error will be transmitted through the digital system.

Test Your Understanding

*16.16 A CMOS inverter is biased at $V_{DD} = 10$ V. The transistor parameters are: $V_{TN} = 2$ V, $V_{TP} = -2$ V, $K_N = 200 \,\mu\text{A/V}^2$, and $K_P = 80 \,\mu\text{A/V}^2$. (a) Sketch the voltage transfer curve. (b) Determine the critical voltages V_{IL} and V_{IH}, and the corresponding output voltages. (c) Calculate the noise margins NM_L and NM_H. (Ans. (b) $V_{IL} = 3.39$ V, $V_{IH} = 4.86$ V (c) $\text{NM}_L = 2.59$ V, $\text{NM}_H = 4.57$ V)

*16.17 Repeat Exercise 16.16 for a CMOS inverter biased at $V_{DD} = 5$ V with transistor parameters of: $V_{TN} = +0.8$ V, $V_{TP} = -2$ V, and $K_N = K_P = 100 \,\mu\text{A/V}^2$. (Ans. (b) $V_{IL} = 1.63$ V, $V_{IH} = 2.18$ V (c) $\text{NM}_L = 1.35$ V, $\text{NM}_H = 2.55$ V)

16.4 CMOS LOGIC CIRCUITS

Large-scale integrated CMOS circuits are used extensively in digital systems, including watches, calculators, and microprocessors. We will look at the basic CMOS NOR and NAND gates, and will then analyze more complex CMOS logic circuits. Since there is no clock signal applied to these logic circuits, they are referred to as **static CMOS logic** circuits.

16.4.1 Basic CMOS NOR and NAND Gates

In the basic or classical CMOS logic circuits, the gates of a PMOS and an NMOS are connected together, and additional PMOS and NMOS transistors are connected in series or parallel to form specific logic circuits. Figure 16.44(a) shows a two-input CMOS NOR gate. The NMOS transistors are in parallel and the PMOS transistors are in series.

If $A = B = $ logic 0, then both M_{NA} and M_{NB} are cut off, and the current in the circuit is zero. The source-to-gate voltage of M_{PA} is V_{DD} but the current is zero; therefore, v_{SD} of M_{PA} is zero. This means that the source-to-gate voltage

A	B	v_O
0	0	V_{DD}
V_{DD}	0	0
0	V_{DD}	0
V_{DD}	V_{DD}	0

(a)   (b)

Figure 16.44   (a) Two-input CMOS NOR logic circuit and (b) truth table

of M_{PB} is also V_{DD}. However, since the current is zero, then v_{SD} of M_{PB} is also zero. The output voltage is therefore $v_O = V_{DD} = $ logic 1.

If the input signals are $A = $ logic $1 = V_{DD}$ and $B = $ logic $0 = 0$ V, then the source-to-gate voltage of M_{PA} is zero, and the current in the circuit is again zero. The gate-to-source voltage of M_{NA} is V_{DD} but the current is zero, so v_{DS} of M_{NA} is zero and $v_O = 0 = $ logic 0. This result also holds for the other two possible input conditions, since at least one PMOS is cut off and at least one NMOS is in a conducting state. The NOR logic function is shown in the truth table of Figure 16.44(b).

A two-input CMOS NAND logic gate is shown in Figure 16.45(a). In this case, the NMOS transistors are in series and the PMOS transistors are in parallel. If $A = B = $ logic 0, the two NMOS devices are cut off and the current in the circuit is zero. The source-to-gate voltage of each PMOS device is V_{DD}, which means that both PMOS transistors are in a conducting state. However, since the current is zero, v_{SD} for both M_{PA} and M_{PB} is zero and $v_O = V_{DD}$. This result applies if at least one input is a logic 0.

If the input signals are $A = B = $ logic $1 = V_{DD}$, then both PMOS transistors are cut off, and the current in the circuit is zero. With $A = $ logic 1, M_{NA} is in a conducting state; however, since the current is zero, then v_{DS} of M_{NA} is zero. This means that the gate-to-source voltage of M_{NB} is also V_{DD} and M_{NB} is also in a conducting state. However, since the current is zero, then v_{DS} of M_{NB} is zero, and $v_O = $ logic $0 = 0$ V. The NAND logic function is shown in the truth table in Figure 16.45(b).

In both the CMOS NOR and NAND logic gates, the current in the circuit is essentially zero when the inputs are in any quiescent state. Only very small reverse-bias pn junction currents exist. The quiescent power dissipation is therefore essentially zero. Again, this is the primary advantage of CMOS circuits.

To obtain symmetrical switching times or propagation delay times for the high-to-low and low-to-high output transitions, the effective conduction parameters of the composite PMOS and composite NMOS devices must be equal. For the CMOS NOR logic gate in Figure 16.44(a), we can write

$$K_{CN} = K_{CP} \tag{16.80}$$

Figure 16.45   (a) Two-input CMOS NAND logic circuit and (b) truth table

where K_{CN} is the effective conduction parameter of the two parallel NMOS devices and K_{CP} is the effective conduction parameter of the two series PMOS transistors.

The effective channel *width* of the parallel NMOS devices is twice the individual width; similarly, the effective channel *length* of the series PMOS devices is twice the individual length. Equation (16.80) is then

$$\frac{k_n'}{2}\left(\frac{2W}{L}\right)_N = \frac{k_p'}{2}\left(\frac{W}{2L}\right)_P \tag{16.81}$$

since $k_n' \approx 2k_p'$, Equation (16.81) becomes

$$2\left(\frac{2W}{L}\right)_N = \left(\frac{W}{2L}\right)_P \tag{16.82(a)}$$

or

$$\left(\frac{W}{L}\right)_P = 8\left(\frac{W}{L}\right)_N \tag{16.82(b)}$$

This equation states that in the two-input CMOS NOR gate, the width-to-length ratio of the PMOS transistors must be approximately eight times that of the NMOS devices in order to provide the current required for symmetrical switching.

Figure 16.46 shows the voltage transfer and current characteristics of the two-input CMOS NOR logic gate from a PSpice analysis. Figure 16.46(a) shows the results when the two inputs are tied together; the circuit then behaves as a composite inverter. When the width-to-length ratios of the PMOS devices are eight times those of the NMOS devices, the transfer characteristics are symmetrical as expected.

Also shown are the voltage transfer and current characteristics if the p-channel width-to-length ratios are only twice as large as those of the n-channel devices. The resulting switching characteristics are not symmetrical and the maximum current is significantly reduced, implying that the switching times would be longer for this case.

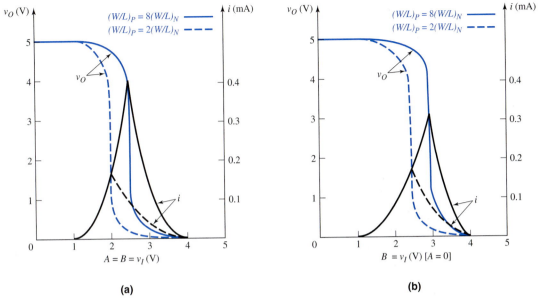

Figure 16.46 Voltage transfer characteristics, two-input CMOS NOR logic circuit for various width-to-length ratios: (a) $A = B = v_I$ and (b) $A = 0$, $B = v_I$

Figure 16.46(b) shows the transfer characteristics for input A held low and input B varied from zero to V_{DD}. This corresponds to only one input changing states while the other input remains constant. In this case, transistor M_{PA} is on and transistor M_{NA} is cut off. The resulting circuit is essentially equivalent to a simple CMOS inverter. When the width-to-length ratios of the PMOS devices are twice as large as those of the NMOS devices, the transfer characteristics are almost symmetrical, as we would expect for the CMOS inverter. When the PMOS width-to-length ratios are increased, the voltage transfer characteristic is no longer symmetrical. However, the maximum current is larger since the PMOS devices are larger, and this shortens the switching times. These results show that we obtain good transfer characteristics when $(W/L)_P = 8(W/L)_N$ for the two-input CMOS NOR circuit.

Test Your Understanding

[Note: In the following Exercises, let $k'_n = 35\,\mu A/V^2$ and $k'_p = 17.5\,\mu A/V^2$.]

16.18 Consider the two-input CMOS NOR gate in Figure 16.44(a). Assume $V_{TN} = 1\,V$, $V_{TP} = -1\,V$, and $V_{DD} = 5\,V$. (a) For $A = $ logic 0 and $(W/L)_P = 8(W/L)_N$, plot the voltage transfer curve v_O versus B. Show all critical voltages. (b) If $A = $ logic 0, determine $(W/L)_P$ and $(W/L)_N$ such that the peak current in the logic circuit is $50\,\mu A$. (Ans. (a) $V_{It} = 2.76\,V$, $V_{OPt} = 3.76\,V$, $V_{ONt} = 1.76\,V$ (b) $(W/L)_N = 0.923$, $(W/L)_P = 8(W/L)_N = 7.38$)

D16.19 In the two-input CMOS NAND gate in Figure 16.45(a), determine the relationship between the (W/L) ratios of the n-channel and p-channel transistors such that the composite conduction parameters of the PMOS and NMOS devices are equal. (Ans. $(W/L)_N = 2(W/L)_P$)

D16.20 Design a three-input CMOS NOR logic gate such that the effective conduction parameters of the composite PMOS and NMOS transistors are equal. Determine $(W/L)_P/(W/L)_N$, where (W/L) is the width-to-length ratio of the individual PMOS and NMOS transistors. (Ans. $(W/L)_P = 18(W/L)_N$)

D16.21 Repeat Exercise 16.20 for a three-input CMOS NAND logic gate. (Ans. $(W/L)_N = 4.5(W/L)_P$)

16.4.2 Complex CMOS Logic Circuits

Just as with NMOS logic designs, we can form complex logic gates in CMOS, which avoids connecting large numbers of NOR, NAND, and inverter gates to implement the logic function. There are formal methods that can be used to implement the logic circuit. However, we can use the knowledge gained in the analysis and design of the NOR and NAND circuits.

Design Example 16.12   Objective: Design a CMOS logic circuit to implement a particular logic function.

Implement the logic function $Y = AB + C(D + E)$ in a CMOS design. The signals A, B, C, D, and E are available.

Design Approach: The general CMOS design is shown in Figure 16.47, in which the inputs are applied to both the PMOS and NMOS networks. We may start the design by considering the NMOS portion of the circuit. To implement a basic OR (NOR) function, the n-channel transistors are in parallel (Figure 16.44) and to implement a basic AND (NAND) function, the n-channel transistors are in series (Figure 16.45). We will consider whether the function or its complement is generated at the end of the design.

Solution: NMOS Design: In the overall function, we note the logic OR between the functions AB and $C(D + E)$, so that the NMOS devices used to implement AB will be in parallel with the NMOS devices used to implement $C(D + E)$. There is a logic AND between the inputs A and B, so that the NMOS devices with these inputs will be in

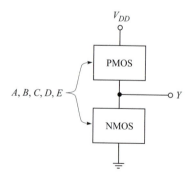

Figure 16.47 General CMOS design

series. Finally, the NMOS devices with the D and E inputs will be in parallel and this combination will be in series with the NMOS device with the C input. The NMOS implementation of the function is shown in Figure 16.48.

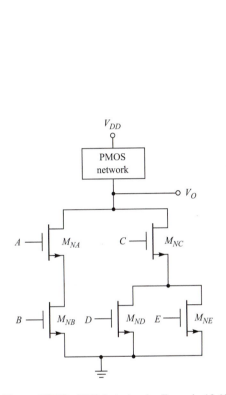

Figure 16.48   NMOS design for Example 16.12

Figure 16.49   Complete CMOS design for Example 16.12

Solution: PMOS Design: The arrangement of the PMOS devices is complementary to that of the NMOS devices. PMOS devices that perform the basic OR function are in series and PMOS devices that perform the basic AND function are in parallel. We then see that the PMOS devices used to implement AB will be in series with the devices used to implement $C(D + E)$. The two PMOS devices with the A and B inputs will be in parallel. The two PMOS devices with the D and E inputs will be in series and in turn will be in parallel with the PMOS device with the C input. The completed circuit is shown in Figure 16.49.

Final Solution: By considering various inputs, we may note that the output signal of the circuit shown in Figure 16.49 is actually the complement of the desired signal. We may then simply add a CMOS inverter to the output to obtain the desired function.

Comment: As mentioned, there are formal ways in which to design circuits. However, in many cases, these circuits can be designed by using the knowledge and intuition gained from previous work. The width-to-length ratios of the various transistors can be determined as we have done in previous examples.

Test Your Understanding

D16.22 Design the width-to-length ratios of the transistors in the static CMOS logic circuit of Figure 16.49 so that the composite conduction parameters are the same. Assume the minimum W/L ratio of an n-channel device is 1 and assume $\mu_n = 2\mu_p$. (Ans. For M_A–M_E: 2, 2, 2, 1, 1. For P_A–P_E: 2, 2, 2, 4, 4.)

D16.23 Design a static CMOS logic circuit that implements the logic function $Y = \overline{(ABC + DE)}$.

16.4.3   Fanout and Propagation Delay Time

Fanout

The term *fanout* refers to the number of load gates of similar design connected to the output of a driver gate. The maximum fanout is the maximum number of load gates that may be connected to the output. Since the CMOS logic gate will be driving other CMOS logic gates, the quiescent currrent required to drive the other CMOS gates is essentially zero. In terms of static characteristics, the maximum fanout is virtually limitless.

However, each additional load gate increases the load capacitance that must be charged and discharged as the driver gate changes state, and this places a practical limit on the maximum allowable number of load gates. Figure 16.50 shows a constant current charging a load capacitance. The voltage across the capacitance is

$$v_O = \frac{1}{C_L} \int I_O dt = \frac{I_O t}{C_L} \tag{16.83}$$

The load capacitance C_L is proportional to the number N of load gates and to the input gate capacitance of each load. The current I_O is proportional to the conduction parameter of the driver transistor. The switching time is therefore

$$t \propto \frac{N(W \cdot L)_L}{\left(\dfrac{W}{L}\right)_D} \tag{16.84}$$

where the gate capacitance is directly proportional to the gate area of the load $(W \cdot L)_L$, and the conduction parameter of the driver transistor is proportional to the width-to-length ratio. Equation (16.84) can be rewritten as

$$t \propto N(L_L L_D)\left(\frac{W_L}{W_D}\right) \tag{16.85}$$

The propagation delay time, which is proportional to the switching time, increases as the fanout increases. The propagation delay time could be reduced by increasing the size of the driver transistor. However, in any given driver logic circuit and load logic circuit, the sizes of the devices are generally fixed. Consequently, the maximum fanout is limited by the maximum acceptable propagation delay time.

Figure 16.50 Constant-current source charging a load capacitor

Propagation delay times are typically measured with a specified load capacitance. The average propagation delay time of a two-input CMOS NOR gate (such as an SN74HC36) is 25 ns, measured with a load capacitance of $C_L = 50\,\text{pF}$. Since the input capacitance is $C_I = 10\,\text{pF}$, a fanout of five would produce a 50 pF load capacitance. A fanout larger than five would increase the load capacitance, and would also increase the propagation delay time above the specified value.

Propagation Delay Time

Although the propagation delay time of the CMOS inverter can be determined by analytical techniques, it can also be determined by computer simulation. This is especially true when more complex CMOS logic circuits are considered. Using the appropriate transistor models in the simulation, the transient response can be produced. Obtaining an accurate transient response depends on using the correct transistor parameters. Some computer simulation problems in the end-of-chapter problems deal with propagation delay times. However, we will not go into detail here.

16.5   CLOCKED CMOS LOGIC CIRCUITS

The CMOS logic circuits considered in the previous section are called static circuits. One characteristic of a static CMOS logic circuit is that the output node always has a low-resistance path to either ground or V_{DD}. This implies that the output voltage is well defined and is never left floating.

Static CMOS logic circuits can be redesigned with an added clock signal while at the same time eliminating many of the PMOS devices. In general, the PMOS devices must be larger than NMOS devices. Eliminating as many PMOS devices as possible reduces the required chip area as well as the input capacitance. The low-power dissipation of the CMOS technology, however, is maintained.

Clocked CMOS circuits are dynamic circuits that generally precharge the output node to a particular level when the clock is at a logic 0. Consider the circuit in Figure 16.51. When the clock signal is low, or CLK = logic 0, M_{N1} is cut off and the current in the circuit is zero. Transistor M_{P1} is in a conducting state, but since the current is zero, then v_{O1} charges to V_{DD}. A high input to the CMOS inverter means that $v_O = 0$. During this phase of the clock signal, the gate of M_{P2} is precharged.

During the next phase, when the clock signal goes high, or CLK = logic 1, transistor M_{P1} cuts off and M_{N1} is biased in a conducting state. If input A = logic 0, then M_{NA} is cut off and there is no discharge path for voltage v_{O1}; therefore, v_{O1} remains charged at $v_{O1} = V_{DD}$. However, if CLK = logic 1 and A = logic 1, then both M_{N1} and M_{NA} are biased in a conducting state, providing a discharge path for voltage v_{O1}. As v_{O1} is pulled low, output signal v_O goes high.

The quiescent power dissipation in this circuit is essentially zero, as it was in the standard CMOS circuits. A small amount of power is required to precharge output v_{O1}, if it had been pulled low during the previous half clock cycle.

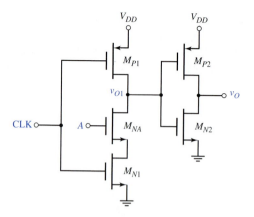

Figure 16.51   Simple clocked CMOS logic circuit

    The single NMOS transistor M_{NA} in Figure 16.51 can be replaced by a more complex NMOS logic circuit. Consider the two circuits in Figure 16.52. When CLK = logic 0, then M_{N1} cuts off and M_{P1} is in its conducting state in both circuits; then, v_{O1} is charged to $v_{O1} = V_{DD}$ and $v_O = 0$. For the circuit in Figure 16.52(a), when CLK = logic 1, voltage v_{O1} is discharged to ground or pulled low only when $A = B =$ logic 1. In this case, v_O goes high. The circuit in Figure 16.52(a) performs the AND function. Similarly, the circuit in Figure 16.52(b) performs the OR function.

    The advantage of the precharge technique is that it avoids the use of extensive pull-up networks: Only one PMOS and one NMOS transistor are required. This leads to an almost 50 percent savings in silicon area for larger circuits, and a reduction in capacitance resulting in higher speed. In addition, the static or quiescent power dissipation is essentially zero, so the circuit maintains the characteristics of CMOS circuits.

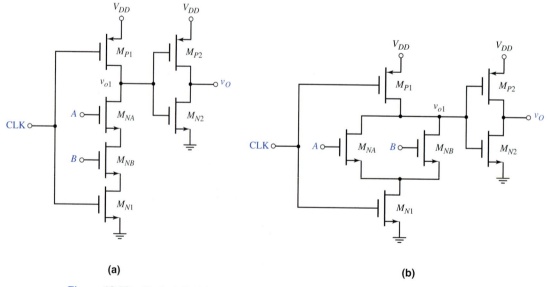

(a)   (b)

Figure 16.52   Clocked CMOS logic circuit: (a) AND function and (b) OR function

The AND and OR logic transistors M_{NA} and M_{NB} in Figures 16.52(a) and 16.52(b) can be replaced by a generalized logic network as indicated in Figure 16.53. The box marked f is an NMOS pull-down network that performs a particular logic function $f(X)$ of n variables, where $X = (x_1, x_2, \ldots, x_n)$. The NMOS circuit is a combination of series–parallel interconnections of n transistors. When the clock signal goes high, the CMOS inverter output is the logic function $f(X)$.

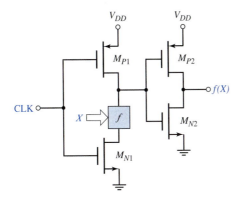

Figure 16.53 Generalized CMOS clocked logic circuit

The set of X inputs to the logic circuits f is derived from the outputs of other CMOS inverters and clocked logic circuits. This means that when CLK = logic 0, the outputs of all CMOS inverters are a logic 0 during the precharge cycle. As a result, all n variables $X = (x_1, x_2, \ldots, x_n)$ are a logic 0 during the precharge cycle. During this time, all NMOS transistors are cut off, which guarantees that output v_{O1} can be precharged to V_{DD}. There can then be only one possible transition at each node during the evaluation phase. The output of the CMOS buffer may change from a 0 to a 1.

An example of a cascaded domino CMOS circuit is shown in Figure 16.54. During the precharge cycle, in which CLK = logic 0, nodes 1 and 3 are charged high and nodes 2 and 4 are low. Also during this time, the inputs A, B, and C are all a logic 0. During the evaluation phase, in which CLK = logic 1, if $A = C = $ logic 1 and $B = $ logic 0, then node 1 remains charged high, $f_1 = $ logic 0, and node 3 discharges through M_{NC} causing f_2 to go high. However, if, during the evaluation phase, $A = B = $ logic 1 and $C = $ logic 0, then node 1 is pulled low causing f_1 to go high, which in turn causes node 3 to go low and forces node 4 high. This chain of actions thus leads to the term **domino circuit**.

Test Your Understanding

D16.24 Design a clocked CMOS domino logic circuit, such as shown in Figure 16.53, to generate an output $f(X) = A \cdot B \cdot C + D \cdot E$.

D16.25 Sketch a clocked CMOS logic circuit that realizes the exclusive OR function.

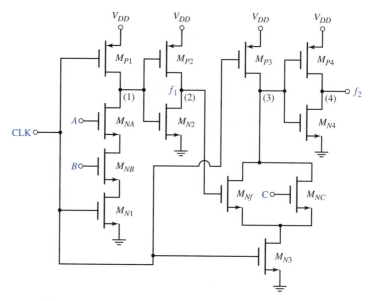

Figure 16.54   Cascaded clocked or domino CMOS logic circuit

16.6   TRANSMISSION GATES

Transistors can act as switches between driving circuits and load circuits. Transistors used to perform this function are called transmission gates. We will examine NMOS and CMOS transmission gates, which can also be configured to perform logic functions.

16.6.1   NMOS Transmission Gate

The NMOS enhancement-mode transistor in Figure 16.55(a) is a transmission gate connected to a load capacitance C_L, which could be the input gate capacitance of a MOS logic circuit. In this circuit, the transistor must be bilateral, which means it must be able to conduct current in either direction. This is a natural feature of MOSFETs. Terminals a and b are assumed to be equivalent, and the bias applied to the transistor determines which terminal acts as the drain and which terminal acts as the source. The substrate must be connected to the most negative potential in the circuit, which is usually ground.

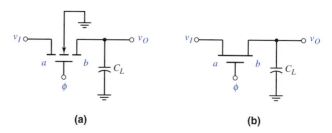

(a)   **(b)**

Figure 16.55   (a) NMOS transmission gate, showing substrate connection, and (b) simplified diagram

Figure 16.55(b) shows a simplified circuit symbol for the **NMOS transmission gate** that is used extensively.

We assume that the NMOS transmission gate is to operate over a voltage range of zero-to-V_{DD}. If the gate voltage ϕ is zero, then the n-channel transistor is cut off and the output is isolated from the input. The transistor is essentially an open switch.

If $\phi = V_{DD}$, $v_I = V_{DD}$, and v_O is initially zero, then terminal a acts as the drain since its bias is V_{DD}, and terminal b acts as the source since its bias is zero. Current enters the drain from the input, charging up the capacitor. The gate-to-source voltage is

$$v_{GS} = \phi - v_O = V_{DD} - v_O \tag{16.86}$$

As the capacitor charges and v_O increases, the gate-to-source voltage decreases. The capacitor stops charging when the current goes to zero. This occurs when the gate-to-source voltage v_{GS} becomes equal to the threshold voltage V_{TN}. The maximum output voltage occurs when $v_{GS} = V_{TN}$, therefore, from Equation (16.86), we have

$$v_{GS}(\text{min}) = V_{TN} = V_{DD} - v_O(\text{max}) \tag{16.87(a)}$$

or

$$v_O(\text{max}) = V_{DD} - V_{TN} \tag{16.87(b)}$$

where V_{TN} is the threshold voltage taking into account the body effect.

Equation (16.87(b)) demonstrates one disadvantage of an NMOS transmission gate. A logic 1 level degrades, or attenuates, as it passes through the transmission gate. However, this may not be a serious problem for many applications.

Figure 16.56 shows the quasi-static output voltage versus input voltage of the NMOS transmission gate. As seen in the figure, when $v_I = V_{DD}$, the output voltage is $v_O = V_{DD} - V_{TN}$ as we have discussed. For input voltages in the range $v_I < V_{DD} - V_{TN}$, the figure demonstrates that $v_O = v_I$. In this range of input voltages, the gate-to-source voltage is still greater than the threshold voltage. However, in steady-state, the current must be zero through the capacitor. In this case, the current becomes zero when the drain-to-source voltage is zero, or when $v_O = v_I$.

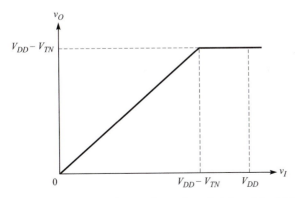

Figure 16.56 Output voltage versus input voltage characteristics of the NMOS transmission gate

Now consider the situation in which $\phi = V_{DD}$, $v_I = 0$, and $v_O = V_{DD} - V_{TN}$ initially. Terminal b then acts as the drain and terminal a acts as the source. The gate-to-source voltage is

$$v_{GS} = \phi - v_I = V_{DD} - 0 = V_{DD} \tag{16.88}$$

The value of v_{GS} is a constant, and the capacitor discharges as current enters the NMOS transistor drain. The capacitor stops discharging when the current goes to zero. Since v_{GS} is a constant at V_{DD}, the drain current goes to zero when the drain-to-source voltage is zero, which means that the capacitor completely discharges to zero. This implies that a logic 0 is transmitted unattenuated through the NMOS transmission gate.

Using an NMOS transmission gate in a MOS circuit may introduce a dynamic condition. Figure 16.57 shows a cross section of the NMOS transistor in the transmission gate configuration. If $v_I = \phi = V_{DD}$, then the load capacitor charges to $v_O = V_{DD} - V_{TN}$. When $\phi = 0$, the NMOS device turns off and the input and output become isolated.

Figure 16.57   NMOS transmission gate with cross section of NMOS transistor

The capacitor voltage reverse biases the pn junction between terminal b and ground. A reverse-biased pn junction leakage current begins to discharge the capacitor, and the circuit does not remain in a static condition. This circuit is now dynamic in that the high output does not remain constant with time.

Example 16.13   Objective: Estimate the rate at which the output voltage v_O in Figure 16.57 decreases with time.

Assume the capacitor is initially charged to $v_O = 4\,\text{V}$. Let $C_L = 1\,\text{pF}$ and assume the reverse-biased pn junction leakage current is a constant at $i_L = 1\,\text{nA}$.

Solution: The voltage across the capacitor can be written as

$$v_O = -\frac{1}{C_L} \int i_L \, dt$$

where the minus sign indicates that the current is leaving the positive terminal of the capacitor. Since i_L is a constant, we have

$$v_O = -\frac{i_L}{C_L} t + K_1$$

where $K_1 = v_O(t = 0) = 4\,\text{V}$ is the initial condition. Therefore,

$$v_O = 4 - \frac{i_L}{C_L} t$$

The rate at which the output voltage decreases is

$$\frac{dv_O}{dt} = -\frac{i_L}{C_L} = -\frac{10^{-9}}{10^{-12}} = -1000\,\text{V/s} \Rightarrow -1\,\text{V/ms}$$

Therefore, in this example, the capacitor would completely discharge in 4 ms.

Comment: Even though the NMOS transmission gate may introduce a dynamic condition into a circuit, this gate is still useful in clocked logic circuits in which a clock signal is periodically applied to the NMOS transistor gate. If, for example, the clock frequency is 25 kHz, the clock pulse period is 40 μs, which means that the output voltage would decay by no more than 1 percent.

Example 16.14   **Objective:** Determine the output of an NMOS inverter driven by a series of NMOS transmission gates.

Consider the circuit shown in Figure 16.58. The NMOS inverter is driven by three NMOS transmission gates in series. Assume the threshold voltages of the n-channel transmission gate transistors and the driver transistor are $V_{TN} = +0.8\,\text{V}$, and the threshold voltage of the load transistor is $V_{TNL} = -1.5\,\text{V}$. Let $K_D/K_L = 3$ for the inverter. Determine v_O for $v_I = 0$ and $v_I = 5\,\text{V}$.

Solution: The three NMOS transmission gates in series act as an AND/NAND function. If $v_I = 0$ and $A = B = C = \text{logic } 1 = 5\,\text{V}$, the gate capacitance to driver M_D becomes completely discharged, which means that $v_{O1} = v_{O2} = v_{O3} = 0$. Driver M_D is cut off and $v_O = 5\,\text{V}$.

Figure 16.58   NMOS inverter driven by three NMOS transmission gates in series

If $v_I = 5\,\text{V}$ and $A = B = C = \text{logic } 1 = 5\,\text{V}$, the three transmission gates are biased in their conducting state, and the gate capacitance of M_D becomes charged. For transistor M_{NA}, the current becomes zero when the gate-to-source voltage is equal to the threshold voltage, or, from Equation (16.87(b)),

$$v_{O1} = V_{DD} - V_{TN} = 5 - 0.8 = 4.2\,\text{V}$$

Transistors M_{NB} and M_{NC} also cut off when the gate-to-source voltages are equal to the threshold voltage; therefore,

$$v_{O2} = v_{O3} = V_{DD} - V_{TN} = 5 - 0.8 = 4.2 \text{ V}$$

This result shows that the drain-to-source voltages of M_{NB} and M_{NC} are also zero. A threshold voltage drop is lost in the first transmission gate, but additional threshold voltage drops are not lost in subsequent NMOS transmission gates in series.

For a voltage of $v_{O3} = 4.2$ V applied to the gate of M_D, the driver is biased in the nonsaturation region and the load is biased in the saturation region. From $i_{DD} = i_{DL}$, we have

$$K_D\left[2(v_{O3} - V_{TN})v_O - v_O^2\right] = K_L[-V_{TNL}]^2$$

The output voltage is found to be

$$v_O = 0.112 \text{ V}$$

If any one of the transmission gate voltages, A or B or C, switches to a logic 0, then v_{O3} will begin to discharge through a reverse-biased pn junction in the transmission gates, which means that v_O will increase with time.

Comment: In this example, the inverter is again in a dynamic condition; that is, when any transmission gate is cut off, the output voltage changes with time. However, this type of circuit can be used in clocked digital systems.

Test Your Understanding

16.26   The threshold voltage of the NMOS transmission gate transistor in Figure 16.55(a) is $V_{TN} = 1$ V. Determine the quiescent output voltage v_O for: (a) $v_I = \phi = 5$ V; (b) $v_I = 3$ V, $\phi = 5$ V; (c) $v_I = 4.2$ V, $\phi = 5$ V; and (d) $v_I = 5$ V, $\phi = 3$ V. (Ans. (a) $v_O = 4$ V (b) $v_O = 3$ V (c) $v_O = 4$ V (d) $v_O = 2$ V)

D16.27   Consider the NMOS inverter with enhancement load driven by an NMOS transmission gate in Figure 16.59. The threshold voltage of each n-channel transistor is $V_{TN} = 2$ V. Neglect the body effect. Design K_D/K_L such that $v_O = 0.5$ V when: (a) $v_I = 8$ V, $\phi = 10$ V, and (b) $v_I = \phi = 8$ V. (Ans. (a) $K_D/K_L = 9.78$ (b) $K_D/K_L = 15$)

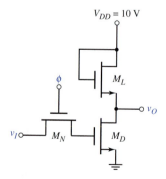

Figure 16.59   Figure for Exercise 16.27

16.6.2 NMOS Pass Networks

As integrated circuit technology advances, one emphasis is on increased circuit density. The maximum number of circuit functions per unit area is determined either by power dissipation density or by the area occupied by transistors and related devices.

One form of NMOS circuit logic that minimizes power dissipation and maximizes device density is called **pass transistor logic**. Pass transistor circuits use minimum-sized transistors, providing high density and high operating speed. The average power dissipation is due only to the switching power consumed by the driver circuits in charging and discharging the pass transistor control gates and driving the pass network inputs.

In this section, we present a few examples of NMOS pass transistor logic circuits. Consider the circuit in Figure 16.60. To determine the output response, we examine the conditions listed in Table 16.1 for the possible states of the input signals A and B. We assume that a logic 1 level is V_{DD} volts. In states 1 and 2, transmission gate M_{N2} is biased in its conducting state. For state 1, $\bar{A} =$ logic 1 is transmitted to the output so $f =$ logic $1'$, where the logic $1'$ level is $(V_{DD} - V_{TN})$. The logic 1 level is attenuated by one threshold voltage drop. For state 2, $A =$ logic 0 is transmitted unattenuated to the output. In states 3 and 4, transmission gate M_{N1} is biased in its conducting state. The $A =$ logic 0 for state 3 is transmitted unattenuated to the output, and $A =$ logic 1 for state 4 is attenuated during transmission; therefore, $f =$ logic $1'$. The output is thus the exclusive-NOR function.

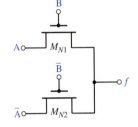

Figure 16.60 Simple NMOS pass logic network

Table 16.1 Input and ouput states for the circuit in Figure 16.60

State	A	B	$\bar{A}$	$\bar{B}$	M_{N1}	M_{N2}	f
1	0	0	1	1	off	on	$1'$
2	1	0	0	1	off	on	0
3	0	1	1	0	on	off	0
4	1	1	0	0	on	off	$1'$

Another example of an NMOS pass transistor logic circuit is shown in Figure 16.61. The ouput response as a function of the input gate controls A and B is shown in Table 16.2. This circuit is a multiplexer; that is, for a specific set of gate controls, the input signals P_i are individually passed to the output. By using both normal and inverted forms of A and B, four inputs can be controlled with just two variables.

A potential problem of NMOS pass transistor logic is that the output may be left floating in a high impedance state and charged high. Consider the circuit shown in Figure 16.62. If, for example, $\bar{B} = C =$ logic 0 and $A =$ logic 1, then $f =$ logic $1'$, which is the logic 1 level attenuated by V_{TN}. When A is switched to logic 0, the output should be low, but there may not be a discharge path to ground, and the output may retain the logic $1'$ stored at the output capacitance.

The NMOS pass network must be designed to avoid a high impedance output by passing a logic 0 whenever a 0 is required at the output. A logic network that performs the logic function $f = A + \bar{B} \cdot C$, as indicated in Figure 16.62, is shown in Figure 16.63. The complementary function $\bar{f} = \bar{A} \cdot (B + \bar{C})$ attached at the output node drives the output to a logic 0 whenever $f = 0$.

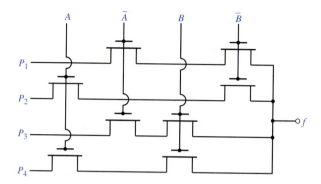

Figure 16.61   NMOS pass logic network example

Table 16.2   Input and output states for the circuit in Figure 16.61

State	A	B	$\bar{A}$	$\bar{B}$	f
1	0	0	1	1	P_1
2	1	0	0	1	P_2
3	0	1	1	0	P_3
4	1	1	0	0	P_4

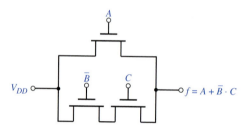

Figure 16.62   NMOS pass logic network with a potential problem

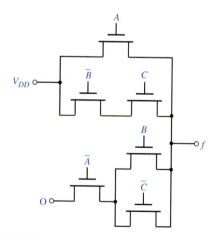

Figure 16.63   NMOS pass logic network with complementary function in parallel

D16.28 Design an NMOS pass network to perform the exclusive-OR function.

16.6.3 CMOS Transmission Gate

A **CMOS transmission gate** is shown in Figure 16.64(a). The parallel combination of NMOS and PMOS transistors, with complementary gate signals, allows the input signal to be transmitted to the output without the threshold voltage attenuation. Both transistors must be bilateral; therefore, the NMOS substrate is connected to the most negative potential in the circuit and the PMOS substrate is connected to the most positive potential (usually, ground and V_{DD}, respectively). Figure 16.64(b) shows a frequently used simplified circuit symbol for the CMOS transmission gate.

(a)

(b)

Figure 16.64 (a) CMOS transmission gate and (b) simplified circuit symbol

We again assume that the transmission gate is to operate over a voltage range of zero-to-V_{DD}. If the control voltages are $\phi = 0$ and $\overline{\phi} = V_{DD}$, then both the NMOS and PMOS transistors are cut off and the output is isolated from the input. In this state, the circuit is essentially an open switch.

If $\phi = V_{DD}$, $\overline{\phi} = 0$, $v_I = V_{DD}$, and v_O is initially zero, then for the NMOS device, terminal a acts as the drain and terminal b acts as the source, whereas for the PMOS device, terminal c acts as the drain and terminal d acts as the source. Current enters the NMOS drain and the PMOS source, as shown in Figure 16.65(a), to charge the load capacitor. The NMOS gate-to-source voltage is

$$v_{GSN} = \phi - v_O = V_{DD} - v_O \qquad \qquad \text{(16.89(a))}$$

and the PMOS source-to-gate voltage is

$$v_{SGP} = v_I - \overline{\phi} = V_{DD} - 0 = V_{DD} \qquad \qquad \text{(16.89(b))}$$

Figure 16.65 Currents and gate–source voltages in CMOS transmission gate for: (a) input high condition and (b) input low condition

As with the NMOS transmission gate, when $v_O = V_{DD} - V_{TN}$, the NMOS transistor cuts off and $i_{DN} = 0$ since $v_{GSN} = V_{TN}$. However, since the source-to-gate voltage of the PMOS device is a constant at $v_{SGP} = V_{DD}$, the PMOS transistor continues to conduct. The drain current i_{DP} goes to zero when the PMOS source-to-drain voltage goes to zero, or $v_{SDP} = 0$. This means that the load capacitor C_L continues to charge through the PMOS device until the output and input voltages are equal, or in this case, $v_O = v_I = 5$ V.

Consider what happens if $\phi = V_{DD}$, $\overline{\phi} = 0$, $v_I = 0$, and $v_O = V_{DD}$ initially. For the NMOS device, terminal a acts as the source and terminal b acts as the drain, whereas for the PMOS device, terminal c acts as the source and terminal d acts as the drain. Current enters the NMOS drain and the PMOS source, as shown in Figure 16.65(b), to discharge the capacitor. The NMOS gate-to-source voltage is

$$v_{GSN} = \phi - v_I = V_{DD} - 0 = V_{DD} \qquad\qquad \textbf{(16.90(a))}$$

and the PMOS source-to-gate voltage is

$$v_{SGP} = v_O - \overline{\phi} = v_O - 0 = v_O \qquad\qquad \textbf{(16.90(b))}$$

When $v_{SGP} = v_O = |V_{TP}|$, the PMOS device cuts off and i_{DP} goes to zero. However, since $v_{GSN} = V_{DD}$, the NMOS transistor continues conducting and capacitor C_L completely discharges to zero.

Using a CMOS transmission gate in a MOS circuit may introduce a dynamic condition. Figure 16.66 shows the CMOS transmission gate with simplified cross sections of the NMOS and PMOS transistors. If $\phi = 0$ and $\overline{\phi} = V_{DD}$, then the input and output are isolated. If $v_O = V_{DD}$, then the NMOS substrate-to-terminal b pn junction is reverse biased and capacitance C_L can discharge, as it did in the NMOS transmission gate. If, however, $v_O = 0$, then the PMOS terminal c-to-substrate pn junction is reverse biased and capacitance C_L can charge to a positive voltage. This circuit is therefore dynamic in that the output high or low conditions do not remain constant with time.

Figure 16.66 CMOS transmission gate showing cross sections of NMOS and PMOS transistors

---

Test Your Understanding

16.29 Consider the CMOS transmission gate in Figure 16.64(a). Assume transistor parameters of $V_{TN} = +0.8$ V and $V_{TP} = -1.2$ V. When $\phi = 5$ V, input v_I varies with time as $v_I = 0.5t$ V for $0 \leq t \leq 10$ s. Let $v_O(t = 0) = 0$ and assume $C_L = 1$ pF. Determine the range of times that the NMOS and PMOS devices are conducting or cut off. (Ans. NMOS conducting, $0 \leq t < 8.4$ s; PMOS conducting, $2.4 < t \leq 10$ s)

16.6.4 CMOS Pass Networks

CMOS transmission gates may also be used in pass network logic design. CMOS pass networks use NMOS transistors to pass 0's, PMOS transistors to pass 1's, and CMOS transmission gates to pass a variable to the output. An example is shown in Figure 16.67. One PMOS transistor is used to transmit a logic 1, while transmission gates are used to transmit a variable that may be either a logic 1 or a logic 0. We can show that for any combination of signals, a logic 1 or logic 0 is definitely passed to the output.

16.7 SEQUENTIAL LOGIC CIRCUITS

In the logic circuits that we have considered in the previous sections, such as NOR and NAND logic gates, the output is determined only by the instantaneous values of the input signals. These circuits are therefore classified as combinational logic circuits.

Another class of circuits is called **sequential logic circuits**. The output depends not only on the inputs, but also on the previous history of its inputs.

Figure 16.67   CMOS pass logic network

This feature gives sequential circuits the property of memory. Shift registers and flip-flops are typical examples of such circuits. We will also briefly consider a full-adder circuit. The characteristic of these circuits is that they store information for a short time until the information is transferred to another part of the system.

In this section, we introduce a basic shift register and the basic concept of a flip-flop. These circuits can become very complex and are usually described with logic diagrams. We will also introduce a CMOS full adder circuit in terms of its logic diagram and then provide the transistor implementation of this logic function. Additional information can be found in more advanced texts.

16.7.1   Dynamic Shift Registers

A **shift register** can be formed from transmission gates and inverters. Figure 16.68 shows a combination of NMOS transmission gates and NMOS depletion-load inverters. The clock signals applied to the gates of the NMOS transmission gates must be complementary, nonoverlapping pulses. The effective capacitances at the gates of M_{D1} and M_{D2} are indicated by the dotted connections to C_{L1} and C_{L2}.

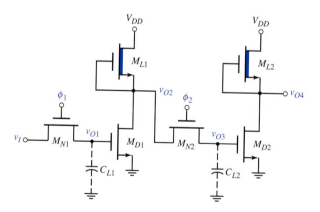

Figure 16.68   Dynamic shift register with NMOS inverters and transmission gates

If, for example, C_{L1} is initially uncharged when $v_{O1} = 0$ and if $v_I = V_{DD}$ when $\phi_1 = V_{DD}$, then a logic $1' = V_{DD} - V_{TN}$ voltage should exist at v_{O1} at the end of clock pulse ϕ_1. The capacitance of C_{L1} charges through M_{N1} and the driving circuit of v_I. The effective RC time constant must be sufficiently small to achieve this charging effect. As v_{O1} goes high, v_{O2} goes low, but the low is not transmitted through M_{N2} as long as ϕ_2 remains low.

Figure 16.69 is used to determine the operation of this circuit and the voltages at various times. For simplicity, we assume that $V_{DD} = 5\,\mathrm{V}$ and $V_{TN} = 1\,\mathrm{V}$ for the NMOS drivers and transmission gate transistors.

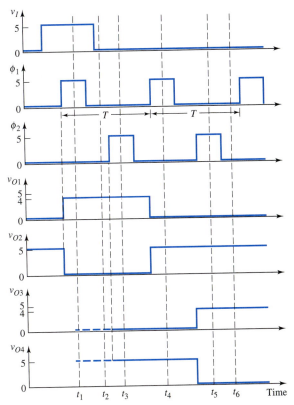

Figure 16.69   NMOS shift register voltages at various times

At $t = t_1$, $v_I = \phi_1 = 5\,\mathrm{V}$, v_{O1} charges to $v_{DD} - V_{TN} = 4\,\mathrm{V}$, and v_{O2} goes low. At this time, M_{N2} is still cut off, which means that the values of v_{O3} and v_{O4} depend on the previous history. At $t = t_2$, ϕ_1 is zero, M_{N1} is cut off, but v_{O1} remains charged. At $t = t_3$, ϕ_2 is high, and the logic 0 at v_{O2} is transmitted to v_{O3}, which forces v_{O4} to $5\,\mathrm{V}$. The input signal $v_I = 5\,\mathrm{V}$ at $t = t_1$ has thus been transmitted to the output; therefore, $v_{O4} = v_I = 5\,\mathrm{V}$ at $t = t_3$. The input signal is transmitted, or *shifted*, from the input to the output during one clock cycle, making this circuit one stage of a shift register.

At $t = t_4$, $v_I = 0$, and $\phi_1 = 5\,\mathrm{V}$, so that $v_{O1} = 0$ and $v_{O2} = 5\,\mathrm{V}$. Since $\phi_2 = 0$, M_{N2} is cut off, and v_{O2} and v_{O3} are isolated. At $t = t_5$, $\phi_2 = 5\,\mathrm{V}$, so that v_{O3} charges to $V_{DD} - V_{TN} = 4\,\mathrm{V}$, and v_{O4} goes low (logic 0). At $t = t_6$, both NMOS transmission gates are cut off, and the two inverters remain in their previous states. It is important that ϕ_1 and ϕ_2 do not overlap, or the signal would propagate through the whole chain at once and we would no longer have a shift register.

In the dynamic condition of NMOS transmission gates, the high output voltage across the output capacitance does not remain constant with time; it discharges through the transmission gate transistor. This same effect applies to the shift register in Figure 16.68. For example, from Figure 16.69, at $t = t_2$, $v_{O1} = 4\,\text{V}$, $\phi_1 = 0$, and M_{N1} is cut off. Voltage v_{O1} will start to decay and v_{O2} will begin to increase. To prevent logic errors from being introduced into the system, the clock signal period T must be small compared to the effective RC discharge time constant. The circuit in Figure 16.68 is therefore called a **dynamic shift register**.

A dynamic shift register formed in a CMOS technology is shown in Figure 16.70. Operation of this circuit is very similar to that of the dynamic NMOS shift register, except for the voltage levels. For example, when $v_I = \phi_1 = V_{DD}$, then $v_{O1} = V_{DD}$ and $v_{O2} = 0$. When ϕ_2 goes high, then v_{O3} goes to zero, $v_{O4} = V_{DD}$, and the input signal is shifted to the output during one clock period.

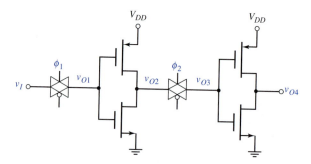

Figure 16.70   CMOS dynamic shift register

16.7.2   R–S Flip-Flop

Flip-flops are bistable circuits usually formed by cross-coupling two NOR gates. Figure 16.71 shows an R–S flip-flop using NMOS NOR logic gates with depletion loads. As shown, M_1, M_2, and M_3 form one NOR gate, and M_4, M_5, and M_6 form the second. The outputs of the two NOR circuits are connected back to the inputs of the opposite NOR gates.

If we assume that $S = \text{logic } 1$ and $R = \text{logic } 0$, then M_1 is biased in its conducting state and output $\overline{Q}$ is forced low. The inputs to both M_4 and M_5 are low, so output Q goes high to a logic $1 = V_{DD}$. Transistor M_2 is then also biased in a conducting state. The two outputs Q and $\overline{Q}$ are complementary and, by definition, the flip-flop is in the set state when $Q = \text{logic } 1$ and $\overline{Q} = \text{logic } 0$.

If S returns to logic 0, then M_1 turns off, but M_2 remains turned on so $\overline{Q}$ remains low and Q remains high. Therefore, when S goes low, nothing in the circuit can force a change and the flip-flop stores this particular logic state.

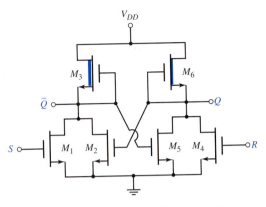

Figure 16.71   NMOS R–S flip-flop

When R = logic 1 and S = logic 0, then M_4 turns on so output Q goes low. With $S = Q$ = logic 0, then both M_1 and M_2 are cut off and $\overline{Q}$ goes high. Transistor M_5 turns on, keeping Q low when R goes low. The flip-flop is now in the reset state.

If both S and R inputs were to go high, then both outputs Q and $\overline{Q}$ would go low. However, this would mean that the outputs would not be complementary. Therefore, a logic 1 at both S and R is considered to be a forbidden or nonallowed condition. If both inputs go high and then return to logic 0, the state of the flip-flop is determined by whichever input goes low last. If both inputs go low simultaneously, then the outputs will flip into one state or the other, as determined by slight imbalances in transistor characteristics.

Figure 16.72 shows an R–S flip-flop using CMOS NOR logic gates. The outputs of the two NOR gates are connected back to the inputs of the opposite NOR gates to form the flip-flop.

If S = logic 1 and R = logic 0, then M_{N1} is turned on, M_{P1} is cut off, and $\overline{Q}$ goes low. With $\overline{Q} = R$ = logic 0, then both M_{N3} and M_{N4} are cut off, both

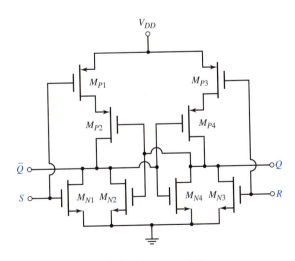

Figure 16.72   CMOS R–S flip-flop

M_{P3} and M_{P4} are biased in a conducting state so that the output Q goes high. With $Q = $ logic 1, M_{N2} is biased on, M_{P2} is biased off, and the flip-flop is in a set condition. When S goes low, M_{N1} turns off, but M_{N2} remains conducting, so the state of the flip-flop does not change.

When $S = $ logic 0 and $R = $ logic 1, then output Q is forced low, output $\overline{Q}$ goes high, and the flip-flop is in a reset condition. Again, a logic 1 at both S and R is considered to be a forbidden or a nonallowed condition, since the resulting outputs are not complementary.

16.7.3   D Flip-Flop

A **D-type flip-flop** is used to provide a delay. The logic bit on the D input is transferred to the output at the next clock pulse. This flip-flop is used in counters and shift registers. The basic circuit is similar to the CMOS dynamic shift register in Figure 16.70, except that additional circuitry makes the D flip-flop a static circuit.

Consider the circuit in Figure 16.73. The CMOS inverter composed of M_{N2} and M_{P2} is driven by a CMOS transmission gate composed of M_{N1} and M_{P1}. A second CMOS inverter, M_{N3} and M_{P3}, is connected in a feedback configuration. If $v_I = $ high, then v_{O1} goes high when the transmission gate is conducting, and output v_O, which is the input to the feedback inverter, goes low.

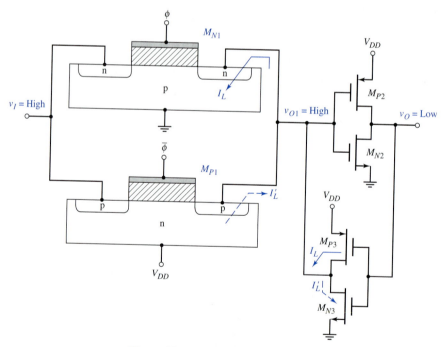

Figure 16.73 CMOS D-type flip-flop

When the CMOS transmission gate turns off, the pn junction in the M_{N1} transmission gate transistor is reverse biased. In this case, however, voltage v_{O1} is not simply across the gate capacitance of inverter M_{N2}–M_{P2}. Transistor M_{P3}

is biased in a conducting state, so the reverse-biased pn junction leakage current I_L is supplied through M_{P3}, as indicated in Figure 16.73. Since this leakage current is small, the source-to-drain voltage of M_{P3} will be small, and v_{O1} will remain biased at essentially V_{DD}. The circuit will therefore remain in this static condition.

Similarly, when v_{O1} is low and v_O is high, the pn junction in the M_{P1} transmission gate transistor is reverse biased and transistor M_{N3} is biased on. Transistor M_{N3} sinks the pn junction leakage current I'_L, and the circuit remains in this static condition until changed by a new input signal through the transmission gate.

The circuit shown in Figure 16.74 is a master–slave configuration of a D flip-flop. When clock pulse ϕ is high, transmission gate TG1 is conducting, and data D goes through the first inverter, which means that $Q' = \overline{D}$. Transmission gate TG2 is off, so data stops at Q'. When clock pulse ϕ goes low, then TG3 turns on, and the master portion of the flip-flip is in a static configuration. Also when ϕ goes low, TG2 turns on, the data are transmitted through the slave portion of the flip-flop, and the output is $Q = \overline{Q'} = D$. The data present when ϕ is high are transferred to the output of the flip-flop during the negative transition of the clock pulse. The various signals in the D flip-flop are shown in Figure 16.75.

Additional circuitry can be added to the D flip-flop in Figure 16.74 to provide a set and reset capability.

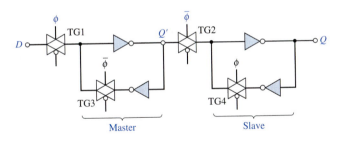

Figure 16.74   CMOS master–slave D flip-flop

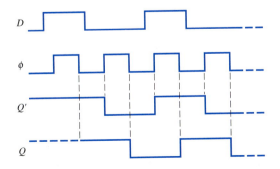

Figure 16.75   D flip-flop signals at various times

16.7.4 CMOS Full-Adder Circuit

One of the most widely used building blocks in arithmetic processing architectures is the one-bit full-adder circuit. We will first consider the logic diagram from the Boolean function and then consider the implementation in a conventional CMOS design.

Assuming that we have two input bits to be added plus a carry signal from a previous stage, the sum-out and carry-out signals are defined by the following two Boolean functions of three input variables A, B, and C.

$$\text{Sum-out} = A \oplus B \oplus C$$
$$= ABC + A\overline{B}\,\overline{C} + \overline{A}\,\overline{B}C + \overline{A}B\overline{C} \qquad \textbf{(16.91(a))}$$

$$\text{Carry-out} = AB + AC + BC \qquad \textbf{(16.91(b))}$$

The logic diagrams for these functions are shown in Figure 16.76. As we have seen previously, the implementation at the transistor level can be done with fewer transistors than would be used if all the NOR and NAND gates were actually connected as shown in the logic diagram.

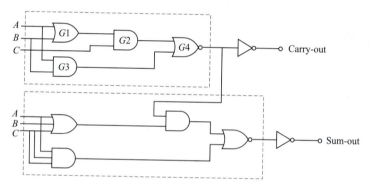

Figure 16.76 Gate configuration of the one-bit full adder

Figure 16.77 is a transistor-level schematic of the one-bit full-adder circuit implemented in a conventional CMOS technology. We can understand the basic design from the logic diagram. For example, we may consider the NMOS portion of the carry-out signal. We see that transistors M_{NA1} and M_{NB1} are in parallel, to perform the basic OR function, and these transistors are in series with transistor M_{NC1}, to perform the basic AND function. These three transistors form the NMOS portion of the design of the two gates labeled G_1 and G_2 in Figure 16.76. We also have transistors M_{NA2} and M_{NB2} in series, to perform the basic AND function of gate G_3. This set of two transistors is in parallel with the previous three transistors, and this configuration performs the basic OR function of gate G_4. This output signal goes through an inverter to become the final carry-out signal.

We can go through the same discussion for the design of the NMOS portion of the sum-out signal. The PMOS design is then the complement of the NMOS design. As mentioned, the total number of transistors in the final design is considerably less than would have occurred if the basic OR and AND gates shown in the logic diagram were actually incorporated in the design.

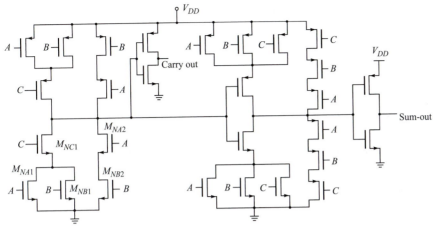

Figure 16.77   Transistor configuration of the CMOS one-bit full adder

16.8   MEMORIES: CLASSIFICATIONS AND ARCHITECTURES

In the previous sections of this chapter, various logic circuits were considered. Combinations of gates can be used to perform logic functions such as addition, multiplication, and multiplexing. In addition to these combinatorial logic functions, digital computers require some method of storing information. Semiconductor circuits form one type of memory, considered in this chapter, and define a class of digital electronic circuits that are just as important as the logic gates.

A memory cell is a circuit, or in some cases just a single device, that can store a bit of information. A systematic arrangement of memory cells constitutes a memory. The memory must also include peripheral circuits to address and write data into the cells as well as detect data that are stored in the cells.

In this section, we define the various types of semiconductor memories, discuss the memory organization, and briefly consider address decoders. In the next section, we analyze in detail some of the basic memory cells and briefly discuss sense amplifiers.

16.8.1   Classifications of Memories

Two basic types of semiconductor memory are considered. The first is the **random access memory** (RAM), a read–write memory, in which each individual cell can be addressed at any particular time. The access time to each cell is virtually the same. Implicit in the definition of the RAM is that both the read and write operations are permissible in each cell with also approximately the same access time. Both static and dynamic RAM cells are considered.

A second class of semiconductor memory is the **read-only memory** (ROM). The set of data in this type of memory is generally considered to be fixed, although in some designs the data can be altered. However, the time required to write new data is considerably longer than the read access time of the memory cell. A ROM may be used, for example, to store the instructions of a system operating program.

A volatile memory is one that loses its data when power is removed from the circuit, while nonvolatile memory retains its data even when power is removed. In general, a random access memory is a volatile memory, while read-only memories are nonvolatile.

Random Access Memories

Two types of RAM are the static RAM (SRAM) and dynamic RAM (DRAM). A static RAM consists of a basic bistable flip-flop circuit that needs only a dc current or voltage applied to retain its memory. Two stable states exist, defined as logic 1 and logic 0. A dynamic RAM is an MOS memory that stores one bit of information as charge on a capacitor. Since the charge on the capacitor decays with a finite time constant (milliseconds), a periodic refresh is needed to restore the charge so that the dynamic RAM does not lose its memory.

The advantage of the SRAM is that this circuit does not need the additional complexity of a refresh cycle and refresh circuitry, but the disadvantage is that this circuit is fairly large. In general, a SRAM requires six transistors. The advantage of a DRAM is that it consists of only one transistor and one capacitor, but the disadvantage is the required refresh circuitry and refresh cycles.

Read-Only Memories

There are two general types of ROM. The first is programmed either by the manufacturer (mask programmable) or by the user (programmable, or PROM). Once the ROM has been programmed by either method, the data in the memory are fixed and cannot be altered. The second type of ROM may be referred to as an alterable ROM in that the data in the ROM may be reprogrammed if desired. This type of ROM may be called an EPROM (erasable programmable ROM), EEPROM (electrically erasable PROM), or flash memory. As mentioned, the data in these memories can be reprogrammed although the time involved is much longer than the read access time. In some cases, the memory chip may actually have to be removed from the circuit during the reprogramming process.

16.8.2   Memory Architecture

The basic memory architecture has the configuration shown in Figure 16.78. The terminal connections may include inputs, outputs, addresses, and read and write controls. The main portion of the memory involves the data storage. A RAM memory will have all of the terminal connections mentioned, whereas a ROM memory will not have the inputs and the write controls.

A typical RAM architecture, shown in Figure 16.79, consists of a matrix of storage bits arranged in an array with 2^M columns and 2^N rows. The array may be square, in which case M and N are equal. This particular array may be only one of several on a single chip. To read data stored in a particular cell within the array, a row address is inputted and decoded to select one of the row lines. All of the cells along this row are activated. A column address is also inputted and decoded to select one of the columns. The one particular memory cell at

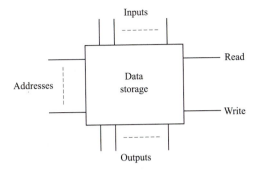

Figure 16.78 Schematic of a basic memory configuration

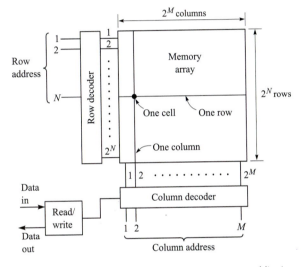

Figure 16.79 Basic random access memory architecture

the intersection of the row and column addressed is then selected. The logic level stored in the cell is routed down a bit line to a sense amplifier.

Control circuits are used to enable or select a particular memory array on a chip and also to select whether data are to be read from or written into the memory cell. Memory chips or arrays are designed to be paralleled so that the memory capacity can be increased. The additional lines needed to address parallel arrays are called **chip select signals**. If a particular chip or array is not selected, then no memory cell is addressed in that particular array. The chip select signal controls the tristate output of the data-in and data-out buffers. In this way, the data-in and data-out lines to and from several arrays may be connected together without interfering with each other.

16.8.3 Address Decoders

The row and column decoders in Figure 16.79 are essential elements in all memories. Access time and power consumption of memories may be largely determined by the decoder design. Figure 16.80 shows a simple decoder with a two-bit input. The decoder uses NAND logic circuits, although the same type

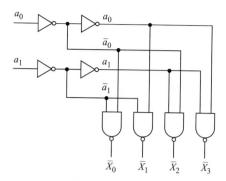

Figure 16.80   Simplified decoder with two-bit input

of decoder may be implemented in NOR gates. The input word goes through input buffers that generate the complement as well as the signal.

Another example of the direct implementation of a decoder is shown in Figure 16.81. Figure 16.81(a) shows a pair of NMOS input buffer-inverters, and Figure 16.81(b) shows a five-input NOR logic address decoder circuit using NMOS enhancement-mode drivers and a depletion load. A pair of input-buffer inverters is required for each input address line. The input signal is then required to drive only an inverter, while the buffer-inverter pair can be designed to drive the remainder of the logic circuits. The output of the NOR decoder goes high only when all inputs are a logic 0. The NOR gate in Figure 16.81(b) would decode the address word 00110 and select the sixth row or column for a read or write operation.

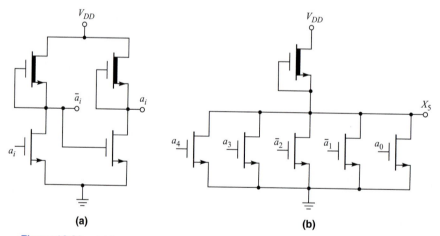

(a)   **(b)**

Figure 16.81   (a) Input buffer-inverter pair; (b) five-input NOR logic address decoder

As the size of the memory increases, the length of the address word must increase. For example, a 64-K (where 1 K = 1024 bits) memory whose cells are arranged in a square array would require an 8-bit word for the row address and another 8-bit word for the column address. As the word size increases, the decoder becomes more complex, and the number of transistors and power dissipation may become large. In addition, the total capacitance of MOS decoder transistors and interconnect lines increase so that propagation delay times may become significant. The number of transistors required to design a decoder may be reduced by using a two-stage decoder using both NOR and NAND gates. These circuits may be found in more advanced textbooks on digital circuits.

Test Your Understanding

16.30 A NOR logic address decoder, such as shown in Figure 16.81(b), is used in both the row and column address decoders in a memory arranged in a square array. Calculate the number of decoder transistors required for a (a) 1-K, (b) 4-K, and (c) 16-K memory. (Ans. 384, 896, 2048 plus buffer transistors.)

16.9 RAM MEMORY CELLS

In this section, we consider two designs of an NMOS static RAM (SRAM), one design of a CMOS static RAM, and one design of a dynamic RAM (DRAM). We also consider examples of sense amplifiers and read/write circuitry. This section is intended to present the basic concepts used in memory cell design. More advanced designs can again be found in advanced texts on digital circuits.

16.9.1 NMOS SRAM Cells

A static RAM cell is designed by cross-coupling the inputs and outputs of two inverters. In the case of an NMOS design, the load devices may be either depletion-mode transistors or polysilicon resistors, as shown in Figure 16.82. In either case, the inputs and outputs of the two inverters are cross-coupled to form a basic flip-flop. If transistor M_1 is turned on, for example, the output Q

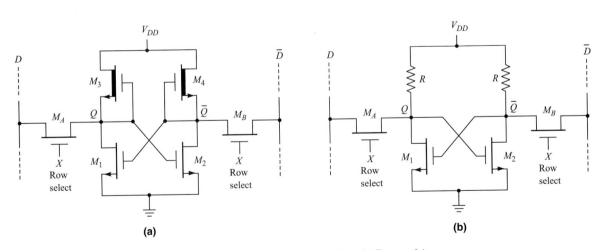

Figure 16.82 Static NMOS RAM cells with (a) depletion loads and (b) polysilicon resistor loads

is low, which means that transistor M_2 is cut off. Since M_2 is cut off, the output $\overline{Q}$ is high, ensuring that M_1 is turned on. Thus, we have a static situation as long as the bias voltage V_{DD} is applied to the circuit.

To access (read or write) the data contained in the memory cell, two NMOS transmission gate transistors, M_A and M_B, connect the memory cell to the complementary bit lines. When the word line signal or row select signal is low, both transmission gate transistors are cut off and the memory cell is isolated or in a standby condition. The data stored in the cell remain stored as long as power is applied to the cell. When the row select or word line signal goes high, the memory cell is then connected to the complementary data lines so that the data in the cell can be read or new data can be written into the cell.

One critical parameter in the design of RAM cells is power dissipation. As we will see in the following example, this is one situation in which incorporating a high-valued resistor as a load device improves the design. A lightly doped polysilicon load resistor is formed by ion implantation, which can accurately dope the polysilicon to produce the designed resistance value.

Example   16.15   **Objective:** Determine the currents, voltages, and power dissipation in two NMOS SRAM cells. The first design uses a depletion-load device and the second design uses a resistor-load device.

Assume the following parameters: $V_{DD} = 3\,\text{V}$ and $k'_n = 60\,\mu\text{A/V}^2$; driver transistors: $V_{TND} = 0.5\,\text{V}$ and $(W/L)_D = 2$; load devices: $V_{TNL} = -1.0\,\text{V}$, $(W/L)_L = 1/2$, and $R = 2\,\text{M}\Omega$.

Solution: With Depletion Load: Assume M_2 is cut off in the circuit in Figure 16.82(a) so that $\overline{Q} = V_{DD} = 3\,\text{V}$. M_1 is on in the nonsaturation region and M_3 is on in the saturation region. The drain current in M_1 and M_3 is then

$$i_D = \frac{k'_n}{2} \cdot \left(\frac{W}{L}\right)_L (V_{GSL} - V_{TNL})^2 = \frac{60}{2} \cdot \left(\frac{1}{2}\right)(0 - (-1))^2$$

or

$$i_D = 15\,\mu\text{A}$$

The power dissipated in the circuit is then

$$P = i_D \cdot V_{DD} = (15)(3) = 45\,\mu\text{W}$$

The logic 0 value of the Q output is found from

$$i_D = \frac{k'_n}{2} \cdot \left(\frac{W}{L}\right)_D [2(V_{GSD} - V_{TND})V_{DSD} - V_{DSD}^2]$$

or

$$15 = \frac{60}{2} \cdot (2)[2(3 - 0.5)Q - Q^2]$$

which yields

$$Q = 50.5\,\text{mV}$$

Solution: With Resistor Load: Again assume M_2 is cut off in the circuit in Figure 16.82(b) so that $\overline{Q} = V_{DD} = 3\,\text{V}$. Again M_1 is on in the nonsaturation region. The drain current is found from

$$\frac{V_{DD} - Q}{R} = \frac{k'_n}{2} \cdot \left(\frac{W}{L}\right)_D \left[2(V_{GSD} - V_{TND})Q - Q^2\right]$$

or

$$\frac{3 - Q}{2} = \frac{60}{2} \cdot (2)\left[2(3 - 0.5)Q - Q^2\right]$$

[Note that dividing by megohms on the left agrees with microamperes on the right.]

We find

$$Q \cong 5\,\text{mV}$$

The drain current is then found:

$$i_D = \frac{V_{DD} - Q}{R} = \frac{3 - 0.005}{2} \cong 1.5\,\mu\text{A}$$

The power dissipated in the circuit is then

$$P = i_D \cdot V_{DD} = (1.5)(3) = 4.5\,\mu\text{W}$$

Comment: We see that the SRAM with the resistive load dissipates 10 times less power than the SRAM with the depletion-load device. Thus, for a given allowed power dissipation per chip, the memory with the resistive load could be 10 times larger than that using the depletion load device.

Since the value of the load resistance R is, in general, very large, the memory must be designed so that the resistor R is not required to be a pull-up device. We will see this type of design later. The resistors can actually be fabricated on top of the NMOS transistors by a double-polysilicon technology, so that the cell with resistor load devices can be very compact, resulting in a high-density memory.

Test Your Understanding

D16.31 A 16-K NMOS static RAM cell using a resistor load is to be designed. Each cell is to be biased at $V_{DD} = 2.5\,\text{V}$. Assume transistor parameters as described in Example 16.15. The entire memory is to dissipate no more than 125 mW in standby. Design the value of R in each cell to meet this specification. (Ans. $R = 0.82\,\text{M}\Omega$)

16.9.2 CMOS SRAM Cells

The basic six-transistor CMOS SRAM cell is shown in Figure 16.83. The inputs and outputs of the two CMOS inverters are cross-coupled so that the circuit will be in one of two static conditions. For example, if $\overline{Q}$ is low, then M_{N1} is cut off so that Q is high, which in turn means that M_{P2} is cut off, ensuring that $\overline{Q}$ remains low. The two NMOS transmission gate transistors again connect the basic memory cell to the complementary data lines.

The traditional advantages of CMOS technology include low static power dissipation, superior noise immunity to either bipolar or NMOS, wide

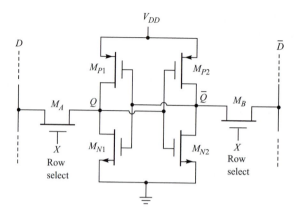

Figure 16.83 A CMOS static RAM cell

operating temperature range, sharp transfer characteristics, and wide voltage supply tolerance.

CMOS is inherently lower power than NMOS, since conducting paths between power and ground do not arise when the circuit is in one logic state or the other. In standard CMOS, the p- and n-channel devices in the memory cell and in the periphery circuits are in series and on at the same time only during switching. Current is, therefore, drawn only during switching. This makes SRAMs and CMOS extremely low power in standby, when there are only surface, junction, and channel leakage currents.

A more complete circuit of the CMOS static RAM is shown in Figure 16.84, which includes PMOS data line pull-up transistors on the complementary bit lines. If all word line signals are zero, then all pass transistors are

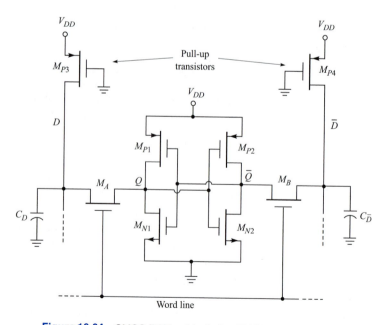

Figure 16.84 CMOS RAM cell including PMOS pull-up transistors

turned off. The two data lines with the relatively large column capacitances are charged up by the column pull-up transistors, M_{P3} and M_{P4}, to the full V_{DD} voltage.

To determine the (W/L) ratios of the transistors in a typical CMOS SRAM cell, two basic requirements must be taken into consideration. First, the read operation should not destroy the information stored in the cell, and second, the cell should allow for the modification of the data stored during a write operation. Consider a read operation in which a logic 0 ($Q = 0$ and $\bar{Q} = V_{DD}$) is stored in the cell. The voltage levels in the cell and on the data lines just prior to the read operation are shown in Figure 16.85. Transistors M_{P1} and M_{N2} are turned off while transistors M_{N1} and M_{P2} are biased in the nonsaturation region.

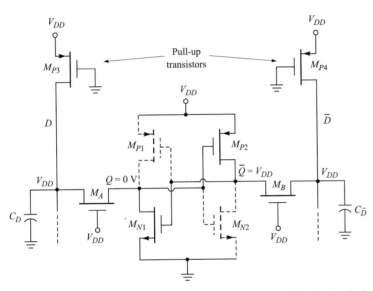

Figure 16.85 Voltage levels and "on" transistors in CMOS RAM cell at the beginning of the read cycle

Immediately after the word select signal is applied to the pass transistors M_A and M_B, the voltage on the $\bar{D}$ data line will not change significantly, since the pass transistor M_B is actually not conducting and no current flows. On the opposite side of the cell, current will flow through M_A and M_{N1} so that the voltage on the D data line will drop and the voltage Q will increase above its initial zero value. The key design point is that Q must not become larger than the threshold voltage of M_{N2}, so that M_{N2} remains cut off during the read phase. This will ensure that there is not a change in the data stored in the cell.

At the initial time the cell is addressed, we can assume that the D bit line remains at approximately V_{DD}, since the line capacitance cannot change instantaneously. The pass transistor M_A is biased in the saturation region and the transistor M_{N1} is biased in the nonsaturation region. Setting the drain currents through M_A and M_{N1} equal, we have

$$K_{nA}(V_{DD} - Q - V_{TN})^2 = K_{n1}[2(V_{DD} - V_{TN})Q - Q^2] \qquad (16.92)$$

Setting $Q = Q_{\max} = V_{TN}$ as our design limit, then from Equation (16.92), we find the relation between the transistor width-to-length ratios to be

$$\frac{(W/L)_{nA}}{(W/L)_{n1}} < \frac{2(V_{DD}V_{TN}) - 3V_{TN}^2}{(V_{DD} - 2V_{TN})^2} \tag{16.93}$$

Assuming that $V_{DD} = 3\,\text{V}$ and $V_{TN} = 0.5\,\text{V}$, we find that $(W/L)_{nA}/(W/L)_{n1} < 0.56$. So the width-to-length of the pass transistor should be approximately one-half that of the NMOS device in the memory cell. By symmetry, the same condition applies to the transistors M_{N2} and M_B.

We now need to consider the write operation. Assume that a logic 0 is stored and we want to write a logic 1 into the memory cell. Figure 16.86 shows the initial voltage levels in the CMOS SRAM cell when the cell is first addressed at the beginning of the write cycle. Transistors M_{P1} and M_{N2} are initially turned off, and M_{N1} and M_{P2} are biased in the nonsaturation region. The cell voltages are $Q = 0$ and $\overline{Q} = V_{DD}$ just before the pass transistors are turned on. The data line D is held at V_{DD} and the complementary data line $\overline{D}$ is forced to a logic 0 value by the write circuitry. We may assume that $\overline{D} = 0\,\text{V}$ for analysis purposes. The voltage Q will remain below the threshold voltage of M_{N2} because of the condition given by Equation (16.93). Consequently, the voltage at Q is not sufficient to switch the state of the memory cell. To switch the state of the cell, the voltage at $\overline{Q}$ must be reduced below the threshold voltage of M_{N1}, so that M_{N1} will turn off. When $\overline{Q} = V_{TN}$, then M_B is biased in the nonsaturation region and M_{P2} is biased in the saturation region. Equating drain currents, we have

$$K_{p2}(V_{DD} + V_{TP})^2 = K_{nB}\left[2(V_{DD} - V_{TN})V_{TN} - V_{TN}^2\right] \tag{16.94(a)}$$

which can be written in the form

$$\frac{K_{p2}}{K_{nB}} < \frac{2(V_{DD}V_{TN}) - 3V_{TN}^2}{(V_{DD} + V_{TP})^2} \tag{16.94(b)}$$

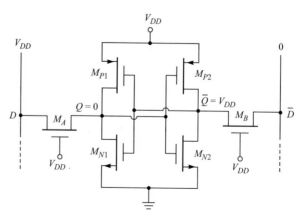

Figure 16.86 Voltage levels in the CMOS RAM at the beginning of a write cycle

Considering the width-to-length ratios, we find

$$\frac{(W/L)_{p2}}{(W/L)_{nB}} < \frac{k_n'}{k_p'} \cdot \frac{2(V_{DD}V_{TN}) - 3V_{TN}^2}{(V_{DD} + V_{TP})^2} \tag{16.95}$$

Assuming that $V_{DD} = 3\,\text{V}$, $V_{TN} = 0.5\,\text{V}$, $V_{TP} = -0.5\,\text{V}$, and $(k_n'/k_p') = (\mu_n/\mu_p)$ $= 2$, we find that $(W/L)_{p2}/(W/L)_{nB} < 0.72$.

From previous results, if we assume that the width-to-length of the pass transistor is one-half that of the NMOS in the memory cell, and if we assume that the width-to-length of the PMOS in the memory cell is 0.7 that of the pass transistor, then the width-to-length of the PMOS in the cell should be approximately 0.35 that of the NMOS in the memory cell.

Test Your Understanding

D16.32   A six-transistor CMOS SRAM cell is biased at $V_{DD} = 2.5\,\text{V}$. The transistor parameters are $V_{TN} = +0.4\,\text{V}$, $V_{TP} = -0.4\,\text{V}$, and $(\mu_n/\mu_p) = 2.5$. Determine the relative width-to-length ratios such that Equations (16.92) through (16.95) are satisfied in terms of read/write requirements.

16.9.3   SRAM Read/Write Circuitry

An example of a read/write circuit at the end of a column is shown in Figure 16.87. We may consider the write portion of the circuit as shown in Figure 16.88(a). We may note that if the column is not selected, then M_3 is cut off and the two data lines are held at their precharged value of V_{DD}. When $X = Y = 1$, then the one-bit cell shown is addressed. If $\overline{W} = 1$ then the write cycle is de-selected and both M_1 and M_2 are cut off. For $\overline{W} = 0$ and $D = 1$, M_1 is cut off and M_2 is turned on so that the $\overline{D}$ data line is pulled low while the D data line remains high. The logic 1 is then written into the cell. For $\overline{W} = 0$ and $D = 0$, the D data line is pulled low and the $\overline{D}$ data line is held high so that logic 0 is written into the cell.

Figure 16.88(b) shows the NMOS cross-coupled sense amplifier that is in the complete circuit of Figure 16.87. This circuit does not generate an output signal, but rather amplifies the small difference in the data bit lines. Suppose that a logic 1 is to be read from the memory cell. When the cell is addressed, the D bit line is high and the $\overline{D}$ bit line voltage begins to decrease. This means that when the M_3 transistor turns on, the M_2 transistor turns on harder than M_1 so that the $\overline{D}$ bit line voltage is pulled low and the M_1 transistor will eventually turn off.

Figure 16.88(c) shows the differential amplifier that senses the output of the memory cell. Note that this sense amplifier is connected to the bit lines through a couple of pass transistors, as seen in Figure 16.87. If the input signal to the pass transistors is also a function of the column select signal, then this configuration enables the use of one main sense amplifier to read the data out of several columns, one at a time. When the clock signal is zero, the M_3 transistor in the differential amplifier is cut off and the common source node

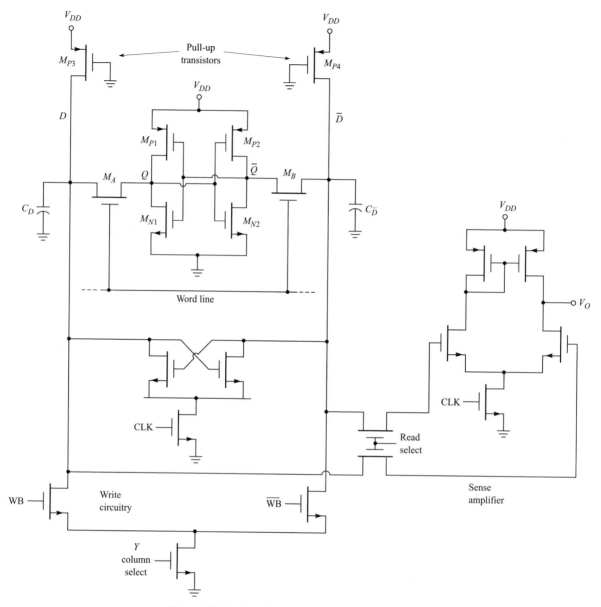

Figure 16.87   Complete circuit diagram of a CMOS RAM cell with write and read circuitry

of M_1 and M_2 is pulled high, which means the output voltage is pulled high. When a memory cell is selected and the clock goes high, M_3 turns on. If a logic 1 level is to be read, then D remains high and the $\overline{D}$ line voltage decreases. This means that the M_2 transistor will turn off and the output voltage remains high. If a logic 0 is to be read, then the D line voltage decreases and $\overline{D}$ remains high. The transistor M_1 will turn off while M_2 is turned on so that the output voltage goes low.

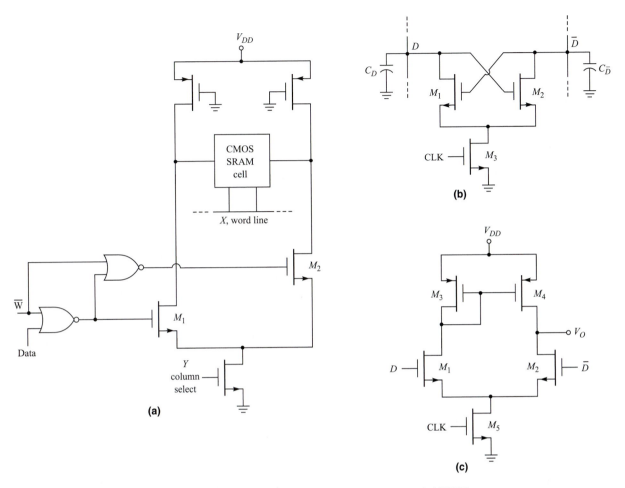

Figure 16.88 (a) Write circuitry associated with CMOS RAM cell; (b) cross-coupled NMOS sense amplifier; (c) CMOS differential sense amplifier

16.9.4 Dynamic RAM (DRAM) Cells

The CMOS RAM cell just considered requires six transistors and five lines connecting each cell, including the power and ground connections. A substantial area, then, is required for each memory cell. If the area per cell could be reduced, then higher-density RAM arrays would be possible.

In a dynamic RAM cell, a bit of data is stored as charge on a capacitor, where the presence or absence of charge determines the value of the stored bit. Data stored as charge on capacitors cannot be retained indefinitely, since leakage currents will eventually remove the stored charge. Thus the name *dynamic* refers to the situation in which a periodic refresh cycle is required to maintain the stored data.

One design of a DRAM cell is the one-transistor cell that includes a pass transistor M_S plus a storage capacitor C_S, shown in Figure 16.89. Binary information is stored in the form of zero charge on C_S (logic 0) and stored charge on C_S (logic 1). The cell is addressed by turning on the pass transistor via the word line signal WL and charges are transferred into or out of C_S on the bit line BL.

Figure 16.89 One-transistor dynamic RAM cell

The storage capacitor is isolated from the rest of the circuit when M_S is off, but the stored charge on C_S decreases because of the leakage current through the pass transistor. This effect was discussed in detail in Section 16.6 during the analysis of the NMOS pass transistor. As a result of this leakage, the cell must be refreshed regularly to restore its original condition.

An example of a sense amplifier to detect the charge stored in the memory cell is shown in Figure 16.90. On one side of the amplifier is a memory cell that either stores a full charge or is empty, depending on the binary value of the data. On the other side of the amplifier is a reference cell with a reference or dummy storage capacitor C_R that is one-half the value of the storage capacitor. The charge on C_R will then be one-half the logic 1 charge on C_S. A cross-coupled dynamic latch circuit is used to detect the small voltage differences and to restore the signal levels. The capacitors C_D and C_{DR} represent the relatively large parasitic bit line and reference bit line capacitances.

In the standby mode, the bit lines on both sides of the sense amplifier are precharged to the same potential. During the read cycle, both the WL and D–WL address signals go high allowing the charges in the cells to be redistributed along the bit lines. After the charge equalization and since the charge in the dummy cell is half the full charge, then $v_1 < v_2$ when the memory cell is empty or a logic 0, and $v_1 > v_2$ when the memory cell is full or a logic 1. The sense amplifier detects and amplifies the voltage difference between the bit lines, and will latch at the logic level stored in the basic memory cell.

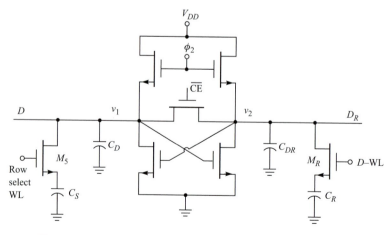

Figure 16.90 Sense amplifier configuration for dynamic RAM cell

Test Your Understanding

16.33 A one-transistor DRAM cell is composed of a 0.05 pF storage capacitor and an NMOS transistor with a 0.5 V threshold voltage. A logic 1 is written into the cell when both the data line and row-select line are raised to 3 V. Sensing circuitry permits the stored charge to decay to 50 percent of its original value. Refresh occurs every 1.5 ms. Determine the maximum allowed leakage current that can exist.

16.10   READ-ONLY MEMORY

We consider several examples of read-only memories in this section. The intent is again to provide an introduction to this type of memory. In the case of EPROMs and EEPROMs, the development effort has been directed toward the characteristics of the basic memory cell.

16.10.1   ROM and PROM Cells

We consider two types of ROMs. The first example is a mask-programmed ROM, in which contacts to devices are selectively included or excluded in the final manufacturing process to obtain the desired memory pattern. Figure 16.91 shows an example of an NMOS 16×1 mask-programmed ROM.

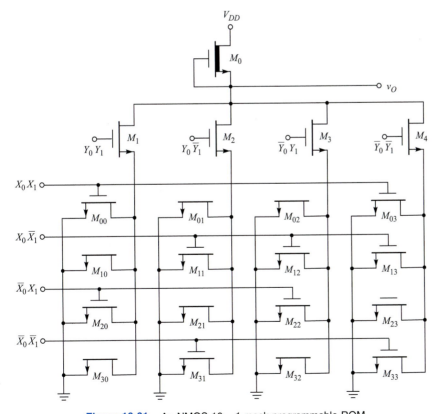

Figure 16.91   An NMOS 16×1 mask-programmable ROM

Enhancement-mode NMOS transistors are fabricated in each of the 16 cell positions (the substrate connections are omitted for clarity). However, gate connections are fabricated only on selected transistors. The transistors M_1–M_4 are column-select transistors and M_0 is a depletion-mode load device.

The inputs X_O, X_1, Y_O, and Y_1 are the row- and column-select signals. If, for example, $X_O = \overline{X}_1 = \overline{Y}_O = Y_1 = 1$, then the M_{12} transistor is addressed. Transistors M_{12} and M_3 turn on with this address, forcing the output to a logic 0. If the address changes, for example, to $\overline{X}_O = X_1 = \overline{Y}_O = \overline{Y}_1 = 1$, then the transistor M_{23} is addressed. However, this transistor does not have a gate connection and consequently never turns on, so the output is a logic 1.

The mask-programmed memory discussed is only a 16×1-bit ROM, while a more useful memory would contain many more bits. Memories can be organized in any desired manner, such as a 2048×8 for a 16-K memory. This ROM is a nonvolatile memory, since the data stored are not lost when power is removed.

The second example of a ROM is a user-programmed ROM. The data pattern is defined by the user after the final manufacture rather than during the manufacture. One specific type is shown schematically in Figure 16.92. A small fuse is in series with each emitter and can be selectively "blown" or left in place by the user. If, for example, the fuse in Q_{00} is left in place and this transistor is addressed by $X_O = X_1 = Y_O = Y_1 = 1$, then Q_{00} turns on, raising the data line voltage at the emitter of Q_{00}. The inverter N_1 is enabled, making the output a logic 0. If the fuse is blown in this transistor, then the input to the inverter is a logic 0, so the output is a logic 1.

The polysilicon fuse in the emitter of an npn bipolar transistor has a fairly low resistance, so with the fuse in place and at low currents, there is very little voltage drop across the fuse. When the current through the fuse is increased to the 20 to 30 mA range, the heating of the polysilicon fuse causes the temperature to increase. The silicon oxidizes, forming an insulator that effectively opens the path between the data line and the emitter. The bipolar ROM circuit with the fuses either in place or "blown" form a permanent ROM that is not alterable and is also nonvolatile.

16.10.2   EPROM and EEPROM Cells

An EPROM transistor is shown in Figure 16.93. The device has a double gate, with gate 1 being a "floating gate" that has no electrical contact. Gate 2 is used for cell selection, taking the role of the single gate of an MOS transistor.

Operation of this EPROM cell relies on being able to store charge on the floating gate. Initially, we assume no charge on the floating gate so that with gate 2, drain, and source grounded, the potential of gate 1 is also zero. As the voltage on gate 2 increases, the gate 1 voltage rises also, but at a lower rate as determined by the capacitive divider. The net effect of this is to effectively raise the threshold voltage of this MOSFET as seen from gate 2. However, when the gate 2 voltage is raised sufficiently (approximately twice the normal threshold voltage), a channel forms. Under these conditions, the device provides a stored logic 0 when used in the NOR array.

To write a logic 1 into this cell, both gate 2 and drain are raised to about 25 V while the source and substrate remain at ground potential. A relatively large drain current flows because of normal device conduction characteristics.

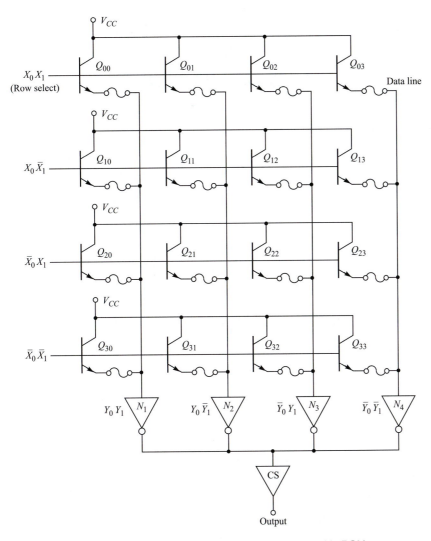

Figure 16.92 A bipolar fuse-linked user-programmable ROM

In addition, the high field in the drain–substrate depletion region results in avalanche breakdown of the drain–substrate junction, with a considerable additional flow of current. The high field in the drain depletion region accelerates electrons to high velocity such that a small fraction traverse the thin oxide and become trapped on gate 1. When the gate 2 and drain potentials are reduced to zero, the negative charge on gate 1 forces its potential to approximately -5 V. If the gate 2 voltage for reading is limited to $+5$ V, then a channel never forms. Thus a logic 1 is stored in the cell.

Gate 1 is completely surrounded by silicon dioxide (SiO_2), an excellent insulator, so charge can be stored for many years. Data can be erased, however, by exposing the cells to strong ultraviolet (UV) light. The UV radiation generates electron–hole pairs in the SiO_2 making the material slightly conductive. The negative charge on the gate can then leak off, restoring the transistor to its original uncharged condition. These EPROMs must be assembled in

Figure 16.93   (a) Cross section of erasable programmable ROM; (b) equivalent circuit

Figure 16.94   (a) Cross section of a floating-gate electrically erasable programmable ROM; (b) charging the floating gate; (c) discharging the floating gate

packages with transparent covers so the silicon chip may be exposed to UV radiation. One disadvantage is that the entire memory must be erased before any reprogramming can be done. In general, reprogramming must also be done on specialized equipment; therefore, the EPROM must be removed from the circuit during this operation.

In the EEPROM, each individual cell can be erased and reprogrammed without disturbing any other cell. The most common form of EEPROM is also a floating gate structure; one example is shown in Figure 16.94(a). The memory transistor is similar to an n-channel MOSFET, but with a physical difference in the gate insulator region. Charge may exist on the floating gate that will alter the threshold voltage of the device. If a net positive charge exists on the floating gate, the n-channel MOSFET is turned on, whereas if zero or negative charge exists on the floating gate, the device is turned off.

The floating gate is capacitively coupled to the control gate with the tunnel oxide thickness less than 200 Å. If 20 V is applied to the control gate while keeping $V_D = 0$, electrons tunnel from the n^+-drain region to the floating gate as demonstrated in Figure 16.94(b). This puts the MOSFET in the enhancement mode with a threshold voltage of approximately 10 V, so the device is effectively off. If zero volts is applied to the control gate and 20 V is applied to the drain terminal, then electrons tunnel from the floating gate to the n^+-drain terminal as demonstrated in Figure 16.94(c). This leaves a net positive charge on the floating gate that puts the device in the depletion mode with a threshold voltage of approximately -2 V, so the device is effectively on. If all voltages are kept to within 5 V during the read cycle, this structure can retain its charge for many years.

16.11   SUMMARY

- In this chapter, NMOS and CMOS digital logic circuits were analyzed and designed. These included basic logic gates, shift registers, flip-flops, and memories.

- The discussion of NMOS logic circuits served as an introduction to the analysis and design of digital logic circuits. Since this technology deals with only one type of transistor (n-channel), the analysis and design is straightforward.

- The NMOS inverter is the basis of NMOS logic circuits. The quasi-static voltage transfer characteristics of NMOS inverters with resistive load, enhancement load, and depletion load were generated. The transfer characteristics were designed to provide appropriate logic 0 values by designing the width-to-length ratios of the transistors. The impact of the body effect on the transfer curves and logic values was analyzed. The noise margin of the NMOS inverter is defined as the point where the magnitude of the voltage gain is unity.

- The basic NMOS NOR and NAND logic gates were analyzed. More sophisticated logic functions can be implemented by combining driver transistors in particular series and parallel combinations. The width-to-length ratios of the driver transistors were designed to produce a composite conduction parameter to produce a specified logic 0 value.

- The CMOS inverter is the basis of the CMOS logic circuits. The quasi-static voltage transfer characteristics were generated. For the CMOS circuit, the quiescent power dissipation is essentially zero when the input is in either logic state. The extremely low static power dissipation is the primary advantage of the CMOS technology. The switching power dissipation is given by $P = fC_L V_{DD}^2$, where f is the switching frequency, C_L is the effective load capacitance, and V_{DD} is the supply voltage. The tendency in CMOS design is toward lower supply voltages on CMOS digital logic circuits because of the squared term in the power equation.

- The basic CMOS NOR and NAND logic gates were analyzed. In the classical CMOS design, the gates of a PMOS and NMOS are connected together. CMOS logic circuits are usually designed to provide equal current drive in the NMOS pull-down and PMOS pull-up portions of the circuit. Transistor width-to-length ratios were designed to provide equal composite conduction parameters in the NMOS and PMOS circuits.

- More sophisticated logic functions can be implemented in the classical CMOS technology. NMOS transistors in series implement the basic AND function and NMOS transistors in parallel implement the basic OR function. The combination of the PMOS transistors is the complement of the NMOS design. By combining transistors in a particular series or parallel combination, more complex logic functions can be implemented.

- Since the mobility of carriers in the PMOS transistor is smaller than that in the NMOS transistor, PMOS devices must be approximately twice as large as NMOS devices to provide the same current drive. Therefore, a savings in chip area as well as reduced capacitance can be achieved by eliminating as many PMOS transistors as possible. Clocked CMOS logic circuits achieve this goal. A generalized NMOS logic circuit is inserted between clocked PMOS and NMOS devices. The advantage of low static power dissipation is maintained.

- Sequential logic circuits are a class of circuits whose output depends not only on the inputs, but is also a function of the previous history of the inputs. Shift registers, flip-flops, and a full one-bit adder were analyzed in this section. Dynamic shift registers are formed with transmission gates and inverters. Both NMOS and CMOS designs were analyzed. A flip-flop can be implemented by cross-coupling two NOR gates. This bistable circuit can remain in either stable state indefinitely, as long as power is applied. A full one-bit CMOS adder was analyzed at both the gate and the transistor level.

- A whole classification of circuits called memories was considered. Typically, an array of memory cells is organized in a square matrix to form a memory. A cell is addressed via row and column decoders and data are read from the cell or written into a cell through data lines.

- A random-access memory (RAM) cell is a circuit or device that can store one bit of information, and whose information can be written (stored) or retrieved (read) with essentially the same access time. A static RAM (SRAM) retains its data as long as power is applied, whereas a dynamic RAM (DRAM) loses its stored data over time by leakage currents. The DRAM data must be refreshed.

- Three SRAM designs were considered. In the two NMOS designs, static power is dissipated in the cell, which limits the size of the memory because of the total chip power limitation. A CMOS SRAM was designed. The primary advantage of essentially no static power dissipation is again the primary advantage of CMOS technology. The size of a CMOS memory is limited primarily by chip area requirements. An example of the peripheral read/write circuitry required was considered.

- Read-only memory (ROM and PROM) contains fixed data that are implemented by the manufacturer (mask programmed) or by the user (user programmed). In both cases, the data cannot be altered. In the case of a mask-programmed ROM, for example, the gates of MOSFETs may be fabricated or may be deliberately left off in a cell depending on whether a logic 1 or logic 0 is to be stored. For a user-programmed ROM, a fuse in a particular memory cell can be left in place or "blown," depending on whether a logic 1 or logic 0 is to be stored.

- Erasable read-only memory (EPROM and EEPROM) cells contain MOSFETs with floating gates. The floating gates can be either charged or left uncharged by the user depending on whether a logic 1 or logic 0 is to be stored. The charge on the floating gate can be altered so that the data in the ROM can be erased and reprogrammed. The writing of new data, however, takes a relatively long time compared to the read access time.

CHECKPOINT

After studying this chapter, the reader should have the ability to:

✔ Analyze the transfer characteristics of NMOS inverters, including the determination of noise margins. (Section 16.1)

✔ Design an NMOS logic circuit to perform a specific logic function. (Section 16.2)

✔ Analyze the transfer characteristics of the CMOS inverter, including the determination of switching power and noise margins. (Section 16.3)

✔ Design a CMOS logic circuit to perform a specific logic function. (Section 16.4)

✔ Design a clocked CMOS logic circuit to perform a specific logic function. (Section 16.5)

✔ Design an NMOS or CMOS pass network to perform a specific logic function. (Section 16.6)

✔ Design an NMOS or CMOS RAM cell and design a simple sense amplifier. (Section 16.9)

REVIEW QUESTIONS

1. Explain qualitatively what is meant by the body effect in an NMOS device and discuss its effect on the threshold voltage of the NMOS transistor.

2. Sketch the quasi-static voltage transfer characteristics of an NMOS inverter with resistive load. Discuss the various intervals in terms of transistor bias. What is the effect on the transfer curve of increasing the transistor W/L ratio?

3. Sketch the quasi-static voltage transfer characteristics of an NMOS inverter with enhancement load. Discuss the various intervals in terms of transistor bias. Why doesn't the output ever reach the V_{DD} value? What is the effect on the transfer curve of changing the transistor W/L ratios?

4. Sketch the quasi-static voltage transfer characteristics of an NMOS inverter with depletion load. Discuss the advantage of the depletion-load inverter compared to the other two NMOS inverter designs. What is the effect on the transfer curve of changing the transistor W/L ratios?

5. Define the noise margin of an NMOS inverter.

6. What is the impact of the body effect on the NMOS inverter voltage transfer characteristics of each of the inverter designs?

7. Sketch an NMOS three-input NOR logic gate. Describe its operation. Discuss the condition under which the maximum logic 0 value is obtained.

8. Sketch an NMOS three-input NAND logic gate. Describe its operation. Discuss the effect of changing the driver transistor W/L ratios.

9. Discuss how more sophisticated (compared to the basic NOR and NAND) logic functions can be implemented in a single NMOS logic circuit.

10. Sketch the quasi-static voltage transfer characteristics of a CMOS inverter. Discuss the various intervals in terms of transistor bias. What is the effect on the transfer

curve of changing the transistor W/L ratios? What is the advantage of the CMOS inverter compared to an NMOS inverter?

11. Sketch the quasi-static current versus input voltage of a CMOS inverter. Discuss the various intervals in terms of transistor bias.

12. Discuss the difference between the static power dissipation and switching power dissipation in a CMOS inverter.

13. Discuss the parameters that affect the switching power dissipation in a CMOS inverter.

14. Define the noise margin in a CMOS inverter.

15. Sketch a CMOS three-input NOR logic gate. Describe its operation. Determine relative transistor W/L ratios to obtain equal NMOS and PMOS composite conduction parameters.

16. Sketch a CMOS three-input NAND logic gate. Describe its operation. Determine relative transistor W/L ratios to obtain equal NMOS and PMOS composite conduction parameters.

17. Discuss how more sophisticated (compared to the basic NOR and NAND) logic functions can be implemented in a single CMOS logic circuit.

18. Discuss the basic principles of a clocked CMOS logic circuit. Discuss why, in general, PMOS transistors must be larger in size than NMOS transistors.

19. Sketch an NMOS transmission gate and describe its operation. If the input and gate voltages are both V_{DD}, determine the maximum output voltage. Why can't the output voltage reach V_{DD}?

20. Consider three NMOS transmission gates in series or in cascade. If the input voltage and each gate voltage is V_{DD}, determine the output voltage. Discuss why three threshold voltage drops are *not* lost between the input and output.

21. Sketch a CMOS transmission gate and describe its operation. For this circuit, discuss why the quasi-static output voltage is always equal to the quasi-static input voltage.

22. Discuss what is meant by pass transistor logic.

23. If an NMOS or CMOS transmission gate is turned off (an open switch), discuss why the output voltage is, in general, not stable.

24. Sketch an NMOS dynamic shift register and describe its operation.

25. Sketch a CMOS R–S flip flop and describe its operation. Why must the input condition $R = S = 1$ be avoided?

26. Describe the basic architecture of a semiconductor random-access memory.

27. Discuss the differences between SRAM and DRAM cells. Discuss advantages and disadvantages of each design.

28. Sketch an NMOS SRAM cell and describe its operation. Discuss any disadvantages of this design.

29. Sketch a CMOS SRAM cell and describe its operation. Discuss any advantages and disadvantages of this design. Describe how the cell is addressed.

30. Describe the voltage levels in a CMOS SRAM cell during a read operation. Describe any limitations in voltage changes in the cell during this read cycle.

31. Describe the voltage levels in a CMOS SRAM cell during a write operation. Assume a logic 0 is initially stored and a logic 1 is to be written into the cell.

32. Sketch a one-transistor DRAM cell and describe its operation. What makes this circuit dynamic?

33. Describe a mask-programmed MOSFET ROM memory.

34. Describe the basic operation of a floating gate MOSFET and how this can be used in an erasable ROM.

PROBLEMS

[Note: In the following problems, unless otherwise stated, assume: $k'_n = 80\,\mu A/V^2$, $k'_p = 35\,\mu A/V^2$, $\lambda = 0$ for all transistors; $V_{TNO} = 0.8\,V$ for all n-channel enhancement-mode transistors; and $V_{TPO} = -0.8\,V$ for all p-channel enhancement-mode transistors. Neglect the body effect unless otherwise stated. The temperature is $300\,°K$.]

Section 16.1   NMOS Inverters

16.1   Consider an NMOS transistor with parameters: $K_n = 0.2\,mA/V^2$, $V_{TNO} = 0.8\,V$, $N_a = 8 \times 10^{15}\,cm^{-3}$, $t_{ox} = 450\,\text{Å}$, and $\phi_{fp} = 0.343\,V$. (a) Determine the change in threshold voltage from $V_{SB} = 1\,V$ and $V_{SB} = 2\,V$. (b) If $V_{GS} = 2.5\,V$ and $V_{DS} = 5\,V$, find the transistor current for $V_{SB} = 0$, $V_{SB} = 1\,V$, and $V_{SB} = 2\,V$.

RD16.2   The load resistor in the NMOS inverter in Figure 16.5(a) is $40\,k\Omega$. The circuit is biased at $V_{DD} = 5\,V$. (a) Redesign the width-to-length ratio of the driver transistor such that $v_O = 0.10\,V$ when $v_I = 5\,V$. (b) Using the results of part (a), find the driver transition point and the maximum power dissipated in the inverter circuit.

16.3   For the circuit in Figure 16.5(a), assume the transistor conduction parameter is $K_n = 50\,\mu A/V^2$. (a) Plot the voltage transfer characteristics for $0 \le v_I \le 5\,V$ and for $R_D = 20\,k\Omega$. Mark the values of v_I and v_O at the transition point. (b) Repeat part (a) for $R_D = 200\,k\Omega$.

RD16.4   Redesign the inverter in Figure 16.5(a) such that the maximum power dissipated is no greater than $1\,mW$, and the output voltage is $0.2\,V$ when the input voltage is $5\,V$. Determine the load resistance R_D and the transistor width-to-length ratio.

16.5   An NMOS inverter with saturated load is shown in Figure 16.8(a). Let $V_{DD} = 10\,V$, $V_{TNL} = V_{TND} = 2\,V$, $K_D = 200\,\mu A/V^2$, and $K_L = 50\,\mu A/V^2$. Calculate the transition point and determine v_O when $v_I = 8\,V$. Sketch the voltage transfer characteristics.

16.6   An NMOS inverter with saturated load is shown in Figure 16.8(a). The bias is $V_{DD} = 3\,V$ and the transistor threshold voltages are $0.5\,V$. (a) Find the ratio K_D/K_L such that $v_O = 0.25\,V$ when $v_I = 3\,V$. (b) Repeat part (a) for $v_I = 2.5\,V$. (c) If $W/L = 1$ for the load transistor, determine the power dissipation in the inverter for parts (a) and (b).

RD16.7   Consider the NMOS inverter with saturated load in Figure 16.8(a). Let $V_{DD} = 3\,V$ and let the threshold voltages be $0.5\,V$. (a) Redesign the circuit such that the power dissipated in the circuit is $400\,\mu W$ and the output voltage is $0.10\,V$ when the input voltage is a logic 1. Determine the driver transition point. (b) Determine the noise margin for this inverter.

16.8   The NMOS inverter with saturated load in Figure 16.8(a) operates with a supply voltage of V_{DD}. The MOSFETs have threshold voltages of $V_{TN} = 0.2V_{DD}$. Determine $(W/L)_D/(W/L)_L$ such that $V_O = 0.08V_{DD}$. Neglect the body effect.

16.9   The enhancement-load transistor in the NMOS inverter in Figure P16.9 has a separate bias applied to the gate. Assume transistor parameters of $K_n = 1\,mA/V^2$ for M_D, $K_n = 0.4\,mA/V^2$ for M_L, and $V_{TN} = 1\,V$ for both transistors. Using the appropriate logic 0 and logic 1 input voltages, determine V_{OH} and V_{OL} for: (a) $V_B = 4\,V$, (b) $V_B = 5\,V$, (c) $V_B = 6\,V$, and (d) $V_B = 7\,V$.

16.10   For the depletion-load NMOS inverter circuit in Figure 16.10(a), assume: $V_{DD} = 5\,V$, $V_{TNL} = -2\,V$, $V_{TND} = 0.8\,V$, $K_L = 100\,\mu A/V^2$, and $K_D = 500\,\mu A/V^2$. (a) Find the transition points for the load and driver transistors. (b) Calculate the

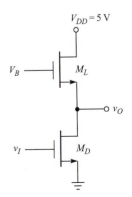

Figure P16.9

value of v_O for $v_I = 5\,\text{V}$. (c) Calculate i_D when $v_I = 5\,\text{V}$. (d) Sketch the voltage transfer characteristics for $0 \le v_I \le 5\,\text{V}$.

16.11 In the depletion-load NMOS inverter circuit in Figure 16.10(a), let $V_{TND} = 0.5\,\text{V}$ and $V_{DD} = 3\,\text{V}$, $K_L = 50\,\mu\text{A/V}^2$, and $K_D = 500\,\mu\text{A/V}^2$. Calculate the value of V_{TNL} such that $v_O = 0.10\,\text{V}$ when $v_I = 3\,\text{V}$.

RD16.12 Consider the NMOS inverter with depletion load in Figure 16.10(a). Let $V_{DD} = 3\,\text{V}$, and assume $V_{TNL} = -1.0\,\text{V}$ and $V_{TND} = 0.5\,\text{V}$. (a) Redesign the circuit such that the maximum power dissipated in the circuit is $150\,\mu\text{W}$ and the minimum output voltage is $0.10\,\text{V}$ when the input voltage is a logic 1. Determine the transition points for the driver and load transistors. (b) Determine the noise margin for this inverter.

D16.13 The NMOS inverter with depletion load is shown in Figure 16.10(a). The bias is $V_{DD} = 2.5\,\text{V}$. The transistor parametes are $V_{TND} = 0.5\,\text{V}$ and $V_{TNL} = -1\,\text{V}$. The width-to-length ratio of the load device is $W/L = 1$. (a) Design the driver transistor such that $v_O = 0.05\,\text{V}$ when the input is a logic 1. (b) What is the power dissipated in the circuit when $v_I = 2.5\,\text{V}$?

16.14 Calculate the power dissipated in each inverter circuit in Figure P16.14 for the following input conditions: (a) Inverter a: (i) $v_I = 0.5\,\text{V}$, (ii) $v_I = 5\,\text{V}$; (b) Inverter b: (i) $v_I = 0.25\,\text{V}$, (ii) $v_I = 4.3\,\text{V}$; (c) Inverter c: (i) $v_I = 0.03\,\text{V}$, (ii) $v_I = 5\,\text{V}$.

Figure P16.14

16.15 For the two inverters in Figure P16.15, assume width-to-length ratios of $W/L = 1$ for the load devices and $W/L = 10$ for the driver devices. Determine the values of v_I and v_{O2} if $v_{O1} = V_{IH}$. What is the value of V_{IH}?

16.16 Consider the circuit in Figure P16.16. The parameters of the driver transistors are $V_{TND} = 0.8\,\text{V}$ and $W/L = 4$, and those of the load transistors are $V_{TNL} = -2\,\text{V}$ and $W/L = 1$. (a) Find the values of v_I and v_{O2} if $v_{O1} = V_{IH}$. (b) Repeat part (a) if $v_{O1} = V_{IL}$. (c) What are the values of V_{IH} and V_{IL} in parts (a) and (b)?

16.17 For the two transistors in the NMOS inverter with saturated load in Figure 16.17(a), assume the parameters are as described in Problem 16.1, except that $K_D = 200\,\mu\text{A/V}^2$ and $K_L = 20\,\mu\text{A/V}^2$. Let $V_{DD} = 5\,\text{V}$. (a) Determine the output voltage when $v_I = 0$ for: (i) neglecting the body effect, and (ii) taking the body effect into account. (b) Compare the results of part (a) with a computer simulation analysis.

Figure P16.15

Figure P16.16

16.18 Consider the NMOS inverter with depletion load in Figure 16.17(b). Assume that the circuit and transistor parameters are the same as those given and determined in Example 16.4. Assume the body effect coefficient for the load transistor is $\gamma = 0.35 \, V^{1/2}$. From a computer simulation, plot the load curve for: (a) neglecting the body effect, and (b) taking the body effect into account.

Section 16.2   NMOS Logic Circuits

16.19 Consider the circuit with a depletion load device shown in Figure P16.19. (a) Let $v_X = 5 \, V$ and $v_Y = 0.20 \, V$. Determine K_D/K_L such that $v_O = 0.20 \, V$. (b) Using the results of part (a), determine v_O when $v_X = v_Y = 5 \, V$. (c) If the width-to-length ratio of the depletion device is $W/L = 1$, determine the power dissipation in the logic circuit for the input conditions listed in parts (a) and (b).

Figure P16.19

Figure P16.20

D16.20 Consider the three-input NOR logic gate in Figure P16.20. The transistor parameters are $V_{TNL} = -1 \, V$ and $V_{TND} = 0.5 \, V$. The maximum value of v_O in its low state is to be $0.1 \, V$. (a) Determine K_D/K_L. (b) The maximum power dissipation in the NOR logic gate is to be $0.1 \, mW$. Determine the width-to-length ratios of the transistors. (c) Determine v_O when $v_X = v_Y = v_Z = 3 \, V$.

16.21 The transistor parameters for the circuit in Figure P16.21 are: $V_{TN} = 0.8 \, V$ for all enhancement-mode devices, $V_{TN} = -2 \, V$ for the depletion-mode devices, and $k'_n = 60 \, \mu A/V^2$ for all devices. The width-to-length ratios of M_{L2} and M_{L3} are 1, and those for

Figure P16.21

Figure P16.22

Figure P16.23

M_{D2}, M_{D3}, and M_{D4} are 8. (a) For $v_X = 5\,\text{V}$, output v_{O1} is 0.15 V, and the power dissipation in this inverter is to be no more than $250\,\mu\text{W}$. Determine $(W/L)_{ML1}$ and $(W/L)_{MD1}$. (b) For $v_X = v_Y = 0$, determine v_{O2}.

16.22 In the NMOS circuit in Figure P16.22, the transistor parameters are: $(W/L)_X = (W/L)_Y = 9$, $(W/L)_L = 1$, and $V_{TN} = 0.8\,\text{V}$ for all transistors. (a) Determine v_O when $v_X = v_Y = 9.2\,\text{V}$. What are the values of v_{GSX}, V_{GSY}, v_{DSX}, and v_{DSY}? (b) Repeat part (a) for $\gamma = 0.5$.

16.23 In the NMOS circuit in Figure P16.23, the transistor parameters are: $(W/L)_X = (W/L)_Y = 4$, $(W/L)_L = 1$, $V_{TNX} = V_{TNY} = 0.8\,\text{V}$, and $V_{TNL} = -1.5\,\text{V}$. (a) Determine v_O when $v_X = v_Y = 5\,\text{V}$. (b) What are the values of v_{GSX}, v_{GSY}, v_{DSX}, and v_{DSY}? Repeat part (a) for $\gamma = 0.5$.

16.24 Find the logic function implemented by the circuit in Figure P16.24.

16.25 Find the logic function implemented by the circuit in Figure P16.25.

D16.26 The Boolean function for a carry-out signal of a one-bit full adder is given by

$$\text{Carry-out} = A \cdot B + A \cdot C + B \cdot C$$

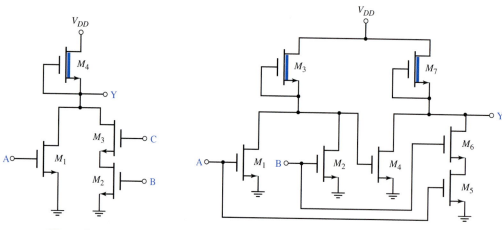

Figure P16.24

Figure P16.25

(a) Design an NMOS logic circuit with depletion load to perform this function. Signals A, B, and C are available. (b) Assume $(W/L)_L = 1$, $V_{DD} = 5$ V, $V_{TNL} = -1.5$ V, and $V_{TND} = 0.8$ V. Determine the W/L ratio of the other transistors such that the maximum logic 0 value in any part of the circuit is 0.2 V.

D16.27 Design an NMOS logic circuit with a depletion load that will sound an alarm in an automobile if the ignition is turned off while the headlights are still on and/or the parking brake has not been set. Separate indicator lights are also to be included showing whether the headlights are on or the parking brake needs to be set. State any assumptions that are made.

Section 16.3   CMOS Inverter

16.28 Consider the CMOS inverter in Figure 16.34. Let $K_P = K_n$, $V_{TN} = +0.8$ V, $V_{TP} = -0.8$ V, and $V_{DD} = 5$ V. (a) Find the transition points for the p-channel and n-channel transistors. (b) Sketch the voltage transfer characteristic, including the appropriate voltage values at the transition points. (c) Find v_O for $v_I = 2$ V and for $v_I = 3$ V.

16.29 For the CMOS inverter in Figure 16.34, let $V_{TN} = +0.4$ V, $V_{TP} = -0.4$ V, $k'_n = 80 \,\mu\text{a/V}^2$, $k'_p = 40 \,\mu\text{A/V}^2$, and $V_{DD} = 3.3$ V. (a) Let $(W/L)_n = 2$ and $(W/L)_p = 4$. (i) Find the transition points for the p-channel and n-channel transistors. (ii) Sketch the voltage transfer characteristics including the appropriate voltage values at the transition points. (iii) Find v_I when $v_O = 0.4$ V and when $v_O = 2.9$ V. (b) For $(W/L)_n = (W/L)_p = 2$, repeat part (a).

16.30 Consider the CMOS inverter pair in Figure P16.30. Let $V_{TN} = 0.8$ V, $V_{TP} = -0.8$ V, and $K_n = K_p$. (a) If $v_{O1} = 0.6$ V, determine v_I and v_{O2}. (b) Determine the range of v_{O2} for which both N_2 and P_2 are biased in the saturation region.

Figure P16.30   **Figure P16.31**

16.31 Consider the series of CMOS inverters in Figure P16.31. The threshold voltages of the n-channel transistors are $V_{TN} = 0.8$ V, and the threshold voltages of the p-channel transistors are $V_{TP} = -0.8$ V. The conduction parameters are all equal. (a) Determine the range of v_{O1} for which both N_1 and P_1 are biased in the saturation region. (b) If $v_{O2} = 0.6$ V, determine the values of v_{O3}, v_{O1}, and v_I.

16.32 For the CMOS inverter in Figure 16.34, (a) calculate and plot the current through the transistors as a function of the input voltage for $0 \le v_I \le 5$ V. Assume $K_n = K_p = 0.1 \,\text{mA/V}^2$, $V_{TN} = 0.8$ V, $V_{TP} = -0.8$ V, and $V_{DD} = 5$ V. (b) Repeat part (a) for $V_{DD} = 15$ V.

16.33 The transistor parameters in the CMOS inverter are: $k_n' = 50\,\mu\text{A/V}^2$, $k_p' = 25\,\mu\text{A/V}^2$, $V_{TN} = 0.8$ V, and $V_{TP} = -0.8$ V. (a) For $(W/L)_n = 2$ and $(W/L)_p = 4$, determine the peak current in the inverter during a switching cycle for $V_{DD} = 5$ V. (b) Repeat part (a) for $(W/L)_n = (W/L)_p = 2$.

16.34 A load capacitor of $0.2\,\text{pF}$ is connected to the output of a CMOS inverter. Determine the power dissipated in the CMOS inverter for a switching frequency of 10 MHz, for inverter parameters described in (a) Problem 16.32 and (b) Problem 16.33.

16.35 A CMOS digital logic circuit contains the equivalent of 2 million CMOS inverters and is biased at $V_{DD} = 5$ V. (a) The equivalent load capacitance of each inverter is $0.4\,\text{pF}$ and each inverter is switching at 150 MHz. Determine the total average power dissipated in the circuit. (b) If the switching frequency is doubled, but the total power dissipated is to remain the same and the load capacitance remains constant, determine the required bias voltage.

16.36 Consider a CMOS inverter. (a) Show that when $v_I \cong V_{DD}$, the resistance of the NMOS device is approximately $1/[k_n'(W/L)_n(V_{DD} - V_{TN})]$, and when $v_I \cong 0$, the resistance of the PMOS device is approximately $1/[k_p'(W/L)_p(V_{DD} + V_{TP})]$. (b) Using the results of part (a), determine the maximum current that the NMOS device can sink such that the output voltage stays below 0.5 V, and determine the maximum current that the PMOS device can source such that the output voltage does not drop more than 0.5 V below V_{DD}.

16.37 Consider the CMOS inverter in Figure 16.34. Let $K_p = K_n$, $V_{TN} = +1.5$ V, $V_{TP} = -1.5$ V, and $V_{DD} = 10$ V. Determine the two values of v_I and the corresponding values of v_O for which $(dv_O/dv_I) = -1$ on the voltage transfer characteristics. What are the noise margins?

16.38 Repeat Problem 16.37 if the CMOS inverter transistor parameters are: $V_{TN} = +1.5$ V, $V_{TP} = -1.5$ V, $K_n = 100\,\mu\text{A/V}^2$, and $K_p = 50\,\mu\text{A/V}^2$. Let $V_{DD} = 10$ V.

Section 16.4   CMOS Logic Circuits

16.39 Consider the three-input CMOS NAND circuit in Figure P16.39. Assume $k_n' = 2k_p'$ and $V_{TN} = |V_{TP}| = 0.8$ V. (a) If $v_A = v_B = 5$ V, determine v_C such that both N_3 and

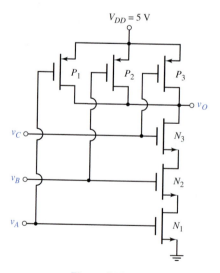

Figure P16.39

P_3 are biased in the saturation region when $(W/L)_p = 2(W/L)_n$. (State any assumptions you make.) (b) If $v_A = v_B = v_C = v_I$, determine the relationship between $(W/L)_p$ and $(W/L)_n$ such that $v_I = 2.5\,\text{V}$ when all transistors are biased in the saturation region. (c) Using the results of part (b) and assuming $v_A = v_B = 5\,\text{V}$, determine v_C such that both N_3 and P_3 are biased in the saturation region. (State any assumptions you make.)

16.40 Consider the circuit in Figure P16.40. (a) The inputs v_X, v_Y, and v_Z listed in the following table are either a logic 0 or a logic 1. These inputs are the outputs from similar-type CMOS logic circuits. The input logic conditions listed are sequential in time. State whether the transistors listed are "on" or "off," and determine the output voltage. (b) What logic function does this circuit implement?

v_X	v_Y	v_Z	N_1	N_2	N_3	N_4	N_5	v_O
1	0	1						
0	0	1						
1	1	0						
1	1	1						

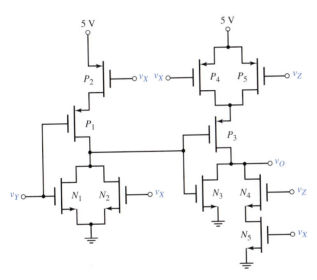

Figure P16.40

D16.41 (a) Given inputs A, B, and C, design a CMOS circuit to implement the logic function $Y = ABC + \overline{ABC}$. (b) For $k'_n = 2k'_p$ and assuming a minimum width-to-length ratio of unity, size the transistors in the design to provide equal composite conduction parameters.

D16.42 (a) Given inputs A, B, C, and D, design a CMOS circuit to implement the logic function $Y = \overline{(A+B)C+D}$. (b) Repeat part (b) of Problem 16.41 for this circuit.

16.43 Determine the logic function implemented by the circuit in Figure P16.43.

D16.44 Consider a five-input CMOS NAND logic gate. Assume that $k'_n = 2k'_p$ and assume that the minimum width-to-length ratio of any single transistor is unity. Design the width-to-length ratio of each transistor such that the composite conduction parameters of the NMOS and PMOS portions of the circuit are equal and such that the composite conduction parameters are equal to those of a CMOS inverter in which $(W/L)_n = 1$ and $(W/L)_p = 2$.

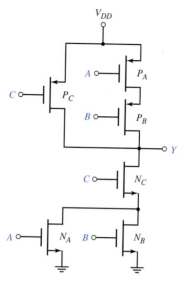

Figure P16.43

*D16.45   (a) Consider a six-input CMOS NOR logic gate whose output is connected to a CMOS inverter, so the output is an OR logic function. Repeat Problem 16.44 for this circuit. (b) Redesign the circuit such that three inputs are connected to one three-input CMOS NOR gate, the other three inputs are connected to another three-input CMOS NOR gate, and the outputs of the NOR gates are connected to a two-input CMOS NAND gate. The output of the NAND gate is still the OR function of the six inputs. Design these logic circuits using the specifications of Problem 16.44. Compare the size of transistors in this design compared to that of part (a). What can be said about the expected propagation delay times of the two circuits?

Section 16.5   Clocked CMOS Logic Circuits

16.46   (a) Figure P16.46 shows a clocked CMOS logic circuit. Make a table showing the state of each transistor ("on" or "off"), and determine the output voltages v_{O1} and v_{O2} for the input logic states listed in the following table. Assume the input conditions are sequential in time from state 1 to state 6. (b) What logic function does the circuit implement?

State	CLK	v_A	v_B	v_C
1	0	0	0	0
2	1	1	0	0
3	0	0	0	0
4	1	0	0	1
5	0	0	0	0
6	1	0	1	1

16.47   (a) For the circuit in Figure P16.47, make a table showing the state of each transistor ("on" or "off"), and determine the output voltages v_{O1}, v_{O2}, and v_{O3} for the input logic states listed in the following table. Assume the input conditions are sequential in time from state 1 to state 6. (b) What logic function does the circuit implement?

Figure P16.46

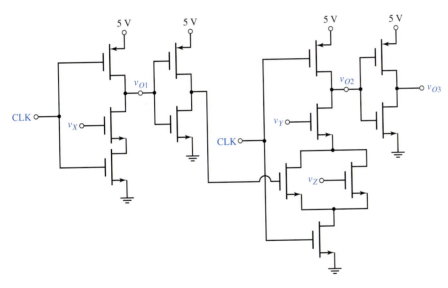

Figure P16.47

State	CLK	v_X	v_Y	v_Z
1	0	0	0	0
2	1	1	1	1
3	0	0	0	0
4	1	0	1	1
5	0	0	0	0
6	1	1	0	1

D16.48 Sketch a clocked CMOS domino logic circuit that realizes the function $Y = ABC + \overline{A}\,\overline{B}\,\overline{C}$.

D16.49 Sketch a clocked CMOS domino logic circuit that realizes the function $Y = (A + B)C + D$.

16.50 Consider the CMOS clocked circuit in Figure 16.52(b). Assume the effective capacitance at the v_{O1} terminal is 25 fF. If the leakage current through the M_{NA} and M_{NB} transistors is $I_{\text{Leakage}} = 2$ pA when these transistors and M_{P1} are cutoff, determine the time for which v_{O1} will decay by 0.5 V.

Section 16.6   Transmission Gates

16.51 The parameters of an NMOS transmission gate are $V_{TN} = 0.8$ V, $K_n = 0.5 \text{ mA/V}^2$, and $C_L = 1$ pF. (a) For a gate voltage of $\phi = 5$ V, determine the quasi-steady-state output voltage for (i) $v_I = 0$, (ii) $v_I = 5$ V, and (iii) $v_I = 2.5$ V. (b) Repeat part (a) for a gate voltage of $\phi = 4$ V.

D16.52 For the circuit in Figure P16.52, the input voltage v_I is either 0.1 V or 5 V. Let $\phi = 5$ V. The threshold voltages are $V_{TN} = -1.5$ V for M_4 and $V_{TN} = 0.8$ V for all other transistors. The width-to-length ratios are 1 for M_2 and M_4 and 10 for M_A and M_B. (a) What are the logic 1 values of v_{O1} and v_{O2}? (b) Design the width-to-length ratios of M_1 and M_3 such that the logic 0 values of v_{O1} and v_{O2} are 0.1 V.

Figure P16.52   **Figure P16.53**

16.53 What is the logic function implemented by the circuit shown in Figure P16.53? Assume that all inputs are either 0 or 5 V.

16.54 Consider the circuit in Figure P16.54. What logic function is implemented by this circuit? Are there any potential problems with this circuit?

16.55 What is the logic function implemented by the circuit in Figure P16.55?

16.56 Consider the circuit in Figure P16.56. Signals ϕ_1 and ϕ_2 are nonoverlapping clock signals. Describe the operation of the circuit and the logic function implemented. Discuss any possible relationship between the width-to-length ratios of the load and driver transistors for "proper" circuit operation.

16.57 The circuit in Figure P16.57 is a form of clocked shift register. Signals ϕ_1 and ϕ_2 are nonoverlapping clock signals. Describe the operation of the circuit. Discuss any possible relationship between the width-to-length ratios of the load and driver transistors for "proper" circuit operation.

Figure P16.54

Figure P16.56

Figure P16.55

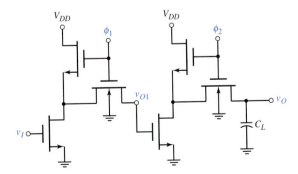

Figure P16.57

Section 16.7   Sequential Logic Circuits

16.58 Consider the NMOS R–S flip-flop in Figure 16.71 biased at $V_{DD} = 5\,\text{V}$. The threshold voltages are $1\,\text{V}$ (enhancement-mode devices) and $-2\,\text{V}$ (depletion-mode devices). The conduction parameters are $K_3 = K_6 = 30\,\mu\text{A/V}^2$, $K_2 = K_5 = 100\,\mu\text{A/V}^2$, and $K_1 = K_4 = 200\,\mu\text{A/V}^2$. If $Q = \text{logic } 0$ and $\overline{Q} = \text{logic } 1$ initially, determine the voltage at S that will cause the flip-flop to change states.

16.59 A CMOS R–S flip-flop is shown in Figure P16.59. Assume $V_{DD} = 5\,\text{V}$, $|V_{TN}| = |V_{TP}| = 1\,\text{V}$, $K_1 = K_2 = K_3 = K_4 \equiv K$, and $K_5 = K_6$. If $Q = \text{logic } 1$ and $\overline{Q} = \text{logic } 0$ initially, determine the relationship between K_5 and K such that the flip-flop changes state when $R = 2.5\,\text{V}$.

D16.60 The CMOS R–S flip-flop in Figure P15.59 is not a fully complementary CMOS design. Design a fully complementary CMOS clocked R–S flip-flop. [Note: the design contains 12 transistors.]

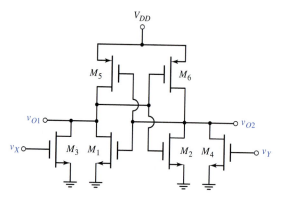

Figure P16.59

D16.61 The circuit in Figure P16.61 is an example of a D flip-flop. (a) Explain the operation of the circuit. Is this a positive- or negative-edge-triggered flip-flop? (b) Redesign the circuit to make this a static flip-flop.

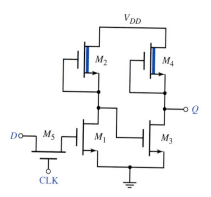

Figure P16.61

16.62 Show that the circuit in Figure P16.62 is a J–K flip-flop.

16.63 Reconsider the circuit shown in Figure P16.40. Show that this circuit is a J–K flip-flop with $J = v_X$, $K = v_Y$, and $\text{CLK} = v_Z$.

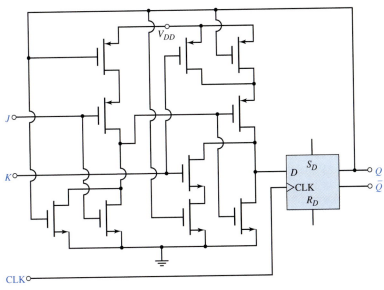

Figure P16.62

Section 16.8   Memories: Classifications and Architectures

16.64 A 64-K memory is organized in a square array and uses the NMOS NOR decoder in Figure 16.81(b) for the row- and column decoders. (a) How many inputs does each decoder require? (b) What input to the row decoder is required to address rows (i) 94 and (ii) 239? (c) What input to the column decoder is required to address columns (i) 39 and (ii) 123?

D16.65 A 1024-bit RAM consists of 128 words of 8 bits each. Design the memory array to minimize the number of row and column address decoder transistors required. How many row and column address lines are necessary?

16.66 Assume that an NMOS address decoder can source $250\,\mu A$ when the output goes high. If the effective capacitance of each memory cell is $C_L = 0.8\,pF$ and the effective capacitance of the address line is $C_{LA} = 5\,pF$, determine the rise time of the address line voltage if $V_{IH} = 2.7\,V$.

Section 16.9   RAM Memory Cells

D16.67 Consider the NMOS RAM cell with resistor load in Figure 16.82(b). Assume parameter values of $k_n' = 35\,\mu A/V^2$, $V_{TN} = 0.7\,V$, $V_{DD} = 5\,V$, and $R = 1\,M\Omega$. (a) Design the width-to-length ratios such that $v_{DS} = 0.1\,V$ for the on transistor. (b) Consider a 16-K memory with the cell described in part (a). Determine the standby current and power of the memory for a standby voltage of $V_{DD} = 2\,V$.

D16.68 A 16-K NMOS RAM, with the cell design shown in Figure 16.82(b), is to dissipate no more than $200\,mW$ in standby when biased at $V_{DD} = 2.5\,V$. Design the width-to-length ratios of the transistors and the resistance value. Assume $V_{TN} = 0.7\,V$ and $k_n' = 35\,\mu A/V^2$.

*16.69 Consider the CMOS RAM cell and data lines in Figure 16.84 biased at $V_{DD} = 5\,V$. Assume transistor parameters $k_n' = 40\,\mu A/V^2$, $k_p' = 20\,\mu A/V^2$, $V_{TN} = 0.8\,V$, $V_{TP} = -0.8\,V$, $W/L = 2$ (M_{N1} and M_{N2}), $W/L = 4$ (M_{P1} and M_{P2}), and $W/L =$

1 (all other transistors). If $Q = 0$ and $\overline{Q} = 1$, determine the voltages at D and $\overline{D}$ a short time after the row has been addressed. Neglect the body effect.

*16.70   Consider the CMOS RAM cell and data lines in Figure 16.84 with circuit and transistor parameters described in Problem 16.69. Assume initially that $Q = 0$ and $\overline{Q} = 1$. Assume the row is selected with $X = 5\,\text{V}$ and assume the data lines, through a write cycle, are at $\overline{D} = 0$ and $D = 4.2\,\text{V}$. Determine the voltages at Q and $\overline{Q}$ a short time after the write cycle voltages are applied.

*16.71   Consider a general sense amplifier configuration shown in Figure 16.90 for a dynamic RAM. Assume that each bit line has a capacitance of $1\,\text{pF}$ and is precharged to $4\,\text{V}$. The storage capacitance is $0.05\,\text{pF}$, the reference capacitance is $0.025\,\text{pF}$, and each are charged to $5\,\text{V}$ for a logic 1 and to $0\,\text{V}$ for a logic 0. The M_S and M_R gate voltages are $5\,\text{V}$ when each cell is addressed and the transistor threshold voltages are $0.5\,\text{V}$. Determine the bit line voltages v_1 and v_2 after the cells are addressed for the case when (a) a logic 1 is stored and (b) a logic 0 is stored.

Section 16.10   Read-Only Memory

D16.72   Design a 4-word $\times$ 4-bit NMOS mask-programmed ROM to produce outputs of 1011, 1111, 0110, and 1001 when rows 1, 2, 3, and 4, respectively, are addressed.

D16.73   Design an NMOS 16×4 mask-programmed ROM that provides the 4-bit product of two 2-bit variables.

D16.74   Design an NMOS mask-programmed ROM that decodes a binary input and produces the output for a seven-segment array. (See Figure 2.40, Chapter 2.) The output is to be high when a particular LED is to be turned on.

COMPUTER SIMULATION PROBLEMS

16.75   The three types of NMOS inverters are shown in Figures 16.5(a), 16.8(a), and 16.10(a). Using PSpice, investigate the voltage transfer characteristics and the current versus input voltage characteristics for the three types of inverters as a function of various width-to-length ratios and as a function of the body effect.

16.76   Again consider the three-types of NMOS inverters. Investigate the propagation delay times and switching characteristics of the three types of inverters using PSpice. Consider a series of inverters as shown in Figure 16.19. Include appropriate transistor capacitance values and assume effective C_T load capacitor values of $0.2\,\text{pF}$. Determine the propagation delay times with and without the body effect. Consider various transistor width-to-length ratios.

16.77   Consider a three-input CMOS NAND logic circuit similar to the two-input circuit shown in Figure 16.45(a). Using PSpice, investigate the voltage transfer characteristics and the current versus input voltage characteristics for various transistor width-to-length ratios and various input conditions similar to the results in Figure 16.46 for the CMOS NOR circuit.

16.78   Investigate the propagation delay times and switching characteristics of the CMOS inverter using PSpice. Set up a series of CMOS inverters similar to the series of inverters shown in Figure 16.19. Include appropriate transistor capacitance values and assume effective C_T load capacitor values of $0.2\,\text{pF}$. Determine the propagation delay times as a function of various transistor width-to-length ratios.

16.79 Consider the dynamic shift register shown in Figure 16.68. Assume appropriate transistor and load capacitance values. Using PSpice, investigate the transient effects in voltages v_{O1}, v_{O2}, v_{O3}, and v_{O4} after the clock signals go to zero.

DESIGN PROBLEMS

*D16.80 Design an NMOS logic circuit that will implement the logic function $Y = (A + (B \cdot C)) \cdot D$.

*D16.81 Design clocked CMOS logic circuits that will implement the logic functions: (a) $Y = [\overline{A + (B \cdot C)}]$, and (b) $Y = [\overline{(A + B) \cdot (C + D)}]$. If the smallest width-to-length ratio is 2, determine the appropriate width-to-length ratios of each transistor in your design.

*D16.82 Design an NMOS pass logic network that implements the logic functions described in Problem 16.81.

*D16.83 Design a clocked CMOS R–S flip-flop such that the output becomes valid on the negative-going edge of a clock signal.

*D16.84 Design a clocked CMOS dynamic shift register in which the output becomes valid on the positive-going edge of a clock signal.

17

Bipolar Digital Circuits

17.0 PREVIEW

In the previous chapter, we presented the basic concepts of MOSFET logic circuits. In this chapter, we discuss the basic principles of bipolar logic circuits. Prior to the emergence of the MOS digital technology, the bipolar digital family of transistor–transistor logic circuits was used extensively. Bipolar digital circuits are now used less frequently because of their relatively large power requirements.

Our study of bipolar digital circuits begins with emitter-coupled logic (ECL). This is the fastest bipolar technology and is used in specialized applications where high speed is required. One price to pay for high speed is a relatively low noise margin. The basis of ECL is a differential amplifier that is operated in the nonlinear region.

A bipolar technology that has a higher noise margin is transistor–transistor logic (TTL). Transistors in this technology are driven between cutoff and saturation. The storage time related to transistors driven into saturation slows the switching speed of TTL compared to that of ECL. Higher speed in TTL is achieved in Schottky TTL circuits. The basic principle of the Schottky transistor is discussed and the transistor is then applied to digital circuits. Low-power Schottky TTL makes a trade-off between speed and power.

The BiCMOS inverter and BiCMOS digital logic circuit are considered. These circuits take advantage of the low-power properties of CMOS and the high-current drive capability of bipolar transistors.

17.1 EMITTER-COUPLED LOGIC (ECL)

The emitter-coupled logic (ECL) circuit is based on the differential amplifier circuit, which we studied in Chapter 11 in the context of linear amplifiers. In digital applications, the diff-amp is driven into its nonlinear region. The transistors are either cut off or in the active region. Saturation is avoided in order to minimize switching times and propagation delay times. ECL circuits have the shortest propagation delay times of any bipolar digital technology.

17.1.1   Differential Amplifier Circuit Revisited

Consider the basic diff-amp circuit in Figure 17.1. For a linear diff-amp, the input voltages are small and both transistors remain biased in the active region at all times. The relationship between collector currents and base–emitter voltages for Q_1 and Q_2 can be written[1]

$$i_{C1} = I_S e^{v_{BE1}/V_T} \tag{17.1(a)}$$

and

$$i_{C2} = I_S e^{v_{BE2}/V_T} \tag{17.1(b)}$$

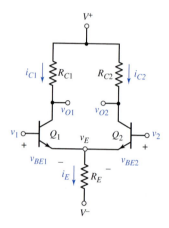

Figure 17.1   Basic differential amplifier circuit

where Q_1 and Q_2 are assumed to be matched and parameter I_S is the same for both devices. The current–voltage transfer curves are shown in Figure 17.2.

In digital applications, the input voltages are large, which means that one transistor remains biased in its active region while the opposite transistor is cut off. For example, if $v_{BE1} = v_{BE2} + 0.12$, then the ratio of i_{C1} and i_{C2} is

$$\frac{i_{C1}}{i_{C2}} = \frac{e^{v_{BE1}/V_T}}{e^{v_{BE2}/V_T}} = e^{(v_{BE1}-v_{BE2})/V_T} = e^{0.12/0.026} = 101 \tag{17.2}$$

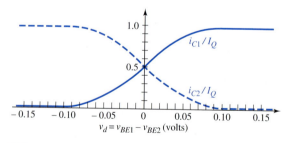

Figure 17.2   Normalized dc transfer characteristics, BJT differential amplifier

---

[1]In most cases in this chapter, total instantaneous current and voltage parameters are used, even though most analyses of logic circuits involve dc calculations.

When the base–emitter voltage of Q_1 is 120 mV greater than the base–emitter voltage of Q_2, the collector current of Q_1 is 100 times that of Q_2; for all practical purposes, Q_1 is on and Q_2 is cut off.

Conversely, if v_1 is less than v_2 by at least 120 mV, then Q_1 is effectively cut off and Q_2 is on. The difference amplifier, when operating as a digital circuit, operates as a current switch. When $v_1 > v_2$ by at least 120 mV, it switches an approximately constant current through R_E to Q_1; when $v_2 > v_1$ by at least 120 mV, the current goes to Q_2.

Example 17.1 **Objective:** Calculate the currents and voltages in the basic differential amplifier circuit used as a digital circuit.

Consider the circuit in Figure 17.1. Assume that $V^+ = -V^- = 5$ V, $R_{C1} = R_{C2} \equiv R_C = 1$ kΩ, $R_E = 2.15$ kΩ, and $v_2 = 0$. In the dc analysis, assume that dc base currents are negligible.

Solution: For $v_1 = 0$, both transistors are on. Assuming a base–emitter turn-on voltage of 0.7 V, then $v_E = -0.7$ V and

$$i_E = \frac{v_E - V^-}{R_E} = \frac{-0.7 - (-5)}{2.15} = 2.0 \text{ mA}$$

Assuming Q_1 and Q_2 are matched, we have $i_{C1} = i_{C2} = i_E/2$ since $v_{BE1} = v_{BE2}$ and $i_{C1} = i_{C2} \equiv i_C = 1$ mA. In this case,

$$v_{O1} = v_{O2} = V_{CC} - i_C R_C = 5 - (1)(1) = 4 \text{ V}$$

Both Q_1 and Q_2 are now biased in the active region.

Now let $v_1 = -1$ V. Since the base voltage of Q_1 is less than the base voltage of Q_2 by more than 120 mV, Q_1 is cut off and Q_2 is on. In this case, $v_E = v_2 - V_{BE}(\text{on}) = -0.7$ V and $i_E = 2$ mA, as before. However, $i_{C1} = 0$ and $i_{C2} = i_E = 2$ mA, so that

$$v_{O1} = V_{CC} = 5 \text{ V}$$

and

$$v_{O2} = V_{CC} - i_{C2} R_C = 5 - (2)(1) = 3 \text{ V}$$

For $v_1 = +1$ V, Q_1 is on and Q_2 is cut off. For this case, $v_E = v_1 - V_{BE}(\text{on}) = 1 - 0.7 = +0.3$ V, the current i_E is

$$i_E = i_{C1} = \frac{v_E - V^-}{R_E} = \frac{0.3 - (-5)}{2.15} = 2.47 \text{ mA}$$

and

$$v_{O1} = V_{CC} - i_{C1} R_C = 5 - (2.47)(1) = 2.53 \text{ V}$$

and

$$v_{O2} = V_{CC} = 5 \text{ V}$$

Comment: For the three conditions given, transistors Q_1 and Q_2 are biased either in cutoff or in the active region. In terms of digital applications, output v_{O2} is in phase with input v_1 and output v_{O1} is 180 degrees out of phase.

When biased on, transistor Q_1 conducts slightly more heavily than Q_2 when it is conducting. To obtain symmetrical complementary outputs, R_{C1} should therefore be slightly smaller than R_{C2}.

Test Your Understanding

D17.1 Consider the differential amplifier circuit in Figure 17.1 biased at $V^+ = 5\,\text{V}$, $V^- = -5\,\text{V}$, and $v_2 = 0$. Assume $V_{BE}(\text{on}) = 0.7\,\text{V}$ and neglect base currents. (a) Design the circuit such that $i_E = 1\,\text{mA}$ and $v_{O1} = v_{O2} = 3.5\,\text{V}$ when $v_1 = 0$. (b) Using the results of part (a), calculate i_E, v_{O1}, and v_{O2} for: (i) $v_1 = +1\,\text{V}$, and (ii) $v_1 = -1\,\text{V}$. (Ans. (a) $R_E = 4.3\,\text{k}\Omega$, $R_{C1} = R_{C2} = 3\,\text{k}\Omega$ (b) (i) $i_E = 1.23\,\text{mA}$, $v_{O1} = 1.31\,\text{V}$, $v_{O2} = 5\,\text{V}$ (ii) $v_{O2} = 2\,\text{V}$, $v_{O1} = 5\,\text{V}$)

17.1.2   Basic ECL Logic Gate

A basic two-input ECL OR/NOR logic circuit is shown in Figure 17.3. The two input transistors, Q_1 and Q_2, are connected in parallel. On the basis of the differential amplifier, if both v_X and v_Y are less than the reference voltage V_R (by at least 120 mV), then both Q_1 and Q_2 are cut off, while the reference transistor Q_R is biased on its active region. In this situation, the output voltage v_{O1} is greater than v_{O2}. If either v_X or v_Y becomes greater than V_R, then Q_R turns off and v_{O2} becomes larger than v_{O1}. The OR logic is at the v_{O2} output and the NOR logic is at the v_{O1} output. An advantage of ECL gates is the availability of complementary outputs, precluding the need for separate inverters to provide the complementary outputs.

One problem with the OR/NOR circuit in Figure 17.3 is that the output voltage levels differ from the required input voltage levels; the output voltages are not compatible with the input voltages. The mismatch arises because ECL

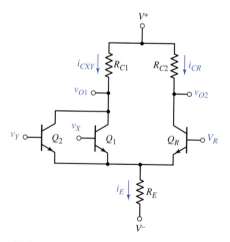

Figure 17.3   Basic two-input ECL OR/NOR logic circuit

circuit transistors operate between their cutoff and active regions, requiring that the base–collector junctions be reverse biased at all times. We see that a logic 1 voltage of the output is $V_{OH} = V^+$. If this voltage were to be applied to either the v_X or v_Y input, then either Q_1 or Q_2 would turn on and the collector voltage v_{O1} would decrease below V^+; the base–collector voltage would then become forward biased and the transistor would go into saturation. Emitter-follower circuits are added to provide outputs that are compatible with the inputs of similar gates.

Test Your Understanding

D17.2 For the ECL logic gate in Figure 17.3, the bias voltages are: $V^+ = 3.5\,\text{V}$, $V^- = -3.5\,\text{V}$, and $V_R = 1.5\,\text{V}$. Assume $V_{BE}(\text{on}) = 0.7\,\text{V}$ and neglect base currents. (a) Determine R_E and R_{C2} such that $i_E = 2\,\text{mA}$ and $v_{O2} = 2\,\text{V}$ when $v_x = v_y = $ logic 0. (b) Find R_{C1} such that $v_{O1} = 2\,\text{V}$ when $v_x = v_y = 2\,\text{V}$. What is i_E? (Ans. (a) $R_E = 2.15\,\text{k}\Omega$, $R_{C2} = 0.75\,\text{k}\Omega$ (b) $i_E = 2.23\,\text{mA}$, $R_{C1} = 0.673\,\text{k}\Omega$)

ECL Logic Gate with Emitter Followers

In the ECL circuit in Figure 17.4, emitter followers are added to the OR/NOR outputs, and supply voltage V^+ is set equal to zero. The ground and power supply voltages are reversed because analyses show that using the collector–emitter voltage as the output results in less noise sensitivity. If the forward current gain of the transistors is on the order of 100, then the dc base currents may be neglected with little error in the calculations.

Figure 17.4 Two-input ECL OR/NOR logic gate with emitter-follower output stages

If either v_X or v_Y is a logic 1 (defined as greater than V_R by at least 120 mV), then the reference transistor Q_R is cut off, $i_{CR} = 0$, and $v_{O2} = 0$. Output transistor Q_3 is biased in the active region, and $v_{OR} = v_{O2} - V_{BE}(\text{on}) = -0.7\,\text{V}$. If both v_X and v_Y are a logic 0 (defined as less than V_R by at least 120 mV), then both Q_1 and Q_2 are cut off, $v_{O1} = 0$, and $v_{\text{NOR}} = 0 - V_{BE}(\text{on}) = -0.7\,\text{V}$. The largest possible voltage that can be achieved at either output is $-0.7\,\text{V}$; therefore, $-0.7\,\text{V}$ is defined as the logic 1 level.

In the following example, we will determine the currents and the logic 0 values in the basic ECL gate.

Example 17.2 **Objective:** Calculate current, resistor, and logic 0 values in the basic ECL logic gate.

Consider the circuit in Figure 17.4. Determine R_{C1} and R_{C2} such that when Q_1, Q_2, and then Q_R are conducting, the B–C voltages are zero.

Solution: Let $v_X = v_Y = -0.7\,\text{V} = \text{logic 1} > V_R$ such that Q_1 and Q_2 are on. We find that

$$v_E = v_X - V_{BE}(\text{on}) = -0.7 - 0.7 = -1.4\,\text{V}$$

and the current is

$$i_E = i_{Cxy} = \frac{v_E - V^-}{R_E} = \frac{-1.4 - (-5.2)}{1.18} = 3.22\,\text{mA}$$

In order for the B–C voltages of Q_1 and Q_2 to be zero, voltage v_{O1} must be $-0.7\,\text{V}$. Therefore

$$R_{C1} = \frac{-v_{O1}}{I_{Cxy}} = \frac{0.7}{3.22} = 0.217\,\text{k}\Omega$$

The NOR output logic 0 value is then

$$v_{\text{NOR}} = v_{O1} - V_{BE}(\text{on}) = -0.70 - 0.7 = -1.40\,\text{V}$$

Input voltages v_X and v_Y are greater than V_R in a logic 1 state and less than V_R in a logic 0 state. If V_R is set at the midpoint between the logic 0 and logic 1 levels, then

$$V_R = \frac{-0.7 - 1.40}{2} = -1.05\,\text{V}$$

When Q_R is on, we have

$$v_E = V_R - V_{BE}(\text{on}) = -1.05 - 0.7 = -1.75\,\text{V}$$

and

$$i_E = i_{CR} = \frac{v_E - V^-}{R_E} = \frac{-1.75 - (-5.2)}{1.18} = 2.92\,\text{mA}$$

For $v_{O2} = -0.7\,\text{V}$, we find that

$$R_{C2} = \frac{-v_{O2}}{i_{C2}} = \frac{0.7}{2.92} = 0.240\,\text{k}\Omega$$

The OR logic 0 value is therefore

$$v_{OR} = v_{O2} - V_{BE}(\text{on}) = -0.7 - 0.7 = -1.40\,\text{V}$$

Comment: For symmetrical complementary outputs, R_{C1} and R_{C2} are not equal. If R_{C1} and R_{C2} become larger than the designed values, transistors Q_1, Q_2, and Q_R will be driven into saturation when they are conducting.

Test Your Understanding

D17.3 Redesign the ECL circuit in Figure 17.4 such that the logic 0 values at the v_{OR} and v_{NOR} terminals are $-1.5\,\text{V}$. The maximum value of i_E is to be $2.5\,\text{mA}$, and the maximum values of i_3 and i_4 are to be $2.5\,\text{mA}$. The bias voltages are as shown. Determine all resistor values and the value of V_R. (Ans. $R_E = 1.52\,\text{k}\Omega$, $R_{C1} = 320\,\Omega$, $V_R = -1.1\,\text{V}$, $R_{C2} = 357\,\Omega$, $R_3 = R_4 = 1.8\,\text{k}\Omega$)

17.4 Using the results of Example 17.2, calculate the power dissipated in the circuit in Figure 17.4; for: (a) $v_x = v_y =$ logic 1, and (b) $v_x = v_y =$ logic 0. (Ans. (a) $P = 45.5\,\text{mW}$ (b) $P = 43.9\,\text{mW}$)

The Reference Circuit

Another circuit is required to provide the reference voltage V_R. Consider the complete two-input ECL OR/NOR logic circuit shown in Figure 17.5. The reference circuit consists of resistors R_1, R_2, and R_5, diodes D_1 and D_2, and transistor Q_5. The reference portion of the circuit can be specifically designed to provide the desired reference voltage.

Figure 17.5 Basic ECL logic gate with reference circuit

Design Example 17.3 **Objective:** Design the reference portion of the ECL circuit.

Consider the circuit in Figure 17.5. The reference voltage V_R is to be $-1.05\,\text{V}$.

Solution: We know that

$$v_{B5} = V_R + V_{BE}(\text{on}) = -1.05 + 0.7 = -0.35\,\text{V} = -i_1 R_1$$

Since there are two unknowns, we will choose one variable. Let $R_1 = 0.25\,\text{k}\Omega$. Then,

$$i_1 = \frac{0.35}{0.25} = 1.40\,\text{mA}$$

Since this current is on the same order of magnitude as other currents in the circuit, the chosen value of R_1 is reasonable. Neglecting base currents, we can now write

$$i_1 = i_2 = \frac{0 - 2V_\gamma - V^-}{R_1 + R_2}$$

where V_γ is the diode turn-on voltage and is assumed to be $V_\gamma = 0.7\,\text{V}$. We then have

$$1.40 = \frac{-1.4 - (-5.2)}{R_1 + R_2}$$

which yields

$$R_1 + R_2 = 2.71\,\text{k}\Omega$$

Since $R_1 = 0.25\,\text{k}\Omega$, resistance R_2 is $R_2 = 2.46\,\text{k}\Omega$.

Also, we know that

$$i_5 = \frac{V_R - V^-}{R_5}$$

If we let $i_5 = i_1 = i_2 = 1.40\,\text{mA}$, then

$$R_5 = \frac{V_R - V^-}{i_5} = \frac{-1.05 - (-5.2)}{1.40} = 2.96\,\text{k}\Omega$$

 Comment: As with any design, there is no unique solution. The design presented will provide the required reference voltage to the base of Q_R.

17.1.3   ECL Logic Circuit Characteristics

In this section, we will determine the power dissipation, fanout, and propagation delay times for the ECL logic gate. We will also examine the advantage of using a negative power supply.

Power Dissipation

Power dissipation is an important characteristic of a logic circuit. The power dissipated in the basic ECL logic gate in Figure 17.5 is given by

$$P_D = (i_{Cxy} + i_{CR} + i_5 + i_1 + i_3 + i_4)(0 - V^-) \tag{17.3}$$

Example　17.4　**Objective:**　Calculate the power dissipated in the ECL logic circuit.
　Consider the circuit in Figure 17.5. Let $v_X = v_Y = -0.7\,\text{V} = \text{logic 1}$.

Solution:　From our previous analysis, we have $i_{Cxy} = 3.22\,\text{mA}$, $i_{CR} = 0$, $i_5 = 1.40\,\text{mA}$, and $i_1 = 1.40\,\text{mA}$, and the output voltages are $v_{OR} = -0.7\,\text{V}$ and $v_{NOR} = -1.40\,\text{V}$. The currents i_3 and i_4 are

$$i_3 = \frac{v_{OR} - V^-}{R_3} = \frac{-0.7 - (-5.2)}{1.5} = 3.0\,\text{mA}$$

and

$$i_4 = \frac{v_{NOR} - V^-}{R_4} = \frac{-1.40 - (-5.2)}{1.5} = 2.53\,\text{mA}$$

The power dissipation is then

$$P_D = (3.22 + 0 + 1.40 + 1.40 + 3.0 + 2.53)(5.2) = 60.0\,\text{mW}$$

Comment:　This power dissipation is significantly larger than that in NMOS and CMOS logic circuits. The advantage of ECL, however, is the short propagation delay times, which can be less than 1 ns.

Propagation Delay Time

The major advantage of ECL circuits is their small propagation delay time, on the order of 1 ns or less. The two reasons for the short propagation delay times are: (1) the transistors are not driven into saturation, which eliminates any charge storage effects; and (2) the logic swing in the ECL logic gate is small (about 0.7 V), which means that the voltages across the output capacitances do not have to change as much as in other logic circuits. Also, the currents in the ECL circuit are relatively large, which means that these capacitances can charge and discharge quickly. However, the trade-offs for the small propagation delay time are higher power dissipation and smaller noise margins.

　ECL circuits are very fast, and they require that special attention be paid to transmission line effects. Improperly designed ECL circuit boards can experience ringing or oscillations. These problems have less to do with the ECL circuits than with the interconnections between the circuits. Care must therefore be taken to terminate the signal lines properly.

Fanout

Figure 17.6 shows the emitter-follower output stage of the OR output of an ECL circuit used to drive the diff-amp input stage of an ECL load circuit. When v_{OR} is a logic 0, input load transistor Q_1' is cut off, effectively eliminating any load current from the driver output stage. With v_{OR} at a logic 1 level, the input load transistor is on and an input base current i_L' exists. (Up to this point, we have neglected dc base currents; however, they are not zero.) The load current must be supplied through Q_3, whose base current is supplied through R_{C2}. As the load current i_L increases with the addition of more load circuits, a voltage drop occurs across R_{C2} and the output voltage decreases. The maximum fanout is determined partially by the maximum amplitude that the output voltage is allowed to drop from its ideal logic 1 value.

Figure 17.6 Output stage of ECL logic gate driving N identical ECL input stages

---

Example 17.5   Objective: Calculate the maximum fanout of an ECL logic gate, based on dc loading effects.

Consider the circuit in Figure 17.6. Assume the current gain of the transistors is $\beta = 50$, which represents a worst-case scenario. Assume that the logic 1 level at the OR output is allowed to decrease by 50 mV at most from a value of -0.70 V to -0.75 V.

Solution: From the figure, we see that

$$i_E' = \frac{v_{OR} - V_{BE}(\text{on}) - V^-}{R_E} = \frac{-0.75 - 0.7 - (-5.2)}{1.18} = 3.18\,\text{mA}$$

The input base current to the load transistor is

$$i_B' = \frac{i_E'}{(1 + \beta)} = \frac{3.18}{51} \Rightarrow 62.4\,\mu\text{A} = i_L'$$

The total load current is therefore $i_L = N i_L'$.

The base current i_{B3} required to produce both the load current i_L and current i_3 is

$$i_{B3} = \frac{i_3 + i_L}{(1 + \beta)} = \frac{0 - v_{B3}}{R_{C2}} = \frac{0 - (v_{OR} + V_{BE}(\text{on}))}{R_{C2}} \qquad (17.4)$$

Also, from the figure we see that

$$i_3 = \frac{v_{OR} - V^-}{R_3} = \frac{-0.75 - (-5.2)}{1.5} = 2.97\,\text{mA}$$

From Equation (17.4), the maximum fanout for this condition is

$$\frac{2.97 + N(0.0624)}{51} = \frac{0 - (-0.75 + 0.7)}{0.24}$$

which yields $N = 122$.

Comment: This maximum fanout is based on dc conditions and is unrealistic. In practice, the maximum fanout for ECL circuits is determined by the propagation delay time. Each load circuit increases the load capacitance by approximately 3 pF. A maximum fanout of about 15 is usually recommended to keep the propagation delay time within specified limits.

The Negative Supply Voltage

In classic ECL circuits, it is common practice to ground the positive terminal of the supply voltage, reducing the noise signals at the output terminal. Figure 17.7(a) shows an emitter-follower output stage with the supply voltage V_{CC} in series with a noise source V_n. The noise signal may be induced by the effect of switching currents interacting with parasitic inductances and capacitances. The output voltage is measured with respect to ground; therefore, if the positive terminal of V_{CC} is grounded, voltage V_o is taken as the output voltage. If the negative terminal of V_{CC} is at ground, then V_o' is the output voltage.

(a)   **(b)**

Figure 17.7 (a) Equivalent circuit, ECL emitter-follower output stage and noise generator, and the (b) small-signal hybrid-π equivalent circuit

To determine the effect of the noise voltage at the output, we assume that Q_R is cut off, and we evaluate the small-signal hybrid-π equivalent circuit shown in Figure 17.7(b).

Example 17.6 **Objective:** Determine the effect of a noise signal on the output of an ECL gate.

Consider the small-signal equivalent circuit in Figure 17.7(b). Let $\beta = 100$. Find V_o and V_o' as a function of V_n.

Solution: From a previous analysis, the quiescent collector current in Q_3 for Q_R in cutoff is 3 mA. Then,

$$r_{\pi 3} = \frac{\beta V_T}{I_{CQ}} = \frac{(100)(0.026)}{3} = 0.867 \, \text{k}\Omega$$

and

$$g_{m3} = \frac{I_{CQ}}{V_T} = \frac{3}{0.026} = 115 \, \text{mA/V}$$

We can also write that

$$V_n = I_{b3}(R_{C2} + r_{\pi3}) + (1 + \beta)I_{b3}R_3$$

which yields

$$I_{b3} = \frac{V_n}{R_{C2} + r_{\pi3} + (1 + \beta)R_3} = \frac{V_n}{0.24 + 0.867 + (101)(1.5)} = \frac{V_n}{152.6}$$

The output voltage V_o is

$$V_o = -I_{b3}(R_{C2} + r_{\pi3}) = -\left(\frac{V_n}{152.6}\right)(0.24 + 0.867) = -0.0073 V_n$$

and output voltage V_o' is

$$V_o' = (1 + \beta)I_{b3}R_3 = (101)\left(\frac{V_n}{152.6}\right)(1.5) = 0.99 V_n$$

Comment: The effect of noise on the collector–emitter output voltage V_o is much less than on output voltage V_o'. It is advantageous, then, to use V_o, which implies that the positive terminal of V_{CC} is grounded. The noise insensitivity gained with a negative power supply may be critical in a logic circuit with a low noise margin.

17.1.4   Voltage Transfer Characteristics

The voltage transfer curve indicates the circuit characterisics during transition between the two logic states. The voltage transfer characteristics can also be used to determine the noise margins.

DC Analysis

A good approximation of the voltage transfer characteristics can be derived from the piecewise linear model of the two input transistors and the reference transistor. Consider the ECL gate in Figure 17.5. If inputs v_X and v_Y are a logic 0, or -1.40 V, then Q_1 and Q_2 are cut off and $v_{\text{NOR}} = -0.7$ V. The reference transistor Q_R is on and, as previously seen, $i_E = i_{C2} = 2.92$ mA, $v_{B3} = -0.70$ V, and $v_{\text{OR}} = -1.40$ V. As long as $v_X = v_Y$ remains less than $V_R - 0.12 = -1.17$ V, the output voltages do not change from these values. During the interval when the inputs are within 120 mV of reference voltage V_R, the output voltage levels vary.

When $v_X = v_Y = V_R + 0.12 = -0.93$ V, then Q_1 and Q_2 are on and Q_R is off. At this point, $i_E = i_{C1} = 3.03$ mA, $v_{B4} = -0.657$ V, and $v_{\text{NOR}} = -1.36$ V. As determined previously, when $v_X = v_Y = -0.7$ V, $v_{\text{NOR}} = -1.40$ V. The voltage transfer curves are shown in Figure 17.8.

Noise Margin

For the ECL gate, we define the threshold logic levels V_{IL} and V_{IH} as the points of discontinuity in the voltage transfer curves. These values are $V_{IL} = -1.17$ V

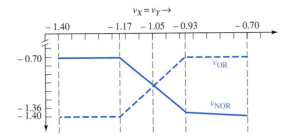

Figure 17.8 ECL OR/NOR logic gate voltage transfer characteristics

and $V_{IH} = -0.93\,\text{V}$. The high logic level is $V_{OH} = -0.7\,\text{V}$ and the low logic value is $V_{OL} = -1.40\,\text{V}$.

The noise margins are defined as

$$\text{NM}_H = V_{OH} - V_{IH} \qquad\qquad\qquad\qquad (17.5(a))$$

and

$$\text{NM}_L = V_{IL} - V_{OL} \qquad\qquad\qquad\qquad (17.5(b))$$

Using the results from Figure 17.8, we find that $\text{NM}_H = 0.23\,\text{V}$ and $\text{NM}_L = 0.23\,\text{V}$. The noise margins for the ECL circuit are considerably lower than those calculated for NMOS and CMOS circuits.

Test Your Understanding

17.5 Consider the ECL circuit in Figure 17.4. Using the results of Example 17.2, plot the voltage transfer characteristics for $-1.40 \le v_x = v_y \le -0.7\,\text{V}$. Find the noise margins NM_H and NM_L. (Ans. $\text{NM}_H = 0.23\,\text{V}$, $\text{NM}_L = 0.23\,\text{V}$)

17.2 MODIFIED ECL CIRCUIT CONFIGURATIONS

The large power dissipation in the basic ECL logic gate makes this circuit impractical for large-scale integrated circuits. Certain modifications can simplify the circuit design and decrease the power consumption, making the ECL more compatible with integrated circuits.

17.2.1 Low-Power ECL

Figure 17.9(a) shows a basic ECL OR/NOR logic gate with reference voltage V_R and a positive voltage supply. We can make the output voltage states compatible with the input voltages, eliminating the need for the emitter-follower output stages. In some applications, both complementary outputs may not be required. If, for example, only the OR output is required, then we can eliminate resistor R_{C1}. Removing this resistor does not reduce the circuit power consumption, but it eliminates one element.

(a)   **(b)**

Figure 17.9   (a) Basic ECL OR/NOR logic gate and (b) modified ECL logic gate

Figure 17.9(b) shows the modified ECL gate. For $v_x = v_y$ logic $1 > V_R$, transistors Q_1 and Q_2 are turned on and Q_R is off. The output voltage is $v_{OR} = V_{CC}$. For $v_x = v_y = $ logic $0 < V_R$, then Q_1 and Q_2 are off and Q_R is on. The currents are

$$i_E = \frac{V_R - V_{BE}(\text{on})}{R_E} \cong i_{CR} \tag{17.6}$$

and the output voltage is

$$v_{OR} = V_{CC} - i_{CR}R_{C2} \tag{17.7}$$

If the resistance values of R_E and R_{C2} vary from one circuit to another because of fabrication tolerances, then current i_E and the logic 0 output voltage will vary from one circuit to another.

To establish a well-defined logic 0 output, we can insert a Schottky diode in parallel with resistor R_C, as shown in Figure 17.10. If the two inputs are a logic 0, then Q_1 and Q_2 are off and Q_R is on. For this condition, we want the Schottky diode to turn on. The output will then be $v_{OR} = V_{CC} - V_\gamma$, where

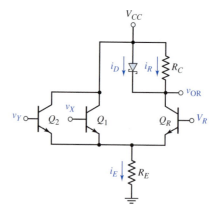

Figure 17.10   Modified ECL logic gate with Schottky diode

V_γ is the turn-on voltage of the Schottky diode. This logic 0 output voltage is a well-defined value. If the diode turns on, then current i_R is limited to $i_R(\max) = V_\gamma/R_C$. Since we must have $i_E > i_R(\max)$, the diode current is $i_D = i_E - i_R(\max)$.

Example 17.7 **Objective:** Analyze the modified ECL logic gate.

Consider the circuit in Figure 17.10 with parameters $V_{CC} = 1.7\,\text{V}$ and $R_E = R_C = 8\,\text{k}\Omega$. Assume the diode and transistor piecewise linear parameters are $V_\gamma = 0.4\,\text{V}$ and $V_{BE}(\text{on}) = 0.7\,\text{V}$.

Solution: The output voltage values are

$$v_{OR} = \text{logic } 1 = V_{CC} = 1.7\,\text{V}$$

and

$$v_{OR} = \text{logic } 0 = V_{CC} - V_\gamma = 1.7 - 0.4 = 1.3\,\text{V}$$

For the output voltages to be compatible with the inputs, the reference voltage V_R must be the average of the logic 1 and logic 0 values, or $V_R = 1.5\,\text{V}$. If $v_x = v_y = \text{logic } 0 = 1.3\,\text{V}$, then Q_R is on. Therefore,

$$i_E = \frac{V_R - V_{BE}(\text{on})}{R_E} = \frac{1.5 - 0.7}{8} \Rightarrow 100\,\mu\text{A}$$

The maximum current in R_C is

$$i_R(\max) = \frac{V_\gamma}{R_C} = \frac{0.4}{8} \Rightarrow 50\,\mu\text{A}$$

and the current through the diode is

$$i_D = i_E - i_R(\max) = 100 - 50 = 50\,\mu\text{A}$$

For $v_x = v_y = \text{logic } 0$, the power dissipation is $P = i_E V_{CC}$, or

$$P = i_E V_{CC} = (100)(1.7) = 170\,\mu\text{W}$$

For $v_x = v_y = \text{logic } 1 = 1.7\,\text{V}$, we have

$$i_E = \frac{v_x - V_{BE}(\text{on})}{R_E} = \frac{1.7 - 0.7}{8} \Rightarrow 125\,\mu\text{A}$$

Therefore, the power dissipation for this condition is

$$P = i_E V_{CC} = (125)(1.7) = 213\,\mu\text{W}$$

Comment: If the resistance values of R_E and R_C were to change by as much as ± 20 percent as a result of manufacturing tolerances, for example, the currents would still be sufficient to turn the Schottky diode on when Q_R is on. This means that the logic 0 output is well defined. Also, the power dissipation in this ECL gate is considerably less than that in the classic ECL OR/NOR logic circuit. The reduced power is a result of fewer components, lower bias voltage, and smaller currents.

When transistor Q_R is off, its collector voltage is 1.7 V and the B–C junction is reverse biased by 0.2 V. When Q_R is conducting, its collector voltage is 1.3 V, the B–C junction is forward biased by 0.2 V, and the transistor is biased slightly in saturation. However, this slight saturation bias does not degrade the switching of Q_R, so the fast-switching characteristic of the ECL circuit is retained.

Test Your Understanding

D17.6   Design the basic ECL logic gate in Figure 17.11 such that the maximum power dissipation is $0.2\,\text{mW}$ and the logic swing is $0.4\,\text{V}$. (Ans. $I_Q = 118\,\mu\text{A}$, $R_C = 3.39\,\text{k}\Omega$, $V_R = 1.5\,\text{V}$)

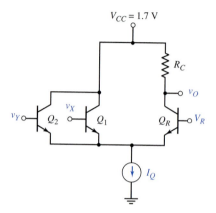

Figure 17.11   Figure for Exercise 17.6

17.2.2   Alternative ECL Gates

In an ECL system, as in all digital systems, a gate is used to drive other logic gates. Connecting load circuits to the basic ECL gate demonstrates changes that can be made to incorporate ECL into integrated circuits more effectively.

Figure 17.12 shows the basic ECL gate with two load circuits. In this configuration, the collectors of Q_2' and Q_2'' are at the same potential, as are the bases of the two transistors. We can therefore replace Q_2' and Q_2'' by a single multiemitter transistor.

In Figure 17.13, the multiemitter transistor Q_O is part of the driver circuit. The operation of the circuit is as follows:

- $v_x = v_y = \text{logic } 1 = 1.7\,\text{V}$: The two input transistors Q_1 and Q_2 are on, Q_R is off, and $v_O = 1.7\,\text{V}$. Since the base voltage of Q_O is higher than the base voltages of Q_R' and Q_R'', then Q_O is conducting, Q_R' and Q_R'' are off, and $v_E' = v_E'' = 1.7 - 0.7 = 1.0\,\text{V}$. The currents i_E' and i_E'' flow through the emitters of Q_O. The output voltages are $v_O' = v_O'' = 1.7\,\text{V}$.
- $v_x = v_y = \text{logic } 0 = 1.3\,\text{V}$: For this case, the two input transistors Q_1 and Q_2 are off, Q_R is on, and $v_O = 1.3\,\text{V}$. The output transistor Q_O is off and both Q_R' and Q_R'' are on. The output voltages are then $v_O' = v_O'' = 1.3\,\text{V}$.

The two load circuits in Figure 17.13 each have only a single input, which limits the circuit functionality. The versatility of the circuit can be further enhanced by making the load transistor Q_R' a multiemitter transistor. This is shown in Figure 17.14. For simplicity, we show only a single input transistor to each of the two driver circuits. The operation of this circuit for various combinations of input voltages is as follows.

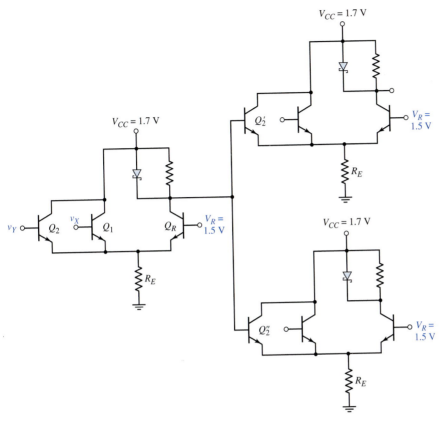

Figure 17.12   Modified ECL logic gate with two load circuits

- $v_1 = v_2 =$ logic $0 = 1.3$ V: The two input transistors Q_1 and Q_2 are off and the two reference transistors Q_{R1} and Q_{R2} are on. This means that $v_{O1} = v_{O2} = 1.3$ V and both output transistors Q_{O1} and Q_{O2} are off. Both emitters of Q'_R are forward biased, currents i_{E1} and i_{E2} flow through Q'_R, and the output voltage is $v'_O =$ logic $0 = 1.3$ V.

- $v_1 = 1.7$ V, $v_2 = 1.3$ V: For this case, Q_1 is on, Q_{R1} is off, Q_2 is off, and Q_{R2} is on. The output voltages are $v_{O1} = 1.7$ V and $v_{O2} = 1.3$ V. This means that Q_{O1} is on and Q_{O2} is off. With Q_{O1} on, current i_{E1} flows through Q_{O1} and no current flows in emitter E_1. With Q_{O2} off, emitter E_2 is forward biased, current i_{E2} flows through Q'_R, and the output voltage is $v'_O =$ logic $0 = 1.3$ V.

- $v_1 = 1.3$ V, $v_2 = 1.7$ V: This case is the complement of the one just discussed. Here, Q_{O1} is off and Q_{O2} is on. This means that i_{E1} flows through emitter E_1 of Q'_R, and i_{E2} flows through Q_{O2}. The output voltage is $v'_O =$ logic $0 = 1.3$ V.

- $v_1 = v_2 = 1.7$ V: The two input transistors Q_1 and Q_2 are on, the two reference transistors Q_{R1} and Q_{R2} are off, and $v_{O1} = v_{O2} = 1.7$ V. This means that both Q_{O1} and Q_{O2} are on and Q'_R is off. Currents i_{E1} and i_{E2} flow through Q_{O1} and Q_{O2}, respectively, and the output voltage is $v'_O =$ logic $1 = 1.7$ V.

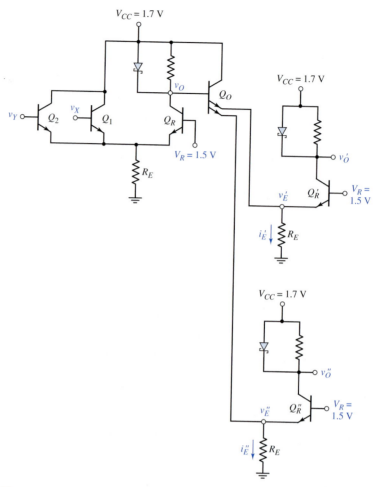

Figure 17.13 Modified ECL logic gate with multiemitter output transistor and two load circuits

These results are summarized in Table 17.1, which shows that this circuit performs the AND logic function. A more complicated or sophisticated logic function can be performed if multiple inputs are used in the driver circuits.

In integrated circuits, resistors R_E are replaced by current sources using transistors. Replacing resistors with transistors in integrated circuits usually results in reduced chip area.

Table 17.1 Summary of results for the ECL circuit in Figure 17.14

$v_1(V)$	$v_2(V)$	$v_o'(V)$
1.3	1.3	1.3
1.7	1.3	1.3
1.3	1.7	1.3
1.7	1.7	1.7

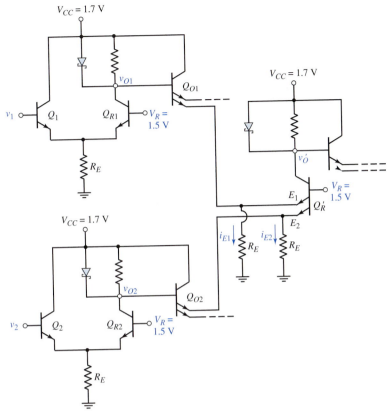

Figure 17.14 Two ECL driver circuits with a multi-input load circuit

17.2.3 Series Gating

Series gating is a bipolar logic circuit technique that allows complex logic functions to be performed with a minimum number of devices and with maximum speed. Series gating is formed by using cascode stages.

Figure 17.15(a) shows the basic emitter-coupled pair, and Figure 17.15(b) shows a cascode stage, also referred to as two-level series gating. Reference voltage V_{R1} is approximately 0.7 V greater than reference voltage V_{R2}. The input voltages v_x and v_y must also be shifted approximately 0.7 V with respect to each other.

As an example, we use the multiemitter load circuit from Figure 17.14 as part of a cascode configuration as shown in Figure 17.16. Transistors Q_{O1}, Q_{O2}, and Q_{O3} represent the output transistors of three ECL driver circuits. We assume a logic 1 level of 2.5 V and a logic 0 level of 2.1 V. The 0.4 V logic swing results from incorporating a Schottky diode in each output stage.

With three input signals, there are eight possible combinations of input states. We will only consider two combinations here:

- $A = B = C = \text{logic } 0 = 2.1 \text{ V}$: In this case, transistors Q_{O1} and Q_{O2} are off and transistor Q_1 is off. This means that current I_Q flows through Q_2 and Q_R, and $v_O = \text{logic } 0 = 2.1 \text{ V}$.

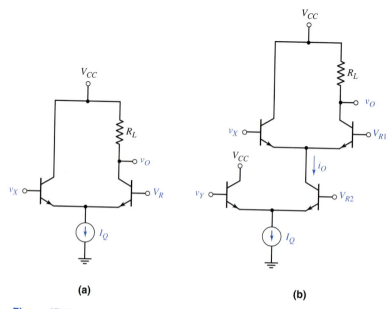

Figure 17.15   (a) Basic emitter-coupled pair and (b) ECL cascode configuration

Figure 17.16   ECL series gating example

$\bullet$ $A = C = 2.1$ V, $B = 2.5$ V: Transistors Q_{O1} and Q_1 are off, Q_{O2} is on, and current I_Q flows through Q_2 and Q_{O2}. Since Q_1 is off, no current is available to flow through Q_R, even though Q_{O1} is off. The output is $v_O = $ logic $1 = 2.5$ V.

For the output voltage v_O to be a logic 1, no current must flow through Q_R. This occurs when both Q_{O1} and Q_{O2} are on, or when a B–E junction of Q_R is turned on but no current is available through Q_1 or Q_2. We can show that this circuit performs the logic function

$$(A \text{ AND } C) \text{ OR } (B \text{ AND } \overline{C}) \tag{17.8}$$

We are now beginning to integrate logic functions into a circuit rather than using separate, distinct logic gates. This reduces the number of devices required, as well as the propagation delay time.

Another example of series gating is shown in Figure 17.17. A negative supply voltage is again used. The operation of the circuit is as follows.

- $v_x = v_y = \text{logic } 0 = -0.4 \text{ V}$: Transistors Q_1, Q_4, and Q_7 are on, current I_Q flows through Q_7 and Q_4, the diode turns on, and the output voltage is -0.4 V.

- $v_x = -0.4 \text{ V}$, $v_y = 0$: Transistors Q_1, Q_4, and Q_6 are on, current I_Q flows through Q_6 and Q_1 to ground, and current I_{Q2} flows through Q_4 and the resistor. The output voltage is $v_O = -R_C I_{Q2} = -(1)(0.05) = -0.05 \text{ V}$. This voltage is not sufficient to turn the Schottky diode on. Although it is not zero volts, the voltage still represents a logic 1.

- $v_x = 0$, $v_y = -0.4 \text{ V}$: Transistors Q_2, Q_3, and Q_7 are on, current I_Q flows through Q_7 and Q_3 to ground, and current I_{Q1} flows through Q_2 and the resistor. Again, $v_O = -0.05 \text{ V} = \text{logic } 1$.

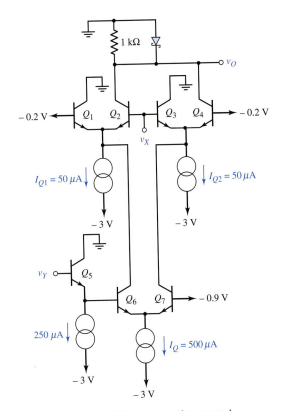

Figure 17.17 ECL series gating example

- $v_x = v_y = \text{logic } 1 \cong 0\,\text{V}$: Transistors Q_2, Q_3, and Q_6 are on, I_Q flows through Q_6, Q_2, and the Schottky diode, and output voltage is $v_O = -0.4\,\text{V} = \text{logic } 0$.

These results are summarized in Table 17.2, in which the logic levels are given. The results show that the circuit performs the exclusive-OR logic function.

Table 17.2 Summary of logic levels for ECL circuit in Figure 17.17

v_x	v_y	v_O
0	0	0
0	1	1
1	0	1
1	1	0

17.2.4 Propagation Delay Time

ECL is the fastest bipolar logic technology. Bipolar technology can produce small, very fast transistors with cutoff frequencies in the range of 3 to 15 GHz. Logic gates that use these transistors are so fast that interconnect line delays tend to dominate the propagation delay times. Minimizing these interconnect delays involves minimizing the metal lengths and using sufficient current drive capability.

Speed is derived from low-signal logic swings, nonsaturating logic, and the ability to drive a load capacitance. Figure 17.18 is the emitter-follower output stage found in many ECL circuits, showing an effective load capacitance. Usually, the emitter-follower current I_Q is two to four times larger than the cell current.

In the pull-down cycle, the current I_Q discharges C_L. The current–voltage relationship of the capacitor is

$$i = C_L \frac{dv_O}{dt} \tag{17.9(a)}$$

or

$$v_O = \frac{1}{C_L} \int i\,dt \tag{17.9(b)}$$

Assuming C_L and $i = I_Q$ are constants, the fall time is

$$\tau_F = (0.8)\frac{C_L V_S}{I_Q} \tag{17.10}$$

where V_S is the logic swing, and the factor (0.8) occurs because τ_F is defined as the time required for the output to swing from 10 percent to 90 percent of its final value.

As an example, if $V_S = 0.4\,\text{V}$ and $I_Q = 250\,\mu\text{A}$, then for a minimum fall time of $\tau_F = 0.8\,\text{ns}$, the maximum load capacitance is $C_L(\text{max}) = 0.625\,\text{pF}$. This calculation shows that the load capacitance must be minimized to realize short propagation delay times.

Figure 17.18 Emitter-follower stage with load capacitance

Test Your Understanding

17.7 Consider the ECL circuit in Figure 17.16. For each of the eight possible combinations of input states, determine the conduction state (on or off) of each transistor. Verify that this circuit performs the logic function given by Equation (17.8).

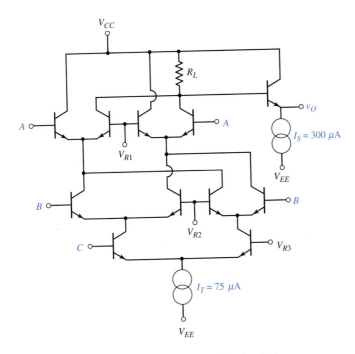

Figure 17.19 Figure for Exercise 17.8

17.8 The ECL circuit in Figure 17.19 is an example of three-level series gating. Determine the logic function that the circuit performs. (Ans. $(A \oplus B) \oplus C$)

17.3 TRANSISTOR–TRANSISTOR LOGIC

The bipolar inverter is the basic circuit from which most bipolar saturated logic circuits are developed, including diode–transistor logic (DTL) and transistor–transistor logic (TTL). However, the basic bipolar inverter suffers from loading effects. Diode–transistor logic combines diode logic (Chapter 2) and the bipolar inverter to minimize loading effects. Transistor–transistor logic, which evolved directly from DTL, provides reduced propagation delay times, as we will show.

In DTL and TTL circuits, bipolar transistors are driven between cutoff and saturation. Since the transistor is being used essentially as a switch, the current gain is not as important as in amplifier circuits. Typically, for transistors used in these circuits, the current gain is assumed to be in the range of 25 to 50. These transistors need not be fabricated to as tight a tolerance as that of high-gain amplifier transistors.

Table 17.3 lists the piecewise linear parameters used in the analysis of bipolar digital circuits, along with their typical values. Also included is the pn junction diode turn-on voltage V_γ. Generally, the B–E voltage increases as the transistor is driven into saturation, since the base current increases. When the transistor is biased in the saturation region, the B–E voltage is $V_{BE}(\text{sat})$, where $V_{BE}(\text{sat}) > V_{BE}(\text{on})$.

Table 17.3 Piecewise linear parameters for a pn junction diode and npn bipolar transistor

Parameter	Value
V_γ	0.7 V
$V_{BE}(\text{on})$	0.7 V
$V_{BE}(\text{sat})$	0.8 V
$V_{CE}(\text{sat})$	0.1 V

17.3.1   Basic Diode–Transistor Logic Gate

The basic diode–transistor logic (DTL) gate is shown in Figure 17.20. The circuit is designed such that the output transistor operates between cutoff and saturation. This provides the maximum output voltage swing, minimizes loading effects, and produces the maximum noise margins. When Q_o is in saturation, the output voltage is $v_O = V_{CE}(\text{sat}) \cong 0.1\,\text{V}$ and is defined as logic 0 for the DTL circuit. As we will see, the basic DTL logic gate shown in Figure 17.20 performs the NAND logic function.

Figure 17.20   Basic diode–transistor logic gate

Basic DTL NAND Circuit Operation

If both input signals v_X and v_Y are at logic 0, then the two input diodes D_X and D_Y are forward biased through resistor R_1 and voltage source V_{CC}. The input diodes conduct, and voltage v_1 is clamped to a value that is one diode drop above the input voltage. If $v_X = v_Y = 0.1\,\text{V}$ and $V_\gamma = 0.7\,\text{V}$, then $v_1 = 0.8\,\text{V}$. Diodes D_1 and D_2 and output transistor Q_o are nonconducting and are off. If D_1 and D_2 were conducting, then voltage v_B would be $-0.6\,\text{V}$ for $V_\gamma = 0.7\,\text{V}$. However, no mechanism exists for v_B to become negative and still have a forward-biased diode current. Thus, the current in D_1 and D_2, the current in Q_o, and the voltage v_B are all zero. Since Q_o is cut off, then the output voltage is $v_O = V_{CC}$. This is the largest possible output voltage and is therefore defined as the logic 1 level. This same condition applies as long as at least one input is at logic 0.

When both v_X and v_Y are at logic 1, which is equal to V_{CC}, both D_X and D_Y are cut off. Diodes D_1 and D_2 become forward biased, output transistor Q_o is driven into saturation, and $v_O = V_{CE}(\text{sat})$, which is the smallest possible output voltage and is defined as the logic 0 level.

This circuit is a two-input DTL NAND logic gate. However, the circuit is not limited to two inputs. Additional input diodes may be included to increase the fan-in.

Example 17.8 **Objective:** Determine the currents and voltages in the DTL logic circuit.

Consider the DTL circuit in Figure 17.20. Assume the transistor parameters are as given in Table 17.3 and let $\beta = 25$.

Solution: Let $v_X = v_Y = \text{logic } 0 = 0.1 \text{ V}$. For this case,

$$v_1 = v_X + V_\gamma = 0.1 + 0.7 = 0.8 \text{ V}$$

and

$$i_1 = \frac{V_{CC} - v_1}{R_1} = \frac{5 - 0.8}{4} = 1.05 \text{ mA}$$

Since diodes D_1 and D_2 and output transistor Q_o are nonconducting, we assume that current i_1 divides evenly between the matched diodes D_X and D_Y. In this case, the currents $i_2 = i_B = i_C = 0$ and the output voltage is $v_O = 5 \text{ V} = \text{logic } 1$.

If $v_X = 0.1 \text{ V}$ and $v_Y = 5 \text{ V}$, or $v_X = 5 \text{ V}$ and $v_Y = 0.1 \text{ V}$, then the output transistor is still cut off and $v_O = 5 \text{ V} = \text{logic } 1$.

If $v_X = v_Y = \text{logic } 1 = 5 \text{ V}$, it is impossible for input diodes D_X and D_Y to be forward biased. In this case, diodes D_1 and D_2 and the output transistor are biased on, which means that, starting at ground potential at the emitter of Q_o, v_1 is

$$v_1 = V_{BE}(\text{sat}) + 2V_\gamma = 0.8 + 2(0.7) = 2.2 \text{ V}$$

Voltage v_1 is clamped at this value and cannot increase. We see that D_X and D_Y are indeed reverse biased and turned off, as assumed.

Currents i_1 and i_2 are

$$i_1 = i_2 = \frac{V_{CC} - v_1}{R_1} = \frac{5 - 2.2}{4} = 0.70 \text{ mA}$$

and current i_R is

$$i_R = \frac{V_{BE}(\text{sat})}{R_B} = \frac{0.8}{10} = 0.08 \text{ mA}$$

The base current into the output transistor is then

$$i_B = i_2 - i_R = 0.70 - 0.08 = 0.62 \text{ mA}$$

Since the circuit is to be designed such that Q_o is driven into saturation, the collector current is

$$i_C = \frac{V_{CC} - V_{CE}(\text{sat})}{R_C} = \frac{5 - 0.1}{4} = 1.23 \text{ mA}$$

Finally, the ratio of collector to base current is

$$\frac{i_C}{i_B} = \frac{1.23}{0.62} = 1.98 < \beta$$

Comment: Since the ratio of the collector current to base current is less than β, the output transistor is biased in the saturation region. Since the output transistor is biased between cutoff and saturation, the maximum swing between logic 0 and logic 1 is obtained.

Test Your Understanding

[Note: In the following exercises, assume the piecewise linear transistor parameters are as listed in Table 17.3.]

17.9 The DTL circuit in Figure 17.20 has new circuit parameters of $R_1 = 6\,\text{k}\Omega$, $R_C = 5\,\text{k}\Omega$, and $R_B = 15\,\text{k}\Omega$. Assume $V_{CC} = 5\,\text{V}$ and $\beta = 25$. Determine i_1, i_2, i_R, i_B, i_{RC}, and v_O for: (a) $v_X = v_Y = 0.1\,\text{V}$, (b) $v_X = 5\,\text{V}$, $v_Y = 0.1\,\text{V}$, and (c) $v_X = v_Y = 5\,\text{V}$. (Ans. (a) $i_1 = 0.7\,\text{mA}$, $i_2 = i_R = i_B = i_{RC} = 0$, $v_O = 5\,\text{V}$ (b) same as part (a) (c) $i_1 = i_2 = 0.467\,\text{mA}$, $i_R = 0.053\,\text{mA}$, $i_B = 0.414\,\text{mA}$, $i_{RC} = 0.98\,\text{mA}$, $v_O = 0.1\,\text{V}$)

17.10 Consider the basic DTL circuit in Figure 17.20 with circuit and transistor parameters given in Example 17.8. Assume no load is connected to the output. Calculate the power dissipated in the circuit for (a) $v_X = v_Y = 5\,\text{V}$ and (b) $v_X = v_Y = 0$.

Minimum β

To ensure that the output transistor is in saturation, the common-emitter current gain β must be at least as large as the ratio of collector current to base current. For example 17.8, the minimum β, or β_{min}, is 1.98. If the common-emitter current gain were less than 1.98, then Q_o would not be driven into saturation, and the currents and voltages in the circuit would have to be recalculated. A current gain greater than 1.98 ensures that Q_o is driven into saturation for the given circuit parameters and for the no-load condition.

Pull-Down Resistor

In the basic DTL NAND logic circuit in Figure 17.20, a resistor R_B is connected between the base of the output transistor and ground. This resistor is called a pull-down resistor, and its purpose is to decrease the output transistor switching time as it goes from saturation to cutoff. As previously discussed, excess minority carriers must be removed from the base before a transistor can be switched to cutoff. This base charge removal produces a current out of the transistor base terminal until the transistor is turned off. Without the pull-down resistor, this reverse base current would be limited to the reverse-bias leakage current in diodes D_1 and D_2, resulting in a relatively long turn-off time. The pull-down resistor provides a path for the reverse base current.

The base charge can be removed more rapidly if the value of R_B is reduced. The larger the reverse base current, the shorter the transistor turn-off time. However, a trade-off must be made in choosing the value of R_B. A small R_B provides faster switching, but lowers the base current to the transistor in the on state by diverting some drive current to ground. A lower base current reduces the circuit drive capability, or maximum fanout.

17.3.2 The Input Transistor of TTL

Figure 17.21(a) shows a basic DTL circuit with one input diode D_X and one offset diode D_1. The structure of these back-to-back diodes is the same as an npn transistor, as indicated in Figure 17.21(b). The base–emitter junction of Q_1 corresponds to input diode D_X and the base–collector junction corresponds to offset diode D_1.

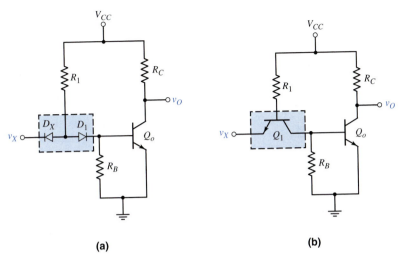

(a)   **(b)**

Figure 17.21   (a) Basic DTL gate and (b) basic TTL gate

In isoplanar integrated circuit technology, the emitter of a bipolar transistor is fabricated in the base region. More emitters can then be added in the same base region to form a multiemitter, multi-input device. Figure 17.22(a) shows a simplified cross section of a three-emitter transistor, which is used as the input device in a TTL circuit. Figure 17.22(b) shows the basic TTL circuit with the multiemitter input transistor.

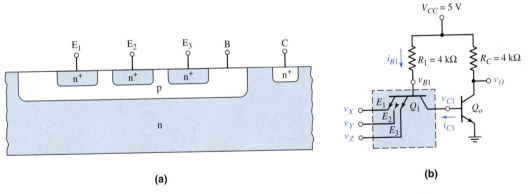

(a)   **(b)**

Figure 17.22   (a) Simplified cross section of three-emitter transistor and (b) TTL circuit with three-emitter input transistor

This circuit performs the same NAND operation as its DTL counterpart. The multiemitter transistor reduces the silicon area required, compared to the DTL input diodes, and it increases the switching speed. Transistor Q_1 assists in pulling output transistor Q_o out of saturation and into cutoff during a low-to-high transition of the output voltage. Pull-down resistor R_B in Figure 17.21(b) is no longer necessary, since the excess minority carriers in the base of Q_o use transistor Q_1 as a path to ground.

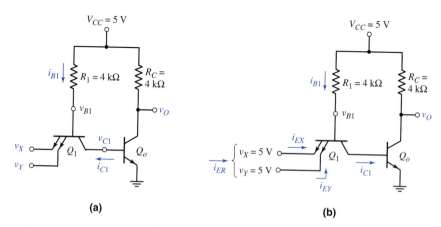

Figure 17.23 TTL circuit (a) with at least one input low and (b) with all inputs high

    The operation of input transistor Q_1 is somewhat unconventional. In Figure 17.23(a), if either or both of the two inputs to Q_1 are in a low state, the base–emitter junction is forward biased through R_1 and V_{CC}. The base current enters Q_1, and the emitter current exits the specific emitter connected to the low input. Transistor action forces the collector current into Q_1, but the only steady-state collector current in this direction is a reverse-bias saturation current out of the base of Q_o. The steady-state collector current of Q_1 is usually much smaller than the base current, implying that Q_1 is biased in saturation.

    If at least one input is low such that Q_1 is biased in saturation, then from Figure 17.23(a), we see that the base voltage of Q_1 is

$$v_{B1} = v_X + V_{BE}(\text{sat}) \tag{17.11}$$

and the base current into Q_1 is

$$i_{B1} = \frac{V_{CC} - v_{B1}}{R_1} \tag{17.12}$$

If the forward current gain of Q_1 is β_F, then Q_1 will be in saturation as long as $i_{C1} < \beta_F i_{B1}$.

    The collector voltage of Q_1 is

$$v_{C1} = v_X + V_{CE}(\text{sat}) \tag{17.13}$$

If both v_X and $V_{CE}(\text{sat})$ are approximately 0.1 V, then v_{C1} is small enough for the output transistor to cut off and $v_O = V_{CC} = \text{logic } 1$.

    If all inputs are high, $v_X = v_Y = 5\,\text{V}$, as shown in Figure 17.23(b), then the base–emitter junctions of the input transistor are reverse biased. Base voltage v_{B1} increases, which forward-biases the B–C junction of Q_1 and drives output transistor Q_o into saturation. Since the B–E junction of Q_1 is reverse biased and the B–C junction is forward biased, Q_1 is biased in the inverse-active mode. In this bias mode, the roles of the emitter and collector are interchanged.

    When input transistor Q_1 is biased in the inverse-active mode, base voltage v_{B1} is

$$v_{B1} = V_{BE}(\text{sat})_{Q_o} + V_{BC}(\text{on})_{Q_1} \tag{17.14}$$

where $V_{BC}(on)$ is the B–C junction turn-on voltage. We assume that the B–C junction turn-on voltage is equal to the B–E junction turn-on voltage. The terminal current relationships for Q_1 are therefore

$$i_{EX} = i_{EY} = \beta_R i_{B1} \tag{17.15}$$

and

$$i_{C1} = i_{B1} + i_{EX} + i_{EY} = (1 + 2\beta_R)i_{B1} \tag{17.16}$$

where β_R is the inverse-active mode current gain of each input emitter of the input transistor.

Since a bipolar transistor is not symmetrical, the inverse and forward current gains are not equal. The inverse current gain is generally quite small, usually less than one. In Figure 17.23(b), the input transistor has a fan-in of two. Transistor Q_1 may be considered as two separate transistors with their bases and collectors connected. For simplicity, when all inputs are high, we assume that current i_{ER} splits evenly between the input emitters.

The inverse-active mode current into the emitters of Q_1 is not desirable, since this is a load current that must be supplied by a driver logic circuit when its output voltage is in its high state. Because of the transistor action, these currents tend to be larger than the reverse saturation currents of DTL circuit input diodes. The major advantage of TTL over DTL is faster switching of the output transistor from saturation to cutoff.

If all inputs are initially high and then at least one input switches to the logic 0 state, 0.1 V, the B–E junction of Q_1 becomes forward biased and base voltage v_{B1} becomes approximately $0.1 + 0.7 = 0.8$ V. Collector voltage v_{C1} is held at 0.8 V as long as output transistor Q_o remains in saturation. At this instant in time, Q_1 is biased in the forward-active mode. A large collector current into Q_1 can exist, which pulls the excess minority carrier charge out of the base of Q_o. A large reverse base current from Q_o will very quickly pull the output transistor out of saturation. In the TTL circuit, the action of the input transistor reduces the propagation delay time compared to that of DTL logic circuits. For example, the propagation delay time is reduced from approximately 40 ns in a DTL NAND gate to approximately 10 ns in an equivalent TTL circuit.

17.3.3   Basic TTL NAND Circuit

We can improve the circuit performance of the simple TTL circuit in Figure 17.23 by adding a second current gain stage. The resulting basic TTL NAND circuit is shown in Figure 17.24. In this circuit, both transistors Q_2 and Q_o are driven into saturation when $v_X = v_Y = $ logic 1. When at least one input switches from high to low, input transistor Q_1 very quickly pulls Q_2 out of saturation and pull-down resistor R_B provides a path for the excess charge in Q_o, which means that the output transistor can turn off fairly quickly.

DC Current–Voltage Analysis

The analysis of the TTL circuit is very similar to that of the DTL circuit, as demonstrated in the following example.

Figure 17.24   TTL circuit with currents and voltages

Example 17.9   **Objective:** Calculate the currents and voltages for the basic TTL NAND circuit.

Consider the TTL circuit in Figure 17.24. Assume the piecewise linear transistor parameters are as listed in Table 17.3. Assume the forward current gain is $\beta_F \equiv \beta = 25$ and the inverse current gain of each input emitter is $\beta_R = 0.1$.

Solution: For $v_X = v_Y = 0.1 \, \text{V}$, Q_1 is biased in saturation and

$$v_{B2} = v_X + v_{CE}(\text{sat}) = 0.1 + 0.1 = 0.2 \, \text{V}$$

which means that Q_2 and Q_o are both cut off. The base voltage v_{B1} is then

$$v_{B1} = v_X + V_{BE}(\text{sat}) = 0.1 + 0.8 = 0.9 \, \text{V}$$

and current i_1 is

$$i_1 = \frac{V_{CC} - v_{B1}}{R_1} = \frac{5 - 0.9}{4} = 1.03 \, \text{mA}$$

This current flows out of the input transistor emitters. Since Q_2 and Q_o are cut off, all other currents are zero and the output voltage is $v_O = 5 \, \text{V}$.

If $v_X = v_Y = 5 \, \text{V}$, then the input transistor is biased in the inverse active mode. The base voltage v_{B1} is

$$v_{B1} = V_{BE}(\text{sat})_{Q_o} + V_{BE}(\text{sat})_{Q_2} + V_{BC}(\text{on})_{Q_1}$$
$$= 0.8 + 0.8 + 0.7 = 2.3 \, \text{V}$$

and the collector voltage v_{C2} is

$$v_{C2} = V_{BE}(\text{sat})_{Q_o} + V_{CE}(\text{sat})_{Q_2} = 0.8 + 0.1 = 0.9 \, \text{V}$$

The currents are

$$i_1 = \frac{V_{CC} - v_{B1}}{R_1} = \frac{5 - 2.3}{4} = 0.675 \, \text{mA}$$

and

$$i_{B2} = (1 + 2\beta_R)i_1 = (1 + 0.2)(0.675) = 0.810 \, \text{mA}$$

Also,

$$i_2 = \frac{V_{CC} - v_{C2}}{R_2} = \frac{5 - 0.9}{1.6} = 2.56 \, \text{mA}$$

which means that

$$i_{E2} = i_2 + i_{B2} = 2.56 + 0.81 = 3.37 \, \text{mA}$$

The current in the pull-down resistor is

$$i_4 = \frac{V_{BE}(\text{sat})}{R_B} = \frac{0.8}{1} = 0.8 \, \text{mA}$$

and the base drive to the output transistor is

$$i_{Bo} = i_{E2} - i_4 = 3.37 - 0.8 = 2.57 \, \text{mA}$$

Current i_3 is

$$i_3 = \frac{V_{CC} - V_{CE}(\text{sat})}{R_C} = \frac{5 - 0.1}{4} = 1.23 \, \text{mA}$$

Comment: As mentioned, the analysis of the basic TTL circuit is essentially the same as that of the DTL circuit. The magnitudes of currents and voltages in the basic TTL circuit are also very similar to the DTL results.

Test Your Understanding

[Note: In the following exercise, assume the piecewise linear transistor parameters are as listed in Table 17.3.]

17.11 The parameters of the TTL NAND circuit in Figure 17.24 are: $R_1 = 6 \, \text{k}\Omega$, $R_2 = 1.5 \, \text{k}\Omega$, $R_B = 1.5 \, \text{k}\Omega$, and $R_C = 2.2 \, \text{k}\Omega$. Assume that $\beta_F \equiv \beta = 20$ and $\beta_R = 0.1$ (for each input emitter). For a no-load condition, determine the base and collector currents in each transistor for: (a) $v_X = v_Y = 0.1 \, \text{V}$, and (b) $v_X = v_Y = 3.6 \, \text{V}$. Prove that Q_2 and Q_o are driven into saturation for $v_X = v_Y = 3.6 \, \text{V}$. (Ans. (a) $i_1 = i_{B1} = 0.683 \, \text{mA}$, $i_{C1} \cong 0$, $i_{B2} = i_{C2} = 0$, $i_{Bo} = i_{Co} = 0$ (b) $i_1 = i_{B1} = 0.45 \, \text{mA}$, $i_{B2} = |i_{C1}| = 0.54 \, \text{mA}$, $i_2 = i_{C2} = 2.73 \, \text{mA}$, $i_{Bo} = 2.74 \, \text{mA}$, $i_3 = i_{Co} = 2.23 \, \text{mA}$)

17.3.4 TTL Output Stages and Fanout

The propagation delay time can be improved by replacing the output collector resistor with a current source.

When the output changes from low to high, the load capacitance must be charged by a current through the collector pull-up resistor. The total load capacitance is composed of the input capacitances of the load circuits and the capacitances of the interconnect lines. The associated RC time constant for a load capacitance of 15 pF and a collector resistance of 4 kΩ is 60 ns, which is large compared to the propagation delay time of a commercial TTL circuit.

Totem-Pole Output Stage

In Figure 17.25, the combination of Q_3, D_1, and Q_o forms an output stage called a totem pole. Transistor Q_2 forms a phase splitter, because the collector

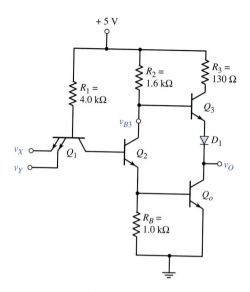

Figure 17.25   TTL circuit with totem-pole output stage

and emitter voltages are 180 degrees out of phase. If $v_X = v_Y = $ logic 1, input transistor Q_1 is biased in the inverse-active mode, and both Q_2 and Q_o are driven into saturation. The voltage at the base of Q_3 is

$$v_{B3} = V_{C2} = V_{BE}(\text{sat})_{Q_o} + V_{CE}(\text{sat})_{Q_2} \qquad (17.17)$$

which is on the order of 0.9 V, and the output voltage is approximately 0.1 V. The difference between the base voltage of Q_3 and the output voltage is not sufficient to turn Q_3 and D_1 on. The pn junction offset voltage associated with D_1 must be included so that Q_3 is cut off when the output is low. For this condition, the saturation output transistor discharges the load capacitance and pulls the output low very quickly.

If $v_X = v_Y = $ logic 0, then Q_2 and Q_o are cut off, and the base voltage to Q_3 goes high. The transistor Q_3 and diode D_1 turn on so that the output load capacitance can be charged and the output goes high. Since Q_3 acts like an emitter follower, the output resistance is small so that the effective RC time constant to charge the load capacitance is now very small.

Fanout

Logic gates are not operated in isolation, but are used to drive other similar type logic gates to implement a complex logic function. Figure 17.26 shows the TTL NAND gate with a totem-pole output stage connected to N identical TTL NAND gates. The maximum fanout is defined as the maximum number of similar-type logic circuits that can be connected to the logic gate output without affecting proper circuit operation. For example, the output transistor Q_o must remain in saturation when the output goes low to its logic 0 value. For a given value of β, there is then a maximum allowable load current, and therefore a maximum allowable number of load circuits that can be connected to the output. As another condition, the output transistor is usually rated for a

Figure 17.26 TTL circuit with totem-pole output stage driving *N* identical TTL stages

maximum collector current. For an output low condition, the current i_{LL} is the load current that Q_o must sink from the load circuits.

Example 17.10 **Objective:** Calculate the maximum fanout for the output low condition.

Let $\beta = 25$ for the output transistor.

Solution: Transistor Q_o to remain in saturation. In Example 17.9, we calculated the base current into Q_o as $i_{Bo} = 2.57\,\text{mA}$. The output voltage is $v_O = 0.1\,\text{V}$ so that $v'_{B1} = 0.1 + 0.8 = 0.9\,\text{V}$. Each individual load current is then

$$i'_{LL1} = i'_1 = \frac{5 - 0.9}{4} = 1.025\,\text{mA}$$

The maximum collector current in Q_o is

$$i_{Co}(\text{max}) = \beta i_{Bo} = Ni'_{LL1}$$

The maximum fanout, N, is then found as

$$N = \frac{\beta i_{Bo}}{i'_{LL1}} = \frac{(25)(2.57)}{1.025} = 62.7$$

The number of load circuits must be an integer, so we round to the next lower integer, or $N = 62$.

With 62 load circuits connected to the output, the collector current would be

$$i_{Co} = Ni'_{LL1} = (62)(1.025) = 63.55\,\text{mA}$$

which is a relatively large value. In most cases, the output transistor has a maximum rated collector current that may limit the maximum fanout.

Solution: Maximum rated output current. If the maximum rated collector current of the output transistor is $i_{Co}(\text{rated}) = 20\,\text{mA}$, then the maximum fanout is determined by

$$i_{Co}(\text{rated}) = N i'_{LL1}$$

or

$$N = \frac{i_{Co}(\text{rated})}{i'_{LL1}} = \frac{20}{1.025} = 19.5 \to 19$$

Comment: In the first solution, the resulting fanout of 62 is not realistic since the output transistor current is excessive. In the second solution, a maximum fanout of 19 is more realistic. However, another limitation in terms of proper circuit operation is propagation delay time. For a large number of load circuits connected to the output, the output load capacitance may be quite large which slows down the switching speed to unacceptably large values. The maximum fanout, then, may be limited by the propagation delay time specification.

Again, Figure 17.26 shows the TTL circuit with N identical load circuits and the inputs in their low state. The input transistor is biased in saturation, and both Q_2 and Q_o are cut off, causing base voltage v_{B3} and the output voltage to go high. The input transistors of the load circuits are biased in the inverse-active mode, and the load currents are supplied through Q_3 and D_1. In this circuit, the input transistors of the load gates are one-input NAND (inverter) gates, to illustrate the worst-case or maximum load current under the high input condition. Since the load current is supplied through Q_3, a base current into Q_3 must be supplied from V_{CC} through R_2. As the load current increases, the base current through R_2 increases, which means that voltage v_{B3} decreases because of the voltage drop across R_2. Assuming the B–E voltage of Q_3 and the diode voltage across D_1 remain essentially constant, the output voltage v_O decreases from its maximum value.

A reasonable fanout of 10 or 15 for the high output condition means that the load current will be small, base current i_{B3} will be very small, and the voltage drop across R_2 will be negligible. The output voltage will then be approximately two diode drops below V_{CC}. For typical TTL circuits, the logic $1 = V_{OH}$ value is on the order of 3.6 V, rather than the 5 V previously determined.

Test Your Understanding

[Note: In the following exercises, assume the piecewise linear transistor parameters are as listed in Table 17.3.]

17.12 (a) For the basic DTL logic circuit, the parameters are as given in Exercise 17.9. Calculate the maximum fanout for the low output condition such that Q_o remains in saturation. (b) Repeat part (a) if the rated collector current of Q_o is $I_{C,\text{rated}} = 15\,\text{mA}$. (Ans. (a) $N = 13$ (b) $N = 13$)

17.13 Consider the TTL circuit shown in Figure 17.24 with parameters as given in Exercise 17.11. Calculate the maximum fanout for the low output. For the low output condition, assume that the output transistor must remain in saturation. (Ans. $N = 76$)

17.14 The TTL circuit shown in Figure 17.25 is redesigned such that $R_1 = 6\,\text{k}\Omega$, $R_2 = 2\,\text{k}\Omega$, $R_3 = 80\,\text{k}\Omega$, and $R_B = 1.5\,\text{k}\Omega$. Assume that $\beta_F \equiv \beta = 20$ and $\beta_R = 0.1$ (for each input emitter). Calculate the fanout for $v_X = v_Y = 3.6\,\text{V}$. For the low output condition, assume that the output transistor must remain in saturation. (Ans. $N = 60$)

Modified Totem-Pole Output Stage

Figure 17.27 shows a modified totem-pole output stage in which transistor Q_4 is used in place of a diode. This has several advantages. First, the transistor pair Q_3 and Q_4 provides greater current gain, which in turn increases the fanout capability of this circuit in its high state. Second, the output impedance in the high state is lower than that of the single transistor, decreasing the switching time. Third, the base–emitter junction of Q_3 fulfills the function of diode D_1; therefore, the diode is no longer needed to provide a voltage offset. In integrated circuits, the fabrication of transistors is no more complex than the fabrication of diodes.

When the output is switched to its low state, resistor R_4 provides a path to ground for the minority carriers that must be pulled out of the base of Q_3 to turn the transistor off. Note that when the output is low, with Q_2 and Q_o in saturation, the voltage at the base of Q_4 is approximately 0.9 V, which is sufficient to bias Q_4 in its active region. However, the voltage at the emitter of Q_4 is only approximately 0.2 V, which means that the current in Q_4 is very small and does not add significantly to the power dissipation.

Figure 17.27 TTL circuit with modified totem-pole output stage

17.3.5   Tristate Output

The output impedances of the totem-pole output TTL logic circuits considered thus far are extremely low when the output voltage is in either the high or low state. In memory circuit applications, situations arise in which the outputs of many TTL circuits must be connected together to form a single output. This creates a serious loading situation, demanding that all other TTL outputs be disabled or put into a high impedance state, as shown symbolically in Figure 17.28. Here, G_1 and G_3 are disconnected from the output; the output voltage v_O then measures only the output of logic gate G_2.

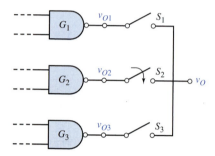

Figure 17.28   Circuit symbolically showing tristate output

The TTL circuit in Figure 17.29 may be used to put the logic output into a high impedance state. When $\bar{D} = 5\,\text{V}$, the state of input transistor Q_1 is controlled by inputs v_X and v_Y. Under these circumstances, diode D_2 is always reverse biased and the circuit function is the NAND function already considered.

Figure 17.29   TTL circuit with tristate output stage

When $\overline{D}$ is driven to a logic 0 state of 0.1 V, the low voltage at the emitter of Q_1 ensures that both Q_2 and Q_o are cut off, and the low voltage applied to D_2 means that D_2 is forward biased. The voltage at the base of Q_4 is approximately 0.8 V, which means that Q_3 is also cut off. In this condition, then, both output transistors Q_3 and Q_o are cut off. The impedance looking back into transistors that are cut off is normally in the megohm range. Therefore, when TTL circuits are paralleled to increase the capability of a digital system, the tristate output stage is either enabled or disabled via the $\overline{D}$ select line. The output stage on only one TTL circuit may be enabled at any one time.

Test Your Understanding

[Note: In the following exercise, assume the piecewise linear transistor parameters are as listed in Table 17.3.]

17.15   For the tristate TTL circuit in Figure 17.29, the parameters are: $R_1 = 6\,\text{k}\Omega$, $R_2 = 2\,\text{k}\Omega$, $R_3 = 100\,\Omega$, $R_4 = 4\,\text{k}\Omega$, and $R_B = 1\,\text{k}\Omega$. Assume that $\beta_F \equiv \beta = 20$ and $\beta_R = 0.1$ (for each input emitter). For $\overline{D} = 0.1$ V, calculate the base and collector currents in each transistor. (Ans. $i_{B1} = 0.683\,\text{mA}$, $|i_{C1}| = i_{B2} = i_{C2} = i_{Bo} = i_{Co} = 0$, $i_{B4} = 1.19\,\mu\text{A}$, $i_{C4} = 23.8\,\mu\text{A}$, $i_{B3} = i_{C3} = 0$)

17.4   SCHOTTKY TRANSISTOR–TRANSISTOR LOGIC

The TTL circuits considered thus far drive the output and phase-splitter transistors between cutoff in the high output stage and saturation in the low output state. The input transistor is driven between saturation and the inverse-active mode. Since the propagation delay time of a TTL gate is a strong function of the storage time of the saturation transistors, a nonsaturation logic circuit would be an advantage. In the Schottky clamped transistor, the transistor is prevented from being driven into deep saturation and has a storage time of only approximately 50 ps.

17.4.1   Schottky Clamped Transistor

The symbol for the Schottky clamped transistor, or simply the Schottky transistor, is shown in Figure 17.30(a); its equivalent configuration is

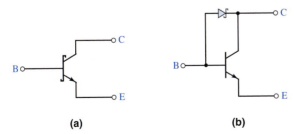

(a)   **(b)**

Figure 17.30   (a) Schottky clamped transistor symbol and (b) Schottky clamped transistor equivalent circuit given in Figure 17.30(b). In this transistor, a Schottky diode is connected between the base and collector of an npn bipolar transistor. Two characteristics of the Schottky diode are: a low turn-on voltage and a fast-switching time. When the transistor is in its active region, the base–collector junction is reverse biased, which means that the Schottky diode is reverse biased and effectively out of the circuit. The Schottky transistor then behaves like a normal npn bipolar transistor. As the Schottky transistor goes into saturation, the base–collector junction becomes forward biased, and the base–collector voltage is effectively clamped at the Schottky diode turn-on voltage, which is normally between 0.3 and 0.4 V. The excess base current is shunted through the diode, and the basic npn transistor is prevented from going deeply into saturation.

Figure 17.31 shows the equivalent circuit of the Schottky transistor with designated currents and voltages. Currents i_C and i_B are the collector and base currents, respectively, of the Schottky transistor, while i_C' and i_B' are the collector and base currents, respectively, of the internal npn transistor.

The three defining equations for the Schottky transistor are

$$i_C' = i_D + i_C \tag{17.18}$$

$$i_B = i_B' + i_D \tag{17.19}$$

and

$$i_C' = \beta i_B' \tag{17.20}$$

Figure 17.31 Schottky clamped transistor equivalent circuit, with currents and voltages

Equation (17.20) is appropriate since the internal transistor is clamped at the edge of saturation. If $i_C < \beta i_B$, then the Schottky diode is forward biased, $i_D > 0$, and the Schottky transistor is said to be in saturation. However, the internal transistor is only driven to the edge of saturation in this case.

Combining Equations (17.19) and (17.20), we find that

$$i_D = i_B - i_B' = i_B - \frac{i_C'}{\beta} \tag{17.21}$$

Substituting this equation into Equation (17.18) yields

$$i_C' = i_B - \frac{i_C'}{\beta} + i_C \tag{17.22(a)}$$

or

$$i_C' = \frac{i_B + i_C}{1 + (1/\beta)} \tag{17.22(b)}$$

Equation (17.22(b)) relates the internal transistor collector current to the external Schottky transistor collector and base currents.

Example 17.11   Objective: Determine the currents in a Schottky transistor.

Consider the Schottky transistor in Figure 17.31 with an input base current of $i_B = 1\,\text{mA}$. Assume that $\beta = 25$. Determine the internal currents in the Schottky transistor for $i_C = 2\,\text{mA}$, and then for $i_C = 20\,\text{mA}$.

Solution: For $i_C = 2\,\text{mA}$, the internal collector current is, from Equation (17.22(b)),

$$i_C' = \frac{1+2}{1+(1/25)} = 2.89\,\text{mA}$$

and the internal base current is

$$i_B' = \frac{i_C'}{\beta} = \frac{2.89}{25} = 0.115\,\text{mA}$$

The Schottky diode current is therefore

$$i_D = i_B - i_B' = 1 - 0.115 = 0.885\,\text{mA}$$

Repeating the calculations for $i_C = 20\,\text{mA}$, we obtain

$$i_C' = 20.2\,\text{mA}$$
$$i_B' = 0.808\,\text{mA}$$
$$i_D = 0.192\,\text{mA}$$

Comment: For a relatively small collector current into the Schottky transistor, the majority of the input base current is shunted through the Schottky diode. As the collector current into the Schottky transistor increases, less current is shunted through the Schottky diode and more current flows into the base of the npn transistor.

Test Your Understanding

[Note: In the following exercise, assume the piecewise linear transistor parameters are as listed in Table 17.3. In addition, assume a Schottky diode turn-on voltage of $V_\gamma(\text{SD}) = 0.3\,\text{V}$.]

17.16 Consider the Schottky clamped transistor in Figure 17.32. Assume $\beta = 10$, $V_{BE}(\text{on}) = 0.7\,\text{V}$, and $V_\gamma(\text{SD}) = 0.3\,\text{V}$. (a) For no load, $i_L = 0$, find the currents i_D, i_B', and i_C'. (b) Determine the maximum load current i_L that the transistor can sink and still remain at the edge of saturation. (Ans. (a) $i_C' = 3.67\,\text{mA}$, $i_B' = 0.367\,\text{mA}$, $i_D = 1.63\,\text{mA}$ (b) $i_L(\text{max}) \cong 18\,\text{mA}$)

Figure 17.32   Figure for Exercise 17.16

Since the internal npn bipolar transistor is not driven deeply into saturation, we assume that the B–E junction voltage remains equal to the turn-on voltage, or $v_{BE} = V_{BE}(\text{on})$. If the Schottky transistor is biased in saturation, then the C–E voltage is

$$v_{CE} = V_{CE}(\text{sat}) = V_{BE}(\text{on}) - V_\gamma(\text{SD}) \tag{17.23}$$

where $V_\gamma(\text{SD})$ is the turn-on voltage of the Schottky diode. Assuming parameter values of $V_{BE}(\text{on}) = 0.7\,\text{V}$ and $V_\gamma(\text{SD}) = 0.3\,\text{V}$, the collector–emitter saturation voltage of a Schottky transistor is $V_{CE}(\text{sat}) = 0.4\,\text{V}$. When the Schottky transistor is at the edge of saturation, then $i_D = 0$, $i_C = \beta i_B$, and $v_{CE} = V_{CE}(\text{sat})$.

17.4.2   Schottky TTL NAND Circuit

Figure 17.33 shows a Schottky TTL NAND circuit in which all of the transistors except Q_3 are Schottky clamped transistors. The connection of Q_4 across the base–collector of Q_3 prevents this junction from becoming forward biased, ensuring that Q_3 never goes into saturation. Another difference between this circuit and the standard TTL circuit is that the pull-down resistor at the base of output transistor Q_o has been replaced by transistor Q_5 and two resistors. This arrangement is called a **squaring network**, since it squares, or sharpens, the voltage transfer characteristics of the circuit.

Device Q_2 is prevented from conducting until the input voltage is large enough to turn on both Q_2 and Q_o simultaneously. Recall that the passive pull-down resistor on the TTL circuit provided a pathway for removing stored charge in the base of the output transistor, when the output transistor was turned off from the saturated state. Transistor Q_5 now provides an active pull-down network that pulls Q_o out of saturation more quickly.

This is one example of a circuit in which the piecewise linear model of a transistor fails to provide an adequate solution for the circuit analysis. With the piecewise linear model, Q_5 would apparently never turn on. However, because of the exponential relationship between collector current and base-emitter voltage, transistor Q_5 does turn on and does help pull Q_o out of saturation during switching.

The two Schottky diodes between the input terminals and ground act as clamps to suppress any ringing that might occur from voltage transitions. The input diodes clamp any negative undershoots at approximately $-0.3\,\text{V}$.

The dc current–voltage analysis of the Schottky TTL circuit in Figure 17.33 is similar to that for the standard TTL circuit. One minor difference is that when the inputs are high and the input transistor is in the inverse-active

Figure 17.33   Schottky TTL NAND logic circuit

mode, the B–C forward bias voltage is 0.3 V, because of the Schottky diode connected between the base and collector junctions.

The major difference between the Schottky circuit and standard TTL circuits is the quantity of excess minority carrier storage in the transistors when they are driven into or near saturation. The internal npn transistor of the Schottky clamped transistor is held at the edge of saturation, and the resulting propagation delay time is on the order of 2 to 5 ns, compared to a nominal 10 to 15 ns for standard TTL circuits.

A slight difference between the Schottky and standard TTL circuits is the value of the output voltage in the logic 0 state. The low output voltage of a standard TTL circuit is in the range of 0.1 to 0.2 V, while the Schottky transistor low output saturation voltage, V_{OL}, is approximately 0.4 V. The output voltage in the logic 1 state is essentially the same for both types of logic circuits.

Test Your Understanding

[Note: In the following exercise, assume the piecewise linear transistor parameters are as listed in Table 17.3. In addition, assume a Schottky diode turn-on voltage of $V_\gamma(\text{SD}) = 0.3$ V.]

17.17 In the Schottky TTL NAND circuit in Figure 17.33, assume $\beta_F \equiv \beta = 25$ and $\beta_R = 0$. For a no-load condition, calculate the power dissipation for: (a) $v_X = v_Y = 0.4$ V, and (b) $v_X = v_Y = 3.6$ V. (Ans. $P = 12.5$ mW (b) $P = 32.1$ mW)

17.4.3 Low-Power Schottky TTL Circuits

The Schottky TTL circuit in Figure 17.33 and the standard TTL circuit dissipate approximately the same power, since voltage and resistance values in the two circuits are similar. The advantage of the Schottky TTL circuit is the reduction in propagation delay time by a factor of 3 to 10.

Propagation delay times depend on the type of transistors (Schottky clamped or regular) used in the circuit, and on the current levels in the circuit. The storage time of a regular transistor is a function of the reverse base current that pulls the transistor out of saturation. Also, the transistor turn-on time depends on the current level charging the base–emitter junction capacitance. A desirable trade-off can therefore be made between current levels (power dissipation) and propagation delay times. Smaller current levels lead to lower power dissipation, but at the expense of increased propagation delay times. This trade-off has been successful in commercial applications, where very short propagation delay times are not always necessary, but reduced power requirements are always an advantage.

A **low-power Schottky** TTL NAND circuit is shown in Figure 17.34. With few exceptions, these circuits do not use the multiemitter input transistor of standard TTL circuits. Most low-power Schottky circuits use a DTL type of input circuit, with Schottky diodes performing the AND function. This circuit is faster than the classic multiemitter input transistor circuit, and the input breakdown voltage is also higher.

Figure 17.34   Low-power Schottky TTL NAND logic circuit

The dc analysis of the low-power Schottky circuit is identical to that of DTL circuits.

Example 17.12   **Objective:** Calculate the power dissipation in a low-power Schottky TTL circuit.

Consider the circuit shown in Figure 17.34. Assume the Schottky diode turn-on voltage is $V_\gamma(\text{SD}) = 0.3\,\text{V}$ and the transistor parameters are: $V_{BE}(\text{on}) = 0.7\,\text{V}$, $V_{CE}(\text{sat}) = 0.4\,\text{V}$, and $\beta = 25$.

Solution: For the low input condition, $v_X = v_Y = 0.4\,\text{V}$ and $v_1 = 0.4 + 0.3 = 0.7\,\text{V}$. Current i_1 is

$$i_1 = \frac{V_{CC} - v_1}{R_1} = \frac{5 - 0.7}{20} = 0.215\,\text{mA}$$

Since Q_2 and Q_o are cut off with a no-load condition, all other currents in the circuit are zero. The power dissipation for the low input condition is therefore

$$P_L = i_1(V_{CC} - v_X) = (0.215) \cdot (5 - 0.4) = 0.989\,\text{mW}$$

For the high input condition, $v_X = v_Y = 3.6\,\text{V}$, voltage v_1 is

$$v_1 = V_{BE}(\text{on})_{Q_o} + V_{BE}(\text{on})_{Q_2} = 0.7 + 0.7 = 1.4\,\text{V}$$

and voltage v_{C2} is

$$v_{C2} = V_{BE}(\text{on})_{Q_o} + V_{CE}(\text{sat})_{Q_2} = 0.7 + 0.4 = 1.1\,\text{V}$$

The currents are then

$$i_1 = \frac{V_{CC} - v_1}{R_1} = \frac{5 - 1.4}{20} = 0.18 \, \text{mA}$$

and

$$i_2 = \frac{V_{CC} - v_{C2}}{R_2} = \frac{5 - 1.1}{8} = 0.488 \, \text{mA}$$

When $v_{C2} = 1.1 \, \text{V}$ and $v_O = 0.4 \, \text{V}$, transistor Q_4 is at the edge of turn-on, however, since there is no voltage drop across R_4, Q_4 has negligible emitter current. For a no-load condition, all other currents are zero. Therefore, the power dissipation for the high input condition is

$$P_H = (i_1 + i_2)V_{CC} = (0.18 + 0.488) \cdot 5 = 3.34 \, \text{mW}$$

Comment: The power dissipation in this low-power Schottky TTL circuit is approximately a factor of five smaller than in the Schottky or standard TTL logic gates. The propagation delay time in the low-power Schottky circuit is approximately 10 ns, which compares closely with the propagation delay time for a standard TTL circuit.

Diodes D_5 and D_6 are called **speedup diodes**. As we showed in the dc analysis, these diodes are reverse biased when the inputs are in either a static logic 0 or a logic 1 mode. When at least one input is in a logic 0 state, the output is high, and Q_3 and Q_4 tend to turn on, supplying any necessary load current. When both inputs are switched to their logic 1 state, Q_2 turns on and v_{C2} decreases, forward biasing D_5 and D_6. Diode D_5 helps to pull charge out of the base of Q_3, turning this transistor off more rapidly. Diode D_6 helps discharge the load capacitance, which means that output voltage v_O switches low more rapidly.

Test Your Understanding

[Note: In the following exercise, assume the piecewise linear transistor parameters are as listed in Table 17.3. In addition, assume a Schottky diode turn-on voltage of $V_\gamma(\text{SD}) = 0.3 \, \text{V}$.]

17.18 Assume the low-power Schottky TTL circuit in Figure 17.34 is redesigned such that $R_1 = 40 \, \text{k}\Omega$ and $R_2 = 12 \, \text{k}\Omega$, and all other circuit parameters remain the same. The transistor and diode parameters are: $V_{BE}(\text{on}) = 0.7 \, \text{V}$, $V_{CE}(\text{sat}) = 0.4 \, \text{V}$, $\beta = 25$, and $V_\gamma(\text{SD}) = 0.3 \, \text{V}$. Assuming no load, determine the base and collector currents in each transistor, and the power dissipation in the gate, for: (a) $v_X = v_Y = 0.4 \, \text{V}$, and (b) $v_X = v_Y = 3.6 \, \text{V}$. (Ans. (a) $i_{B2} = i_{C2} = i_{Bo} = i_{Co} = i_{B5} = i_{C5} = 0$, $i_{B3} = i_{C3} = i_{B4} = i_{C4} = 0$, $P = 497 \, \mu\text{W}$ (b) $i_{B2} = 90 \, \mu\text{A}$, $i_{C2} = 325 \, \mu\text{A}$, $i_{B4} = i_{C4} = i_{B3} = i_{C3} = 0$, $i_{B5} \cong i_{C5} \cong 0$, $i_{Bo} = 415 \, \mu\text{A}$, $i_{Co} = 0$, $P = 2.08 \, \text{mW}$)

17.4.4   Advanced Schottky TTL Circuits

The advanced low-power Schottky circuit possesses the lowest speed–power product with a propagation delay time short enough to accommodate a large number of digital applications, while still maintaining the low power dissipation of the low-power Schottky family of logic circuits. The major modification lies in the design of the input circuitry. Consider the circuit shown in Figure 17.35. The input circuit contains a pnp transistor Q_1, a current amplification transistor Q_2, and a Schottky diode D_2 from the base of Q_3 to the input. Diode D_2 provides a low-impedance path to ground when the input makes a high-to-low transition. This enhances the inverter switching time. The current driver transistor Q_1 provides a faster transition when the input goes from low to high than if a Schottky diode input stage were used. Transistor Q_1 provides the switch element that steers current from R_1 either to Q_2 or the input source.

Figure 17.35 Advanced low-power Schottky (ALS) inverter gate

When $v_X = 0.4\,\text{V}$, the E–B junction of Q_1 is forward biased, and Q_1 is biased in its active region. The base voltage of Q_2 is approximately 1.1 V; Q_2, Q_3, and Q_5 are cut off; and the output voltage goes high. Most of the current through R_1 goes to ground through Q_1, so very little current sinking is required of the driver output transistor. When $v_X = 3.6\,\text{V}$, transistors Q_2, Q_3, and Q_5 turn on, the voltage at the base of Q_2 is clamped at approximately 2.1 V, the E–B junction of Q_1 is reverse biased, and Q_1 is cut off.

With fast switching circuits, inductances, capacitances, and signal delays may introduce problems requiring the use of transmission line theory. Clamping diodes D_1 and D_4 at the input and output terminals clamp any negative-going switching transients that result from ringing signals on the interconnect lines.

Test Your Understanding

[Note: In the following exercise, assume the piecewise linear transistor parameters are as listed in Table 17.3. In addition, assume a Schottky diode turn-on voltage of $V_\gamma(\text{SD}) = 0.3\,\text{V}$.]

17.19 Consider the advanced low-power Schottky circuit shown in Figure 17.35. Let $V_{CC} = 5\,\text{V}$. Determine the current in R_1 for: (a) $v_X = 0.4\,\text{V}$, and (b) $v_X = 3.6\,\text{V}$. (Ans. (a) $i_1 = 97.5\,\mu\text{A}$ (b) $i_1 = 72.5\,\mu\text{A}$)

17.5   BiCMOS DIGITAL CIRCUITS

As we have discussed previously, BiCMOS technology combines bipolar and CMOS circuits on one IC chip. This technology combines the high-input-impedance, low-power characteristics of CMOS with the high-current drive characteristics of bipolar circuits. If the CMOS circuit has to drive a few other similar CMOS logic circuits, the current drive capability is not a problem. However, if a circuit has to drive a relatively large capacitive load, bipolar circuits are preferable because of the relatively large transconductance of BJTs.

We consider a BiCMOS inverter circuit and then a simple example of a BiCMOS digital circuit. This section is intended only to introduce this technology.

17.5.1   BiCMOS Inverter

Several BiCMOS inverter configurations have been proposed. In each case, npn bipolar transistors are used as output devices and are driven by a quasi-CMOS inverter configuration. The simplest BiCMOS inverter is shown in Figure 17.36(a). The output stage of the npn transistors is similar to the totem-pole output stage of the TTL circuits that were considered in Section 17.3.

When the input voltage v_I of the BiCMOS inverter in Figure 17.36(a) is low, the transistors M_N and Q_2 are cut off. The transistor M_P is turned on and provides base current to Q_1 so that Q_1 turns on and supplies current to the load capacitance. The load capacitance charges and the output voltage goes high. As the output voltage goes high, the output current will normally become very small, so that M_N is driven into its nonsaturation region and the drain-to-source voltage will become essentially zero. The transistor Q_1 will essentially cut off and the output voltage will charge to a maximum value of approximately $v_O(\text{max}) = V_{DD} - V_{BE}(\text{on})$.

When the input voltage v_I goes high, M_P turns off, eliminating any bias current to Q_1, so Q_1 is also off. The two transistors M_N and Q_2 turn on and provide a discharge path for the load capacitance so the output voltage goes low. In steady state, the load current will normally be very small, so M_N will be biased in the nonsaturation region. The drain-to-source voltage will become essentially zero. The transistor Q_2 will be essentially off and the

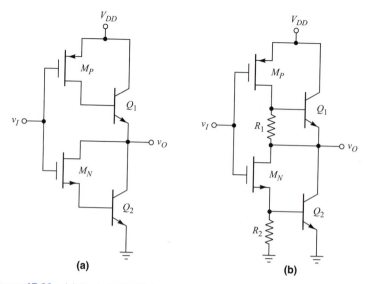

Figure 17.36   (a) Basic BiCMOS inverter. (b) Improved version of BiCMOS inverter

output voltage will discharge to a minimum value of approximately $v_O(\text{min}) \cong V_{BE}(\text{on})$.

One serious disadvantage of the inverter in Figure 17.36(a) is that there is no path through which base charge from the npn transistors can be removed when they are turning off. Thus, the turn-off time of the two npn transistors can be relatively long. A solution to this problem is to include pull-down resistors, as shown in the circuit in Figure 17.36(b). Now, when the npn transistors are being turned off, the stored base charge can be removed to ground through R_1 or R_2. An added advantage of this circuit is, that when v_I goes high and the output goes low, the very small output current through M_N and R_2 means the output voltage is pulled to ground potential. Also, as v_I goes low and the output goes high, the very small load current means that the output is pulled up to essentially V_{DD} through the resistor R_1. We may note that the two npn output transistors are never on at the same time.

Other circuit designs incorporate other transistors that aid in turning transistors off and increasing switching speed. However, these two examples have demonstrated the basic principle used in BiCMOS inverter circuit designs.

17.5.2   BiCMOS Logic Circuit

In BiCMOS logic circuits, the logic function is implemented by the CMOS portion of the circuit and the bipolar transistors again act as a buffered output stage providing the necessary current drive. One example of a BiCMOS logic circuit is shown in Figure 17.37. This is a two-input NOR gate. As seen in the figure, the CMOS configuration is the same as the basic CMOS NOR logic gate considered previously. The two npn output transistors and the R_1 and R_2 resistors have the same configuration and purpose as was seen in the BiCMOS inverter.

Other BiCMOS logic circuits are designed in a manner similar to that shown for the BiCMOS NOR gate.

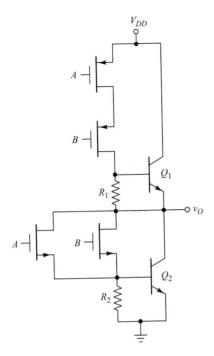

V_{DD}

A

B

Q_1

R_1

v_O

A B

Q_2

R_2

Figure 17.37   Two-input BiCMOS NOR circuit

17.6 SUMMARY

- This chapter presented the analysis and design of bipolar digital logic circuits, which were historically the first logic gate technology used in digital systems.

- Emitter coupled logic (ECL) is used in specialized high-speed applications. The basic ECL gate is the same as the differential amplifier, but transistors are switched between cutoff and the active region. Avoiding driving transistors into saturation keeps the propagation delay time to a minimum. The classical ECL gate uses the diff-amp configuration in conjunction with emitter-follower output stages and a reference voltage circuit. Both NOR and OR output are available. Although the propagation delay of this logic gate is short, on the order of a nanosecond, the power dissipated in the circuit is rather large.

- Transistor–transistor logic (TTL) was introduced by discussing Diode–transistor logic (DTL). The analysis of the DTL circuit introduced saturating bipolar logic circuits and their characteristics.

- The input transistor of the TTL circuit is driven between saturation and the inverse active mode. This transistor reduces the switching time by quickly pulling charge out of the base of a saturated transistor. The totem-pole output stage was introduced in order to increase the switching speed of the output stage. The maximum fanout was determined by specifying that the output transistor was to remain biased in the saturation region and also by specifying a maximum collector current in the output transistor. Maximum fanout is also a function of the specified propagation delay time.

- Schottky TTL was introduced. The Schottky clamped transistor has a Schottky diode between base and collector of an npn transistor. When the transistor starts into saturation, this diode turns on and clamps the forward-bias base–collector voltage to approximately 0.3 V, thus preventing the transistor from being driven deep into saturation. This effect substantially reduces the turn-off time of the transistor. The propagation delay time of Schottky TTL, then, is shorter than that of regular TTL.

- Low-power Schottky TTL has the same basic configuration as the DTL circuit. Resistor values are increased so as to reduce the currents, which in turn reduce the power dissipated per circuit. However, since current is reduced, the time to charge and discharge circuit and load capacitances is increased and propagation time increases. The trade-off is between power dissipation and propagation delay time.
- BiCMOS circuits incorporate the best characteristics of both the CMOS and bipolar technologies. Two examples of a BiCMOS inverter were discussed. A basic CMOS inverter drives a bipolar output stage. Thus, the high input impedance and low power dissipation of the CMOS design is coupled with the high-current drive capability of a bipolar output stage. An example of a BiCMOS NOR logic circuit was considered.

CHECKPOINT

After studying this chapter, the reader should have the ability to:

✔ Analyze and design a basic ECL OR/NOR logic gate. (Section 17.1)

✔ Analyze and design modified, lower-power ECL logic gates. (Section 17.2)

✔ Describe the operation and characteristics of the input transistor of a TTL logic circuit. (Section 17.3)

✔ Analyze and design a TTL NAND logic gate. (Section 17.3)

✔ Describe the operation and characteristics of a Schottky transistor, and analyze and design a Schottky TTL logic circuit. (Section 17.4)

REVIEW QUESTIONS

1. Sketch a basic bipolar differential amplifier circuit and sketch the dc transfer characteristics. Explain how the circuit is used in a digital application.

2. Why must emitter-follower output stages be added to the diff-amp to make this circuit a practical logic gate? Explain the operation of the circuit in terms of the reference voltage.

3. Sketch the voltage transfer characteristics of the basic ECL circuit. Describe the noise margins.

4. Sketch a modified ECL circuit in which a Schottky diode is incorporated in the collector portion of the circuit. Explain the purpose of the Schottky diode.

5. Explain the concept of series gating for ECL circuits. What are the advantages of this configuration?

6. Sketch a diode–transistor NAND circuit and explain the operation of the circuit. Explain the concept of minimum β and the purpose of the pull-down resistor.

7. Explain the operation and purpose of the input transistor in a TTL circuit.

8. Sketch a basic TTL NAND circuit and explain its operation.

9. Sketch a totem-pole output stage and explain its operation and the advantages of incorporating this circuit in the TTL circuit.

10. Explain how maximum fanout can be based on maintaining the output transistor in saturation when the output is low.

11. Explain how maximum fanout can be based on a maximum rated collector current in the output transistor when the output is low.

12. Explain the operation of a Schottky clamped transistor. What are its advantages?

13. What is the primary advantage of a Schottky TTL NAND gate compared to a regular TTL NAND gate.

14. Sketch a low-power Schottky TTL NAND circuit. What are the primary differences between this circuit and the regular DTL circuit considered earlier in the chapter?

15. Sketch a basic BiCMOS inverter and explain its operation. Explain the advantages of this inverter compared to a simple CMOS inverter.

16. Sketch a BiCMOS NAND logic circuit and explain its operation.

PROBLEMS

[Note: In the following problems, assume the transistor and diode parameters are as listed in Table 17.3 and $T = 300\,^\circ$K, unless otherwise stated.]

Section 17.1   Emitter-Coupled Logic (ECL)

17.1 For the differential amplifier in Figure P17.1, neglect base currents. (a) For $v_I = -1.5\,\text{V}$, calculate i_E, v_{O1}, and v_{O2}. (b) For $v_I = 1.0\,\text{V}$, calculate i_E and v_{O2}. (c) Determine R_{C1} such that the logic 0 level at v_{O1} is the same as the logic 0 value at v_{O2}.

Figure P17.1   **Figure P17.2**

17.2 Consider the circuit in Figure P17.2. (a) Determine R_{C2} such that $v_2 = -1\,\text{V}$ when Q_2 is on and Q_1 is off. (b) For $v_{in} = -0.7\,\text{V}$, determine R_{C1} such that $v_1 = -1\,\text{V}$. (c) For $v_{in} = -0.7\,\text{V}$, find v_{O1} and v_{O2}, and for $v_{in} = -1.7\,\text{V}$, find v_{O1} and v_{O2}. (d) Find the power dissipated in the circuit for (i) $v_{in} = -0.7\,\text{V}$ and for (ii) $v_{in} = -1.7\,\text{V}$.

17.3 Consider the ECL logic circuit in Figure P17.3. Neglect base currents. (a) Determine the reference voltage V_R. (b) Find the logic 0 and logic 1 voltage values at each output v_{O1} and v_{O2}. Assume that inputs v_X and v_Y have the same values as the logic levels at v_{O1} and v_{O2}.

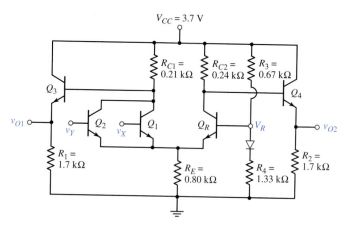

Figure P17.3

17.4 Consider the circuit in Figure P17.4. Neglect base currents. Calculate all resistor values such that the following specifications are satisfied: logic $1 = 1.0$ V and logic $0 = 0$ V; V_R is the average of logic 1 and logic 0; $i_E = 1.0$ mA when Q_R is on; $i_1 = i_2 = 1.0$ mA; $i_3 = 3.0$ mA when $v_{OR} = $ logic 1; and $i_4 = 3.0$ mA when v_{NOR} = logic 0.

Figure P17.4

Figure P17.5

17.5 In the ECL circuit in Figure P17.5, the outputs have a logic swing of 0.60 V, which is symmetrical about the reference voltage. Neglect base currents. The maximum emitter current for all transistors is 5.0 mA. Assume the input logic voltages v_I are compatible with the output logic voltages. Calculate the resistances of R_{C1}, R_{C2}, R_E, R_2 and R_3.

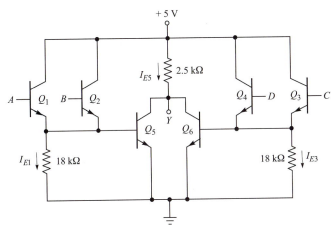

Figure P17.6

17.6 For the circuit in Figure P17.6, complete the following table. What logic function does the circuit perform?

A	B	C	D	I_{E1}	I_{E3}	I_{E5}	Y
0	0	0	0				
5 V	0	0	0				
5 V	0	5 V	0				
5 V	5 V	5 V	5 V				

17.7 Consider the ECL circuit in Figure P17.7. The input voltages A and B are compatible with the output voltages v_{O1} and v_{O2}. (a) Determine the reference voltage V_R. (b) Determine the logic 0 and logic 1 levels at the outputs v_{O1} and v_{O2}. (c) Determine the voltage V_E for $A = B = $ logic 0 and for $A = B = $ logic 1. (d) Determine the total power dissipated in the circuit for $A = B = $ logic 0 and for $A = B = $ logic 1.

Figure P17.7

17.8 A positive-voltage-supply ECL logic gate is shown in Figure P17.8. Neglect base currents. (a) What logic function is performed by this circuit. (b) What are the logic 1 and logic 0 values of v_2 at the output? (c) When $v_1 = $ logic 0 for one of the three inputs, determine i_{E1}, i_{E2}, i_{C3}, i_{C2}, and v_2. (d) Repeat part (c) when $v_1 = $ logic 1 for all three inputs.

Figure P17.8

Section 17.2   Modified ECL Circuit Configurations

D17.9 In the circuit in Figure P17.9, the output voltages v_{O1} and v_{O2} are compatible with the input voltages v_X and v_Y. Neglect base currents. (a) Design an appropriate value of V_R. State the reason for your selection. (b) Determine R_{C1} such that when Q_1 is on, the current in R_{C1} is the same as the current in D_1. (c) Determine R_{C2} such that when Q_2 is on, the current in R_{C2} is the same as the current in D_2. (d) Calculate the power dissipated in the circuit when $v_X = $ logic 0 and $v_Y = $ logic 1.

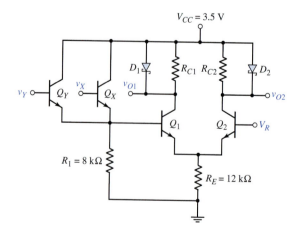

Figure P17.9

17.10 Consider the circuit in Figure P17.10. Neglect base currents. (a) What are the logic 1 and logic 0 voltage levels at the output terminals v_{O1} and v_{O2}? (b) When $v_X = v_Y = $ logic 0, the current i_E is to be 0.8 mA. Determine R_E. (c) Using the results

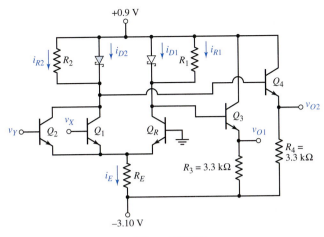

Figure P17.10

of part (b), determine R_1 such that $i_{D1} = R_{R1}$ when Q_R is conducting. (d) If $R_1 = R_2$, determine i_{R2} and i_{D2} for Q_1 and Q_2 conducting. (e) For $v_X = v_Y = $ logic 1, calculate the power dissipated in the circuit.

17.11 For the circuit in Figure P17.11, assume transistor and diode parameters of $V_{BE}(\text{on}) = 0.7\,\text{V}$ and $V_\gamma = 0.4\,\text{V}$. Neglect base currents. Find i_1, i_2, i_3, i_4, i_D, and v_O for: (a) $v_X = v_Y = -0.4\,\text{V}$, (b) $v_X = 0$, $v_Y = -0.4\,\text{V}$, (c) $v_X = -0.4\,\text{V}$, $v_Y = 0$, (d) $v_X = v_Y = 0$.

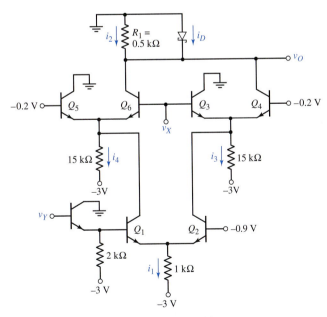

Figure P17.11

17.12 Assume the inputs A, B, C, and D to the circuit in Figure P17.12 are either 0 or 2.5 V. Let the B–E turn-on voltage be 0.7 V for both the npn and pnp transistors. Assume $\beta = 120$ for the npn devices and $\beta = 50$ for the pnp devices. (a) Determine the voltage at Y for: (i) $A = B = C = D = 0$, and (ii) $A = C = 0$, $B = D = 2.5\,\text{V}$. (b) What logic function does this circuit implement?

Part III Digital Electronics

17.13 The input and output voltage levels for the circuit in Figure P17.13 are compatible. (a) What are the logic 0 and logic 1 voltage levels? (b) What are the logic functions implemented by this circuit at v_{O1}, v_{O2}, and v_{O3}?

17.14 Consider the circuit in Figure P17.14. (a) Explain the operation of the circuit. Demonstrate that the circuit functions as a clocked D flip-flop. (b) Neglecting base currents, if $i_{DC} = 50\,\mu\text{A}$, calculate the maximum power dissipated in the circuit.

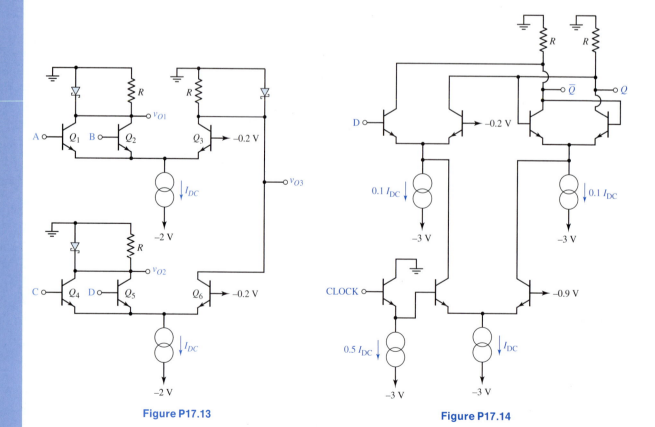

Figure P17.13

Figure P17.14

Section 17.3   Transistor–Transistor Logic

17.15   In Figure P17.15, the transistor current gain is $\beta = 20$. Find the currents and voltages i_1, i_3, i_4, and v' for the input conditions: (i) $v_X = v_Y = 0.10\,\text{V}$, and (ii) $v_X = v_Y = 5\,\text{V}$.

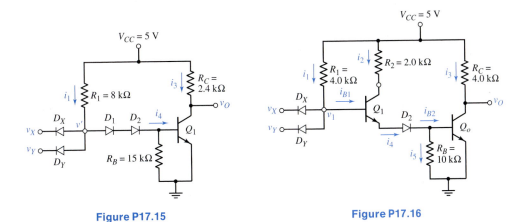

Figure P17.15   **Figure P17.16**

17.16   Figure P17.16 shows an improved version of the DTL circuit. One offset diode is replaced by transistor Q_1, providing increased current drive to Q_o. Assume $\beta = 20$ for both transistors. (a) For $v_X = v_Y = 5\,\text{V}$, determine the currents and voltages listed in the figure. (b) Calculate the maximum fanout for the low output condition.

17.17   For the modified DTL circuit in Figure P17.17, calculate the indicated currents in the figure for $v_X = v_Y = 5\,\text{V}$.

17.18   For the transistors in the TTL circuit in Figure P17.18, the parameters are $\beta_F = 20$ and $\beta_R = 0$. (a) Determine the currents $i_1, i_2, i_3, i_4, i_{B2}$, and i_{B3} for the following input conditions: (i) $v_X = v_Y = 0.1\,\text{V}$, and (ii) $v_X = v_Y = 5\,\text{V}$. (b) Show that for $v_X = v_Y = 5\,\text{V}$, transistors Q_2 and Q_3 are biased in saturation.

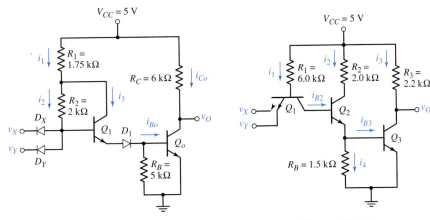

Figure P17.17   **Figure P17.18**

17.19 Reconsider the circuit in Figure P17.15. (a) Calculate the maximum fanout for the output low condition for the condition that Q_1 remains in saturation. (b) If the maximum collector current in Q_1 is limited to 5 mA, determine the maximum fanout for the low output condition.

17.20 In the TTL circuit in Figure P17.20, the transistor parameters are $\beta_F = 20$ and $\beta_R = 0.10$ (for each input emitter). (a) Calculate the maximum fanout for $v_X = v_Y = 5$ V. (b) Calculate the maximum fanout for $v_X = v_Y = 0.1$ V. (Assume v_O is allowed to decrease by 0.10 V from the no-load condition.)

Figure P17.20   **Figure P17.21**

17.21 For the TTL circuit in Figure P17.21, assume parameters of $\beta_F = 50$, $\beta_R = 0.1$, $V_{BE}(\text{on}) = 0.7$ V, $V_{BE}(\text{sat}) = 0.8$ V, and $V_{CE}(\text{sat}) = 0.1$ V. Determine the power dissipated in the circuit (no load condition) for (a) $V_{in} = 0.1$ V and (b) $V_{in} = 5$ V.

17.22 Consider the basic TTL logic gate in Figure P17.22 with a fanout of 5. Assume transistor parameters of $\beta_F = 50$ and $\beta_R = 0.5$ (for each input emitter). Calculate the base and collector currents in each transistor for: (a) $v_X = v_Y = v_Z = 0.1$ V, and (b) $v_X = v_Y = v_Z = 5$ V.

Figure P17.22

17.23   For the transistors in the TTL circuit in Figure P17.23, the parameters are $\beta_F = 100$ and $\beta_R = 0.3$ (for each input emitter). (a) For $v_X = v_Y = v_Z = 2.8$ V, determine i_{B1}, i_{B2}, and i_{B3}. (b) For $v_X = v_Y = v_Z = 0.1$ V, determine i_{B1} and i_{B4} for a fanout of 5.

17.24   A low-power TTL logic gate with an active pnp pull-up device is shown in Figure P17.24. The transistor parameters are $\beta_F = 100$ and $\beta_R = 0.2$ (for each input emitter). Assume a fanout of 5. (a) For $v_X = v_Y = v_Z = 0.1$ V, determine i_{B1}, i_{B2}, i_{B3}, i_{C2}, and i_{C3}. (b) Repeat part (a) for $v_X = v_Y = v_Z = 2$ V.

Figure P17.23   Figure P17.24

Section 17.4   Schottky Transistor–Transistor Logic

17.25   Consider the Schottky transistor circuit in Figure P17.25. Assume parameter values of $\beta = 50$, $V_{BE}(\text{on}) = 0.7$ V, and $V_\gamma = 0.3$ V for the Schottky diode. (a) Determine I_B, I_D, I_C, and V_{CE}. (b) Remove the Schottky diode and repeat part (a) assuming additional parameter values of $V_{BE}(\text{sat}) = 0.8$ V and $V_{CE}(\text{sat}) = 0.1$ V.

Figure P17.25

17.26   Consider the Schottky TTL circuit in Figure 17.33. The transistor parameters are $\beta_F = 30$ and $\beta_R = 0.1$ (for each emitter). (a) Determine all base currents, collector currents, and node voltages for $v_X = v_Y = 0.4$ V. (b) Repeat part (a) for $v_X = v_Y = 3.6$ V.

17.27 A modified Schottky TTL NAND gate is shown in Figure P17.27. The current gain of all transistors is $\beta = 50$. (a) With all inputs high and only one load connected, Q_2 is biased in saturation and $i_{B2} = i_{C2} = 0.5\,\text{mA}$. Determine the values of R_{B1} and R_{C1}. (b) With all inputs at logic 0 and with one load circuit, calculate v_{B1}, v_{C2}, and all base and collector currents. (c) With all inputs at logic 1 and with one load circuit, calculate v_{B1}, v_{C1}, and all base and collector currents. (d) Determine the maximum fanout for a low output state.

Figure P17.27   **Figure P17.28**

17.28 A low-power Schottky TTL logic circuit is shown in Figure P17.28. Assume a transistor current gain of $\beta = 30$ for all transistors. (a) Calculate the maximum fanout for $v_X = v_Y = 3.6\,\text{V}$. (b) Using the results of part (a), determine the power dissipated in the circuit for $v_X = v_Y = 3.6\,\text{V}$.

17.29 For all transistors in the circuit in Figure 17.35, the current gain is $\beta = 50$. (a) Calculate the power dissipation in the circuit when the input is at logic 0. (b) Repeat part (a) when the input is at logic 1. (c) Calculate the output short-circuit current. (Assume the input is a logic 0 and the output is inadvertently shorted to ground.)

Section 17.5   BiCMOS Digital Circuits

17.30 Consider the basic BiCMOS inverter in Figure 17.36(a). Assume circuit and transistor parameters of $V_{DD} = 5\,\text{V}$, $K_n = K_p = 0.1\,\text{mA/V}^2$, $V_{TN} = +0.8\,\text{V}$, $V_{TP} = -0.8\,\text{V}$, and $\beta = 50$. (a) For $v_I = v_O = 2.5\,\text{V}$, determine the current in each transistor. (b) If the current calculated for Q_1 were charging a 15 pF load capacitance, how long would it take to charge the capacitance from 0 to 5 V? (c) Repeat part (b) for the current in the transistor M_P.

17.31 Repeat Problem 17.30 for the BiCMOS inverter shown in Figure 17.36(b).

COMPUTER SIMULATION PROBLEMS

17.32 Consider the modified ECL logic circuit in Figure 17.17. Using PSpice, generate the voltage transfer characteristics and determine the power dissipation. Investigate the transfer characteristics at several temperatures.

17.33 Using PSpice, generate the voltage transfer characteristics of the DTL logic circuit shown in Figure 17.20.

17.34 Repeat Problem 17.32 for the TTL logic circuit in Figure 17.27. In addition, investigate the propagation delay time of this TTL circuit for one load circuit and for five load circuits connected to the output.

17.35 Repeat Problem 17.34 for the low-power Schottky TTL NAND logic circuit shown in Figure 17.34.

DESIGN PROBLEMS

*D17.36 Design an ECL R–S flip-flop.

*D17.37 Design an ECL series gating logic circuit, similar to the one shown in Figure 17.16, that will implement the logic functions: (a) $Y = [\overline{A + (B \cdot C)}]$, and (b) $Y = [\overline{(A + B) \cdot (C + D)}]$.

*D17.38 Design a clocked D flip-flop, using a modified ECL circuit design, such that the output becomes valid on the negative-going edge of the clock signal.

*D17.39 Design a low-power Schottky TTL exclusive-OR logic circuit.

*D17.40 Design a TTL R–S flip-flop.

APPENDIX

A

Physical Constants and Conversion Factors

General Constants and Conversion Factors

Angstrom	Å	$1\,\text{Å} = 10^{-4}\,\mu\text{m} = 10^{-8}\,\text{cm} = 10^{-10}\,\text{m}$
Boltzmann's constant	k	$k = 1.38 \times 10^{-23}\,\text{J/K} = 8.6 \times 10^{-5}\,\text{eV/K}$
Electron–volt	eV	$1\,\text{eV} = 1.6 \times 10^{-19}\,\text{J}$
Electronic charge	e or q	$q = 1.6 \times 10^{-19}\,\text{C}$
Micron	μm	$1\,\mu\text{m} = 10^{-4}\,\text{cm} = 10^{-6}\,\text{m}$
Mil		$1\,\text{mil} = 0.001\,\text{in.} = 25.4\,\mu\text{m}$
Nanometer	nm	$1\,\text{nm} = 10^{-9}\,\text{m} = 10^{-3}\,\mu\text{m} = 10\,\text{Å}$
Permittivity of free space	ε_o	$\varepsilon_o = 8.85 \times 10^{-14}\,\text{F/cm}$
Permeability of free space	μ_o	$\mu_o = 4\pi \times 10^{-9}\,\text{H/cm}$
Planck's constant	h	$h = 6.625 \times 10^{-34}\,\text{J–s}$
Thermal voltage	V_T	$V_T = kT/q \cong 0.026\,\text{V at } 300\,^\circ\text{K}$
Velocity of light in free space	c	$c = 2.998 \times 10^{10}\,\text{cm/s}$

Semiconductor Constants

	Si	Ge	GaAs	SiO$_2$
Relative dielectric constant	11.7	16.0	13.1	3.9
Bandgap energy, E_g(eV)	1.1	0.66	1.4	
Intrinsic carrier concentration, n_i (cm^{-3} at 300 $^\circ$K)	1.5×10^{10}	2.4×10^{13}	1.8×10^6	

B

Introduction to PSpice

B.0 PREVIEW

Several computer software packages enhance electronic analysis and design. SPICE, an acronym for **S**imulation **P**rogram with **I**ntegrated **C**ircuit **E**mphasis, is by far the most widely used computer simulation program for elecronic circuits. The program was first developed by the University of California at Berkeley in the mid-1970s. The original version was used on mainframe computers, but many upgrades have been developed, including versions written for the personal computer. These programs are generally referred to as PSpice (the prefix P denoting the personal computer). Relatively simple and inexpensive PSpice versions, generally referred to as student versions, are available.

The 8.0 student version from MicroSim Corporation was used in this text. More sophisticated programs included in SPICE, such as a Monte Carlo analysis, are not usually available in the student versions. However, this version is adequate for conducting basic PSpice analyses of transistor circuits. As mentioned in the Preface, the computer simulation should be used in conjunction with hand analyses and to fine-tune a circuit design.

Electronic circuit design generally begins by systematically combining various subcircuits, using relatively simple mathematical models of transistors. These models enable the designer to determine if the circuit can potentially meet the required specifications. However, a complex IC design generally requires a computer analysis that incorporates sophisticated device models. This prefabrication phase of the design process is important because any changes in the IC design after fabrication are expensive. A computer simulation can minimize design errors.

This appendix is intended to provide a basic description of PSpice. A few examples are included to illustrate various simulation analyses. The references listed in Appendix E will provide much more comprehensive descriptions of PSpice, as well as more detailed model parameters of diodes and transistors.

B.1 INTRODUCTION

There are three major programs to this version of PSpice: *Schematics*, *PSpice*, and *Probe*. Schematics is the program that lets you draw the circuit on the screen. PSpice is the program that analyzes the circuit created in Schematics and generates voltages and currents. The combination of Schematics and PSpice eliminates the need to create a netlist before an analysis can be performed. Probe is a graphics program that generates plots of specified circuit parameters such as currents and voltages.

The description in this appendix assumes that the software has already been installed.

B.2   DRAWING THE CIRCUIT

To begin, open the Schematics program. A blank page may appear or the page may have a grid that looks like engineering paper. At the top of the page is a menu bar. Drawing the circuit begins by selecting components from a library. Resistors, inductors, capacitors, and power supplies are available. In addition, a large number of standard transistors, op-amps, and digital components are available.

The mouse is an important tool in drawing the circuit. A single click selects an item, either a menu item or a device in the circuit. A double click with the left mouse button performs an action, such as editing a selection or ending an operation. To drag a selected item, click on the item with the left mouse button, and then, holding the button down, drag the item to a new location. Release the button when the item has been placed.

The steps in drawing a circuit are as follows:

1. A component is chosen from the **Get New Part** menu. Drag the component to the drawing board and place it in an appropriate position.

2. The component may be rotated or flipped by using the **Edit** menu to place the item in the proper orientation.

3. Components can be wired together by choosing **Wire** from the menu. The cursor will change to a pencil shape. Click the left mouse button with the pencil on one terminal of a device and drag the pencil to the terminal of another device. Double click to end this mode of operation.

4. Components can be relabeled by clicking on the item label (such as R, L, or Q). An **Edit Preference Designator** box will appear. Type in the new label and click on the OK.

5. The attributes of the items can be changed by clicking on the item value (such as 1K, 10μF, etc.). A **Set Attribute Value** box will appear. Type in the new value and click on the OK button.

6. Be sure to include a ground connection in the circuit.

7. Save the schematic.

B.3   TYPE OF ANALYSIS

The **Setup** command from the **Analysis** menu allows you to choose the type of circuit analysis to be performed. The most common types of simulations are dc bias point, dc sweep analysis, ac sweep analysis, and transient analysis.

The *dc bias point analysis* calculates all the dc nodal voltages and also calculates all electronic device quiescent values. This analysis includes determining transistor quiescent currents and voltages. As part of this analysis, the small signal parameters are determined for the electronic devices.

The *dc sweep analysis* involves allowing the voltage of a particular source to vary over a range of values with a given increment. The current through a particular component or the voltage at a given node can then be measured as the source voltage changes. This analysis can be used in diode or transistor circuits to determine the "proper" dc voltages that need to be applied.

The *ac sweep analysis* performs a frequency analysis of the circuit by varying the input signal frequency over a range of values with a given increment. A linear, decade, or octave frequency scale can be chosen. This analysis can be used to determine the bandwidth of an amplifier.

The *transient analysis* determines the circuit response as a function of time. The start and end times as well as the time increment can be chosen. This analysis can be used to determine propagation delay times in digital circuits, for example.

B.4   DISPLAYING RESULTS OF SIMULATION

Probe is the program that allows the simulation results to be graphically displayed. A voltage level or current level marker is placed at the point in the circuit where the voltage or current is to be measured. To use Probe, select Run Probe from the **Analysis** menu. From the Probe setup options, Probe can be automatically run after a simulation. Probe will open with an initial graph in which the axes are automatically set.

B.5   EXAMPLE ANALYSES

The following three examples illustrate the various types of analyses.

Example   B.1   **Objective:** Determine the dc operating point and the dc transfer characteristics of a diode circuit.

The dc bias voltages will be determined for the circuit in Figure B.1 for an input voltage of 3 V, and then the output voltage will be measured as the input voltage is swept between −2 and +6 V. Standard IN4002 diodes are used in the circuit.

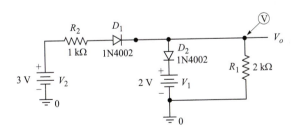

Figure B.1   Diode circuit for Example B.1

DC Analysis: The results of the dc analysis with the input voltage set at 3 V show that the output voltage is 1.625 V, which means that the diode D_2 is reverse biased. Listed in Table B.1 are the quiescent currents and voltages of the two diodes. As indicated, the current and voltage of the diode D_2 are for a reverse-biased diode.

Appendix B

Table B.1 Quiescent diode parameters for Example B.1

NAME	D_D1	D_D2
MODEL	D1N4002	D1N4002
ID	8.13E-04	-1.42E-08
VD	5.62E-01	-3.75E-01
REQ	6.31E+01	4.35E+09

DC Voltage Sweep: The dc sweep analysis was chosen from the **Setup** command in the **Analysis** menu. The input voltage V_2 was set to sweep from -2 to $+6$ V. A voltage level marker was placed at the output node, as shown in the figure, to measure the output voltage. The Probe pogram was set to run automatically after the simulation.

Figure B.2 shows the analysis results. The output voltage begins to increase when the input voltage is approximately 0.4 V, indicating that the diode D_1 has begun to conduct. When the input voltage reaches approximately 4.5 V, the output voltage tends to reach a maximum value, indicating that diode D_2 has turned on. Since the output voltage is not exactly a constant, this result shows that the voltage across the diode does increase slightly as the current through the diode increases.

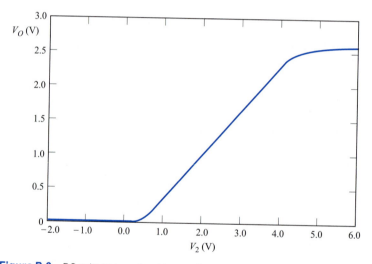

Figure B.2 DC voltage transfer characteristics of the diode circuit in Example B.1

Example B.2 **Objective:** Determine the input resistance and small-signal voltage gain versus frequency of a common-emitter amplifier.

This analysis is an example of a steady-state sinusoidal frequency analysis.

A common-emitter circuit is shown in Figure B.3. A standard 2N3904 npn bipolar transistor is used in the circuit. A 10 mV, 1 kHz ac signal is initially applied at the input. The input coupling capacitor is 1 μF, the output load capacitor is 15 pF, and the emitter-bypass capacitor is 1 kF, which means that it is essentially a short circuit to all signal currents and voltages.

DC Analysis: A dc analysis was initially performed to ensure that the bipolar transistor was biased in the forward active region. The model parameters of the 2N3904

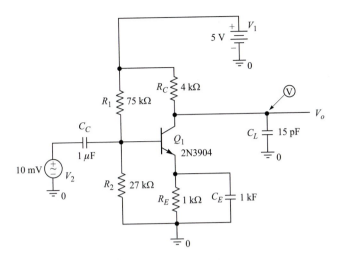

Figure B.3   Figure for Example B.2

transistor and the quiescent characteristics of the transistor are listed in Table B.2. The quiescent collector current is 0.577 mA and the quiescent collector-emitter voltage is 2.11 V, which means that the transistor is indeed biased in the forward active region.

Input Resistance: A current level marker was placed at the node of the input voltage source. With a 1 kHz, 10 mV input signal applied, the input current was measured to be

Table B.2   Model parameters and quiescent characteristics of the transistor in Example B.2

Model parameters		Quiescent characteristics	
	Q2N3904		
	NPN	NAME	Q_Q1
IS	6.734000E−15	MODEL	Q2N3904
BF	416.4	IB	4.59E−06
NF	1	IC	5.77E−04
VAF	74.03	VBE	6.51E−01
IKF	.06678	VBC	−1.46E+00
ISE	6.734000E−15	VCE	2.11E+00
NE	1.259	BETADC	1.26E+02
BR	.7371	GM	2.21E−02
NR	1	RPI	6.58E+03
RB	10	RX	1.00E+01
RC	1	RO	1.31E+05
CJE	4.493000E−12	CBE	1.31E−11
MJE	.2593	CBC	2.61E−12
CJC	3.638000E−12	CJS	0.00E+00
MJC	.3085	BETAAC	1.46E+02
TF	301.200000E−12	CBX	0.00E+00
XTF	2	FT	2.25E+08
VTF	4		
ITF	.4		
TR	239.500000E−09		
XTB	1.5		

2.03 µA. The input resistance is then found to be 4.93 kΩ. This agrees very well with calculated values of $R_1 \| R_2 \| r_\pi$. The value of r_π is given in Table B.2.

AC Sweep Analysis: The frequency of the input signal source was swept from 1 Hz to 100 MHz with 100 data points calculated per decade of frequency. The magnitude of the output voltage, plotted on a log scale, is shown in Figure B.4(a) for the case when a 15 pF capacitor is included in the output. The lower corner frequency, which is a function of the coupling capacitor, is approximately 30 Hz, and the upper corner frequency, which is a function of the load capacitor, is approximately 30 MHz. The midband voltage gain is $(0.85 \text{ V})/(0.01 \text{ V}) = 85$.

The frequency response for the case when the load capacitance is set equal to zero is shown in Figure B.4(b). The upper corner frequency is now a result of the transistor capacitances and the effective Miller capacitance. The transistor capacitances were determined for this transistor during the dc analysis and are listed in Table B.2.

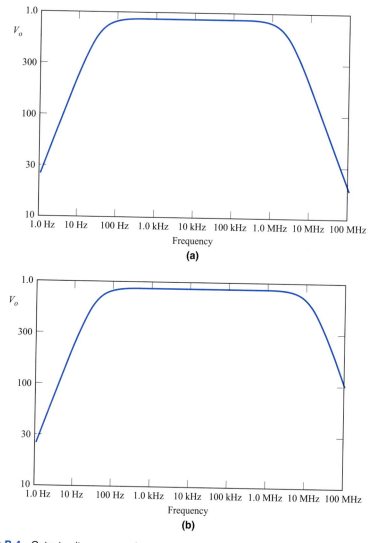

Figure B.4 Output voltage versus frequency for the circuit in Example B.2: (a) load capacitance is 15 pF and (b) load capacitance is zero

Example B.3 **Objective:** Determine the transient response of cascaded CMOS inverters.

A series of three CMOS inverters is shown in Figure B.5. The input voltage is a 5 V pulse lasting 400 ns. Capacitances are shown at the output of each inverter. These capacitors model the transistor capacitances as well as any interconnect capacitance. The capacitance values are larger than typical IC capacitance values, but are used to illustrate this type of analysis.

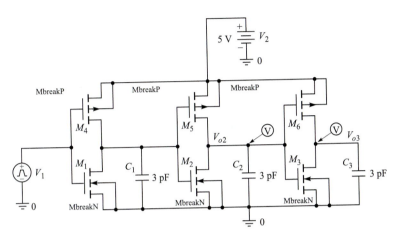

Figure B.5 CMOS inverter circuit in Example B.3

The voltages at the outputs of the second and third inverters, V_{o2} and V_{o3}, were measured as a function of time. These curves are shown in Figure B.6. This type of measurement is useful in determining propagation delay times. At the midpoint voltage of 2.5 V, there is a delay between the voltage of the third inverter compared to that of the second inverter. These time delays are referred to as propagation delay times and are important parameters in digital circuits.

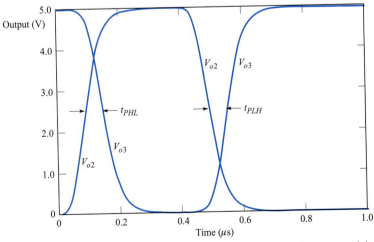

Figure B.6 Voltage versus time at the outputs of the second and third inverters of the circuit for Example B.

C

Selected Manufacturers' Data Sheets

This appendix contains data sheets representative of transistors and op-amps. This appendix is not meant as a substitute for the appropriate data books. In some cases, therefore, only selected information is presented. These data sheets are provided courtesy of National Semiconductor.

CONTENTS

National Semiconductor

2N2222
2N2222A

TO–18

PN2222
PN2222A

TO–92

MMBT2222
MMBT2222A

TO–236
(SOT–23)

MPQ2222

TO–116

NPN General Purpose Amplifier

Electrical Characteristics $T_A = 25\ °C$ unless otherwise noted

Symbol	Parameter		Min	Max	Units
OFF CHARACTERISTICS					
$V_{(BR)CEO}$	Collector-Emitter Breakdown Voltage (Note 1) ($I_C = 10$ mA, $I_B = 0$)	2222 2222A	30 40		V
$V_{(BR)CBO}$	Collector-Base Breakdown Voltage ($I_C = 10\ \mu A$, $I_E = 0$)	2222 2222A	60 75		V
$V_{(BR)EBO}$	Emitter Base Breakdown Voltage ($I_E = 10\ \mu A$, $I_C = 0$)	2222 2222A	5.0 6.0		V
I_{CEX}	Collector Cutoff Current ($V_{CE} = 60$ V, $V_{EB(OFF)} = 3.0$ V)	2222A		10	nA
I_{CBO}	Collector Cutoff Current ($V_{CB} = 50$ V, $I_E = 0$) ($V_{CB} = 60$ V, $I_E = 0$) ($V_{CB} = 50$ V, $I_E = 0$, $T_A = 150\ °C$) ($V_{CB} = 60$ V, $I_E = 0$, $T_A = 150\ °C$)	2222 2222A 222 2222A		0.01 0.01 10 10	μA
I_{EBO}	Emitter Cutoff Current ($V_{EB} = 3.0$ V, $I_C = 0$)	2222A		10	nA
I_{BL}	Base Cutoff Current ($V_{CE} = 60$ V, $V_{EB(OFF)} = 3.0$)	2222A		20	nA
ON CHARACTERISTICS					
h_{FE}	DC Current Gain ($I_C = 0.1$ mA, $V_{CE} = 10$ V) ($I_C = 1.0$ mA, $V_{CE} = 10$ V) ($I_C = 10$ mA, $V_{CE} = 10$ V) ($I_C = 10$ mA, $V_{CE} = 10$ V, $T_A = -55\ °C$) ($I_C = 150$ mA, $V_{CE} = 10$ V) (Note 1) ($I_C = 150$ mA, $V_{CE} = 1.0$ V) (Note 1) ($I_C = 500$ mA, $V_{CE} = 10$ V) (Note 1)	2222 2222A	35 50 75 35 100 50 30 40	300	

Note 1: Pulse Test: Pulse Width $\leq 300\ \mu$s, Duty Cycle $\leq 2.0\%$.

2N2222/PN2222/MMBT2222/MPQ2222/2N2222A/PN2222A/MMBT2222A NPN General Purpose Amplifier

2222/PN2222/MMBT2222/MPQ2222/2N2222A/PN2222A/MMBT2222A NPN General Purpose Amplifier

NPN General Purpose Amplifier (Continued)

Electrical Characteristics $T_A = 25$ °C unless otherwise noted (Continued)

Symbol	Parameter		Min	Max	Units
ON CHARACTERISTICS (Continued)					
$V_{CE\,(sat)}$	Collector-Emitter Saturation Voltage (Note 1) ($I_C = 150$ mA, $I_B = 15$ mA) ($I_C = 500$ mA, $I_B = 50$ mA)	2222 2222A 2222 2222A		0.4 0.3 1.6 1.0	V
$V_{BE\,(sat)}$	Base-Emitter Saturation Voltage (Note 1) ($I_C = 150$ mA, $I_B = 15$ mA) ($I_C = 500$ mA, $I_B = 50$ mA)	2222 2222A 2222 2222A	0.6 0.6	1.3 1.2 2.6 2.0	V
SMALL-SIGNAL CHARACTERISTICS					
f_T	Current Gain—Bandwidth Product (Note 3) ($I_C = 20$ mA, $V_{CE} = 20$ V, f = 100 MHz)	2222 2222A	250 300		MHz
C_{obo}	Output Capacitance (Note 3) ($V_{CB} = 10$ V, $I_E = 0$, f = 100 kHz)			8.0	pF
C_{ibo}	Input Capacitance (Note 3) ($V_{EB} = 0.5$ V, $I_C = 0$, f = 100 kHz)	2222 2222A		30 25	pF
$rb'C_C$	Collector Base Time Constant ($I_E = 20$ mA, $V_{CB} = 20$ V, f = 31.8 MHz)	2222A		150	ps
NF	Noise Figure ($I_C = 100\,\mu$A, $V_{CE} = 10$V, $R_S = 1.0$ kΩ, f = 1.0 kHz)	2222A		4.0	dB
$Re(h_{ie})$	Real Part of Common-Emitter High Frequency Input Impedance ($I_C = 20$ mA, $V_{CE} = 20$ V, f = 300 MHz)			60	Ω
SWITCHING CHARACTERISTICS					
t_D	Delay Time	($V_{CC} = 30$ V, $V_{BE(OFF)} = 0.5$ V, $I_C = 150$ mA, $I_{B1} = 15$ mA) except MPQ2222		10	ns
t_R	Rise Time			25	ns
t_S	Storage Time	($V_{CC} = 30$ V, $I_C = 150$ mA, $I_{B1} = I_{B2} = 15$ mA) except MPQ2222		225	ns
t_F	Fall Time			60	ns

Note 1: Pulse Test: Pulse Width < 300 μs, Duty Cycle $\leq$ 2.0%.
Note 2: For characteristics curves, see Process 19.
Note 3: f_T is defined as the frequency at which h_{fe} extrapolates to unity.

National Semiconductor

2N2907
2N2907A

TO–18

PN2907
PN2907A

TO–92

MMBT2907
MMBT2907A

TO–236
(SOT–23)

MPQ2907

TO–116

PNP General Purpose Amplifier

Electrical Characteristics $T_A = 25\ °C$ unless otherwise noted

Symbol	Parameter		Min	Max	Units
OFF CHARACTERISTICS					
$V_{(BR)CEO}$	Collector-Emitter Breakdown Voltage (Note 1) $(I_C = 10\ mAdc, I_B = 0)$	2907 2907A	40 60		Vdc
$V_{(BR)CBO}$	Collector-Base Breakdown Voltage $(I_C = 10\ \mu Adc, I_E = 0)$		60		Vdc
$V_{(BR)EBO}$	Emitter Base Breakdown Voltage $(I_E = 10\ \mu Adc, I_C = 0)$		5.0		Vdc
I_{CEX}	Collector Cutoff Current $(V_{CE} = 30\ Vdc, V_{BE} = 0.5\ Vdc)$			50	nAdc
I_{CBO}	Collector Cutoff Current $(V_{CB} = 50\ Vdc, I_E = 0)$ $(V_{CB} = 50\ Vdc, I_E = 0, T_A = 150\ °C)$	2907 2907A 2907 2907A		0.020 0.010 20 10	μAdc
I_B	Base Cutoff Current $(V_{CE} = 30\ Vdc, V_{EB} = 0.5\ Vdc)$			50	nAdc

PNP General Purpose Amplifier (Continued)

Electrical Characteristics $T_A = 25\ °C$ unless otherwise noted (Continued)

Symbol	Parameter		Min	Max	Units
ON CHARACTERISTICS					
h_{FE}	DC Current Gain				
	($I_C = 0.1$ mAdc, $V_{CE} = 10$ Vdc)	2907	35		
		2907A	75		
	($I_C = 1.0$ mAdc, $V_{CE} = 10$ Vdc)	2907	50		
		2907A	100		
	($I_C = 10$ mAdc, $V_{CE} = 10$ Vdc)	2907	75		
		2907A	100	300	
	($I_C = 150$ mAdc, $V_{CE} = 10$ Vdc) (Note 1)		100		
	($I_C = 500$ mAdc, $V_{CE} = 10$ Vdc) (Note 1)	2907	30		
		2907A	50		
$V_{CE\ (sat)}$	Collector-Emitter Saturation Voltage (Note 1)				
	($I_C = 150$ mAdc, $I_B = 15$ mAdc)			0.4	Vdc
	($I_C = 500$ mAdc, $I_B = 50$ mAdc)			1.6	
$V_{BE\ (sat)}$	Base-Emitter Saturation Voltage				
	($I_C = 150$ mAdc, $I_B = 15$ mAdc), (Note 1)			1.3	Vdc
	($I_C = 500$ mAdc, $I_B = 50$ mAdc)			2.6	
SMALL-SIGNAL CHARACTERISTICS					
f_T	Current Gain—Bandwidth Product				
	($I_C = 50$ mAdc, $V_{CE} = 20$ Vdc, $f = 100$ MHz)		200		MHz
C_{obo}	Output Capacitance				
	($V_{CB} = 10$ Vdc, $I_E = 0$, $f = 100$ kHz)			8.0	pF
C_{ibo}	Input Capacitance				
	($V_{EB} = 2.0$ Vdc, $I_C = 0$, $f = 100$ kHz)			30	pF
SWITCHING CHARACTERISTICS					
t_{on}	Turn-On Time			45	ns
t_d	Delay Time	($V_{CC} = 30$ Vdc, $I_C = 150$ mAdc, $I_{B1} = 15$ mAdc) Except MPQ2907		10	ns
t_r	Rise Time			40	ns
t_{off}	Turn-Off Time			100	ns
t_s	Storage Time	($V_{CC} = 6.0$ Vdc, $I_C = 150$ mAdc, $I_{B1} = I_{B2} = 15$ mAdc) Except MPQ2907		80	ns
t_f	Fall Time			30	ns

Note 1: Pulse Test: Pulse Width $\leq 300\ \mu$s, Duty Cycle $\leq 2.0\%$.

NDS9410
Single N-Channel Enhancement Mode Field Effect Transistor

General Description

These N-channel enhancement mode power field effect transistors are produced using National's proprietary, high cell density, DMOS technology. This very high density process has been especially tailored to minimize on-state resistance, provide superior switching performance, and withstand high energy pulses in the avalanche and commutation modes. These devices are particularly suited fo low voltage applications such as laptop computer power management and other battery powered circuits where fast switching, low in-line power loss, and resistance to transients are needed.

Features

• 7.0A, 30V. $R_{DS(ON)}$ = 0.03 Ω

• Rugged internal source-drain diode can eliminate the need for an external Zener diode transient suppressor

• High density cell design (3.8 million/in^2) for extremely low $R_{DS(ON)}$

• High power and current handling capability in a widely used surface mount package

• Critical DC electrical parameters specified at elevated temperature

8-SOIC

ABSOLUTE MAXIMUM RATINGS T_C = 25 °C unless otherwise noted

Symbol	Parameter	NDS9410	Units
V_{DSS}	Drain-Source Voltage	30	V
V_{DGR}	Drain-Gate Voltage ($R_{GS} \leq$ 1 MΩ)	30	V
V_{GSS}	Gate-Source Voltage	±20	V
I_D	Drain Current – Continuous @ T_A = 25 °C	±7.0	A
	– Continuous @ T_A = 70 °C	±5.8	A
	– Pulsed	±20	A
P_D	Maximum Power Dissipation @ T_A = 25 °C	2.5 (Note 1)	W
T_jT_{STG}	Operating and Storage Temperature Range	−55 to 150	°C
THERMAL CHARACTERISTICS			
$R_{\theta JA}(t)$	Thermal Resistance, Junction-to-Ambient (Pulse = 10 seconds)	50 (Note 1)	°C/W
$R_{\theta JA}$	Thermal Resistance, Junction-to-Ambient (Steady-State)	100 (Note 2)	°C/W

ELECTRICAL CHARACTERISTICS (T_C = 25 °C unless otherwise noted)

Symbol	Parameter	Conditions		Min	Typ	Max	Units
OFF CHARACTERISTICS							
BV_{DSS}	Drain–Source Breakdown Voltage	V_{GS} = 0 V, I_D = 250 μA		30			V
I_{DSS}	Zero Gate Voltage Drain Current	V_{DS} = 24 V, V_{GS} = 0 V				2	μA
			T_C = 125 °C			25	μA
I_{GSSF}	Gate–Body Leakage, Forward	V_{GS} = 20 V, V_{DS} = 0 V				100	nA
I_{GSSR}	Gate–Body Leakage, Reverse	V_{GS} = –20 V, V_{DS} = 0 V				–100	nA
ON CHARACTERISTICS (Note 3)							
$V_{GS(h)}$	Gate Threshold Voltage	V_{DS} = V_{GS}, I_D = 250 μA		1	1.4	3	V
			T_C = 125 °C	0.7	1	2.2	V
$R_{DS(ON)}$	Static Drain–Source On-Resistance	V_{GS} = 10 V, I_D = 7.0 A			0.022	0.03	Ω
			T_C = 125 °C		0.033	0.045	Ω
		V_{GS} = 4.5 V, I_D = 3.5 A			0.031	0.05	Ω
			T_C = 125 °C		0.045	0.075	Ω
$I_{D(ON)}$	On-State Drain Current	V_{GS} = 10 V, V_{DS} = 5 V		20			A
		V_{GS} = 2.7 V, V_{DS} = 2.7 V			7.7		A
g_{FS}	Forward Transconductance	V_{DS} = 15 V, I_D = 7.0 V			15		S
DYNAMIC CHARTACTERISTICS							
C_{ISS}	Input Capacitance	V_{DS} = 15 V, V_{GS} = 0 V, f = 1.0 MHz			1250		pF
C_{OSS}	Output Capacitance				610		pF
C_{RSS}	Reverse Transfer Capacitance				260		pF
SWITCHING CHARACTERISTICS (Note 3)							
$t_{D(ON)}$	Turn-On Delay Time	V_{DD} = 25 V, I_D = 1 A, V_{GEN} = 10 V, R_{GEN} = 6 Ω			10	30	ns
t_r	Turn-On Rise Time				15	60	ns
$t_{D(OFF)}$	Turn-Off Delay Time				70	150	ns
t_r	Turn-Off Fall Time				50	140	ns
Q_g	Total Gate Charge	V_{DS} = 15 V, I_D = 2.0 A, V_{GS} = 10 V			41	50	nC
Q_{gs}	Gate-Source Charge				2.8		nC
Q_{gd}	Gate-Drain Charge				12		nC
DRAIN-SOURCE DIODE CHARACTERISTICS AND MAXIMUM RATINGS							
I_S	Maximum Continuous Drain–Source Diode Forward Current					2.2	A
V_{SD}	Drain-Source Diode Forward Voltage	V_{GS} = 0 V, I_S = 2.0 A (Note 3)			0.76	1.1	V
t_{rr}	Reverse Recovery Time	V_{GS} = 0 V, I_S = 2 A, dI_S/dt = 100 A/μs			100		ns

Notes:
1. Maximum power dissipation and thermal resistance based on an assumption that a 10 second pulse is equivalent to steady-state and using a single-sided maximum-copper mounting board.
2. Junction-to-ambient thermal resistance based on steady-state conditions in still air using mounting board with minimum heat dissipation characteristics.
3. Pulse Test: Pulse Width $\leq$ 300 μs, Duty Cycle $\leq$ 2.0%.

TYPICAL ELECTRICAL CHARACTERISTICS

Figure 1. On-Region Characteristics

Figure 2. On-Resistance Variation with Drain Current and Gate Voltage

Figure 3. On-Resistance Variation with Temperature

Figure 4. On-Resistance Variation with Drain Current and Temperature

Figure 5. Transfer Characteristics

Figure 6. Gate Threshold Variation with Temperature

National Semiconductor

LM741 Operation Amplifier

General Description

The LM741 series are general purpose operational amplifiers which feature improved performance over industry standards like the LM709. They are direct, plug-in replacements for the 709C, LM201, MC1439 and 748 in most applications. The amplifiers offer many features which make their application nearly foolproof: overload protection on the input and output, no latch-up when the common mode range is exceeded as well as freedom from osciliations.

The LM741C/LM741E are identical to the LM741/LM741A except that the LM741C/LM741E have their performance guaranteed over a 0 °C to +70 °C temperature range, instead of −55 °C to +125 °C.

Schematic Diagram – (See Figure 13.3 in text)

Offset Nulling Circuit

LM741

OUTPUT

10 kΩ

V−

Absolute Maximum Ratings

	LM741A	LM741E	LM741	LM741C
Supply Voltage	±22 V	±22 V	±22 V	±18 V
Power Dissipation	500 mW	500 mW	500 mW	500 mW
Differential Input Voltage	±30 V	±30 V	±30 V	±30 V
Input Voltage (Note 2)	±15 V	±15 V	±15 V	±15 V
Output Short Circuit Duration	Continuous	Continuous	Continuous	Continuous
Operating Temperature Range	−55 °C to +125 °C	0 °C to +70 °C	−55 °C to +125 °C	0 °C to +70 °C
Storage Temperature Range	−65 °C to +150 °C	−65 °C to +150 °C	−65 °C to +150 °C	−65 °C to +150 °C
Junction Temperature	150 °C	100 °C	150 °C	100 °C

Electrical Characteristics

Parameter	Conditions	LM741A/LM741E			LM741			LM741C			Units
		Min	Typ	Max	Min	Typ	Max	Min	Typ	Max	
Input Offset Voltage	$T_A = 25$ °C $R_S \leq 10$ kΩ $R_S \leq 50$ Ω		0.8	3.0		1.0	5.0		2.0	6.0	mV mV
	$T_{AMIN} \leq T_A \leq T_{AMAX}$ $R_S \leq 50$ Ω $R_S \leq 10$ kΩ			4.0			6.0			7.5	mV mV
Average Input Offset Voltage Drift				15							μV/°C
Input Offset Voltage Adjustment Range	$T_A = 25$ °C, $V_S = \pm 20$ V	±10				±15			±15		mV
Input Offset Current	$T_A = 25$ °C		3.0	30		20	200		20	200	nA
	$T_{AMIN} \leq T_A \leq T_{AMAX}$			70		85	500			300	nA
Average Input Offset Current Drift				0.5							nA/°C
Input Bias Current	$T_A = 25$ °C		30	80		80	500		80	500	nA
	$T_{AMIN} \leq T_A \leq T_{AMAX}$			0.210			1.5			0.8	μA
Input Resistance	$T_A = 25$ °C, $V_S = \pm 20$ V	1.0	6.0		0.3	2.0		0.3	2.0		MΩ
	$T_{AMIN} \leq T_A \leq T_{AMAX}$, $V_S = \pm 20$ V	0.5									MΩ
Input Voltage Range	$T_A = 25$ °C							±12	±13		V
	$T_{AMIN} \leq T_A \leq T_{AMAX}$				±12	±13					V
Large Signal Voltage Gain	$T_A = 25$ °C, $R_L \geq 2$ kΩ $V_S = \pm 20$ V, $V_O = \pm 15$ V $V_S = \pm 15$ V, $V_O = \pm 10$ V	50			50	200		20	200		V/mV V/mV
	$T_{AMIN} \leq T_A \leq T_{AMAX}$, $R_L \geq 2$ kΩ $V_S = \pm 20$ V, $V_O = \pm 15$ V $V_S = \pm 15$ V, $V_O = \pm 10$ V $V_S = \pm 5$ V, $V_O = \pm 2$ V	32 10			25			15			V/mV V/mV V/mV

Electrical Characteristics (Continued)

Parameter	Conditions	LM741A/LM741E			LM741			LM741C			Units
		Min	Typ	Max	Min	Typ	Max	Min	Typ	Max	
Output Voltage Swing	$V_S = \pm 20$ V $R_L \geq 10$ kΩ $R_L \geq 2$ kΩ	±16 ±15									V V
	$V_S = \pm 15$ V $R_L \geq 10$ kΩ $R_L \geq 2$ kΩ				±12 ±10	±14 ±13		±12 ±10	±14 ±13		V V
Output Short Circuit Current	$T_A = 25$ °C $T_{AMIN} \leq T_A \leq T_{AMAX}$	10 10	25	35 40		25			25		mA mA
Common-Mode Rejection Ratio	$T_{AMIN} \leq T_A \leq T_{AMAX}$ $R_S \leq 10$ kΩ, $V_{CM} = \pm 12$ V $R_S \leq 50$ Ω, $V_{CM} = \pm 12$ V	80	95		70	90		70	90		dB dB
Supply Voltage Rejection Ratio	$T_{AMIN} \leq T_A \leq T_{AMAX}$ $V_S = \pm 20$ V to $V_S = \pm 5$ V $R_S \leq 50$ Ω $R_S \leq 10$ Ω	86	96		77	96		77	96		dB dB
Transient Response Rise Time Overshoot	$T_A = 25$ °C, Unity Gain		0.25 6.0	0.8 20		0.3 5			0.3 5		μs %
Bandwidth	$T_A = 25$ °C	0.437	1.5								MHz
Slew Rate	$T_A = 25$ °C, Unity Gain	0.3	0.7			0.5			0.5		V/μs
Supply Current	$T_A = 25$ °C					1.7	2.8		1.7	2.8	mA
Power Consumption	$T_A = 25$ °C $V_S = \pm 20$ V $V_S = \pm 15$ V		80	150		50	85		50	85	mW mW
LM741A	$V_S = \pm 20$ V $T_A = T_{AMIN}$ $T_A = T_{AMAX}$			165 135							mW mW
LM741E	$V_S = \pm 20$ V $T_A = T_{AMIN}$ $T_A = T_{AMAX}$			150 150							mW mW
LM741	$V_S = \pm 15$ V $T_A = T_{AMIN}$ $T_A = T_{AMAX}$					60 45	100 75				mW mW

Note 2: For supply voltages less than ±15 V, the absolute maximum input voltage is equal to the supply voltage.

A P P E N D I X

D

Standard Resistor and Capacitor Values

In this appendix, we list standard component values, which are used for selecting resistor and capacitor values in designing discrete electronic circuits and systems. Low-power carbon and film resistors with 2 percent to 20 percent tolerances have a standard set of values and a standard color-band marking scheme. These tabulated values may vary from one manufacturer to another, so the tables should be considered as typical.

D.1   CARBON RESISTORS

Standard resistor values are listed in Table D.1. The lightface type indicates 2 percent and 5 percent tolerance values; the boldface type indicates 10% tolerance resistor values.

Discrete carbon resistors have a standard color-band marking scheme, which makes it easy to recognize resistor values in a circuit or a parts bin, without having to search for a printed legend. The color bands start at one end of the resistor, as shown in Figure D.1. Two digits and a multiplier digit determine the resistor value. The additional color bands indicate the tolerance and reliability. The digit and multiplier color-code is given in Table D.2.

For example, the first three color bands of a $4.7\,\text{k}\Omega$ resistor are yellow, violet, and red. The first two digits are 47 and the multiplier is 100. The first three color bands on a $150\,\text{k}\Omega$ resistor are brown, green, and yellow.

Ten percent tolerance carbon resistors are available in the following power ratings: $\frac{1}{4}$, $\frac{1}{2}$, 1, and 2 W.

Table D.1 Standard resistance values ($\times 10^{n}$)

10	16	**27**	43	**68**
11	**18**	30	**47**	75
12	20	**33**	51	**82**
13	**22**	36	**56**	91
15	24	**39**	62	**100**

1st digit
2nd digit
Multiplier
Reliability (optional)
Tolerance
Red   2%
Gold   5%
Silver 10%
None   20%

Figure D.1 Color-band notation of low-power carbon-composition resistors

Table D.2 Digit and multiplier color code

Digit	Color	Multiplier	Number of zeros
	Silver	0.01	−2
	Gold	0.1	−1
0	Black	1	0
1	Brown	10	1
2	Red	100	2
3	Orange	1 k	3
4	Yellow	10 k	4
5	Green	100 k	5
6	Blue	1 M	6
7	Violet	10 M	7
8	Gray		
9	White		

D.2   PRECISION RESISTORS (ONE PERCENT TOLERANCE)

Metal-film precision resistors can have tolerance levels in the $\frac{1}{2}$ percent to 1 percent range. These resistors use a four-digit code printed on the resistor body, rather than the color-band scheme. The first three digits denote a value, and the last digit is the multiplier for the number of zeros. For example, 2503 denotes a 250 kΩ resistor, and 2000 denotes a 200 Ω resistor. If the resistor's value is too small to be described in this way, an R is used to indicate the decimal point; for example, 37R5 is a 37.5 Ω resistor, and 10R0 is a 10.0 Ω resistor.

The standard values typically range from 10 Ω to 301 kΩ. Standard values in each decade are given in Table D.3.

Table D.3   Standard precision resistance values

100	140	196	274	383	536	750
102	143	200	280	392	549	768
105	147	205	287	402	562	787
107	150	210	294	412	576	806
110	154	215	301	422	590	825
113	158	221	309	432	604	845
115	162	226	316	442	619	866
118	165	232	324	453	634	887
121	169	237	332	464	649	909
124	174	243	340	475	665	931
127	178	249	348	487	681	953
130	182	255	357	499	698	976
133	187	261	365	511	715	
137	191	267	374	523	732	

One percent resistors are often used in applications that require excellent stability and accuracy; a small adjustable trimmer resistor may be connected in series to the 1 percent resistor to set a precise resistance value. It is important to realize that 1 percent resistors are only guaranteed to be within 1 percent of their rated value under a specified set of conditions. Resistance variations due to temperature or humidity changes, and operation at full rated power can exceed the 1 percent tolerance.

D.3   CAPACITORS

Typical capacitor values for 10 percent tolerance capacitors from one manufacturer are listed in Table D.4. The range of capacitance values for the ceramic-disk capacitor is approximately 10 pF to 1 µF.

Table D.4   Ceramic-disk capacitors

3.3	30	200	600	2700
5	39	220	680	3000
6	47	240	750	3300
6.8	50	250	800	3900
7.5	51	270	820	4000
8	56	300	910	4300
10	68	330	1000	4700
12	75	350	1200	5000
15	82	360	1300	5600
18	91	390	1500	6800
20	100	400	1600	7500
22	120	470	1800	8200
24	130	500	2000	
25	150	510	2200	
27	180	560	2500	

Tantalum capacitors $(\times 10^n)$ (to 330 µF)

0.0047	0.010	0.022
0.0056	0.012	0.027
0.0068	0.015	0.033
0.0082	0.018	0.039

APPENDIX

E

Reading List

GENERAL ELECTRONICS TEXTS

Burns, S. G.; and P. R. Bond. *Principles of Electronic Circuits*. 2nd ed. Boston: PWS Publishing Co., 1997.

Colclaser, R. A.; D. A. Neamen; and C. F. Hawkins. *Electronic Circuit Analysis: Basic Principles*. New York: John Wiley and Sons, Inc., 1984.

Gaussi, M. S. *Electronic Devices and Circuits: Discrete and Integrated*. New York: Holt, Rinehart, and Winston, 1985.

Hambley, A. R. *Electronics*. New York: Macmillan Publishing Co., 1994.

Hayt, W. H., Jr.; and G. W. Neudeck. *Electronic Circuit Analysis and Design*. 2nd ed. Boston: Houghton Mifflin Co., 1984.

Horenstein, M. N. *Microelectronic Circuits and Devices*. 2nd ed. Englewood Cliffs, NJ: Prentice Hall, Inc., 1995.

Horowitz, P.; and W. Hill. *The Art of Electronics*. 2nd ed. New York: Cambridge University Press, 1989.

Jaeger, R. C. *Microelectronic Circuit Design*. New York: McGraw-Hill Companies, Inc., 1997.

Malik, N. R. *Electronic Circuits: Analysis, Simulation, and Design*. Englewood Cliffs, NJ: Prentice Hall, Inc., 1995.

Mauro, R. *Engineering Electronics*. Englewood Cliffs, NJ: Prentice Hall, Inc., 1989.

Millman, J.; and A. Graybel. *Microelectronics*. 2nd ed. New York: McGraw-Hill Book Co., 1987.

Mitchell, F. H., Jr.; and F. H. Mitchell, Sr. *Introduction to Electronics Design*. 2nd ed. Englewood Cliffs, NJ: Prentice-Hall, Inc., 1992.

Rashid, M. H. *Microelectronic Circuits: Analysis and Design*. Boston: PWS Publishing Co., 1999.

Roden, M. S.; and G. L. Carpenter. *Electronic Design: From Concept to Reality*. 3rd ed. Burbank, CA: Discovery Press, 1997.

Sedra, A. S.; and K. C. Smith. *Microelectronic Circuits*. 4th ed. New York: Oxford University Press, 1998.

LINEAR CIRCUIT THEORY

Alexander, C. K.; and M. N. O. Sadiku. *Fundamentals of Electric Circuits*. Boston: McGraw-Hill Companies, Inc., 2000.

Bode, H. W. *Network Analysis and Feedback Amplifier Design*. Princeton, NJ; D. Van Nostrand Co., 1945.

Hayt, W. H., Jr.; and J. E. Kemmerley. *Engineering Circuit Analysis*. 4th ed. New York: McGraw-Hill Book Co., 1986.

Irwin, J. D.; and C-H. Wu. *Basic Engineering Circuit Analysis*. 6th ed. Upper Saddle River, NJ: Prentice-Hall, Inc., 1999.

Johnson, D. E.; J. L. Hillburn; J. R. Johnson; and P. D. Scott. *Basic Electric Circuit Analysis*. 5th ed. Englewood Cliffs, NJ: Prentice Hall, Inc., 1995.

Nilsson, J. W.; and S. A. Riedel. *Electric Circuits*. 6th ed. Upper Saddle River, NJ: Prentice-Hall, Inc., 2000.

SEMICONDUCTOR DEVICES

Neamen, D. A. *Semiconductor Physics and Devices: Basic Principles*. 2nd ed. Homewood, IL: Richard D. Irwin, Inc., 1997.

Streetman, B. G. *Solid State Electronic Devices*. 4th ed. Englewood Cliffs, NJ: Prentice Hall, Inc., 1995.

ANALOG INTEGRATED CIRCUITS

Allen, P. E.; and D. R. Hoberg. *CMOS Analog Circuit Design*. New York: Holt, Rinehart, and Winston, 1987.

Geiger, R. L.; P. E. Allen; and N. R. Strader. *VLSI Design Techniques for Analog and Digital Circuits*. New York: McGraw-Hill Publishing Co., 1990.

Gray, P. R.; and R. G. Meyer. *Analysis and Design of Analog Integrated Circuits*. 3rd ed. New York: John Wiley and Sons, Inc., 1993.

Johns, D. A.; and K. Martin. *Analog Integrated Circuit Design*. New York: John Wiley and Sons, Inc., 1997.

Laker, K. R.; and W. M. C. Sansen. *Design of Analog Integrated Circuits and Systems*. New York: McGraw-Hill, Inc., 1994.

Northrop, R. B. *Analog Electronic Circuits*. Reading, MA: Addison-Wesley Publishing Co., 1990.

Soclof, S. *Design and Applications of Analog Integrated Circuits*. Englewood Cliffs, NJ: Prentice Hall, Inc., 1991.

Solomon, J. E. "The Monolithic Op-Amp: A Tutorial Study," *IEEE Journal of Solid-State Circuits* SC-9, No. 6 (December 1974), pp. 314–32.

Widlar, R. J. "Design Techniques for Monolithhic Operational Amplifiers," *IEEE Journal of Solid-State Circuits* SC-4 (August 1969), pp. 184–91.

OPERATIONAL AMPLIFIER CIRCUITS

Barna, A.; and D. I. Porat. *Operational Amplifiers*. 2nd ed. New York: John Wiley and Sons, Inc., 1989.

Berlin, H. M. *Op-Amp Circuits and Principles*. Carmel, IN: SAMS, A division of Macmillan Computer Publishing, 1991.

Coughlin, R. F.; and F. F. Driscoll. *Operational Amplifiers and Linear Integrated Circuits*. Englewood Cliffs, NJ: Prentice Hall, Inc., 1977.

Graeme, J. G.; G. E. Tobey; and L. P. Huelsman. *Operational Amplifiers: Design and Applications*. New York: McGraw-Hill Book Co., 1971.

Helms, H. *Operational Amplifiers 1987 Source Book*. Englewood Cliffs, NJ: Prentice Hall, Inc., 1987.

DIGITAL CIRCUITS AND DEVICES

CMOS/NMOS Integrated Circuits, RCA Solid State, 1980.

DeMassa, T. A.; and Z. Ciccone. *Digital Integrated Circuits*. New York: John Wiley and Sons, Inc., 1996.

Glasford, G. M. *Digital Electronic Circuits*. Englewood Cliffs, NJ: Prentice Hall, Inc., 1988.

Hauser, J. R., "Noise Margin Criteria for Digital Logic Circuits." *IEEE Transactions on Education* 36, No. 4 (November 1993), pp. 363–68.

Hodges, D. A.; and H. G. Jackson. *Analysis and Design of Digital Integrated Circuits*. New York: McGraw-Hill Book Co., 1983.

Kang, S-M.; and Y. Leblebici. *CMOS Digital Integrated Circuits: Analysis and Design*. 2nd ed. Boston: WCB/McGraw-Hill, 1999.

Lohstroh, J. "Static and Dynamic Noise Margins of Logic Circuits," *IEEE Journal of Solid-State Circuits* SC-14, No. 3 (June 1979), pp. 591–98.

Mead, C.; and L. Conway. *Introduction to VLSI Systems*. Reading, MA: Addison-Wesley Publishing Co., Inc., 1980.

Mukherjee, A. *Introduction to nMOS and CMOS VLSI Systems Design*. Englewood Cliffs, NJ: Prentice Hall, Inc., 1986.

Prince, B. *Semiconductor Memories; A Handbook of Design, Manufacture and Applications*. 2nd ed. New York: John Wiley and Sons, Inc., 1991.

Wang, N. *Digital MOS Integrated Circuits*. Englewood Cliffs, NJ: Prentice Hall, Inc., 1989.

Wilson, G. R. "Advances in Bipolar VLSI." *Proceedings of the IEEE* 78, No. 11 (November 1990), pp. 1707–19.

SPICE AND PSPICE REFERENCES

Banzhap, W. *Computer-Aided Circuit Analysis Using PSpice*. 2nd ed. Englewood Cliffs, NJ: Prentice Hall, Inc., 1992.

Brown, W. L.; and A. Y. J. Szeto. "Verifying Spice Results with Hand Calculations: Handling Common Discrepanices." *IEEE Transactions on Education*, 37, No. 4 (November 1994), pp. 358–68.

Goody, R. W. *MicroSim PSpice for Windows: Volume I: DC, AC, and Devices and Circuits*. 2nd ed. Upper Saddle River, NJ: Prentice-Hall, Inc., 1998.

Goody, R. W. *MicroSim PSice for Windows: Volume II: Operational Amplifiers and Digital Circuits*. 2nd ed. Upper Saddle River, NJ: Prentice-Hall, 1998.

Herniter, M. E. *Schematic Capture with MicroSim PSpice*. 3rd ed. Upper Saddle River, NJ: Prentice-Hall, 1998.

Meares, L. G.: and C. E Hymowitz. *Simulating with Spice*. San Pedro, CA: Intusoft, 1988.

MicroSim Staff. *PSpice User's Manual Version 4.03*. Irvine, CA: MicroSim Corporation, 1990.

Natarajan, S. "An Effective Approach to Obtain Model Parameters for BJTs and FETs from Data Books." *IEEE Transactions on Education* 35, No. 2 (May 1992), pp. 164–69.

Rashid, M. H. *SPICE for Circuits and Electronics Using PSpice*. Englewood Cliffs, NJ: Prentice Hall, Inc., 1990.

Roberts, G. W.; and A. S. Sedra. *SPICE for Microelectronic Circuits*. 3rd ed. New York: Saunders College Publishing, 1992.

Thorpe, T. W. *Computerized Circuit Analysis with SPICE*. New York: John Wiley and Sons, Inc., 1992.

Tuinenga, P. W. *SPICE: A Guide to Circuit Simulation and Analysis Using PSpice*. 2nd ed. Englewood Cliffs, NJ: Prentice Hall, Inc., 1992.

Chapter 1

1.1 (a) (i) 1.9×10^9 cm^{-3},
         (ii) 8.71×10^{10} cm^{-3}
    (b) (i) 1.34×10^5 cm^{-3},
         (ii) 1.63×10^7 cm^{-3}

1.3 (a) n-type; $n_o = 5 \times 10^{15}$ cm^{-3},
         $p_o = 4.5 \times 10^4$ cm^{-3}
    (b) n-type; $n_o = 5 \times 10^{15}$ cm^{-3},
         $p_o = 6.48 \times 10^{-4}$ cm^{-3}

1.6 (a) Add donors, $N_d = 7 \times 10^{15}$ cm^{-3},
    (b) $T = 324$ K

1.10 (a) $p_o = 10^{17}$ cm^{-3}, $n_o = 3.24 \times 10^{-5}$ cm^{-3},
     (b) $n = n_o + \delta n \cong 10^{15}$ cm^{-3}; $p = p_o + \delta p = 1.01 \times 10^{17}$ cm^{-3}

1.13 For $N_d = 10^{16}$ cm^{-3} and $N_a = 10^{15}$ cm^{-3},
     $V_{bi} = 0.637$ V; For $N_d = 10^{16}$ cm^{-3} and
     $N_a = 10^{18}$ cm^{-3}, $V_{bi} = 0.817$ V

1.17 (a) $f_o = 8.38$ MHz;
     (b) $f_o = 13.2$ MHz

1.20 (a) 0.430 V;   (b)   0.549 V

1.23 2.83×10^3

1.25 (a) $I_D = 0.145$ μA, $V_D = 0.046$ V
     (b) $I_D = -30$ nA, $V_D = -1.2$ V

1.27 $V_I = 1.81$ V

1.29 (a) $I_D = 0.0267$ mA, $V_D = 0.7$ V
     (b) $V_D = 0.45$ V, $I_D = 0$

1.32 $I_{D1} = 0.65$ mA, $I_{D2} = 1.30$ mA,
     $R_1 = 2.35$ kΩ

1.35 (a) and (b) $v_d = 1.30$ mV (peak-to-peak)

1.37 0.599 V; 0.299 V

1.41 (a) $V_O = 5.685$ V,
     (b) $\Delta V_O = 0.039$ V,
     (c) $V_O = 5.658$ V

1.43 (a) 6.921 V;   (b) -0.13 V

Chapter 2

2.4 (a) 6.06,   (b)   1.58; PIV $= 25.7$ V for
    (a), PIV $= 100.7$ V for (b)

2.8 $R = 1.19$ Ω, 32.25%, 18 W

2.11 3.04 V

2.13 (a) $I_L = 26.3$ mA, $I_I = 45.0$ mA,
         $I_Z = 18.8$ mA
     (b) $R_L = 2$ kΩ
     (c) $R_L = 585$ Ω

2.15 (a) $\Delta V_O = 0.815$ V,   (b)   4.08%

2.18 $R_i = 18.2$ Ω, $C = 9900$ μF

2.21 (a) $v_O = v_I$ for $0 \le v_I \le 5.7$ V;
         $$v_O = \frac{v_I}{2.5} + 3.42 \text{ for } 5.7 \le v_I \le 15 \text{ V}$$
     (b) $i_D = 0$ for $0 \le v_I \le 5.7$ V;
         $$i_D = \frac{0.6v_I - 3.42}{1 \text{ k}\Omega} \text{ for}$$
         $5.7 \le v_I \le 15$ V

2.33 (a) $I = I_{D1} = I_{D2} = 0$, $V_O = 10$ V
     (b) $I_{D1} = 0$, $I = I_{D2} = 0.94$ mA,
         $V_O = 1.07$ V

(c) $I_{D1} = 0$, $I = I_{D2} = 0.44\,\text{mA}$,
     $V_O = 5.82\,\text{V}$
(d) $I = 0.964\,\text{mA}$,
     $I_{D1} = I_{D2} = 0.482\,\text{mA}$,
     $V_O = 0.842\,\text{V}$

2.35 (a) $V_1 = 6.9\,\text{V}$, $V_2 = -0.6\,\text{V}$,
           $I_{D1} = 1.25\,\text{mA}$, $I_{D2} = 0$,
           $I_{D3} = 0.95\,\text{mA}$
      (b) $V_1 = 4.4\,\text{V}$, $V_2 = -0.3\,\text{V}$,
           $I_{D1} = 0.833\,\text{mA}$,
           $I_{D2} = 0.107\,\text{mA}$, $I_{D3} = 0$
      (c) $V_1 = 4.4\,\text{V}$, $V_2 = -0.6\,\text{V}$,
           $R_1 = 10\,\text{k}\Omega$, $R_2 = 5\,\text{k}\Omega$,
           $R_3 = 2.93\,\text{k}\Omega$

2.38 (a) $I_{D1} = 0.86\,\text{mA}$, $V_O = 0$;
      (b) $I_D = 0$, $V_O = -3.57\,\text{V}$

2.40 (a) $I_D = 0$, $V_D = -2.5\,\text{V}$;
      (b) $I_D = 0.19\,\text{mA}$, $V_D = 0.6\,\text{V}$

2.43 (a) $V_{O1} = V_{O2} = 5\,\text{V}$;
      (b) $V_{O1} = 0.6\,\text{V}$, $V_{O2} = 1.2\,\text{V}$;
      (c) $V_{O1} = 0.6\,\text{V}$, $V_{O2} = 1.2\,\text{V}$

2.47 $V_{PS} = 2.6\,\text{V}$

Chapter 3

3.1   (a) $\beta_F = 85$, $\alpha_F = 0.9884$,
           $i_E = 516\,\mu\text{A}$
      (b) $\beta_F = 53$, $\alpha_F + 0.9815$,
           $i_E = 2.70\,\text{mA}$

3.4 $i_C = 1.85\,\text{mA}$, $i_B = 0.0154\,\text{mA}$,
      $i_E = 1.865\,\text{mA}$

3.7 $I_{S1} = 1.69 \times 10^{-13}\,\text{A}$, $I_{S2} = 6.94 \times 10^{-15}\,\text{A}$,
      $I_{S1}/I_{S2} = 24.35$

3.10 60.6

3.14 $R_B = 120\,\text{k}\Omega$, $I_{CQ} = 1.05\,\text{mA}$,
      $R_C = 2.38\,\text{k}\Omega$

3.16 (a) $I_E = 0$, $V_C = 6\,\text{V}$,
      (b) $I_E = 0.3\,\text{mA}$, $V_C = 3\,\text{V}$,
      (c) $I_E = 1.3\,\text{mA}$, $V_C = 1.5\,\text{V}$

3.19 $V_B = 1.19\,\text{V}$, $I_E = 0.49\,\text{mA}$

3.22 $V_E = -0.7\,\text{V}$, $V_C = 2.84\,\text{V}$

3.25 $I_{E1} = I_{E2} = 0.5\,\text{mA}$, $V_{C1} = V_{C2} = 3\,\text{V}$

3.28 3.97 V

3.31 $R_1 = 338\,\text{k}\Omega$, $R_2 = 58.7\,\text{k}\Omega$, $R_C = 6.49\,\text{k}\Omega$

3.34 (a) $I_{BQ} = 0.0624\,\text{mA}$, $I_{CQ} = 4.68\,\text{mA}$,
           $V_{CEQ} = 5.22\,\text{V}$,
      (b) $I_{BQ} = 0.0326\,\text{mA}$, $I_{CQ} = 4.89\,\text{mA}$,
           $V_{CEQ} = 4.41\,\text{V}$

3.37 $I_{CQ} = 4.41\,\text{mA}$, $V_{ECQ} = 6\,\text{V}$, $R_C = 1.26\,\text{k}\Omega$

3.40 $I_{CQ} = 2.73\,\text{mA}$, $V_{CEQ} = 6\,\text{V}$,
      $R_1 = 23.2\,\text{k}\Omega$, $R_2 = 2.83\,\text{k}\Omega$

3.43 (a) $I_{BQ} = 0.0214\,\text{mA}$, $I_{CQ} = 1.60\,\text{mA}$,
           $V_{ECQ} = 15.2\,\text{V}$,
      (b) $I_{BQ} = 0.0161\,\text{mA}$, $I_{CQ} = 1.61\,\text{mA}$,
           $V_{ECQ} = 15.1\,\text{V}$

3.46 $R_E = 4.90\,\text{k}\Omega$, $R_1 = 72.4\,\text{k}\Omega$,
      $R_2 = 50.9\,\text{k}\Omega$, designed using $\beta = 60$.

3.49 (a) $R_{TH} = 6.67\,\text{k}\Omega$, $V_{TH} = 1.67\,\text{V}$,
      (b) $I_{BQ} = 0.593\,\text{mA}$, $I_{CQ} = 3.56\,\text{mA}$,
           $V_E = 2.76\,\text{V}$, $V_C = -2.17\,\text{V}$

3.52 (a) $R_{TH} = 54.7\,\text{k}\Omega$, $V_{TH} = -3.03\,\text{V}$,
      (b) $I_{CQ} = 0.227\,\text{mA}$, $V_{CEQ} = 7.51\,\text{V}$

3.55 $I_{E2} = 3.6\,\text{mA}$, $I_{B2} = 0.0444\,\text{mA}$,
      $I_{C2} = 3.56\,\text{mA}$, $I_{E1} = 0.259\,\text{mA}$,
      $I_{B1} = 0.0032\,\text{mA}$, $I_{C1} = 0.256\,\text{mA}$

Chapter 4

4.1   (a) $g_m = 76.9\,\text{mA/V}$, $r_\pi = 2.34\,\text{k}\Omega$,
           $r_o = 75\,\text{k}\Omega$,
      (b) $g_m = 19.2\,\text{mA/V}$, $r_\pi = 9.36\,\text{k}\Omega$,
           $r_o = 300\,\text{k}\Omega$

4.4 $41.5 \leq g_m \leq 50.8\,\text{mA/V}$,
      $1.58 \leq r_\pi \leq 2.89\,\text{k}\Omega$

4.7   (a) $V_B = -0.0347\,\text{V}$, $V_E = -0.735\,\text{V}$,
      (b) $R_C = 6.43\,\text{k}\Omega$, (c) $A_V = -83.7$,
      (d) $A_V = -74.9$

4.10 (a) $I_{CQ} = 1.19\,\text{mA}$, $V_{ECQ} = 8.42\,\text{V}$,
      (b) $A_V = -1.94$,
      (c) $1.76 \leq |A_V| \leq 2.14$

4.13 (a) $R_E = 11.0\,\text{k}\Omega$, (b) $R_C = 3.71\,\text{k}\Omega$,
      (c) $A_V = -43.9$, (d) $R_i = 4.81\,\text{k}\Omega$

4.16 (a) $39.0 \leq |A_V| \leq 43.2$,

(b) $1.64 \le R_i \le 2.13\,\text{k}\Omega$,

(c) $3.70 \le R_o \le 3.85\,\text{k}\Omega$

4.19 $r_e = r_\pi \left\| \left(\dfrac{1}{g_m} \right) \right\| r_o$

4.25 3.24 V peak-to-peak

4.28 0.342 mA peak-to-peak

4.31 Δi_C (peak-to-peak) = 1.29 mA,
Δv_{CE} (peak-to-peak) = 2.58 V

4.33 (a) $I_{CQ} = 2.09\,\text{mA}$, $V_{CEQ} = 3.69\,\text{V}$,

(c) $A_V = 0.988$,

(d) $R_{ib} = 122\,\text{k}\Omega$, $R_o = 12.2\,\Omega$

4.37 (a) $I_{CQ} = 0.650\,\text{mA}$, $V_{ECQ} = 3.01\,\text{V}$,

(c) $A_V = 0.977$, $A_i = 4.61$,

(d) $R_{ib} = 88.2\,\text{k}\Omega$, $R_o = 38.7\,\Omega$,

(e) $4.21 \le A_i \le 5.05$

4.40 (a) $V_B = 0.0617\,\text{V}$, $V_E = 0.762\,\text{V}$;

(b) $g_m = 19\,\text{mA/V}$, $r_\pi = 4.21\,\text{k}\Omega$,
$r_o = 304\,\text{k}\Omega$;

(c) $A_v = 0.906$,
$A_i = 14.8$; (d) $A_v = 0.728$,
$A_i = 14.8$

4.44 (a) $I_{CQ} = 1.46\,\text{mA}$, $V_{CEQ} = 7.75\,\text{V}$;

(b) $R_m = 1.93\,\text{k}\Omega$; (c) $A_v = 26.0$

4.47 (a) $V_E = 0.5\,\text{V}$, $V_B = 1.20\,\text{V}$,
$V_C = 1.70\,\text{V}$,

(b) $A_V = 9.36$, (c) $R_i = 49.5\,\Omega$

4.50 (a) $g_{m1} = 42.7\,\text{mA/V}$, $r_{\pi1} = 2.34\,\text{k}\Omega$,
$g_{m2} = 48.5\,\text{mA/V}$, $r_{\pi2} = 2.06\,\text{k}\Omega$,
$r_{o1} = r_{o2} = \infty$,

(b) $A_{V1} = -85.4$, $A_{V2} = -97$,

(c) $A_V = 3890$

4.53 (a) $I_{C1} = 12.8\,\mu\text{A}$, $V_{CE1} = 5.11\,\text{V}$,
$I_{C2} = 1.29\,\text{mA}$, $V_{CE2} = 5.81\,\text{V}$,

(b) $A_V = -55.2$,

(c) $R_{is} = 74.4\,\text{k}\Omega$, $R_o = 2.2\,\text{k}\Omega$

4.56 (a) $P_{RE} = 1.66\,\text{mW}$, $P_{RC} = 13.0\,\text{mW}$,
$P_Q = 7.0\,\text{mW}$,

(b) $\overline{P}_{RL} = 1.44\,\text{mW}$

4.59 (a) $P_{RC} = 7.02\,\text{mW}$, $P_Q = 2.65\,\text{mW}$,

(b) $\overline{P}_{RL} = 0.290\,\text{mW}$,
$\overline{P}_{RC} = 0.0289\,\text{mW}$, $\overline{P}_Q = 2.33\,\text{mW}$

Chapter 5

5.1 (a) 3.06 mA, (b) 2.81 mA

5.4 $W/L = 9.375$

5.7 7.21 µm

5.10 (a) $V_{SD}(\text{sat}) = 1\,\text{V}$, $I_D = 0.12\,\text{mA}$

(b) $V_{SD}(\text{sat}) = 2\,\text{V}$, $I_D = 0.48\,\text{mA}$

(c) $V_{SD}(\text{sat}) = 3\,\text{V}$, $I_D = 1.08\,\text{mA}$

5.13 781 kΩ, 63.7 kΩ, 100 V

5.16 $1.24\,\text{V}^{1/2}$

5.19 $V_{GS} = 2.05\,\text{V}$, $I_D = 0.775\,\text{mA}$,
$V_{DS} = 5.35\,\text{V}$

5.22 $V_S = 2.21\,\text{V}$, $V_{SD} = 5.21\,\text{V}$

5.25 For example, let $W/L = 10$, then
$V_{SG} = 4\,\text{V}$, $R_S = 5\,\text{k}\Omega$, $R_D = 7.5\,\text{k}\Omega$,
$R_1 = 100\,\text{k}\Omega$, $R_2 = 150\,\text{k}\Omega$

5.28 $R_D = 5\,\text{k}\Omega$, $R_S = 2.36\,\text{k}\Omega$

5.31 $R_D = 4\,\text{k}\Omega$. Let $W/L = 10$, then
$R_S = 3.94\,\text{k}\Omega$

5.34 $(W/L)_1 = 3.23$

5.37 20.3

5.40 $I_D = 0.49\,\text{mA}$, $W/L = 0.731$

5.43 $V_{DS} > V_{DS}(\text{sat}) = -V_P$, $I_D = I_{DSS}$

5.46 $V_{DD} \le -2.5\,\text{V}$, $V_S = -1.06\,\text{V}$

5.49 $V_{GSQ} = -1.17\,\text{V}$, $I_{DQ} = 5.85\,\text{mA}$,
$V_{DSQ} = 7.13\,\text{V}$

5.52 $R_D = 0.9\,\text{k}\Omega$, $R_1 = 8.6\,\text{k}\Omega$,
$R_2 = 91.4\,\text{k}\Omega$

5.55 $R_D = 1.75\,\text{k}\Omega$, $I_D = I_{DSS} = 4\,\text{mA}$

5.58 $128\,\mu\text{A/V}^2$

Chapter 6

6.1 (a) 12.5, (b) 2.21 V

6.4 0.833 mA

6.6   (a) $R_D = 8\,\text{k}\Omega$, W/L = 11.6,
       (b) $g_m = 0.835\,\text{mA/V}$, $r_o = 133\,\text{k}\Omega$,
       (c) -6.3

6.10 $2.1\,\text{mA/V}$

6.13   (b) -2.88,   (c)   2.76 V peak-to-peak

6.16   (a) $R_S = 0.5\,\text{k}\Omega$, $I_{DQ} = 1.0\,\text{mA}$,
       (b) -1.33

6.19 $K_n = 0.202\,\text{mA/V}^2$, $V_{TN} = -2.65\,\text{V}$,
       $R_D = 1.23\,\text{k}\Omega$, $R_S = 0.10\,\text{k}\Omega$,
       $R_1 = 529\,\text{k}\Omega$, $R_2 = 123\,\text{k}\Omega$

6.23   (No load) $A_V = 0.995$, $R_o = 0.249\,\text{k}\Omega$;
       (With load) $A_V = 0.905$, $R_o = 0.226\,\text{k}\Omega$

6.26   (a)   47.0,   (b)   3.13 mA

6.30   (a) $100\,\Omega$,   (b) $100\,\Omega$

6.33   (a) $I_{DQ} = 0.365\,\text{mA}$, $V_{DSQ} = 4.53\,\text{V}$,
       (b) $g_m = 2.09\,\text{mA/V}$, $r_o = \infty$,
       (c) $A_V = 4.64$

6.36   (a) $R_S = 2.26\,\text{k}\Omega$, $R_D = 1.07\,\text{k}\Omega$,
       (b) $A_V = 4.74$

6.39 $0.936\,\text{k}\Omega$

6.42   (a)   0.731,   (b) $0.40\,\text{k}\Omega$

6.45   (a) $R_1 = 545\,\text{k}\Omega$, $R_2 = 1.50\,\text{M}\Omega$,
       (b) $I_{DQ1} = 0.269\,\text{mA}$, $I_{DQ2} = 0.5\,\text{mA}$,
       $V_{DSQ1} = 4.62\,\text{V}$,
       (c) $A_V = 0.714$, $R_o = 1.25\,\text{k}\Omega$

6.48   (a) $R_1 = 38.8\,\text{k}\Omega$, $R_2 = 35\,\text{k}\Omega$,
       $R_3 = 26.2\,\text{k}\Omega$, $R_D = 0.6\,\text{k}\Omega$,
       (b) $A_V = -5.36$

6.53   (a) $I_{DQ} = 1.42\,\text{mA}$, $V_{SDQ} = 2.9\,\text{V}$,
       (b) $A_V = 0.844$, $A_i = 4.18$,
       (c)   5.8 V peak-to-peak

Chapter 7

7.1   (c) $v_o(t) = 1 - e^{-t/R_1 C_1}$

7.5   (a) $\tau_S = (R_i + R_P)C_S = 0.40\,\text{s}$,
       $\tau_P = (R_i \| R_P)C_P = 0.375\,\mu\text{s}$
       (b) $f_L = 0.398\,\text{Hz}$, $f_H = 424\,\text{kHz}$,
       $|T|_{\max} = 7.5\,\text{k}\Omega$

7.8   (a) $|A_V| = 159$;
       (b) τ_S (open circuit) $= 5.31\,\text{ms}$, τ_P
       (short circuit) $= 0.332\,\mu\text{s}$;
       (c) $C_C = 0.932\,\mu\text{F}$, $C_L = 55.3\,\text{pF}$

7.11   (a)   959 Hz,   (b) $|A_V| = 6.70$

7.13   (a) $R_S = 2.59\,\text{k}\Omega$, $R_D = 4.41\,\text{k}\Omega$,
       (c) $1.89\,\mu\text{F}$

7.16   (a) $I_{DQ} = 1.8\,\text{mA}$, $V_{SDQ} = 5.68\,\text{V}$,
       $g_m = 2.68\,\text{mA/V}$, $r_o = \infty$
       (b)   For C_{C1}, $\tau_{S1} = 2.28\,\text{ms}$; For C_{C2},
       $\tau_{S2} = 51.2\,\text{ms}$
       (c) C_{C2} dominates; $f_{3-\text{dB}} = 3.1\,\text{Hz}$

7.19 $C_C = 456\,\mu\text{F}$

7.22   (a) $C_E = 57.2\,\mu\text{F}$,
       (b) $f_B \approx 199.7\,\text{Hz}$, $f_A = 0.556\,\text{Hz}$

7.25 $C_L = 121\,\text{pF}$

7.31 $f_T = 511\,\text{MHz}$, $f_\beta = 4.26\,\text{MHz}$

7.33   (a) $f_\beta = 13.3\,\text{MHz}$,   (b) $f = 199\,\text{MHz}$

7.37 $f_L = 540\,\text{Hz}$, $f_H = 344\,\text{kHz}$

7.41   (a) $r_S = 198\,\Omega$,   (b)   12%

7.44   (a) $C_\pi = 2.21\,\text{pF}$, $C_M = 27.7\,\text{pF}$,
       (b) $f_H = 3.06\,\text{MHz}$, $A_V = -19.5$

7.47   (a) $f_H = 10.4\,\text{MHz}$,   (b) $C_M = 18.2\,\text{pF}$,
       (c) $A_V = -4.66$

7.50 $f_{P\mu} = 17.9\,\text{MHz}$, $A_V = 0.863$

Chapter 8

8.2   (a) $R_L = 7.2\,\Omega$, $R_B = 1.12\,\text{k}\Omega$,
       (b) $V_P = 26\,\text{mV}$

8.5   (b)   9.38, 30, 39.4, 10.8, 7.16 W,
       (c)   Yes

8.9 $T_{\text{dev}} = 136\,°\text{C}$, $T_{\text{case}} = 101\,°\text{C}$, $T_{\text{snk}} = 85\,°\text{C}$

8.11 $P_D = 10\,\text{W}$

8.13   (a) $I_Q = 9.8\,\text{mA}$, $R = 949\,\Omega$,
       $i_{E1}(\max) = 19.6\,\text{mA}$,
       $i_{E1}(\min) = 0$, $i_L(\max) = 9.8\,\text{mA}$,
       $i_L(\min) = -9.8\,\text{mA}$,
       (b)   16.3%

8.16 (a) $v_o(\text{max}) = 8\,\text{V}$, $i_L = 1.6\,\text{mA}$,
$v_I = 10\,\text{V}$,
(b) 62.7%

8.19 (a) $V_{BB} = 5\,\text{V}$, $P = 5\,\text{mW}$,
(b) $v_o(\text{max}) = 8\,\text{V}$, $i_L = i_{Dn} = 8\,\text{mA}$,
$i_{Dp} = 0$, $v_I = 9.5\,\text{V}$, $P_L = 64\,\text{mW}$,
$P_{Mn} = 16\,\text{mW}$, $P_{Mp} = 0$

8.21 (a) $200\,\mu\text{A/V}^2$

8.24 $\overline{P}_L(\text{max}) = 112.5\,\text{mW}$, $R_1 = 40.4\,\text{k}\Omega$,
$R_2 = 13.3\,\text{k}\Omega$

8.29 (a) Set $V_e = 0.9V_{CC} = aV_P$, then
$a = 2.86$;
(b) $P_Q = 4.95\,\text{W}$

8.33 $R_i = 46.4\,\text{k}\Omega$

Chapter 9

9.2 (a) $A_v = -10$, $R_i = 10\,\text{k}\Omega$;
(b) $A_v = -5$, $R_i = 10\,\text{k}\Omega$;
(c) $A_v = -20$, $R_i = 5\,\text{k}\Omega$

9.5 $R_2 = 1\,\text{M}\Omega$, $R_1 = 33.3\,\text{k}\Omega$

9.7 (a) $v_O = -150\sin\omega t\,(\text{mV})$;
(b) $i_2 = 10\sin\omega t\,(\mu\text{A})$,
$i_L = -37.5\sin\omega t\,(\mu\text{A})$,
$i_O = -47.5\sin\omega t\,(\mu\text{A})$

9.11 (a) $450\,\text{k}\Omega$, (b) $4.95\,\text{M}\Omega$

9.15 (a) $-1.996\,\text{V}$, (b) $+1.996\,\text{V}$

9.19 (a) $v_O = -2.667\,\text{V}$,
(b) $v_{I3} = 0.525\,\text{V}$

9.20 $R_1 = 20\,\text{k}\Omega$, $R_2 = 160\,\text{k}\Omega$, $R_F = 80\,\text{k}\Omega$

9.24 (a) $R_F = 10\,\text{k}\Omega$,
(b) $v_O = 0.3125\,\text{V}$, $4.6875\,\text{V}$

9.27 $R_1 = 8\,\text{k}\Omega$, $R_2 = 72\,\text{k}\Omega$

9.31 $A_v = 5$

9.34 $v_{O1} = -v_{O2} = \left(1 + \dfrac{R_2}{R_1}\right)\cdot v_I$

9.37 (b) $R_S \geq 1.1\,\text{k}\Omega$

9.40 $R_2 = R_3 = 1\,\text{k}\Omega$, Set $R_1 = R_F = 1\,\text{k}\Omega$

9.43 $v_{12} = 2.5\,\text{V}$

9.47 $R_{if} = 1.52\,\text{k}\Omega$, Potentiometer $\approx 300\,\text{k}\Omega$

9.50 (a) $f = 31.8\,\text{Hz}$, Phase $= -90°$,
(b) $f = 15.9\,\text{Hz}$, $159\,\text{Hz}$

9.53 (a) $A_v = -\dfrac{R_2}{R_1}\cdot\dfrac{j\omega R_1 C_1}{1 + j\omega R_1 C_1}$,
(b) $A_v = -\dfrac{R_2}{R_1}$, (c) $f = \dfrac{1}{2\pi R_1 C_1}$

Chapter 10

10.2 $I_{C1} = I_{C2} = 962\,\mu\text{A}$, $I_{B1} = I_{B2} = 19.2\,\mu\text{A}$

10.5 $I_{\text{REF}} = 0.54\,\text{mA}$, $R_1 = 7.96\,\text{k}\Omega$

10.7 (a) $R_1 = 9.3\,\text{k}\Omega$, (b) $I_O = 2\,\text{mA}$,
(c) $R_{C2} = 4.65\,\text{k}\Omega$

10.11 (a) $I_2 = 1.0\,\text{mA}$, $I_3 = 1.5\,\text{mA}$;
(b) $I_1 = 0.25\,\text{mA}$, $I_3 = 0.75\,\text{mA}$;
(c) $I_1 = 0.167\,\text{mA}$, $I_2 = 0.333\,\text{mA}$

10.14 $I_{\text{REF}} = 0.500392\,\text{mA}$, $R_1 = 17.19\,\text{k}\Omega$

10.17 $I_{\text{REF}} = 1.00053\,\text{mA}$, $R_1 = 9.295\,\text{k}\Omega$

10.20 (a) $0.466\,\text{mA}$, (b) $400\,\Omega$

10.23 $R_O = 12.8\,\text{M}\Omega$, 0.936%

10.25 $V_{BE1} = 0.681\,\text{V}$, $I_{\text{REF}} = 0.482\,\text{mA}$,
$I_O = 8.7\,\mu\text{A}$, $V_{BE2} = 0.577\,\text{V}$

10.29 $R_{E2} = 10.17\,\text{k}\Omega$, $R_{E3} = 2.44\,\text{k}\Omega$,
$R_1 = 18.6\,\text{k}\Omega$,
$V_{BE2} = 0.598\,\text{V}$, $V_{BE3} = 0.6268\,\text{V}$

10.32 (a) $I_{O1} = 4.64\,\text{mA}$, $I_{O2} = 2.32\,\text{mA}$,
$I_{O3} = 6.96\,\text{mA}$,
(b) $R_{C1} \simeq 2\,\text{k}\Omega$, $R_{C2} \simeq 4\,\text{k}\Omega$, $R_{C3} \simeq 1.34\,\text{k}\Omega$

10.35 (a) $0.475 \leq I_O \leq 0.525\,\text{mA}$,
(b) $0.451 \leq I_O \leq 0.551\,\text{mA}$

10.38 $(W/L)_2 = 3.125$, $(W/L)_1 = 1.25$,
$(W/L)_3 = 0.2$

10.41 $(W/L)_2 = 20$, $(W/L)_1 = 60$,
$(W/L)_3 = 0.986$

10.44 $I_{\text{REF}} = I_O = 0.74\,\text{mA}$, $V_{DS2}(\text{sat}) = 0.86\,\text{V}$

10.47 (a) $I_{\text{REF}} = 80\,\mu\text{A}$, $I_O \cong 80\,\mu\text{A}$;
(b) From a PSpice analysis,
$\Delta I_O = 0.052\,\mu\text{A}$

10.50 $I_{REF} = 89.5\,\mu A$, $I_1 = 17.9\,\mu A$,
       $I_2 = 112\,\mu A$, $I_3 = 71.6\,\mu A$, $I_4 = 358\,\mu A$

10.52 (a) $i_D = 2.5\,mA$, (b) $i_D = 3\,mA$,
       (c) $i_D = 3.5\,mA$

10.56 (a) $V_{BE} = 0.5208\,V$,
       (b) $R_1 = 8.96\,k\Omega$,
       (c) $V_I = 4.462\,V$,
       (d) $A_v = -1846$

10.60 $A_v = -4447$

Chapter 11

11.1   (a) $R_E = 2\,k\Omega$, $R_c = 4\,k\Omega$;
       (c) $v_{CM}(max) = 4\,V$, $v_{CM}(min) = -8\,V$

11.5   (a) (i) $0\,V$, (ii) $-0.1\,V$
       (b) (i) $0.768\,V$, (ii) $0.758\,V$

11.8   (a) $R_E = 62\,k\Omega$,
       (b) $A_d = 71.0$, $A_{cm} = -0.398$,
       $CMMR_{dB} = 45\,dB$
       (c) $R_{id} = 70.4\,k\Omega$, $R_{icm} = 6.28\,M\Omega$

11.18 (a) $R_{id} = 46.8\,k\Omega$,
       (b) $R_{icm} = 43.1\,M\Omega$

11.22 (a) $R_D = 47.5\,k\Omega$, $R_1 = 73.75\,k\Omega$,
       $I_Q = I_1 = 240\,\mu A$
       (b) $\Delta I_Q \cong 13\,\mu A$

11.25 (a) $v_d = 1\,V$, (b) $v_{d,max} = 1.58\,V$

11.30 $v_{cm}(max) = 6\,V$

11.36 (a) $I_Q = 1\,mA$, $R_D = 6\,k\Omega$,
       (b) $g_f(max) = 0.25\,mA/V$,
       (c) $A_d = 1.5$

11.41 (a) $A_d = 2307$, (b) $R_L = 150\,k\Omega$

11.44 (a) $I_O = 2\,\mu A$, (b) $A_d = 1923$,
       (c) $A_d = 641$

11.48 (a) $V_{DS3} = V_{DS4} = 2\,V$,
       $V_{DS1} = V_{DS2} = 10\,V$,
       (b) $A_d = 80$, (c) $R_o = 400\,k\Omega$

11.56 (a) $A_d = 88.9$

11.62 $R_i \cong 1.05\,M\Omega$, $R_o = 0.472\,k\Omega$,
       $A_v = -438$

11.66 (a) $R_{C1} = 80\,k\Omega$, $R_{C2} = 20\,k\Omega$;
       (b) $A_{d1} = -69.6$, $A_d = -5352$

Chapter 12

12.1   (a) 1.249×10^{-2},
       (b) -0.016%, 79.987,
       (c) $\beta = 1.15 \times 10^{-2}$, -1.6%, 78.72

12.4 $A = 4999$

12.6   (a) $f_H = 8\,kHz$, (b) $f_H = 40\,kHz$

12.8 1000

12.11 (a) $R_i(max) = 10^5\,k\Omega$, $R_i(min) = 1\,\Omega$
       (b) $R_o(max) = 10^4\,k\Omega$,
       $R_o(min) = 0.1\,\Omega$

12.14 $R_{if} \cong 500\,M\Omega$, $R_{of} = 0.0219\,\Omega$

12.18 $R_{if} = 10^6\,k\Omega$, $R_{of} = 5.04\,M\Omega$

12.22 (a) $I_{C1} = I_{C2} = 0.5\,mA$,
       $I_{C3} = 2\,mA$, $v_O = 0$;
       (b) $A_{vf} = 5.68$

12.26 $A_{vf} = 45.4$

12.30 (a) $r_{\pi1} = 15.8\,k\Omega$, $g_{m1} = 7.62\,mA/V$,
       $r_{\pi2} = 2.28\,k\Omega$, $g_{m2} = 52.7\,mA/V$
       (b) $A_{if} = 8.63$; (c) $R_{if} = 45.1\,\Omega$

12.34 $A_{if} = 5.33$

12.38 $A_{gf} = 98.06\,mA/V$

12.41 (a) $A_v = -3.41$,
       (b) $A_{zf} = -85.0\,V/mA$,
       (c) $R_{if} = 14.9\,k\Omega$,
       (d) $R_{of} = 4.88\,k\Omega$

12.45 $R_F = 27.2\,k\Omega$

12.49 $T = 84.45$

12.51 (a) $f_{180} \cong 1.05 \times 10^4\,Hz$,
       (b) $\beta = 4.42 \times 10^{-4}$

12.55 (c) For $\beta = 0.005$, system is stable.
       Phase margin $= 14°$;
       For $\beta = 0.05$, system is unstable.

12.60 $\beta = 0.01428$

12.65 $f_{PD} = 555\,Hz$

Chapter 13

13.5 56.4 V

13.9 $I_{C2} = 10.28\,\mu\text{A}$, $I_{C9} = 17.13\,\mu\text{A}$,
$I_{B9} = 1.713\,\mu\text{A}$, $I_{B4} = 0.9345\,\mu\text{A}$,
$I_{C4} = 9.345\,\mu\text{A}$

13.12 $I_{C14} = 21.8\,\text{mA}$, $I_{C15} = 0.071\,\text{mA}$

13.14 $R_1 = 30.32\,\text{k}\Omega$, $R_2 = 33.96\,\text{k}\Omega$

13.18 $R_{id} = 2.095\,\text{M}\Omega$

13.22 (a) $I_{\text{REF}} = I_Q = I_{D7} = 89.2\,\mu\text{A}$,
      (b) $A_d = 141$, $A_{v2} = 141$, $A_v = 19{,}881$

13.26 $R_o = 1.26\,\text{M}\Omega$

13.38 $-15 \le v_{CM} \le 11.6\,\text{V}$

13.41 $A_d = 10.38$, $|A_{v2}| = 1917$, $|A| = 19{,}895$

13.43 $I_{DSS} = 0.8\,\text{mA}$

Chapter 14

14.1 (a) $A_{CL} = -4.52$, $R_{if} = 90.8\,\Omega$
      (b) $A_{CL} = -4.92$, $R_{if} = 98.9\,\Omega$
      (c) $A_{CL} = -4.965$, $R_{if} = 99.8\,\Omega$

14.5 (a) $A_v \approx 1$, (b) $R_{of} = 0.02\,\Omega$

14.8 (a) $R_{if} = 99.1\,\Omega$, (b) $R_{of} = 18.4\,\Omega$,
      (c) $A_{CL} = 0.65$, (d) 0.65

14.11 $f_{3-\text{dB}} = 40\,\text{Hz}$, $f_T = 2\,\text{MHz}$

14.14 $f_{\text{max}} = 159\,\text{kHz}$

14.18 6.37 V

14.22 10^3 s

14.26 $i_{C1}/i_{C2} = 1.0155$

14.31 (a) $v_{O1} = v_{O2} = 0.5\,\text{V}$, $v_{O3} = -0.3\,\text{V}$,
       (b) $R_A = 8.33\,\text{k}\Omega$, $R_B = 10\,\text{k}\Omega$,
       (c) $v_{O1} = v_{O2} = 0.1\,\text{V}$, $v_{O3} = -0.14\,\text{V}$

14.34 (a) $R_2 = 22.48\,\text{M}\Omega$, (b) $R_1 = 6\,\text{k}\Omega$

14.37 For (a) $v_O = 9\,\text{mV}$,
       for (b) $v_O = -1.0815\,\text{V}$

14.40 (a) Circuit a: $v_O = 0$, Circuit b:
       $v_O = -0.975\,\text{V}$

       (b) Circuit a: $v_O = -0.010\,\text{V}$, Circuit
       b: $v_O = -1.18\,\text{V}$

       (c) Circuit a: $I_B \to v_O = 0$,
       $I_{OS} \to v_O = 0.0125\,\text{V}$
       Circuit b: $I_B \to v_O = -1.365\,\text{V}$,
       $I_{OS} \to v_O = -1.62\,\text{V}$

14.42 $\text{CMRR}_{\text{dB}} = 37.5\,\text{dB}$

Chapter 15

15.6 $N = 5$

15.10 (b) $|A_v|_{\text{max}} = 28.3$, $f_o = 5.305\,\text{kHz}$,
       $f_1 = 5.315\,\text{kHz}$, $f_2 = 5.296\,\text{kHz}$

15.13 (b) $R_2 = 524\,\text{k}\Omega$, $C_1 = 0.0732\,\mu\text{F}$,
       $C_2 = 66.3\,\text{pF}$

15.15 (a) $10\,\text{M}\Omega$, (b) $1\,\text{M}\Omega$, (c) $333\,\text{k}\Omega$

15.18 (a) $\tau = 60\,\mu\text{s}$, (b) $\Delta v_O = 0.167\,\text{V}$,
       (c) $N = 78$

15.21 $R = 8.12\,\text{k}\Omega$, $R_2 = 236\,\text{k}\Omega$

15.25 (a) $f_o = \dfrac{1}{2\pi\sqrt{R_A R_B C_A C_B}}$,

       (b) $\dfrac{R_2}{R_1} = \dfrac{R_A}{R_B} + \dfrac{C_B}{C_A}$

15.29 $f_o = \dfrac{1}{2\pi}\sqrt{\dfrac{C_1 + C_2}{C_1 C_2 L}}$, $\dfrac{C_1}{C_2} = g_m R_L$

15.34 $V_{\text{REF}} = -5\,\text{V}$, $R_F = 10\,\text{k}\Omega$,
       $R_{VAR} = 40\,\text{k}\Omega$

15.36 (a) $V_{TH} = 2\,\text{V}$, $V_{TL} = -2\,\text{V}$

15.40 (b) $R_2 = 190\,\text{k}\Omega$, $V_{\text{REF}} = 1.579\,\text{V}$

15.42 (a) $R_2 = 12.6\,\text{k}\Omega$, (b) $R = 3.02\,\text{k}\Omega$

15.47 (b) Duty cycle $= 50\%$, $f_o = 257\,\text{Hz}$

15.52 $T = 3.80\,\text{ms}$, recovery time $\approx 2\,\text{ms}$

15.56 $627\,\text{Hz} \le f \le 4.81\,\text{kHz}$,
       $52.2 \le \text{d.c.} \le 66.7\%$

15.60 $\dfrac{R_2}{R_1} = 14$, $\dfrac{R_4}{R_3} = 15$, Bias voltage $=$
       $\pm 12\,V$, $I_P = 2\,\text{A}$,
       Peak output voltage $= \pm 10\,\text{V}$

Chapter 16

16.1   (a) $V_{SB} = 1\,\text{V} \Rightarrow \Delta V_{TN} = 0.315\,\text{V}$,
           $V_{SB} = 2\,\text{V} \Rightarrow \Delta V_{TN} = 0.544\,\text{V}$
        (b) $I_D = 0.578, 0.384, 0.267\,\text{mA}$

16.5 $v_{it} = 4.67\,\text{V}, v_{Ot} = 2.67\,\text{V}, v_O = 1.09\,\text{V}$

16.8   5.79

16.11 $V_{TNL} = -2.88\,\text{V}$

16.14   (a)   (i)   0,   (ii)   1.16 mW;
        (b)   (i)   0,   (ii)   825 μW;
        (c)   (i)   0,   (ii)   200 μW

16.17   (a)   (i)   4.2 V,   (ii)   3.4 V

16.20   (a)   1.82;
        (b) $(W/L)_L = 0.444$,
           $(W/L)_D = 0.808$;
        (c)   0.0369 V

16.24 $\overline{(B \cdot C) + A}$

16.28   (a) $V_{It} = 2.5\,\text{V}, V_{ONt} = 1.7\,\text{V}$,
           $V_{OPt} = 3.3\,\text{V}$;
        (c)   4.64 V, 0.356 V

16.33   (a)   144.5 μA,   (b)   99.4 μA

16.37 $V_{IL} = 4.125\,\text{V}, V_{OHU} = 9.125\,\text{V}$,
        $V_{IH} = 5.875\,\text{V}, V_{OLU} = 0.875\,\text{V}$
        $NM_L = 3.25\,\text{V}, NM_H = 3.25\,\text{V}$

16.39   (a)   2.5 V,   (b) $\left(\dfrac{W}{L}\right)_n = 4.5\left(\dfrac{W}{L}\right)_p$,
        (c)   1.65 V

16.43 $\overline{(A\ \text{OR}\ B)\ \text{AND}\ C}$

16.46   (b) $v_{O2} = (v_A\ \text{OR}\ v_B)\ \text{AND}\ v_C$

16.50   6.25 ms

16.55   Exclusive-OR function

16.67   (a) $(W/L) = 0.329$;
        (b)   32.8 mA, 65.6 mW

Chapter 17

17.1   (a) $i_E = 0.56\,\text{mA}, v_{O1} = 3.5\,\text{V}$,
           $v_{O2} = 2.38\,\text{V}$
        (b) $i_E = 0.76\,\text{mA}, v_{O2} = 3.5\,\text{V}$
        (e)   For $v_{O1} = 2.38\,\text{V}, R_{C1} = 1.47\,\text{k}\Omega$

17.4 $R_5 = 2.8\,\text{k}\Omega, R_E = 2.1\,\text{k}\Omega, R_2 = 2.1\,\text{k}\Omega$,
        $R_1 = 0.5\,\text{k}\Omega, R_3 = 1.1\,\text{k}\Omega$,
        $R_4 = 0.767\,\text{k}\Omega$,
        $R_{C2} = 1\,\text{k}\Omega, R_{C1} = 0.808\,\text{k}\Omega$

17.8   (a)   AND logic function;
        (b)   Logic $0 = 0\,\text{V}$, Logic $1 = 1.8\,\text{V}$;
        (c) $i_{E1} = 1.65\,\text{mA}, i_{C3} = 0$,
           $i_{C2} = i_{E3} = 3\,\text{mA}, V_2 = 0$
        (f) $i_{E1} = 0.962\,\text{mA}, i_{C2} = 0$,
           $i_{C3} = i_{E2} = 2.25\,\text{mA}, V_2 = 1.8\,\text{V}$

17.10   (a)   Logic $1 = +0.2\,\text{V}$, Logic $0 = -0.2\,\text{V}$,
        (b) $R_E = 3\,\text{k}\Omega$,   (c) $R_1 = 1\,\text{k}\Omega$,
        (d) $i_{R2} = 0.4\,\text{mA}, i_{D2} = 0.467\,\text{mA}$,
        (e)   10.98 mW

17.13   (a)   Logic $1 = 0\,\text{V}$, Logic $0 = -0.4\,\text{V}$,
        (b) $v_{O1} : A + B, v_{O2} : C + D$,
           $v_{O3} : (A + B) \cdot (C + D)$

17.15   (a)   (i) $v^1 = 0.8\,\text{V}, i_1 = 0.525\,\text{mA}$,
              $i_3 = i_4 = 0, v_O = 5\,\text{V}$
           (ii) $v^1 = 2.2\,\text{V}, i_1 = 0.35\,\text{mA}$,
              $i_3 = 2.04\,\text{mA}, i_4 = 0.297\,\text{mA}$
        (b)   7,   (c)   5

17.18   (a)   (i) $i_1 = 0.683\,\text{mA}$,
              $i_{B2} = i_2 = i_4 = i_{B3} = i_3 = 0$,
           (ii) $i_1 = i_{B2} = 0.45\,\text{mA}$,
              $i_2 = 2.05\,\text{mA}, i_4 = 0.533\,\text{mA}$,
              $i_{B3} = 1.97\,\text{mA}, i_3 = 2.23\,\text{mA}$

17.22   (a) $i_{B1} = 1.05\,\text{mA}$, other currents $= 0$,
        (b) $i_{B1} = 0.926\,\text{mA}, i_{B2} = 1.59\,\text{mA}$,
           $i_{C2} = 2.05\,\text{mA}$,
           $i_{B3} = 2.64\,\text{mA}, i_{C3} = 7.29\,\text{mA}$

17.26   (a) $v_{B1} = 1.1\,\text{V}, i_{B1} = 1.39\,\text{mA}$,
           $v_{B2} = 0.8\,\text{V}, i_{B4} = 0.0394\,\text{mA}$,
           $i_{C4} = 1.18\,\text{mA}, v_{B4} = 4.97\,\text{V}$,
           all other currents $= 0$
        (b) $v_{B1} = 1.7\,\text{V}, v_{B2} = 1.4\,\text{V}$,
           $v_{BO} = 0.7\,\text{V}, v_{C2} = 1.1\,\text{V}$
           $i_{B1} = 1.18\,\text{mA}, i_{B2} = 1.42\,\text{mA}$,
           $i_{B4} = 0.00369\,\text{mA}$,
           $i_{C2} = 5.13\,\text{mA}, i_{BO} = 6.55\,\text{mA}$

17.29   (a) $i_{E1} = 0.0975\,\text{mA}, P = 0.4875\,\text{mW}$;
        (b) $P = 1.98\,\text{mW}$;
        (c) $i_{Sc} \cong 78\,\text{mA}$

T

Transmission gates—*Cont.*
  NMOS pass networks, 1062–1065
  NMOS transmission gate, 1058–1062
Transresistance (shunt-shunt) amplifiers, 768–778
  discrete circuit representation, 770–778
  op-amp circuit representation, 768–770
Trigger comparator, 966
Triode region, 249–251
TTL. *See* Transistor-transistor logic
TTL NAND circuit, Schottky, 1152–1153
Tuinenga, P. W., 1202
Turn-off time, 22
Turn-on time, 22
Turn-on voltage, 27
Turns ratio, transformer, 51, 53
Two-diode circuits, 76–77
Two-input bipolar NOR logic circuit, 133
Two-pole active filters, 926
Two-pole amplifiers, Bode plot of, 785–789
Two-pole Butterworth filters
  high-pass, 929–931
  low-pass, 927–929
Two-port equivalent circuits, 189
Two-sided output, 647
Two-terminal MOS structure, 244–246
Two-transistor current sources, 578–583
  current relationships, 579–580, 590
  mismatched resistors, 583
  MOSFET, 598–603
  output resistance, 580–583

U

Undefined range, 1020
Unity-gain bandwidth, 421
  in FETs, 428–430
Upper corner frequency, 395

V

Valence electrons, 7–8, 11
Valentine, R. J., 2
Varactor diodes, 15
V_{BE} multiplier, class-AB biasing using, 501–504
Virtual ground, 525, 527
Virtual short concept, 536, 543
VMOS (vertical MOS) process, 476
Voltage
  breakdown, 20–21, 110–113
  forward-bias, 21

Voltage—*Cont.*
  offset, 892–906
  peak inverse, 51
  peak reverse, 21
Voltage divider biasing, 190–191
  and bias stability, 140–145
Voltage doubler circuit, for rectification, 63–64
Voltage feedback ratio, 182
Voltage follower, for noninverting amplifiers, 537–539
Voltage gain, 691–695
  in BJT active load circuits, 614–616
  closed-loop, 740
  in MOSFET active load circuits, 618
  small-signal, 173, 205–207
Voltage limitations, input and output, 873–875
Voltage regulators, 64, 923, 978–986
  applications and design, 978–986
  described, 978
  foldback characteristic, 985
  output resistance and load regulation, 978–980
  positive, 982–986
  simple series-pass, 980–982
Voltage (series-shunt) amplifiers, 749–755
  discrete circuit representation, 752–755
  op-amp circuit representation, 749–752
Voltage-to-current converter, 540–543
Voltage transfer characteristics, ECL, 1124–1125
Voltage transfer curve, CMOS inverter, 1037–1041
Voltages, sinusoidal, 170

W

Wang, N., 1201
Wide-swing current mirrors, 606–607
Widlar, R. J., 1200
Widlar current sources, 589–595, 695
Wien-bridge oscillator, 941–945
Wilson, G. R., 1201
Wilson current mirror, 605–606
Wilson current source, 587–588
  for MOSFETs, 605–606
Wu, C-H., 1200

Z

Zener diode circuits, 21, 40, 64–68
  ideal voltage reference circuit in, 64–67